A Competitive Approach to

MODERN ALGEBRA

Useful for

- Undergraduate and Post Graduate Courses for All Indian Universities.
- UGC-CSIR-JRF, NET, SET, GATE and All Entrance Examinations for Admission in M.Sc., M.Phil. and Ph.D. programmes.

Dr. Sudhir K. Pundir

M.Sc., M.Phil, NET, Ph.D.

Associate Professor
Department of Mathematics
S.D. (P.G.) College,
Muzaffarnagar (U.P.)

CBS

CBS Publishers & Distributors Pvt. Ltd.

New Delhi • Bengaluru • Chennai • Kochi • Kolkata • Mumbai
Bhubaneswar • Hyderabad • Jharkhand • Nagpur • Patna • Pune • Uttarakhand • Dhaka

ISBN: 978-81-239-2533-2

First Edition: 2015
Reprint: 2020

Published by **Satish Kumar Jain** and produced by **Varun Jain** for
CBS Publishers & Distributors Pvt. Ltd.,
4819/XI Prahlad Street, 24 Ansari Road, Daryaganj, New Delhi - 110002
delhi@cbspd.com, cbspubs@airtelmail.in • www.cbspd.com
Ph.: 23289259, 23266861, 23266867 • Fax: 011-23243014

Corporate Office: 204 FIE, Industrial Area, Patparganj, Delhi - 110 092
Ph: 49344934 • Fax: 011-49344935
E-mail: publishing@cbspd.com • publicity@cbspd.com

Branches:
• *Bengaluru:* 2975, 17th Cross, K.R. Road, Bansankari 2nd Stage,
  Bengaluru - 70 • Ph: +91-80-26771678/79 • Fax: +91-80-26771680
  E-mail: cbsbng@gmail.com, bangalore@cbspd.com
• *Chennai:* No. 7, Subbaraya Street, Shenoy Nagar, Chennai - 600030
  Ph: +91-44-26681266, 26680620 • Fax: +91-44-42032115
  E-mail: chennai@cbspd.com
• *Kochi:* Ashana House, 39/1904, A.M. Thomas Road, Valanjambalam,
  Ernakulum, Kochi • Ph: +91-484-4059061-65
  Fax: +91-484-4059065 • E-mail: cochin@cbspd.com
• *Kolkata:* 6-B, Ground Floor, Rameshwar Shaw Road, Kolkata - 700014
  Ph: +91-33-22891126/7/8 • E-mail: kolkata@cbspd.com
• *Mumbai:* 83-C, Dr. E. Moses Road, Worli, Mumbai - 400018
  Ph: +91-9833017933, 022-24902340/41 • E-mail: mumbai@cbspd.com

Representatives:

• Hyderabad: 0-9885175004	• Nagpur: 0-9021734563
• Patna: 0-9334159340	• Pune: 0-9623451994
• Jharkhand: 0-9811541605	• Uttarakhand: 0-9716462459

Printed at:
India Binding House, Noida, UP (India)

Preface

The book entitled 'A Competitive Approach to **Modern Algebra'** is meant for UG and PG students of all Indian Universities. Besides, it will also be very useful for those students preparing for various competitive examinations like CSIR-JRF/NET, SET and various entrance examinations for admission in M.Sc., MPhil and Ph.D programmes.

Special and conscious efforts have been made to keep the writing style simple. Students who are tired of complex concepts and abstract presentations styles, will find this book simple and straight forward. It is a collection and compilation work from various sources and has been endeavoured to include as much as information could be possible. The book's objective is to provide a conceptual understanding of the fundamentals of modern algebra. Different concepts have been explained with the help of examples. A large number of problems with solutions have been provided to assist one get a firm grip on the ideas developed. There is plenty of scope in the form of exercise for the reader to try and solve the problem on his own. To make the book self-contained and competition oriented a chapter review of basic terms, results and questions has been given at the end of each chapter. Also, at the end of each section, graded examples, illustrations and facts have been given with the name of competition corner, which help the students to grasp the thing better. These include problems on the entire section and are carefully selected to represent variety.

I express my gratitude to the authors and publishers of various books I consulted.

I wish to sincerely thank **Sh S.K. Jain**, Managing Director, CBS Publishers and Distributors, New Delhi for his encouragement and help in bringing out this publication on a present nice form.

My special thanks to Sh. Y.N. Arjuna, Senior Director Publishing, editorial and publicity, CBS Publishers & Distributors, New Delhi, whose encouragement and unstinted support enabled me to complete my book. Sh. Sunil Dutt, Promotion and development manager, CBS Publishers and Distributors deserve special mention for their kind help and support. Mr. Peeyush Goel, M/s Dreamshapers also deserve special mention for nice type setting.

I must also record my appreciation due to my wife Dr. Rimple, daughter Rijuta and son Shreesh for their understanding and love during the long period that I have taken to complete this book.

Above all I am thankful to The Almighty God, without whose grace nothing is possible for any one.

Readers are welcomed to point out errors, if any and send their valuable suggestions for improving the quality of the book.

Dr. Sudhir K. Pundir
email : skpundir05@yahoo.co.in

Contents

1

Basic Set Theoretic Concepts

■ Outlines

- Number system ;
- Ordered pairs;
- Types of relations and functions;
- Sets and its type;
- Relation;
- Some particular functions;
- Operations on sets;
- Functions;
- Algebra of functions;

1.1 INTRODUCTION

The concept of set is fundamental in all branches of mathematics. It was developed by German mathematician George Cantor. This chapter introduces the notations and terminology of set theory. Classical set theory, also termed as crisp set theory, is fundamental to the study of pure mathematics.

1.2 NUMBER SYSTEM

The number system plays a key role in mathematics. The real number system **R** is one of the most important and beautiful mathematical system. There are different ways of introducing the real number system, but the most common way is to start with Peano's Axioms for the natural numbers. The axioms for natural numbers, discovered by the Italian Mathematician Peano are:

(i)   1 is a natural number

(ii)   Each natural number n has a successor $(n+1)$.

(iii)   Two natural numbers are equal if their successors are equal.

(iv)   Except 1, each natural number is a successor of a natural number.

(v)   Any set of natural numbers which contains 1 and the successor $(k+1)$ of every natural number (k) whenever it contains k in the set **N** of natural numbers.

REMARKS

☞ Axiom (v) is commonly known as the axiom of induction or principle of finite induction.

☞ The above axioms completely define the set of natural numbers.

Definition: *The numbers* 1, 2, 3, ... *are called natural numbers.* We represent the set of natural numbers by **N**.

i.e.,   **N** = {1,2,3, ...}

The Peano's axioms can be used to extend the set **N** of natural numbers to another large system, known as the set of integers.

Definition: *The numbers* ... , –3, –2, –1, 0, 1, 2, 3... *are called integers. We represent the set of integers by* **Z**.

i.e., $\mathbf{Z} = \{ \, \dots -3, -2, -1, 0, 1, 2, 3, \dots \}$

Integers are used to define the rational numbers.

Definition: *Any number of the form p/q, where p,q∈* **Z**, *q ≠ 0 and p,q have no common factor (except ± 1) is called a rational number.*

The set of rational numbers is denoted by **Q**.

∴ $$\mathbf{Q} = \left\{ \frac{p}{q} ; \, p, q \in \mathbf{Z}, q \neq 0 \right\}$$

REMARK

☞ The set of rational numbers consists of integers and fractions.

Definition: *Any number which is not rational, is called an irrational number.*

For example, $\sqrt{2}, \sqrt{3}$ etc. It should be noted that every rational number can be expressed as a terminating or recurring decimal whereas every irrational number can be expressed as a non-terminating infinite decimal.

1.2.1 Real Number

A number which is either rational or irrational is called a real number. The set of real numbers is denoted by **R**.

1.2.2 Integral Powers of a Real Number

Let $a \in \mathbf{R}$, and n be any positive integer then we can define $a^n = a.a.a \dots n$ times.

In particular $a = a$

$$a^2 = a.a$$

$$a^3 = a.a.a = a^2.a \text{ and so on.}$$

Also, if n is any negative integer, then we have $x^{-n} = (x^n)^{-1} = (x^{-1})^n$

✎ General Definitions

(i) A real number a is called positive, if $a > 0$ and the set of all positive real numbers, denoted by $\mathbf{R}^+$, is given by $\mathbf{R}^+ = \{x : x \in \mathbf{R}, x > 0\}$

(ii) A real number a is called negative if $a < 0$ and the set of all negative real numbers, denoted by $\mathbf{R}^-$, is given by $\mathbf{R}^- = \{x : x \in \mathbf{R}, x < 0\}$

1.3 INTERVAL

A subset S of **R** *is called an interval if a, b ∈ S, x ∈* **R** *such that a < x < b implies x ∈ S.* There are following four type of intervals.

(i) $a \circ\!\!-\!\!\!-\!\!\!-\!\!\!-\!\!\!-\!\!\!-\!\!\!-\!\!\!-\!\!\!-\!\!\circ b \Rightarrow \;]a, b[= \{x : a < x < b\}$

(ii) $a \bullet\!\!-\!\!\!-\!\!\!-\!\!\!-\!\!\!-\!\!\!-\!\!\!-\!\!\!-\!\!\!-\!\!\bullet b \Rightarrow \;[a, b] = \{x : a \leq x \leq b\}$

(iii) $a \circ\!\!-\!\!\!-\!\!\!-\!\!\!-\!\!\!-\!\!\!-\!\!\!-\!\!\!-\!\!\bullet \;\; \Rightarrow \;]a, b] = \{x : a < x \leq b\}$

(iv) $a \bullet\!\!-\!\!\!-\!\!\!-\!\!\!-\!\!\!-\!\!\!-\!\!\!-\!\!\!-\!\!\circ b \Rightarrow \;[a, b[= \{x : a \leq x < b\}$

Observations

▶ The set $]a, b[$ in which the end points are not included, is called an open interval.
▶ The set $[a, b]$ also contains both its end points, is called a closed interval.
▶ The sets $[a, b[$ and $]a, b]$ are called half open (or half closed) intervals or semi-open (or semi-closed) as they contain only one end point.

Apart from the four types of intervals listed above; there are a few more types: These are

(i) $]a, \infty[$ $=$ $\{x : a < x\}$ (open right ray)
(ii) $[a, \infty[$ $=$ $\{x : a \le x\}$ (closed right ray)
(iii) $]-\infty, b[$ $=$ $\{x : x < b\}$ (open left ray)
(iv) $]\infty, b]$ $=$ $\{x : x \le b\}$ (closed left ray)
(v) $]-\infty, \infty[$ $=$ (open interval)

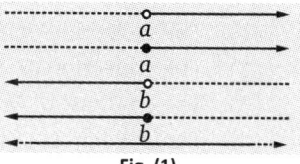

Fig. (1)

REMARKS

☞ If S is any interval and if c and d are two elements of S, then all numbers lying between c and d are also elements of S.

☞ The proper use of a bracket, for example, parenthesis' for open and square brackets for closed, itself specifies the interval. As such, to emphasize the nature of an interval, we shall drop the used 'description' and shall simply express the interval by using the appropriate brackets.

1.3.1 Length of an Interval

The number $b - a$ is called length of the intervals $]a, b[$, $[a, b[$, $]a, b]$ *and* $[a, b]$. If the length of the interval is finite, the interval is said to be finite and if the length is infinite, then it is known as infinite interval.

1.3.2 Absolute Value of a Real Number

The absolute value of a real number a denoted by $|a|$ is the real number $a, -a$ or 0 according as a is positive, negative or zero, *i.e.*,

$$|a| = \begin{cases} a & if \quad a \ge 0 \\ -a & if \quad a < 0 \end{cases}$$

From the above definition, it is clear that

(i) $|a| = \max\{a, -a\}$ (ii) $-|a| = \min\{a, -a\}$ (iii) $|a| \ge a \ge -|a|$

1.3.3 Some Useful Results

(i) $|xy| = |x| \cdot |y|$ (ii) $|x+y| \le |x| + |y|$
(iii) $|x-y| \ge ||x| - |y||$ (iv) $|x-y| \le |x| + |y|$

(v) $\left|\dfrac{x}{y}\right| = \dfrac{|x|}{|y|}$

(vi) If $\epsilon > 0$, then $|x - y| < \epsilon \Leftrightarrow y - \epsilon < x < y + \epsilon$

1.4 CONCEPT OF SETS

The theory of sets is one of the most important tools of pure mathematics. Pure mathematics is the study of sets equipped with assigned structures, known as mathematical systems. In this section, we shall study some fundamental concept of set theory.

Definition: 'A set is a well defined collection of objects'.

The objects of a set are called the elements or members of that set and their membership is defined by certain conditions.

The sets are usually denoted by the capital letters of English alphabets: Say $A, B, C, ...$, X, Y, Z.

For example :

(1) The collection of the letters a, b, c, d, ...

(2) The collection of all natural numbers denoted by **N**.

(3) The students of M.Sc., Mathematics in C.C.S. University, Meerut.

(4) The collection of vowels in English alphabet. This set containing only five elements, namely a, e, i, o, u.

(5) The collection of all states in Indian union.

If S is a set, an object a in the collection S is called an element of S. This fact is expressed in symbol as $a \in S$ (read as a is in S or a belongs to S). If a is not in S, we write $a \notin S$. For example, $4 \in \mathbf{R}$, but $\sqrt{-2} \notin \mathbf{R}$.

Here, Greek letter $\in$ denotes 'belongs to'. It is the abbreviation of the Greek word meaning 'is'.

REMARKS

☞ By the term 'well defined' we mean that we are given a collection of objects, with certain definite property, so that we are able to determine whether a given object belongs to our collection or not. Thus, every collection of objects is not a set.

☞ Set and aggregate both have the same meaning.

☞ The elements of a set must be distinguished from one another. The collection of sand particles does not form a set.

☞ The collection of rich persons of a city is not a set. However the collection of those persons of city whose wealth exceeds, a fixed amount, say ruppes fifty thousands, is a set.

☞ The order is not preserved in case of a set, whereas order is necessarily preserved in case of sequence. That is to say, each of the sets {1,2,3}, {3,2,1}, {1,3,2} denotes the same sets.

☞ The repetition of an element does not change the nature of a set, i.e., each of the sets {1,2,3}, {1,2,2,3}, {1,3,3,2} denotes the same sets.

1.4.1. Representation of a Set

There are two ways of representing a set:

(1) Roster or tabulation method

(2) Set-builder or rule method

(1) Roster Method: In this method, the elements of the set are listed within braces, and separated by comma.

For example:

(1) A= {1,2,3,4,5,6}

(2) The set of vowels of English alphabet may be represent as {a, e, i, o, u}.

(3) The set of a natural numbers from 1 to 100 may be written as **N**= {1,2,3, ..., 100}. We use three dots in the middle to include the missing elements.

(4) The set of positive integers, which is a non-ending set may be written as $Z^+ = \{1,2,3,4,5, ...\}$. The three dots in the end means that the elements continue in the same manner.

(5) The set of prime number is written as $P = \{2,3,5,7,11,13,17,19....\}$

(2) Set-Builder Method: In this method, we first try to find a property which characterizes the elements of a set, that is, a property P, which all the elements of the set possess and which no other objects possess. Then, we describe the set as $\{x : x \text{ has property } P\}$.

This is to be read as "the set of all x such that x has property P".

For example:

(1) The set of all integers can be written as $Z = \{x : x \text{ is an integer}\}$

(2) The set $A = \{1, 2, 3, 4, 5\}$ can be written as $A = \{x \in N : x \leq 5\}$.

(3) The set of complex numbers can be written as $C = \{a+ib : a, b \in R\}$

(4) The set $A = \{1,8,27, ...\}$ can be written as $A = \{x^3 : x \in Z^+\}$.

Solved Examples

Example 1. *Use the Roster method to identify each set:*

    (a) The set of possible integers greater than 8 and less than 14.

    (b) The set of numbers whose elements are the first five positive odd integers.

    (c) The set of even positive integers.

    (d) The set of even positive integers that are divisible by 10.

    (e) The set of all vowels in English alphabets which precedes r.

Solution.   (a) $\{9, 10, 11, 12, 13\}$ (b) $\{1,3,5,7,9\}$

          (c) $\{2, 4, 6, 8, 10 ...\}$ (d) $\{10,20,30,40,50 ...\}$

          (e) $\{a, e, i, o\}$

Example 2. *Use the set-builder method, identify the following sets :*

    (a) $A = \{1,3,5,7,9,...\}$ *(b)* $B = \left\{1, \dfrac{1}{4}, \dfrac{1}{9}, \dfrac{1}{16}, \dfrac{1}{25}, ...\right\}$

    (c) $C = \{0,1, 2, 3, ...\}$ *(d)* $D = \left\{\dfrac{1}{2}, \dfrac{2}{3}, \dfrac{3}{4}, \dfrac{4}{5}, ...\right\}$

Solution.   (a) The set of odd positive integers.

          (b) Here, elements of the set B are the reciprocals of the squares of the natural numbers.

So, the set $B = \left\{\dfrac{1}{n^2} : n \in N\right\}$

          (c) The set of whole numbers.

          (d) Here, each element in the given set has the denominator one more than the numerator. Hence,

$$D = \left\{x : x = \dfrac{n}{n+1} : n \in N\right\}$$

Example 3. *Write the set* $\left\{\dfrac{1}{2}, \dfrac{2}{5}, \dfrac{3}{10}, \dfrac{5}{26}, ...\right\}$ *in the set-builder form.*

Solution. We observe that each element in the given set has the denominator one more than the square of the numerator. Also, the numerator begins with 1. Hence, in the set builder form, the given set can be written as

$$\left\{x : x = \dfrac{n}{n^2 + 1} : n \in \mathbf{N}\right\}$$

EXERCISE 1.1

1. Which of the following collections are sets?

   (i) All mathematics students in your college.

   (ii) All poor hockey players in the college.

   (iii) All odd numbers less than 20.

   (iv) The collection of good teachers in your college.

   (v) All successful and rich people in your city.

   (vi) The people in your immediate family (father, mother, sister, brother).

2. Write the members of each of following sets by the Roster method.

   (i) $\{x : x$ is an odd whole number less than 14$\}$

   (ii) $\{x : x^2 < 36$ and $x \in \mathbf{N}\}$

   (iii) $\{x :$ squares of all whole numbers less than 8$\}$

   (iv) $\{x : x$ is a prime number, $10 < x < 20\}$

   (v) $\{x: x$ is a composite number less than 20$\}$

3. Rewrite the following sets using set-builder method.

   (i) $A = \{2, 4, 6, 8, ...\}$

   (ii) $B = \left\{1, \dfrac{1}{2}, \dfrac{1}{3}, \dfrac{1}{4}, ...\right\}$

   (iii) $C = \{0, 3, 6, 9, 12, ...\}$

   (iv) $D = \{0, 4, 6, 8, 10, ...\}$

4. List the elements of the following sets.

   (i) $A = \{x : x^2 \leq 16 : x \in \mathbf{Z}\}$

   (ii) $B = \{x : 1 \leq x \leq 5$ and $x \in \mathbf{N}\}$

   (iii) $C = \{x : x \in \mathbf{N}$ and x is a proper factor of 15$\}$

   (iv) $D = \{x : x$ is a month of year having 31 days$\}$

5. Use the appropriate symbols $\in$ or $\notin$ to fill in the blanks below:

   (i) 12 ... the set of all numbers dividing 84.

   (ii) K ... the set of all vowels of the English alphabets.

   (iii) $\dfrac{1}{2}$ .. the set of natural number.

   (iv) India ... the set of members of UNO.

   (v) $\sqrt{2}$ .... The set of rational number

   (vi) 15 ... the set of multiples of 3.

Answers

1. (i), (iii), (vi)

2. (i) $\{1, 3, 5, 7, 9, 11, 13\}$ (ii) $\{1, 2, 3, 4, 5\}$ (iii) $\{0, 1, 4, 9, 16, 25, 36, 49\}$
(iv) $\{11, 13, 17, 19\}$ (v) $\{1, 4, 6, 8, 9, 10, 12, 14, 15, 16, 18\}$

3. (i) $A = \{x : x = 2n : n \in \mathbf{N}\}$ (ii) $\{1/n : n \in \mathbf{N}\}$
(iii) $\{x : x = 3n, n$ is the whole number$\}$ (iv) $\{x : x = 2n, n$ is the whole number$\}$

4. (i) $\{-4, -3, -2, -1, 0, 1, 2, 3, 4\}$ (ii) $\{1, 2, 3, 4, 5\}$ (iii) $\{3, 5\}$
(iv) $\{$Jan, March, May, July, August, October, December$\}$

5. (i) $\in$ (ii) $\notin$ (iii) $\notin$ (iv) $\in$ (v) $\notin$ (vi) $\in$

1.5 TYPE OF SETS

(i) **Empty Set:** *A set containing no element is called empty set and is denoted by the symbol* ϕ.

For example:

(1) $\phi = \{x : x$ is a negative integer whose square is $-1\}$

(2) $\phi = \{x : x$ is a natural number lying between 2 and 3$\}$

(3) $\phi = \{$the set of such persons, who never die$\}$

(4) $\phi = \{x : x$ is a real number, $x^2 < 0\}$

(5) $\phi = \{x : x$ is an even prime number greater than five$\}$

(6) $\phi = \{$the set of real numbers which are solution of equation $x^2 + 1 = 0\}$

(7) $\phi = \{x : x$ is a straight ling passing through three distinct points on a circle$\}$

REMARKS

☞ The empty set is also known as null set or void set.

☞ The Roster method, the empty set is denoted by {}.

☞ To describe the null set, we can use any property, which is not true for any element.

☞ It is wrong to use the expression 'an empty' or 'a null set' as there is one and only one empty set through, it may have many-many descriptions. We shall always call 'The empty or the null set.'

☞ A set consisting of at least one element is called a non-empty or non-void set.

☞ $\{\phi\}$ is not a null set.

(ii) **Singleton Set:** *Set containing only one element is a singleton set.* The set {a} is a singleton set.

REMARKS

☞ {0} is not a null set, since it contains 0 as its member. It is a singleton set.

☞ A room containing only one man is not same thing as a man. In a similar way, the singleton set {a} is not the same thing as the element a.

(iii) **Finite Set:** *A set is said to be finite if it consists of only finite number of elements.* Here, the process of counting the different elements comes to an end.

For example:

(1) Set of natural numbers less than 50.

(2) Set of all persons in a city.

(3) Set of English alphabets.

(4) Set of all persons on the earth.

(iv) **Infinite Set:** *A set which is not finite, i.e., it contains infinite number of elements.* Here, process of counting the different elements never comes to an end.

For example:

(1) Set of natural numbers **N** = {1,2,3, ...}

(2) Set of all points of plane.

(3) Set of all even integers.

(4) Set of rational numbers lying between two integers.

(v) **Equal Sets:** *Two sets are said to be equal if they contain exactly the same elements.*

For example: Let

$$A = \{x : x \text{ is a letter in the word 'Area'}\}, \ i.e., \ A = \{a, r, e\}$$

and $\quad B = \{y : y \text{ is a letter in the word 'ear'}\}, \ i.e., \quad B = \{a, r, e\}$

Here A and B are equal sets.

1.5.1 Cardinal Number of a Set

The number of distinct elements contained in a finite set A is called cardinal number of A and is denoted by $n(A)$.

1.5.2 Equivalent Sets

Two finite sets are said to be equivalent if they have the same cardinal number.

REMARKS

☞ Equivalent sets are not always equal but equal sets are always equivalent.
☞ The number of distinct elements in a finite set is also called the order of the set. If the order of a set is zero, the set is empty.
☞ If the order of a set is one, the set is singleton.
☞ The order of an infinite set is never defined.

1.6 SUBSET

Let A and B be two sets. *The set A is said to be a subset of the set B if every element of A is also an element of B.* Symbolically, we write $A \subseteq B$.

When A is subset of B, it means that 'A is contained in B' or 'B contains A'. Here B is called superset of A and is written as $B \supset A$.

REMARKS

☞ Every set is a subset of itself.
☞ Empty set is a subset of every set.
☞ If A is not a subset of B, we write $A \nsubseteq B$.
☞ An element cannot be a subset of a set, only a set can be subset of a set.

1.6.1 Proper Subset

We know that for A to be a subset of B all that is needed is that every element of A is in B. It is possible that every element of B may or may not be in A. If it so happens that every element of B is also in A, then we will have $B \subset A$. Obviously, then A and B are the same set, so that we have $A \subset B$ and $B \subset A \iff A = B$.

If every element of A is in B, but every element of B is not in A, *i.e.*, if $A \subset B$ and $B \not\subset A$, then A is said to be a proper subset of B.

For example:

(1) $\{a, b\}$ is a proper subset of $\{a, b, c\}$.
(2) Set of natural numbers **N** is a proper subset of set **Z** of integers.

REMARKS

☞ Here, it follows that every element of A is an element of B and B contains at least one element which does not belong to A.
☞ If the subset is not proper, it is called **improper subset.** $A \subseteq A$ and $\phi \subset A$ are improper subsets.

1.6.2 Number of Subsets of a Set

If A is a set contains n distinct element such that $0 < r \leq n$. If we consider those subsets of A that have r elements each, then we know that the number of ways in which r elements can be choose out of n elements is ^nC_r. Therefore, the number of subsets of A having r elements each is ^nC_r.

Hence, the total number of subsets of A is equal to

$$^nC_0 + {}^nC_1 + {}^nC_2 + ... + {}^nC_n = (1+1)^n = 2^n$$

For example:

(1) If a set A has one element, then it has $2^1 = 2$ subsets.

(2) If a set A has two elements, then it has $2^2 = 4$ subsets.

REMARKS

☞ The number of proper subsets of a set with n elements is $2^n - 2$.

☞ The collection of all possible subsets of a given set A is called power set. It is denoted by $P(A)$. For example : If A = {1,2,3} then the power set $P(A)$ = { ϕ, {1}, {2}, {3}, {1,2}, {1,3}, {2,3}, {1,2,3}}.

☞ $P(\phi) = \{\phi\}$

☞ The power set of any given set is always non-empty.

1.7 UNIVERSAL SET

In any discussion , we are given particular set and we consider different subsets of the given set. This given set is called Universal Set. It is denoted by U.

For Example:

(1) The universal set is of real numbers **R**, while considering the set of natural numbers, whole numbers, integers and rational numbers.

(2) The set of alphabets is the universal set from which the letters of any word may be chosen to form a set.

(3) In geometry, we discuss set of lines, triangles and circles, then the universal set is the plane, in which the lines, triangles and circles lie.

REMARKS

☞ Universal set is a super set of each of the given sets.

☞ The universal set is not unique.

1.7.1 Complement of a Set

Let U be the universal set and the set $A \subseteq U$. Complement of set A with respect to the universal set U is the set of all those elements of U which are not the elements of A and is denoted by A' or A^c,

i.e., $A' = \{x : x \in U \text{ and } x \notin A\}$

For example:

(i) If U= {1, 2, 3, 4, 5, 6, 7, 8, 9, 11} and A= {1, 2, 3}

then A'= {4, 5, 6, 7, 8, 9, 11}.

REMARKS

☞ Complement of the universal set is the null set and *vice-versa*.

☞ $(A')' = A$

☞ If $A \subseteq B$, then $B' \subseteq A'$.

☞ $x \in A' \Leftrightarrow x \notin A$

Solved Examples

Based on the following Results

▶ A set containing no element is called empty set.

▶ Set containing finite number of elements is called finite otherwise infinite.

▶ The number of distinct elements contained in a finite set is called its cardinality.

▶ Two finite sets are said to be equivalent if they have same cardinality.

▶ A set A is said to be subset of a set B if every elements of A belongs to B.

▶ Total number of subsets of a set A of n elements is 2^n.

▶ $A' = \{x : x \in A' \text{ and } x \notin A\}$

Example 1. *Let $A = \{1,2,3\}$, then find $P(A)$.*

Solution. Since $A = \{1, 2, 3\}$ then,

$$P(A) = \{\phi, \{1\}, \{2\}, \{3\}, \{1,2\}, \{1,3\}, \{2,3\}, \{1,2,3\}\}$$

Example 2. *Let $A = \{a,b,c,d\}$, $B = \{a,b,c\}$ and $C = \{b,d\}$, find all sets X such that*

(i) $X \subset B$ *and* $X \subset C$ (ii) $X \subset A$ *and* $X \not\subset B$

Solution. (i) Here, we have

$$P(B) = \{\phi, \{a\}, \{b\}, \{c\}, \{a, b\}, \{a,c\}, \{b, c\}, \{a, b, c\}\}.$$

And $P(C) = \{\phi, \{b\}, \{d\}, \{b, d\}\}$, then $X \subset B$ and $X \subset C$ implies

$$X \in P(B) \text{ and } X \in P(C)$$

$$X = \{\phi, \{b\}\}$$

(ii) Here, we have, $X \subset A$ and $X \not\subset B$, which implies that

$$X \in P(A) \text{ and } X \notin P(B)$$

Therefore

$$X = \{\{d\}, \{a,b,d\}, \{b,c,d\}, \{a,c,d\}, \{a,d\}, \{b,d\}, \{c,d\}, \{a,b,c,d\}\}$$

Example 3. *Write down all the subsets of the following sets.*

(i) $\{a\}$ (ii) $\{a,b\}$ (iii) $\{a,b,c\}$ (iv) ϕ

Solution. (i) Let $A = \{a\}$. Since A contains only one element, therefore, the total number of subsets is $2^1 = 2$, which are given by ϕ and $\{a\}$.

(ii) Here, total number of subsets, $= 2^2 = 4$, which are given by ϕ, $\{a\}, \{b\}, \{a, b\}$

(iii) Here, total number of subsets $= 2^3 = 8$, given by

$$\phi, \{a\}, \{b\}, \{c\}, \{a,b\}, \{a,c\}, \{b,c\}, \{a,b,c\}$$

(iv) since ϕ contains no element therefore the number of subsets $= 2^0 = 1$. The only subset is ϕ.

Example 4. *Which of the following sets are empty. Also, give the reason.*

(i) $A = \{x : x \neq x,$ *is a real number*$\}$.

(ii) $B = \{x : x + 4 = 4\}$

(iii) $C = \{x : x^3 - 3 = 0$ *and x is rational number*$\}$

Solution. (i) Here, $A = \{x : x \neq x, x$ is a real number$\}$. Since $x \neq x$ is not true

$\Rightarrow \quad A = \phi$

(ii) $B = \{x : x + 4 = 4\} = \{x : x = 0\} = \{0\}$

$\Rightarrow \quad B$ has one element 0, therefore $B \neq \phi$.

(iii) Since there is no rational number whose square is 3, so $x^3 - 3 = 0$ is not satisfied for any rational numbers. Therefore, C is an empty set.

Example 5. *Which of the following sets are finite and which are infinite.*

(i) *The set of natural numbers divisible by 2.*

(ii) *The set of natural numbers less then 8.*

(iii) *The set of integers whose square is even.*

(iv) *The set of integers greater than* −18.

(v) *The set of lines passing through a point.*

(vi) *The set of points of a plane at a fixed distance from a given point in the plane.*

(vii) *The set of points common to two given parallel lines.*

(viii) *The set of the roots of a polynomial of nth degree.*

Solution. (i) The given set is $\{2, 4, 6, 8, ...\}$. It has an infinite number of elements, therefore it is an infinite set.

(ii) The given set is $\{1,2,3,4,5,6,7\}$. It has seven elements, *i.e.*, finite number of elements. Hence, it is a finite set.

(iii) The given set is $\{..., -8, -6, -4, -2, 0, 2, 4, 6, 8,...\}$. It has infinite number of elements, therefore it is an infinite set.

(iv) Here, the given set is $\{-17, -16, ..., 0, 1, 2 ...\}$. It has infinite number of elements therefore, it is an infinite set.

(v) Since infinite number of lines can pass throught a fixed point, therefore the given set is an infinite set.

(vi) Since the points in a plane at a fixed distance from a given point in the plane lie on a circle with the given point as center and the number of points on a circle is infinite. Therefore, the given set is an infinite set.

(vii) Since two parallel lines cannot meet anywhere, therefore, the set of points common to two given parallel lines is empty, therefore the given set cannot be infinite. Hence, it is a finite set.

(viii) Since, a polynomial of nth degree always have atmost n roots.

Therefore, the given set is always a finite set.

Example 6. *Which of the following sets are equivalent* ϕ ,$\{0\}$ *and* $\{\phi\}$.

Solution. Since ϕ has no element. Also, $\{0\}$ and $\{\phi\}$, each contains one element namely 0 and ϕ respectively. Hence, $\{0\}$ and $\{\phi\}$ are equivalent.

Example 7. *Which of the following sets are equal ?*

$$A = \{1,2,3\}, B = \{2,3,4\}, C = \{3,2,1\}, D = \{2,3,5\}$$

Solution.   Since $1 \in A$ but $1 \notin B$, therefore $A \neq B$. A and C have exactly the same element, therefore $A = C$.

Also,

$$1 \in C \quad \text{but} \quad 1 \notin D \quad \Rightarrow \quad C \neq D$$
$$4 \in B \quad \text{but} \quad 4 \notin C \quad \Rightarrow \quad B \neq C$$
$$4 \in B \quad \text{but} \quad 4 \notin C \quad \Rightarrow \quad B \neq C$$
$$1 \in A \quad \text{but} \quad 1 \notin D \quad \Rightarrow \quad A \neq D$$

Hence, only A and C are equal sets.

EXERCISE 1.2

1. Fill in the blanks:
   (i) A set which contains no element is called ... set.
   (ii) If $A = \{1,2,3\}$ and $B = \{3,2,1\}$ then they are said to be ...
   (iii) If $A = \{a, b, c\}$ and $B = \{c, d, e\}$ then they are said to be ...
   (iv) If every element of a set B is also an element of A, then B is said to be ... of A.
   (v) The empty set is a ... of every set.
   (vi) Every set is a .... of itself.
   (vii) The set **Z** of integers is a ... of set of natural numbers **N**.

2. Which of the followings sets are equal?
   (i) $A = \{1,2,3\}$
   (ii) $B = \{1,2,2,3\}$
   (iii) $C = (x \in \mathbf{R} : x^3 - 6x^2 + 11x - 6 = 0)$

3. Which of the following sets are equivalent to the set $\{4,7,11,17,20\}$?
   (i) $\{5,1,2,3,4\}$
   (ii) {all odd numbers less then 10}
   (iii) {the months of a year of 30 days}
   (iv) {all the prime numbers which lie between 10 and 25}.

4. Which of the following sets are finite and which are infinite ?
   (i) $\{x \in \mathbf{N} : x > 10\}$
   (ii) $\{x \in \mathbf{N} : x < 100\}$
   (iii) $\{x \in \mathbf{R} : 1 \leq x \leq 2\}$
   (iv) Set of vowels in English alphabets.
   (v) The set of prime numbers less than 100.
   (vi) The set of multiple of 8.

5. Which of the following statements are true? Give the reason.
   (i) For any two sets A and B either $A \subseteq B$ or $B \subseteq A$
   (ii) Every subset of a finite set is finite.
   (iii) A subset of an infinite set may be finite.
   (iv) Every set has a proper subset.
   (v) A set containing n elements have 2^n subsets.
   (vi) If $A = \{1,2,3,4,5,6\}$ and $B = \{$whole numbers less than 6$\}$, then $A = B$.
   (vii) The empty set has no proper subset.

6. Examine which of the following sets are empty?
   (i) The set of tigers in your class.
   (ii) The set of triangles having three equal sides.
   (iii) The set of all numbers which, when added to zero, yield sum greater than the original.
   (iv) The set of odd numbers which are divisible by 2.
   (v) The set of men, who never die.

7. Which of the following statements are true?
   (i) If $x \in A$ and $A \subset B$, then $x \in B$
   (ii) If $A \subset B$ and $B \subset C$, then $A \subset C$
   (iii) If $A \not\subset B$ and $B \not\subset C$, then $A \not\subset C$
   (iv) If $x \in A$ and $A \not\subset B$, then $x \in B$
   (v) If $A \subset B$ and $x \notin B$, then $x \notin A$

8. Are the following sets, *i.e.*, (A and B) are equal.
   (i) $A = \{x : x$ is a letter of the word 'LITTLE'$\}$
       $B = \{x : x$ is a letter in the word 'TITLE'$\}$
   (ii) $A = \{x : x$ is a letter in the word 'FOLLOW'$\}$
       $B = \{x : x$ is a letter in the word 'WOLF'$\}$
   (iii) $A = \{x : x$ is a letter in the word 'LOYAL$\}$
       $B = \{x : x$ is a letter In the word 'ALLOY'$\}$

9. Write down all possible subsets of each of the following sets.
   (i) $\{a\}$ (ii) $\{0,1\}$ (iii) $\{a, b, c\}$

(iv) {1, {1}}　　　(v) φ

(ii) If $A \subseteq B$ and $B \subseteq C$, then $A \subseteq C$

10. Which of the following statements are true?

(iii) If $a \in B$ and $B \subseteq C$, then $a \in C$

(i) {a, ϕ} $\in$ {$a, \{a, \phi\}$}

Answers

1. (i) Empty (ii) equal (iii) equivalent (iv) subset (v) subset
(vi) subset (vii) super set.

2. $A = B = C$　　　　**3.** (i), (ii), (iv)

4. (ii), (iv), (v) are finite sets and (i), (iii), (vi) are infinite.

5. (i) F　(ii) T　(iii) T　(iv) F　(v) T　(vi) F　(vii) T

6. (i), (iii), (iv), (v)

7. (i), (ii), (v)

8. (i) Equal, (ii) Equal, (iii) Equal

9. (i) φ , {a}; (ii) φ ,{0},{1},{0,1}; (iii) φ ,{a},{b},{c},
{a, b},{b,c},{a,c}, {a,b,c}　　　(iv) {1}; {1}, {{1}}, {1,{1}};

10. all are true

1.8 VENN DIAGRAMS

A set can be represented by closed figures like circles, triangles, rectangles, etc. The point in the interior of the figure represents the elements of the set. Such a representations is called a Venn diagram. In Venn diagram, the universal set is usually represented by a rectangular region and its subset by closed bounded regions inside the rectangular region. For example, if A is a subset of B, i.e., $A \subset B$. It's Venn diagram is shown in Fig. (2).

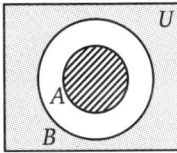

Fig. (2)

REMARKS

☞ The diagrams drawn to represent sets are called Venn diagram or Venn-Euler diagrams, after the name of British mathematician **Venn.**

☞ If A and B are two sets, which are not equal, but have common elements, then to represent A and B, we draw two intersecting circles.

☞ Two disjoint sets are represented by two non-intersecting circles.

☞ Venn diagrams are to be used for clarity and are no substitute for precise proof.

1.9 OPERATIONS ON SETS

1.9.1 Union and Intersection Operations

(i) Union of Two sets

Let A and B be two sets. Then Union of A and B, denoted by $A \cup B$ is the set of all those elements, which either belongs to A or B or to both A and B.

It should be noted that the common elements are to be taken only once.

Symbolically: $A \cup B = \{x : x \in A$ or $x \in B\}$ It is shown in the adjoining figure 3.

For example:

(1) Let $A = \{3,4,5,6,7\}$ and $B= \{5,6,7,8,9\}$
Then $A \cup B = \{3, 4, 5, 6, 7, 8, 9\}$

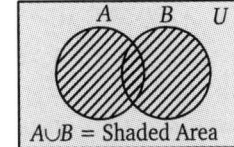

$A \cup B$ = Shaded Area

Fig. (3)

(ii) Let $A = \{x : x = 2n, n = 1, 2, 3, ...\} = \{2, 4, 6, 8, ...\}$

$B = \{x : x = 3n, n = 1, 2, 3, ...\} = \{3, 6, 9, 12, ...\}$

Then $A \cup B = \{x : x$ is multiple of 2 or a multiple of 3$\}$
$= \{2, 3, 4, 6, 8, 9, 10, 12, ...\}$

(iii) Let A = set of even natural numbers = $\{2, 4, 6, 8,...\}$

and B = set of natural numbers = $\{1, 2, 3, 4, 5,...\}$

Then $A \cup B = \{1, 2, 3, 4,...\}$

REMARKS

☞ $x \in (A \cup B) \Leftrightarrow x \in A$ or $x \in B$.

☞ $x \notin (A \cup B) \Leftrightarrow x \notin A$ and $x \notin B$

☞ $A \cup B = B \cup A$, *i.e.*, union of sets is commutative.

☞ $A \cup A' = U$ and $A \cup U = U$

☞ $A \cup \phi = A$

☞ If $A,B,C,D, ...,Z$ is a finite family of sets, then their union is denoted by $A \cup B \cup C \cup D... \cup Z$.

☞ $(A \cup B) \cup C = A \cup (B \cup C)$, *i.e.*, a union of sets is associative.

(ii) Intersection of Two sets

Let A and B be two sets. Then intersection of A and B, denoted by $A \cap B$ is the set of all those elements, which belongs to both A and B.

Symbolically: $A \cap B = \{x : x \in A$ and $x \in B \}$ It is shown in the adjoining figure 4.

For example:

(1) Let $A = \{2, 4, 6, 8, 10\}$ and $B = \{1, 2, 3, 4, 5\}$

Then $A \cap B = \{2, 4\}$

(2) If $A = \{x : x = 3n, n \in \mathbf{Z}\}$

$B = \{x : x = 4n, n \in \mathbf{Z}\}$

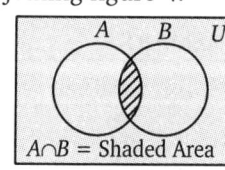

$A \cap B$ = Shaded Area

Fig. (4)

Then $A \cap B = \{x : x$ is multiple of 3 and x is a multiple of 4$\}$
$= \{x : x$ is multiple of 3 and 4 both)
$= \{x : x = 12n, n \in \mathbf{Z}\}$

REMARKS

☞ $x \in (A \cap B) \Leftrightarrow x \in A$ and $x \in B$.

☞ $x \notin (A \cap B) \Leftrightarrow x \notin A$ or $x \notin B$

☞ $A = A \cap A$, *i.e.*, intersection of sets is idempotent.

☞ $A \cap \phi = \phi$

☞ $A \cap U = A$, where U is a universal set.

☞ $A \cap B = B \cap A$, *i.e.*, intersection of sets is commutative.

☞ $(A \cap B) \cap C = A \cap (B \cap C)$ intersection of sets is associative.

☞ If $A,B,C,D, ...,Z$ is a finite family of sets, then their intersection is denoted by $A \cap B \cap C... \cap Z$.

(iii) Distributive Property of Union and Intersection

(i) $A \cup (B \cap C) = (A \cup B) \cap (A \cup C)$ (ii) $A \cap (B \cup C) = (A \cap B) \cup (A \cap C)$

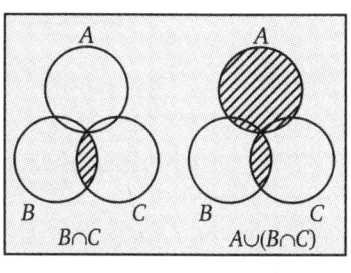
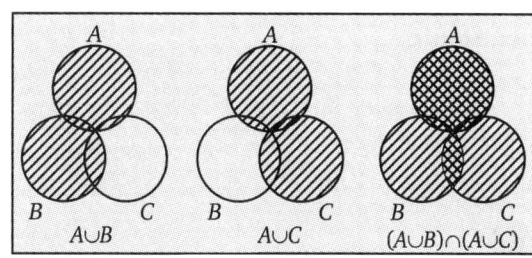

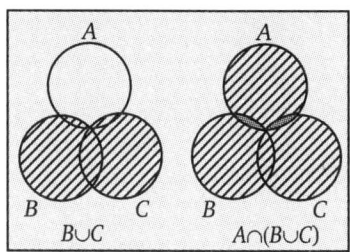
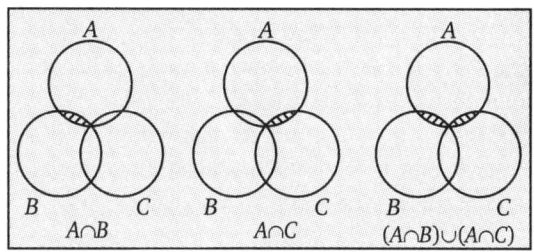

Fig. (5)

1.9.2. Disjoint Sets

When two sets have no common elements, they are called disjoint sets. Thus, if $A \cap B = \phi$, then A and B are disjoint. It is shown in the adjoining figure 6.

For example:

(i) If $A = \{2, 4, 6, 8\}$ and $B = \{1, 3, 5, 7, 9\}$

Then, $A \cap B = \phi$

(ii) If $\qquad A$ = Boys in school

$\qquad\qquad B$ = Girls in school

Then, $A \cap B = \phi$

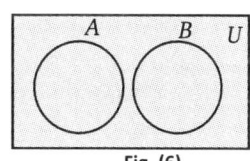

Fig. (6)

REMARKS

☞ If $A \cap B \neq \phi$, , then A and B are said to be intersecting or overlapping sets.

☞ A family of sets is said to be pairwise disjoint family of sets if and only if any two sets of this family are disjoint. For example, classes of A_2, A_3, A_5 and A_7 defined as

$A_2 = \{2, 2^2, 2^3, ...\}$; $A_3 = \{3, 3^2, 3^3, ...\}$; $A_5 = \{5, 5^2, 5^3, ...\}$ and $A_7 = \{7, 7^2, 7^3, ...\}$ are pairwise disjoint.

☞ $\phi \cap A = \phi$, *i.e.*, null set is disjoint from every subset.

1.9.3 Difference of Two Sets

If A and B are two sets, then the set of all elements which belong to A but do not belong to B is called the difference of sets A and B and is denoted by $A \sim B$. The set of all elements which belong to B but do not belong to A is called the difference of sets B and A and is denoted by $B \sim A$.

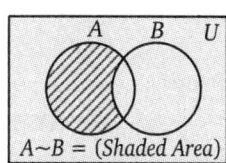

$A \sim B$ = (Shaded Area)

Fig. (7)

Therefore,

$A \sim B = \{x : x \in A \ and \ x \notin B\} = A \cap B'$

And $B \sim A = \{x : x \notin A \ and \ x \in B\} = B \cap A'$

For example:

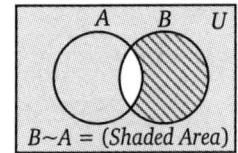

Let $A = \{1, 2, 3, 4, 5\}$

And $B = \{-1, 0, 1, 2\}$

Then, $A \sim B = \{3, 4, 5\}$

And $B \sim A = \{-1, 0\}$

$B \sim A = (Shaded\ Area)$

Fig. (8)

REMARKS

☞ $x \in (A - B) \Leftrightarrow x \in A$ and $x \notin B$.

☞ $x \notin (A - B) \Leftrightarrow x \notin A$ and $x \in B$

☞ $A - B \neq B \sim A$, i.e., , difference of two sets is not commutative.

☞ $A \subset B$ then $A \sim B = \phi$

☞ The sets $A \sim B$, $A \cap B$ and $B \sim A$ are mutually disjoint.

☞ Difference of a set with the universal set is known as complementation.

☞ $A \sim B$ is a subset of A and $B \sim A$ is a subset of B.

1.9.4 Symmetric Difference of Two Sets

If A and B are two sets, then the symmetric difference of two sets A and B is denoted by $A\Delta B$ is given by $A\Delta B = (A \sim B) \cup (B \sim A)$

Symbolically:

$A \Delta B = \{x : (x \in A \ and \ x \notin B) \ or \ (x \in B \ and \ x \notin A)\}$

For example:

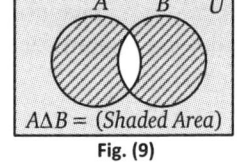

$A\Delta B = (Shaded\ Area)$

Fig. (9)

(i) If $A = \{1,2,3,4,5,6,7,8\}$ and $B = \{1,3,5,6,7,8,9\}$

Then $A \sim B = \{2, 4\}$ and $B \sim A = \{9\}$

and $A \Delta B = \{2, 4, 9\}$

Equivalent Sets: Two finite sets A and B are said to be equivalent if their cardinal numbers are same , i.e., $n(A) = n(B)$.

1.9.5 Law of Excluded Middle and Law of Contradiction

Two special properties of set operations are known as the excluded middle axioms and law of contradiction. The excluded middle axioms are very important because they are the only set operations described here that are not valid for both classical sets and fuzzy sets. Let A be any subset of universal set X. Then , we define.

(i)   Axiom of the excluded middle: $A \cup A' = U$

(ii) Axiom of the contradiction: $A \cap A' = \phi$

Theorem 1.

    (i) $A \cup \phi = A$ (ii) $A \cap \phi = \phi$ (iii) $A \cup A = A$

    (iv) $A \cap A = A$ (v) $A \cup B = B \cup A$ (vi) $A \cap B = B \cap A$

Proof.

    (i) Let x be an arbitrary element of $A \cup \phi$.

      i.e., $x \in A \cup \phi$

Then, by definition $x \in A \cup B \Leftrightarrow x \in A$ or $x \in B$

i.e., $x \in A \cup \phi$ $\Rightarrow x \in A$ or $x \in \phi$

$\Leftrightarrow$ $x \in A$ $(\because \phi$ is a null set $\Rightarrow x \notin \phi)$

Therefore, $A \cup \phi = A$

(ii) Let x be an arbitrary element of $A \cap \phi$.

     $x \in A \cap \phi \Leftrightarrow x \in A$ and $x \in \phi$ $(\because \phi$ is a null set$)$

     Therefore, $A \cap \phi = \phi$

(iii) Let x be an arbitrary element of $A \cup A$,

     $x \in A \cup A \Leftrightarrow x \in A$ or $x \in A$ (Repeated statement)

            $\Leftrightarrow x \in A$

     Therefore, $A \cup A = A$

(iv) Let x be an arbitrary element of $A \cap A$,

     $x \in A \cap A$ $\Leftrightarrow x \in A$ and $x \in A$ (Repeated statement)

            $\Leftrightarrow x \in A$

     Therefore, $A \cap A = A$

(v) Let x be an arbitrary element of $A \cup B$,

     $x \in A \cup B$ $\Leftrightarrow x \in A$ or $x \in A$ (Writing in reverse order)

     $\Leftrightarrow x \in B$ or $x \in A \Leftrightarrow x \in B \cup A$

     Therefore, $A \cup B = B \cup A$

(vi) Let x be an arbitrary element of $A \cap B$

     $x \in A \cap B$ $\Leftrightarrow x \in A$ and $x \in A$ (Writing in reverse order)

     $\Leftrightarrow x \in B$ and $x \in A \Leftrightarrow x \in B \cap A$

     Therefore, $A \cap B = B \cap A$

Theorem 2. *For any three sets A, B and C*

     (i) $A \cup (B \cup C) = (A \cup B) \cup C$ (ii) $A \cap (B \cap C) = (A \cap B) \cap C$

     (iii) $A \cup (B \cap C) = (A \cup B) \cap (A \cup C)$ (iv) $A \cap (B \cup C) = (A \cap B) \cup (A \cap C)$

Proof.

(i) Let x be an arbitrary element of $A \cup (B \cup C)$, then

         $x \in A \cup (B \cup C)$

   $\Leftrightarrow$ $x \in A$ or $x \in (B \cup C) \Leftrightarrow x \in A$ or $(x \in B$ or $x \in C)$

   $\Leftrightarrow$ $(x \in A$ or $x \in B)$ or $x \in C$ (By associativity)

   $\Leftrightarrow$ $x \in (A \cup B)$ or $x \in C \Leftrightarrow x \in A \cup (B \cup C)$

     Therefore, $A \cup (B \cup C) = (A \cup B) \cup C$

(ii) Let x be an arbitrary element of $A \cap (B \cap C)$, then

         $x \in A \cap (B \cap C)$

   $\Leftrightarrow$ $x \in A$ and $x \in (B \cap C) \Leftrightarrow x \in A$ and $(x \in B$ and $x \in C)$

   $\Leftrightarrow$ $(x \in A$ and $x \in B)$ and $x \in C$ (By associativity)

$\Leftrightarrow \quad x \in (A \cap B) \text{ and } x \in C \Leftrightarrow x \in (A \cap B) \cap C$

Therefore, $A \cap (B \cap C) = (A \cap B) \cap C$

(iii) Let x be an arbitrary element of $A \cup (B \cap C)$, then

$$x \in A \cup (B \cap C)$$

$\Leftrightarrow \quad x \in A \text{ or } x \in (B \cap C) \Leftrightarrow x \in A \text{ or } (x \in B \text{ and } x \in C)$

$\Leftrightarrow \quad (x \in A \text{ or } x \in B) \text{ and } (x \in A \text{ or } x \in C) \Leftrightarrow x \in (A \cup B) \text{ and } x \in (A \cup C)$

$\Leftrightarrow \quad x \in (A \cup B) \cap (A \cup C)$

Therefore, $A \cup (B \cap C) = (A \cup C) \cap (A \cup C)$

(iv) Let x be an arbitrary element of $A \cap (B \cup C)$, then

$$x \in A \cap (B \cup C)$$

$\Leftrightarrow \quad x \in A \text{ and } x \in (B \cup C) \Leftrightarrow x \in A \text{ and } (x \in B \text{ or } x \in C)$

$\Leftrightarrow \quad (x \in A \text{ and } x \in B) \text{ or } (x \in A \text{ and } x \in C) \Leftrightarrow x \in (A \cap B) \text{ or } x \in (A \cap C)$

Therefore, $A \cap (B \cup C) = (A \cap B) \cup (A \cap C)$

Theorem 3.

(i) $(A')' = A$ 　　　　　　(ii) $A \cup A' = U$, *where U is the universal set.*

(iii) $A \cap A' = \phi$ 　　　　　(iv) $(A \cup B)' = A' \cap B'$ *(De' Morgan's Law)*

(v) $(A \cap B)' = A' \cup B'$ *(De' Morgan's Law)*

Proof.

(i) Let x be an arbitrary element of $(A')'$,

$$x \in (A')' \Leftrightarrow x \notin A' \Leftrightarrow x \in A$$

Therefore, 　　　　$(A')' = A$

(ii) Let x be an arbitrary element of $(A \cup A')$,

$$x \in (A \cup A') \qquad\qquad \Leftrightarrow x \in A \text{ or } x \in A' \Leftrightarrow x \in A \text{ or } x \in U - A$$

$\Leftrightarrow \quad x \in A \text{ or } (x \in U, x \notin A) \Leftrightarrow x \in U$

Therefore, 　　　　$A \cup A' = U$

(iii) Let x be an arbitrary element of $(A \cap A')$,

$$x \in (A \cap A') \qquad\qquad \Leftrightarrow x \in A \text{ and } x \in A' \text{ but if } x \in A \text{ then } x \notin A'$$

Therefore, 　　　$A \cap A' = \phi$

(iv) Let x be an arbitrary element of $(A \cup B)'$,

$$x \in (A \cup B)' \qquad\qquad \Leftrightarrow x \notin (A \cup B) \qquad \Leftrightarrow x \notin A \text{ and } x \notin B$$

$\Leftrightarrow \quad x \in A' \text{ and } x \in B' \Leftrightarrow x \in A' \cap B'$

Therefore, 　　　$(A \cup B)' = A' \cap B'$

(v) Let x be an arbitrary element of $(A \cap B)'$,

$$x \in (A \cap B)' \qquad\qquad \Leftrightarrow x \notin (A \cap B) \qquad \Leftrightarrow x \notin A \text{ or } x \notin B$$

$\Leftrightarrow \quad x \in A' \text{ or } x \in B' \Leftrightarrow x \in A' \cup B'$

Therefore, 　　　$(A \cap B)' = A' \cup B'$

Solved Examples

Based on the following Results

- ▶ $A \cup B = \{x : x \in A \text{ or } x \in B\}$
- ▶ $x \notin A \cap B \Leftrightarrow x \notin A \text{ or } x \notin B$
- ▶ $x \in A - B \Leftrightarrow x \in A \text{ and } x \notin B$
- ▶ $A \cup \phi = A$
- ▶ $(A')' = A$
- ▶ $A \cap A' = \phi$
- ▶ $(A \cap B)' = A' \cup B'$

- ▶ $A \cap B = \{x : x \in A \text{ and } x \in B\}$
- ▶ $x \notin A \cup B \Leftrightarrow x \notin A \text{ and } x \notin B$
- ▶ $A \triangle B = (A \sim B) \cup (B \sim A)$
- ▶ $A \cap \phi = \phi$
- ▶ $A \cup A' = U$
- ▶ $(A \cup B)' = (A' \cap B')$

Example 1. *Show that (i) $A \subset (A \cup B)$, (ii) $(A \cap B) \subset A$.*

Solution :   (i) Let $x \in A$ be arbitrary then $x \in A$ certainly but may or may not belong to B.

$$\Rightarrow \qquad x \in A \cup B \qquad \Rightarrow \qquad A \subset A \cup B$$

(ii) Let $\qquad x \in A \cap B \qquad$ where x is arbitrary

$$x \in A \cap B \qquad \Rightarrow \qquad x \in A \text{ and } x \in B$$

In particular, $\qquad x \in A \cap B \qquad \Rightarrow \qquad x \in A$

Therefore, $\qquad (A \cap B) \subset A$

REMARK

☞ Similarly we can show that (i) $B \subset (A \cup B)$ and (ii) $A \cap B \subset B$.

Example 2. *Let A and B be two sets, if $A \cap X = B \cap X = \phi$ and $A \cup X = B \cup X$ for some set X, prove that A = B.*

Solution.   Given that $A \cup X = B \cup X$

$$\Rightarrow \quad A \cap (A \cup X) = A \cap (B \cup X) \qquad \text{(taking intersection by } A \text{ on both sides)}$$

$$\Rightarrow \quad A = A \cap (B \cup X) \qquad\qquad (\because A \cap (A \cup X) = A)$$

$$\Rightarrow \quad A = (A \cap B) \cup (A \cap X) \qquad \text{(By distributive law)}$$

$$\Rightarrow \quad A = (A \cap B) \cup \phi \quad \Rightarrow \quad A = A \cap B$$

$$\Rightarrow \quad A \subset (A \cap B) \qquad \Rightarrow \quad A \subseteq B \qquad\qquad\qquad ...(1)$$

Again consider, $A \cup X = B \cup X$

$$\Rightarrow \quad B \cap (A \cup X) = B \cap (B \cup X) \qquad \text{(taking intersection with } B)$$

$$\Rightarrow \quad B \cap (A \cup X) = B$$

$$\Rightarrow \quad (B \cap A) \cup (B \cap X) = B \qquad\qquad \text{(By distributive law)}$$

$$\Rightarrow \quad (B \cap A) \cup \phi = B \qquad\qquad\qquad \text{(Given } B \cap X = \phi)$$

$$\Rightarrow \quad (B \cap A) = B \qquad\qquad\qquad\qquad (\because A \cap B = B \cap A)$$

$$\Rightarrow \quad A \cap B = B \qquad \Rightarrow \quad B \subset A \cap B \quad \Rightarrow \quad B \subseteq A \qquad ...(2)$$

Hence, (1) and (2) gives $A \subseteq B$ and $B \subseteq A$.

$$\Rightarrow \qquad A = B$$

Example 3. *For any two sets A and B, show that*

(i) $P (A \cap B) = P(A) \cap P(B)$, (ii) $P(A) \cup P(B) \subset P (A \cup B)$

Solution :   (i) Let $X \in P(A \cap B)$ $\Rightarrow X \subseteq A \cap B$

$\Rightarrow$ $X \subseteq A$ and $X \subseteq B \Rightarrow X \in P(A)$ and $X \in P(B)$

$\Rightarrow$ $X \in P(A) \cap P(B)$

Therefore, $P(A \cap B) \subset P(A) \cap P(B)$ ...(1)

Now, let $X \in P(A) \cap P(B) \Rightarrow X \in P(A)$ and $X \in P(B)$

$\Rightarrow$ $X \subseteq A$ and $X \subseteq B \Rightarrow X \subseteq A \cap B$

$\Rightarrow$ $X \in P(A \cap B)$

Therefore, $P(A) \cap P(B) \subset P(A \cap B)$ ...(2)

From (1) and (2), we conclude that

$$P(A \cap B) = P(A) \cap P(B)$$

(ii) Let $X \in P(A) \cup P(B)$ $\Rightarrow X \in P(A)$ or $X \in P(B)$

$\Rightarrow$ $X \subseteq A$ or $X \subseteq B$ $\Rightarrow X \subseteq A \cup B$

$\Rightarrow$ $X \in P(A \cup B)$

Therefore, $P(A) \cup P(B) \subset P(A \cup B)$

REMARK

☞ Converse of the result (ii) is not necessarily true. For example, let $A = \{1,2\}$ and $B = \{3,5,6\}$, then we find that $X = \{1,2,3,5\}$ which is a subset of $A \cup B$. Therefore, $X \in P(A \cup B)$. But $X \notin P(A), X \notin P(B)$. So,

$$X \notin P(A) \cup P(B) \Rightarrow P(A \cup B) \not\subset P(A) \cup P(B)$$

1.9.6 Some More Results

1. If A and B are any two sets, then

(i) $A - B = A \cap B'$

(ii) $A - B = A \Leftrightarrow A \cap B = \phi$

(ii) $(A - B) \cup B = A \cup B$

(iv) $A \subset B \Leftrightarrow B' \subset A'$

(v) $(A - B) \cup (B - A) = (A \cup B) - (A \cap B)$

2. If A and B are any two sets, then

(i) $A - (B \cap C) = (A - B) \cup (A - C)$

(ii) $A - (B \cup C) = (A - B) \cap (A - C)$

(ii) $A \cap (B - C) = (A \cap B) - (A \cap C)$

EXERCISE 1.3

1. Let $A = \{a, b\}$, $B = \{a, b, c\}$. Is $A \subset B$? Find $A \cup B$ and $A \cap B$.

2. If $A = \{1,2,3,4\}$, $B = \{2,4,6,8\}$, $C = \{3,4,5,6\}$ and universal set $U = \{1,2,3,4,...9\}$. Verify that $A \cap (B \cup C) = (A \cap B) \cup (A \cap C)$.

3. If A, B, C are subsets of a set X, then show that $A \subseteq B$ and $B \subseteq C \Rightarrow A \subseteq C$.

4. Find the union of the following sets:

(i) $A = \{x : x$ is an even integer$\}$,
   $B = \{x : x$ is an odd integer$\}$.

(ii) $A = \{x : x$ is a multiple of 2$\}$,
   $B = \{x : x$ is a multiple of 3$\}$.

(iii) $A = \{x : x$ is a rational number $\}$,
   $B = \{x : x$ is an irrational number$\}$.

(iv) $A = \{x : x$ is a negative integer$\}$,
   $B = \{x : x$ is a non- negative integer$\}$

5. Find the intersection of the following sets.

(i) $A = \{x : x$ is an even integer$\}$,
   $B = \{x : x$ is an odd integer$\}$

(ii) $A = \{x : x$ is a rational number $\}$,
   $B = \{x : x$ is an irrational number$\}$.

(iii) $A = \{x : x$ is a multiple of 5$\}$,
   $B = \{x : x$ is a multiple of 2$\}$

(iv) $A = \{x : x$ is a rational number $\}$,
   $B = \{x : x$ is a real number$\}$

6. If $A = \{1,2,3,4\}$, $B = \{2,4,6,8\}$ and $C = \{3,4,5,6\}$, find

(i) $(A \cup B) \cap C$

(ii) $A \cup (B \cap C)$

7. Write T for true and F for false statement.

(i) $A \subset (A \cup B)$ (T/F)

(ii) $(A \cup B) \subset B$ (T/F)

(iii) $(A \cap B) \subset A$ (T/F)

(iv) $A \cup A = A$ and $A \cap A = A$ (T/F)

(v) If $A \cap B = \phi$, then $A \cap \phi = B$ (T/F)

(vii) If A and B are disjoint sets, then intersection of their union and intersection is the null set.   (T/F)

(viii) If A is the proper subset of U, then the union of A and A' is U.   (T/F)

(ix) $U' = \phi$ and $\phi' = U$ (T/F)

(x) $(A \cup B)' = A' \cap B'$ (T/F)

(xi) $A \cap A'$ is always empty   (T/F)

(xii) $(A \cap B)' = A' \cup B'$ (T/F)

8. If $A = \{1, 2, 3, 4, 5, 6, 7, 8\}$ and $B = \{1, 3, 5, 6, 7, 8, 9\}$, then show that
$$A \, \Delta \, B = \{2, 4, 9\}$$

9. Let $A = \{x : x \in \mathbf{N}\}$,

$B = \{x : x = 2n : n \in \mathbf{N}\}$,

$C = \{x : x = 2n{-}1 : n \in \mathbf{N}\}$

and $D = \{x : x$ is a prime natural number$\}$. Find

(i) $A \cap B$ (ii) $A \cap C$

(iii) $A \cap D$ (iv) $B \cap C$

(v) $B \cap D$ (vi) $C \cap D$

10. For any two sets A and B, prove that $P(A) = P(B)$ implies that $A = B$

11. For any two sets A and B, show that

(i) $A \cup (A \cap B) = A$

(ii) $A \cap (A \cup B) = A$

(iii) $(A \cup B) \cap (A \cap B') = A$

(iv) $A' \cup B = U \Rightarrow A \subset B$

(v) $A \subset B \Leftrightarrow B' \subset A'$

(vi) $B \subset A \Leftrightarrow A \cap B = B$

12. Let $A = \{1, 2, 3, 4\}$, $B = \{2, 3, 4, 5\}$ and $C = \{4, 5, 6, 7\}$. Verify that

(i) $A \cup (B \cap C) = (A \cup B) \cap (A \cup C)$

(ii) $A \cap (B \cup C) = (A \cap B) \cup (A \cap C)$

(iii) $A \cap (B - C) = (A \cap B) - (A \cap C)$

(iv) $A - (B \cup C) = (A - B) \cap (A - C)$

(v) $A - (B \cap C) = (A - B) \cup (A - C)$

13. Show that

(i) If a sets has only even element, then it has 2 subsets.

(ii) If $B \subset A$ and B has one element less than that of A, show that A has twice as many subset as B has.

(iii) A set with 2 element has 2^2 subsets, a set with 3 elements has 2^3 subsets and so on.

14. If $X = \{4^n - 3n - 1 : n \in \mathbf{N}\}$ and $Y = \{9(n-1) : n \in \mathbf{N}\}$, show that $X \subset Y$.

15. Show that $A - B$, $A \cap B$ and $B - A$ are pairwise disjoint.

16. Show that $A \cup B \subseteq A \cap B$ implies that $A = B$.

═══════════════ **Answers** ═══════════════

1. (i) Yes. $\{a, b, c\}$, $\{a, b\}$;

4. (i) $A \cup B = \{x : x$ is non-zero integer$\}$ (ii) $A \cup B = \{x : x$ is a multiple of 2 or 3$\}$

(iii) $A \cup B = \{x : x$ is a real number$\}$ (iv) $A \cup B = \{x : x$ is an integer$\}$

5. (i) ϕ (ii) ϕ (iii) ϕ (iv) $\{x : x$ is a rational number$\}$

6. (i) $\{3, 4, 6\}$,   (ii) $\{1, 2, 3, 4, 6\}$

7. (i) T   (ii) F   (iii) T   (iv) T   (v) F   (vi) T   (vii) T

(viii) T   (ix) T   (x) T   (ix) T   (x) T   (xi) T   (xii) T

9. (i) B   (ii) C   (iii) D   (iv) ϕ (v) 2   (vi) $D - \{2\}$

1.10 SOME RESULTS ON VENN DIAGRAMS

If A is a finite set, then $n(A) = $ No. of elements in the set A.

The following results may be remembered for direct application :

(i) $n(A \cup B) = n(A) + n(B) - n(A \cap B)$

(ii) $n(A \cup B) = n(A) + n(B)$, provided A and B are disjoint, i.e., if $n(A \cap B) = 0$

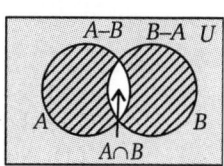

Fig. (10)

(iii)　$n(A \cap B') = n(A) - n(A \cap B)$

(iv)　$n(B \cap A') = n(B) - n(A \cap B)$

(v)　$n(A \cup B) = n(A \cap B') + n(B \cap A') + n(A \cap B)$

(vi)　$n(A \triangle B) = n(A) + n(B) - 2n(A \cap B)$

(vii)　$n(A' \cup B') = n[(A \cap B)'] = n(U) - n(A \cap B)$

(viii)　$n(A' \cap B') = n[(A \cup B)'] = n(U) - n(A \cup B)$

(ix)　$n(A - B) = n(A) - n(A \cap B) \Rightarrow n(A - B) + n(A \cap B) = n(A)$

(x)　$n(A \cup B \cup C) = n(A) + n(B) + n(C) - n(A \cap B) - n(B \cap C)$
$$- n(A \cap C) + n(A \cap B \cap C)$$

Solved Examples

Example 1. *In a group of athletic teams in a school, 21 are in the basket ball, 26 in the hockey team and 29 in the football team. If 14 play hockey and basket ball, 12 play football and basket ball, 15 play hockey and football and 8 play all the three games. Find (i) how many players are there in all (ii) how many play football only.*

Solution. Let A, B and C denote the set of players, who play basket ball, hockey and football respectively. Then, according to question, we have

$$n(A) = 21, n(B) = 26, n(C) = 29$$
$$n(A \cap B) = 14, n(A \cap C) = 12, n(B \cap C) = 15 \text{ and } n(A \cap B \cap C) = 8$$

Therefore, $\quad n(A \cap B \cap C) = [n(A) + n(B) + n(C) + n(A \cap B \cap C)]$
$$- [n(A \cap B) + n(A \cap C) + n(B \cap C)]$$
$$= [21 + 26 + 29 + 8] - [14 + 12 + 15] = 43$$

Hence, the total number of players is 43. Now, the number of players playing football only is $[29 - (7+8+4)] = 10$.

Example 2. *In a canteen, out of 123 students, 42 students buy ice-cream, 36 buy burst and 10 buy cakes, 15 students buy ice-cream and 11 buy ice-cream and buns but no cakes. Draw Venn diagram to illustrate the above information and find (i) how many students buy nothing at all (ii) how many students buy at least two items. (iii) how many students buy all three items.*

Solution : Define the sets A, B and C such that

$\qquad A = $ Set of students who buy cakes

$\qquad B = $ Set of students who buy ice-cream

$\qquad C = $ Set of students who buy buns

According to question, we have,

$\qquad n(A) = 10; \ n(B) = 42; n(C) = 36; n(B \cap C) = 15;$

$\qquad n(A \cap B) = 10; \ n[(A \cap C) - B] = 4;$

$n[(B \cap C) - A] = 11$ and $n[A - B \cup C] = 10$

Now we have $n(B \cup C) = n(B) + n(C) - n(B \cap C)$
$$= 42 + 36 - 15 = 63$$
$n(B \cup C) - n(B) = 63 - 42 = 21$

and $\quad n(B \cup C) - n(C) = 63 - 36 = 27$

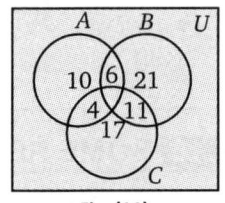

Fig. (11)

The above distribution of the students can be illustrated by Venn diagram (Figure 11). Now, total number of students buying something.

$$= 10+6+21+4+4+11+17 = 73$$

(i) Number of students who did not buy anything = 123 – 73 = 50

(ii) Number of students buying at least two items = 6+4+4+11 = 25

And (iii) Number of students buying all three items = 4

EXERCISE 1.4

1. Out of 80 students who secured first class marks in Mathematics or in Physics, 50 obtained first class marks in Mathematics, 10 in both Physics and Mathematics. How many students secured first class marks in Physics only?

2. The Mathematics club in a school held an open house on three afternoons 115, 110 and 135 students attended both the first, second and third afternoons respectively. 25 attended just the first, 30 attended both the first and second days, 80 attended both the first and third day, and 60 attended both the second and third day. How many attended (i) all three days (ii) just the second day (iii) just the third day?

3. In a school of 250 pupils, 100 are girls, and 200 pupils stay at school for lunch. If 40 girls go home for lunch. Find the number of boys who go home for lunch.

4. In a class of 150 students, the following results were obtained in a certain examination. 45 students failed in Maths; 50 students failed in Physics, 48 students failed in Chemistry, 35 failed in both Maths and Chemistry, 25 failed in the three subject. Find the number of students who have failed in at least one subject.

Answers

1. 30   **2.** 20, 30, 15   **3.** 10   **4.** 71

1.11 ORDERED PAIR

Sometimes, there are situations in which order is very important. Some results may be affected by order and other are not.

Definition: *An ordered pair is a pair of entries whose components occur in a specific order. It is written by listing the two components in the specific order, separating them by a comma and enclosing the pair in parentheses.*

Symbolically: If A and B are two sets, then by ordered pair of elements, we must mean a pair (a,b): $a \in A$, $b \in B$ in that order.

REMARKS

☞ It may be noted that (a, b) is not the same as $\{a, b\}$. The former denotes an ordered pair whereas the latter denotes a set.

☞ $(a, b) \neq (b, a)$ unless $a = b$.

☞ Ordered pair may have the same first and second components, *i.e.*, two elements of an ordered pair need not be distinct.

☞ Two ordered pairs are said to be equal when both the first components are equal and their second components are also equal.

1.11.1 Cartesian Product of Two Sets

The set of all ordered pairs of elements (a,b), $a \in A$, $b \in B$ is called the cartesian product of two sets A and B. It is denoted by $A \times B$.

Symbolically: $A \times B = \{(a, b) : a \in A, b \in B\}$

For example :

If $A = \{2, 3\}$ and $B = \{4,5,6\}$, then

$$A \times B = \{(2, 4), (2, 5), (2, 6), (3, 4), (3, 5), (3, 6)\}$$

REMARKS

☞ $A \times B = \phi \Leftrightarrow A = \phi$ or $B = \phi$

☞ If A and B are finite sets, then $n(A \times B) = n(A) . n(B)$

☞ If either A or B is infinite set, then $A \times B$ is an infinite set.

1.11.2 Ordered Triplet

If A, B, C are three sets, then by ordered triple product of elements, we mean a triplet $(a, b, c) : a \in A, b \in B, c \in C$ in that order.

This is also called ordered 3-tuple.

The set of all ordered triplets $(a, b, c): a \in A, b \in B, c \in C$ is also called the cartesian triple product of three sets A, B and C and is denoted by $(A \times B \times C)$

Symbolically: $A \times B \times C = \{(a, b, c): a \in A, b \in B, c \in C\}$

REMARK

☞ In general, the cartesian product on n sets $A_1, A_2, ..., A_n$ is a ordered n tuples $(a_1, a_2, ..., a_n)$, where

$a_1 \in A_1, a_2 \in A_2, ..., a_n \in A_n$. It is denoted by $A_1 \times A_2 ... \times A_n$ or briefly by $\prod_{i=1}^{n} A_i$ where $\prod$ stands for the product.

Solved Examples

Example 1. *If $A = \{1, 2\}$ and $B = \{a, b, c\}$, find the value of $A \times B$, $B \times A$, $A \times A$, $B \times B$.*

Solution. We have $A = \{1, 2\}$ and $B = \{a, b, c\}$.

Therefore,

$$A \times B = \{(1, a), (1, b), (1, c), (2, a), (2, b), (2, c)\}$$
$$B \times A = \{(a, 1), (a, 2), (b, 1), (b, 2), (c, 1), (c, 2),\}$$
$$A \times A = \{(1, 1), (1, 2), (2, 1), (2, 2)\}$$
$$B \times B = \{(a, a), (a, b), (a, c), (b, a) (b, b), (b, c), (c, a), (c, b), (c, c)\}$$

Example 2. *If $A = \{1, 2, 3\}$, $B = \{a, b, c, d\}$ and $C = \{-1, -2\}$, find $A \times B$, $B \times A$ and $C \times (B \cup C)$.*

Solution. Given that $A = \{1, 2, 3\}, B = \{a, b, c, d\}$ and $C = \{-1, -2\}$.

Therefore,

$$A \times B = \{(1, a), (1, b), (1, c), (1, d), (2, a), (2, b), (2, c), (2, d),$$
$$(3, a), (3, b), (3, c), (3, d)\}$$
$$B \times A = \{(a, 1), (b, 1), (c, 1), (d, 1), (a, 2), (b, 2), (c, 2), (d, 2), (a, 3),$$
$$(b, 3), (c, 3), (d, 3)\}$$

Also, $B \cup C = \{a, b, c, d, -1, -2\}$

Therefore,

$$C \times (B \cup C) = \{(-1, a), (-1, b), (-1, c), (-1, d), (-1, -1), (-1, -2), (-2, a),$$
$$(-2, b), (-2, c), (-2, d), (-2, -1), (-2, -2)\}$$

Example 3. *Find the values of a and b if $(4a-2, b+4) = (2a, 4)$.*

Solution. Since we know that two ordered pairs (a_1, b_1) and (a_2, b_2) are said to be equal if $a_1 = a_1$ and $b_1 = b_2$. Therefore, for the equality of two given ordered pairs, we have

$$4a - 2 = 2a \text{ and } b + 4 = 4$$

Therefore, $4a - 2a = 2 \implies a = 1$ and $b + 4 = 4 \implies b = 0$

Example 4. *If $A = \{1, 2, 3, 4\}$ and $B = \{4, 5\}$, represent $A \times B$, $B \times A$ and $B \times B$ pictorially and find their values.*

Solution. Given $A = \{1, 2, 3, 4\}$ and $B = \{4, 5\}$

$A \times B = \{(1, 4), (1, 5), (2, 4), (2, 5), (3, 4), (3, 5), (4, 4), (4, 5)\}$

$B \times A = \{(4, 1), (5, 1), (4, 2), (5, 2), (4, 3), (5, 3), (4, 4), (5, 4)\}$

And $B \times B = \{(4, 4), (4, 5), (5, 4), (5, 5)\}$

Pictorially, $A \times B$, $B \times B$ and $B \times A$ can be represented as shown in figure 12.

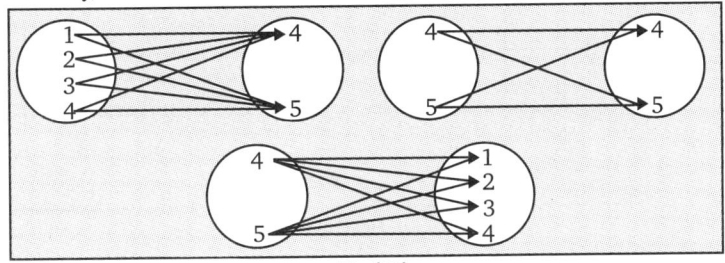

Fig. (12)

Example 5. *Let $A = \{1, 2, 3, 4\}$ and $B = \{5, 7, 9\}$. Determine (i) $A \times B$, (ii) $B \times A$. Also represent $A \times B$ and $B \times A$ graphically.*

Solution. (i) Given $A = \{1, 2, 3, 4\}$ and $B = \{5, 7, 9\}$. Then,

$A \times B = \{(1, 5), (1, 7), (1, 9), (2, 5), (2, 7), (2, 9)\ (3, 5), (3, 7), (3, 9),$

$(4, 5), (4, 7), (4, 9)\}$

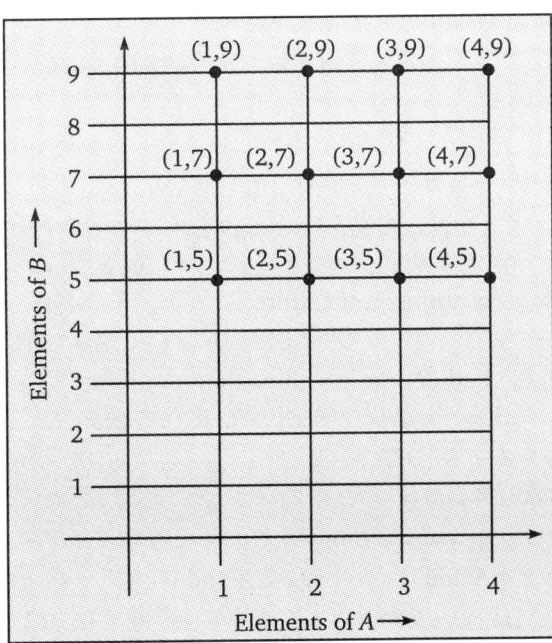

Fig. (13) : $A \times B$

Graphically, it can be represented as shown in Figure 13.

Now, $B \times A = \{(5, 1), (5, 2), (5, 3)\ (5, 4)\ (7, 1), (7, 2), (7, 3)\ (7, 4)\ (9, 1),$
$(9, 2), (9\ 3)\ (9, 4)\}$

Graphically, it can be represented as shown in Figure 14.

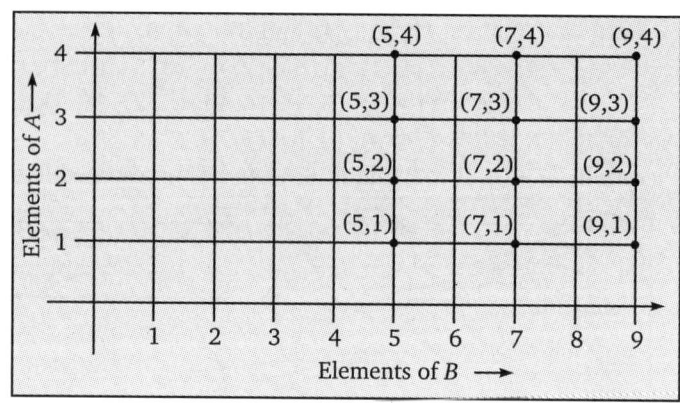

Fig. (14) : B × A

Theorem 1. *For any three subsets A, B and C , we have.*

(i) $A \times (B \cap C) = (A \times B) \cap (A \times C)$ (ii) $A \times (B \cup C) = (A \times B) \cup (A \times C)$

Proof. (i) Let $(x, y) \in A \times (B \cap C)$

$\Rightarrow$ Then, $x \in A$ and $y \in (B \cap C)$

$\Rightarrow$ $x \in A$ and $y \in B$ and $y \in C$ $\Rightarrow x \in A, y \in B$ and $x \in A, y \in C$

$\Rightarrow$ $(x, y) \in A \times B$ and $(x, y) \in (A \times C) \Rightarrow (x, y) \in (A \times B) \cap (A \times C)$

But (x, y) is arbitrary, therefore

$$A \times (B \cap C) \subset (A \times B)\ (A \times C) \qquad \qquad \dots (1)$$

Conversely ,

If $(x, y) \in (A \times B) \cap (A \times C)$

Then, $(x, y) \in A \times B$ and $(x, y) \in A \times C$

$\Rightarrow$ $x \in A, y \in B$ and $x \in A, y \in C \Rightarrow x \in A, y \in B$ and $y \in C$

$\Rightarrow$ $x \in A$ and $y \in (B \cap C)$ $\Rightarrow (x, y) \in A \times (B \cap C)$

But (x, y) is arbitrary, therefore

$$(A \times B) \cap (A \times C) \subseteq A \times (B \cap C) \qquad \qquad \dots (2)$$

From (1) and (2), we conclude that

$$A \times (B \cap C) = (A \times B) \cap (A \times C)$$

(ii) $(x, y) \in A \times (B \cup C)$

Then, $x \in A$ and $y \in (B \cup C)$

$\Rightarrow$ $x \in A$ and $y \in B$ or $y \in C$

$\Rightarrow$ $(x \in A$ and $y \in B)$ or $(x \in A$ and $y \in C)$

$\Rightarrow$ $\{(x, y) \in (A \times B)\}$ or $\{(x, y) \in (A \times C)\}$

$\Rightarrow$ $(x, y) \in (A \times B) \cup (A \times C)$

Since (x, y) is arbitrary, therefore
$$A \times (B \cup C) \subseteq (A \times B) \cup (A \times C) \quad \dots (1)$$

Conversely,

If $\qquad\qquad (x, y) \in (A \times B) \cup (A \times C)$

Then, $\qquad\qquad (x, y) \in (A \times B)$ or $(x, y) \in (A \times C)$

$\Rightarrow \qquad (x \in A$ and $y \in B)$ or $(x \in A$ and $y \in C)$

$\Rightarrow \qquad x \in A$ and $(y \in B$ or $y \in C) \Rightarrow (x, y) \in A \times (B \cup C)$

But (x, y) is arbitrary, therefore
$$(A \times B) \cup (A \times C) \subseteq A \times (B \cup C) \qquad \dots (2)$$

From (1) and (2), we conclude that
$$A \times (B \cup C) = (A \times B) \cup (A \times C).$$

Theorem 2. *For any sets A, B, C, D we have* $(A \times B) \cap (C \times D) = (A \cap C) \times (B \cap D)$

Proof. $\qquad$ If $(a, b) \in (A \times B) \cap (C \times D)$, then

$\Rightarrow \qquad (a, b) \in (A \times B)$ and $(a, b) \in (C \times D)$

$\Rightarrow \qquad (a \in A$ and $b \in B)$ and $(a \in C$ and $(b \in D)$

$\Rightarrow \qquad (a \in A$ and $a \in C)$ and $(b \in B$ and $b \in D)$

$\Rightarrow \qquad a \in (A \cap C)$ and $b \in (B \cap D) \Rightarrow (a, b) \in (A \cap C) \times (B \cap D)$

Since (a, b) is arbitrary, therefore
$$(A \times B) \cap (C \times D) \subseteq (A \cap C) \times (B \cap D) \qquad \dots (1)$$

Now, let $(a, b) \in (A \cap C) \times (B \cap D)$

$\Rightarrow \qquad a \in (A \cap C)$ and $b \in (B \cap D) \Rightarrow (a \in A$ and $a \in C)$ and $(b \in B$ and $b \in D)$

$\Rightarrow \qquad (a \in A$ and $b \in B)$ and $(a \in C$ and $b \in D)$

$\Rightarrow \qquad (a, b) \in (A \times B) \cap (C \times D)$

Since, (a, b) is arbitrary, therefore
$$(A \cap C) \times (B \cap D) \subseteq (A \times B) \cap (C \times D) \qquad \dots (2)$$

From (1) and (2), we conclude that
$$(A \times B) \cap (C \times D) = (A \cap C) \times (B \cap D)$$

REMARKS

☞ $(A \times B) \cap (B \times A) = (A \cap B) \times (B \cap A)$

☞ $A \times (B' \cup C')' = A \times (B \cap C) = (A \times B) \cap (A \times C)$

☞ $A \times (B' \cap C')' = A \times (B \cup C) = (A \times B) \cup (A \times C)$

Theorem 3. *If A and B are two non-empty sets having n elements in common, then $A \times B$ and $B \times A$ have n^2 elements in common.*

Proof. $\qquad$ We know that $\quad (A \times B) \cap (C \times D) = (A \cap C) \times (B \cap D)$

$$(A \times B) \cap (B \times A) = (A \cap B) \times (B \cap A)$$

$$(A \times B) \cap (B \times A) = (A \cap B) \times (A \cap B)$$

Since $(A \times B)$ has n elements, therefore $(A \cap B) \times (B \cap A)$ has n^2 elements.

$$(A \times B) \cap (B \times A) = (A \cap B) \times (B \cap A) \text{ has } n^2 \text{ elements.}$$

Hence, $(A \times B)$ and $(B \times A)$ have n^2 elements in common.

REMARKS

☞ For any three sets, A, B, C, we have $A \times (B - C) = (A \times B) - (A \times C)$.
☞ If A and B are any two non-empty sets, then $A \times B = B \times A$ iff $A = B$.
☞ If $A \subseteq B$, then $A \times A \subseteq (A \times B) \cap (B \times A)$
☞ If $A \subseteq B$, then $A \times C \subseteq B \times C$ for any set C.
☞ If $A \subseteq B$ and $C \subseteq D$, then $A \times C \subseteq B \times D$.
☞ $A \times B = A \times C \Rightarrow B = C$

EXERCISE 1.5

1. If $A = \{a, b, c\}$, $B = \{d\}$, $C = \{2\}$, then verify
   (i) $A \times (B \cup C) = (A \times B) \cup (A \times C)$
   (ii) $A \times (B \cap C) = (A \times B) \cap (A \times C)$
   (iii) $A \times (B - C) = (A \times B) - (A \times C)$
   (iv) $(A \cap B) \times C = (A \times C) \cap (B \times C)$

2. If $A = \{2, 3\}$, $B = \{1, 2, 3\}$, $C = \{2, 3, 4\}$, show that $A \times A = (B \times B) \cap (C \times C)$.

3. If $A = \{1, 2, 3\}$, $B = \{4, 5\}$ and $C = \{1, 2, 3, 4, 5\}$, then show that $(C \times B) - (A \times B) = B \times B$.

4. The ordered pairs $(2,7)$, $(4, 8)$ and $(5, 9)$ and among nine elements of the set $A \times B$. Determine the other six elements of $A \times B$.

5. Let $A = \{2, 3, 5, 7\}$, $B = \{1, 12, 13, 15\}$. How many elements are there in $A \times B$? In $B \times A$? Is $A \times B = B \times A$? Is $n(A \times B) = n(B \times A)$?

6. Let A and B be two sets. Show that the sets $A \times B$ and $B \times A$ have an element in common if and only if the sets A and B have an element in common.

7. Some elements of $A \times B$ are (a, x), (a, y), (d, z). If $A : \{a, b, c\ d\}$, find the remaining elements of $A \times B$ such that $n(A \times B)$ is least.

8. If A and B are two sets having 3 elements in common. If $n(A) = 5$, $n(B) = 4$, find $n(A \times B)$ and $n\{(A \times B) \cap (B \times A)\}$.

9. The ordered pairs $(1, 1)$, $(2, 2)$ and $(3, 3)$ are among the elements in the set $A \times B$. If A and B have 3 elements each, how many elements in all does the set $A \times B$ have? Also find the remaining elements.

10. If A and B are two sets such that $n(A) = 3$ and $n(B) = 2$. If $(x, 1)$, $(y, 2)$, $(z, 1)$ are in $A \times B$, find A and B, where x, y, z are distinct.

11. Write 'T' for true and 'F' for false statement:
    (a) If $A = (a, b)$ and $B = (b, a)$, then $A \times B = \{(a, b)\ (b, a)\}$ **(T/F)**
    (b) $\{(a, x), (a, y), (b, x), (b, y)\}$ is product set. **(T/F)**
    (c) If $n(A) = x$ and $n(B) = y$ and $A \cap B = \phi$, then $n(A \times B) = xy$ **(T/F)**
    (d) If A and B are non-empty sets, then $A \times B$ is a non-empty set of ordered pairs (x, y) such that $x \in A$ and $y \in A$. **(T/F)**

12. (a) If $A = \{1, 2, 3\}$, $B = \{4, 5\}$ and $C = \{1, 2, 3, 4, 5\}$. Find
    (i) $A \times B$, (ii) $C \times B$, (iii) $B \times B$
    (b) If $A = \{1, 2, 3, 4\}$ and $B = \{5, 7, 9\}$, find $(A \times B) \cap (A \cap B)$.

═══ **Answers** ═══

4. $(2, 8), (2, 9), (4, 7), (4, 9), (5, 7), (5, 8)$ 5. 16, 16, No, yes
7. $(a, y), (a, 2), (b, x), (b, y), (b, z), (c, x), (c, z), (d, x), (d, y)$ 8. 20, 9
9. 9, $(1, 2), (1, 3), (2, 1), (2, 3), (3, 1), (3, 2)$
10. (i) $A = \{x, y, z\}$, $B = \{1, 2\}$,   (ii) (a) F   (b) T,   (c) T   (d) F
12. (a) (i) $A \times B = (1, 4), (1, 5), (2, 4), (2, 5), (3, 4), (3, 5)$
    (ii) $C \times B = \{(1, 4), (1, 5), (2, 4), (2, 5), (3, 4), (3, 5), (4, 4), (4, 5), (5, 4), (5, 5), \}$
    (iii) $B \times B = \{(4, 4), (4, 5), (5, 5)\}$ (b) ϕ

1.12 RELATION

Let us take two sets of natural numbers N_1 and N_2. We define R as a relation between them such that N_1 is a square of N_2. Then we can write 1R1, 2R4, 3R9, ...

In terms of ordered pair, we can write

$$R = \{(1, 1), (2, 4), (3, 9), (4, 16), ...\} = \{(x, y : x, y \in \mathbf{N} \text{ and } y = x^2\}$$

The relation from set $\mathbf{N}$ to $\mathbf{N}$ is a subset of $\mathbf{N} \times \mathbf{N}$ such that $y = x^2$.

Definition: *Let A and B be two sets. Then a relation R from A to B is a subset of $A \times B$.*

Symbolically: R is a relation from A to $B \Leftrightarrow R \subseteq A \times B$.

REMARKS

☞ If R is a relation from A to B, then A is called the domain and B the range of R.

☞ If R is a relation from a non-empty set A to a non-empty set B and if $(a, b) \in R$, then we write aRb, read as "a is related to b by the relation R." On the other hand, if $(a, b) \notin R$, we write $a\cancel{R}b$ and say that 'a is not related to b by the relation R'.

☞ In particular, any subset $A \times A$ defined a relation in A, known as Binary relation.

■ Illustrations

(i)   If $a, b \in \mathbf{N}$ and R is defined as "a is divisor of b" then R is relation on $\mathbf{N}$.
      The subset $\mathbf{N} \times \mathbf{N}$, which corresponds to the relation R is $S = \{(n, r): n \in \mathbf{N}, r \in \mathbf{N}\}$
      Here, it is clear that $(1, 3), (2, 4), (3, 9)$ $(4, 8), (4, 4)$, are in S, whereas $(2, 3), (4, 5)$, $(5, 6)$ are not in S.

(ii) If R is a relation from set $A = \{1,2,3\}$ to the set $B = \{-1, -2\}$ defined by $x + y = 0$, then $R = \{(1, -1), (2, -2)\}$
      Here, domain of R is $\{1, 2\}$ and Range $= \{-1, -2\}$.

(iii) If $A = \{a, b, c, d, e\}$ and $B = \{f, g, h, i\}$ and let $R = \{(a, g), (a, i), (d, h), (e, f)\}$ by a relation from A to B then
      Domain of $R = \{a, d, e\}$ and Range of $R = \{g, i, h, f\}$

(iv) If $a, b \in \mathbf{R}$, the set of real numbers and R is "$|a - b|$ is a rational number" then R is a relation on $\mathbf{R}$. The subset S of $\mathbf{R} \times \mathbf{R}$ which corresponds to the relation is
      $$S = \{(a, b + a): a \in \mathbf{R}, b \in \mathbf{Q}\}$$
      It is observed that $\left(1, 2\frac{1}{2}\right), \left(\pi, \pi - \frac{1}{2}\right)$ belongs to S, while $(\sqrt{2}, \pi + \sqrt{2}) \notin S$.

(v)   If $A = \{2, 3, 4\}$ and $B = \{a, b, c\}$, then $R = \{(2, b), (3, c), (2, a), (4, a)\}$ being a subset of $A \times B$, is a relation from $A \times B$. Here $(2, b), (3, c), (2, a), (4, a) \in R$, so we may write $2Rb, 3Rc, 2Ra, 4Ra$. But $(3, b) \notin R$ therefore, $3 \cancel{R} b$.

(vi) If $a, b \in \mathbf{N}$ and R is defined by "$a - b$ is divisible by a number $n \in \mathbf{N}$", then R is a relation on $\mathbf{N}$. The subset S of $\mathbf{N} \times \mathbf{N}$ corresponding to the relation by
      $$S = \{n, n + rm : n \in \mathbf{N}, r \in \mathbf{N}\}$$
      Here, $m = 3$, $(2, 8), (5, 11) \in S$ [$\because 2 - 8 = 6$, which is divisible by 3]
      While $(3, 8) \in S$ [$\because 3 - 8 = 5$, which is not divisible by 3]

1.12.1 Total number of Relations

Let A and B be two non-empty finite sets consisting p and q elements respectively, then $A \times B$ consists of $p\,q$ ordered pairs. Therefore, total number of subset of $A \times B$ is 2^{pq}.

REMARKS

☞ For a non-empty set A, $\phi \in A \times A$, therefore it is a relation on A, called void or empty relation on A.

☞ The void relation ϕ and the universal relation $A \times B$ are called trivial relations from A to B.

☞ The void and universal relation on set A respectively the smallest and the largest relation on A.

1.12.2 Identity Relation

Let A be a set. The identity relation on A is the relation $I_A = \{(x, x) : x \in A\}$ on A.

For example : If $A = \{a, b, c\}$ then the relations $I_A = \{(a, a), (b, b), (c, c)\}$ is the identity relation. $R = \{(a, a), (b, b)\}$ is not an identity relation as $(c, c) \notin R$.

1.12.3 Inverse of a Relation

Let A, B be two non-empty sets and R be a relation from a set A to B and let (x,y), number of the subset D of $A \times B$ corresponding to the relation R from A to B.

To the relation R from the set A to the set B, there corresponds a relation from the set B to the set A called the inverse of the relation, denoted by R^{-1} such that the subset $B \times A$ corresponding to the relation R^{-1} is $= \{(y, x): (x, y) \in D\}$.

i.e.,
$$yR^{-1}x \Leftrightarrow xRy$$

For example:

(i) Let $A = \{a, b, c\}$ and $B = \{1,2,3\}$ be two sets and let $R = \{(a, 1), (a, 2), (b, 1), (b, 2)\}$ be a relation from A to B then $R^{-1} = \{(1, a), (2, a), (1, b), (2, b)\}$

(ii) If $A = \{1, 2, 3\}$, $B = \{5, 6, 7\}$ and let $R = \{(1, 5), (2, 5), (2, 7)\}$ be a relation from A to B.

Then $R^{-1} = \{(5, 1), (5, 2), (7, 2)\}$ which is a relation from B to A.

Also, Domain $(R) = \{1, 2\} = $ Range (R^{-1})

And, Range $(R) = \{5, 7\} = $ Domain (R^{-1})

(iii) The inverse of the relation "*is less than*" In **R** "*is greater than*".

REMARK

☞ Sometimes, the inverse of a relation coincides with the relation itself.
For example, the inverse of the relation "perpendicular to" in the set of straight lines coincides with itself.

1.13 CLASSIFICATION OF RELATIONS

(a) Reflexive Relation: Let R be a relation on a set A.

"*A relation R is said to be reflexive if* $(x, x) \in R \; \forall \; x \in A$"

i.e.,
$$x\,R\,x \; \forall \; x \in A$$

For example :

(i) In a set of integers, a relation R defined by $x\,R\,y$ iff $x - y$ is divisible by 4, then R is a reflexive relation because $x - x = 0$ which is a divisible by 4.

(ii) The universal relation on a non-empty set A is reflexive.

(iii) The relation "is less than," *i.e.,* '<' in the set of rational number is not reflexive, because no member have the relation is less than to itself.

(iv) The relation "is a factor of" in the set of rational number is reflexive, since every rational number is a factor of itself.

(v) The relation "is less than or equal to." *i.e.,* $\leq$ in the set of natural number is reflexive.
$$n \leq n \ \forall \ n \in \mathbf{N}$$

(b) Symmetric Relation. *A relation R on a set A is said to be symmetric if*
$$(y, x) \in R \text{ whenever } (x, y) \in R \ \forall \ x, y \in R$$
i.e., $\qquad x \, R \, y \Leftrightarrow y \, R \, x \ \forall \ x, y \in R$

For example:
(i) Let l_1, l_2 be two lines such that l_1 is perpendicular to l_2, *i.e.,* $l_1 \perp l_2$. Then $l_1 \perp l_2 \Rightarrow l_2 \perp l_1$. Therefore the relation $\perp$ is symmetric.

(ii) The identity and the universal relation on a non-empty set are symmetric relations.

(iii) Consider the set $\mathbf{N}$ of natural numbers and the relation 'is less than'. This relation is not symmetric. Since if $2 < 3$ then $3 \not< 2$.

Let $A = \{1, 2, 3\}$ and relations R_1 and R_2 defined by
$$R_1 = \{(1, 2), (1, 3), (3, 1), (2, 1)\} \text{ and } R_2 = \{(1, 2), (2, 3), (3, 1)\}$$
Then R_1 is a symmetric relation, but R_2 is not symmetric.

(c) Transitive Relation: *A relation R on a set A is said to be transitive iff* $(x,y) \in R$ *and* $(y, z) \in R \Rightarrow (x, z) \in R \ \forall \ x, y, z \in A$, *i.e.,* $x \, R \, y, y \, R \, z \Rightarrow x R z$.

For example:
(i) Let a, b, c be three numbers such that a is a factor of b and b is a factor of c, then obviously a is a factor of c. Therefore, 'is a factor of' is a transitive relation.

(ii) If l_1, l_2, l_3 are three lines such that $l_1 \perp l_2$ and $l_2 \perp l_3$ then it is obvious that l_1 is parallel to l_3. Therefore the relation " $\perp$ " is not transitive.

(iii) The identity and universal relation on a non-empty set are transitive.

(iv) Let l_1, l_2, l_3 be three straight lines, such that l_1 is parallel to l_2 and l_2 is parallel to l_3 then it is clear that l_1 is parallel to l_3. Therefore, 'is parallel to' is a transitive relation.

(d) Anti-symmetric Relation. *A relation R on a non-empty set A is said to be an anti-symmetric relation iff* $(x, y) \in R$ *and* $(y, x) \in R \Rightarrow x = y \ \forall \ x, y \in R$

REMARKS

☞ The identity relation R on a set A is an anti – symmetric relation.
☞ If $(x, y) \in R$ and $(y, x) \notin R$, then it may be noted that $x = y$.
☞ The universal relation on a set A containing at least two elements is not anti – symmetric.

1.13.1 Equivalence Relations

A relation R on a set E is said to be equivalence if it is

(i) Reflexive,   (ii) Symmetric and   (iii) Tansitive

For example :
(1) In a set of integers, a relation R is defined by $x \, R \, y$ if and only if $x - y$ is divisible by 4. Then R is an equivalence relation. Since
   (a) For $x \, R \, x$, $x - x = 0$ is divisible by 4. Therefore, it is reflexive.
   (b) For $x \, R \, y$. Let $x - y = 4m$ so $y - x = 4m$, which is also divisible by 4. Therefore, it is symmetric.

(c) For $x \, R \, y$, let $x - y = 4m$; for $y \, R \, z$, let $y - z = 4n$. By adding these two equations, we get $x - z = -4(m + n)$,

which is divisible by 4. Therefore it is transitive.

(2) Let R be a relation on the set of all lines in a plane L defined by $(l_1, l_2) \in R$ if and only if line l_1 is parallel to l_2, then R is an equivalence relation because

(a) For each line $l \in L$, we have l is parallel to l.

$\Rightarrow lRl \Rightarrow R$ is reflexive.

(b) Let $l_1, l_2 \in L$ such that $(l_1, l_2) \in R$, then

$\Rightarrow (l_1, l_2) \in R \Rightarrow l_1$ is parallel to $l_2 \Rightarrow l$ is symmetric.

(c) Let $l_1, l_2, l_3 \in L$ such that (l_1, l_2) and $(l_2, l_3) \in R$, then obviously $(l_1, l_3) \in R$ because if l_1 is parallel to l_2 and l_2 is parallel to l_3, then l_3 should be parallel to l_1.

1.13.2 Congruence Modulo 'm'

Let m be an arbitrary but fixed integer. If $x - y$ is divisible by m, then two integers x and y are said to be congruence modulo m of one another.

Symbolically: $x \equiv y \pmod{m}$ if $x - y$ divisible by m.

For example: $32 \equiv 2 \pmod{3}$, as $32 - 2 = 30$ which is divisible by 3.

1.13.3 Composition of Relations

Let R_1 and R_2 be two relations from sets A to B and B to C respectively, then we can define a relation $R_1 \, o \, R_2$ from A to C, such that $(x, z) \in R_1 \, o \, R_2$ if and only if there exist $y \in Y$ such that $(x, y) \in R_1$ and $(y, z) \in R_2$.

This relation is called composition of R_1 and R_2.

REMARKS

☞ $R_1 o R_2 \# R_2 o R_1$

☞ $(R_2 o R_1)^{-1} = R_1^{-1} o R_2^{-1}$

For example : Let A, B, C be three sets such that

$A = \{-1, -2\}, B = \{p, q, r\}$ and $C = \{\alpha, \beta, \gamma\}$

Also, $\qquad R_1 = \{(-1, p), (-1, r), (-2, q)\}$ is a relation from A and B and

$\qquad R_2 = \{(p, \alpha), (q, \beta), (r, \gamma)\}$ and is a relation from set to B to C.

Then $R_2 o R_1$ is a relation from A to C given by

$\qquad R_2 o R_1 = \{(-1, \alpha), (-1, \gamma), (-z, \beta)\}$

Theorem 4. *The intersection of two equivalence relations on a set is an equivalence relation.*

Proof. Let R_1, R_2 be two equivalence relation on a set A. To show $(R_1 \cap R_2)$ also an equivalence relation.

(i) Let $a \in A$ be arbitrary.

Since R_1 and R_2 both are reflexive on A.

$\therefore (a, a) \in R_1$ and $(a, a) \in R_2 \Rightarrow (a, a) \in R_1 \cap R_2$

Therefore, $(R_1 \cap R_2)$ is reflexive.

(ii) Let $a, b \in A$ such that $(a, b) \in R_1 \cap R_2$

$$(a, b) \in R_1 \cap R_2 \Rightarrow (a, b) \in R_1 \text{ and } (a, b) \in R_2$$

Also, R_1 and R_2 both are symmetric on A.

Therefore, $(b, a) \in R_1$ and $(b, a) \in R_2 \Rightarrow (b, a) \in R_1 \cap R_2 \Rightarrow (R_1 \cap R_2)$ is symmetric on A.

(iii) Let $a, b, c \in A$ such that $(a, b) \in R_1 \cap R_2$, $(b, c) \in R_1 \cap R_2$

Then, $(a, b) \in R_1 \cap R_2$ and $(b, c) \in R_1 \cap R_2$

$\Rightarrow \quad \{(a, b) \in R_1 \text{ and } (a, b) \in R_2 \text{ and } \{(b, c) \in R_1 \text{ and } (b, c) \in R_2\}$

$\Rightarrow \quad \{(a, b) \in R_1, (b, c) \in R_1\} \text{ and } \{(a, b) \in R_2, (b, c) \in R_2\}$

$\Rightarrow \quad (a, c) \in R_1 \text{ and } (a, c) \in R_2 \qquad [\because R_1 \text{ and } R_2 \text{ both are transitive.}]$

$\Rightarrow \quad (a, c) \in R_1 \cap R_2$

Therefore, $(R_1 \cap R_2)$ is transitive on A.

From (i), (ii) and (iii), we have $R_1 \cap R_2$ is reflexive, symmetric and transitive, and hence $R_1 \cap R_2$ is an equivalence relation.

REMARK

☞ The union of two equivalence relations on a set is not necessarily an equivalence relation.

Theorem 5. *If R is an equivalence relation, then R^{-1} is also an equivalence relation.*

Proof.   Let R be an equivalence relation on a sct A. Then by definition of relation on a set, we have

$$R \subseteq A \times A \Rightarrow R^{-1} \subseteq A \times A$$

Therefore, R^{-1} is a relation on A.

Now, to show R^{-1} is an equivalence relation.

(i) Let $a \in A$, then $(a, a) \in R$ ($\because R$ is an equivalence relation and hence reflexive)

$\Rightarrow \quad (a, a) \in R^{-1}$

Thus, $(a, a) \in R^{-1} \forall a \in R \Rightarrow R^{-1}$ is reflexive on A.

(ii) Let $(a, b) \in R^{-1}$, then $(a, b) \in R^{-1} \Rightarrow (b, a) \in R$

$\Rightarrow \quad (a, b) \in R \qquad\qquad\qquad (\because R \text{ is symmetric})$

$\Rightarrow \quad (b, a) \in R^{-1}$

Therefore R^{-1} is symmetric .

(iii) Let $(a, b) \in R^{-1}$ and $(b, c) \in R^{-1}$ then $(a, b) \in R^{-1} \Rightarrow (b, a) \in R$

and $(b, c) \in R^{-1} \Rightarrow (c, b) \in R$

Now, $(c, b) \in R$ and $(b, a) \in R$

$(c, a) \in R \qquad\qquad\qquad\qquad (\because R \text{ is transitive})$

$(a, c) \in R^{-1}$

Therefore R^{-1} is transitive .

From (i), (ii) and (iii), we conclude that R^{-1} is an equivalence relation.

Solved Examples

Based on the following Results

▶ If $n(A) = p$, $n(B) = q$ then total number of subsets of $A \times B = 2^{pq}$.

▶ **Reflexive Relation:** xRx, $\forall x \in A$

▶ **Symmetric relation:** $xRy \Leftrightarrow yRx \ \forall \ x, y \in R$

▶ **Transitive relation:** xRy, $yRz \Rightarrow xRz$

▶ **Anti-symmetric relation:** $xRy \Rightarrow yRx \Leftrightarrow x = y$

▶ **Equivalence relation:** Reflexive, symmetric and transitive **(RST).**

▶ **Partrial ordered relation:** Reflexive, anti-symmetric and transitive **(RAT)**

▶ R is equivalence $\Rightarrow R^{-1}$ is equivalence.

▶ Intersection of two equivalence relations on a set is again equivalence.

Example 1. *Let **Z** be the set of integers. Define a relation R on **Z** such that x R y holds if and only if x − y is divisible by 5, x ∈ **Z**, y ∈ **Z**. Show that it is an equivalence relation.*

Solution : (i) For each $x \in$ **Z**, $x - x$ i.e., 0 is divisible by 5.

Therefore, for all $x \in$ **Z** , $x R x \Rightarrow x$ is reflexive.

(ii) Let $\quad x R y \Rightarrow x - y$ is divisible by 5.

$\Rightarrow y - x$ is divisible by 5.

Thus $\quad xRy = yRx$

Therefore R is symmetric.

(iii) Let us suppose xRy and yRz, then $(x - y)$ and $(y - z)$ are both divisible by 5. Hence, 5 is also a divisor of $(x - y) + (y - z)$.

5 is a divisor of $(x - z)$.

Therefore, xRy, $yRz \Rightarrow xRz \Rightarrow R$ is transitive.

From (i), (ii) and (iii), we conclude that R is an equivalence relation.

Example 2. *Let **N×N** be the set of ordered pairs of natural numbers. Also, let R be the relation in **N×N**, defined by (a, b) R (c, d) if and only if a+d = b+c. Show that R is an equivalence relation.*

Solution : (i) For all $(a, b) \in$ **N×N**, we have $a+b = b+a$, i.e., $(a, b) R (b, a)$.

Therefore, R is reflexive.

(ii) Let $(a, b) R (c, d)$, then, by definition of R

$$(a+d) = (b+c) \text{ or } (c+b) = (d+a)$$

$(c, d) R (a, b) \Rightarrow R$ is symmetric.

(iii) Let us suppose $(a, b) R (c, d)$ and $(c, d) R (e, f)$, then

$$a + d = b+c \text{ and } c+f = d+e$$

$\Rightarrow \quad (a + d) + (c + f) = (b + c) + (d + e) \quad \Rightarrow \ a + f = b + e$

$\Rightarrow \quad (a, b)R(e, f)$

Therefore, R is transitive.

Hence, from (i), (ii) and (iii), we conclude that R is an equivalence relation.

Example 3. *If R is the relation for natural number defined by x+4y = 20. Find the domain and range of the relation R.*

Solution. Let $x + 4y = 20$ $\Rightarrow$ $y = \dfrac{20 - x}{4}$

For $x = 4$, $y = 4$ and for $x = 8$, $y = 3$.

For $x = 16$, $y = 1$ and for $x = 12$, $y = 2$

Therefore, Domain = {4, 8, 12, 16} and range = {4, 3, 2, 1}

Example 4. *A relation R defined on the set of integers Z, as follows*

$$(x, y) \in R \Rightarrow x^2 + y^2 = 25$$

Express R and R⁻¹ as the sets of ordered pairs and hence find their respective domains.

Solution. Since $(x, y) \in R \Leftrightarrow x^2 + y^2 = 25$ $\Rightarrow$ $y = \pm\sqrt{25 - x^2}$

If $x = 0$ $\Rightarrow$ $y = 5$.

Therefore, $(0, 5) \in R$ and $(0, -5) \in R$

Now, $x = 3$ $\Rightarrow$ $y = \sqrt{25 - 9} = \pm 4$

$(3, 4) \in R$, $(-3, 4) \in R$, $(3, -4) \in R$ and $(-3, -4) \in R$

$x = \pm 4$ $\Rightarrow$ $y = \pm 3$

Therefore, $(4, 3) \in R$, $(-4, 3) \in R$, $(4, -3) \in R$ and $(-4, -3) \in R$

$x = \pm 5$ $\Rightarrow$ $y = \sqrt{25 - 25} = 0$ $\therefore$ $(5, 0) \in R$ and $(-5, 0) \in R$

Here, it is clear that for any other integral value of x, y is not an integer. Therefore,

$R =$ {(0, 5), (0, –5), (3, 4), (–3, 4), (3, –4), (–3, – 4), (4, 3), (–4, 3), (4, –3), (– 4, –3), (5, 0), (–5, 0)}

and $R^{-1} =$ {(5, 0), (–5, 0), (4, 3), (4, –3), (– 4, 3), (– 4, –3), (3, 4), (3, –4), (–3, 4), (–3, – 4), (0, 5), (0, –5)}

Also, Domain $(R) =$ {(0, 3, –3, 4, – 4, 5, –5} = domain of (R^{-1}).

Example 5. *Consider the set A = {a, b, c}. Give an example of a relation R on A which is*

(i) *reflexive and symmetric but not transitive.*

(ii) *symmetric and transitive, but not reflexive.*

(iii) *reflexive and transitive, but not symmetric.*

Solution. (i) Given $A = \{a, b, c\}$

Let $R = \{(a, a), (a, b), (b, a), (b, c), (c, b), (b, b), (c, c)\}$ on A.

Clearly, R is reflexive and symmetric but not transitive.

(ii) Let $R = \{(a, a), (a, b), (b, a), (b, b)\}$ on A.

Here, R is symmetric and transitive but not reflexive.

(iii) Let $R = \{(a, a), (b, b), (c, c), (a, b)\}$ on A.

Here, R is reflexive, transitive but not symmetric.

Example 6. *If R is a relation on N×N, show that the relation **R** defined by (a, b) R (c, d) if and only if ad = bc is an equivalence relation.*

Solution. (i) Since $ab = ba$ $\forall$ $a, b \in \mathbf{N}$.

Therefore, $(a, b) R (a, b) \forall a, b \in \mathbf{N} \Rightarrow R$ is reflexive.

(ii) We have $(a, b) R (c, d)$ iff $ad = bc$ $\forall$ $a, b, c, d \in \mathbf{N}$

Now, $(c, d) R (a, b)$ iff $cb = da$ $\forall$ $a, b, c, d \in \mathbf{N} \Rightarrow R$ is symmetric.

(iii) We have $(a, b) R (c, d)$ iff $ad = bc \ \forall \ a, b, c, d \in \mathbf{N}$

Therefore, $(a, b) R (c, d), (c, d) R (e, f) \Rightarrow (a, b) R (e, f) \ \forall \ a, b, c, d \in \mathbf{N}$

Using $(a, d), (c, f) = (b, c)(d, e)$

$\Rightarrow$ $(a, f) = (b, e) \Rightarrow R$ is transitive

Hence, from (i), (ii) and (iii), we conclude that R is an equivalence relation.

Example 7. *Let R_1 and R_2 be two relations on a set A, where $A = \{1, 2, 3, 5\}$ such that*

$$R_1 = \{(1, 1), (1, 2), (1, 5), (2, 1), (2, 5)\}$$

and $R_2 = \{(3, 3), (3, 2), (2, 3), (1, 2), (2, 1)\}$

Then, which of the following statement is false :

(i) $R_1 \cup R_2$ is symmetric   *(ii) $R_1 \cap R_2$ is transitive*

(iii) $R_1 \cap R_2$ is symmetric   *(iv) $R_1 \cup R_2$ is transitive.*

Solution.   (i) As $(1, 2) \in R_1$, also $(2, 1) \in R_1$, therefore, it is symmetric and as $(1, 2) \in R_2$, also $(2, 1) \in R_2 \Rightarrow R_2$ is symmetric.

Now, $R_1 \cup R_2 = \{(1, 1), (1, 2), (1, 5), (2, 1), (2, 5), (3, 3), (3, 2), (2, 3)\}$

In $R_1 \cup R_2$, as $(1, 2) \in R_1 \cup R_2$, also $(2, 1) \in R_1 \cup R_2 \Rightarrow R_1 \cup R_2$ is symmetric

Therefore, (i) is true.

(ii) We have $R_1 \cap R_2 = \{(1, 2), (2, 1)\}$

$\Rightarrow$ $(1, 1)$ should also belong to $R_1 \cap R_2$.

But in this case $(1, 1) \notin R_1 \cap R_2$. Hence, $R_1 \cap R_2$ is not transitive.

Therefore, (ii) is false.

(iii) We have, $R_1 \cap R_2 = \{(1, 2), (2,1)\}$

$(1, 2) \in R_1 \cap R_2$ and also $(2, 1) \in R_1 \cap R_2$.

Therefore, (iii) is true.

(iv) In $R_1 \cup R_2$, $(1, 2) \in R_1 \cup R_2$

and $(2, 5) \in R_1 \cup R_2$, also $(1, 5) \in R_1 \cup R_2$

$\Rightarrow$ $R_1 \cup R_2$ is transitive

Therefore, (iv) is true.

Example 8. *If A be the set of all triangles in a plane and $R = \{(a, b) : \Delta a = \Delta b\}$, i.e.,*

$aRb \Leftrightarrow$ Area of triangle $a =$ Area of triangle b, then show that R is an equivalence relation.

Solution.   (i) Since, for all $a \in A$ we have $\Delta a = \Delta a$

Therefore, $aRa \Rightarrow R$ is reflexive.

(ii) For any $a, b \in A$, we have $(a, b) \in R \quad \Rightarrow \Delta a = \Delta b$

$\Rightarrow$ $\Delta b = \Delta a \Rightarrow \quad (b, a) \in R$

Therefore, $(b, a) \in R$, i.e., $bRa \Rightarrow R$ is symmetric.

(iii) For all $a, b, c \in A$, we have $(a, b) \in R, (b, c) \in R$

$$\Delta a = \Delta b \text{ and } \Delta b = \Delta c \quad \Rightarrow \quad \Delta a = \Delta c \quad \Rightarrow \quad (a, c) \in R$$

Therefore, R is transitive.

Hence, from (i), (ii) and (iii), we conclude that R is an equivalence relation.

Example 9. *Let **Z** be a set of non-zero integers and a relation R defined by $xRy \Rightarrow x^y = y^x \ \forall \ x,$ $y \in \textbf{Z}$, then show that R is not an equivalence relation on **Z**.*

Solution. (i) Let $x \in \textbf{Z}$, then $x^x = x^x, \ \forall \ x \in \textbf{Z}$

     $\Rightarrow$ $xRx, \ \forall \ x \in \textbf{Z}$

     Therefore, R is reflexive.

   (ii) Let $x, y \in \textbf{Z}$, such that xRy, i.e., $x^y = y^x$

     $\Rightarrow$ $x^y = y^x \Rightarrow y^x = x^y$

     Therefore, $xRy \Rightarrow yRx, \ \forall \ x, y \in \textbf{Z}$

     $\Rightarrow$ R is symmetric.

   (iii) Let $x, y, z \in \textbf{Z}$ such that xRy and yRz

     i.e., $x^y = x^y$ and $y^z = z^y$ which does not give $x^z = z^x$

     $\Rightarrow$ R is not transitive.

     Hence, we conclude that R is not an equivalence relation.

Example 10. *Let $A = \textbf{R} \times \textbf{R}$ (**R** is the set of real numbers) and define the following relation on $A : (a, b) R (c, d)$ iff $a^2 + b^2 = c^2 + d^2$*

 (i) verify that R is an equivalence relation on A.

 (ii) describe geometrically what the equivalence classes are for this reason.

Solution. (i) we have $(a, b)R(c, d)$ $\Rightarrow \ a^2 + b^2 = c^2 + d^2$

    $\Rightarrow$ $c^2 + d^2 = a^2 + b^2 \ \Rightarrow \ (c, d)R(a, b)$ ...(1)

    $\Rightarrow$ R is symmetric.

    Now, $(a, b)R(c, d)$ and $(c, d)R(x, y) \Rightarrow a^2 + b^2 = c^2 + d^2$

    and $c^2 + d^2 = x^2 + y^2$

    $\Rightarrow$ $a^2 + b^2 = x^2 + y^2 \ \Rightarrow \ (a, b)R(x, y)$ ...(2)

    $\Rightarrow$ R is transitive.

    Again $(a, b)R(a, b) \Leftrightarrow a^2 + b^2 = a^2 + b^2$ ...(3)

    $\Rightarrow$ R is reflexive.

    Hence, from (1), (2) and (3), we conclude that R is an equivalence relation.

   (ii) For any point (a, b), the sum $a^2 + b^2$ is the square of the distance from the origin. The equivalence classes are, therefore, the set of points in the plane which have the same distance from the origin. Hence, the equivalence classes are concentric circles centered at the origin.

Example 11. *Let R be the binary relation defined as $R = \{(a, b) \in R^2 : a - b \leq 3\}$. Determine whether R is reflexive, symmetric, anti symmetric and transitive.*

Solution.   We have $(a, b) \in R^2 : a - b \leq 3$.

    $\Rightarrow$ $(a, a) \in R^2 : a - a \leq 3$ i.e., $0 \leq 3$, which is true. So, R is reflexive.

    In a similar way, we can easily show that R is neither symmetric, anti symmetric nor transitive.

1.13.4 Relations other than Equivalence

Let R be a given relation on the set X. Then R is

(i) non-reflexive if $\exists x$, such that $(x, x) \notin R$.

(ii) anti-reflexive or reflexive if $i_x \cap R = \phi$ (where i_x is the identity relation on X or $\forall x \in X : (x, x) \notin R$

(iii) non-symmetrical if for some $(x, y) \in R$, we have $(y, x) \notin R$

(iv) anti-symmetric if $R \cap R^{-1} = i$, *i.e.*, $(x, y) \in R$ and $(y, x) \in R \Rightarrow x = y$

(v) asymmetric if $R \cap R^{-1} = \phi$, *i.e.*, $(x, y) \in R \Rightarrow (y, x) \notin R$

(vi) non-transitive if $R \circ R \not\subset R$

(vii) anti-transitive if $(R \circ R) \cap R = \phi$

(viii) A reflexive and symmetric, but not transitive relation is called a tolerance relation.

(ix) A non-symmetric transitive relation is called an ordered relation.

(x) A reflexive, anti-symmetric and transitive relation is called partial-ordered relation.

EXERCISE 1.6

1. If R is the relation 'is less than' from $A = \{1, 2, 3, 4, 5\}$ to $B = \{1, 4, 5\}$, find the set of ordered pairs corresponding to R. Also find R^{-1}.

2. A relation R defined from a set $A = \{2, 3, 4, 5\}$ to a set $B = \{3, 6, 7, 10\}$ as follows :

$(x, y) \in R \Rightarrow x$ divides y. Write R as a set of ordered pairs and determine the domain and range of R. Also find R^{-1}.

3. Find the domain and range of $A = \{1, 2, 3, 4, 5, 6\}$ when the relation are defined as

(i) xR_1y if and only if $x - y > 0$

(ii) xR_2y if and only if $x + y < 0$

4. Two sets A and B are given by $A = \{1, 2, 8, 9\}$ and $B = \{2, 3, 4, 6, 7\}$ and if R is the relation form A to B given by $\{(1,2), (1,3) ,(2,4), (2,6)\}$, then which of the following statement is true?

(i) Domain (R) = Range (R^{-1}) and Range (R) = Domain (R^{-1})

(ii) Domain (R) = Domain (R^{-1}) and Range (R) = Range (R^{-1})

(iii) Domain (R) = Range (R^{-1}) and Range (R) = Domain (R^{-1})

(iv) Domain (R) = Range (R)

5. If R is a relation on a set A, then which of the following statement is not true?

(i) If R is reflexive then R^{-1} is reflexive.

(ii) If R is symmetric then R^{-1} is symmetric.

(iii) If R is transitive, then R^{-1} is transitive.

(iv) None of these

6. Find the domain and range of the following relations:

(i) $R = \{(x + 1, x + 5)\} : x \in \{0, 1, 2, 3, 4, 5\}$

(ii) $R = \{(x, x^3) : x$ is a prime number, less than 10$\}$

(iii) $R = \{(a, b) : a \in \mathbf{N}, a < 5, b = 4\}$

(iv) $R = \{(a, b) : b = |a - l|, a \in \mathbf{Z}$, and $|a| \leq 3\}$

7. Let R_1 be the relation defined on the set of reals **R** such as $(a, b) \in R_1$ if and only if $1 + ab > 0$ for all $a, b \in \mathbf{R}$. Show that R_1 is reflexive, symmetric but not transitive.

8. Let R be relation on $\mathbf{N} \times \mathbf{N}$, defined by $(a, b) R (c, d)$ if and only if $ad (b + c) = bc (a + d)$. Show that R is an equivalence relation.

9. Show that the relation 'congruence modulo m' on the set of integers is an equivalence relation.

10. Let R_1 be a relation on the set of reals defined by $R_1 = \{(a, b) \in R \times R : a^2 + b^2 = 1\}$ Show that R_1 is not an equivalence relation on R.

11. In a set L of all straight lines in a plane, discuss which of the following two relations are equivalence relations L.

(i) $R_1 = \{(x, y): x, y \in L$ and x is parallel to $y\}$

(ii) $R_2 = \{(x, y): x, y \in L$ and x is perpendicular to $y\}$.

12. Show that the relation $R = \{(a, b) : a - b = $ even integer $\forall a, b \in \mathbf{Z}\}$, *i.e.,* $aRb \Leftrightarrow a - b = $ even integer, is an equivalence relation.

13. Show that the relation R in $\mathbf{N}$, the set of natural numbers, defined by xRy if $x^2 - 4xy + 3y^2 = 0$, $(x, y \in \mathbf{N})$ is reflexive, not symmetric and not transitive.

14. For the given relation R on a set S, determine which are equivalence relations:

   (i) S is the set of all rational numbers, aRb if and only if $a = b$

   (ii) S is the set of all real numbers iff

      (a) $|a| = |b|$ (b) $a \geq b$

   (iii) S is the set of all triangles in a plane, aRb iff a is congruent to b.

   (iv) S is the set of all triangles in a plane, aRb iff a and b have equal perimeters.

15. An integer m is said to be related to another integer n if m is a multiple of n. Show that this relation is reflexive and transitive but not symmetric.

16. Let R be a relation defined on the set of natural number $\mathbf{N}$ as $R = \{(x, y): x, y \in \mathbf{N}, 2x + y = 41\}$. Find the domain and range of R.

17. Let O be the origin. Define a relation between two points P and Q in a plane if $PO = OQ$. Show that the relation is an equivalence relation.

18. Given the relation $R = \{(1, 2), (2, 3)\}$ on the set of natural number $\mathbf{N}$, add a minimum of ordered pairs so that the enlarged relation is symmetric, transitive and reflexive.

19. Let $\mathbf{N}$ denote the set of all natural numbers and R be the relation on $\mathbf{N} \times \mathbf{N}$ defined by $(a, b) R (c, d) \Leftrightarrow ad (b + c) = bc (a + d)$. Show that R is an equivalence relation.

20. Show that the relation, which is symmetric and transitive, is not necessarily reflexive.

Answers

1. $aRb = \{(1, 4), (1, 5), (2, 4), (3, 4), (2, 5), (3, 5), (4, 5)\}$,
   $R^{-1} = \{(4, 1), (5, 1), (4, 2), (5, 2), (4, 3), (5, 3), (5, 4)\}$
2. Domain $(R) = \{2, 3, 5\}$, Range $(R) = \{3, 6, 10\}$, $R^{-1} = \{(6, 2), (10, 2), (3, 3), (6, 3), (10, 5)\}$
3. (i) $\{2, 3, 4, 5, 6\}$, $\{1, 2, 3, 4, 5\}$, (ii) ϕ, ϕ 4. (iii) 5. (iv) 6. (i) Domain $(R) = \{1, 2, 3, 4, 5, 6\}$, Range $(R) = \{5, 6, 7, 8, 9, 10\}$ (ii) Domain $(R) = \{2, 3, 5, 7\}$, Range $(R) = \{8, 27, 125, 243\}$ (iii) Domain $(R) = \{1, 2, 3, 4\}$, Range $(R) = \{4\}$ (iv) Domain $(R) = \{0, -1, -2, -3, 1, 2, 3\}$, Range $(R) = \{1, 2, 3, 4, 0, 1, 2\}$ 11. R_1 = Equivalence relation, R_2 = Not equivalence 14. (i), (ii) 16. Domain $(R) = \{1, 2, ..., 19, 20\}$, Range $(R) = \{39, 37, 35, ..., 5, 3, 1\}$ 18. $\{(1, 2), (2, 1), (2, 3), (3, 2), (1, 3), (3, 1), (1, 1), (2, 2), (3, 3), (4, 4), ...\}$

1.14 FUNCTIONS

Definition: *Let A and B be two sets, then a rule or corresponding, which associates each element of A to a unique element to B, is called a function from set A to set B.*

If a general element of set A is denoted by x, and of set B is denoted by y, then we say that y is a function of x if, for every $x \in A$, one and only one value of $y \in B$ can be determined.

Symbolically: If f is a function from a set A to a set B, then we write $f : A \to B$, read as f is a function from A to B or f maps A to B.

1.14.1 Range and Domain of a Function

Let an element $y \in B$ be corresponded by an element $x \subset A$, then y is called the image of x and is denoted by $f(x)$. Here, x is defined as the pre-image of y.

The set A is called the domain and the set B is called the co-domain of the function f.

The set of all f-images of the elements of A, is called image set or the range of f and is denoted by

$$f(A) \quad \text{or} \quad \{f(x) : x \in A\}$$

Evidently, $f(A) \subseteq B$.

Thus, a mapping $f : A \to B$ is the set of ordered pairs $\{(a, b) : a \in A, b \in B\}$, so that no two ordered pairs have the same finite element.

$$f = \{(a, b): a \in A, b \in B, b = f(x) \; \forall \; a \in A\}$$

For example: Let $A = \{-2, -1, 0, 1, 2\}$ and B is the set of natural numbers for every $x \in A$, $f(x) \in B$ and $f(x) = x^2$.

Here, A is the domain and B is the co-domain.

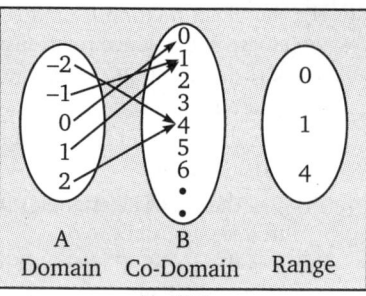

$f(a)$ is the value of the function $f(x)$, when x takes the value a, *i.e.,* when x is replaced by a.

The elements of the co-domain which is equal to $f(x)$ form the range.

Fig. (15)

When $x = -2$, $f(-2) = (-2)^2 = 4$

When $x = -1$, $f(-1) = 1$

When $x = 0$, $f(0) = 0$

When $x = 1$, $f(1) = 1$

When $x = 2$, $f(2) = 4$.

Which can be illustrated in the figure (15).

REMARKS

☞ If $f : A \to B$ then a single element in A cannot have more than one image in B. However, two or more elements in A may have the same image in B.

☞ Every element in A must have its image in B, but every element in B may not have it pre-image in A.

☞ To each element x in A, there exists a unique element y in B such that $y = f(x)$.

☞ The unique element y of B is called the value of f at x (the image of f under x), and written as $y = f(x)$.

☞ The range of f consist of those elements in B which appear as the image of at least one element in A.

☞ Range of a function is the image of its domain.

☞ Range is a subset of co-domain.

1.15 TYPE OF FUNCTIONS

(a) One-One function: *A function f from A to B, i.e., $f : A \to B$ is said to be one-one (or injective) iff distinct elements of A have distinct images.*

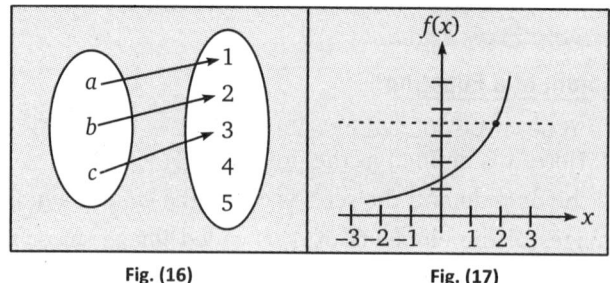

Fig. (16)   Fig. (17)

Symbolically: f is one-one if for $x_1, x_2 \in A$, we have

$$x_1 \neq x_2 \quad \Rightarrow \quad f(x_1) \neq f(x_2) \ \forall \ x_1, x_2 \in A$$

or $$f(x_1) = f(x_2) \Rightarrow \quad x_1 = x_2 \ \forall \ x_1, x_2 \in A$$

It is also called Univalent function.

Graphically, a function is one-one if and only if no line parallel to x-axis meets the graph of the function at more than one point.

(b) Many-One Function: *A function $f : A \rightarrow B$ is called many-one, if at least one element of co-domain B has two or more than two pre-images in domain A.*

Symbolically: f is many-one if for $x_1, x_2 \in A$, we have $x_1 \neq x_2 \Rightarrow f(x_1) = f(x_2)$

This can be illustrated in the following figures.

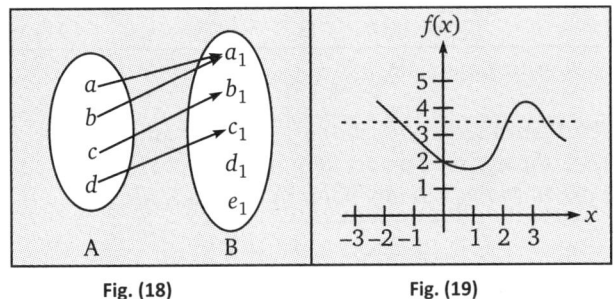

Fig. (18)   Fig. (19)

Graphically, a function is many-one if and only if a line parallel to *x*-axis meets the graph of the function at more than one point.

REMARK

☞ One-many function does not exist.

(c) Onto function: *A function $f : A \rightarrow B$ is called an onto function, if there is no element of B which is not an image of some element of A,* i.e., every element of B appears as the image of at least one element of A. This is illustrated in Figure 20.

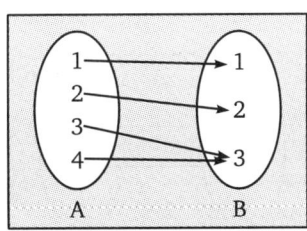

Fig. (20): Onto Function

REMARKS

☞ In an onto function, Range = Co-domain
☞ Onto function is also called surjective.

(d) Into function: *A function $f : A \rightarrow B$ is called an into function,* i.e., *if there is at least one element of set B which has no pre-image in the set A. This is illustrated in Figure 21.*

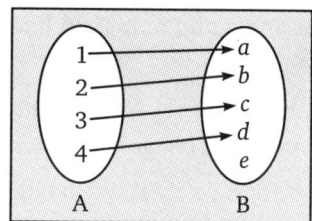

Fig. (21): Into Function

REMARK

☞ In an into function, Range $\subset$ Co-domain.

(e) One-One Into Function: *A function* $f : A \rightarrow B$ *is called a one-one into function, if it is both one-one and into, i.e., the different points in A are joined to different points in B and there are some points in B which are not joined to any point in A. This is illustrated in* Figure 22.

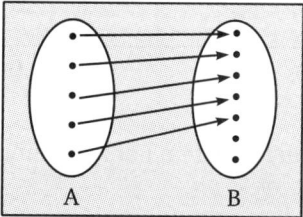

Fig. (22): One-One Into Function

Symbolically : One-one into function is defined as

(i) Range $\subset$ Co-domain.

(ii) $f(x_1) \neq f(x_2) \Rightarrow x_1 \neq x_2.$

(f) One-One Onto Function: *A function* $f : A \rightarrow B$ *is both one-one and onto, i.e., the different points in A are joined to different points in B and no point in B is left vacant. This is illustrated in Figure* 23.

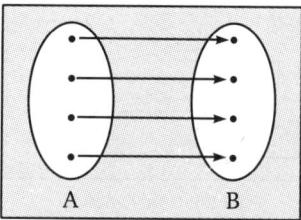

Fig. (23): One-one Onto Function

REMARKS

☞ One-one onto mapping is also known as bijective or one-to-one.
☞ For a one-one onto function

Range = Co-domain, and $x_1 \neq x_2 \Rightarrow f(x_1) \neq f(x_2)$ or $f(x_1) = f(x_2) \Rightarrow x_1 = x_2$

(g) Many-One Into Function: *A function f : A → B which is both many-one and into function is called a many-one into function, i.e., two or more points in A are joined to same points in B and there are some point in B which are not joined to any point in A. Therefore, for many-one into function.*

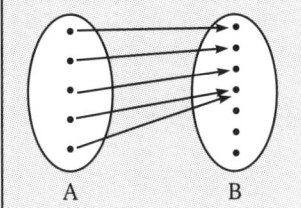

Fig. (24): Many-One Into Function

(i) Range ⊂ Co-domain.

(ii) $x_1 \neq x_2$

$\Rightarrow f(x_1) = f(x_2)$

(h) Many-One Onto Function: *If function f : A → B is both many-one and onto function is called a many one onto function, i.e., in B one point is join   ed to at least one point in A and two or more points in A are joined to some points in B. Therefore, for many-one onto function.*

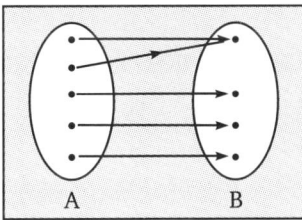

Fig. (25): Many-One Onto Function

(i) Range = Co-domain.

(ii) $x_1 \neq x_2$

$\Rightarrow f(x_1) = f(x_2)$

WORKING PROCEDURE

1. **For checking the Injectivity (One-One) of the function**

   Let x and y be two arbitrary elements in the domain of f.

   Step 1. Take $f(x) = f(y)$

   Step 2. If we get $x = y$, after solving $f(x) = f(y)$. Then, $f : A \to B$ is one-one.

2. **For checking the surjectivity (Onto) of a function**

   Step 1. Take an arbitrary element y in the co-domain.

   Step 2. Put $f(x) = f(y)$

   Step 3. Solve $f(x) = y$ for x and obtain x in terms of y.

   Step 4. Get the equation of the form $x = g(y)$

   Step 5. If $x = g(y)$ belongs to domain f, for all values of y, then f is onto.

Solved Examples

Based on the following Results

▶ For a function $f : A \to B$, A = domain, B = co-domain.

▶ **For one-one function:** $x_1 \neq x_2 \Rightarrow f(x_1) \neq f(x_2) \, \forall \, x_1, x_2 \in A$

▶ or $f(x) = f(x_2) \Rightarrow x_1 = x_2 \, \forall \, x_1, x_2 \in A$

▶ **For many-one function:** $x_1 \neq x_2 \Rightarrow f(x_1) = f(x_2)$, $x_1, x_2 \in A$

▶ **For onto function:** Range = co-domain

▶ **For into function:** Range $\subset$ co-domain

▶ **For one-one into function:** (i) Range $\subseteq$ co-domain
  (ii) $f(x_1) \neq f(x_2) \Rightarrow x_1 \neq x_2$.

▶ **For one-one onto function:** (i) Range = codomain
  (ii) $x_1 \neq x_2 \Rightarrow f(x_1) \neq f(x_2)$ or $f(x_1) = f(x_2) \Rightarrow x_1 = x_2$

▶ **For many-one into function:** (i) Range = co-domain   (ii) $x_1 = x_2 \Rightarrow f(x_1) = f(x_2)$

▶ **For many-one onto function:** (i) Range = co-domain (ii) $x_1 \neq x_2 \Rightarrow f(x_1) = f(x_2)$

Example 1. *Let $f : \mathbf{R} \to \mathbf{R}$ be a function defined by*

$$f(x) = \begin{cases} 3x - 1 \text{ when } & x > 3 \\ x^2 - 1 \text{ when } -2 \leq x \leq 3 \\ x + 3 \quad \text{ when } & x < -2 \end{cases}$$

Find (i) f(2), (ii) f(4), (iii) f(–1), (iv) f(–3)

Solution. (i) $f(2) = (2)^2 - 2 = 4 - 2 = 2$

(ii) $f(4) = 3(4) - 1 = 12 - 1 = 11$

(iii) $f(-1) = (-1)^2 - 2 = 1 - 2 = -1$

(iv) $f(-3) = 2(-3) + 3 = -6 + 3 = -3$

Example 2. *For $y = +\sqrt{x}$, say whether it is a function or not. If it is a function, find its domain and range.*

Solution. Here we have $y = +\sqrt{x}$ ...(1)

Since y is real if $x \geq 0$ and is unique and finite for each $x \geq 0$.
Therefore, (1) is a function with domain $[0, \infty[$.
Again from (1), $y \geq 0 \, \forall \, x \geq 0$
Hence, range $= [0, \infty [$

Example 3. *Find the domain of $f(x) = \dfrac{x^3 - x^2 + 4x + 2}{3x + 11}$*

Solution. Since f is defined for all real values of x except when $3x + 11 = 0$

i.e., when, $x = -\dfrac{11}{3}$

Hence, domain of $f = \mathbf{R} - \left\{ -\dfrac{11}{3} \right\}$

Example 4. *Let $f : \mathbf{N} - \{1\} \to \mathbf{N}$ be defined by $f(n) =$ the highest prime factor of n. Show that f is neither one-one nor onto. Also, find the range f.*

Solution. Since we have

$f(6) =$ the highest prime factor of $6 = 3$

$f(9) =$ the highest prime factor of $9 = 3$

$f(12) =$ the highest prime factor of $12 = 3$

Therefore, f is a many-one function.

Clearly, image of any $n \in \mathbf{N} - \{1\}$ is the largest prime number that divides n. So the range of f consists of prime number only. Consequently, range of $f \neq \mathbf{N}$ (Co-domain)

$\Rightarrow$ f is not onto function.

Hence, f is neither one-one nor onto. The range of f is the set of all prime numbers.

Example 5. *Let $A = \{1, 2\}$. Find all one-to-one function from A to A.*

Solution. Let $f : A \to A$ be a one-one function.

Then, for $f(1)$, there are two choices, *i.e.*, 1 or 2.

Let us first suppose $f(1) = 1$.

As $f : A \to A$ is one-one, $f(2) = 2$

Therefore, we have $f(1) = 1, f(2) = 2$

Now, let $f(1) = 2$

Since, $f : A \to A$ is one-one, therefore $f(2) = 1$.

Therefore, we have $f(1) = 2$ and $f(2) = 1$.

Hence, we have two one-one function say f and g form A and A given by $f(1) = 1$, $f(2) = 2$ and $f(2) = 1$ and $f(1) = 2$.

Example 6. *Let $\{x \in \mathbf{R} : -1 \leq x \leq 1\} = B$. Show that $f : A \to B$ given by $f(x) = x\,|x|$ is one-one and onto.*

Solution. Let x, y be any two elements in A, then

$$x \neq y \Rightarrow x|x| \neq y|y| \Rightarrow f(x) \neq f(y).$$

Therefore, f is one-one.

Since, range of $f = f(A) = B$ so $f : A \to B$ is onto mapping. Hence f is one-one and onto.

Example 7. *Find the domain and range of the function.*

$$f(x) = -\sqrt{-5 - 6x - x^2}$$

Solution. Given that, $f(x) = -\sqrt{-5 - 6x - x^2}$

For f to be real, $-5 - 6x - x^2 \geq 0$ $\Rightarrow$ $x^2 + 6x + 5 \leq 0$

$\Rightarrow$ $x^2 + 6x \leq -5$ $\Rightarrow$ $x^2 + 6x + 9 \leq -5 + 9$

$\Rightarrow$ $(x + 3)^2 \leq 4$ $\Rightarrow$ $|x + 3|^2 \leq 4$

$\Rightarrow$ $|x + 3| \leq 2$ $\Rightarrow$ $-2 \leq x + 3 \leq 2$

$\Rightarrow$ $-2 - 3 \leq x \leq 2 - 3$ $\Rightarrow$ $-5 \leq x \leq -1$

Therefore, domain of $f(x) = [-5, -1]$

To find the range of $f(x)$, put $y = f(x)$

Therefore, $f(x) = -\sqrt{-5 - 6x - x^2}, y \leq 0$

$$\Rightarrow \quad y^2 = -5 - 6x - x^2 \qquad\qquad \Rightarrow \quad x^2 + 6x + (y^2 + 5) = 0$$

For real x, discriminant ≥ 0, $\qquad$ *i.e.,* $(6)^2 - 4 \times 1 \times (y^2 + 5) \geq 0$

$$\Rightarrow \qquad 36 - 4y^2 - 20 \geq 0 \qquad\qquad \Rightarrow \qquad\qquad -4y^2 \geq -16$$

$$\Rightarrow \qquad\qquad y^2 \leq 4 \qquad\qquad \Rightarrow \qquad\qquad |y|^2 \leq 4$$

$$\Rightarrow \qquad\qquad |y| \leq 2 \qquad\qquad i.e., \qquad\qquad -2 \leq y \leq 2$$

But $y \leq 0$ therefore, $-2 \leq y \leq 0$.

Hence, Range of $f = [-2, 0]$

Example 8. *For a finite set A, if $f : A \to A$ is a one-one function, show that f is onto.*

Solution. Let $A = \{a_1, a_2, ..., a_n\}$ be a finite set.

Since $f : A \to A$ is one-one function, therefore $f(a_1), f(a_2), ..., f(a_n)$ are distinct elements of the set A, but A has only n elements. Therefore,

$$A = \{f(a_1), f(a_2), ..., f(a_n)\}$$

$\Rightarrow \qquad$ Co-domain = Range

Hence, every element in A (co-domain) has its pre-image in the domain A.

$\Rightarrow \quad f : A \to A$ is onto.

REMARK

☞ For a finite set A, if $f : A \to A$ is onto function, then f is one-one.

Example 9. *If $f : \mathbf{R} \to \mathbf{R}$ be a function defined by $f(x) = 4x^3 - 7$, show that the function f is bijective.*

Solution. Given that $f(x) = 4x^3 - 7$; $x \in \mathbf{R}$

(i) **f is one-one :** Let $x_1, x_2 \in \mathbf{R}$

Now, $\qquad f(x_1) = f(x_2)$

$$\Rightarrow \qquad 4x_1^3 - 7 = 4x_2^3 - 7 \qquad\qquad \Rightarrow \quad 4x_1^3 = 4x_2^3$$

$$\Rightarrow \qquad\qquad x_1^3 = x_2^3 \qquad\qquad \Rightarrow \quad x_1^3 - x_2^3 = 0$$

$$\Rightarrow \qquad (x_1 - x_2)(x_1^2 + x_1 x_2 + x_2^2) = 0$$

$$\Rightarrow \qquad (x_1 - x_2)\left[\left(x_1 + \frac{x_2}{2}\right)^2 + \frac{3x_2^2}{4}\right] \qquad \left\{\because \left[\left(x_1 + \frac{x_2}{2}\right)^2 + \frac{3x_2^2}{4} \neq 0\right]\right\}$$

$$\Rightarrow \qquad (x_1 - x_2) = 0 \qquad\qquad \Rightarrow \quad x_1 = x_2$$

Therefore, f is one-one.

(ii) **f is onto :** Let $c \in \mathbf{R}$

$$f(x) = c \quad \Rightarrow 4x^3 - 7 = c \qquad \Rightarrow x = \left(\frac{c+7}{4}\right)^{1/3}$$

Now, $\left(\dfrac{c+7}{4}\right)^{1/3} \in \mathbf{R}$ and $f\left\{\left(\dfrac{c+7}{4}\right)^{1/3}\right\} = 4\left[\left(\dfrac{c+7}{4}\right)^{1/3}\right]^3 - 7 = c + 7 - 7 = c$

Which implies that c is the image of $\left(\dfrac{c+7}{4}\right)^{1/3}$

Therefore, f is onto. Hence, f is bijective function.

Example 10. *Let A and B be two sets. Prove that $f : A \times B \rightarrow B \times A$ difined by $f(a, b) = (b, a)$ is one-one and onto.*

Solution.   (i)   **f is one-one :** Let (a_1, b_1) and $(a_2, b_2) \in A \times B$ such that

$$f(a_1, b_1) = f(a_2, b_2)$$
$$\Rightarrow \qquad (b_1, a_1) = (b_2, a_2)$$
$$\Rightarrow \qquad b_1 = b_2 \text{ and } a_1 = a_2$$

Therefore, $(a_1, b_1) = (a_2, b_2)$

Thus, $f(a_1, b_1) = f(a_2, b_2)$

$\Rightarrow \qquad (a_1, b_1) = (a_2, b_2) \; \forall \; (a_1, b_1), (a_2, b_2) \in A \times B$

$\Rightarrow \qquad f$ is one-one.

(ii)   **f is onto :** Let $(b, a) \in B \times A$ such that $b \in B$ and $a \in A$.

$\Rightarrow \qquad\qquad (a, b) \in A \times B$

Therefore, for all $(b, a) \in B \times A$, there exist $(a, b) \in A \times B$ such that $f(a, b) = (b, a)$

$\Rightarrow \qquad f$ is onto. Hence f is one-one and onto.

EXERCISE 1.7

1. Let $A = \{-2, -1, 0, 1, 2\}$ and $f : A \rightarrow \mathbf{Z}$ given by $f(x) = x^2 - 2x - 3$. Find :

    (i) the range of f,

    (ii) pre-image of 6, –3 and 5.

2. Find the domain and range of the following function

$$f(x) = \sqrt{(x-1)(3-x)}$$

3. Find the range of the following function

$$f(x) = \frac{1}{(2x-3)(x+1)}$$

4. Find the domain and range of the following functions :

    (i) $f(x) = \dfrac{x^2 - 1}{x - 1}$ (ii) $y = -|x|$

    (iii) $f(x) = \dfrac{|x-1|}{x-1}$ (iv) $y = \sqrt{x-3}$

5. If $A = \{-1, 0, 2, 5, 6, 11\}$,

    $B = \{-2, -1, 0, 18, 25, 108\}$

    and $f(x) = x^2 - x - 2$, find $f(A)$.

6. Let A be the set of two positive integers. Let $f : A \rightarrow \mathbf{Z}^+$, set of positive integers be defined by $f(n) = p$, where p is the highest prime factor of n. If range of $f = \{3\}$, find A.

7. Find the domain for which the function $f(x) = 2x^2 - 1$ and $g(x) = 1 - 3x$ are equal.

8. Let $f_1 : \mathbf{R} \rightarrow \mathbf{R}$ and $f_2 : \mathbf{C} \rightarrow \mathbf{C}$ be two functions defined as $f_1(x) = x^3$ and $f_2(x) = x^3$. Show that they are not equal.

9. Let $A = \{p, q, r, s\}$ and $B = \{1, 2, 3\}$. Which of the following relations from A to B not a funcion?

    (i) $R_1 = \{(p, 1), (q, 2), (r, 1), (s, 2)\}$

    (ii) $R_2 = \{(p, 1), (q, 1), (r, 1), (s, 1)\}$

    (iii) $R_3 = \{(p, 1), (q, 2), (r, 2), (s, 3)\}$

    (iv) $R_4 = \{(p, 2), (q, 3), (r, 2), (s, 2)\}$

10. Write the following relations as sets of ordered pairs and find which of them are functions :

    (i) $\{(x, y) : y = 3x, x \in (1, 2, 3),$
          $y \in (3, 6, 9, 12)\}$

    (ii) $\{(x, y) : y > x + 1, x = 1, 2$ and
          $y = 2, 4, 6\}$

    (iii) $\{(x, y) : x + y = 3$
          $x, y \in (0, 1, 2, 3)\}$

11. Express the following functions as sets of ordered pairs, and find their range :

    (i) $f_1 : A \rightarrow \mathbf{R} : f_1(x) = x^2 + 1$
        where $A = \{-1, 0, 2, 4\}$

    (ii) $f_2 : A \rightarrow \mathbf{N} : f_2(x) = 2x$
        where $A = \{x : x \in \mathbf{N}, x \leq 10\}$

12. Let $f : \mathbf{R} \rightarrow \mathbf{R}$ be a function such that $f(x) = 2^x$. Determine :

    (i) range of f

    (ii) $\{x : f(x) = 1\}$

    (iii) whether $f(x + y) = f(x) \cdot f(y)$ holds

13. Let $f : \mathbf{R}^+ \rightarrow \mathbf{R}$, be a function such that $f(x) = \log x$. Determine :

    (i) the image set of domain of f

    (ii) $\{x : f(x) = -2\}$

(iii) whether $f(xy) = f(x) + f(y)$ holds

14. Give an example of a map which is :

    (i) one-to-one but not onto

(ii) not one to one, but onto

(iii) neither one-to-one nor onto

=========================**Answers**=========================

1. (i) $f(A) = \{-4, -3, 0, 5\}$, (ii) ϕ, $\{1, 2\}$, -2 **2.** Domain $= [1, 3]$, Range $= [-1, 1]$ **3.** $\left]-\infty, \dfrac{-8}{25}\right] \cup [0, \infty[$

4. (i) $\mathbf{R} - \{1\}$, $\mathbf{R} - \{2\}$, (ii) $\mathbf{R} : \mathbf{R} - \mathbf{R}^+$, (iii) $\mathbf{R} - \{1\}$, $\{-1, 1\}$, (iv) $[3, \infty[$, $[0, \infty]$ **5.** $f(A) = \{1, -2, 18,$ $28, 108\}$ **6.** $A = \{3, 6\}$ or $(3, 9)$ or $[3, 12]$ etc. **7.** $(-2, 1/2)$ **9.** (iii) **10.** (i) $\{(1, 3), (2, 6), (3, 9)\}$, function,   (ii) $\{(1, 4), (1, 6), (3, 4), (3, 6)\}$, not function (iii) $\{(0, 3), (1, 2), (2, 1), (3, 0)\}$, function **11.** (i) $f_1 = \{x, f(x) : x \in A\} = \{(-1, 2), (0, 1), (2, 5), (4, 17)\}$ (ii) $f_2 = \{(x, g(x)) : x \in A\} = \{(1,2),(2,4),$ $(3, 6), ..., (10, 20)\}$ **12.** (i) Range of $f = \mathbf{R}^+$, the set of positive real numbers, (ii) $(x : f(x) = 1) = \{0\}$, (iii) $f(x+y) = f(x) \cdot f(y)$ holds for all $x, y \in \mathbf{R}$ **14.** (i) $n \rightarrow n^2 : \mathbf{N} \rightarrow \mathbf{N}$ (ii) $n \rightarrow |n| : \mathbf{Z} \rightarrow \mathbf{N} \cup \{0\}$ (iii) $n \rightarrow |n|^2 : \mathbf{Z} \rightarrow \mathbf{N} \cup \{0\}$

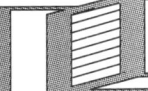 **Chapter Review:** *A competitive Approach*

Selected Terms and Results

Terms

- **Interval :** A subset S of $\mathbf{R}$ is called an interval if $a, b \in S$; $x \in \mathbf{R}$ such that $a < x < b$ implies $x \in S$.

- **Set:** A set is a well defined collection of objects.

- **Empty set :** A set containing no element is called empty set.

- **Finite set :** A set is said to be finite if it consists of only finite number of elements. Otherwise it is said to be infinite.

- **Equivalent set :** Two finite sets are said to be equivalent if they have the same cardinal number.

- **Union of two sets :** The union of two sets A and B denoted by $A \cup B$ is the set of those elements which either belong to A or B or to both.

- **Intersection of two sets :** The intersection of two sets A and B, denoted by $A \cap B$ is the set of all those elements which belongs to both A and B.

- **Disjoint sets :** When two sets have no common elements, they are called disjoint sets.

- **Ordered pair :** An ordered pair is a pair of entries whose components occur in a specific order.

- **Cartesian product of two sets :** The set of all ordered pairs of elements (a, b) : $a \in A$, $b \in B$ is called the cartesian product of two sets A and B.

- **Relation :** Let A and B be two sets. Then a relation R from A to B is a subset of $A \times B$.

- **Reflexive Relation :** A relation R is said to be reflexive if $(x, x) \in R \ \forall \ x \in A$.

- **Symmetric Relation :** A relation R on a set A is said to be symmetric if $(y, x) \in R$ whenever $(x, y) \in R$.

- **Transitive Relation :** A relation R on a set A is said to be transitive if $(x, y) \in R$, $(y, z) \in R$ $\Rightarrow (x, z) \in R$.

- **Anti-symmetric Relation :** A relation R on a non-empty set A is said to be an anti-symmetric iff $(x, y) \in R$ and $(y, x) \in R$ $\Rightarrow x = y \ \forall \ x, y \in R$.

- **Equivalence Relation :** A relation R on a set E is said to be equivalence if it is
  (i) reflexive
  (ii) anti-symmetric and
  (iii) transitive

- **Partial Ordered Relation :** A relation R on a set E is said to be partial ordered relation if it is
  (i) reflexive
  (ii) anti-symmetric and
  (iii) transitive

- **Function :** Let A and B be two sets, then the rule or correspondence which associates each element of A to a unique element of B is called a function or mapping.

- **Range and domain of a function:** Let an element $y \in B$ be corresponded by an element $x \in A$, then y is called the image of x and is denoted by $f(x)$. The set A is called the domain and the set B is called the co-domain of the function f.

- **One-One function :** A function $f : A \rightarrow B$ is said to be one-one iff distinct elements of A have distinct images.

- **Onto function :** A function $f : A \rightarrow B$ is called an onto function if there is no element of B which is not the image of some element A, i.e., every element of B appears as the image of at least one element of A.

- **Into function :** A function $f : A \rightarrow B$ is called an into function, if there is at least one element of set B which has no pre-image in the set A.

- **Even function :** A function $f : A \rightarrow B$ is said to be an even function if $f(-x) = f(x) \forall x \in A$.

- **Odd function :** A function $f : A \rightarrow B$ is said to be an odd function if $f(-x) = -f(x) \forall x \in A$.

Results

- Total number of subsets of a set A is equal to 2^n where n is the number of elements of A.
- The number of proper subsets of a set with n elements is $2^n - 2$.
- Union of sets is commutative, associative and idempotent.
- Difference of two sets is not commutative.
- Difference of a set with the universal set is called complementation.
- The identity and universal relations on a non-empty sets are transitive.
- The intersection of two equivalence relations on a set is equivalence relation.
- The union of two equivalence relations on a set is not necessarily an equivalence relation.
- If gof and fog both exist, they may not be equal.
- The composition of function is associative but not commutative.
- The composition of any function with the identity function is the function itself.
- The inverse of bijective function is unique.
- The inverse of bijective function is again bijective.

Review Questions and Project Work

1. Define union, intersection, difference and symmetric difference of two sets.
2. Define the power set of a set
3. How many element does the power set of a set S with n elements have?
4. Define what it mean for a function from the set of positive integers to the set of positive integer to be one to one.
5. Define the inverse of a function.
6. Define the floor and ceiling functions from the set of real numbers to the set of integers.
7. Let $f(n)$ be the function from the set of integers to the set of integers such that $f(n) = n^2 + 1$. What are the domain, co-domain and range of this function.
8. Give an example of a function from the set of positive integers to the set of positive integers

that is :
   (a) both one-one and onto.
   (b) one-one but not onto.
   (c) neither one-one nor onto.
   (d) not one-one but is onto.
9. When the empty set the power set of a set?
10. (a) Define what is means for two sets to be equal.
    (b) Describe the ways to show that two sets are equal.
11. Let A and B be sets in a finite universal set U. List the following in order of increasing size :
    (a) $|A|, |A \cup B|, |A \cap B|, |U|, |\phi|$
    (b) $|A - B|, |A \oplus B|, |A| + |B|, |A \cup B|, |\phi|$
12. Research where the concept of a function first arose and describe how this concept was first used.

Objective type Questions

Fill in the Blanks

1. A relation R on a set A is symmetric iff $R =$ _____.
2. Let R be an anti-symmetric relation on a set A such that $(a, b) \in R$ and $(b, a) \in R$. Then _____.
3. Let R be a relation on a set A such that $R = R^{-1}$. Then R is _____.
4. Let $A = \{1, 2, 3\}$, then the smallest equivalence relation on A is _____.
5. Let A be a finite set. Then the smallest equivalence relation on A is the _____
   relation on A.
6. The void relation on a set is _____ and _____ but not _____.
7. Let R be a relation defined by $R = \{(4, 5), (1, 4), (4, 6), (7, 6), (3, 7)\}$ on N. Then $R \circ R^{-1} =$ _____.
8. Let $R = \{(a, a), (b, c), (a, b)\}$ be a relation on a set $A = \{a, b, c\}$. Then the minimum number of ordered pairs which when added to R make it transitive is _____.

True/False

Write 'T' for true and F for false statement

1. A binary relation is a set. **(T/F)**
2. A void set defines a relation. **(T/F)**
3. The total number of relations from a set containing m elements to a finite set containing n elements is 2^{mn}. **(T/F)**
4. Every relation is a function. **(T/F)**
5. Every function is a relation. **(T/F)**

6. The total number of bijections from a set containing n elements to a set containing n elements is n^n. **(T/F)**
7. Every equivalence relation is symmetric. **(T/F)**

8. Every symmetric relation is equivalence. **(T/F)**
9. Every anti-symmetric relation is symmetric. **(T/F)**
10. The composition of functions is commutative. **(T/F)**
11. Reflexivity is redundant in the definition of an equivalence relation on a set A, because by symmetry $(a, b) \in R \Rightarrow (b, a) \in R$ and by transitivity $(a, b) \in R$ and $(b, a) \in R \Rightarrow (a, a) \in R$ **(T/F)**
12. The relation $R = \{(1, 2), (1, 3)\}$ is a transitive relation on a set $A = \{1, 2, 3\}$. **(T/F)**
13. The identity relation on a finite set A is the smallest equivalence relation on A. **(T/F)**

Multiple choice Questions

Choose the most appropriate one

1. Let R_1 and R_2 be two equivalence relation on a set. Consider the following assertion :
   (i) $R_1 \cup R_2$ is an equivalence relation.
   (ii) $R_1 \cap R_2$ is an equivalence relation.
   Which of the following is correct?
   (a) Both assertions are true.
   (b) Assertion (i) is true but assertion (ii) is not true.
   (c) Assertion (ii) is true but assertion (i) is not true.
   (d) Neither (i) nor (ii) is true.

2. The 'subset' relation on a set of set is :
   (a) a partial ordering
   (b) an equivalence relation
   (c) transitive and symmetric only
   (d) transitive and anti-symmetric only

3. Let R be a symmetric and transitive relation on a set A, then :
   (a) R is reflexive and hence an equivalence relation.
   (b) R is reflexive and hence a partial order.
   (c) R is not reflexive and hence is not an equivalence relation .
   (d) None of the above

4. The number of equivalence relations of the set $\{1, 2, 3, 4\}$ is :
   (a) 4   (b) 15
   (c) 16   (d) 24

5. Suppose A is a finite set with n elements. The number of elements in the large equivalence relation of A is :

   (a) 1   (b) n
   (c) $n + 1$ (d) n^2

6. The binary relation $S = \phi$ on the set $A = \{1, 2, 3\}$ is :
   (a) neither reflexive nor symmetric
   (b) symmetric and reflexive
   (c) transitive and reflexive
   (d) transitive and symmetric

7. Let $f(x) = x^2 + x$ and $g(x) = x + 1$ then fog is :
   (a) $x^2 + 3x + 2$ (b) $x^2 + x + 1$
   (c) $(x+1)^2 + (x+1)$ (d) None of these

8. Let A and B be sets with cardinalities m and n respectively. The number of one-to-one mapping from A to B where $m < n$ is :
   (a) m^n (b) ^nP_m
   (c) ^mC_n (d) ^nC_m

9. The number of functions from m element set to n element set is :
   (a) $m + n$ (b) m^n
   (c) n^m (d) $m * n$

10. _____ is an unordered collection of elements where an element can occur as a member more than once :
    (a) Multiset   (b) Ordered set
    (c) Set   (d) None of these

11. The number of substrings of all lengths that can be formed from a character string of length $n = $ _____
    (a) n (b) n^2
    (c) $\dfrac{n(n-1)}{1}$ (d) $\dfrac{n(n+1)}{2}$

12. In a room containing 28 females, there are 18 females who speak English, 15 females speak French and 22 speak German. 9 females speak both English and French, 11 Females speak both French and German whereas 13 speak both German and English. How many females speak all the three languages?

   (a) 9
   (b) 8
   (c) 7
   (d) 6

13. Consider the following statements :

   S_1 : There exist infinite set A, B and C such that $A \cap (B \cap C)$ is finite.

   S_2 : There exist two irrational numbers x and y such that $(x + y)$ is rational.

   Which of the following is True about S_1 and S_2?

   (a) Only S_1 is correct.
   (b) only S_2 is correct.
   (c) Both S_1 and S_2 are correct.
   (d) None of the S_1 and S_2 is correct.

14. The power set 2^S of the set $S = \{3, (1, 4), 5\}$ is:

   (a) $\{5, 3, 1, 4, (1, 3, 5), (1, 4, 5), (3, 4), \phi\}$
   (b) $\{5, 3 (1, 4), 5\}$
   (c) $\{5, (3), [3, (1,4)], (3, 5), \phi\}$
   (d) None of the above.

15. Let A be a finite set of size n, the number of elements in the power set of $A \times A$ is :

   (a) 2^n
   (b) 2^{n^2}
   (c) $(2^n)^2$
   (d) None of these

16. Let S be an infinite set and $S_1, S_2, S_3, \ldots S_n$ be the sets such that $S_1 \cup S_2 \cup S_3 \cup \ldots \cup S_n = S$. Then :

   (a) at least one of the set S_i is a finite set.
   (b) not more than one of the set S_i can be finite.
   (c) at least one of the sets S_i is an infinite set.
   (d) None of the above

17. The number of elements in the power set $P(S)$ of the set $S = \{\{\phi\}, 1, \{2, 3\}\}$ is :

   (a) 2
   (b) 4
   (c) 8
   (d) None of these

18. Let A and B be sets and A' and B' denote the complements of the sets A and B. The set $(A - B) \cup (B - A) \cup (A \cap B)$ is equal to :

   (a) $A \cup B$
   (b) $A' \cup B'$
   (c) $A \cap B$
   (d) $A' \cap B'$

19. Let $P(S)$ denote the power set of set S which of the following is always TRUE?

   (a) $P(P(S)) = P(S)$
   (b) $P(S) \cap S = P(S)$
   (c) $P(S) \cap P(P(S)) = (\phi)$
   (d) $S \in P(S)$

===== Answers =====

Fill in the Blanks

1. $R = R$ 2. $a = b$ 3. Many-One into 4. Symmetric 5. $\{(1, 1), (2, 2), (3, 3)\}$ 6. The identity relation on A.
7. Symmetric, transitive but not reflexive 8. $\{(5, 5), (4, 4), (6, 5), (6, 6), (5, 6), (7, 7)\}$

True/ False

1. T	2. T	3. F	4. F	5. T	6. F	7. T	8. F	9. F
10. F	11. F	12. T	13. T					

Multiple choice questions

1. (c)	2. (a)	3. (d)	4. (b)	5. (d)	6. (d)	7. (a)	8. (b)	9. (c)
10. (a)	11. (d)	12. (d)	13. (c)	14. (d)	15. (b)	16. (c)	17. (c)	18. (a)
19. (c)								

● ● ●

Chapter

2 Groups

2.1 INTRODUCTION

It is known that binary operation on a set X is a mapping form $X \times X$ to X. It is usually denoted by means of a symbol such as $+, \times, ., *, o, \oplus$. If we denote a binary operation on a set X by $*$, then the result of $*$ on the elements x, y of X is expressed by $x * y$.

If $f : X \times X \to X$ be a binary composition in X and $x, y \in X$. Then $f(x, y)$ is called the composition of x and y under the composition f.

For Example

(1) Addition and multiplication are binary operations in the following sets :

   (i) The set **N** of natural numbers.

   (ii) The set **Z** of all integers.

   (iii) The set **Q** of all rational numbers.

   (iv) The set **R** of real numbers.

(2) Subtraction is not a binary composition in **N** because for given $m, n \in N$, $m-n$ may not be an element of **N**, *e.g.*, $2 - 3$.

REMARK

☞ The adjective binary is used because our rule combines two elements at a time.

2.1.1 *n*-Ray Operations

Let X be any non empty set A mapping $f : X^n \to X$ is called an n-ray operations for $n = 1, 2, \ldots$ on the set X.

For $n = 1$, such an operation is called a unary operation. Therefore any function form X to it-self is an binary operation on X.

For $n = 2$, n-ray operation is called binary operations.

For $n = 3$ *i.e.*, a function from $X \times X \times X$ into X is called 3-ray (or ternary) operations.

2.2 GROUPOID

Let S be any non-empty set and $*$ be any binary operation. Then algebraic structure $(S, *)$ is said to be groupoid if closure property is satisfied in S. We shall however, use the phrases

'S is a groupoid with respect to * ' and 'S is a groupoid' in case of × is taken for granted. This convection of using S for the structure as well as the set underlying the structure will be followed throughout and should cause no confusion.

For Example: (**N**, +), where **N** is the set of natural numbers and + is the operations on **N**, is a groupoid. It is also note that (**N**, –) is not a groupoid.

2.2.1 Commutative Groupoid

Definition. *A groupoid (S, *) is said to be commutative groupoid if * is a commutative operations.*

For Example

(1) (**N**, +), (**Z**, ×) are both commutative groupoids.

(2) (**Q**, –) and (**R**~{0}, ÷) are not commutative groupoids because neither subtraction is in **Q** nor division in the set of non-zero real numbers is commutative operations.

(3) Let S be the set of all square matrices of order 3 over **R** and let ⊕ and ⊗ denote respectively the matrix addition and matrix multiplication. Then (S, ⊕) is commutative but (S, ⊗) is not a commutative groupoid.

2.3 SEMIGROUP

Definition. *A groupoid (S, *) is said to be semigroup if * is associative, i.e., if S is any non-empty set and * is a binary composition on S such that*

$$a*(b*c) = (a*b)*c, \forall\, a, b, c \in S$$

Then (S, *) a semigroup.

For Example

(i) (**N**, +), (**N**, .)(**Z**, ×) and (**R**, +) are the examples of semigroup.

(ii) Let* be defined by $a* b = |a - b|, \forall\, a, b \in \mathbf{R}$

Then clearly, (**R**, *) is a commutative groupoid but not a semigroup.

2.3.1 Relation between Groupoid, Commutative Groupoid and Semigroup

Let us denote the set of all semigroups by S and C denote the set of all commutative groupoid. Then following diagram shows the relation between the set of groupoids, commutative groupoid and semigroup.

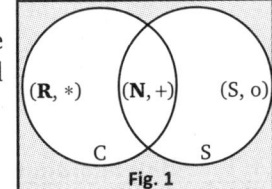

Fig. 1

2.4 MONOID (SEMIGROUP WITH IDENTITY)

Definition. *A semigroup (S, *) is said to be a monoid if S has an identity element with respect to *, i.e., (S, *) is a monoid if S is a non-empty set and * is a binary composition on S such that*

   (i) $a*(b*c) = (a*b)*c, \forall\, a, b, c \in S$ (Associativity)

and (ii) $\exists\, e \in S$ such that $a * e = e * a = a \;\forall\, a \in S.$ (Existence of Identity)

For example

(1) The semigroup (**N**, ×), (**Q**, ×), (**R**, ×), are all monoids. 1 being the identity element for each.

(2) (**N**, +) is a semigroup which is not a monoid.

(3) (**Q**, +), (**R**, +) are both monoids, 0 being the identity element for each.

(4) (P(X), ∩) is a monoid, X being the identity element.

(5) (P(X), ∪) is a monoid, φ being the identity element.

(6) Let S be the set of all square matrices of order 4 over **R** and let ⊕ and ⊗ denote respectively the operations of matrix addition and matrix multiplication. Then both (S, ⊕) and (S, ⊗) are monoids. The identity of (S, ⊕) is the null matrix of under 4 and that of (S, ⊗) the unit matrix of order 4.

2.4.1 Relation between the Groupoid, Commutative groupoid, Semigroup and Monoid

Let M denote the set of all monoids. Then following diagram shows the relation between these algebraic structure.

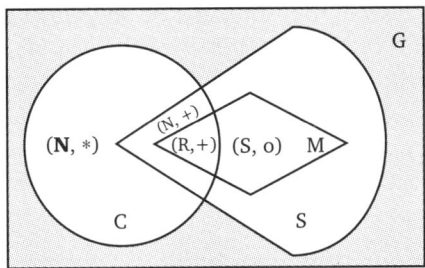

Fig. 2

2.4.2 Monoid with Inverses

Definition. *A monoids (S, ∗) in which every element has an inverse is known as a group.*

For Example

(1) (**R, +**) is a group.

(2) Let S denotes the set of real functions and ∘ is the operations of composite of functions. Then (S, ∘) is not a group because only the bijective functions have an inverse.

■ Illustrations

(i) Algebraic structure (G, .) is a monoid in each of the following cases, where '.' denote multiplication of numbers.

   (a) $G = \{2n : n \in \mathbf{Z}\}$

   (b) G = {The set of all positive rational numbers}.

   (c) $G = \{1, -1, i, -i\}$

   (d) $G = \{z : z \in \mathbf{C}. \ |z| = 1\}$

   (e) $G = \{a + b\sqrt{3} : a, b \in \mathbf{Q}, a^2 + b^2 \neq 0\}$

(ii) Algebraic structure (G, +) is a monoid in each of the following cases, where + denote addition of numbers.

   (a) $G = \{a + b\sqrt{3} : a \in \mathbf{Q}, b \in \mathbf{Q}\}$

   (b) $G = \{3n : n \in \mathbf{Z}\}$

   (c) $G = \{a + ib : a \in \mathbf{Z}, b \in \mathbf{Z}\}$

   (d) $G = \{a + ib; a \in \mathbf{Q}, b \in \mathbf{Q}\}$

(iii) Following are the examples of monoids :

   (a) The set of all real numbers of the form $x + y\sqrt{3}$ where $x, y \in \mathbf{Q}$, with respect to addition.

(b) The set of all non-zero complex number with respect to the multiplication.

(c) The set of real numbers **R** with respect to $*$ where $a*b = a+b - ab$, $\forall a, b \in \mathbf{R}$.

(d) The set **Z** with respect to $*$ where $a * b = a+b +1$, $\forall a, b \in \mathbf{Z}$.

(e) The set **Z** with respect to $*$ where $a * b = a+b -1$, $\forall a, b \in \mathbf{Z}$.

(f) The set of all matrices of the form $\begin{pmatrix} a & 0 \\ 0 & 0 \end{pmatrix}$ where $a \in \mathbf{R}$ is a monoid with respect to the matrix multiplication.

(g) The set of square symmetric matrices of order n over **C**.

2.5 Subsemigroup

Definition. *Let $(S, *)$ be a semigroup and let T be any subset of S. If the set T is closed under the operation $*$ (i.e., $a * b \in T$, $\forall a, b \in T$) defined on S. Then $(T, *)$ is called a subsemigroup of $(S, *)$.*

For Example

Let E denote the set of even positive integers. Then $(E, .)$ where $.$ is the multiplication of integers is a subsemigroup of semigroup $(N, .)$.

REMARKS

☞ The associative property holds in any subset of a semigroup. Thus a subsemigroup $(T, *)$ of a semigroup $(S, *)$ in itself a semigroup.

☞ If $(S, *)$ is a semigroup then $(S, *)$ itself is a subsemigroup of $(S, *)$. Such a subsemigroup is called trival sub semigroup.

2.5.1 Direct Product of Subsemigroups

Definition. *Let $(S, *)$ and $(T, \circ)$ be two semigroups. Then direct product of $(S, *)$ and $(T, \circ)$ is the system $(S \times T, \oplus)$ where the operation $\oplus$ on $S \times T$ is defined as follows*

$$(a, b) \oplus (c, d) = (a*c, b \circ d); (a, b), (c, d) \in S \times T$$

Theorem 1. *The direct product of any two semigroups is a semigroup.*

Step outlines : To make the proof easier, we shall proceed as follows:

Step 1. *Define $\oplus$ by $(a, b) \oplus (c, d) = (a \times c, b \circ d)$*

Step 2. *Show that $\oplus$ is associative on $S \times T$*

Proof. Let $(S, *)$ and $(T, \circ)$ be two semigroups. To show that $(S \times T, \oplus)$ where $\oplus$ is defined by $(a, b) \oplus (c, d) = (a*c, b \text{ o } d)$ for all $(a, b), (c, d) \in S \times T$, is a semigroup.

Let $a, c \in S \Rightarrow a*c \in S$ [$\because S$ is a semigroup.]

Similarly

$b, d \in T \Rightarrow b \circ d \in T$ [$\because T$ is a semigroup.]

We have to show that $\oplus$ is associative on $S \times T$.

Let $(a, b), (c, d)$ and (e, f) be any three elements in $S \times T$.

Then $((a, b) \oplus (c, d) \oplus (e, f)) = (a*c, b \circ d) \oplus (e, f)$

$$= ((a*c)* e, (b \circ d) \circ f)$$

$$= ((a*(c, e), b \circ (d \circ f))[\because * \text{ and } \circ \text{ both are associative.}]$$

$$= (a, b) \oplus ((c*e, d \circ f)$$

$$= (a, b) \oplus ((c, d) \oplus (e, f))$$

$\Rightarrow$ $\oplus$ is associative. Hence, $(S \times T, \oplus)$ is a semigroup.

Theorem 2. *The product of two commutative semigroups is again a commutative semigroup.*

Proof.   In Theorem 1, we have already proved that if $(S, *)$ and $(T, \circ)$ are semigroups. Then their product is again a semigroup. It remains to prove that $S \times T$ is commutative.

Let (a, b) and (c, d) be any two elements in $S \times T$. Then

$$(a, b) \oplus (c, d) = (a \times c, b \circ d)$$

$$= (c \times a, d \circ b) \qquad [\because * \text{ and } \circ \text{ both are commutative}]$$

$$= (c, d) \oplus (a, b)$$

$\Rightarrow$ $\oplus$ is commutative on $S \times T$.

Hence, $(S \times T, \oplus)$ is a commutative semigroup.

2.5.2 Morphisms of Semigroups

Definition 1. *Let $(S, *)$ and $(T, \circ)$ be any two semigroups. A mapping $f : S \rightarrow T$ is called semigroup homomorphism if $f(a \times b) = f(a) \circ f(b)$, $\forall$ $a, b \in S$ (structure preserving property).*

REMARK

☞ A homomorphism of a semigroup $(S, *)$ into itself is called endomorphism, *i.e.*, endomorphism is a mapping $f : S \rightarrow S$ such that $f(a * b) = f(a) * f(b)$; $\forall a, b \in S$.

Definition 2. *Let $(S, *)$ and (T, o) be any two semigroups. A semigroup homomorphism mapping $f : S \rightarrow T$ is called*

(a) *epimorphism if f is onto also.*

(b) *monomorphism is f is one-one also.*

(c) *isomorphism if f is one-one and onto (i.e., bijective)*

Definition 3. *A semigroup isomorphism from a semigroup $(S, *)$ onto itself is called semigroup automorphism, i.e., a semigroup automorphism is a mapping from a semigroup $(S, *)$ onto itself which is homomorphism and bijective.*

Definition 4. *Two semigroups $(S, *)$ and $(T, \circ)$ are said to be isomorphic if there exists a semigroup isomorphism from S to T (or from T to S).*

REMARK

☞ If two semigroups $(S, *)$ and $(T, \circ)$ are isomorphic then they are structurally identical in the sense that they can differ only in the nature of their elements and the involved operations.

Definition 5. *An element a in a semigroup $(S, *)$ is called an idempotent if $a^2 = a$.*

Theorem 1. *Let $f : S \rightarrow T$ be an onto mapping from a semigroup $(S, *)$ to an algebraic structure $(T, \circ)$ where $\circ$ is a binary operation on T. If f is a semigroup homomorphism then $(T, \circ)$ is semigroup.*

Proof.   We have to prove that $(T, \circ)$ is a semigroup. For this we shall prove that $\circ$ is an associative operation on T.

Let x, y, z be any three elements in T. Now, since f is onto mapping, there exist a, b, c in S such that $x = f(a)$

$$y = f(b) \text{ and } z = f(c)$$

Consider

$$(x \circ y) \circ z = (f(a) \circ f(b)) \circ f(c)$$

$$= f(a*b) \circ f(c) \qquad [\because f \text{ is a homomorphism.}]$$

$$= f((a*b) * c) \qquad [\because f \text{ is a homomorphism.}]$$

$$= f(a*(b*c)) \qquad [\because * \text{ is associative.}]$$

$$= f(a) \circ f(b*c) \qquad [\because f \text{ is a homomorphism.}]$$

$$= x \circ (y \circ z) \qquad [\because f \text{ is a homomorphism.}]$$

$\Rightarrow$ $\circ$ is associative.

Hence, $(T, \circ)$ is a semigroup.

Theorem 2. *Let f be a semigroup homomorphism from semigroup $(S, *)$ to semigroup $(T, \circ)$. If $a \in S$ is an idempotent element then $f(a)$ is idempotent in $(T, \circ)$.*

Proof.   Let $(S, *)$ and $(T, \circ)$ be two semigroups and let $f: S \to T$ be a homomorphism. Let $a \in S$ be idempotent element.

Then, by definition

$$a^2 = a$$

Now $$[f(a)]^2 = f(a) \circ f(a)$$

$$= f(a* a) \qquad [\because f \text{ is a homomorphism}]$$

$$= f(a^2)$$

$$= f(a) \qquad [\because a^2 = a]$$

$\Rightarrow$ $f(a)$ is an idempotent element in $(T, \circ)$

Theorem 3. *Every finite semigroup has an idempotent element.*

Proof.   Let S be a finite semigroup on S and x be any element in S.

Consider positive integral powers of x.

$$x, x^2, x^3 \ldots$$

Since S is finite, therefore all these powers of x cannot be distinct. Thus, there exist positive integers r and s such that $x^r = x^s$.

Let $k = r - s$. Then $k \geq 1$.

Also, we can write

$$x^r = x^s \text{ as}$$

Now, $$x^{s+k} = x^s \qquad \qquad \ldots(1)$$

$$x^{2s+k} = x^s . x^{s+k}$$

$$= x^s . x^s = x^{2s} \qquad \text{[using (1)]}$$

Similarly $$x^{3s+k} = x^{3s}.$$

Then we can write by the principle of mathematical induction

$$x^{ms+k} = x^{ms} \text{ for any } m \in \mathbf{N} \qquad \ldots (2)$$

Further $$x^{ms+2k} = x^{ms+k} . x^k$$

$$= x^{ms} . x^k \qquad \text{[using (2)]}$$

$$=x^{ms+k}$$

$$= x^{ms} \qquad\qquad .... (3)$$

and $\qquad x^{ms+3k} = x^{ms+2k} . x^k$

$$= x^{ms} . x^k$$

$$= x^{ms+k}$$

$$= x^{ms}$$

Again using principle of induction, we have

$$x^{ms+nk} = x^{ms} \text{ for } n\in \mathbf{N}$$

for $m = k$ and $n = s$, we get

$$x^{ks+sk} = x^{ks}$$

$\Rightarrow \qquad\qquad x^{2ks} = x^{ks}$

$\Rightarrow \qquad\qquad a^2 = a \qquad\qquad$ [for $a = x^{ks}$]

Hence, every finite semigroup has an idempotent element.

Theorem 4. *If f is a homomorphism from a commutative semigroup (S, $*$) onto a semigroup (T, $\circ$) then (T, $\circ$) is also commutative.*

Proof. *Let $f : S \to T$ be a homomorphism.*

Let x and y be any two elements of T. Now, since f is onto, therefore, there exist elements a and b in S such that

$$x = f(a) \text{ and } y = f(b)$$

Consider

$x \circ y = f (a) \circ f(b)$

$\qquad = f(a*b) \qquad\qquad$ [$\because f$ is a homomorphism.]

$\qquad = f (b * a) \qquad\qquad$ [$\because (S, *)$ is commutative.]

$\qquad = f (b) \circ f(a)$

$\qquad = y \circ x$

Hence, (T, $\circ$) is commutative.

Theorem 5. *Let f be a homomorphism from a semigroup (S, $*$) to semigroup (T, $\circ$). If H is a subsemigroup of (S, $*$). Then*

$$f(H) = \{t\in T : T = f(h), \text{ for some } h\in H\} \text{ is a subsemigroup of } (T, \circ)$$

Proof. Let x and y be any two elements of $f(H)$. By definition of $f(H)$, there must exist a and b in H such that

$$x = f(a) \text{ and } y = f(b)$$

Now, $\qquad x \circ y = f(a) \circ f(b)$

$\qquad\qquad = f(a*b)$

$\qquad\qquad = f(c) \quad \text{where } c = a*b\in H$

$\Rightarrow \qquad\qquad f(c) \in f(H)$

$\Rightarrow f(H)$ is closed under the operator $\circ$ of T.

Hence, $f(H)$ is a subsemigroup of (T, $\circ$)

2.6 COMMUTATIVE MONOID

Definition: *A monoid* $(M, *)$ *is said to be commutative if the operator* $*$ *is commutative on M.*

2.6.1 Cyclic Monoid

Definition. *A monoid* $(M, *)$ *is said to be cyclic if there exists an element* $a \in M$ *such that every element of M can be written as some integral powers of a. In such a case a is called the generator of the cyclic monoid.*

Theorem 1. *Every cyclic monoid is a commutative monoid.*

Proof.   Let $(M, *)$ be a cyclic monoid generated by an element a of M. Further let x, y be any two elements of M.

Then $x = a^m$ and $y = a^n$ for some $m, n \in \mathbf{N}$

Consider

$$x * y = a^m * a^n$$
$$= a^{m+n}$$
$$= a^{n+m}$$
$$= a^n * a^m$$
$$= y * a$$

Hence, $(M, *)$ is commutative.

Theorem 2. *The identity element of a monoid is unique.*

Proof.   Let $(M, *)$ be a monoid. Let if possible e and e' be two identity elements in $(M, *)$

Now, since e is identity element therefore

$$e * e' = e' * e = e \qquad \qquad \text{... (1)}$$

Further, since e' is the identity element , therefore

$$e' * e = e * e' = e \qquad \qquad \text{... (2)}$$

From (1) and (2), we conclude that

$$e = e'$$

Hence, the identity element of a monoid is unique.

2.6.2 Submonoid

Let $(M, *)$ be a monoid with identity element e and let T be a non-empty subset of M. If T is closed under the operation $*$ and identity $e \in T$, then $(T, *)$ is known as submonoid of $(M, *)$.

REMARK

☞ Since associativity holds in a monoid $(M, *)$, therefore it holds in any subset of monoid M. Hence, a submonoid of a monoid is also a monoid.

Theorem 1. *The intersection of two submonoids of a monoid* $(M, *)$ *is again a submonoid of* $(M, *)$.

Proof.   Let $(T_1, *)$ and $(T_2, *)$ be two submonoids of a monoid $(M, *)$. Also, let e be the identity element in M. Using the definition of monoid, we have

$$e \in T_1, e \in T_2, \text{ which implies } T_1 \cap T_2 \neq \phi$$

Firstly, we shall prove that $T_1 \cap T_2$ is closed with respect to $*$. Let $x, y \in T_1 \cap T_2$

$\Rightarrow \qquad x, y \in T_1$ and $x, y \in T_2$

$\Rightarrow \qquad x * y \in T_1$ and $x * y \in T_2$

$\Rightarrow \qquad x * y \in T_1 \cap T_2$

$\Rightarrow \quad T_1 \cap T_2$ is closed with respect to $*$. Finaly, we show that $e \in T_1 \cap T_2$.

Since T_1 and T_2 are submonoids, thus T_1 and T_2 both contains identity element.

Now $e \in T_1$ and $e \in T_2 \Rightarrow e \in T_1 \cap T_2$

$\Rightarrow T_1 \cap T_2$ contains identity element e of monoid $(M, *)$.

Hence, $(T_1 \cap T_2, *)$ is a submonoid.

Theorem 2. *If $(M, *)$ is a commutative monoid, then the set of all idempotent element of M forms a submomoids.*

Proof. *Let S be the set of all idempotent elements of M such that*

$$S = \{ x \in M : x^2 = x \}$$

It is known that the identity element $e \in M$ is always idempotent. Thus, we have $e \in S$. Now to show that S is closed with respect to $*$.

Let $a \in S$, $b \in S$ Then

$$a^2 = a, \ b^2 = b \qquad\qquad\qquad \text{... (1)}$$

Consider

$$(a * b)^2 = (a * b) * (a * b)$$
$$= a * (b * a) * b \qquad\qquad [\because * \text{ is associative}]$$
$$= a * (a * b) * b \qquad\qquad [\because * \text{ is commutative}]$$
$$= (a * a)(b * b)$$
$$= a^2 * b^2 = a * b \qquad\qquad [\text{Using (1)}]$$

$\Rightarrow a * b$ is idempotent element of M. Hence, $a \times b \subset S$.

Hence, S is a submonoid.

2.7 MORPHISMS OF MONOIDS

Definition 1. *Let $(M, *)$ and $(T, \circ)$ be two monoids with e and e' as identity elements respectively. Then a mapping*

$$f : M \rightarrow T$$

is said to be a monoid homonorphism if for any two elements a, b $\in$ M

$$f(a * b) = f(a) \circ f(b) \qquad\qquad [\text{Structure preserving property}]$$

and $\qquad\qquad f(e) = e'$

Definition 2. *A homomorphism of a monoid into itself is called a monoid endomorphism.*

Definition 3. *A homomorphism $f : M \rightarrow T$ is called*

       (i) *monoid epimorphism if f is onto also.*

       (ii) *monoid monomorphism if f is one-one also.*

       (iii) *monoid isomorphism if f one-one and onto (bijective) also.*

Definition 4. *Two monoids $(M, *)$ and $(T, \circ)$ are said to be isomorphic monoids if there* *exists a monoid isomorphism from M to T.*

      Definition 5. *An isomorphism of a monoid onto itself is called a monoid automorphism. Thus a mapping*

$$f : (M, *) \to (M, *)$$

is said to be a monoid automorphism of M if

(i) *f is bijective.*

(ii) $f(a * b) = f(a) * f(b); \forall a, b \in M$

(iii) $f(e) = e$

Theorem 1. *Let (M, *) and (T, ∘) be two monoids with identities e and e' respectively. If f is onto from M onto T such that*

$$f(a*b) = f(a) \circ f(b); \forall a, b \in M \qquad\qquad Then\ f(e) = e'$$

Proof.   Define a map $f : M \xrightarrow{\text{onto}} T$

Let y be any element of T. Since, f is onto, therefore there exists an element $x \in M$ such that

$$f(x) = y$$

Now $y = f(x) = f(x*e)$

$$= f(x) \circ f(e)$$

$$= y \circ f(e) \qquad\qquad\qquad [\because f \text{ is onto.}]$$

Similarly

$$y = f(x) = f(e*x)$$

$$= f(e) \circ f(x)$$

$$= f(e) \circ y$$

$\Rightarrow$ $f(e) \circ y = y \circ f(e) = y$

$\Rightarrow$ $f(e)$ is the identity for T.

$\Rightarrow$ $e' = f(e)$ [$\because$ Identity element of a monoid is always unique.]

Theorem 2. *Let (M, *) be a monoid with identity element e and (T, ∘) be any algebraic structure. If a mapping f : M → T is onto and satisfying*

$$f(a*b) = f(a) \circ f(b) \ \forall a, b \in M$$

then (T, ∘) is a monoid with f(e) as its identity element.

Proof.   Firstly, we shall prove that ∘ is associative. Let x, y and z be any three elements in T. Now, since f is onto then there exist a, b and c in M such that

$$x = f(a),\ y = f(b),\ z = f(c)$$

Consider

$$x \circ (y \circ z) = f(a) \circ (f(b) \circ f(c))$$

$$= f(a) \circ f(b*c) \qquad\qquad [\because f(b*c) = f(b) \circ f(c)]$$

$$= f(a*(b*c))$$

$$= f((a*b)*c) \qquad\qquad [\text{By associativity}]$$

$$= f(a*b) \circ f(c)$$

$$= (f(a) \circ f(b)) \circ f(c)$$

$$= (x \circ y) \circ z$$

$\Rightarrow \circ$ is associative in T.

Further, we shall show that $f(e)$ is the identity for T. Let x be any element of T. Now, since f is onto, there exists an element $a \in M$ such that $f(a) = x$.

Consider

$$x = f(a)$$
$$= f(a * e)$$
$$= f(a) \circ f(e)$$
$$= x \circ f(e)$$

Similarly, we can prove that

$$x = f(a) = f(e*a) = f(e) \circ f(a) = f(e) \circ x$$

Therefore

$$x = x \circ f(e) = f(e) \circ x$$

Thus, we conclude that $f(e)$ is the identity element for $(T, \circ)$.

Hence, $(T, \circ)$ is a monoid with $f(e)$ as its identity element.

REMARKS

☞ In view of the Theorem-1, an onto mapping f from a monoid $(M, *)$ onto monoid $(T, \circ)$ is called homomorphism if for any two elements $a, b \in M$

$$f(a*b) = f(a) \circ f(b)$$

This is because $f(e) = e'$ is a consequence of f being onto and satisfying

$$f(a * b) = f(a) \circ f(b) ; \forall a, b \in M.$$

☞ If f is a homomorphism from a commutative monoid $(M, *)$ onto a monoid $(T, \circ)$, then T is also commutative.

2.8 CONGRUENCE RELATION AND QUOTIENT SEMIGROUPS

Definition. *An equivalence relation R on the semigroup $(S, *)$ is called a congruence relation if*

$$a \, R \, a' \text{ and } bRb' \Rightarrow (a * b) \, R \, (a' * b')$$

Theorem 1. *Let R be a congruence relation on a semigroup $(S, *)$. Then quotient set S/R is a semigroup with respect to the operation $\oplus$ defined by*

$$(a) \oplus (b) = (a*b), \forall a, b \in S$$

where $[a]$ denote the equivalence class of element a in S corresponding to the relation R.

Proof. Let $(S, *)$ be a semigroup and R be a congruence relation on S. Then, for any $a \in S$, the equivalence class $[a]$ is a set contataining all those elements of S which are related to a under the relation R. Now, for $a, b \in S$, define an operation $\oplus$ on S/R such that

$$[a] \oplus [b] = [a * b]$$

Firstly, we shall prove that operation $\oplus$ is well defined on S/R.

Now, suppose that $\quad [a] = [a']$ and $[b] = [b']$

Since $[a] = [a']$ and $[b] = [b']$ Therefore $a\,R\,a'$ and bRb'

Now, since R is a congruence relation, therefore

$$a\,R\,a' \text{ and } b\,R\,b' \Rightarrow a * bRa' * b'$$
$$\Rightarrow (a * b) = (a' * b')$$
$$\Rightarrow \oplus \text{ is well defined on } S/R$$
$$\Rightarrow \oplus \text{ is binary operation on } S/R$$

Further, we shall show that $\oplus$ is associative.

For $a, b, c \in S$, we have

$$[a] \oplus [b] \oplus [c] = [a] \oplus [b * c]$$
$$= [a * (b * c)]$$
$$= [(a * b) * c]$$
$$= [a * b] \oplus [c]$$
$$= ([a] \oplus [b]) \oplus [c] \qquad [\because * \text{ is associative.}]$$

Therefore, $\oplus$ is associative. Hence S/R is a semigroup.

Definition. *The semigroup defined in above theorem $(S/R, \oplus)$ is called quotient semigroup or factor semigroup of semigroup $(S, *)$ by the congruence relation R.*

REMARK

☞ The operator $\oplus$ on S/R is called binary relation.

Theorem 2. *Let R be a congruence relation on the monoid $(M, *)$ Then $(M/R, \oplus)$ is a monoid, where $\oplus$ on M/R is defined by*

$$[a] + [b] = [a * b]$$

Proof. Let $(M, *)$ be a monoid with e as identity. Using Theorem-1, we can say that $(M/R, \oplus)$ is a semigroup. We have to show that $(M/R, \oplus)$ has identity element. For this we shall prove that $[e]$ is the identity element in $(M/R, \oplus)$. Let $[a]$ be any element in M/R where $a \in M$. Then we have

$$[a] \oplus [e] = [a * e]$$
$$= [a]$$
$$= [e * a]$$
$$= [e] \oplus [a]$$

Hence, $[e]$ is the identity in M/R. Therefore $(M/R, \oplus)$ is a monoid.

Theorem 3. *Let $(S, *)$ be a semigroup and let R be a congruence relation on $(S, *)$. Then there exists a homomorphism from $(S, *)$ onto quotient semigroup $(S/R, \oplus)$.*

Proof. Define a function

$$f : S \rightarrow S/R$$

such that $\qquad\qquad f(a) = [a]$, the equivalence class of $a \in S$ under R.

To show, f is onto and homomorphism.

(i) f is onto

Let $[a]$ be any arbitary element in S/R. Then $a \in S$ and $f(a) = [a]$

$\Rightarrow f$ is onto.

(ii) *f* is homomorphism.

Let $a \in S$, $b \in S$, Then we have

$$f(a * b) = [a * b]$$
$$= [a] \oplus [b], \qquad \text{(by definition of } \oplus)$$
$$= f(a) \oplus f(b)$$

$\Rightarrow$ Structure preserving property is satisfied by *f*.

$\Rightarrow$ *f* is a homomorphism.

REMARK

☞ The mapping $f : S \to S/R$ defined by $f(a) = [a]$ is called natural homomorphism.

Theorem 4. *Let $f : S \to T$ be a semigroup homomorphism from the semigroup $(S, *)$ to the semigroup $(T, \circ)$. Then the relation R on $(S, *)$ defined by a R b if and only if $f(a) = f(b)$ for a, $b \in S$ is a congruence relation on $(S, *)$ is an equivalence relation.*

Proof.   Recall that a relation is said to be equivalence if it is reflexive, symmetric and transitive.

(i) R is reflexive: Since $f(a) = f(a)$

$\Rightarrow$ $a \, R \, a; \, \forall a \in S$

(ii) R is symmetric: Let aRb then $f(a) = f(b)$ which can easily be written as

$$f(b) = f(a)$$

$\Rightarrow$ $b \, R \, a$

(iii) R is transitive: Let us suppose that $a \, R \, a$ and $b \, R \, c$

Now

$$a \, R \, b \Rightarrow f(a) = f(b)$$
$$b \, R \, c \Rightarrow f(b) = f(c)$$

Therefore, $a \, R \, b$ and $b \, R \, c \Rightarrow f(a) = f(c)$

$$\Rightarrow a \, R \, c \Rightarrow R \text{ is transitive}$$

Hence, *R* is an equivalence relation.

Further, it remains to prove that *R* is a congruence relation.

Suppose that

$$a \, R \, a' \text{ and } b \, R \, b'$$

Then $a \, R \, a' \Rightarrow f(a) = f(a')$

and $b \, R \, b' \Rightarrow f(b) = f(b')$

Now, $f(a * b) = f(a) \circ f(b)$ $[\because f \text{ is a homomorphism.}]$

$$= f(a') \circ f(b')$$
$$= f(a' * b')$$

$\Rightarrow$ $(a * b) \, R \, (a' * b')$

Therefore

$$a \, R \, a' \text{ and } b \, R \, b' \Rightarrow (a * b) \, R \, (a' * b')$$

Hence *R* is a congruence relation on $(S, *)$.

Theorem 5. **(Fundamental Theorem of Homomorphism of Semigroups)**

Let $(S, *)$ and $(T, \circ)$ be two semigroups. If $f : S \to T$ is a semigroup homomorphism, then semigroup $(T, \circ)$ is isomorphic to some quotient semigroup of $(S, *)$.

Step Outlines: To make the proof easier, we shall proceed as follows:

Step 1. *Define a homomorphism $f : S \to T$ from semigroup $(S, \circ)$ to $(T, \circ)$ and a cengruence relation R on $(S, *)$ by aRb iff $f(a) = f(b) \forall$ a, b$\in$S*

Step 2. *Show that $(T, \circ)$ is isomorphic to quotient semigroup $(S/R, \oplus)$ of $(S, *)$ by a mapping $\psi : S/R \to T$ defined by $\psi([a]) = f(a) \forall a \in S$*

Proof. Let $f : S \to T$ be a homomorphism of the semigroup $(S, *)$ onto the semigroup $(T, \circ)$.

Let R be the congruence relation on $(S, *)$ corrosponding to the homomorphism f. Then, R is defined by

$$a \, R \, b \text{ iff } f(a) = f(b) \, ; \, \forall a, b \in S$$

To show that $(T, \circ)$ is isomorphic to quotient semigroup $(S/R, \oplus)$ of $(S, *)$.

If $a \in S$, then $[a] \in S/R$ and $f(a) \in T$

Now, define a mapping

$$\psi : S/R \to T$$

by setting $\qquad \psi([a]) = f(a)$ for all $a \in S$

Firstly, we shall prove that ψ is a semigroup isomorphism. For this first of all we shall prove that ψ is well defined.

Let us suppose $a, a' \in S$

and $\qquad\qquad\qquad [a] = [a']$

Now $\qquad\qquad\qquad [a] = [a'] \quad \Rightarrow a \, R \, a'$

$\Rightarrow \qquad\qquad\qquad f[a] = f[a']$

$\Rightarrow \qquad\qquad\qquad \psi([a]) = \psi([a'])$

$\Rightarrow \psi$ is well defined

(i) $\quad \Psi$ **is one-one.** Let us suppose that $\psi([a]) = \psi([b])$ then, we have

$$f[a] = f[b]$$

$\Rightarrow \qquad\qquad a \, R \, a$

$\Rightarrow \qquad\qquad [a] = [b]$

$\Rightarrow \psi$ is one-one.

(ii) $\quad \Psi$ **is onto.** Let b be any arbitrary element of T. Now, since f is onto, then there exists an element a in S such that

$$b = f[a]$$

Now, $\quad \psi([a]) = f[a] = b$

$\Rightarrow \psi$ is onto.

(iii) $\quad \Psi$ **is a homomorphism.**

Let $[a]$, $[b]$ be any two element in S/R.

$$\psi \left([a] \oplus [b]\right) = \psi([a * b])$$
$$= f(a * b)$$
$$= f[a] \circ f(b) \qquad [\because f \text{ is a homomorphism.}]$$
$$= \psi([a]) \circ \psi[(b)]$$

Hence, we conclude that ψ is an isomorphism of S/R onto T. Therefore, $(S/R, \oplus)$ and $(T, \circ)$ are isomorphic.

Solved Examples

Based on the following Results

▶ An algebraic structure $(S, *)$ is said to be groupoid if S is closed w.r.t. $*$.

▶ A groupoid $(S, *)$ is said to be semigroup if $*$ is associative.

▶ A semigroup $(S, *)$ is said to be monoid if S has an identity element.

▶ A monoid $(S, *)$ in which every element has an inverse is known as a group.

Example 1. *Let $\mathbf{R}$ be the set of real numbers and operation $*$ on $\mathbf{R}$ be defined as follows*

$$a * b = |a - b|; a, b \in \mathbf{R}$$

Then $*$ is a commutative binary operation on $\mathbf{R}$ but $(\mathbf{R}, *)$ is not a semigroup.

Solution. By definition of an absolute value of a number we have

$|a - b|$ is always positive real number.

and $a * b \in \mathbf{R}$; $a, b \in \mathbf{R}$

$\Rightarrow *$ is a binary operation on $\mathbf{R}$.

Now, since

$$|a - b| = |b - a| \quad ; \forall \, a, b \in \mathbf{R}$$

We have $\qquad a * b = b * a ; \qquad \forall \, a, b \in \mathbf{R}$

Therefore, $*$ is a commutative operation.

Also, $*$ is not associative because

$$1 * (2 * 3) = 1 * |2 - 3| = 1 * 1 = |1 - 1| = 0$$

but $\qquad (1 * 2) * 3 = |1 - 2| * 3 = 1 * 3 = |1 - 3| = 2$

$\Rightarrow \qquad 1 * (2 * 3) \neq (1 * 2) * 3$

Example 2. *Show that there exists a semigroup homomorphism from the semigroup $(\mathbf{N}, +)$ of natural numbers under addition to the semigroup $(\{0, 1, 2, 3\}, +_4)$ where $+_4$ denotes the operation of addition modulo 4 on the set $\{0, 1, 2, 3\}$).*

Solution. Define a mapping $f : \mathbf{N} \rightarrow \{0, 1, 2, 3\}$

such that $\qquad f(a) = a \pmod 4; \forall a \in \mathbf{N}$

$\qquad\qquad\qquad$ = The remainder r, $0 \leq r < 4$ when a is divided by 4.

Now, for any $a, b \in \mathbf{N}$, let $f(a) = i$ and $f(b) = j$. Then

$$f(a + b) = (a + b) \pmod 4$$

$$= (i+j) \pmod 4$$
$$= i +_4 j$$
$$= f(a) +_4 f(b)$$

Hence, f is a homomorphism.

Example 3. *Show that monoids* $(\mathbf{Z}, +)$ *and* $(E, +)$ *are isomorphic where* $\mathbf{Z}$ *is the set of integers and E is the set of even integers.*

Solution. Define a mapping

$$f : \mathbf{Z} \to E$$

such that $\qquad f(x) = 2x \; ; \; \forall x \in \mathbf{Z}$

We shall prove that f is an isomorphism.

(i) f is one-one

Let $x, y \in \mathbf{Z}$

Now, suppose that

$$f(x) = f(y)$$
$$\Rightarrow \qquad 2x = 2y$$
$$\Rightarrow \qquad x = y$$

Therefore, $f(x) = f(y)$

$$\Rightarrow \qquad x = y \quad \Rightarrow \quad f \text{ is one-one.}$$

(ii) f is onto.

Let y be any arbitrary element of E. Then y is an even integer. Thus $y/2$ is an integer.

Let $x = y/2$, then $x \in \mathbf{Z}$ such that

$$f(x) = 2x = 2.\frac{y}{2} = y$$
$$\Rightarrow \qquad f \text{ is onto.}$$

(iii) f is a homomorphism.

Let $x, y \in \mathbf{Z}$, then

$$f(x+y) = 2(x+y)$$
$$= 2x + 2y$$
$$= f(x) + f(y)$$

Therefore, f is a homomorphism. Hence, monoids $(\mathbf{Z}, +)$ and $(E, +)$ are isomorphic.

Example 4. *Let A be a non-empty set of symbols and let A^* be the free semigroup generated by A under the operation of concatenation. Show that the functions*

$$f : A^* \to \mathbf{N} \text{ defined by } f(\alpha) = l(\alpha), \text{ the length of } \alpha,$$

where $l(\alpha)$ denoted the number of symbols in α (each symbols is counted as many times as it appear in α) is a homomorphism and if R is the relation induced by f, then show that $(A^*/R, \oplus)$ is isomorphic to $(\mathbf{N}, +)$

Solution. Let α, β be any two elements (words) in A^*

Then, we have

$$f(\alpha, \beta) = l(\alpha.\beta)$$

$$= l(\alpha) + l(\beta)$$
$$= f(\alpha) + f(\beta)$$
$\Rightarrow$ f is a homomorphism.

By Theorem 4 (before the solved examples), f induces a congruence relation on A^* defined by

$$\alpha \, R \, \beta \text{ is any only if } l(\alpha) = l(\beta)$$

Now, since f is onto, then by fundamental theorem of semigroup homomorphism $(A^*/\mathbf{R}, \oplus)$ is isomorphic to $(\mathbf{N}, +)$.

EXERCISE 2.1

1. Let $S = \{1, 2, 3, 6, 12\}$ and the operation $*$ be defined as follows

   $a * b$ = greatest common divisor of a and b.

   Show that $(S, *)$ is a semigroup.

2. Show that the set $\mathbf{N}$ of natural numbers is a semigroup under the operation $*$ defined by $x * y = \max \{x, y\}$

3. Show that every finite semigroup has an idempotent element.

4. Show that the non-empty intersection of two sub-semigroups of a semigroup is again a sub-semigroup.

5. Show that the union of two subsemigroups need not be a subemigroup.

6. Show that the set of all semigroups endomorphism of a semigroup is a semigroup under the operation of composition of functions.

7. Let $(S, *)$ and $(T, *)$ be two semigroups. Show that $S \times T$ and $T \times S$ are isomorphic semigroups.

8. Show that the monoids $(\{0, 1, 2, 3\}, +_4)$ and $(\{1, 3, 7, 9\}, \times_{10})$ are isomorphic.

9. Show that the set of all invertible elements of a monoid forms a monoid under the same operation as that of the monoid.

10. Show that the intersection of two congruence relations on a semigroup is a congruence relation.

2.9 INTRODUCTION TO GROUPS

The concept of a group is of fundamental importance in the study of algebra. The objective of studying groups is to classify all groups upto isomorphism which in practice means finding necessary and sufficient conditions for two groups to be isomorphic. The theory of groups, an important part in present mathematics, started early nineteenth century in connection with the solution of algebraic equations. Originally, a group was the set of all permutations of the roots of an algebraic equation, which has the property that combinations of any two of these permutations again belong to the set. Later, the idea was generalised to the concept of an abstract group. An abstract group is essentially the study of a set with an operation defined and Group Theory has many useful applications both within and outside mathematics.

2.10 GROUP

Let G be a non-empty set and $*$ be a binary operation defined on it, then the structure $(G, *)$ is said to be a group if the following axioms are satisfied:

(i) **Closure Property.** $a * b \in G$; $\forall \, a, b \in G$.

(ii) **Associativity.** *The opration $*$ is associative on G, i.e.,*

$$a * (b * c) = (a * b) * c; \; \forall \, a, b, c \in G$$

(iii) Existence of identity. *There exists an element e* $\in G$ *such that*

$$a*e = e*a = a; \; \forall a \in G$$

e is called identity of $*$ *in G.*

(iv) Existence of inverse. *For each element* $a \in G$, *there exist an element* $b \in G$ *such that*

$$a * b = b * a = e$$

The element b is called the inverse of element a with respect to $*$ *and we write* $b = a^{-1}$.

REMARKS

☞ When we say $*$ is a binary operation defined on a non-empty set G, it implies that G is closed for the binary operation $*$, *i.e.*,

$a \in G, b \in G \Rightarrow a * b \in G \; \forall \, a, b \in G.$

☞ A group is not simply a set, but it is an algebraic structure.

☞ Because of the associativity, the parenthesis can be dropped in products of more than two elements of a group and instead of writing $a * (b * c)$ or $(a * b) * c$ we may simply write $a * b * c$. The associative law can be extended to any finite number of elements.

☞ We know that '·' is a binary operation on G we must have $a \cdot b \in G \; \forall \, a, b \in G$. Hence, in our defnition of a group there is no necessity of mentioning the closure axioms. We mentioned it to emphasize the fact that while showing the group postulates in a problem, one should not forget the closure axions.

2.11 ABELIAN OR COMMUTATIVE GROUP

A group $(G, *)$ is said to be abelian or commutative if $a * b = b * a; \; \forall \, a, b \in G$. The group which is not abelian is called non-abelian or non-commutative.

REMARKS

☞ An abelian group under addition is sometimes called a 'module'.

☞ The commutative group is also known as Abelian group after the name of famous mathematician Abel.

☞ The smallest group for a given composition is the set $\{e\}$, containing identity element only .

☞ A group consisting the identity element only, is called a trivial group, other are called non-trivial groups.

2.12 FINITE AND INFINITE GROUPS

If a group contains a finite number of elements, it is called a **finite group.** If the number of elements in a group is infinite, it is called an **infinite group.**

Order of a Group. *The number of elements in a finite group is called the order of the group.* It is denoted by $o(G)$.

An infinite group is called a group of infinite order.

◾ Illustrations

(1) The set **Z** of integers is an infinite abelian group with respect to the operation of addition but **Z** is not a group with respect to the multiplication.

(2) Let $G = \{1\}$, then G is an abelian group of order 1 with resepect to multilpication.

(3) Let $G=\{0\}$, then G is an abelian group of order 1 with respect to addition.

(4) Let $G = \{1, -1\}$, then G is an abelian group of order 2 with respect to multiplication.

2.12.1 General Properties of Groups

Here, we shall discuss some important properties of groups.

Theorem 1. *Let $(G, *)$ be a group, then*

(i) *the identity element is unique.*

(ii) *every element of G has unique inverse in G.*

Proof. (i) Let, if possible e_1 and e_2 be two distinct identities of a group G. Then, by difinition of identity, we have

$$e_1 * e_2 = e_1 \qquad \text{(since } e_2 \text{ is identity.)}$$

and $\qquad e_1 * e_2 = e_2 \qquad \text{(since } e_1 \text{ is identity.)}$

Hence, it follows that $e_1 = e_2$

$\Rightarrow \qquad$ Identity is unique.

(ii) If possible, let any element $a \in G$ have two inverses say b and c, then, we have

$$a * b = e = b * a$$

and $\qquad a * c = e = c * a.$

Therefore, $\qquad b = b*e = b* (a*c) = (b*a) * c \quad$ (By associativity)

$$= e * c = c$$

$\Rightarrow \qquad b = c$

Hence, every element of a group has unique inverse.

REMARKS

☞ The identity element has its own inverse.

Theorem 2. *If $(G, *)$ is a group, then*

(i) $(a^{-1})^{-1} = a; \ \forall a \in G.$

(ii) $(a * b)^{-1} = b^{-1} * a^{-1}; \ \forall a, b \in G \qquad$ *(Reversal rule)*

Proof. (i) For each element $a \in G$, there exist an element $b \in G$ such that

$$a*b = b*a = e$$

From the symmetry of this result, we have

$$a^{-1} = b \qquad \qquad \text{... (1)}$$

and $\qquad \qquad b^{-1} = a \qquad \qquad \text{... (2)}$

Putting the value of b in equation (2), we get

$$(a^{-1})^{-1} = a$$

(ii) For all $a, b \in G$ we have

$$(a*b) * (b^{-1} * a^{-1}) = a*(b * b^{-1}) * a^{-1} \qquad \text{(by associativity)}$$

$$= a*(e) * a^{-1} = (a * e) * a^{-1} \quad \text{(by associativity)}$$

$$= a * a^{-1} = e$$

Similarly, we can easily show that

$$(b^{-1} * a^{-1}) * (a * b) = e$$

Thus, $\quad (b^{-1} * a^{-1}) * (a * b) = e$

Thus, $(a * b) * (b^{-1} * a^{-1}) = e = (b^{-1} * a^{-1}) * (a * b)$

Hence, it follows that

$$(a * b)^{-1} = b^{-1} * a^{-1}$$

REMARKS

☞ The above reversal law can be generalised as follows:

If $a_1, a_2, \ldots a_n$, are elements of a group G, then
$$(a_1 * a_2 * \ldots * a_n)^{-1} = a_n^{-1} * a_{n-1}^{-1} * a_{n-2}^{-1} * \ldots * a_2^{-1} * a_1^{-1}$$

☞ In additive composition, above result can be started as follows:

(i) $-(-a) = a \ \forall \ a \in G$

(ii) $-(a+b) = (-b) + (-a) \ \forall \ a, b \in G.$

Theorem 3. *If a, b, c are three elements of a group (G, *) then*

$$a * c = b * c \Rightarrow a = b \qquad \qquad \text{(Right cancellation law)}$$

$$c * a = c * b \Rightarrow a = b \qquad \qquad \text{(Left cancellation law)}$$

Proof. (i) $a * c = b * c \Rightarrow \quad (a * c) * c^{-1} = (b * c) * c^{-1} \qquad (\because c^{-1} \in G)$

$$\Rightarrow \quad a * (c * c^{-1}) = b * (c * c^{-1}) \qquad \text{(By Associativity)}$$

$$\Rightarrow \qquad \qquad a * e = b * e \Rightarrow a = b$$

Now, $c * a = c * b \Rightarrow c^{-1} * (c * a) = c^{-1} * (c * b)$

$$\Rightarrow \quad (c^{-1} * c) * a = (c^{-1} * c) * b$$

$$\Rightarrow \qquad \quad e * a = e * b \Rightarrow a = b$$

Theorem 4. *In a group G, the equation a*x = b and y*a = b where a, b∈ G have unique solution in G.*

Proof. Let $\qquad \qquad \qquad a * x = b$

$$\Rightarrow \qquad \qquad a^{-1} * (a * x) = a^{-1} * b \qquad \qquad (\because a^{-1} \in G)$$

$$\Rightarrow \qquad \qquad (a^{-1} * a) * x) = a^{-1} * b \qquad \qquad \text{(By associativity)}$$

$$\Rightarrow \qquad \qquad e * x = a^{-1} * b$$

$$\Rightarrow \qquad \qquad x = a^{-1} * b \in G \qquad \qquad (\because a^{-1}, b \in G \Rightarrow a^{-1} * b \in G)$$

Therefore, the equation $a * x = b$ has a solution $x = a^{-1} * b$ in G.

Similarly it can be proved that the equation $y * a = b$ has a solution $y = b * a^{-1}$ in G.

Uniqueness. Let if possible x_1 and x_2 be any two solutions of the equation

$$a * x = b$$

so that

$$a * x_1 = b \text{ and } a * x_2 = b$$

$$\Rightarrow \qquad \qquad a * x_1 = a * x_2$$

$$\Rightarrow \qquad \qquad x_1 = x_2 \qquad \qquad \text{(By left cancellation law)}$$

Therefore, the solution of the equation $a * x = b$ is unique. Similarly, it can be proved that the equation $y * a = b$ has a unique solution.

Hence, the given equations have unique solutions in G.

REMARKS

☞ With the help of the above theorem, we can define the group alternatively as follows:
" A set G with a binary composition ∗ is a group iff"
   (i) the composition ∗ is associative
   (ii) the equations $ax = b$ and $ya = b$ have unique solutions in G.

Theorem 5. *The left identity is also the right identity.*

Proof. Let e be the left identity of a group G and let $a \in G$ be any element.

Then
$$ea = a \qquad\qquad\qquad \text{... (1)}$$

To prove that e is also the right identity, it is sufficient to show that
$$ea = a \qquad\qquad\qquad \text{... (2)}$$

Let a^{-1} be the inverse of a , then
$$a^{-1} a = e \qquad\qquad\qquad \text{... (3)}$$

Now, $a^{-1} (ae) = (a^{-1}a)e = e.e = e = a^{-1}a$

$\Rightarrow$ $a^{-1} (ae) = a^{-1}a$

$\Rightarrow$ $ae = a.$ (By left cancellation law)

Hence, we have the left identity is also the right identity of G.

Theorem 6. *The left inverse of an element is also its right inverse.*

Proof. Let a^{-1} be the left inverse of an element a of a group G, so that
$$a^{-1}a = e \qquad\qquad\qquad \text{... (1)}$$

where e is the identity of G.

To prove that aa^{-1} is also the right inverse of a, it is sufficient to show that
$$aa^{-1} = e \qquad\qquad\qquad \text{...(2)}$$

By associativity, we have
$$a^{-1}(aa^{-1}) = (a^{-1}a)\, a^{-1} = ea^{-1} \qquad\qquad \text{[Using (1)]}$$
$$= a^{-1} = a^{-1}e$$

$\Rightarrow$ $aa^{-1} = e$ (By left cancellation law)

Hence, from (2), we can say that the left inverse of an element is also the right inverse of that element.

2.12.2 Definition of a Group Based on Left Axioms

A non-empty set G with a binary operation ∗ is a group iff the following properties are satisfied:

(i) Associative law. The binary operation ∗ must be associative *i.e.,*
$$a * (b * c) = (a * b) * c; \ \forall\, a, b, c \in G.$$

(ii) Existence of left identity. There exists an element $e \in G$ (called the left identity) such that
$$e * a = a \, ; \forall\, a \in G.$$

(iii) **Existence of left inverse.** For each element $a \in G$, there exists an element $b \in G$ such that

$$b * a = e$$

b is called the left inverse of a and is denoted by a^{-1}.

2.12.3 Definition of a Group Based on Right Axioms

A non-empty set G with a binary operation $*$ is said to be group iff the following properties are satisfied :

(i) **Associative law.** The binary operation $*$ must be associative, *i.e.*,

$$a * (b * c) = (a * b) * c ; \quad \forall\, a, b, c \in G.$$

(ii) **Existence of right identity.** There exists an element $e \in G$ (called the right identity) such that

$$a * e = a ; \quad \forall\, a \in G.$$

(iii) **Existence of right inverse.** For each element $a \in G$, there exists an element $b \in G$ such that $\quad a * b = e$

b is called the right inverse of a and is denoted by a^{-1} .

Theorem 7. *A finite set G, with a binary operation $*$ which is associative, is a group iff the cancellation laws hold.*

Proof. If G is a group, then cancellation laws always hold (Theorem 3). Conversely, if the cancellation laws hold, we have to show that G is a group.

Let $G = (a_1, a_2, ..., a_n)$ and let a, b be any two elements of G.

Consider $\quad a * a, a * a_2, .., a * a_n$

By closure property each are belongs to G.

Further, these element are all distincts, since

$$\text{for } i \neq j \qquad a * a_i = a * a_j \Rightarrow a_i = a_j, \text{ by left cancellation law}$$

Therefore, $a * a_1, a * a_2$ are n distinct elements of G, so that

$$G = \{a * a_1, a * a_2, ..., a * a_n\}$$

Now, $b \in G$, so one of these elements must coincide with b, *i.e.*,

$$a * a_r = b \text{ for some } a_r \in G.$$

Therefore, the equation $a * x = b$ has a unique solution in G. Again considering

$$a_1 * a, a_2 * a, ... a_n * a$$

and proceeding in the similar way, we have

$$y * a = b$$

has a unique solution in G, which implies that G is a group.

REMARKS

☞ The result of the above theorem may not hold when G is an infinite group. For example, in the set **N** of the numbers, for the binary operation $+$, associative and cancellation laws also hold. But (**N**, $+$) is not a group.

Solved Examples

▶ An algebraic structure $(G, *)$ is said to be group if it satisfying the folllowing four properties: (i) Closure property (ii) Associativity (iii) Existence of identity (iv) Existence of inverse

▶ A group G is said to be abelian if $a*b = b*a \ \forall \ ab \in G$.

▶ $(a^{-1})^{-1} = a \ \forall \ a \in G$.

▶ $(a*b)^{-1} = b^{-1}*a^{-1} \ \forall \ a, b \in G$ (Reversal rule)

▶ A finite set G, with binary operation $*$, which is associative is a group if and only if the cancellation laws hold.

Example 1. *Show that the set **Z** of integers (positive or negative integers including 0) with additive unary operation is an infinite abelian group.*

Solution. Let us apply the group-axioms to all integers :

   (i) Closure property. Closure property is satisifed because the sum of any two integers is an integer.

   (ii) Associativity. The associative property satisfied, because if a, b, c are any three integers, then
$$(a + b) + c = a + (b + c)$$

   (iii) Existence of identity. The axiom on identity is satisfied, because 0 is the identity element in the set **Z** such that
$$a + 0 = a \ \forall \ a \in \mathbf{Z}$$

   (iv) Existence of inverse. The axiom on inverse is satisfied, because the inverse of any integer a is the integer $-a$ such that $a+(-a)=(-a)+a=0$, the identity element.

   (v) Commutativity. Since, we know that $a+b = b+a \ \forall a, b \in \mathbf{Z}$, the commutative law is satisfied.

   Also, the number of elements in **Z** is infinite.

   Hence, the set **Z** is an infinite abelian group with additive binary operation.

Example 2. *Show that the set $\{1, -1, i, -i\}$ is an abelian finite group of order 4 under multiplication.*

Solution.   **(i) Closure property.** Closure property is satisfied as
$$1(-1) = -1, 1.i = i, i(-i)=1, 1(-i)=-i \ \text{etc.}$$

   (ii) Associativity. Associative property is satisfied as
$$(1 . i)(-i) = 1. \{i(-i)\} = 1, (1, i). (-i) = 1, \text{etc.}$$

   (iii) Existence of identity. Axioms on identity is satisfied, 1 being the multiplicative identity.

   (iv) Existence of inverse. Axiom on inverse is satisfied since the inverse of each element of the set exists
$$1 \cdot 1 = e = 1, (-1)(-1) = e = 1, i(-i) = e = 1, (-i)(i) = e = 1$$

   (v) Commutativity : The commutative law is also satisfied as
$$1(-1) = (-1) \cdot 1, (-1)i = i(-1) \ \text{etc.}$$

   Finally, since, there are four elements in the given set, hence it is a group of order 4.

Example 3. *Show that the set of all positive rational numbers forms an abelian group under the composition defined by* $a*b = \dfrac{(ab)}{2}$.

Solution. Let $\mathbf{Q}^+$ be the set of all positive rational numbers. To show $(\mathbf{Q}^+, *)$ is a group.

    (i) Closure Property. For every $a, b \in \mathbf{Q}^+$, $ab/2 \in \mathbf{Q}^+$

        $\Rightarrow$ $\mathbf{Q}^+$ is a closed under the composition $*$.

    (ii) Associativity. Let $a, b, c \in \mathbf{Q}^+$ then

$$(a*b)*c = \left(\frac{ab}{2}\right)*c = \frac{[ab/2]\cdot c}{2} = \frac{[ab/2]}{2} = a*\left(\frac{ab}{2}\right) = a*(b*c)$$

    (iii) Existence of Identity. An element e will be the identity if $e \in \mathbf{Q}^+$ and if

$$e*a = a = a*e; \ \forall \ a \in \mathbf{Q}^+$$

    Now, $e*a = a\dfrac{(ea)}{2} = a \Rightarrow \left(\dfrac{a}{2}\right)(e-2) = 0$

    $\Rightarrow$ $e = 2$

    Since, $a \in \mathbf{Q}^+ \Rightarrow a \neq 0$

    Therefore, $2*a = (2a)/2 = a = a*2; \ \forall \ a \in \mathbf{Q}^+$ is the identity element.

    (iv) Existence of Inverse. Let $a \in \mathbf{Q}^+$ be the inverse of a, then we must have

$$b*a = e = 2$$

    $\Rightarrow$ $\dfrac{(ba)}{2} = 2 \Rightarrow b = \dfrac{4}{a}$

    Now, $a \in \mathbf{Q}^+ \Rightarrow 4/a \in \mathbf{Q}^+$

    We have $(4/a)*a = \{(4/a)\cdot a\}/2 = 2 = a*(4/a)$

    $\Rightarrow$ $4/a$ is the inverse of a.

    $\Rightarrow$ Inverse of each element of $\mathbf{Q}^+$ exists.

    (v) Commutativity. Let $a, b \in \mathbf{Q}^+ \Rightarrow a*b = (ab)/2 = (ba)/2 = b*a$

    Hence $(\mathbf{Q}^+, *)$ is an abelian group.

Example 4. *Show that the set* $\mathbf{Z}$ *of all integers form a group with respect to binary operation* $*$ *defined by*

$$a*b = a + b + 1; \ \forall \ a, b \in \mathbf{Z}$$

is an abelian group.

Solution.   **(i) Closure property.** Let $a, b \in \mathbf{Z}$

    $\Rightarrow$ $a + b + 1 \in \mathbf{Z} \Rightarrow a*b \in \mathbf{Z}$

    $\Rightarrow$ $\mathbf{Z}$ is closed with respect to $*$.

    (ii) Associativity. If $a, b, c \in \mathbf{Z}$, then

$$(a*b)*c = (a+b+1)*c = (a+b+1)+c+1 = a+b+c+2$$

    Also, $a*(b*c) = a*(b+c+1) = a+(b+c+1) = a+b+c+2$

    $\Rightarrow$ $(a*b)*c = a*(b*c); \ \forall \ a, b, c \in \mathbf{Z}$

    (iii) Existence of Identity. An element $e \in \mathbf{Z}$ will be the identity if

$$e*a = a; \ \forall \ a \in \mathbf{Z}$$

    Now, $e*a = e + a + 1$

$$e + a + 1 = a \quad \Rightarrow \quad e = -1$$

Since, $-1 \in \mathbf{Z}$ and we have for any $a \in \mathbf{Z}$

$$(-1) * a = -1 + a + 1 = a \quad \Rightarrow \quad -1 \text{ is the identity element.}$$

(iv) Existence of inverse. If $a \in \mathbf{Z}$, then $b \in \mathbf{Z}$ will be the inverse of a if

$$b * a = -1 \qquad (\because -1 \text{ is the identity element})$$

Now, $\qquad b * a = -1 \quad \Rightarrow \quad b + a + 1 = -1 \quad \Rightarrow \quad b = -2 - a.$

Also $\qquad a \in \mathbf{Z} \quad \Rightarrow \quad -2 - a \in \mathbf{Z}$

and $\quad (-2 - a) * a = (-2 - a) + a + 1 = -1$, identity element.

$\therefore (-2 - a)$ is the inverse of a.

(v) Commutativity. Since $\qquad a * b = a + b + 1 = b + a + 1 = b * a$

$\Rightarrow$ Commutativity satisfied.

Hence, $\mathbf{Z}$ is an inifinite abelian group under the given composition.

EXERCISE 2.2

1. Show that following are groups :
   (i) Set of all even integers (including zero) under addition.
   (ii) Set of all non-zero rational numbers with respect to binary operation of multiplication.
   (iii) The set of all real numbers with respect to addition.
   (iv) The set $\mathbf{C}$ of all non-zero complex numbers with respect to multiplication.

2. Does the set of all odd integers form a group with respect to addition ?

3. Show that the set of positive rational numbers does not form a group with respect to the binary operation $*$ defined by $a * b = a/b$.

4. Show that the set $A = \{a + b\sqrt{2} : a, b \in \mathbf{Q}\}$ is a group with respect to addition.

5. Show that the set of all $n \times n$ non-singular matrices having their elements as rational (real or complex) number is an infinite non-abelian group with respect to matrices multiplication.

6. Show that the matrices $A = \begin{bmatrix} \cos\alpha & -\sin\alpha \\ \sin\alpha & \cos\alpha \end{bmatrix}$

   where α is a real number, form a group with respect to matrix multiplication.

7. Show that the four matrices $\begin{bmatrix} 1 & 0 \\ 0 & 1 \end{bmatrix}, \begin{bmatrix} -1 & 0 \\ 0 & 1 \end{bmatrix},$ $\begin{bmatrix} 1 & 0 \\ 0 & -1 \end{bmatrix}, \begin{bmatrix} -1 & 0 \\ 0 & -1 \end{bmatrix}$ forms a group with respect to matrix multiplication.

8. Show that the set of all n, n^{th} roots of unity forms a finite abelian group of order n with respect to multiplication.

9. Show that the set $\mathbf{Z}$ of all integers is an abelian group with operation defined by $a * b = a + b + 2.$

10. Show that the set $\mathbf{Q}$ of all rational numbers, other than 1 with operation $*$, defined by $a * b = a + b + 2$ forms a group under binary operation $*$.

11. Show that the set $G = (1, \omega, \omega^2)$, where ω is an imaginary cube root of unity is a group with respect to multiplication.

12. Show that the set of complex numbers $\mathbf{C}$ with the condition $|z| = 1$ forms a group with respect to the operation of multiplication of complex numbers.

13. Show that the set of four transformation f_1, f_2, f_3, f_4 on the set of complex numbers denoted by $f_1(z) = z, f_2(z) = -z, f_3(z) = 1/z,$ $f_4(z) = -1/z$ forms a finite abelian group with respect to the composite composition.

14. Show that the set V of all vectors (defined as directed line segment) froms an infinite abelian group with respect to vector addition.

15. Show that $\mathbf{Q}$, the set of all rational numbers without 1 by the operation defined by $a * b = a + b - ab$ is an infinte abelian group.

2.13 INTEGRAL POWERS OF AN ELEMENT

Let G be a group with respect to multiplication. If $a \in G$, then aa is denoted by a^2, aaa is denoted by a^3 and so on. We have

$$aaa \ldots n \text{ times} = a^n , n \in \mathbf{Z}^+$$

But by closure property $a^2, a^3, \ldots, a^n \in G$

Also, if e is the identity element in G, we define $a^\circ = e$

If n is a positive integer, we define

$$a^{-n} = (a^n)^{-1} \in G \text{ since } a^n = a \cdot a \cdot a \ldots n \text{ times} \in G.$$

Further,　　　　$$(a^n)^{-1} = (aa \ldots n \text{ times})^{-1} = a^{-1}a^{-1} \ldots n \text{ times} = (a^{-1})^n$$

Thus,　　　　$$a^{-n} = (a^n)^{-1} = (a^{-1})^n$$

REMARKS

☞ For additive groups we write na instead of a^n. Thus, if n is a positive integer, we write

$$na = a + a + \ldots \text{ upto } n \text{ terms.}$$

☞ For an arbitrary element a of a multiplicative group G and for arbitrary constant m and n, it is easy to verify that

(i) $a^m a^n = a^{m+n}$　　　　(ii) $(a^m)^n = a^{mn}$

(iii) $e^n = e$, where e is the identity of G.

2.14 ORDER OF AN ELEMENT OF A GROUP

Let G be a group under multiplication. Let e be the identity element in G. Suppose a is any element of G then the least positive integer n, if exist, such that $a^n = e$ is said to be order of an element $a \in G$, and can be written as

$$o(a) = n.$$

In case, such a positive integer n does not exist, we say that element a is of infinite or zero order.

REMARKS

☞ If G is an additive group, we write na in place of a^n.

☞ If m is a positive integer such that $a^n = e$ the $0(a) \leq m$.

☞ Identity element e in a group G, is the only element whose order is one.

☞ The order of an element of an infinte group may be finite or infinte.

　　For example : In the multiplicative group $\mathbf{Q_0}$ of non-zero rational number $1, -1 \in \mathbf{Q_0}$ such that $0(1) = 1$ and $0(-1) = 2$. The order of any other element of this group is infinite.

Solved Examples

Example 1.　*Consider the multiplicative group $G = \{1, -1, i, -i\}$ of cube roots of unity. Find the order of each element of G.*

Solution.　Since 1 is the identity element, therefore o(1) = 1

　　　　Also,　　　　$(-1)^2 = 1$　　$\Rightarrow$　　$o(-1) = 2$

　　　　　　　　　$(i)^4 = 1$　　$\Rightarrow$　　$o(i) = 4$

　　　　　　　　　$(-i)^4 = 1$　　$\Rightarrow$　　$o(-i) = 4$

Example 2. *Consider the additive group* $\mathbf{Z} = \{..., -3, -2, -1, 0, 1, 2, 3, ...\}$ *of all integers. Show that* 0 *is the only element of finite order.*

Solution. If a be any non-zero integers, then there exists no positive integer n such that

$$na = (a + a + ... + n \text{ times}) = 0$$

$\Rightarrow$ $o(a)$ is infinite.

Hence, in $\mathbf{Z}$ the identity 0 is the only element of finite order.

Theorem 1. *The order of every element of a finite group is finite.*

Proof. Let G be a finite multiplicative group and $a \in G$.

Consider all positive integral powers of a, i.e.,

$$a, a^2, a^3, ..., a^s, ..., a^r, ...$$

By closure property, these all are elements of G.

Since, G is finite, therefore all the integral powers of a cannot be distinct elements of G.

Suppose that $\qquad a^r = a^s$, where $r > s$ $\qquad\qquad\qquad$ ...(1)

Then, $\qquad\qquad\qquad a^r = a^s \Rightarrow \quad a^r a^{-s} = a^s \cdot a^{-s}$

$\Rightarrow \qquad\qquad\qquad a^{r-s} = a^0 = e$

$\Rightarrow \qquad\qquad\qquad a^m = e$, where $m = r - s > 0$

Thus, there exists a positive integer m such that $a^m = e$. Now since every set of positive integers has a least member, it follows that the set of all positive integers m such that $a^m = e$ has a least member say n.

Thus, $\qquad\qquad o(a) = n$, $\qquad$ which is finite.

Hence, the order of every element of the finite group G is finite.

Theorem 2. *If the element a of a group G is of order n, then $a^m = e$. iff n is a divisor of m.*

Proof. Let $o(a) = n$ and $a^m = e$, for some positive integer m then $m \geq n$. If $m = n$ then n is a divisor of m.

If $m \geq n$, then by division algorithm, there exists two integers q and r such that $m = nq + r$, where $0 \leq r \leq n$.

Therefore, $\qquad\qquad a^m = e \quad \Rightarrow \quad a^{nq+r} = e$

$\Rightarrow \qquad\qquad a^{nq} \cdot a^r = e \quad \Rightarrow \qquad a^r = e \qquad\qquad [\because a^{nq} = (a^n)^q = e^q = e]$

Thus, $\qquad\qquad a^r = e$, where $0 < r < n$ or $r = 0$ $\qquad$ (By division algorithm)

Now, since $\qquad o(a) = n$, n is the least positive integer such that $a^n = e$. Hence, it follows that $\qquad r = 0$.

Therefore, $\qquad\qquad m = nq \Rightarrow \quad n$ is a divisor of m.

Conversely, Let n be a divisor of m, so that

$$m = nq, \text{ for } q \in \mathbf{Z}^+$$

Hence, $\qquad\qquad a^m = a^{nq} = (a^n)^q = e^q = e$

Theorem 3. *The order of an element of a group is the same as that of its inverse.*

Proof. Let G be a group under multiplication and a is any element of G.

Suppose that

$$o(a) = m \text{ and } o(a^{-1}) = n$$

Now, $\qquad o(a) = m \qquad \Rightarrow \quad a^m = e \ \Rightarrow \ (a^m)^{-1} = e^{-1} = e$

$\Rightarrow \qquad (a^{-1})^m = e, \quad$ since $\quad (a^m)^{-1} = e^{-1} = e$

$\Rightarrow \qquad o(a^{-1}) \leq m \qquad \Rightarrow \qquad n \leq m \qquad \qquad \qquad$ ...(1)

Again $\qquad o(a^{-1}) = n \qquad \Rightarrow (a^{-1})^n = e \ \Rightarrow \ (a^n)^{-1} = e$

$\Rightarrow \qquad ((a^n)^{-1})^{-1} = e^{-1} \qquad \Rightarrow \qquad a^n = e$

$\Rightarrow \qquad o(a) \leq n \qquad \qquad \Rightarrow \qquad m \leq n \qquad \qquad \qquad$ ...(2)

From (1) and (2), we conclude that

$$m = n, \qquad i.e., \ o(a) = o(a^{-1})$$

Theorem 4. *The order of any integral power of an element a cannot exceed that order of a.*

Proof. Let a^r be any integral power of a and let $o(a) = n$. Now, $o(a) = n \Rightarrow a^n = e$

$\Rightarrow \qquad (a^n)^r = e^r \qquad \Rightarrow \qquad a^{nr} = e$

$\Rightarrow \qquad (a^r)^n = e \qquad \Rightarrow o(a^r) \leq n$

$\Rightarrow \quad o(a^r)$ cannot exceeds the $o(a)$.

Theorem 5. *If a and b are any two elements of a group G, then $o(a) = o(b^{-1}ab)$.*

Proof. Let $o(a) = m$, hence m is the least positive integer such that $a^m = e$.

Now, $\qquad (b^{-1}ab)^2 = (b^{-1}ab)(b^{-1}ab) = b^{-1}a(bb^1)ab \qquad$ (By associativity)

$\qquad \qquad \qquad = b^{-1}aeab \qquad \qquad \qquad \qquad (\because bb^{-1} = e)$

$\qquad \qquad \qquad = b^{-1}a^2b \qquad \qquad \qquad \qquad \ (\because ae = a)$

Similarly $\qquad (b^{-1}ab)^3 = b^{-1}a^3b$

$\qquad \qquad \qquad \vdots \qquad$ ... and so on.

$\qquad (b^{-1}ab)^m = (b^{-1}ab)(b^{-1}ab) \ ... $ to m factors

$\qquad \qquad \qquad = b^{-1}abb^{-1}ab \ ... \ b^{-1}ab \qquad \qquad$ (By associativity)

$\qquad \qquad \qquad = b^{-1}a(bb^{-1})a(bb^{-1}) \ ... \ (bb^{-1})ab \qquad$ (By associativity)

$\qquad \qquad \qquad = b^{-1}a^mb = b^{-1}eb = b^{-1}b = e \qquad \quad (\because a^m = e)$

Thus, we have $(b^{-1}ab)^m = b^{-1}a^mb = e$

Now, since, m is the least positive integer such that $a^m = e$, it follows that m is the least positivie integer such that

$$(b^{-1}ab)^m = e$$

Hence, $\qquad o(b^{-1}ab) = m$

Theorem 6. *For any two elements a, b of a group G, $o(ab) = o(ba)$.*

Proof. We have

$$ba = e(ba) = (a^{-1}a)(ba) = a^{-1}(ab)(a)$$

Hence, by Theorem 5

$$o[a^{-1}(ab)a] = o(ab) \quad \Rightarrow \quad o(ba) = o(ab)$$

Theorem 7. *The order of any integral power of an element of a group is a divisor of the order of that element.*

Proof. Let a be any element of a group G with $o(a) = n$.

$\Rightarrow \qquad \qquad \qquad a^n = e$

Let a^r be any integral power of a such that $o(a^r) = m$

Then $\qquad (a^r)^n = a^{rn} = (a^n)^r = e^r = e$

Thus, $\qquad (a^r)^n = e \implies o(a^r)$ divides n

Hence, $o(a^r)$ is a divisor of $o(a)$.

REMARK

☞ The order of any integral power of an element of a group cannot exceeds the order of the element.

Theorem 8. *If a is an element of order n and p is prime to n, then a^p is also of order n.*

Proof. Let r be the order of a^p.

Now $\qquad o(a) = n \qquad\qquad \implies \qquad a^n = e$

$\implies \qquad (a^n)^p = e^p = e \qquad \implies (a^n)^p = e$

$\implies \qquad o(a^n) \le n \qquad\qquad \implies \qquad r \le n \qquad\qquad$ ...(1)

Now, since p and n are relatively prime, there exists integers x and y such that

$$px + ny = 1$$

$\therefore \qquad\qquad a = a^1 = a^{px+ny} = a^{px}\cdot a^{ny} = a^{px} = a^{px}(a^n)^y = a^{px}e^y$

$\qquad\qquad\qquad = a^{px}\cdot e = a^{px} = (a^p)^x$

Also, $\qquad a^r = [(a^n)^x]^r = (a^p)^{rx} = [(a^p)^r]^x = e^x \quad [\because o(a^p) = r \implies (a^p)^r = e]$

$\qquad\qquad = e$

$\therefore \qquad\qquad o(a) \le r \qquad\qquad \implies \qquad n \le r \qquad\qquad$ ...(2)

From (1) and (2), we conclude that

$$n = r$$

Solved Examples

Based on the following Results

▶ Let G be a group under multiplication, e the identity of G and $a \in G$, then least positive integer n if exist, such that $a^n = e$ is said to be order of a. It can be written as $o(a) = n$.

▶ If the element a of a group G is of order n then $a^m = e$ iff n is a divisor of m.

▶ The order of an element of a group is the same as that of its inverse.

▶ The order of any integral power of an element of a group is a divisor of the order of that element.

Example 1. *For any two elements a and b of a group G, show that G is abelian iff $(ab)^2 = a^2b^2$*

Solution. Let us first suppose that G be abelian so that $ab = ba \; \forall \; a, b \in G$.

Consider $\qquad (ab)^2 = (ab)(ab) = a(ba)b \qquad\qquad$ (by associativity)

$\qquad\qquad\qquad = a(ab)b \qquad\qquad\qquad$ (by commutativity)

$\qquad\qquad\qquad = (aa)(bb) \qquad\qquad\qquad$ (by associativity)

$\qquad\qquad\qquad = a^2\cdot b^2$

Thus, $\qquad (ab)^2 = a^2b^2 \; \forall \; a, b \in G$

Conversly, Let $(ab)^2 = a^2b^2 \; \forall \; a, b \in G$

To show $\qquad\qquad ab = ba$

Consider $\qquad (ab)^2 = a^2b^2$

$\implies \qquad\qquad (ab)(ab) = (aa)(bb)$

$$\Rightarrow \qquad\qquad a(ba)b = a(ab)b \qquad\qquad \text{(By associativity)}$$
$$\Rightarrow \qquad\qquad ab = ba \qquad\qquad \text{(By left and right concellation law)}$$

Thus, we have $\quad ab = ba \; \forall \; a, b \in G.$

Hence, G is abelian.

Example 2. *Show that if G is an abelian group then for all $a, b \in G$ and all integers n, $(ab)^n = a^n b^n$.*

Solution. (i) Let $n = 0$

Then $\qquad\qquad (ab)^0 = e$

Also, $\qquad\qquad a^0 b^0 = e \cdot e = e$

$\therefore \qquad\qquad (ab)^0 = a^0 b^0$

(ii) Let $n > 0$

If $n = 1$, then $(ab)^1 = ab = a^1 b^1$

Let us suppose that result is true for $n = r$

i.e., $\qquad\qquad (ab)^r = a^r b^r$

Then $\qquad (ab)^{r+1} = (ab)^r . ab = a^r b^r ab = a^r ab^r b \qquad (\because ab^r = b^r a)$
$$= a^{r+1} b^{r+1}$$

Then, by mathematical induction for all $n > 0$, $(ab)^n = a^n \cdot b^n$

(iii) Let $n > 0$

Let $n = -r$, where r is a positive integer.

Then $\quad (ab)^n = (ab)^{-r} = [(ab)^r]^{-1} = [a^r b^r]^{-1} = [b^r a^r]^{-1} \quad (\because a^m b^m = b^m a^m)$
$$= [a^r]^{-1} [b^r]^{-1} \qquad\qquad [\because (ab)^{-1} = b^{-1} a^{-1}]$$
$$= a^{-r} b^{-r} = a^n b^n$$

Example 3. *If number of elements in a group G is less than or equal to four, then group must be abelain.*

Solution. Case(i) When $o(G) = 1$, $G = \{e\}$, where e is the identity element.

$\Rightarrow \quad G$ is abelian, since $e \cdot e = e$

Case (ii) When $o(G) = 2$, Let $G = \{e, a\}$ where $a \neq e$

In this case $ae = ea = a$ (by closedness of G) and hence G is abelian.

Case (iii) When $o(G) = 3$, Let $G = \{e, a, b\}$, then $a \neq b \neq e$

In this case we prepare the composition table as follows :

$\bullet$	e	a	b
e	e	a	b
a	a	a^2	ab
b	b	ba	b^2

Since, in any row of the composition table of a group, each element appears only once.

So, $\qquad\qquad\qquad a^2 = e$ or $ab = e$ $\qquad\qquad\qquad$ ...(1)

and $\qquad\qquad\qquad b^2 = e$ or $ba = e$ $\qquad\qquad\qquad$ ...(2)

Now, $\quad a^2 = e \implies ab = b$ since each element in second row occurs only once.

$\Rightarrow \qquad\qquad\qquad ab = eb \implies \qquad a = e$

This is not possible, since $a \neq e$. Thus, $a \neq e \Rightarrow ab = e$

Similarly $b^2 = e \Rightarrow b = a$, by the last row of the composition table.

$\Rightarrow \qquad\qquad ba = ea \Rightarrow \qquad b = e$

Hence, G is abelian.

Case (iv)   Let $G = \{e, a, b, c\}$ be a group of order 4, the identity element e has its own inverse. Since G is a group of even order, there must be at least one more element in G which has its own inverse.

Let $a^{-1} = a$. If $b^{-1} = b$ and $c^{-1} = c$, then in G, every element has its own inverse.

$\Rightarrow \qquad\qquad bc = e = cb$

Also, $\qquad\qquad a^{-1} = a \Rightarrow \qquad a^2 = e$

Now, $\qquad\qquad ab = b \quad$ or $\quad ab = c$

$\qquad\qquad$ ($\because$ each element in the second row occurs only once.)

Since, $\qquad\qquad ab = b \Rightarrow \qquad a = e$, which is not possible so $ab = c$

Then, $\qquad\qquad ac = b$

Now, we prepare the composition table as follows :

$\bullet$	e	a	b	c
e	e	a	b	c
a	a	e	c	b
b	b	c	a	e
c	c	b	e	a

From the above table we conclude that each row is identical to corresponding columns.

$\Rightarrow$ G is commutative.

Hence, G is abelian.

Example 4. *If G is a group of even order, then show that there exists an element a, other than the identity e, such that $a^2 = e$*

Solution. Let $o(G) = 2r$, where r is any positive integer.

Since, we know that, in a group every element possesses a unique inverse and $e^{-1} = e$. The remaining $(2r-1)$ elements should therefore be divided into pairs in such a way that each pair consists of two distinct elements, which are inverse of each other. But this is not possible, since $(2r-1)$ is odd.

Hence, $\exists$ an element $a \in G$, such that,

$$a^{-1} = a, \text{ where } a \neq e$$

But $\qquad\qquad a = a^{-1} \Rightarrow a^2 = a \cdot a = a^{-1}a = e$

Thus, there exists $a \in G$ such tha $a \neq e$ and $a^2 = e$.

Example 5. *In a group, if $ba = a^m b^n$, prove that the elements $a^m b^{n-2}$, $a^{m-2}b^n$, ab^{-1} have the same order.*

Solution. We can write

$$a^m b^{n-2} = a^m b^n b^{-2} = bab^{-2} \qquad\qquad (\because ba = a^m b^n)$$

$$= bab^{-1}b^{-1} = (b^{-1})^{-1}(ab^{-1})b^{-1}$$

In a group, we know that $o(a) = o(x^{-1}ax)$, where a, x are any two elements of the group.

$\therefore$ $o(a^m b^{n-2}) = o[(b^{-1})^{-1}(ab^{-1})b^{-1}] = o(ab^{-1})$ ...(1)

Also, $a^{m-2}b^n = a^{-2}a^m b^n = a^{-2}ba = a^{-2}ba^{2-1} = a^{-2}ba^{-1}a^2$

$= (a^2)^{-1}(ba^{-1})a^2$

$\therefore$ $o(a^{m-2}b^n) = o[(a^2)^{-1}(ba^{-1})a^2] = o(ba^{-1})$

$= o[(ba^{-1})^{-1}]$ $[\because o(a^{-1}) = o(a)]$

$= o[(a^{-1})^{-1}b^{-1}] = o(ab^{-1})$ ...(2)

Hence, from (1) and (2) we get $o(a^m b^{n-2}) = o(ab^{-1}) = o(a^{m-2}b^n)$

Example 6. *If G is a finite abelian group with elements $a_1, a_2, ..., a_n$ show that $a_1 \cdot a_1 ... a_n$ is an element whose square is an identity.*

Solution. We know that

$$(a_1 a_2 ... a_n)^2 = (a_1 a_2 ... a_n)(a_1 a_2 ... a_n)$$ ...(1)

Now, each element in a group has a unique inverse. Thus each of $a_1, a_2, ..., a_n$ is the inverse of exactly one of them. Hence, associating each of $a_1, a_2, ..., a_n$ with its inverse, the relation (1) reduces to

$$(a_1 . a_2 . .... a_n)^2 = (a_1 a_1^{-1})(a_2 a_2^{-1})...(a_n a_n^{-1})$$

$$= e \cdot e \cdot e \cdot ... n \text{ times} = e$$

Example 7. *In any group G if $a^5 = e$, $aba^{-1} = b^2$ for $a, b \in G$. Find $o(b)$.*

Solution. We have

$$(aba^{-1})^2 = aba^{-1}aba^{-1} = ab(a^{-1}a)ba^{-1}$$

$$= ab(e)ba^{-1} = ab^2a^{-1} = aaba^{-1}a^{-1} \qquad (\because aba^{-1} = b^2)$$

$$= a^2ba^{-2}$$

Therefore, $(aba^{-1})^4 = [\{(aba^{-1})\}^2]^2 = (a^2ba^{-2})^2 = a^2ba^{-2}a^2ba^{-2}$

$$= a^2b^2a^{-2} = a^2aba^{-1}a^{-2} = a^3ba^{-3}$$

$\Rightarrow$ $(aba^{-1})^8 = [(aba^{-1})^4]^2 = (a^3ba^{-3})^2 = a^3ba^{-3}a^3ba^{-3} = a^3b^2a^{-3}$

$$= a^3aba^{-1}a^{-3} = a^4ba^{-4}$$

and $(aba^{-1})^{16} = a^4aba^{-4}a^{-4} = a^5ba^{-5} = ebe \qquad (\because a^5 = e \Rightarrow a^{-5} = e)$

$$= b$$

Hence, $(aba^{-1})^{16} = b$

$\because$ $(b^2)^{16} = b$

$\Rightarrow$ $b^{32} = b \quad \Rightarrow \quad b^{31} = e$

Since, $b^m = e \quad \Rightarrow \quad o(b)|m$, therefore $o(b)|31$

but 31 is prime integer, therefore $o(b) = 1$ or 31

Therefore if $b = e$, then $o(b) = 1$ and if $b \neq e$ then $o(b) = 31$.

Example 8. *In a group G if $xy^2 = y^3x$ and $yx^2 = x^3y$, show that $x = y = e$, where e is the identiy of G.*

Solution. It is given that $xy^2 = y^3x \qquad \Rightarrow x^2y^2 = xy^3x$

$\Rightarrow$ $x^2y = xy^3xy^{-1} = xy^2yxy^{-1} = y^3xyxy^{-1}$ ...(1)

Further $\qquad yx^2 = x^3y \qquad \Rightarrow \quad yx^2 = xx^2y$

$\Rightarrow \qquad\qquad yx^2 = xy^3xyxy^{-1}$ $\qquad\qquad$ [Using (1)]

$\Rightarrow \qquad\qquad x^2 = y^{-1}xy^3xyxy^{-1}$

$\Rightarrow \qquad\qquad x^2y = y^{-1}xy^3xyx$ $\qquad\qquad$ ...(2)

Using (1) and (2), we get

$\qquad\qquad y^3xyxy^{-1} = y^{-1}xy^3xyx$

$\Rightarrow \qquad\qquad y^4xyx = xy^3xyxy = xy^2yxyxy = y^3xyxyxy$

$\Rightarrow \qquad\qquad yxyx = xyxyxy$

$\Rightarrow \qquad\qquad (yx)^2 = (xy)^3$ $\qquad\qquad$ ...(3)

Now, since the given relation is symmetrical in x and y, so interchanging x and y in (3), we get

$\qquad\qquad (xy)^2 = (yx)^3$ $\qquad\qquad$ ...(4)

Hence, from (3) and (4), we have

$\qquad\qquad (xy)^2 = (yx)^3 = (yx)^2(yx) = (xy)^3(yx)$

Cancelling $(xy)^2$ from both sides, we get

$\qquad\qquad e = (xy)(yx) = xy^2x$

$\Rightarrow \qquad\qquad x^{-2} = y^2$

Now, $\qquad\qquad xy^2 = y^3x \Rightarrow xx^{-2} = yx^{-2}x$

$\Rightarrow \qquad\qquad x^{-1} = yx^{-1} \Rightarrow \quad y = e$

Further, $\qquad\qquad yx^2 = x^3y$

$\Rightarrow \qquad\qquad ex^2 = x^3e$

$\Rightarrow \qquad\qquad x^2 = x^3$

$\Rightarrow \qquad\qquad x = e$

EXERCISE 2.3

1. If a and b any two elements of a group G, then show that $(bab^{-1})^n = ba^nb^{-1}$ for any $n \in \mathbf{Z}$.

2. Show that if for every element a in a group G, $a^2 = e$, then G is abelian.

3. Show that if every element of a group G has its own inverse, then G is abelian.

4. If G is group such that $(ab)^p = a^pb^p$ for three consecutive integers $p \,\forall\, a, b \in G$, show that G is abelian.

5. Show that a group G is abelian if every element of G except the identity element is of order 2.

6. Find the order of each element in the multiplicative group $G = [1, \omega, \omega^2$, where ω is the cube root of unity].

7. If the element a, b and ab of a group are each of order 2 show that $ab = ba$.

8. Show that in a group G, we have (i) $ab = e \Rightarrow a = b^{-1}$ and $b = a^{-1}$ (ii) $ab = a$ or $ba = a \Rightarrow b = e$.

9. If a is an element of a group, prove that the integral powers of a form a multiplicative group.

10. Show that a group G is abelian iff $(ab)^{-1} = a^{-1}b^{-1} \,\forall\, a, b \in G$.

11. If in a group G, the elements a and b commutes, then prove that
    (i) a^{-1} and b^{-1} also commute.
    (ii) a^{-1} and b also commute.
    (iii) a and b^{-1} also commute.

Hints to selected problems

1. Consider the three cases : $n = 0, n > 0, n < 0$

2. Prove the result for three consecutive integral values of m, i.e., for $n - 1$, n, and $n+1$

3. $o(1) = 1, o(\omega^3) = 1$. Also　$(\omega^2)^3 = \omega^6 = (\omega^3)^2 = 1 \Rightarrow o(\omega^2) = 3$

Answers

6. $o(1) = 1, (\omega) = 3, (\omega^2) = 3$

2.15 MODULO SYSTEMS

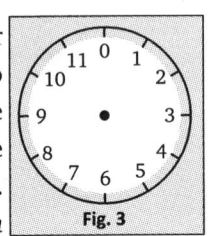

Fig. 3

In fig. (3), the dial is similar to an ordinary clock, with 0 substituted for 12. The number 13 is equvalent to 1, 14 is equivalent to 2. 23 is equivalnet to 11, 30 is equivalent to 6. Thus all the positive integers are equivalent to one or the other of the integers, 0, 1, 2, ... with modulo 12. Similarly, negative integers can be represented by the same numbers. Thus, $-2 \equiv 10, 20 \equiv -8 \equiv 4$. Here is the integers are divided into 12 classes. We can similarly divide into m classes ($m \in \mathbf{Z}^+$) and we write mod m. This system of representing numbers is called modulo system.

2.15.1 Addition Modulo m

Let a and b be any two integers and m is a fixed positive integers then
$$a +_m b = r, 0 \leq r < m$$
where r is the least non-negative remainder when $a + b$ is divide by m.

For Example　　　　$8 +_3 8 = 1$

Since,　　　　$8 + 8 = 16 = 3(5) + 1,$

i.e., 1 is the least non-negative remainder, when $8 + 8$ is divided by 3.

2.15.2 Multiplication modulo p

The multiplication modulo p is written as $a \times_p b$ where a and b are any integers and p is fixed.

Then, by definition, we have $a \times_p b = r, 0 \leq r < p$, where r is the least non-negative remainder when $a \cdot b$ is divided by p.

For Example　　　　$4 \times_3 5 = 2$

Since　　　　$4 \times 5 = 20 = 3(6) + 2$

Definition. *Let $n > 0$ be a fixed integer. If a and b are two integers, we define* $a \equiv b \pmod{n}$ *if $n \mid (a - b)$, i.e., if n divides $(a - b)$.*

The relation is referred to as " congruent modulo n", n is called the modulus of the relation and we read $a \equiv b$ (mod n) as "a is congruent to b modulo n."

REMARK

☞ $a \times_p b = ab \pmod{p}$

2.16 DIVISION ALGORITHM

Let $a, b \in \mathbf{Z}$ with $b \neq 0$, then we can divide a by b to get a non-negative remainder r which is smaller than b. In other words, If $a, 0 \neq b \in \mathbf{Z}$ then there exists integers q and r such that
$$a = qb + r, \text{ where } 0 \leq r < |b|$$

2.16.1 Divisibility in the Set of Integers

Let $a, b \in \mathbf{Z}$ and $b \neq 0$, then we say that a is divisible by b if $a = km$, where k is some integer.

REMARKS

☞ If a is divisible by b, then we also say that b is divisor of a and can be written as $b|a$ read as 'b is a divisor of a'.

☞ If $a|b$, $b|c$ then $a|c$.

☞ If $a|b$, $a|d$ then $a|(b+d)$.

☞ If $a|b$ then $a|bc$ where c is any integer.

2.16.2 Greatest Common Divisor.

Let a and b be any two integers then a positive integer c is said to be the greatest common divisor of a and b if

(i) $c|a$ and $c|b$

(ii) whenever $d|a$ and $d|b$, then $d|c$

The g.c.d. of a and b can be written as (a, b).

REMARKS

☞ If a and b are two integers, not both 0 then they have unique greatest common divisor say c. Also, we can find integers x and y such that
$$c = xa + yb$$
☞ Two integers a and b are said to be relatively prime if their greatest common divisor is 1.
For Example (i) 1 and 4, (ii) 2 and 3.

2.16.3 Prime Integers

An integer p is said to be prime integer if $p \neq 0$, $p \neq \pm 1$ and the only divisors of p are $\pm 1, \pm p$.

For Example

$$\pm 2, \pm 3, \pm 5, \ldots \text{ are prime integers.}$$

Theorem 1. *The set $\{0, 1, 2, \ldots, (n-1)\}$ of n elements is a group under addition modulo n.*

Proof.   Let $G = \{0, 1, 2, \ldots (n-1)\}$

(i) **Closure property.** Let $h, k \in G$, then by definition of addition modulo n $h+k \equiv r \pmod{n}$, where $0 \leq r < n$.

Thus, sum of any two elements in G is congruent modulo n to some one of the n integers in G. Hence, G is closed under addition modulo n.

(ii) **Associativity.** Since, the associative law of addition holds for the set of integers $\mathbf{Z}$. It holds for all S which is a subset of $\mathbf{Z}$.

(iii) **Existence of inverse.** Addittive inverse of any $h \in G$. is $(n-h) \in G$.
Since, $h+(n-h) = n \equiv 0 \pmod{n}$.

(iv) **Commutativity.** Here, $h+k = k+h$, for every, h, k in G.
Hence all the group axioms are satisfied, so that given set is a commutative group under addition modulo n.

Theorem 2. *If p is a prime number, then the set {1,2,3, ... , (p−1)} is a group under multiplication modulo p.*

Proof.   Let us denote the given set by G as

$$G = \{1,2,3, ..., (p-1)\}$$

(i) **Closure property.** Let $h, k \in G$. Since p is prime and does not divide any element in G. We have

$$h \cdot k \equiv r \pmod{p}, \text{ where } 0 < r < p.$$

Thus, product of any two elements in G is again in G. Therefore, G is closed under multiplication modulo p.

(ii) **Associativity.** Since the associative law of multiplication holds for the set of integer **Z**, it does hold for the given set which is a subset of **Z**. Thus, if $h, k, l \in$ **G**, we certainly have

$$(h. k). l = h . (k. l)$$

(iii) **Existence of identity.** The identity element is 1, since for any $h \in G$, $1. h = h$.

(iv) **Existence of inverse.** Let $a \in G$, then x be the inverse of a if

$$ax \equiv 1 \pmod{p}$$

Since, p is prime, this just has one solution in G.

Thus, every element in G has a unique inverse in G.

Therefore, all the group axioms are satisfied. Moreover hk or $h . k = h = kh$. For every $h, k \in G$. For every $h, k \in G$. Hence, the given set is an abelian group under multiplication modulo n.

Theorem 3. *The relation, congruence modulo n, is an equivalence relation on the set of integers. Also this equivalence relation has n distinct equivalence classes.*

Step outlines: To prove the desired result, we must verify the following conditions:

Step 1. **Reflexive:** $a \equiv a \pmod{n}$ ...(i)

Step 2. **Symmetric:** $a \equiv b \pmod{n}, b \equiv a \pmod{n}$ ...(ii)

Step 3. **Transitive:** $a \equiv a \pmod{n}, b \equiv c \pmod{n} \Rightarrow a \equiv c \pmod{n}$ ...(iii)

Proof.   Since, $n \mid 0$, we surely have $n \mid (a - a)$ and so $a \equiv a \pmod{n}$ for every integer a

$\Rightarrow$ (i) is verified.

Also, if $a \equiv b \pmod{n}$, then $n \mid (a - b)$ and so $n \mid -(a - b)$, i.e., $n \mid (b - a)$; therefore $b \equiv a \pmod{n} \Rightarrow$ (ii) is verified.

Finally $a \equiv b \pmod{n} \Rightarrow n \mid (a - b)$
$$b \equiv c \pmod{n} \Rightarrow n \mid (b - c)$$
$$\Rightarrow n \mid (a - b) + (b - c)$$
$$\Rightarrow n \mid (a - c)$$
$$\Rightarrow n \equiv c \bmod n$$

$\Rightarrow$ (iii) is verified.

Now we have an equivalence relation defined on the set of integers Z; we can consider equivalence classes.

Let the equivalence class under the equivalence relation of congruence modulo n of $k \in \mathbf{Z}$ be denoted by $[k]$, then

$$[k] = \{a \in \mathbf{Z} ; a \equiv k \ (\text{mod } n)\}$$

Thus, $[k]$ is the set of integers which are congruent to k modulo n, we call it the congruence class (mod n) of k.

Now, let a be any integer then by divison algorithm $a = qn + r$ where $0 \le r < n$. Then $a - r = qn$ and hence $a \equiv r$ (mod n). Since, $0 \le r < n$, it follows that every integer is congruent modulo n to some one of the n integers 0, 1, 2, ... , $n-1$. Morever, since each of these integers is less than n, no two of them can be congruent to each other modulo n. Furthermore $a \equiv r$ (mod n) implies that $a \in [r]$ and so $[a] = [r]$. Consequently there are exactly n distinct congruence classes (mod n), namely.

$$[0], [1], [2], ..., [n-1]$$

2.17 RESIDUE CLASSES MODULO n

The equivalence relation $\equiv$ (mod n), i,e., congruence modulo n decomposes the set $\mathbf{Z}$ into disjoint classes, called equivalence classes. These equivalence classes are called residue calsses modulo n. The set of these equivalence classes is called the quotient set of $\mathbf{Z}$ and is denoted by $\mathbf{Z}_n$.

Thus $\mathbf{Z}_n = \{[0], [1], [2], ... [n-1]\}$

2.17.1 Addition of Residue Classes

Let $[a]$ and $[b]$ be two residue calsse modulo m, their sum is defined by

$$[a] + [b] = [r]$$

where r is the least non-negative remainder when $a + b$ is divided by m.

For example.

If $[1], [2], ... [5], [6], [7]$ are residue classes modulo 8.

$$[2] + [5] = [7] \text{ for } 2 + 5 = 7 < 8$$
$$[1] + [5] = [7] \text{ for } 1 + 5 = 6 < 8$$
$$[6] + [7] = [5] \text{ for } 6 + 7 > 8 \text{ and } 6 + 7 = 13 = 1 \times 8 + 5$$

where r is he least non-negative remainder when $a + b$ is divided by m.

REMARK

☞ The addition of residue classes can also be defined by

$$[a] + [b] = \begin{cases} [a + b] & \text{if } a + b < m \\ [r] & \text{if } a + b \ge m \end{cases}$$

where r is the least non-negative remainder *when* $a + b$ is divided by m.

2.17.2 Multiplication of Residue Classes

Let $[a]$ and $[b]$ be redidue classes modulo m, then their product is defined by

$$[a] \cdot [b] = [r]$$

where r is the least non-negative remainder when ab is divided by m.

For example.

If $[1], [2], ... [6], [7]$ are residue classes modulo 8, then

(i) $[2] \cdot [5] = [2]$ for $2 \times 5 > 8$ and $2 \times 5 = 10 = 1 \times 8 + 2$

(ii) $[1] \cdot [5] = [5]$ for $1 \times 5 < 8$.

REMARK

☞ The multiplication of residue classes can also be defined by

$$[a][b] = \begin{cases} [ab] & \text{iff } ab < m \\ [r] & \text{iff } ab \geq m \end{cases}$$

where r is the least non-negative remainder when ab is divided by m.

Theorem 1. *The set of residue classes modulo m is a group with respect to addition of residue classes.*

Proof. Let $\mathbf{Z}_m$ denote the set of all residue classes modulo m so that

$$\mathbf{Z}_m = \{[0], [1], [2] \, ... [m-1]\}$$

Let $[a], [b], \in \mathbf{Z}_m$ be arbitrary.

To prove $(\mathbf{Z}_m, +)$ is a group.

(i) Closure property. $[a], [b] \in \mathbf{Z}_m \Rightarrow [a] + [b] \in \mathbf{Z}_m$

For $[a] + [b] = [r]$, where r is the least non-negative remainder when $a + b$ is divided by m.

$$\Rightarrow \qquad 0 \leq r < m \Rightarrow [r] \in \mathbf{Z}_m \Rightarrow [a] + [b] \in \mathbf{Z}_m$$

(ii) Associativity. Consider $([a] + [b]) + [c] = [(a+b)+c]$, when $(a+b)+c$ is reduced by m.

$$= [a+(b+c)] \qquad\qquad [\because (a+b) + c = a + (b+c)]$$
$$= [a] + [b+ c] = [a]+([b]+[c])$$

(iii) Existence of identity. There exists an identity element $[0] \in \mathbf{Z}_m$

for $[0]+[a] = [a] = [a] + [0]$.

(iv) Existence of inverse. Every element $[a] \in \mathbf{Z}_m$ has its inverse $[m-a] \in \mathbf{Z}_m$

For $[a] + [m-a] = [0] = [m-a] + [a]$

Thus, all group axioms are satisfied and hence $(\mathbf{Z}_m, +)$ is a group.

Theorem 2. *The set of all non-zero residue classes modulo a prime number p is a group with respect to multiplication of residue classes.*

Proof. Let G be a set of all non-zero residue classes modulo a prime integer p so that

$$G = \{[1], [2], [3] \, ... [p-1]\}$$

The divisors of p are only ± 1 and $\pm p$.

To prove that $(G, *)$ is a group.

Let $[a], [b], [c] \in G$

(i) Closure property. $[a].[b] = [r]$, where r is least non-negative remainder when ab is divided by p.

$$\Rightarrow \qquad 0 \leq r < p$$

$[a]$ and $[b] \in G \Rightarrow 0 < a < p, 0 < b < p$

$\Rightarrow$ Neither a nor b is divisible by p, also p is prime.

$\Rightarrow$ Remainder $r \neq 0$. But $0 \leq r < p$

$$\Rightarrow \qquad 0 < r < p \Rightarrow [r] \in G \Rightarrow [a] [b] \in G$$

(iii) Associativity. $([a] . [b]). [c] = [a]. ([b]. [c]).$

For L.H.S. $= [a . b].[c]$, where ab is reduced by p

$$= [a(bc)] \qquad\qquad\qquad [\because (ab)c = a(bc)]$$

$$=[a]\,[bc] = [a]\,([b][c]) = \text{R.H.S.}$$

(iv) Existence of identity. There exists an element $[a] \in \mathbf{G}$ such that

$$[a]\,.\,[1] = [a.1] = [a] = [1]\,.\,[a]$$

(v) Existenc of inverse. Every element $[a] \in \mathbf{G}$ has its inverse $[a']\in \mathbf{G}$ where

$$[a']\,[a] = [aa'] = [a'a] = [1] \quad aa' \equiv 1 \ (\text{mod } p)$$

Thus, every element $[a] \in G$ has its inverse $[a'] \in G$ such that $aa' \equiv 1 (\text{mod } p)$.

Since $ax \equiv 1$ (mod p) has a solution x if p is prime. Hence $[a']$ exists.

Thus, all the group axioms are satisfied and hence $(G, *)$ is a group.

Solved Examples
Based on the following Results

▶ Let a and b be any two integers and m be a fixed positive integers, then

$$a +_m b = r,\ 0 \le r < m$$

where r is the least non-negative remainder when $a+b$ is divided by m.

▶ The multiplication modulo p is written as $a \times_p b$ where a and b are any integers and p is fixed. Then $a \times_p b = r,\ 0 \le r < p$, where r is the least non-negative remainder when $a.b$ is divided by p.

▶ **Division Algorithm** If $a/b\ (b \ne 0) \in \mathbf{Z}$ then there exist integers q and r such that $a = qb+r,\ 0 \le r < |b|$

Example 1. *Show that the set* $G = \{1, 2, 3, 4, 5, 6\}$ *is a finite abelian group of order 6 with respect to multiplication modulo 7.*

Solution. Let us prepare the composition table as follows:

$\times_7$	1	2	3	4	5	6
1	1	2	3	4	5	6
2	2	4	6	1	3	5
3	3	6	2	5	1	4
4	4	1	5	2	6	3
5	5	3	1	6	4	2
6	6	5	4	3	2	1

Here we have the following obervations:

(i) Closure property. We see that all the entries in the composition table are elements of the set G. Therefore G is closed with respect to multiplication modulo 7.

(ii) Associativity. Associative law always holds in modulo system.

(iii) Existence of identity. We have $1 \in G$. Let $a \in G$. Then from the composition table we see that

$$1 \times_7 a = a = a \times_7 1.$$

$\therefore$ 1 is the identity element.

(iv) Existence of inverse. From table we have the inverses of 1, 2, 3, 4, 5, 6 which are given by 1, 4, 5, 2, 3, 6 respectively.

(v) **Commutativity.** Since each row is identical to corresponding column therefore the composition is commutative.

Also, the set G has six element $\Rightarrow o(G) = 6$

Hence, $(G, \times_7)$ is a finite abelian group of order 6.

2.18 TRANSFORMATION, PERMUTATION AND PERMUTATION GROUPS

(i) **Transformation.** A one-one mapping of a non-empty set S onto itself is called a transformation.

(ii) **Permutations.** A one-one mapping of a finite non-empty set S onto itself is called a permutation.

If the set S consists of n distinct elements, then a one-one mapping of S onto itself is called a permutation of degree n or a permutations of n-symbols.

Notation. An easy way of denoting a permutation f is write down the elements of the domain in a row and then write the image of each element directly below it. In this manner we would write the permutations as follows

Let $S = \begin{pmatrix} a_1, a_2, \ldots a_i, \ldots a_n \\ b_1, b_2, \ldots b_i, \ldots b_n \end{pmatrix}$

so that in the first row all the element of S are written in a certain order and

$$f(a_1) = b_1, f(a_2) = b_2, \ldots f(a_i) = b_i, \ldots f(a_n) = b_n$$

REMARKS

☞ It is clear that each $b_i \in S$, $i = 1, 2, \ldots, n$.

☞ It is immaterial in which order, the elements of S are written in the first row, but the image of element a_i must be written under a_i. Thus, the interchange of columns does not change the permutation.

For Example

$$\begin{pmatrix} 1 & 2 & 3 & 4 \\ 2 & 3 & 4 & 1 \end{pmatrix} = \begin{pmatrix} 2 & 1 & 3 & 4 \\ 3 & 2 & 4 & 1 \end{pmatrix} = \begin{pmatrix} 1 & 4 & 3 & 2 \\ 2 & 1 & 4 & 3 \end{pmatrix}.$$

2.18.1 Equality of Permutations

Two permutations f and g of a set S are said to be equal if

$$f(a) = g(a) \ \forall \ a \in S$$

For example. If $f = \begin{pmatrix} 1 & 2 & 3 \\ 3 & 1 & 2 \end{pmatrix}$ and $\begin{pmatrix} 1 & 2 & 3 & 4 & 5 \\ 2 & 3 & 4 & 5 & 1 \end{pmatrix}$ are two permutations of degree 3

then we have $f = g$ since $f(1) = g(1) = 3, f(2) = g(2) = 1, f(3) = g(3) = 2$.

2.18.2 Identity Permutations.

A permutations on the set S is called the identity permutation if it maps each element of S onto itself.

Thus, $I(a) = a, \forall a \in S$

For Example

$$I = \begin{pmatrix} 1 & 2 & 3 & \ldots & n \\ 1 & 2 & 3 & \ldots & n \end{pmatrix}$$ is the identity permutation of degree n.

2.18.3 Total Number of Distinct Permutations

Let S be a set conssisting of n distinct elements. Then the elements of S can be permuted in $n!$ distinct ways, *i.e*, $n!$ distinct arrangement of the elements belonging to S are possible. If P_n be the set consisting of all permutations of degree n, then it contains $n!$ distinct permutations of degree n.

This set P_n is called the symmetric set of permutations of degree $n!$ and is denoted by

$$P_n = \{f : f \text{ is a permutation of degree } n\}.$$

2.18.4 Inverse Permutations

Since a permutation is a one-one onto mapping and hence it is invertible, *i.e.*, every permutation f on a set $P = \{a_1, a_2, \dots a_n\}$ has a unique inverse permutation denoted by f^{-1}.

For Example

If $f = \begin{pmatrix} a_1 & a_2 & \dots & a_n \\ b_1 & b_2 & \dots & b_n \end{pmatrix}$, then $f^{-1} = \begin{pmatrix} b_1 & b_2 & \dots & b_n \\ a_1 & a_2 & \dots & a_n \end{pmatrix}$.

2.18.5 Product or Composite of Permutations

The product or composite of two permutations f and g of the same degree, denoted by $f.g$ is obtained by first carrying out the operation defined by mapping f and then by mapping g.

WORKING PROCEDURE

Arrange the first row of f so that it is identical with the second row of g (also, arrange the second row of f accordingly). Then fg is the permutations whose first row is the first row of g and whose second row is the second row of f after the rearrangement.

Let $f, g \in P_n$, so that

$$f = \begin{pmatrix} a_1 & a_2 & \dots & a_n \\ b_1 & b_2 & \dots & b_n \end{pmatrix}, g = \begin{pmatrix} b_1 & b_2 & \dots & b_n \\ c_1 & c_2 & \dots & c_n \end{pmatrix}$$

where, $a_i, b_i, c_i \in S, i = 1, 2, \dots, n$.

Then,

$$fg = \begin{pmatrix} a_1 & a_2 & \dots & a_n \\ b_1 & b_2 & \dots & b_n \end{pmatrix} \begin{pmatrix} b_1 & b_2 & \dots & b_n \\ c_1 & c_2 & \dots & c_n \end{pmatrix} = \begin{pmatrix} a_1 & a_2 & \dots & a_n \\ c_1 & c_2 & \dots & c_n \end{pmatrix} \in P_n$$

For Example

Let $f = \begin{pmatrix} 1 & 2 & 3 & 4 \\ 2 & 3 & 4 & 1 \end{pmatrix}, g = \begin{pmatrix} 1 & 2 & 3 & 4 \\ 3 & 4 & 2 & 1 \end{pmatrix}$ be two permutations of degree 4, then

$$fg = \begin{pmatrix} 1 & 2 & 3 & 4 \\ 2 & 3 & 4 & 1 \end{pmatrix} \begin{pmatrix} 1 & 2 & 3 & 4 \\ 3 & 4 & 2 & 1 \end{pmatrix} = \begin{pmatrix} 1 & 2 & 3 & 4 \\ 2 & 3 & 4 & 1 \end{pmatrix} \begin{pmatrix} 2 & 3 & 4 & 1 \\ 4 & 2 & 1 & 3 \end{pmatrix} = \begin{pmatrix} 1 & 2 & 3 & 4 \\ 4 & 2 & 1 & 3 \end{pmatrix}$$

and

$$gf = \begin{pmatrix} 1 & 2 & 3 & 4 \\ 3 & 4 & 2 & 1 \end{pmatrix} \begin{pmatrix} 1 & 2 & 3 & 4 \\ 2 & 3 & 4 & 1 \end{pmatrix} = \begin{pmatrix} 1 & 2 & 3 & 4 \\ 3 & 4 & 2 & 1 \end{pmatrix} \begin{pmatrix} 3 & 4 & 2 & 1 \\ 4 & 1 & 3 & 2 \end{pmatrix} = \begin{pmatrix} 1 & 2 & 3 & 4 \\ 4 & 1 & 3 & 2 \end{pmatrix}$$

Here, we observe that $fg \neq gf$. Thus, the product of permutations is not commutative.

2.19 CYCLIC PERMUTATION

Let f be a permutation of degree n on a set having n distinct elements and let it be possible to arrange m elements of the set S in a row in such a way that the f-image of each element in the row is the element which follows and the f-image of the last element is the first element and the remaining $n - m$ elements of the set S are left unchanged by f. Then, f is called a cyclic permutation or a cycle of length m or an m-cycle.

REMARKS

☞ The following have the same meaning: cyclic permutation, circular permutations or cycles.

☞ The number of distinct objects permuted by a cycle is known as the length of the cycle.
   For Example

(i) $\begin{pmatrix} 1 & 2 & 3 \\ 2 & 3 & 1 \end{pmatrix}$ is a cyclic permutation of length 3.

(ii) $\begin{pmatrix} 1 & 2 & 3 & 4 & 5 \\ 2 & 3 & 4 & 5 & 1 \end{pmatrix}$ is a cyclic permutation of length 5.

2.19.1 Symbol for Cyclic Permutation

We denote the cyclic permutations $\begin{pmatrix} a_1 & a_2 & a_3...a_n & a_n \\ a_2 & a_3 & a_4...a_{n-1} & a_1 \end{pmatrix}$ by the symbols $(a_1, a_2, ..., a_n)$, which means that each member in the bracket is replaced by its successor on the right and the last member is replaced by the first one. Thus the cyclic permutation $(1, 4, 2, 6)$ is expressible as

$$\begin{pmatrix} 1 & 4 & 2 & 6 \\ 4 & 2 & 6 & 1 \end{pmatrix}$$

It is interpreted as 1 replaced by 4, 4 replaced by 2, 2 replaced by 6 and 6 replaced by 1.

REMARKS

☞ The length of the cycle (4 5 6) is 3, where as the degree of the permutation $\begin{pmatrix} 1 & 2 & 3 & 4 & 5 & 6 \\ 1 & 2 & 3 & 5 & 6 & 4 \end{pmatrix}$ is 6.

☞ A cycle does not changes by changing the places of its elements in cyclic order. Thus, (1 2 3 4 5) = (2 3 4 5 1) = (3 4 5 1 2)

2.19.2 Cycle of Length 1

A cycle of length 1 means that the image of the element involved is the same element and the missing element are unchanged. Thus, all the element are unchanged. Hence, every cycle of length one represents the identity permutation.

2.19.3 Transposition.

A cycle of length two is called a transposition. Thus, the cycle (9 6) or (1 2) is a transposition. If the cycle (1 2) is a permutation of degree 3 on three symbols 1, 2, 3 then the corresponding permutation is $\begin{pmatrix} 1 & 2 & 3 \\ 2 & 1 & 3 \end{pmatrix}$.

2.19.4 Disjoint Cycles.

Two cycles are said to be disjoint, if they have no elements in common.

For Example

(i) (1 2) and (3 4) are disjoint cycles.

(ii) (1 5 3) and (8 9 10) are disjoint cycles.

2.19.5 Multiplication of Cycles

We multiply cycles by multiplying the permutations represented by them.

For Example

Let us suppose that (2 3 4) and (5 3 1 2) represents permutations of degree 6 on six symbols 1, 2, 3, 4, 5, 6. Then, we have

$$\begin{pmatrix} 2 & 3 & 4 \\ 3 & 4 & 2 \end{pmatrix}\begin{pmatrix} 5 & 3 & 1 & 2 \\ 3 & 1 & 2 & 5 \end{pmatrix} = \begin{pmatrix} 1 & 2 & 3 & 4 & 5 & 6 \\ 1 & 3 & 4 & 2 & 5 & 6 \end{pmatrix}\begin{pmatrix} 5 & 3 & 1 & 2 & 4 & 6 \\ 3 & 1 & 2 & 5 & 4 & 6 \end{pmatrix}$$

$$= \begin{pmatrix} 1 & 2 & 3 & 4 & 5 & 6 \\ 1 & 3 & 4 & 2 & 5 & 6 \end{pmatrix}\begin{pmatrix} 1 & 3 & 4 & 2 & 5 & 6 \\ 2 & 1 & 4 & 5 & 3 & 6 \end{pmatrix}$$

$$= \begin{pmatrix} 1 & 2 & 3 & 4 & 5 & 6 \\ 2 & 1 & 4 & 5 & 3 & 6 \end{pmatrix} = (1,2)(3,4,5),(6)$$

($\because$ a cycle of length 1 represents the identity permutations.)

REMARK

☞ It is easy to verify that the product of the disjoint cycles is commutative. Therefore, if f and g are any two cycles, then we have

$$(fg)^{-1} = g^{-1}f^{-1}. \text{ Also, } (fgh)^{-1} = h^{-1}g^{-1}f^{-1}$$

Theorem 1. *Every permutation can be expressed as a product of disjoint cycles.*

Proof. Let f be a given permutation of degree n, defined on the set $S = \{a_1, a_2, \dots a_n\}$.

All the cycles of length 1 is given by the invariant element.

Now select an element, which is non-invariant and construct a row, starting with this element and writing each element after its image under f.

As the number of elements in S is finite, after a finite number of steps, we get an element whose image under f is the one with which we started. This row is a cycle.

Now, we choose an element of S which is not contained in the above cycles and get another cycle, as above.

Proceeding in the same way, each and every element of S is included in one or the other cycle. Obviously, these cycles have no element in common and hence they are disjoint.

Hence, the permutation f can be expressed as a product of disjoint cycles.

Theorem 2. *Every permutations can be expressed as a product of transpositions.*

Proof. By theorem 1 every permutation can be expressed as a product of disjoint cycle. Consider the cycle $(a_1, a_2, \dots a_n)$ of length n, where $n > 1$, then, we see

$$(a_1, a_2, \dots a_n) = (a_1, a_2)(a_1, a_3)(a_1, a_4)\dots(a_1, a_n)$$

For Example

$$(a_1, a_2, a_3) = (a_1, a_2)(a_1, a_3).$$

Also $n = 1$

$$(a_1) = (a_1, a_2)(a_2, a_1).$$

i.e., the identity permutation can also be expressed as a product of transpositions.

    $\Rightarrow$ Every cycle can be expressed as a product of transpositions. Hence, it follows that every permutations can be expressed as a product of transpositions.

REMARK

☞ For any manner of expressing a given permutations as a product of transpositions, the number of transpositions is either even or a odd.

2.20 EVEN AND ODD PERMUTATIONS

A permutation is said to be even or odd according as it can be expressed as a product of even or odd number of transpositions.

There is another easy way of determining whether a permutation is even or odd.

Let $\begin{pmatrix} a_1 & a_2 & a_3...a_n & a_n \\ a_2 & a_3 & a_4...a_{n-1} & a_1 \end{pmatrix}$ be a permutation of degree n. The pair $(i . j)$ is said to be regular if $i - j$ and $a_i - a_j$ both have the same sign. Otherwise irregular. Thus for irregularity of any pair (i, j), $(i, -j)$ and $(a_i, -a_j)$ are of opposite signs. The number of irregular pairs denotes number of inversions.

A permutation of a set of integers onto itself is even or odd according as it contains even or odd number of inversions.

For Example :

(1) $\begin{pmatrix} 1 & 2 & 3 \\ 3 & 1 & 2 \end{pmatrix}$ $\Rightarrow$ No inversion

(2) $\begin{pmatrix} 1 & 2 & 3 \\ 3 & 1 & 2 \end{pmatrix}$ $\Rightarrow$ Two inversion

(3) $\begin{pmatrix} 1 & 2 & 3 \\ 3 & 2 & 1 \end{pmatrix}$ $\Rightarrow$ Three inversion

(4) $\begin{pmatrix} 1 & 2 & 3 & 4 & 5 \\ 5 & 3 & 2 & 4 & 1 \end{pmatrix}$ $\Rightarrow$ Eight inversion

(5) $\begin{pmatrix} 1 & 2 & 3 \\ 2 & 1 & 3 \end{pmatrix}$ $\Rightarrow$ One inversion

Hence, (1), (2), (4) $\Rightarrow$ Even permutations

        (3) and (5) $\Rightarrow$ odd permutations

REMARKS

☞ Every transposition is always an odd permutations.

☞ Identity permutation is an even permutation.

☞ The product of two odd permutations is an even permutation.

☞ The product of two even permutations is an even permutations.

☞ The product of an odd and an even permutation is an odd permutation.

☞ The inverse of an even permutation is even.

☞ The inverse of an odd permutation is odd.

☞ A cycle of length n is expressible as the product of $(n - 1)$ transpositions.

☞ A permutation cannot be both odd and even.

Theorem 1. *Of the n! permutation of n symbols, $\dfrac{n!}{2}$ are even permutation and $\dfrac{n!}{2}$ are odd permutations.*

Proof. Let P_n be the set of all permutations on n distinct symbols. The P_n contains $n!$ distinct permutations of degree n. Also P_n is a group with respect to permutation multiplication as compositions.

Out of these $n!$ permutation, let the even permutation be $E_1, E_2, \dots Ep$ and the odd permutation be $O_1, O_2, \dots, Oq$ so that $p+q = n!$

Let $t \in P_n$ be arbitrary such that t is a transposition so that t is an odd permutation. Let t be operated on each of E_i $(i = 1, 2, \dots, p)$ and similarly on each $O_i (i = 1, 2, \dots, q)$

Now, since $(P_n, \, . \,)$ is a group and therefore $tE_i \in P_m$ and $tO_j \in P_n$ for
$$1 \le i \le p, \ 1 \le j \le q$$

The permutations tE_i for $i = 1, 2 \dots, p$ are all permutation

($\because$ transposition is an odd permutation and product of even and odd permutation is an odd permutation). Also, these permutation are all distinct.

For
$$tE_i = tE_j \Rightarrow E_i = E_j$$
$\therefore$
$$tE_i \ne tE_j \ \text{if} \ \ E_i \ne E_j$$

Similarly, the permutations tO_j $(j = 1, 2, \dots, q)$ are all distinct even permutations.

Since, a permutation cannot be both even and odd, we have the even permutations $E_1, E_2, \dots, E_p$ are equal to q even permutations.

Similarly, q odd permutations $O_1, O_2, \dots, O_q$ are equal to p odd permutation tE_1, $tE_2, tE_3, \dots, tE_p$. Consequently $p = q$. Also $p+q = n!$

$\therefore$
$$p = q = \frac{n!}{2}$$

REMARK

☞ If A_n is the set of all even permutations of degree n then $A_n \subset P_n$ and A_n contatins $\dfrac{n!}{2}$ elements.

The set A_n is called an alternating set of permutations.

Theorem 2. *The set A_n of all even permutations of degree n forms a finite non- abelian group of order $\dfrac{n!}{2}$ with respect to permutation multiplication as composition.*

Proof. Let A_n be set of all even permutations of degree n. Let $f, g, h \in A_n$ be arbitrary.

To show that $(A_n, .)$ is a finite non-abelian group of order $\dfrac{n!}{2}$.

(i) Closure property. Let $f, g, h \in A_n \Rightarrow f$ and g are even permutations.
$\Rightarrow \quad f \cdot g$ is an even permutation.
$\qquad \qquad (\because$ product of two even permutations is even permutation.)
$\Rightarrow \qquad \qquad f \cdot g \in A_n$

(ii) **Associativity.** Let $f, g, h \in A_n \Rightarrow f, g, h$ are expressible as

$$f = \begin{pmatrix} a_1 \ldots a_n \\ b_1 \ldots b_n \end{pmatrix}, g = \begin{pmatrix} b_1 \ldots b_n \\ c_1 \ldots c_n \end{pmatrix}, h = \begin{pmatrix} c_1 \ldots c_n \\ d_1 \ldots d_n \end{pmatrix}$$

where the element $b_1, b_2, \ldots, b_n, c_1, c_2, \ldots, c_n, d_1, d_2, \ldots, d_n$, are simply different arrangement of the same n elements $a_1, a_2, \ldots, a_n$.

Then
$$f = \begin{pmatrix} a_1 \ldots a_n \\ b_1 \ldots b_n \end{pmatrix} \begin{pmatrix} b_1 \ldots b_n \\ c_1 \ldots c_n \end{pmatrix} = \begin{pmatrix} a_1 \ldots a_n \\ c_1 \ldots c_n \end{pmatrix}$$

$$(fg)h = \begin{pmatrix} a_1 \ldots a_n \\ c_1 \ldots c_n \end{pmatrix} \begin{pmatrix} c_1 \ldots c_n \\ d_1 \ldots d_n \end{pmatrix} = \begin{pmatrix} a_1 \ldots a_n \\ d_1 \ldots d_n \end{pmatrix} \qquad \ldots(1)$$

Now
$$gh = \begin{pmatrix} b_1 \ldots b_n \\ c_1 \ldots c_n \end{pmatrix} \begin{pmatrix} c_1 \ldots c_n \\ d_1 \ldots d_n \end{pmatrix} = \begin{pmatrix} b_1 \ldots b_n \\ d_1 \ldots d_n \end{pmatrix}$$

$$f(gh) = \begin{pmatrix} a_1 \ldots a_n \\ b_1 \ldots b_n \end{pmatrix} \begin{pmatrix} b_1 \ldots b_n \\ d_1 \ldots d_n \end{pmatrix} = \begin{pmatrix} a_1 \ldots a_n \\ d_1 \ldots d_n \end{pmatrix} \qquad \ldots(2)$$

$\therefore$

Now from (1) and (2), we have $(fg)h = f(gh)$

(iii) **Existence of identity.** Let I be the identity permutation of degree n.

Then $\qquad fI = If ; f \in A_n$

(iv) **Existence of inverse.** Let f^{-1} denote the inverse of f. Then $f^{-1}f = I = $ an identity permutation.

Also f is an even permutation.

$\Rightarrow \qquad f^{-1}$ is an even permutation.

$\therefore \qquad f^{-1} \in A_n$

(v) **Commutativity.** The product of permutations is not commutative. Also the set of permutations of degree n contains $n!$ permutations out of which $\dfrac{n!}{2}$ are even and $\dfrac{n!}{2}$ are odd.

$\Rightarrow (A_n, .)$ is a non-abelian group of order $\dfrac{n!}{2}$.

REMARK

☞ The set of all odd permutations is not a group with respect to permutation multiplication as composition. Because closure property is not satisfied, since product of two odd permutations is an even permutations.

Solved Examples
Based on the following Results

▶ A one-one mapping of a finite set onto itself is called a permutation.

▶ A permutation on the set S is called the identity permutation if it maps each element of S onto itself.

▶ If $f = \begin{pmatrix} a_1 & a_2 \ldots a_n \\ b_1 & b_2 \ldots b_n \end{pmatrix}$ then $f^{-1} = \begin{pmatrix} b_1 & b_2 \ldots b_n \\ a_1 & a_2 \ldots a_n \end{pmatrix}$

▶ A cycle of length 2 is called transposition.

▶ Every permutation can be expressed as a product of disjoint cycles.

▶ A permutation is said to be even or odd according as it can be expresed as a product of even or odd number of permutations.

Example 1. *Express the permutation* $f = \begin{pmatrix} 1 & 2 & 3 & 4 & 5 & 6 \\ 1 & 6 & 5 & 3 & 4 & 2 \end{pmatrix}$ *as a product of disjoint cycles.*

Solution. Clearly, we have

$$f = \begin{pmatrix} 1 & 2 & 3 & 4 & 5 & 6 \\ 1 & 6 & 5 & 3 & 4 & 2 \end{pmatrix} = \begin{pmatrix} 1 & 2 & 6 & 3 & 5 & 4 \\ 1 & 6 & 2 & 5 & 4 & 3 \end{pmatrix}$$

$$= (1)\,(2\ 6)\,(3\ 5\ 4)$$

Example 2. *Express the permutation* $f = \begin{pmatrix} 1 & 2 & 3 & 4 & 5 & 6 & 7 \\ 1 & 3 & 2 & 6 & 4 & 5 & 7 \end{pmatrix}$ *as a product of transpositions.*

Solution. $f = \begin{pmatrix} 1 & 2 & 3 & 4 & 5 & 6 & 7 \\ 1 & 3 & 2 & 6 & 4 & 5 & 7 \end{pmatrix} = (1)\,(2\ 3)\,(4\ 5\ 6)\,(7)$

$$= (1\ 2)\,(2\ 1)\,(2\ 3)\,(4\ 6)\,(4\ 5)\,(7\ 1)\,(1\ 7).$$

Since, $(1) = (1\ 2)\,(2\ 1)$ and $(4\ 6\ 5) = (4\ 6)\,(4\ 5),\ (7) = (7\ 1)\,(1\ 7)$

Example 3. *Decompose the permutation* $f = \begin{pmatrix} 1 & 2 & 3 & 4 & 5 & 6 & 7 & 8 \\ 3 & 1 & 4 & 7 & 2 & 5 & 8 & 6 \end{pmatrix}$ *into transposition.*

Hence, show that f is an odd permutations.

Solution. Here, we have

$$f = \begin{pmatrix} 1 & 3 & 4 & 7 & 8 & 6 & 5 & 2 \\ 3 & 4 & 7 & 8 & 6 & 5 & 2 & 1 \end{pmatrix} = (1\ 3\ 4\ 7\ 8\ 6\ 5\ 2)$$

$$= (1\ 3)\,(1\ 4)\,(1\ 7)\,(1\ 8)\,(1\ 6)\,(1\ 5)\,(1\ 2)$$

Hence, f is an odd permutation.

Example 4. *Show that four permutations* $I,\ (ab),\ (cd),\ (ab)(cd)$ *on four symbols a, b, c, d form a finite abelian group with respect to the permutation multiplication.*

Solution. Let $I = f_1,\ (ab) = f_2,\ (cd) = f_3$ and $f_4 = (ab)(cd)$

New let $G = \{f_1, f_2, f_3, f_4\}$

Since, f_2 and f_3 are transposition $f_2 \circ f_2 = I$ and $f_3 \circ f_3, = I = f_1$

Also, $f_2 \circ f_2 = f_3 \circ f_2 = f_4$

Further, $f_2 \circ f_2 = (ab)\,(ab)\,(cd) = I\,(cd) = (cd) = f_3,$

Similarly, by computing the various product, we get the following compositions table:

•	f_1	f_2	f_3	f_4
f_1	f_1	f_2	f_3	f_4
f_2	f_2	f_1	f_4	f_3
f_3	f_3	f_4	f_1	f_2
f_4	f_4	f_3	f_2	f_1

From the table, it is clear that:

(i) All the entries in the composition table are elements of the given set G. Hence, closure property is satisfied.

(ii) The multiplication of permutations is commutative.

(iii) f_1 is the identity element.

(iv) Each element of G possesses inverse, for

$$f_1^{-1} = f_1 \qquad\qquad \because f_1 \circ f_1 = f_1 = 1$$
$$f_2^{-1} = f_2 \qquad\qquad \because f_2 \circ f_2 = f_1 = 1$$
$$f_3^{-1} = f_3 \qquad\qquad \because f_3 \circ f_3 = f_1 = 1$$
$$f_4^{-1} = f_4 \qquad\qquad \because f_4 \circ f_4 = f_1 = 1$$

(v) Each row is identical to corresponding column. Therefore, the composition is commutative.

Hence, the given set is a finite abelian group of order 4 with respect to permutation multiplication.

EXERCISE 2.4

1. Find the order of each element of the group $[\{0. 1, 2, 3, 4\}, +_5]$.

2. Show that the set $G = \{0, 1, 2, 3, 4, 5\}$ is a finite abelian group of order 6 with respect to addition modulo 6.

3. Show that $G = \{1\ 5\ 7\ 11\}$ is a group multiplication modulo 12.

4. Show that the set P_3 of all permutation on three symbols 1, 2, 3 is a finite non-abelian group of order 6 with respect to permutation multiplication as composition.

5. Show that the set A_3 of three permutations (a) (a b c), (a c b) on three symbols a, b, c forms a finite abelian group with respect to the permutation multiplication.

6. If $f = \begin{pmatrix} 1 & 2 & 3 \\ 3 & 1 & 2 \end{pmatrix}$ and $g = \begin{pmatrix} 1 & 2 & 3 \\ 1 & 3 & 2 \end{pmatrix}$ then find fg and gf.

7. Find the inverse of the following permutation:

(a) $\begin{pmatrix} 1 & 2 & 3 & 4 \\ 1 & 3 & 4 & 2 \end{pmatrix}$ (b) $\begin{pmatrix} 1 & 2 & 3 & 4 \\ 3 & 4 & 1 & 2 \end{pmatrix}$

8. Show that if S has more than two elements of G. Then the symmetric group S_n is not abelian.

9. Show that a cycle contataining an odd number of symbols is an even permutation where as a cycle containing an even number of symbols is an odd permutation.

10. If $f = (1, 2, 3, 4, 5, 6)$, then show that $f^3 = (1\ 4)\ (2\ 5)\ (6\ 7)$

11. Examine whether the following permutation is even or odd: $\begin{pmatrix} 1 & 2 & 3 & 4 & 5 & 6 & 7 \\ 6 & 5 & 2 & 4 & 3 & 1 & 7 \end{pmatrix}$

Answers

1. o (0) $= 1$ and order of other element is 5.

6. $fg = \begin{pmatrix} 1 & 2 & 3 \\ 2 & 1 & 3 \end{pmatrix}$ and $gf = \begin{pmatrix} 1 & 2 & 3 \\ 3 & 2 & 1 \end{pmatrix}$

7. (a) $\begin{pmatrix} 1 & 2 & 3 & 4 \\ 1 & 4 & 2 & 3 \end{pmatrix}$ (b) $\begin{pmatrix} 1 & 2 & 3 & 4 \\ 3 & 4 & 1 & 2 \end{pmatrix}$

2.21 SUBGROUPS OF A GROUP

Let $(G, *)$ be a group and H is any subset of G such that $H \neq \phi$. Then by properties of a group $\forall\ a, b \in H \Rightarrow a \cdot b \in G$ for $H \subset G$.

$$\Rightarrow \qquad a * b \in G \Rightarrow a * b \in H \text{ or } a * b \notin H$$

If $a * b \in H$, they we say that H is stable for the composition in G and the composition in G has induced a composition in H. Now, there are two possibilities.

(i) H is itself a group with respect to the operation $*$.

(ii) H is not a group with respect to $*$.

Complex of a Group

Any non-empty subset H of G is called a complex of the group G.

Subgroup

A non-empty subset H of a group G is calld a subgroup of G if H itself a group with respect to the operation defined in G.

REMARKS

☞ The two subgroups, G and $\{e\}$ of the group G are called Improper (or trivial) subgroups of G. Any subgroups other than these two subgroups is called a proper (or non-trivial) subgroups.

☞ It is clear that, if H is a subgroup of G and K is a subgroup of H, then K is subgroup of G.

☞ Every subgroup of G is a complex of G, but every complex is not always a subgroup.

■ Illustrations

(i) $(\{1, -1\}, .)$ is a subgroup of $(\{1, -1, i, -i\}, .)$

(ii) $(\mathbf{Z}, +)$ is a subgroup of $(\mathbf{Q}, +)$.

(iii) $(\mathbf{Q}, +)$ is a subgroup of $(\mathbf{R}, +)$.

(iv) The set of all non-singular matrices with real elements whose determinant value is 1, is a subgroup of multiplicative group of all non-singular $n \times n$ matrices.

(v) The multiplicative group of positive rational numbers is a subgroup of the multiplicative group of all non-zero rational numbers.

REMARKS

☞ If H and K are two complexes of a group G, then
$$HK = \{x \in G : x = hk, h \in H, k \in K \}$$
Also, HK is a complex consisting of the elements of G, which obtained on multiplying each number of H to each member of K.

☞ Multiplication of complexes is associative.

☞ If H is be any complex of G, then inverse of H, i.e., H^{-1} is the complex of G consisting the inverse of the elements of H. i.e., $H^{-1} = \{h^{-1} : h \in H\}$

Theorem 1. *Let H be a subgroup of G. Then*

    (i) identity of H is same as that of G.

    (ii) the inverse of an element $a \in H$ is the same as the inverse of the same element a regard as an element of the group G.

    (iii) the order of an element $a \in H$ is the same as the order of the same element regarded as an element of the group G.

Proof.   Let H be a subgroup of G.

    (i) Let e and e' be identities in H and G respectively. To show $e = e'$

Let $\quad\quad\quad a \in H$

Now, $\quad\quad\quad a \in H \quad\quad \Rightarrow \quad ae = a \quad$ ($\because$ e is the identity of H.)..(1)

Also $\quad\quad\quad a \in H \quad\quad \Rightarrow \quad ae \in a \quad$ ($\because$ $H \subset G$)

$\quad\quad\quad\quad ae' = a \quad\quad\quad\quad\quad\quad$ (e' is the identity in G.) $\quad\quad$ ...(2)

From (1) and (2), we have $ae = ae' \Rightarrow e = e'$.

(ii) Let e be the identity of H as well as G. Let $a \in G$ be arbitrary. Let b be the inverse of a in H. Let c be the inverse of a in G.

To show that $\quad\quad b = c$

By assumption $\quad ba = e = ca$ in G, we have $ba = ca \Rightarrow b = c$.

(iii) Let e be the identity of H as well as G.

Let $a \in H$ be the arbitrary. Let n be the order of a in H and m be the order of a in G.

To show $\quad\quad\quad m = n$

Since e is the identity in $\quad H = a^n = e$

Since e is the identity in $\quad H \Rightarrow a^m = e$, where n and m are positive integers.

Therefore, $\quad\quad a^n = a^m$

$\Rightarrow \quad\quad\quad a^{m-n} = e = a^0 \Rightarrow a^{m-n} = a^0$

$\Rightarrow \quad\quad\quad m - n = 0 \quad\quad \Rightarrow \quad\quad m = n$

Theorem 2. *Let H be any complex of group G, then $(HK)^{-1} = K^{-1}H^{-1}$.*

Proof. $\quad$ Let x be any arbitrary element of $(HK)^{-1}$. Then

$$x = (hk)^{-1}, h \in H, k \in K$$
$$= k^{-1}h^{-1} \in K^{-1}H^{-1}$$
$$(HK)^{-1} \subseteq K^{-1}H^{-1} \quad\quad\quad\quad\quad ...(1)$$

Now, let y be any arbitrary element of $K^{-1}H^{-1}$.

Then $\quad\quad\quad\quad y = k^{-1}h^{-1} = (hk)^{-1} \in (HK)^{-1}, k \in K, h \in H$

$\therefore \quad\quad\quad\quad K^{-1}H^{-1} \subseteq (HK)^{-1} \quad\quad\quad\quad\quad ...(2)$

Now from (1) and (2), we conclude that $(HK)^{-1} = K^{-1}H^{-1}$.

Theorem 3. *If H is any subgroup of G, then $H^{-1} = H$. Also, show that converse is not true.*

Proof. $\quad$ Let $h^{-1} \in H$. Then $h \in H$.

Since, H is subgroup of G, therefore, $h \in H \Rightarrow h^{-1} \in H$.

Thus, $\quad\quad h^{-1} \in H^{-1} \Rightarrow h^{-1} \in H$

$\Rightarrow \quad\quad\quad H^{-1} \subseteq H \quad\quad\quad\quad\quad\quad\quad ...(1)$

Again $\quad\quad h \in H \Rightarrow h^{-1} \in H \Rightarrow (h^{-1})^{-1} \in H^{-1} \Rightarrow h \in H^{-1}$

$\therefore \quad\quad\quad H \subseteq H^{-1}$

Now, from (1) and (2), we have

$$H = H^{-1} \quad\quad\quad\quad\quad\quad ...(2)$$

Now, to show converse is not true.

i.e., If H is a complex of a group G and $H^{-1} = H$, then it is not necessary that H is a subgroup of G.

For Example

$H = \{-1\}$ is a complex of the multiplicative group $G = \{-1, 1\}$.

Also $H^{-1} = \{-1\}$ ($\because -1$ is the inverse of -1). But $H = \{-1\}$ is not a subgroup of G.

We have $(-1)(-1) = 1 \notin H$, i.e., H is not closed with respect of multiplication.

2.21.1 Criterion for a Subset to be a Subgroup

It is not necessary for an arbitrary subset of H of a group G to be a subgroup. It would be interesting to find out criteria by which H can be identified as a subgroup or otherwise.

Theorem 4. *A non-empty subset H of a group G is a subgroup of G if and only if*

    (i) *$a, b \in H \Rightarrow ab \in H$*

    (ii) *$a \in H \Rightarrow a^{-1} \in H$, where a^{-1} is the inverse of $a \in G$.*

Proof.   Let H be a subgroup of G, then H must be closed with respect to multiplication, i.e., the composition in G.

Therefore $a \in H, b \in H \Rightarrow ab \in H$.

Conversely. Suppose H is a subset of a group G such that, the given conditions, holds. In order to show that H is a subgroup, all that is needed is to verify that the identity element $e \in H$ and that the associative law holds for elements of H.

If $a \in H$, then by (ii), $a^{-1} \in H$ and so by (i) we see that $e = aa^{-1} \in H$, again since associative law does holds in G, it holds in H, which is a subset of G, Hence, H is a subgroup of G.

Theorem 5. *Let H be a non-empty subset of a group G. Then H is a subgroup of G iff*

$$a, b \in H \Rightarrow ab^{-1} \in H, \text{ where } b^{-1} \text{ is the inverse of } b \text{ in } G.$$

Proof.   **Necessary condition.** Let us first suppose H is a subgroup of G and $a, b \in H$. Since H is a group, each element of H must have its inverse in H. Thus if $b \in H \Rightarrow b^{-1} \in H$ and then by closure property $ab^{-1} \in H$. This proves the necessary condition.

Condition is sufficient. Conversely, let H be a subset of G for which $a, b \in H$ implies $ab^{-1} \in H$. To show that H is a subgroup of G, we must verify that H is closed, the existence of identity element and each element of H has an inverse in H and the associative law holds for all elements of H.

Let $b = a$, then we see that $a \in H \Rightarrow aa^{-1} \in H \Rightarrow e \in H$

$\Rightarrow$ Identity element of G also belongs to H.

Now, for the elements e and b of H, we have $eb^{-1} \in H$ and so $b^{-1} \in H$, since b is arbitrary element of H, we see that for any $b \in H$, $b^{-1} \in H$.

Now, $a, b \in H \Rightarrow a, b^{-1} \in H$

$\Rightarrow$ $a(b^{-1})^{-1} \in H$

$\Rightarrow$ $ab \in H$

$\Rightarrow$ H is closed.

Finally, since the associative law does hold for H, it also holds for H which is a subset of G. Hence, $(H, .)$ is a subgroup.

REMARKS

☞ In case of additive composition, the condition of the above theorem becomes
$$a, b \in H \Rightarrow a - b \in H$$

☞ To show that a non-empty subset H of a group G is a subgroup of G, we should take any two arbitrary elements $a, b \in H$ and try to show that $ab^{-1} \in H$.

Theorem 6. *The necessary and sufficient condition of a non-empty subset H of a group G to be a subgroup is $HH^{-1} \subset H$.*

Proof. Let H be a non-empty subset of a group G such that H is a subgroup of G.

To show that $\qquad HH^{-1} \subset H$

Let $x \in HH^{-1} \qquad \Rightarrow \quad \exists\, a, b \in H \ \ s.t. \ \ x = ab^{-1} \qquad (\because H$ is a subgroup.)

$\Rightarrow \qquad\qquad a, b^{-1} \in H \ s.t. \ \ x = ab^{-1}$

$\Rightarrow \qquad\qquad ab^{-1} \in H$

$\Rightarrow \qquad\qquad x \in H$

$\Rightarrow \qquad\qquad HH^{-1} \subset H \qquad\qquad (\because x$ is arbitrary)

Conversely. Suppose that H is a non-empty subset of a group G such that $HH^{-1} \subset H$.

To show that H is a subgroup of G. For this we shall show that
$$a, b \in H \Rightarrow ab^{-1} \in H$$

Now, $\qquad a, b \in H \Rightarrow ab^{-1} \in HH^{-1}$ (by definition of HH^{-1})

$\qquad\qquad \Rightarrow \quad ab^{-1} \in H$ for $HH^{-1} \subset H$.

Theorem 7. *A necessary and sufficient condition of a non empty subset H of a group G to be a subgroup is that is $HH^{-1} = H$.*

Proof. Let H be a non-empty subset of a group G such that H is a subgroup of H so that $(H, .)$ is a group.

To show that $\qquad\qquad HH^{-1} = H$

Let $x \in HH^{-1} \qquad \Rightarrow \qquad \exists\, a \in H, b^{-1} \in H^{-1}, \ s.t. \ \ x = ab^{-1}$

$\Rightarrow \qquad\qquad a, b \in H \ s.t. \ \ x = ab^{-1}$

$\Rightarrow \qquad\qquad a \cdot b^{-1} \in H \ \ s.t. \ \ x = ab^{-1}$

$\Rightarrow \qquad\qquad a \cdot b^{-1} \in H \ \ s.t. \ \ x = ab^{-1} \ \Rightarrow x \in H.$

Therefore, $\qquad\qquad HH^{-1} \subset H \qquad\qquad\qquad\qquad ...(1)$

Now let

$$x \in H \quad \Rightarrow x \in H, e \in H.$$

For $(H, .)$ is a group and e is the identify for G.

$\Rightarrow \qquad\qquad xe^{-1} \in HH^{-1} \Rightarrow xe \in HH^{-1} \qquad (\because e^{-1} = e)$

$\Rightarrow \qquad\qquad x \in HH^{-1}$

$\Rightarrow \qquad\qquad H \subset HH^{-1} \qquad\qquad\qquad\qquad ...(2)$

Now , from (1) and (2), we conclude that
$$HH^{-1} = H$$

Conversely, Let H be a non-empty subset of a group G such that $HH^{-1} = H$.

To show that H is a subgroup of G, it is sufficient to show that

$$a, b \in H \quad \Rightarrow \quad ab^{-1} \in H$$

Now, $a, b \in H \quad \Rightarrow \quad a \in H, b^{-1} \in H^{-1}$

$$\Rightarrow \quad ab^{-1} \in H H^{-1} = H$$

$$\Rightarrow \quad ab^{-1} \in H$$

Theorem 8. *If H, K are subgroups of a group G, then HK is a subgroup of G iff $HK = KH$.*

Proof. Let H and K be two subgroups of a group G so that

$$H H^{-1} = H, K K^{-1} = K \qquad \qquad \dots (1)$$

and $\qquad \qquad K^{-1} = K, H^{-1} = H \qquad \qquad \dots (2)$

Step I. Let HK be a subgroup of G so that

$$(HK)^{-1} = HK \qquad \qquad \dots(3)$$

To show $\qquad \qquad HK = KH$

(3) $\qquad \Rightarrow \qquad K^{-1} H^{-1} = HK \qquad \qquad \dots. (4)$

Using (2), we have $\qquad \qquad HK = KH$

Step II. Let $HK = KH$

To show that HK is a subgroup of G. For this we have to prove,

$$HK(HK)^{-1} = HK \qquad \qquad \dots(5)$$

Consider $\qquad (HK) (HK)^{-1} = (HK)(K^{-1}H^{-1}) = H(KK^{-1}) (H^{-1})$

$$\text{(By associatively)}$$

$$= HKH^{-1} \qquad \qquad \text{[by (1)]}$$

$$= KHH^{-1} \qquad \qquad \text{[by (4)]}$$

$$= K(HH^{-1}) = KH = HK \qquad \qquad \text{[by (4)]}$$

Hence, the theorem is proved.

Theorem 9. *If H and K are subgroups of an abeliam group G, then HK is a subgroup of G.*

Proof. Let us suppose H and K be subgroups of an abelian group G, so that

$$H^{-1} = H, K^{-1} = K \qquad \qquad \dots (1)$$

$$HH^{-1} = H, K K^{-1} = K \qquad \qquad \dots (2)$$

To show that (HK) is a subgroup of G, we have to show that

$$(HK)(HK)^{-1} = HK$$

Now, since G is abelian $\Rightarrow \quad ab = ba \,\, \forall \, a, b \in G$

$\therefore \quad h \in H, \,\, k \in K \Rightarrow \quad hk = kh \Rightarrow HK = KH \qquad \qquad \dots (3)$

Consider $\qquad (HK) (HK)^{-1} = (HK) (K^{-1} H^{-1}) = H(KK^{-1}) H^{-1}$

$$= HKH^{-1} = (HK)H^{-1} = (KH)^{-1} = K (HH^{-1})$$

$$= KH = HK$$

Theorem 10. *The necessary and sufficient condition for a non-empty finite subset H of a group G, with respect to multiplication to be a subgroup is that H must be closed with respect to multiplication, i.e., $a \in H, b \in H \Rightarrow ab \in H$.*

Proof. **Necessary Condition.** Let us suppose H is a subgroup of G. Then H must be closed with respect to multiplication. Therefore,

$$a \in H, b \in H \Rightarrow ab \in H$$

Sufficient condition. Here, it is a given that H is closed with respect to multiplication, *i.e.,* $a \in H, b \in H \Rightarrow ab \in H$

$\Rightarrow$ Closure property is satisfied.

Associativity. For all $a, b, c \in H \Rightarrow a, b, c \in G$

$\Rightarrow$ $\qquad\qquad\qquad a(bc) = (ab)\,c$ $\qquad\qquad\qquad$ (by associativity in G)

Existence of identity. Let e be the identyty in G. By the given condition, we have

$a \in H, a \in H \qquad \Rightarrow \qquad a^2 \in h = a.a \in H$

$\qquad\qquad\qquad \Rightarrow \qquad a^3 = a^2 . a \in H \Rightarrow a^4 = a^3 .a \in H$

Proceeding in this way, we see that all the elements

$$a, a^2, a^3, a^4, \dots a^r, \dots a^s$$

belong to H if $a \in H$.

But H is a finite set. Consequently all these elements are not distinct, *i.e.,* there must be repetition in this collection of elements, *i.e.,* $a^r = a^s$, where $r, s \in \mathbf{N}$ and $r > s$.

$\Rightarrow$ $\qquad\qquad\qquad a^{r-s} = a^0 \qquad \Rightarrow a^{r-s} = e$

Also $(r - s)$ is the positive integer

$\Rightarrow$ $\qquad\qquad\qquad e = a^{r-s} \in H \Rightarrow e \in H$

Existence of inverse.

$$r - s \geq 1 \Rightarrow r - s - 1 \geq 0 \ a^{\,r-s-1} \in H$$

$\Rightarrow$ $\qquad\qquad a^{\,r-s} . a^{-1} \in H \Rightarrow a^{-1} \in H$

Thus $\qquad\qquad\qquad a \in H \Rightarrow a^{-1} \in H, \forall \, a \in H$

This shows that every element of H is invertible. Hence , H is subroup of G.

REMARKS

☞ The above theorem can also be stated as " A necessary and sufficient condition for a non-empty finite subset H of a group G to be a subgroup is that H must be closed .

☞ The criterion given in the above theorem is valid only for finite subsets of a group G. It is not valid for infnite subsets of an infinite group G.

2.22 UNION AND INTERSECTION OF SUBGROUPS

Theorem 1. *The intersection of any two subgroups of a group G is a subgroup of G.*

Proof. Let H_1 and H_2 be two subgroups of a group G. To show that $H_1 \cap H_2$ is a subgroup of G.

Let $\qquad\qquad a, b \in H_1 \cap H_2 \Rightarrow a, b \in H_1$ and $a, b \in H_2$

Now, since H_1 and H_2 are subgroups, then

$$a, b \in H_1 \Rightarrow ab^{-1} \in H_1 \text{ and } a, b \in H_2 \Rightarrow ab^{-1} \in H_2$$

Finally, $ab^{-1} \in H_1$ and $ab^{-1} \in H_2 \Rightarrow ab^{-1} \in H_1 \cap H_2$. Hence $H_1 \cap H_2$ is a subgroup of G.

Theorem 2. *An arbitrary intersection of subgroups of a group G is a subgroup of G.*

Proof. Let H_r be the collection of subgroups for $r \in \mathbf{N}$.

Let $H = \bigcap\limits_{r=1}^{\infty} H_r$

To show that H is a subgroup of G.

$$a, b \in H \Rightarrow a \in \bigcap\limits_{r=1}^{\infty} H_r \text{ and } b \in \bigcap\limits_{r=1}^{\infty} H_r$$

$$\Rightarrow a \in H_r, b \in H_r \;\; \forall r \in N \Rightarrow ab^{-1} \in H_r \; \forall r$$

$$\Rightarrow ab^{-1} \in \bigcap\limits_{r=1}^{\infty} H_r = H$$

$$\Rightarrow ab^{-1} \in H$$

Thus, we have proved that $a, b \in H \Rightarrow ab^{-1} \in H$.

This declares that H is a subgroup of G.

REMARK

☞ $H_1 \cap H_2$ is a largest subgroup of G which is contained in H_1 as well as H_2.

Theorem 3. *The union of two subgroups of a group G is a subgroup of G iff one is contained in the other.*

Proof. Let H_1 and H_2 be subgroups of a group G. Let us first suppose

$$H_1 \subset H_2 \text{ or } H_2 \subset H_1.$$

To show that $H_1 \cup H_2$ is a subgroup of G

$$H_1 \subset H_2 \Rightarrow H_1 \cup H_2 = H_2$$

Also, H_2 is a subgroup of $G \Rightarrow H_1 \cup H_2$ is a subgroup of G. Again

$$H_2 \subset H_1 \Rightarrow H_1 \cup H_2 = H_1$$

Also, H_1 is a subgroup of $G \Rightarrow H_1 \cup H_2$ is a subgroup of G.

Hence, $H_1 \cup H_2$ is a subgroup of G, in both cases.

Conversely, Suppose that H_1 and H_2 are subgroups of a group G such that $H_1 \cup H_2$ is a subgroup of G. To prove either $H_1 \subset H_2$ or $H_2 \cup H_1$.

Suppose the contrary. Then $H_1 \not\subset H_2$ or $H_2 \not\subset H_1$

$$H_1 \not\subset H_2 \Rightarrow \exists\, a \in H_1 \text{ s. t. } a \notin H_2$$

and $\qquad\qquad H_2 \not\subset H_1 \Rightarrow \exists\, a \in H_2 \text{ s. t. } a \notin H_1$

Now $a, b \in H_1 \cup H_2$ and $H_1 \cup H_2$ is a subgroup of G

$$\Rightarrow \qquad\qquad ab \in H_1 \cup H_2$$

This implies $ab \in H_1$ or $ab \in H_2$

$$a \in H_1, ab \in H_1 \Rightarrow a^{-1}(ab) \in H_1 \qquad\qquad (\because H_1 \text{ is a subgroup.})$$

$$\Rightarrow \qquad (a^{-1}a)b \in H_1 \Rightarrow eb \in H_1 \Rightarrow b \in H_1$$

Which is a contradiction $\qquad\qquad (\because b \notin H_1)$

Also, $\qquad b \in H_2, ab \in H_2 \Rightarrow (ab)b^{-1} \in H_2$

$$\Rightarrow \qquad\qquad\qquad a \in H_2 \qquad\qquad (\text{For } (ab)b^{-1} = a(bb^{-1}) = ae = a)$$

Again, we get a contradiction $\qquad\qquad (\because a \notin H_1)$

Also, $b \in H_2, ab \in H_2 \Rightarrow (ab)b^{-1} \in H_2$

$\Rightarrow \qquad\qquad\qquad a \in H_2$ $(\because (ab)b^{-1} = a(bb^{-1}) = ae = a)$

Again, we get a contradiction. $(\because a \notin H_1)$

Hence, our assumption is wrong.

Consequently $H_1 \subset H_2$ or $H_2 \subset H_1$

REMARK

☞ The union of two subgroups is not necessarily a subgroup.

For Example

Let H_1 and H_2 be two subgroups of the group $(\mathbf{Z}, +)$ where
$$H_1 = \{2n : n \in \mathbf{Z}\}, H_2 = \{5n : n \in \mathbf{Z}\}.$$
Then, $H_1 \cup H_2 = [x : x = 2n \text{ or } 5n \text{ where } n \in \mathbf{Z}]$

$$= [0, \pm 2, \pm 4, \pm 6, \dots, 0, \pm 5, \pm 10, \pm 15, \dots]$$

$$6, 15 \in H_1 \cup H_2 \Rightarrow 6 + 15 = 21 \notin H_1 \cup H_2$$

$\Rightarrow$ Closure property is not satisfied.

Hence, $H_1 \cup H_2$ is not a subgroups of $(\mathbf{Z}, +)$.

Solved Examples

Based on the following Results

▶ The necessary and sufficient candition for a non-empty subset H of a group G to be a subgroup is $a, b \in H \Rightarrow ab^{-1} \in H$.

▶ If H and K are subgroups of a group G, then HK is a subgroup of G if and only if $HK = KH$.

▶ The intersection of any two subgroups of a group G is a sub group of G.

▶ The union of two subgroups of a group G is a subgroup of G if and only if one is contained in the other.

Example 1. *Is* **Z** *a subgroup of* $(\mathbf{Q}, +)$?

Solution.   Let $\mathbf{Z} \subset \mathbf{Q}$ and the inverse of $b \in \mathbf{Q}$ is $-b$.

Also, $a, b \in \mathbf{Z} \Rightarrow a + (-b) = a - b \in \mathbf{Z}$

Therefore **Z** is a subgroup of **Q**, under addition.

Example 2. *Let G be the additive group of integers and* $H = \{nl : n$ *is a fixed integer and* $l \in \mathbf{Z}\}$. *Show that* H *is a subgroup of* G.

Solution.   Here, we have $H \subseteq G$.

Let $a = nh$ and $b = nk$ be any two elements of H with $h, k \in \mathbf{Z}$. Then $a + b = n(n + k)$ certainly belongs to H.

Thus $a, b \in H$ implies that $a + b \in H$.

Also, $-a = n(-h)$, the additive inverse of a is in H. Thus $a \in H$ implies that $-a \in H$. Hence, H is a subgroup of G.

Example 3. *If G is a group, then show that the set Z, defined by*

$$Z = \{xz = zx : x \in G, z \in Z\}$$

(it is called centre of the group) is a subgroup of G.

Solution.   Let $z_1, z_2 \in Z$, then

$$z_1 x = x z_1, \; z_2 x = x z_2, \; \forall x \in G \qquad \qquad \ldots (1)$$

Now, $$x z_1 = z_1 x = z_1 (z_2^{-1} z_2 x), \forall \, x \in G$$

$$= z_1 z_2^{-1} (z_2 x) = z_1 z_2^{-1} (x \, z_2)$$

$$\therefore \quad (x z_1) z_2^{-1} = z_1 z_2^{-1} (x z_2) z_2^{-1} \qquad \qquad \text{[From (1)]}$$

or $$x (z_1 z_2^{-1}) = z_1 z_2^{-1} x (z_2 z_2^{-1}) = (z_1 z_2^{-1}) x, \forall \, x \in G$$

Therefore, $z_1 z_2^{-1} \in Z$. Hence, Z is a subgroup of G.

REMARK

☞ The subgroup defined in above example is known as centre of the group. Thus "If G is a group, then the subgroup Z, defined by $Z = \{xz = zx : x \in G, z \in Z\}$ is said to be centre of the group.

Example 4.   *If a is any element of a group G, then show that $\{a^n : n \in \mathbf{Z}\}$ is a subgroup of G.*

Solution.   Let a be any arbitrary element of a group G. Define

$$H = [a^n : n \in \mathbf{Z}].$$

To show H is a subgroup of G.

For any

$$
\begin{aligned}
h_1, h_2 \in H & \Rightarrow & h_1 = a^x, h_2 = a^y & \quad \text{where } x, y \in \mathbf{Z}. \\
& \Rightarrow & h_1 h_2^{-1} = a^{x-y}, & \quad \text{where } x - y \in \mathbf{Z}. \\
& \Rightarrow & h_1 . h_2^{-1} \in H & \\
& \Rightarrow & H \text{ is a subgroup of } G. &
\end{aligned}
$$

Example 5.   *If a is a fixed element of a group G, then prove that the set*

$$N(a) = \{x \in G : xa = ax\}$$

is a subgroup of G.

Solution.   Let $x, y \in N(a)$, then $xa = ax, ya = ay.$

Now $$ya = ay \quad \Rightarrow y^{-1}(ya)y^{-1} = y^{-1}(ay)y^{-1}$$

$$\Rightarrow \qquad \qquad ay^{-1} = y^{-1}a \quad \Rightarrow \qquad \qquad y^{-1} \in N(a)$$

Also $$(xy^{-1})a = x(y^{-1}) = x(ay^{-1})$$

$$= (xa)y^{-1} = (ax)y^{-1} = a(xy^{-1})$$

$$\Rightarrow \qquad \qquad xy^{-1} \in N(a), \text{ where } x, y \in N(a)$$

Hence, $N(a)$ is a subgroup of G.

REMARK

☞ The subgroup define in example (5) is called the Normalizer $N(a)$ of an element $a \in G$. i.e., $N(a) = \{x \in G : xa = ax\}$ i.e., $N(a)$ is the set of all those elements which commutes with a.

2.23 COSETS

Let H be a subgroup of a group (G,.). Let $a \in G$ be arbitrary. We define

$$aH = \{ah : h \in H\} \text{ and } Ha = \{ha : h \in H\}$$

that aH is called left coset of H in G gnerated by a, and Ha is called right coset of H in G generated by a.

REMARKS

☞ If e the identity of G, then $e \in H$ is also identity for H.

Then $\qquad\qquad a = ae \in aH, a = ea \in Ha$

This gives that any left or right cosets of H in G is not-empty.

☞ Since $He = H = eH$, hence H itself is right is well as left cosets.

☞ If the group $(G,.)$ is abelian, then the right coset of H in G generated by a is defined as $aH = Ha$, $\forall a \in H$

☞ If the composition in G is additive, then the right coset of H in G generated by a is defined by $H + a = \{h + a : h \in H\}$ and $a + H = \{a + H; h \in H\}$

2.23.1 Index of a Subgroup of a Group

If H is a subgroup of a group G, then number of distinct right (or left) cosets of H in G is called index of H in G and is denoted by $[G : H]$ or by $i_G(H) = o(G)/o(H)$

2.23.2 Relation of Congruence Modulo a Subgroup in a Group

Let H be a subgroup of a group G and $a, b \in G$ be arbitrary, we define

$$a \equiv b \bmod H \text{ iff } \qquad ab^{-1} \in H.$$

The symbol $a \equiv b \pmod{H}$ is read as a is congruent to b modulo H.

REMARK

☞ $a \equiv b \pmod{H}$ *iff* $ab^{-1} \in H$ or $Ha = Hb$.

☞ $a \in Hb \Leftrightarrow ab^{-1} \in H \Leftrightarrow Ha = Hb$

Theorem 1. *Let $a \in G$ be arbitrary and let H be a subgroup of a group G. Then $Ha = H = aH \Leftrightarrow a \in H$*

Proof. Let H be a subgroup of G and let $a \in G$ be arbitrary.

Step I. To show that

$$Ha = H \Leftrightarrow a \in H.$$

Let us first suppose $\quad Ha = H$. To show $a \in H$.

$$e \in H, a \in H \Rightarrow ea \in Ha \Rightarrow a \in Ha$$

$\Rightarrow \qquad\qquad a \in H, \text{ for } H = Ha$

Now, let $a \in H$. To show $Ha = H$.

Let $\quad xa \in Ha \qquad\qquad \Rightarrow \quad x \in H$

$\qquad\qquad\qquad\qquad\qquad\quad \Rightarrow \quad x \in H, a \in H$

$\qquad\qquad\qquad\qquad\qquad\quad \Rightarrow \quad xa \in H \qquad\qquad\qquad (\because H \text{ is a subgroup})$

Thus, any $\quad xa \in Ha \qquad \Rightarrow \quad xa \in H$

This prove that $\qquad Ha = H.$ $\qquad\qquad\qquad\qquad\qquad\qquad\qquad$ ... (1)

Now, let $\qquad\qquad a \in H \Rightarrow a^{-1} \in H \qquad\qquad (\because H \text{ is a subgroup})$

For any $\quad y \in H, a^{-1} \in H \Rightarrow ya^{-1} \in H$

$\qquad\qquad\qquad\qquad \Rightarrow (ya^{-1})a \in Ha$

$\qquad\qquad\qquad\qquad \Rightarrow y \in Ha \text{ for } ya^{-1}a = ye = y$

Thus any $y \in H$ $\Rightarrow y \in Ha$

               $\Rightarrow H \subset Ha$ ... (2)

Now from (1) and (2) we get $H = Ha$.

Step II. To show $aH = H \Leftrightarrow a \in H$.

We can prove step II by making the parallel arguments as in step I.

Theorem 2. *If a and b are arbitrary distinct elements of a group G and H is any subgroup of G, then*

$$Ha = Hb \Leftrightarrow ab^{-1} \in H$$

$$Ha = Hb \Leftrightarrow b^{-1}a \in H$$

Proof. Let a and b be arbitrary elements of a group G such that $a \neq b$

Let e be the identity of $G \Rightarrow e \in H$. Firstly, we shall show that

$$Ha = Hb \Leftrightarrow ab^{-1} \in H$$

$$Ha = Hb \Leftrightarrow (Ha)b^{-1} = (Hb)b^{-1}$$

$\Rightarrow$ $H(ab^{-1}) = H(bb^{-1}) = He = H$

$\Rightarrow$ $H(ab^{-1}) = H \Rightarrow ab^{-1} \in H$ (By previous theorem)

Conversely $ab^{-1} \in H \Rightarrow H(ab^{-1}) = H$

                       $\Rightarrow H(ab^{-1})(b) = Hb \Rightarrow (Ha)(b^{-1}b) = Hb$

                       $\Rightarrow (Ha) e = Hb \Rightarrow Ha = Hb$

Therefore , we have

$$Ha = Hb \Leftrightarrow ab^{-1} \in H$$

Theorem 3. *Any two left cosets of a subgroup are either disjoint or identical.*

Proof. Let aH and bH be any two left cosets of H. To show if aH and bH have an element in common. i.e., If $aH \cap bH$ is not the empty set, then they are identical, i.e., $aH = bH$.

Let $aH \cap bH \neq \phi$ and let c be any element of $aH \cap bH$ then there exist elements $h_1, h_2 \in H$ such that $c = ah_1$ and $c = bh_2$, it follows that

$$ah_1 = bh_2 \text{ so } a = bh_1 (h_1)^{-1} \qquad \text{... (1)}$$

Now, let ah be any element of aH. Then

$$ah = bh_2 (h_1)^{-1} h \qquad \text{[Using (1)]}$$

Now, since H is a subgroup, $h_2(h_1)^{-1} h \in H$ and so $ah \in bH$.

This shows that every $ah \in aH$ is also in bH. Therefore $aH \subseteq bH$.

Similary, we can show that $bH \subseteq aH$.

Therefore, we have $aH = bH$.

Hence, we have shown that any left cosets which are not disjoint are identical.

REMARK

☞ In other words, we can state this theorem as "Any two left (right) cosets either coincide or have no element in common. In symbols

either $aH = bH$ or $aH \cap bH = \Phi$

similarly, either $Ha = Hb$ or $Ha \cap Hb = \Phi$

Theorem 4. *There exists a one-one correspondence between a subgroup H of G and a left coset aH of H in G.*

Proof. Consider a map $f : H \to aH$, defined by $f(h) = ah$ for every $h \in H$.

The mapping f is clearly onto on aH, because for each $ah \in aH$ there exists $h \in H$ such that $f(h) = ah$.

Now, let $h_1, h_2 \in H$ such that $f(h_1) = f(h_2)$, i.e., $ah_1 = ah_2$

$\Rightarrow \qquad\qquad h_1 = h_2$

$\Rightarrow$ f is one-one.

Thus, f maps H such that f is one-one and onto.

Hence, there is one-to-one correspondence between H and aH.

Theorem 5. *If H is a subgroup of G, then there exist one-one correspondence between the set of left cosets of H in G and the set of right cosets of H in G.*

Proof. Let H be a subgroup of a group G.

Also, suppose

$$L = \{aH : a \in G \}$$

and $$R = \{Ha : a \in G\},$$

Now define a map $f : L \to R$ such that

$$f(aH) = Ha^{-1}, \ \forall \ a \in G.$$

To show f is one-one and onto.

(i) f is one-one :

For $\quad f(aH) = f(bH); aH, bH \in L$

$\Rightarrow \qquad Ha^{-1} = Hb^{-1} \qquad\qquad \Rightarrow \qquad Ha^{-1}a = Hb^{-1}a$

$\Rightarrow \qquad H = Hb^{-1}a \qquad\qquad \Rightarrow \qquad b^{-1}a \in H$

$\Rightarrow \qquad (b^{-1}a)^{-1} \in H \qquad\qquad \Rightarrow \qquad a^{-1}b \in H \qquad \Rightarrow a^{-1}bH = H$

$\Rightarrow \qquad a(a^{-1}bH) = aH \qquad\qquad\qquad (\because Ha = H = aH \Leftrightarrow a = H)$

$\Rightarrow \qquad\qquad bH = aH$ for $aa^{-1}b = eb = b$

$\Rightarrow \qquad\qquad aH = bH$

$\Rightarrow \qquad f$ is one-one.

(ii) f is onto :

For any $Ha \in R, \exists a^{-1}H \in L$ such that

$$f(a^{-1}H) = H(a^{-1})^{-1} = Ha$$

Hence, there exists a one-one correspondence between the set of left cosets of H in G and the set of right cosets of H in G.

REMARKS

☞ This theorem can also be stated as follows : "If H is a subgroup of G, the number of distinct left cosets of H in G is equal to the number of distinct right cosets of H in G."

☞ The two right cosets Ha and Hb are distinct iff the two left cosets $a^{-1}H$ and $b^{-1}H$ are distinct.

☞ If H is a subgroup of a group G, there is a one-one correspondence between any two right cosets of H in G.

Theorem 6. (Lagrange's Theorem) *The order of each subgroup of a finite group is a divisor (factor) of the group.*

Steps outlines : To make the proof easier, we shall proceed as follows :

Step 1. *Assume a subgroup H of G such that $o(G) = n$ and $o(H) = m$.*

Step 2. *Prove that m is a divisor of n, i.e., $n = mp$ for some $p \in \mathbf{N}$.*

Proof. Let H be a subgroup of a finite group G and let

$$o(G) = n \text{ and } o(H) = m$$

To show m is a divisor of n.

For this we have to show that $n = mp$ for some $p \in \mathbf{N}$.

Let Ha be any right coset of H in G.

Then $o(Ha) = m \Rightarrow \exists m$ distinct elements $h_1, h_2, ..., h_m \in H$

$\Rightarrow \exists m$ distinct elements $h_1 a, h_2 a, ..., h_m a \in Ha$, for any map from H into Ha is one-one onto, we have

$$o(Ha) = m = o(H), a \in G.$$

$\Rightarrow$ Every right coset of H in G has m distinct elements.

Since, G finite and therefore, number of distinct right cosets of H in G will be finite say p. Also, any two right cosets of H in G will be either identical or disjoint. Hence, p disjoint right cosets of H in G will contains mp distinct elements.

$\therefore$ $$G = H \cup Ha \cup Hb \cup Hc \cup ... \text{ } p \text{ times}$$

$$o(G) = o(H) + o(Ha) + o(Hb) + ... = m + m + ... + p \text{ times} = mp$$

Hence, order of the subgroup of a finite group is a divisor of the group.

REMARKS

☞ The converse of the Lagrange's theorem is not true, *i.e.,* if G is a finte group of order n and m is any divisor of n then it is not necessary that G must have a subgroup of order m.

For Example: Consider the symmetric group P_4 of permutation of degree 4. Then $o(P_4) = 4! = 24$. Let A_4 be the alternating group of even permutation of degree 4. Then $o(A_4) = 24/2 = 12$. There exists no subgroup H of A_4 such that $o(H) = 6$, though 6 is a divisor of 12.

☞ The Lagrange's theorem has important applications in group theory. If G is a group of order 8, then there will not exist subgroups of G of order 3, 5, 6, 7. The only subgroup of G may be of order 2 and 4. Since, 2 and 4 are divisors of 8.

Theorem 7. *The order of every element of a finite group G is a divisor of the order of the group, i.e., $o(a) | o(G)$.*

Proof. Let G be a finite group of order n and let $a \in G$ be arbitrary, such that $o(a) = m$.

To show m is a divisor of n.

Define $$H = \{a^p : p \in \mathbf{Z}\}$$

$$o(a) = m$$

$\Rightarrow$ m is the least positive integer such that $a^m = e$.

$$H = \{a^{-2}, a^{-1}, a^0, a^1, a^2, ..., a^m = e\}$$

Let $x, y \in H$ $\Rightarrow \exists \, p, q \in \mathbf{Z}$ such that

$$xy^{-1} = a^{p-q} = a^r \text{ where } p - q = r \in \mathbf{Z}$$

$\Rightarrow \qquad\qquad\qquad xy^{-1} = a^r \in H$

$\Rightarrow \quad H$ is a subgroup of G.

Now, to show $\qquad o(H) = m, \quad i.e., H$ contains m distinct elements.

$$a, a^2, a^3, ..., a^m = e = a^0$$

Let $r, s \in \mathbf{Z}^+$ such that $s > 0, 1 \le r \le m, 1 \le s \le m$

Now, $a^r = a^s \Rightarrow \quad a^{r-s} = e \Rightarrow o(a) \le r - s < m \Rightarrow o(a) < m$

which is a contradiction.

$\therefore \qquad\qquad\qquad a^r \ne a^s$ if $r \ne s$

$\Rightarrow \quad a, a^2, a^3, ..., a^m$ are distinct elements of H.

$\Rightarrow \qquad\qquad\qquad o(H) = m = o(a)$

Then, by Lagrange's theorem, we have m is a divisor of n.

Hence, $o(a)$ is a divisor of $o(G)$.

Theorem 8. *Let G be a finite group of order n and $a \in G$ then $a^n = e$.*

Proof. Let G be a finite group of order n and let $a \in G$ be an element of order m so that $a^m = e$.

To show $\qquad a^m = e$. $\qquad$ Let $\quad H = \{a^p : p \in \mathbf{Z}\}$.

Then, by previous theorem, H is a subgroup of order m.

Using Lagrange's theorem, we have m is a divisor of n.

$\Rightarrow \quad \exists\, p \in \mathbf{N}$ such that $n/m = p \Rightarrow n = mp$

Now, $\qquad a^n = a^{mp} = (a^m)^p = (e)^p = e \Rightarrow a^n = e$

Theorem 9 (Fermat's Theorem). *If a is any integer and p is prime, then $a^p \equiv a \pmod p$*

Proof. If $a \ne 0$. Let G be a multiplicative group of residue classes modulo p, then G contains $(p-1)$ distinct elements namely

$$[1], [2], ..., [p-1]$$

$\Rightarrow \qquad\qquad o(G) = p - 1$ and $[e] = 1$

Then, by theorem 8, we have $[a^{p-1}] = [1] \Rightarrow a^{p-1} \equiv 1 \pmod p$

Now, since p is prime, hence, $a^{p-1}a \equiv 1. a \pmod p \Rightarrow a^p \equiv a \pmod p$.

2.23.3 EULER'S Φ FUNCTION

For any positive integer n, the Euler's Φ function is defined as follows :

$\qquad\qquad\qquad \Phi(1) = 1$ and for $n > 1$, $\Phi(n) =$ The number of positive integers less than n and relatively prime to n.

For example

$\Phi(6) = 2$, since the positive integer less than 6 and relatively prime to 6 are 5 and 1 and their number is 2.

Theorem 10 (Euler's Theorem). *If n is positive integer and a is any integer relatively prime to n, then $a^{\Phi(n)} \equiv 1 \pmod n$*

Proof. Let us suppose $[x]$ denote the residue class of the set of integer mod n, for any integer-n.

Now , the residue classes G is group of order $\Phi(n)$ with identiy element, residue class $[1]$.

Now, we have

$$[a] \in G \Rightarrow [a]^{0(G)} = [1] \qquad \Rightarrow [a]^{\Phi(n)} = [1]$$
$$\Rightarrow [a][a] \text{ ... up to } \Phi(n) \text{ times} = [1]$$
$$\Rightarrow [a \cdot a \text{ ... up to } \Phi(n) \text{ times}] = [1]$$
$$\Rightarrow [a^{\Phi(n)}] = [1] \qquad \Rightarrow a^{\Phi(n)} = 1 \pmod{n}$$

Theorem 11. *Let H and K be finite subgroups of a groups G, then* $o(HK) = \dfrac{o(H)o(K)}{o(H \cap K)}$.

Proof. Let G be group and H and K be two subgroups of G. Clearly $H \cap K$ is a subset of K also. Let $D = H \cap K$, then D is a subgroup of G and $D \subseteq K$. Therefore D is a subgroup of K also. Since, K is finite therefore the number of distinct right cosets is finte. Let it be n. Then by Lagrange's theorem, we have

$$n = \frac{o(K)}{o(D)} \qquad \qquad \text{...(1)}$$

Let $Dk_1, Dk_2, ..., Dk_n$ are some distinct right cosets of D in K, then elements in K can be written as

$$K = Dk_1 \cup Dk_2 \cup ... \cup Dk_n = \bigcup_{i=1}^{n} Dk_i$$

We should see that $k_1, k_2, ..., k_n$ are some distinct elements in K.

Then, $\qquad HK = H\left(\bigcup_{i=1}^{n} Dk_i\right) = \bigcup_{i=1}^{n} HDk_i = \bigcup_{i=1}^{m} Hk_i \quad (\because D \subseteq H \Rightarrow HD = D)$

$$= Hk_1 \cup Hk_2 \cup ... \cup Hk_m$$

Now, we shall prove that the cosets $Hk_1, Hk_2, ..., Hk_n$ are pairwise distinct, we have

$$Hk_i = Kk_j \Rightarrow k_i k_j^{-1} \in H \qquad \qquad (\because k_i \, k_j \in K \Rightarrow k_i k_j^{-1} \in K)$$

$$\Rightarrow \qquad k_i k_j^{-1} \in H \cap K$$

$$\Rightarrow \qquad k_i k_j^{-1} \in D \quad \Rightarrow \quad Dk_i = Dk_j$$

$$\Rightarrow \qquad k_i = k_j \qquad \qquad (\because Dk_i, ..., Dk_j \text{ are distinct cosets.})$$

Therefore, $Hk_1, Hk_2, ..., Hk_n$ are distinct right cosets and therefore they are pairwise distinct elements. Also, number of elements in each of them is equal to $o(H)$, *i.e.,* the number of elements in H. Now from (1).

The number of elements in $HK = n \times o(H)$

$$\Rightarrow \qquad o(HK) = n \times o(H) = \frac{o(K)}{o(D)} \cdot o(H) = \frac{o(K)o(H)}{o(H \cap K)}$$

Corollary. Let H and K be subgroups of a finite group G and let $o(H) > \sqrt{o(G)}, o(K) > \sqrt{o(G)}$. Then $H \cap K \neq \{e\}$.

Proof. We know that $HK \subseteq G \qquad \Rightarrow o(HK) \leq o(G) \qquad \qquad \text{...(1)}$

But $\qquad o(HK) = \dfrac{o(H)o(K)}{o(H \cap K)} \qquad \qquad \text{...(2)}$

Using (1) and (2) we have

$$o(G) \geq \frac{o(H)o(K)}{o(H \cap K)} \qquad \text{...(3)}$$

But $\quad \dfrac{o(H)o(K)}{o(H \cap K)} > \dfrac{\sqrt{o(G)} \cdot \sqrt{o(G)}}{o(H \cap K)} \qquad \text{...(4)}$

From (3) and (4), we have

$$o(G) > \frac{o(G)}{o(H \cap K)}$$

$\Rightarrow \qquad o(H \cap K) > 1$

$\Rightarrow \qquad H \cap K \neq \{e\}$

Theorem 12 (Cayley's Theorem) *Every finite group G is isomorphic to a permutation group.*

Steps outline : To prove this theorem, we shall proceed as follows :

Step 1. *Find a set G' of permutation for forming a group under permutation multiplication isomorphic to G.*

Step 2. *Prove that G' is a group under the composition of permutation multiplication.*

Step 3. *Define a mapping* $\phi : G \to G'$ *and prove that* $f : G \to G'$ *is an isomorphism.*

Proof.   Let G be a finite group of order n such that $G = \{a_1, a_2, ..., a_n\}$

Let $a \in G$. Define a map $f_a : G \to G$ given by

$$f_a(x) = ax \ \forall \ x \in G.$$

f_a **is one-one.**

Let $\qquad\qquad f_a(x_1) = f_a(x_2) : x_1, x_2 \in G$

$\Rightarrow \qquad\qquad ax_1 = ax_2 \ \Rightarrow \ x_1 = x_2$

f_a **is onto.**

$f_a : G \to G$ is one-one and G is finite, therefore f is onto. Thus, f_a is one-one map of a finite set G onto itself. It means that f_a is a permutation of degree n.

Here $\qquad\qquad f_a = \begin{pmatrix} a_1 & a_2 & ... & a_n \\ aa_1 & aa_2 & ... & aa_n \end{pmatrix}$

The elements $aa_1, aa_2, ..., aa_n$ are all distinct elements of G.

Write $\qquad G' = \{f_a : a \in G\}$ then G' is a set of permutations of degree n.

Now, we claim that $(G', .)$ is a group, where $(.)$ denotes permutations multiplications.

Let $a, b, c \in G$ be arbitrary and e be the identity in G.

Let a^{-1} denote inverse of a in G so that $a^{-1}a = aa^{-1} = e$

(i) Closure property. $f_a, f_b \in G' \ \Rightarrow \ f_a f_b \in G'$

$\qquad$ Consider $\quad (f_a f_b)(x) = f_a[f_b(x)] = f_a(bx)$

$\qquad\qquad\qquad\qquad\qquad\qquad = a(bx) = (ab)x, \qquad$ (By associativity in G)

$\qquad\qquad\qquad\qquad\qquad\qquad = f_{ab}(x)$

$\Rightarrow \qquad\qquad f_a f_b = f_{ab} \qquad\qquad\qquad\qquad\qquad \text{...(1)}$

$\qquad\qquad\qquad a, b \in G \qquad \Rightarrow \qquad f_{ab} \in G' \ \Rightarrow \ f_a f_b \in G'$

(ii) Associativity. Let $a, b, c \in G$

$\Rightarrow \qquad\qquad (ab)c = a(bc)$

$\Rightarrow \qquad\qquad f_{(ab)} f_c = f_a f_{bc} \ \Rightarrow \ (f_a \ f_b) f_c = f_a (f_b \ f_c)$

(iii) Existence of identity. $a, e \in G \Rightarrow f_e \in G'$ and $ae = ea = a$

$\Rightarrow \qquad\qquad f_{ae} = f_{ea} = f_a \Rightarrow \qquad f_a f_e = f_e f_a = f_a$

$\Rightarrow \quad f_e \in G'$ is the identity element of G'.

(iv) Existence of inverse. $a \in G \Rightarrow a^{-1} \in G \Rightarrow f_a, f_{a^{-1}} \in G'$.

Also, $\qquad\qquad aa^{-1} = a^{-1}a = e$

$\qquad\qquad f_{aa^{-1}} = f_{a^{-1}a} = f_e$ or $f_a f_{a^{-1}} = f_{a^{-1}a} = f_e$

$\Rightarrow \quad f_{a^{-1}} \in G'$ is the inverse of $f_a \in G$.

(v) Order of G'. For $\quad o(G) = n, G = \{ f_a : a \in G \} \Rightarrow o(G') = n$

Therefore, we have $(G', .)$ is a finite group of order n.

We claim that $(G, .) \cong (G', .)$.

Now define a map $g : G \to G'$ such that $g(x) = f_x \ \forall \ x \in G$

(i) g is one-one. For $g(x_1) = g(x_2)$; $x_1, x_2 \in G$

$\Rightarrow \qquad\qquad f_{x_1} = f_{x_2}$

$\Rightarrow \qquad\qquad f_{x_1}(x) = f_{x_2}(x) \ \forall \ x \in G$

$\Rightarrow \qquad\qquad x_1 x = x_2 x$

$\Rightarrow \qquad\qquad x_1 = x_2$

(ii) g is onto. For any $f_a \in G' \Rightarrow a \in G$ such that $g(a) = f_a$.

(iii) g preserves composition in G and G'.

For $\quad g(x_1 x_2) = f_{x_1 x_2}$ $\qquad\qquad$ (where $x_1, x_2 \in G \Rightarrow x_1 x_2 \in G$)

Hence, G is an isomorphism of G onto G' and hence $G \cong G'$.

REMARKS

☞ Cayley's theorem can also be stated as :

Any finite group of order n is isomorphic to a subgroup of the symmetric group s_n.

☞ Cayley's theorem is holds, even if G is not finite. In this case the word permutation should not exist in the statement of the theorem. In that case we state as "Every group is isomorphic to a group of one-one onto function."

2.23.4 Regular Permutation Group

The permutation group to which G is isomorphic is called a regular permutation group.

Solved Examples
Based on the following Results

▶ $aH = \{ah : h \in H\}$, $Ha = \{ha : h \in H\}$

▶ Any two left cosets of a subgroup are either disjoint or identical.

▶ The order of each subgroup of a finite group is a divisor of the group. **[Lagrange's theorem]**

▶ If a is any integer and p is prime, then $a^p \equiv a (\bmod\, p)$. **[Fermat's theorem]**

▶ Every finite group G is isomorphic to a permutation group. **[Cayley's theorem]**

Example 1. If G is group and $a \in G$, then show that the set $H = \{a^n : n \in \mathbf{Z}\}$ is a subgroup of G and it is the smallest subgroup of G which contains the element a.

Solution. Clearly, H is non-empty subset of G.

Let $x, y \in H$, then $x = a^p, y = a^q$, where $p, q \in \mathbf{Z}$

Thereore; $(a^p)(a^q)^{-1} = a^p a^{-q} = a^{p-q} \in H$

$\Rightarrow$ H is a non-empty subset of G and $x, y \in H \Rightarrow xy^{-1} \in H$. Therefore, H is a subgroup of G.

Now, if K is any subgroup of G which contain a, they by closure property in K, $a^n \in K$ for every integer n. Also every integral power of a belong to K, i.e., $H \subseteq K$. Hence, H is the smallest subgroup of G which contain a.

REMARK

☞ If G is a group $a \in G$, then the subgroup $H = \{a^n : n \in \mathbf{Z}\}$ is called the subgroup of G, generated by a.

Example 2. *Prove that those elements of a group G which commute with the square of a given elements b of G form a subgroup of G.*

Solution.   Let $H = \{x \in G : xb^2 = b^2x\}$

Since, $b^2 \in G$ and $eb^2 = b^2e$, so $e \in H \Rightarrow H$ is a non-empty subset of G.

Let $x_1, x_2 \in H$ so that $x_1b^2 = b^2x_1$ and $x_2b^2 = b^2x_2$

$$\begin{aligned}
\text{Now, } (x_1x_2)b^2 &= x_1(x_2b^2) &&\text{(By associativity)} \\
&= x_1(b^2x_2) &&(\because x_2b^2 = b^2x_2) \\
&= (x_1b^2)x_2 &&\text{(By associativity)} \\
&= (b^2x_1)x_2 = b^2(x_1x_2) &&(\because x_1b^2 = b^2x_1)
\end{aligned}$$

$\Rightarrow$ $x_1, x_2 \in H$ $\Rightarrow$ $x_1 \cdot x_2 \in H$ ...(1)

Also, $x_1 \in H$ $\Rightarrow$ $x_1b^2 = b^2x_1$

$\Rightarrow$ $x_1^{-1}(x_1b^2)x_1^{-1} = x_1^{-1}(b^2x_1)x_1^{-1}$

$\Rightarrow$ $(x_1^{-1}x_1)(b^2x_1^{-1}) = (x_1^{-1}b^2)(x_1x_1^{-1})$

$\Rightarrow$ $e(b^2x_1^{-1}) = (x_1^{-1}b^2)e$

$\Rightarrow$ $b^2x_1^{-1} = x_1^{-1}b^2$

Thus, $x \in H \Rightarrow x_1^{-1} \in H$ ...(2)

Hence, from (1) and (2), we conclude that H is a subgroup of G.

Example 3. *If H is a subgroup of group G and $T = \{x \in G : xH = Hx\}$, show that T is a subgroup of G.*

Solution.   Since $e \in G$ and $eH = He \Rightarrow e \in T$

$\Rightarrow$ T is a non-empty subset of G.

Let $x_1, x_2 \in T$ so that $x_1H = Hx_1$, $x_2H = Hx_2$

Now, $x_2 \in T$ $\Rightarrow$ $x_2H = Hx_2$

$\Rightarrow$ $x_2^{-1}(x_2H)x_2^{-1} = x_2^{-1}(Hx_2)x_2^{-1}$

$\Rightarrow$ $x_2^{-1}x_2(Hx_2^{-1}) = (x_2^{-1}H)(x_2x_2^{-1})$

$\Rightarrow$ $e(Hx_2^{-1}) = (x_2^{-1}H)e$

$\Rightarrow$ $Hx_2^{-1} = x_2^{-1}H$

$\Rightarrow x_2^{-1} \in T$

Thus, $x_2T \Rightarrow x_2^{-1} \in T$

Also, $(x_1x_2^{-1})H = x_1(x_2^{-1}H) = x_1(Hx_2^{-1}) = (x_1H)x_2^{-1} = (Hx_1)x_2^{-1} = H(x_1x_2^{-1})$

$$\Rightarrow \qquad x_1 x_2^{-1} \in T$$

Thus, T is a non-empty subset of G and $x_1, x_2 \in T \Rightarrow x_1 x_2^{-1} \in T$.

Hence, T is a subgroup of G.

Example 4. *Find the regular permutation group isomorphic to the multiplicative group* $G = \{1, -1, i, -i\}$.

Solution. We know by Cayley's theorem that the regular permutation group G' isomorphic to G consist the following four permutations :

$$f_1 = \begin{pmatrix} 1 & -1 & i & -i \\ 1 \cdot 1 & 1 \cdot (-1) & 1 \cdot i & (1) \cdot (-i) \end{pmatrix} = \begin{pmatrix} 1 & -1 & i & -i \\ 1 & -1 & i & -i \end{pmatrix} = I$$

$$f_2 = \begin{pmatrix} 1 & -1 & i & -i \\ (-1) \cdot 1 & (-1) \cdot (-1) & (-1) \cdot i & (-1) \cdot (-i) \end{pmatrix} = \begin{pmatrix} 1 & -1 & i & -i \\ -1 & 1 & -i & i \end{pmatrix} = (1, -1)(i, -i)$$

$$f_3 = \begin{pmatrix} 1 & -1 & i & -i \\ i \cdot 1 & i \cdot (-1) & 1 \cdot i & i \cdot (-i) \end{pmatrix} = \begin{pmatrix} 1 & -1 & i & -i \\ i & -i & -1 & 1 \end{pmatrix} = (1, i, -1, -i)$$

$$f_4 = \begin{pmatrix} 1 & -1 & i & -i \\ (-i) \cdot 1 & (-i) \cdot (-1) & (-i) \cdot i & (-i) \cdot (-i) \end{pmatrix} = \begin{pmatrix} 1 & -1 & i & -i \\ -i & i & 1 & -1 \end{pmatrix} = (1, -i, -1, i)$$

EXERCISE 2.7

1. Let G be a additive group of integers. Then show that the set of all multiples of integers by a fixed integer m is a subgroup of G.

2. Show that the integral multiples of 5 form a subgroup of the additive group of integers.

3. Show that the 24 permutations on 4 symbols form a group with respect to permutation multiplication.

4. Use Lagrange's theorem to show that any group of prime order can have no proper subgroups.

5. Find the regular permutation group isomorphic to the group $(1, \omega, \omega^2)$ with respect to multiplication.

6. If a finite group G contains an element of even order, show that G must also be of even order.

7. If a finite group possesses an element of order 2, show tha it possesses an odd number of such elements.

8. If $H \subseteq K$ are two subgroups of a finite group G, prove that $[G : H] = [G : K][K : H]$.

9. Show that the set of inverses of the elements of a right coset is a left coset.

10. Show that the intersection of two subgroups, each of finite index, is again of finite index.

2.24 CYCLIC GROUPS

If a group G contain an element a such that every element $x \in G$ is of the form a^m, where $m \in \mathbf{Z}$, then G is said to be cyclic group and G is generated by a i.e., a is the generator of G, and we write $G = \{a\}$.

REMARKS

☞ It G is a cyclic group generated by a then, since G is closed under multiplication, then $a^k \in G \; \forall \, k \in \mathbf{Z}^+$. Also, since the inverse of a^k is a^{-k} we see that $a^{-k} \in G \; \forall \, k \in \mathbf{Z}^+$. Also a^0 is the identity e of G. Then

$$G = \{a\} = \{a^k : k \in \mathbf{Z}\}$$

For example

The multiplicative group $G = \{1, -1, i, -i\}$ is a cyclic group with generator i, because

$$(i)^1 = i, \; (i)^2 = -1, \; (i)^3 = -i \text{ and } (i)^4 = 1$$

$\Rightarrow$ each element of G can be expressed as some integral power of i.

$\Rightarrow$ G is cyclic, generated by i.

REMARKS

☞ A cyclic group always have at least two generators. *i.e.,* if a is a generator of G then a^{-1} is also the generator of G.

☞ The multiplicative group of n, n^{th} roots of unity is cyclic with generator $e^{2\pi/n}$.

☞ Let n be a positive integer. We construct a group G of order n as follows :

Suppose that G consists of all symbols. a^i, $i = 0, 1, 2, ..., n-1$, where we insists that $a^0 = a^n = e$, $a^i a^j = a^{i+j}$ if $i+j \leq n$ and $a^i a^j = a^{i+j-n}$. If $i+j > n$. Then we may easily verify that this is a cyclic group of order n and having the form $G = \{a\} = \{e, a, a^2, ..., a^{n-1}, ...\}$

☞ The additive group of integers $\{..., -3, -2, -1, 0, 1, 2, 3, ...\}$ is a cyclic group with generator 1 and -1.

2.24.1 Properties of Cyclic Groups

Theorem 1. *Every cyclic group is necessarily abelian but the converse is not necessarily true.*

Proof. Let $G = \{a\}$ is a cyclic group generator by an element $a \in G$.

Let x and y be any two elements of G.

Then, $x = a^m$ and $y = a^n$, for some integers m and n.

Now, $xy = a^m a^n = a^{m+n} = a^{n+m} = a^n \cdot a^m = yx$

$\Rightarrow$ $xy = yx \ \forall \ x, y \in G$

$\Rightarrow$ G is abelian.

Conversely, we have to show that an abelian group is not always a cyclic groups. It is illustrated by the following example :

The set $\mathbf{R_0}$ of all non-zero real numbers is an abelian group with respect to multiplication.

If $a \in \mathbf{R_0}$, then $\{a^n : n \in \mathbf{Z}\}$ is a countable subset of $\mathbf{R_0}$ and so it cannot be equal to the uncountable set $\mathbf{R_0}$.

$\Rightarrow$ All the elements of $\mathbf{R_0}$ cannot be expressed as some integral power of a single element of $\mathbf{R_0}$

$\Rightarrow$ $(\mathbf{R_0}, .)$ is not cyclic group.

Theorem 2. *If the generator of a cyclic group G is of infinite order (or of zero order), then G is isomorphic to the additive group of integers.*

Proof. Let a be the generator of the cyclic group $G = \{a\}$. Let $o(a) = \infty \Rightarrow a^n \neq e$ for any n.

To show $(G, .) \cong (\mathbf{Z}, +)$.

Firstly we shall show that any two powers of a cannot be equal. Let if possible,

$$a^m = a^n \text{ for } m \neq n$$

$$a^m = a^n \ \Rightarrow \ a^m = ea^n \Rightarrow a^{m-n} = e$$

$\Rightarrow$ $o(a) \leq m - n = a$ finite number

$\Rightarrow$ $o(a)$ is finite, a contradiction.

$\Rightarrow$ $a^m \neq a^n$ for $m \neq n$

$\Rightarrow$ G contains an infinite number of distinct elements.

$$G = \{a^0 = e, a^{\pm 1}, a^{\pm 2}, a^{\pm 3}, ...\}$$

and $\mathbf{Z} = \{(0), \pm 1, \pm 2, \pm 3, \pm ...\}$

(i) Now define a map $f : G \to \mathbf{Z}$ such that f **is one-one.**

Let $f(a^m) = f(a^n) : a^m, a^n \in G$. Now $m = n \Rightarrow a^m = a^n$

i.e., $f(a^m) = f(a^n) \Rightarrow a^m = a^n$

(ii) f is onto.

Since, $o(G) = \infty = o(\mathbf{Z})$ and f is one-one.

$\Rightarrow f$ is onto.

(iii) f preserves composition in G and $\mathbf{Z}$.

Let $a^m, a^n \in G$, then

$$f(a^m \cdot a^n) = f(a^{m+n}) = m+n = f(a^m)+f(a^n)$$

i.e., $f(a^m \cdot a^n) = f(a^m) + f(a^n)$

$\Rightarrow f$ preserves compositions in G and $\mathbf{Z}$.

Hence, f is an isomorphism in G and $\mathbf{Z}$.

and $(G, .) \cong (\mathbf{Z}, +)$

REMARKS

☞ Cyclic group is infinite $\Rightarrow$ order of its generator is infinite.

☞ The above theorem can also be stated as "Every infinite group is isomorphic to the additive group of integers."

In a Similar way we can prove the following theorem :

"A cyclic group of finite order n is isomorphic to the additive group of residue classes modulo n."

Theorem 3. *The order of a cyclic group is equal to the order of any generator of the group.*

Proof.   Let a be the generator of a group $G = \{a\}$. Let $o(a) = $ finite $= n$

$\Rightarrow$ $a^n = e, a^r \neq e$ for $o < r < n$

To show $o(a) = o(G) = n$.

Step I. Firstly we shall show that G contain n elements.

The elements of the cyclic group G are given below :

$$a, a^2, a^3, ..., a^n = e = a^0$$

Let if possible, G contains an element a^m besides these elements, where $m > n$.

Then by division algorithm, we have

$$m = nq+r, 0 \leq r \leq n \text{ and } q, r \in N$$
$$a^m = a^{nq+r} = a^{nq} \cdot a^r = (a^n)^q \cdot a^r = e^q \cdot a^r = e \cdot a^r = a^r$$

∴ $a^m = a^r, 0 \leq r \leq n$

∵ a^r is already contained in the set of n elements and so a^m is also contained.

$\Rightarrow$ G contains n elements.

Step II. Now to show that any two elements of G are not equal. For this we have to show that $a^r \neq a^s$ where $r \neq s$, $0 < r < n$, $0 < s < n$

Let $r < s < n$

Then $s - r > 0$

$$a^r = a^s \Rightarrow ea^s = a^s \Rightarrow a^{s-r} = e$$

$\Rightarrow$ $o\ (a) \leq s - r$ and $s - r < n$

$\Rightarrow$ $o\ (a) < n$

which is a contradiction. Hence, $a^r \neq a^s$ where $r \neq s$

Thus, we have shown that G contains n distinct elements and hence $o(G) = n$.

Theorem 4. *A cyclic group G with a generator of finite order n, is isomorphic to the multiplicative group of n, n^{th} roots of unity.*

Proof. Since we know that, if G be a cyclic group with a generator of finite order n, then G contains exactly n distinct elements namely.

$$a, a^2, a^3, ..., a^n = e$$

Now, let G' be the multiplicatiive group of n, n^{th} roots of unity so that

$$G' = \{x : x \text{ is a solution of } x^n = 1\}$$

$$= \{x : x \text{ is a solution of } x^{1/n} = e^{2\pi i/n}\}$$

$$= \{e^{2\pi r i/n} : r = 0, 1, 2, ... (n-1)\}.$$

Also, $G = \{a^r : r = 1, 2, 3, ..., n\} = \{a^r : r = 0, 1, 2, 3, ..., (n+1)\}$

For $a^n = e = a^0$

Consider a map $f : G \to G'$ such that $f(a^r) = e^{2\pi r i/n}, \forall a^r \in G$

Now $f(a^r) = f(a^s) \Rightarrow e^{2\pi s i/n} = e^{2\pi r i/n} \Rightarrow r = s \Rightarrow a^r = a^s$

$\Rightarrow$ f is onto.

Now $f(a^r.a^s) = f(a^{r+s}) = e^{2\pi(r+s)i/n} = f(a^r).f(a^s)$

i.e., $f(a^r.a^s) = f(a^r).f(a^s), \forall a^r, a^s \in G$

$\Rightarrow$ f preserves composition in G and G'.

$\Rightarrow$ f is an isomorphism.

Hence, $(G, .) \cong (G', .)$.

Theorem 5. *Every isomorphic image of a cyclic group is cyclic.*

Proof. Let a be the generator of a cyclic group G so that $G = \{a\}$. Let G' be the isomoprphic image of G so that there exists an isomorphism $f : G \to G'$.

To show G' is cyclic.

Since G contains the elements a, a^2, a^3, a^4, ... therefore, the elements of G' are

$$f(a), f(a^2), f(a^3)....$$

Let $a^n \in G$ be arbitrary, then $f(a^n) \in G$. By the property of isomorphism

$$f(a^n) = f(a.a.a...n \text{ times})$$

$$= f(a).f(a).f(a)...n \text{ times} = [f(a)]^n$$

$\Rightarrow$ $f(a^n) = [f(a)]^n$

$\Rightarrow$ every element of G' is expresible as some integral power of $f(a)$.

Hence, G' is cyclic with generator $f(a)$.

Theorem 6. *A finite group of order n contining an element of order n must be cyclic.*

Proof. Let G be a finite group of order n. Let $a \in G$ such that $o(a) = n$. Then

$$o(G) = n = o(a)$$

To show G is cyclic.

Let H be a cyclic group generated by a then

$$o(H) = o(a) = n$$

$\Rightarrow$ H can be expressed as $\{H = a^r : r = 1, 2, 3, ..., n\}$.

Since, G is a group, then $\Rightarrow a \in G \Rightarrow a^r \in G$ for every integral value of r.

Thus $H \subseteq G$.

Moreover

$$o(G) = n = o(H)$$

$\therefore$ $H = G$, but H is cyclic. Hence, G is cyclic.

Theorem 7. *If cyclic group G is generated by an element a of order n, then a^m is a generator of G iff $(m, n) = 1$, i.e., the greatest common divisor of m and n is 1.*

Proof. Let G be a cyclic group generated by an element a of order n so that $a^n = e$. Let us first suppose the g.c.d. of m and n is 1.

To show that a^m is a also a generator of G. Let H be a cyclic group generated by a^m so that $H = \{a^m\}$.

For any $a \in G$, we have $a^r \in G$ for every integral value of $r \Rightarrow H \subset G$.

Now, since H.C.F. of m and n is $1 \Rightarrow \exists\, x, y \in \mathbf{Z}$ such that $mx + ny = 1$.

$\Rightarrow$ $a^1 = a^{mx+ny} = a^{mx} \cdot a^{ny}$

$\Rightarrow$ $a^1 = a^{mx}(a^n)^y = a^{mx} \cdot e^y = a^{mx} \cdot e = a^{mx}$

$\Rightarrow$ $a = a^{mx} = (a^m)^x \Rightarrow a = (a^m)^x$

$\Rightarrow$ Every element of G can be expressed as some integral power of $a^m \in H$

$\Rightarrow$ $G \subset H$

Now $H \subset G$, $G \subset H \Rightarrow G = H$

Also, $H = G$, a^m is the generator of H.

Hence, a^m is also a generator of G.

Conversely, let G be a cyclic group generated by an element a of order n. Also, suppose that a^m is the generator of G.

To show that m and n are relatively prime, i.e., g.c.d of m and n is 1.

Let d be the H.C.F. of m and n then $m|d, n|d$

To show that $d = 1$.

Let if possible.

$$d \neq 1,\ i.e.,\ d > 1$$
$$(a^m)^{n/d} = (a^n)^{m/d} = (e)^{m/d} = e$$

Then $(a^m)^{n/d} = e \Rightarrow o(a^m) \le n/d < n$

$\Rightarrow o(a^m) \le n$, which is a contradiction.

$\Rightarrow d = 1$.

Theorem 8. *Every group of prime order is cyclic.*

Proof. Let G be a finite group of order p, where p is prime. To show G is cyclic. Since G is a group of prime order $\Rightarrow G$ must contains at least 2 elements.

$\qquad\qquad\qquad\qquad\qquad$ ($\because$ 2 is the least positive prime integer.)

$\Rightarrow \quad$ There must exists an element $a \in G : a \neq e$

$\because \quad a \neq e \ \Rightarrow o(a) \geq 2$

Let o $(a) = m$. Then $H = \{a\}$ is a cyclic group of G and $o(H) = o(a) = m$. Then, by Lagrange's theorem m must be a divisor of p. But p is prime and $m \geq 2$. Hence $m = p$.

Therefore, $H = G$.

Since, H is cyclic, therefore G is cyclic with generator a.

Theorem 9. *Every subgroup of a cyclic group is cyclic.*

Proof. Let $G = \{a\}$ be a cyclic group generated by a . If $H = G$ or $\{e\}$, then obviously H is cyclic.

Now, let H be a proper subgroup of G, such that H contains the element or integral power of a.

If $a^s \in H \Rightarrow a^{-s} \in H$. Therefore, H contain elements which are positive as well as negative integral power of a. Let k be the least positive integer such that $a^k \in H$. To show $H = \{a^m\}$.

Let $a^t \in H$. Then, by division algorithm, there exist $q, r \in \mathbf{Z}$ such that $t = kq + r, 0 \leq r \leq k$.

Now $a^k \in H \Rightarrow (a^k)^q \in H \Rightarrow a^{kq} \in H \Rightarrow (a^{kq})^{-1} \in H \Rightarrow a^{-kq} \in H$

Also, $a^t \in H, a^{-kq} \in H \Rightarrow a^t a^{-kq} \in H \Rightarrow a^{t-kq} \in H \Rightarrow a^r \in H$

$\because \quad k$ is the least positive integer such that $a^k \in H$ and $0 \leq r < k$.

$\Rightarrow \quad r$ must be equal to 0.

$\Rightarrow \qquad\qquad t = kq$

$\therefore \qquad\qquad a^t = a^{kq} = (a^k)^q$

$\Rightarrow \quad$ every element $a^t \in H$ is of the form $(a^k)^q$

Hence, H is cyclic with generator a^m.

Solved Examples

Example 1. *How many generators are there of the cyclic group of order 8 ?*

Solution. Let us suppose that the cyclic group G of order 8 generated by an element a then $o(a) = 8$

Clearly, $G = \{a, a^2, a^3, a^4, a^5, a^6, a^7, a^8 = e\}$

Now from theorem 7, we know that an element a^m is also a generator of G, if m is less than 8 and relatively prime to 8.

Such numbers are 1, 3, 5 and 7.

Hence, a, a^3, a^5 and a^7 are generators of G.

$\Rightarrow \quad$ There are four generators of G.

Example 2. *Show that the group G = [{1, −1, i, −i},.] is cyclic.*

Solution.   Let $G = \{1, -1, i, -1\}$

To show G is cyclic.

If there exists an element $a \in G$ such that $o(a) = 4 = o(G)$. Then G will be cyclic group with generator a.

Evidently,   (i)$^1 = i$, $i^2 = -1$, $i^3 = -i$, $i^4 = 1$.

Here identity element of G is 1.

Thus $i^4 = 1$, $i^r \neq 1$ for any $r < 4$.

$\Rightarrow$ $o(i) = 4 = o(G)$

$\Rightarrow$ i is the generator of G.

Thus G is expressible as $G = \{i, i^2, i^3, i^4\}$

Hence, G is cyclic.

Example 3. *Give an example of a finite abelian group which is not cyclic.*

Solution.   Define G as the set of four real matrices

$$I = \begin{bmatrix} 1 & 0 \\ 0 & 1 \end{bmatrix}, A = \begin{bmatrix} -1 & 0 \\ 0 & 1 \end{bmatrix}, B = \begin{bmatrix} 1 & 0 \\ 0 & -1 \end{bmatrix}, C = \begin{bmatrix} -1 & 0 \\ 0 & -1 \end{bmatrix}$$

We can easily verify that G is an abelian group w.r.t. multiplication of matrices. The identity element of this group is the identity matrix I. Now, we have to find the order of each element of G.

Clearly $o(I) = 1$, being the identity element.

$$A^2 = \begin{bmatrix} -1 & 0 \\ 0 & 1 \end{bmatrix}\begin{bmatrix} -1 & 0 \\ 0 & 1 \end{bmatrix} = \begin{bmatrix} 1 & 0 \\ 0 & 1 \end{bmatrix} = I \Rightarrow o(A) = 2$$

Now, since

$$B^2 = \begin{bmatrix} 1 & 0 \\ 0 & -1 \end{bmatrix}\begin{bmatrix} 1 & 0 \\ 0 & -1 \end{bmatrix} = \begin{bmatrix} 1 & 0 \\ 0 & 1 \end{bmatrix} = I \Rightarrow o(B) = 2$$

and $$C^2 = \begin{bmatrix} -1 & 0 \\ 0 & -1 \end{bmatrix}\begin{bmatrix} -1 & 0 \\ 0 & -1 \end{bmatrix} = \begin{bmatrix} 1 & 0 \\ 0 & 1 \end{bmatrix} = I \Rightarrow o(C) = 2$$

Since, G is a group of order 4 and G contains no element of order 4 therefore G is not abelian. Hence, G is a finite abelian non-cyclic group.

EXERCISE 2.8

1. If ω is the cube roots of unity, show that the set $\{1, \omega, \omega^2\}$ is a cyclic group of order 3 with respect to multiplication.

2. Show that the group $[\{1, 2, 3, 4, 5, 6\}, \times_7]$ is cyclic. How many generators are there?

3. Show that the two cyclic groups of same order are isomorphic.

4. Show that every finite group of composite order possesses proper subgroups.

5. Show that the set U_n of n, n^{th} complex roots of units forms a cyclic group with respect to multiplication.

6. Show that every finite group of order 6 must be abelian.

7. Show that the group $[\{1, 2, 3, 4\}, \times_5]$ is cyclic.

8. How many generators are there of the cyclic group of order 10 ?

9. Show that the residue classes [1], [2], [3], [4], [5], [6] with mod 7 form a multiplicative cyclic group. Find the number of generators.

10. Let G be the set of four matrices $\begin{bmatrix} 0 & 0 \\ 0 & 0 \end{bmatrix}, \begin{bmatrix} 1 & 1 \\ 0 & 0 \end{bmatrix}, \begin{bmatrix} 0 & 0 \\ 1 & 1 \end{bmatrix}, \begin{bmatrix} 1 & 1 \\ 1 & 1 \end{bmatrix}$ where 0 and 1 are the elements of the set $Z = \{0,1\}$ modulo 2.

Show that G is abelian, non-cyclic group under matrix addition.

11. Find the order of each element in the cyclic group $G = \{a, a^2, a^3, a^4, a^5, a^6 = e\}$

Hints to Selected Problems

5. $x^n - 1 = 0 \Rightarrow x = (1)^{1/n} = (\cos 2\pi r + i \sin 2\pi r)^{1/n} = \cos\dfrac{2\pi r}{n} + i \sin\dfrac{2\pi r}{n}$

Let $\omega = \cos\dfrac{2\pi}{n} + i \sin\dfrac{2\pi}{n} \Rightarrow \omega^2 = \cos\dfrac{4\pi}{n} + i \sin\dfrac{4\pi}{n}$

Therefore $U_n = \{1, \omega, \omega^2, ..., \omega^{n-1}\}$. Then show that $(U_n, \bullet)$ is a group.

8. Since 1, 3, 7, 9 are relatively prime to p. Thus a, a^3, a^7 and a^9 are generators of G.

Answers

2. Two, *i.e.*, 3 and 5 **8.** Four, *i.e.*, a, a^3, a^7, a^9

16. $o(a) = 6$, $o(a^2) = 3$, $o(a^3) = 2$, $o(a^4) = 3$, $o(a^5) = 6$ and $o(a^6) = 1$.

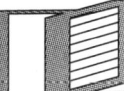

Chapter Review: | *A competitive Approach*

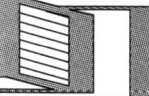

Selected Terms and Results

Terms

- **Groupoid :** A non-empty set S with binary operation $*$.
- **Semigroup:** A groupoid which is associative.
- **Monoid :** Semigroup with identity.
- **Group:** Monoid with inverse of each element.
- **Abelian group:** A group which is commutative.
- **Order of an element :** $o(a) = n \Rightarrow n$ is the least positive integer such that $a^n = e$.
- **Division algorithm :** If $a \neq 0$, $b \in \mathbf{Z}$ then there exists integers q and r such that $a = qb+r$ where $0 \leq r < |b|$
- **Transformation:** A one-one mapping f of a non-empty set S onto itself is called a transformation.
- **Permutation :** A one-one mapping f of a finite non-empty set S onto itself is called a permutation.
- **Identity permutation :** A permutation on the set S is called the identity permutation if it maps each element of S onto itself.
- **Transposition :** A cycle of length 2 is called transposition.

- **Even and odd permutation :** A permutation is said to be even or odd according as it can be expressed as a product of even or odd number of transposition.
- **Complex of a group :** A non-empty subset H of a group G is called complex.
- **Subgroup :** A non-empty subset H of a group G is called a subgroup of G if H itself is a group with respect to the operation defined in G.
- **Cosets :** Left coset $aH = \{ah : h \in H\}$
  Right coset $Ha = \{ha : h \in H\}$
- **Index of a subgroup in a group :** The number of distinct right or left cosets of subgroup H in group G is called index of H in G.
- **Regular permutation group :** The permutation group to which the group G is isomorphic is called a regular permutation group.
- **Cyclic group :** If a group G contain an element a such that every element $x \in G$ can be expressed in some integral power of a, then it is called cyclic group.

Results

- The direct product of any two semigroups is a semigroup.
- The product of two commutative semigroups is again commutative semigroup.
- Every finite semigroup has an idempotent element.
- Every cyclic monoid is a commutative monoid.
- The identity element of a group is unique.
- The inverse of each element of a group is unique.
- A finite set G with a binary operation $*$ which is associative is a group if and only if the cancellation laws hold.
- The identity element e in a group G is the only element whose order is 1.

- If the element a of a group G is of order n, then $a^m = e$ iff n is a divisor of m.
- The order of an element of a group is the same as that of its inverse.
- The relation 'Congruence modulo n' is an equivalence relation.
- The set of residue classes modulo m is a group w.r.t. addition of residue classes.
- Every permutation can be expressed as a product of disjoint cycles.
- Every permutation can be expressed as a product of transpositions.
- Of the $n!$ permutation of n symbol, $n!/2$ are even permutation and $n!/2$ are odd permutation.

- Let H be a non-empty subset of a group G. Then H is a subgroup of G iff a, $b \in H \Rightarrow ab^{-1} \in H$.

- A necessary and sufficient condition for a non-empty subset H of a group G to be a subgroup is that $HH^{-1} = H$.

- If H and K are subgroups of an abelian group G, then HK is a subgroup of G.

- If H and K are subgroups of a group G, then HK is a subgroup of G iff $HK = KH$.

- An arbitrary intersection of subgroups of a group G is a subgroup of G.

- The union of two subgroups of a group G is a subgroup of G iff one is contained in the other for $a \in G$.

- If H be a subgroup of G then for $a \in G$, $Ha = aH = H$, $\Leftrightarrow a \in H$.

- Any two left cosets of subgroups are either disjoint or identical.

- The order of each subgroup of a finite group is a divisor of the group. (Lagrange's theorem).

- The order of each element of a finite group G is a divisor of the order of the group.

- If G is a finite group of order n and $a \in G$, then $a^n = e$.

- If a is any integer and p is prime, then $a^p \equiv a \pmod{p}$. (Fermat's theorem).

- If n is a positive integer and a is any integer relatively prime to n then $a^{\phi(n)} \equiv 1 \pmod{n}$. (Euler's theorem)

- $O(HK) = \dfrac{O(H)O(K)}{O(H \cap K)}$

- Every finite group is isomorphic to a permutation group. (Cayley's theorem)

- Every cyclic group is necessarily abelian but converse is not true.

- The order of a cyclic group is equal to the order of any generator of the group.

- Every isomorphic image of a cyclic group is cyclic.

- A finite group of order n containing an element of order n must be cyclic.

- Every group of prime order is cyclic.

- Every subgroup of a cyclic group is cyclic.

Review Questions and Project Work

1. Show that the set of all rational number x such that $o < x \leq 1$ does not form a group w.r.t. ordinary multiplication.

2. Show that the set of vectors defined as directed line segments does not form a group (i) with respect to dot product (ii) with respect to cross product.

3. Show that set of all rational numbers of the form $2^a 3^b$ (a, $b \in \mathbf{Z}$) is a group with respect to multiplication of rationals.

4. Distinguish between an abelian group and a non-abelian group by giving some suitable examples.

5. Show that the set G of all square matrices $[a_{ij}]_{n \times n}$ such that det $[a_{ij}] = \pm 1$ is a group under matrix multiplication. Show also that those matrices in G for which det $[a_{ij}] = 1$ form a group.

6. Forming the composition table for the multiplicative group $\{e, a, b\}$ of order 3 show that every group of order 3 must be abelian.

7. Show that if S has more than two elements, then the symmetric group S_n is not abelian

8. Prove that a cycle containing an odd number of symbols is an even permutation where as a cycle containing an even number of symbols is an odd permutation.

9. Distinguish between the order of a group and the order of an element of a group.

10. If in a group G, the element a and b commutes, then prove that
    (a) a^{-1} and b^{-1} also commute.
    (b) a^{-1} and b also commute.
    (c) a and b^{-1} also commute.

11. If a group has an even number of elements, prove that at least one element other than identity must equal to its inverse.

12. Show that the elements of finite order in any commutative group G form a subgroup of G.

13. If H and K are subgroups of a finite group G give an example to show that $O(HK)$ need not divide $o(G)$.

14. Prove that a non-Commutative group has at least 6 elements.

15. If a finite group possesses an element of order 2 prove that it possessses an odd number of such elements.

Objective type Questions

Fill in the Blanks

1. A group G is said to be abelian if _____.

2. The number of element in a finite group G is said to be _____ group.

3. The identity of a group is _____.

4. The inverse of each elements of a group G is _____.

5. If G = {1,–1} then G is a group w.r.t. _____.

6. A set which does not contain zero, never be a group under _____.

7. A set which contain zero , never be a group under _____.

8. A one-one mapping of a finite set onto itself is called _____.

9. The identity permutation is always an _____ permutation.

10. A non-empty subset H of G is called _____.

11. Every group of prime order is _____.

12. A group containing less than six elements is always _____.

13. The necessary and sufficient condition to be a non-empty subset H of G to be subgroup is _____ . $\forall a, b \in H$.

14. The product of two even permutations is an _____ permutation.

15. A finite set G with a binary operation which is associative is a group if and only if _____ holds.

16. The left inverse of an element is also _____ inverse.

17. A group in which identity element e is the only element of finite order is called _____ .

18. A group in which every element is of finite order is called _____ group.

19. The order of every element of a finite group is _____ .

20. For any $a, b \in G$, $(ab)^2 = a^2 b^2$ then G must be _____ .

21. The set **N** of natural numbers is a _____ w.r.t. addition.

22. The set **N** of natural numbers is a _____ w.r.t. multiplication.

23. For every $a, b \in G$, $(ab)^{-1} =$ _____

24. The set of all n, nth roots of unity forms a finite abelian group of order w.r.t. _____.

25. A non-empty set G equipped with one or more binary operations is called an _____ .

True/False

Write 'T' for true and 'F' for false statement

1. The product of two odd permutations is an odd permutation. **(T/F)**

2. If for every element a in a group , $a^2 = e$, then G is abelian. **(T/F)**

3. If each element having its own inverse then group is abelian. **(T/F)**

4. If * be a commutative composition in a set S then $a*(b*c) = (c*b)*a \ \forall \ a,b,c \in S$ **(T/F)**

5. Every composition in a set consisting of 6 elements is commutative. **(T/F)**

6. A semigroup may have more than one identity element. **(T/F)**

7. The identity element of a group does not possesses the inverse. **(T/F)**

8. A finite semigroup in which the cancellation law hold is necessarily a group. **(T/F)**

9. No subgroup of a non-abelian group can be abelian. **(T/F)**

10. Every subgroup of a cyclic group is cyclic **(T/F)**

11. The union of two subgroups of a group is again a subgroup. **(T/F)**

12. All subgroups of an infinite group are infinite. **(T/F)**

13. The index of each subgroup of a finite group is finite. **(T/F)**

14. A cycle of length 2 is called transposition. **(T/F)**

15. The identity of a subgroup is the same as that of the group. **(T/F)**

16. An infinite group is said to be of infinite order. **(T/F)**

17. There may exists infinite groups in which the cancellation laws do not holds good. **(T/F)**

18. No group has an element of order 1. **(T/F)**

19. There exist at least one positive integer n such that there is no abelian group of order n. **(T/F)**

20. The set of all odd integers forms a group w.r.t. addition. **(T/F)**

21. The order of 1 in the group {(0, 1, 2, 3, 4, 5). $+_5$} is 1. **(T/F)**

22. For every a, b in a group G, $(ab)^{-1} = (b^{-1} a^{-1})$ **(T/F)**

23. A group of order 121 is abelian   (T/F)
24. The associative law hold in every group   (T/F)
25. There may be a group in which the cancellation laws do not hold   (T/F)
26. Every group is a subgroup of itself.   (T/F)
27. In a cyclic group, every element is a generator.   (T/F)
28. Every set of numbers that is a group under addition is also a group under multiplication.   (T/F)
29. A subgroup may be defined as a subset of the group .   (T/F)
30. Every subset of every group is a subgroup under the induced operation.   (T/F)

Multiple Choice Questions

Choose the most apprapriate one.

1. The groupoid form the following is :
   (a) Natural number w.r.t. addition
   (b) {1, 2, 3} w.r.t +
   (c) [–1, –2, –3] w.r.t +
   (d) none of the above
2. The semigroup form the following:
   (a) (**Z**, ±)   (b) (**Z**, •)
   (c) (**N**, +)   (d) all are true
3. A monoid is one which satisfy :
   (a) only closure axioms
   (b) only closure and associativity
   (c) closure, associativity and identity axioms
   (d) none of the above
4. A semigroup is a monoid if :
   (a) identity axiom is satisfied
   (b) inverse axiom is satisfied
   (c) both (a) and (b) are true
   (d) none of the above
5. A subgroup of an abelian group is :
   (a) not abelian
   (b) necessarily abelian
   (c) may be abelian
   (d) none of the above
6. Let $G = \{1, \omega, \omega^2\}$ be cyclic then generaters of G are :
   (a) 1 and ω (b) ω and ω^2
   (c) 1 and ω^2 (d) none of these
7. The number of identity element in a group is:
   (a) 1   (b) 2
   (c) 3   (d) none of these
8. The smallest group for a binary operation $*$ is:
   (a) ({e}, $*$)   (b) {e, $*$}
   (c) {ϕ, $*$}   (d) none of these
9. Which of the following is not a group?
   (a) (**R**, ×)   (b) (**N**, +)
   (c) (**Z**, ×)   (d) none of these
10. Which of the following is a group?
    (a) ((1, ω), ×)   (b) ((1, ω, ω^2), ×)
    (c) ((1, –1),+)   (d) none of these

11. Set of all integers is a group under :
    (a) addition   (b) multiplication
    (c) division   (d) none of these
12. Which of the following is/are group ?
    (a) {0, 1, 2, 3(mod 4) w.r.t. addition }
    (b) {0, 1, 2, 3(mod 4) w.r.t. multiplication }
    (c) {[0], [1], [2], [3],}
    (d) all are true.
13. If $\mathbf{Q}^+$ be the set of all positive rational numbers and binary operation is defined by $a*b = \dfrac{ab}{2}, a,b \in \mathbf{Q}$. Then identity of **Q** is :
    (a) 1   (b) 2
    (c) 3   (d) none of these
14. In the set of rational number **Q**, operation $*$ is defined by $a*b = a+b - ab$. Then inverse of a is :
    (a) $\dfrac{a}{a-1}$ (b) $\dfrac{a}{1-c}$
    (c) $1-a$ (d) none of these
15. A semigroup G following both the cancellation laws is a group if G is :
    (a) finite   (b) infinite
    (c) abelian   (d) none of these
16. Set of all natural numbers is :
    (a) a group under multiplication
    (b) a group under addition
    (c) a cyclic group
    (d) none of the above
17. The set of integers under the operation of multiplication is :
    (a) not a group   (b) is a group
    (c) may be a group   (d) none of these
18. Which one of the following forms a group?
    (a) The set of integers w.r.t. multiplication
    (b) The set of integers w.r.t. addition
    (c) The set of rational numbers w.r.t. multiplication
    (d) none of the above

19. The set $G = \{1, -1, i, -i\}$ w.r.t multiplication is:
 (a) an abelian group
 (b) a non-abelian group
 (c) not a group
 (d) none of the above

20. In a group $\{(1, \omega, \omega^2), \times\}$ where ω is the cube root of unity is , the inverse of ω^2 is :
 (a) 1 (b) ω
 (c) ω^2 (d) none of these

21. A group $(G, *)$ is said to be abelian if $*$ is :
 (a) associative (b) commutative
 (c) both (a) & (b) (d) none of these

22. Let $(\mathbf{Z}, *)$ be an algebraic system such that $a * b = a + b + 1$, $a, b \in \mathbf{Z}$ then the identity element of the set of integers $\mathbf{Z}$ is :
 (a) 0 (b) 1
 (c) -1 (d) 2

23. If H, K are two subgroups of a group G, then HK is a subgroup of G if :
 (a) $HK = KH$ (b) $HK = e$
 (c) $KH = e$ (d) none of these

24. Every group of prime order is :
 (a) abelian (b) cyclic
 (c) both (a) and (b) are true
 (d) none of the above

25. The number of generators of the cyclic group G of order 8 is :
 (a) 2 (b) 4
 (c) 6 (d) 8

26. If G is a finite group of order n and $a \in G$, then a^n is equal to :
 (a) e (b) $1/a$
 (c) a^{-1} (d) none of these

27. The order of $a.b$ in a group is equal to the order of :
 (a) a (b) b
 (c) ba (d) none of these

28. The set A_n of all even permutation of degree n forms a finite group w.r.t. permutations multiplications, then order of this group is :
 (a) $\dfrac{n}{2}$ (b) $n!$
 (c) $\dfrac{n!}{2}$ (d) none of these

29. The set $G = [1, 2, 3, 4, 5, 6]$ is a finite abelian group of order 6 w.r.t. multiplication modulo :
 (a) 2 (b) 3
 (c) 6 (d) 7

30. If the element a, b of a group G commute and $o(a) = n$, $o(b) = m$ such that $(m, n) = 1$, then $O(ab) =$
 (a) m (b) n
 (c) mn (d) none of these

31. The identity element of the group $(\mathbf{Z}, *)$ where $a \times b = a + b + 1$ is :
 (a) -1 (b) 1
 (c) 2 (d) none of these

32. $(aob)^{-1} =$
 (a) $a^{-1}ob^{-1}$ (b) $b^{-1}oa^{-1}$
 (c) ab (d) none of these

33. For three elements a, b, c in a group G, the values of $(abc)^{-1} =$
 (a) $a^{-1}b^{-1}c^{-1}$
 (b) $c^{-1}b^{-1}a^{-1}$
 (c) both (a) and (b) are true
 (d) none of the above

34. The identity element of the group $(\mathbf{Z}, +)$ is :
 (a) 1 (b) 0
 (c) 2 (d) 3

35. If a is an element of a group G, then $(a^{-1})^{-1} =$
 (a) a^{-1} (b) a
 (c) 1 (d) none of these

36. A finite semigroup is a group if and only if :
 (a) commutativity holds
 (b) associativity holds
 (c) reversal law holds
 (d) none of these

37. The identity element of the multiplicative group $\{2^n : n \in \mathbf{Z}\}$ is :
 (a) 0 (b) 1
 (c) 1/2 (d) none of these

38. If a is an element of a group $(G, *)$ such that $a * a = e$ then $a^{-1} =$
 (a) e (b) a
 (c) $1/a$ (d) none of these

39. Which one of the following is true ?
 (a) Multiplication of integers is distributive over addition.
 (b) Addition of integers is distributive over addition.
 (c) Both (a) and (b) are true.
 (d) None of the above

40. The value of $5 \times_7 6 \times_7 8 =$
 (a) 1 (b) 2
 (c) 3 (d) 5

41. If $o(a) = o(b) = o(ab) = 2$ then group is:
 (a) abelian (b) not abelian
 (c) may be abelian (d) none of these

42. If H and K are two subgroups of a group G then which one of the following is a subgroup of G?
 (a) $H \cap K$ (b) HK
 (c) $H \cup K$ (d) none of these

43. If $(G, *)$ is a group and for all a, $b \in G$, $b^{-1} * a^{-1} * b * a = e$ then G is :
  (a) abelian   (b) not a group
  (c) not a semigroup   (d) none of these

44. If $o(a) = n$, $a \in G$ then $a^m = e$ if :
  (a) $n \mid m$ (b) $m \mid n$
  (c) $n \nmid m$ (d) none of these

45. If G is a group of even order and $a \neq e$, if $a^2 = e$ then G is :
  (a) abelian   (b) not abelian
  (c) may be abelian   (d) none of these

46. Four fourth roots of unity *i.e.*, $\{1, -1, i, -i\}$ forms a group with respect to :
  (a) addition   (b) subtraction
  (c) multiplication   (d) all are true

47. The set of rational numbers of the form $m/2^n$, m, $n \in \mathbf{Z}$ is a group under :
  (a) addition   (b) multiplication
  (c) division   (d) none of these

48. The number of generators in cyclic group of order 10 are :
  (a) 1   (b) 2
  (c) 3   (d) 4

49. If b and c are the inverse of some element $a \in G$, then :
  (a) $b = c$ (b) $b \neq c$
  (c) $b = mc$, for some m (d) none of these

50. If G is *a* finite group then for every $a \in G$, the order of a is :

  (a) finite   (b) zero
  (c) infinite   (d) none of these

51. In the additive group of integers, the order of identity element is :
  (a) 0   (b) 1
  (c) 2   (d) none of these

52. If order of group G is p^2, p is prime, then G is:
  (a) abelian   (b) not abelian
  (c) may be abelian   (d) none of these

53. The inverse of an odd permutation is :
  (a) odd permutation   (b) even permutation
  (c) may be even   (d) none of these

54. The inverse of an even permutation is
  (a) odd permutation
  (b) even permutation
  (c) may be even permutation
  (d) none of the above

55. The identity permutation is:
  (a) even permutation (b) odd permutation
  (c) either even or odd (d) none of these

56. Which one of the following is not true ?
  (a) The product of two even permutations is even permutataion.
  (b) The product of two odd permutations is even permutataion.
  (c) The product of one even and other odd permutation is odd permutation.
  (d) all are true.

═══════════════════ **Answers** ═══════════════════

Fill in the Blanks

  1. $ab = ba \ \forall \ a, b \in G$ **2.** order **3.** unique **4.** unique **5.** multiplication **6.** addition **7.** multiplication
  8. permutaion **9.** even **10.** complex **11.** cycle **12.** abelian **13.** $ab^{-1} \in H$ **14.** even **15.** cancellation laws
  16. right **17.** torsion free **18.** torsion or periodic **19.** finite **20.** abelian **21.** semigroup **22.** monoid
  23. $b^{-1}a^{-1}$ **24.** n **25.** algebraic structure.

True or False

  1. F   **2.** T   **3.** T   **4.** T   **5.** F   **6.** F   **7.** F   **8.** T   **9.** F   **10.** T   **11.** F
  12. F   **13.** T   **14.** T   **15.** T   **16.** T   **17.** F   **18.** F   **19.** F   **20.** F   **21.** F   **22.** F
  23. T   **24.** T   **25.** F   **26.** T   **27.** T   **28.** F   **29.** F   **30.** F

Multiple Choice Questions

  1. (a)   **2.** (d)   **3.** (c)   **4.** (a)   **5.** (b)   **6.** (b)   **7.** (a)   **8.** (a)   **9.** (b)   **10.** (b)   **11.** (a)
  12. (d) **13.** (b) **14.** (a) **15.** (a) **16.** (d) **17.** (a) **18.** (b) **19.** (a) **20.** (b) **21.** (b) **22.** (c)
  23. (a) **24.** (c) **25.** (b) **26.** (a) **27.** (c) **28.** (c) **29.** (d) **30.** (c) **31.** (a) **32.** (b) **33.** (b)
  34. (b) **35.** (b) **36.** (d) **37.** (b) **38.** (a) **39.** (a) **40.** (b) **41.** (a) **42.** (a) **43.** (a) **44.** (a)
  45. (a) **46.** (c) **47.** (a) **48.** (d) **49.** (a) **50.** (a) **51.** (b) **52.** (a) **53.** (a) **54.** (b) **55.** (a)
  56. (d)

● ● ●

3 Normal Subgroups and Homomorphism

◾ Outlines

- Normal subgroups;   • Quotient groups;   • Homomorphism;
- Fundamental theorem of homomorphism

3.1 INTRODUCTION

Let H be a subgroup of an abelian group G, then $Hx = xH$, $\forall\, x \in G$.

Sometimes, it is possible that G is a non - abelian group possesses a subgroup H such that $Hx = xH$, $x \in G$. Such subgroups whose left and right cosets coincide are called normal subgroup.

Definition. *A subgroup H of a group G is said to be a normal subgroup of G if for every $x \in G$ and for every $h \in H$, $xhx^{-1} \in H$.*

REMARKS

☞ Galois had given special attention first times, on such types of groups.

☞ A normal subgroup is also called as an invariant subgroup or a self conjugate subgroup or a normal divisor of the group.

☞ Every group G possesses at least two normal subgroups, namely G itself and the subgroup consisting of the identity element e alone, i.e., $\{e\}$. These are called improper normal subgroups of G.

☞ From above definition we just conclude that H is a normal subgroup of G if and only if $xHx^{-1} \subseteq H\ \forall\, x \in G$.

3.1.1 Simple Groups

A group $G \neq \{e\}$ is known as simple group if it has no proper normal subgroup.

3.1.2 Hamiltonian Groups

A non-abelian group, each of whose subgroup is normal, is said to be Hamiltonian group.

Solved Examples

Example 1. *Show that every subgroup of an abelian group is normal.*

Solution. Let H be any subgroup of an abelian group G, then we have

$$Hx = xH \;\forall\, x \in G \qquad\qquad (\because G \text{ is abelian}).$$

Hence, H is normal subgroup of G.

Example 2. *Show that the centre Z of group G is a normal subgroup of G.*

Solution. Recall the definition of the centre of Z, we have

$$Z = \{z \in G : xz = zx,\ x \in G\}$$

We have proved earlier that Z is a subgroup of G.

Now, to show Z is normal in G.

For all $x \in G$ and $z \in Z$, we have

$$xzx^{-1} = zxx^{-1} \qquad (\because xz = zx)$$

$$= ze = z \in Z$$

$$\Rightarrow \qquad xzx^{-1} \in Z$$

Hence, Z is normal subgroup of G.

Example 3. *If G is a group and H is a subgroup of index 2 in G, show that H is a normal subgroup of G.*

Solution. Since index of H in G is 2, there are two distinct left (right) cosets of H is G. Also we have $H = eH = He$, so H is itself a left as well as right coset of H in G.

Let $a \in G$. If $a \in H$ then

$$aH = Ha = H$$

Again for $a \in G$, $a \notin H$ the left cosets aH is different from H and likewise the right coset Ha is different from H. Since there are only two distinct left (right) cosets of H in G, the decomposition of G into left cosets with respect to H consist of H and aH for $a \in G$, $a \notin H$, therefore

$$G = H \cup aH$$

Also, decomposing G into right cosets with respect to H, we have

$$G = H \cup Ha$$

Now the left cosets H and aH as well as the right cosets H and Ha have no element in common (since they are disjoint, it follows from the above relation), we must have

$$Ha = aH$$

Now, since a is arbitrary, therefore, every left cosets of H is also a right coset of H. Hence, H is a normal subgroup of G.

Example 4. *If H is the only subgroup of order $o(H)$ in a finite group G, show that H is the normal subgroup of G.*

Solution. Let $o(H) = m$. If $x \in G$, then xHx^{-1} is also a subgroup of G.

Let $$H = \{h_1, h_2, h_3, ..., h_m\}$$

Then $$xHx^{-1} = \{xh_1x^{-1}, xh_2x^{-1}, ..., xh_mx^{-1}, \}$$

The number of distinct elements in xHx^{-1} is m, because

$$xh_ix^{-1} = xh_jx^{-1} \quad \Rightarrow h_i = h_j$$

$$\therefore \qquad o(xHx^{-1}) = o(H) = m$$

Since H is the only subgroup of G of order m, we must have

$$xHx^{-1} = H \ \forall \ x \in G$$

Hence, H is normal subgroup of G.

Theorem 1. *A subgroup H of a group G is normal iff $xHx^{-1} = H \ \forall \ x \in G$.*

Proof. Let us first suppose

$$x H x^{-1} = H \ \forall \ x \in G$$

Then $$xHx^{-1} \subseteq H, \ \forall \ x \in G$$

$$\Rightarrow \quad H \text{ is normal subgroup of } G.$$

Conversely, let H be a normal subgroup of G, then
$$xHx^{-1} \subseteq H, \forall\, h \in H \text{ and for all } x \in G$$
$\Rightarrow$ $xHx^{-1} \subseteq H, \forall\, x \in G$ ... (1)

Also, let h be any element of H, then for every $x \in G$, we have
$$h = ehe = xx^{-1}hxx^{-1} = x(x^{-1}hx)x^{-1} \in xHx^{-1}$$

Since $x^{-1}hx = x^{-1}h(x^{-1})^{-1} \in H$, $x^{-1} \in G$ and H being normal.

Therefore, $H \subset xHx^{-1}, \forall x \in G.$ ... (2)

From (1) and (2), we conclude that
$$xHx^{-1} = H.$$

Theorem 2.   *A subgroup H of a group G is a normal subgroup of G if and only if every left coset of H in G is a right coset of H in G.*

Proof.   Let us first suppose H is a normal subgroup of G. Then, by previous theorem
$$xHx^{-1} = H \;\forall\, x \in G$$
$\Rightarrow$ $(xHx^{-1})x = Hx, \;\forall\, x \in G$
$\Rightarrow$ $(xH)\,x^{-1}x = Hx, \;\forall\, x \in G$
$\Rightarrow$ $xH = Hx, \;\forall\, x \in G$
$\Rightarrow$ each left coset xH is the right coset Hx

Conversely, suppose each left coset of H in G is a right cost of H in G.
$\Rightarrow$ $xH = Hy$ for some $x, y \in G.$

Since $e \in H$, so $x = xe \in xH$

Thus, $x \in xH = Hy$

Now $x \in Hy \Rightarrow xy^{-1} \in H \Rightarrow Hx = Hy$

Therefore, we have
$$xH = Hx \;\forall x \in G$$
$\Rightarrow$ $xHx^{-1} = Hxx^{-1} \;\forall x \in G$
$\Rightarrow$ $xH^{-1} = H \;\forall x \in G$

Hence, H is a normal subgroup of G.

Theorem 3.   *A subgroup H of a group G is a normal subgroup of G if and only if the product of two right cosets of H in G in again a right coset of H in G.*

Proof.   Let H be a normal subgroup of a group G and x, y be any two elements of G.
$\Rightarrow$ Hx, Hy are any two right cosets of H in G.

we have
$$(Hx)\,(Hy) = H\,(xH)y \qquad \text{(By associativity)}$$
$$= H\,(Hx)y \qquad (\because H \text{ is normal} \Rightarrow Hx = xH)$$
$$= (HH)xy \qquad \text{(By associativity)}$$
$$= Hxy \qquad (\because HH = H)$$

Now, $x, y \in G \Rightarrow xy \in G \Rightarrow Hxy$ is a right coset of H in G, thus the product of two right cosets Hx and Hy is the right coset Hxy.

Conversely. Let H be a subgroup of G such that the product of two right cosets of H in

G is again right cosets of H in G.

i.e., $\qquad (Hx)\,(Hy) = Hxy,\ \forall x,\, y \in G.$

Let $x \in G$, $h \in H$.

Then $\qquad\qquad xhx^{-1} = (ex)\,(hx^{-1}) \in (Hx)(Hx^{-1}) \qquad\qquad [\because ex = x]$

Since $\qquad\qquad e \in H = Hxx^{-1} = He = H.$

Thus, we have $a \in Z$. Therefore, H is normal subgroup of G.

REMARK

☞ If H is a normal subgroup of G, then the formula $(Hx)\,(Hy) = Hxy\ \forall x,\, y \in G$ is highly suggestive. We shall use this binary operation to show that the collection of right cosets is a group.

Theorem 4. *The intersection of any two normal subgroups of a group G is a normal subgroup of G.*

Proof. Let H and K be any two normal subgroups of G. Since the intersection of any two subgroups of a group G is also a subgroup of G.

$\Rightarrow \quad H \cap K$ is also a subgroup of G.

Let $x \in G$ and $h \in H \cap K$, then $h \in H$ and $h \in K$.

Now, since H is a normal subgroup of G, so

$$x \in G,\ h \in H \Rightarrow xhx^{-1} \in H$$

Also, K is a normal subgroup of G, so

$$x \in G,\ h \in k \Rightarrow xhx^{-1} \in K$$

Thus, we have $\quad xhx^{-1} \in H \cap K\ \forall\, x \in G$ and for all $h \in H \cap K$.

Hence, $H \cap K$ is a normal subgroup of a G.

REMARKS

☞ This result can be extended for the arbitrary collection of normal subgroups as follows; " The intersection of an arbitrary collection of normal subgroups of a group G is itself a normal subgroup of G.

Solved Examples
Based on the following Results

▶ A subgroup H of a group G is said to be normal if any one of the following conditions is satisfied :
  (i) for every $x \in G$, $h \in H$, $xhx^{-1} \in H$
  (ii) for every $x \in G$, $xHx^{-1} \in H$
  (iii) for every $x \in G$, $h \in H$, each left coset of H in G is a right coset of H in G.
  (iv) If the product of two right cosets of H in G is again a right cosets of H in G.

Example 1. *Show that a normal subgroup is commutative with every complex.*

Solution. Let N be a normal subgroup and H be a complex of G.

Since N is normal , therefore

$$Nx = xN, x \in G$$

$\Rightarrow \quad N$ commutes with every element of the group.

Let $\qquad\qquad H = \{h_1, h_2, \dots\} \subset G.$

If we replace x by $h_1, h_2, \dots$ we see that N commutes with every element of H.

Hence, we have $\qquad NH = HN.$

Example 2. *If N is a normal subgroup of G and H is any subgroup of G, show that NH is a subgoup of G.*

Solution. Since N is a normal subgroup of G and H is any subgorup of G.

Then $\qquad\qquad\qquad\qquad NH = HN$ $\qquad\qquad\qquad\qquad$ (By previous examples)

Here, N and H are two subgroups of G such that $NH = HN$

$\Rightarrow\quad NH$ is also a subgroup of G.

Example 3. *If N and M are normal subgroups of G, then NM is also a normal subgroup of G.*

Solution. By example (1), we have NM is a subgroup of G.

Let $x \in G$ and $n \in N$, $m \in M$.

Now $\qquad\qquad x(nm)x^{-1} = x(nx^{-1}xm)^{-1} = (xnx^{-1})(xmx^{-1}) \in NM.$

Since, N is normal in G, $xnx^{-1} \in N$ and since M is normal in G, $xmx^{-1} \in M$

Thus, NM is a subgroup of G and

$$x(nm)x^{-1} \in NM, \forall\, x \in G \text{ for all } nm \in NM$$

$\Rightarrow NM$ in a normal subgroup of G.

Example 4. *Let G be a group of order 2p, where p is prime. Show that G has a normal subgroup of order p.*

Solution. Here, we have

$O(G) = 2p$, where p is prime.

Now, since G is a finite group of composite order, so it possesses proper subgroup, say H of G.

Then by Lagrange's theorem, we have $\dfrac{o(G)}{o(H)}$

Therefore, either $o(H) = 2$ or $o(H) = p$

If $o(H) = p$, then index of H in $G = [G : H] = \dfrac{o(G)}{o(H)} = \dfrac{2p}{p} = 2$

$\Rightarrow\quad H$ is a subgroup of index 2 in G.

Hence, H is a normal subgroup of G.

EXERCISE 3.1

1. If H is a normal subgroup of G and K is a subgroup of G such that $H \subseteq K \subseteq G$, show that H is a normal subgroup of K.

2. If H is a subgroup of G and N is a normal subgroup of $G.$, show that $H \cap N$ is a normal subgroup of H.

3. If N and M are two normal subgroups of G such that $N \cap M = \{e\}$, show that every element of N commutes with every element of M.

4. If H is the only subgroup of finite order in a group G, then show that H is a normal subgroup of G.

5. If a cyclic subgroup N of G is normal in G, then show that every subgroup of N is normal in G.

6. Show that a subgroup H of a group G is normal if and only if the set G/H of all its left cosets is closed under multiplication.

7. If H is a subgroup of G and N is a normal subgroup of G, then show that
   (i) HN is a subgroup of G.
   (ii) N is a normal subgroup of HN.

8. If a cyclic subgroup of a group G is normal in G, then show that every subgroup of G is normal in G.

3.2 CONJUGATE ELEMENT

Let G be a group. An element $a \in G$ is called conjugate to an element $b \in G$ if and only if $a = x^{-1} b x$, for some $x \in G$.

If $a = x^{-1} b x$, then somtimes, we say that a is transform of b by x.

REMARKS

☞ Here, the element x is not unique.

☞ If a conjugate to b then we write $a \sim b$ and this relation is known as relation of conjugacy.

☞ The relation of conjugacy is an equivalence relation.

3.3 NORMALIZER OF AN ELEMENT

Definition. *The normalizer of an element $a \in G$ is the set of all those element of G which commutes with a and is denoted by $N(a)$, i.e. $N(a) = \{x \in G : ax = xa\}$.*

REMARKS

☞ $N(a)$ is a subgroup of G.

☞ In general, $N(a)$ is not a normal subgroup of G.

☞ $N(e) = G$ $\qquad\qquad$ [∵ $ex = xe$, $\forall x \in G$]

☞ $N(e) = G$ iff G is abelian.

3.3.1 Centralizer

Let A be a non-empty subset of a group G. The centralizer $C(A)$ of A, in G is defined as $C(A) = \{x \in G : ax = xa, \forall a \in A\}$

REMARKS

☞ $C(A)$ is a subgroup of G.

☞ $C(A) \subset N(A)$.

☞ The abelian part of a group is defined as the centre of the group.

3.3.2 Commutator

Let G be a group and x, $y \in G$. The element $x^{-1} y^{-1} xy$ is called the commutator of x and y taken in this order.

REMARKS

☞ The inverse of a commutator is a commutator.

☞ The product of two commutators is not necessarily a commutator.

3.3.3 Conjugate Class

Since an equivalence relation defined on a set decomposes the set into mutually disjoint equivalence classes, hence the relation of conjugacy, defined on a group G decomposes G into mutually disjoint equivalence classes known as classes of conjugate element.

Let C_a denote the equivalence class determined by an element $a \in G$. then

$$C_a = \{x \in G : x \approx a\} = \{x \in G : x = y^{-1} ay \text{ for some } y \in G\}$$

$$= \{y^{-1}ay : y \in G\}$$

Here, C_a is defined as conjugate class of a in G. Also C_a is the class element conjugate to a.

3.3.4 Self Conjugate element

An element a of a group G is called self conjugate element iff $C_a = \{a\}$

i.e., iff C_a consists of the single element a.

i.e., iff $\qquad\qquad\qquad x^{-1}\,ax = a, \forall\, x \in G.$

$\Rightarrow$ iff $\qquad\qquad\qquad ax = xa, \forall\, x \in G$

REMARKS

☞ A self conjugate element is one which commutes with every element of the group.

☞ The transform of a self conjugate element remains the same.

☞ Self conjugate element is also called invariant element.

3.3.5 Conjugate subgroups

Transformation of an element by another element gives us a conjugate element. Similarly, transformation of a subgroup by another subgroup gives a conjugate to A if there exists an element $x \in G$ such that $B = x^{-1}Ax$. The symbol "$B \subseteq A$" is read as B is conjugate to A.

Theorem 1. *The normalizer $N(a)$ of $a \in G$ is a subgroup of G.*

Proof. By definition, we have

$$N(a) = \{x \in G : ax = xa\}$$

Let $x_1, x_2 \in N(a)$, then $\qquad ax_1 = x_1a,\; ax_2 = x_2a$

To show $x_1x_2^{-1} \in N(a)$.

Firstly we shall show that $\quad x_2^{-1} \in N(a)$

Consider

$$ax_2 = x_2a$$

$\Rightarrow \qquad\qquad x_2^{-1}(ax_2)x_2^{-1} = x_2^{-1}(x_2a)x_2^{-1}$

$\Rightarrow \qquad\qquad x_2^{-1}a = ax_2^{-1} \Rightarrow x_2^{-1} \in N(a).$

Now to show $\qquad x_1x_2^{-1} \in N(a)$.

Consider

$$a(x_1x_2^{-1}) = (ax_1)x_2^{-1} = (x_1a)x_2^{-1} = x_1(ax_2^{-1})$$

$$= x_1(x_2^{-1}a) = (x_1x_2^{-1})a$$

$$x_1x_2^{-1} \in N(a)$$

$\therefore$

i.e., $\qquad\qquad x_1, x_2 \in N(a) \Rightarrow x_1x_2^{-1} \in N(a)$

Hence, $N(a)$ is a subgroup of G.

Theorem 2. *Let G be a finite group and $a \in G$. Let N(a) denotes the normalizer of a and G and C_a the conjugate class of a in G. Then* $o(C_a) = \dfrac{o(G)}{o(N(a))}$ *i.e., number of distinct elements conjugate to a in G is the index of the normalizer of a in G.*

Proof. Let G be a finite group and $a \in G$.

To show $o(C_a) = \dfrac{o(G)}{o(N(a))}$.

By definition, we have

$$N(a) = \{x \in G; \ ax = xa\} \quad \text{and} \quad C_a = \{x^{-1} ax : x \in G\}$$

Write $M = \{N(a)x : x \in G\}$

$= $ set of right cosets of $N(a)$ in G

Define a map $f : M \to C_a$ such that $f[N(a). x] = x^{-1} ax \ \forall \ x \in G$. Obviously, f is well defined.

To show f is one-one and onto.

(i) f is one-one.

Consider $f[N(a).x] = f[N(a).y]; x, y \in G$

$\Rightarrow \qquad x^{-1}ax = y^{-1}ay$

$\Rightarrow \qquad x(x^{-1}ax)y^{-1} = x(y^{-1}ay)y^{-1}$

$\Rightarrow \qquad (xx^{-1})(axy^{-1}) = (xy^{-1}a)(yy^{-1})$

$\Rightarrow \qquad e(axy^{-1}) = (xy^{-1}a)e$

$\Rightarrow \qquad axy^{-1} = xy^{-1}a \Rightarrow \qquad a(xy^{-1}) = (xy^{-1})a$

Therefore, $\qquad xy^{-1} \in N(a) \Rightarrow \qquad a(xy^{-1}) = (xy^{-1})a$

$\Rightarrow \qquad xy^{-1} \in N(a) \Rightarrow \qquad N(a)xy^{-1} = N(a)$

$\Rightarrow \qquad N(a).x = N(a).y$

$\Rightarrow \qquad$ f is one-one.

(ii) f is onto.

For given any $x^{-1} ax \in C_a \ \exists N(a). x \in M$ such that

$$f[N(a).x] = x^{-1} ax$$

$\Rightarrow$ f is onto.

Hence $o(C_a) = o(M) = $ number of distinct right cosets of $N(a)$ in G.

$= $ index of $N(a)$ in G

$= \dfrac{o(G)}{o(N(a))}$ by definition of index.

Theorem 3. *If G is a finite group, then* $o(G) = \sum \dfrac{o(G)}{o(N(a))}$ *where this sum is taken over one element of each conjugate class.*

Proof. Since, we know that, the relation of conjugacy is an equivalence relation on the set G.

$\Rightarrow$ This relation partitions the group G into disjoint equivalence classes.

Let C_a, C_b, C_c respectively denote conjugate classes of elements a, b, $c \in G$.

Also let

$$o(C_a) = C_a, \; o(C_b) = C_b, \; o(C_c) = C_c.$$

Then
$$G = C_a \cup C_b \cup C_c \; ....$$

$\Rightarrow$
$$o(G) = o(C_a) + o(C_b) + o(C_c) + ...$$

$$(\because C_a, C_b, C_c \text{ are mutually disjoint.})$$

$$= \sum C_a,$$

where the sum runs over each elements of conjugate class.

But
$$C_a = \frac{o(G)}{o(N(a))}$$
[By theorem 2]

$\therefore$
$$o(G) = \sum \frac{o(G)}{oN(a)}$$
... (1)

REMARK

☞ The above equation (1) is known as class equation of G.

Theorem 4. *Let Z denote the centre of a group G. Then show that $a \in Z$ iff $N(a) = G$. Also, show that if G is finite, then $a \in Z$ iff $o(N(a)) = o(G)$.*

Proof. Let Z denote the centre of a group G.

Let us first suppose $a \in Z$. To show $N(a) = G$.

Since, we know that $N(a)$ is a subgroup of G and

$$N(a) = \{x \in G : ax = xa\}$$
$$Z = \{x \in G : zx = xz \; \forall \, x \in G\}$$

Let $y \in G$ and $a \in Z$, then

$\Rightarrow$ $ay = ya$ (By definition of Z)

$\Rightarrow$ $y \in N(a)$ $\Rightarrow G \subset N(a)$

But $N(a) \subset G$ $\Rightarrow G = N(a)$

Conversely, let $G = N(a)$, where $a \in G$

To show that $a \in G$.

         $N(a) = G$ $\Rightarrow G \subset N(a)$

$\therefore$ $y \in G \Rightarrow y \in N(a)$ $\Rightarrow ay = ya$

i.e., $ay = ya, \, y \in G$ $\Rightarrow a \in Z$

Now, let G be finite. To show $a \in Z \Leftrightarrow o(G) = o(N(a))$.

We have $a \in Z \Leftrightarrow N(a) = G$

Also, G is finite $\Leftrightarrow$ $o(N(a)) = o(G)$

Theorem 5. *If G is a finite group and Z is its centre, then class equation of G is expressible as*

$$o(G) = o(Z) + \sum_{a \notin Z} \left\{ \frac{o(G)}{o(N(a))} \right\}$$

where the summation runs over one element a in each conjugate class.

Proof. Let a be an element of a conjugate class. The class equation of G is

$$o(G) = \sum \frac{o(G)}{oN(a)} \qquad \text{... (1)}$$

We know that

$$a \in Z \Leftrightarrow o(N(a)) = o(G) \Rightarrow \frac{o(G)}{o(N(a))} = 1$$

or $\qquad a \in Z \Rightarrow \dfrac{o(G)}{o(N(a))} = 1$

Now (1) becomes $\qquad o(G) = o(Z) + \sum_{a \notin Z} \left\{ \dfrac{o(G)}{o(N(a))} \right\}$

Theorem 6. *If G is group of order p^n, then its centre $Z \neq \{e\}$, p being a prime number.*

Proof. Let p be a prime number and G be a group of order p^n.

To show $\qquad\qquad Z \neq \{e\}$, *i.e.,* $o(Z) > 1$.

Let $a \in G$.

p is prime $\Rightarrow |p| > 1$ and ± 1 and $\pm p$ are only divisors of p.

By Lagrange's theorem, $o(N(a))$ is a divisor of $o(G) = p^n$, so we can write

$o(N(a)) = p^m$, $m \in \mathbf{Z}^+$ and $m \leq n$

By class equation of G, we have

$$o(G) = \sum \frac{o(G)}{o(N(a))}$$

i.e., $\qquad\qquad P^n = \sum_{m \leq n} \dfrac{p^n}{p^m} \qquad \text{... (1)}$

Next, suppose that $\qquad o(Z) = q \qquad \text{...(2)}$

We know that $\qquad a \in Z \Leftrightarrow o(N(a)) = o(G)$

i.e., $\qquad a \in Z \Leftrightarrow p^m = p^n \Leftrightarrow n = m$

$\Rightarrow \exists\, q$ elements in Z, *i.e.,* in G such that

$$n = m, i.e., \frac{p^n}{p^m} = 1$$

Now (1) becomes

$$p^n = q + \sum_{m < n} \left(\frac{p^n}{p^m} \right)$$

or $\qquad\qquad q = p^n - \sum_{m<n} \left(\dfrac{p^n}{p^m} \right) \qquad \text{... (3)}$

$\Rightarrow p$ is a divisor of R.H.S. of (3). Hence, p is a divisor of q and so there exists positive integer k such that $\left(\dfrac{p}{q}\right) = k$ or $q = kp$.

But $p > 1$ $q > 1$. Hence, $o(Z) > 1$.

Theorem 7. *A group of order p^2 is abelian, where p is prime number.*

Proof. Let G be a group of order p^2. We wish to show that G is abelian. For this, we shall show that $Z = G$. (Then G will be abelian, since the centre Z is abelian).

Since p is prime, then by theorem (6) $Z \neq \{e\}$. Therefore, $o(Z) > 1$. Again since Z is a subgroup of G. Then, by Lagrange's theorem $o(Z) > 1$. Again since Z is a subgroup of G. Then, by Lagrange's theorem $o(Z)\,|\,o(G)$, i.e., $o(Z)\,|\,p^2$

$\Rightarrow$ either $o(Z) = p$ or $o(Z) = p^2$

if $o(Z) = p^2$, then $Z = G$ and the result follows.

Now, suppose $o(Z) = p$, so $o(Z) < o(G)$. Then there must exist an element which is in G but is not in Z. Let $a \in G$ and $a \notin Z$. Since a commutes with itself, $a \in N(a)$. Also, $N(a)$ is a subgroup of G. Let $x \in Z$ so $xa = ax \Rightarrow x$ commutes with $a \Rightarrow x \in N(a)$

Thus $x \in Z \Rightarrow x \in N(a) \Rightarrow Z \subseteq N(a)$

Now $a \notin Z$, $a \in N(a)$, $Z \subseteq N(a)$
$\Rightarrow o(N(a)) > o(Z)$
$\Rightarrow o(N(a)) > p$
$\Rightarrow o(N(a)) > p^2$ ($\because$ $o(N(a))\,/p^2$)
$\Rightarrow N(a) = G \Rightarrow a \in Z$

which is a contradiction.

Thus $o(Z) = p$ is not possible

$\Rightarrow$ $o(Z) = p^2$

Hence $Z = G$

$\Rightarrow$ G is abelian.

Application. A group of order 9 is abelian.

For, let $p = 3 \Rightarrow p^2 = 3^2 = 9$

$\because$ $o(G) = p^2$, is prime $\Rightarrow G$ is abelian

$\therefore$ $o(G) = 3^2 = 9 \Rightarrow G$ is abelian

EXERCISE 3.2

1. Show that if p is prime, then any group G of order $2p$ has a normal subgroup of order $2p$.

2. If an abelian group is simple, show that order of G is prime

3. Show that two elements are conjugate iff they can be put in the form xy and yx respectively, where x and y are suitable elements of G.

4. Show that the normalizer $N(A)$ of a subgroup A of group G is a subgroup of G.

5. If a cyclic subgroup N of G is normal in G., then show that every subgoup of N is normal in G.

6. Show that the group of order 121 is abelian.

3.4 QUOTIENT GROUP

Let H be a normal subgroup of G.

Denote the set of all right cosets of H in G by G/H, i.e., $G/H = \{H\,a: a \in G\}$.

Let $Ha, Hb \in G/H$. Now define the operation, multiplication on G/H as follows:
$$Ha. Hb = Hab$$
It can be easily shown that G/H is a group with respect to this operation.

This group G/H is called quotient group.

Definition. *If G is a group and H is a normal subgroup of G. then the set G/H of all cosets of H in G is a group with respect to multiplication of cosets, is called quotient group or factor group of G by H.*

REMARK

☞ The identity elements of the quotient group G/H is H.

Theorem 1. *The set of all cosets of a normal subgroup is a group w.r.t. multiplication of complexes as compositions.*

Proof. Let H be a normal subgroup of G, so that
$$Hx = xH, \forall\, x \in G$$
Write $\qquad\qquad G/H = \{Hx : x \in G\}$

Let e be the identity of G. It is obvious that
$$(Ha)\,(Hb) = Hab \qquad\qquad ...(1)$$

Now to show G/H is a group w.r.t. multiplication of complexes as composition.

Let $a, b, c \in G$

(i) **Closure property.** Let $Ha, Hb \in G/H$
$$\Rightarrow \qquad\qquad a, b \in G \Rightarrow a.b \in G$$
$$\Rightarrow \qquad\qquad Hab \in G/H \Rightarrow Ha \cdot Hb \in G/H$$

(ii) **Associativity.** Since $(ab)c = a(bc)$ [By associativity in G]
$$\therefore \qquad\qquad H\,(ab)c = H\,a(bc)$$
using (1),
$$H(ab).Hc = Ha. H\,(bc)$$
or $\qquad (HaHb)Hc = Ha.(Hb.Hc)$

(iii) **Existence of Identity.**
For $ae = ea = a$ when $Hae = Hea = Ha$
Also, $\qquad\qquad Ha. He = He. Ha = Ha$
$$\Rightarrow \qquad\qquad He = H \in G/H, \text{ is the identity element.}$$

(iv) **Existence of inverse.**
We have $a \in G \Rightarrow a^{-1} \in G \Rightarrow Ha^{-1} \in G/H$ and $aa^{-1} = a^{-1}a = e$
$$\therefore \qquad\qquad H(aa^{-1}) = H(a^{-1}a) = He = H$$
Using (1),
$$Ha. Ha^{-1} = Ha^{-1}. Ha = H$$
$$\Rightarrow \qquad\qquad (Ha)^{-1} = Ha^{-1} \in G/H$$

Hence, G/H is group with respect to the operation of multiplication of complexes.

Theorem 2. *A subgroup H of a group G is normal iff the set G/H of all its left cosets is closed under multiplication.*

Proof. Let H be a subgroup of a group G.

Define $\qquad\qquad\qquad$ $G/H = \{aH: a \in G\}$

Let us first suppose H be normal in G so that

$$Ha = aH \ \forall a \in G. \qquad\qquad ... (1)$$

To show G/H is closed under multiplication.

Let $aH, bH \in G/H$ so that $\quad a, b \in G$

$a, b \in G \qquad\qquad \Rightarrow \qquad\qquad ab \in G$

$\qquad\qquad\qquad \Rightarrow \qquad\qquad (ab)H \in G/H \qquad\qquad\qquad ... (2)$

Consider $\quad (aH)(bH) = a(Hb)H = a(bH)H$

$\qquad\qquad\qquad = (ab)(HH) = (ab)(H) \qquad\qquad (\because HH = H)$

$\therefore \qquad\qquad (aH)(bH) = abH \in G/H \qquad\qquad\qquad\qquad (\text{By } (2))$

$\Rightarrow \qquad\qquad (aH)(bH) \in G/H \Rightarrow G/H$ is closed

Conversely, let G/H be closed under multiplication of left cosets of H. To show H is normal in G.

Let $a \in G$, then aH and $a^{-1}H$ both are left cosets of H in G.

Also, G/H is closed under multiplication.

$\Rightarrow \qquad\qquad (aH)(a^{-1}H) \in G/H$

Now, Since H is a subgroup of $G \Rightarrow e \in H$

$\Rightarrow \qquad\qquad (ae)(a^{-1}e) = aa^{-1} = e$ is an element of $(aH)(a^{-1}H)$.

Since, we know that any two left cosets are either identical or disjoint.

Consequently, $\qquad\qquad H = (aH)(a^{-1}H), \ \forall a \in G$

$\Rightarrow \qquad\qquad (ah)(a^{-1}h_1) \in (aH)(a^{-1}H) = H$

$\Rightarrow \qquad\qquad (aha^{-1})h_1 \in H.h, \ h_1 \in H$

$\Rightarrow \qquad\qquad (aha^{-1}h_1)h_1^{-1} \in Hh_1^{-1} = H \qquad\qquad (\text{For } h_1 \in H \Rightarrow Hh_1 = H = Hh_1^{-1})$

$\Rightarrow \qquad\qquad aha^1 \in H, \text{ for } h_1h_1^{-1} = e$

Thus $aha^1 \in H \ \forall h \in H$ and $a \in G$

Hence, H is normal in G.

Theorem 3. *Every quotient group of a cyclic group is cyclic, but converse is not true.*

Proof. $\quad$ Let H be a subgroup of a cyclic group so that H is also cyclic. To show G/H is cyclic.

Let a be the generator of G. Also, let m be the least positive integer such that $a^m \in H$, then a^m is a generator of H.

Since H is cyclic, therefore, H is an abelian subgroup of G.

$\Rightarrow H$ is a normal subgroup of G.

$\Rightarrow G/H$ will contain the elements of the form Ha^n, where $a^n \in G$.

Also, $\qquad\qquad (Ha)^2 = Ha.Ha = Haa = Ha^2 \in G/H$.

In general $\qquad (Ha)^n = Ha^n \in G/H$.

Thus the general element Ha^n of G/H is expressible as $(Ha)^n$.

$\Rightarrow \quad G/H$ is cyclic with generator Ha.

Converse of the above theorem is not true. For example, P_3/A_3 is cyclic, while P_3 is not cyclic. P_3 and A_3 being symmetric group and alternating group of permutation of degree 3.

Theorem 4. *Every quotient group of an abelian group G is abelian but converse is not true.*

Proof. Let H be a subgroup of an abelian group G and hence H is a normal subgroup of G.

To show G/H is abelian.

Let $Ha, Hb \in G/H$, then

$$(Ha)(Hb) = Hab = Hba$$
$$= (Hb)(Ha)$$
$$\Rightarrow \qquad (Ha)(Hb) = (Hb)(Ha)$$
$$\Rightarrow \qquad G/H \text{ is abelian.}$$

Theorem 5. *Let H be a normal subgroup of a finite group G. Then $o(G/H) = \dfrac{o(G)}{o(H)}$, i.e., $o(G/H)$ is the index of H in G.*

Proof. By definition

$$o(G/H) = \text{number of distinct right (or left) cosets of } H \text{ in } G$$
$$= \frac{\text{Number of distinct element in } G}{\text{Number of distinct element in } H}$$
$$= \frac{o(G)}{o(H)}.$$

Solved Examples

Example 1. *If $G = \{a\}$ is a cyclic group of order 8, then find the quotient groups corresponding to the subgroup generated by a^2 and a^4 respectively.*

Solution. Let $\qquad G = \{a, a^2, a^3, a^4, a^5, a^6, a^7, a^8 = e\}$

and $\qquad H_1 = \{a^2, a^4, a^6, a^8 = e\}$

$\qquad H_2 = \{a^4, a^8 = e\}$

Since, G is abelian, therefore the subgroup H_1 and H_2 are normal in G.

$$o(G/H_1) = \frac{8}{4} = 2, O(G/H) = \frac{8}{2} = 4$$

$$G/H_1 = \{H_1, H_1a\}, \text{where } H_1a = \{a^3, a^5, a^7, a\}$$

$$H_1a^3 = H_1a, H_1a^2 = H_1a^4 = H_1a^6 = H_1a^8 = H_1, \text{ etc.}$$

$$G/H_2 = \{H_2, H_2a, H_2a^2, H_2a^3\}$$

EXERCISE 3.3

1. Let N_1 and N_2 be two normal subgroups of G, then show that $G/N_1 = G/N_2$ if and only if $N_1 = N_2$.

2. If P is a prime number and G is a non-abelian group of order p^3, show that centre Z has exactly p elements.

3. If N is a normal subgroup of a group G and $a \in G$ such that $o(a) = n$, then show that Na in G/H is a divisor of n.

4. Let a be an arbitrary element of a group G, show that the element $x, y \in G$ give rise to the same conjugate of a iff they belong to the same right coset of the normalizer of a in G.

3.5 HOMOMORPHISM OF A GROUP

Let $G = \{a, b, c, ...\}$ be a group with respect to the binary operation $\circ$ and

$G' = \{a', b', c'\}$ be another group with respect to $*$.

A mapping f of G onto G' to be homomorphism if for all $a, b \in G$

$f(a \circ b) = f(a) * f(b)$

REMARKS

☞ The word 'homomorphism' is derived from the two Greek words 'homos' meaning 'link' and 'morphic' meaning 'form'.

☞ A homomorphism f from G and G' carries the product $x \circ y$ in G to the product $f(x) * f(y)$ in G'.

☞ $f(G)$, if exists, is called homomorphic image of G.

☞ The relation of homomorphism is expressed by writting $G = G'$. If G is homomorphic to G', then there may exist more than one homomorphism of G and G'.

Some More Definitions

(i) **Epimorphism.** A homomorphism f which is also onto is called an epimorphism.

(ii) **Monomorphism.** A homomorphism f which is also one-one is called a monomorphism.

(iii) **Endomorphism.** A homomorphism of a group into itself is called an endomorphism.

(iv) **Natural homomorphism.** Let H be a normal subgroup of group G then the map $f: G \to G/H$ such that $f(x) = Hx$ is a homomorphism. This homomorphism is called natural or canonical homomorphism.

For example

(1) If $f(x) = e, \forall x \in G$, this is homomorphism.

(2) If G be a group of all real numbers under addition, i.e, $a * b$ for $a, b \in G$ is the real number $a+b$ and G' be the multiplicative group of non-zero real numbers and

$$f: G \to G' \text{ defined by } f(a) = 2^a \text{ is a homomorphism}$$
$$[\because f(a+b) = 2^{a+b} = 2^a \cdot 2^b = f(a) \cdot f(b)]$$

(3) If G be the group of integers under addition and $G' = G$ and let $f(x) = 2x \; \forall x \in G$, then

$$f(x+y) = 2(x+y) = 2x+2y = f(x)+f(y)$$

$\Rightarrow f$ is a homomorphism.

(4) If G is the group of integers under addition and G'_n is the group of integers under addition modulo n and let f be defined as $f(x) = $ remainder of x on division by n, then it is a homomorphism.

(5) Let G be a group and let e be the identity element of G. Then the mapping $f: G \to G$ defined by $f(a) = e, \forall a \in G$ is an endomorphism of G.

3.5.1 Kernel of a Homomorphism.

Let f be a homomorphism of a group G into a group G', then the set K of all those elements of G, which are mapped by f onto the identity e' of G' is called the kernel of a homomorphism.

3.5.2 Isomorphism

A mapping f from a group (G, o) into $(G', *)$ is called an isomorphism if f is one-one and

onto and
$$f(a \circ b) = f(a) \circ f(b), \forall a, b \in G$$

WORKING PROCEDURE

Define a function $f : G \to G'$. To show ϕ is isomorphism, use the following steps:

Step 1. *show that f is one-one.*

Step 2. *show that f is onto.*

Step 3. *show that it satisfy structure preserving property.*

REMARKS

☞ The word 'isomorphism' is derived from the Greek word 'iso' meaning 'equal'.

☞ An isomorphism of a group G onto itself is called an automorphism of G.

3.5.3 Inner Automorphism

Let a be a fixed element of a group G. Then the map
$$f_a : G \to G \text{ such that } f_a(x) = a^{-1} xa \ \forall \ x \in G$$
is an automorphism of G, known as inner automorphism.

3.5.4 Automorphism

An isomorphic mapping of a group G onto itself is called an automorphism of G.

Therefore $f : G \to G$ is an automorphism if

(i) f is one-one

(ii) f is onto

(iii) $f(ab) = f(a) . f(b) \ \forall a, b \in G$

REMARK

☞ An automorphism, which is not inner is called outer automorphism.

<u>**Theorem 1.**</u> If $f : G \to G'$ be a homomorphism of group and e, e' be the identities in G and G' respectively. Then

(i) $f(e) = e'$

(ii) $f(a^{-1}) = [f(a)]^{-1}$ where $a \in G$.

(iii) If the order of an element $x \in G$ is finite, then the order of $f(x)$ is a divisor of the order of x.

Proof.   (i) We have
$$f(a)e' = f(a) = f(ae) \qquad\qquad (\because a = ae)$$
$$= f(a) . f(e)$$
$$\Rightarrow \qquad\qquad e' = f(e)$$

(ii) If $a \in G$, then
$$aa^{-1} = a^{-1} . a = e$$
Thus
$$f(a^{-1}) f(a) = f(a^{-1}a) = f(e) = e'$$
and
$$f(a) f(a^{-1}) = f(aa^{-1}) = f(e) = e'$$
$$f(a^{-1}) = [f(a)]^{-1}$$

(iii) Let $x \in G$ and $x^m = e$. Then
$$f(x^m) = f(e) \Rightarrow f(x.x.x \ ... \ m \text{ times}) = e'$$
$$\Rightarrow \qquad (f(x). f(x) \ ... \ m \text{ times}) = e'$$
$$\Rightarrow \qquad (f(x). f(x) \ ... \ m \text{ times}) = e'$$

$$\Rightarrow \qquad [f(x)^m] = e'$$

$$\Rightarrow \qquad o(f(x)\,|\,m \qquad\qquad (\because x^m = e \Leftrightarrow o(x)/m)$$

$$\Rightarrow \qquad o(f(x))\,|\,o(x)$$

Thus, the order of $f(x)$ is divisor of the order of x.

Theorem 2. *If $f : G \to G'$ is an isomormhism of groups, then the order of an element $a \in G$ is equal to order of the f-image of a, i.e., $o(a) = o[f(a)]$*

Proof.   Let $f : G \to G'$ be an isomorphism. Let $a \in G$ be arbitrary such that

$$o(a) = n \text{ and } o[f(a)] = m$$

To show that $\qquad o(a) = [f(a)], \text{ ie., } n = m.$

Let e and e' be identities in G and G' respectively, then we have $e' = f(e)$.

Now, $\qquad\qquad o(a) = n \Rightarrow a^n = e \Rightarrow f(a^n) = f(e) = e'$

$$\Rightarrow \qquad\qquad f(a.a \ldots n \text{ times}) = e'$$

$$\Rightarrow \qquad\qquad (f(a).f(a) \ldots n \text{ times}) = e'$$

$$\Rightarrow \qquad\qquad [f(a)]^n = e' \Rightarrow 0\,[f(a)] \le n \quad (\because a^n = e \Rightarrow o(a) \le n)$$

$$\Rightarrow \qquad\qquad m \le n$$

Also, $\qquad\qquad o[f(a)] = m \Rightarrow [f(a)]^m = e'$

$$\Rightarrow \qquad\qquad (f(a).f(a) \ldots m \text{ times}) = e'$$

$$\Rightarrow \qquad\qquad f(a.a \ldots m \text{ times}) = e'$$

$$\Rightarrow \qquad\qquad f(a^m) = e' = f(e)$$

$$\Rightarrow \qquad\qquad a^m = e \qquad\qquad\qquad (\because f \text{ is one-one.})$$

$$\Rightarrow \qquad\qquad o(a) \le m \qquad\qquad\qquad \ldots (2)$$

$$\Rightarrow \qquad\qquad n \le m$$

From (1) and (2), we conclude that $n = m$.

Theorem 3. *The relation of isomorphism in the set of all groups is an equivalence relation.*

Proof.   Let G, G', G'' be three groups. Then

(i) Reflexive. Consider the map $f : G \to G$ given by $f(x) = x \ \forall \ x \in G$.

Evidently f is one-one and onto

Also, $\qquad\qquad f(xy) = xy, \ \forall x, y \in G.$

$$= f(x)\,f(y)$$

$\therefore f$ preserves the composition in G.

$$\Rightarrow \qquad\qquad G \cong G.$$

$$\Rightarrow \qquad\qquad \cong \text{ is Reflexive.}$$

(ii) Symmetric. $\qquad G \cong G' \Rightarrow \ G' \cong G$

For $\ G \cong G' \Rightarrow \exists$ an isomorphism $f : G \xrightarrow{\text{onto}} G'$

$$\Rightarrow \qquad f \text{ is one-one onto and } f(xy) = f(x)\,f(y) \ \forall \ x, y, \in G$$

$$\Rightarrow \qquad f^{-1} : G' \to G \text{ is one-one and onto.}$$

Let f^{-1} peserves compositions in G and G'.

Also, we have f^{-1} is one-one and onto.

$$\therefore \qquad f^{-1} : G' \to G \text{ is an isomorphism.}$$

$$\Rightarrow \qquad\qquad G' \cong G. \Rightarrow \ \cong \text{ is symmetric.}$$

(iii) Transitive. For this, we shall prove that

$$G \cong G', \ G' \cong G'' \Rightarrow G \cong G''.$$

For $G \cong G'$, $G \cong G'' \; \exists$ isomorphisms $f : G \xrightarrow{\text{onto}} G'$, $g : G' \rightarrow G''$

$\Rightarrow$ f and g are one-one onto maps and $f(xy) = f(x) \cdot f(y)$; $x, y \in G$

$\qquad g(x'y') = g(x')g(y')$; $x', y' \in G'$

$\Rightarrow$ $gf : G \rightarrow G''$ is one-one and onto

and $(gf)(xy) = g[f(xy)]$; $x, y \in G$

$\qquad\qquad\quad = g[f(x)f(y)] = g[f(x)]g[f(y)]$

$\qquad\qquad\quad = (gf)(x)(gf)(y)$, *i.e.*, gf is order preserving.

$\Rightarrow$ $gf : G \rightarrow G''$ is an isomorpohism.

$\Rightarrow$ $\qquad\qquad G \cong G'' \; \Rightarrow \; \cong$ is transitive.

Thus, the relation satisfies all the conditions of an equivalence relation and hence the given relation is an equivalence relation.

REMARKS

☞ The equivalence relation decomposes the family of groups into disjoint classes such that groups of one equivalence class are isomorphic to each other and groups of different classes are not isomorphic.

☞ When two groups are isomorphic, we consider the groups of the same type from the point of view of their abstract structure, although the elements as well as the compositions might be entirely different.

Solved Examples

Based on the following Results

▶ A mapping $f : (G, o) \rightarrow (G', *)$ is said to be a homomorphism if $f(aob) = f(a) * f(b)$

▶ A mapping f from a group (G, o) and $(G', *)$ is called an isomorphism if f is one-one onto and $f(aob) = f(a)of(b) \; \forall \; a, b \in G.$

▶ An isomorphism $f : G \rightarrow G$ is called an automorphism.

Example 1. *Let G be a group and e be the identity element of G. Then show that mapping $f : G \rightarrow G$ defined by $f(a) = e$, $\forall \; a \in G$ is an endomorphism of G.*

Solution.   Let a and b be any two elements of G then

$\qquad\qquad f(a) = e = f(b)$

Now $f(ab) = e = ee = f(a) \cdot f(b)$

$\Rightarrow$ f is a homomorphism of G into G. Therefore, f is an endomorphism of G.

Example 2. *Show that the multiplicative group $\{1, -1, i, -i\}$ is isomorphic to the group of residue classes modulo 4 under addition of residue classes.*

Solution.   Let $G = \{1, -1, i, -i\}$ and $G' = \{[0], [1], [2], [3]\}$

To show that $(G', *) \cong (G, +)$, where f denote a map

$\qquad\qquad f : G \rightarrow G'$ such that $o(a) = o[f(a)]$, $\forall \; a \in G$

If we show that f is an isomorphism, then result will follow, 1 and [0] are identities in G and G' respectively. Also, order of identiy element in every group is one.

Hence, $o(1) = 1 = o([0])$

For elements of $G : o(a) = n$ $\qquad \Rightarrow \qquad a^n = e = 1$

$$(-1)^1 = -1, (-1)^2 = 1 = e \quad \Rightarrow o(-1) = 2$$
$$(i)^4 = 1 = e \qquad\qquad \Rightarrow \quad o(i) = 4$$
$$(-i)^4 = 1 = e \qquad\qquad \Rightarrow o(-i) = 4$$

Now for element of G', $o([1]) = n \Rightarrow n[a] = [0]$

$$1[1] = [1], 2[1] = [2], 3[1] = [3], 4[1] = [0] = e \Rightarrow o[(1)] = 4$$

Similarly,

$$o([2]) = 2 \text{ and } o([3]) = 4$$

Therefore, order of elements $1, -1, i, -i \in G$ are respectively 1, 2, 4, 4 and order of elements $[0], [1], [2], [3] \in G'$ are respectively given by 1, 4, 2, 4. Clearly

$$o(1) = 1 = o([1]), o(-1) = 2 = o([2])$$
$$o(i) = 4 = o([1]), \ o(-i) = 4 = o([3])$$

This implies

$$f(1) = [0], f(-1) = [2], f(i) = [1], f(-i) = [3]$$

$\Rightarrow \quad f$ is one-one onto.

Also, $\quad f(1 \cdot i) = f(i) = [1] = [0] + [1] = f(1) + f(i)$

Similarly,

$$f\{i(-i)\} = f(1) = [0] = [1] + [3] = f(i) + f(-i)$$

$\Rightarrow \quad f$ preserves composition in G and G'.

Hence, f is an isomorphism.

Example 3. *Show that the additive group of integers $G = \{..., -3, -2, -1, 0, 1, 2, 3, ...\}$ is isomorphic to the additive group $G' = \{..., -3m, -2m, -1m, 0, 1m, 2m, 3m, ...\}$ where m is any fixed integer not equal to zero.*

Solution. Let $a \in G$. Then we have $ma \in G$.

Define a map $f : G \rightarrow G'$ such that $f(a) = ma, \ \forall \ a \in G$

   (i) f is one-one. Let $a_1, a_2 \in G$. Then $f(a_1) = f(a_2)$ [By definition of f]

$$\Rightarrow \qquad\qquad ma_1 = ma_2$$
$$\Rightarrow \qquad\qquad a_1 = a_2$$

$\Rightarrow \quad f$ is one-one.

   (ii) f is onto. Since, m is any fixed integer. Therefore for any $a \in G$ there exists only one elemennt ma in G', which implies that G and G' both have the same number of elements.

Also, f is one-one. Hence, f is onto.

   (iii) Structure preserving property.

Let $\qquad a_1, a_2 \in G$

then, $\quad f(a_1 + a_2) = m(a_1 + a_2) = ma_1 + ma_2 = f(a_1) + f(a_2)$

$\Rightarrow f$ preserves composition in G and G'.

Finally, from (i), (ii) and (iii) we conclude that f is an isomorphism.

Hence, $\qquad\qquad G \cong G'$

Example 4. *Show that the multiplicative group G of n, nth roots of unity is isomorphic to the group $G' = (\{0, 1, 2, ..., n-1\}, +_n)$.*

Solution. Let the element of G be $e^{2r\pi i/n}$, where $r = 0, 1, 2, ..., n-1$.

Define a map $f : G \to G'$ such that $f(e^{2r\pi i/n}) = r$

(i) **f is one-one.** If $0 \le r_1 \le n-1$ and $0 \le r_2 \le n-1$

then, $\quad f(e^{2r_1\pi i/n}) = r_1$ and $f(e^{2r_2\pi i/n}) = r_2$

We have $f(e^{2r_1\pi i/n}) = f(e^{2r_2\pi i/n}) \Rightarrow r_1 = r_2$

$\Rightarrow \qquad\qquad e^{2r_1\pi i/n} = e^{2r_2\pi i/n}$

$\Rightarrow f$ is one-one.

(ii) **f is onto.** Since, the number of elements in G is equal to the number of elements in G', therefore f is one-one $\Rightarrow f$ is onto.

(iii) **Strucutre preserving property.** Consider

$$f(e^{2r_1\pi i/n} \cdot e^{2r_2\pi i/n}) = f[e^{2(r_1+r_2)\pi i/n}] = f[e^{2q\pi i/n}]$$

(where q is least non-negative remainder when $r_1 + r_2$ is divided by n.)

$$= q = r_1 +_n r_2 = f(e^{2r_1\pi i/n}) +_n f(e^{2r_2\pi i/n})$$

$\Rightarrow f$ preserves cmposition. Hence, $G \cong G'$.

REMARKS

☞ In above example if we take $n = 6$, we get 6th roots of unity as z_0, z_1, z_2, z_3, z_4 and z_5, where
$z_j = \cos(2\pi j/6) + i\sin(2\pi j/6), j = 0, 1, ..., 5$.

☞ Here we observe that all these roots lies on the unit circle. [*i.e.,* the circle of radius one with centre $(0, 0)$]. They form the vertices of a regular hexagon.

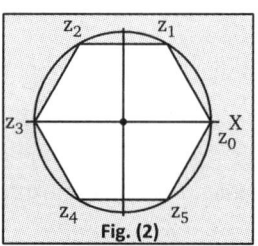

Fig. (2)

EXERCISE 3.4

1. If **R** is the additive group of real numbers and **R$^+$** is the multiplicative group of positive real numbers, show that the mapping $f : \mathbf{R} \to \mathbf{R}^+$ defined by $f(x) = e^x, \forall x \in \mathbf{R}$ is an isomorphism of **R** onto **R$^+$**.

2. Let **R$^+$** be the multiplicative group of all positive real numbers and **R** is the additive group of all real numbers. Show that the mapping $f : \mathbf{R}^+ \to \mathbf{R}$ defined by $f(x) = \log x, \forall x \in \mathbf{R}^+$ is an isomorphism.

3. Let G be the group of all ordered paris (a, b) of real numbers with the binary operation defined by $(a, b) + (c, d) = (a+c, b+d)$ and G' be the additive group of all numbers. Then show that the mapping $f : G \to G'$ defined by $f(a, b) = a \; \forall \; (a, b) \in G$ is a homomorphisms of G onto G'.

4. Show that the mapping $f : G \to G'$, defined by $f(x) = 2x, \forall x \in G$ is an isomorphism of G onto G', where G is the additive group of integers and G' is the additive group of even integers with zero.

5. Show that the multiplicative group $G = \{1, \omega, \omega^2\}$ is isomorphic to the group G' of residue classes (mod 3) under addition of residue classes.

6. If f is an isomorphism of a group G onto a group G', show f^{-1} is an isomorphism of G' onto G.

7. Show that the mapping $x \to x^{-1}$ of G onto G is an isomorphism iff G is abelian, where $x \in G$.

8. Show that the product of two isomorphisms is again an isomorphism.

9. Show that the group G of four transformations f_1, f_2, f_3, f_4 defined by $f_1(z) = z, f_2(z) = -z, \quad f_3(z) = 1/z, f_4(z) = -1/z$ with composite composition is isomorphic to the permutaion group G' of degree 4 consisting of the permutations $I, (ab), (cd), (ab)(cd)$.

10. Let a be a fixed element of a group G. Define a map $f : G \to G$ such that $f(x) = axa^{-1}, \forall x \in G$. Show that f is an isomorphism of G onto itself.

Some More Theorems

Theorem 1. *A homomorphism f of a group G into a group G' is an isomorphism iff $\ker f = \{e\}$.*

Proof.   Let f be a homomorphism of a group G into a group G'. Suppose f is one-one.

Let $x \in \ker(f)$.

Then $f(x) = e'$ (where e' is the identity of G')

Also $f(x) = e' \Rightarrow f(x) = f(e) \Rightarrow x = e$

          $\operatorname{Ker} f = \{e\}$.

Conversely. Let $\operatorname{Ker} f = \{e\}$.

For $x, y \in G$ be such that $f(x) = f(y)$

Then $f(x)f(y)^{-1} = e' \Rightarrow f(xy^{-1}) = f(x)f(y^{-1}) = e'$

                      $\Rightarrow xy^{-1} \in \ker f = \{e\}$

                      $\Rightarrow xy^{-1} = e \Rightarrow x = y$.

Hence, f is one-one $\Rightarrow f$ is an isomorphism.

Theorem 2. *If f is a homomorphism of group G into a group G' with Kernel K, then K is a normal subgroup of G.*

Proof.   Let e, e' be the identities of G and G' respectively. Let K be the Kernel of f. Then

$$K = \{x \in G : f(x) = e'\}$$

Since $f(e) = e \Rightarrow$ at least $e \in K$.

$\Rightarrow$ $K \neq \phi$

Let $a, b \in K$, then $f(a) = e', f(b) = e'$

Therefore, we have

$$f(ab^{-1}) = f(a)f(b^{-1}) = f(a)\{f(b)\}^{-1} = e'(e')^{-1} = e'e' = e'$$

$\Rightarrow$ $ab^{-1} \in K$

$\Rightarrow$ K is a subgroup of G.

Let g be any element of G and k be any element of K then

         $f(k) = e'$

Therefore

$$f(gkg^{-1}) = f(g)f(k)f(g^{-1})$$
$$= f(g)e'[f(g)]^{-1}$$
$$= f(g)[f(g)]^{-1} = e'$$

Now $g \in G, k \in K \Rightarrow gkg^{-1} \in K$

Hence, K is a normal subgroup of G.

Theorem 3. *If f is a homomorphism of a group G into a group G' with Kernel K. Let $a \in G$ be such that $f(a) \in G'$. Then the set of all those elements of G which have the image a' in G' is the coset Ka of K in G.*

Proof.   Let e and e' be the identities of G and G' respectively.

Let $a \in G$ and $f(a) = a' \in G'$

Let $f^{-1}(a') = \{x \in G : f(x) = a'\}$.

Now to show
$$f^{-1}(a') = Ka$$
Let $y \in Ka$. then for $k \in K, y = Ka$

We have $\qquad f(y) = f(ka) = f(k)f(a) = e'f(a)$ as $k \in K \Rightarrow f(k) = e'$
$$= f(a) = a'$$
$\therefore \qquad\qquad\qquad y \in f^{-1}(a')$

Thus, $\qquad\qquad y \in Ka \qquad\qquad \Rightarrow y \in f^{-1}(a')$

$\therefore \qquad\qquad\quad Ka \subseteq f^{-1}(a') \qquad\qquad\qquad\qquad\qquad$ ...(1)

Now, let z be any element of $f^{-1}(a')$

Then $\qquad\qquad f(z) = a'$

$\therefore \qquad\quad f(za^{-1}) = f(z)f(a^{-1}) = f(z)\{f(a)\}^{-1} = a'(a')^{-1} = e'$

$\therefore \qquad\qquad\quad za^{-1} \in K$

$\Rightarrow \qquad\qquad (za^{-1})a \in Ka \qquad\qquad \Rightarrow z \in Ka$

Thus, $\qquad\qquad z \in f^{-1}(a') \qquad\quad \Rightarrow z \in Ka.$

$\therefore \qquad\qquad f^{-1}(a') \subseteq Ka \qquad\qquad\qquad\qquad\qquad$ ...(2)

From (1) and (2), we conclude that
$$f^{-1}(a') = Ka.$$

Theorem 4. *Let G be a group and N is a normal subgroup of G. Let f be a mapping from G to G/N defined by*
$$f(x) = Nx; \ \forall \ x \in G.$$
Then f is a homomorphism of G onto G/N and Kernel f = N.

Proof. Define a mapping
$$f: G \to G/N \text{ such that } f(x) = Nx, \ \forall \ x \in G.$$
Let $Nx \in G/N$, then $x \in G$.

We have $\qquad\qquad f(x) = Nx \ \Rightarrow \ f$ is onto G/N.

Let $a, b \in G$, then $f(ab) = Nab = (Na)(Nb) = f(a).f(b) \qquad\qquad (\because N \text{ is normal.})$

$\therefore \quad f$ is a homomorphism of G onto G/N.

Thus, every quotient group of a group is a homomorphic image of the group.

Let K be the Kernel of f. The identity of the quotient group G/N is the coset N.

So, $\qquad\qquad\qquad K = \{y \in G : f(y) = N\}$

To show $K = N$.

Let $k \in K$ then $f(k) = N$, *i.e.,* identity of G/N

By definition of f, $f(k) = Nk$

Now, $\qquad\qquad\qquad Nk = N \ \Rightarrow \ k \in N \ \Rightarrow \ K \subseteq N. \qquad\qquad\qquad$ ...(1)

Now let $n \in N$, then $Nn = N$

We have $\qquad\qquad f(n) = Nn = N$

$\therefore \qquad\qquad\qquad N \subseteq K \qquad\qquad\qquad\qquad\qquad\qquad$ ...(2)

From (1) and (2), we conclude that
$$N = K$$

Theorem 5 (Fundamental theorem on homomorphism of groups). *Every homomorphic image of a group G is isomorphic to some quotient group of G.*

> **Step Outline :** To make the proof easier, use the following steps :
>
> **Step 1.** *Define a mapping* $\phi : G/K \rightarrow G'$ *and show that it is well defined.*
>
> **Step 2.** *Show that* ϕ *is one-one.*
>
> **Step 3.** *Show that* ϕ *is onto.*

Proof.   If G' is a homomorphic image of G and f is the homomorphism of G onto G'.

Let $K = \operatorname{Ker} f$

If $a \in G$, then $Ka \in G/K$ and $f(a) \in G'$.

Define a mapping $\phi : G/K \rightarrow G'$ such that $\phi(Ka) = f(a), \forall \ a \in G$.

Step 1 : ϕ *is well defined.* Now we shall show that ϕ is well defined.

i.e., if $a, b \in G$ and $Ka = Kb$, then

$$\phi(Ka) = \phi(Kb)$$

Then $Ka = Kb$ $\Rightarrow$ $ab^{-1} \in K$

$\Rightarrow$ $f(ab^{-1}) = e'$ $\Rightarrow$ $f(a)f(b^{-1}) = e'$

$\Rightarrow$ $f(a)[f(b)]^{-1} = e'$ $\Rightarrow$ $f(a)[f(b)]^{-1} \cdot f(b) = e'f(b)$

$\Rightarrow$ $f(a)e' = f(b)$ $\Rightarrow$ $f(a) = f(b)$

$\Rightarrow$ $\phi(Ka) = \phi(Kb)$

$\Rightarrow$ ϕ is well defined.

Step 2 : ϕ *is one-one.* We have

$$\phi(Ka) = \phi(Kb) \quad \Rightarrow \quad f(a) = f(b)$$

$\Rightarrow$ $f(a)[f(b)]^{-1} = f(b)[f(b)]^{-1} = e'$

$\therefore$ $f(a) \cdot f(b^{-1}) = e' \Rightarrow f(ab^{-1}) = e'$

$\Rightarrow$ $ab^{-1} \in K$ $(\because K \text{ is Kernel.})$

$\Rightarrow$ $Ka = Kb$

$\therefore$ ϕ is one-one.

Step 3 : ϕ *is onto.* Let $y \in G'$, then $y = f(a)$ for some $a \in G$ as f is onto on G'.

Now $Ka \in G/K$ and we have $f(Ka) = f(a) = y$

$\Rightarrow$ ϕ is onto.

Therefore, ϕ is an isomorphism of G/K onto G'. Hence $G/K \cong G'$.

REMARKS

☞ According to famous Mathematician **Jacobian**, the theorem can be stated as follows : "Any factor group of G is a homomorphic image of G and conversely every homomorphic image of group G is isomorphic to a quotient group of G."

☞ The fundamental theorem on homomorphism of groups tells us how to find all possible homomorphic image of a given group G.

☞ Except for isomorphism, these homomorphic images must be expressible, in the form of G/N, where N is normal in G.

☞ For any normal subgroup N of G, G/N is a homomorphic image of G. Thus we have a one-one correspondence between the normal subgroups of G and the homomorphic image of G.

Theorem 6 (First law of isomorphism). *Let f be a homomorphism of a group G onto a group G′ and H = Ker f, K′ is any normal subgroup of G′ and*

$$K = \{x \in G : f(x) \in K'\} = f^{-1}(K')$$

then K is a normal subgroup of G containing H and G/K ≅ G′K′.

Step Outline : To make the proof easier, use the following steps :

Step 1. *Show that g : G → G′/K′ defined by g(x) = K′f(x) is a homomorphism.*

Step 2. *Show that Ker(g) = K*

Step 3. *Use fundamenetal theorem of homomorphism to show G/K ≈ G′/K′*

Proof. Defined a map $g : G \to G'/K'$ such that $g(x) = K'f(x)$, $x \in G$.

Let $a, b \in G$, then

$$g(ab) = K'f(ab) = K'f(a)f(b) \qquad (\because f \text{ is a homomorphism.})$$

⇒ g is a homomorphism.

Now let $K'a' \in G'/K'$, $a' \in G'$.

∵ f is onto, $\exists\ a \in G'$ such that $f(a) = a'$. Then

$$g(a) = K'f(a) = K'a'$$

⇒ g is a homomorphism of G onto G′/K′.

Then by fundamental theorem of homomorphism, we have

$$G/\mathrm{Ker} \cdot g \cong G'/K'$$

Now, to show Ker · g = K.

Now as K′ is the identiy of G′/K′.

$$x \in \mathrm{Ker}\ g \Leftrightarrow g(x) = K',$$
$$\Leftrightarrow K'f(x) = K' \Leftrightarrow f(x) \in K$$
$$\Leftrightarrow x \in K \qquad \text{(By definition of } K\text{)}$$

This gives $x \in \mathrm{Ker}.g \Leftrightarrow x \in K$, hence, Ker. g= K

⇒ K is a normal subgroup of G and $G/K \cong G'/K'$

Now $x \in H$ ⇒ $f(x) = e' \in K'$ ⇒ $g(x) = K'f(x) = K'$

⇒ $x \in \mathrm{Ker}.g = K$ ⇒ $H \subseteq K$

Hence, K is a normal subgroup of G containing H.

and $G/K \cong G'/K'$

Theorem 7 (Second law of isomorphism). *If H be a normal subgroup of a group G and K be a normal subgroup of G containing H, then G/K ≅ (G/H)/(K/H)*

Step Outline : To make the proof easier, use the following steps :

Step 1. *Using first law of isomorphism to show that (G/H)/(K/H) is a quotient group.*

Step 2. *Define a map f : G/H→G/K such that f(Hx) = Kx ∀ x ∈ H and show that f is well defined and homomorphism.*

Step 3. *Show that Ker·f = K/H.*

Step 4. *Use fundamental theorem of homomorphism.*

Proof. Let G be a group such that H is a subgroup and K is normal subgroup of G, containing H.

Then by first law of isomorphism, we have K/H is a normal subgroup of the quotient group G/H.

$\Rightarrow$ $(G/H)/(K/H)$ is a quotient group.

Define a map $f : G/H \to G/K$ such that $f(Hx) = Kx,\ \forall\ x \in G$

Now, firstly we shall show that f is well defined.

Let $Hx = Hy$, where $x, y \in G$

$\Rightarrow\ \ xy^{-1} \in H \quad \Rightarrow\ \ xy^{-1} \in H \subset K \Rightarrow \quad xy^{-1} \in K$

$\Rightarrow\ \ Kx = Ky \quad \Rightarrow\ \ f(Hx) = f(Hy)$

Therefore, f is well defined.

Now, to show f is a homomorphism of G/H onto G/K. For this, we shall show that it satisfies sturcture preserving property.

Consider $f[(Hx)(Hy)] = f(Hxy) = Kxy = (Kx)(Ky) = f(Hx)f(Hy)$

$\Rightarrow$ Structure preserving property is satisfied.

Also, we have f is onto because $Kx \in G/K$.

$\Rightarrow\ \ \exists\ Hx \in G/H$ such that $f(Hx) = Kx$

Therefore, f is a homomorphism of G/H onto G/K.

Now,   to show that kernel of $f = K/H$

Since the identity element of the quotient group G/K is K.

Therefore, if $Hx \in G/H$, then $Hx \in$ Ker.f

$\qquad \Leftrightarrow\ f(Hx) = K\ \Leftrightarrow\ Kx = K\ \Leftrightarrow\ x \in K\ \Leftrightarrow\ Hx \in K/H$

$\Rightarrow$ The Ker. $f = K/H$

$\Rightarrow$ f is a homomorphism of G/K onto G/K, with kernel K/H.

Now by fundamental theorem on homomorphism, since every homomorphic image is isomorphic to some quotient group.

Therefore, $G/K \cong (G/K)/(K/H)$.

Theorem 8 (Third law of isomorphism). *Let G be a group and H be any subgroup of G. If N is any normal subgroup of G, then $H/N \cong H/(H \cap N)$.*

Step Outline : To make the proof easier, use the following steps :

Step 1. *Define quotient group $H/(H \cap N)$*

Step 2. *Show that HN/N is a quotient group.*

Step 3. *Define a map $f : H \to HN/N$ by $f(x) = Nx,\ \forall\ x \in H$ and show that f is well defined and homomorphism onto.*

Step 4. *Show that Ker. $f = H \cap N$ and then use fundamental theorem on homomorphism.*

Proof.   Let G be a group such that H be a subgroup of G and N, a normal subgroup of G.

Then obviously we have $H \cap N$ is a normal subgroup of H.

$\Rightarrow$ $H/(H \cap N)$ is a quotient group.

Now HN is a subgroup of G ($\because$ H and N both are subgroups of G) and N is a subset of H/N.

$\Rightarrow$ N is normal in HN.

$\Rightarrow$ HN/N is a quotient group.

Now define a map $f : H \to HN/N$ such that $f(x) = Nx \ \forall \ x \in H$. Firstly, we shall show that the mapping f is well defined.

Since $H \subset HN$

Therefore $x \in H \ \Rightarrow \ x \in HN$

$\Rightarrow$ $Nx \in HN/N$

$\Rightarrow$ f is well defiend.

Now to show f is a homomorphism.

Consider $f(xy) = Nxy = Nx \cdot Ny = f(x) \cdot f(y)$

$\Rightarrow$ Structure preserving property is satisfied.

$\Rightarrow$ f is a homomorphism.

Now to show f is onto.

Let $Hx \in HN/N$, where $x \in HN$

Since $x \in HN$, let $x = hn, \ h \in H, \ n \in N$

Also, since N is normal is $G \ \Rightarrow \ HN = NH$.

Now $f(h_1) = Nh_1 = (Nn_1)h_1$ $(\because Nn_1 = N)$

$\qquad\qquad = N(n_1 h_1) = N(hn) = nx$

$\Rightarrow$ f is onto.

Therefore, we have shown that f is a homomorphism of H onto HN/N. Now, to show Ker. $f = H \cap N$

Let $h \in H$, then

$\qquad\qquad h = $ Ker. $f \ \Rightarrow \ f(h) = N$ $(\because N$ is the identiy of $HN/N.)$

$\Rightarrow$ $Nh = N$ $\Rightarrow$ $h \in N \ \Rightarrow \ h \in H \cap N$

Also, if $h \in H \cap N$, then $h \in H$ and $h \in N$

$\qquad\qquad f(h) = Hh = N$

Therefore, $h \in H \cap N \ \Rightarrow \ h \in$ Ker. f

$\qquad\qquad\qquad \Rightarrow$ if $h \in H$, then $h \in$ Ker. $f \ \Leftrightarrow \ h \in H \cap N$

$\qquad\qquad\qquad \Rightarrow$ Ker. $f = H \cap N$

Then, by fundamental theorem on homomorphism, we have

$\qquad HN/N \cong H/(H \cap N)$

Solved Examples

Example 1. *If* **R** *be the additive group of real numbers and* **R**$^+$ *the multiplicative group of positive real numbers, show that following mapping are isomorphism.*

(i) $f : \mathbf{R} \to \mathbf{R}^+$ *s.t.* $f(x) = e^x, x \in \mathbf{R}$

(ii) $f : \mathbf{R}^+ \to \mathbf{R}$ *s.t.* $f(x) = \log x, x \in \mathbf{R}$

Solution. (i) Since $x = y \ \Rightarrow \ e^x = e^y \ \forall \ x, y \in \mathbf{R} \ \Rightarrow \ f$ is one-one. Now, to every positive real number $x \in \mathbf{R}^+$, there is a real number $\log x \in \mathbf{R}$ such that $f(\log x) = e^{\log x}$ $= x.$

$\Rightarrow f$ is onto.

Also, $f(x+y) = e^{x+y} = e^x \cdot e^y = f(x) \cdot f(y)$.

Hence, f is an isomorphism of $\mathbf{R}$ onto $\mathbf{R}^+$.

(ii) Since any two distinct real positive numbers have two different logarithms as real numbers.

$\Rightarrow f$ is one-one.

Also, every real number has a unique antilogarithms as a positive real number

i.e., $a = b \Rightarrow \log a = \log b, \forall a, b \in \mathbf{R}^+$

and for all $x \in \mathbf{R}, x = \log e^x = f(e^x)$, where $e^x \in \mathbf{R}^+$

$\Rightarrow f$ is onto.

Also for all $x, y \in \mathbf{R}^+$

$$f(xy) = \log(xy) = \log x + \log y = f(x) + f(y)$$

Hence, f is an isomorphism.

Example 2. *If n is any given positive integer, show that the mapping $f : C_0 \to C_0$ defined by $f(z) = z^n$ is an endomorphism of the multiplicative group of non-zero complex numbers. What is the kernel of this endormorphism.*

Solution. Let $z_1 \cdot z_2 \in C_0$. Then $f(z_1) = z_1^n$ and $f(z_2) = z_2^n$

We have $f(z_1 z_2) = (z_1 z_2)^n = z_1^n z_2^n = f(z_1)f(z_2)$

$\Rightarrow$ f is an endomorphism.

Since, the identity of C_0 is 1. The kernel of f consist of the n, n^{th} roots of unity

i.e.,   Kernel of $f = (e^{2r\pi i/n}, r = 0, 1, 2, ..., n-1)$

$$f(e^{2r\pi i/n}) = (e^{2r\pi i/n})^n = e^{2r\pi i} = \cos 2r\pi + i\sin 2r\pi = 1 + 0i = 1,$$

which is the identity of C_0.

Also, $f(z) = 1 \Rightarrow z^n = 1 \Rightarrow z = (1)^{1/n} \Rightarrow z$ is an nth root of unity.

Hence, kernel of f consist n, n^{th} roots of unity.

EXERCISE 3.5

1. Consider the cyclic group $G = \{a, a^2, a^3, ..., a^{12} = e\}$ and its subgroup $G' = \{a^2, a^4, a^8, a^{12}\}$. Show that $f : G \to G'$ defined by $f(a^n) = a^{2n}$ is a homomorphism.

2. If f is a homomorphism of G onto G' and g is a homomorphism of G' onto G'', show that the composite mapping $g \circ f$ is a homomorphism of G onto G''. Also show that the kernel of f is a subgroup of $g \circ f$.

3. Show that the mapping $f : \mathbf{C} \to \mathbf{R}$ such that $f(x+iy) = x$ is a homomorphism of the additive group of complex numbers onto the additive group of real numbers. Also show that the kernal of f consists of all complex numbers whose real part is zero.

4. Show that any two conjugate classes of a group are either disjoint or identical.

5. Let $\mathbf{R}$ be the additive group of real numbers and $\mathbf{C}$, the multiplicative group of complex numbers of absolute value unity show that the mapping $x \to e^{ix}$ is a homomorphism of $\mathbf{R}$ onto $\mathbf{C}$.

6. Show that the multiplicative group G consisting of the three cube roots of unity $1, \omega, \omega^2$ is isomorphic to the group G of residues classes (mod 3) under addition.

7. If N is a normal subgroup of G, having the prime index p. Show that G/N is cyclic.

8. Let C_0 and R_0 be the multiplicative groups of non-zero complex numbers and of non-zero real number respectively. Then show that the mapping $f : C_0 \to R_0$ defined by $f(z) = z$, $\forall z \in C_0$ is a homomorphism of C_0 into R_0. Also show that the kernel of f is the multiplicative subgroup of complex numbers whose modulus is one.

 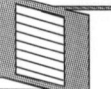

Chapter Review: *A competitive Approach*

Selected Terms and Results

Terms

- **Normal subgroup :** A subgroup H of a group G is said to be a normal subgroup of G if for every $x \in G$ and for every $h \in H$, $xHx^{-1} \in H$.

- **Simple group :** A group $G \neq \{e\}$ is known as simple group if it has no proper normal subgroup.

- **Hamiltonian group :** A non-abelian group, each of whose subgroups is normal is said to be Hamiltonian group.

- **Conjugate element :** An elemnt of a group G is called conjugate to an element $b \in G$ if and only if $a = x^{-1}bx$ for some $x \in G$.

- **Normalizer :** $N(a) = \{x \in G : ax = xa\}$.

- **Centralizer :** $C(A) = \{x \in G : ax = xa \ \forall \ a \in A\}; A \subseteq G$.

- **Commutator :** Let G be a group and $x, y \in G$. The elemnt $x^{-1}y^{-1}xy$ is called the commutator of x and y taken in the order.

- **Self conjugate element :** An element of a group G is called self conjugate element iff $ax = xa \ \forall \ x \in G$.

- **Quotient group :** If G is a group and H, a normal subgroup of G then the set G/H of all cosets of H in G is a group w.r.t. multiplication of cosets is called quotient group.

- **Homomorphism :** A mapping $f : (G, o) \to (G', *)$ is said to be homomorphism if $f(aob) = f(a)*f(b) \ \forall \ a, b \in G$.

- **Kernel :** Let f be a homomorphism of a group G onto G', then the set K of all those elements of G which are mapped under f onto the identity e' of G'.

- **Isomorphism :** A homomorphism which is one-one and onto.

- **Automorphism :** An isomorphism $f : G \to G$.

- **Epimorphism :** A homomorphism onto.

- **Monomorphism :** A homomorphism into.

- **Inner automorphism :** Let $a \in G$ be fixed. Then the mapping $f_a : G \to G$ such that $f_a(x) = a^{-1}xa \ \forall x \in G$ is an automorphism of G is called an inner automorphism.

Results

- Every subgroup of an abelian group is normal.

- The centre of a group is a normal subgroup of G.

- If G is a group and H is a subgroup of index 2 in G then H is a normal subgroup of G.

- If H is the only subgroup of $o(H)$ in a finite group G then H is the normal subgroup of G.

- A subgroup H of a group G is a normal subgroup of G if and only if every left cosets of H in G is a right cosets of H in G.

- A subgroup H of a group G is a normal subgroup of G if and only if the product of two right cosets of H in G is again a right cosets of H in G.

- The intersection of any two normal subgroups of a group G is a normal subgroup of G.

- A normal subgroup is commutative with every complex.

- If N is a normal subgroup of G and H is any subgroup of G then NH is a subgroup of G.

- If G be a group of order $2p$, where p is prime, then G has a normal subgroup of order p.

- If H is a normal subgroup of G and K is a subgroup of G such that $H \subseteq K \subseteq G$, then H is a normal subgroup of K.

- If N and M are normal subgroups of G, then NM is also a normal subgroup of G.

- If N and M are two normal subgroups of G such that $N \cap M = \{e\}$ then every element of N commutes with every element of M.

- If H is the only subgroup of finite order in a group G then H is a normal subgroup of G.

- If a cyclic subgroup N of G is normal in G then every subgroup of N is normal in G.

- A subgroup H of G is normal if and only if the set G/H of all its left cosets is closed under multiplication.

- If H is a subgroup of G and N is a normal

subgroup of G then

(i) HN is a subgroup of G

(ii) N is a normal subgroup of HN.

- A self conjugate element commutes with every element of the group.
- The normalizer $N(a)$ of $a \in G$ is a subgroup of G.
- A group of order p^2 is abelian, where p is a prime number.
- The set of all cosets of a normal subgroup is a group w.r.t. multiplication of complexes as composition.
- If p is prime, then any group of order $2p$ has a normal subgroup of order p.
- If an abelian group is simple, then order of G is prime.
- Every quotient group of a cyclic group is cyclic but converse is not true.
- Every quotient group of an abelian group is abelian but converse is not true.
- Let H be a normal subgroup of a finite group G, then $o(G/H) = o(G)/o(H)$.
- A homomorphism f of a group G into G' is one-one iff Ker(f) = {e}.
- If f is a homomorphism of group G into a group G' with Kernel K then K is a normal subgroup of G.
- Every homomorphic image of a group G is isomorphic to some quotient group G. (Fundamental theorem of homomorphism)
- If f is a homomorphism of a group G onto a group G' and H = Ker(f), then K' is a normal subgroup of G' and $K = \{x \in G : f(x) \in K'\} = f^{-1}(K')$. (First theorem on isomorphism)
- If H be a normal subgroup of a group G and K is a normal subgroup of G containing H then $G/K \cong (G/H)/(K/H)$. (Second law of isomorphism)
- Let G be a group and H be any subgroup of G. If N is any normal subgroup of G then $HN/N \cong H/(H \cap N)$. (Third law of isomorphism)
- Any two conjugate classes of a group are either disjoint or identical.

Review Questions and Project Work

1. Show that a subgroup of index 2 is always a normal subgroup.
2. If M and N are two normal subgroups of G and $M \cap N = \{e\}$, show that for every $n \in N$, $m \in M$, $nm = mn$.
3. Show that in a group G, the normalizer of an element is not necessarily normal subgroup of G.
4. Show that a homomorphism from a simple group is either trivial or one to one.
5. Let G be a group and H be any subgroup of G, show that following statement are equivalents :

(i) H is normal subgroup of G.

(ii) H is the kernel of a homomorphism of G.

(iii) Every left coset of H in G is a right cosets of H in G.

6. Show that the group of all automorphism of a cyclic group G of order r is isomorphic to the group of integers less than and relatively prime to r under multiplication modulo r.
7. Show that a simple group has no non-trivial homomorphic image.
8. Give an example of a group in which the inner automorphism corresponding to any two elements are the same.

Objective type Questions

Fill in the Blanks

1. All group of prime order is _____ .
2. Every subgroup of an abelian group is _____ .
3. The normalizer of an element $a \in G$ is a _____ of G.
4. The abelian part of a group is _____ of group.
5. If every left cosets of H in G is also the right cosets then group is said to be _____ .
6. Every subgroup of index 2 is _____ .
7. Intersection of two normal subgroups again _____ .
8. If N is a normal subgroup of G and H is any subgroup of G, then HN is a _____ of G.
9. The identity element of the quotient group G/H is _____ .
10. The monomorphism is also known as _____ .

11. Every homomorphism image of a group G is isomorphic to some _____ group of G.

12. The index of each subgroup of a finite group is _____ .

13. H is a normal subgroup of G iff $xHx^{-1} =$ _____ $\forall x \in G$.

14. The kernel of a homomorphism is a _____ subgroup of a group G.

15. Every homomorphic image of an abelian group is _____ .

16. If the product of every pair of right cosets of H in G is a right cosets of H in G then it is _____ .

17. The fundamenetal theorem on homomorphism is also known as _____ law of isomorphism.

18. A homomorphism of a group G onto itself is called _____ .

19. Any subgroup H of a group G is normal if $xH =$ _____ .

20. If G is a finite group and it is a normal subgroup of G then $o(G/H)$ is _____ .

True/False

1. It makes sense to speak of the quotient group G/N if and only if N is normal subgroup of the group G. **(T/F)**

2. Every subgroup of an abelian group G is normal. **(T/F)**

3. An inner automorphism of an abelian group must be just the identity mapping. **(T/F)**

4. Every quotient group of a finite group is again of finite order. **(T/F)**

5. Every quotient group of a torsion group is a torsion group. **(T/F)**

6. Every factor group of a torsion free group is torsion free. **(T/F)**

7. Every factor group of an abelian group is abelian. **(T/F)**

8. Every factor group of non-abelian group is non-abelian. **(T/F)**

9. A_n is a normal subgroup of S_n. **(T/F)**

10. Every isomorphism is also a homomorphism. **(T/F)**

11. Every homomorphism is an isomorphism. **(T/F)**

12. A homomorphism is an isomorphism of the domain with the image iff the kernel consist of the group of the identity element alone. **(T/F)**

13. The image of a group of six elements under some homomorphism may have four elements. **(T/F)**

14. The image of a group of six elements under a homomorphism may have twelve elements. **(T/F)**

15. There is a homomorphism of some group of six elements into some group of twelve elements. **(T/F)**

16. There is a homomorphism of some groups of six elements into some group of ten elements. **(T/F)**

17. All homomorphism of a group of prime order are in some sense trivial. **(T/F)**

18. It is not possible to have a homomorphism of some infinite groups into some finite group. **(T/F)**

19. A group of order 169 is abelian. **(T/F)**

20. Union of two normal subgroups is again normal.

Multiple Choice Questions

Choose the most appropriate one :

1. Let G and G' be two groups; $f : G \rightarrow G'$ a homomorphism of G into G' and K, the kernel of this homomorphism then K is :
   (a) a non-empty subset of G
   (b) a subgroup of G
   (c) a normal subgroup of G
   (d) all are true.

2. A subgroup H of a group G is normal if :
   (a) $xhx^{-1} \in H$ (b) $xHx^{-1} = H$
   (c) $xH = Hx$ (d) all are true.

3. Every group of prime order is :
   (a) cyclic   (b) non-cyclic

   (c) non-abelian   (d) none of these

4. The normal subgroups of each group G are :
   (a) e and G (b) only e
   (c) only G (d) none of these

5. Of which order of the following subgroups is always a normal subgroup :
   (a) 3   (b) 2
   (c) 4   (d) 1

6. A subgroup of a group G need not be normal if G is :
   (a) abelian   (b) cyclic
   (c) non-abelian   (d) of prime order

7. Every subgroup of an abelian group :
   (a) is not abelian
   (b) is abelian
   (c) may or may not be abelian
   (d) none of the above

8. Every quotient group of a cyclic group :
   (a) is cyclic
   (b) not cyclic
   (c) may or may not be cyclic
   (d) none of the above

9. If $f : G \rightarrow G'$ is homomorphism of groups and e, e' are respectively identities of G and G'. Then:
   (a) $f(e) = e$
   (b) $f(e) = e'$
   (c) $f(e) = 1$
   (d) none of these

10. G and G' are isomorphic groups. If G is cyclic then :
   (a) G' must be cyclic
   (b) G' may be cyclic
   (c) G' should not be cyclic
   (d) G and G' have same number of elements

11. A subgroup H of a group G is normal if for all $g \in G$, and $h \in H$:
   (a) $ghg^{-1} = H$
   (b) $ghg^{-1} \in H$
   (c) $g^{-1}h^{-1}g \in H$
   (d) none of these

12. Of which order, a subgroup is always a normal subgroup :
   (a) 3
   (b) 2
   (c) 4
   (d) 1

13. A subgroup of a group G need not be normal if G is :
   (a) abelian
   (b) cyclic
   (c) non-alelian
   (d) none of these

14. If G is a group and K is a normal subgroup of G, then the elements of G/K are :
   (a) elements of G
   (b) elements of K
   (c) cosets of K
   (d) none of these

15. The normalizer of an element of a group is :
   (a) not a group
   (b) not a subgroup
   (c) a subgroup of the group
   (d) none of the above

16. If G is a finite group and N is a normal subgroup of G then $o(G/N)$ is :
   (a) $o(G)$
   (b) $o(N)$
   (c) $o(G)/o(N)$
   (d) none of these

17. If any homomorphism is onto, it can be :
   (a) into also
   (b) not into
   (c) may be into
   (d) none of these

18. If $f : G \rightarrow G'$ is homomorphism of groups and e, e' are respectively identities of G and G', then:
   (a) $f(e) = e'$
   (b) $f(e) = e$
   (c) $f(e) = 1$
   (d) none of these

19. If G is an abelian group, then the mapping defined by $f : G \rightarrow G'$, $f(x) = x^5 \ \forall \ x \in G$:
   (a) isomorphic mapping
   (b) not isomorphic mapping
   (c) indefinite
   (d) none of the above

20. G and G' are isomorphic groups. If G is cyclic, then :
   (a) G and G' have the same number of elements.
   (b) G' must be cyclic.
   (c) G' may be cyclic.
   (d) none of the above

21. Every infinite cyclic group is isomorphic to the group :
   (a) $(\mathbf{Z}, +)$
   (b) $(\mathbf{Q}, +)$
   (c) $(\mathbf{R}, +)$
   (d) none of these

22. If G and H are two groups then $f : G \rightarrow H$ is injective only if :
   (a) f is one-one.
   (b) $\text{Ker}(f) = \{e\}$
   (c) $f(g_1) = f(g_2) \Rightarrow g_1 = g_2$
   (d) all are true.

23. If $f : (\mathbf{Z}, +) \rightarrow (\mathbf{Z}, +)$ is a group homomorphism such that $f(a) = 1$ and $f(b) = 2$, $\forall \ a, b \in \mathbf{Z}$ then value of $f(a+b) =$
   (a) 1
   (b) 2
   (c) 3
   (d) none of these

24. The number of abelian groups of order 20 upto isomorphism is :
   (a) 2
   (b) 1
   (c) 3
   (d) none of these

25. The number of element of order 11 in a group of order 33 are :
   (a) 20
   (b) 10
   (c) 0
   (d) none of these

26. The smallest odd order for a group to have a non-abelian subgroup is :
   (a) 11
   (b) 21
   (c) 23
   (d) none of these

27. The smallest order for a group to have a non-abelian proper subgroup is :
   (a) 14
   (b) 24
   (c) 24
   (d) none of these

28. Any subgroup of a group of rationals is :
   (a) normal, cyclic and finitely generated

(b) cyclic, finitely generated but not abelian

(c) cyclic, abelian but not finitely generated

(d) none of the above

29. Let G be a group such that $a^2 = e$ for each $a \in G$ where e is the identity of G then G is :

(a) cyclic   (b) abelian

(c) finite   (d) none of these

30. If a non-trivial group has only one generator, then the number of elements in the group is :

(a) 2   (b) 3

(c) an odd prime   (d) none of these

Answers

Fill in the Blanks

1. Simple **2.** Abelian **3.** Subgroup **4.** Centre **5.** Normal **6.** Normal **7.** Normal **8.** Subgroup **9.** H **10.** Injective mapping **11.** Quotient **12.** Finite **13.** H **14.** Normal **15.** Abelian **16.** Normal **17.** First **18.** Endomorphism **19.** Hx **20.** $o(G)/o(H)$.

True/False

1. T	**2.** T	**3.** T	**4.** T	**5.** T	**6.** F	**7.** T	**8.** F	**9.** T
10. T	**11.** F	**12.** T	**13.** F	**14.** F	**15.** T	**16.** T	**17.** T	**18.** F
19. T	**20.** F							

Multiple Choice Questions

1. (d)	**2.** (d)	**3.** (a)	**4.** (a)	**5.** (b)	**6.** (b)	**7.** (b)	**8.** (a)	**9.** (b)
10. (d)	**11.** (b)	**12.** (b)	**13.** (b)	**14.** (d)	**15.** (c)	**16.** (c)	**17.** (b)	**18.** (a)
19. (a)	**20.** (a)	**21.** (a)	**22.** (d)	**23.** (c)	**24.** (a)	**25.** (b)	**26.** (b)	**27.** (b)
28. (a)	**29.** (b)	**30.** (a)						

COMPETITION CORNER
for JRF, NET/SET, GATE Aspirants

Some Fascinating Facts

1. 1 is the left identity for the system $(\mathbf{R}, .)$.
2. 0 is the left identity for the system $(\mathbf{C}, +)$.
3. The system $(\mathbf{N}, +)$ does not have a left identity.
4. No element of $(\mathbf{Z}, \cdot)$ different from 1 and -1 has a left inverse with respect to 1. Each of 1 and -1 has its own left inverse with respect to 1.
5. An algebraic system has always a unique identity, if at all. The left/right identities need not be unique. Similarly left/right inverses are not necessarily unique. When both sided they are unique. Even when not both sided, they may be unique.
6. The word isomorphism means 'of the same form or pattern.'
7. The identity mapping of a group onto itself is always an automorphism.
8. The composite of two isomorphisms is an isomorphism.
9. The inverse of an isomorphism is again an isomorphism.
10. All groups of order one are isomorphic.
11. All groups of order two are isomorphic.
12. The set of all automorphism of a group forms a group with respect to composite of mapping.
13. Every non-abelian group has an automorphism other than the identity automorphism.
14. The subgroup generated by the identity element of any group is the trivial subgroup consisting of the identity element alone.

15. The subgroups generated by different elements of a group need not be different.
16. Every non-trivial subgroup of an infinite cyclic group is infinite and cyclic.
17. Every non-trivial subgroup of an infinite cyclic group is isomorphic to the group itself.
18. A cyclic group of order p has $p-1$ generators if p is prime.
19. A cyclic group of order n has $\phi(n)$ generators, $\phi(n)$ being the number of integers less than n and relatively prime to n.
20. Union of two normal subgroups need not be considered for being a normal subgroup because the union of two subgroups need not be a subgroup at all.
21. If G is abelian then G/N is abelian.
22. If G is non-abelian, G/N may be abelian.
23. Quotient groups of non-isomorphic groups by a common subgroups may be isomorphic.
24. A homomorphism is injective if and only if its kernel contains only the identity element.
25. An injective homomorphism is called a monomorphism and a surjective homomorphism is called an epimorphism.
26. For $n > 1$, the alternating group A_n has order $n!/2$.
27. The alternating group A_n is a subgroup of symmetric group S_n.
28. A permutation with odd order must be an even permutation.

Some Important illustrations

1. The semigroups $(\mathbf{N}, \cdot)$, $(\mathbf{Q}, \cdot)$, $(\mathbf{R}, \cdot)$ are all monoids.
2. The algebraic strucutre $(\mathbf{N}, +)$ is a semigroup which is not a monoid.
3. Algebraic strucutre $(\mathbf{Q}, +)$, $(\mathbf{R}, +)$ are both monoids.
4. The algebraic structure $(P(x), \cap)$ is a monoid, x being the identity element.
5. $(P(x), \cup)$ and $(P(x), \Delta)$ are both monoids, ϕ being the identity.
6. Following sets are monoid with respect to multiplication.
   (i) The set of positive rational numbers.

   (ii) $G = \{1, -1, i, -i\}$
   (iii) $G = \{z : z \in \mathbf{C} : |z| = 1\}$
   (iv) $G = \{x + y\sqrt{2} : x, y \in \mathbf{Q}, x^2 + y^2 \neq 0\}$
7. Following sets are monoids with respect to addition:
   (i) $G = \{x + y\sqrt{2} : x, y \in \mathbf{Q}\}$
   (ii) $G = \{x + iy : x, y \in \mathbf{Z}\}$
   (iii) $G = \{x + iy : x, y \in \mathbf{Q}\}$
8. Following sets are group with respect to multiplication :
   (i) The set of positive real numbers.
   (ii) The set of positive rational numbers.
   (iii) $\{z : z \in \mathbf{C} : |z| = 1\}$

    (iv) $\{x+iy : x, y \in \mathbf{Q}, x^2+y^2 \neq 0\}$

    (v) $\{x+y\sqrt{2} : x, y \in \mathbf{Q}, x^2+y^2 \neq 0\}$

9. Following sets are group with respect to addition :

    (i) $G = \{x+iy : x, y \in \mathbf{Z}\}$

    (ii) $G = \{x+y\sqrt{2} : x, y \in \mathbf{Q}\}$

    (ii) $G = \{x+iy : x, y \in \mathbf{Q}\}$

10. The set $\mathbf{R}$ with respect to $*$ where $a*b = a+b-ab \;\forall\; a, b \in \mathbf{R}$ is a group.

11. The set $\mathbf{R}$ with respect to $*$ where $a*b = \min\{a, b\} \;\forall\; a, b \in \mathbf{R}$ is a group.

12. Following algebraic structures are groups :

    (i) $(\{0, 4, 8, 12\}, +_{16})$

    (ii) $(\{0, 3, 6, 9\}, +_{12})$

    (iii) $(\{0, 2, 4, 6, 8\}, +_{10})$

    (iv) $(\{1, 2\}, \times_3)$

    (v) $(\{1, 2, ..., 6\}, \times_7)$

    (vi) $(\{1, 2, ..., 10\}, \times_{11})$

13. The set of real symmetric matrices of order n is an abelian group with respect to matrix addtion.

14. The set of all non-singular matrices of order n with real entries is a group with respect to matrix multiplication.

15. The set of all n-rowed square matrices over $\mathbf{Q}$ with determinant ± 1 is a group with respect to matrix multiplication.

16. The set of all n-rowed skew-symmetric matrices over $\mathbf{R}$ or $\mathbf{C}$ is an abelian group with respect to matrix addition.

17. Each of the following set of matrices is a group w.r.t. matrix multiplication :

    (i) The set of matrices of the form

$$\begin{pmatrix} 1 & b \\ 0 & a^{-1} \end{pmatrix}, a \neq 0, a, b \in \mathbf{R}$$

    (ii) The set of matrices of the form

$$\begin{pmatrix} 1 & b \\ 0 & 1 \end{pmatrix}, b \in \mathbf{R}$$

    (iii) The set of matrices of the form

$$\begin{pmatrix} a & a-a^{-1} \\ 0 & a^{-1} \end{pmatrix}, a \neq 0, a \in \mathbf{R}$$

    (iv) The set of matrices of the form

$$\begin{pmatrix} a & b \\ c & d \end{pmatrix}, ad - bc \neq 0, a, b, c, d \in \mathbf{R}$$

18. Let p be any prime and G be the set of all 2×2 matrices $\begin{pmatrix} a & b \\ c & d \end{pmatrix}$ where a, b, c, d are integers modulo p and $ad - bc \neq 0$. Then G is a group of order $(p-1)^2 p(p+1)$.

An Important Table

S.No.	Algebraic structure	Form of the element	Group	Abelian Group	Identity	Inverse
1.	$(\mathbf{Z}, +)$	x	Yes	Yes	0	$-x$
2.	$(\mathbf{Q}^+, \cdot)$	p/q	Yes	Yes	1	q/p
3.	$(\mathbf{Z}_{n}, +_n)$	x	Yes	Yes	0	$n - x$
4.	$(\cup_n, \times_n)$	x, g.c.d.$(x, n) = 1$	Yes	Yes	1	solution of the equation $xy \equiv 1 \bmod x$
5.	$(\{GL(2, F), \times\})$	$\begin{bmatrix} a & b \\ c & d \end{bmatrix}$	Yes	No	$\begin{bmatrix} 1 & 0 \\ 0 & 1 \end{bmatrix}$	$\begin{bmatrix} \dfrac{d}{ad-bc} & \dfrac{-b}{ad-bc} \\ \dfrac{-c}{ad-bc} & \dfrac{a}{ad-bc} \end{bmatrix}$
6.	$(\{SL(2, F)\}, \times)$	$\begin{bmatrix} a & b \\ c & d \end{bmatrix}$	Yes	No	$\begin{bmatrix} 1 & 0 \\ 0 & 1 \end{bmatrix}$	$\begin{bmatrix} d & -b \\ -c & a \end{bmatrix}$

Symbols —

1. $\mathbf{Z}_n$ = group $\{0, 1, \ldots (n - 1)\}$ under addition modulo n.
2. $GL(2, F)$ = 2×2 matrices of non-zero determinant with coefficient from the field K.
3. $SL(2, F)$ = group of 2×2 matrices over F with determinant 1.
4. U_n = group of units modulo n.

Self Assessment Test

1. Let G be a group with the following property : If a, b and c belong to G and $ab = ca$ then $b = c$. Show that G is abelian.

2. Prove that a group G is abelian if and only if $(ab)^{-1} = a^{-1}b^{-1} \ \forall \ ab \in G$.

3. Show that the number of elements x of a finite group G such that $x^3 = e$ is odd. Show that the number of elements x of G such that $x^2 \neq e$ is even.

4. Prove that if G is a group with the property that the square of every element is the identity, then G is abelian.

5. Show that in a finite group, the number of non-identity elements that satisfy the equation $x^5 = e$ is a multiple of 4.

6. Show that an abelian group with two elements of order 2 must have a subgroup of order 4.

7. Show that a group of order 6 cannot have a subgroup of order 4.

8. Suppose that G is a group that has exactly eight elements of order 3. How many subgroups of order 3 does G have ?

9. Let G be a finite group with more than one element. Show that G has an element of prime order.

10. Let us suppose that a cyclic group G has exactly three subgroups : G itself, $\{e\}$ and a subgroup of order 7. Find the order of G.

11. Let G be a finite group. Show that there exist a fixed positive integer n such that $a^n = e \ \forall$ $a \in G$.

12. Let G be an abelian group and n be a fixed positive integer. Let $G^n = \{a^n : a \in G\}$. Prove that G^n is a subgroup of G.

13. If p is an odd prime show that there is no group that has exactly p elements of order p.

14. If G is a group that has exactly one non-trivial proper subgroup. Show that G is a cyclic group of order p^2, where p is prime.

15. Prove that **Z** under addition is not isomporphic to **Q** under addition.

16. Prove that **Q** under addition is not isomorphic to $\mathbf{R}^+$ under multiplication.

17. Let G be an abelian group with an odd number of elements, show that the product of all of the elements of G is the identity.

18. Show that a non-abelian group of order 10 have five elements of order 2.

19. Show that a group or order 12 must have an element of order 2.

20. Let G be a finite group and if a normal subgroup of G. Prove that the order of the element gH in G/H must divide the order of g in G.

21. If N is a normal subgroup of G and $o(G/H) = m$, prove that $x^m \in N \ \forall \ x \in G$.

22. Let G be a group. If $H = \{a^2 : a \in G\}$ is a subgroup of G, prove that it is a normal subgroup of G.

23. Show that alternating group A_5 cannot have a normal subgroup of order 2.

24. If H and K are distinct subgroups of index 2. Prove that $H \cap K$ is a normal subgroup of G of index 4 and that $G/ (H \cap K)$ is not cyclic.

25. If $G = H \times K$ and N is a normal subgroup of H. Prove that N is normal in G.

26. Let G be a finite group and p is prime. If $p^2 > o(G)$, show that any subgroup of order p is normal in G.

● ● ●

Chapter 4

Structure Theory of Groups
(Some Miscellaneous Concepts)

■ Outlines

- Isotropy group;
- Normal and subnormal series;
- Maximal subgroup and composition series;
- Ascending and descending subnormal chain;   • Sylow's p-group

4.1 INTRODUCTION

In this chapter we shall discuss some miscellaneous topics of group theory. The main emphasis will be on obtaining structure theorems of some depth for certain classes of abelian groups and for various classes of groups that share some desirable properties with abelian groups.

4.2 OPERATIONS OF A GROUP ON A SET

Definition. *Let G be a group and S be any set. An operation or an action of G on S is a homomorphism*

$$\pi : G \to Perm \ (S)$$

of G into the group of permutations of S.

Let us call S a G-set. We denote the permutation associated with an element $x \in G$ by π_x. Therefore, the homomorphism is denoted by $x \to \pi_x$. Given $s \in S$, the image of S under the permutations π_x is $\pi_x \ (s)$. Using such type of operation, we get a mapping.

$$\pi : G \times S \to S$$

which to each pair (x, s) with $x \in G$, $s \in S$ associates the element $\pi_x(s)$.

Further, if we are given a mapping from $G \times S \to S$ denoted by $(x, s) \to xs$ satisfying the following two properties:

(i)   For all $x, y \in G$ and $s \in S$, $x(y \ s) = (xy)s$

(ii) If e is the identity of G, then $es = s \ \forall \in S$

Then for each $x \in G$, the mapping $s \to xs$ is a permutation of S, which is denoted by $\pi_x(s)$. Then $x \to \pi_x$ is a homomorphism of G into perm (S). Hence, an operation of G on S could also be defined as a mapping $G \times S \to S$ satisfying the above two properties.

4.3 REPRESENTATIONS OF G AS A GROUP OF PERMUTATIONS

(1) Conjunction. For each $x \in G$, let $f_x : G \to G$ be the mapping such that $f_x(y) = xyx^{-1}$. We can easily verified that the association $x \to f_x$ is a homomorphism $G \to Aut \ (G)$ and therefore this mapping gives operations of G on itself, called conjunction.

The kernel of the homomorphism $x \to f_x$ is a normal sugroup of G, consists of all $x \in G$ which consists of all $x \in G$ which comⁿ ute with every element of G. Thus kernel is called the centre of G.

REMARKS

☞ G also operates by conjuction on the set of subsets of G. Indeed, let S be the subset of G and let $A \in S$ be a subset of G. Then $x A x^{-1}$ is also a subset of G which may be denoted by $f_x(A)$.

☞ If A is a subgroup of G, then $x A x^{-1}$ is also a subgroup, so that G operator on the set of subgroups by conjunctions.

☞ If A and B are two subsets of G. Then, they are conjugate if there exists $x \in G$ such that $B = x A x^{-1}$.

(2) Translation. Let G be a group. For each $x \in G$, we define the translation $T_x : G \to G$ by $T_x(y) = xy$.

Then, the mapping $(x, y) \to xy = T_x(y)$ define and operation of G on itself.

In a similar way, G operate by translation on the set of subsets. For if A is a subset of G, then $xA = T_x(A)$ is also a subset. If H is is a subgroup of G, then $T_x(H) = xH$ is in general not a subgroup but a coset of H. We denote the set of left cosets of H by G/H. Therefore, even though H need not be normal, G/H is a G-set.

REMARK

☞ T_x is not a group homomorphism, it is only a permutation of G.

(3) Morphism of G-Set.

Let S and S' be two G-sets and $f : S \to S'$ is a mapping. Then f is said to be a morphism of G-set or a G-map if, $\qquad f(xs) = x f(x)$ for all $x \in G$ and $s \in S$.

4.4 ISOTROPY GROUP

Let S be any set and $s \in S$. The set of element $x \in G$ such that $xs = s$ is obviously a subgroup of G, called the isotropy group in G and denoted by G_s.

When G operates on itself by conjuction, then the isotropy group of an element is none other that the normalizer of this element. Also, when G operates on the set of subgroups by conjunction, the isotropy group of a subgroup is again its normalizer.

Further Discussion

Let G operator on a set S, let s, s' be elements of S any y be an element of G such that $ys = s'$, then

$$Gs' = yG_s y^{-1}$$

Here, it is clear that $yG_s y^{-1}$ leaves s' fixed. Conversely, if $x's = s'$ then $x'ys = ys$, therefore $y^{-1}x'y \in G_s$ and $x' \in yG_s y^{-1}$. Hence, the isotropy groups of s and s' are conjugate.

Definition 1. *An action of G is said to be faithful if $K = \{e\}$, i.e., the kernel of $G \to perm$ (s) is trivial.*

Definition 2. *A fixed point of G is an element $s \in S$ such that $xs = s$ for all $x \in G$ or $G = G_s$.*

Definition 3. *Let G operates on a set S and $s \in S$. The subset of S consisting of all element x s (with $x \in G$) is denoted by Gs and is called the orbit of s under G.*

REMARKS

☞ If x and y are in the same coset of the subgroup $H = Gs$. Then $xs = ys$ and conversely.

☞ If G is a group operator on a set S and $s \in S$ then the order of the orbit Gs is equal to the index $(G: Gs)$.

☞ When G operates by conjuction on the set of subgroups and H is a subgroup, then the number of conjugate subgroups to H is equal to the index of the normalizer of H.

4.5 ISOTROPY SUBGROUPS

Let S be a G-set. Let $s \in S$ and $g \in G$. Then it is known that $gs = s$.
Further we let

$$S_g = \{s \in S : gs = s\}$$

and
$$G_s = \{g \in G : gs = s\}$$

Then, we can easily shown that G_s is a subgroup of G for each $s \in S$.

Definition. *Let S be a G-set and $x \in S$. Then subgroup G_x is known as isotropy subgroup of x.*

Theorem 1. *Let S be a G-set, for $x_1, x_2 \in S$, let $x_1 \sim x_2$ if and only if there exists $g \in G$ such that $gx_1 = x_2$. Then $\sim$ is an equivalence relaton on S.*

Proof.   (i) $\sim$ **is Reflexive.** For each $x \in S$. we have

$$ex = x$$

Therefore $x \sim x$ and $\sim$ is reflexive.

(ii) $\sim$ **is Symmetric.** Let $x_1 \sim x_2 \Rightarrow gx_1 = x_2$ for same $g \in G$

Then $g^{-1} x_2 = g^{-1} (gx_1) = (g^{-1}g) x_1 = ex_1 = x_1$

$\Rightarrow \qquad x_2 \sim x_1$

$\Rightarrow \qquad \sim$ is symmetric.

(iii) $\sim$ **is Transitive.** If $x_1 \sim x_2$ and $x_2 \sim x_3$

Then clearly, we have

$$g_1 x_1 = x_2 \text{ and } g_2 x_2 = x_3 \text{ ; for some } g_1, g_2 \in G$$

Then $\qquad (g_1 g_2) x_2 = g_2(g_1 x_1) = g_2 x_2 = x_3$

$\Rightarrow \qquad x_1 \sim x_3$. Hence , $\sim$ is Transitive.

Theorem 2. *Let S be a G-set and $x \in S$ Then $|Gx| = (G : G_x)$. $|G|$ is finite, then $|G_x|$ is a divisor of G.*

Proof.   Define a onc-to-onc mapping ψ from G_x onto the collection of left cosets of G_x in G. Let $x_1 \in G_x$. Then, there exists $g_1 \in G$ such that $g_1 x = x_1$. We define $\psi(x_1)$ to be the left coset $g_1 G_x$ of G_x. To show that this mapping ψ is well defined independent of the choice of $g_1 \in G$ such that $g_1 x = x_1$.

Suppose that $\qquad\qquad g_1' x = x_1$. Then $g_1 x = g_1' x$. so $g_1^{-1}(g x) = g_1^{-1} (g_1' x)$

Thus, we can find $\qquad\quad x = (g_1^{-1} g_1')x \Rightarrow g_1^{-1} g_1' \in G_x$

$\Rightarrow \qquad\qquad\qquad g_1^{-1} \in g_1 G_x \text{ and } g_1 G_x = g_1' G_x$

$\Rightarrow \quad \psi$ is well defined.

ψ **is one-one.** Let $x_1, x_2 \in G_x$ and $\psi(x_1) = \psi(x_2)$

Then, there exist $g_1, g_2 \in G$ such that

$$x_1 = g_1 x, x_2 = g_2 x, \text{ and } g_2 \in g_1 G_x$$

$\Rightarrow \qquad\qquad\qquad g_2 = g_1 g \text{ for some } g \in G_x$

Therefore $\qquad\qquad x_2 = g_2 x = g_1 (gx) = g_1 x = x_1$

$\Rightarrow \quad \psi$ is one-one.

Now, we want to show that each left coset of G_x in G is of the form $\psi(x_1)$ for some $x_1 \in G_x$. Let $g_1 G_x$ be a left coset.

Then, if $g_1 x_1 = x_1$, we have $g_1 G_x = \psi\,(x_1)$

Therefore ψ maps G_x one to one onto the collection of the right cosets so $|G_x| = (G : G_x)$

Finally, if G is finite, then the equation $|G| = |G_x|\,(G : G_x)$, shows that
$$|G_x| = (G : G_x), \text{ is a divisor of } |G|.$$

4.6 APPLICATIONS OF G-SETS TO COUNTING

4.6.1 Burnside's Formula

Let G be a finite group and X a finite G-set. If r be the number of orbits in X under G, then
$$r.\,|G| = \sum_{g \in G} |X_g| \qquad\qquad \dots (1)$$

Proof. Let N be the number of pairs such that
$$gx = x$$

Then for each $g \in G$, there are $|X_g|$ pairs having g as first number.

Therefore.
$$N = \sum_{g \in G} |X_g| \qquad\qquad \dots (2)$$

Also, for each $x \in X$, there are $|G_x|$ pairs having x as second member.

Therefore, we also have
$$N = \sum_{x \in X} |G_x|$$

It is known that $\qquad\qquad |G_x| = (G : G_x)$.

Also $\qquad\qquad (G : G_x) = |G| / |G_x|$

$$|G_x| = \frac{|G|}{|G_x|}$$

$\Rightarrow$

Then, we get
$$N = \sum_{x \in X} \frac{|G|}{|G_x|} = |G| \left(\sum_{x \in X} \frac{1}{|G.x|} \right) \qquad\qquad \dots (3)$$

Further, $\dfrac{1}{|G.x|}$ has the same value for all x in the same orbits. Let o be the orbit, then we have
$$\sum_{x \in X} \frac{1}{|G.x|} = \sum_{x \in X} \frac{1}{|o|} = 1 \qquad\qquad \dots (4)$$

Using (3) and (4) we get

$N = |G|$ (number of orbits in X under G)
$$= |G|.\,r \qquad\qquad \dots (5)$$

On comparing (5) and (2) we get
$$r.\,|G| = \sum_{g \in G} |X_g|$$

REMARK

☞ If G is a finite group and X is a finite G-set, then

$$\text{number of orbits in } X \text{ under } G = \frac{1}{|G|} \sum_{g \in G} |X_g|$$

4.7 MAXIMAL SUBGROUPS AND COMPOSITION SERIES

Definition 1. *A normal subgroup N of a group G is said to be maximal if there exists no proper normal subgroup K of G which properly contains N, i.e., a normal subgroup N of a group G is said to be maximal if and only if there exists no normal subgroup K of G such that $N \subset K \subset G$.*

Definition 2. *Let G be a group. Then a finite sequence of its subgroups*

$$G = H_1, H_2, \dots H_n = \{e\}$$

is called a composition series for G if each H_i except H_1 is a maximal normal subgroup of H_{i-1}.

Definition 3. *The quotient groups G/H_1, H_1/H_2, ..., H_{n-1}/H_n which are necessarily simple are known as composition factor groups or composition quotient groups of the defined composition series.*

Theorem 1.   *A normal subgroup H of a group G is maximal if and only if the group G/H is simple.*

Proof.   Let if possible, H is maximal in G and G/H is not simple. Now, since, G/H is not simple, there exists a proper normal subgroup K/H of G/H.

Therefore, we conclude that K is a normal subgroup of G containing H, which is a contradiction, because by our hypothesis H is maximal in G. Hence G/H is simple.

Conversely, let us suppose that G/H is simple. Let H be not maximal in G.

Now, since H is not maximal in G, there exists a normal subgroup K of G which properly contains H. Thus K/H is a normal subgroup of G/H which is a contradiction, because G/H is simple. Hence, H must be maximal in G.

4.8 SERIES OF GROUPS : NORMAL AND SUBNORMAL SERIES

In this sections, we shall discuss the notion of a series of a group G, which gives in right into the structure of a group G. Then results hold for both abelian and non-abelian group.

4.9 SUBNORMAL SERIES

A subnormal series of a group G is a finite sequence $H_0, H_1, ..., H_n$ of subgroups of G such that $H_i \subseteq H_{i+1}$ and H_i is normal subgroup of H_{i+1} with $H_0 = \{e\}$ and $H_n = G$.

REMARKS

☞ A subnormal series is begin with G itself and ending with the unit subgroup.

☞ In a subnormal series, every subgroup H_i is a proper normal subgroup of H_{i-1}.

☞ Every group has subnormal series that passes through a given normal subgroup of G.

☞ A subnormal series is also known as subinvariant series.

4.9.1 Normal series

A normal series of G is a finite sequence $H_0, H_1, ..., H_n$ of normal subgroup of G such that

$$H_i \subseteq H_{i+1}, H_0 = \{e\} \text{ and } H_n = G.$$

REMARKS

☞ For abelian group, the notions of subnormal and normal series coincides, since every subgroup of an abelian group is normal.

☞ A normal series is always subnormal, but the converse is not necessarily true.

For example

1. Let $G = \{\mathbf{Z}\}$, group of integers under addition, then $\{0\} < 8\mathbf{Z} < 4\mathbf{Z} < \mathbf{Z}$ and $\{0\} < 9\mathbf{Z} < 3\mathbf{Z}$ both are subnormal series of $\mathbf{Z}$.

2. Consider the group D_4 (Dihedral group) of symmetries of the square then the series
$$\{P_0\} < \{P_0, \mu_1\} < \{P_0, P_2 \mu_1, \mu_2\} < D_4$$
is a subnormal series of D_4. It should be noted that this series is not a normal series since $\{P_0, \mu_1\}$ is not normal in D_4.

Definition 1. *A subnormal (normal) series* $\{K_j\}$ *is a refinement of subnormal (normal) series* $\{H_i\}$ *of a group G if* $\{H_i\} \subseteq \{K_j\}$, *i.e., if each* H_i *is one of the* K_j.

In other words, we can say that a subnormal series

$$G \supset K_1 \supset K_2 \supset ... \supset K_n = \{e\} \qquad \qquad ... (1)$$

is called refinement of the subnormal series

$$G \supset H_1 \supset H_2 \supset H_3 .... \supset H_m = \{e\} \qquad \qquad ... (2)$$

if every subgroup H_i coincides with one of the subgroup K_j, i.e., all the subgroup H_i that occur in (2) also occur in (1) or $\{H_i\} \subset \{K_j\}$.

For example

1. The series $\{0\} < 72 \, \mathbf{Z} < 24 \, \mathbf{Z} < 8\mathbf{Z} < 4\mathbf{Z} < \mathbf{Z}$ is a refinement of the series
$$\{0\} < 72 \, \mathbf{Z} < 8\mathbf{Z} < \mathbf{Z}.$$

REMARKS

☞ Particularly, every subnormal series is a refinement of itself.

☞ The length of the subnormal series (2) also occur in (1) or $\{H_i\} \subset \{K_j\}$

Definition 2. *Let G a group. The factor groups (quotient groups)*
$$G/H_1. \ H_1./ H_2 ,..., \ H_{m-1}/\{e\}$$
are called the factors of the subnormal series
$$G = H_0 \supset H_1 \supset H_2 .... \supset H_m = \{e\} \text{ of } G.$$

Definition 3. *The number of factors in a subnormal series*
$$G = H_0 \supset H_1 \supset H_2 .... \supset H_m = \{e\}.$$
is called the length of the series.

REMARK

☞ The length of the series defined above is m.

Definition 4. *Two subnormal (normal) series* $\{H_i\}$ *and* $\{K_j\}$ *of the same group G are isomorphic if there is a one-to-one correspondence between the collections of factor groups* $\{H_{i+1}/H_i\}$ *and* $\{K_{j+1}/K_j\}$ *such that corresponding factors groups are isomorphic.*

For exmple.

1. If $G = \mathbf{Z}_{15}$, group of integers module 15 then, the two series of G
$$G \supset \{5\} \supset \{0\} \text{ and } G \supset \{3\} \supset \{0\}$$
are isomorphic because $\mathbf{Z}_{15}/5$ and $(3)/\{0\}$ are isomorphic to $\mathbf{Z}_5$ and $\mathbf{Z}_{15}/(3)$ is isomorphic to $(5)/\{0\}$.

2. If G be a cyclic group of order 6 and $g = (a)$, $a^6 = 1$, then the subnormal series
$$G \supset \{a^2\} \supset \{1\} \text{ and } G \supset \{a^3\} \supset \{1\}$$
are isomorphic, since their factors are, one cyclic group if order two and one of order three.

REMARK

☞ Two isomorphic subnormal (normal) series must have the same number of groups.

4.9.2 Zassenhaus Lemma

Let H and K be subgroups of a group G and let H^* and K^* be normal subgroups of H and K respectively.

Then

(i) $H^*(H \cap K^*)$ *is a normal subgroup of* $H^*(H \cap K)$.

(ii) $K^*(H^* \cap K)$ *is a normal subgroup of* $H^*(H \cap K)$.

(iii) $H^*(H \cap K)/H^*(H \cap K^*) \cong K^*(H \cap K)/K^*(H^* \cap K)$
$$\approx (H \cap K)/[H^* \cap K (H \cap K^*)]$$

Proof. Let us define
$$C = H \cap K \text{ and } D = (H \cap K^*)(K \cap H^*)$$

Then clearly, we have
$$D \subseteq C$$

Now, since K^* is normal in K and since C is a subgroup of K.

Then $\quad C \cap K^* = H \cap K \cap K^* = H \cap K^* \qquad [\because K^\wedge \subset K]$

is a normal subgroup of C, i.e., $H \cap K^*$ is normal in $H \cap K$.

It is known that product of two normal subgroups is also normal.

Therefore $H^*(H \cap K^*)$ is normal in $H^*(H \cap K)$.

On interchanging H and K, we have $K^*(K \cap H)$ is normal in $K^*(K \cap H)$ and then D is normal in C.

Thus, we get a factor group of D in C say F such that
$$F = \frac{C}{D} = \frac{H \cap K}{(H \cap K^*)(K \cap H^*)}$$

Now, since H^* is normal subgroup of H, therefore
$$H^*(H \cap K) = H^* C \text{ is a subgroup}$$

Therefore, if

$a * c \in A * C$, then $a^* \in H$ and $c \in C$

Then we have
$$F = \{Dc : c \in C\}$$

If $a*c$ has another representation in the same form, the $a*_1 \in H$ and $c_1 \in C$, we have

$$a*c \Rightarrow a_1^* c_1 \Rightarrow a_1^{*-1} a^* = c_1 c^{-1}$$

But

$$H^* \cap C = H^* \cap H \cap K \subseteq H^* \cap K$$

$$\Rightarrow \qquad C_1 C^{-1} \in H^* \cap C \subset H^* \cap K \subset D$$

$$\Rightarrow \qquad a_1^{*-1} a^* \subseteq D \Rightarrow (a_1^{*-1} a^*) C \in Dc$$

$$\Rightarrow \qquad C_1 \in Dc \Rightarrow C_1 \in F$$

In this way we get a simple valued mapping of the group $H*C$ into the group F, and since every element $c \in C$ is mapped onto its cosets Dc, therefore, the single valued mapping $H*C$ is onto the whole group F. Also this mapping is homomorphic.

Now, since H^* is normal in $H*C$, we have

$$(a_1^* c_1)(a_2^* c_2) = a_3^*(C_1 C_2) \text{ for some } a_3^* \in H^*$$

To prove kernel of this homomorphism is $H^*(H \cap K^*)$

Since

$$H \cap K^* \subset D$$

If an element $a*c$ is mapped by above homomorphism then $c \in D$.

But

$$D = (H \cap K^*)(K \cap H^*), \text{ then we have}$$

$$C = u.v, \text{ where } u \in (H \cap K^*)$$

and

$$v \in (H \cap K^*)$$

and therefore,

$$a*c = a*(uv) = (a* u)v = a_1^* v, \text{ for some } a_1^* \in H^*$$

$$\Rightarrow \qquad a*c = a_1^* v \in H^*(H \cap K^*)$$

Therefore, the kernel of the homomorphism is therefore the subgroup $H^* (H \cap K^*)$. So, by theorem on homomorphism, we have

$$\frac{H^*(H \cap K)}{H^*(H \cap K^*)} \cong \frac{(H \cap K)}{(H \cap K^*)(K \cap H^*)} \qquad \text{... (1)}$$

Similarly, we may get

$$\frac{K^*(K \cap H)}{K^*(K \cap H^*)} \cong \frac{K \cap H}{(K \cap H^*)(H \cap K^*)} \qquad \text{... (2)}$$

From (1) and (2) we conclude that

$$\frac{H^*(H \cap K)}{H^*(H \cap K^*)} \cong \frac{K^*(K \cap H)}{K^*(K \cap H^*)}$$

4.9.3 Schrierer Theorem

Any two subnormal series of an arbitrary group have isomorphic refinements.

Proof. Let G be a group. Define.

$$\{e\} = H_0 \subset H_1 \subset H_2 \subset \dots \subset H_n = G \qquad \text{... (1)}$$

$$\{e\} = K_0 \subset K_1 \subset K_2 \dots \subset K_m = G \qquad \text{... (2)}$$

be two subnormal series for G, which form the chain of groups for i, $0 \le i \le n-1$ such that

$$H_i = H_i (H_{i+1} \cap K_0) \subseteq H_i(H_{i+1} \cap K_1) \subseteq H_i(H_{i+1} \cap K_m) = H_{i+1}$$

Thus insert $m-1$, not necessarily distinct groups between H_i and H_{i+1}. If we do this for each i, where $0 \le i \le n-1$ and let

$$H_{ij} = H_i (H_{i+1} \cap Kj)$$

Then, we get the chain of groups

$$\{e\} = H_{0,0} \subseteq H_{0,1} \subseteq H_{0,2} \subseteq \dots \subseteq H_{0,m-1} = H_{1,0}$$
$$\subseteq H_{1,1} \subseteq H_{1,2} \subseteq \dots \subseteq H_{1, m-1} \subseteq H_{2,0}$$
$$\subseteq H_{2,1} \subseteq H_{2,2} \subseteq \dots \subseteq H_{2, m-1} \subseteq H_{3,0}$$
$$\subseteq \dots$$
$$\subseteq H_{n-1,1} \subseteq H_{n-1, 2} \subseteq \dots \subseteq H_{n-1, m-1} \subseteq H_{n-1,m}$$
$$= G \qquad \qquad \dots (3)$$

The above chain contains $nm+1$, not necessarily distinct groups and $H_{i,0} = H_i$ for each i.

By Zassenhaus lemma, (3) is a subnormal series, *i.e.*, each group is normal in the following group. This chain refines the series (1).

By symmetry, we get

$$K_{j, i} = K_j (K_{j+1} \cap H_i) \text{ for } 0 \le j \le m-1$$

and $0 \le i \le n$

This gives a subnormal chain

$$\{e\} = K_{0,0} \subseteq K_{0,1} \subseteq K_{0,2} \subseteq \dots \subseteq K_{0, n-1} \subseteq K_{1,0}$$
$$\subseteq K_{1,1} \subseteq K_{1,2} \subseteq \dots \subseteq K_{1, n-1} \subseteq K_{2,0}$$
$$\subseteq K_{2,1} \subseteq K_{2,2} \subseteq \dots \subseteq K_{2, n-1} \subseteq K_{3,0}$$
$$\subseteq \dots$$
$$\subseteq K_{m-1,1} \subseteq K_{m-1, 2} \subseteq \dots \subseteq K_{m-1, n-1} \subseteq K_{m-1, n} \qquad \dots (4)$$
$$= G$$

This chain contain $mn+1$, not necessarily distinct groups and $K_{j, 0} = K_j$ for each j. This chain refines series (2).

Further, by Zassenhaus lemma, we have

$$H_i(H_{i+1} \cap K_{j+1})/H_i(H_{i+1} \cap K_j) \simeq K_j (K_{j+1} \cap H_{i+1})/K_j(K_{j+1} \cap H_i)$$

or $$H_{ij+1}/H_{ij} \simeq K_{j, i+1}/K_{j,i} \qquad \dots (5)$$

for $$0 \le i \le n-1 \text{ and } 0 \le j \le m-1$$

The isomorphisms of (5) gives a one-to one correspondence of isomorphic factor groups between the subnormal chain (3) and (4).

For verification, note that $H_{i, 0} = H_i$ and $H_{i,m} = H_{i+1}$ while $K_{j,0} = K_j$ and $K_{j,n} = K_{j+1}$.

Each chain in (3) and (4) contains a rectangular array of m n symbols $\subseteq$. Each $\subseteq$ gives rise to a factor group. The factor group arising from the rth row of $\subseteq$'s in chain (3).

Corresponding to the factor group arising from the rth column of $\subseteq$'s in chain (4). Deleting repeated groups from the chain in (3) and (4), we get subnormal series of distinct groups that are isomorphic refinements of chain (1) and (2).

REMARK

☞ For normal series, where all H_i and K_j are normal in G we observe that all the groups $H_{i, j}$ and $K_{j, i}$ formed above are also normal in G. Then for Schreirer theorem of normal series, we apply the same proof as above.

4.10 COMPOSITION SERIES

Let G be a group. Then a finite sequence of its subgroups

$$G = H_1, H_2, H_3, \ldots, H_n = \{e\} \qquad \ldots (1)$$

is known as composition series for G if each H_i except H_1 is a maximal normal subgroup of H_{i-1}.

Definition 1. *A subnormal series that has no refinement other than itself is known as composition series.*

For Example :

1. $G = P_3 = \{I, (1\ 2), (2\ 3), (3\ 1), (1\ 2\ 3), (1\ 3\ 2)\}$ and let $H_2 = \{I, (123), (132)\}$ then F, H_2, $\{I\}$ is a composition series for G.

2. Let G be a cyclic group of order 6 generated by a, *i.e.*,

$$G = \{a, a^2, a^3, a^4, a^5, a^6, = e\}$$

   Then

$$G, H_2 = \{e, a^3\}, \{e\}$$
$$G, N_2 = \{e, a^2, a^4\}, \{e\}$$

   are two different compositon series for G.

3. Let G be a cyclic group generated by a of order 12.

   Then $\{a\}, \{a^2\}, \{a^4\}, \{e\}$ and $\{a\}, \{a^3\}, \{a^6\}, \{e\}$ are two different composition series for G.

4. If $G = S_3$ (symmetric group of degree 3), then

$$G \supset \{e, (1\ 2\ 3). (1\ 3\ 2)\} \supset \{e\}$$

   is a compositon series of G.

5. If G = Quarterian group, then $G \supset \{e, e', i, i\} \supset \{e, e'\} \supset \{e\}$ is a compositon series for G.

REMARK

☞ Principal series also known as 'chief series'.

Theorem 1. *There exists at least one compositon series for every finite group G.*

Proof.   (i) Let first suppose G is simple, then G, $\{e\}$ is a composition series.

    (ii) Let us suppose G is not simple, then there exists a proper normal subgroup H of G. If H is maximal in G and $\{e\}$ is maximal in H, then G, H, $\{e\}$ is a composition series.

Suppose H is not maximal in G, but $\{e\}$ is maximal in H. Then, there exists a normal subgroup K of G such that $H \subset K \subset G$. If K is maximal in G and H is maximal in K then G, H, $\{e\}$ is a composition series.

Further suppose that H is maximal in G but $\{e\}$ is not maximal in H. Then there exists a normal subgroup J of H such that

$$\{e\} \subset J \subset H.$$

If $\{e\}$ is maximal in J and J is maximal in H, then $G, H, J, \{e\}$ is a composition series.

Further, suppose that H is not maximal in G and $\{e\}$ is not maximal in H. Then there exists a normal subgroup L of G such that $H \subset L \subset G$. Also, there exists a normal subgroup N of H such that $\{e\} \subset N \subset H$. Therefore $\{e\} \subset N \subset H \subset L \subset G$. If L is maximal in G, H is maximal in L, N is maximal in H and $\{e\}$ is maximal in N, then $G, L, H, N, \{e\}$ is a composition series.

Finally, since G is finite there are only a finite number of subgroups. Hence, we get a composition series.

Theorem 2. *If group G has a compositon series, then any two composition series of G are isomorphic.*

Proof. Let us suppose

$$\{e\} = G_m \subset G_{m-1} \subset .... \subset G_2 \subset G_1 \subset G_0 = G$$

and $\qquad \{e\} = H_n \subset H_{n-1} \subset .... \subset H_2 \subset H_1 \subset H_0 = G$

be two compositon series of G.

Since they both are subnormal series of G. Therefore by Shreirer's Theorem, both series have isomorphic refinements so that $m = n$. Also, since all the factors groups are simple. Therefore, both series have maximal length $m = n$ so that neither series has any further refinements. Hence, both composition series are isomorphic.

Theorem 3. *If G has a compositon series and H is a proper normal subgroup of G then there exists a composition series containing H.*

Proof. It is given that H is a proper normal subgroup of G.

Therefore $\qquad\qquad\qquad H \neq \{e\} \text{ or } H \neq G.$

Then $\qquad\qquad\qquad\qquad \{e\} \subset H \subset G$ $\qquad\qquad\qquad$ ... (1)

is both a subnormal and normal series.

Further, since G has a composition series and let the series be

$$g = G_0 \supset G_1 \supset G_{i....} \supset G_m = \{e\}$$

Then by Schreirer's theorem, there is a refinements of (1), which is isomorphic to a refinement of (2) but (2) is a composition series so it does not have any further refinement. Hence, series (1) is the required composition series.

Theorem 4. *If G is a commutative group having a composition series then G is finite.*

Proof. It is known that if a group G has no proper subgroup then it must be cyclic group of prime order, *i.e.*, a simple abelian group is a cyclic group of prime order.

Now, let $\{e\} = G_m \subset G_{m-1} \subset .... G_2 \subset G_1 \subset G_0 = G$

be a composition series of G. Here each G_i is normal because every subgroup of an abelian group is normal.

Then the quotient group $G_{m-1}/G_m \cong G_{m-1}/\{e\} \cong G_m$ is simple and abelian.

Thus, the order of G_{m-1} must be prime.

Let $o(G_{m-1}) = P_{m-1}$, a prime number.

Also G_{m-2}/G_{m-1} must be prime. Let $o(G_{m-2}/G_{m-1}) = P_{m-2}$.

Thus, the number of cosets of G_{m-1} in G_{m-2} are P_{m-2} but G_{m-1} has P_{m-1} element.

Therefore, $\quad\quad o(G_{m-2}) = P_{n-1} \cdot P_{n-2}$.

Continuing in the same manner, we get

$$o(G) = P_0 \cdot P_1 \cdots P_{m-1} \cdot P_{m-2} = \text{finite}$$

Hence, G is finite.

Theorem 5. **(Jordan-Holder theorem for finite groups).** *Two composition series of a finite group G are isomorphic.*

Proof. Let G be a group of finite order n.

Also, let

$$\{e\} = G_r \subset G_{r-1} \subset \ldots \subset G_2 \subset G_1 \subset G_0 = G \quad\quad \ldots (1)$$
$$\{e\} = H_s \subset H_{s-1} \subset \ldots \subset H_2 \subset H_1 \subset H_0 = G \quad\quad \ldots (2)$$

be two composition series of G. We shall prove this theorem by mathematical induction on order of G.

If n = 2, then $G = \{e\}$. in this case, theorem is trivially true.

If n = 2, then $G = G_0 \supset \{e\}$. is the only compositon series of G.

Now, suppose that $o(G) = n > 2$ and let above result be true for all groups of order less than n.

Then, we have the follwing two cases:

Case I. If $G_1 = H_1$

Then $\quad\quad \{e\} = G_r \subset G_{r-1} \subset \ldots \subset G_2 \subset G_1 \quad\quad \ldots (3)$

and $\quad\quad \{e\} = H_s \subset H_{s-1} \subset \ldots \subset H_2 \subset H_1 = G \quad\quad \ldots (4)$

be two composition series of G_1 and $o(G_1) < n$. Then by induction hypothesis, the series (3) and (4) are isomorphic. Therefore, the length of (3) is equal to the length of (4) which implies $r = s$.

Also $G/G_1 \cong G/H_1$

Hence, series (1) and (2) are isomorphic.

Case II. If $G_1 \neq H_1$

Then G_1 and H_1 are maximal normal subgroups of G.

It is known that product of two normal subgroups is again a normal subgroup, therefore $G_1 H_1$ is a normal subgroup of G containing G_1 as well as H_1 and G_1 and H_1 being maximal normal subgroup.

Therefore $\quad\quad G_1 H_1 = G$

Further let $N = G_1 \cap H_1 = G_1$ then clearly N is a maximal normal subgroup of G. But $N \subset G_1$ and $N \subset H_1$ because $G_1 \neq H_1$. Therefore $G_1 \cap H_1$ is a maximal normal subgroup of G_1 as well as H_1.

Let $\quad N = N_0 \supset N_1 \supset N_2 \supset \ldots \supset N_t = \{e\}$

be a composition series of N and consider two compositon series

$$G = G_0 \supset G_1 \supset N = N_0 \supset N_1 \supset N_2 \supset \ldots \supset N_t = \{e\} \quad\quad \ldots (5)$$

and $\quad\quad G = H_0 \supset H_2 \supset N = N_0 \supset N_1 \supset N_2 \supset \ldots \supset N_t = \{e\} \quad\quad \ldots (6)$

We claim that the series (5) and (6) are compositon series of G. For this, we shall prove that G_1/N and H_1/N both are simple.

Now, $\qquad G_1/N = G_1/G_1 \cap H_1 \cong G_1 H_1/H_1 = G_1/H_1$

But G/H_1 is simple so that G_1/N is simple.

Also, $\qquad H_1/N = H_1/G_1 \cap H_1 \cong G_1 H_1/G_1 = G/G_1$

Similarly, we can prove that

$$H_1/N \cong G/G_1$$

Thus, the series (5) and (6) each having the length $t+2$ and having all the factor groups isomorphic. Hence, the series (5) and (6) are isomorphic.

Now, using case I, series (1) and (5) are isomophic so that $r = t+2$

Similarly, series (2) and (6) are isomorphic so that $s = t+2$

Hence, $r = s$.

Finally since the series (5) and (6) are isomorphic, the series (1) and (2) are isomorphic. Hence, the theorem.

Solved Examples

Example 1. *Find isomorphic refinement of the subnormal series*

$$\{0\} \subset 8Z \subset 4Z \subset Z \text{ and } \{0\} \subset 9Z \subset Z.$$

Solution. Consider two refinements

$$\{0\} \subset 72\,Z \subset 8Z \subset 4Z \subset Z \qquad \qquad \text{... (1)}$$

and $\qquad \{0\} \subset 72Z \subset 81Z \subset 9Z \subset Z. \qquad \qquad \text{...(2)}$

of the given series $\{0\} \subset 8Z \subset 4Z \subset Z$ and $\{0\} \subset 9Z \subset Z.$ respectively.

Hence, both series (1) and (2) have same length. Thus having four factor groups. These factors groups are given by

$$Z/4\,Z \cong Z_4 \cong 18\,Z/72Z$$

$$4\,Z/8\,Z \cong Z_2 \cong 9\,Z/18Z$$

$$8\,Z/72\,Z \cong Z_9 \cong Z/9Z$$

and $\qquad 72\,Z/\{0\} \cong 72\,Z$

Hence, we conclude that refinements (1) and (2) are the required isomorphic refinements of the given subnormal series.

Example 2. *Show that the group of integers **Z** has no compositon series.*

Solution. Let us suppose

$$\mathbf{Z} = G_0 \supset G_1 \supset G_2 \supset ... \supset G_{n-1} \supset G_n = \{0\}$$

be a subnormal series of Z and G_{n-1} must be of the form rZ for some positive integer r.

Now, since $\dfrac{G_{n-1}}{G_n} \cong rZ$ and rZ is infinite cyclic group having many non-trivial proper normal subgroups of $\mathbf{Z}$ therefore $\dfrac{G_{n-1}}{G_n}$ is not simple . Hence, $\mathbf{Z}$ has no compositon series.

Example 3. *Find all composition series of $\mathbf{Z}_{60}$.*

Solution. If is known that $\mathbf{Z}_{60}$ represent the group of integers under addition modulo 60.

Clearly, the sets (2), (3), (4), (5), (6), (10), (12), (15), (20), (30) are all subsets of $\mathbf{Z}_{60}$.

Hence, all composition series of Z_{60} are given as below:

$\{0\} \subset (12) \subset (4) \subset (2) \subset Z_{60};$ $\{0\} \subset (12) \subset (6) \subset (2) \subset Z_{60}$

$\{0\} \subset (12) \subset (6) \subset (3) \subset Z_{60};$ $\{0\} \subset (20) \subset (4) \subset (2) \subset Z_{60}$

$\{0\} \subset (20) \subset (10) \subset (2) \subset Z_{60};$ $\{0\} \subset (30) \subset (6) \subset (2) \subset Z_{60}$

$\{0\} \subset (30) \subset (10) \subset (2) \subset Z_{60};$ $\{0\} \subset (30) \subset (6) \subset (3) \subset Z_{60}$

$\{0\} \subset (30) \subset (15) \subset (3) \subset Z_{60};$ $\{0\} \subset (20) \subset (10) \subset (5) \subset Z_{60}$

$\{0\} \subset (30) \subset (10) \subset (5) \subset Z_{60};$ $\{0\} \subset (30) \subset (15) \subset (5) \subset Z_{60}$

Example 4. *Show that the series* $\{e\} \subset A_n \subset S_n$ *of* S_n *is a composition series of* S_n *where* S_n *is a symmetric group and* A_n *is an alternating group.*

Solution. It is known that the factor group and $A_n/\{e\}$ is isomorphic to A_n, which is simple for all $n \geq 5$ and the factor group S_n/A_n is isomorphic to Z_2, which is also simple. Hence the series $\{e\} \subset A_n \subset S_n$ is a composition series of S_n.

4.11 ASCENDING AND DESCENDING SUBNORMAL CHAIN

 (i) **Accessible subgroup.** Let G be a group. A subgroup H of G is said to be accessible subgroup if it occur in a subnormal series of G.

 (ii) **Descending subnormal chain.** A descending sequence of subgroups of a group G given by

$$G = H_0 \supset H_1 \supset H_2 \supset \dots \supset H_n \supset \dots.$$

is called a decending subnormal chain of G if every subgroup $H_n (n \in N)$ is a proper normal subgroup of H_{n-1}.

For Example

If $G = \{a\}$ is an infinite cyclic group, then following sequence of subgroup of G given by

$$G \supset (a^2) \supset (a^4) \supset (a^6) \supset \dots. \supset (a^{2n}) \supset \dots$$

is an infinite descending subnormal chain of G.

 (iii) **Ascending subnormal chain.** Let G be a group. An ascending sequence of subgroups of G given by

$$\{e\} \supset F_1 \supset F_2 \supset \dots. \supset F_n \supset \dots.$$

is known as an ascending subnormal chain of G if every subgroup $F_n (n \in N)$ is a proper normal subgroup of F_{n+1} and if all the subgroups F_n are accessible in G.

For Example

If G is an additive group of rational numbers, then all the subgroups of G form a ascending subnormal chain of G.

Theorem 1. *A group G has a composition series if and only if all ascending and all descending subnormal chain break off.*

Proof. Let G be group. Firstly, let us suppose that G has a composition series and k be its composition length.

If G has an infinite descending subnormal chain

$$G \supset H_0 \supset H_1 \supset H_2 \supset \dots. \supset H_n \supset \dots \qquad \dots (1)$$

then for $n \geq k$ the subnormal series

$$G \supset H_1 \supset H_2 \supset \dots. \supset H_n \supset \{e\}$$

consisting of the first n terms of (1) and unit subgroup which contradict the Schrierer's theorem. Further, assume that G has an infinite ascending subnormal chain given by

$$\{e\} \subset F_1 \subset F_2 \subset .... \subset F_n \subset .... \qquad ... (2)$$

Then for $n \geq k$ we may construct any subnormal series of G containing F_n as follows

$$G \supset G_1 \supset G_2 \supset ... \supset G_{s-1} \supset F_n \supset ... \supset \{e\} \text{ for } s \geq 1$$

Now, since F_n is assumed to accessible subgroup so that such a series exists. But then the series

$$G \supset G_1 \supset ... \supset G_{s-1} \supset F_n \supset F_{n-1} \supset ... \supset F_2 \supset F_1 \supset \{e\}$$

is subnormal series and its length is greater then K, which is also a contradiction. Therefore, both subnormal chain break off.

Conversely, suppose that all ascending and descending subnormal chain of a group G break off. Then it follows that every accessible subgroup H of $G \neq \{e\}$ must have at least one proper maximal normal subgroup. Now if every proper subgroup of H are contained in a larger proper normal subgroup, then we get an infinite ascending chain of normal subgroup of H, and this would be an ascending subnormal chain of G.

Now, to construct the composition series of G, we proceed as follows:

Take a proper maximal normal subgroup H_1 of G. If $H_0 = G, H_1, H_2, ..., H_n$ have been choosen in such a way that each is proper normal subgroup of the proceeding one, then H_n is obviously accessible in G.

If $H_n \neq \{e\}$, we take as H_{n+1} one of the maximal normal subgroups of H_n. Using this, we must arrive at the unit subgroup $\{e\}$ after a finite number of steps and therefore, we get a compostion series of G.

EXERCISE 4.1

1. Let $G = \sigma(30)$ (a cyclic group of order 30) and x being its generator. Then show that
   $$G \supset (x^5) \supset (x^{10}) \supset \{1\}$$
   and $G \supset (x^2) \supset (x^6) \supset \{1\}$ are isomorphic.

2. Show that the series $S_n \supset A_n \supset \{e\}$ of S_n is a composition series of S_n where S_n is a symmetric group and A_n is an alternating group.

3. Show that a finite cyclic group of prime order has only one composition series.

4. Find all composition series of
   (i) $Z_5 \times Z_5$ (ii) $S_3 \times Z_2$

4.12 SYLOW'S p-SUBGROUP

4.12.1 p-group

Let p be a fixed prime number. Then a group G is said to be p-group is every element of G has a order, a power of p.

4.12.2 p-subgroup

Let p be a prime number. Then a subgroup H of a group G is said to be p-subgroup if every element of H is a power of p.

Theorem 1. *Let G be a finite group. Then G is a p-group if and only if $o(G) = p^n$.*

Proof.   Let G be a p-group and q be any prime dividing $o(G)$ Then, by Cauchy's theorem we can say that there exists $x \in G$ such that $o(x) = q$. But $o(x) = p^r$ as G is a p-group

Therefore $q = p^r$

$$\Rightarrow \qquad\qquad\qquad q = p$$

$\Rightarrow$ p is the only prime dividing $o(G)$.

$$\Rightarrow \qquad\qquad\qquad o(G) = p^n$$

Conversely, let $o(G) = p^n$, where p is prime.

Further, let $x \in G$. then $o(x)/o(G) = p^n$

$$\Rightarrow \qquad\qquad\qquad o(x) = p^r$$

Thus every element of G has order which is some power of p.

Hence, G is a p-group.

Theorem 2. **(Converse of the Lagrange's Theorem).** *Let G be an abelian group of order n. Then for every divisor m of n, G has a subgroup of order m.*

Proof. Let G be an abelian group of finite order. To show, for every divisor m of n, group G has a subgroup of order m.

We prove this result by induction.

For $n = 1$, we have $\qquad\qquad G = \{e\}$

$\Rightarrow \qquad\qquad$ Result is true for $n = 1$

Now, suppose that results is true for all groups with order less then $o(G)$.

Let $\qquad\qquad\qquad o(G) = n,\ m/n,\ m > 1$

Let p be a prime dividing m, therefore $p/n = o\ (G)$

Then by Cauchy's theorem of finite groups there exists $x \in G$ such that

$$o(x) = p$$

Let K be a cyclic group generated by x.

Then we have

$$o(K) = o(x) = p$$

[$\therefore$ order of each subgroup of a cyclic group is equal to the order of its generator.]

Now, since G is abelian, K is normal in G.

Consider $\quad \dfrac{G}{K}$ such that $o\left(\dfrac{G}{K}\right) = \dfrac{n}{p} < n$

Also $\dfrac{G}{K}$ is abelian.

Let $m | p m_1$.

Then, $\qquad\qquad\qquad m = p m_1 | o(G) = o(G/K) o\ (K)$

$\Rightarrow \qquad\qquad\qquad m_1 | o(G/K)$

Then, by induction hypothesis $\exists$ subgroup H/K of G/K such that

$$o(H/K) = m_1,\ H \subseteq G$$

Thus, $\qquad\qquad\qquad o(H) = o(K) m_1 = p m_1 = m$

$\Rightarrow \quad$ Result is true in this case also.

Hence, by induction theorem is true for all n.

4.12.3 Sylow p- Subgroups

Definition. *Let G be a group and p be a prime number such that p^n divides order of G and p^{n+1} does not divide it. Then, a subgroup H of G such that $o(H) = p^n$ is known as sylow p-subgroup of G or p-sylow subgroup of G.*

4.13 SYLOW'S GENERAL THEOREMS

Theorem 1. **(Sylow's first Theorem).** *Let G be a group and p be any prime number and m be a positive integer such that p^m divides o(G). Then, there exists a subgroup H of G such that $o(H) = p^m$.*

Proof. We shall prove this theorem by induction.

When $o(G) = 1$, then result is true.

Further, assume that result is true for all groups with order less than $o(G)$. Let

$$p^m \mid o(G).$$

If K is a subgorup of G such that $K \neq G$ and $p^m \mid o(K)$

Then, by induction $H \subseteq K$ such that $o(H) = p^m$.

Then $H \subseteq K \Rightarrow H \subseteq G.$

Therefore, result holds in this case.

Now, assume p^m does not divide order of any proper subgroup of G.

Consider class equation of G.

$$o(G) = o[Z(G)] + \sum_{a \notin Z(G)} \frac{o(G)}{o(N(a))}$$

$$a \notin Z(G) \Rightarrow N(a) \notin G$$

$$\Rightarrow p^m \nmid o(N(a))$$

But we have

$$p^m \mid o(G) \Rightarrow p^m \left| \frac{o(G)}{o N(a)} \cdot o(N(a)) \right.$$

$$\Rightarrow p^m \left| \frac{o(G)}{o(N(a))} \right. , \text{ for all } a \notin Z(G) \text{ as } p^m \nmid o(N(a))$$

$$\Rightarrow p \left| \sum_{a \notin Z(G)} \frac{o(G)}{o(N(a))} \right.$$

$$\Rightarrow p \left| o(G) - \sum_{a \notin Z(G)} \frac{o(G)}{o(N(a))} = o(Z(G)) \right.$$

$$\Rightarrow \exists x \in Z(G) \text{ such that } o(x) = p$$

Let $K = \{x\} \subseteq Z(G)$

$\Rightarrow$ K is normal in G.

Now $o(G/K) < o(G)$ and $p^m \mid o(G) = (o(G) \mid o(K)) \cdot o(K)$

$p^m \nmid o(K)$ and therefore $p^{m-1} \mid p^m \mid o(G/K)$

By induction hypothesis there exists a subgroup H/K of G/K such that

$$o(H/K) = p^{m-1}$$

Therefore $o(H) = p^m$ Also, $H/K \subseteq G/K \Rightarrow H \subseteq G$. Thus, result is true in this case also.

Hence, by mathematical induction, theorem is proved.

REMARK

☞ If p is prime such that $p^n \mid o(G)$ and $p^{n+1} \nmid o(G)$, then $\exists$ a p-Sylow subgroup of G.

Theorem 2. **(Sylow's Second Theorem).** *Any two Sylow p-subgroup of a finite group G are conjugate in G.*

Proof. Let G be a group and P, Q be Sylow's p-subgrops of G such that

$$O(P) = p^n = o(Q)$$

where $$p^{n+1} \nmid o(G)$$

Let us suppose P and Q are not conjugate in G.

i.e., $$P \ne gQg^{-1} \text{ for any } g \in G$$

It is known that

$$o(PxQ) = \frac{o(P)o(Q)}{o(P \cap xQx^{-1})}$$

Now, since

$$P \cap (x \, Q \, x^{-1}) \subseteq P$$
$$o \, (P \cap (x \, Q \, x^{-1})) = P^m , \, m \le n$$

If $m = n$, then we have

$$P \cap (x \, Q \, x^{-1}) = P$$
$$\Rightarrow \qquad\qquad P \subseteq (x \, Q \, x^{-1})$$
$$\Rightarrow \qquad\qquad P = (x \, Q \, x^{-1}) \qquad\qquad (\because o(x \, Q \, x^{-1}) = o(Q) = o(P))$$

which is a contradiction

Therefore, we get $m < n$ and hence

$$o(pxQ) = p^{2n-m}, \, m < n, \, \forall \, x \in G.$$

which implies

$$o(pxQ) = p^{n+1} (p^{n-m+1}) = \text{multiple of } p^{n+1}$$

Therefore $$o(G) = \sum_x o(PxQ) = \text{multiple of } p^{n+1}$$

$$\Rightarrow \qquad\qquad p^{n+1} \mid \text{RHS} \Rightarrow p^{n+1} \mid o(G)$$

which is a contradiction. Hence, $P = g \, Q \, g^{-1}$, for some $g \in G$.

REMARK

☞ Let P be a Sylow p-subgroup of G. Then the number of sylow p- subgroup of G is equal to $\dfrac{o(G)}{o(N(p))}$.

Theorem 3 (Sylow's third Theorem). *The number of Sylow p-subgroups of G is of the form $1+kp$ where $(1+kp) \mid o(G)$, k being a non-negative integer.*

Proof.   Let G be a group and P be a Sylow p-subgroup of G.

Let $O(P) = p^n$

Now $G = \bigcup_x PxP$

$$= \bigcup_{x \in N(P)} PxP = \bigcup_{x \notin N(P)} PxP$$

$$x \in N(P) \Rightarrow Px = xP$$

$\Rightarrow$ $Px = PxP$ $[\because PP = P]$

Thus $\bigcup_{x \notin N(P)} PxP = \bigcup_{x \in N(P)} P(x) = N(P)$

Since $P \leq N(P)$ and union of disjoint right sets equal the set, therefore

$$x \notin N(P) \Rightarrow Px = xP$$

$\Rightarrow$ $xPx^{-1} \neq P$

$\Rightarrow$ $o(P \cap xPx^{-1}) = p^m, \quad m < n$ [By Sylow's second theorem]

$\Rightarrow$ $o(PxP) = p^{2n-m}, \ m < n$

Thererfore $o(G) = o(N(P)) + \sum_{x \notin N(P)} o(PxP)$

$$= o(N(P)) + \sum_{x \notin N(P)} p^{2n-m}$$

Thus $\dfrac{o(G)}{o(N(P))} = 1 + \sum \dfrac{p^{2n-m}}{o(N(P))} = 1 + \dfrac{p^{n+1} \cdot t}{o(N(P))}$, where t is an integer.

Further, since $\text{LHS} = \text{integer}$

$$\dfrac{p^{n+1} \cdot t}{o(N(P))} = r \text{ , an integer}$$

Therefore, $p^{n+1} \cdot t = r \cdot o\,(N(P))$

Also $P \leq N(P)$

$\therefore$ $o(P) \mid O(N(P))$

$\Rightarrow$ $p^n \mid O(N(P))$

$\Rightarrow$ $o(N(P)) = p^n \cdot u$

Thus $p^{n+1} \cdot t = r \cdot o\,(N(P))$

$\Rightarrow$ $pt = r \cdot u$

$\Rightarrow$ $p \mid ru$

If $p \mid u$, then $p^{n+1} \mid o(N(P)) \mid o(G)$

$\Rightarrow$ $p^{n+1} \mid o(G)$, which is a contradiction.

Therefore, $\quad p \mid r \Rightarrow \dfrac{r}{p} =$ integer

$$\dfrac{t}{u} = \text{integer } k = \dfrac{r}{p}$$

$\Rightarrow$

Thus $\quad \dfrac{o(G)}{o(N(P))} = 1 + \dfrac{p^{n+1}.t}{o(N(P))} = 1 + \dfrac{p.t}{u} = 1 + kp$

Then, we get

$$\dfrac{o(G)}{o(N(P))} = \text{number of Sylow } p\text{-subgroup of } G.$$

Thus, the number of Sylow p-subgroups is of the form $1 + kp = \dfrac{o(G)}{o(N(P))}$

Hence, $\quad (1+kp) \mid o(G).$

REMARK

☞ If $o(G) = p^n . q$, $(p\,q) = 1$ then the number of Sylow p-subgroup is $1 + kp$ where $1+kp \mid p^n q$ which implies $(1 + kp) \mid q$ as $(1 + kp, p^n) = 1$.

Theorem 4. *If P is the only Sylow p-subgroup of G, then P is normal in G and conversely.*

Proof. By Sylow's third theorem, we have

$$\dfrac{o(G)}{o(N(P))} = 1$$

$\Rightarrow \qquad\qquad o(G) = o(N(P))$

Since $\qquad\qquad N(P) \subseteq G$

$\Rightarrow \qquad\qquad N(P) = G$

Therefore, P is normal in G.

Conversely, if Sylow p-subgroup P is normal in G, then

$$N(P) = G$$

$\Rightarrow \qquad\qquad o(N(P)) = o(G)$

$\Rightarrow \qquad\qquad \dfrac{o(G)}{o(N(P))} = 1$

$\Rightarrow \quad$ The number of Sylow p-subgroup of G is 1.

Hence, P is the only Sylow p-subgroup of G.

Theorem 5. *Let P be a Sylow p-subgroup of G. Let $x \in N(P)$ such that $o(x) = p^i$. Then $x \in P$.*

Proof. Let G be a group and P be a Sylow's p-subgroup of G.

Let us suppose $\qquad O(P) = p^n, p^{n+1} \nmid O(G)$

Now $\qquad\qquad (px)^{p^i} = P.x^{p^i} = P . e = P$

$\Rightarrow \qquad\qquad o(px) \mid p^i$

$\Rightarrow$ $o(Px) = p^j, j \geq 0$

Let $j > 0, \bar{k} = <px> \leq \dfrac{N(P)}{P}$ such that $o(\bar{K}) = p^j$

Further, Since $(\bar{K}) \subseteq \dfrac{N(P)}{P}, \bar{k} = \dfrac{k}{p},$ where $k \subseteq N(P)$

$$p^j = o(\bar{k}) = \dfrac{o(k)}{o(P)} = \dfrac{o(k)}{p^n}$$

$\Rightarrow$ $o(k) = p^{n+j}, j > 0$

But $o(k) | o(N(P)) | o(G)$

$\Rightarrow$ $p^{n+j} | o(G), j > 0$, which is a contradiction.

Thus, $j = 0, o(Px) = p^j = 1$

Hence, $Px = P \Rightarrow x \in P$

REMARKS

☞ Every p-subgroup of a finite group G is contained in some Sylow's p-subgroup of G.

☞ If G is a finite Group and P is a p-subgroup of G, then P is a Sylow p-subgroup of G if and only if no p-subgroup of G properly contains P.

Solved Examples

Example 1. Let $o(G) = pq$ where p and q are distinct primes, $p < q$, $p \nmid q$. Show that that G is cyclic.

Solution. It is known that the number of Sylow p-subgroup is $1 + kp$ and $1 + kp | q$ which implies $1 + kp = 1$ or q.

Now, $1 + kp = 1 \Rightarrow$ Sylow p-subgroup is unique.

Therefore, Sylow p-subgroup H is normal in G.

Further, $1 + kp = 1 \Rightarrow kp = q - 1$, i.e., $p | q - 1$, which is a contradiction.

Therefore, $1 + kp \neq q \Rightarrow$ Sylow p-subgroup is normal.

The number of Sylow q-subgroups is $1 + k_1 q$ and $1 + k_1 q | p \Rightarrow 1 + k_1 q = 1$ or p.

If $1 + k_1 q = p$ then $k_1 q = p - 1$

$\Rightarrow$ $q | p - 1 \Rightarrow q \leq p - 1 < p \Rightarrow q < p$, which is a contradiction.

If $1 + k_1 q = 1$, then Sylow q-subgroup K is normal in G.

We have $o(H) = p, o(K) = q, H \cap K = \{e\}$. H and K both are normal to G.

So, $hk = kh, \forall h \in H, k \in K$

Let $H = \{a\}$ and $K = \{b\}$ [∵ Groups of prime order are cyclic.]

               $o(a) = o(H) = p, o(b) = o(K) = q$

Now $ab = ba$ $(o(a), o(b)) = (p, q) = 1$

Thus $o(ab) = o(a) \, o(b) = pq = o(G)$

Hence, G is cyclic.

Example 2. *Let $o(G) = 30$, then show that*

(i) *either Sylow-3 subgroup or Sylow-5-subgroup is normal in G.*

(ii) *G has a normal subgroup of order 15.*

(iii) *both Sylow-3- subgroup and Sylow-5 subgroup are normal in G.*

Solution. (i) It is given that $o(G) = 30 = 2 \times 3 \times 5$

It is known that the number of Sylow 3-subgroup is $1+3k$ and $(1+3k)\,|\,10$

$\Rightarrow$ $\qquad\qquad k = 0$ or 3

If $k = 0$, the Sylow 3-subgroup is normal.

If $k \neq 0$**,** then $k = 3$, then we have 10 Sylow 3- subgroups H_i each of order 3.

Then, we have 20 elements of order 3.

Because for $i \neq j$, $o(H_i \cap H_j)\,|\,o(H_i) = 3$, which gives

$\qquad o(H_i \cap H_j) = 1$ only.

Therefore, 20 elements are different.

Also, each H_i has one element e of order 1 and other two of order 3 and

$$a \in H_i \Rightarrow o(a)\,|\,o(H_i) = 3 \Rightarrow o(a) = 1, 3.$$

Further, the number of Sylow 5-subgroup is $1+5k_1$

and $\qquad (1+5k_1)\,|\,6 \Rightarrow k_1 = 0, 1.$

If $k_1 = 0$, the Sylow 5-subgroup is normal.

If $k_1 \neq 0$, then $k_1 = 1$, then we have 6 Sylow 5-subgroups each of order 5 and proceed same as above, we get 24 elements of order 5. But we have already counted 20 elements of order 3. Thus, we have more than 44 elements in G, which is a contradiction.

Thus the only possibility is that either $k = 0$ or $k_1 = 0$.

Hence, either Sylow 3-subgroup or Sylow 5-subgroup is normal in G.

(ii) Let us suppose that H be a Sylow 3-subgroup of order 3 and K be a Sylow 5-subgroup of order 5. Then by (i), either H is normal in G or K is normal in G.

In both the cases, $HK \subset G$, $o(HK) = 15$, because $o(H \cap K)$ divide $o(H) = 3$ and $o(K) = 5$, which gives $o(H \cap K) = 1$

$$\left[\because o(HK) = \frac{o(H)o(K)}{o(H \cap K)} \right]$$

Now, since index of HK in G is 2, hence HK is normal in G.

(iii) Let us suppose that H is normal in G, K is not normal in G. Then by (i), G has 6 Sylow 5-subgroups and therefore 24 elements of order 5. But $o(HK) = 15$, therefore HK is cyclic.

Thus , HK has 8 elements of order 15.

$\Rightarrow$ $\qquad G$ has 32 (= 4×8) elements, which is a contradiction.

$\Rightarrow$ $\qquad K$ is normal in G.

Hence, both H and K are normal in G.

Example 3. *Let G be a group of order 231. Show that Sylow 11-subgroup of G is contained in the centre of G.*

Solution. It is given that $o(G) = 231 = 3 \times 7 \times 11$.

It is known that Sylow 11-subgroup of G is $1 + 11k$ and $(1 + 11k) | 21$. Clearly, then $k = 0$.

Therefore, Sylow 11-subgroup H of G is normal in G.

Further, Sylow 7-subgroup K of G is normal in G.

$$o(H) = 1, o(K) = 7.$$

Now, $o(G/K) = 33 = 3 \times 11$ and $3 \nmid (11-1)$,

so G/K is cyclic and therefore G/K is abelian.

But G' is smallest subgroup of G such that G/G' is abelian, where G' is the commutator subgroup of G.

So $G' \subseteq K \Rightarrow o(G') = 1$ or 7.

If $o(G') = 1$, then $G' = \{e\} \Rightarrow x^{-1}y^{-1}xy = e \Rightarrow xy = yx, \forall x, y \in G$

$\Rightarrow \quad G$ is abelian.

$\Rightarrow \quad G = Z(G)$

$\Rightarrow \quad H \subset Z(G)$

If $o(G') = 7$, then $G' = K$.

Then, clearly $H \cap K = \{e\}$, because $o(H \cap K)$ divides $o(H) = 11$ and $o(K) = 7$.

Let $x \in H$, $y \in G$, then $x^{-1}y^{-1}xy \in G' = K$.

Now, $x^{-1}y^{-1}xy = x^{-1}(y^{-1}xy) \in H$, H is normal in G.

Thus, $x^{-1}y^{-1}xy \in H \cap K = \{e\}$

$\Rightarrow \quad xy = yx \, ; \, y \in G, \, x \in H$

Example 4. *If G is a finite non-abelian simple group and $H \subseteq G$, show that $[G : H] \geq 5$.*

Solution. Let $[G : H] = n$.

Let us denote the set of all left cosets of H in G by L .

Then, $o(L) = n$

Define $f : G \to A(L) = S_n$ such that $f(g) = Tg$, where $Tg : L \to L$ such that $Tg(xH) = gxH$.

It can be easily shown that Tg is one-one and onto.

Thus, $Tg \in A(L)$ and f is a homomorphism.

Now, let $g \in \text{Ker } f \Rightarrow f(g) = Tg = I$

$\Rightarrow \quad gxh = xH \, ; \, \forall \, x \in G$

$\Rightarrow \quad gH = H \Rightarrow g \in H$

$\Rightarrow \quad \ker f \subseteq H$

We know that Ker. f is normal in G and G has no non-trivial normal subgroup

so Ker. $f = \{e\}$ or G.

But Ker. $f = G \Rightarrow H = G$, which is not true.

$\therefore \quad \ker. f = \{e\}$

Therefore, G is isomorphic to a subgroup of S_n.

For $n = 4$, G is a subgroup of S_4.

Now, since G is simple, $G \neq S_4$ and $o(G)$ must be divisible by at least two primes, because of $o(G)$ is divisible by one prime only, then G has non-trivial centre as normal subgroup.

Thus, $\qquad o(G) = 2^2 \times 3$ or $o(G) = 2 \times 3$.

If $o(G) = 2^2 \times 3$, the Sylow 3-subgroup or Sylow 2 subgroup is normal, therefore,

$$o(G) \neq 2^2 \times 3.$$

If $o(G) = 2 \times 3$ and G is non-abelian, then $G \cong S_3$.

Therefore, G has normal subgroup A_3.

In both the cases, we get a contradiction. Thus $n \neq 4$.

If $n = 3$, then G is a subgroup of S_3. Since, G is simple, $G \neq S_3$, therefore, $G \subset S_3 \Rightarrow$ $o(G) = 1, 2$ or 3, which is not possible (because G is non-abelian) and it is known that group of order less than or equal to 4 is always abelian.

$\Rightarrow \qquad\qquad n \neq 1, 2.$

Thus, $n > 4$. Hence, $[G : H] \geq 5$.

Example 5. *Show that there is no simple group of order* 144.

Solution. Let G be a group of order 144. We can write

$$144 = 2^4 \times 3^2$$

Suppose G is simple.

It is known that the number of Sylow 3-subgroup of G is $1+3k$ and $(1+3k)\,|\,16$

$\Rightarrow \qquad\qquad k = 0, 1, 5.$

If $k = 0$, then Sylow 3-subgroup is unique and normal, which is not possible.

If $k = 1$, then there exist 4 Sylow 3-subgroups of G and if P is any one of these, then as

$$\frac{o(G)}{o(N(P))} = 4 = \text{number of Sylow 3-subgroups}$$

We find $N(P)$ as a subgroup of G with index, which is not possible.

If $k = 5$, then there exist 16 Sylow 3-subgroups each of order 9 in G. Let H_1, H_2 be any Sylow 3-subgroups.

Now, $\qquad H_1 \cap H_2 \subseteq H_1$

$o(H_1 \cap H_2)\,|\,9 \Rightarrow o(H_1 \cap H_2) = 1, 3$ or 9.

If $\qquad o(H_1 \cap H_2) = 9$, then $o(H_1 \cap H_2) = o(H_1) = o(H_2)$

$\Rightarrow \qquad\qquad H_1 = H_1 \cap H_2 = H_2$, which is a contradiction.

If $\qquad o(H_1 \cap H_2) = 3$, then $H_1 \cap H_2$ is normal in H_1 and H_2.

Further, since $N(H_1 \cap H_2)$ is the largest subgroup of G in which $H_1 \cap H_2$ is normal.

Now, $\qquad H_1 \subseteq N(H_1 \cap H_2),\ H_2 \subseteq N(H_1 \cap H_2)$

$\Rightarrow \qquad\qquad H_1 H_2 \subseteq N(H_1 \cap H_2) \subseteq G$

Also, $\qquad\qquad o(H_1 H_2) = \dfrac{o(H_1)o(H_2)}{o(H_1 \cap H_2)} = 27$

Since $\qquad o(N(H_1 \cap H_2)) \geq 27$ and divides $o(G) = 144$, therefore

$o(N(H_1 \cap H_2)) = 36, 48$, 72 or 144.

But then $[G : N(H_1 \cap H_2)] = 4, 3, 2$ or 1, which is not possible.

Thus, $o(H_1 \cap H_2) = 1$

Thus, we conclude that any 2 Sylow 3-subgroups of G intersects trivially. This gives 128 elements of order 3^i ($i = 1$ or 2).

Since Sylow 2-subgroup is of order 16 and not normal, there are at least 16 elements of order $2^i = (i = 1, 2, 3$ or 4) and are identity element, therefore, we get 145 elements in G, which is a contradiction. Hence, G is a simple group.

Example 6. *Let G be a finite group and H be normal in G. If p is a prime dividing o(G) such that* $([G : H], p) = 1$, *show that H contains every Sylow p-subgroup of G.*

Solution. Suppose $[G : H] = m$.

Thus, $(m, p) = 1$

$\Rightarrow$ $p \nmid m \Rightarrow p^i \nmid m, i > 0$

Further, let $p^n \mid o(G), p^{n+1} \nmid o(G)$

Then, we can write

$$o(G) = \frac{o(G)}{o(H)}.o(H) = [G : H] \, o(H) = m.o(H)$$

Now, since $p^n | o(G), p^n | m$, implies $p^n | o(H)$, then by Sylow's first theorem, there exist $K \subseteq H$ such that $o(K) = p^n$. Thus K is a Sylow p-subgroup of G. If P is any p-subgroup of G, then

$$P = g K g^{-1}, g \in G$$

$\Rightarrow$ $K = g^{-1} P g$

But $K \subseteq H \Rightarrow g^{-1} P G \subseteq H$

$\Rightarrow$ $P \subseteq g H g^{-1} = H$, which is normal in G.

Hence, H contains all Sylow p-subgroups of G.

Example 7. *Let G be a finite group and $H \subseteq G$. If p is a prime dividing o(G). Let P be a Sylow p-subgroup of H contained in some Sylow p-subgroup S of G, show that $P = S \cap H$.*

Solution. It is given that

$$P \subseteq S, P \subseteq H$$

$\Rightarrow$ $P \subseteq S \cap H$

Also, we have $S \cap H \subseteq S \Rightarrow S \cap H$ is a p-subgroup.

Now, since $S \cap H \subseteq H$ and $S \cap H$ is a p-subgroup of H, therefore

$$P \subseteq S \cap H \subseteq H$$

Since P is Sylow p-subgroup of H, there is no p-subgroup of H properly containing P, therefore,

$$P = S \cap H$$

Example 8. *Let G be a group of order pqr, $p < q < r$, where p, q, r are primes. Show that some Sylow subgroup of G is normal. Also, show that G is not simple.*

Solution. Let us suppose that no Sylow subgroup of G is normal. Then, the number of Sylow p-subgroup of G is $1 + kp$ and $(1+kp)|qr$ which implies

$$1 + kp = q, r \text{ or } qr \ (\geq q)$$

and the number of Sylow q-subgroups of G is $(1+k'q)|pr$

$$\Rightarrow 1 + k'q = p, r \text{ or } pr$$

If $1 + k'q = p$, then $q | (p - 1) \Rightarrow q < p$, which is a contradiction.

Thus, the number of Sylow q-subgroups of G is r or pr $(\geq r)$

and, the number of Sylow r-subgroups of G is $(1 + k^n r) | pq$

$$1 + k^n r = p, q \text{ or } pq$$

If $1 + k^n r = p$ or q, then $r | (p - 1)$ or $r | q - 1 \Rightarrow r < p$ or $r < q$

Which is a contradiction.

Thus, the number of Sylow-r-subgroups of G is pq.

Further, Sylow p-subgroups give at least $q(p-1)$ elements of order p and Sylow q-subgroups give $r(q-1)$ elements of order q and Sylow r-subgroups give $pq(r-1)$ elements of order r.

Thus, $o(G) = pqr \geq q(p - 1) + r(q - 1) + pq(r - 1) + 1$

$$0 \geq rq - q - r + 1 = (q - 1)(r - 1)$$

$\Rightarrow$ $(q - 1)(r - 1) \leq 0$, which is a contradiction.

$\Rightarrow$ Some Sylow subgroup of G is normal.

Hence, G is not a simple group.

Example 9.   *Let G be a non-abelian group of order 12 in which Sylow 3-subgroup is normal. Show that G has an element of order 6.*

Solution.   It is known that Sylow 3-subgroup is normal, thus it will be unique. Let it be denoted by H. Then clearly $o(H) = 3$. Since 3 is prime and we know that every group of prime order is cyclic, therefore, H is cyclic. Let $H = \{a\}$, then

$$o(a) = o(a^{-1}) = 3 \quad [\because \text{ order of a cyclic group is equal to the order of its generator.}]$$

Let $C'(a)$ is the conjugate class of a in G, then we have

$$o[C'(a)] = \frac{o(G)}{o(N(a))} = 1 \text{ or } 2$$

It is known that

$$C'(a) = [g^{-1}ag : g \in G]$$

and $o(g^{-1}a\,g) = o(a) = 3$

and as a, a^{-1} are the only elements of order 3, either

$$C'(a) = [a] \text{ or } C'|(a) = [a, a^{-1}]$$

Therefore, $o[N(a)] = 6$ or 12

$\Rightarrow$ $\exists\, b \in N(a)$ such that $o(b) = 2$.

$\Rightarrow$ $ab = ba$ and $o(ab) = 6$

Hence, G has an element of order 6.

Example 10.   *Show that there is no simple group of order 120.*

Solution.   Let G be a group of order 120. Then

$$o(G) = 120 = 2^3 . 3 . 5$$

Now, the number of Sylow 2-subgroups of G is of the form $1 + 2k$ such that $1 + 2k\,|o(G)$, *i.e.*,$|\ 1 + 2k | 15$ as $1 + 2k$ is 1, 3 or 15.

If $1+2k = 1$, then there is a unique Sylow 2-subgroup of G of order 8. Therefore, it is normal in G.

If $1+2k = 3$, then there are 3 Sylow 2-subgroups of G of order 8. Let, these subgroups be H_1, H_2 and H_3.

Since $$o(H_1) = o(H_2) = o(H_3) = 2^3.$$

Therefore, the center of either H_1, H_2 or H_3 will have 2 elements and then centre of a group is normal subgroup. Therefore, each H_1, H_2, H_3 contains a normal subgroup of order 2.

If $1+2k = 15$, then there are 15 Sylow 2-subgroup of order 8. Using the same argument, each subgroup will contain a normal subgroup.

$\Rightarrow$ G has a proper normal subgroup. Hence, G is not simple.

Example 11. *Let G be a group of order* 108. *Show that there exists a normal subgroup of order* 27 *or* 9.

Solution. Let G be a group of order 108, *i.e.*, $o(G) = 108 = 2^2.3^3$.

The numbe of Sylow 3-subgroup of G is of the form $1+3k$ such that $1+3k \mid o(G)$, *i.e.*, $1+ 3k \mid 2^2.3^3$.

Now, since $1+ 3k$ is relatively prime to 3 so that $1+3k \mid 2^2$.

Since the factor of 4 are 1, 2 and 4, so $1+3k = 1$ or 4.

If $1+3k = 1$, then there is a unique Sylow 3-subgroup of G of order 27. which is normal in G.

If $1+3k = 4$, then there are 4 Sylow 3-subgroups each of order 27. Let H and K be two such distinct Sylow 3-subgroups.

Then, we have

$$o(HK) = \frac{o(H)o(K)}{o(H \cap K)} = \frac{27 \times 27}{o(H \cap K)}$$

But $$o(HK) \le o(G) = 108$$

Thus, $$\frac{27 \times 27}{o(H \cap K)} \le 108$$

$\Rightarrow$ $$o(H \cap K) > \frac{27 \times 27}{108}$$

$\Rightarrow$ $$o(H \cap K) > \frac{27}{4}$$

Since $o(H) = o(K) = 27$, therefore, $o(H \cap K) = 9$

We know that $H \cap K \subseteq H$ and $H \cap K \subseteq K$, therefore $H \cap K$ is a subgroup of H as well as K. Then, subgroups of order p^{n-1} are normal in a group of order p^n.

Now, since $H \cap K$ is normal in H and K, therefore, $H \subset N(H \cap K)$ and $K \subset N(H \cap K)$, then $HK \subset N(H \cap K)$, where $N(H \cap K)$ is a normalizer of $H \cap K$.

Also, $$o(HK) = \frac{o(H)o(K)}{o(H \cap K)} = \frac{27 \times 27}{9} = 81$$

and $N(H \cap K)$ is a subgroup of G, $o(N(H \cap K)) = 108$.

Therefore, $N(H \cap K) = G$. Hence, $H \cap K$ is a normal subgroup of order 9.

Example 12. *Find all non-abelian groups of order 8.*

Solution. Let G be a group of order 8. Since $o(a) | o(G) \, \forall a \in G$, so $o(a) = 1, 2, 4$ or 8. If for some $a \in G$, $o(a) = 8$, G is cyclic. Therefore, there is no element in G of order 8. If each non-identity element is of order 2, then G is abelian, therefore $\exists \, a \in G$ such that $o(a) = 4$.

Now, let $\qquad\qquad H = \{a\}. \, o(H) = o(a) = 4.$

So, H is normal is G as index of H in G is 2.

Now, let $\qquad\qquad G = H \cup Hb, \, b \notin H.$

Then, $b^2 \in H$ as $b^2 \notin H \Rightarrow H, Hb, Hb^2$ are distinct right coset of H in G, which is a contradiction.

If $b^2 = a$, then $\qquad o(b^2) = \dfrac{o(b)}{(2, o(b))} = \dfrac{o(b)}{2}$

$\Rightarrow \qquad\qquad\qquad o(b) = 2.o(a) = 8$, a contradiction.

Similarly, if $b^2 = a^3$, then $b^2 = a^{-1} \Rightarrow o(b^2) = o(a^{-1}) = 4$

$\Rightarrow \qquad\qquad\qquad o(b) = 8$

Thus, $\qquad\qquad\qquad b^2 = a$ or a^2.

Since H is normal in G, $b^{-1} \, ab \in H$.

But, $\qquad\qquad\qquad o(b^{-1}ab) = o(a) = 4$

$\Rightarrow \qquad\qquad\qquad b^{-1} \, ab = a$ or a^3.

If $\qquad\qquad\qquad b^{-1}ab = a$, then $ab = ba$.

Since $G = H \cup Hb$, $G = \{e, a, a^2, a^3, ab, a^2b, a^3b, b\}$, and $ab = ba$ would imply G is abelian.

Therefore, $\qquad\qquad b^{-1}ab = a^3.$

Hence, we have two non-abelian groups G of order 8 given by

(i) G generated by a, b, such that $a^4 = e, b^2 = e, b^{-1}ab = a^3 = a^{-1}$

(ii) G generated by a, b, such that $a^4 = e, -b^2 = e, b^{-1}ab = a^3 = a^{-1}$

REMARKS

☞ The group defined in (i) is the dihedral group of order 8 and (ii) is the quaternion group or order 8.

☞ Althrough the dehedral group and quaternian groups have same number of elements, they are not isomorphic.

Example 13. *Find all non-abelian groups of order 6.*

Solution. Let G be a non-abelian group of order 6. Then, by Cauchy's theorem, there exists $a, b \in G$ such that

$$o(a) = 3, o(b) = 2$$

Let $H = \{a\}$, then clearly, we have $o(H) = o(a) = 3$.

Also, since index of H in G is 2, thus H is normal in G. If $b \in H$, then

$o(b) | o(H) \Rightarrow 2 | 3$, which is a contradiction, so $b \notin H$.

$\Rightarrow \quad H$ and Hb are distinct right cosets of H in G.

Thus, $\qquad\qquad\qquad G = H \cup Hb = \{e, a, a^2, b, ab, a^2b\}.$

Now, since H is normal in G, therefore $b^{-1} ab \in H$.

$\Rightarrow \qquad\qquad b^{-1} ab = a \text{ or } a^2.$

If $b^{-1} ab = a$, then $\quad ab = ba \Rightarrow G$ is abelian, which is not possible.

Then, $\qquad\qquad b^{-1} ab = a^2 = a^{-1}$.

Thus, there is only one non-abelian group G of order 6 given by

$$G = \{e, a, a^2b, ab, a^2b, a^3 = e = b^2, b^{-1} ab = a^{-1}\}$$

Example 14. *Let G be a group of $n \times n$ invertible matrices over the integers modulo p, where p is prime. Find a p-Sylow subgroup of G.*

Solution. Let G be a group of $n \times n$ invertible matrices. Let $A \in G$. Since A is invertible, rows of A are linearly independent over the field F of integers mudulo p. Now, since first row of A is linearly independent, it is non-zero. It can be chosen in $(p^n - 1)$ ways. Second row should not be $\alpha(\alpha \in F)$ times the first row. Therefore, second row can be chosen in $p^n - p$ ways. Further, third row should not be α-times first row plus β times second row $(\alpha, \beta \in F)$. Therefore, third row can be chosen in $(p^n - p^2)$ ways. In this way, last n^{th} row can be chosen in $p^n - p^{n-1}$ ways.

Thus, $\qquad\qquad o(G) = (p^n - 1)(p^n - p) \dots (p^n - p^{n-1})$

$$= p^{1+1+\dots+(n-1)} [(p^n - 1)(p^{n-1} - 1)\dots(p-1)]$$

$$= p^{\frac{n(n-1)}{2}} [(p^n - 1)\dots(p-1)]$$

Further, since $(p, (p^i - 1)) = 1$, order of Sylow p-subgroup of G is $p^{\frac{n(n-1)}{2}}$

Now, let

$$P = \left\{ \begin{bmatrix} 1 & \cdots & \cdots & \cdots & \cdots \\ & 1 & \cdots & \cdots & \cdots \\ & & 1 & \cdots & \cdots \\ O & & & 1 & \cdots \end{bmatrix} \{1 \text{ entries above diagonal from } F\} \right\}$$

Since $1 \in P$, therefore, $P \neq \phi$

Also, we have $A, B \in P \Rightarrow A = \begin{bmatrix} 1 & \cdots & \cdots & \cdots \\ & 1 & \cdots & \cdots \\ & & 1 & \cdots \\ O & & \cdots & 1 \end{bmatrix}, B = \begin{bmatrix} 1 & \cdots & \cdots & \cdots & \cdots \\ & 1 & \cdots & \cdots & \cdots \\ & & 1 & \cdots & \cdots \\ O & & & 1 & \cdots \end{bmatrix}$

which implies $\qquad\qquad AB = \begin{bmatrix} 1 & \cdots & \cdots & \cdots & \cdots \\ & 1 & \cdots & \cdots & \cdots \\ & & 1 & \cdots & \cdots \\ O & & & 1 & \cdots \end{bmatrix} \in P$

Therefore, $\qquad\qquad P \subseteq G$.

Let $A \in P$, the first row in A can be chosen in p^{n-1} ways, second row in p^{n-2} ways and in this way $(n-1)^{\text{th}}$ row in p ways and last row is fixed.

Therefore, $\qquad o(P) = p^{n-1} p^{n-1} \dots p^1$

$$= p^{1+2+\dots+(n-1)} = p^{\frac{n(n-1)}{2}}$$

Hence, P is Sylow p-subgroup of G.

EXERCISE 4.2

1. Show that there are no simple group of order 63, 56 and 36.

2. Show that any group of order 15 is cyclic.

3. If H is a normal subgroup of a finite group G and if the index of H is 1 is prime to p, show that H contains every Sylow p-subgroup of G.

4. Show that there are only two non-abelian groups of order 8.

5. Let $o(G) = p^m . n$, where g.c.d. $(p,n) = 1$ and $p > 1$. Prove that G has a normal Sylow p-subgroup.

6. Show that a group of order 200 must contain a normal Sylow subgroup.

7. Let p be a prime dividing $o(G)$ and $(ab)^p = a^p b^p$; for all $a, b \in G$. Show that Sylow p-subgroup P is normal in G.

8. Let G be a finite group and p is the smallest prime divisor of $o(G)$. Show that a subgroup H of index p in G is normal in G.

9. If G is a finite group and p is the prime number dividing $o(H)$, where $H \subseteq G$, then show that the number of Sylow p-subgroup of H is less than or equal to the number of Sylow p-subgroups of G.

10. If in a finite group G, there are atmost d elements of order d for every $d \mid o(G)$, then G is cyclic.

11. If G is a group of order 385, show that Sylow 11-subgroup is normal in G and Sylow 7-subgroup is in the centre of G.

12. If $o(G) = p^n q$, $p > q$ both are primes, show that G contins a unique normal subgroup of index q.

13. Find a Sylow 2-subgroup and a Sylow 3-subgroup of S_6.

14. Show that 21 is the smallest possible odd integer that can be the order of a non-abelian group.

15. Show that there are no simple groups of order $p^r m$, where p is a prime, r is a positive integer and $m < p$.

16. Let G be a finite group and P be a normal p-subgroup of G. Show that P is contained in every Sylow p-subgroup of G.

4.14 NILPOTENT AND SOLVABLE GROUPS

4.14.1 Nilpotent Group

A finite group G is said to be nilpotent if it has a normal series

$$\{e\} \subset Z_1(G) \subset Z_2(G) \subset \dots \subset Z_k(G) \subset Z_{k+1}(G) \dots$$

such that $Z_s(G) = G$ for some index s and the first such index s is known as the "class of nilpotency" of G.

4.12.2 Solvable Group

A group G is said to be solvable if it has a finite decreasing sequence of subgroups.

$$G = G_0 \supset G_1 \supset G_2 \supset \dots \supset G_k = \{e\}$$

such that G_{i+1} is a normal subgroup of G_i and factor group $G_i \mid G_{i+1}$ is abelian.

For example

S_3, the symmetric group of order 3 is solvable, because S_3 has a subnormal series

$$\{e\} \subset A_3 \subset S_3$$

such that S_3/A_3 and $S_3/\{e\}$ are both abelian being of order 2 and 3 respectively.

REMARK

☞ A solvable group is also known as matacyclic group.

4.14.3 Commutator Subgroup of a Group

Let G be a group and $a, b \in G$. Then it is known that $aba^{-1}b^{-1}$ is called as the commutator of the ordered pair (a, b).

Let $U = \{aba^{-1}b^{-1} : a, b \in G\}$. If G' is the subgroup of G generated by U, then G' is called the commutator subgroup of G.

REMARK

☞ If G' is the subgroup of G generated by U, then G' is the smallest subgroup of G containing U. Thus, the commutator subgroup of G containing the set of all commutators in G.

Theorem 1. *Every abelian group is solvable.*

Proof.   Let G be an abelian group. Let us write $N_0 = G$ and $N_1 = \{e\}$.

Then, we have,
$$G = N_0 \supseteq N_1 = \{e\}$$

which is a solvable series for G.

Obviously $N_1 = \{e\}$ is a normal subgroup of $N_0 = G$, because if a is any element of G, then $a^{-1} \, ea = a^{-1}a = e \in \{e\}$. Now, since G is abelian, the quotient group $N_0/N_1 = G(e)$ is also abelian, because every quotient group of an abelian group is abelian. Hence, G is solvable.

Theorem 2. *Every subgroup of a solvable group is solvable.*

Proof.   Let G be a solvable group and H be any subgroup of G. Also, let
$$G = G_0 \supseteq G_1 \supseteq G_2 \supseteq \dots \supseteq G_n = \{e\}$$

be a solvable series for G.

We have to prove that
$$H = H_0 \supseteq (H \cap G_1) \supseteq (H \cap G_2) \supseteq \dots \supseteq (H \cap G_n) = (e) \qquad \dots (1)$$

is a solvable series for H.

Now, since for $i = 0, 1, \dots, n-1$, G_{i+1} is normal in G_i, therefore $H_{i+1} = H \cap G_{i+1}$ is normal in $H_i = H \cap G_{i+1}$.

Define a mapping $f : H \to G_i/G_{i+1}$ such that
$$f(x) = xG_{i+1} \quad \forall x \in H_i$$

Since $H_i \subseteq G_i$, therefore $x \in H_i$ implies $x \in G_i$. Thus, the coset xG_{i+1} is an element of the quotient group G_i/G_{i+1} and therefore the mapping f is well defined.

Further, if $x, y \in H_i$, then
$$f(xy) = (xy) \, G_{i+1}$$
$$= (xG_{i+1}) \, (yG_{i+1}) \qquad [\because x, y \in G_i \text{ and } G_{i+1} \text{ is normal is } G_i.]$$
$$= f(x) \cdot f(y)$$

Thus, the mapping f is a homomorhpism of H_i into G_i/ G_{i+1}.

Also, $x \in \ker (f)$

$\Leftrightarrow$ $f(x) = G_{i+1}$ $[\because$ Identity of G_i/G_{i+1} is $G_{i+1}.]$

$\Leftrightarrow$ $xG_{i+1} = G_{i+1}$

$$\Leftrightarrow \qquad\qquad x \in G_{i+1}$$

$$\Leftrightarrow \qquad\qquad x \in H \cap G_{i+1}, \text{ since } x \in H_i \subseteq H.$$

Therefore, $\text{Ker } f = H \cap G_{i+1} = H_{i+1}$

Then, by fundamental theorem on homomorphism of groups, we have

$$H_i / H_{i+1} \cong f(H_i)$$

But $f(H_i)$ is a subgroup of G_i/G_{i+1} which is abelian because $f(H_i)$ is also abelian. Thus, H_i / H_{i+1} is also abelian, because it is isomorphic to $f(H_i)$. Therefore (1) is a solvable series for H and hence H is a solvable group.

Theorem 3. *If G is a group and N is a normal subgroup of G such that both N and G/N are solvable. Then G is solvable.*

Proof.   Let G be a group and N is a normal subgroup of G such that N and G/N both are solvable. We have to prove that G is solvable.

The identity of G/N is N.

Let the solvable series for G/N be given by

$$G/N = G_0/N \supseteq G_1/N \supseteq \ldots \supseteq G_{m-1}/N \supseteq G_m/N/=(N) \qquad \ldots (1)$$

where each G_i is a subgroup of G containing N. Now, since G_{i+1}/N is normal in G_i/N therefore, each G_{i+1} is normal in G_i. Also,

$$(G_i/N)(G_{i+1}/N) \cong G_i/G_{i+1}.$$

It is known that each quotient group $(G_i/N)|(G_{i+1}/N)$ is abelian because (1) is solvable series for G/N. Thus each quotient group G_i/G_{i+1} is also abelian because it is isomorphic to an abelian group. Also, $G_m/N = (N)$ implies $G_m = N$. It is also given that N is solvable. Let

$$N = N_0 \supseteq N_1 \supseteq N_2 \supseteq \ldots \supseteq N_t = \{e\}$$

be a solvable series for N. Then

$$G = G_0 \supseteq G_1 \supseteq G_2 \supseteq \ldots \supseteq G_{m-1} \supseteq N \supseteq N_1 \supseteq N_2 \supseteq \ldots \supseteq N_t = \{e\}$$

is a solvable series for G. Hence, G is a solvable series.

Theorem 4. *Let G' be the commutator subgroup of a group G. Then G is abelian if and only if $G' = \{e\}$, where e is the identity of G.*

Proof.   Let G be a group. Define $U = [aba^{-1}b^{-1} : a, b \in G]$

If G' is the commutator subgroup of G, then G' is the subgroup of G gnerated by U. Therefore, G' is the smallest subgroup of G containing U.

Now, suppose G is abelian. We have to prove that $G' = \{e\}$

If G is abelian, then we have

$$aba^{-1}b^{-1} = abb^{-1}a^{-1} = aa^{-1} = e \ \forall \ a, b \in G$$

Then, U consists only one element namely e. Then $\{e\}$ is the smallest subgroup of G containing e.

$$\Rightarrow \qquad\qquad G' = \{e\}$$

Conversely, let us suppose $G' = \{e\}$

We have to prove that G is abelian.

Let a, b be any two elements of G. Then $aba^{-1}b^{-1} \in U$

$$\Rightarrow \qquad\qquad aba^{-1}b^{-1} \in G'$$

But G' contains only one element, namely, e.

Therefore, $\quad\quad\quad aba^{-1}b^{-1}=e$

$\Rightarrow \quad\quad\quad (ab)(ba)^{-1}= e \quad\quad\quad\quad\quad [\because (ab)^{-1}=b^{-1}a^{-1}]$

$\Rightarrow \quad\quad\quad\quad ab = [(ba)^{-1}]^{-1}$

$\Rightarrow \quad\quad\quad\quad ab = ba$

Hence, G is abelian.

Theorem 5. *Let G be a group and G' be the commutator subgroup of G. Then*

    (i) G' is normal in G.

    (ii) G/G' is abelian.

    (iii) If N is any normal subgroup of G, then G/N is abelian if and only if $G' \subseteq N$.

    (iv) If H is a subgroup of G, such that $G' \subseteq H$, then H is a normal subgroup of G.

Proof.   Let G be a group and G' be the commutator subgroup of G. Define

$$U = [aba^{-1}b^{-1}: a, \, b \in G]$$

By definition of commutator subgroup, G' is the smallest subgroup of G containing U.

(i)   Let $x \in G$ and $c \in G'$.

       Then, consider,

$$xcx^{-1} = (xcx^{-1}) \, c^{-1}c = (xcx^{-1}c^{-1})c \in G'$$

$\Rightarrow \quad\quad\quad xcx^{-1} \in G'$

$\Rightarrow \quad\quad G'$ is normal.

(ii)   Since G' is normal in G, we can define quotient group G/G'. Let $a, \, b \in G$. Then $G'a, \, G'b$ are any two elements of G/G'. We have

$$aba^{-1}b^{-1} \in U \quad\quad \Rightarrow \quad\quad aba^{-1}b^{-1} \in G'$$

$\Rightarrow \quad\quad (ab)(ba)^{-1} \in G' \quad \Rightarrow \quad\quad G'(ab) = G'(ba)$

$\Rightarrow \quad\quad (G'a)(G'b) = (G'b)(G'a)$

$\Rightarrow \quad\quad G/G'$ is abelian.

(iii)   Let N be any normal subgroup of G. Let $a, \, b$ be any two elements of G. Then, $Na, \, Nb$ are any two elements of G/N. Now, let G/N be abelian. Then

$$(Na)(Nb) = (Nb)(Na)$$

$\Rightarrow \quad\quad\quad Nab = Nba$

$\Rightarrow \quad\quad\quad (ab)(ba)^{-1} \in N$

$\Rightarrow \quad\quad\quad aba^{-1}b^{-1} \in N \quad\quad\quad\quad\quad [\, (ba)^{-1} = a^{-1}b^{-1} \,]$

$\Rightarrow \quad\quad\quad\quad U \subset N$

$\Rightarrow \quad\quad\quad N$ is a subgroup of G containing U, but G' is the smallest subgroup of G containing U.

$\Rightarrow \quad\quad\quad G'$ must contained in N, i.e., $G' \subseteq N$.

       Conversely, let $G' \subseteq N$. Then $U \subseteq N$ because $U \subseteq G'$

$\Rightarrow \quad\quad\quad\quad aba^{-1}b^{-1} \in N$

$\Rightarrow \quad\quad\quad (ab)(ba)^{-1} \in N$

$\Rightarrow \quad\quad\quad\quad Nab = Nba$

$$\Rightarrow \qquad (Na)\,(Nb) = (Nb)\,(Na)$$

$$\Rightarrow \qquad G/N \text{ is abelian}$$

(iv) Let g be any element of G and let h be any element of H. Then

$$ghg^{-1} = (ghg^{-1})\,(h^{-1}h) = (ghg^{-1}h^{-1})\,h \in H$$

(because $h \in H$ and $ghg^{-1}h^{-1} \in G'$ and $G' \subseteq H \Rightarrow ghg^{-1}h^{-1} \in H$)

$$\Rightarrow \qquad ghg^{-1} \in H$$

$$\Rightarrow \qquad H \text{ is a normal subgroup of } G.$$

Theorem 6. *A group G is solvable if and only if $G(k) = \{e\}$ for some integer k.*

Proof.   Let us first suppose $G(k) = \{e\}$. We have to prove that G is solvable.

Let us write $\qquad N_0 = G, N_1 = G', N_2 = G^{(2)}, \dots N_K = G^{(k)} = e$

Then, we have

$$G = N_0 \supseteq N_1 \supseteq N_2 \supseteq \dots N_k = \{e\} \qquad \qquad \dots(1)$$

We want to prove that (1) is solvable for G. We have already proved in previous theorem that $G^{(i)} = (G^{(i-1)})'$ is a normal subgroup if $G^{(i-1)}$ for each i. Thus, N_i is a normal subgroup if N_{i-1} for each i.

Further, $\qquad \dfrac{N_{i-1}}{N_i} = \dfrac{G^{(i-1)}}{G^{(i)}} = \dfrac{G^{(i-1)}}{(G^{(i-1)})'}$

Again , by previous theorem $\dfrac{G^{(i-1)}}{(G^{(i-1)})'}$ is abelian.

Thus, N_{i-1}/N_i is abelian for each i.

$\Rightarrow$ Series (1) is solvable for G and hence G is solvable. Conversely, let G be a solvable group. We have to prove that for some integer k, $G^{(k)} = \{e\}$.

Let $\qquad \qquad G = N_0 \supseteq N_1 \supseteq N_2 \supseteq \dots \supseteq N_k = \{e\} \qquad \qquad \dots (2)$

be a solvable series for G, where each N_i is a normal subgroup of N_{i-1} and N_{i-1}/N_i is an abelian group. Again by previous theorem, the commutator subgroup N'_{i-1} of N_{i-1} must be contained in N_i.

Therefore

$$N_1 \supseteq N'_0 = G'$$

$$N_2 \supseteq N'_1 \supseteq (G')^2 = G^{(2)} \qquad \qquad [\because N_1 \supseteq G' \Rightarrow N'_1 \supseteq (G')']$$

$$N_3 \supseteq N'_2 \supseteq (G^{(2)})' = G^{(3)}$$

$$\dots \quad \dots \quad \dots \quad \dots \quad \dots \quad \dots$$

$$N_1 \supseteq G^{(i)}$$

$$\dots \quad \dots \quad \dots \quad \dots \quad \dots$$

$$N_k \supseteq G^{(k)}$$

Clearly, we see that $(e) = N_k \supseteq G^{(k)}$

$$\Rightarrow \qquad \qquad G^{(k)} \subseteq (e)$$

But $(e) \subseteq G^{(k)}$

$\Rightarrow \quad G^{(k)} = (e)$

Theorem 7. *Every homomorphic image of a solvable group is solvable.*

Proof. Let G be a solvable group and f be the homomorphism from G to G^*.

We have to prove that G^* is solvable.

Clearly, $(G^*)^{(k)}$ is the image of $(G)^{(k)}$ under f, i.e., $f(G^k) = (G^*)^{(k)}$.

Now, since G is solvable, thus by previous theorem, we have $(G)^{(k)} = (e)$ for some integer k.

Then, $(G^*)^{(k)} = f(G^{(k)}) = f(e) = (e^*)$

Where e^* is the identity of G^*. Then again by previous theorem

$(G^*)^{(k)} = (e^*)$

$\Rightarrow \quad G^*$ is solvable.

Hence, every homomorphic image of a solvable group is solvable.

Theorem 8. *Let $G \neq \{e\}$ be a solvable group, then G contains a normal abelian subgroup $H \neq (e)$.*

Proof. If G is abelian, then we can take $H = G$. In this case, proof is complete. Further, consider the case when the group G is non-abelian. Since G is solvable group, therefore, there exists a positive integer k such that $G^{(k)} = (e)$. Since G is non-abelian, therefore, we cannot take $k = 1$. If k is the least positive integer such that $G^{(k)} = (e)$, let us take $H = G^{(k-1)}$. Then, clearly H is a normal subgroup of G. Further

$$H' = (G^{(k-1)})' = G^{(k)} = (e)$$

Hence, H is abelian.

Theorem 9. *If H is a normal subgroup of a solvable group G, then the quotient group G/H is solvable.*

Proof. By definition of solvable series, G has a subnormal series

$$G = G_0 \supset G_1 \supset G_2 \supset \dots \supset G_k = \{e\} \qquad \dots (1)$$

such that G_{i+1} is normal in G_i, for all i and G_i / G_{i-1} are abelian. Consider a series

$$G/H = G_0/H \supset \overline{H}_1/H \supset \overline{H}_2/H \supset \dots \supset \overline{H}_k/H = \{\overline{e}\} \qquad \dots (2)$$

of G/H, where $\overline{H}_i = G_i H$ for all $i = 0, 1, 2, \dots, k-1$.

We claim that series (2) is solvable series of G/H.

First, we shall prove that $\overline{H}_{i+1}/H$ is a normal subgroup of $\overline{H}_i/H$. For this, we shall show that $\overline{H}_{i+1}$ is normal in $\overline{H}_i$.

Let $a \in \overline{H}_i = G_i H$, then $a = gh$ for some $g \in G_i$, $h \in H$.

Now, since G_{i+1} is normal in G_i, then for $g \in G_i$, we must have

$$gG_{i+1} = G_{i+1} g$$

Further,

$$a\overline{H}_{i+1} = aG_{i+1} H$$
$$= ghG_{i+1} H$$
$$= ghHG_{i+1} = gHG_{i+1}$$
$$= gHG_{i+1} = gG_{i+1}H = G_{i+1}H = G_{i+1}gH$$

$$= G_{i+1}gH$$
$$= G_{i+1}ghH$$
$$= G_{i+1}Hga \qquad\qquad [\because H \text{ is normal in } G_i.]$$
$$= G_{i+1}Ha = H = \bar{H}_{i+1}a$$

$\Rightarrow \qquad\qquad a\bar{H}_{i+1} = \bar{H}_{i+1}\,a$

Therefore, H_{i+1} is normal in H_i.

Now, it remains to prove that $(\bar{H}_i / H) | (\bar{H}_{i+1} / H)$ is abelian.

We know that $\bar{H} / \bar{H}_{i+1} \cong \dfrac{\bar{H}_i / H}{\bar{H}_{i+1} / H}$

Further, define a mapping

$$f : G_i \to \bar{H}_i / \bar{H}_{i+1}$$

such that $\qquad f(x) = \bar{H}_{i+1}x\,;\ x \in G_i$

Then, clearly f is a homomorphism.

For some $\qquad y \in \bar{H}_i = G_iH = HG_i$, we have $y = hg$, for some $g \in G_i$ and $h \in H$.

Then, $\qquad\qquad \bar{H}_{i+1}\, y = G_{i+1}\,Hy$
$$= G_{i+1}H\,hg$$
$$= G_{i+1}H\,g$$
$$= \bar{H}_{i+1}\, g = f(g)$$

$\Rightarrow \quad f$ is onto.

Now, since $\qquad \ker(f) = [x \in G : f(x) = \bar{H}_{i+1} = G_{i+1}H]$

If $\quad x \in G_{i+1} \Rightarrow \quad x \in G_i \qquad\qquad\qquad [\because G_{i+1} \text{ is a subset of } G_i.]$

which implies that

$$f(x) = \bar{H}_{i+1}x$$
$\Rightarrow \qquad\qquad f(x) = G_{i+1}Hx$
$$= HG_{i+1}x = HG_{i+1} = G_{i+1}H$$
$\Rightarrow \qquad\qquad x \in \text{Ker}(f)$

Therfore, $\quad G_{i+1} \subseteq \text{Ker}(f)$.

Finally, define a map

$$\phi : G_i / G_{i+1} \to \bar{H}_i / \bar{H}_{i+1}$$

such that

$$\phi\,(G_{i+1}x) = \bar{H}_{i+1}x = G_{i+1}Hx,\ \forall\, x \in G_i$$

Clearly ϕ is onto. Also, $\bar{H}_i / \bar{H}_{i+1}$ is homomorphic image of abelian group G_i / G_{i+1}. Therefore, $\bar{H}_i / \bar{H}_{i+1}$ is abelian, thus $(\bar{H}_i / H) | (\bar{H}_{i+1} / H)$ is abelian.

Hence, the series (2) for G/H is solvable, which implies that G/H is solvable.

Theorem 10. *Let G be a solvable group and* $f : G \to \bar{G}$ *is an epimorphism, then* $\bar{G}$ *is solvable.*

Proof.   It is given that G is solvable, therefore it has a subnormal series.

$$G = G_0 \supset G_1 \supset G_2 \supset... \supset G_k = \{e\} \qquad\qquad ...(1)$$

such that G_i/G_{i+1} is abelian, where k is some positive integer.

Further, since $\bar{G}$ is homomorphic image of G such that $\bar{G} = f(G)$, so we may define

$$\Rightarrow \qquad \bar{G}_i = f(G_i)$$

Thus, we have

$$f(G) = \bar{G} = \bar{G}_0 \supset \bar{G}_1 \supset \bar{G}_2 \supset \dots \supset \bar{G}_k = \{e\} \qquad \dots(2)$$

is a subnormal series of $\bar{G}$.

Now, we shall prove that $\dfrac{\bar{G}_i}{\bar{G}_{i+1}}$ is abelian.

For this, let us define

$$\phi : G_i \rightarrow \bar{G}_i / \bar{G}_{i+1} \text{ such that}$$

$$\phi(x) = f(x)\bar{G}_{i+1}; \text{ for all } x \in G_i$$

Then, clearly ϕ is homomorphism.

Now, since $\bar{G}_{i+1}$ is normal in G_i, then $y\,\bar{G}_{i+1} = \bar{G}_{i+1}y$ for some $y \in \bar{G}_i$. But we have $\bar{G}_i = f(h_i)$, therefore, $y = f(x)$ for some $x \in G_i$.

$$\Rightarrow \qquad y\,\bar{G}_{i+1} = f(x)\bar{G}_{i+1}; \text{ for some } x \in G_i.$$

$$= \phi(x); \text{ for some } x \in G_i.$$

$\Rightarrow \quad \phi$ is onto.

Also, $\bar{G}_{i+1} \subset \text{Ker}(\phi)$ and therefore introduce a homomorphism

$$\psi : G_i / G_{i+1} \rightarrow \bar{G}_i / \bar{G}_{i+1}$$

such that $\quad \psi(G_{i+1}x) = \bar{G}_{i+1}x$ for some $x \in G_i$.

Clearly, ψ is onto, therefore $\bar{G}_i / \bar{G}_{i+1}$ is homomorphic image of an abelian group G_i / G_{i+1} and so $\bar{G}_i / \bar{G}_{i+1}$ is abelian, which shows that subnormal series (2) is a solvable series of $\bar{G}$. Hence, $\bar{G}$ is solvable.

Theorem 11. *Let H be a normal subgroup of a group G. If both H and G/H are solvable, then G is solvable.*

Proof. It is given that H is solvable, therefore, by definition, there is a subnormal series

$$H = H_0 \supset H_1 \supset H_2 \supset \dots \supset H_k = \{e\} \qquad \dots(1)$$

such that H_i / H_{i+1} is abelian. Also, G/H is solvable. Therefore, there exists a subnormal series for G/H with abelian factor.

$$G / H = \bar{G}_0 \supset \bar{G}_1 \supset \bar{G}_2 \supset \dots \supset \bar{G}_s = \{\bar{e}\} \qquad \dots(2)$$

where each G_i is a subgroup of G containg H and $\bar{G}_i = G_i / H$.

Further, since each G_{i+1}/H is normal in G_i / H so that G_{i+1} is normal in G_i.

Also, since

$$\frac{\bar{G}_i}{\bar{G}_{i+1}} = \frac{G_i / H}{G_{i+1} / H} \text{ is abelian}$$

and $\qquad \dfrac{\bar{G}_i}{\bar{G}_{i+1}} = \dfrac{G_i / H}{G_{i+1} / H}$ for all $i = 0, 1, 2,\dots,s-1$, therefore G_i / G_{i+1} is abelian.

Using (2), we have

$$\bar{G}_s = e \quad \Rightarrow \quad \frac{G_s}{H} = \bar{e}$$
$$\Rightarrow \quad G_s = H$$

Therefore, we obtain a subnormal series.

$$G = G_0 \supset G_1 \supset G_2 \supset ... \supset G_{s-1} \supset G_s = H = H_0 \supset H_1 \supset H_2 \supset ... \supset H_k = \{e\}$$

for G such that G_i/G_{i+1} and H_i/H_{i+1} are abelian. Hence, G is solvable.

Theorem 12. *Product of solvable groups is solvable.*

Proof.   Let G be a group and H, K be two solvable groups such that G = HK. We have to prove that G is solvable. Clearly, G/H is isomorphic to K and K is solvable so that G/H is solvable. Therefore, we have H is solvable and G/H is also solvable, then using above theorem, we can say that G is solvable. Hence, HK is solvable, meaning thereby product of solvable groups is solvable.

Theorem 13. *Any finite p-group is solvable.*

Proof.   Let G be the given p-group. We have to prove that G is solvable.

Since G is a p-group, therefore order of G is of the form p^n; $n > 0$.

If $n = 1$, then $G \cong Z/pZ$ and Z/pZ is abelian and we know that every abelian group is solvable, therefore G is solvable. Now, suppose that this result is true for every p-group of order p^k with $k < n$ and let $o(G) = p^n$. Then clearly we have the centre $Z(G) = e$ is an abelian subgroup of G, so that it is solvable.

It is also known that $Z(G)$ is a normal subgroup of G, therefore the quotient $G/Z(G)$ is p-group, because

$$o(G/Z(G)) = \frac{o(G)}{o(Z(G))} = \frac{p^n}{p^k} = p^{n-k}$$

Then by induction, $G/Z(G)$ is solvable.

Therefore, $Z(G)$ and $G/Z(G)$ are solvable. Hence, by Theorem 11, we can say G is abelian.

Theorem 14. *Let G be a finite group. Then G is solvable if and only if each factor in a composition series for G is of prime order.*

Proof.   Let us first suppose G is solvable, then by definition G has a subnormal series with abelian factors given by

$$G = G_0 \supset G_1 \supset G_2 \supset ... \supset G_k = \{e\} \qquad \qquad ...(1)$$

Now, since G is finite, so that the series (1) can be refined to a composition series with abelian factors. Since each factor in a composition sereis is a simple group and we know that the only simple finite abelian groups are cyclic groups of prime order. Also, the composition factors are determined by G to within isomorphism, therefore, in any composition series in G, the order of each factor must be prime.

Conversely, if the order of each factor in a composition series for G is prime, then each factor must be cyclic and therefore abelian. Hence, G itself is solvable.

Theorem 15. *A group G of order pq for distinct primes p and q is solvable.*

Proof.   It is given that p and q are distinct, therefore, we may assume that $p > q$. Let t be the number of p-Sylow subgroups of G, then $t = 1 + kp$ and t divides pq.

Now, since t is relatively prime to p, therefore t must divide q, but we have assume that $p > q$. Therefore only possibility is that $t = 1$. Thus, G has a unique Sylow p-subgroup which is normal in G. Let this p-subgroup be H.

Further, since every p-subgroup is solvable, therefore H is solvable and also H is normal in G. So the quotient group G/H is solvable. Hence, by theorem 11, G is solvable.

Theorem 16. *A group G of order pqr for distinct primes p, q, and r is solvable.*

Proof. Without loss of any generality, we may assume that $p > q > r$.

Further, suppose that G does not have any normal subgroups. Then, using Sylow's theorem, we can find

(i) $t_1 = 1 + k_1q$, $t_1 > 1$ and t_1 divides pqr so that $t_1 = qr$.

(ii) $t_2 = 1 + k_2q$, $t_2 > 1$ and t_2 divides pqr so that $t_2 = pr$.

(iii) $t_3 = 1 + k_3q$, $t_3 > 1$ and t_3 divides pqr so that $t_3 = pq$.

where t_1, t_2, t_3 are the numbers of p-Sylow, q-Sylow and r-Sylow subgroups respectively. It is known that all Sylow subgroups for distinct primes p, q and r have trivial intersection, thus we can find

$$o(G) \geq qr(p-1) + rp\ (q-1) + pq\ (r-1) + 1 > pqr$$

which is a contradiction because $o(G) = pqr$. Therefore, G must have a normal subgroup of prime order so that it is solvable. Let this solvable subgroup be H. Therefore, G/H is solvable. Then by Theorem 11, G is solvable.

Theorem 17. *A group G of order p^2q for distinct primes p, q is solvable.*

Proof. Without loss of any generality, we may assume that $p > q$. Further, let t_1 be the number of p-Sylow subgroup of G. Then by Sylow's theorem, we have $t_1 = 1+kp$ and t_1 divides p^2q. Since t_1 is relatively prime to p, so t_1 must divide q. But $p>q$ then we only have $t_1 = 1$. Therefore, G has a normal subgroup H of order p^2. Therefore, being a p-subgroup, H is solvable and the quotient group G/H of order q is solvable. Hence, by Theorem 11, G is solvable. Now, consider the case when $p < q$ and assume that G has no normal subgroups. Then, by Sylow's theorem, we find

(i) $t_2 = 1 + k_1p$, $t_2 > 1$ and t_2 divides p^2q so that $t_2 = q$

(ii) $t_3 = 1 + k_2q$, $t_3 > 1$ and t_3 divides p^2q so that $t_3 = q^2$

where t_2 and t_3 are the number of p-Sylow and q-Sylow subgroups respectively.

Since, we know that ϕ Sylow p-subgroup and q-subgroup have trivial intersection. Thus, we can find $o(G) \geq q(p^2 - 1) + p^2\ (q - 1) + 1 > p^2q$ which is a contradiction, because $o(G) = p^2q$.

Thus, G must have a normal subgroup H (say) of order p^2 or q. Also, H is solvable. Further, quotient group G/H is of order q or p^2 so it is also solvable. Hence, by Theorem 11, G is solvable.

EXERCISE 4.3

1. Show that the symmetric group S_n of n symbols is not solvable for $n \geq 5$.

2. Show that S_3 and S_4 are solvable but not nilpotent.

3. Show that every finite p-group is nilpotent.

4. Show the product of two nilpotent groups is nilpotent.

Chapter Review: *A competitive Approach*

Selected Terms and Results

Terms

- **Operation or action :** Let G be a group and S be any set. An operation or an action of G on S is a homomorphism $\pi : G \to$ Perm (s) of G into the group of permutations of S.

- **Conjunction:** Let G be a group. For each $x \in G$, let $f_x : G \to G$ be the mapping such that $f_x(y) = xyx^{-1}$. Clearly $x \to f_x$ is a homomorphism $G \to$ Aut (G) and therefore this mapping gives an operations of G on itself, called the conjunction.

- **Translation :** Let G be a group. Then for each $x \in G$, translation $T_x : G \to G$ is defined by $T_x(y) = xy$.

- **Morphism of G-set :** Let S and S' be two G-sets and $f : S \to S'$ is a mapping. Then f is said to be morphism of G-sets or a G-map if $f(xs) = xf(s)$ for all $x \in G, s \in S$.

- **Isotropy group :** Let S be any set and $s \in S$. The set of element $x \in G$ such that $xs = s$ is obviously a subgroup of G, called the isotropy group in G and denoted by G_s.

- **Faithful action :** An action of G is said to be faithful if $k = \{e\}$, *i.e.*, the kernal of $G \to$ Perm(s) is trivial.

- **Fixed point :** A fixed point of G is an element $s \in S$ such that $xs = s \; \forall \; x \in G$.

- **Isotropy subgroup :** Let S be a G-set, $s \in S$ and $g \in G$. Then subgroup $G_s = \{g \in G : gs = s\}$ is called isotropy subgroup of s.

- **Composition series :** Let G be a group. Then a finite sequence of its subgroups $G = H_1, H_2, ..., H_n = \{e\}$ is called a composition series for G if each H_i except H_1 is a maximal normal subgroup of H_{i-1}.

- **Composition factor group :** The quotient group $H_1/H_2 ...H_{n-1}/H_n$ which are necessarily simple are known as composition factor group.

- **Subnormal series :** A subnormal series of G is a finite sequence $H_0, H_1, H_2.., H_n$ of subgroups of G such that $H_i \subseteq H_{i+1}$ and H_i is a normal subgroup of H_{i+1} with $H_0 = \{e\}$ and $H_n = G$.

- **Normal series:** A normal series of G is a finite sequence of normal subgroup of G such that $H_i \subseteq H_{i+1}, H_0 = \{e\}$ and $H_n = G$.

- **Composition series :** A subnormal series $G = G_0 \supset G_1 \supset G_2 \supset ... \supset G_m = \{e\}$ is said to be composition series of G if all the factor of the series G_i / G_{i+1} are simple groups.

- **Principal series:** A normal series $G = G_0 \supset G_1 \supset G_2 \supset ... \supset G_m = \{e\}$ of a group G is said to be principal series if all the factor groups G_{i-1}/G_i are simple.

- **Accessible subgroup :** A subgroup H of a group G is said to be a accessible subgroup if it occur in a subnormal series in G.

- **Descending subnormal chain :** A descending sequence of subgroups of a group G given by $G = H_0 \supset H_1 \supset H_2 \supset ... \supset H_n \supset ...$ is called a descending subnormal chain of G if every subgroup H_n ($n \in N$) is a proper normal subgroup of H_{n-1}.

- **Ascending subnormal chain :** A ascending sequence of subgroups of a group G given by $\{e\} \subset F_1 \subset F_2 \subset ... \subset F_n \subset ...$ is called an ascending subnormal chain of G if every subgroup F_n is a proper normal subgroup of F_{n+1} and if all the subgroups F_n are accessible in G.

- **p-group :** Let p be a fixed prime number then a group G is said to be p-group if every element of G has a order, a power of p.

- **p-subgroup :** Let p be a prime number then a subgroup H of a group G is said to be p-subgroup if every element of H is a power of p.

- **Sylow p-subgroup :** Let G be a group and p be a prime number such that p^n divides order of G and p^{n+1} does not divide it. Then a subgroup H of G such that $o(H) = p^n$ is called a sylow p-subgroup of G or p-sylow subgroup of G.

- **Nilpotent group :** A finite group G is said to be nilpotent if it has a normal series $\{e\} \subset Z_1(G) \subset Z_2(G) \subset ... \subset Z_k(G)$ such that $Z_s(G) = G$ for some index s. Further, the first such index s is known as the "class of Nilpotency" of G.

- **Solvable group :** A group G is said to be solvable it it has a finite decreasing sequence of subgroups $G = G_0 \supset G_1 \supset G_2 \supset ... \supset G_k = \{e\}$ such that G_{i+1} is a normal subgroup of G_i and every factor group G_i/G_{i+1} is abelian.

- **Commutator subgroup of a group:** Define $U = \{aba^{-1}b^{-1} : a, b \in G\}$. If G' is the subgroup of G generated by U, then G' is called the commutator subgroup of G.

Results

- A normal subgroup H of a group G is maximal if and only if the group G/H is simple.

- A subnormal series is also known as subvariant series.

- A normal series is always subnormal but converse is not necessarily true.

- Every subnormal series is a refinement of itself.

- Two isomorphic subnormal (normal) series must have the same number of groups.

- Any two subnormal series of an arbitrary group have isomorphic refinement **(Schrier's theorem)**.

- A subnormal series that has no refinement other than itself is known as composition series.

- The common length of the composition series of a group is known as composition length.

- There exists at least one composition series for every finite group.

- Any two composition series of a group are isomorphic.

- Two composition series of a finite group are isomorphic **(Jordon-Holder's Theorem)**.

- The group of integers has no composition series.

- A subgroup of a group G is said to be accessible if it occur in a subnormal series of G.

- A group G has a composition series iff all ascending and descending subnormal chain break-off.

- A group G is a p-group iff $o(G) = p^n$.

- Every p-subgroup of a finite group G is contained in some Sylow's p-subgroup of G.

- A solvable group is also known as metacyclic group.

- Every abelian group is solvable.

- Every subgroup of a solvable group is solvable.

Review Questions and Project Work

1. If G is a finite non-trivial p-group, show that G has a normal subgroup of index p.

2. If H is a normal subgroup of order p^k of a finite group G, show that H is contained in every Sylow p-subgroup of G.

3. Show that the groups of order 12, 28, 56 and 200 must contain a normal Sylow subgroup.

4. If a group of order p^n contains exactly one subgroup each of order $p, p^2, ..., p^{n-1}$. Show that G is cyclic.

5. If in a finite group G, there are atmost of elements of order d for every $d|o(G)$, then show that G is cyclic.

6. If G is a finite group and p is a prime dividing $o(H)$ where $H \subseteq G$, then show that the number of Sylow p-subgroup of H is less than or equal to the number of Sylow p-subgroups of G.

7. If G is a finite group and p is the smallest prime divisor of $o(G)$, show that a subgroup H of index p in G is normal in G.

8. Let N be a normal subgroup of a group G. If both N and G/N are solvable, then show that G is also solvable.

Objective type Questions

Fill in the blanks

1. A normal subgroup N of a group G is called _____ normal subgroups.

2. A subgroup is said to be _____ if G has no proper normal subgroup.

3. Every _____ group has a composition series.

4. Any two composition series of a finite group are _____ .

5. A finite group is solvable if and only if its composition factors are _____ groups of prime order.

True/False

Write 'T' for true and 'F' for false statement

1. An abelian group G has a composition series if and only if G is finite. **(T/F)**

2. A cyclic group which has exactly one composition series is a p-group. **(T/F)**

3. Jordan-Holder theorem implies the fundamental theorem of arithmetic. **(T/F)**

4. A group G is solvable if and only if $G^{(k)} \neq \{e\}$ **(T/F)**

5. Every homomorphic image of a solvable group is solvable. **(T/F)**

Multiple choice Questions

Choose the most appropriate one.

1. Any subgroup of a solvable group is :
   (a) solvable
   (b) may be solvable
   (c) not solvable
   (d) none of these

2. If H is a normal subgroup of a solvable group G, then G/H is :
   (a) solvable
   (b) may be solvable
   (c) not solvable
   (d) none of these

3. Let N be a normal subgroup of G, then G is solvable if :
   (a) N is solvable
   (b) G/H is solvable
   (c) both N and G/N are solvable
   (d) none of the above

4. If two groups H and K are solvable then $H \times K$ is :
   (a) not solvable
   (b) solvable
   (c) may be solvable
   (d) none of these

5. Every abelian group of order greater than 1 is nilpotent group of class :
   (a) 0
   (b) 1
   (c) 2
   (d) 3

6. If G is a nilpotent group. Then :
   (a) every subgroup of G is nilpotent.
   (b) every homomorphic image of G are nilpotent.

   (c) both (a) and (b) are true.
   (d) none of the above

7. Which is not true ?
   (a) Every finite p-group is nilpotent.
   (b) A group of order p^n (p is prime) is nilpotent.
   (c) Both (a) and (b) are true.
   (d) None of the above

8. Which is not true :
   (a) A group of order pq for distinct primes p and q is solvable.
   (b) A group of order pqr for distinct primes p, q and r is solvable.
   (c) A group of order p^2q for distinct primes p and q is solvable.
   (d) None of the above

9. If G' be the commutator subgroup of a group G. Then G is abelian if and only if :
   (a) $G' \neq \{e\}$
   (b) $G' = \{e\}$
   (c) $G = \{e\}$
   (d) None of these

10. If G be a group of order 108, then $\exists$ a normal subgroup of order :
    (a) 27
    (b) 9
    (c) both (a) and (b) are true.
    (d) none of these

=== **Answers** ===

Fill in the Blanks

1. maximal   **2.** simple   **3.** finite   **4.** equivalent   **5.** cyclic

True/ False

1. T   **2.** T   **3.** T   **4.** F   **5.** T

Multiple choice questions

1. (a)   **2.** (a)   **3.** (c)   **4.** (b)   **5.** (b)   **6.** (c)   **7.** (d)   **8.** (d)   **9.** (b)

10. (c)

● ● ●

5 Rings

Outlines

- Ring;
- Field;
- Integral domain;
- Subring;
- Subfield;
- Polynomial Rings;
- Ideals;
- Homomorphism of Rings;
- Embedding of Rings;

5.1 INTRODUCTION

In previous chapters, our work has been concerned with sets on which a single binary operation has been defined. The study of sets on which two binary operations have been defined should be of great importance. Algebraic sutucture of this type are introduced in this chapter. The most general algebraic structure with two binary operations that we shall study is called a ring.

In this chapter, we shall take up the study of general rings and their homomorphisms, showing how these are associated with ideals.

5.2 RING

Definition. *Let R be a non-empty set. An algebraic structure* $(R, +, .)$ *together with two binary operations addition and multiplication for all a, $b \in R$ is called a ring if this structure satisfies following properties:*

(1) Addition Axioms

(i) *Closed under addition.*
$$a+b \in R, \ \forall a, b \in R$$

(ii) *Associative under addition.*
$$a+(b+c) = (a+b)+c, \ \forall a,b,c \in R$$

(iii) *Existence of identity.* If $0+a=a=a+0, \ \forall a \in R$
then 0 is an additive identity of *R*.

(iv) *Existence of inverse.* There exists an element $-a \in R$ such that
$$-a +a=0 = a+(-a) \ \forall a \in R$$
then *–a* is an additive inverse of *a*.

(v) *Commutative under addition.* $a+b = b + a \ \forall a, b \in R$

(2) Multiplication Axioms

(i) *Closed under multiplication.* $a.b \in R \ \forall a, b \in R$

(ii) *Associative under multiplication.* $a.(b.c) = (a.b).c \ \forall a, b \in R$

(iii) *Distributive laws.* (Distribution of '.' over '+')
$$a.(b+c) = a.b+a.c \ \forall a, b, c \in R \qquad \text{(Left distributive)}$$

and $\qquad\qquad\qquad (b+c).a = b.a+c.a \qquad\qquad\qquad$ (Right distributive)

OR

An algebraic structure $(R, +, .)$ is said to be a ring provided $a+b \in R$, $a.b \in R$ for all a, $b \in R$ and satisfies following properties:

   (i) $(R, +)$ is an abelian group.

   (ii) Multiplication is associative.

   (iii) Multiplication is distributive over addition.

5.2.1 Ring with unity

A ring R is said to be a ring with unity if there exists the multiplicative identity, *i.e.*, $1 \in R$ such that $\qquad\qquad\qquad 1.a = a = a.1, \ \forall a \in R$

5.2.2 Commutative ring

A ring R is said to be a commutative ring if $a.b = b.a$, $\forall a$, $b \in R$

5.2.3 Special classes of rings

1. **Null ring (or zero ring).** A set R having a single element 0 with two binary operations, addition and multiplication defined by $0+0 = 0$ and $0.0 = 0$ is called a null ring.

   We conclude that

   (i) The zero of a ring $(R, +, \cdot)$ can not have a multiplicative inverse unless R is the zero ring $(\{0\}, + \cdot)$

   (ii) The additive identity 0 and the multiplicative identity 1 of a ring $(R, +, \cdot)$ are not equal unless R is the zero ring.

2. **Ring of integers.** The set **Z** of all integers with respect to addition and multiplication forms a ring. This ring is called a ring of integers.

3. **Ring of real numbers.** The set **R** of all real numbers with two binary operations addition and multiplication forms a ring which is called a ring of real numbers.

4. **Ring of rational number.** The set **Q** of all rational numbers forms a commutative ring under addition and multiplication. This ring is called a ring of rational numbers.

5. **Ring of matrices.** The set M of all matrices of order $n \times n$ whose elements are integers, real , complex number forms a non-commutative ring with unit element with respect to matrix addition and matrix multiplication, is called a ring of matrices.

5.3 ELEMENTARY PROPERTIES OF A RING

For all a, b, c in a ring R, we have

     (i) $a0 = 0a = 0$

    (ii) $a(-b) = -(ab) = (-a)b$

   (iii) $(-a)(-b) = ab$

   (iv) $a(b - c) = ab - ac$

    (v) $(b - c)a = ba - ca$

Proof.   (i) $\qquad\qquad\qquad a0 = a(0+0) \qquad\qquad\qquad\qquad$ ($\because 0+0 = 0$)

	$= a0+a0$	(By left distributive law)
or	$0+a0 = a0 + a0$	(By the property of identity)
		($\because 0+a0 = a0$ as $a0 \in R$)
$\therefore$	$0 = a0$	(By right cancellation law $\because R$ is a group)

Similarly,

	$0a = (0+0)a$	($\because 0+0=0$)
	$= 0a+0a$	(By right distributive law)
or	$0a +0 =0a+0a$	(By the property of identity)
$\therefore$	$0 = 0a$	(By left cancellation law)
Hence	$a0 = 0a = 0.$	

(ii)

	$a0 = 0$	[From (i)]
or	$a[(-b) +b]= 0$	(By the property of inverse in R)
or	$a(-b)+ab = 0$	(By left distributive law)
or	$a(-b) = -(ab)$	($\because a+b = 0 \Rightarrow a=-b$)
Similarly,	$0b = 0$	[Using (i)]
or	$[(-a) + a]b = 0$	(By the property of inverse in R)
or	$(-a)b+ab = 0$	(By right distributive law)
or	$(-a)b = -(ab)$	($\because a+b = 0 \Rightarrow a=-b$)
Hence	$(-a)b = -(ab) = a(-b)$	

(iii)

	$(-a)(-b) = (ab)$	
Since we have $a(-b) = -(ab)$		
$\therefore$	$(-a)(-b) = - [(-a)b]$	
	$= - [-(ab)]$	$[\because (-a)b= -(ab)$]
	$= (ab)$	
Hence	$(- a)(-b) = (ab)$	

(iv)

	$a(b - c) =ab - ac$	
Now	$a(b - c) = a[b +(- c)]$	
	$= ab +a(- c)$	(By left distributive law)
	$=ab - ac$	$[\because a(-c)= -(ab)]$
Hence	$a(b-c) = ab - ac.$	

(v)

	$(b - c)a = ba - ac$	
Now	$(b - c)a =[b +(- c)]a$	
	$=ba +(- c)]a$	(By right distributive law)
	$=ba - ca$	$[\because (-c)a= -(ca)]$
Hence	$(b-c)a = ba - ca.$	

5.4 RING WITH AND WITHOUT ZERO DIVISORS

5.4.1 Zero Divisors

By elementary property of a ring we know that if 0 is the additive identity in R (a ring), then $a0 = 0 = 0a$, $\forall a \in R$. But in some rings it is possible that $ab=0$ when neigher $a =0$ nor $b = 0$. Such type of elements a and b are called zero divisors.

5.4.2 Ring with zero Divisors

A ring R is said to be ring with zero divisor if there exist non-zero elements a, b in R such that $ab = 0$, that is, if $a \neq 0$, $b \neq 0$ but $ab = 0$.

For example :

The set M of all matrices of order 2×2 having their elements as integers forms a ring with zero divisors under addition and multiplication of matrices, that is,

$$A = \begin{bmatrix} 1 & 1 \\ 1 & 1 \end{bmatrix}, B = \begin{bmatrix} -1 & -1 \\ 1 & 1 \end{bmatrix}, O = \begin{bmatrix} 0 & 0 \\ 0 & 0 \end{bmatrix}, \text{ then } AB = 0 \text{ but } A \neq 0 \text{ and } B \neq 0.$$

5.4.3 Ring without zero divisors

A ring R is said to be ring without zero divisor if $ab=0$, then either $a =0$ or $b=0$.

For Example

The ring of integers is a ring without zero divisors.

REMARKS

☞ Even though the existence of multiplicative inverses implies the absence of zero divisors, the absence of zero divisors does not guarantee the existence of multiplicative inverses, *i.e.*, in a ring, zero divisors and multiplicative inverses for all the elements (non-zero) cannot exist together.

5.5 CANCELLATION LAWS IN A RING

Let R be a ring and a, b, $c \in R$. For $a \neq 0$, if $ab= ac$ implies $b = c$, and for $a \neq 0$, if $ba= ca$ implies $b = c$. Then first is known as left cancellation law while second is known as right cancellation law.

Theorem 1. *A ring R is without zero divisors if and only if the cancellations laws holds in R.*

Proof. Let us first suppose that the ring R is without zero divisors, then we have to show that the cancellations laws hold in R.

Let $a \in R$ and $a \neq 0$ and we have

$$ab = ac$$
$\Rightarrow \qquad ab - ac = 0$
$\Rightarrow \qquad a(b - c) = 0$ 　　　　　　　　　　(By left distributive law)
$\Rightarrow \qquad b - c = 0$ 　　　　　　　　　　($\because R$ is of without zero divisor.)
$\Rightarrow \qquad b = c$

Similarly, 　　　　　$ba = ca$
$\Rightarrow \qquad ba - ca = 0$
$\Rightarrow \qquad (b - c) a = 0$ 　　　　　　　　　　(By left distributive law)
$\Rightarrow \qquad b - c = 0$ 　　　　　　　　　　($\because R$ is of without zero divisor.)
$\Rightarrow \qquad b = c$

Conversely, suppose cancellation laws hold in R, then we have to show that R is without zero divisor. Let us assume R is with zero divisor, that is

$$ab = 0, \text{ with } a \neq 0, b \neq 0$$
$\therefore \qquad ab = a0$ 　　　　　　　　　　($\because a0 = 0$)
$\Rightarrow \qquad b =0$ 　　　　　　　　　　(By left cancellation law and $a \neq 0$)

Which is a contradiction.

Similarly, let is take $\qquad ba = 0$

$\therefore \qquad\qquad\qquad ba = 0a$

$\Rightarrow \qquad\qquad\qquad b = 0 \qquad\qquad$ (By right cancellation law)

This gives again contradiction. Hence R is without zero divisors.

5.6 FIELD, INTEGRAL DOMAIN AND SKEW FIELD

5.6.1 Integral Domain

A commutative ring R with unit element having no zero divisors is called an integral domain.

For Example :

1. The ring of integers $(\mathbf{Z}, +,.)$ is an integral domain.

2. Let $S = \{a+b\sqrt{2} : a, b \in \mathbf{R}\}$ then $(S, +, .)$ is an integral domain.

5.6.2 Field

A commutative ring $\mathbf{R}$ with unit element having at least two elements is called a field if every non-zero element of R possesses multiplicative inverse.

For Example

1. The ring of rational numbers $(\mathbf{Q}, '+', '.')$ is a field.

2. Let $S = (\{0, 1,2,3,4\}, +_5, \times_5)$, then S is a finite field.

REMARK

☞ An element is a multiplicative inverse of b if $ab = 1 = ba$ for all $a \neq 0, b \neq 0$ in R.

5.6.3 Skew field

A ring R with unit element having at least two elements is called a skew field if every non-zero elements of R possesses their multiplicative inverse. The skew field is also known as division ring.

REMARKS

☞ We can define the field as a commutative division ring.

☞ Every field is also a divisor ring.

☞ Every integral domain satiesfies the cancellation laws.

☞ Every division ring satisfies the cancellation laws.

☞ If R is a ring with unity and without zero divisors, then the only solutions of $a^2 = a, a \in R$ are 0 and 1.

☞ A ring satisfies the left (right) cancellation law iff it has no left (right) divisors of zero.

☞ It should be noted that every ring satisfies the cancellation laws with respect to addition, when we use the phrase 'the ring R satisfies the cancellation laws', we are referring to the fact that the ring R satisfies the cancellation laws with respect to multiplication.

☞ You must be cautious while applying the cancellation laws in a ring $ab = ac \Rightarrow b = c$ only when $a \neq 0$. Compare this situation with that in R, where division by a is allowed only when $a \neq 0$.

Theorem 1. *Every field is an integral domain.*

Step outlines. To make the proof easier, use the following steps :

Step 1. *Use the definition of field.*

Step 2. *Show that F has no zero divisors.*

Proof. Let F be a field. We have to show that F is an integral domain. Since F is a field so it is commutative ring with unit element. Therefore in order to show F to be an integral domain, we only have to show that F has no zero divisors.

Let $a \in F$ and $a \neq 0$, then a^{-1} exists in F. We have $ab = 0$

$\Rightarrow \qquad\qquad a^{-1}(ab) = a^{-1}\,0$

$\Rightarrow \qquad\qquad (a^{-1}a)b = a^{-1}\,0$ \hfill (By associative law)

$\Rightarrow \qquad\qquad 1b = a^{-1}\,0$ \hfill $(\because a^{-1}a = 1)$

$\Rightarrow \qquad\qquad 1b = 0$ \hfill (By elementary property of ring)

$\Rightarrow \qquad\qquad b = 0$ \hfill $(\because 1.a = a = a.1)$

Similarly, let $b \in F$ and $b \neq 0$

$\qquad\qquad\qquad ab = 0$

$\Rightarrow \qquad\qquad (ab)b^{-1} = 0b^{-1}$

$\Rightarrow \qquad\qquad a(bb^{-1}) = 0$ \hfill (By associative law and $0b^{-1} = 0 = b^{-1}0$)

$\Rightarrow \qquad\qquad a1 = 0$ \hfill $(\because bb^{-1} = 1)$

$\Rightarrow \qquad\qquad a = 0$ \hfill $(\because a.1 = a = 1.a)$

Thus we obained that in F, $ab = 0$, then either $a = 0$ or $b = 0$ which implies F is without zero divisors. Hence F is an integral domain.

Theorem 2. *Disprove that every integral domain is a filed.*

Proof. To disprove that every integral domain is a field we shall give an example. Since the ring of integers is an integral domain but it is not field, because if $a \in \mathbf{Z}$ and $a \neq 0$, then $a^{-1} \in \mathbf{Z}.$

Theorem 3. *A field has no zero divisors.*

Proof. Since field is a commutative ring with unit element in which every non-zero element possesses its multiplicative inverse. Let $a \in F$ and $a \neq 0$ (a field), then a^{-1} exists in F, is also non-zero. Let us assume

$\qquad\qquad\qquad a^{-1} = 0$

$\Rightarrow \qquad\qquad aa^{-1} = a0$

$\Rightarrow \qquad\qquad 1 = 0$

This gives a contradiction. \hfill $(\because aa^{-1} = 1, a0 = 0)$

Thus $a^{-1} \neq 0$ and $aa^{-1} = 1 \neq 0$

Therefore in a field F, product of two non-zero elements is again a non-zero element. Hence F has no zero divisors.

REMARK

☞ A skew-field has no zero divisors.

Theorem 4. *Every finite integral domain is a field.*

Step outlines. To make the proof easier, use the following steps :

Step 1. *Use the definition of integral domain.*

Step 2. *Show that for each $a \in D$, $a \neq 0$, a^{-1} exists.*

Proof. Let D be a finite integral domain. Therefore by definition of integral domain we have that D is a commutative ring with unit element having no zero divisors. Let D = $\{a_1, a_2, ..., a_n\}$. In order to show that D is a field we only have to show that D has multiplicative inverse for every non-zero element in D. For this purpose let $a \neq 0$ be any arbitrary element of D and consider the set

$$D_1 = \{aa_1, aa_2, ..., aa_n\}$$

D_1 has n distinct products. For this let us suppose $aa_i = aa_j$ for $i \neq j$

$\Rightarrow \qquad a(a_i - a_j) = 0$ \hfill (By left distributive law)

Since D has no zero divisors and $a \neq 0$, then

$$a_i - a_j = 0 \quad \text{for } i \neq j$$

$\Rightarrow \qquad a_i = a_j \quad \text{for } i \neq j$

$\Rightarrow$ This is a contradiction, because D has n distinct elements $a_1, a_2, ..., a_n$.

Consequently D_1 has n distinct products. But the elements of D_1 are the elements of D placed in some order. Further D has unit element, that is ,

$$1 \in D \Rightarrow 1 \in D_1.$$

This implies that there exists an element b in D such that $ab = 1$.

But D is commutative. Therefore $ab = 1 = ba$.

Thus a^{-1} exists in D. Hence D is a field.

Theorem 5. *A finite commutative ring without zero divisors is a field.*

Step outlines. To make the proof easier, use the following steps :

Step 1. *Show that R has unit element.*

Step 2. *Show that every non-zero element of R has its multiplicative inverse.*

Proof. Let R be a finite commutative ring without zero divisor. We have to show that R is a field. In order to show R is a field, we only have to show that R has unit element and every non-zero element of R has its multiplicative inverse. Since R is a finite set so let us assume

$$R = \{a_1, a_2, ..., a_n\}$$

has n distinct elements. Let $a \neq 0$ be any non-zero arbitrary element of R. Then consider the set

$$R_1 = \{aa_1, aa_2, ..., aa_n\}$$

This set R_1 will have n distinct products.

Let if possible, $\qquad aa_i = aa_j$ for $i \neq j$

$\Rightarrow \qquad aa_i - aa_j = 0$

$\Rightarrow \qquad a(a_i - a_j) = 0$ \hfill (By left distributive law)

Since R has no zero divisors and $a \neq 0$, then

$$a_i - a_j = 0 \quad \text{for } i \neq j$$

or $\qquad a_i = a_j \quad \text{for } i \neq j.$

This is a contradiction, because R has distinct elements. Consequently R_1 has n distinct product and these n products are the elements of R which are placed in some order. Since $a \in R$ and $a \neq 0$, so we have

$$aa_k = a \quad \text{for some } k$$

or $\qquad aa_k = a.1$ \hfill $(\because a.1 = a)$

$\Rightarrow \qquad a(a_k - 1) = 0$

$\Rightarrow$ $a_k = 1$ ($\because$ R is without zero divisor.)

Thus there exists an element $1 \in R$. Now we have to show that this element 1 is a multiplicative identity element of R. For this purpose let a_m be any arbitrary element of R then

$$a_m = a a_r = a_r a \text{ for some } a_r \in R$$
$$1 . a_m = 1 . (a a_r) = (1 . a) a_r \qquad \text{(By associative law)}$$
$$= a a_r$$
$$= a_m$$

Similarly, $a_m . 1 = a_m$

Thus $a_m . 1 = a_m = 1 . a_m$. Consequently 1 is a multiplicative identity of R.

Therefore there exists $b \in R$ such that $ab = 1 = ba$. Hence a^{-1} exists in R and hence R is a field.

REMARK

☞ If the coefficient of a polynomial form an integral domain, we can solve a polynomial equation in which the polynomial can be factored into linear factors in the useful fashion by setting each factor equal to zero.

Solved Examples
Based on the following Results

▶ An algebraic structure $(R, +, \cdot)$ is said to be ring if it is
(i) abelian group under addition
(ii) semigroup under multiplication
(iii) multiplication is distributive over addition

▶ **Ring with zero divisors.** $ab = 0$ when neither $a = 0$ nor $b = 0$

▶ **Ring without zero divisors.** $ab = 0$ when either $a = 0$ or $b = 0$

▶ A commutative ring with unity without zero divisors is called an integral domain.

▶ A commutative ring with unity having at least two elements is called a field if every non-zero elements of R possesses multiplicative inverse.

Example 1. *Show that the set $R = \{0,1,2,3,4,5\}$ forms a commutative ring with respect to binary operations '$+_6$' and '$\times_6$'.*

Solution. First we shall prove that $(R, '+_6')$ is an abelian group. Now forming a compositioin table for '$+_6$'

$+_6$	0	1	2	3	4	5
0	0	1	2	3	4	5
1	1	2	3	4	5	0
2	2	3	4	5	0	1
3	3	4	5	0	1	2
4	4	5	0	1	2	3
5	5	0	1	2	3	4

From composition table it is clear that R is closed under '$+_6$'.

(i) **Associativity.** If $a, b, c \in R$, then we have to show $a +_6 (b +_6 c) = (a +_6 b) +_6 c$
$a +_6 (b +_6 c) = a$ remainder when $a + (b + c)$ is divided by 6

$$= a \text{ remainder when } (a + b) + c \text{ is divided by } 6$$
$$= (a +_6 b) +_6 c$$

(ii) Existence of identity. Since $0 \in R$, then $0 +_6 a = a = a +_6 0$ for $a \in R$. Thus 0 is the identity which exists in R.

(iii) Existence of inverse. From composition table it is observed that $1 +_6 5 = 0$, $2 +_6 4 = 0$, $3 +_6 3 = 0$. Hence inverse of every element exists in R.

(iv) Commutatively. If $a, b \in R$, then we have to show that
$$a +_6 b = b +_6 a$$
$$a +_6 b = a \text{ remainder when } (a+b) \text{ is divided by } 6$$
$$= a \text{ remainder when } (b+a) \text{ is divided by } 6 = b +_6 a$$

Thus $(R, '+_6')$ is an abelian group.

Next we form a composition table for '$\times_6$' as follows :

$\times_6$	0	1	2	3	4	5
0	0	0	0	0	0	0
1	0	1	2	3	4	5
2	0	2	4	0	2	4
3	0	3	0	3	0	3
4	0	4	2	0	4	2
5	0	5	4	3	2	1

From this composition table it is observed that R is closed w.r.t. '$\times_6$'. Also '$\times_6$' is associative. For this we have
$$a \times_6 (b \times_6 c) = (a \times_6 b) \times_6 c, \text{ for all } a, b, c \in R$$

Now $\quad a \times_6 (b \times_6 c) = a \text{ remainder when } a \times (b \times c) \text{ is divided by } 6$
$$= a \text{ remainder when } (a \times b) \times c \text{ is divided by } 6$$
$$= (a \times_6 b) \times_6 c$$
$\therefore \qquad a \times_6 (b \times_6 c) = (a \times_6 b) \times_6 c$

(v) Distributive laws. If $a, b, c \in R$, then
$$a \times_6 (b \times_6 c) = (a \times_6 b) \times_6 (a \times_6 c)$$
and $\quad (b \times_6 c) \times_6 a = (b \times_6 a) +_6 (c \times_6 a)$
Now $\quad a \times_6 (b \times_6 c) = a \text{ remainder when } a(b+c) \text{ is divided by } 6.$
$$= a \text{ remainder when } ab+ac \text{ is divided by } 6.$$
$$= (a \times_6 b) +_6 (a \times_6 c)$$
Similarly, $(b \times_6 c) \times_6 a = (b \times_6 a) +_6 (c \times_6 a).$

(vi) Commutativity. If $a, b \in R$, then
$$(a \times_6 b) = (b \times_6 a)$$
Now $\qquad a \times_6 b = a \text{ remainder when } a \times b \text{ is divided by } 6.$
$$= a \text{ remainder when } b \times a \text{ is divided by } 6.$$
$$= b \times_6 a$$
$\therefore \qquad a \times_6 b = b \times_6 a$

Hence $(R +_6, \times_6)$ is a commutative ring.

Example 2. *If R is a ring such that $a^2 = a$ for all $a \in R$ prove that*
(i) $a + a = 0, \forall a \in R$

(ii) $a + b = 0 \Rightarrow a = b$

(iii) *R is a commutative ring.*

Solution. (i) Since $\qquad a \in R \Rightarrow a + a \in R$

Now $\qquad (a + a)^2 = (a + a)$ $\qquad (\because a^2 = a)$

$\Rightarrow \qquad (a + a)(a + a) = a + a$

$\Rightarrow \quad (a + a)a + (a + a)a = a + a$ $\qquad$ (By left distributive law)

$\Rightarrow (a^2 + a^2) + (a^2 + a^2) = a + a$ $\qquad$ (By right distributive law)

$\Rightarrow \qquad (a + a) + (a + a) = a + a$ $\qquad (\because a^2 = a)$

$\Rightarrow \qquad (a + a) + (a + a) = (a + a) + 0$ $\qquad (\because a + 0 = a)$

$\Rightarrow \qquad\qquad a + a = 0$ $\qquad$ (By left cancellation law)

(ii) $\qquad\qquad a + b = 0$ $\qquad$ (given)

$\Rightarrow \qquad\qquad a + b = a + a$ $\qquad (\because a + a = 0)$

$\Rightarrow \qquad\qquad\qquad a = b$ $\qquad$ (By left cancellation law)

(iii) If $\qquad\qquad a, b \in R \Rightarrow a + b \in R$

Now $\qquad\qquad (a + b)^2 = (a + b)$ $\qquad (\because a^2 = a)$

$\Rightarrow \qquad (a + b)(a + b) = a + b$

$\Rightarrow \quad (a + b)a + (a + b)b = a + b$ $\qquad$ (By left distributive law)

$\Rightarrow \qquad a^2 + ba + ab + b^2 = a + b$ $\qquad$ (By right distributive law)

$\Rightarrow \qquad a + ba + ab + b = a + b$ $\qquad (\because a^2 = a)$

$\Rightarrow \qquad\quad a + ba + ab = a$ $\qquad$ (By right cancellation law)

$\Rightarrow \qquad\qquad ba + ab = 0$ $\qquad$ (By left cancellation law)

$\Rightarrow \qquad\qquad\quad ba = ab$ $\qquad (\because a + b = 0 \Rightarrow a = b)$

Hence, R is a commutative ring.

REMARK

☞ If $a^2 = a, \forall\, a \in R$ (Ring), then R is called a boolean ring.

Example 1. *Show that the set of numbers of the form* $a + b\sqrt{2}$, *where a and b are rational numbers, is a field with respect to addition and multiplication.*

Solution. Let $R = \{a + b\sqrt{2} : a, b \in \mathbf{Q}\}$. First we shall show that $(R, +)$ is an abelian group.

(i) Closure property. Let $a_1 + b_1\sqrt{2}, a_2 + b_2\sqrt{2} \in R$

$(a_1 + b_1\sqrt{2}) + (a_2 + b_2\sqrt{2}) = (a_1 + a_2) + (b_1 + b_2)\sqrt{2}$ $\qquad$ (By associative law)

Since $\qquad\qquad a_1, a_2 \in \mathbf{Q} \Rightarrow a_1 + a_2 \in \mathbf{Q}$ and $b_1, b_2 \in \mathbf{Q} \Rightarrow b_1 + b_2 \in \mathbf{Q}$

$\therefore \qquad (a_1 + a_2) + (b_1 + b_2)\sqrt{2} \in R$

Thus R is closed under addition.

(ii) Associative property. Since $a + b\sqrt{2}$ is a real number and addition in real numbers is associative therefore associative law holds in **R**.

(iii) Existence of identity. Since $0 \in \mathbf{Q}$, then $0 + 0.\sqrt{2} \in \mathbf{R}$.

Now $\quad (a + b\sqrt{2}) + (0 + 0.\sqrt{2}) = (a + 0) + (b + 0)\sqrt{2}$

$\qquad\qquad\qquad\qquad\qquad = a + b\sqrt{2}$ $\qquad \left(\because \begin{array}{l} a + 0 = a \\ b + 0 = b \end{array} \right)$

Thus 0 is an additive identity of **R**.

(iv) Existence of inverse. Since $a \in Q \Rightarrow -a \in Q$

If $a + b\sqrt{2} \in R$, then $(-a) + (-b)\sqrt{2} \in R$,

$$(a + b\sqrt{2}) + [(-a) + (-b)\sqrt{2}] = (a-a) + (b-b)\sqrt{2}$$
$$= 0 + 0.\sqrt{2}$$

Thus $(-a) + (-b)\sqrt{2}$ is an additive inverse of $a + b\sqrt{2}$ in **R**.

(v) Addition is always commutative.

Hence $(R '+')$ is an abelian group.

(vi) Closed under multiplication.

If $a_1 + b_1\sqrt{2}, a_2 + b_2\sqrt{2} \in R$, then

$$(a_1 + b_1\sqrt{2}).(a_2 + b_2\sqrt{2}) = (a_1 a_2 + 2b_1 b_2) + (a_1 b_2 + b_1 a_2)\sqrt{2}$$

Since $a_1, a_2, b_1, b_2 \in Q$, then $a_1 a_2 + 2b_1 b_2 \in Q$ and $a_1 b_2 + b_1 a_2 \in Q$

Thus $(a_1 a_2 + 2b_1 b_2) + (a_1 b_2 + b_1 a_2)\sqrt{2} \in$ **R.**

Hence **R** is closed under multiplication.

(vii) Multiplication is always distributive over addition.

(viii) Multiplication is always commutative.

Hence $(R '+', '.')$ is an abelian ring. In order to prove that R is a field we have to prove that unity belong to R and every non-zero elements of R possesses its multiplicative inverse. For this purpose we have

$$1 \in Q \text{ then } 1 + 0.\sqrt{2} \in R \text{ then}$$
$$(a + b\sqrt{2}).(1 + 0.\sqrt{2}) = a + b\sqrt{2}$$

Thus $1 + 0.\sqrt{2} \in R$ is a multiplicative identity. Further let

$$a + b\sqrt{2} \neq 0 \text{ and } a + b\sqrt{2} \in R \text{ either } a \neq 0 \text{ or } b \neq 0.$$

Then, we have

$$\frac{1}{a + b\sqrt{2}} = \frac{a - b\sqrt{2}}{a^2 - 2b^2} = \left(\frac{a}{a^2 - 2b^2}\right) + \left(\frac{-b}{a^2 - 2b^2}\right)\sqrt{2}.$$

Since $a, b \in Q$, then we can have $a^2 = 2b^2$ only if $a = 0$ and $b = 0$ but this is not possible because either $a \neq 0$ or $b \neq 0$. Thus $\dfrac{a}{a^2 - 2b^2}$ and $-\dfrac{b}{a^2 - 2b^2}$ are both rational numbers and hence, we have

$$\frac{1}{a + b\sqrt{2}} \in R \text{ and } (a + b\sqrt{2}).\frac{1}{a + b\sqrt{2}} = 1 = 1 = 0.\sqrt{2}.$$

Hence $\dfrac{1}{a + b\sqrt{2}}$ is a multiplicative inverse of non-zero element $(a + b\sqrt{2})$.

Consequently, $(\mathbf{R}, +, .)$ is a field.

Example 4. *Give an example of a skew field which is not a field.*

Solution. Let **M** be the set of all matrices of order 2×2 of the form

$$\begin{bmatrix} a + ib & c + id \\ -c + id & a - ib \end{bmatrix}$$

where $a, b, c, d \in \mathbf{R}$ (the set of real numbers)

First we have to show that the set M forms a ring with respect to addition and multiplication of matrix.

(i) Closure axiom. Let $A, B \in M$, where

$$A = \begin{bmatrix} a_1 + ib_1 & c_1 + id_1 \\ -c_1 + id_1 & a_1 - ib_1 \end{bmatrix}, \ B = \begin{bmatrix} a_2 + ib_2 & c_2 + id_2 \\ -c_2 + id_2 & a_2 - ib_2 \end{bmatrix},$$

and $a_1, b_1, c_1, d_1; a_2, b_2, c_2, d_2$ are all reals.

Now $$A + B = \begin{bmatrix} (a_1 + a_2) + i(b_1 + b_2) & (c_1 + c_2) + i(d_1 + d_2) \\ -(c_1 + c_2) + i(d_1 + d_2) & (a_1 + a_2) - i(b_1 + b_2) \end{bmatrix}.$$

Obeviously $A + B$ is of the form as given in M. Therefore $A + B \in M$. Also

$$AB = \begin{bmatrix} a_1 + ib_1 & c_1 + id_1 \\ -c_1 + id_1 & a_1 - ib_1 \end{bmatrix} \begin{bmatrix} a_2 + ib_2 & c_2 + id_2 \\ -c_2 + id_2 & a_2 - ib_2 \end{bmatrix}$$

$$AB = \begin{bmatrix} (a_1a_2 - b_1b_2 - c_1c_2 - d_1d_2) & (a_1c_2 - b_1d_2 - c_1a_2 + d_1b_2) \\ +i(a_1b_2 + b_1a_2 + c_1d_2 - c_2d_1) & +i(a_1d_2 + b_1c_2 + d_1a_2 - c_1b_2) \\ & \\ -(c_1a_2 + d_1b_2 + a_1c_2 - b_1d_2) & (-c_1c_2 - d_1d_2 + a_1a_2 - b_1b_2) \\ +i(d_1a_2 + a_1d_2 - c_1b_2 + b_1c_2) & +i(c_1d_2 - d_1c_2 + a_1b_2 + b_1a_2) \end{bmatrix}$$

Thus $AB \in M$. Hence M is closed under addition and multiplication of matrices.

(ii) Matrix addition and multiplication are always associative.

(iii) Existence of identity. The null matrix

$$O = \begin{bmatrix} 0 & 0 \\ 0 & 0 \end{bmatrix} = \begin{bmatrix} 0 + i.0 & 0 + i.0 \\ -0 + i.0 & 0 - i.0 \end{bmatrix}$$

is the additive identity for M. That is,

$$AO = \begin{bmatrix} a + ib & c + id \\ -c + id & a - ib \end{bmatrix} \begin{bmatrix} 0 + i.0 & 0 + i.0 \\ -0 + i.0 & 0 - i.0 \end{bmatrix}$$

$$= \begin{bmatrix} (a + 0) + i(b + 0) & (c + 0) + i(d + 0) \\ -(c + 0) + i(d + 0) & (a + 0) - i(b + 0) \end{bmatrix}$$

$$= \begin{bmatrix} a + ib & c + id \\ -c + id & a - ib \end{bmatrix} = A.$$

(iv) Existence of inverse. Let $A \in M$, that is $A = \begin{bmatrix} a + ib & c + id \\ -c + id & a - ib \end{bmatrix}$

Then $$-A = \begin{bmatrix} (-a) + i(-b) & (-c) + i(-d) \\ -(-c) + i(-d) & (-a) - i(-b) \end{bmatrix} \in M$$

and $$(-A) + A = \begin{bmatrix} 0 + i.0 & 0 + i.0 \\ -0 + i.0 & 0 - i.0 \end{bmatrix} = O.$$

Hence, $(-A)$ is the additive inverse of A.

(v) Matrix addition is commutative.

Matrix multiplication is always distributive over matrix addition. Hence $(M, +, .)$ is a ring.

(vi) Existence of multiplicative identity. The unit matrix

$$I = \begin{bmatrix} 1 & 0 \\ 0 & 1 \end{bmatrix} = \begin{bmatrix} 1+0.i & 0+i.0 \\ -0+0.i & 1-i.0 \end{bmatrix} \in M$$

is the multiplicative identity of M.

(vii) Existence of multiplicative inverse. Let A be any non-zero element of M and let

$$A = \begin{bmatrix} a+ib & c+id \\ -c+id & a-ib \end{bmatrix} \neq O = \begin{bmatrix} 0 & 0 \\ 0 & 0 \end{bmatrix}.$$

This implies that a, b, c, d are not all zero, and

$$|A| = \begin{bmatrix} a+ib & c+id \\ -c+id & a-ib \end{bmatrix} = a^2 + b^2 + c^2 + d^2 \neq 0.$$

Thus matrix A is non-singular and therefore A^{-1} exist.

Now $\qquad A^{-1} = \dfrac{AdjA}{|(A)|} = \dfrac{1}{a^2+b^2+c^2+d^2}\begin{bmatrix} a-ib & -c+id \\ c-id & a+ib \end{bmatrix} \in M.$

$\therefore \qquad A^{-1}$ exist and belongs to M.

Hence M is a skew field. Now we have to show that M is not a field. For this, let $A, B \in M$ then

$$A = \begin{bmatrix} a_1+ib_1 & c_1+id_1 \\ -c_1+id_1 & a_1-ib_1 \end{bmatrix}, B = \begin{bmatrix} a_2+ib_2 & c_2+id_2 \\ -c_2+id_2 & a_2-ib_2 \end{bmatrix}$$

Now $\qquad AB = \begin{bmatrix} a_1+ib_1 & c_1+id_1 \\ -c_1+id_1 & a_1-ib_1 \end{bmatrix}\begin{bmatrix} a_2+ib_2 & c_2+id_2 \\ -c_2+id_2 & a_2-ib_2 \end{bmatrix}$

$$= \begin{bmatrix} (a_1a_2-b_1b_2-c_1c_2-d_1d_2) & (a_1c_2-b_1d_2+c_1a_2+d_1b_2) \\ +i(a_1b_2+b_1a_2+c_1d_2-c_1d_1) & +i(a_1d_2+b_1c_2+d_1a_2-c_1b_2) \\ & \\ -(a_1c_2-b_1d_2+c_1a_2+d_1b_2) & (a_1a_2-b_1b_2-c_1c_2-d_1d_2) \\ +i(a_1d_2+b_1c_2+d_1a_2-c_1b_2) & -i(a_1b_2+b_1a_2+c_1d_2-c_2d_1) \end{bmatrix}$$

and $\qquad BA = \begin{bmatrix} (a_1a_2-b_1b_2-c_1c_2-d_1d_2) & (c_1a_2-d_1b_2+a_1c_2+b_1d_2) \\ +i(a_1b_2+a_2b_1+c_1d_2+c_2d_1) & +i(c_1b_2+a_1d_2+a_1d_2-b_1c_2) \\ & \\ -(a_1c_2+b_1d_2+c_1a_2-d_1b_2) & (-c_1c_2-d_1d_2+a_1a_2-b_1b_2) \\ +i(b_1d_2-b_1c_2+a_2d_1+c_1b_2) & -i(d_1c_2-c_1d_2+a_1b_2+a_2b_1) \end{bmatrix}$

Obviously $AB \neq BA$. Hence M is not a field.

Example 5: *Prove that the set of residue classes modulo p is a commutative ring with respect to addition and multiplication of residue classes. Further show that the ring of residue classes modulo p is a field iff p is a prime.*

Solution. Let I_p be the set of residue classes modulo p. That is,

$$I_p = \{[0], [1], [2], \ldots, [p-1]\}$$

Obviously I_p has p distinct elements. Now we shall have to show that $(I_p, '+', '\cdot')$ forms a commutative ring.

(i) **Closure axiom.** Let $[a]. [b] \in I_p$, then

$$[a] + [b] = [a+b] \qquad \text{(By addition of residue classes)}$$
and $\quad [a] . [b] = [ab] \qquad \text{(By multiplication of residue classes)}$

Since $[a+b]$ and $[ab]$ are residue classs modulo p. Then I_p is closed under addition and multiplication.

(ii) **Associative for addition.** Let $[a], [b], [c] \in I_p$, then

$$[a] + ([b]+[c]) = [a]+[b+c]$$
$$= [a+ (b+c)] = [(a+b)+c] = [a+b]+[c] = ([a]+[b])+[c]$$
$$\therefore \ [a] + ([b]+[c]) = ([a]+[b])+[c]$$

Therefore addition is associative in I_p.

(iii) **Existence of identity.** Since $[0] \in I_p$, and for all $[a] \in I_p$

$$[0]+[a] = [0+a] = [a].$$

Thus $[0]$ is an additive identity.

(iv) **Existence of inverse.** Let $[a] \in I_p$, then $[-a] \in I_p$

Now $\quad [a]+[-a] = [a-a] = [0].$

$\therefore \qquad [-a]$ is an additive inverse of $[a]$.

(v) **Commutativity under addition and multiplication.**

Let $[a], [b] \in I_p$, we have

$$[a] + [b] = [a+b] = [b+a] = [b] + [a]$$
and $\quad [a][b] = [ab] = [ba] = [b][a].$

Thus I_p is commutative under addition and multiplicaiton.

(vi) **Distributivity.** Let $[a], [b], [c] \in I_p$, then

$$[a] . ([b]+[c]) = [a].[b+c] = [a . (b+c)]$$
$$= [ab + ac] = [ab] +[ac]$$
$$= ([a][b]) + ([a][c])$$

Similarly,

$$([b] + [c]). [a] = [b][a] + [c][a].$$

Hençe I_p is a commutative ring.

Next we have to show that I_p is a field iff p is prime. Suppose I_p is a field. Then I_p is an integral domain, therefore, I_p is without zero divisors. We have to prove that p is a prime. Let us suppose p is not prime. That is $p = mn$, where $1 < m < p, 1 < n < p$. Then

$$[mn] = [p]$$
$\Rightarrow \qquad\qquad [mn] = [0] \qquad\qquad\qquad\qquad (\because [p] = [0])$

$\Rightarrow$ either $[m] = [0]$ or $[n] = 0$.

But $1 < m < p$ so $[m] \neq [0]$. Similarly, $[n] \neq 0$. Therefore neither $[m] = 0$ nor $[n] = 0$. Thus our assumption that p is not prime is wrong. Hence p is prime.

Conversely, suppose p is prime. We have to show I_p is field. Since I_p has p element which is finite. Therefore we only have to show that I_p is an integral domain. For this, consider

$$[a], [b] \in I_p, \text{ If } [a][b] = [0]$$

$\Rightarrow$ $[ab] = [0] \Rightarrow$ Either p divides a or p divides b.

$\Rightarrow$ Either $[a] = 0$ or $[b] = 0$

$\Rightarrow$ I_p is without zero divisors.

Thus I_p is an integral domain and since we know that every finite integral domain is a field. Hence I_p is a field.

Example 6. *If a, b, c, d are elements of a ring R, then evaluate* $(a+b)(c+d)$.

Solution. Since $a, b, c, d \in R$ so that $ac, ad, bc, bd \in R$. We have

$$(a+b)(c+d) = a(c+d) + b(c+d) \quad \text{(By right distributive law)}$$
$$= ac + ad + bc + bd \quad \text{(By left distributive law)}$$

Example 7. *Show that the set* $J[i] = \{a+ib : a,b \in \mathbf{Z}\}$ *of Gaussian integers forms a ring under ordinary addition and multiplication of complex numbers. Is it an integral domain? Is it a field ?*

Solution. Let $a + ib$ and $c + id$ be any two elements of $J[i]$.

(i) Closure property.

$$(a+ib) + (c+id) = (a+c) + i(b+d) \in J[i] \quad (\because a+c, b+d \in I)$$

Also, $(a+ib)(c+id) = (ac - bd) + i(ad+bc) \in J[i] \quad (\because ac - bd, ad+bc \in I)$

Therefore $J[i]$ is closed under addition and multiplication of complex numbers.

(ii) Associative property. In complex numbers both the addition and multiplication are associative.

(iii) Additive inverse. Let $a + ib \in J[i], i.e., a, b \in \mathbf{Z}$ so that $(-a), (-b) \in \mathbf{Z}$, therefore, $(-a) + i(-b) \in J[i]$. Thus $(-a) + i(-b)$ is an additive inverse of $a + ib$.

(iv) Distributive property. In complex numbers multiplication is distributive over addition.

(v) Multiplicative identity. Since $1, 0 \in \mathbf{Z}$ so that $1+i.0 \in J[i]$ thus $1+i.0$ is the multiplicative identity.

(vi) Commutative property. Let $a + ib, c + id \in J[i]$.

We have

$$(a+ib)(c+id) = (ac - bd) + i(ad+bc) = (ca - bd) + i(da+cb)$$

$$(\because \text{ Integers are commutative w.r.t., multiplication})$$

$$= (c+id)(a+ib).$$

Thus $J[i]$ is commutative w.r.t. multiplication.

Hence, the set of Gaussian integers is a commutative ring with unity for given composition.

Also, the product of two non-zero complex numbers cannot be zero, so that $J[i]$ has no zero divisors. Therefore, $J[i]$ is an integral domain. But $J[i]$ is not a field because the multiplicative inverse of $a + ib$ will be $\left(\dfrac{a}{a^2 + b^2}\right) + i\left(\dfrac{b}{a^2 + b^2}\right)$

which is not always a Gaussian integer as $\dfrac{a}{a^2 + b^2}$ and $\dfrac{b}{a^2 + b^2}$ are not necessarily integers.

Example 8. *Prove that the set M of 2×2 matrices over the field of real numbers is a ring with respect to matrix addition and multiplication. Is it a commutative ring with unity element ? The zero element? Does this ring possess zero divisors ?*

Solution.   **(i) Closure property.** Let A, $B \in M$. Then $A + B \in M$ and $AB \in M$, therefore M is closed with respect to addition and multiplication of matrices.

   (ii) Associative property. Let A, B, $C \in M$. Then $A + B \in M$ and $AB \in M$, therefore M is closed with respect to addition and multiplication of matrices.

  (iii) Associative property. Let A, B, $C \in M$. Then

$$A + (B+C) = (A+B)+C, \ \forall \, A, B, C \in M$$

and $\qquad\qquad A(BC) = (AB)C, \ \forall \, A, B, C \in M.$

Thus, both addition and multiplication of matrices are associative.

  (iv) Additive Identity. Let $O = \begin{bmatrix} 0 & 0 \\ 0 & 0 \end{bmatrix}$ be a null matrix of order 2×2, then $O \in M$ and

$$O+A = A+O = A, \ \forall \, A \in M$$

$\therefore$ O is the additive identity of M.

  (v) Additive inverse. If $A \in M$, then $-A$ will be a matrix of order 2×2, so that $-A \in M$ also,

$$A + (-A) = O, \ \forall \, A \in M$$

$\therefore$ $-A$ is the additive inverse of A.

  (vi) Commutative property. Let A, $B \in M$, then

$$A+B = B+A, \ \forall \, A, B \in M.$$

$\therefore$ Addition is commutative in M.

 (vii) Distributive property. Let A, B, $C \in M$, then

$$A(B+C) = AB+AC$$

and $\qquad\qquad (B+C)A = BA + CA, \ \forall \, A, B, C \in M.$

$\therefore$ Multiplication of matrices is distributive over addition.

Hence, M forms a ring with respect to the given compsotions.

If I is the unit matrix of order 2×2, *i.e.*,

$$I = \begin{bmatrix} 1 & 0 \\ 0 & 1 \end{bmatrix}$$

then $I \in M$ and also $AI = A = IA \qquad A \in M.$

$\therefore$ M has unit element.

But multiplication of matrices is not in general a commutative composition.

For Example $A = \begin{bmatrix} 1 & 2 \\ -1 & 3 \end{bmatrix}, \ B = \begin{bmatrix} 2 & -1 \\ -1 & 2 \end{bmatrix}$

Then $\qquad AB = \begin{bmatrix} 0 & 3 \\ -5 & 7 \end{bmatrix}$ and $BA = \begin{bmatrix} 3 & 1 \\ -3 & 4 \end{bmatrix}$

∴ $\qquad\qquad AB \neq BA$

Therefore M is non-commutative ring but has unity. The zero element of M is the null matrix O. The ring possesses zero divisors, because the product of two non-zero elements of the ring is equal to the zero element of the ring.

For Example $\qquad A = \begin{bmatrix} 0 & 1 \\ 0 & 2 \end{bmatrix}, \ B = \begin{bmatrix} 2 & 3 \\ 0 & 0 \end{bmatrix}$

Then $\qquad\qquad AB = \begin{bmatrix} 0 & 1 \\ 0 & 2 \end{bmatrix} \cdot \begin{bmatrix} 2 & 3 \\ 0 & 0 \end{bmatrix} = \begin{bmatrix} 0 & 0 \\ 0 & 0 \end{bmatrix} = O.$

Example 9. *Show that the set R of all real valued continuous functions defined in the closed interval [0,1] is a commutative ring with unity with respect to the addition and multiplication of functions defined pointwise as follows :*

$$(f+g)\ (x) = f(x) + g(x) \text{ and } (f\,g)\ (x) = f(x)\ g(x) \in [0, 1]$$

where f, g are any two members of R.

Solution. **(i) Closure property.** Since $f(x)$ is a real valued continuous function defined in a closed interval [0, 1]. This means that $f(x)$ is a real number. We know that the sum and product of two real numbers is again real number and also the product and sum of two continuous functions is continuous. Therefore, the set R is closed under given composition of functions.

(ii) Associative property. Let f, g, h be any elements of R. Then for every $x \in [0, 1]$, we have

$$[(f+g) + h]\ (x) = (f+g)(x) + h(x)$$
$$= [f(x)+g(x)] + h(x)$$
$$= f(x)+[g(x) + h(x)]$$

$\qquad\qquad$ (∵ Addition is associative for real numbers.)

$$= f(x) +(g+h)(x)$$
$$= [f+(g+h)](x)$$

∴ $\qquad (f+g)+h = f+(g+h) \qquad$ (By definition of equality of functions)

Also, $\qquad [(fg)h](x) = (fg)(x). \, h(x)$
$$= [f(x)\ g(x)]\ h(x)$$
$$= f(x)[g(x)\ h(x)]$$

$\qquad\qquad$ (∵ Multiplication is associative for real numbers.)

$$= f(x)(gh)(x)$$
$$= [f(gh)](x)$$

∴ $\qquad\qquad (fg)h = f(gh) \qquad$ (By eqaulity of functions)

Thus R is associative under the given compositions of functions.

(iii) Existence of identity. Define a function e by $e(x) = 0$, $\forall\ x \in [0, 1]$. Then $e \in R$ and also if $f \in R$,

then $\qquad (e+f)(x) = e(x) + f(x) = 0 + f(x) = f(x)$

$$\therefore \qquad\qquad e+f=f, \qquad \forall f \in R$$

Thus e is the additive indentity for addition.

(iv) Existence of inverse. Let $f \in R$, then define a function
$$(-f)(x) = -f(x) = -f(x), \forall x \in [0, 1], \text{ then } -f \in R \text{ and also,}$$
$$(-f+f)(x) = -f(x) + f(x)$$
$$= 0$$
$$= e(x)$$
$$\therefore \qquad\qquad -f + f = e, \ \forall f \in R$$

Thus $-f$ is the additive inverse of f.

(v) Commutative property. Let $f, g \in R$.

Then
$$(f+g)(x) = f(x) + g(x)$$
$$= g(x) + f(x)$$
$$(\because \text{ Addition is commutative for real numbers.})$$
$$= (g+f)(x)$$
$$\therefore \qquad\qquad f+g = g+f$$

Thus, R is commutative under addition.

Then
$$(fg)(x) = f(x) \, g(x)$$
$$= g(x) \, f(x)$$
$$(\because \text{ Multiplication is commutative for real numbers.})$$
$$= (gf)(x)$$
$$\therefore \qquad\qquad fg = gf \qquad\qquad\qquad \text{(By equality of functions)}$$

Thus, R is commutative under multiplication of functions.

(vi) Distributive property. Let $f, g, h \in R$. Then
$$[f(g+h)](x) = f(x)[(g+h)(x)]$$
$$= f(x) \, [g(x) + h(x)]$$
$$= f(x) \, g(x) + f(x) \, h(x)$$
$$(\because \text{ Multiplication is distributive w.r.t. addition for real numbers.})$$
$$= (fg)(x) + (fh)(x)$$
$$= (fg+fh)(x)$$
$$\therefore \qquad\qquad f(g+h) = fg + fh$$
Similarly, $\qquad (g+h)f = gf + hf$

Hence, R is a ring with respect to given compositions.

(vii) Existence of multiplication identity. Define a function i by $i(x) = 1$, $\forall x \in [0, 1]$, then $i \in R$. If $f \in R$, then
$$(if)(x) = i(x) \, f(x)$$
$$= 1. \, f(x) = f(x)$$
$$\therefore \qquad\qquad if = f$$
Similarly, $\qquad fi = f$
$$\therefore \qquad\qquad if = f = fi, \ \forall f \in R$$

This function i is the multiplicative identity. Hence, R is a ring with unit element.

Example 10. *Let p be a prime number. Prove that the set of integers I_p given by $I_p = \{0,1,2,3,...,p-1\}$ forms a field with respect to addition and multiplication modulo p.*

Solution. First we shall show that $\{I_p, +_p\}$ is an abelian group.

    (i) Closure property. Let $a, b \in I_p$. Then

$$a +_p b \equiv a+b \ (mod \ p)$$
$$a +_p b \in I_p \ \forall \ a, b \in I_p$$

Thus, I_p is closed under addition modulo p.

    (ii) Associative property. Let $a, b, c \in I_p$, then

$$(a +_p b) +_p c = (a+b) +_p c \qquad [\because a +_p b = a+b \ (mod \ p)]$$
$$= \text{least non-negative remainder when } (a+b) +_p c \text{ is divided by } p$$
$$= \text{least non-negative remainder when } a + (b+c) \text{ is divided by } p$$
$$= a +_p (b+c) = a +_p (b +_p c)$$
$$\therefore \qquad (a +_p b) +_p c = a +_p (b +_p c) \ \forall \ a, b, c \in I_p$$

    (iii) Existence of identity.

Since $0 \in I_p$ and if $a \in I_p$ then $0 +_p a = 0 + a (mod \ p) = a (mod \ p), \ \forall \ a \in I_p$

$\therefore$ 0 is the identity for addition modulo p.

    (iv) Existence of inverse. If $a \in I_p$, then $p - a \in I_p$ since

$$a +_p (p - a) \equiv a + p - a (mod \ p)$$
$$\equiv p (mod \ p) \equiv 0 (mod \ p)$$

Thus, $(p - a)$ is the inverse of a for addition modulo p.

    (v) Commutative property. Let $a, b \in I_p$. Then

$$a +_p b = \text{least non-negative remainder when } a+b \text{ is divided by } p$$
$$= \text{least non-negative remainder when } b+a \text{ is divided by } p$$
$$= b +_p a$$
$$\therefore \qquad a +_p b = b +_p a \ \forall \ a, b \in I_p$$

Hence, $\{I_p, +_p\}$ is an abelian group.

    (vi) Also, I_p is closed with respect to $'\times_p'$ and $'\times_p'$ is associative as well as commutative.

    (vii) Distributive law. Let $a, b, c \in I_p$, then

$$a \times_p (b +_p c) = a \times_p (b+c) \qquad\qquad [\because b +_p c \equiv b+c \ (mod \ p)]$$
$$= \text{least non-negative remainder when } a(b+c) \text{ is divided by } p$$
$$= \text{least non-negative remainder when } ab+bc \text{ is divided by } p$$
$$= (ab) +_p (ac)$$
$$= (a \times_p b) +_p (ac) \qquad\qquad [\because ab \equiv a \times_p b \ (mod \ p)]$$
$$= (a \times_p b) +_p (a \times_p c)$$

Similarly,

$$(b +_p a) \times_p a = (b \times_p a) +_p (c \times_p a)$$

    (viii) $1 \in I_p$, which is the identity elment for $'\times_p'$.

    (ix) The zero element of the ring $\{I_p, +_p, \times_p\}$ is 0.

    (x) Let a be any non zero element of I_p where, $1 \le a \le p -1$. Consider the following $p -1$ products :

$$1 \times_p a, 2 \times_p a, 3 \times_p a,..., (p-1) \times_p a$$

all elements of I_p and also no two of these can be equal.

Let if possible, for $i \neq j$

$\Rightarrow$ $\qquad\qquad 1 \leq i \leq p-1, \; 1 \leq j \leq p-1, i \times_p a = j \times_p a$

$\Rightarrow$ ia and ja leave the same least non-negative remainder when divided by p.

$\Rightarrow$ $\qquad (ia - ja)$ is divisible by p

$\Rightarrow$ $\qquad (i-j)a$ is divisible by p

$\Rightarrow$ $\qquad$ either $p\,|\,(i-j)$ or $p\,|\,a$.

But p is neither a divisor of $i - j$ nor a divisor of 'a' because both $i - j$ and a are positive integers less than p.

$\Rightarrow$ $\qquad\qquad i \times_p a \neq j \times_p a$ for $i \neq j$

Thus, $1 \times_p a, 2 \times_p a, ..., (p-1) \times_p a$ are $p-1$ distinct elements of I_p and one of these elements must be equal to 1. Now, $b \times_p a = 1 = a \times_p b$ by commutativity. Thus b is the inverse of a. Therefore, every non-zero element of I_p has its multiplicative inverse.

Hence $\{I_p, +_p, '\times_p')$ is a field.

EXERCISE 5.1

1. Show that the set **Q** of all rational numbers is a commutative ring with unity under the addition and multiplication.

2. Show that the set set **C** of all complex numbers is a commutative ring under the addition and multilpication of complex numbers.

3. Show that the set **M** of all matrices of order 2×2 forms a ring with zero divisors under the addition and multiplication of matrices. The elements of matrices being integers.

4. Let C be the set of ordered pairs (a, b) of real numbers. The addition and multiplication are defined as

$$(a, b) + (c, d) = (a+c, b+d)$$
$$(a, b).(c, d) = (ac - bd, bc + ad)$$

Prove that **C** is a field.

5. Define integral domain with respect to addition and multiplication.

6. Prove that the set of integers $R=\{0,1,2,3,4\}$ forms a field under addition and multiplication modulo 5.

7. Prove that the set $\{0,1,2,\}$ (mod 3) is a field with respect to addition and multiplication.

8. If two binary operations * and o on the set of integer **Z** are defined as follows :

$$a * b = a + b - 1, \; aob = a+b - ab$$

Show that $(\mathbf{Z}, *, o)$ is a field.

9. Prove that the set $R=\{0,2,4,6,8\}$ (mod 10) is a ring with unity with respect to addition and multiplication.

10. Prove that a ring R is commutative if and only if $(a+b)^2 = a^2 + 2ab + b^2$ $a, b \in R$.

11. If **R** is a ring with unity element 1, then $a = -a = a(-1) \; \forall \; a \in R$ and $(-1)(-1) = 1$.

12. Prove that the set **Z**(2) of numbers of the form $a + \sqrt{2}b$, with a as integers is an integral domain with respect to addition and multiplication. Is it a field ?

13. Give an example of a ring which is not and integral domain.

14. If R is a ring satisfying all the conditions for a ring with unity with the possible exception of $a+b = b+a$, prove that the axiom of $a+b = b+a$ must hold in R that R is thus a ring.

15. Prove that the set R = $\{a+b. \; 3^{1/3} + c \; .a^{1/3};$ $a, b, c \in \mathbf{Q}\}$ is a ring under addition and multiplication.

16. Let p be a prime number. Prove the set of integers $I_p = \{0,1,2,3, ... \; p-1\}$ forms a field with respect to addition and multiplication modulo p.

17. In a ring R prove that
    (i) $(a+b)^2 = a^2 + ab + b^2, \; \forall \; a, b \in R$
    (ii) $-(-a) = a, a \in R$.

18. Show that the set of all matrices of the form

$$\begin{bmatrix} 0 & a \\ 0 & b \end{bmatrix}$$ where a and b are real numbers forms

a ring with respect to matrix addition and matrix multiplication. Is it a commutative ring ?

1. $Q = \left\{ \dfrac{a}{b} \ ; \ b \neq 0, \ a, b \in \mathbf{R} \right\}$, then prove that $(\mathbf{Q}, '+', '.')$ is a commutative ring with unity.

3. We have to prove that $(M, '+', '.')$ is a ring with zero divisors :
   (i) $A, B \in M, A+B \in M$
   (ii) Matrix addition is always associative.
   (iii) Since $O = \begin{bmatrix} 0 & 0 \\ 0 & 0 \end{bmatrix} \in M$ such that for $A \in M$.
         $$-A + A = O$$
         $\therefore$ O is the additive identity of M.
   (iv) There exists a unique matrix $-A$ corresponding to $A \in M$ such that
         $$A+A = O = A + (-A)$$
   (v) Matrix addition is always commutative.
   (vi) Since the product of two 2×2 matrix is again a 2×2 matrix so M is closed under matrix multiplication.
   (vii) Matrix multiplication is always associative.
   (viii) For $A, B, C \in M$, we have $A.(B+C) = A.B + A.C$ and $(B+C).A = B.A + C.A$.
         Hence, $(M, '+', '.')$ forms a ring.

         Also let $A = \begin{bmatrix} 0 & a \\ 0 & b \end{bmatrix}$ and $B = \begin{bmatrix} c & d \\ 0 & 0 \end{bmatrix}$ which are non-zero matrices.

         But $AB = O$, this implies that M is of with zero divisors.

6. The composition table for $+_5$ and $\times_5$ are given by

$+_5$	0	1	2	3	4
0	0	1	2	3	4
1	1	2	3	4	0
2	2	3	4	0	1
3	3	4	0	1	2
4	4	0	1	2	3

$\times_5$	0	1	2	3	4
0	0	0	0	0	0
1	0	1	2	3	4
2	0	2	4	1	3
3	0	3	1	4	2
4	0	4	3	2	1

From above tables we can easily verify the required results.

8. We have to prove that $(\mathbf{Z}, *, .)$ is a commutative ring with unit elements.
   (i) Since $a * b = a + b - 1 \in \mathbf{Z}$ so $\mathbf{Z}$ is closed under '*'.
   (ii) For $a, b, c \in \mathbf{Z}$, we have
         $$(a * b) * c = (a+b-1) * c = a+b-1 + c -1$$
         $$= a+b+c - 2$$
         and $$a * (b * c) = a *(b+c-1) = a+b + c -1 -1$$
         $$= a+b+c - 2$$
         $\therefore$ $$(a * b) * c = a * (b * c), \ \forall \ a, b, c \in \mathbf{Z}.$$
         Thus '*' is associative over $\mathbf{Z}$.
   (iii) Identity element for '*' is $e = 1 \in \mathbf{Z}$.
   (iv) Inverse element of a is $2 - a \in \mathbf{Z}$.
   (v) Also, $a * b = b * a \ \forall \ a, b \in \mathbf{Z}$.
   (vi) $a \circ b = a + b - ab \in \mathbf{Z}$ so $\mathbf{Z}$ is closed under '$\circ$'.
   (vii) Also $(a \circ b) \circ c = a \circ (b \circ c), \ \forall \ a, b, c \in \mathbf{Z}$.

(viii) $a \circ b = b \circ a,$ $\forall\, a, b \in \mathbf{Z}$

(ix) Unit element for the operation '$\circ$' is 0 which belongs to **Z**.

Hence, $(\mathbf{Z}, \,'*', \,'\circ')$ is a commutative ring with unity.

12. $\mathbf{Z}(\sqrt{2}) = \{a+b\sqrt{2}, a, b \in \mathbf{Z}\}$, then we have to prove that this set form an integral domain. For this first we prove that $\mathbf{Z}(\sqrt{2})$ forms a commutative ring with unity, which is a similar as Hint number 8.

Let $a + b\sqrt{2}$ and $c + d\sqrt{2}$ be any two elements of $\mathbf{Z}(\sqrt{2})$,

and $(a+b\sqrt{2})(c+d\sqrt{2}) = 0 + 0.\sqrt{2}$

$\Rightarrow$ $ac+2bd = 0$ and $bc+ad = 0$

These equations are possible only when either $a=0, b=0$ or $c=0, d=0$.

Thus when $(a+b\sqrt{2})(c+d\sqrt{2})=0+0.\sqrt{2}$, then either $(a+b\sqrt{2}) = 0$ or $(c+d\sqrt{2}) = 0$, this shows that $\mathbf{Z}(\sqrt{2})$ is of without zero divisors. Also $\mathbf{Z}(\sqrt{2})$ is not a field, because for $5+3\sqrt{2} \in \mathbf{Z}(\sqrt{2})$, $\dfrac{1}{5+3\sqrt{2}} = \dfrac{5}{7}+\dfrac{3}{7}\sqrt{3} \notin \mathbf{Z}(\sqrt{2})$. That is $5 +3\sqrt{2}$ has no multiplication inverse.

Answers

12. No **18.** No

5.7 SUBRINGS : RING WITHIN RINGS

Definition. *Let S be any non-empty subset of a ring R. Then S is said to be a subring of R if S itself is a ring for the same operations as defined on R.*

5.7.1 Improper Subrings

Let R be a ring. Then $\{0\}$ and R itself are the subrings of R. These subrings are called *improper subrings* or *trivial subrings*.

5.7.2 Proper Subrings

Let R be a ring. Then other subrings of R except $\{0\}$ and R called *proper subrings*.

For Example :

(1) The ring of rational numbers is a proper subring of the ring of real numbers.

(2) The ring of even integers is a subring of the ring of integers.

5.8 PROPERTIES OF SUBRINGS

Theorem 1. *The necessary and sufficient condition for a non-empty subset S of a ring R to be a subring is*

 (i) $a \in S, b \in S \Rightarrow a - b \in S$ *(ii) $a \in S, b \in S \Rightarrow ab \in S$*

Proof. Suppose S is a subring of a ring R, then we have to show that S is closed under addition and multiplication.

If $b \in S \Rightarrow -b \in S$ ($\because$ S is a group under multiplication.)

$\therefore$ $a \in S, -b \in S$

$\Rightarrow$ $a + (-b) \in S$ ($\because$ S is closed under addition.)

$\Rightarrow$ $a - b \in S$

Hence, $a \in S, b \in S \Rightarrow a - b \in S$ and $a \in S, b \in S \Rightarrow ab \in S$

Conversely, suppose S is any non-empty subset of a ring R and

 (i) $a \in S, b \in S \Rightarrow a - b \in S$

(ii) $a \in S, b \in S \Rightarrow ab \in S$

Then we have to show that S is a subring of R.

Now, from (i) we have

$$a \in S, a \in S$$

$$\Rightarrow \qquad a - a \in S \Rightarrow 0 \in S.$$

∴   Additive identity exists in S.

Again from (i) we have

$$0 \in S, a \in S$$

$$\Rightarrow \qquad 0 - a \in S \Rightarrow -a \in S$$

∴   Additive inverse exists in S.

From (i), we have

$$a \in S, \ -b \in S$$

$$\Rightarrow \qquad a - (-b) \in S \Rightarrow a + b \in S$$

∴ S is closed under addition

and form (ii) S is closed under multiplication. Further since addition and multiplication are always associative and multiplication is distributive over addition. So, S is a ring and hence S is a subring of R.

Theorem 2. *The necessary and sufficient condition for a non-empty subset S of a ring R to be a subring of R are*

$$(i) \ S + (-S) = S \qquad\qquad (ii) \ SS \subseteq S$$

Proof.   **The conditions are necessary.** Let us suppose S is a subring of R, then we shall have to show that $S + (-S) = S$ and $SS \subseteq S$.

Let $a \in S$

∴ $a = a + 0$ (By the definition of additive identity is S)

But $a + 0 \in S + (-S)$ ($\because 0 \in -S$)

$\Rightarrow$ $a \in S + (-S)$

$\Rightarrow$ $S \subseteq S + (-S)$ ...(1)

Now, let $a + (-b)$ be any arbitrary element of $S + (-S)$, then we have

$$a + (-b) \in S + (-S) \Rightarrow a \in S, -b \in (-S)$$

$\Rightarrow$ $a \in S, b \in S$

$\Rightarrow$ $a - b \in S$ ($\because S$ is a subring.)

∴ $S + (-S) \subseteq S$

Thus, from (1) and (2), we have

$$S + (-S) = S \qquad\qquad\qquad ...(2)$$

Further, since S is a subring so that it is closed under multiplication, that is,

if $a \in S, b \in S \Rightarrow ab \in S$

But $ab \subseteq SS$

∴ $SS \subseteq S$

Hence, the necessary conditions.

Sufficient conditions. Now suppose,

(i) $S + (-S) = S$ (ii) $SS \subseteq S$

Then we shall have to show that S is subring of R. To show S to be subring, we have only to prove $a - b$, $ab \in S$. For this we proceed as follows :

If $a \in S, b \in S \Rightarrow a \in S, -b \in (-S)$

$\Rightarrow$ $a + (-b) \in S + (-S)$

$\Rightarrow$ $a + (-b) \in S$ $(\because S + (-S) = S)$

$\Rightarrow$ $a - b \in S$

Also, $a \in S , b \in S \Rightarrow ab \in S$

Therefore, if $a - b \in S$, $ab \in S$ then S is a subring of R.

Hence, proved the theorem.

Theorem 3. *The intersection of two subrings is again a subring.*

Proof.   Let S_1 and S_2 be two subrings of a ring R. Then we have to show that $S_1 \cap S_2$ is a subring of R. Let $a, b \in S_1 \cap S_2 \Rightarrow a, b \in S_1$ and $a, b \in S_2$ since S_1 and S_2 are subring of R, then $a - b \in S_1$, $a - b \in S_2$ and $ab \in S_1$, $ab \in S_2$.

Therefore we have

If $a - b \in S_1$ and $a - b \in S_2$, then $a - b \in S_1 \cap S_2$

And if $ab \in S_1$ and $ab \in S_2$, then $ab \in S_1 \cap S_2$

$\therefore$ $a \in S_1 \cap S_2, b \in S_1 \cap S_2 \Rightarrow a - b \in S_1 \cap S_2$

and $a \in S_1 \cap S_2, b \in S_1 \cap S_2 \Rightarrow ab \in S_1 \cap S_2$

Hence $S_1 \cap S_2$ is a subring of R.

Generallisation. Further, we can also prove that arbitrary intersection of subrings of a ring are again a subring.

Let $a, b \in \bigcup_{i \in \Lambda} S_i \Rightarrow a, b \in S_i, \forall i \in \Lambda.$

Since each S_i is a subring of R so we have

$$a - b \in S_i \ \forall i \in \Lambda \text{ and } ab \in S_i, \forall i \in \Lambda.$$

Therefore we obtain

$$a - b \in \bigcap_{i \in \Lambda} S_i \text{ and } ab \in \bigcap_{i \in \Lambda} S_i .$$

Hence $\bigcap_{i \in \Lambda} S_i$ is a subring of R.

REMARK

☞ The union of two subrings is a subring iff f one is contained in the other.

Theorem 4. *The intersection of the collection of all the subrings containing a given subset M of a ring R is the smallest subring containing M.*

Proof.   Let $\{S_\alpha : \alpha \in \Lambda\}$ be the collection of all the subring containing a given subset M of a ring R. That is, $M \subseteq S_\alpha$ for each α. Since R is itself a subrings containing M so that the collection S_α is non-empty. Further, the intersection of all subrings containing M is a subring of R containing M. Now we have to show that $\cap S_\alpha$ is a smallest subring of R containing M.

5.9 SUBFIELD : FIELD WITHIN FIELD

Definition. *Let K be a non-empty subset of a field F. Then K is a subfield of F if K itself is a field under the same operations as defined on F.*

For Example

(1) The field of real numbers is a subfield of the field of complex numbers.

(2) The field of real numbers is a subfield of the field of real numbers.

5.9.1 Necessary and Sufficient Conditions for a Subset to be Subfield

Theorem 1. *The necessary and sufficient condition for a non-empty subset K of a field F to be a subfield of F are*

$$\text{(i) } a \in K, b \in K \Rightarrow a - b \in K$$

$$\text{(ii) } a \in K, b \in K \Rightarrow ab^{-1} \in K$$

Proof.   Suppose K is a subfield of F. Then

If $b \in K \Rightarrow -b \in K$

Now $a \in K, \; -b \in K$

$\Rightarrow$ $a + (-b) \in K$ ($\because$ K is closed under addition.)

$\Rightarrow$ $a - b \in K$

and if $b \in K \Rightarrow b^{-1} \in K$ ($\because$ K is a field.)

Now $a \in K, b^{-1} \in K$

$\Rightarrow$ $ab^{-1} \in K$ ($\because$ K is closed under multiplication.)

Conversely, Suppose K is a non-empty subset of F and

(i) $a \in K, b \in K \Rightarrow a - b \in K$

(ii) $a \in K, b \in K \Rightarrow ab^{-1} \in K$

Then we have to show that K is a subfield of F. For this, we have from (i)

$$a \in K, a \in K$$

$\Rightarrow$ $a - a \in K$

$\Rightarrow$ $0 \in K$

$\therefore$ Additive identity exists in K.

Also from (i), we have $0 \in K, a \in K$

$\Rightarrow$ $0 - a \in K$

$\Rightarrow$ $- a \in K$

$\therefore$ Additive inverse exists in K.

Also, from (ii), we have $a \in K, a \in K$

$\Rightarrow$ $aa^{-1} \in K$ (By (i))

$\Rightarrow$ $1 \in K$

$\therefore$ Multiplicative identity exists in K.

Again from (ii), we have $1 \in K, a \in K$

$\Rightarrow$ $1 . a^{-1} \in K$

$\Rightarrow$ $a^{-1} \in K$

$\therefore$ Multiplicative inverse exists in K.

Further since, we have addition and multiplication are always associative and commutative and also multiplication is always distributive over addition. Hence, K is a field and hence K is a subfield of F.

5.10 CHARACTERISTIC OF A RING

Let R be a ring with zero element '0' and suppose there exists a positive integer n such that $na = a + a + a + \ldots + a$ (upto n terms) $= 0$, $\forall\ a \in R$. Then the smallest such positive integer n is called the characteristic of a ring.

If there does not exist such positive integer n such that $na = 0$ $\forall\ a \in R$ and $a \neq 0$. Then the ring R is said to be of characteristic zero or infinite.

For Example :

(1) The ring of integers is of characteristic zero and the ring of rational number is also of characteristic zero.

(2) If $I_6 = \{0,1,2,3,4,5,6\}$, then the ring $\{I_6,\ '+_6',\ '\times_6'\}$ is of characteristic 6, because $6a = 0 \ \forall\ a \in I_6$.

REMARK

☞ When we say that the characteristic of R is n, we mean that n is the least positive integer for which the relation $na = 0$ is universally true in the ring.

Theorem 1. The characteristic, n of a ring with unity is zero or > 0 according as the unity element 1 regarded as a member of the additive group of the ring has the order of zero or n.

Proof. Let R be a ring with unity element 1 and if order of 1 is zero, it is obvious that the characteristic of the ring is zero. Let us suppose that the order of 1 is finite say n, then

$$1 + 1 + 1 + \ldots + 1\ (n\ \text{times}) = 0$$

or $\qquad n \cdot 1 = 0 \qquad \qquad \ldots(1)$

Let a be any non-zero element of R, then we have

$$na = a + a + a + \ldots + a(n\ \text{times})$$
$$= 1.a + 1.a + 1.a + \ldots + .a \qquad (\because R \text{ is with unity.})$$
$$= [1 + 1 + 1 + \ldots + 1\ (n\ \text{times})] \cdot a$$
$$= (n \cdot 1)\,a$$
$$= 0 \cdot a = 0 \qquad \qquad \text{[Using (1)]}$$

$\Rightarrow \qquad$ order of $a \leq n$.

Hence, the characteristic of R is n.

Theorem 2. The characteristic of an integral domain is zero or $n > 0$ according as the order of any non-zero element of the integral domain regarded as a member of additive group of the integral domain either zero or n.

Proof. Let D be an integral domain and let a be any non-zero element of D.

If the order of a is zero, then obviously the characteristic of D is zero.

Suppose that the order of $n > 0$, then we have

$$na = 0. \qquad \qquad \ldots(1)$$

Furthermore, suppose b is any other non-zero element of D.

Mutliplying both sides of (1) by b, we have

$$(na)b = 0.\,b$$

$$\Rightarrow \qquad [a + a + a + ... + a(n \text{ times})] b = 0 \qquad\qquad (\because\ 0.b = 0)$$
$$\Rightarrow \qquad ab + ab + ab + ... + ab(n \text{ times})] = 0 \qquad \text{(By right distributive law)}$$
$$\Rightarrow \qquad a[b + b + b + ... + b(n \text{ times})] = 0 \qquad \text{(By left distributive law)}$$
$$\Rightarrow \qquad a(nb) = 0$$
$$\Rightarrow \qquad nb = 0 \qquad\qquad (\because\ a \neq 0) \qquad\qquad ...(2)$$

But $0(a) = n \Rightarrow n$ is the least positive integer such that $na = 0$. Also, $n.0 = 0$.

Thus (1) and (2) implies that n is the least positive integers such that $nx = 0, \forall\, x \in D$.

Hence, the characteristic of D is n.

Theorem 3. *The characteristic of an integral domain is either zero or a prime number.*

Step Outlines : To make the proof easier, use the following steps :

Step 1. *Let $o(a) = p$ then characteristic of D is p.*

Step 2. *Show that p is prime.*

Proof. Let D be an integral domain and let a be any non-zero element of D. If the order of a is zero, then the characteristic of D is zero.

Let us suppose that the order of a is p, then the characteristic of D is p.

Now we shall show that p is prime. Let if possible p is not prime, then

$$p = p_1 p_2 \text{ with } p_1 < p, p_2 < p \text{ and } p_1 \neq 1, p_2 \neq 1$$

Since $a \neq 0$ so that $a^2 \neq 0$ and the characteristic of D is p, then

$$p(a^2) = 0$$
$$\Rightarrow \qquad p_1 p_2 (a^2) = 0 \qquad\qquad (\because p = p_1 p_2)$$
$$\Rightarrow \qquad (p_1 a)(p_2 a) = 0$$
$$\Rightarrow \qquad \text{either } p_1 a = 0 \text{ or } p_2 a = 0 \qquad [\because D \text{ has no zero divisor.}]$$
$$\Rightarrow \quad \text{the characteristic of } D \text{ is either } p_1 \text{ or } p_2.$$

But the characteristic of D is p, which is the least positive integer, therefore neither $p_1 a = 0$ nor $p_2 a = 0$, this implies that p is a prime number. Hence the theorem is proved.

Theorem 4. *Each non-zero element of integral domain D, regarded as a member of the additive group D is of the same order.*

Proof. Let a be any non-zero element of D and let $o(a) = n$ (say).

Suppose b is any other non-zero element of D and $o(b) = m$ (say).

Since $\qquad\qquad\qquad o(a) = n$

$$\Rightarrow \qquad\qquad\qquad na = 0$$
$$\Rightarrow \qquad\qquad\qquad nb = 0 \qquad\qquad (\because nx = 0 \,\forall\, x \in D)$$
$$\Rightarrow \qquad\qquad\qquad o(b) \leq n$$
$$\therefore \qquad\qquad\qquad m \leq n \qquad\qquad ...(1)$$

Also, $\qquad\qquad\qquad o(a) = m$

$$\Rightarrow \qquad\qquad\qquad mb = 0$$
$$\Rightarrow \qquad\qquad a(mb) = a.0 = 0$$
$$\Rightarrow \qquad a[b + b + b + ... + b(m \text{ times})] = 0$$
$$\Rightarrow \qquad ab + ab + ab + ... + ab(m \text{ times}) = 0$$
$$\qquad\qquad [a + a + a + ... + (m \text{ times})]b = 0$$
$$\qquad\qquad (ma)b = 0 \qquad\qquad (b \neq 0 \text{ and } D \text{ has no zero divisor.})$$
$$\qquad\qquad (ma) = 0$$
$$\qquad\qquad o(a) \leq m.$$

$$n \le m. \qquad\qquad ...(2)$$

From (1) and (2), we have

$$m = n$$
$$o(b) = o(a)$$

Also, if $o(a)$ is zero, then $o(b)$ will also be zero. Hence the theorem is proved.

5.11 CHARACTERISTIC OF A FIELD

Since every field is an integral domain, therefore, the characteristic of a field F is zero or $n > 0$ according as any non-zero element (in particular the unit element 1) of F is order zero or n.

In order to find the characteristic of field F, we should find the order of the unit element 1 of F when regarded as a member of the additive group of F. Thus if the order of 1 is zero, then the characteristic of F is zero and if the order of 1 is finite say n, then F is a characteristic of n.

For Example:

(1) The characteristic of the field of real numbers is 0.

(2) If $F = \{\mathbf{Z}_7, '+_7', '\times_7'\}$ be a field, where $\mathbf{Z}_7 = \{0,1,2,3,4,5,6\}$, then the characteristic of $\mathbf{Z}_7$ is 7.

5.12 ORDERED INTEGRAL DOMAIN

Definition (1). *Let* $(R, +, \cdot)$ *be an integral domain. This integral is said to be ordered if* R *contains a subset* $\mathbf{R}^+$ *(the set of positive elements of* R*) such that*

(i) R^+ is closed under addition and multiplication as define on R.

(ii)   There is one and only one $a = 0$, $a \in R^+$, $-a \in R^-$ for all $a \in R$ holds.

For Example

The integral domain of integers is ordered.

Definition (2). *A field F is said to be ordered if it is ordered as an integral domain.*

5.13 ORDERED RELATIONS IN AN INTEGRAL DOMAIN

Definition. *Let D be an integral domain and* D_+ *be the set of positive elements of D. Then we define two relations 'less than'* $(<)$ *and 'greater than'* $(>)$ *on D as follows :*

(i) $a > b$ when $a - b \in D_+$

(ii) $a < b$ when $b - a \in D_+$

Thefore $a > b$ *if and only if* $b < a$.

Theorem 1. *The set of complex numbers is not ordered integral domain.*

Proof.   Let $\mathbf{C}$ be the set of complex numbers. Then $(\mathbf{C}, '+', '\cdot')$ is an integral domain and let $\mathbf{C}^+$ be the set of the positive elements of $\mathbf{C}$. We have to show that $\mathbf{C}$ is not ordered. Obviously $i \in \mathbf{C}$ and $i \ne 0$, then we have to show that $i \notin \mathbf{C}^+$, $-i \notin \mathbf{C}^+$. For this let us suppose either $i \in \mathbf{C}^+$ o r $-i \in \mathbf{C}^+$

If $i \in \mathbf{C}^+$

$\Rightarrow$ $i \cdot i \in \mathbf{C}^+$ ($\because \mathbf{C}^+$ is closed under multiplication.)

$\Rightarrow$ $i^2 \in \mathbf{C}^+ \;\Rightarrow\; -1 \in \mathbf{C}^+$

$\Rightarrow$ This gives a contradiction because $\mathbf{C}^+$ contains all positive elements.

$\therefore$ $i \notin \mathbf{C}^+$

If $\qquad -i \in \mathbf{C}^+ \Rightarrow (-i).(-i) \in \mathbf{C}^+$

$\Rightarrow \qquad i^2 \in \mathbf{C}^+ \Rightarrow -1 \in \mathbf{C}^+$

$\Rightarrow$ This gives again a contradiction.

$\therefore \qquad -i \notin \mathbf{C}^+$

Hence $i \neq 0, i \notin \mathbf{C}^+$, $-i \notin \mathbf{C}^+$. Consequently we can say that one and only one of $\qquad i = 0, i \in \mathbf{C}^+$, $-i \in \mathbf{C}^+$ does not hold, and hence $\mathbf{C}$ is not ordered.

Theorem 2.   *The field $\{\mathbf{Z}_p, '+_p' '\times_p'\}$ where p is a prime and $\mathbf{z}_p = \{0,1,2,3, ...p-1\}$ is not ordered.*

Proof.   Let us suppose that $\mathbf{Z}_p$ is an ordered field and let S be the set of positive elements of $\mathbf{Z}_p$. The zero elements of $\mathbf{Z}_p$ is 0.

Since $1 \neq 0$ and the additive inverse of 1 is $p-1$. By definition of an ordered field we have, either $1 \in S$ or $p-1 \in S$.

Also, S is closed with respect to $'+_p'$, therefore

$$1 \in S \Rightarrow 1 +_p 1 +_p \ ....+_p 1 \ (\text{upto } p-1 \text{ times})$$

$$\Rightarrow p-1 \in S$$

This gives a contradiction of the principle of trichotomy. Similarly, if it is assumed that $p-1 \in S$. We can show that $1 \in S$, which is again a contradiction. Hence, the field $\mathbf{Z}_p$ is not an ordered field.

5.14 POLYNOMIAL RINGS

Definition. *The expression of the form*

$$f(x) = a_0 + a_1 x + a_2 x^2 + ... + a_n x^n, \ a_n \neq 0$$

is called a polynomial of degree n and the variable x is called indeterminate.

Definition. *Let R be a ring and x be an indeterminate which does not belong to R. Then a polynomial of the form*

$$f(x) = a_0 + a_1 x + a_2 x^2 + ... + a_n x^n, \ a_n \neq 0$$

where $a_0, a_1, a_2, ..., a_n...$ all are in R with finite and non-zero is called a polynomial over a ring R.

5.15 SET OF ALL POLYNOMIALS OVER A RING

Let R be a ring and x be an indeterminate, then the set of all polynomial of the form $f(x) = a_0 + a_1 x + a_2 x^2 + ... + a_n x^n + ...,$ where $a_0, a_1, a_2,..., a_n...$ all are in R with finite number of non-zero elements is called the set of polynomials over R and it is denoted by $R[x]$.

5.15.1 Zero Polynomial

A polynomial over a ring R of the type

$$a_0 + a_1 x + a_2 x^2 + ... + a_n x^n + ...$$

is called a zero polynomial if $a_n = 0 \ \forall \ n$.

5.15.2 Equality, Sum and Product of Polynomials

(i)   **Equality.** Let R be a ring and let

$$f(x) = a_0 + a_1 x + a_2 x^2 + ... \quad \text{and} \quad g(x) = b_0 + b_1 x + b_2 x^2 + ..$$

be two polynomials over R. Then they are said to be equal if $a_n = b_n, \ \forall n$.

(ii) Sum of polynomials. Let R be a ring and x be indeterminate and let

$$f(x) = a_0 + a_1x + a_2x^2 + ... \quad \text{and} \quad g(x) = b_0 + b_1x + b_2x^2 + ..$$

be two polynomials over R, i.e., $f(x), g(x) \in R[x]$.

Then $\qquad\qquad\qquad f(x) + g(x) = c_0 + c_1x + c_2x^2 + ...$

is said to be the **sum** of $f(x)$ and $g(x)$ if $c_n = a_n + b_n$, $\forall n$.

(iii) Product of polynomials. Let R be a ring and x be an indeterminate and let $f(x)$, $g(x) \in R[x]$, where,

$$f(x) = a_0 + a_1x + a_2x^2 + ... \quad \text{and} \quad g(x) = b_0 + b_1x + b_2x^2 + ..$$

Then the polynomial $f(x). g(x) = c_0 + c_1x + c_2x^2 + ..$

is said to be the product of $f(x)$ and $g(x)$ if

$$c_n = a_0b_n + a_1b_{n-1} + a_2b_{n-2} + ... + a_nb_0$$

$$= \sum_{i+j=n} a_ib_i \quad \text{for all } b \text{ belongs to the set of all non-negative integers.}$$

(iv) Degree of a polynomial. Let $f(x) \in R[x]$ is of the form

$$f(x) = a_0 + a_1x + a_2x^2 + ... + a_nx^n + ...$$

Then the degree of $f(x)$ is n if and only if $a_n \neq 0$ and $a_m = 0$, $m > n$.

Definition. *If a polynomial $f(x)$ is of degree n, then the term a_nx^n is called the leading term, a_n is called the leading coefficient and a_0 is called the constant term.*

Theorem 1. *The set $R[x]$ of all polynomial over an arbitrary ring R forms a ring with respect to addition and multiplication of polynomials.*

Proof. Since by definition of addition (sum) and multiplication (product) of polynomials $f(x), g(x) \in R[x]$ we have $[f(x)+g(x)]$ and $[f(x) . g(x)] \in R[x]$ and hence $R[x]$ is closed under addition and multiplication.

(i) Associativity. Let $f(x), g(x), h(x) \in R[x]$, i.e.,

$$f(x) = a_0 + a_1x + a_2x^2 + ... = \sum_{k=0} a_kx^k$$

$$g(x) = b_0 + b_1x + b_2x^2 + ... = \sum_{k=0} b_kx^k$$

and $\qquad\qquad h(x) = c_0 + c_1x + c_2x^2 + ... = \sum_{k=0} c_kx^k$

we have $f(x)+[g(x)+ h(x)] = \sum_{k=0} a_kx^k + \sum_{k=0} (b_k +c_k)x^k$

$$= \sum_{k=0} [a_k + (b_k +c_k)]x^k = \sum_{k=0} [(a_k +b_k)+c_k]x^k$$

$$= \sum_{k=0} (a_k +b_k)x^k + \sum_{k=0} c_k x^k$$

$$= [f(x) + g(x)]+ h(x)$$

(ii) Existence of identify. Let $0(x)$ be the zero polynomial of $R[x]$ that is,

$$0(x) = 0 + 0 . x + 0. x^2 + ...$$

Let $f(x) = a_0 + a_1x + a_2x^2 +... \in R[x]$, we have

$$f(x) = f(x)+0(x)=(a_0+0)+(a_1+ 0)x+(a_2+0)x^2+ ..$$
$$= a_0 + a_1x + a_2x^2 +... = f(x)$$

$\therefore$ $0(x)$ is the additive identity of $R[x]$.

(iii) Existence of additive inverse. Let $f(x) \in R[x]$ and let $-f(x)$ be a polynomial of $R[x]$ defined as $-f(x) =(-a_0) + (-a_1)x + (-a_2)x^2 +...$. Then we have

$$-f(x) + f(x) = (-a_0+a_0) + (-a_1+a_1)x + (-a_2+ a_2)x^2 +...$$
$$= 0 + 0.x + 0.x^2 +....$$
$$= 0(x)$$

$\therefore$ $-f(x)$ is the additive inverse of $f(x)$ and exists in $R[x]$.

(iv) Commutativity for addition. Let $f(x), g(x) \in R[x]$, i.e.,

$$f(x) = a_0 + a_1x + a_2x^2 +.... \text{ and } g(x) = b_0 + b_1x + b_2x^2 +..$$

We have $f(x) + g(x) = (a_0+b_0) + (a_1+b_1)x +(a_2+ b_2)x^2 +...$
$$= (b_0+a_0) + (b_1+a_1)x +(b_2+ a_2)x^2 +...$$
$$= g(x) + f(x)$$

Thus $(R[x] . +)$ is an abelian group.

(v) Associativity for multiplication. Let $f(x), g(x), h(x) \in R[x]$, i.e.,

$$f(x) = a_0 + a_1x + a_2x^2 +... = \sum_{i=0} a_ix^i$$

$$g(x) = b_0 + b_1x + b_2x^2 +... = \sum_{i=0} b_jx^j$$

and $h(x) = c_0 + c_1x + c_2x^2 +... = \sum_{i=0} c_kx^k$

We have $f(x)[g(x)h(x)] = \sum_{i=0} a_ix^i \left[\sum_{j=0} b_jx^j . \sum_{k=0} c_kx^k \right]$

$$= \sum_{i=0} a_ix^i \left[\sum_{l=0} d_lx^l \right] \text{ where } d_l = \sum_{j+k=l} b_jc_k$$

$$= \sum_{m=0} e_mx^m \qquad \text{where } e_m = \sum_{i+l=m} a_id_l$$

$\therefore$ e_m is the coefficient of x^m in $f(x)[g(x)h(x)]$

$$e_m = \sum_{i+l=m} a_id_l = \sum_{i+l=m} a_i \left[\sum_{j+k=l} b_jc_k \right] = \sum_{i+j+k=m} a_ib_jc_k$$

Similarly, the coefficient of x^m in $[f(x)g(x)]h(x)$ is e_m which is equal to

$$= \sum_{i+j+k=m} a_ib_jc_k$$

Thus $f(x)[g(x)h(x)] = [f(x)g(x)]h(x)$

(vi) Distributivity of multiplication over addition.

Let
$$f(x) = \sum_{i=0} a_i x^i; \ g(x) = \sum_{j=0} b_j x^j;$$

and
$$h(x) = \sum_{k=0} c_k x^k \text{ be the element of } R[x], \text{ we have}$$

Now,
$$f(x)[g(x) + h(x)] = \sum_{i=0} a_i x^i \left[\sum_{j=0} b_j x^j + \sum_{k=0} c_k x^k \right]$$

$$= \sum_{i=0} a_i x^i \left[\sum_{l=0} d_l x^l \right], \text{ where } d_l = b_l + c_l$$

$$= \sum_{m=0} l_m x^m$$

where
$$l_m = \sum_{i+l=m} a_i d_l$$

$\therefore$ l_m is the coefficient of x^m in $f(x)[g(x) + h(x)]$

i.e.,
$$l_m = \sum_{i+l=m} a_i d_i = \sum_{i+l=m} a_i (b_l + c_l)$$

or
$$l_m = \sum_{i+l=m} a_i b_l + \sum_{i+l=m} a_i c_l$$

$$= \text{Coefficient of } x^m \text{ in } [f(x)+g(x)]+\text{coefficient of } x^m \text{ in } [f(x)h(x)]$$
$$= \text{Coefficient of } x^m \text{ in } [f(x) \ g(x)+f(x)h(x)]$$
$\therefore$ $\quad f(x) \ [g(x)+h(x)] = f(x) \ g(x) +h(x)f(x)$

Similarly, we can prove that
$$[g(x)+h(x)]f(x) = g(x)f(x)+h(x)f(x)$$
Hence $R[x]$ forms a ring. This ring is called *polynomial ring*.

Solved Examples

Example 1. *Show that the set of all matrices of order 2×2 of the form* $\begin{bmatrix} a & 0 \\ b & c \end{bmatrix}$ *where a,b,c are integers is a subring of the ring M of all matrices of order 2×2 with integral elements.*

Solution. Since M is a ring of all matrices of order 2×2 and let M_1 be the set of all matrices of the type $\begin{bmatrix} a & 0 \\ b & c \end{bmatrix}$, where $a, b, c, \in \mathbf{Z}$ (set of integers). We have to show that M_1 is a subring of M. For this purpose, let A, B be any two elements of M_1, i.e.,

$$A = \begin{bmatrix} a_1 & 0 \\ b_1 & c_1 \end{bmatrix}, B = \begin{bmatrix} a_2 & 0 \\ b_2 & c_2 \end{bmatrix}, a_1, b_1, c_1, a_2, b_2, c_2 \in \mathbf{Z}$$

Then, we have

$$A - B = \begin{bmatrix} a_1 - a_2 & 0 \\ b_1 - b_2 & c_1 - c_2 \end{bmatrix}$$

since $a_1 - a_2$, $b_1 - b_2$, $c_1 - c_2$, are all in **Z**. Thus $A - B \in M_1$. Also, we have

$$AB = \begin{bmatrix} a_1 & 0 \\ b_1 & c_1 \end{bmatrix} \begin{bmatrix} a_2 & 0 \\ b_2 & c_2 \end{bmatrix} = \begin{bmatrix} a_1 a_2 & 0 \\ b_1 a_2 + c_1 b_2 & c_1 c_2 \end{bmatrix}$$

It is the type of $\begin{bmatrix} a & 0 \\ b & c \end{bmatrix}$.

$\therefore$ $AB \in M$

Hence M_1 is a subring of M.

Example 2. *Let R be a ring of integer and let S = {mx : x∈**Z**}, m being a fixed integer. Show that S is a subring of R.*

Solution.   Let $a, b \in S$. Then

	$a = mx$	(For some $x \in$ **Z**)
and	$b = my$	(For some $y \in$ **Z**)
Now	$a - b = mx - my = m(x - y)$	
$\therefore$	$a - b \in S$	$(x - y \in$ **Z**)
Now	$ab = (mx)(my) = m(mxy)$	
$\therefore$	$ab \in S$	$(\because mxy \in$ **Z**)

Hence S is subring of R.

Example 3. *Add and multiply the following polynomials over ring* $(\mathbf{Z}_5, +_5, \times_5,)$.
$$f(x) = 3 + 4x + 2x^2, \quad g(x) = 1 + 3x + 4x^2 + 2x^3.$$

Solution.   We have $f(x) + g(x) = 3 + 4x + 2x^2 + 1 + 3x + 4x^2 + 2x^3$

$$= 4 + 7x + 6x^2 + 2x^3$$
$$= 4 + 2x + x^2 + 2x^3 \qquad [\because 7 \equiv 2 \pmod 5, 6 \equiv 1 \pmod 5]$$

and $f(x)\,g(x) = (3 + 4x + 2x^2)(1 + 3x + 4x^2 + 2x^3)$

$$= 3 + 13x + 26x^2 + 28x^3 + 16x^4 + 4x^5$$
$$= 3 + 3x + x^2 + 3x^3 + x^4 + 4x^5$$
$$[\because 13 \equiv 3 \pmod 5, 28 \equiv 3 \pmod 5, 16 \equiv 1 \pmod 5]$$

EXERCISE 5.2

1. Show that the set of matrices of the form $\begin{bmatrix} a & b \\ 0 & c \end{bmatrix}$ is a subring of the ring of all matrices of order 2×2 with integral elements.

2. Let M be the ring of 2×2 matrices with real elements. Show that the set of matrices of the type $\begin{bmatrix} 0 & a \\ 0 & b \end{bmatrix}$ with real elements is a subring of R.

3. Show that the set of even integers forms a subring of the ring of integers.

4. If $(R, +, .)$ is a ring, show that the set $Z(R) = \{c \in R : xy = yx, \forall y \in R\}$ is a subring of R.

5. Give an example to show that the union of two subrings is not necessarily a subring.

6. If D is an integral domain, then show that the polynomial ring $D[x]$ is also an integral domain.

7. Show that the set of integers is a subring of the ring of rational numbers.

8. Let x, y be commutative elements of a ring R of characteristic 2, show that

$$(x+y)^2 = x^2+y^2 = (x-y)^2.$$

9. Let R be a non-zero ring such that for all $x \in R$, $x^2 = x$. Prove that R is a commutative ring of characteristic 2.

Hints to the Selected Problems

1. Let $M = \left\{ \begin{bmatrix} a & b \\ 0 & c \end{bmatrix} : a, b \in \mathbf{Z} \right\}$, then we have to prove that M is a subring of the ring of all matrices of order 2×2 with integral elements. Let $A = \begin{bmatrix} a_1 & b_1 \\ 0 & c_1 \end{bmatrix}$, $B = \begin{bmatrix} a_2 & b_2 \\ 0 & c_2 \end{bmatrix}$ be any two elements of M.

Now $A - B = \begin{bmatrix} a_1 - a_2 & b_1 - b_2 \\ 0 & c_1 - c_2 \end{bmatrix} \in M$ as $a_1 - a_2, b_1 - b_2$ and $c_1 - c_2$ are the integers. Similarly $AB \in M$.

Hence M is a subring.

3. Let $E = \{2x : x \in \mathbf{Z}\}$, then we have to prove that E form a subring of ring of integers. For $a_1, b_1 \in E$, i.e., $a_1 = 2x$, $b_1 = 2y$ for x, $y \in \mathbf{Z}$.

$$a_1 - b_1 = 2x - 2y = 2(x - y) \in E \text{ as } x - y \in \mathbf{Z}.$$

Also $a_1 b_1 = (2x)(2y) = 2(2xy) \in E$ as $2xy \in \mathbf{Z}$. Hence E forms a subring.

5. Let $(\mathbf{Z}, +, .)$ be a ring of integers and let $R_1 = 2\mathbf{Z}$, $R_2 = 3\mathbf{Z}$, then $R_1 \cup R_2$ is not a subring. But if $R_3 = 4\mathbf{Z}$, then $R_1 \cup R_3$ is a subring.

8. Since characteristic $(R) = 2$, i.e., $1 + 1 = 0$ or $2 = 0$. Then for x, $y \in R$

$$(x+y)^2 = (x+y)(x+y) = (x+y)x + (x+y)y = x^2 + yx + xy + y^2 = x^2 + 2xy + y^2$$

But $2 \simeq 0$. $\therefore (x + y)^2 = x^2 + y^2$, similarly $(x - y)^2 = x^2 + y^2$.

5.16 IDEALS

Definition.

An additive subgroup S of a ring R is said to be a *left ideal* of R if $a \in S$, $r \in R \Rightarrow ra \in S$.

Definition.

An additive subgroup S of a ring R is said to be a *right ideal* of R if $a \in S$, $r \in R$ $ar \in S$.

Definition.

A non-empty subset S of a ring R is said to be an ideal of R if it is both left and right ideal. That is, an additive subgroup S of R is an ideal of R if $a \in S$, $r \in R \Rightarrow ra \in S$, $ar \in S$.

REMARKS

☞ In particular, the subset $\{0\}$ consisting of 0 alone and the ring R itself are ideals in R. These two ideals $\{0\}$ and R are called the *improper* ideals of R; all other ideals of R are called *proper.*

☞ A two-sided ideal is called simply an ideal in this section.

<u>**Theorem 1.**</u> *The necessary and sufficient conditions for a non-empty subset S of a rings R to be an ideal of R are*

(i) $a \in S$, $b \in S \Rightarrow a - b \in S$ (ii) $a \in S$, $r \in R \Rightarrow ra \in S$, or $ar \in S$.

<u>**Proof.**</u>   **The conditions are necessary.** Let S be an ideal of R, then by definition of an ideal, S is additive subgroup of R and $ar \in S$ and $ra \in S$ for all $a \in S$ and $r \in S$.

But the necessary and sufficient condition for a non-empty subset S to be a subgroup of an additive group R is

$$a \in S, b \in S \Rightarrow a - b \in S$$

Hence, $a \in S$, $b \in S \Rightarrow a - b \in S$ and

$a \in S$, $r \in R \Rightarrow ra \in S$ and $ar \in S$

The conditions are sufficient. Let S be any non-empty subset of a ring R such that

(i) $\quad a \in S, b \in S \Rightarrow a - b \in S$

(ii) $\quad a \in S, r \in R \Rightarrow ra \in S$ and $ar \in S$

If a and b be any two elements of S, then from condition (i), we have

$$a \in S, b \in S \Rightarrow a - b \in S$$

And from condition (ii), we have

$$a \in S, b \in R \Rightarrow ab \in S.$$

Thus, $\quad a \in S, b \in S \Rightarrow a - b \in S$ and $ab \in S$ implying that S is an additive subgroup of R because the condition $a \in S, b \in S \Rightarrow a - b \in S$ and $ab \in S$ is a necessary and sufficient condition for a non-empty subset S to be a subgroup. Therefore, S is an additive subgroup of R and $ra \in S$ and $ar \in S$ for all $a \in S$ and $r \in R$. Hence, S is an ideal of R.

Theorem 2. *The intersection of any two ideals of a ring R is again an ideal of R.*

Proof. Let S_1 and S_2 be any two ideals of a ring R, then we have to show that $S_1 \cap S_2$ is also an ideal of R.

Since S_1 and S_2 are ideals of R so that S_1 and S_2 both are additive subgroups of R. But intersection of two subgroups is also a subgroup of R, thus $S_1 \cap S_2$ is also an additive subgroup of R.

Now, $\quad a \in S_1 \cap S_2 \Rightarrow a \in S_1$ and $a \in S_2$

$\therefore \quad a \in S_1, r \in R \Rightarrow ra \in S_1$ and $ar \in S_1 \qquad (\because S_1 \text{ being an ideal})$

and $\quad a \in S_2, r \in R \Rightarrow ra \in S_2$ and $ar \in S_2 \qquad (\because S_2 \text{ being an ideal})$

Thus, $\quad a \in S_1 \cap S_2, r \in R \Rightarrow ra \in S_1 \cap S_2$ and $ar \in S_1 \cap S_2.$

Hence, $S_1 \cap S_2$ is an additive subgroup of R and $ra \in S_1 \cap S_2$ and $ar \in S_1 \cap S_2$ for all $a \in S_1 \cap S_2$ and $r \in R$. Hence, $S_1 \cap S_2$ is an ideal of R.

Theorem 3. *If S is any ideal of a ring R and T any subring of R, then S is an ideal of $S + T$.*

Proof. Since S is an ideal of R so that

$$a \in S, b \in S \Rightarrow a - b \in S$$

and $\quad a \in S, b \in R \Rightarrow ab \in S$

Now, first we shall show that $S + T$ is a subring of R.

Let $a + \alpha$ and $b + \beta$ be any two elements of the set $S + T$. That is,

$$a + \alpha \in S + T \Rightarrow a \in S \text{ and } \alpha \in T$$

and $\quad b + \beta \in S + T \Rightarrow b \in S$ and $\beta \in T$

But T is a subring of R so that

$$\alpha \in T, \beta \in T \Rightarrow \alpha - \beta \in T$$

and $\quad \alpha \in T, \beta \in T \Rightarrow \alpha\beta \in T$

$\therefore \quad (a + \alpha) - (b + \beta) = (a - b) + (\alpha - \beta) \in S + T \qquad (\because a - b \in S \text{ and } \alpha - \beta \in T)$

and $\quad (a + \alpha)(b + \beta) = (a + \alpha)b + (a + \alpha)\beta$

$$= ab + \alpha b + a\beta + \alpha\beta.$$

Further, since S is an ideal of R, therefore

$$b \in S, \alpha \in R \Rightarrow \alpha b \in S \text{ and } a \in S, \beta \in R \Rightarrow a\beta \in S$$

so that $ab + \alpha b + a\beta + \alpha\beta \in S \qquad (\because S \text{ is closed under addition.})$

$\therefore \quad (a + \alpha)(b + \beta) \in S + T$

Hence, $S + T$ is a subring of R.

Next, since S is an ideal of R so that

$$a = a + 0$$

$$\therefore \qquad a = a + 0 \in S + T \qquad\qquad (\because 0 \in T)$$

$$\Rightarrow \qquad a \in S + T$$

$$\Rightarrow \qquad S \subseteq S + T$$

But $S + T$ is contained in R. Hence, S is an ideal of $S + T$.

Theorem 4. *The intersection of all ideals of a ring R containing a non-empty subset M of R is at the smallest ideal of R containing M.*

Proof. Let $(S_\alpha : \alpha \in \Lambda)$ be the collection of all ideals of a ring R containing a non-empty subset M of R. Since each S_α is an ideal of R so that S_α is an additive subgroups for each $\alpha \in \Lambda$.

But arbitrary intersection of subgroups is again a subgroup. Therefore, $\cap S_\alpha$ for each $\alpha \in \Lambda$ is an additive subgroup of R containing M.

Let A be an any element of $\cap S_\alpha$ and $r \in R$, $\alpha \in \Lambda$.

$$\Rightarrow \qquad a \in S_\alpha \text{ for each } \alpha \in \Lambda \text{ and } r \in R$$

$$\Rightarrow \qquad ra \in S_\alpha \text{ and } ar \in S_\alpha \text{ for each } \alpha \in \Lambda \qquad (\because S_\alpha \text{ is an ideal of } R \text{ for each } \alpha.)$$

$$ra \in \underset{\alpha \in \Lambda}{\cap} S_\alpha \text{ and } ar \in \underset{\alpha \in \Lambda}{\cap} S_\alpha$$

Hence, $\cap S_\alpha$ is an ideal of R containing M.

Let S be any other ideal of R containing M, therefore $\cap S_\alpha \subseteq S$. This implies that $\underset{\alpha \in \Lambda}{\cap} S_\alpha$ is the smallest ideal of R containing M.

REMARK

☞ This smallest ideal $\underset{\alpha \in \Lambda}{\cap} S_\alpha$ containing M is also called an ideal generated by M.

Theorem 5. *A field has no proper ideal.*

Proof. If F be a field, then we shall prove that its only ideals are $\{0\}$ and F itself. Let S be an ideal of F and if $S = \{0\}$, then in this case the theorem is obviously proved. Therefore, let $S \neq \{0\}$ and a be an arbitrary non-zero element of S, then

$$a \in S \Rightarrow a \in F \qquad\qquad (\because S \subseteq F)$$

$$\Rightarrow a^{-1} \in F \qquad\qquad (\because F \text{ is a field})$$

Since S being an ideal of F so that

$$a \in S, a^{-1} \in F \Rightarrow aa^{-1} \in S \Rightarrow 1 \in S \qquad\qquad (\because aa^{-1} = 1)$$

Now, let x be any element of F, then

$$1 \in S, x \in F \Rightarrow 1 . x \in S \Rightarrow x \in S$$

$$\therefore \qquad F \subseteq S$$

But $S \subseteq F$ because S being an ideal of R. Hence, $S = F$ and hence only ideals of F are $\{0\}$ and F itself.

Theorem 6. *If R is a commutative ring and $a \in R$, then the set $Ra = \{ra : r \in R\}$ is an ideal of R.*

Proof. In order to prove that Ra is an ideal of R, we must show that it is an additive subgroup of R and if $u \in R$ and $r \in R$, then ru is also in Ra. (Here we only need to check that $ru \in Ra$ because R is commutative *i.e.*, $ru = ur$).

Now if $u, v \in Ra$, then $u = r_1 a$ and $v = r_2 a$, for some $r_1, r_2 \in R$. Thus

$$u - v = r_1 a - r_2 a = (r_1 - r_2) a \in Ra \qquad (\because r_1 - r_2 \in R)$$

Hence, Ra is an additive subgroup of R.

Moreover, if r is any arbitrary element of R and $u \in Ra$, then

$$ru = r(r_1 a) \qquad (\because u = r_1 a)$$
$$= (rr_1) a \in Ra \qquad (\because rr_1 \in R)$$

Hence, Ra is an ideal of R.

Theorem 7. *Let R be a commutative ring with unit element whose only ideals are $\{0\}$ and R itself. Then R is a field.*

Proof. In order to prove that R is a field we must show that every non-zero element of R has its multiplicative inverse in R.

Let $a \in R$ and $a \neq 0$ and consider a set $Ra = \{ra : r \in R \}$ then Ra is an ideal of R (By Theorem 6). But R is a commutative ring with unit element having only ideals $\{0\}$ and R itself. Therefore, either $Ra = 0$ or $R = Ra$. Since $1 \in R$ so that $0 \neq a \neq 1$. $a \in Ra$, thus $Ra \neq \{0\}$ and hence only possibility is that $Ra = R$. This implies that every element in R is a multiple of 'a' by some element of R. But $1 \in R$ so that it is also a multiple of a. That is, there exists an element $b \in R$ such that $ba = 1$. But R is commutative and every non-zero element in R has its multiplicative inverse in R. Hence R is a field.

Theorem 8. *Let S_1 and S_2 be any two ideals of R, then ideal generated by $S_1 \cup S_2$ is the set $S_1 + S_2$ obtained by adding each element of S_1 with each element of S_2.*

Proof. Firstly, we shall show that the set $S_1 + S_2 = \{a + b : a \in S_1, b \in S_2\}$ is an ideal of R. Since $0 = 0 + 0 \in S_1 + S_2$ so $S_1 + S_2$ is non-empty.

Now, if $u, v \in S_1 + S_2$ then $u = a_1 + b_1$, $v = a_2 + b_2$ such that $a_1, a_2 \in S_1$ and $b_1, b_2 \in S_2$. Thus,

$$u - v = (a_1 + b_1) - (a_2 + b_2)$$
$$= (a_1 - b_2) + (b_1 - a_2) \in S_1 + S_2$$

$(\because S_1$ and S_2 being an ideal of R so that $a_1 - b_2 \in S_1$ and $b_1 - a_2 \in S_2)$

Hence, $S_1 + S_2$ is an additive subgroup of R.

Moreover, if r is any arbitrary element of R and $u \in S_1 + S_2$ such that $u = a_1 + b_2$ for some $a_1 \in S_1$ and $b_1 \in S_2$, then

$$ru = r(a_1 + b_1) = ra_1 + rb_1 \in S_1 + S_2$$
$$(\because S_1 \text{ and } S_2 \text{ being an ideals implies } ra_1 \in S_1 \text{ and } rb_1 \in S_2)$$

Also, $$ur = (a_1 + b_1)r$$
$$= a_1 r + b_1 r \in S_1 + S_2 \qquad (\because a_1 r \in S_1 \text{ and } b_1 r \in S_2)$$

Therefoere, $S_1 + S_2$ is an additive subgroup of R and $ru \in S_1 + S_2$

Hence, $S_1 + S_2$ is an ideal of R.

Let a be any arbitrary element of S_1 and since $0 \in S_2$ so that

$$a = a + 0 \in S_1 + S_2$$

Thus, $$S_1 \subseteq S_1 + S_2$$

Similarly, $\qquad\qquad S_2 \subseteq S_1 + S_2$

$\therefore \qquad\qquad\qquad S_2 \cup S_1 \subseteq S_1 + S_2$

Thus $S_1 + S_2$ is an ideal of R containing $S_1 \cup S_2$.

Let S be any ideal of R containing $S_1 \cup S_2$. If $u \in S_1 + S_2$ then $u = a_1 + b_1$ such that $a_1 \in S_1$ and $b_1 \in S_2$. Now

$$a \in S_1 \cup S_2 \text{ and } b \in S_1 \cup S_2 \qquad (\because S_1 \subseteq S_1 \cup S_2, S_2 \subseteq S_1 \cup S_2)$$

Therefore $\qquad\quad a, b \in S \qquad\qquad\qquad\qquad\qquad\qquad (\because S_1 \cup S_2 \subseteq S)$

$\Rightarrow \qquad\qquad\quad a + b \in S \qquad\qquad\qquad\qquad\qquad (\because S \text{ is an additive subgroup.})$

Thus, $\qquad\quad S_1 + S_2 \subseteq S.$

Hence, $\qquad\quad S_1 + S_2 = \{S_1 \cup S_2\}$ or $S_1 + S_2 = (S_1 \cup S_2)$

Theorem 9. *Let S_1 and S_2 be two ideals of a ring R, then $S_1 \cup S_2$ is an ideal of R if and only if either $S_1 \subseteq S_2$ or $S_2 \subseteq S_1$.*

Proof. Since S_1 and S_2 are ideals of R and let us first suppose either $S_1 \subseteq S_2$ or $S_2 \subseteq S_1$. If $S_1 \subseteq S_2$ then $S_1 \cup S_2 = S_2$ and S_2 is given to be an ideal so that $S_1 \cup S_2$ is an ideal of R. Similarly, if $S_2 \subseteq S_1$ then $S_1 \cup S_2 = S_1$ and S_1 is given to be an ideal so that $S_1 \cup S_2$ is an ideal of R.

Conversely, suppose $S_1 \cup S_2$ is an ideal of R and let if possible neither $S_1 \subseteq S_2$ nor $S_2 \subseteq S_1$. Then there exists an element a in S_1 and b in S_2 but not in S_1, therefore, $a, b \in S_1 \cup S_2$. But $S_1 \cup S_2$ is an ideal of R so $a - b \in S_1 \cup S_2$. This implies either $a - b \in S_1$ or $a - b \in S_2$. If $a - b \in S_1$ and $a \in S_1$, then $b = a - (a-b) \in S_1$ because S_1 being an ideal. This gives a conteradiction because $b \notin S_1$. Similarly, if $a - b \in S_2$, then $a = b + (a-b) \in S_2$ which is against the fact that $a \notin S_2$. Hence our assumption is wrong. Consequently either $S_1 \subseteq S_2$ or $S_2 \subseteq S_1$.

Solved Examples
Based on the following Results

▶ An additive subgroup S of a ring R is said to be a left ideal of R if $a \in S, r \in R \Rightarrow ra \in S$ and if $a \in S, r \in R \Rightarrow ar \in S$ then it is called right ideal.

▶ The necessary and sufficient condition for a non-empty subset S of a ring R to be an ideal of R are
  (i) $a \in S, b \in S \Rightarrow a - b \in S$
  (ii) $a \in S, r \in R \Rightarrow ra \in S$ and $ar \in S$

Example 1. *If S is an ideal of R and $1 \in S$, prove that $S = R$.*

Solution. Since S is an ideal of R so that $\quad S \subseteq R.$

Let a be any arbitrary element of R, S being an ideal of R and $1 \in S$, then

$$1 \in S, a \in R \Rightarrow a.1 \in S$$

$$\Rightarrow a \in S.$$

Thus, $\qquad\qquad\qquad\qquad R \subseteq S.$ But $S \subseteq R$

Hence, $\qquad\qquad\qquad\qquad S = R$

Example 2. *If S is an ideal of R, let $r(S)=\{x\in R : xa=0, \forall a\in S\}$. Prove that $r(S)$ is an ideal of R.*

Solution. Since $0\in R$ such that $0a=0$, it follows that $0\in r(S)$. Thus $r(S)$ is non-empty. Now if u,v are any elements of $r(S)$, then $ua=0$ and $va=0$ where, $a\in S$, then

$$(u-v)a = ua - va = 0 - 0 = 0, \forall a\in S$$

Thus, $u-v\in r(S)$. Hence, $r(S)$ is an addititve subgroup of R.

Let r_1 be any arbitrary elemet of R and $u\in r(S)$, then $ya=0$, $\forall a\in S$. Since S is an ideal of R, therefore,

$$a\in S, r_1\in R \Rightarrow ar_1\in S \text{ and } r_1a\in S$$

Then, $$(r_1u)a=r_1(ua)=r_10=0$$

i.e., $$(r_1u) \in r(S)$$

Also $$(ur_1)a=u(r_1a)=0 \quad (\because u\in r(S) \text{ and } r_1a\in S \text{ so } u(r_1a)=0)$$

i.e., $$ur_1 \in r(S).$$

Thus, $r(S)$ is an additive subgroup of R and $r_1u\in r(S)$ and $ur_1\in S$. Hence, $r(S)$ is an ideal of R.

Example 3. *Let R be the ring of all matrices of order 2×2 over integers and let S be the set of all 2×2 matrices of the form $\begin{pmatrix} a_1 & 0 \\ b_1 & 0 \end{pmatrix}$, where a,b are integers, then S is a left ideal but not right ideal of R.*

Solution. Let $A, B\in S$, then

$$A=\begin{pmatrix} a_1 & 0 \\ b_1 & 0 \end{pmatrix}, B=\begin{pmatrix} a_2 & 0 \\ b_2 & 0 \end{pmatrix}$$

Now, $$A-B=\begin{pmatrix} a_1 & 0 \\ b_1 & 0 \end{pmatrix} - \begin{pmatrix} a_2 & 0 \\ b_2 & 0 \end{pmatrix} = \begin{pmatrix} a_1-a_2 & 0 \\ b_1-b_2 & 0 \end{pmatrix} \in S$$

$$(\because a_1, a_2, b_1, b_2 \text{ are integers so are } a_1-a_2 \text{ and } b_1-b_2.)$$

Thus, S is an additive subgroup of R.

Moreover, if $X=\begin{pmatrix} a & b \\ c & d \end{pmatrix}$ be an arbitrary element of R and let $A=\begin{pmatrix} p & 0 \\ q & 0 \end{pmatrix}$ be any element of S, then

$$XA=\begin{pmatrix} a & b \\ c & d \end{pmatrix}\begin{pmatrix} p & 0 \\ q & 0 \end{pmatrix} = \begin{pmatrix} ap+bq & 0 \\ cp+dq & 0 \end{pmatrix} \in S \quad (\because ap+bq, cp+dq \text{ are integers.})$$

$\therefore A\in S, X\in R \Rightarrow XA\in S$

Hence, S is a left ideal of R.

Also, if $A=\begin{pmatrix} 1 & 0 \\ 1 & 0 \end{pmatrix} \in S$ and $X = \begin{pmatrix} 1 & 3 \\ 0 & 1 \end{pmatrix} \in R$,

Then $AX=\begin{pmatrix} 1 & 0 \\ 1 & 0 \end{pmatrix} \begin{pmatrix} 1 & 3 \\ 0 & 1 \end{pmatrix} \neq S$

Thus, $AX \neq S$. Hence, S is not a right ideal of R.

Example 4. *If S and T are ideals of a ring R, let ST be the set of all elements that can be written as finite sums of element of the form ab where, $a\in S$ and $b\in T$. Prove that ST is an ideal.*

Solution. Since $0\in S$ and $0\in T$ and $0=0$. $0\in ST$, so that ST is non-empty.

Let x, y be any two element of ST, then

$$x = \sum_{i=1}^{n} a_i b_i \;, y = \sum_{j=1}^{m} c_j d_j \qquad a_i c_j \in S \text{ and } b_i d_j \in T$$

Now, $$x - y = \sum_{i=1}^{n} a_i b_i - \sum_{j=1}^{m} c_j d_j = \sum_{k=1}^{m+n} e_k f_k$$

where $e_k = a_k, \forall\, k = 1, 2, ..., n,\ e_{n+1} = -c_t, \forall\, t = 1, 2, ..., m;$
$$f_k = b_k, \forall\, k = 1, 2, ..., n,\ f_{n+1} = d_t, \forall\, t = 1, 2, ..., m.$$
Since $e_k \in S, f_k \in T\ \forall\ k = 1, 2, 3, ..., n$, therefore, $x - y \in ST$. Thus ST is an additive subgroup of R.

Moreover, if r be any arbitrary element of R and $x \in ST$, i.e., $x = \sum\limits_{i=1}^{n} a_i b_i$, then

$$xr = \left(\sum_{i=1}^{n} a_i b_i\right) r = \sum_{i=1}^{n} (a_i b_i)\, r = \sum_{i=1}^{n} a_i (b_i r) \in ST$$

$$(\because T \text{ being an ideal of } R \text{ so that } b_i \in T,\ r \in R \Rightarrow b_i r \in T)$$

Thus, $\qquad\qquad xr \in ST$
Similarly, $\qquad\quad rx \in ST$
Hence, ST is an additive subgroup of R and $xr \in ST$ and $rx \in ST$ and therefore ST is an ideal of R.

Example 5. *The set of all nilpotent elements of R is an ideal of R.*

Solution. It is known that an element $a \in R$ is said to be nilpotent if $a^n = 0$ for some positive integer $n \geq 1$.

Let I be the set of all nilpotent elements of R. That is, if $a, b \in I$, then $a^n = 0$ and $b^m = 0$ for some $n \geq 1$ and $m \geq 1$.

Now, $$(a+b)^{m+n} = \sum_{i=j+m+n} a^i b^j \qquad \text{(By Binomial theorem)}$$

$$= 0$$
Thus, $\qquad\qquad\qquad a + b \in I.$
Next, a be any element of I and $r \in R$, then
$$(ar)^m = a^m r^m = 0.r^m$$
and $\qquad\qquad\qquad (ra)^m = r^m a^m = 0.$
Thus, $ar, ra \in I$. Hence, I is an ideal of R.

5.17 HOMOMORPHISM OF RINGS

In studying group, we have seen that for group a homomorphism was defined as a mapping ϕ such that $\phi\,(a*b) = \phi(a) * \phi(b)$. But a ring has two operations so we can define homomorphism of rings as follows:

Definition. *Let R and R' be two rings with two binary operations addition and multiplication, then a mapping ϕ from the ring R into R' is said to be a homomorphism if*

(i) $\phi(a+b) = \phi(a) + \phi(b)$

(ii) $\phi(ab) = \phi(a)\phi(b)$ *for all $a, b \in R$*

REMARKS

☞ Here both rings have same binary operations, in case both have the operations as $\{R,'+','.'\}$ and $\{R,'*','o'\}$, then above condition (i) and (ii) can be written as

(i) $\phi(a+b)=\phi(a) * \phi(b)$

(ii) $\phi(ab)=\phi(a) \circ \phi(b)$.

☞ If ϕ is surjective, then R' is said to be a homomorphic image of R.

☞ If ϕ is bijective, it is known as isomorphism, then R and R' are called isomorphic.

☞ If the ring R is identical with R', then ϕ is known as endomorphism.

☞ If R and R' are identical and ϕ is bijective, then ϕ is known as ring automorphism.

5.18 THEOREMS ON HOMOMORPHISM

Theorem 1. *If ϕ is a homomorphism of R into R', then*

(i) $\phi(0)=0', 0 \in R, 0' \in R'$

(ii) $\phi(-a)=-\phi(a), \forall a \in R$

Proof. (i) Since $0 \in R$ and a be any arbitrary element of R, then

$$a+0= a$$

$\Rightarrow \qquad \phi(a+0)=\phi(a)$

$\Rightarrow \qquad \phi(a)+\phi(0)=\phi(a)$ $\qquad\qquad$ ($\because \phi$ is homomorphism.)

Also, $\qquad 0+a=a$

$\Rightarrow \qquad \phi(0+a)=\phi(a)$

$\Rightarrow \qquad \phi(0)+\phi(a)=\phi(a)$

$\Rightarrow \qquad \phi(a)+\phi(0)=\phi(a)= \phi(0)+\phi(a) \; \forall \; \phi(a) \in R'.$

This show that $\phi(0)$ is the zero element of R', if $0'$ is the zero element of R', then

$$\phi(0)=0'.$$

(ii) Let a be an arbitrary element of R, then $-a \in R$ therefore,

$$a+(-a)= 0$$

$\Rightarrow \qquad \phi[a+(-a)]=\phi(a)$

$\Rightarrow \qquad \phi(a)+\phi(-a)=\phi(0)$ $\qquad\qquad$ ($\because \phi$ is homomorphism.)

Also, $\qquad (-a)+a=0$

$\Rightarrow \qquad \phi(-a)+\phi(a)=\phi(0)$

Thus $\qquad \phi(a)+\phi(-a)=\phi(0)= \phi(-a)+\phi(a) \; \forall \; \phi(a) \in R'.$

Hence $\qquad \phi(a)=- \phi(a).$

Theorem 2. *Every homomorphic image of a commutative ring is a commutative ring.*

Proof. Let ϕ be a homomorphism from a commutative ring R onto a ring R', then we have to show that R' is a commutative ring. Let a', b' be any elements of R'. Since ϕ is onto, then there exist two elements a and b of R such that $a'= \phi(a)$ and $b'=\phi(b)$.

Now, $\qquad a'b'= \phi(a)\phi(b)$

$\qquad\qquad\quad = \phi(ab)$ $\qquad\qquad$ ($\because \phi$ is homomorphism.)

$\qquad\qquad\quad =\phi(ba)$ $\qquad\qquad$ ($\because R$ is commutative.)

$\qquad\qquad\quad =\phi(b)\phi(a)$ $\qquad\qquad$ ($\because \phi$ is homomorphism.)

$=b'a'$.

Hence, R' is a commutative ring.

Theorem 3. *If ϕ is a homomorphism of a R into a ring R', then $\phi(R)$ is a subring of R'.*

Proof. Let $\phi(a)$ and $\phi(b)$ be any arbitrary elements of $\phi(R)$ for some $a, b \in R$. Since the ring R is itself a subring so that $a-b \in R$ and $ab \in R$, therefore,

$$\phi(a)-\phi(b) = \phi(a) +\phi(-b) \qquad\qquad [\because \phi(-b) =\phi(b)]$$
$$= \phi[a+(-b)] \in \phi(R) \qquad (\because \phi \text{ is homomorphism and } a-b \in R)$$

Now $\qquad\qquad \phi(a)\phi(b) = \phi(ab) \in \phi(R) \qquad\qquad\qquad (\because ab \in R)$

Thus, $\qquad\quad \phi(a)-\phi(b) \in \phi(R)$ and $\phi(a)\phi(b) \in \phi(R)$.

Hence, $\phi(R)$ is subring of R'.

5.19 KERNEL OF A RING HOMOMORPHISM

Definition. *Let R and R' be two rings and ϕ be a homomorphism from a ring R into a ring R' then the set of all those elements of R' which are mapped onto the zero element of R', is said to be* ***kernel*** *of ϕ and is usually denoted by ker (ϕ).*

i.e., If 0' is the zero element of R, then $\ker(\phi) = \{a \in R: \phi(a) = 0'\}$

Theorem 1. *If ϕ is a homomorphism of a ring R into a ring R' with kernel S, then S is an ideal of R.*

Proof. By definition of kernel of a homomorphism, we have

$S = \{a \in R: \phi(a) = 0'$, where $0'$ is the zero element of $\phi'\}$.

Now we shall show that S is an ideal.

Let x, y be any arbitrary element of S, then $x, y \in R$

and $\qquad\qquad \phi(x) = 0'$, $\phi(y) = 0'$

Now $\qquad\qquad \phi(x-y) = \phi[x+(-y)]$
$$= \phi(x)+\phi(-y) \qquad\qquad (\because \phi \text{ is homomorphism.})$$
$$= \phi(x)-\phi(y) \qquad\qquad [\because \phi(-y) = -\phi(y)]$$
$$=0'- 0'=0'$$

Thus, $x-y \in S$, therefore, S is an additive subgroup of R.

Moreover, if r be any arbitrary element of R and $a \in S$, then
$$\phi(ra)= \phi(r)\phi(a) \qquad\qquad (\because a \text{ is homomorphism.})$$
$$= \phi(r).0' \qquad\qquad (\because \phi(a)=0')$$
$$=0'$$

Thus, $ra \in S$. Similarly $ar \in S$.

Hence, S is an additive subgroup of R and $ra \in S$ and $ar \in S$ for all $r \in R, a \in S$. Consequently, S is an ideal of R.

Theorem 2. *If ϕ is a homomorphism of a ring R into a ring R' and let S be an ideal of R, then ϕ (S) is an ideal of ϕ (R).*

Proof. Let $\phi(a)$ and $\phi(b)$ be any arbitrary elements of $\phi(S)$, where $a, b \in S$ and S being an ideal of R, then $a-b \in S$ and $ra, ar \in S$ for all $r \in R$

Now $\qquad \phi(a) - \phi(b) = \phi(a)+ \phi(-b) \qquad\qquad [\because \phi(-b)=- \phi(b)]$
$$= \phi(a + (-b)) \qquad\qquad (\because \phi \text{ is homomorphism.})$$
$$= \phi(a - b) \in \phi(S) \qquad\qquad (\because a - b \in S)$$

Thus, $\qquad \phi(a) - \phi(b) \in \phi(S)$, so $\phi(S)$ is an additve subgroup of $\phi(R)$.

Moreover, if $\phi(r)$ be any arbitrary element of $\phi(R)$ and $\phi(a) \in \phi(S)$ then

$$\phi(r)\phi(a) = \phi(ra) \in \phi(S) \qquad (\because \phi \text{ is homomorphism and } ra \in S)$$

Similarly, $\qquad \phi(a)\phi(r) \in \phi(S)$.

Thus $\phi(S)$ is an additive subgroup of $\phi(R)$ and $\phi(r) \in \phi(a) \subseteq \phi(S)$ and $\phi(a)\phi(r) \in \phi(S)$.

Hence, $\phi(S)$ is an ideal of $\phi(R)$.

5.20 ISOMORPHISMS AND QUOTIENT RINGS

Definition 1. *A homomorphism ϕ from a ring R into a ring R' is said to be* isomorphism *if*

(i) ϕ *is one to one.*

(ii) ϕ *is onto.*

Definition 2. *Let R be a ring and S be any ideal of R, then the algebraic structure $\{R/S, '+'.'\}$ where $R/S = \{a+S : a \in R\}$ and the operations '+' and '.', on R/S defined by*

$$(a+S) + (b+S) = (a+b) + S$$

and $\qquad (a+S).(b+S) = a.b + S$

for all $a, b \in R$, forms a ring. This ring is called the quotient ring of R with respect to the ideal S. The quotient ring is also known as residue class ring.

REMARKS

☞ Zero element of R/S is S.

☞ Unit element of R/S is $S+1$.

☞ $S+a = S+b \Rightarrow a - b \in S$

Theorem 1. *Let S be an ideal of a ring R and let T be an ideal of R, containing S, then*

$$R/T \cong \frac{R/S}{T/S}$$

Proof. Since S is an ideal of R so R/S is meaningful and T being an ideal of R, R/T is also meaningful. Also T being an ideal of R containing S, *i.e.*,

$$S \subseteq T \subseteq R$$

Thus S is an ideal of T so T/S has a meaning. Now we shall show that T/S is an ideal of R/S. Let $S+a$ be any arbitrary element of T/S, then

$$S+a \in T/S$$

$\Rightarrow \qquad\qquad a \in T$

$\Rightarrow \qquad\qquad a \in R$

$\Rightarrow \qquad\qquad S + a \in R/S$

Thus, $\qquad\qquad T/S \subseteq R/S$

Again let $S+a$ and $S+b$ be any elements of T/S, then

$$a \in T \text{ and } b \in T$$

$\Rightarrow \qquad\qquad a - b \in T \qquad\qquad\qquad (\because T \text{ being an ideal of } R)$

$\Rightarrow \qquad\qquad (a-b)+S \in T/S$

$\Rightarrow \qquad (S+a) - (S+b) \in T/S$

$\Rightarrow \qquad T/S$ is an additive subgroup of R/S.

Moreover, if $S+r$ be any arbitrary element of R/S for $r\in R$ and $S+a$ be any element of T/S, then

$$(S+r)(S+a) = S+ra \in T/S$$

$$[\because T \text{ being an ideal so that } a\in T, r\in R \Rightarrow ra\in T \text{ and } ar\in T]$$

Similarly, $\quad (S+a)(S+r) \in T/S$.

Thus, T/S is an additive subgroup of R/S and $(S+r)(S+a) \in T/S$ and $(S+a)(S+r) \in T/S$.

Consequently T/S is an ideal of R/S. Hence $\dfrac{(R/S)}{(T/S)}$ has a meaning.

Now consider a mapping

$$\phi : R/S \to R/T$$

defined by

$$\phi(S+a)=T+a, \ \forall a\in R$$

First we show that ϕ is well defined.

Let $S+a$ and $S+b$ be any element of R/S and suppose

$$S+a= S+b$$

$\Rightarrow \qquad\qquad a-b \in S$

and $\qquad\qquad a-b \in T$

$\Rightarrow \qquad\qquad T + a = T + b$

$\Rightarrow \qquad\qquad \phi(S+a) = \phi(S+b)$

Thus ϕ is well defined, now we shall prove that ϕ is onto and preserves both the compositions.

   (i) **ϕ is onto.** Since corresponding to the element $T+a$ of R/T there exists an element $S+a$ of R/S such that

$$\phi(S+a)= T+a$$

(ii) **ϕ preserves both the compositions.** Let $S+a$ and $S+b$ be any arbitrary element of R/S, then

$$\phi[(S+a)(S+b)]= \phi[S+(a+b)]$$
$$= T+(a+b)= (T+a)+ (T+b)=\phi(S+a)+\phi(S+b)$$

Also, $\ \phi[(S+a)(S+b)] = \phi[S+(ab)]$
$$= T+(ab)= (T+a)+ (T+b)$$
$$= \phi(S+a)\phi(S+b)$$

Thus, ϕ is homomorphism of R/S onto R/T i.e., is the homomorphic image of R/S. Finally, we show that kernel of ϕ is T/S.

By definition of kernel, we have

$$\text{Kernel } \phi =\{(S+a)\in R/S: \phi(S+a)=T, \text{ where } T \text{ is zero element of } R/T\}.$$
$$=\{(S+a)\in R/S: (T+a)\in T\} \qquad\qquad (\because \ \phi(S+a)=T+a)$$
$$=\{S+a\in R/S: a\in T\}= T/S$$

Therefore, ϕ is the homomorphic image of R/S with kernel T/S. Hence by fundamental theorem of homomorphism, we have

$$R/T \cong \frac{R/S}{T/S}$$

i.e., R/T is isomorphic to the quotient ring of R/S with respect to the ideal T/S.

Theorem 2. *If S be an ideal of a ring R and T be any subring of R. Then*
$$(S+T)/S \cong T/S(S\cap T).$$

Proof. Here first, we shall show that $S+T$ is a subring of R and S is an ideal of $S+T$.

Let a_1+b_1 and a_2+b_2 be any two arbitrary elements of $S+T$ such that $a_1,a_2 \in S$ and $b_1,b_2 \in T$.

Now, $\quad (a_1+b_1) - (a_2+b_2) = (a_1-a_2) + (b_1-b_2) \in S+T$

$$(\because a_1,a_2 \in S \text{ and } b_1,b_2 \in T)$$

and $\qquad (a_1+b_1)(a_2+b_2) = (a_1+b_1)a_2 + (a_1+b_1)b_2 \qquad$ (By distributive law)

$$= a_1a_2+b_1a_2+a_1b_2+b_1b_2 \qquad \text{(By distributive law)}$$

Since S being an ideal so that a_1a_2, b_1a_2, $a_1b_2 \in S$ and T being a subring of R so $b_1b_2 \in T$. Thus,

$$a_1a_2+b_1a_2+a_1b_2+b_1b_2 \in S+T$$

Also $\qquad (S+T) \subseteq R$. Hence, $S+T$ is a subring of R.

Since $S \subseteq (S+T) \subseteq R$ so that S is an ideal of $S+T$. Therefore, $(S+T)/S$ has a meaning. Now consider a mapping

$$\phi : T \to (S+T)/S$$

defined by $\qquad \phi(a) = S+a, \forall a \in T.$

ϕ **is onto.** Corresponding to each element $S+a$ of $(S+T)/S$ there must exists an element a of T such that $\phi(a) = S+a$. Thus ϕ is onto.

ϕ **preserves both the compositions.** Let a, b be any two arbitrary elements of T, then

$$\phi(a+b) = S + (a+b)$$
$$= (S+a) + (S+b)$$
$$= \phi(a) + \phi(b)$$

Also, $\qquad \phi(ab) = S + (ab)$

$$= (S+a)(S+b) = \phi(a)\,\phi(b)$$

Hence, $(S+T)/S$ is the homomorphic image of T.

Now, we shall show that kernel of ϕ is $S\cap T$. Therefore, by definition of kernel, kernel $\phi = \{a \in T: T(a) = S, \text{ the zero element of } (S+T)/S\}$.

Let $x \in$ kernel ϕ

$\Rightarrow \qquad x \in T$ such that $\phi(x) = S$

$\Rightarrow \qquad x \in T$ such that $S+x = S$

$\Rightarrow \qquad x \in T$ and $x \in S \qquad\qquad (\because S+x = S \text{ so } x \in S)$

$\Rightarrow \qquad x \in S \cap T \qquad\qquad\qquad\qquad\qquad \dots (1)$

$\therefore \qquad\quad \phi \subseteq S \cap T$

Also, if $\quad y \in S \cap T$

$\qquad\qquad y \in S \quad$ and $\quad y \in T$

$\qquad\qquad y \in S \quad$ and $\quad \phi(y) = S+y$

$\qquad\qquad\qquad \phi(y) = S \qquad\qquad\qquad (\because y \in S \Leftrightarrow S+y = S)$

$\qquad\qquad\qquad y \in$ kernel ϕ

$\therefore \qquad\qquad S \subseteq T \text{ kernel } \phi \qquad\qquad\qquad\qquad \dots(2)$

Hence, from (1) and (2), we conclude that
$$\text{kernel } \phi = S \cap T.$$
Therefore, $(S+T)/S$ is the homomorphic image of T with kernel $S \cap T$, then by fundamental theorem of homomorphism on rings, we have
$$(S+T)/S \cong T/(S \cap T)$$
i.e., $(S+T)/S$ is isomorphic to $T/(S \cap T)$.

REMARK

☞ Fundamental theorem of homomorphism states that "every homomorphic image of a ring is isomorphic to some quotient ring."

5.21 PRINCIPAL IDEAL

Definition. *An ideal S of a ring R is said to be principal ideal if it is generated by a single element of S, i.e., if $a \in S$ then $S = \{a\}$.*

An ideal S of R is said to be principal ideal if there exists an element $a \in S$ such that an ideal T of R containing a and also contains S.

If R is a ring with unity, then the ideal generated by 1, i.e., $\{1\}$ is the ring R itself, because $r.1 = r$, $\forall r \in R$. Therefore, the ring R itself is called the unit ideal.

Also, the ideal generated by the zero element of R, i.e., $\{0\}$ is called null ideal, because $\{0\}$ consists of the zero element alone.

REMARKS

☞ Every ring R has at least one principal ideal, namely (0).
☞ Every ring with unit element has at least two principal ideals, namely (0) and (1).

Theorem 1. *If a is an element in a commutative ring R with unity, then the set $S = \{ra : r \in R\}$ is a principal ideal of R generated by the element a, i.e., $S = \{a\}$.*

Step Outline : To make the proof easier, use the following steps :

Step 1. *Show that S is an additive subgroup of R.*

Step 2. *Show that for $r \in R$, $u \in S$ and $ur \in S$*

Proof. Since $a \in R$ and $1 \in R$, then by definition of S, $1.a = a \in S$.

Now we shall prove that S is an additive subgroup of R.

Let u and v be any two elements of S, then
$$u = r_1 a \text{ and } v = r_2 a \text{ for some } r_1, r_2 \in R$$

Consider
$$\begin{aligned} u - v &= r_1 a - r_2 a \\ &= (r_1 - r_2) a \qquad \text{(By right distributive law)} \end{aligned}$$

$\therefore \qquad\qquad u - v \in S$ as $r_1 - r_2 \in R$

Thus S is an additive subgroup of R.

Now we shall show that for $r \in R$, $u \in S \Rightarrow ru \in S$ and $ur \in S$.

But R is a commutative ring so it is sufficient to prove that $ru \in S$.

Consider,
$$\begin{aligned} ru &= r(r_1 a) \qquad\qquad\qquad (\because u = r_1 a) \\ &= (rr_1)a \qquad\qquad\quad \text{(By associative law)} \end{aligned}$$

$\therefore \qquad\qquad ru \in S$ as $rr_1 \in R$

Hence, S is an ideal of R.

Now we shall prove that $S = (a)$.

Let ra be any element of S and let T be an ideal of R containing a, then

$$a \in T, \ r \in T \Rightarrow ra \in T$$

$$\therefore \quad S \subseteq T$$

Hence, S is a principal ideal of R generated by the element a.

Theorem 2. *Let S be an ideal of a commutative ring R. Let a be an element of S such that $x \in S \Rightarrow x = ya$ for some $y \in R$. Then S is a principal ideal of R generated by a.*

Proof. Since S is an ideal of a commutative ring R and $a \in S$. We shall prove that S is a principal ideal.

Let $x \in S$. Then $x = ya$ for some $y \in R$.

Again, let T be an ideal of R containing a, then

$$a \in T, \ y \in R \Rightarrow ya \in T \Rightarrow x \in T$$

Thus, $\quad x \in S \Rightarrow x \in T$

$$\therefore \quad S \subseteq T$$

Hence, S is a principal ideal of R generated by an element a.

5.22 PRINCIPAL IDEAL RING

Definition: *A commutative ring R with unity having no zero divisor is called a* principal ideal ring *if every ideal of R is a principal ideal.*

Theorem 1. *The ring of integers is a principal ideal ring.*

Step Outline : To make the proof easier, proceed as follows :

Prove that every ideal of $\mathbf{Z}$ is a principal ideal by using division algorithm and Zorn's lemma.

Proof. Let $\{\mathbf{Z}, +, .\}$ be a ring of integers. Obviously $\mathbf{Z}$ is a commutative ring with unit element having no zero divisors.

In order to prove that $\mathbf{Z}$ is a principal ideal ring, we shall prove that every ideal of $\mathbf{Z}$ is a principal ideal.

Let S be any ideal of $\mathbf{Z}$ and if $S = \{0\}$ then obviously S is a principal ideal as it is generated by single element 0.

So assume that $S \neq \{0\}$, therefore, S contains at least one non-zero integer, let it be 'a'. Since S is an additive subgroup of $\mathbf{Z}$ so $a \in S \Rightarrow -a \in S$. This implies that S contains at least one positive integer as one of a and $-a$ is positive.

Let S_+ denote the set of all positive integers of S, obviously $S_+ \neq \phi$, therefore, by well ordering principal (Zorn's lemma), S_+ must possess a least positive integer. Let this least positive integer be m. we will now show that $S = \{m\}$.

Let n be any integer in S, then by division algorithm, there exist integers q and r such that

$$n = mq + r, \quad \text{where } 0 \leq r \leq m$$

Now $\qquad m \in S, \ q \in \mathbf{Z} \Rightarrow mq \in S \qquad\qquad [\because S \text{ is an ideal.}]$

and $\qquad n \in S, \ mq \in S \Rightarrow n - mq \in S \quad [\because S \text{ is an additive subgroup of } \mathbf{Z}.]$

$$\Rightarrow r \in S$$

But $0 \leq r \leq m$ and m is assumed to be least positive integers of S.

Therefore, $r = 0$ so that $\quad n = mq$.

Thus, $n \in S \quad \Rightarrow \qquad n = mq$ for some $q \in \mathbf{Z}$.

Hence, S is a principal ideal of $\mathbf{Z}$ generated by the element m. Since S is an arbitrary ideal of $\mathbf{Z}$ therefore, every ideal of $\mathbf{Z}$ is a principal ideal. Hence, $\mathbf{Z}$ is a principal ideal ring.

Theorem 2. *Every field is a principal ideal ring.*

Proof. Since a field has no proper ideals, therefore its only ideal are $\{0\}$ and the field itself. The ideal $\{0\}$ is a principal ideal and the field itself is also principal ideal as it is generated by 1. Hence, both these ideals of field are principal. Therefore, every field is a principal ideal ring.

5.23 DIVISIBILITY IN AN INTEGRAL DOMAIN

Definition. *Let a be a non-zero element of a commutative ring R. Then a divides $b \in R$, if there exists an element $c \in R$ such that $b = ca$.*

We shall use the symbol $a|b$ to represents the fact that "a divides b".

In partiular, if $a \neq 0$ and $0 = a.0$, then $a|0$ this implies that every non-zero element of R is a divisor of its zero element.

Theorem 1. *If R is a commutative ring, then*

    (i) $a|b$ and $b|c \Rightarrow a|c$, i.e., relation of divisibility in R is transitive relation.

    (ii) $a|b$ and $a|c \Rightarrow a|(b + c)$

    (iii) $a|b \Rightarrow a|bx$, for all $x \in R$

Proof.   (i) $a|b \Rightarrow b = am$, for some $m \in R$

        and $b|c \Rightarrow \qquad c = bn$, for some $n \in R$

        $\therefore \qquad\qquad\qquad c = bm = (am)n = a(mn)$

        $\Rightarrow \qquad\qquad\qquad a|c$ as $mn \in R$

    (ii) $a|b \Rightarrow b = am$ for some $m \in R$

        and $\qquad\qquad a|c \Rightarrow c = an$ for some $n \in R$

        $\therefore \qquad\qquad\quad b+c = am + an = a(m+n)$

        $\Rightarrow \quad a|(b+c)$ as $m+n \in R$

    (iii) $a|b \Rightarrow \qquad\qquad b = an$, for some $n \in R$

        $\Rightarrow \qquad\qquad bx = anx$, for all $n \in R$ $a|bx$, for all $nx \in R$.

5.24 UNITS AND ASSOCIATES

5.24.1 Units

Let R be a commutative ring with unity, *i.e.*, $1 \in R$. Then an element a of R is said to be unit in R if there exists an element $b \in R$ such that $ab = 1$.

In other words, we can say that units of R are those elements of R which possess multiplicative inverse.

For Example:

1. In the ring of integers the only units are 1 and -1 as these are the only invertible elements of the ring of integers.

2. Every non-zero element in a field is a unit.

REMARKS

☞ If a is a unit in a ring R, then a^{-1} is also a unit in R. Also the product of two units is again unit, as if a, b are units of R, then $(ab)^{-1} = b^{-1}\,a^{-1} \in R$.

☞ Do not confuse a units with unit element of R.

☞ There may be more than one units in a ring but the unit element is always unique.

☞ The set of all units in R forms a group under multiplication.

Solved Examples

Example 1. *Find all the units of the integral domain of Gaussian integers.*

Solution. Let $J[i] = \{a+ib : a,\ b \in \mathbf{Z}\}$ be the ring of Gaussian integers. The element $1+0.\,i$ is the unity element of $J[i]$.

Let $x+iy$ be a unit of $J[i]$ and $x_1 + iy_1$ be its inverse. Then we have

$$(x+iy)\,(x_1 - iy_1) = 1 + 0.\,i$$

or $\quad (xx_1 - yy_1) + i(xy_1 + x_1 y) = 1 + 0.i$

Separating the real and imaginary parts, we get

$$xx_1 - yy_1 = 1 \qquad\qquad \text{... (1)}$$

and $\qquad\qquad xy_1 + x_1 y = 0 \qquad\qquad \text{... (2)}$

Squaring and adding (1) and (2), we get

or $\quad (xx_1 - yy_1)^2 + (xy_1 + x_1 y)^2 = 1$

or $\quad x^2 x_1^2 + y^2 y_1^2 + x^2 y_1^2 + x_1^2 y^2 = 1$

Now, the product of two positive integers can be equal to 1 if and only if each of them is 1.

$\therefore \qquad\qquad x^2 + y^2 = 1$ and $x_1^2 + y_1^2 = 1$

If $x^2 + y^2 = 1$, then $x^2 = 0,\ y^2 = 1$ or $x^2 = 1,\ y^2 = 0$

$$x = 0,\ y = \pm 1 \text{ and } x = \pm 1,\ y = 0$$

Thus the only units of the integral domain of Gaussian integers are $0 \pm i$ and $\pm 1 \pm 0.i$ or $1,\ i,\ -1$.

5.24.2 Associates

Let R be a commutative ring with unity. Then an element a of R is said to be an associate of $b \in R$ if $a = ub$ for some unit u of R.

If a is an associate of b, then it can be denoted by $a \sim b$.

For Example :

(1) Since the ring of integers has only two units namely 1 and -1. Therefore, if a is any non-zero integer, then it has exactly two associates namely $1.\,a$ and $(-1)\,.\,a$, *i.e.*, a and $-a$.

(2) The units of Gaussian integers $J[i]$ are $1,\ -1,\ i$ and $-i$. Therefore, if $a+ib$ be any non-zero element of $J[i]$, then it has exactly four associates namely,

$$1.\,(a+ib) = a+ib,\ -1(a+ib) = -a-ib,\ i(a+ib) = -b+ai$$

and $\qquad\qquad -i(a+ib) = b - ai.$

Theorem 1. *Let D be an integral domain with unity 1. Then, two non-zero elements a,b∈D are associates if and only if a|b and b|a.*

Proof. Let us first suppose that two non-zero elements a, $b \in D$ are associates of each other in D. Then $a = bu$ for some unit u in D.

Now $$a = bu \Rightarrow b|a$$

Also, $$a = bu \Rightarrow au^{-1} = buu^{-1}$$
$$\Rightarrow au^{-1} = b \qquad (\because u^{-1} \text{ us a unit of } u.)$$
$$\Rightarrow a|b$$

∴ a and b are associates $\Rightarrow a|b$ and $b|a$.

Conversely, suppose $a|b$ and $b|a$. Then we shall prove that a and b associates. Now
$$a|b \Rightarrow b = ax \text{ for some } x \in D$$
Also $$b|a \Rightarrow a = by \text{ for some } y \in D$$
∴ $$b = ax = (by)x = b(yx)$$
or $$1. b = b(yx) \text{ or} \qquad b(1 - yx) = 0$$
$$\Rightarrow \qquad 1 - yx = 0 \qquad (\because b \neq 0 \text{ and } D \text{ has no zero divisors.})$$
$$\Rightarrow \qquad yx = 1$$

∴ both x and y are units in D.

Thus, $a = by$ where y is a unit in D. Hence, a and b are associates.

5.25 PRIME IDEALS

Definition: *If R be a commutative ring, then an ideal S≠ R is said to be* prime ideal *if ab∈S implies that either a∈S or b∈S for all a, b∈R.*

For Example :

In a ring of integers **Z**, the ideal $S = \{3n : n \in \mathbf{Z}\}$ is prime, because if $ab \in S$ this implies $3|ab$ $\Rightarrow$ either $3|a$ or $3|b$, which gives that either $a \in S$ or $b \in S$.

REMARK

☞ In a ring of integers **Z**, every ideal $(p) = \{pn : n \in \mathbf{Z}, \text{where } p \text{ is prime}\}$ is a prime ideal of **Z**.

Theorem 1. *Let R be a commutative ring with unity and let S ≠ R be an ideal of R. Then R|S is an integral domain if and only if S is a prime ideal of R.*

Proof. Since R is a commutative ring with unity so that R/S is commutative ring with unity. Therefore, in order to show that R/S is an integral domain, we shall show that R/S has no zero divisors.

Let us suppose that S is a prime ideal of R and let a, $b \in R$, then $S+a$ and $S+b$ be any arbitrary elements of R/S.

If $$(S+a)(S+b) = S \qquad (\because \text{ zero element of } R/S \text{ is } S.)$$
$$\Rightarrow \qquad S+(ab) = S$$
$$\Rightarrow \qquad \text{either } ab \in S \text{ or } b \in S \qquad (\because S \text{ is prime ideal of } R.)$$
$$\Rightarrow \qquad \text{either } S+a = S \text{ or } S+b = S$$
$$\Rightarrow \qquad R/S \text{ has no zero divisors.}$$
$$\Rightarrow \qquad R/S \text{ is an integral domain.}$$

Conversely, R/S is an integral domain so that it has no zero divisors.

Therefore, if $S+a$, $S+b \in R/S \ \forall \ a, b \in R$ then,

$$(S+a) \ (S+b) = S$$
$$\Rightarrow \qquad \text{either } S+a = S \text{ or } S+b = S$$
$$\Rightarrow \qquad \text{either } a \in S \text{ or } b \in S$$

But $\qquad (S+a) \ (S+b) = S \Rightarrow S + ab = S \Rightarrow ab \in S$

Thus, if $ab \in S$, then either $a \in S$ or $b \in S$. Hence, S is a prime ideal of R.

Theorem 2. *Let S be a prime ideal of R and suppose that a product a_1, a_2, $a_3 \dots a_n$, of element of R, belongs to S. Then at least one of a_i, $i = 1, 2, 3, \dots, n$, belong to S.*

Proof. We shall prove this result by induction hypothesis on n. If $n = 1$ or 2, then

$$a_1 a_2 \in S \Rightarrow \text{either } a_1 \in S \text{ or } a_2 \in S \qquad (\because S \text{ is prime ideal.})$$

Thus in this case theorem is proved. Now it will therefore, be supposed that the theorem has been proved for products with $n-1$ elements. That is, if

$a_1 a_2 a_3 \dots a_{n-1} \in S$, then at least one of the a_i, $i = 1, 2, 3, \dots, n-1$ belongs to S. Now we shall prove that the theorem, for the products $a_1 a_2 a_3 \dots a_n \in S$. Then

$$a_1 a_2 a_3 \dots a_n \in S \ \Rightarrow \ (a_1 a_2 a_3 \dots a_{n-1}) \ a_n \in S$$
$$\Rightarrow \text{either } a_1 a_2 \dots a_{n-1} \in S \text{ or } a_n \in S \qquad (\because S \text{ is prime ideal.})$$

But $a_1 a_2 a_3 \dots a_{n-1} \in S$ is obtained by the pervious step of the induction hypothesis. Hence, the theorem is proved for the product $a_1 a_2 a_3 \dots a_n$ of elements of R.

Theorem 3. *Let $\{S_i : i \in \Lambda\}$ be a non-empty family of prime ideals of a ring R and suppose if i, j $\in \Lambda$, then either $S_i \subseteq S_j$ or $S_j \subseteq S_i$. Then both $\bigcup_i S_i$ and $\bigcap_i S_i$ are prime ideals of R.*

Proof. Let us suppose $S_1 = \bigcup_i S_i$ and since each S_i is proper ideal so that S_1 is proper ideal.

If $ab \in S_1$ and suppose $a \notin S_i$. Then there exists $i \in \Lambda$ such that $ab \in S_i$. Since $a \neq S_i$ then $b \in S_i$ because S_i is prime ideal.

But $S_i \subseteq S_1$ so that $b \in S_1$. Hence S_1 is a prime ideal of R.

Next, let us suppose $S_2 = \bigcap_i S_i$. Clearly, S_2 is proper ideal. Assume that $xy \in S_2$ and

$x \notin S_2$, then there exists $i \in \Lambda$ so that $x \notin S_i$ but S_i is prime ideal, therefore, $y \in S_i$. Finally, let j be an arbitrary element of Λ. If $S_1 \subseteq S_j$ then $y \in S_j$. On the other hand if $S_j \subseteq S_i$, then $xy \in S_j$, while $x \in S_j$ because $xy \in S_2 \subseteq S_j$. Again S_j is prime so that $y \in S_j$. Thus in any way $y \in S_j$ and therefore, $y \in S_2$. Accordingly S_2 is prime ideal of R.

Theorem 4. *If $S_1 S_2 \dots S_n$ or $S_1 \cap S_2 \cap \dots \cap S_n$ is contained in a prime ideal S, then for $i(1 \leq i \leq n) S_i$ is contained in S.*

Proof. Since $S_1 S_2 \dots S_n \subseteq S$. Assuming that number S_i is not contained in S. For each i choose $a_i \in S_i$ so that $a_i \notin S_i$. Then $a_1 a_2 \dots a_n \in S_1 S_2 \dots S_n$. But $S_1 S_2 \dots S_n \subseteq S$, therefore, $a_1 a_2 \dots a_n \in S$ and S is prime ideal so that at least one of a_i belong to S. This is against our assumption. Hence, for some $i(1 \leq i \leq n) S_i$ is contained in S.

Theorem 5. *Let S_1, S_2, $\dots$, S_n be prime ideals of a ring R and let S be ideal of R which is not wholly contained in any single one of them. Then S contains an element a which does not belong to any S_i.*

Proof. Assuming that no S_i is contained in any of the remaining ideals. For if $S_1 \subseteq S_2$ then it is sufficient to prove the theorem for S and the reduces system $S_2 S_3 \dots S_n$. Since no S_i is contained in any of the remaining so that S_i does not contain any of the ideals of S, S_1, $S_2, \dots, S_{i-1}$, $S_{i+1}, \dots, S_n$ therefore, by above theorem 4, it cannot contian their product. It is therefore possible to choose a_i so that

$$a_i \in SS_1 S_2 ...... S_{i-1} S_{i+1} ... S_n \subseteq S \cap S_1 \cap S_2 \cap ..... \cap S_{i-1} \cap S_{i+1} \cap ..... \cap S_n$$

and $a_i \notin S_i$. Let such an element be chosen for each value of i and put

$$a = a_1 + a_2 + ... + a_n$$

Clearly 'a' belongs to S_1. However, since

$$a_i = a - a_2 - ... a_n.$$

and all of $a_2, a_3, ... a_n \in S_1$, a does not belong to S_1. This is against our assumption that $a_1 \in S_1$.

Similarly, 'a' does not belong to any of $S_2, S_3, ... S_n$. Hence the theorem is proved.

5.26 MAXIMAL IDEALS

Definition. *A proper ideal S of a ring R is said to be maximal ideal of R if it is not strictly contained by any other proper ideal of R.*

Or

A proper ideal S of a ring R is said to be maximal ideal of R if T is any other proper ideal of R such that either S=T or T=R.

Or

A proper ideal S of a ring R is said to be maximal ideal of R if there exists no proper ideal between S and R.

For Example :

If R is a ring of even integers, then the ideal $(4) = \{4m : m \in \mathbf{Z}\}$ is maximal ideal of R because $2 \notin (4)$ and $(4) \neq R$.

Theorem 1. *If S be an ideal of a ring $\mathbf{Z}$ of all integers. Then S is maximal if and only if it is generated by some prime integers.*

Proof. Let the ideal S generated by some positive integer p. Since the ring of integers is a principal ideal ring, then every ideal of $\mathbf{Z}$ is principal ideal so that S is also principal ideal generated by p, *i.e.,* $S=(p)$.

Now suppose p is prime, then we will show that S is maximal.

Let T be an ideal of Z containing S so that T is also a principal ideal and let $T=(q)$.

Since $S \subseteq T$ and $p \in S$ so that $p \in T$

But $\qquad\qquad\qquad T=(q)$, theefore, $p=aq$ for some $a \in Z$.

Since p is prime, then $\qquad p=aq$ either $q=1$ or $q=p$.

If $q=1$, then $\qquad\qquad T=(q)=(1)=\mathbf{Z}$. Also if $q = p$,

then $\qquad\qquad\qquad\qquad T=(q)=(p)=S$

Hence the ideal S generated by a prime, is maximal ideal.

Conversely, S is a maximal ideal generated by p, then we will show that p is prime.

Let if possible p is not prime so that it is composite, therefore assume that $p=mn$, where $m \neq 1$ and $n \neq 1$.

Let T be an ideal of Z generated by m, *i.e.,* $T = (m)$, then obviously $S \subseteq T \subseteq \mathbf{Z}$.

But S being a maximal ideal of $\mathbf{Z}$, so that either $S=T$ or $T=\mathbf{Z}$.

If $T=\mathbf{Z}$, then $\qquad\qquad (m)= (1)$

$\Rightarrow \qquad\qquad\qquad\qquad m=1$

This gives a contradiction.

Now if $T=S$, then $\quad\quad (m)=(p)$

$\Rightarrow\quad\quad\quad\quad m = xp$ for some $x \in \mathbf{Z}$

$\Rightarrow\quad\quad\quad\quad m = xnm \quad\quad\quad\quad\quad\quad (\because p=mn)$

$\Rightarrow\quad\quad\quad\quad xn = 1 \quad\quad\quad\quad\quad\quad\quad (\because p \neq 0)$

$\Rightarrow\quad\quad\quad\quad n = 1$

This gives again a contradiction.

Thus the contradiction arises against the assumption that p is not prime.

Hence, p is prime.

REMARK

☞ An ideal generated by a single element is called principal ideal and the ring whose every ideal is a principal ideal called *principal ideal ring*.

Theorem 2. *An ideal S of a commutative ring R with unity is maximal if and only if the residue class ring R/S is a field.*

Proof. Since R is a commutative ring with unity so that R/S is also commutative ring with unity. In order to show that R/S is a field, we only have to show that every non-zero element of R/S has its multiplicative inverse in R/S.

Let S be a maximal ideal of R and let $S + a$ be any non-zero element of R/S, i.e.,

$$S + a \neq a + S.$$

Let (a) be an ideal of R. Since the sum of two ideals is again an ideal of R so that $S+(a)$ and $(a)+S$ are again ideals.

Since S is maximal, so that we must have

$$S+(a)=R$$

But $1 \in R$, then for some $b \in S$ such that

$$b+ \alpha a = 1 \text{ for some } \alpha \in R$$

$\Rightarrow\quad\quad\quad\quad b = 1 - \alpha a$

$\Rightarrow\quad\quad\quad\quad (1 - \alpha a) \in S \quad\quad\quad\quad\quad (\because b \in S)$

$\Rightarrow\quad\quad\quad\quad S + \alpha a = S+1$

$\Rightarrow\quad\quad\quad\quad (S + \alpha)(S + a) = S+1$

Similarly, $\quad\quad (S + a)(S + \alpha) = S+1 \quad\quad\quad (\because R/S \text{ is commutative.})$

Therefore, $\quad\quad (S+a)^{-1}= (S + \alpha) \in R/S$

Hence, every non-zero element of R/S has its multiplicative inverse and hence R/S is a field.

Conversely, let R/S be a field. Now we shall have to show that S is maximal.

Let T be any ideal of R properly containing S and let a be any arbitrary element of R not in S, then $\quad\quad a \neq S \Rightarrow S+a \neq S$

Thus $S+a$ a non-zero element of R/S. Now b be any element of T not in S, then $S+b$ is again a non-zero element of R/S. Since R/S is a field so that

$$(S+b) \in R/S \Rightarrow (S+b)^{-1} \in R/S$$

Now $(S+a) \in R/S, (S+b) \in R/S \Rightarrow (S+a)(S+b)^{-1} \in R/S$

$$\Rightarrow (S+a)(S+b^{-1}) \in R/S$$
$$\Rightarrow S+ab^{-1} \in R/S$$
$$\Rightarrow ab^{-1} \in R$$

Since T is an ideal of R, then

$$b \in T, \ ab^{-1} \in R \Rightarrow ab^{-1}b \in T$$
$$\Rightarrow \qquad a.1 \in T \Rightarrow a \in T$$
$$\therefore \qquad a \in R \Rightarrow a \in T$$

Thus $R \subseteq T$. But $T \subseteq R$. Therefore, $T = R$. Hence, there is no ideal of R between S and R. Hence S is a maximal ideal of R.

REMARK

☞ A ring R is a field if and only if its zero ideal is a maximal ideal.

Theorem 3. *Let R be a commutative ring with unity. Then every maximal ideal of R is a prime.*

Proof. Let S be any maximal ideal of a commutative ring R with unity. Then by previous theorem, R/S is a field and since every field is an integral domain so R/S is also an integral domain. Further, since we have any ideal S of a commutative ring R with unity is prime if and only if R/S is an integral domain. Therefore, S is a prime ideal of R.

Theorem 4. *Let S be a proper ideal of a ring R. Then S is contained in a maximal ideal of R.*

Proof. Let A be a multiplicatively closed subset of a ring R (multiplicatively closed means whenever a and b belong to A, then product ab also belongs to A) and does not contain the zero element. Now let $\{S_\alpha : \alpha \in \Lambda\}$ be the collection of all ideals of **R** such that $S_\alpha \cap S = \phi$ for each $\alpha \in A$ and if S_i and S_j belong to the collection such that $S_i \subseteq S_j$, then $\{S_\alpha : \alpha \in \Lambda\}$, together with the relation $\subseteq$ (inclusion), is a non-empty inductive system. Since $\{S_\alpha : \alpha \in \Lambda\}$ consists of the proper ideals of R partially ordered by inclusion, then in particular it contains S. But $\{S_\alpha : \alpha \in \Lambda\}$ is an inductive system, therefore, there exists a maximal element of $\{S_\alpha : \alpha \in \Lambda\}$. Let this maximal element be T such that $S \subseteq T$. The fact that T is a maximal in $\{S_\alpha : \alpha \in \Lambda\}$ means simply that T is a maximal ideal of R. Hence, if S is a proper ideal of R, then S will be contained in a maximal ideal of R.

5.27 EMBEDDING OF RINGS

Here we shall discuss particular kinds of a subring of a given ring. We now consider the problem of extending a ring to another ring which has special properties or bears a special relationship to the given ring.

Since a non-empty subset S of a given ring R is said to be a subring of R if S forms a ring under the binary compositions of R. Thus a subring inherits its composition from the parent ring.

If a ring is given, can we find a super ring R so that S is a subring of R? The answer in general is not yes because subring is defined only when a ring is given. But it is of course, possible to find a super ring R, so that the starting S is isomorphic to a subring of R. Then we can say that the ring S as a subring of the ring R. This phenomena to find such super ring R of a ring S so that S is a subring of R, is called *embedding of rings*.

Definition. *Let R and R' be two rings. A monomorphism (or one-one homomorphism) ϕ from R and R' is called an embedding (or imbedding) mapping.*

The three problems we shall consider here are:

(i)   Embedding any ring in a ring with identity.

(ii) Embedding of an integral domain in a quotient field.

(iii) Embedding any ring in a ring endomorphisms.

Theorem 1. *Every ring can be embeded in a ring with identity.*

Proof.   Let R be a ring and $\bar{R} = R \in \mathbf{Z} = \{(r, n): r \in R, n \in \mathbf{Z}\}$.

Define addition and multiplication in $\bar{R}$ by

$$(r, n) + (s, m) = (r+s, n+m)$$

and $$(r, n).(s, m) = (rs, nm)$$

for all $$(r, n), (s, m) \in \bar{R} = R \times \mathbf{Z}.$$

It is easy to check that $\bar{R}$ forms a ring with respect to these operations, and $(0,1)$ serves as multiplicative identity for $\bar{R}$.

Now define a mapping $\theta : R \to \bar{R}$ by

$$\theta(r) = (r, 0), \forall\, r \in R$$

Obviously θ is a homomorphism and θ is also one-one as

$$\theta(r) = \theta(s) \Rightarrow (r, 0) = (s, 0)$$

$\Rightarrow$ $r = s$

Hence, R is embeded in a ring $\bar{R}$ which has unity element namely $(0,1)$.

Theorem 2. *An integral domain can be embeded into a field.*

Proof.   Let D be an integral domain and let,

$$S = \{(a, b): a, b \in D, b \neq 0\}$$

Now define a relation $\sim$ on S by

$$(a, b) \sim (c, d) \Leftrightarrow ad = bc.$$

It is easy to check that $\sim$ is an equivalence relation on S. This relation partitions S into disjoint equivalence classes. Let equivlence class of (a, b) be denoted by $[a, b]$.

Let F be the set of all these equivalence classes. Define addition and multiplication on F as follows:

$$[a, b] + [c, d] = [ad+bc, bd]$$

and $$[a, b].[c, d] = [ac, bd].$$

Now we shall show that F forms a field under these operations.

First of all we shall show that the operations of addition and multiplication are well defined.

Let $[a, b] = [a_1, b_1]$ and $[c, d] = [c_1, d_1]$

Then $(a, b) \sim (a_1, b_1)$ and $(c, d) \sim (c_1, d_1)$

$\Rightarrow$ $ab_1 = ba_1$ and $cd_1 = dc_1$

$\Rightarrow$ $(ab_1)dd_1 = (ba_1)dd_1$ and $(cd_1)bb_1 = (dc_1)bb_1$

$\Rightarrow$ $ab_1dd_1 + cd_1bb_1 = ba_1dd_1 + dc_1bb_1$ ...(1)

Now we shall show that

$$[ad+bc, bd] = [a_1d_1 + b_1c_1, b_1d_1].$$

This equality will hold if

$$(ad + bc, bd) \sim (a_1 d_1 + b_1 c_1, b_1 d_1)$$

or $\qquad (ad + bc)(b_1 d_1) = bd(a_1 d_1 + b_1 c_1)$

or $\qquad adb_1 d_1 + bcb_1 d_1 = bda_1 d_1 + bd\, b_1 c_1$

This is true, by equation (1).

Hence, addition is well defined.

Next, let $\qquad [a, b] = [a_1, b_1] \qquad$ and $\qquad [c, d] = [c_1, d_1]$

Then $\qquad ab_1 = ba_1 \qquad$ and $\qquad cd_1 = dc_1$

or $\qquad ab_1 cd_1 = ba_1 dc_1 \qquad\qquad\qquad\qquad\qquad$ ... (2)

Now $\qquad [a, b] \cdot [c, d] = [a_1, b_1] \cdot [c_1 \cdot d_1]$

or $\qquad (ac, bd) = (a_1 c_1, b_1 d_1)$

or $\qquad acb_1 d_1 = bda_1 c_1$

This is true by equation (2).

Hence, multiplication is well defined.

(i) Associative property.

Let $\quad [a, b], [c, d]$ and $[e, f] \in F$. Then

$$\begin{aligned}
[a, b] + ([c, d] + [e, f]) &= [a, b] + [cf + de, df] \\
&= [a(df) + b(cf + de), b(df)] \\
&= [(ad)f + (bc)f + (bd)e, (bdf)] \\
&= [(ad + bc)f + (bd)e, (bd)f] \\
&= [(ad + bc, bd) + e, f] \\
&= ([a, d] + [c, d]) + [e, f].
\end{aligned}$$

$\therefore \qquad$ Addition is associative in F.

Commutative property.

(ii) Let $\qquad [a, b], [c, d] \in F$, then

$$\begin{aligned}
[a, b] + [c, d] &= [ad + bc, bd] \\
&= [bc + ad, db] = [cd + da, db] \\
&= [c, d] + [a, b] \; \forall [a, b], [c, d] \in F
\end{aligned}$$

$\therefore \qquad$ Addition is commutative in F.

(iii) Existence of additive identity.

Let $\quad [a, b] \in F$, then

$$[a, b] + [0, b] = [ab + b.\, 0,\, b^2] = [ab, b^2] = [a, b] \text{ as } (ab, b^2) \sim (a, b)$$

Thus, $[0, b]$ the additive identity, since $[0, x] = [0, y]$ for any non-zero element $x, y \in D$.

(iv) Existence of additive Inverse.

Let $\quad [a, b] \in F$, be any element, then $[-a, b] \in F$

Also, $\quad [a, b] + [-a, b] = [ab + b\,(-a), b^2]$

$$\begin{aligned}
&= [ab - ba, b^2] = [0, b^2] \\
&= [0, b] \qquad\qquad\qquad\qquad\qquad\qquad [(0, b^2) \sim (0, b)]
\end{aligned}$$

Thus $[-a, b]$ is the additive inverse of $[a, b]$.

Therefore, $\{F,'+'\}$ forms an abelian group.

Now we shall verify that multiplication is commutative, associated and is distributive over addition.

(v) Associative Property.

Let $[a,b],[c,d]$ and $[e, f]$ be any three elements of F. Then we have

$$[a, b].([c, d].[e,f]) = [a, b].[ce, df]$$
$$= [a(ce), b(df)]$$
$$= [(ac)e,(bd)f]$$
$$= [ac, bd].[e, f]$$
$$= ([a, b].\ [c, d]).[e, f]$$

Thus multiplication is associative in F.

(vi) Commutative Property.

Let $[a, b],[c, d]$ be any two elements of F, then we have

$$[a,b].[c,d] = [ac, bd]$$
$$= [ca, db]=[c, d].[a, b]$$

Thus multiplication is commutative in F.

(vii) Existence of Multiplicative Identity.

Let x be any non-zero element of D and let $[a, b] \in F$, then we have

$$[a,b].[x,x]= [ax,bx]= [a, b] \text{ as } (ax, bx) \sim (a, b)$$

Thus $[x, x]$ acts as a multiplicative identity.

(viii) Existence of Multiplicative Inverse.

Let $[a, b]$ be any non-zero element of F, then $a \neq 0$ and $b \neq 0$.

Therefore, $[b, a] \in F$

Also, $\qquad [a, b].[b, a] = [ab, ba] = [x, x] \text{ as } (ab, ba) \sim (x, x)$

$\therefore$ Every non-zero element possesses multiplicative inverse.

Hence $\{F, +,.\}$ is a field.

We now show that D can be embeded into F as follows:

Define a mapping

$$\theta: D \to F$$

by $\qquad\qquad\qquad \theta(a) = [ax, x]$, where x is a non-zero element of D.

First we shall show that that θ is well defined.

Let $a, b \in D$ and if $\quad a = b$, then

$$ax = bx \qquad\qquad\qquad\qquad\qquad (x \neq 0)$$

$\Rightarrow \qquad\qquad\qquad xax = xbx \Rightarrow axx = xbx$

$\Rightarrow \qquad\qquad\qquad [ax, x] = [bx, x] \Rightarrow \theta(a) = \theta(b)$

Now to show that θ is one-one:

Let $\theta(a) = \theta(b)$ for $\quad a, b \in D$, then

$$[ax, x] = [bx, x]$$

$$\Rightarrow \qquad (ax, x) \sim (bx, x) \Rightarrow axx = xbx$$

$$\Rightarrow \qquad xxa = xxb \Rightarrow x^2a = x^2b$$

$$\Rightarrow \qquad x^2(a-b) = 0$$

$$\Rightarrow \qquad a-b = 0 \qquad\qquad (\because x^2 \neq 0 \text{ as } x \neq 0)$$

$$\Rightarrow \qquad a = b$$

$\therefore \quad \theta$ is one-one mapping.

Now to show that θ is homomorphism:

Let $a, b \in D$, then we have

$$\theta(a+b) = [(a+b)x, x] = [ax+bx, x]$$

$$= [ax^2+bx^2, x^2] \qquad [\because (ax+bx, x) \sim (ax^2+bx^2, x^2)]$$

$$= [axx+xbx, x^2]$$

$$= [ax, x] + [bx, x]$$

$$= \theta(a) + \theta(b)$$

Also, $\qquad \theta(ab) = [ab, x] = [abx^2, x^2] \qquad [\because (abx, x) \sim (abx^2, x^2)]$

$$= [ax, x] = [bx, x]$$

$$= \theta(a).\theta(b)$$

Hence, θ is one-one homomorphism, therefore θ is required mapping which embeds the integral domain D into a field F.

Definition. *The above field F is called the* field of quotients of the integral of domain D.

REMARKS

☞ Sometimes, we use the notation $a|b$ to denote $[a, b]$.

☞ For instance, $\qquad [a, b] + [c, d] = [ad+bc, bd]$

$$\frac{a}{b} + \frac{c}{d} = \frac{ad+bc}{bd}$$

and $[a,b].[c,d] = [ac,bd]$, i.e., $\qquad \frac{a}{b}.\frac{c}{d} = \frac{ac}{bd}$

☞ Each element of F can be expressed as a quotient of two elements of D, that is why the field F is called field of quotient of integral domain.

Solved Examples
Based on the following Results

▶ Let R be a commutative ring, then an ideal $S+R$ is said to be prime ideal if $a, b \in S \Rightarrow$ either $a \in S$ or $b \in S \; \forall a,b \in R$.

▶ A proper ideal S of a ring R is said to be maximal ideal of R if there exists no proper ideal between S and R.

▶ For a commutative ring with unity, every maximal ideal of R is a prime ideal.

Example 1. *If E is a ring of even integers, then show that an ideal (4) generated by 4 is a maximal ideal of E.*

Solution. Let $S=(4)$ and let T be an ideal of E such that $S \subseteq T$. Then there exists $a \in T$ such that

$a{\neq}(4){=}S$ implying that a is an even integer not divisible by 4. So that we have
$$a=4m+2, \text{ for some integer } m$$
Now, $\quad\quad\quad 2=a-4m{\in}T$ because $4m{\in}(4){=}S$

and $S \subseteq T \Rightarrow 4m{\in}T$ and $a{\in}T$. Thus T has every even integral multiple of 2. Hence, $T=E$, consequently, $S=(4)$ is maximal ideal of E.

Example 2. *Show that in a commutative ring without unity a maximal integral need not be prime.*

Solution. Let E, the ring of even integers be a commutative ring without unity and an ideal (4) of E is maximal. But it is not prime because
$$2.2{\in}(4) \text{ and } 2{\notin}(4)$$

Example 3. *Let R be the ring of all real valued continuous functions on the closed interval $[0,1]$.*

Let $M = \left\{ f \in R\left(\dfrac{1}{3}\right) = 0 \right\}$, show that M is a maximal ideal of R.

Solution. First we shall show that M is an ideal of R. Since the continuous functions $\omega(x){=}0$, $\forall x{\in}[0,1]$ belongs to M, so that M is non-empty.

Let $f(x), g(x){\in}M$, then $f\left(\dfrac{1}{3}\right) =0, g\left(\dfrac{1}{3}\right) =0$ now consider $h(x) = f(x) - g(x)$, then

$$h\left(\frac{1}{3}\right) = f\left(\frac{1}{3}\right) - g\left(\frac{1}{3}\right) = 0 - 0 = 0$$

$$\therefore \quad\quad\quad h(x){=}0{\in}M, \text{ i.e., } f(x){=} g(x){\in}M$$

Let $r(x)$ be any arbitrary element of R and $f(x){\in}M$ such that $f\left(\dfrac{1}{3}\right) =0$, then suppose
$$t(x){=} r(x)f(x){\in}M$$

$$\Rightarrow \quad\quad\quad t\left(\frac{1}{3}\right) = r\left(\frac{1}{3}\right) f\left(\frac{1}{3}\right) {=}r\left(\frac{1}{3}\right).0{=}0$$

$\therefore t(x){\in}M$, i.e., $r(x)f(x){\in}M$.

Similarly, $\quad\quad\quad f(x) = r(x) \in M$.

Hence M is an ideal of R. Now we will show that M is maximal. Obviously, $M{\neq}R$ because $i(x){\in}R$ given by $i(x){=}1$, which does not belong to M.

Let N be an ideal of R such that $M{\subseteq}N$, then there exists $\lambda(x){\in}N$ and $\lambda(x){\notin}M$, therefore, $\lambda\left(\dfrac{1}{3}\right){\neq}0$ so assume $\lambda\left(\dfrac{1}{3}\right){=}c$, where $c \neq 0$. Now consider $\mu(x){=}R$ given by

$$\mu(x) {=}\lambda(x){-}\beta(x), \text{ where } \beta(x){=} c, \forall x{\in}[0,1]. \text{ Then}$$

$$\mu\left(\frac{1}{3}\right) {=}\lambda\left(\frac{1}{3}\right){-}\beta\left(\frac{1}{3}\right) {=}c - c{=}0$$

$$\therefore \quad\quad\quad \mu(x){\in} M$$
$$\Rightarrow \quad\quad\quad \mu(x){\in} N \quad\quad\quad\quad\quad\quad (\because M{\subseteq}N)$$
$$\therefore \quad\quad\quad \beta(x){=}\lambda(x) - \mu(x){\in}N$$

Let us define $\gamma(x){\in}R$ by $\gamma(x){=}\dfrac{1}{c}$, $\forall x{\in}[0,1]$ and N is an ideal of R, then

$$\beta(x) \in N, \gamma(x) \in R \Rightarrow \beta(x)\gamma(x) \in N$$

$\Rightarrow$ c. $\dfrac{1}{c} \in N$

$\Rightarrow$ $1 \in N$

$\Rightarrow$ $i(x) \in N$ $\hspace{2cm}$ ($\because i(x)=1$)

But $i(x)$ is the unit of R. Hence, $N=R$. Consequently M is a maximal ideal.

Example 4. *If Z_{12} is a ring under addition and multiplication modulo 12, then find its all prime and maximal ideals.*

Solution : Since $Z_{12} = \{0,1,2,3,4,5,6,7,8,9,10,11\}$ it is commutative ring with unity.

Let $S_1 = \{0,2,4,6,8,10\}$ and $S_2 = \{0,3,6,9\}$ be two subsets of Z_{12}. Obviously, S_1 and S_2 are both ideals of Z_{12}.

For S_1: If $ab \in S_1$, then either $a \in S_1$ and $b \in S_1$

i.e.,
$$10 \in S_1 \Rightarrow 2.5 \in S_1 \Rightarrow 2 \in S_1$$
$$8 \in S_1 \Rightarrow 2.4 \in S_1 \Rightarrow 2 \in S_1 \text{ and } 4 \in S_1$$
$$6 \in S_1 \Rightarrow 2.3 \in S_1 \Rightarrow 2 \in S_1$$
$$4 \in S_1 \Rightarrow 2.2 \in S_1 \Rightarrow 2 \in S_1$$
$$2 \in S_1 \Rightarrow 2.1 \in S_1 \Rightarrow 2 \in S_1$$
$$0 \in S_1 \Rightarrow 0.0 \in S_1 \Rightarrow 0 \in S_1$$

Thus S_1 is a prime ideal of Z_{12}.

For S_2: If $ab \in S_2$, then either $a \in S_2$ and $b \in S_2$,

i.e.,
$$9 \in S_2 \Rightarrow 3.3 \in S_2 \Rightarrow 3 \in S_2$$
$$6 \in S_2 \Rightarrow 2.3 \in S_2 \Rightarrow 3 \in S_2$$
$$3 \in S_2 \Rightarrow 3.1 \in S_2 \Rightarrow 3 \in S_2$$
$$0 \in S_2 \Rightarrow 0.0 \in S_2 \Rightarrow 0 \in S_2$$

Thus S_2 is a prime ideal of Z_{12}.

Further since Z_{12} is commutative ring with unity so that S_1, S_2 are also maximal ideals.

Example 5. *Prove that if K is any field which contains D, then K contains a subfield isomorphic to F, where F is the field of quotients of the integral domain D.*

Solution. Define a mapping

$$\theta : F \to K$$

by $\hspace{2cm} \theta([a,b]) = ab^{-1}$

Since $a, b \in D \subseteq K$ and $b \neq 0$, therefore b^{-1} exists in K. First we shall show that θ is well defined.

Let $[a_1, b_1], [a_2, b_2] \in F$ and $[a_1, b_1] = [a_2, b_2]$

$\Rightarrow$ $\hspace{2cm} (a_1 b_1) \sim (a_2 b_2)$

$\Rightarrow$ $\hspace{2cm} a_1 b_2 = a_2 b_1$

$\Rightarrow$ $\hspace{2cm} a_1 b_1^{-1} = a_2 b_2^{-1}$

$\Rightarrow$ $\hspace{2cm} \theta([a_1 b_1]) = \theta([a_2 b_2])$

$\therefore$ $\hspace{2cm} \theta$ is well defined.

Now to show that θ is one-one.

Let $[a_1,b_1],[a_2,b_2] \in F$ then

$$\theta([a_1, b_1]) = \theta([a_2, b_2])$$

$\Rightarrow \qquad a_1 b_1^{-1} = a_2 b_2^{-1}$

$\Rightarrow \qquad a_1 b_2 = a_2 b_1$

$\Rightarrow \qquad a_1 b_2 = b_1 a_2$

$\Rightarrow \qquad [a_1,b_1] = [a_2,b_2]$

$\therefore \qquad \theta$ is one-one.

Now to show that θ is homomorphism.

Let $[a_1, b_1]$ and $[a_2, b_2]$ be any two elements of F, then

$$\theta([a_1, b_1] + [a_2, b_2]) = \theta([a_1 b_2 + b_1 a_2, b_1 b_2]) = (a_1 b_2 + b_1 a_2)(b_1 b_2)^{-1}$$

$$= (a_1 b_2 + b_1 a_2) b_2^{-1} b_1^{-1}$$

$$= (a_1 b_2) b_2^{-1} b_1^{-1} + (b_1 a_2) b_2^{-1} b_1^{-1}$$

$$= a_1 (b_2^{-1} b_2^{-1}) b_1^{-1} + a_2 (b_1 b_2^{-1}) b_1^{-1}$$

$$= a_1 b_1^{-1} + a_2 b_2^{-1} (b_1 b_1^{-1})$$

$$= a_1 b_1^{-1} + a_2 b_2^{-1}$$

$$= \theta([a_1,b_1]) + \theta([a_2, b_2])$$

Also, $\theta([a_1,b_1].[a_2, b_2]) = \theta([a_1 a_2, b_1 b_2]) = a_1 a_2 (b_1 b_2)^{-1}$

$$= a_1 a_2 (b_2^{-1} b_1^{-1}) = (a_1 b_1^{-1})(a_2 b_2^{-1})$$

$$= \theta([a_1,b_1]). \theta([a_2, b_2])$$

$\therefore \qquad \theta$ is homomorphism.

Thus F will be isomorphic to $\theta(F)$ which will be a subfield of K. Hence, the quotient field is the smallest field containing D.

Example 6. If D_1 and D_2 be two isomorphic integral domain then show that their respective field of quotients F_1 and F_2 are also isomorphic.

Solution : Let $\phi : D_1 \rightarrow D_2$ be the given isomorphism. The fields of quotients F_1 and F_2 are given by

$$F_1 = \{[a, b]: a, b \in D_1, b \neq 0\}$$

and $\qquad F_2 = \{[x, y]: x, y \in D_2, y \neq 0\}$

Now Define a mapping

$$\theta : F_1 \rightarrow F_2$$

by $\qquad \theta([a, b]) = [\phi(a), \phi(b)]$

since $\qquad [a, b] \in F_1 \Rightarrow a, b \in D_1, b \neq 0$

$\Rightarrow \qquad \phi(a), \phi(b) \in D_2$ with $\phi(b) \neq 0$

$\Rightarrow \qquad [\phi(a), \phi(b)] \in F_2$

We shall show that θ is well defined and one-one:

Let $[a_1,b_1], [a_2,b_2]$ be any two elements of F_1, then we have

$$\theta([a_1, b_1]) = \theta([a_2,b_2])$$

$\Leftrightarrow \qquad [\phi(a_1),\phi(b_1)] = [\phi(a_2),\phi(b_2)]$

$\Leftrightarrow \qquad [\phi(a_1),\phi(b_1)] \sim [\phi(a_2),\phi(b_2)]$

$$\Leftrightarrow \qquad \phi(a_1)\phi\,(b_2)=\phi(b_1)\,\phi(a_2)$$

$$\Leftrightarrow \qquad \phi(a_1b_2)=\phi(b_1a_2) \qquad\qquad\qquad (\phi \text{ is a homomorphism.})$$

$$\Leftrightarrow \qquad a_1b_2=b_1a_2 \qquad\qquad\qquad\qquad\qquad (\phi \text{ is one-one})$$

$$\Leftrightarrow \qquad (a_1,b_1)\sim(a_2,b_2)$$

$$\Leftrightarrow \qquad [a_1,b_1]=[a_2,b_2]$$

$\therefore$ θ is well defined and one-one mapping.

Now to show that θ is onto.

Let $[x, y]$ be any element of F_2, then $x,\ y\in D_2$. Since ϕ is onto from D_1 and D_2, therefore, there exists $a,\ b\in D_1$ such that

$$\phi(a)=x \text{ and } \phi(b)=y$$

Now $\qquad\qquad \theta([a,\ b])=[\phi(a),\phi(b)]=[x, y]$

$\therefore$ θ is onto.

Now to show that θ is homomorphism:

$$\begin{aligned}
\theta([a_1,b_1]+[a_2,b_2]) &= \theta([a_1b_2+b_1a_2,b_1b_2])\\
&= [\phi(a_1b_2+b_1a_2),\phi(b_1b_2)]\\
&= [\phi(a_1b_2)+\phi(b_1a_2),\phi(b_1)\phi(b_2)]\\
&= [\phi(a_1)\phi(b_2)+\phi(b_1)\phi(a_2),\phi(b_1)\phi(b_2)]\\
&= [\phi(a_1),\phi(b_1)]+[\phi(a_2),\phi(b_2)]\\
&= \theta([a_1,\ b_1)]+\theta([a_2,\ b_2)]
\end{aligned}$$

Also $\quad \begin{aligned}[t]
\theta([a_1,b_1].[a_2,b_2]) &= \theta[a_1a_2+b_1b_2] = [\phi(a_1,a_2),\phi(b_1,b_2)]\\
&= [\phi(a_1)\phi(a_2),\phi(b_1)\phi(b_2)]\\
&= [\phi(a_1),\phi(b_1)].[\phi(a_2),\phi(b_2)]\\
&= \theta([a_1,b_1)].\theta([a_2,b_2)]
\end{aligned}$

Thus θ is homomorphism.

Consequently θ is an isomorphism.

Example 7. *Find the field of the quotient of the integral domain* $J[i]=\{a+ib:a,b\in\mathbf{Z}\}$

Solution. Let F be the field of quotients of $J[i]$. Then by definition

$$F = \left\{ \frac{a+ib}{c+id} : a,b,c,d \in \mathbf{Z} \text{ and } c+id \neq 0 \right\}$$

$$= \left\{ \frac{(a+ib)(c-id)}{c^2+d^2} ; a,b,c,d\in\mathbf{Z} \right\}$$

$$= \left\{ \left(\frac{ac+bd}{c^2+d^2} + i\frac{bc-ad}{c^2+d^2} \right) : a,b,c,d \in \mathbf{Z} \right\}$$

$\therefore$ $F\subseteq F_1$, where $F_1=\{x+iy : x, y\in\mathbf{Q}\}$

Now let $x+iy$ be any element of F_1, then $x,y\in\mathbf{Q}$, then we have,

$$x+iy = \frac{p_1}{q_1} + \frac{p_2}{q_2} \text{ , with } p_1,q_1,p_2,q_2\in\mathbf{Z} \text{ and } q_1\neq 0,\ q_2 \neq 0$$

$$= \frac{p_1 q_2 + i p_2 q_1}{q_1 q_2}$$

$$= \frac{p_1 q_2 + i p_2 q_1}{q_1 q_2 + i.0} \in F \text{ as } (p_1 q_2 + i p_2 q_1) \text{ and } (q_1 q_2 + i.0) \in J[i]$$

∴ $\quad\quad\quad (x+iy) \in F_1 \Rightarrow x+iy \in F$

∴ $\quad\quad\quad\quad\quad F_1 \subseteq F_2$

Now Define a mapping $\theta : J[i] \rightarrow F_1$

by $\quad\quad\quad\quad \theta(a+ib) = a+ib$

This mapping will be the embeddng mapping.

Hence $\quad\quad\quad F_1 = \{x+iy : x, y \in \mathbf{Q}\}$ is the field of quotients of $J[i]$.

EXERCISE 5.3

1. If in a ring R, $x^2=x, \forall x \in R$, then show that R is commutative ring of characteristic 2.

2. Let R be a commutative ring with unity and $a,b \in R$ show that $S=\{xa+yb : x, y \in R\}$ is an ideal of R containing a and b.

3. If S is an ideal of R let $[R:S]=\{x \in R: rx \in S, \forall s \in R\}$. Prove that $[R : S]$ is an ideal of R and that it contains S.

4. If R is a ring and $a \in R$ let $r(a)=\{x \in R : ax=0\}$. Prove that $r(a)$ is a right ideal of R.

5. Let p be a prime integer and let $\mathbf{Z_p}=\{0,1,2,3, \ldots\ldots p-1\}$. Show that the field $(\mathbf{Z_p}, +_\mathbf{p}, \times_\mathbf{p})$ is not ordered.

6. If S_1 and S_2 are two ideals of a ring R, then the set of all elements of the form $b_1 b_2 + c_1 c_2 + \ldots\ldots + I_1 I_2$ where $b_1, c_1, \ldots\ldots, I_1 \in S_1$ and $b_2, c_2, \ldots\ldots, I_2 \in S_2$ is an ideal of R.

7. Show that the set of all integers $\mathbf{Z}$ is not an ideal of a ring $\mathbf{Q}$ of all rational numbers.

8. Let M be a ring of 2×2 matrices over integers and let $K = \left\{ \begin{pmatrix} a & b \\ 0 & 0 \end{pmatrix} : a, b \in \mathbf{Z} \right\}$. Then K is a right ideal of M but not a left ideal of M.

9. Let R be a commutative ring. Prove that the set of all nilpotent elements of R is an ideal of R.

10. (i) If S is a left ideal and T is a right ideal of a ring R, then show that $S+T$ need not be even a one-sided ideal of R.

    (ii) If S is a left ideal and T is a right ideal of a ring R, then show that $S \cap T$ need not be even a one-sided ideal of R.

11. Prove that in a ring of integers $\mathbf{Z}$, an ideal (n) is maximal if and only if n is prime, where $(n)=\{na:a \in \mathbf{Z}\}$

12. Give an example of a ring R without unity which has no maximal ideals and in which $xy=0 \forall x, y \in R$.

13. Find a maximal ideal of $\mathbf{Z}+\mathbf{Z}$ where $\mathbf{Z}$ is a ring of integers.

14. Find a prime ideal of $\mathbf{Z}+\mathbf{Z}$ which is not maximal.

15. Let R be a ring of all real-valued continuous functions on the closed integral $[0,1]$. If S is a maximal ideal of R, prove that there exist a real number $y \in [0,1]$, such that $$S=S_y=\{f(x) \in R : f(y)=0\}.$$

16. Show that the field of reals can be embeded into the field of complex numbers.

17. Find a field of quotient of the integral domain $\mathbf{Z}[\sqrt{2}] = \{a + b\sqrt{2} : a,b \in \mathbf{Z}\}$

18. Show that the field of quotients of a finite integral domain is the integral domain itself.

19. Let D be an integral domain, and $a, b \in D$ be such that $a^n=b^n$, $a^m=b^m$ for relatively prime positive integers m, n. Prove that $a=b$.

20. Show that the result given in Q.19 may fail to hold in case D is not an integral domain.

21. Give an example of an embedding mapping in which unity is not mapped to unity.

22. If $R = \{a + b\sqrt{3} : a,b \in \mathbf{Z}\}$. Show that R is an integral domain with unity. Obtain its field of quotients.

23. Determine the field of quotients of the following integral domain:

(i) The domain of all rational number of the form $\dfrac{m}{10^n}$, $m, n \in \mathbf{Z}$.

(ii) The domain of all complex number $a+bi$, where $a, b \in \mathbf{Z}$.

24. Give an example to show that non-isomorphic integral domains may have isomorphic field of quotients.

Hints to the Selected Problems

1. In a ring R, if $x^2=x$, then $x+x=0$, $\forall \ x \in R$ and $x+y=0 \Rightarrow x=y$. Using these two results we can easily prove that R is commutative *i.e.*, $xy=yx$.

2. Since $S=\{xa+yb : x,y \in R\}$. If $x=1, y = 0$, then $a \in S$, also $x=0, y=1$, then $a \in S$. Thus, $a,b \in S$. Now we have to prove that S is an ideal of R.

(i) If $u,v \in S$, then $u=x_1a+y_1b$, $v=x_2a+y_2b$, for $x_1, x_2, y_1, y_2 \in R$

$\therefore \qquad u-v=(x_1a+y_1b) - (x_2a+y_2b) = (x_1-x_2)a - (y_1-y_2)b \in S$.

$\therefore$ S is an additive subgroup of S.

(ii) For $u=xa+yb \in S$ and for $r \in R$

$\qquad ru = r(xa+yb) = (rx)a + (ry)b \in S$ as $rx, ry \in R$.

Also, R is commutative so $ur \in S$.

Hence, S is an ideal of R containing a and b.

4. $\qquad r(a) = \{x \in R : ax=0\}$.

For, $u,v \in r(a)$, *i.e.*, $u,v \in R$ such that $au=0, av=0$,

$\qquad a(u-v)=au-av=0-0=0$

$\therefore \qquad u-v \in r(a)$.

Also, for $u \in r(a)$, and $r_1 \in R$,

$\qquad a(ur_1)=(au)r_1=0.r_1=0$ as $au=0$.

Hence $r(a)$ is a right ideal of R.

7. Since the product of a rational number and an integer is not necessarily an integer. For example, $2 \in \mathbf{Z}$,

$$\left(\frac{1}{3}\right) \in \mathbf{Q} \text{ but } 2\left(\frac{1}{3}\right) = \frac{2}{3} \notin \mathbf{Z}.$$

Hence the set of an integer is not an ideal of ring of all rational numbers.

9. An element a of R is said to be nilpotent if for some positive integer n such that $a^n=0, a \in R$.

10. (ii) Consider the ring R of 2×2 matrices whose elements are real numbers.

Let S be the set of 2×2 matrices of the form $\begin{bmatrix} a & b \\ 0 & 0 \end{bmatrix}$, where a,b are real numbers. Clearly S forms a left ideal.

Let T be the set of 2×2 matrices of the form $\begin{bmatrix} a & b \\ 0 & 0 \end{bmatrix}$, where a,b are real numbers. Clearly T forms a right ideal.

Then $S \cap T$ be the set of 2×2 matrices of the form $\begin{bmatrix} a & 0 \\ 0 & 0 \end{bmatrix}$, where a is real number. Clearly this set is neither left nor right ideal or R.

13. $2\mathbf{Z}+\mathbf{Z}$ is a maximal ideal of $\mathbf{Z}+\mathbf{Z}$.

14. $\mathbf{Z}+\{0\}$ is a prime ideal which is not maximal.

18. Suppose R is a finite integral domain. Then R is a field. It means that the smallest field containing R is R itself. Also the quotient field of R is the smallest field containing R.

Answers

13. $2\mathbf{Z}+\mathbf{Z}$ **14.** $\mathbf{Z}+\{0\}$ **15.** Field of quotients of $\mathbf{Z}\sqrt{2} = \{x+y\sqrt{2} : x, y \in \mathbf{Q}\}$

22. Fields of quotisions of $R = \{x+y\sqrt{3} : x,y \in \mathbf{Q}\}$ **23.** (i) $\mathbf{Q}$ (ii) $\{c+di : c,d \in \mathbf{Q}\}$ **12.** No. **18.** No.

5.20 CONCEPTS OF DIVISIBILITY IN A RING

Definition1. *Let R be a commutative ring and $a,b \in R$, $a \neq 0$. Then we say $a|b$ (a divides b) if $\exists\ c \in \mathbf{R}$ such that $b=ac$. Here,a is called a factor of b.*

Definition2. *Let R be a commutative ring and $a,b \in R$. Then an element $d \in R$ is called greatest common divisior (g.c.d.) or highest common factor of a and b. if*

(i) *$d|a$ and $d|b$*

(ii) *whenever $c|a$, $c|b$ then $c|b$*

In this case we write $d=g.c.d.(a.b)$. It is sometimes denoeted only by (a, b).

SOME GENERAL RESULTS

(i) *If $a|b$, $b|c$ then $a|c$.*

(ii) *If $a|b$, $a|c$ then $a|b \pm c$.*

(iii) *If $a|b$ then $a|bx$ for all $x \in R$.*

(iv) *If R has unity then $1|x$ for all $x \in R$ and if a is a unit then $a|x$ for all $x \in R$.*

Definition 3. *Let R be a commutative ring. A non-zero element $l \in R$ is called least common multiple (l.c.m.) of two non-zero elements $a,b \in R$ if*

(i) *$a|l,b|l$*

(ii) *If $a|x,b|x$ then $l|x$.*

we denote l by l.c.m. $(a,b) = [a,b]$.

Definition 4. *Let R be a commutative ring with unity. Then $a,b \in R$ are called associates if $b=ua$ for some unit u in R.*

REMARKS

☞ A pair of element in a ring may not have an *l.c.m.* and a pair could have more than one *l.c.m.*

☞ By a **unit**, we mean an element which has multiplication inverse.

Theorem 1. *Let R be an integral domain with unity and $a,b \in R$ be non-zero elements then $a|b$ and $b|a$ if and only if a and b are associates.*

Proof. Let R be an integral domain with unity and $a,b \in R$ be non-zero elements of R . Let us first suppose $a|b$ and $b|a$.

If $a|b$, then $b = xa$, for some $x \in R$

         $b|a \Rightarrow$ $a = yb$ for some $y \in R$

Therefore,

$$b = xa=x(yb)$$

$\Rightarrow$ $b(1-xy)= 0$

$\Rightarrow$ $1-xy = 0$

$\Rightarrow$ $xy =1$

$\Rightarrow$ y is a unit in R and $a=yb$.

Conversely, if a, b are associates then there exists a unit u such that $a= bu$ (so $au^{-1}=b$)

$\Rightarrow$ $b|a$ and $a|b$.

Theorem 2. *Let R be an integral domain with unity. If $d_1=g.c.d.(a,b)$ in R then d_2 is also a g.c.d. (a,b) if and only if d_1 and d_2 are associates.*

Proof. Let us first suppose d_1 and d_2 both are g.c.d.(a,b) Then, by definition, we can write

$$d_1|a, d_1|b$$

and $$d_2|a, d_2|b$$

By definition we get $d_1|d_2$, which gives that d_1 and d_2 are associates.

Conversely, let us suppose $d_1=$ g.c.d.(a,b) and d_2 be an associates of d_1. Then

$$u\, d_2= d_1, \text{ for some unit } u.$$

$\Rightarrow$ $d_2|d_1$ and as $d_2|a$, $d_2|b$, we find $d_2|a$ and $d_2|b$

Now, let $x|a,x|b$ then $x|d_1$ as d_1 is g.c.d.(a,b)

Also, as $$d_2= d_1u^{-1}$$

$$d_1|d_2$$

$\Rightarrow$ $$x|d_2$$

Hence, $$d_2=g.c.d.(a,b)$$

REMARKS

☞ In a similar manner we can prove that "If $l_1= l.c.m.(a,b)$ in R then l_2 is also an l.c.m. (a,b) iff l_1 and l_2 are associates."

Theorem 3. *Let R be an integral domain with unity. If $g.c.d.(a,b)=d$ for $a,b \in R$. Then c.d and g.c.d. (ca,cb) are associates.*

Proof. Let us suppose that g.c.d.$(ca,cb)=d'$

Now, since $d|a$ Therefore $a=dk$, for some k $\in R$

$\Rightarrow$ $ac=dkc=cdk$

$\Rightarrow$ $cd|ca.$

Similarly $cd|cb$ $\Rightarrow$ $cd|d'$ $\Rightarrow cdt$

Further $d'|ca$ $\Rightarrow$ $ca=d's$

$\Rightarrow$ $ca=d's=cdts$

$\Rightarrow$ $a=(dt)s \Rightarrow dt|a$

Similarly $dt|b$

$\Rightarrow$ $dt|d$ $\Rightarrow$ $d= dtp \Rightarrow d(1-tp)=0$

$\Rightarrow$ $1= tp$

$\Rightarrow$ t is a unit.

$\Rightarrow$ g.c.d.$(ca,cb)= d'=cdt$

Hence, cd and d' are associates.

Theorem 4. *In an integral domain if there exists a greatest common divisior of any two elements, then it is unique apart from the distinction between associates.*

Proof. Let D be any integral domain and a, b be any non-zero elements of D such that g.c.d. of a and b exists. Let if possible d_1 and d_2 be the g.c.d. of a and b then when d_1 is g.c.d.,we have

$$d_1|a \text{ and } d_1|b$$

Further $d_2|a$ and $d_2|b$ then $d_2|d_1$.

Also when d_2 is g.c.d., then we have $d_2|a$ and $d_2|b$

Also, $d_1|a$ and $d_1|b$, then $d_1|d_2$.

Now, $d_2|d_1$ and $d_1|d_2$ which shows that d_1 and d_2 are associates.

Theorem 5. *In every principal ideal domain (PID) each pair of non-zero elements surely has a greatest common divisior.*

Proof. Let us suppose that D be any principal ideal domain (PID) and a,b be any non-zero elements of D. Also let (a) and (b) denote the principal ideal generated by a and b respectively.

It is known that linear sum of two ideals is again an ideal thus $(a)+(b)$ is an ideal in D. Also, every ideal of D is principal ideal, therefore there exists an element $d \in D$ such that

$$(a)+(b)=(d)$$

We have to prove that d is the g.c.d. of a and b.

Now, since $(a)+(b)=(d)$

$\Rightarrow$ $(a) \subset (d)$ and $(b) \subset (d)$

Also $(a) \subset (d) \Rightarrow$ $a = dx_1$,   for some $x_1 \in D$

$\Rightarrow$ $d|a$

and $(b) \subset (d) \Rightarrow$ $b=dx_2$,   for some $x_2 \in D$

$\Rightarrow$ $d|b$

Thus, we conclude that d is the common divisor of a and b . Further, suppose that c is also a common divisor of a and b, so that $c \mid a$ and $c \mid b$. Thus $a = cy_1$ and $b = cy_2$, for $y_1, y_2 \in D$. Therefore

$x \in (a) \Rightarrow$ $x = az_1$, for some $z_1 \in D$

$\Rightarrow$ $x = (cy_1)z_1$

$= c(y_1z_1)$, where $y_1z_1 \in D$

$\Rightarrow$ $x \in (c) \Rightarrow (a) \subset (c)$

Similarly, we can show that

$$(b) \subset (c)$$

Therefore $(a) \subset (c)$ and $(b) \subset (c)$ implies that

$\Rightarrow$ $(a)+(b) \subset (c)$

$\Rightarrow$ $(d) \subset (c)$

$\Rightarrow$ $d \in (c)$

$\Rightarrow$ $d = cz$, for some $z \in D$

$\Rightarrow$ $c|d$

Therefore, $d \mid a$ and $d \mid b$ and whenever $c \mid a$, $c \mid b$ then $c \mid d$.

Hence, d is the g.c.d. of a and b.

Theorem 6. *In an integral domain if there exists a lowest common multiple of any two elements, then it is unique apart from the distinction between associates.*

Proof. Let us suppose D be an integral domain and a, b be any two non-zero elements of D such that least common multiple of a and b exists. Let if possible l_1 and l_2 be two l.c.m. of a and b .

If l_1 is the l.c.m. of a and b then we have

$$a \mid l_1 \text{ and } b \mid l_1$$

If l_2 is the l.c.m. of a and b, then we have

$$a|l_2 \text{ and } b|l_2$$

Therefore, we have $l_1 |l_2$. Similarly, we may get $l_2 |l_1$.

Thus, $l_1| l_2$ and $l_2 |l_1$ implies that l_1 and l_2 are associates.

Theorem 7. *In every principal ideal domain, each pair of elements surely has a common multiple.*

Proof. Let D be a principal ideal domain and a, b be any two non-zero elements of D. Further suppose that (a) and (b) denotes the principal ideals of D generated by a and b respectively.

It is known that intersection of two ideals is again an ideal therefore $(a) \cap (b)$ is also an ideal of D. Further it is also known that $(a) \cap (b)$ is a principal ideal. Therefore, there exists an element $l \in D$ such that

$$(a) \cap (b) = (l)$$

We have to show that l is the lowest common multiple.

We have $\qquad\qquad (a) \cap (b) = (l)$

$\Rightarrow \quad (l) \subset (a) \qquad$ and $\qquad (l) \subset (b)$

$\Rightarrow \quad l \in (a) \qquad$ and $\qquad l \in (b)$

$\Rightarrow \quad l = ax_1 \qquad$ and $\qquad l = bx_2$, for some $x_1, x_2 \in D$

$\Rightarrow \quad a \mid l \qquad$ and $\qquad b \mid l$

$\Rightarrow \quad l$ is a common multiple of a and b.

If m is a common multiple of a and b , then $a \mid m$ and $b \mid m$.

Thus, $m = ay_1 \quad$ and $\qquad m = by_2$, for some $y_1, y_2 \in D$.

Now $\quad x \in (m) \qquad \Rightarrow \qquad x = mz_1$, for some $z_1 \in D$

$\qquad\qquad\qquad\quad \Rightarrow \qquad x = (ay_1)\, z_1 \Rightarrow x = a(y_1, z_1),\, y_1, z_1 \in D.$

Thus $\qquad\qquad\qquad\quad (m) \subset (a)$

Similarly, we can prove that

$$(m) \subset (b) .$$

Now $(m) \subset (a) \quad$ and $\qquad (m) \subset (b)$ implies that

$$(m) \subset (a) \cap (b) = (l)$$

$\Rightarrow \quad (m) \subset (l) \qquad \Rightarrow \qquad m \in (l)$

$\Rightarrow \quad m = l\,y$, for some $y \in D \Rightarrow l \mid m$.

Thus $a \mid l$, $b|l$ and if $m \mid a$ and $m \mid b$, then $l \mid m$.

Hence, l is the least common multiple of a and b .

5.29 PRIME AND IRREDUCIBLE ELEMENTS

Definition 1. *Let R be a commutative ring with unity. An element $p \in R$ is called prime element if*

(i) *$p \neq 0$, p is not a unit.*

(ii) *for any $a, b, c \in R$ if $p|ab$ then $p| a$ and $p| b$.*

Definition 2. *Let R be a commutative ring with unity. An element $p \in R$ is called an irreducible element if*

(i) *p ≠ 0, p is not a unit.*

(ii) *whenever p = ab, then one of a and b must be a unit.*

For example, in the ring $(\mathbf{Z}, +, .)$ of integers, every prime number is a prime element as well as irreducible element.

Solved Examples

Example 1. *Consider the ring* $\mathbf{Z}(\sqrt{-5}) = \{a + (\sqrt{-5})b : a, b \in \mathbf{Z}]$ *under the operation defined by*

$$(a + \sqrt{-5}b) + (c + \sqrt{-5}d) = (a + c) + \sqrt{-5}(b + d)$$

$$(a + \sqrt{-5}b).(c + \sqrt{-5}d) = (ac - 5bd) + \sqrt{-5}(ad + bc)$$

Show that $\sqrt{-5}$ *is a prime element. Further, show that 3 is an irreducible element which is not prime.*

Solution.   Clearly we have $\qquad\qquad \sqrt{-5} \neq 0$

Further, it is also a unit because if it is not a unit, then there exists $a + \sqrt{-5}b$ such that

$$\sqrt{-5}(a + \sqrt{-5}b) = 1 \implies \sqrt{-5} = 1 + 5b$$

which is not possible because RHS is an integers whereas it is not an integers.

Now suppose $\sqrt{-5}$ divides $(a + \sqrt{-5}b)(c + \sqrt{-5}d)$ then there exists $(x + \sqrt{-5}y)$ such that

$$\sqrt{-5}(x + \sqrt{-5}y) = (a + \sqrt{-5}b)(c + \sqrt{-5}d)$$

On comparision, we get

$$-5y = ac - 5bd$$

$$\implies \qquad\qquad 5(bd - y) = ac \implies 5 | ac$$

But 5 is a prime number, therefore either $5 | a$ or $5 | c$

If $5 | a$ then $(\sqrt{-5})(\sqrt{-5}) | a$

$$\implies \qquad\qquad \sqrt{-5} | a$$

$$\implies \qquad\qquad \sqrt{-5} | (a + b\sqrt{-5})$$

Similarly, if $5 | c$, then $\qquad \sqrt{-5} | c + \sqrt{-5}d.$

Hence $\sqrt{-5}$ is a prime element.

Further, we shall show that 3 is an irreducible element which is not prime. Let us suppose

$$3 = (a + \sqrt{-5}b)(c + \sqrt{-5}d), a, b, c, d \in \mathbf{Z}$$

$$\implies \qquad \overline{3} = (a - \sqrt{-5}b)(c - \sqrt{-5}d)$$

$$\implies \qquad 3.\overline{3} = (a^2 + 5b^2)(c^2 + 5d^2)$$

$$\implies \qquad 9 = (a^2 + 5b^2)(c^2 + 5d^2)$$

$$\implies \qquad a^2 + 5b^2 = 1, 3 \text{ or } 9$$

Hence, $a^2 + 5b^2 = 3$ is not possible as $a, b \in \mathbf{Z}$.

If $a^2 + 5b^2 = 1$ then $a = \pm 1$ and $b = 0$.

If $a^2 + 5b^2 = 9$ then $a^2 + 5d^2 = 1$, then $c = \pm 1$ and $d = 0$.

Therefore, if $a^2 + 5b^2 = 1$, then $a^2 + \sqrt{-5}b = \pm 1$, which is a unit and if $a^2 + 5b^2 = 9$, then $c + \sqrt{-5}\, d = \pm 1$, which is also a unit.

Thus, 3 is an irreducible element of $\mathbf{Z}(\sqrt{-5}\,)$.

Further, we have

$$(2+\sqrt{-5}\,)(2-\sqrt{-5}\,) = 9$$

and therefore

$$3\,|\,(2+\sqrt{-5}\,)(2-\sqrt{-5}\,)$$

we have to show that it does not divide any one of these.

Suppose $3\,|\,(2+\sqrt{-5}\,)$ in $\mathbf{Z}(\sqrt{-5}\,)$

Then $\qquad\qquad (2+\sqrt{-5}\,) = 3(a+\sqrt{-5}\,b),\ a,\ b \in \mathbf{Z}.$

$\Rightarrow \qquad\qquad\qquad 2-\sqrt{-5} = 3(a-\sqrt{-5}\,b)$

$\Rightarrow \qquad\qquad\qquad\qquad 9 = 9(a^2+5b^2)$

$\Rightarrow \qquad\qquad\qquad\qquad 1 = a^2+5b^2,\ i.e.,\ a = \pm 1,\ b = 0$

$\Rightarrow \qquad\qquad\qquad 2+\sqrt{-5} = \pm 3$, which is not possible.

Therefore $3 \nmid (2+\sqrt{-5})$. Similarly, we can prove that $3 \nmid (2-\sqrt{-5})$. ·

Hence, 3 is a not a prime element of $\mathbf{Z}(\sqrt{-5}\,)$.

Example 2. *Find all the units of* $\mathbf{Z}(\sqrt{-5}\,)$.

Solution : Let us suppose $a+\sqrt{-5}\ b$ is unit in $\mathbf{Z}(\sqrt{-5}\,)$

Then $\qquad\qquad (a+\sqrt{-5}\,b)\ (c+\sqrt{-5}\,b) = 1+\sqrt{-5}\,.0$, for some $c,\ d \in \mathbf{Z}$.

Therefore $\qquad (a-\sqrt{-5}\,b)\ (c+\sqrt{-5}\,b) = I = 1$

$\Rightarrow \qquad\qquad (a^2+5b^2)\ (c^2+5b^2) = 1$ in $\mathbf{Z}$

$\Rightarrow \qquad\qquad\qquad (a^2+5b^2) = 1 \ \Rightarrow a = \pm 1,\ b = 0$

So $\qquad\qquad\qquad a+\sqrt{-5}\,b = \pm 1$ are the units.

Example 3. *In the domain $J(i)$ of Gaussian integers, show that $(1+i)$ is prime.*

Solution. Let if possible $(1+ i)$ is not prime. Then we can write

$$1+i = (a+ib)(c+id),\ \text{for some integers } a,\ b,\ c,\ d.$$

Taking conjugate of both the sides, we get

$$1-i = (a-ib)(c-id)$$

$\Rightarrow \qquad\qquad\qquad |1-i|^2 = |(a-ib)(c-id)|^2$

$\Rightarrow \qquad\qquad\qquad 2 = (a^2+b^2)\ (c^2+d^2)$

$\Rightarrow$ either $\qquad\qquad a^2+b^2 = 1$ or $(a^2+b^2) = 2$

If $(a^2+b^2) = 2$, then $\qquad c^2+d^2=1$

Therefore $\qquad c^2+d^2 \Rightarrow (c+id)(c-id) = 1$

$\Rightarrow \quad c+id$ is a unit.

Similarly if $(a^2+b^2) = 1$, then $(a + ib)(a-ib)=1$

$\Rightarrow \quad (a+ib)$ is a unit.

Therefore, if $1+i = (a + ib)(c + id)$ then either $(a + ib)$ is a unit or $(c+ ib)$ is a unit, which is not possible. Hence, $(1+i)$ is a prime in $J(i)$.

Theorem 1. *In a Principal Ideal Domain (PID), an element is prime if and only if it is irreducible.*

Proof. Let D be a principal domain and $p \in D$ be a prime element. We have to prove that if $p=ab$. Then a or b is a unit.

Let $p = ab$ then $p\,|\,ab$

$\Rightarrow \quad p|a$ or $p|b$, where p is prime.

If $p|a$ then $a = px$, for some x. Then $p = ab = (px)b$

$\Rightarrow \qquad\qquad p(1-xb) = 0$

Since $p \neq 0$ therefore $1 - xb = 0$

$\Rightarrow \qquad\qquad xb = 1$

$\Rightarrow \quad b$ is a unit.

Similarly, we can prove that if $p|b$ then a will be a unit.

Conversely, suppose that p is irreducible element and $p|ab$. We have to show that $p|a$ or $p|b$.

Let if possible $p \nmid a$. Since p and a are the elements of PID, they have a g.c.d., say d. To show d is a unit.

Now $d|p$ and $d|a$ [By definition of g.c.d.]

$\Rightarrow \quad \exists\, u$ and v such that $p = du,\ a = dv$

If d is not a unit, then as p is irreducible and $p = du$, u will be a unit.

$\Rightarrow \quad u^{-1}$ exists

$\Rightarrow \qquad\qquad\qquad pu^{-1} = d$

Therefore, $\qquad\qquad a = pu^{-1}v \Rightarrow \quad p|a$, which is a contradiction.

Therefore, d is a unit.

It is also known that the g.c.d., d can be expressed as

$$d = \lambda a + \mu p$$

which gives

$$dd^{-1} = d^{-1}\lambda\, a + d^{-1}\,\mu p$$
$$b.1 = \lambda\, d^{-1}\, ab + \mu d^{-1}\, bp$$

But $p|ab,\ p|\mu d^{-1}\, bp$, therefore

$$p\,|\,(ab\lambda d^{-1} + \mu d^{-1}\, bp)$$
$$\Rightarrow \qquad\qquad p|b$$

REMARK

☞ In an integral domain with unity, every prime element is irreducible. Converse is not true.

Theorem 2.   *Let D be a PID which is not a field then an ideal $A = (a_0)$ is a maximal ideal if and only if a_0 is an irreducible element.*

Proof.   Let $A = (a_0)$ be a maximal ideal.

   (i) We have to prove that $a_0 \neq 0$.

     Let if possible $a_0 = 0$ then since D is not a field, there exists at least one $b(\neq 0)$ such that b^{-1} does not exist. Let $B = (b)$ and as $a_0 = 0$.

$$A = (0) \text{ and}$$
$$(0) \subseteq B \subseteq D \Rightarrow A \subseteq B \subseteq D$$

     Now, $B \neq A$ as $b \in B$, $b \neq 0$ and $A = (0)$

$$B \neq R \text{ as } 1 \in D, \text{ but } 1 \notin B$$

     because if $1 \in B = (b)$ then exist some x such that $1 = bx$, which show that b is invertible, which is not so.

     Hence $a_0 \neq 0$.

(ii) Now, we have to show that a_0 is not a unit.

Let if possible a_0 is a unit, then $a_0 a_0^{-1} = 1$

$$a_0 \in A, a_0^{-1} \in D \quad \Rightarrow \quad a_0 a_0^{-1} \in A$$

$\Rightarrow \qquad\qquad 1 \in A$

$\Rightarrow \qquad\qquad A = D$

which is not possible because A is maximal ideal.

Hence, a_0 is not a unit.

(iii) Let $a_0 = bc$ for some $b, c \in D$. We have to show that b or c is a unit.

Let $B = (b)$.

Since $a_0 = bc$, $a_0 \in B$

$\Rightarrow \qquad$ all multiplies of a_0 are in B.

$\Rightarrow \qquad A \subseteq B.$

But since A is maximal, therefore either $B = D$ or $B = A$

If $B = D$, then $\qquad 1 \in B = (b)$ as $1 \in D$

$\Rightarrow \qquad\qquad 1 = x\,b$ for some x

$\Rightarrow \qquad b$ is a unit. If $\;\; B = A$, then $B \subseteq A = (a_0)$

$\Rightarrow \qquad\qquad b = ya_0,$ for some y

$\Rightarrow \qquad\qquad a_0 = bc = ya_0 c$

$\Rightarrow \qquad\qquad a_0 - ya_0 c = 0$

$\Rightarrow \qquad\qquad a_0 (1 - yc) = 0$

$\Rightarrow \qquad\qquad 1 - yc = 0 \qquad\qquad [\because a_0 \neq 0]$

$\Rightarrow \qquad\qquad yc = 1$

$\Rightarrow \qquad c$ is a unit.

Conversely, let a_0 be an irreducible element. We have to show that $A(a_0)$ is maximal.

Let I be any other ideal such that $A \subseteq I \subseteq D$.

Now, since D is a PID, therefore I is generated by some element say x. So $x \notin A$ as if $x \in A$, then $(x) \subseteq A$, i.e., $\qquad I \subseteq A$ but $A \subseteq I$.

$\Rightarrow \qquad\qquad A = I$, which is not possible. Thus $x \notin A$.

Further $\qquad\qquad A = (a_0) \subseteq I.$

$\Rightarrow \qquad\qquad a_0 = xy,$ for some y

a_0 is irreducible $\qquad \Rightarrow x$ or y is a unit.

If y is a unit, then $yy^{-1} = 1$ and $a_0 = xy$

$\Rightarrow \qquad\qquad a_0 y^{-1} = x$

But $\qquad\qquad a_0 \in A, y^{-1} \in D \Rightarrow a_0 y^{-1} \in A$

$\Rightarrow \qquad x \in A$, which is not true

Therefore, y is not a unit. Thus x is a unit and $xx^{-1} = 1$.

Now $x \in I, x^{-1} \in D$, I is an ideal, therefore

$$x\, x^{-1} \in I \Rightarrow 1 \in I \Rightarrow I = D$$

Hence, A is maximal ideal of D.

5.30 METHOD OF FINDING THE G.C.D. OF ANY TWO MEMBERS OF F(X) WHERE F(X) IS AN EUCLIDEAN DOMAIN

Let f_1, f_2 be any two members of $F(x)$.

Step 1. Divide f_1 by f_2 to get $f_1 = f_2 g_1 + f_3$, $\deg f_3 < \deg f_2$

Step 2. Divide f_2 by f_3 to get $f_2 = f_3 g_2 + f_4$, $\deg f_4 < \deg f_3$

Continuing this process, we get

$$f_{n-1} = f_n g_{n-1} + 0$$

Step 3. Then we claim $\qquad f_n = \text{g.c.d. } (f_1, f_2)$

Consider the ideal (f_1, f_2) generated by f_1 and f_2

$$(f_1, f_2) = \{ g f_1 + h f_2 : g, h \in F(x) \}$$

Let $g f_1 + h f_2$ be any member of this ideal, then

$$g f_1 + h f_2 = g(f_2 g_1 + f_3) + h f_2$$
$$= f_2(g g_1 + h) + g f_3 \in (f_2, f_3)$$
$$\Rightarrow \qquad (f_1, f_2) \subseteq (f_2, f_3)$$

Similarly, we can show that

$$(f_2, f_3) \subseteq (f_1, f_2)$$
$$\Rightarrow \qquad (f_1, f_2) = (f_2, f_3)$$

Continuing this process, finally we get

$$(f_1, f_2) = (f_2, f_3) = \ldots = (f_n, 0) = (f_n)$$

Then we get the g.c.d. $(f_1, f_2) = (f_n)$.

Solved Examples

Example 1. *Find g.c.d. of*

    (i) 9, 15; 7, 10 *in* **Z**   *(ii)* $11 + 7i$, $18 - i$ *in* **Z** *(i)*

    (iii) $x^4 + x^3 + 2x^2 + x + 1$, $x^3 - 1$; $x^2 + 1$, $x^6 + x^3 + x + 1$ *in* **Q**(x)

Solution.   (i) We can write

$$15 = 9 \times 1 + 6 \qquad\qquad 10 = 7 \times 1 + 3$$
$$9 = 6 \times 1 + 3 \qquad\qquad 7 = 3 \times 2 + 1$$
$$6 = 3 \times 2 + 0 \qquad\qquad 3 = 1 \times 3 + 0$$

i.e.,   g.c.d. (9, 15) = 3   g.c.d. (7, 10) = 1

    (ii) Dividing $18 - i$ by $11 + 7i$, we get

$$\frac{18 - i}{11 + 7i} = \frac{(18 - i)(11 - 7i)}{(11 + 7i)(11 - 7i)} = \frac{191}{170} - \frac{137}{170} i$$

$$= \left(1 + \frac{21}{170}\right) - \left(1 - \frac{33}{170}\right) i = (1 - i) + \left(\frac{21}{170} + \frac{33i}{170}\right)$$

Therefore, $18 - i = (11 + 7i)(1 - i) + 3i$ ... (1)

Further, dividing $11+7i$ by $3i$, we get

$$\frac{11+7i}{3i} = \frac{(11+7i)(-3i)}{3i(-3i)} = \frac{21}{9} - \frac{33}{9}i$$

$$= \frac{7}{3} - \frac{11}{3}i = (2-3i) + \left(\frac{1}{3} - \frac{2i}{3}\right)$$

$\Rightarrow \qquad\qquad 11+7i = 3i(2-3i) + (2+i)$ ... (2)

Also $\qquad\qquad \dfrac{3i}{2+i} = \dfrac{3i(2-i)}{(2+i)(2-i)} = i + \left(\dfrac{3}{5} + \dfrac{1}{5}i\right)$

$\Rightarrow \qquad\qquad 3i = i(2+i) + \left(\dfrac{3}{5} + \dfrac{1}{5}i\right)(2+i)$ ... (3)

Now, dividing $(2+i)$ by $1+i$, we get

$$\frac{2+i}{1+i} = \frac{2+i}{1+i}\frac{1-i}{1-i} = \frac{3}{2} - \frac{1}{2}i = 1 + \left(\frac{1}{2} - \frac{1}{2}i\right)$$

$\because \qquad\qquad 2+i = (1+i)1 + 1$ ... (4)

Again dividing $1+i$, by 1 we have $(1+i) = 1(1+i) + 0$

Hence, required g.c.d. $(11+7i, 18 - i) = 1$

(iii) We have

$$
\begin{array}{r}
x+1 \\
x^3 - 1 \overline{\smash{\big)}\, x^4 + x^3 + 2x^2 + x + 1} \\
\underline{x^4 - x } \\
-+ \\
\overline{ x^3 + 2x^2 + 2x + 1} \\
\underline{x^3 - 1} \\
-+ \\
\overline{ 2x^2 + 2x + 2}
\end{array}
$$

Thus, we can write

$$x^4 + x^3 + 2x^2 + x + 1 = (x^3 - 1)(x+1) + (2x^2 + 2x + 2)$$

Similarly, $\qquad x^3 - 1 = (2x^2 + 2x + 2)\left(\dfrac{x}{2} - \dfrac{1}{2}\right) + 0$

Therefore, required g.c.d. is given by $2x^2 + 2x + 2$.

Now, we can write

$x^6 + x^3 + x + 1 = (x^2 + 1)x^2 + (x+1)$ [Dividing $x^6 + x^3 + x + 1$ by $x^2 + 1$]

$x^2 + 1 = (x+1)(x-1) + 2$ [Dividing $x^2 + 1$ by $x+1$]

$x+1 = 2\left(\dfrac{1}{2}x\right) + 1$ [Dividing $x+1$ by 2]

$2 = 1 \times 2 + 0$ [Dividing 2 by 1]

Hence, required g.c.d is 1.

5.31 EUCLIDEAN RING (OR EUCLIDEAN DOMAIN)

Definition. *Let R be an integral domain. Then this integral domain R is said to be a Euclidean Ring if for every non-zero element $a \in R$ there exists a non-negative integer $d(a)$ such that*

(i) for all non-zero elements

$$a, b \in R, d(a) \le d(ab) \text{ (or } d(b) \le d(ab)).$$

(ii) for each $a, b \in R$ with $b \ne 0$, there exist $q, r \in R$ such that $a = bq + r$ where either $r = 0$ or $d(r) < d(b)$.

REMARKS

☞ The second property of above definition is known as division algorithm.

☞ d is called the Euclidean valuation or Euclidean norm function.

☞ We do not assign a value to $d(0)$.

Solved Examples

Example 1. *Show that the ring $\mathbf{Z}$ of all integers is an Euclidean ring.*

Solution. Defing a mapping d from a set of all integers $\mathbf{Z}$ into the set $\mathbf{N}$ of all non-negative integers as d: $\mathbf{Z} \to \mathbf{N}$ given by $d(a) = |a|$; $a \in \mathbf{Z}$ with $a \ne 0$.

Since $a \ne 0$ is any element of $\mathbf{Z}$, then $|a|$ is also a non-negative integer, thus the mapping d assigns every non-zero integer to a non-negative integer of $\mathbf{Z}$.

(i) If a and b any two non-zero elements of $\mathbf{Z}$, then

$$d(a) = |ab| = |a||b| \ge |a| \qquad [\because |b| \ge a \text{ for all non-zero } b \in \mathbf{Z}]$$

$$\Rightarrow \quad d(ab) \ge d(a)$$

$$\therefore \quad d(a) \le d(ab); a, b \in \mathbf{Z}$$

(ii) If a and b are any two non-zero elements of $\mathbf{Z}$, then by division algorithm there exist $q, r \in \mathbf{Z}$ such that

$$a = bq + r \text{ where, } 0 \le r < |b|.$$

So that, $\quad 0 \le r < |b| \Rightarrow$ either $r = 0$ or $0 < r < |b|$

$$\Rightarrow \text{ either } r = 0 \text{ or } |r| < |b| (\because 0 < r \Rightarrow |r| = r)$$

$$\Rightarrow \text{ either } r = 0 \text{ or } d|r| < d(b).$$

Hence, the mapping d is an Euclidean valuation on $\mathbf{Z}$ and accordingly $\mathbf{Z}$ is a Euclidean ring.

Example 2. *Show that the ring of Gaussian integers is an Euclidean ring.*

Solution. Let $J\{i\}$ be a Gaussian integers. Then we have,

$$J[i] = \{a + ib : a, b \in \mathbf{Z}\}$$

Define a mapping $d : J[i] \to \mathbf{N}$ given by $d(a + ib) = a^2 + b^2$, for all non-zero elements of $J[i]$.

If $a + ib$ is a non-zero element of $J[i]$, then $a^2 + b^2$ is a non-negative integers. Thus the mapping 'd' assigns every non-zero element of $J[i]$ to a non-negative integer.

(i) If $(a_1 + ib_1)$ and $(a_2 + ib_2)$ be any two non-zero elements of $J[i]$. Then

$$d[(a_1 + ib_1)(a_2 + ib_2)] = d[(a_1 a_2 - b_1 b_2) + i(a_1 b_2 + a_2 b_1)]$$

$$= (a_1 a_2 - b_1 b_2)^2 + (a_1 b_2 + a_2 b_1)^2$$

$$= a_1^2 a_2^2 + b_1^2 b_2^2 + a_1^2 b_2^2 + a_2^2 b_1^2$$

$$= (a_1^2 + b_1^2)(a_2^2 + b_2^2)$$

since $\quad (a_1^2 + b_1^2)(a_2^2 + b_2^2) \geq (a_1^2 + b_1^2) \qquad\qquad [\because (a_2^2 + b_2^2) \geq 1]$

so that $d[a_1 + ib_1)(a_2 + ib_2)] \geq (a_1^2 + b_1^2)$

or $\qquad d[(a_1 + ib_1)(a_2+ib_2)] \geq d(a_1+b_1)$

$\therefore \qquad\qquad d(a_1+ib_1) \leq d[(a_1+ib_1)(a_2+ib_2)]$

(ii) If (a_1+ib_1) and (a_2+ib_2) are any two elements of $J[i]$. Then

$$\frac{a_1 + ib_1}{a_2 + ib_2} = \frac{(a_1 + ib_1)(a_2 - ib_2)}{(a_2 + ib_2)(a_2 - ib_2)}$$

$$= \frac{(a_1 a_2 + b_1 b_1) + i(a_2 b_1 - a_1 b_2)}{(a_2^2 + b_2^2)}$$

$$= \frac{(a_1 a_2 + b_1 b_1)}{(a_2^2 + b_2^2)} + i\frac{(a_2 b_1 - a_1 b_2)}{(a_2^2 + b_2^2)} = \alpha + i\beta \text{(say)}$$

where, $\qquad \alpha = \dfrac{(a_1 a_2 + b_1 b_1)}{(a_2^2 + b_2^2)}$ and $\beta = \dfrac{(a_2 b_1 + b_1 b_2)}{(a_2^2 + b_2^2)}$

Here $\alpha + i\beta$ is not necessarily a Gaussian integer.

But $a_2 + ib_2$ is a non-zero Gaussian integer, therefore division by $a_2 + ib_2$ is possible. Now choose the integers α' and β' such that

$$|\alpha - \alpha'| \leq \frac{1}{2} \text{ and } |\beta - \beta'| \leq \frac{1}{2}$$

Since α' and β' are integers, therefore $\alpha' + i\beta'$ is a Gaussian integer. Consider

$$(a_1 + ib_1) - (\alpha' + i\beta')(a_2 + ib_2) = \left[\frac{a_1 + ib_1}{a_2 + ib_2} - (\alpha' + i\beta')\right](a_2 + ib_2)$$

$$= [(\alpha+i\beta) - (\alpha'+i\beta')](a_2+ib_2) \qquad\qquad \dots (1)$$

Further since $(\alpha+i\beta)$, $(\alpha'+i\beta')$ and (a_2+ib_2) each being a Gaussian integer, therefore, $(a_2+ib_2)-(\alpha'+i\beta')(a_2+ib_2)$ is a Gaussian integer and so $[(\alpha+i\beta) - (\alpha'+i\beta')](a_2+ib_2)]$

Now equation (1) can be written as

$$(a_1+ib_1) = (\alpha' + i\beta')(a_2+ib_2) + [(\alpha+i\beta)-(\alpha'+i\beta')](a_2+ib_2)$$

where, $(\alpha' + i\beta')$ and $[(\alpha+i\beta)-(\alpha'+i\beta')][(a_2+ib_2)$ are Gaussian integers.

Furthermore either $[(\alpha+i\beta)-(\alpha'+i\beta')](a_2+ib_2)=0$

or $d\{[(\alpha+i\beta)-(\alpha'+i\beta')](a_2+ib_2)\}+i\{b_2(\alpha-\alpha')+i\beta(\beta-\beta')](a_2+ib_2)\}$

$$= d[\{(\alpha-\alpha')a_2 - b_2(\beta-\beta')\} + i\{\alpha-\alpha')+i(\beta-\beta')(a_2+ib_2)\}]$$

$$= [a_2(\alpha-\alpha') - b_2(\beta-\beta')]^2 + [b_2(\alpha-\alpha') + a_2(\beta-\beta')]^2$$

[by definition of d- mapping]

$$= (a_2^2 + b_2^2)[(\alpha - \alpha')^2 + (\beta - \beta')^2]$$

$$= (a_2^2 + b_2^2)\left[\left(\frac{1}{2}\right)^2 + \left(\frac{1}{2}\right)^2\right] \quad \left[\because |\alpha - \alpha'| \le \frac{1}{2}, |\beta - \beta'| \le \frac{1}{2}\right]$$

$$= \frac{1}{2}(a_2^2 + b_2^2) < (a_2^2 + b_2^2) = d(a_2 + ib_2)$$

Thus the mapping d is an Euclidean evaluation. Hence the ring of Gaussian integers is an Euclidean ring.

Example 3. *Show that every field is an Euclidean ring.*

Solution. Let F be an arbitrary field. Then we shall show that F is an Euclidean ring.
Define a mapping.

$$d : F \to \mathbf{N}$$

given by $d(a) = 0$, $\forall$ non-zero $a \in F$,

(i) If a and b are any two non-zero elements of F, then ab is also a non-zero element of F, therefore

$$d(a) = 0 \text{ and } d(ab) = 0$$

so $\qquad d(a) \le d(ab)$

(ii) If a and b are any two elements of F with $b \ne 0$, then we can write

$$a = a(b^{-1} b) \qquad\qquad\qquad [\because a = a.1]$$

or $\qquad a = a(b^{-1} b) + 0$ or $a = (ab^{-1}) b + r$ where $r = 0$

or $\qquad a = qb + r$, where $r = 0$ and $q = ab^{-1}$

Hence, F is an Euclidean ring.

Example 4. *Show that the ring of polynomials over a field of reals is a Euclidean ring.*

Solution. Let $F(x)$ be the ring of polynomials over a field F. Then we shall show that $F(x)$ is an Euclidan ring.
Define a mapping

$$d : F[x] \to \mathbf{N}$$

by $\qquad d[f(x)] = \deg. f(x)$, $\forall$ all non-zero $f(x) \in F(x)$

(i) Let $f(x)$ and $g(x)$ be any two non-zero polynomials of $F[x]$. Then we have

$$d[f(x) g(x)] = \deg. [f(x)g(x)] = \deg. f(x) + \deg. g(x)$$

Now, $\deg. f(x) + \deg. g(x) \ge \deg. f(x) \qquad [\because \deg g(x) \ge 0]$

$$d[f(x)g(x)] \ge \deg. f(x) = d[(f(x)]$$

(ii) Let $f(x)$ and $g(x)$ be any two non-zero polynomials of $F(x)$ with $g(x) \ne 0$, then there exist $q(x)$ and $r(x)$ such that

$$f(x) = q(x)g(x) + r(x) \qquad \text{[By division algorithm]}$$

where, either $\qquad r(x) = 0$ or $\deg r(x) < \deg g(x)$

or either $\qquad r(x) = 0$ or $\deg r(x) < d[g(x)]$

Thus the mapping d is an Euclidean evaluation.

Therefore, $F[x]$ is an Euclidean ring.

5.31.1 Properties of Euclidean Rings

Theorem 1. *Every Euclidean ring is a principal ideal ring.*

Proof.   Let R be an Euclidean ring. Then we have to show that every ideal of R is a principal ideal.

Let S be an arbitrary ideal of R. If S just consists of the zero element, *i.e.*, S is a null ideal or $S = (0)$ so it is generated by 0, thus S is a principal ideal.

Now we may assume that $S \neq 0$, therefore there exists a non-zero element in S. Let b be a non-zero element in S such that $d(b)$ is a minimal, this implies that there is no element c is S such that $d(c) < d(b)$. We shall show that S is generated by b, *i.e.*, $S = (b)$.

Let a be any element of S and S be an Euclidean ring, therefore, there exist q and r in R such that

$$a = bq + r$$

where, either $r = 0$ or $d(r) < d(b)$

Since S is an ideal so that,

$$q \in R \text{ and } b \in S \quad \Rightarrow \quad bq \in S.$$

Also $q \in R$ and $bq \in S \Rightarrow a - bq \in S \Rightarrow r \in S.$

But we have either $r = 0$ or $d(r) < d(b)$ therefore $d(r) < d(b)$ contradicts that $d(b)$ is generated by b, *i.e.*, $S = (b)$, so that S is a principal ideal and since S is an arbitrary ideal of R. Hence R is a principal ideal ring.

$$\therefore \qquad\qquad a = bq$$

This shows that every elements of S can be expressed as the multiple of b. Thus S is generated by b, *i.e.*, $S = (b)$, so that S is a principal ideal and since S is an arbitrary ideal of R. Hence, R is a principal ideal ring.

Theorem 2. *An Euclidean ring possesses a unit element.*

Proof.   Let R be an Euclidean ring so it is a principal ideal ring. Since R itself is an ideal of R so it is principal ideal. Let $R = (u_0)$ for $u_0 \in R$. Thus every element in R is a multiple of u_0. In particular, $u_0 = u_0c$, for some $c \in R$.

Let a be any element of R, then

$$a = xu_0, \text{ for some } x \in R.$$

Now, $ac = (xu_0)c = x(u_0c)$ [$\because R$ is associative.]

Also R is a commutative ring.

$$\therefore \qquad\qquad ac = a = ca, \forall a \in R. \text{ Hence } c \text{ is unit element of } R.$$

Theorem 3. *Let R be an Euclidean ring. Then any two non-zero elements a and b in R have a greatest common divisor d. Moreover, $d = \lambda a + \mu b$ for some $\lambda, \mu \in R$.*

Proof.   Consider the set

$$S = \{ra + sb \; ; \; r, s \in R\}.$$

We claim that S is an ideal of R.

Let x and y be any two elements of S, then

$$x = r_1a + s_1b \text{ and } y = r_2a + s_2b, \text{ for some } r_1, r_2, s_1, s_2 \in R.$$

Now, $x - y = (r_1a + s_1b) - (r_2a + s_2b)$

$$= (r_1 - r_2)\, a + (s_1 - s_2)b \in S \qquad [\because r_1 - r_2 \in S, s_1 - s_2 \in S]$$

Thus S is an additive subgroup of R.

Also,   for any $u \in R$,

$$ux = u(r_1 a + s_1 b) = (ur_1)a + (us_1)\, b \in S \qquad [\because ur_1, us_1 \in R]$$

R is commutative ring so that $ux = xu \in S$.

Hence, S is an ideal of R and R is a Euclidean ring, so S is a principal ideal. Therefore, there exists an element $d \in S$ such that every element of S is a multiple of d.

Since $d \in S$ so that

$$d = \lambda a + \mu b \text{ for some } \lambda, \mu \in R.$$

We know that every Euclidean ring possesses a unit element 1.

$\therefore$ \qquad\qquad\qquad $a = 1.\, a + 0.\, b \in S$

and \qquad\qquad\qquad $b = 0.\, a + 1.\, b \in S.$

$\therefore$ \qquad\qquad\qquad $ab \in S$ and both are multiple of d.

$\Rightarrow$ \quad $d\,|\,a$ and $d\,|\,b$.

Now suppose that $c\,|\,a$ and $c\,|\,b$, then $c\,|\,\lambda a$ and $c\,|\,\mu b$ for some $\lambda, \mu \in R$.

$$c\,|\,\lambda a + b\mu \Rightarrow c\,|\,d \qquad\qquad [\because d = \lambda a + \mu b]$$

Hence, d is a greatest common divisor of a and b.

Theorem 4. *Let R be an Euclidean ring and let a, b, c be three non-zero elements of R such that a and b are relatively prime and $a\,|\,bc$, then $a\,|\,c$.*

Proof. Since a and b are relatively prime, therefore their greatest common divisor is 1. Then by above theorem, we have

$$1 = \lambda a + \mu b, \text{ for some } \lambda, \mu \in R$$

$\Rightarrow$ \qquad\qquad $c = (\lambda a + \mu b)\, c = (\lambda a)c + (\mu b)c$ \qquad [By right distributive law]

$\therefore$ \qquad\qquad\qquad $c = \lambda(ac) + \mu(bc)$ \hfill ...(1)

But \qquad\qquad $a\,|\,bc \Rightarrow a\,|\,\mu\,(bc)$ also $a\,|\,\lambda\,(ac)$.

$\therefore$ \qquad\qquad $a\,|\,\{\lambda\,(ac) + \mu(bc)\} \Rightarrow a\,|\,c$ \qquad [From equation (1)]

Theorem 5. *Let p be a prime element of an Euclidean ring R and let a, b be non-zero elements of R such that $p\,|\,ab$ then either $p\,|\,a$ or $p\,|\,b$.*

Proof. Suppose that p does not divide a, so that p and a are relatively prime, and $p\,|\,ab$, then by above theorem, we have $p\,|\,b$.

Now suppose that p does not divide b so they are relatively prime, p being a prime and $p\,|\,ab$, then $p\,|\,a$.

REMARK

☞ If p is a prime element in the Euclidean ring R and $p\,|\,a_1.\, a_2.\ \dots\ a_n$, then p divides at least one of $a_1, a_2, \dots, a_n$.

Theorem 6. *Let R be an Euclidean ring and let d be the Euclidean evaluation of R. Let a and b two non-zero elements in R, then*

  (i) *if b is a unit in R,* \quad $d(ab) = d(a)$

  (ii) *if b is not a unit in R, $d(ab) > d(a)$*

Proof. Since R is a Euclidean ring so by definition of Euclidean ring we must have

$$d(a) \leq d(ab) \hfill ...(1)$$

for any two non-zero elements a and b of R.

(i) If b is a unit in R, then by definition it has multiplicative inverse. Therefore b^{-1} exists in R, then

$$d(a) = d[a(bb^{-1})] \qquad [\because bb^{-1} = 1]$$
$$= d[(ab)b^{-1}] \qquad \text{[By associative law]}$$

Since $\qquad d[(ab)b^{-1}] \geq d(ab)$

$$d(a) \geq d(ab) \qquad \qquad ...(2)$$

From equation (1) and (2), we have

$$d(ab) = d(a).$$

(ii) If b is not a unit in R, then we have

$$0 \neq a \in R, 0 \neq b \in R \Rightarrow 0 \neq ab \in R$$

Since R is a Euclidean ring so that for two non-zero elements a and ab in R there exist q and r in R such that

$$a = q(ab) + r$$

where, either $r = 0$ or $d(r) < d(ab)$.

Now, if $r = 0$, then

$$a = q(ab) \Rightarrow a - q(ab) = 0$$
$$\Rightarrow \qquad a\,(1 - qb) = 0$$
$$\Rightarrow \qquad 1 - qb = 0 \quad [a \neq 0, 1 - qb \in r \text{ and } R \text{ is without zero divisors.}]$$
$$\Rightarrow \qquad qb = 1$$
$$\Rightarrow \qquad b \text{ has a multiplicative inverse in } R.$$
$$\Rightarrow \qquad b \text{ is a unit in } R.$$

But, by given hypothesis b is not a unit in R so we must have $r \neq 0$.

Consequently, we must have

$$d(r) < d(ab) \Rightarrow d[a - q(ab)] < d(ab)$$
$$\Rightarrow \qquad d[a\,(1 - qb)] < d(ab)$$
$$\Rightarrow \qquad d(a) \leq d[a(1 - qb)] < d(ab) \Rightarrow d(a) < d(ab)$$

Theorem 7. *The necessary and sufficient condition that non-zero element a in the Euclidean ring R is a unit is that $d(a) = d(1)$.*

Proof. Suppose that a non-zero element a in a Euclidean ring R is a unit.

They by the definition of Euclidean ring, we have

$$d(a) = d(1. a) \geq d(1) \quad \text{or} \quad d(a) \geq d(1) \qquad ...(1)$$

Since a is a unit in R so that it has a multiplicative inverse, *i.e.*, a^{-1} exists in R, we have

$$1 = aa^{-1} \Rightarrow d(1) = d(aa^{-1}) \geq d(a)$$
$$\Rightarrow \qquad d(1) \geq d(a) \qquad \qquad ...(2)$$

From equation (1) and (2), we get

$$d(a) = d(1)$$

Conversely, suppose that $d(a) = d(1)$, then we shall show that a is a unit in R.

Assume that a is not a unit in R, then by previous theorem, we have

$$d(1. a) > d(1) \Rightarrow d(a) > d(1)$$

which gives a contradiction. Hence, a is a unit in R.

Theorem 8. *Let R be a Euclidean ring. Then every non-zero element in R is either a unit in R or it is expressible as the product of a finite number of prime elements of R.*

Proof. Let a be a non-zero element in R. If a is a unit in R, then theorem follows. So let a be non-unit element in R, then we have to show that a can be expressed as the product of a finite number of prime element of R. We shall prove the result by induction on $d(a)$.

Since a is not a unit in R, then by induction we have $d(x) < d(a)$ for all non-zero element $x \in R$. That is, the theorem is true for all $x \in R$ such that $d(x) < d(a)$.

Then we shall show theorem is true for 'a' also.

If a is prime element in R, then there is nothing to prove. Further, suppose that a is not a prime, then we have $a = bc$, where, neither b nor c is a unit in R.

Again b and c both are not units in R, therefore

$$d(b) < d(bc) \text{ and } d(c) < d(bc)$$

But $$d(a) = d(bc)$$

∴ $$d(b) < d(a) \text{ and } d(c) < d(a)$$

Thus by induction hypothesis both b and c can be written as a product of a finite number of prime elements in R.

Let $$b = p_1 \cdot p_2 \cdot \text{ ... } \cdot p_m \text{ and } c = q_1 \cdot q_2 \cdot \text{ ... } \cdot q_n$$

where, each p_i and q_j are prime elements in R. Then

$$a = bc = (p_1 \cdot p_2 \cdot \text{ ... } \cdot p_m)(q_1 \cdot q_2 \cdot \text{ ... } \cdot q_n)$$

$$= p_1 \cdot p_2 \cdot \text{ ... } \cdot p_m q_1 \cdot q_2 \cdot \text{ ... } \cdot q_n.$$

Thus, in this way a is expressible as the product of a finite number of prime element in R.

Theorem 9 (Unique Factorization Theorem). *Let R be a Euclidean ring and $a \neq 0$ a non-unit in R. Suppose $a = p_1 \cdot p_2 \cdot \text{ ... } \cdot p_m = q_1 \cdot q_2 \cdot \text{ ... } \cdot q_m$, where, each p_i and q_i are prime elements of R. Then $m = n$ and each p_i $(1 \leq i \leq m)$ is an associate of some q_i $(1 \leq i \leq n)$ and each q_i is an associative of some p_i.*

Proof. Since we have

$$a = p_1 \cdot p_2 \cdot \text{ ... } \cdot p_m = q_1 \cdot q_2 \cdot \text{ ... } \cdot q_n \qquad \text{...(1)}$$

Also $p_1 | p_1 p_2 \cdots p_m \Rightarrow p_1 | q_1 q_2 \cdots q_n$ [From (1)]

$\Rightarrow \quad p_1$ must divide at least one of $q_1, q_2 \cdots q_n$.   [∵ R is commutative.]

Now by left cancellation law, we have

$$p_1 \cdot p_2 \cdot \text{ ... } \cdot p_m = u_1 q_2 q_3 \cdots, q_n \qquad \text{...(2)}$$

Repeat the above argument on (2) with p_2, p_3 and so on.

In case $m < n$, after m steps, the left side becomes 1 and the right side reduces to a product of some units in R and certain numbers of $q's$, but $q'_j s$ are not units in R, so that the product of some units are some $q'_j s$ cannot be equal to 1, which shows that $m < n$. Therefore, we obtain

$$m \geq n \qquad \text{...(3)}$$

Now interchanging the roles of $p'_i s$, and $q'_j s$, we get

$$n \geq m \qquad \text{...(4)}$$

From (3) and (4), we obtain $m = n$.

In above process, we have also showed that every p_i has some q_j as an associate and conversely.

REMARKS

☞ Combining theorem 8 and 9, we state that non-zero elements in a Euclidean ring R can be uniquely expressed (upto associated) as a product of prime elements or is a unit in R.

☞ From this result, we can say that every Euclidean ring is a unique factorization domain.

Theorem 10. *An ideal S of the Euclidean ring R is maximal if and only if S is generated by some prime element of R.*

Proof. Let $S = (a)$ be an ideal generated by an element a of an Euclidean ring.

Since every Euclidean ring is a principal ideal ring, therefore S is a principal ideal. Suppose S is maximal ideal, then we shall show that a is prime element in R.

Now assuming that a is not prime, so it is a composite number, therefore

$$a = bc \qquad \qquad ...(1)$$

where b and c are non-zero units in R.

From (1), we have $\qquad b \mid a$.

For some $x \in R$, $ax \in (a)$, then

$$ax = (bc)x = b(x) \in (b)$$
$$(a) \subset (b) \subset R \Rightarrow S \subset (b) \subset R$$
$$\Rightarrow \qquad \text{either } S = (b) \text{ or } S = R \qquad [\because S \text{ is maximal.}]$$

If $S = (b)$, then

$$(a) = (b) \qquad \Rightarrow (b) \subseteq (a)$$
$$\Rightarrow \quad b \in (a) \Rightarrow b = ay \text{ for some } y \in R$$
$$\Rightarrow \quad b = (bc)y$$
$$\Rightarrow \quad b = b(cy) \Rightarrow b - b(cy) = 0 \qquad [\text{from (1)}]$$
$$\Rightarrow \quad b(1 - cy) = 0$$
$$\Rightarrow \quad b(1 - cy) = 0$$
$$\Rightarrow \qquad 1 - cy = 0$$
$$[\because \ 0 \neq b \in R \text{ and } R \text{ is without zero divisor.}]$$
$$\Rightarrow \qquad cy = 1 \Rightarrow c \text{ is a unit in } R.$$

This gives a contradiction.

Again, if $S = R$, then

$$(a) = R \ \Rightarrow \ 1 \in (a)$$
$$\Rightarrow \qquad 1 = az \text{ for some } z \in R.$$
$$\Rightarrow \quad a \text{ is a unit in } R.$$

which is again a contradiction because a is not a unit in R.

Here contradiction arises by assuming that a is not prime. Hence, a is a prime element in R.

Conversely, suppose $S = (a)$ and a is prime element in R. Then we shall show that

S is maximal.

Let *T* be an ideal of R such that $S \subset T \subset R$.

Since the Euclidean ring *R* is a principal ideal ring, so that *T* is a principal ideal.

Let $T = (b)$, for some $b \in R$

But $\qquad S \subset T \qquad \Rightarrow \quad (a) \subset (b) \Rightarrow a = bx$ for some $x \in R$

$\qquad\qquad\qquad\qquad\qquad \Rightarrow \quad b$ is either a unit in R or an associate of *a*.

$\qquad\qquad\qquad\qquad\qquad\qquad\qquad\qquad\qquad$ [$\because$ *a* is prime element in R.]

$\qquad\qquad\qquad\qquad\qquad \Rightarrow$ either $(b) = R$ or $(b) = (a)$

$\qquad\qquad\qquad\qquad\qquad \Rightarrow$ either $T = R$ or $T = S$

Hence, *S* is a maximal ideal of *R*.

EXERCISE 5.3

1. In a Euclidean ring, prove that any two greatest common divisors of *a* and *b* are associates.

2. Prove that any two elements in the Euclidean ring *R* have a least common multiple in R.

3. Let *F* be a field. Show that $F(x)$ is a Euclidean ring. Also show that the Euclidean valuation *d* satisfies the additional condition :

$$d(f+g) \leq \max. [d(f), d(g)].$$

4. Show that $\mathbf{Z}(\sqrt{n})$ is an Euclidean domain for $n = -1, -2, 2, 3$ where $\mathbf{Z}[n] = [a + b\sqrt{n} : a, b \in \mathbf{Z}]$

5. Let *D* be an integral domain given by $D = \mathbf{Z}(i\sqrt{3}) = [a + ib\sqrt{3} : a, b \in \mathbf{Z}]$ and $d: D \rightarrow \mathbf{N}$ be the map defined by $d(a + ib\sqrt{3}) = a^2 + 3b^2$. Show that *d* is not a Euclidean valuation on *D*.

5.32 UNIQUE FACTORIZATION DOMAIN

Definition. *An integral domain D is said to be unique factorization domain (UFD) if every non-zero element of D is either a unit or it is expressible as the product of finite number of prime elements in D and this factorization apart from order and associates is unique.*

or

An integral domain D with unity is said to be UFD if

(i) *Every non-zero, non-unit element a of D can be expressed as a product of finite number of irreducible elements of D, and*

(ii) *If $a = p_1 p_2 \ldots p_n$, and*

$\qquad a = q_1 q_2 \ldots q_m$

where p_i and q_j are irreducible in D then $m = n$ and each p_i is an associates of q_j.

For Example :

(1) The ring (**Z**, +, .) of integers is a UFD. Clearly it is an integral domain with unity. If $n \in \mathbf{Z}$ be any non-zero non-unit element of **Z** then if $n > 0$, we can write

$$n = p_1^{\alpha_1} p_2^{\alpha_2} \ldots p_m^{\alpha_m}, \text{ where } p_i \text{ are primes.}$$

$\Rightarrow \qquad n = (p_1 p_1 \ldots p_1)(p_2 p_2 \ldots p_2) \ldots (p_r p_r \ldots p_r)$

i.e., *n* is a product of prime element of **Z**. Also by fundamental theorem of arithmetic, this representation of *n* is unique.

If $n < 0$, let $n = (-m)$, $m > 0$. Then it can be expressed as a product of primes in **Z**, i.e.,

$$m = q_1 q_2 \ldots q_k$$

then $(-m) = n = (-q_1)(q_2)...(q_k)$

(2)   A field $(F, +, .)$ is always a UFD, because it contains no non-zero, non-unit elements.

(3) $\mathbf{Z}[\sqrt{-5}]$ is an integral domain but not a UFD.

5.32.1 Properties of Unique Factorization Domain

Theorem 1. *If a, b are two arbitrary element of a unique factorization domain (UFD) D and p is a prime element of D then $p|ab \Rightarrow$ either $p|a$ or $p|b$.*

Proof.   Let D be a unique factorization domain, and $a \in D, b \in D$.

By definition we can write

$$a = p_1 \cdot p_2 ... p_r \quad \text{and} \quad b = q_1 \cdot q_2 ... q_s$$

where each p_i and q_j are prime elements of D.

Therefore $ab = p_1 \cdot p_2 ... p_r \, q_1 \cdot q_2 ... q_s$

Now, since $p|ab \Rightarrow p| \, p_1 \cdot p_2 ... p_r \, q_1 \cdot q_2 ... q_s$

$\Rightarrow p$ must be one of the primes $p_1 \cdot p_2 ... p_r \, q_1 \cdot q_2 ... q_s$

$\Rightarrow$ either $p|a$ or $p|b$.

Theorem 2. *In a UFD D, an element is prime if it is irreducible.*

Proof.   Let D be unique factorization domain and $a \in D$ be irreducible. We have to prove that a is prime.

Since a is irreducible, then a is non-zero and non-unit.

Let $a|bc$ then $bc = ak$, for some k.

If b is a unit, then $c = a \, kb^{-1} = a(kb^{-1}) \Rightarrow a|c$.

If c is a unit, then similarly $a|b$.

If b, c are non-units and if k is unit, then $bc = ak$

$$\Rightarrow a = n(ck^{-1}).$$

Now, since a is irreducible, therefore either b or ck^{-1} is a unit, but in this case b is not unit, therefore ck^{-1} is a unit, which implies c is a unit which is again not true. Thus, k is a not a unit.

Now, we can express

$$b = p_1 p_2 ... p_m$$
$$c = q_1 q_2 ... q_n$$
$$k = r_1 r_2 ... r_t$$

as product of irreducible. Therefore, $bc = ak$ becomes

$$p_1 p_2 ... p_m \, q_1 q_2 ... q_n = a \, r_1 r_2 ... r_t = x \text{ (say)}$$

Then x is an element having two representations as product of irreducible elements. Further, by definition of UFD each element in one representation is an associate of some element in the other.

Therefore, a is an associate of some p_i or some q_j.

$\Rightarrow$ $ua = p_i$ or $ua = q_j$, for some unit u

$\Rightarrow$ $a \mid p_i$ or $a \mid q_j$

$\Rightarrow$ $a \mid b$ or $a \mid c$ $[p_i|b, q_j|c]$

Hence, a is a prime element.

Theorem 3. *Any two elements of a UFD possesses greatest common divisor as well as least common multiple.*

Proof. Let D be a unique factorization domain, and a, b be any two elements of D. We can write

$$a = p_1^{\alpha_1} p_2^{\alpha_2} \dots p_r^{\alpha_r} \qquad \dots (1)$$

and

$$b = p_1^{\beta_1} p_2^{\beta_2} \dots p_r^{\beta_r} \qquad \dots (2)$$

We have arrange both the expressions (1) and (2) in such a way that the same prime factors appear in both by employing the integer zero as index in any case, if necessary.

We have $p_1, p_2, \dots, p_r$ are all distinct prime and $\alpha_1, \alpha_2, \dots, \alpha_r, \beta_1, \beta_2, \dots \beta_r$ are all non-negative integers.

Let us define

$$d_i = \max\{\alpha_i, \beta_i\} \text{ and } l_i = \min\{\alpha_i, \beta_i\}$$

then $p_1^{l_1} p_2^{l_2} \dots p_r^{l_r}$ and $p_1^{d_1} p_2^{d_2} \dots p_r^{d_r}$ are the required greatest common divisor and least common multiple of a and b respectively.

Theorem 4. *If D is an integral domain with unity in which every non-zero, non-unit element is a finite product of irreducible element and every irreducible element is prime, then D is a UFD.*

Proof. Let D be an integral domain. To prove D is a *UFD*. For this, we shall prove that if $a \in D$ be a non-zero non-unit element and

$$a = p_1 p_2 \dots p_m = q_1 q_2 \dots q_n$$

where p_i and q_j are irreducible elements then $m = n$ and each p_i is an associate of some q_j.

Now, we shall use induction on n.

For n = 1, In this case $a = p_1 p_2 \dots p_m = q_1$ and since q_1 is irreducible, some p_i is a unit. But each p_i being irreducible cannot be a unit.

Therefore $\qquad\qquad m = 1$

$\Rightarrow \qquad\qquad qa = p_1 = q_1$

$\Rightarrow \qquad$ result is true for $n = 1$.

Now, suppose that result is true for $n - 1$.

Further, let $\quad a = p_1 p_2 \dots p_m = q_1 q_2 \dots q_n$

Then $\qquad\qquad p_1 p_2 \dots p_m = (q_1 q_2 \dots q_n)$

$\Rightarrow \qquad\qquad q_1 \mid p_1 p_2 \dots p_m$

Now, since q_1 is irreducible, it is prime

$\Rightarrow \qquad\qquad q_1 \mid p_i;$ for some i.

Let us take $i = 1$, then

$$q_1 \mid p_i \Rightarrow p_1 = q_1 u_1$$

But p_1 is irreducible q_1 or u_1 is a unit.

As q_1 is not a unit (being irreducible), u_1 will be a unit and therefore p_1, q_1 are associates.

Now $(q_1u_1)p_2p_3 \cdots p_m = q_1q_2 \cdots q_n$

$\Rightarrow \qquad (u_1p_1)p_3 \cdots p_m = q_1q_3 \cdots q_n$

$\Rightarrow \qquad p_2'p_3 \cdots p_m = q_2q_3 \cdots q_n$ ...(1)

when $p_2' = u_1p_2$ is irreducible.

Hence, RHS of (1) contains $n-1$ elements and result being true for $n-1$, we find

$$m-1 = n-1 \Rightarrow m = n.$$

Similarly, we can prove that q_2 is an associate of p_2 and so finally we get q_i will be an associate of p_i.

Thus, D is a UFD.

REMARK

☞ Any integral domain D with unity is a UFD if and only if every non-zero, non-unity element is a finite product of irreducible element and every irreducible element is prime.

Theorem 5. *An integral domain with unity is a UFD iff every non-zero, non-unit element is product of primes.*

Proof. Let D be a unique factorization domain. Then by definition of UFD every non-zero, non-unit element is finite product of irreducibles and we know that every irreducible element is prime.

Conversely, let $a \in D$ be any non-zero, non-unit element. Then

$$a = p_1p_2 \cdots p_n,$$

where p_i are prime elements for all i. Now, since D is an integral domain, therefore, prime elements are irreducible. Thus each p_i is irreducible. Now, we show that every irreducible element of D is a prime element. Let $x \in D$ be any irreducible element. Then $x \neq 0$, non unit. Therefore, $x = q_1q_2 \cdots q_m$, where q_j are primes. Suppose $m > 1$. Now since x is irreducible, either q_1 or $(q_2q_3 \cdots q_m)$ is a unit. But q_1 is a prime and therefore cannot be a unit. Therefore $(q_2q_3 \cdots q_m)$ is a unit which gives that q_2 is a unit which is not possible because q_2 is a prime. Thus $m = 1$ or that x is prime. Then by above remark D is a UFD.

Theorem 6. *If in a UFD, a and b are relatively prime then $a \mid bc \Rightarrow a \mid c$.*

Proof. Let D be a UFD.

Also, let $a \mid bc$ then there exist r such that $bc = ar$.

Now, if a is a unit, then $c = (aa^{-1})c = a(a^{-1}c) \Rightarrow a \mid c$

If b is a unit, then $ba = ar \Rightarrow c = b^{-1}ar$

$$\Rightarrow c = a(b^{-1}r) \Rightarrow a \mid c$$

If c is a unit, then $bc = ar$

$\Rightarrow \qquad b = ar\, c^{-1} \Rightarrow a \mid b$

$\Rightarrow \qquad$ g.c.d. $\{a,b\} = a$

It is given that a and b are relatively prime, therefore g.c.d.(a,b) will be a unit

$\Rightarrow \qquad a$ is a unit .

$\Rightarrow \qquad a \mid c.$

If r is a unit, then

$$bc = ar$$

$\Rightarrow$ $bcr^{-1} = a$ $\Rightarrow b|a$

$\Rightarrow$ g.c.d. $(a, b) = b$ $\Rightarrow b$ is a unit.

$\Rightarrow$ $a \mid c$.

Now, suppose that none of a,b,c,r are units.

If b = 0, g.c.d.$(a, b) = a$, a unit which is not true. Therefore, $b \neq 0$.

If c = 0, then $c = a .0 \Rightarrow a \mid c$

Therefore, assuming that $b \neq 0$, $c \neq 0$, we will get

$$a \neq 0, \ r \neq 0 \ [\because bc = ar]$$

Since a,b,c,r being non-zero, non-unit in a UFD, therefore, we can write

$$a = a_1 a_2 ... a_m$$
$$b = b_1 b_2 ... b_n$$
$$c = c_1 c_2 ... c_t$$
$$r = r_1 r_2 ... r_k$$

as product of irreducible elements.

Therefore $(b_1 b_2 ... b_n)(c_1 c_2 ... c_t) = (a_1 a_2 ... a_m)(r_1 r_2 ... r_k) = x$(say). Clearly, x has two representation as product of irreducible element.

Therefore by definition of a UFD, these representation should have same number of elements and each element on one side will be associate of an element on the other. Therefore, $n + 1 = m + k$ and each a_i is an associate of some, b_i or q.

If a_i is an associate of some b_j, then

$$b_j = a_i u \text{ for a unit } u.$$

$\Rightarrow$ $a_i \mid b_j$ and as $b_j \mid b$, we get $a_i \mid b$

$\Rightarrow$ $a_i \mid$ g.c.d.$(a,b) = 1$ as $a_i \mid a$

$\Rightarrow$ a_i is a unit, which is not true because a_i is irreducible.

Therefore, each a_i has to be an associates of some c_i.

$\Rightarrow$ $a_i = c_i u_i$, for unit u_i

which gives

$$(b_1 b_2 ... b_n)(c_1 c_2 ... c_t) = (c_1 u_1 c_2 u_2 ... c_m u_m)(r_1 r_2 ... r_k)$$

$\Rightarrow b(c_{m+1} c_{m+2} ... c_t) = (u_1 u_2 ... u_m) r$

$\Rightarrow$ $b(c_{m+1} c_{m+2} ... c_t)(u_1 u_2 ... u_m)^{-1} = r$

$\Rightarrow$ $b \mid r \Rightarrow r = bd$ for some d.

$\Rightarrow$ $bc = ar = abd$

$\Rightarrow$ $b(c-ad) = 0 \Rightarrow c = ad \Rightarrow a|c.$

Theorem 7. *In a principal ideal domain, every ascending chain of ideals must terminate after a finite number of steps.*

Proof.   Let D be a principal ideal domain. Suppose $A_1 \subseteq A_2 \subseteq A_3 \ ...$ be the given chain.

Suppose, $A = \cup A_i$

Then, clearly A is an ideal of D. Now, since D is a principal ideal domain, therefore A is a principal ideal. Let $A = (a)$.

Then, $\qquad a \in (a) = A = \cup A_i$

$\Rightarrow \qquad\qquad a \in A_i,$ for some i

$\Rightarrow \qquad$ all multiples of a are in A_i.

$\Rightarrow \qquad\qquad\quad (A) \subseteq a_i$

$\Rightarrow \qquad\qquad\quad A \subseteq A_i,$ for each $t \geq i$

$\Rightarrow \qquad\qquad\quad A \subseteq A_i \subseteq A_t \subseteq A$

$\Rightarrow \qquad\qquad\quad A_i = A$

Hence, we conclude that every ascending chain of ideals must terminate after a finite number of steps.

Theorem 8. *In a FID, every non-zero non-unit element is divisible by an irreducible element.*

Proof. Let D be a principal ideal domain and $a \in D$ be a non-zero, non-unit element.

Let $\quad I_1 = (a)$.

If I_1 is a maximal ideal, then a is irreducible and as $a \mid a$.

In this case, result is proved.

Now, suppose I_1 is not maximal, then there exists some ideal $\quad I_2 \neq D$ such that $I_1 \subseteq I_2 \subseteq D$.

Let $\quad I_2 = (p_2)$.

If I_2 is maximal, then p_2 will be irreducible and as $p_2 \mid a$, the result is proved.

Further, suppose that I_2 is not maximal, then there exists an ideal I_3 such that $I_1 \subseteq I_2 \subseteq I_3 \subseteq D$. Proceeding in the same way, we get an ascending chain of ideals in D, which must terminate after a finite number of steps (Theorem-7), say at $I_n = (p_n)$.

In this case, I_n will be maximal and p_n will be irreducible with $p_n \mid a$.

Theorem 9. *A principal ideal domain is a unique factorization domain.*

Proof. Let D be a principal ideal domain and $a \in D$ be any non-zero non-unit element. If a is irreducible, then as $a = a$, we may express a as finite product of irreducibles. If a is not irreducible, then by theorem 8, a is divisible by some irreducible element p_1.

$$P_1 \mid a \Rightarrow a = a_1 p_1, \text{ for some } a_1.$$

Now, if a_1 is irreducible, we can express a_1 as a product of finite number of irreducible elements. Suppose a_1 is not irreducible, then clearly a_1 is a non-zero, non-unit element. Then again by Theorem 7, there exists an irreducible element p_2 such that $p_2 \mid a_1$.

$\Rightarrow \qquad\qquad\qquad a_1 = p_2 a_2$ for some a_2.

If a_2 is irreducible, then

$$a = a_1 p_1 = p_2 p_1 a_2$$

If a_2 is not irreducible, then continuing same as above.

Now, consider the ideals (a), (a_1), (a_2), ...

Then, $\qquad\qquad\qquad (a) \subseteq (a_1) \subseteq (a_2) \subseteq \dots$

Now, $\qquad\qquad\qquad x \in (a) \Rightarrow x = ar = p_1 a_1 r \in (a_1),$ etc.

Therefore, we get an ascending chain of ideals which must terminate after a finite number of steps. Therefore, we get some irreducible elements a_n such that

$$a = p_1 p_2 \dots p_n a_n$$

⇒ *a* is expressed as a product of finite number of irreducible elements.

Now, it remains to prove that if *a* has more than two such representation, then the number of elements is same in both and each element in one representation is an associate of an element in that order. Finally since we know that an integral domain *D* with unity in which every non-zero, non-unit element is a finite product of irreducible element, then *D* is a unique factorization domain. Hence *D* is a unique factorization domain.

5.33 POLYNOMIAL RINGS OVER UNIQUE FACTORIZATION DOMAIN

Let *D* be a unique factorization domain. By definition we have *D* is an integral domain with unity. Thus $D(x)$ is also an integral domain with unity. The only units in $D(x)$ are units of *D*. Consider a polynimial $p(x) = a(x) b(x)$ with $a(x), b(x) \in R(x)$, then one of $a(x)$ or $b(x)$ is a unit of $D(x)$, i.e., a unit in *D*.

Definition 1. *Let* $f(x) = a_0 + a_1 x + a_2 x^2 + ... + a_n x^n$ *be a polynimial over a UFD D. Then the content of* $f(x)$ *denoted by* $C(f)$ *is defined as the greatest common divisor of the coefficients* $a_0, a_1, ... a_n$ *of* $f(x)$.

REMARK

☞ The content of $f(x)$ is unique within units of *D*. Therefore, if C_1 and C_2 are two contants of $f(x)$, then we must have $C_1 = uC_2$, where *u* is some unit in *D*.

Definition 2. (*Primitive polynomials*). *Let D be a unique factorization domain. Then a polynomial* $f(x) = a_0 + a_1 x + a_2 x^2 + ... + a_n x^n \in D(x)$ *is said to be primitive if the greatest common divisor of its coefficients* $a_0, a_1, ..., a_n$ *is a unit in D, i.e., a polynomial* $f(x)$ *is said to be primitive if its content is 1.*

REMARKS

☞ If $f(x)$ is a monic polynomial in *D*, then $f(x)$ is always primitive.
☞ Every irreducible polynomial of positive degree belonging to $D(x)$ is necessarily primitive.
☞ An irreducible polynomial of zero degree may not be primitive.
☞ A primitivie polynomial may not be irreducible.

Theorem 1. *Let D be a unique factorization domain. Then every non-zero member* $f(x)$ *of* $D(x)$ *can be written as* $f(x) = gf_1(x)$, *where* $g = C(f)$ *and* $f_1(x)$ *is primitive. Also, decomposition of* $f(x)$ *as an element of D by a primitive polynomial in* $D(x)$ *is unique apart from its distinction between associates.*

Proof. Let *D* be a unique factorization domain and let

$$f(x) = a_0 + a_1 x + ... + a_n x^n \in D(x)$$

Now, since *D* is a unique factorization domain, thus the elements a_0, $a_1, ..., a_n \in D$ must possess a greatest common divisor. Let $g \in D$ be the g.c.d. of these elements.

Then, 　　　　　　$g = C(f)$

Further, let 　　　$a_i = gb_i$, where $i = 0, 1, ..., n$.

Then, 　　　　　　$f(x) = gb_0 + gb_1 x + ... + gb_n x^n$

　　　　　　　　　$= g[b_0 + b_1 x + ... + b_n x^n]$

Since g is the g.c.d. of $a_0, a_1, \ldots\ldots a_n$, thus the elements $b_0, b_1, \ldots, b_n$ can have no common factor other than units of D. So the polynomial

$$f_1(x) = b_0 + b_1x + \ldots + b_nx^n$$

is a primitive member of $D(x)$. Therefore, we have $f(x) = gf_1(x)$, where $g \in D$ and $f_1(x) \in D(x)$ is primitive.

Now, we shall prove the uniqueness.

Let if possible

$$f(x) = hf_2(x), \text{ where } h \in D \text{ and } f_2(x) \in D(x) \text{ is primitive.}$$

Then, $gf_1(x) = hf_2(x)$ ......(1)

Now, since $f_1(x)$ and $f_2(x)$ are both primitive, then the content of the polynomial on the LHS of (1) is g and the content of RHS of (1) is h. But we know that content of a polynomial is unique upto associates.

Thus, g and h are associates.

Therefore, $g = hu$, where u is some unit in D.

$\Rightarrow$ $huf_1(x) = hf_2(x)$

$\Rightarrow$ $uf_1(x) = f_2(x)$

Hence, $f_1(x)$ and $f_2(x)$ are associates.

Theorem 2. *Let D be a UFD. Then the product of two primitive polynomials in $D(x)$ is again a primitive polynomial in $D(x)$.*

Proof.   Consider two primitive polynomials

$$f(x) = a_0 + a_1x + \ldots\ldots + a_nx^n$$

and $$g(x) = b_0 + b_1x + \ldots\ldots + b_nx^n \text{ in } D(x).$$

Define $$h(x) = f(x).g(x) = c_0 + c_0x + \ldots\ldots + c_{m+n}x^{m+n}$$

Let us suppose $h(x)$ is not primitive. Then all the coefficients of $h(x)$ must be divisible by some prime element p of D. Since $f(x)$ is primitive, therefore, the prime element p must not divide some coefficients of $f(x)$. Let a_i be the first coefficient of $f(x)$, which p does not divide. Similarly, let b_j be the first coefficient of $g(x)$ which p does not divide.

Further, in $f(x).g(x)$, the coefficient of x^{i+j} is

$$c_{i+j} = a_i b_j + (a_{i-1}b_{j+1} + a_{i-2}b_{j+2} + \ldots. + a_0 b_{i+j})$$
$$+ (a_{i+1}b_{j-1} + a_{i+2}b_{j-2} + \ldots + a_{i+j}b_0)$$

Using this relation, we obtain

$$a_i b_j = c_{i+j} - \{(a_{i-1}b_{j+1} + a_{i-2}b_{j+2} \ldots\ldots + a_0b_{i+j})$$
$$+ (a_{i+1}b_{j-1} + a_{i+2}b_{j-2} + \ldots + a_{i+j}b_0)\}$$ ....(1)

By our choice of a_i, p is a divisor of each of the elements $a_0, a_1, \ldots, a_{i-1}$.

Therefore,

$$p \,|\, (a_{i-1}b_{j+1} + a_{i-2}b_{j+2} + \ldots\ldots + a_0b_{i+1})$$

In a similar manner, by our choice of b_j, p is a divisor of each of the elements $b_0, b_1, \ldots, b_{j-1}$. Therefore, $p \,|\, (a_{i+1}b_{j-1} + a_{i+2}b_{j-2} + \ldots. + a_{i+j}b_0)$

Also, we have assumed that $p \,|\, c_{i+j}$.

Therefore, from (1), we get

$$p \mid a_i \, b_j$$

$\Rightarrow \quad p \mid a_i$ or $p \mid b_j$, since p is a prime element of D, which is not possible, because by our assumption, p is not a divisor of a_i and not a divisor of b_j. Hence, $h(x)$ must be primitive.

Theorem 3. *If D is a unique factorization domain and if $f(x)$, $g(x)$ are in $D(x)$, then*

$$C(fg) = C(f)C(g)$$

Proof. We can write the polynomial $f(x)$ in $D(x)$ as $f(x) = a\,f_1(x)$, where, $a = C(f)$ and $f_1(x)$ is primitive. Similarly, the polynomial $g(x)$ can be written as $g(x) = bg_1(x)$, where $b = C(g)$ and $g_1(x)$ is primitive.

Then, $\qquad\qquad f(x)g(x) = abf_1(x)g_1(x)$ ......(1)

Since $f_1(x)$ and $g_1(x)$ both are primitive and it is known that product of two primitives is again primitive, so $f_1(x)\,g_1(x)$ is primitive.

Then, from (1), the content of $f(x)\,g(x)$ is either ab or some associates of ab. Therefore, the content of $f(x).g(x)$ is ab. Then

$$C(fg) = ab = C(f)C(g)$$

5.34 FIELD OF QUOTIENTS OF A UNIQUE FACTORIZATION DOMAIN

Let D be a unique factorization domain, then D is necessarily an integral domain. So D has a field of quotients.

Notations. In this section, we shall denote the field of quotients of D by F. Also, we can consider $D(x)$ to be a subring of $F(x)$.

Theorem 1. *Let D be an integral domain and F be its field of quotients. Then any element $f(x)$ in $F(x)$ can be written as*

$$f(x) = \frac{f_0(x)}{a}, \text{where } f_0(x) \in D(x) \text{ and } a \in D.$$

Proof. Let F be the field of quotients of an integral domain D.

Then, $\qquad F = \left\{ \dfrac{p}{q} : p \in D, 0 \neq q \in D \right\}$

Now, let $f(x)$ be an element of $F(x)$. Further let

$$f(x) = \frac{a_0}{b_0} + \frac{a_1}{b_1}x + ... + \frac{a_n}{b_n}x^n, \text{where } a_0, a_1, a_2, ...a_n \in D$$

and $b_0, b_1, ... b_n$ are non-zero elements of D.

Further, since $b_0, b_1, ... b_n$ are also non-zero elements of F, therefore each of them must be inversible. Also, $b_0, b_1, ... b_n$ is a non-zero element of F, so it is also inversible. Therefore, we can write

$$f(x) = \frac{b_0 b_1 ... b_n}{b_0 b_1 ... b_n} \left(\frac{a_0}{b_0} + \frac{a_1}{b_1}x + ... + \frac{a_n}{b_n}x^n \right)$$

$$= \frac{(a_0 b_1 b_2 ... b_n) + (b_0 a_1 b_2 ... b_n)x + ... + (b_0 b_1 ... b_{n-1}a_n)x^n}{b_0 b_1 ... b_n} = \frac{f_0(x)}{a}$$

where, $f_0(x) = (a_0b_1b_2 \dots b_n) + (b_0a_1b_2 \dots b_n)x + \dots + (b_0b_1 \dots b_{n-1}a_n)x^n$ in $D(x)$

and $\quad a = b_0b_1 \dots \dots b_n$ in D.

Theorem 2 (Gauss Lemma). *Let F be the field of quotients of a unique factorization domain D. If the primitive polynomial $f(x) \in D(x)$ can be factored as the product of two polynomials having coefficients in F, then it can be factored as the product of two polynomials having coefficient in D.*

Proof.   Let D be a unique factorization domain and F be its field of quotients.

Let $f(x) \in D(x)$ be primitive.

Let us write $f(x) = g(x) h(x)$, where $g(x)$ and $h(x)$ have coefficients in F.

Now, since $g(x), h(x) \in F(x)$, thus using previous theorem, we can write

$$g(x) = \frac{g_0(x)}{a}, h(x) = \frac{h_0(x)}{b}$$

where, $a, b \in D$ and $g(x), h(x) \in D(x)$.

Also, we have $g_0(x) = \alpha g_1(x)$, $h_0(x) = \beta h_1(x)$, where, $\alpha = C(g_0)$, $\beta = C(h_0)$ and $g_1(x)$ and $h_1(x)$ are primitive members of $D(x)$.

Now, $\quad f(x) = \dfrac{\alpha\beta}{ab}g_1(x)h_1(x)$

$\Rightarrow \quad abf(x) = \alpha\beta \, g_1(x)h_1(x)$ ... (1)

Further, since $g_1(x)$ and $h_1(x)$ are both primitive members of $D(x)$, so $g_1(x).h_1(x)$ is also a primitive member of $D(x)$. Thus, from (1), we conclude that $f(x)$ and $g_1(x)$ $h_1(x)$ are associates in $D(x)$.

Therefore, $f_1(x) = ug_1(x)h_1(x)$, where u is a unit in $D(x)$ and so a unit of D.

Now, $\quad\quad\quad u \in D, g_1(x) \in D(x)$

$\Rightarrow \quad\quad ug_1(x) \in D(x)$.

Also, $\quad\quad h_1(x) \in D(x)$.

Hence, $f(x)$ can be factored as the product of two polynomials with coefficients in D.

Theorem 3. *Let F be the field of quotients of a unique factorization domain D. If $f(x) \in D(x)$ is both primitive and irreducible as an element of $D(x)$, then it is irreducible as an element of $F(x)$. Conversely, if the primitive element $f(x)$ in $D(x)$ is irreducible as an element of $F(x)$, it is also irreducible as an element of $D(x)$.*

Proof.   Let $f(x) \in D(x)$ be primitive. Let us suppose $f(x)$ is irreducible in $D(x)$ but it is reducible in $F(x)$. Now, since F is a field and $f(x)$ is irreducible in $F(x)$, thus we must have

$$f(x) = g(x)h(x)$$

where $g(x), h(x) \in F(x)$ and are of positive degree.

Further, we can write

$$g(x) = \frac{g_0(x)}{a}, h(x) = \frac{h_0(x)}{b} \quad\quad\quad \text{[Using Theorem 1]}$$

where $a, b \in D$ and $g_0(x), h_0(x) \in D(x)$

Also, $\quad\quad g_0(x) = \alpha g_1(x), h_0(x) = \beta h_1(x)$

where, $\alpha = C(g_0)$, $\beta = C(h_0)$ and $g_1(x)$, $h_1(x)$ are primitive in $D(x)$.

Then, $$f(x) = \frac{\alpha\beta}{ab} g_1(x)h_1(x)$$

which gives

$$abf(x) = \alpha\beta g_1(x)h_1(x) \qquad \qquad ...(1)$$

Now, since $g_1(x)$ and $h_1(x)$ are both primitive in $D(x)$, then $g_1(x)h_1(x)$ is also a primitives in $D(x)$.

Thus, from (1), we conclude that $f(x)$ and $g_1(x)h_1(x)$ are associates in $D(x)$.

Therefore, we can write

$$f(x) = ug_1(x)\, h_1(x)$$

where u is a unit in $D(x)$ and therefore a unit in D. Now, let

$$ug_1(x) = g_2(x)$$

Then, we have

$$f(x) = g_2(x)h_1(x), \text{ where } g_2(x), h_1(x) \in D(x).$$

Also, $\deg(g_2(x)) = \deg(g(x))$

and $\deg(h_1 x)) = \deg(h(x))$

Therefore, $\deg(g_2(x)) > 0$, $\deg(h_1(x)) > 0$

$\Rightarrow$ neither $g_2(x)$ nor $h_1(x)$ is a unit in $D(x)$.

$\Rightarrow$ $f(x) = g_2(x)h_1(x)$ is a proper factorization of $f(x)$ in $D(x)$, which is a contradiction, because we have assumed that $f(x)$ is irreducible in $D(x)$.

Hence, $f(x)$ must be irreducible in $F(x)$.

Conversely, let us suppose $f(x)$ is a primitive member of $D(x)$ and is irreducible as an element of $F(x)$. We have to prove that $f(x)$ is also irreducible as an element of $D(x)$. Let

$$f(x) = g(x)\, h(x), \text{ where } g(x), h(x) \in D(x)$$

Then, $f(x)$ will be irreducible in $D(x)$ if one of $g(x)$ or $h(x)$ is a unit in $D(x)$, i.e., a unit in D.

Now, $g(x)$, $h(x) \in D(x)$ can be treated as $g(x). h(x) \in F(x)$.

Now, since $f(x)$ is irreducible as an element of $F(x)$. Thus, one of $g(x)$ or $h(x)$ must be of degree 0. Now, suppose $\deg g(x) = 0$. In this case $g(x)$ is a constant polynomial. Further, let $g(x) = K \in D$, then $f(x) = K\, h(x)$.

Since $f(x)$ is a primitive member of $D(x)$, therefore $C(f)$ is a unit in D. If K is not a unit in D, then content of $Kh(x)$ cannot be a unit in D and so it cannot be equal to $C(f)$. Thus, K must be a unit in D. Hence, $f(x)$ is irreducible in $D(x)$.

Theorem 4. *Let F be a field of quotients of a unique factorization domain D. If $f_1(x)$, $f_2(x)$ are two primitive members of $D(x)$ and are associates in $F(x)$, then they are also associates in $D(x)$.*

Proof. It is given that $f_1(x), f_2(x)$ are associates in $F(x)$. Thus, we have

$$f_1(x) = Kf_2(x), \text{ where } 0 \neq K \in F, \text{ i.e., } K \text{ is a unit in } F(x).$$

We know that the only units of $F(x)$ are the non-zero elements of F. Since F is the field of quotient of D, thus

$$0 \neq K \in F \qquad \Rightarrow \qquad K = \frac{g}{h} : g, h \neq 0 \in D.$$

Thus, $\quad f_1(x) = \frac{g}{h} f_2(x) \quad \Rightarrow \quad hf_1(x) = gf_2(x)$

Finally, since h, $g \in D$ and $f_1(x)$, $f_2(x)$ are primitive members of $D(x)$, hence $f_1(x)$, $f_2(x)$ are associates in $D(x)$.

5.35 EIENSTEIN'S CRITERION OF IRREDUCIBILITY

Theorem 1. *Let F be the field of quotients of a unique factorization domain D.*

If $f(x) = a_0 + a_1x + a_2x^2 + ... + a_nx^n \in D(x)$ and p is a prime element of D such that $p|a_0, p|a_1, p|a_2, ..., p|a_{n-1}$, whereas p is not a divisor of a_n and p^2 is not a divisor of a_0, then $f(x)$ is irreducible in $F(x)$.

Proof. Let us assume that $f(x)$ is a primitive. Let if possible $f(x)$ is reducible in $F(x)$. Then $f(x)$ can be factored as the product of two polynomials of positive degree in $F(x)$. Then by Gauss lemma, $f(x)$ can be factored as the product of two polynomials of positive degree in $D(x)$. Since we have assumed that $f(x)$ is reducible in $F(x)$, then

$$f(x) = a_0 + a_1x + ... + a_nx^n$$
$$= (b_0 + b_1x + ... + b_rx^r)(c_0 + c_1x + ... + c_sx^s) \qquad ...(1)$$

where b's and c's are elements of D and $r > 0, s > 0$.

Now, from (1), we have $a_0 = b_0 c_0$

Since, p is prime element of D, therefore, $p|a_0 \Rightarrow p|b_0$ or $p|c_0$.

Since p^2 is not a divisor of a_0, therefore p cannot divide both b_0 and c_0.

Now, suppose that $p|b_0$ and p is not a divisor of c_0.

If p is a divisor of all the coefficients $b_0, b_1 .... b_r$, then from (1), we can say that p is a divisor of all the coefficients of $f(x)$. But p is not a divisor of a_n, so not all the coefficients of $b_0, b_1,.... b_r$, cannot be divisible by p. Let b_k, where $k \leq r$ be the first b which is not divisible by p. Then, each of $b_0, b_1 .... b_{k-1}$ is divisible by p and b_k is not divisible by p.

Also, $k < n$, since $r < n$. Then, from (1), we have

$$a_k = b_k c_0 + b_{k-1}c_1 + b_{k-2}c_2 + ... + b_0 c_k$$
$$\Rightarrow \qquad b_k c_0 = a_k - b_{k-1}c_1 - b_{k-2}c_2 - ... b_0 c_k \qquad ... (2)$$

Since $k < n$, thus $p|a_k$. Also, $p|b_{k-1}, b_{k-2},..., b_0$

Then, from (2), we have $p|b_k c_0$

$\Rightarrow \quad p|b_k$ or $p | c_0$, since p is a prime element of D which is not possible, because we have assumed that p is neither a divisor of b_k nor a divisor of c_0. Hence, $f(x)$ must be irreducible in $F(x)$.

REMARKS

☞ Any polynomial of degree 1 is irreducible over a field F.

☞ If $f(x) \in F(x)$ is any polynomial of degree greater than one and $f(a) = 0$ for some $a \in F$, then $f(x)$ is reducible over F.

☞ Let $f(x) \in F(x)$ be a polynomial of degree 2 or 3, then $f(x)$ is reducible implies there exists $a \in F$ such that $f(a) = 0$.

☞ A polynomial $f(x) \in F(x)$ of degree 2 or 3 is reducible if and only if there exists $a \in F$ such that $f(a) = 0$.

☞ An irreducible polynomial need not be an irreducible element over $\mathbf{Z}$.

☞ If F is a field then every irreducible polynomial of $F(x)$ is irreducible element of $F(x)$ and conversely.

☞ If D is a UFD and if $p(x)$ is a primitive polynomials in $D(x)$, then it can be factored in a unique way as the product of irreducible elements in $D(x)$.

☞ The polynomial ring $D(x)$ over a *UFD*, D is itself a *UFD*.

☞ If D is a UFD, then so is $D[x_1, ..., x_n]$.

☞ If F is a field, then $F[x_1, x_2, ..., x_n]$ is a UFD.

Solved Examples

Example 1. *Show that $2x+1$ is a unit in $\mathbf{Z}_4(x)$.*

Solution. We can write $(2x+1)(0x+1) = 0\,x^2 + 0x+1 = 1 \qquad [\because 4 = 0 \text{ in } \mathbf{Z}_4]$

$\Rightarrow \quad (2x+1)$ is a unit in $\mathbf{Z}_4(x)$

Example 2. *Show that $\mathbf{Q}(x)/\mathbf{Z}$, where $I = [x^2 - 5x+6]$ is not a field.*

Solution. We can write $x^2 - 5x+6 = (x-2)(x-3)$

$\Rightarrow \quad$ It is not irreducible polynomial over $\mathbf{Q}$.

Therefore I is not a maximal ideal of $\mathbf{Q}(x)$. Hence, $\mathbf{Q}(x)/\mathbf{Z}$ is not a field.

Example 3. *Show that $f(x) = x^3 - 9$ is reducible in $\mathbf{Z}_{11}$.*

Solution . We can write $4 \oplus 4 \oplus 4 = 9$ in $\mathbf{Z}_{11}$.

Therefore, we find $(x - 4)$ is a factor of $x^3 - 9$.

By actual division, we can find $x^3 - 9 = (x - 4)(x^2+4x+5)$ in $\mathbf{Z}_{11}$.

Hence, $x^3 - 9$ is reducible.

Example 4. *Show that $\mathbf{Z}_5(x)$ is a UFD. Is x^2+2x+3 will be reducible over $\mathbf{Z}_5(x)$?*

Solution. Since 5 is a prime, therefore $\mathbf{Z}_5$ is a field.

$\Rightarrow \quad \mathbf{Z}_5$ is a field $\Rightarrow \mathbf{Z}_5[x]$ is a UFD. Further, x^2+2x+3 will be reducible over $\mathbf{Z}_5$ as if it has a root in $\mathbf{Z}_5$, but none of the element in $\mathbf{Z}_5$ is a root of $x^2 +2x +3$. Hence, it is an irreducible polynomial over $\mathbf{Z}_5$.

Example 5. *Find g.c.d. $(2, x)$ in $\mathbf{Z}(x)$.*

Solution. Here, we have $2 = 2 + 0.x + 0.x^2 + ...; x = 0 +1.x +0.x^2 + ...$

We have $1|2$ and $1|x$. Suppose $f|2$ and $f|x$. We have to show that $f|1$.

Now $f|2 \Rightarrow 2 = fg,$ for some g.

$\Rightarrow \deg 2 = \deg f + \deg g \Rightarrow 0 = \deg f + \deg g$

$\Rightarrow \deg f = 0 \Rightarrow f$ is a constant polynomial.

Let $\quad f = a_0 + 0.x + 0.x^2 + ...$

Again, $f|x \Rightarrow a_0|x \Rightarrow x = a_0\, h(x)$

$$\deg x = 0 + \deg h \Rightarrow \deg h = 1$$

Let $h(x) = b_0 + b_1 x + 0. x^2 + 0. x^3 + \dots$

Then, $x = a_0 h = a_0(b_0 + b_1 x) = a_0 b_0 + a_0 b_1 x$

$\Rightarrow \quad 1 = a_0 b_1 = f(x) b_1 \Rightarrow f \mid 1$. Hence, g. c. d. $(2, x) = 1$

Example 6. *If p is a prime number, show that the polynomial $x^n - p$ is irreducible over the field of rational numbers.*

Solution . We can write $f(x) = x^n - p = -p + 0x + 0x^2 + \dots + 0x^{n-1} + 1. x^n$

$\Rightarrow \quad f(x)$ is a polynomial with integer coefficient. Also, p is prime.

We observe that p divide each of the coefficients of $f(x)$ except the coefficient 1 of the last term x^n. Also, p^2 is not a divisor of the constant term $-p$. Hence, by Eisenstein's criterion of irreduciblity, $f(x)$ is irreducible over the field of rational numbers.

Example 7. *Show that the polynomial $x^2 - 3$ is irreducible over the field of rational numbers.*

Solution. Let $f(x) = x^2 - 3 = -3 + 0 x + 1. x^2$

Clearly, $f(x)$ is a polynomial with integer coefficients. Also, 3 is a prime number such that 3 divides each of the coefficients of $f(x)$ except the coefficient 1 of x^2. Also, 3^2 is not a divisor of the constant term -3. Hence, by Eisenstein's criterion of irreduciblility, $f(x)$ is irreducible over **Q**.

Example 8. *Show that the polynomial $1 + x + \dots + x^{p-1}$, where p is prime, is irreducible over the field of rational numbers.*

Solution . Let $f(x) = 1 + x + x^2 + \dots + x^{p-1}$

$\Rightarrow \quad (x-1)f(x) = (x-1)(1 + x + \dots + x^{p-1}) \Rightarrow (x-1)f(x) = x^p - 1$

Putting $x - 1 = y \quad \Rightarrow \quad y + 1 = x$, we get

$$yf(y+1) = (y+1)^p - 1$$
$$= y^p + {}^pC_1 y^{p-1} + {}^pC_2 y^{p-2} + \dots + {}^pC_{p-1} y + 1 - 1$$

[Using Binomial Expansion]

$$= y^p + {}^pC_1 y^{p-1} + {}^pC_2 y^{p-2} + \dots + {}^pC_{p-2} y + {}^pC_{p-1}$$
$$= y[y^{p-1} + {}^pC_1 y^{p-2} + {}^pC_2 y^{p-3} + \dots + {}^pC_{p-2} y + {}^pC_{p-1}]$$

It is known that ${}^pC_r = \dfrac{p(p-1)(p-2)\dots(p-r+1)}{r!} ; 1 \le r \le p - 1$

Clearly ${}^pC_r$ is divisible by p for each $1 \le r \le p - 1$, where p is prime.

Further, $f(y+1)$ is a polynomial with integer coefficients.

Also, p is a prime number such that p divides each of the coefficients of $f(y+1)$ except the coefficent of y^{p-1} which is 1. Also, p^2 is not a divisor of the constant term which is equal to ${}^pC_{p-1} = p$. Thus, by Eisenstein's criterion of irreducibility $f(y+1)$ is irreducible over the field of rational numbers. Hence, $f(x)$ is irreducible over the field or irrational numbers.

EXERCISE 5.4

1. Show that the following polynomials are irreducible over **Q**.
   (i) $x^4+x^3+x^2+x+1$ (ii) $8x^3-6x+1$

2. Examine whether the polynomial
   x^3+3x^2+x-4 is irreducible over.
   (i) the field of integers modulo 5
   (ii) the field of integers modulo 7.

3. Show that in a principal ideal domain, every ideal is contained in a maximal ideal.

4. Let $p(x)$ be an irreducible polynomial over a field F, show that the ideal generated by $p(x)$ in $F(x)$ is a maximal ideal.

5. Show that x^2+1 and x^2+x+4 are irreducible over the field of integers.

6. If P is a prime ideal of $D(x)$, show that $P \cap D$ is a prime ideal of D.

7. Show that in a UFD D, every non-zero prime ideal ($\neq D$) contains a prime element.

8. Factorise x^2+x+5 in $F[x]$, where F is the field of integers modulo 11.

9. Find the g.c.d. of
   (i) $2x^2+x^3-6x^2+7x-2,\ 2x^3-7x^2+8x-2$
      in $\mathbf{Q}(x)$
   (ii) $10+11i$ and $8+i$ in $\mathbf{Z}[i]$.

10. Let K be a field of quotients of a UFD D, then show that primitive polynomials $f, g \in D[x]$ are associated in $K[x]$.

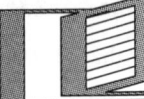

Chapter Review: *A competitive Approach*

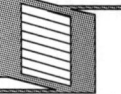

Selected Terms and Results

Terms

- **Ring :** An algebraic structure $(R, +, \cdot)$ is said to be ring if
  (i) $(R, +)$ is an abelian group.
  (ii) $(R, \cdot)$ is a semigroup.
  (iii) multiplication is distributive over addition.

- **Commutative Ring:** A ring which is commutative under multiplication.

- **Null ring:** A ring having a single element 0 with two binary operations addition and multiplication.

- **Ring with unity:** If $1 \in R$.

- **Ring with zero divisors:** A ring R is said to be ring with zero divisor if there exist two non-zero elements a, b in R such that $ab = 0$

- **Ring without zero divisors:** A ring R is said to be ring without zero divisors if $ab = 0$ implies either $a = 0$ or $b = 0$.

- **Integral domain:** A commutative ring with unity without zero divisors is called an integral domain.

- **Field:** A commutative ring R with unity having at least two elements is called a field if every non-zero element in R has a multiplicative inverse.

- **Skew field or division ring:** A ring R with unity having at least two elements is called a skew field if every non-zero elements of R having the multiplicative inverse.

- **Subring:** A non-empty subset S of a ring R is said to be subring if S itself is a ring.

- **Subfield:** A non-empty subset K of a field F is said to be subfield if K itself is a field.

- **Characteristic of a ring:** Let R be a ring and $a \in R$. If there exists a positive integer n such that $na = a + a + \ldots + a$ (upto n times) $= 0 \forall a \in R$, then the smallest such positive integer n is called the characteristic of a ring.

- **Left ideal:** An additive subgroup S of a ring R is said to be a left ideal of R if $a \in S, r \in R \Rightarrow ra \in S$

- **Right ideal:** An additive subgroup S of a ring R is said to be a right ideal of R if $a \in S, r \in R \Rightarrow ar \in S$.

- **Homomorphism of Rings:** Let R and R' be two rings with two binary operation addition and multiplication, then a mapping $\phi : R \to R'$ is said to be a homomorphism if
  (i) $\phi(a+b) = \phi(a) + \phi(b)$
  (ii) $\phi(ab) = \phi(a) \, \phi(b) \; \forall a, b \in R$

- **Quotient ring:** Let R be a ring and S be any ideal of R, then the algebraic structure $(R/S, +, \cdot)$ where $R/S = [a+S : a \in R]$ and the operations '+' and '.' on R/S defined by
  $$(a+S) + (b+S) = (a+b)+S$$
  and $(a+S).(b+S) = ab + S \qquad \forall \, a, b \in R$
  forms a ring. This ring is called the quotient ring or residue class ring.

- **Principal Ideal:** An ideal generated by a single element.

- **Principal Ideal Ring:** A Commutative ring R with unity having no zero divisor is called a principal ideal ring if every ideal of R is a principal ideal.

- **Units:** Let R be a commutative ring with unity, *i.e.*, $1 \in R$. Then an element a of R is said to be unit in R if there exist an element $b \in R$ such that $ab = 1$.

- **Associates:** Let R be a commutative ring with unity. Then an element a of R is said to be an associate of $b \in R$ if $a = ub$ for some unit u in R.

- **Prime ideal:** Let R be a commutative ring, then an ideal $S \subseteq R$ is said to be prime ideal if $ab \in S$ applies either $a \in S$ or $b \in S$ for all a, $b \in R$.

- **Maximal ideal:** A proper ideal S or R is said to be maximal ideal of R if it is not strictly contained by any other proper ideal of R.

- **Prime element:** Let R be a commutative ring with unity. An element $p \in R$ is called prime element if (i) $p \neq 0$, p is not a unit (ii) for any a, b, $c \in R$ if $p|ab$ then $p|a$ and $p|b$.

- **Irreducible element:** Let R be a commutative ring with unity. An element $p \in R$ is called an irreducible element if
  (i) $p \neq 0$, p is not a unit.
  (ii) whenever $p = ab$, then one of a or b must be a unit.

- **Euclidean Ring or Euclidean Domain:** Let R be an integral domain then it is said to be an Euclidean ring if for every non-zero element $a \in R$ there exists a non-negative integer $d(a)$ such that

  (i) for all non-zero elements $a, b \in R$,
  $d(a) \le d(ab)$ (or $d(b) \le d(ab)$)
  (ii) for every a, $b \in R$ with $b \neq 0$, there exist q, $r \in R$ such that $a = bq + r$ where either $r = 0$ or $d(r) < d(b)$

- **Unique Factorization Domain:** An integral domain is said to be unique factorization domain (UFD) if every non-zero element of D is either a unit or it is expressible as the product of finite number of prime elements in D and this factorization is apart from order and associates is unique.

Results

- A ring R is without zero divisor if and only if the cancellation laws hold in R.

- Every field is an integral domain.

- Every finite integral domain is a field.

- The characteristic of an integral domain is either zero or a prime number.

- The characteristic of the field of real numbers is zero.

- The set of complex numbers is not ordered integral domain.

- A field has no proper ideals.

- An ideal generated by a single element is known as principal ideal.

- Every homomorphic image of a commutative ring is commutative.

- A commutative ring R with unity having no zero divisors is called a principal ideal ring if every ideal of R is a principal ideal.

- The ring of integers is a principal ideal ring.

- Every field is a principal ideal ring.

- A proper ideal S of a ring R is said to be maximal ideal of R if there exists no proper ideal between S and R.

- A ring R is a field if and only if its zero ideal is a maximal ideal.

- An element which have multiplicative inverse is known as unit.

- In an integral domain if there exists a g.c.d. of any two elements, then it is unique apart from the distinction between associates.

- In every PID, each pair of non-zero elements surely has a g.c.d.

- In every PID, each pair of non-zero elements surely has a common multiple.

- In a PID, an element is prime if and only if it is irreducible.

- In an integral domain with unity, every prime element is irreducible. Converse is not necessarily true.

- The ring of integers is an Euclidean ring.

- Every field is an Euclidean ring.

- Every Euclidean ring is a principal ideal ring.

- The ring of polynomials over a field of reals is a Euclidean ring.

- An ideal of S of the Euclidan ring R is maximal if and only if S is generated by some prime element of R.

- An integral domain with unity is a UFD iff every non-zero non-unit element is product of primes.

- In a PID, every ascending chain of ideals must terminate after a finite number of steps.

- In a PID, every non-zero non-unit element is divisible by an irreducible element.

- A PID is a UFD.

- An irreducible polynomial of zero degree may not be primitive.

- A primitive polyomial may not be irreducible.

- The product of two polynomials of UFD, is again a primitive polynomial.

- Any polynomial of degree 1 is irreducible.

Review Questions and Project Work

1. Define the addition and mulitplication table for a ring with exactly one element.

2. If a ring R has more than one element and has a unity 1, show that $1 \neq 0$.

3. Show that in a ring with unity, a proper left zero divisor cannot have a multiplicative inverse.

4. Prove that with respect to addition and multiplication, the order of every non-zero element of the integral domain D is same although that element has been taken from the additive group D.

5. If R is a commutative ring with two or more elements and with the property that for $a, b \in R$, $a \neq 0$ there exists $x \in R$ such that $a.x = b$, prove that R is a field.

6. Prove that the quotient field of a finite integral domain coincides with itself.

7. Show that if F is a unique factoriztion domain then $F(x)$ is also a unique factorization domain.

8. Let R be a unique factorization domain. Then show that every prime element in R generates a prime ideal.

9. Let I be an ideal of a ring R. If R is an integral domain, is $R|I$ also an integral domain? Justify your answer.

10. Give an example of a ring in which some prime ideal is not a maximal ideal.

Objective type Questions

Fill in the blanks

1. If $(R, +,.)$ is a ring, then $a.(b + c) = $ ..... $\forall a, b\ c, \in R$.

2. If $(R, +,.)$ is a commutative ring, then $ab = $ ..... $\forall a, b, \in R$.

3. If $(R,+,.)$ is a ring, then $(R, +)$ is ...... group.

4. If $(R, +,.)$ is a ring and $a, b \in R$, then $a = $ .....

5. A ring R is said to be ring with unity if $a = $ .... $\forall a, b, \in R$ and $1 \in R$.

6. In a ring R, $a(-b) = -(ab)$ .... $\forall a, b, \in R$.

7. A ring R is with zero divisors if the product of two non-zero elements of R is ....

8. The set of integers is a ring ....

9. An integral domain is a commutative ring with .... having no zero divisors.

10. A commutative division ring is called ....

11. Every field is an ....

12. If $(F, + ,.)$ is a field and $a \in F$ and $a \neq 0$, the multiplicative inverse of a is ....

13. A skew field is not...........

14. Every finiie integral domain is a...........

15. If R is a ring such that $a^2 = a$ $\forall a \in R$, then R is ....

16. Every field has a multiplicative ....

17. Every ring has at least......... elements.

18. In a ring R if $a^2 = a$ $\forall a \in R$, then R is a ....

19. The ring of integers is .... but it is not a field.

20. An algebraic structure $\{(0,1,2), \text{'}+_3\text{'}, \text{'}\times_3\text{'}$ is a ....

21. A non-empty subset S of a ring R is a subring of R if $a-b \in S$ and .... $\in S$ $\forall a$.

22. If $S \subseteq R$ is a subring of a ring R, then $S+$ ...... $=$ S and $SS \subseteq$ ....

23. The intersection of two subrings is a ....

24. If S_1 and S_2 be two subrings of R and either $S_1 \subseteq S_2$ or $S_2 \subseteq S_1$, then .... is a subring of R.

25. A non-empty subset K of F (Field) is a subfield of F if $a -b \in K$, ... $\in K$ $\forall a, b \in K$.

26. The field of complex numbers is not.........

27. The zero divisors do not exist in a ......

28. The ring $(\{0, 1, 2, 3, ... p-1\}, +_5, \times_5)$ is a field if p is ....

29. The algebraic structure $(\{0, 1, 2, 3, 4\}) +_5,$ $\times_5)$ is a field, then multiplicative inverse of 3 is .....

30. In a ring $R - (-a) = \dots \forall a \in R$

True/False

Write 'T' for true and 'F' for false statement

1. Every field is also a ring.   **(T/F)**

2. In a ring $(R, +,.), (R, +)$ is an abelian group.   **(T/F)**

3. If R is a ring, then for all $a, b \in R,$ $(-a)(-b) = ab$ **(T/F)**

4. The set of even integers is not a ring. **(T/F)**

5. A ring is commutative ring if $ab = ba \ \forall a, b \in R$ **(T/F)**

6. The-structure $(\{0, 1, 2, 3, 4, 5\}, +_6, \times_6)$ is a commutative ring.   **(T/F)**

7. In a ring R, there exist $a \neq 0, b \neq 0$ such that $ab = 0$, then ring is without zero divisors.   **(T/F)**

8. The ring of integers is a ring with zero divisors.   **(T/F)**

9. Every commutative division ring is a field.   **(T/F)**

10. In a field every non-zero element has a multiplicative inverse.   **(T/F)**

11. Every integral domain is a field.   **(T/F)**

12. Every field is an integral domain.   **(T/F)**

13. Every skew field has zero divisors.   **(T/F)**

14. In a ring R, $a^2 = a \ \forall a \in R$, if $a+b=0$, then $a=b$.   **(T/F)**

15. If $F = \{0, 1\}$, then $(F, +_2, \times_2)$ is an integral domain.   **(T/F)**

16. If $R = \{0, 2, 4, 6, 8\}$, then $(R, +_{10}, \times_{10})$ is a field.   **(T/F)**

17. Every field is also a division ring.   **(T/F)**

18. The union of two subrings is always a surbing.   **(T/F)**

19. The field of real numbers is a subfield or rational numbers.   **(T/F)**

20. A finite commutative ring without zero divisors is a field.   **(T/F)**

21. The algebraic structure $(\{[0], [1], [2], \dots \dots [p-1]\} +_p, \times_p)$ is a field if p is a prime.   **(T/F)**

22. If $f(x) = 1+2x$ and $g(x) = 3+2x+6x^2$ are polynomials over the ring $(I_5, +_5, \times_5)$ then the leading coefficient of $f(x) \ g(x)$ is 3. **(T/F)**

23. If the ring $(R, +,.)$ is a commutative ring, then $(a+b)^2 = a^2 + ab + b^2.$ **(T/F)**

24. The ring of 2×2 matrices over the field of real numbers possesses zero divisors.   **(T/F)**

25. The ring of 2×2 matrices of a field of real numbers is a non-commutative ring.   **(T/F)**

26. The set $\{mx : x \in \mathbf{Z}; m$ being fixed integer$\}$ has zero divisors if m is not prime.   **(T/F)**

27. A divisor of zero is a commutative ring with unity can have no multiplicative inverse.   **(T/F)**

28. In a commutative ring R, if $ab = 0$ then we must have $ba = 0$ **(T/F)**

29. The ring $(\{0, 1, 2, 3, 4, 5\}, +_6, \times_6)$ possesses no zero divisor.   **(T/F)**

30. In a ring R, if $a \neq 0 \in R$ and $ab = ac$ then $b=c$.   **(T/F)**

31. A subset of any field is field.   **(T/F)**

32. The ring $(\{0,1, 2, 3, 4\}, +_5, \times_5)$ is a field.   **(T/F)**

Multiple choice Questions

Choose the most appropriate one :

1. In a commutative ring R, ab is equal to:
    (a) b (b) a
    (c) ba (d) 0

2. In a ring $(R, +,.)$ if $a, b \in R$, then $a. 0$ equals:
    (a) 0 (b) a
    (c) 1 (d) none of these

3. In a ring R if $a^2 = a \forall a \in R$, and $a = b$, then $a + b$ equals:
   (a) 1
   (b) a
   (c) b
   (d) 0

4. If S_1 and S_2 are two subrings of R and either $S_1 \subseteq S_2$ or $S_2 \subseteq S_1$, then which is a subring?
   (a) $S_1 - S_2$
   (b) S_1 / S_2
   (c) $S_1 \cup S_2$
   (d) None of these

5. The zero divisor do not exist in:
   (a) Field
   (b) Ring
   (c) Group
   (d) Vector space

6. The ring $(\{0, 1, 2, ...p -1\}, +_p, \times_p)$, is a field if p equals:
   (a) 9
   (b) 31
   (c) 33
   (d) 15

7. In the field $(\{0, 1, 2, 3, 4\}, +_5, \times_5)$, the multipicative inverse of 4 is:
   (a) 3
   (b) 1
   (c) 2
   (d) 4

8. Every field has at least ...... elements:
   (a) 2
   (b) 3
   (c) 1
   (d) 0

9. If a ring R is with zero divisor, then there exist $a \neq 0, b \neq 0$ in R such that ab equals:
   (a) 1
   (b) 3
   (c) 0
   (d) none of these

10. Every division ring is a field if it is:
    (a) associative
    (b) Commutative
    (c) closed
    (d) distributive

11. A ring is commutative if:
    (a) $a + b = b + a$
    (b) $a - b = b - a$
    (c) $ab = ba$
    (d) none of these

12. Every finite integral domain is a:
    (a) Field
    (b) Ring with zero divisors
    (c) Group
    (d) none of the above

13. The ring of integers is not a:
    (a) Commutative ring
    (b) Integral domain
    (c) Field
    (d) none of the above

14. A non-empty subset K of a field F is a subfield of F if $a - b \in K$, and for $a, b \neq 0 \in K$:

(a) $ab \in K$
(b) $a + b \in K$
(c) $b^{-1} \in K$
(d) $ab^{-1} \in K$

15. The leading coefficient of $f(x)g(x)$, where $f(x) = 2 + 5x + 3x^2$, $g(x) = 1 + 4x + 2x^3$ over the ring $(I_6, +_6, \times_6)$ is:
    (a) 3
    (b) 4
    (c) 2
    (d) 6

16. How many binary operations are there in a ring ?
    (a) 1
    (b) 2
    (c) 3
    (d) 4

17. A ring R is said to be commutative ring if :
    (a) addition is commutative
    (b) multiplication is commutative
    (c) there must eixsts a multiplicative identity
    (d) none of the above

18. If R is a ring with zero divisors and $a, b \in R$, then which of the following is true?
    (a) $a \cdot b = 0, a \neq 0, b \neq 0$
    (b) $a \cdot b = 0, a = 0, b = 0$
    (c) $a \cdot b \neq 0, a = 0, b = 0$
    (d) none of the above

19. If R is a ring without zero divisors and for $a, b \in R$ if $ab = 0$, then which of the following is true?
    (a) $a = 0$ or $b = 0$
    (b) $a = 0$ and $b = 0$
    (c) $a = 0, b \neq 0$
    (d) None of these

20. Which one of the following is not an integral domain ?
    (a) **Z**
    (b) 2**Z**
    (c) 3**Z**
    (d) Division ring

21. Which of the following ring is not an integral domain ?
    (a) (**Z**, +, .)
    (b) (**R**, +, .)
    (c) (**C**, +, .)
    (d) None of these

22. If R is a ring and $a \in R$. Let $T = \{x \in R : ax = 0\}$ then T is a :
    (a) right ideal of R
    (b) left ideal of R
    (c) not an ideal
    (d) none of these

23. Which of the following integral domain is not ordered ?
    (a) **Z**
    (b) **R**
    (c) **Q**
    (d) **C**

24. If R is a commutative ring and S be an ideal of R then R/S is a ring of modulo class is an integral domain if S is a :
    (a) prime ideal
    (b) maximal ideal
    (c) subring
    (d) none of these

Answers

Fill in the Blanks

1. $a.b+a.c$ **2.** ba **3.** abelian **4.** 0 **5.** 1. a **6.** $(-a).b$ **7.** zero **8.** with unity **9.** unity **10.** Field **11.** Integral domain **12.** a^{-1} **13.** Commutative **14.** Field **15.** Commutative **16.** Identity **17.** Two **18.** Boolean ring **19.** Integral domain **20.** Field **21.** ab **22.** $(-S)$, S **23.** Subring **24.** $S_1 \cup S_2$ **25.** ab^{-1} **26.** Ordered **27.** Field **28.** Prime **29.** 2 **30.** a

True/ False

1. T	**2.** T	**3.** F	**4.** F	**5.** T	**6.** T	**7.** F	**8.** F	**9.** T
10. T	**11.** F	**12.** T	**13.** F	**14.** T	**15.** T	**16.** F	**17.** F	**18.** F
19. F	**20.** T	**21.** T	**22.** F	**23.** F	**24.** T	**25.** T	**26.** F	**27.** T
28. T	**29.** F	**30.** T	**31.** F	**32.** T				

Multiple choice questions

1. (d)	**2.** (a)	**3.** (d)	**4.** (c)	**5.** (a)	**6.** (b)	**7.** (d)	**8.** (a)	**9.** (c)
10. (b)	**11.** (c)	**12.** (a)	**13.** (c)	**14.** (d)	**15.** (b)	**16.** (b)	**17.** (b)	**18.** (a)
19. (a)	**20.** (d)	**21.** (d)	**22.** (b)	**23.** (d)	**24.** (a)			

COMPETITION CORNER
for JRF, NET/SET, GATE Aspirants

Some Fascinating Facts

1. Every field is an integral domain.
2. Every field is a division ring.
3. Every commutative division ring is a field and hence also an integral domain.
4. A division ring need not be an integral domain, nor an integral domain need be a division ring.
5. Every field is division ring, integral domain, a ring with unity.
6. Every field (integral domain) is a commutative ring.
7. Every non-zero element of every field (division ring) is a unit (*i.e.*, has a multiplicative inverse)
8. In a ring, zero divisors and multiplicative inverse for all the non-zero elements cannot exist together.
9. Multiplication in a field is commutative.
10. Multiplication in an integral domain is commutative.
11. Multiplication in a Boolean ring is commutative.
12. If 0 and 1 are the additive and the multiplicative identities of a ring R then $0 \neq 1$.
13. The unity of a ring is a unit of the ring.
14. If a and b are the element of a field F, then the equation $ax + b = 0$ has a solution.
15. Every field is a skew field.
16. The multiplicative identity of a subring may be different from that of the parent ring.
17. There is a ring in which the equation $x^2 = -1$ has infinitely many solutions.
18. The ring $2\mathbf{Z}$ and $3\mathbf{Z}$ are isomorphic.

19. $\mathbf{Z}$ is an ideal of $\mathbf{Q}$.
20. Every ideal of a ring is a subring of the ring.
21. $\mathbf{Q}$ is its own prime subfield.
22. Let f be a homomorphism from a ring R to a ring S. Let A be a subring of R and B is an ideal of S then
    - for any $r \in R$ and any positive integer n, $f(nr) = nf(r)$ and $f(r^n) = [f(r)]^n$
    - $f(A) = \{f(a) : a \in A\}$ is a subring of S.
    - If A is an ideal and f is onto S then $f(A)$ is an ideal.
    - $f^{-1}(B) = \{r \in R : f(r) \in B\}$ is an ideal of R.
    - If R is commutative then $f(R)$ is commutative.
    - f is an isomorphism if and only if f is onto then $f(1)$ is the unity of S.
    - f is an isomorphsim if and only if f is onto and $Ker\ (f) = \{r \in R : f(r) = 0\} = \{0\}$
    - If f is an isomorphism from R onto S then f^{-1} is an isomorphism from S onto R.
23. If R is a commutative ring and I is an ideal of R then $R|I$ has no zero divisors if and only if I is a prime ideal.
24. A non-zero homomorphic image $D|I$ of an integral domain D is a field if and only if I is a maximal ideal of D.
25. An ideal I of a commutative ring R with unity is maximal if and only if $R|I$ is a field.

Some Examples of Rings

1. The set of integers under adddition and multiplication is a commutative ring with unity.
2. The set $Z_n = \{0, 1, ..., n-1\}$ under addition and multiplication modulo n is a commutative ring with unity.
3. The set $Z(x)$ of all polynomials in the variable x with integer coefficients under adddition and multiplication is a commutattive ring with unity $f(x) = 1$.
4. The set of even integers under addition and multiplication is a commutative ring without unity.

5. The set $M_2(Z)$ of 2×2 matrices with integer entries is a non-commutative ring with unity $\begin{bmatrix} 1 & 0 \\ 0 & 1 \end{bmatrix}$.
6. The set of all continuous real valued functions of a real variable whose graphs pass through the point $(1, 0)$ is a commutative ring without unity.
7. The set of all matrices of the form $\begin{pmatrix} p+iq & r+is \\ -r+is & p-iq \end{pmatrix}$ where p, q, r, s are even integers is a non-commutative ring without

unity and without zeo divisors with respect to matrix addition and multiplication.

8. The set of all matrices of the form $\begin{pmatrix} 0 & x \\ 0 & y \end{pmatrix}$,

$x, y \in \mathbf{Q}$ is a non-commutative ring having zero divisors with respect to matrix addition and multiplication.

9. Let $(G, +)$ be an abelian group. Let R be the set of all homomorphism of G to G. Let $\oplus$ and o be defined on R as follows :

for all $f, g \in R$

$$(f \oplus g)(x) = f(x) + g(x) \ \forall \ x \in G$$
$$fog = \text{composition of } f \text{ and } g$$

Then $(R, \oplus, o)$ is a non-commutative ring with unity.

Some Examples of Integral Domain

1. The algebraic structures $(\mathbf{Q}, +, .)$, $(\mathbf{R}, +, .)$ and $(\mathbf{C}, +, .)$ are all integral domains.

2. The set $\mathbf{Z}(\sqrt{2})$ of all real numbers of the form $a+b\sqrt{2}$ with a and b as integers is an integral domain.

3. The set of Gaussian integers form an integral domain w.r.t. ordinary addition and multiplication of complex numbes.

4. The ring of rational numbers of the type $\dfrac{a}{p^n}$,

where p is a fixed prime, a any integer greater than or equal to zero is an integral domain.

5. Each of the following is an integral domain for the usual addition and multiplication :

  (i) $\{a + b\sqrt{-3} : a, b \in \mathbf{Z}\}$

  (ii) $\{a + b\sqrt{-5} : a, b \in \mathbf{Z}\}$

  (iii) $\{a + b\sqrt{-7} : a, b \in \mathbf{Z}\}$

6. The ring $\mathbf{Z}[x]$ of polynomials with integral coefficients is an integral domain.

7. The ring $\mathbf{Z}_p$ of integers molulo p (prime) is an integral domain.

An Important Table

S. No.	Algebraic structure	Form of the element	Ring	Commutative ring	Integral domain	Field
1.	$(\mathbf{Z}, +, .)$	x	yes	yes	yes	no
2.	$(\mathbf{Z_n}, +, .)$ n is not prime.	x	yes	yes	no	no
3.	$(\mathbf{Z_n}, +, .)$ n is prime.	x	yes	yes	yes	yes
4.	$(\mathbf{Z}(x), +, .)$	Polynomial type	yes	yes	yes	no
5.	$(M_2(\mathbf{Z}), +, .)$	$\begin{bmatrix} a & b \\ c & d \end{bmatrix}$	yes	no	no	no
6.	$(M_2[2\mathbf{Z}], +, .)$	$\begin{bmatrix} 2a & 2b \\ 2c & 2d \end{bmatrix}$	yes	no	no	no
7.	$(n\mathbf{Z}, +, .)$	nx	yes	no	no	no
8.	$(\mathbf{Z}[i], +, .)$	$x+iy$	yes	yes	yes	no
9.	$(\mathbf{Z}[\sqrt{2}], +, .)$	$x+\sqrt{2}\,y; x, y \in \mathbf{Z}$	yes	yes	yes	no
10.	$(\mathbf{Q}[\sqrt{2}], +, .)$	$x+\sqrt{2}\,y; x, y \in \mathbf{Q}$	yes	yes	yes	yes

Self Assessment Test

1. Prove that a ring can have atmost one unity.
2. Describe all the subrings of the ring of integers.
3. Prove that the centre of a ring is a subring.
4. Show that a unit of ring divides every element of the ring.
5. Show that every subgroup of $\mathbf{Z}_n$ under addition is also a subring of $\mathbf{Z}_n$.
6. Let $M_2(\mathbf{Z})$ be ring of all 2×2 matrices over the integers and let $\mathbf{R}$ $\left\{ \begin{bmatrix} a & a-b \\ a-b & b \end{bmatrix} : a, b \in \mathbf{Z} \right\}$.

   Prove or disprove that R is a subring of $M_2(\mathbf{Z})$.
7. Show that $2\mathbf{Z} \cup 3\mathbf{Z}$ is not a subring of $\mathbf{Z}$.
8. Let us suppose there is a positive even integer n such that $a^n = a$ for all elements a of some ring R. Show that $-a = a \ \forall \ a \in R$.
9. Show that every non-zero element of $\mathbf{Z}_n$ is a unit or a zero divisor.
10. Show that 0 is the only nilpotent element in an integral domain.
11. Show that the set of idempotent of a commutative ring is closed under multiplication.
12. Find all elements of a ring that are both units and idempotents.
13. Let R be a ring with unity. If the product of any pair of non-zero elements of R is non-zero, prove that $ab = 1 \Rightarrow ba = 1$.
14. For any two elements a and b of an integral domain, if $a^m = b^m$ and $a^n = b^n$, where m and n are positive integers that are relatively prime, show that $a = b$.
15. Let R be a commutative ring without zero divisors. Show that all the non-zero elements of R have the same additive order.
16. Let D be an integral domain and f is a non-constant function from D to the non-negative integers such that $f(xy) = f(x)f(y)$. If x is a unit in D. Show that $f(x) = 1$.
17. Let $S = \{x+iy; \ x, y \in \mathbf{Z}, y \text{ is even}\}$. Show that S is a subring of $\mathbf{Z}(i)$ but not an ideal of $\mathbf{Z}(i)$.
18. If I_1 and I_2 are two ideals of a commutative ring R with unity and $I_1 + I_2 = R$, show that $I_1 \cap I_2 = I_1 I_2$.
19. Let p be prime. Show that $I = \{(bx, y) : x, y \in \mathbf{Z}\}$ is a maximal ideal of $\mathbf{Z} \oplus \mathbf{Z}$.
20. Let $R = \{0, 2, 4, 6, 8, 10\}$ be a ring under addition and multiplication modulo 12. Show that the characteristic of R is 6.
21. Show that the set of all units in a ring $(\mathbf{R}, +, .)$ forms a group with respect to multiplication $(.)$.
22. Prove that a ring R is commutative iff $(a+b)^2 = a^2 + 2ab + b^2 \ \forall \ a, b \in R$.
23. Prove that in a commutative ring of characteristic 2
$$(a+b)^2 = a^2 + b^2, \ a, b \in R.$$
24. Prove that in a ring with unity and without zero divisors, the only idempotents are the unity and zero.
25. Prove that if $a \neq 0$ and a is idempotent in a ring, then a is not nilpotent.

• • •

6 Structure Theory of Rings
(Some Miscellaneous Concepts)

■ Outlines

- Radicals ;
- Ring with ORE Conditions:
- Grobner Basis;
- Noetherian Ring;
- Ring of Fraction;
- Artinian Ring

6.1 INTRODUCTION

In this chapter we shall discuss some advanced topic of ring theroy. The topics include ring of fraction ORE condition, primary decomposition of ideals, Grobner basis for ideals, Wedderburn's theorem on finite division ring etc.

6.2 PRIMARY DECOMPOSITION OF IDEALS

Definition 1. *Let R be a commutative ring and I be an ideal of R. The radical of I is said to be the ideal $\cap$ P, where the intersection is taken over all primes ideals P which contain I. Usually, it is denoted by Rad. I.*

For example :

(1) In any integral domain, zero ideal is prime. Therefore Rad. (0) =0.

(2) In the ring of intergers **Z,**

$$\text{Rad. } (12) = (2) \cap (3) = 6$$
$$\text{Rad. } (4) = (2) = (2) \text{ Rad. } (32)$$

REMARKS

☞ Radical of *I* is also known as prime radical of *R*.

☞ The set of primes containing *I* is empty. Then Rad. *I = R*.

Definition 2. *Let R be a commutative ring. In ideal I(R) of R is said to be primary ideal, if for any a, b $\in$ R, a.b $\in$ I and a $\in$ I $\Rightarrow$ $b^n \in$ I, for some n $\geq$ 1.*

REMARKS

☞ Every prime ideal is clearly primary.

☞ If *p* is a prime integer and $n \geq 2$, a positive integer, then $(p)^n = (p^n)$ is a primary ideal in **Z,** which is not prime ideal. Therefore, a power p^n of a prime ideal *P* need not be primary.

Definition 3. *Let R be a commutative ring and I be a primary ideal of R such that P = Rad.I. Then, I is said to belong to p or to be p-primary or I is primary to P or associated prime ideal of I.*

REMARK

☞ Every prime ideal is primary and belongs to itself.

Definition 4. *Let R be a commutative ring. An ideal I of R has a primary decomposition if* $I = I_1 \cap I_2 \cap ... \cap I_n$, *where each* I_i *is primary.*

If no I_i contains $I_1 \cap I_2 \cap ... \cap I_{i-1} \cap I_{i+1} \cap .... \cap I_n$ and the radicals of the I_i are all distinct, then the primary decompositions is said to be *reduced* (or *irredudant*).

Definition 5. *Let R be a commutative ring. An ideal I of R has a normal decomposition if* $I = I_1 \cap I_2 \cap ... \cap I_n$ *with each* I_i, P_i *-primary ideal.*

Theorem 1. *The radical of an ideal is also an ideal.*

Proof.   Let R be a commutative ring and I be any ideal of R. We have to show that Rad. I is also an ideal.

Let us suppose that $a, b \in$ Rad. I.

By definition, there exist positive integers m and n such that $a^m, b^n \in I$. Further , $(a+b)^{m+n}$ can be written as a sum of terms of the form $a^\mu b^\nu$ where μ, ν are non-negative integers such that $\mu + \nu = m + n$.

For such μ, ν, either $\mu \geq m$ or $\nu \geq n$. Therefore in any way $a^\mu b^\nu \in I$

$\Rightarrow$ $\qquad\qquad\qquad (a+b)^{m+n} \in I$

$\Rightarrow$ $\qquad\qquad\qquad (a+b) \in$ Rad. I.

Now, let r be any arbitrary element of R, then $(ra)^m = r^m a^m \in I$. Because I is an ideal of R, thus $ra \in$ Rad.I.

In a similar way we can show that $ar \in$ Rad. I. Hence, Rad. I is an ideal of R.

Theorem 2. *Let R be a commutative ring and I be an ideal of R.*

Then,   Rad. I = $\{r \in R : r^n \in I$, for some positive integer n$\}$.

Proof.   If Rad. $I = R$, then the set

$\qquad\qquad \{r \in R : r^n \in I\} \subset$ Rad. I $\qquad\qquad\qquad$ ...(1)

Thus, assumimg Rad. $I \neq R$. Now, if $r^n \in I$ and P is any prime ideal containing I, then $r^n \in P, r \in P$.

Therefore

$\qquad\qquad\qquad \{r \in R : r^n \in I\} \subset$ Rad. I

Now, let $t \in R$ and $t^n \notin I$ for all $n > 0$, then the set

$\qquad\qquad\qquad S = \{t^n + x : n \in \mathbf{Z}^+; x \in I\}$

is a multiplicative set such that $S \cap I = \phi$.

Then, there is a prime ideal P disjoint from S contains I

Therefore, $\qquad\qquad\qquad t \notin P$

$\Rightarrow$ $\qquad\qquad\qquad t \notin$ Rad. I

$\Rightarrow$ $\qquad\qquad\qquad t \notin \{r \in R : r^n \in I\}$

$\Rightarrow$ $\qquad\qquad$ Rad. $I \subset \{r \in R : r^n \in I\}$ $\qquad\qquad\qquad$ ...(2)

From (1) and (2) we conclude that

$\qquad\qquad\qquad$ Rad. $I = \{r \in R : r^n \in I\}$.

Theorem 3. *Let R be a ring. If I is an ideal of R. Then Rad (Rad I) = Rad I.*

Proof. By definition of Rad I, we have

$$a \in \text{Rad. } I \text{ if and only if } a^m \in I, m \in \mathbf{Z}^+$$

Thus, we have $\qquad I \subseteq \text{Rad } I$

$\Rightarrow \qquad\qquad \text{Rad } (I) \subseteq \text{Rad(Rad } I) \qquad\qquad\qquad\qquad\qquad$ ...(1)

$$[\text{By using } A \subseteq B \Rightarrow \text{Rad } A \subseteq \text{Rad } B]$$

Now, it remains to prove that Rad(Rad I) $\subseteq$ Rad I

Let $a \in \text{Rad (Rad } I)$

$\Rightarrow \quad \exists \, m \in \mathbf{Z}^+$, such that $a^m \in \text{Rad } (I)$

$\Rightarrow \qquad\qquad\qquad (a^m)^n = a^{mn} \in I, n \in \mathbf{Z}^+$

$\Rightarrow \qquad\qquad\qquad\qquad a \in \text{Rad } I$

$\qquad\qquad\qquad \text{Rad(Rad } I) \subseteq \text{Rad } I \qquad\qquad\qquad\qquad\qquad$ ...(2)

From (1) and (2), we conclude that

$$\text{Rad(Rad } I) = \text{Rad } I.$$

Theorem 4. *Let R be a ring and $I_1, I_2, ..., I_m$ be ideals of R*

Then,   *Rad $(I_1, I_2, ...I_m) =$ Rad $(I_1 \cap I_2 \cap ... \cap I_m)$*

                       $=$ Rad $I_1 \cap$ Rad $I_2 \cap ... \cap$ Rad I_m.

Proof. Clearly, we have

$$\text{Rad } (I_1, I_2, ...I_m) \subseteq \text{Rad } (I_1 \cap I_2 \cap ... \cap I_m) \subseteq \text{Rad } I_1 \cap \text{Rad } I_2 \cap ... \cap \text{Rad } I_m.$$

Further, suppose that $\qquad a \in \text{Rad } I_1 \cap \text{Rad } I_2 \cap ... \cap \text{Rad } I_m.$

Then for each i, $1 \leq i \leq m$, we may choose an integer n_i such that $a^{n_i} \in I_i$.

Now putting

$$n = n_1 + n_2 + ... + n_m$$

$\Rightarrow \qquad\qquad a^n = a^{n_1} . a^{n_2} .... a^{n_m} \in (I_1, I_2, ...I_m)$

$\Rightarrow \qquad\qquad a \in \text{Rad } (I_1, I_2, ...I_m)$

Then, we have

$$\text{Rad } I_1 \cap \text{Rad } I_2 \cap ... \cap \text{Rad } I_m \subseteq \text{Rad } (I_1 \cap I_2 \cap ... \cap I_m) \subseteq \text{Rad } (I_1, I_2, ...I_m)$$

Hence, we conclude that

$$\text{Rad } (I_1, I_2, ...I_m) = \text{Rad } (I_1 \cap I_2 \cap ... \cap I_m)$$

$$= \text{Rad } I_1 \cap \text{Rad } I_2 \cap ... \cap \text{Rad } I_m .$$

REMARK

☞ In a similar way, we may prove that the following result :

If I is an ideal of ring R and m is a positive integer, then Rad $(I^m) = $ Rad I.

Theorem 5. *The radical of a primary ideal is a prime ideal.*

Proof. Let R be a commutative ring and I be any ideal of R.

We have to show that Rad I is a prime ideal of R.

Now suppose $\qquad\qquad ab \in \text{Rad } I \text{ and } a \notin \text{Rad } I$

Then, we have

$$a^n b^n = (ab)^n \in I, \text{ for some positive integer } n.$$

Now, since $a \notin \text{Rad } I$, $a^n \notin I$ and I is a primary ideal then there exists a positive integer m such that

$$(b^n)^m \in I$$

$$\Rightarrow \qquad b \in \text{Rad } I.$$

Hence, I is prime ideal of R.

Theorem 6. *Let R be a commutative ring and I, P be ideals in R. Then, I is primary for P (i.e., P-primary) if and only if*

(i) $I \subset P \subset \text{Rad } I$

(ii) $ab \in I$ and $a \notin I$, then $b \in P$.

Proof. Let us first suppose (i) and (ii) holds. We have to show that I is primary for P. If $ab \in I$ with $a \notin I$, then $b \in P \subset \text{Rad } I$, $b^n \in I \Rightarrow I$ is primary.

To show I is primary for P. We shall show that $P = \text{Rad } I$.

Clearly, from (i), we have $P \subseteq \text{Rad } I$.

If $b \in \text{Rad } I$, and n is the least positive integer such that $b^n \in I$.

If $n = 1$, $b \in I \subset P$. If $n > 1$, then $b^{n-1}b = b^n \in I$ with $b^{n-1} \notin I$, by the minimality of n.

Also, by (ii) $\qquad b \in P$

$$\Rightarrow \qquad b \in \text{Rad } I \Rightarrow b \in P \text{ whenever Rad } I \subset P. \text{ Hence } P = \text{Rad } I.$$

Conversely, let I be primray for P, then $P = \text{Rad. } I$.

Now, since $I \subseteq \text{Rad } I$ so that $I \subset P \subset \text{Rad } I$.

Also $\qquad I \subset \text{Rad } I \Rightarrow I \subset P \subset \text{Rad } I$. Since I is primary so if $ab \in I$ and $a \notin I$, then $b^n \in I$ for some positive integer. In particular $b \in I$ and $I \subset \text{Rad } I$.

Thus, $\qquad b \in \text{Rad } I = P$

$$\Rightarrow \qquad b \in P$$

Hence, we conclude that if $ab \in I$ and $a \notin I$, then $b \in P$.

Theorem 7. *Let R be a commutative ring and $I_1, I_2, ..., I_n$ be prime ideals in R, all of which are primary for the prime ideal P, then $\bigcap_{i=1}^{n} I_i$ is also a primary ideal belonging to P.*

Proof. Let us write $I = \bigcap_{i=1}^{n} f_i$

Then $\text{Rad } I = \text{Rad}\left(\bigcap_{i=1}^{n} I_i\right) = \bigcap_{i=1}^{n} I_i \text{ Rad } f_i$

Now, since for each i, I_i is P-primary so that $P = \text{Rad } I_i$.

and therefore $\text{Rad } I = \bigcap_{i=1}^{n} P = P$

$\Rightarrow \quad I$ is P-primary.

$\Rightarrow \quad \bigcap_{i=1}^{n} f_i$ is also primary ideal belonging to P.

Theorem 8. *Let R be a ring and I be a maximal ideal of R, then all its powers I ($n \geq 1$) are primary ideals.*

Proof. Since we have $I^n \subseteq I = \text{Rad } I^n$

Now, suppose that $ab \in I_n$ with $a \notin I$. Then, the ideal (I, a) generated by I and a must be the ring R because I is maximal. Therefore, the identify element $1 \in (I, a)$.

Thus, for some $m \in I$ and $r \in R$, we must have

$$1 = m + ra$$

Now, $m^n \in I^n$, then

$$1 = (m + ra)^n = m^n + r'a, \text{ for } r' \in R$$

$$\Rightarrow \qquad b = bm^n + r'(ab) \in I^n$$

Therefore $I^n \subseteq I = \text{Rad } I^n$ and if $ab \in I^n$ with $a \notin I$ then $b \in I^n$. Hence I^n are primary ideals.

REMARK

☞ Let I be an ideal of a ring R. Rad I is maximal ideal of R, then I is primary.

Theorem 9. *Let R be commutative ring and I be an ideal of R. If I has a primary decomposition, then I has a reduced primary decomposition.*

Proof. As per given, I has a primary decomposition, therefore $I = I_1 \cap I_2 \cap ... \cap I_n$ with each I_i and some I_i contains

$$I_1 \cap I_2 \cap ... \cap I_{i-1} \cap I_{i+1} \cap ... \cap I_n$$

then $I = I_1 \cap I_2 \cap ... \cap I_{i-1} \cap I_{i+1} \cap ... \cap I_n$ is also a primary decomposition. Now eliminating the superfluous I_i we have

$$I = I_1 \cap I_2 \cap ... \cap I_k$$

with no I_i containing the intersection of the other I_j.

Further let $P_1, P_2, ... P_r$ be the distinct prime ideals in the set $\{\text{Rad } I_1, \text{Rad } I_2, ..., \text{Rad } I_k\}$ and let I'_1 ($1 \leq i \leq r$) be the intersection of all I's that belong to the primary P_i. Thus, clearly no I'_1 contains the intersection of all the other I_j^i. Therefore, we have

$$I = \bigcap_{i=1}^{k} I_i = \bigcap_{i=1}^{r} I'_i, I \text{ is a reduced decomposition.}$$

Theorem 10. *If I_1 and I_2 are primary ideals that have the same radical β. Then $I_1 \cap I_2$ is primary.*

Proof. It is known that

$$\text{Rad } (I_1 \cap I_2) = \text{Rad } (I_1) \cap \text{Rad } (I_2) \qquad \qquad ...(1)$$

Also, it is given that Rad $(I_1) = \text{Rad } (I_2) = \beta$

Therefore, from (1) we have

$$\text{Rad } (I_1 \cap I_2) = \beta$$

Further, let a be a non-divisor modulo $(I_1 \cap I_2)$. Then we have $ab \neq 0 \pmod{I_1 \cap I_2}$ such that $\qquad \qquad ab \equiv 0 \pmod{I_1 \cap I_2}$

Now, since $b \neq 0 \pmod{I_1 \cap I_2}$,

We can suppose that $b \neq 0 \pmod{I_1}$.

Then, $ab \equiv 0 \pmod{I_1}$ gives $a \equiv 0 \pmod{\text{Rad } (f_1)}$

$$\Rightarrow \qquad \qquad \alpha = \beta.$$

Hence, $I_1 \cap I_2$ is primary.

REMARKS

☞ Let D be a prime ideal and let β' be a prime ideal contains D then $\beta = \beta' = \text{Rad}(D)$.

☞ Let D be primary, β its associated prime and let S be any ideal not contains in β then $(D : S) = D$.

Theorem 11 (First Uniqueness Theorem). *Let* $\beta = D_1 \cap D_2 \cap .... \cap D_r = D'_1 \cap D'_2 \cap .... \cap D'_s$ *be two irredundant intersections into prime ideals whose associated primes are distinct. Then* $r = s$ *and the sets of primes of two decompositions are identical.*

Proof. Let
$$\beta_1 = \text{Rad}(D_i)$$
$$\beta'_1 = \text{Rad}(D'_i)$$

Then there exists ideals in the set $\beta_1, \beta_2,..., \beta_r, \beta'_1, \beta'_2,..., \beta'_i$ that are not contain properly in any of the ideals of this collection. We may suppose that β_1 has this property. Firstly, we shall prove that β_1 is also in the set $\beta'_1, \beta'_2,..., \beta'_s$. If not, then $\beta_1 \not\leq \beta'_i$ for $i = 1, 2,..., s$. Thus using above remark, we have

$$D_1 \not\leq \beta'_i.$$

Also $\qquad\qquad D'_i : D_1 = D'_i$

Therefore,

$$\begin{aligned}
\beta : D_1 &= (D'_1 \cap D'_2 \cap ... \cap D'_s) : D_1 \\
&= D'_1 : D_1 \cap D'_2 : D_1 \cap .... \cap D'_s : D_1 \\
&= D'_1 \cap D'_2 \cap ... \cap D'_s = \beta
\end{aligned}$$

Similarly, $\qquad\qquad D_j : D_1 = D_j \text{ if } j > 1$

Therefore,

$$\begin{aligned}
\beta = \beta : D_1 &= (D_1 \cap D_2 \cap ... \cap D_r) : D_1 \\
&= D_2 \cap D_3 \cap ... \cap D_r
\end{aligned}$$

which contradicts the fact that the first decomposition is irredundant. Now, suppose that
$$\beta_1 = \beta'_1 .$$
The ideal $D_1 \cap D'_1$ is primary. The ideal $D_1 \cap D'_1$ is primary with β_1 as associated prime.

Therefore

$$D_j : (D_1 \cap D'_1) = D_j, \quad \text{for } j > 1$$
and $\qquad D'_i : (D_1 \cap D'_1) = D'_i, \text{ for } i > 1.$

Therefore,

$$\begin{aligned}
\beta : (D_1 \cap D'_1) &= D_2 \cap D_3 \cap ... \cap D_r \\
&= D'_2 \cap D'_3 \cap ... \cap D'_s.
\end{aligned}$$

and those are two irredundant decompositions of $\beta = (D_1 \cap D_2)$ satisfying the conditions of the theorem. We shall use induction to conclude that the sets of prime ideals $\beta_2, \beta_3,..., \beta_r$ coincide with the set $\beta'_2, \beta'_3,..., \beta'_s$.

Theorem 12. *Let* S_1 *and* S_2 *are ideals in a Noetherian ring* S_1 *such that* $S_2 = S_1$ *if and only if* I_2 *is not contained in any of the associated prime of* S_1.

Proof. Let $S_1 = D_1 \cap D_2 \cap ... \cap D_r$ be an irredundant decompositions of D into primary ideals let $\beta_i = \text{Rad}.(D_i)$ and assume that $S_2 \subseteq \beta_i$.

Then, we have

$$S_1 : S_2 = (D_1 \cap D_2 \cap ... \cap D_r) : S_2$$
$$= D_1 : S_2 \cap D_2 : S_2 \cap ... \cap D_r : S_2$$
$$= D_1 \cap D_2 \cap ... \cap D_r = S_1$$

On the other hand, suppose that $S_2 \subseteq \beta_i$ for some i, say $S_2 \nsubseteq \beta_1$. Then, there exists an integer m such that

$$S_2^m \subseteq D_1.$$

Therefore,

$$S_2^m (D_2 \cap ... \cap D_r) \subseteq S_2^m \cap D_2 \cap ... \cap D_r \subseteq b_1.$$

Now, let n be the smallest integer such that

$$S_2^n (D_2 \cap ... \cap D_r) \nsubseteq S_1$$

Since $S_1 = D_1 \cap ... \cap D_r$ is irredundant, $n \geq 1$.

Therefore

$$S_2^{n-1} (D_2 \cap ... \cap D_r) \nsubseteq S_1$$

On the other hand,

$$S_2^{n-1} (D_2 \cap ... \cap D_r) \subseteq S_1 : S_2$$

Hence, $S_1 : S_2 \subseteq S_1.$

Theorem 13. *If S_1 and S_2 are ideals in a Noetherian ring, then $S_1 : S_2 = S_1$ if and only if no assiciated prime of S_1 is contained in any F, the associated prime of S_1 .*

Proof.   Suppose we have a decomposition of S_2 as an irredundant intersection

$$D'_1 \cap D'_2 \cap .... \cap D'_s$$

with associated prime .

$$\beta'_1, \beta'_2,..., \beta'_s.$$

Then if $S_2 \leq \beta_1, D'_1 ... D'_r .... D'_s \leq \beta_1$ therefore one of D'_j and consequently one of the β_j is contained in β_1. Conversely, it is clear that, if $\beta'_j \leq \beta_1$, then $S_2 \leq \beta_j$.

Theorem 14 (Second Uniqueness Theorem).

Let $I = D_1 \cap D_2 \cap .... \cap D_r = D'_1 \cap D'_2 \cap .... \cap D'_r$ be two decompositions of I that satisfy the conditions of the first uniqueness theorem.

Let $S = D_{i1} \cap D_{i2} \cap .... \cap D_{ik}$ be an isolated component in the first decomposition and let S' be the isolated component of the second decomposition that has the same set of associates prime as S. Then $S = S'$.

Proof.   We can write

$$I = S \cap D = S' \cap D'$$

where D and D' are respectively the intersection of the D_i and D'_i that not contains S and S'. Then, the associated prime of $D \cap D'$ are contained in none of the associated prime of S. Thus $S : (D \cap D') = S$

Similarly

$$S' : (D \cap D') = S'$$

Therefore

$$I : (D \cap D') = (S : (D \cap D')) \cap (D : (D \cap D')) = S$$

and
$$I : (D \cap D') = (S' : (D \cap D')) \cap (D' : (D \cap D')) = S'$$

Hence,
$$S = S'.$$

Solved Examples

Example 1. *Find a reduced primary decomposition of an ideal (12600) in **Z**.*

Solution. We have
$$12600 = 2^4 . 3^2 . 5^2 . 7$$
$$= (2^4) \cap (3^2) \cap (5^2) \cap (7)$$
$$= (16) \cap (9) \cap (25) \cap (7)$$

Now, Rad $(16) = (4)$, Rad $(9) = 3$, Rad $(25) = 5$

Rad $(7) = 7$

Therefore, the radicals of all ideals are distinct and no ideal contains in the intersection of other ideals. Also, the ideals (16), (9), (25) and (7) are all primary.

Hence, $(12600) = (16) \cap (9) \cap (25) \cap (7)$

is the reduced primary decomposition.

Example 2. *Let F be a field and I, the ideal (x^2, xy) in $F(x, y)$. Show that I has a reduced primary decomposition as $I = (x) \cap (x^2, xy, y^2)$.*

Solution. We know that (x^2, xy) consists of those polynomials which have x as a factor and which do not posses linear terms. The non-zero element of (x^2, xy, y^2) are precisely the polynomials each of whose terms are of degree greater then equal to two. Thus, the intersection $(x) \cap (x^2, xy, y^2)$ contains the zero polynomials together, with all polynomials of degree ≥ 2 which have x as a factor. Therefore, we have

$$I = (x^2, xy) = (x) \cap (x^2, xy, y^2) \qquad \qquad ...(1)$$

Also, (x) and (x^2, xy, y^2) are primary ideals.

Hence, I has a reduced primary decomposition of (1).

EXERCISE 6.1

1. If **Z** is a ring of integers, then show that the ideal (x^2, xy, y^2) of **Z**$[x, y]$ is a primary ideal.
2. In the ring of integers, show that following are primary decomposition
   (i) $(4, 2x, x^2) = (4, x) \cap (2, x^2)$
   (ii) $(9, 3x+3) = (3) \cap (9, x+1)$
3. In the ring of integers, the ideals (x^2, y), (x^2, y^2), (x^3, y^3) ...(x^i, y^i)... are all ideals

belonging to the prime ideals (x, y)
4. Show that the radical of (x, y^3) in $F(x, y)$ is (x, y).
5. Show that $(x^2, 2x, 4)$ is a primary ideal of **Z**$[x]$ but is expressible in the form $(x^2, 2) \cap (x, 4)$
6. Show that the intersection of two primary ideals with common radicals is a primary ideal with the radical.

6.3 GRÖBNER BASES FOR IDEALS

6.3.1 Algebraic Varieties and Ideals

Let F be a field, we know that $F(x_1, x_2,..., x_n)$ is the ring of polynomial in n indeterminants $x_1, x_2,..., x_n$ with coefficients in F. Let F^n be the Cartesian product $F \times F \times .... F$ for n factors. Also, we denote $(a_1, a_2,..., a_n)$ of F^n by a in bold type.

Definition 1. *Let S be a finite subset of F(x). The algebraic variety V(S) in F^n is the set of all common zero in F^n of the polynomials in S.*

For example: Let $S = \{2x+y-2\} \subset R\{x, y\}$. The algebraic variety $V(S)$ in R^2 is the line in with x intercept 1 and y-intercept 2.

Definition 2. *Let I be an ideal in a commutative ring R with unity. A subset $(b_1, b_2,..., b_r)$ of I is a basis for I if I = $<b_1, b_2,..., b_r>$*

Theorem 1. *Let $f_1, f_2,..., f_r \in F(x)$. The set of common zeros in F^n of the polynomials f_i for $i = 1, 2, ..., r$ is the same as the set of common zeros in F^n of all the polynomials f_i for $i = 1, 2,..., r$ is the same as the set of common zero in F^n of the polynomials in the entire ideal I = $<f_1, f_2,..., f_r>$*

Proof. Let

$$f = c_1 f_1 + c_2 f_2 + ... + c_r f_r \qquad ...(1)$$

by any element of I.

Further let $a \in F^n$ be a common zero of $f_1 . f_2,...,$ and f_r.

From (1), we obtain

$$f(a) = c_1(a)f_1(a) + c_2(a)f_2(a) + ... + c_r(a)f_r(a)$$
$$= c_1(a).0 + c_2(a).0 + ... + c_r(a).0 = 0$$

which shows that a is also a zero of every polynomials f in I. Also, a zero of every polynomials in I will be a zero of each f_i because each f_i belongs to I.

For an ideal I in $F(x)$, we let $V(I)$ be the set of all common zero of all elements of I.

Hence, $V(\{f_1, f_2,..., f_r\}) = V(<f_1, f_2,..., f_r>)$

6.3.2 Gröbner Bases

We consider the problem of finding a nice basis for ideal for an ideal I in $F(x) = F(x_1, x_2,..., x_n)$.

6.3.3 Properties for a Ordering of Power Products

(1) $1 < P$ for all power products $P \neq \phi$.

(2) For any two power products P_i and P_j exactly one of $P_i < P_j$, $P_i = P_j$, $P_j < P_i$ hods.

(3) If $P_i < P_j$ and $P_j < P_k$ then $P_i < P_k$.

(4) If $P_i < P_j$ then $PP_i < PP_j$, for any power product P.

Let us consider a power product in $F(x)$ to be an expression

$$P = x_1^{m_1} \cdot x_2^{m_2} ..... x_n^{m_n} \text{ where all } m_i \geq 0 \text{ in } \mathbf{Z}.$$

Hence, all x_i are present perhaps some with some exponent 0. Therefore, in $F(x, y, z)$ we must write xz^2 as xyz^2 to be a power product. Now, we want to describe a total ordering $<$ on the set of all power products so that we know just what it means to say that $P_i < P_j$ for two power products, providing us with a notion of relative size for power products. We can then try to change an ideal basis in a systematic way to create one with polynomials having terms $a_i P_i$ with as "small" power products P_i as possible. We denote by 1 the power product with all exponents 0 and require that an ordering has been describe and that $P_i \neq P_j$ and P_i divides P_j so that $P_j = PP_i$ where $1 < P$. Using property (4), we have $1P_i < PP_i P_j$ therefore $P_i | P_j$. Therefore P_i divides P_j implies that $P_i < P_j$.

In $F(x)$ with x, the only indeterminate, there is only one power product ordering, we must have $1 < x$. Multiplying repeatedly by x and again using property (4) we have $x < x^2$, $x^2 < x^3$ etc.

Then property (3) shows that $1 < x < x^2 < x^3 < \ldots$ is the only possible order. There are a number of possible ordering for power products in $F(x)$ with n indeterminate. We consider just one, the lexicographical order (lex). In lex, we have

$$x_1^{s_1} x_2^{s_2} \ldots x_n^{s_n} < x_1^{t_1} x_2^{t_2} \ldots x_n^{t_n} \qquad \ldots(1)$$

if and only if $s_i < t_i$ for the first subscript i, reading from left to right, such that $s_i \neq t_i$. Therefore in $F(x, y)$, if we write power products in the order $x^n y^m$, we have $y = x^0 y^1 < x^1 y^0 = x$ and $xy < xy^2$. Using lex, the order of n indeterminate is given by $1 < x_n < x_{n-1} < \ldots < x_2 < x_1$.

For the two-indeterminate case with $y < x$, the total lex term order schematically is

$$1 < y < y^2 < y^3 < \ldots < x < xy < xy^2 < xy^3 < \ldots < x^2 < x^2 y < x^2 y^2 < \ldots$$

REMARKS

☞ An ordering of power products P induces an obvious ordering of terms aP of a polynomial $F(x)$, known as **term order.** Now, given an ordering of power products, we consider every polynomial f in $F(x)$ to be written in decreasing order of terms, so that the leading term has the highest order. We denote by $1t(f)$, the leading term of f and by $1p(f)$, the power product of the leading term.

☞ If f and g are polynomials in $F(x)$ such that $1\,p(g)$ divides $1\,p(f)$, then we can execute a division of f by g to obtain

$$f(x) = g(x)\,g(x) + r(x)$$

where $\qquad\qquad 1p(x) < 1p(f).$

☞ The above properties guarantee that step by step process for modifying a finite ideal basis that does not increase the size of any maximal power product in a basis element and replace at least one by something smaller at each step will terminate in a finite number of steps.

Definition. *A set* $\{g_1, g_2, \ldots, g_r\}$ *of non-zero polynomials in* $F(x_1, x_2, \ldots, x_n)$ *with term ordering* $<$, *is a Gröbner basis for the ideal* $I = <g_1, g_2, \ldots, g_f>$ *if and only if for each non-zero* $f \in I$, *there exists some* i *where* $1 \leq i \leq r$ *such that* $1\,p(g_i)$ *divides* $1p(f)$.

REMARKS

☞ Any basis $\{g_1, g_2, \ldots, g_r\}$ can fail to be a Gröbner basis for $I = <g_1, g_2, \ldots, g_r>$ because when we form an element $c_1 g_1 + c_2 g_2 + \ldots + c_r g_r$ in I, we see that $1p(g_i)$ is a divisor of $1p(c_i g_i)$ for $i = 1, 2, \ldots, r$.

☞ A basis $G = \{g_1, g_2, \ldots, g_r\}$ is a Gröbner basis for the ideal $< g_1, g_2, \ldots, g_r >$ if and only if, for all $i \neq j$, the polynomials $S(g_i, g_r)$ can be reduced to zero by repeatedly dividing remainders by elements of G.

Solved Examples

Example 1. By division, reduce the basis $\{xy^2, y^2 - y\}$ for the ideal $I = <xy^2, y^2 - y>$ in $\mathbf{R}[xy]$ to one with smaller maximum term size, assuming the order lex with $y < x$.

Solution. Clearly, we have y^2 divides xy^2

Now

$$y^2 - y \,\overline{\smash{\big)}\, xy^2} \quad \overset{\textstyle x}{}$$

$$\begin{array}{r} x \\ y^2 - y \,\overline{\smash{\big)}\, xy^2} \\ \underline{xy^2 - xy} \\ {}^{-}\quad {}^{+} \\ \hline xy \end{array}$$

Since, y^2 does not divide xy, we cannot continue this division. Also, note that $1p(xy) = xy$ is not less than $1p(y^2 - y) = y^2$. However, we do have $1p(xy) < 1p(xy^2)$. Hence, our new basis for I is $\{xy, y^2 - y\}$.

6.4 RING OF FRACTIONS

Definition 1. *A non-empty subset S of a ring R is said to be multiplicatively closed if $a, b \in S \Rightarrow a.b \in S$.*

Definition 2. *Let R be a ring and S be a multiplicative subset of R. Then the set of equivalence classes of $R \times S$ defined by R_s forms a ring under addition and multiplication. This ring R_s is called ring of fractions.*

Theorem 1. *If R is a non-zero ring with a no zero divisors and $0 \notin S$, then R_s is an integral domain.*

Proof.   Let R be a non-zero ring with no zero divisors and $0 \notin S$,

It is known that

$$r/s = 0/s \text{ if and only if } r = 0 \text{ in } R$$

Now, let r/s, r'/s' be any two elements of a ring R_s such that

$$(r/s)(r'/s') = 0 \Rightarrow rr'/ss' = 0$$

$\Rightarrow \qquad rr'/ss' = 0$ in R_s if and only if $rr' = 0$ in R.

But R has no zero divisors, thus if $rr' = 0$, then either $r = 0$ or $r' = 0$.

$\Rightarrow \qquad$ either $r/s = 0$ or $r'/s' = 0$

Hence, R_s is an integral domain.

Theorem 2. *Let R be a non-zero ring with no zero divisors and S is the set of all non-zero elements of R, then R_s is a field.*

Proof.   It is given that R_s is a commutative ring with unity.

To prove, R_s is a field. By definition we know that a commutative ring with unity whose every non-zero element having a multiplicative inverse.

Therefore, it remains to prove that every non-zero element of R_s has multiplicative inverse.

Let $r \neq 0$ in R. Then r/β be any non-zero element of R_s for some $s \in S$.

Then, multiplicative inverse of r/s is s/r. Since, $r \neq 0$ belongs to S

$\Rightarrow \qquad s/r \in R_s$

Hence, R_s is a field.

Theorem 3. *Let R be a ring with unity and S be a multiplicative subset containing unity of R, then there exists a natural homomorphism $\phi : R \rightarrow R_S$ defined by $\phi(a) = a/1$.*

Proof.   Let R be a ring with unity and S be a multiplicative subset containing unity of R. Let $a, b \in R$ and $\phi(a) = a/1$, $\phi(b) = b/1$

Then, we have

$$\phi(a + b) = (a + b)/1 = a/1 + b/1 = \phi(a) + \phi(b)$$

and $\qquad \phi(ab) = (ab)/1 = (a/1)(b/1) = \phi(a)\,\phi(b)$

Hence, ϕ is a natural homomorphism.

Theorem 4. *Let R be a commutative ring and S be a multiplicative subset of R. A mapping $\phi_s : R \to R_s$ given by*

$$\phi_s(r) = rs / s \qquad s \in S$$

is a well defined homomorphism such that $\phi_s(s)$ is a unit in R_s for every $s \in S$.

Proof. If $r_1, r_2 \in R$, then assume that

$$r_1 = r_2 \Rightarrow r_1 s/s = r_2 s/s, \quad s \in S$$

$\Rightarrow \qquad \phi_s\,(r_1) = \phi_s\,(r_2)$

$\Rightarrow \quad \phi_s$ is is well defined.

Further, we shall prove that ϕ_s is a homomorphism. Let $r_1, r_2 \in R$, then

$$\phi_s\,(r_1 + r_2) = (r_1 + r_2)s/s \text{ for any } s \in S$$
$$= (r_1 s + r_2 s)/s$$
$$= r_1 s/s + r_2 s/s$$
$$= \phi_s\,(r_1) + \phi_s\,(r_2)$$

Also $\qquad \phi_s\,(r_1 r_2) = (r_1 r_2)s/s \, , s \in S$
$$= (r_1 s/s)(r_2 s/s)$$
$$= \phi_s\,(r_1)\,\phi_s\,(r_2)$$

$\Rightarrow \quad \phi_s$ is a homomorphism.

Finally, for each $s \in S$, $s/s^2 \in R_s$ and s/s^2 is a multiplicative inverse of s^2/s, but $\phi_s\,(s) = s^2/s$. Hence, for each $s \in S$, $\phi_s\,(s)$ is a unit in R_s.

Theorem 5. *Let R be a commutative ring and S be any multiplicative subset of R. If I be an ideal of R, then $I_s = \{a/s : a \in I, s \in S\}$ is an ideal of R_s.*

Proof. Let us suppose

$$a/s, a'/s', \in I_s \text{ such that } a, a' \in I, s, s' \in S.$$

Now, since I is an ideal of R, so that $as', a's \in I$.

Therefore, $\qquad as' - a's \in I$

Consider $a/s - a'/s' = \dfrac{as' - a's}{ss'} \in I_s$ [By using $(as' - a's) \in I$ and $ss' \in S$]

Also, let r/s be any element of R_s and $a/s' \in I_s$, then

$$(r/s)(a/s) = ra/ss' \in I,$$

Further, since R is commutative, therefore, R_s is commutative

$\Rightarrow \qquad (a/s')(r/s) \in I,$

Hence, I_s is an ideal of R_s.

Theorem 6. *Let R be a commutative ring and S be any multiplicative subset of R. If I is an ideal of R then $I_s = R_s$ if and only if $S \cap I = \phi$.*

Proof. Let us first suppose that $S \cap I \neq \phi$. We have to show that $I_s = R_s$

If $\qquad s \in S \cap I \Rightarrow s \in S$ and $s \in I.$

We want to show that unity of R_s belongs to I_s. Since $I_s = \{a\,/\,s : a \in I, s \in S\}$ is an ideal of R_s. But $s \in I$ so that $S/s \in I_s$ because s/s is the unity of R_s

therefore $\qquad I_s = R_s$.

Conversely, suppose that $I_s = R_s$, then we have to prove that $S \cap I \neq \phi$.

If $I_s = R_s$ then $\phi_s^{-1}(I_s) = R$ because $\phi_s : R \to R_s$ given by $\phi_s(r) = rs/s$ for any $s \in S$. Thus, $\qquad\qquad \phi_s(1) = a/s$, for some $a \in I, s \in S$.

But $\phi_s(1) = 1.s/s$, thus we have $s^2 s_1 = ass_1$; for some $s_1 \in S$.

Therefore $s^2 s_1 \in S$ and $ass_1 \in I$. Hence $S \cap I \neq \phi$.

Theorem 7. *Let R be a commutative ring and S be any multiplicative subset of R. Also, let R' be any commutative ring with unity. If $f : R \to R'$ is a homomorphism of rings such that $f(s)$ is a unit in R', $\forall\ s \in S$, then there exists a unique homomorphism of rings $\overline{f} : R_s \to R'$, such that $\overline{f}(\phi_s) = f$ where $\phi_s : R \to R_s$ given by $\phi_s(r) = rs/s$, for any $s \in S$ is a homomorphism with $\phi_s(s)$ as units in R_s.*

Proof. Firstly we shall prove that the mapping

$$\overline{f} : R_s \to R'$$

given by $\overline{f}\ (r/s) = f(r)f(s)^{-1}$ is a well defined homomorphism of ring such that

$$\overline{f}(\phi_s) = f$$

Let r/s, and $r\,'/\,s'$ be any two elements of R_s and assuming

$$r/s = r'/s' \text{ which implies } s_1(rs' - r's) = 0, s_1 \in S$$

$\Rightarrow \quad f(s_1)f\{(rs') - (r's)\} = f(0)$ $\qquad\qquad\qquad$ [$\because f$ is a homomorphism.]

$\Rightarrow \quad f(s_1)f\{(rs') - (r's)\} = 0$

$\Rightarrow \qquad\quad f\{(rs' - r's)\} = 0$ $\qquad\qquad\qquad\qquad$ [$\because f(s_1)$ is a unit in R'.]

$\Rightarrow \qquad\qquad\quad f(rs') = f(r's)$

$\Rightarrow \qquad\qquad\quad f(r)f(s') = f(r')f(s)$

$\Rightarrow \qquad\quad f(r)f(s)^{-1} = f(r')f(s')^{-1} = \overline{f}\ (r/s) = \overline{f}\ (r'/s')$

$\Rightarrow \qquad \overline{f}$ is well defined

If $r/s, r'/s' \in R_s$ then we have

$$\overline{f}\ (r/s + r'/s') = \overline{f}\ \left(\frac{rs' + r's}{ss'}\right) = f(rs' + r's)f(ss')^{-1}$$

$$= [f(rs') + f(r's)]f(ss')^{-1}$$

$$= f(rs')f(ss')^{-1} + f(rs')f(ss')^{-1}$$

$$= f(r)f(s')\ [f\ (ss')^{-1} + f(r')f(s)\,(f(ss')]^{-1}$$

$$= f(r)f(s')\ (f\ (ss'))^{-1} + f(r')\ f(s)\ f(ss')^{-1}$$

$$= f(r)f(s')(f(s)f(s'))^{-1} + f(r')f(s)\,(f(s) + f(s'))^{-1}$$

$$= f(r)f(s')f(s')^{-1}f\ (s)^{-1} + f\ (r')f(s)\ f(s')^{-1}f\ (s)^{-1}$$

$$= f(r)((f\ (s')\ f\ (s')^{-1}\ f(s)^{-1} + f(r')\,(f(s)\ f(s)^{-1}\ ((f\ (s'))^{-1}$$

$$= f(r)f(s^{-1}) + f(r')\,(f(s'))^{-1}$$

$$= f(r/s) + f(r'/s')$$

Further

$$\bar{f}\ (r/s.\ r'/s') = \bar{f}\left(\frac{rr'}{ss'}\right) = f(rr')f(ss')^{-1}$$

$$= f(rr')(f(ss'))^{-1}$$
$$= f(r)f(r'))(f(s)f(s'))^{-1} \qquad [\because f \text{ is a homomorphism.}]$$
$$= f(r)f(r')f(s')^{-1}f(s)^{-1})$$
$$= (f(r)f(s)^{-1})(f(r')f(s')^{-1}) \qquad [\text{By commutativity of } R'.]$$
$$= f(r/s)f(r'/s')$$

$\Rightarrow$ $\bar{f}$ is a homomorphism of rings.

Now, it remains to prove that $\bar{f}$ is unique homomorphism. Let if possible $g : R_s \to R'$ be another homomorphism such that $g(\phi_s) = f$, then every $s \in S$, $g(\phi_s(s))$ is a unit in R'. But $g(\phi_s(s))^{-1} = g(\phi_s(s))^{-1}$ for every $s \in S$.

Now for each $s \in S$,

$$\phi_s(s) = s^2/s$$

Therefore

$$\phi_s(s^{-1}) = s^2/s \in R_s$$

Then for each $\qquad r/s \in R_s$
$$g(r/s) = g(\phi_s(r)\phi_s(s^{-1}))$$
$$= g\ (\phi_s(r))\ g(\phi_s(s^{-1})) \qquad [\because g \text{ is a homomorphism.}]$$
$$= g\ (\phi_s(r))\ g(\phi_s(s))^{-1}$$
$$= f(r)f(s)^{-1}$$
$$= \bar{f}\ (r/s)$$

$\Rightarrow$ $\qquad\qquad g = \bar{f}$

Hence, $\bar{f}$ is a unique homomorphism.

Theorem 8. *If I and J be two ideals of a commutative ring R and S be a multiplicative subset of R then*

(i) $(I +J)s= I_s + J_s$
(ii) $(IJ)_s = (I_s)\ (J_s)$
(iii) $(I \cap J)_s = I_s \cap J_s$
where $I_s=\{a/s : a \in I,\ s \in S\}$ is an ideal of R_s.

Proof. Let I, J be the ideals of R. Then we know that
$I + J, IJ, I \cap J$ are also ideals of R

(i) Let $a/s \in (I + J)_s$ such that $a \in (I + J), s \in S$
Now $a \in (I + J) \Rightarrow a =$ (some element of I, some element of J)
$$= a_1+a_2, a_1 \in I, a_2 \in J$$
$$a/s = (a_1+a_2)/s = a_1/s + a_2/s$$
But $a_1/s \in I_s$ and $a_2/s \in J_s$, therefore
$$a/s = a_1/s + a_2/s \in I_s + J_s$$
Thus $\qquad (I + J) \subseteq I_s+J_s$ ...(1)

Further, let b/s be any element of $(I_s + J_s)$, then

$$b/s = \text{(some element of } I_s + \text{ some element of } J_s)$$
$$= b_1/s + b_2/s, \; b_1 \in I, \; b_2 \in J, \; s \in S$$
$$= \{b_1 + b_2\}/s$$

$\Rightarrow \qquad\qquad b = b_1 + b_2$

Now, $\qquad\qquad b_1 \in I, \; b_2 \in J \Rightarrow b_1 + b_2 \in I + J$

and $\qquad\qquad b \in I + J \quad \Rightarrow b/s \in (I+J)_s$

$\Rightarrow \qquad\qquad (I + J)_s \supseteq I_s + J_s$ ...(2)

From (1) and (2) we conclude that

$$(I + J)_s = I_s + J_s$$

(ii) If $a \in IJ$ then we can write $a = \sum\limits_{i=1}^{m} a_i b_i$ such that $a_i \in I, \; b_i \in J$

Let a/s be any element of $(I J)_s$ so that

$$a/s = \left(\sum_{i=1}^{m} a_i b_i \right) / s, \; s \in S$$
$$= \sum_{i=1}^{m} (a_i / s)(b_i / s)$$

But $\qquad\qquad a_i /s \in I_s, \; b_i /s \in J_s$, therefore

$$a/s = \sum_{i=1}^{m} (a_i / s)(b_i / s) \in (I_s)(J_s)$$

Therefore

$$(IJ)_s \subseteq (I_s)(J_s)$$ ...(3)

In a similar way

$$\sum_{i=1}^{m} (a_i / s)(b_i / s) = \left(\sum_{i=1}^{m} a_i b_i \right) / s, \; s \in (IJ)_s$$

But $\sum\limits_{i=1}^{m} (a_i / s)(b_i / s) \in (I_s)(J_s)$

$\Rightarrow \qquad\qquad (IJ)_s \subseteq (I_s)(J_s)$ ...(4)

from (3) and (4) we conclude that

$$(IJ)_s = (I_s)(J_s)$$

(iii) We have to prove that

$$(I \cap J)_s = I_s \cap J_s$$

Let $a|s \in (I \cap J)_s$

$\Leftrightarrow \qquad\qquad a \in (I \cap S) \text{ and } s \in S$

$\Leftrightarrow \qquad\qquad a \in I \qquad \text{and} \quad a \in J \,; s \in S$

$\Leftrightarrow \qquad\qquad a/s \in I_s \qquad \text{and } a/s \in J_s$

$\Leftrightarrow \qquad\qquad a/s \in I_s \cap J_s$

Hence, $\qquad (I \cap J)_s = I_s \cap J_s$

Theorem 9. *Let R be a commutative ring, P, a prime ideal and $S = R - P$. Then, R_s contains a unique maximal ideal $P_S = [a/s : a \in P, s \notin P]$.*

Proof. Let $a/s, a'/s' \in P_S$ such that $a, a' \in P, s, s' \notin P$.

Then, we have

$$a/s - a'/s' = (as' - a's) \mid ss' \in P_S$$

Since P is a prime ideal.

$\Rightarrow$ $as' - a's \in P$.

Let r/s be any element of R_S and $a_1/s_1 \in P_S$, then

$$(a_1/s_1)(r/s) = a_1 r/s_1 s \in P_S$$

and $(r/s)(a_1/s_1) = ra_1/ss_1 \in P_S \Rightarrow P_S$ is an ideal of R_S.

Further, let M be a maximal ideal of R_S, then M is prime when $M = Q_S$ for some prime ideal Q of R with $Q \subset P$. But $Q \subset P$ implies $Q_S \subset P_S$. Since $P_S \neq R_S$, we must have $Q_S = P_S$.

Hence, P_S is the unique maximal ideal in R_S.

Theorem 10. *Let R be a commutative principal ideal ring with unity and S be a multiplicative subset of R such that $1 \in S$. Then R_S is also a commutative principal ideal ring with unity.*

Proof. Let R be a commutative principal ideal ring with unity and S be a multiplicative subset of R. Also, let I be an ideal of R_S.

Define $I^C = \{a \in R : a/s \in I, s \in S\}$

We want to prove that I^C is an ideal of R.

Let $a, b \in I^C, r \in R$. Then, there exist elements $s_1, s_2 \in S$ such that

$$a/s_1, b/s_2 \in I.$$

Therefore, $a \mid s_1 s_2 = (a \mid s_1)(1 \mid s_2) \in I$ [$\because$ I is an ideal of R_S]

and $b \mid s_1 s_2 \in I.$

Now, since I is an ideal of R_S so that $(a - b) / s_1 s_2 \in I$ and hence $a - b \in I^C$. If r/s_1, be any element of R_S and $a/s_1, r_1/1 \in I$, then $a_r/s_1 \in I$. Therefore $a_r \in I^C$.

$\Rightarrow$ I^C is an ideal of R.

Now, since R is a principal ideal ring so that $I^C = (a)$, for some $a \in R$.

$\Rightarrow$ $I = (a/s), s \in S.$

Therefore, I is a principal ideal of R_S. Hence, R_S is a principal ideal ring.

Theorem 11. *Let R be a unique factorisation domain (UFD) and S be a multiplicative subset of R containing the unity of R, then R_S is also a UFD.*

Proof. Firstly, we shall prove that if $a \in R$ and it is irreducible in R, then $a/1$ is irreducible in R_S.

Let if possible $a/1$ is reducible, so that

$$a/1 = (b/s_1) (c/s_1)$$ ...(1)

where $b \mid s_1$ and $c \mid s_1$ are non-unity in R_S.

Then, from (1), we have

$$as_1 s_2 = bc$$

$\Rightarrow$ either a/b or a/c $\qquad$ [$\because$ a is irreducible.]

If a/b, so $ax = b$ for some $x \in R$, then

$$a/1 = (b/s_1)(c/s_1)$$

$\Rightarrow \qquad 1 = (x/s_1)(c/s_1)$

which is a contradiction because (c/s_2) is not a unit.

$\Rightarrow a/1$ is irreducible in R_S.

Further, we shall show that every non-unit of R_S is a finite product of irreducible factors. Let $a|s \in R_S$ is not a unit of R_S. Since, R is a UFD, therefore if $a \in R$, then

$$a = a_1 a_2 \,....\, a_k$$

where a_i's are all irreducible elements in R. Then

$$a/s = (1/s)(a_1/1)\,(a_2/1)\,....(a_k/1)$$

is a product of irreducible elements in R_S.

Now, we shall prove that every irreducible element in R_S is prime.

Let a/s be any irreducible element in R_S, therefore, a is irreducible in R, but R is UFD, thus a is prime in R. We have to prove that $a/1$ is prime in R_S.

Let if possible $(a/1)|(b/s_1)(c/s_1)$

$\Rightarrow \qquad (a/1)(d/s_3) = (b/s_1)$, for some d/s_3 in R_S

$\Rightarrow \qquad ads_1s_2 = bcs_3$

$\Rightarrow \qquad a/bcs_3$

$\Rightarrow \qquad a/b$ or a/c or a/s_3

If a/s_3, then $(a/s)/1 \Rightarrow (a/s)$ is a unit in R_S, which is a contradiction. Thus, we conclude that either a/b or a/c.

$\Rightarrow \qquad a|s$ divides either (b/s_1) or (c/s_2).

Therefore a/s is prime in R_S.

Hence, R_S is a unique factorisation domain.

Theorem 12. *Let R be a commutative Noetherian ring and let S be a multiplicative subset of R. Then, R_S is Noetherian.*

Proof. As per given, S is a multiplicative subset of a commutative Noetherian ring, then there is one-to-one correspondence between the set of prime ideals of R which are disjoint from S and the set of prime ideals of R_S.

Let $I_{15} \subseteq I_{25} \subseteq I_{35} \subseteq ....$be an ascending chain of left ideals of R_S and let $I_i = \phi_S^{-1}(I_{iS})$, then I_i form an ascending chain of left ideals of R, i.e.,

$$I_1 \subseteq I_2 \subseteq I_3 \subseteq \,......\subseteq I_n \subseteq \,......$$

But R is Noetherian so that there exists some positive integer m such that

$$I_n = I_m, \text{ for all } m > n.$$

Now, since $\qquad \phi_S(I_i) = I_{i5}$ for all i. Therefore, we have

$$I_{ns} = I_{ms}, \quad \text{ for all } m > n.$$

Hence, R_S is Noetherian ring.

Theorem 13. *Let R be a commutative ring with unity and S be a multiplicative subset of R with identity and let I be an ideal of R. Then,*

(i) $I \subset \phi_S^{-1}(I_S)$

(ii) If $I = \phi_S^{-1}(J)$ for some ideal J in R_S, then $I_S = J$, i.e., every ideal in R_S is of the form I_S, for some ideal I in R.

(iii) If P is a prime ideal in R and $S \cap P = \phi$, then P_S is a prime ideal in R_S and $\phi_S^{-1}(P_S) = P$.

Proof.

(i) Let $a \in I$, then $as \in I$, for some $s \in S$. Therefore, $\phi_s(a) = a/s \in I_S$.

When $\qquad a \in \phi_S^{-1}(I_S)$

Hence, $\qquad I \subset \phi_S^{-1}(I_S)$

(ii) We have $\qquad I = \phi_S^{-1}(J)$ and every element of I_S is of the form r/s with $\phi_S(r) \in J$. Thus

$$r/s = (1/s)(rs/s) \qquad\qquad [\because 1 \in R]$$
$$= (1/s)\phi_S(r) \in J$$
$$\Rightarrow \qquad I_S \subset J \qquad\qquad ...(1)$$

If $r/s \in J$, then

$$\phi_S(r) = rs/s = (r/s)(s^2/s) \in J$$
$$\Rightarrow \qquad r \in \phi_S^{-1}(J) = I$$
$$\Rightarrow \qquad r \in I.$$

Therefore, $r/s \in I_S$. Then, we have

$$J \subset I_S \qquad\qquad ...(2)$$

From (1) and (2), we conclude that

$$I_S = J.$$

(iii) If P is a prime ideal of R and $S \cap P = \phi$, then

$$P_S = [a/s : a \in P, s \notin P]$$

Let $\quad a/s, a'/s' \in P_S$, then $a/s - a'/s' = \dfrac{as' - a's}{ss'} \in P_S$

because $s \notin P$, $s' \notin P \Rightarrow ss' \in P$, as P is prime ideal and $(as' - a's) \in P$.

Also, if r/s be any element of R_S and $a_1/s_1 \in P_S$, then
$$(a_1/s_1)(r/s) = a_1 r/s_1 s \in P_S$$
and $(r/s)(a_1/s_1) = r a_1/ss_1 \in P_S$

Therefore, P_S is an ideal of R_S such that $P_S \neq R_S$.

Hence, $\qquad P \subset \phi_S^{-1}(P_S) \qquad\qquad$ [By result (i)]

Conversely, let $r \in \phi_S^{-1}(P_S)$, then $\phi_S(r) \in P_S$. Therefore
$$\phi_S(r) = rs/s = a/s_1, \text{ with some } a \in P, s, s_1 \in S$$
Therefore, $s_2 ss_1 r = s_2 sa \in P$, for some $s_2 \in S$

Now, since $s_2 ss \in S$ and $S \cap P = \phi$, then $r \in P$

$$\Rightarrow \qquad \phi_S^{-1}(P_S) \subset P$$

Hence, $\quad \phi_S^{-1}(P_S) = P$.

Theorem 14. *Let R be a commutative ring with unity and S be a multiplicative subset of R. Then there is one-to-one correspondence between the prime ideals of R which are disjoint from S and the set of prime ideals of R_S.*

Proof. Let U and V be the set of prime ideals of R and R_S respectively. Then, using previous theorem, the assignment $P \to P_S$ defines an injecture mapping from $U \to V$. We have only to prove that it is surjective also.

Let J be a prime ideal of R_S and let

$$P = \phi_S^{-1}(J)$$

Since $\qquad\qquad P_S = J$ [Using (ii) part of previous theorem]

Then, it is sufficient to prove that P is prime.

If $ab \in P$, then

$$\phi_S(a)\, \phi_S(b) = \phi_S(ab) \in J.$$

Since $\qquad\qquad P = \phi_S^{-1}(J).$

Now, since J is prime ideal of R_S, then either $\phi_S(a) \in J$ or $\phi_S(b) \in J$. Therefore, either $a \in \phi_S^{-1}(J) = P$ or $b \in \phi_S^{-1}(J) = P$.

Hence, P is prime.

6.5 RING WITH 'ORE' CONDITIONS

Definition 1. *An integral domain R is said to be a right (left) ORE domain if for every pair of non-zero elements, $a, b \in R$, there exist non-zero elements x, y of R such that $ax = by$ $(xa = yb)$.*

Definition 2. *Let R be an integral domain and let I be a non-zero right ideals of R, the set of mapping $f : I \to R$ denoted by $\text{Hom}_R (I, R)$ is known as set of an R-linear mappings if*

(i) $f(x+y) = f(x) + f(y)$, *for all $x, y \in I$*

(ii) $f(xr) = f(x) \cdot r$, *for all $x \in I, r \in R$*

Theorem 1. *An integral domain is a right ORE domain if and only if every pair of non-zero right ideals has a non-zero intersection.*

Proof. Let us suppose R is a right ORE domain and let I_1 and I_2 be any two non-zero ideals of R.

Let $0 \neq a \in I_1$ and $0 \neq b \in I_2$

$\Rightarrow \qquad\qquad a, b \in R$

Since R is right ORE domain, then there exist non-zero elements, $x, y \in R$ such that

$$ax = by.$$

Now, since I_1 and I_2 are right ideals of R.

Then, we have

$\qquad ax \in I_1$ and $\qquad by \in I_r$, for $x, y \in R$.

But $ax = by$, $ax \in I_1$, $by \in I_2$ for $x, y \in R$. But $ax = by$. Therefore

$$ax = by \in I_1 \cap I_2$$

$\Rightarrow \qquad\qquad I_1 \cap I_2 \neq [0].$

Conversely, suppose that I_1 and I_2 are any non-zero right ideals of R such that
$$I_1 \cap I_2 \neq [0].$$
For $0 \neq a \in I_1$, and $0 \neq x \in R$, $ax \in I_1$ and for $0 \neq b \in I_2$ and $0 \neq y \in R$, $by \in I_2$.

Let if possible, $\quad I_1 \cap I_2 = [0].$

If $t \in I_1 \cap I_2$, then clearly $t = 0$.

and $t = ax \in I_1$ and $\quad t = by \in I_2$.

$\Rightarrow \quad\quad\quad\quad\quad ax = by = 0,$

which gives a contradiction, because R is an integral domain, therefore it is without zero divisors and we take
$$a \neq 0, b \neq 0, x \neq 0, y \neq 0.$$
Hence, R is right ORE domain.

Theorem 2. *If R is a right ORE domain, then*

     (i) *I_0 is closed under intersection.*

     (ii) *If $f : I \to R$ is an R-linear mapping, then $f^{-1}(X) \in I_0$, $X \in C$.*

Proof.

     (i) Let R is a right ORE domain. Also, let I_1, and $I_2 \in I_0$, then clearly $I_1 \cap I_2$ is also a zero right ideal.

        Therefore, $\quad I_1 \cap I_2 \in I_0$

     (ii) First of all, we shall prove that $f^{-1}(X)$ is a right ideal of R

        Let $a, b \in f^{-1}(X)$, then $f(a), f(b) \in X$.

        Now, since X is a non-zero ideal of R, therefore
$$f(a) - f(b) \in X$$

$\Rightarrow \quad\quad\quad f(a - b) \in X \quad\quad\quad\quad\quad\quad\quad\quad\quad [\because f \text{ is linear.}]$

$\Rightarrow \quad\quad\quad a - b \in f^{-1}(X)$

        Further, $f(a) \in X$, $r \in R \Rightarrow f(a).r \in X$

$\Rightarrow \quad\quad\quad\quad f(ar) \in X \Rightarrow ar \in f^{-1}(X)$

$\Rightarrow \quad\quad f^{-1}(X)$ is a non-zero right ideal.

        We have to show that $f^{-1}(X) \neq 0$. If $f(I) = 0$, then $f(I) \subset X$, so that $I \subset f^{-1}(X)$.

        If $f(I) \neq 0$, then $f(I) \cap X \neq [0]$, because R is a right ORE domain. Thus, there exists a non-zero element $a \in I$ such that $f(a) \in X$.

$\Rightarrow \quad\quad\quad\quad a \in f^{-1}(X).$

        Hence, $\quad f^{-1}(X) \in I_0$.

6.6 EQUIVALENCE RELATION AND EQUIVALENCE CLASS WITH ORE CONDITION

Definition 1. *Let R be a right ORE domain and $H = \bigcup\limits_{I \in I_0} \text{Hom}(I, R)$. An equivalence relation $\sim$ is defined on H by $f \sim g$ if $f = g$ on some $X \in I_0$.*

Definition 2. *The equivalence relation makes the partition of H into disjoint classes. Let the equivalence class determined by $f \in H$ be denoted by $[f]$.*

Now, let $Q = H$ and $\sim$ denote the set of equivalence classes. This set Q forms a division ring under the addition and multiplication of the members of Q given by
$$[f] + [g] = [f + g]$$

where, $f + g : \text{Dom}(f) \cap \text{Dom}(g) \rightarrow R$

and $[fg] = [f].[g]$

where, $fg : g^{-1} (\text{Dom}(f)) \rightarrow R$

Theorem 1. *Let R be a right ORE domain. Then* $Q = \bigcup_{I \in I_0} \text{Hom}(I,R)/ \sim$ *is a division ring.*

Proof.   Let R be a right ORE domain and $[f]$ be a non-zero element in Q, where $f : I \rightarrow R$ is a homomorphism.

If Kernel $[f] \neq 0$, then kernel $[f] \in I_0$, because kernel $[f]$ is a non-zero right ideal. If f is a zero mapping on kernel $[f]$, then $[f] = [0]$, which is a contradiction, because $[f]$ is a non-zero element of Q.

Therefore,   kernel $[f] = (0)$.

Further, define a mapping

$$g : I_m(f) \rightarrow R$$

by $g(f(x)) = x, \; \forall \, x \in \text{Dom}(f)$

Since,   kernel $[f] = (0)$, therefore, g is well defined.

From definition of g, we have

$$g(f(x)) = x = I(x)$$

or $(gf)(x) = I(x)$

Thus, gf is the identity mapping I on domain(s), so that

$$[gf] = [I]$$

$\Rightarrow$ $[g] . [f] = [I]$

which shows that $[f]$ has a left inverse.

Hence, Q is a division ring.

Theorem 2. *Let R be an integral domain. Then, R is a right ORE domain if and only if there exists a division ring Q such that*

   (i) R is a subring of Q.

   (ii) Every element of Q is of the form ab^{-1}, *for some a,b $\in R$.*

Proof.   (i) Let us first suppose R is a right ORE domain, then

$$Q = \bigcup_{I \in I_0} \text{Hom}(I,R)/\sim$$

is a division ring.

Firstly, we have to show that R is a subring of Q.

Define a mapping $f_a : R \rightarrow Q$ given by $Q^*(x) = ax$, where $a^* \in \text{Hom}(R, R)$. Now, for $a^*, b^* \in \text{Hom}.\{R, R)$ and $x \in R$, we have

$$(a + b)^* (x) = (a + b) x = ax + bx$$

$$= a^*(x) + b^*(x) = (a^* + b^*)x$$

Therefore,

$$(a + b)^* = a^* + b^* \text{ which implies}$$

$$[(a + b)^*] = [a^* + b^*] = [a^*] + [b^*]$$

Further, $(ab)^*(x) = (ab)x = a(bx)$

$$= a(b^*(x)) = a^*(b^*(x)) = (a^*b^*)(x)$$

$\Rightarrow \qquad (ab)^* = a^*b^*$

$\Rightarrow \qquad [(ab)^*] = [a^*b^*] = [a^*][b^*]$

Therefore, a^* is a homomorphism.

Now, to show that a^* is one-one.

Let $\qquad\qquad [a^*] = [0]$, then $a^* = 0$, on some $I \in I_0$

$\Rightarrow \qquad\qquad a^* I = 0$

$\Rightarrow \qquad\qquad aI = 0$

$\Rightarrow \qquad\qquad a = 0$ [By definition of integral domain]

$\Rightarrow \qquad a^*$ is one one.

Therefore, R can be embedded in a division ring through this mapping. Hence, R can be regarded as a subring of Q.

(ii) We have to show that every element of Q is of the form ab^{-1}, for some $a, b \in R$.

Let $q \in Q$ be arbitrary.

Then, by definition of Q, there exists $I \in I_0$ such that $q(I) \subset R$.

If b is any non-zero element of I, then there exists an element $a \in R$ such that

$\qquad\qquad q(b) = a$

$\Rightarrow \qquad\qquad qb = a$

$\Rightarrow \qquad\qquad q = ab^{-1}$

Since, q is arbitrary, therefore, we conclude that every element of Q is of the form ab^{-1}.

Conversely, suppose that (i) and (ii) are given. To show R is a right ORE domain.

Let a and b be any two non-zero elements of R.

Then, we can write $\quad a^{-1}b \in Q$.

Therefore $a^{-1}b = xy^{-1}$ for some non-zero elements x and y of R.

Thus $\qquad\qquad by = ax$ or $ax = by$

Hence, R is a right ORE domain.

EXERCISE 6.2

1. If R is a principal ideal domain, then show that R_S is also a principal ideal domain.

2. If R is an integral domain with unity in which each right ideal is of the form $aR, a \in R$, show that R is a right ORE domain.

6.7 WEDDERBURN'S THEOREM ON FINITE DIVISION RING

Theorem 1. *Let R be a ring and let $a \in R$. Let T_a be the mapping of R into itself defined by $xT_a = xa - ax$. Then*

$$xT_a^m = xa^m - \max a^{m-1} + \frac{m(m-1)}{2}a^2xa^{m-2} - \frac{m(m-1)(m-2)}{3!}a^3xa^{m-3} + \dots$$

Proof.   By definition $\qquad xT_a = xa - ax$.

Consider $\qquad xT_a{}^2 = (xT_a)\,T_a$

$$= (xa - ax)\,T_a$$

$$= (xa - ax)a - a(xa - ax) = xa^2 - 2axa + a^2x$$

Similarly

$$xT_a{}^3 = (xT_a{}^2)T_a = (xa^2 - 2axa + a^2x)a - a(xa^2 - 2axa + a^2x)$$

$$= xa^3 - 3axa^2 + 3a^2xa - a^3x$$

Continuing in this way and by using the method of induction, we get the required result.

Theorem 2. *If R is a ring such that px = 0 for all x ∈ R, where p is a prime number, then*

$$xT_a^{p^m} = xa^{p^m} - a^{p^m}x.$$

Proof.   Using previous theorem, we have

$$xT_a{}^2 = xa^2 - a^2x, \text{ since } 2axa = 0$$

Therefore,

$$xT_a{}^4 = (xa^2 - a^2x)a^2 - a^2(xa^2 - a^2x)$$

$$= xa^4 - a^4x$$

and so on

$$xT_a{}^{2m} = xa^{2m} - a^{2m}x$$

If p is an odd prime, then again by previous theorem, we have

$$xT_a^p = xa^p - paxa^{p-1} + \frac{p(p-1)}{2}a^2xa^{p-2} + \ldots - a^px$$

and since

$$p \mid \frac{p(p-1)\ldots(p-i+1)}{i!} \text{ for } i < p.$$

Therefore, all the middle terms drop out. Thus, we have

$$xT_a^p = xa^p - a^px$$

Now, $\qquad xT_a^{p^2} = x(T_a^p)^p = xT_a^{p^2}$

and so on for the higher powers of p.

Theorem 3.(Wedderburn's Theorem on Finite Division Ring). *A finite division ring is necessarily a commutative field.*

Proof.   Let D be a division ring and **Z** be its centre. We may assume that any division ring having fewer elements then D is a commutative field.

If $a, b \in D$ are such that $\qquad b^ta = ab^t$, but $ba \neq ab$, then $b^t \in$ **Z.**

Consider $\qquad N(b^t) = \{x \in D : b^tx = xb^t\}$. $N(b^t)$ is a subdivision ring of D, if it were not D, by our hypothesis, it would be commutative. However, both a and b are in $N(s^t)$ and these do not commute, therefore $N(s^t)$ is not commutative so must be all of D, therefore $b^t \in$ **Z** .

Since every non-zero element in D has finite order, so some positive power of it falls in **Z.** Given $w \in D$, let the order of w relative to **Z** be the smallest positive integer $m(w)$ such that $w^{m(w)} \in$ **Z.** Take an element a in D but not in **Z** having minimal possible order relative to **Z,** and let this order be r. We claim that r is a prime number, let if possible $r = r_1 r_2$ with $1 < r_1 < r$, then a^{r_1} is not in **Z,** yet $(a^{r_1})^{r_2} = a^r \in$ **Z** implying that a^{r_1} has an order relative to **Z** smaller than that of a. Now, there is an $x \in D$ such that $xax^{-1} = a^i \neq a$, therefore $x^2 a x^{-2} = x(xax^{-1})x^{-1} = xa^i x^{-1} = (xax^{-1})^i = (a^i)^i = a^{i^2}$. Similarly, we get $x^{r-1}ax^{-(r-1)}$. Since, r is a prime number, then by Fermat little theorem

$$i^{r-1} = 1 + u_0 r$$

Thus $$a^{i^{r-1}} = a^{1+u_0 r} = aa^{u_0 r} = \lambda a \text{, where } \lambda = a^{u_0 r} \in \textbf{Z}$$

Thus, $$x^{r-1}a = \lambda ax^{r-1}.$$

Since, $x \notin$ **Z** , by the minimal nature of r, xr^{-1} cannot be in **Z.** Now, since $xa \neq ax$, $x^{r-1}a \neq ax^{r-1}$ and then $\lambda \neq 1$. Let $b = x^{r-1}$. Therefore, $bab^{-1} = \lambda a$.

$$\Rightarrow \qquad\qquad \lambda^r a^r = (bab^{-1})^r = ba^r b^{-1} = a^r \qquad\qquad [\because a^r \in Z]$$

$$\Rightarrow \qquad\qquad \lambda^r = 1$$

Further, we claim that if $y \in D$, then whenever $y^r = 1$, we have $y = \lambda^i$ for some i for in the field $\textbf{Z}(y)$, there are atmost r roots of the polynomials $u^r - 1$, the elements $1, \lambda, \lambda^2, ..., \lambda^{r-1}$ in **Z** are all distinct, since λ is of prime order r and they already account for r roots of $u^r - 1$ in $\textbf{Z}(y)$ in consequence of which $y = \lambda^i$.

Now, since $\lambda^r = 1$, $b^r = \lambda^r b^r = (\lambda b)^r = (a^{-1}ba)^r = a^{-1}b^r a$.

$$\Rightarrow \qquad\qquad ab^r = b^r a$$

Since a commutes with b^r but does not commute with b, then b^r must be in **Z.** But we know that the multiplicative group of non-zero elements of **Z** is cyclic; let $\gamma \in Z$ be a generator. Therefore

$$a^r = \gamma^j, b^r = \gamma^k$$

If $j = sr$, then $a^2 = \gamma^{sr}$, whence $(a \mid \gamma^s)r = 1$.

$$\Rightarrow \qquad\qquad a \mid \gamma^s = \lambda^i$$

$$\Rightarrow \qquad a \in Z \text{, which contradicts the fact that } a \notin \textbf{Z} .$$

Thus, $r \nmid j$. Similarly, we can prove that $r \nmid k$.

Further, let $$a_1 = a^k \qquad \text{and} \qquad b_1 = b^j$$

$$\Rightarrow \qquad\qquad ba = \lambda ab$$

$$\Rightarrow \qquad\qquad a_1 b_1 = \mu b_1 a_1 \text{, where, } \mu = \lambda^{-jk} \in \textbf{Z}$$

Since r is prime, which is the order of λ does not divide j or k.

$$\lambda^{jk} \neq 1$$

$$\Rightarrow \qquad\qquad \mu \neq 1 \qquad\qquad \Rightarrow \qquad \mu^r = 1$$

Thus, we conclude that there exists two elements a_1 , b_1 such that

(i) $a_1^r = b_1^r = \alpha \in$ **Z**

(ii) $a_1 b_1 = \mu b_1 a_1$, $\mu \neq 1$, $\mu \in$ **Z**

(iii) $\mu^r = 1$.

Now, we compute $(a_1^{-1}b_1)^r$;

Consider $(a_1^{-1}b_1)^2 = a_1^{-1}b_1a_1^{-1}b_1 = a_1^{-1}(b_1a_1^{-1})b = a_1^{-1}(\mu a_1^{-1}b_1)b_1$

$$= \mu a_1^{-1}b_1^2$$

Similarly,

$$(a_1^{-1}b_1)^3 = \mu^{1+2}a_1^{-3}b_1^3$$

Continuing in this way, we get

$$(a_1^{-1}b_1)^r = \mu^{1+2+...+(r-1)}a_1^{-r}b_1^r = \mu^{1+2+..+(r-1)} = \mu^{r(r-1)/2}$$

where, $(a_1^{-1}b_1)^r = 1$

Being a solution of $\gamma^r = 1$

$a_1^{-1}b_1 = \lambda^i$, so that $b_1 = \lambda^i a_1$

but then $\mu b_1 a_1 = a_1 b_1 = b_1 a_1$, which contradict the fact that $\mu \neq 1$.
Therefore, if r is an odd prime number, the theorem is proved.
Now, consider the case of $r = 2$.

In this case, we have two elements $a_1, b_1 \in D$ such that $a_1^2 = b_1^2 = \alpha \in Z$

$$a_1 b_1 = \mu b_1 a_1, \text{ where } \mu^2 = 1 \text{ and } \mu \neq 1.$$

Therefore, $\mu = -1$ and $a_1 b_1 = -b_1 a_1 \neq b_1 a_1$

$\Rightarrow$ Characteristic of D is not 2.

Then, we can find elements $\zeta, \eta \in Z$ such that $1 + \zeta^2 - \alpha \eta^2 = 0$.

Consider $(a_1 + \zeta b_1 + \eta a_1 b_1)^2 = \alpha(1 + \zeta^2 - \alpha \eta^2) = 0$

$\Rightarrow$ $a_1 + \zeta b_1 + \eta a_1 b_1 = 0$ [Being a division ring]

$\Rightarrow$ $0 \neq 2a^2 = a_1(a_1 + \zeta b_1 + \eta a_1 b_1) + (a_1 + \zeta b_1 + \eta a_1 b_1)a_1 = 0$

which is a contradiction. Hence, the theorem.

Theorem 4. (Jacobian Theorem). *Let D be a division ring such that for every $a \in D$, there exists a positive integer $n(a) > 1$ depending on a such that $a^{n(a)} = a$. Then, D is a commutative field.*

Proof. Let D be a division ring and $a \neq 0$ is in D. Then $a^n = a$ and $(2a)^m = 2a$ for some integers, $n, m > 1$.

Let $s = (n-1)(m-1) + 1; s > 1$

$\Rightarrow$ $a^s = a$ and $(2a)^s = 2a$

But $(2a)^s = 2^s a^s = 2^s a$

When $2^s a = 2a$ from which we get $(2^s - 2) a = 0$.

$\Rightarrow$ D has a characteristic $p > 0$.

If $P \subset Z$ is the field having p-elements. Since a is algebraic over P, $P(a)$ has a finite number of elements, in fact p^h elements for some integer h.

Therefore, since $a \in P(a)$, $a^{p^h} = a$.

Thus, if $a \neq Z$, there exists a $b \in D$ such that

$$bab^{-1} = a^\mu \neq a \qquad \qquad ...(1)$$

Similarly, $b^{p^k} = b$ for some integer $k > 1$.

Now, let, $W = \left\{ x \in D : x = \sum_{i=1}^{p^h} \sum_{j=1}^{p^k} p_{ij} a^j b^j, \text{ where } p_{ij} \in P \right\}$

Then, W is finite and is closed under addition. By (1), it is closed under multiplication. Therefore, if W is a finite ring and being a subring of the division ring D, it itself must be a division ring. Therefore, W is a finite division ring. Also, by previous theorem, it is commutative. But a and b are both in W, therefore, $ab = ba$, which is a contradiction because $a^\mu b = ba$.

Hence, the theorem.

REMARK

☞ Jacobian's theorem actually holds for any ring R satisfying $a^{n(a)} = a$ for every $a \in R$, not just for division ring.

EXERCISE 6.3

1. If R is a finite ring in which $x^n = x$ for all $x \in R$, where $n > 1$. Show that R is commutative.

2. If R is a finite ring in which $x^2 = 0$ implies that $x = 0$. Show that R is commutative.

3. If $t > 1$ is an integer and $(t^m - 1) \mid (t^n - 1)$, prove that $m \mid n$.

4. Show that any finite subring of a division ring is a division ring.

5. If D is a division ring, show that its dimension over its centre cannot be 2.

6.8 NOETHERIAN RINGS

It is known that every ascending chain of ideals become stationary after a finite number of terms. In this chapter, we have to discuss some of the properties of rings satisfying similar conditions for one sided ideals. The abstract development of this theory on the basis only of the ascending chain conditions (a.c.c) and commutativity was initiated by *Emmy Noether.*

Definition 1 *(Ascending Chain Conditions). A ring R satisfies the ascending chain conditions (a.c.c.) for ideals if given any sequence of ideals $I_1, I_2, ...$ of R with $I_1 \subseteq I_2 \subseteq I_3 ... \subseteq I_n \subseteq ...,$ there exists an integer m such that $I_m = I_n, \forall\, m \geq n.$*

Definition 2. *A commutative ring with unity in which every strictly ascending chain of right (left) ideals is finite is called a right (left) Noetherian.*

or

A ring in which for every infinite ascending chain $I_1 \subseteq I_2 \subseteq ...$ of right (left) ideals there exists a positive integers n such that $I_m = I_n \forall\, m \geq n$.

Definition 3. *A commutative ring with unity in which ascending chain condition (a.c.c.) holds for right as well as left ideals is called a Noetherian ring.*

For example:

(1) The ring of integers **Z** is a Noetherian ring since it is a principal ideal domain (PID) and every FID is a Noetherian ring.

(2) Every finite ring is clearly both right and left Northerian.

(3) In a division ring D, the only right ideals of D are (0) and D itself. D is a right Noetherian. Similarly, D is a left Noetherian.

Definition 4. *If in a ring R, every non-empty set of right (left) ideals, partially ordered by* *inclusion relation has a maximal element, we say that maximum conditions holds for right (left) ideals of R.*

REMARKS

☞ A is said to be maximal element of F if and only if for each B in F, $A \subseteq B \Rightarrow A = B$.

☞ A set may possesses more than one maximal element.

Definition 5. *A commutative ring R with unity is said to be Noetherian ring if R satisfies the ascending chain conditions for ideals or R holds maximum condition for ideals.*

Definition 6. *A right ideal I of R is said to be finitely generated if it is generated by a finite subset of R.*

REMARK

☞ If generating set S consists of n elements $a_1, a_2, ..., a_n$, we write $I = <a_1, a_2, ..., a_n>$.

6.9 BASIC PROPERTIES OF NOETHERIAN RINGS

Theorem 1. *For any ring R, the following statements are equivalent*

    (1) *R is Noetherian.*

    (2) *Maximum condition holds for ideals of R.*

    (3) *Every ideal of R is finitely generated.*

Proof. (1)⇒(2) Let S be a non-empty set of ideals of R and let I_1 be an ideal of S. If It is not a maximal element of S, we can find an ideal I_2 of R such that $I_2 \in S$ and $I_1 \subseteq I_2$. If S has no maximal element, the above process can be continued indefinitely.

Then, we get an infinite strictly ascending chain

$$I_1 \subseteq I_2 \subseteq I_3 ...$$

of ideals of R. This contradicts the fact that R is Noetherian. Hence, S has a maximal element. Therefore, (1) ⇒ (2) .

(2)⇒(3) Let I be an ideal of R and $S = [J : J$ is finitely generated ideal of R such that $J \subseteq I \}$.

Clearly, $(0) \in S$, therefore S is a non-empty set of ideals. We have assumed that S has a maximal element. Let us suppose M is a maximal element of S. Then, clearly M is finitely generated. We have to show that $I = M$.

Since $M \in S, M \subseteq I$, let if possible $I \neq M$, then there exists an element $a \in I$, $a \notin M$. Consider the ideal $M + <a>_r$.

Here $M + <a>_r$ is finitely generated ideal of R generated by the set $F \cup [a]$ where $< F > = M$.

Now, since $a \in I, M + <a>_r \subseteq I$

$\Rightarrow$ $M + <a>_r \in S$

But, $a \notin M \Rightarrow M \subset M + <a>_r$

$\Rightarrow$ M is not maximal.

$\Rightarrow$ $I = M$

$\Rightarrow$ I is finitely generated.

Hence,   (2) ⇒ (3).

(3)⟹(1). Let $I_1 \subseteq I_2 \subseteq I_3 \subseteq \ldots$ be an infinite ascending chain of ideals of R. Consider $I = \bigcup_i I_i$. Then being the union of ideals, I is an ideal of R. Now, since we have assumed that each ideal is finitely generated. Let $F = \{a_1, a_2, \ldots, a_n\}$ be a generating set of I. Then, we can write $I = <a_1, a_2, \ldots, a_n>$. Now, for each $j \in \{1, 2, 3, \ldots, n\}$, $a_j \in I$ implies the existence of some positive integer ij such that $a_j \in I_{ij}$. Now, let $k = \max \{i_1, i_2, \ldots, i_n\}$. Then $a_j \in I_k$, $\forall j = 1, 2, 3, \ldots, n$. This gives that $I \subseteq I_k$, but $I_k \subseteq I$. Therefore $I = I_k$. Thus, we conclude that for all $m \geq k$, $I_m \subseteq I = I_k$ gives $I_m = I_k$. Hence, R is Noetherian.

REMARK

☞ As the result for left Noetherian rings are same as the results for right Noetherian, thus in this section, we shall confine ourselves to right Noetherian rings only.

Definition. *An ideal I of a ring R is said to be right Noetherian if every strictly ascending chain of right ideals of R, which are contained in I, is finite.*

REMARK

☞ Every ideal of a right Noetherian ring is right Noetherian.

Theorem 2. *A homomorphic image of a right Noetherian ring is right Naetherian.*

Proof. Let R be a Noetherian ring and S be any homomorphic image of R. Then by fundamental theorem of homomorphism, we can say that $S = R / I$ for some ideal I of R. We have to show that R/I is right Noetherian.

Let $J_1 \subseteq J_2 \subseteq J_3 \subseteq \ldots$ be an ascending chain of right ideals of R/I. Each J_i is of the form K_i/I, where K_i is the right ideal of R containing I. Further, $J_i \subseteq J_{i+1} \Rightarrow K_i \subseteq K_{i+1}$.

Therefore, we get an ascending chain $K_1 \subseteq K_2 \subseteq K_3 \subseteq \ldots$ of right ideals of R. But, since R is right Noetherian there exists a positive integer n such that $K_m = K_n$, for all $m \geq n$. Thus $J_m = J_n$, $\forall m \geq n$. Hence, R/I is right Noetherian.

Theorem 3.(Modular Law). *If L, M, N are right ideals of a ring R such that $M \subseteq L$, then*
$$L \cap (M + N) = M + (L \cap N).$$

Proof. As per given, M and $L \cap N$ both are contained in L. Therefore,
$$M + L \cap N \subseteq L.$$
Also, $\qquad\qquad L \cap N \subseteq N \Rightarrow M + L \cap N \subseteq M + N$
$$\Rightarrow \qquad M + (L \cap N) \subseteq L \cap (M + N)$$
Let $x \in L \cap (M + N)$, then $x \in L$ and $x \in M + N$, i.e., $x = y + z$, $y \in M$, $z \in N$.
Now, $z = x - y \in L$ because $M \subseteq L \Rightarrow z \in L \cap N$.
$$\therefore \qquad\qquad x \in M + L \cap N$$
$$\Rightarrow \qquad L \cap (M + N) \subseteq M + L \cap N$$
Hence, $\qquad L \cap (M + N) = M + (L \cap N).$

Theorem 4. *Let I be an ideal of R, then R is right Noetherian if and only if both I and R/I are, right Noetherian.*

Proof. Let R be right Noetherian and I be an ideal of R. Then clearly I is right Noetherian. Since, we know that every homomorphic image of a right Noetherian ring is right Noetherian, therefore R / I is also right Noetherian. Conversely, let us suppose that I and R / I both are right Noetherian. We have to prove that R is right Noetherian.

Let $J_1 \subseteq J_2 \subseteq ...$ be an ascending chain of right ideals of R. Then $J_1 \cap I \subseteq J_2 \cap I \subseteq J_3 \cap I \subseteq ...$ is an ascending chain of right ideals of R contained in I. Now, since I is a right Noetherian there exists a positive integer n such that $J_m \cap I = J_n \cap I, \ \forall \ m \geq n$.

Also, $(J_1 + I)/I \subseteq (J_2 + I)/I \subseteq (J_3 + I)/I \subseteq ...$ is an ascending chain of right ideals of R/I. Since R/I is right Noetherian, therefore, there exists a positive integer t such that $(J_m + I)/I = (J_i + I)/I$ for all $m \geq t$. If $r = \max\{n, t\}$, then $J_m \cap I = J_r \cap I$ and $(J_m + I)/I = (J_r + I)/I, \ \forall \ m \geq r$. But then we have

$$J_m + I = J_r + I \text{ for all } m \geq r$$

We want to prove that

$$J_m = J_r \ \forall \ m \geq r$$

Now, for all $m \geq r$, $J_m = J_m \cap (J_m + I) = J_m \cap (J_r + I)$

$$\hspace{5cm} [\because J_m + I = J_r + I; \ \forall \ m \geq r]$$
$$= J_r + J_m \cap I \hspace{2cm} [\because J_r \leq J_m \ \forall \ m \geq r \text{ and by modular law}]$$
$$= J_r + J_r \cap I \hspace{2.5cm} [\because J_m \cap I = J_r \cap I]$$
$$= J_r \hspace{4cm} [\because J_r \cap I = J_r]$$

Therefore, $\quad J_m = J_r \ \forall \ m \geq r$. Hence, R is right Noetherian.

REMARK

☞ The nil radical of $I = \sqrt{I} = \cap \ [P : I \subseteq P, P \text{ is a prime ideal}]$.

Theorem 5. *If I is an ideal of the Noetherian ring R then I contains some power of its nil radical that is if $(\sqrt{I})^n \subseteq I$ for some positive integer.*

Proof. Let R be a Noetherian ring and I be an ideal of R. Also, $\sqrt{I}$ being an ideal of R so that $\sqrt{I}$ is finitely generated, say $\sqrt{I} = <a_1, a_2,..., a_k>$. Now, since each $a_i \in \sqrt{I}$, there exists positive integer n_i for which $a_i^{n_i} \in I$.

Let us take

$$n = n_1 + n_2 + ... + n_k$$

Therefore, a generated system for $(\sqrt{I})^n$ is given by the products

$$a_1^{m_1} a_2^{m_2} ... a_k^{m_k}, \quad \text{where } m_i \in \mathbb{Z}$$

and

$$n = m_1 + m_2 + ... + m_k$$

But if $\ m_1 + m_2 + ... + m_k = n_1 + n_2 + ... + n_k$

Then, we must have $\quad m_i \geq n_i \quad$ for $i = 1, 2,..., m$.

$$\Rightarrow \hspace{3cm} a_i^{m_i} \in I$$

$$\Rightarrow \hspace{2.5cm} a_1^{m_1} a_2^{m_2} ... a_k^{m_k} \in I$$

Hence, all the generator of $(\sqrt{I})^n$ lie in I. Therefore $(\sqrt{I})^n \subseteq I$.

Theorem 6. *Let Q be a primary ideal of the Noetherian ring R and I and J be ideals with $IJ \subseteq Q$. Then either $I \subseteq Q$ or else $(\sqrt{I})^n \subseteq Q$ for some positive integer n.*

Proof. Since $IJ \subseteq Q$

$$\Rightarrow \quad I \subseteq Q \text{ or there exists a positive integer } k \text{ for which } J^m \subseteq Q.$$

Now, since R is Noetherian, therefore $(\sqrt{J})^k \subseteq J$ (Using Theorem 5) for some positive integer k, then we have

$$(\sqrt{J})^{mk} \subseteq J^m \subseteq Q$$

If $n = mk$, then we have

$$(J)^n \subseteq Q$$

Theorem 7. (Hilbert-Basis Theorem). *If R is a right Noetherian ring with unity, then $R[x]$, the rings of polynomial over R is right Noetherian.*

Proof. Let I be a non-zero right ideal of $R[x]$. We have to prove that I is finitely generated.

For each integer $k \geq 0$, define $I_k = [a \in R : a \neq 0$ and there exists a polynomial

$$a_0 + a_1 x + ... + a_{k-1}x^{k-1} + ax^k \in I] \cap (0).$$

Clearly I_k is a right ideal of R and $I_k \subseteq I_{k+1}$ for all integers $k \geq 0$. Now, since R is right Noetherian, there exists a positive integer n such that $I_m = I_n$, $\forall\ m \geq n$. Also, each I_i being a right ideal of right Noetherian ring R is finitely generated. Let

$$I_i = \langle a_{i1}, a_{i2}, ..., a_{im} \rangle$$

for all $i = 0, 1, 2, ..., n$ where a_{ij} is the leading coefficient of a polynomial $f_{ij} \in I$ of degree i. We want to prove that I is generated by $m_0 + m_1 + m_2 + ... + m_n$ polynomials $f_{01}, f_{02}, ..., f_{0m_0}, f_{11}, f_{12}, ..., f_{1m_1}, ..., f_{n1}, f_{n2}, ..., f_{nm_n}$.

Let $\quad J = \langle f_{01}, f_{02}, ..., f_{n1}, f_{n2}, ..., f_{nm_n} \rangle$ by the choice of $f_{ij} \in I, J \subseteq I$

Let $f \neq 0 \in R[x]$ be such that $f \in I$. Then, we have

$$f = C_0 + C_1 x + C_2 x^2 ..... + C_s x^s, C_s \neq 0$$

Now, we shall apply induction on s.

For $s = 0, f = C_0 \in I_0$ and by definition of f_{ij}'s

$$a_{01} = f_{01}, \quad a_{02} = f_{02}, ..., a_{0m_0} = f_{0m_0}$$

are elements of I_0 which generates I_0 so $I_0 \subseteq J$, we get $C_0 \in J$

$$\Rightarrow \qquad f \in J.$$

Now, suppose that all non-zero polynomials in I of degree less than s belong to J and let $\deg f = s$.

Let $s < n$. The leading coefficient C_s of f belongs to I_s.

But $I_s = I_n$ as $I_m = I_n\ \forall\ m \geq n$. This implies that $C_s \in I_n$, i.e.,

$$C_s = a_{n1}b_1 + a_{n2}b_2 + a_{n3}b_3 + ... + a_{nm_n} b_{m_n} \text{ for some } b_1, b_2 ... b_{m_n} \in R \qquad ...(1)$$

Now, the polynomial $g = f - (f_{n_1}b_1 + f_{n_2}b_2 + ... + f_{nm_n} b_{m_n})x^{s-n}$ is zero or of degree less than s as the coefficient of x^s in g is equal to $C_s - (a_{n_1}b_1 + a_{n_2}b_2 + ... + a_{n_m} b_{m_n})$ which is zero by (1).

If $g = 0$, then $g \in J$, if $g \neq 0$, then by induction hypothesis, $g \in J$. Therefore, in this case

$$f = g + (f_{n_1}b_n + f_{n_2}b_2 + ... + f_{nm_n} b_{m_n})x^{s-n} \in J$$

If $\quad s \leq n$, then $\quad C_s \in I_s \Rightarrow C_s = a_{s_1}d_1 + a_{s_2}d_2 + ... + a_{s m_s} d_{m_s})$ for some $d_1, d_2, ..., d_{m_s} \in R$.

Now, the polynomial $h = f - (f_{s_1}d_1 + f_{s_2}d_2 + ... + f_{s m_s} d_{m_s})$ is either zero or of degree less than s as the coefficient of x^s in h is equal $C_s - (a_{s_1}d_1 + a_{s_2}d_2 + ... + a_{s m_s} d_{m_s})$

$$\Rightarrow \quad h \in J, \text{ i.e., } f \in J.$$

Therefore, in each case, every non-zero polynomial f which is in *I* is also in *J*.

This gives that $I \leq J$, which implies that $I = J$.

$\Rightarrow$ *I* is finitely generated.

Hence, $R[x]$ is right Noetherian ring.

6.10 DECOMPOSITION OF IDEALS IN NOETHERIAN RINGS

Definition 1. *An ideal I of a ring R is said to be irreducible if it can not be expressed as a finite intersection of ideals of R properly containing I.*

If *I* is not irreducible, then it is said to be reducible.

Definition 2. *The quotient of two ideals I and J denoted by I : J is defined as the set* $\{b \in R : ab \in I \text{ for some } a \in J\}$, *i.e.,* $I : J = \{b \in R : ab \in I, \text{ for some } a \in J\}$

Theorem 1. *Every ideal in a Noetherian ring R is a finite intersection of irreducible ideals.*

Proof. Let *R* be a Noetherian. ring and *S* be the family of all ideals of *R* which are not finite intersection of irreducible ideals. We have to show that $S = \phi$. Let if possible $S \neq \phi$. Since *R* is Noetherian, thus there exists an ideal *I* of *R* which is maximal in *S*.

Now, since $I \in S$ and *I* is not irreducible. Therefore

$$I = J \cap K$$

where *J* and *K* are ideals of *R* strictly containing *I*. Since *I* is maximal, therefore *J* and *K* both are finite intersection of irreducible ideals and hence *I* is one also which contradicts the fact that $I \in S$.

Thus, $S = \phi$.

Hence, every ideal in a Noetherian ring is a finite intersection of irreducible ideals.

Theorem 2. *Every irreducible ideal in a Noetherian ring is primary.*

Proof. Let *R* be a Noetherian ring and *I* be any ideal of *R*. Here, it is enough to show that any ideal of *R* which is not primary is necessarily reducible. Let *I* be not primary, then there exist elements *a, b* in *R* such that $ab \in I$ with $b \notin I$ and $a^k \notin I$ for all positive integers *k*.

Now, $I : (a) \subseteq I : (a^2) \subseteq ... \subseteq I : (a^k) \subseteq ...$

For an ascending chain of ideals of *R*, indeed if $ra^k \in I$ for $r \in R$, then certainly $ra_{k+1} \in I$. But since *R* is Noetherian, we can find an integer *n* such that

$$I : (a^n) = I : (a^{n+1})$$

We have to show that *I* can be expressed as $I = (I, a^n) \cap (I, b)$

Clearly $I \subseteq (I, a^n)$ and $I \subseteq (I, b)$ so that

$$I \subseteq (I, a^n) \cap (I, b) \qquad\qquad ...(1)$$

Further, let $x \in (I, a^n) \cap (I, b)$, then

$$x \in (I, a^n) \Rightarrow x = \text{some element of } I + a, \text{ multiple of } a^n$$
$$= \alpha + \beta a^n, \ \alpha \in I, \beta \in R$$

and $x \in (I, b) \Rightarrow$ $x = \text{some element of } I + a, \text{ multiple of } b.$
$$= \alpha' + \beta'b, \alpha' \in I, \beta' \in R$$

Thus, $x = \alpha + \beta a^n = \alpha' + \beta'b, \ \alpha, \alpha' \in I, \beta, \beta' \in R$

Thus, the product

$$\beta a^{n+1} = (x - \alpha)a = (\alpha' + \beta'b - \alpha)a$$
$$= (\alpha' - \alpha)a + \beta'(ba) \in I$$

$\Rightarrow$ $\qquad\qquad \beta \in I : (a^{n+1})$

But $\qquad\qquad I : (a^{n+1}) = I : (a^n)$ so that $\beta \in I: (a^n)$, therefore

$$\beta a^n \in I$$

$\Rightarrow$ $\qquad\qquad x = \alpha + \beta a^n \in I$

Therefore,

$$(I, a^n) \cap (I, b) \subseteq I \qquad\qquad\qquad ...(2)$$

From (1) and (2), we conclude that

$$I = (I, a^n) \cap (I, b)$$

Hence, I is reducible.

Theorem 3. *Every ideal of a Noetherian ring can be represented as a finite intersection of primary ideals.*

Proof. Let R be a Noetherian ring and I be any ideal of a Noetherian ring and if I is irreducible, then using previous theorem, it is also primary. In this case theorem is obvious. Now, let I is not irreducible and assume that the property stated in the theorem is valid for all ideals in R which contains I. Since I is now reducible, therefore, we can write

$$I = I_1 \cap I_2$$

for ideals I_1 and I_2 properly containing I. Both I_1 and I_2 have a representation as a finite intersection of primary ideals. Thus, the combined intersection is a representation of I.

Theorem 4. *A commutative ring R with unity is Noetherian if every prime ideal of R is finitely generated.*

Proof. Let R be a commutative ring with unity such that every ideal of R is finitely generated. To prove R is Noetherian. Let if possible R is not Noetherian and let S be the collection of ideals of R which are not finitely generated is non-empty. Then, by Zorn's lemma, S must contain a maximal element say I. Now, by our hypothesis, I cannot be a prime ideal of R, therefore, there exist elements $a, b \in R$, which are not in I such that $a, b \in I$.

Now, both the ideals (I, b) and $I : b$ property contain I; in particular $a \in I: (b)$. Since I is maximal in S, their ideals are finitely generated.

Now, suppose that

$$(I, b) = (c_1, c_2, ..., c_n)$$

and $\qquad\qquad I : (b) = (d_1, d_2, ..., d_m)$

Then, $\qquad\qquad c_1 = a_i + br_i, \quad a_i \in I, r_i \in R, i = 1, 2, ..., n$

so that

$$(I, b) = \{a_1 a_2 ... a_n, b\}$$

Consider an ideal J generated by the elements a_i and bd_j, i.e.,

$$J = (a_1, a_2, ..., a_n, bd_1, bd_2, ..., bd_m)$$

Now, since $bd_j \in I$ for all j, therefore

$$J \subseteq I \qquad\qquad\qquad ...(1)$$

Further, let x be any arbitrary element of I, $x \in (I, b)$. Then, x can be written as

$$x = a_1y_1 + a_2y_2 + \ldots + a_ny_n + by \quad \forall \, y_i, y \in R$$

Since each $a_i \in I$, therefore $y \in I$: (b). Thus, we can find $z_i \in R$ such that

$$y = d_1z_1 + d_2z_2 + \ldots + d_mz_m$$

Then,

$$x = a_1y_1 + a_2y_2 + \ldots + a_ny_n + b(d_1z_1 + d_2z_2 + \ldots + d_mz_m)$$
$$= a_1y_1 + a_2y_2 + \ldots + a_ny_n + (bd_1)z_1 + (bd_2)z_2 + \ldots + (bd_m)z_m$$

$$\Rightarrow \qquad x \in J$$

$$\Rightarrow \qquad I \subseteq J \qquad\qquad\qquad \ldots(2)$$

From (1) and (2), we conclude that

$$I = J = (a_1, a_2, \ldots, a_n, bd_1, \ldots, bd_m)$$

Thus, I is finitely generated, which is not possible, because $I \in S$. Hence, R is Noetherian.

6.11 ARTINIAN RINGS

Definition 1. *A ring in which every strictly descending chain of right (left) ideals is finite, is called a right (left) Artinian ring.*

or

A ring in which for every infinite descending chain of right (left) ideals $I_1 \supseteq I_2 \supseteq I_3 \supseteq \ldots$, there exists a positive integer n such that $I_m = I_n$, $\forall \, m \geq n$.

Definition 2. *A right (left) Artinian ring is also known as a ring with descending chain condition (DCC) on right (left) ideals.*

For example:

(1) Every division ring D is right (as well as left) Artinian because its only right (left) ideals are (0) and D itself.

(2) Every finite ring is both right and left Artinian.

Definition 3. *If in a ring R, every non-empty collection of all right (left) ideals of R, partially ordered by inclusion relation has a minimal element, then we say that a minimum condition holds for right (left) ideals of R.*

REMARKS

☞ A right (left) Artinian ring is also known as a ring with descending chain condition (DCC).

☞ By a minimal element of a partially ordered non-empty set S, under the relation $\leq$, we mean an element $A \in S$ such that there exists no $B \in S$ satisfying $B \subseteq A$, i.e., A is minimal element of S if and only if $B \subseteq A \Rightarrow B = A$.

☞ A set may possess more than one minimal element.

6.12 BASIC PROPERTIES OF ARTINIAN RINGS

Theorem 1. *For a ring R, the following statements are equivalent:*

    (i) *R is right Artinian.*

    (ii) *Minimum conditions hold for right ideals of R.*

Proof. (i) $\Rightarrow$ (ii) Let S be a non-empty set of right ideals of R and I be an element of S. If I_1 is not a minimal element, we can find another right ideal I_2 of R such that $I_1 \supseteq I_2$. Now, if S has no minimal element, this process can be repeated indefinitely. We get an infinite strictly descending chain $I_1 \supseteq I_2 \supseteq I_3 \supseteq \ldots$

of right ideals of R, which is a contradiction, because R is a right Artinian. Hence, S has a minimal element.

(ii) ⇒ (i) Let $I_1 \supseteq I_2 \supseteq I_3 \supseteq \dots$ be a descending chain of right ideals of R. Consider $S = \{I_i : i = 1, 2, 3, \dots\}$. Then S is non-empty because $I_i \in S$. By our assumption S has a minimal element, say I_n, for some positive integer n. Now, for all $m \geq n, I_m \subseteq I_n$. If $I_m \neq I_n$, then since I_n is a minimal element of S, $I_m \notin F$, which is not possible. Thus, $I_m = I_n \ \forall \ m \geq n$. Hence, R is right Artinian.

REMARKS

☞ An ideal I of R is said to be right Artinian if every strictly descending chain of right ideals of R, which are contained in I_i is finite.

☞ Every ideal I of a right Artinian ring is right Artinian.

☞ A homomorphic image of right Artinian ring is right Artinian.

☞ If I is an ideal of a ring R, then R is right Artinian if and only if both I and R/I are right Artinian.

Theorem 2. *A right Artinian ring having more than one elements and having no proper zero divisors is a division ring.*

Proof. Let R be a right Artinian ring having at least two elements and having no proper zero-divisors.

Let $a \neq 0$ and $a \in R$. Now, consider the descending chain $< a >_r \supseteq < a^2 >_r \supseteq < a^3 >_r \supseteq \dots$ of right ideals of R. Now, since R is right Artinian, there exists a positive integer n such that $<a^m>_r = < a^n >_r; \forall \ m \geq n$.

In particular, $< a^{n+1} >_r = < a_n >_r$

$\Rightarrow \qquad a_n \in < a^{n+1} >_r$

$\Rightarrow \qquad a^n = a^{n+1}. r + ma^{n+1}$, for some $r \in R, m \in \mathbf{Z}$

$\Rightarrow \qquad a^n = a^n (ar + ma) = a^n e$, where $e = ar + ma \in R$

$\Rightarrow \qquad a = ae \qquad\qquad [\because a^{n-1} \neq 0 \text{ is cancelled.}]$

$\Rightarrow \qquad ae = ae^2$

$\Rightarrow \qquad a (e - e^2) = 0, \ i.e., \ e = e^2 \qquad\qquad [\because a \neq 0]$

and R is without proper divisors.

Also, $\qquad\qquad e \neq 0$, as $a \neq 0$.

Further, for any $x \in R, (xe - x) e = 0$

$\Rightarrow \qquad\qquad xe = x$

and $\qquad e (ex - x) = 0$

$\Rightarrow \qquad\qquad ex = x$

Therefore, e is the unity of R. Thus, $< a^n >_r = a^nR = < a^{n+1} >_r = a^{n+1}. R$

Now, $\qquad\qquad a^ne \in a^nR$

$\Rightarrow \qquad\qquad a^ne = a^{n+1} b$; for some $b \in R$

$\Rightarrow \qquad\qquad ab = e$

Hence, R is a division ring.

REMARKS

☞ An Artinian integral domain with at least two elements is a field.

☞ In an Artinian commutative ring R with unity, every prime ideal different from R is maximal.

Theorem 3. *In a right Artinian ring , every nil right ideal is nilpotent.*

Proof. Let R be a right Artinian ring and I be a nil right ideal of R. Now, for the descending chain of right ideals $I \supseteq I^2 \supseteq I^3 \supseteq ...$, there exists a positive integer n such that $I^m = I^n$; $\forall \ m \geq n$. Particularly, $I^{2n} = I^n$. We claim that $I^n = (0)$.

Let if possible $I^n \neq (0)$. Then $I^{2n} \neq (0)$.

Let $F = \{A : A \text{ is a right ideal of } R \text{ such that } AI^n \neq (0)\}$.

Clearly, F is non-empty, because $I^n \in F$.

Now, since R is right Artinian, F has a minimal element.

Let K be a minimal element of F, then $KI^n \neq (0)$.

$\Rightarrow$ there exists $k \in K$ such that $kI^n \neq (0)$.

But $kI^n \subseteq K$, because K is a right ideal of R.

Let $KI^n \subseteq K$. Also, $(kI^n) \ I^n = kI^{2n} = kI^n \neq (0)$.

$\Rightarrow \ kI^n \in F$ as kI^n is a right ideal of R, which is not possible, because K is minimal.

Therefore, $kI^n = K$. Now, since $k \in K$, there exists $a \in I^n$ such that $ka = k$.

But since I is a nil right ideal and $I^n \subseteq I$, a is nilpotent.

Thus, there exists a positive integer t such that $a^t = 0$.

Then, $\qquad k = ka = ka^2 = ka^3 = ... = ka^t = 0$

$\Rightarrow \qquad kI^n = (0)$, which is a contradiction.

Thus, $\qquad I^n = (0)$. Hence, I is a nilpotent right ideal.

Solved Examples

Example 1. *Let S be an infinite set and let R be the set of all subsets of S and $(R, \Delta, \cap)$ is a commutative ring with unity. Then, show that R is not Noetherian.*

Solution. We have to show that the ring R is not Noetherian. For this, it is sufficient to prove that there exists an ideal of R which is not finitely generated. Let C be the subset of R consisting of all finite subsets of S. Then C is an ideal of R.

Let us suppose, C is finitely generated. Then $C = < C_1, C_2,..., C_t >$, where for each $i = 1, 2,..., t$, C_i is a finite subset of S. Let $A \in C$, then

$$A = (C_1 \cap D_1)\Delta(C_2 \cap D_2)\Delta...\Delta(C_t \cap D_t)$$

for some subsets $D_1, D_2,..., D_t$ of S.

Then, for each $i, y \in A$, $y \in \{C_1 \cap D_1)\Delta...\Delta(C_t \cap D_t)$

$\Rightarrow \qquad y \in C_i \cap D_i$ at least for some i, $1 \leq i \leq t$.

Therefore, each element of A is in some C_t.

Since S is an infinite set and $\bigcup\limits_{i=1}^{t} C_t$ is a finite subset of S, there exists an $x \in S$,

$x \notin \bigcup\limits_{i=1}^{i} C_i$ *i.e.,* $x \in S$ and $x \notin C_i$ $\forall$ $i = 1, 2,..., t$. Now consider the finite subset $[x]$ of S. By suitable choice of C, $[x] \in C$, which is not possible. Therefore, C is not finitely generated. Hence, R is not Noetherian.

Example 2. *Let* $R = \left[\begin{pmatrix} a & b \\ 0 & c \end{pmatrix} : a \in \mathbf{Z}, b, c \in \mathbf{Q} \right]$

be a ring under matrix addition and multiplication. Show that R is right Noetherian but not left Noetherian.

Solution. For any non-negative integer k, consider the set

$$A_k = \left[\begin{pmatrix} 0 & m \,|\, 2^k \\ 0 & 0 \end{pmatrix} : m \in \mathbf{Z} \right]$$

Then, A_k is a left ideal of R. Also $A_k \subseteq A_{k+1}$ because $m \,|\, 2^k = \dfrac{2m}{2^{k+1}}$ and $\begin{pmatrix} 1 & 1 \,|\, 2^{k+1} \\ 0 & 0 \end{pmatrix} \notin A_k$

Then, we get a non-terminating strictly ascending chain $A_0 \subseteq A_1 \subseteq A_2...$ of left ideals of R. Therefore, R is not left Noetherian. We have to show that R is right Noetherian. For this purpose, it is sufficient to prove that each non-zero right ideal of R is finitely generated.

Let A [$\neq 0$] be a right ideal of R and let $\alpha e_{11} + \beta e_{12} + \gamma e_{22} \in A$, where e_{ij} denotes the matrix with 1 in $(i, j)^{\text{th}}$ position and zero elsewhere and $\alpha \in \mathbf{Z}$, $\beta, \gamma \in \mathbf{Q}$.

Also, $e_{ij} \, e_{kl} = e_{il}$ if $j = k$ and $e_{ij} \, e_{kl} = 0$ if $j \neq k$.

Then, there are following two cases arise.

Case I. If $\alpha \neq 0$. Let δ be the least positive integer such that $\delta e_{11} + b e_{12} + c e_{22} \in A$ for some $b, c \in Q$. We claim that either A is generated by matrices $\delta e_{11}, e_{12}, e_{22}$ or by δe_{11} and e_{12}.

Now, $(\delta e_{11} + b e_{12} + c e_{22}) e_{12} \in A$

$\Rightarrow \qquad\qquad\qquad \delta e_{12} \in A$

$\Rightarrow \qquad\qquad \delta e_{12}(1 \,|\, \delta) \, e_{22} = e_{12} \in A$

$\Rightarrow \qquad\qquad\qquad b e_{12} = e_{12} \, (b e_{22}) \in A$

Further, $\quad \delta e_{11} + b e_{12} + c e_{22} \in A \Rightarrow (\delta e_{11} + b e_{12} + c e_{22}) e_{11} \in A$

$\Rightarrow \delta e_{11} \in A$

Therefore, we get δe_{11} and e_{12}.

In the other case, $c e_{22} (1 \,|\, c) e_{22} \in A$ and A is then generated by matrices δe_{11}, e_{12} and e_{22}.

Case II. If $\alpha = 0$. Here, we have $b e_{12} + c e_{22} \in A$.

If all the elements of A are of the type $\lambda (b e_{12} + c e_{22})$ for some $\lambda \in \mathbf{Q}$, then A is generated by simple matrix $b e_{12} + c e_{22}$. Otherwise there exists $b_1, c_1 \in \mathbf{Q}$ such that

$$b_1 e_{12} + c_1 e_{22} \in A$$

but

$$b_1 c \neq b c_1.$$

Then, $(bc_1 - b_1c) e_{12} \in A$.

$\Rightarrow \quad (bc_1 - b_1c) e_{12} [1/ (bc_1 - b_1c)] e_{22} \in A$

$\Rightarrow \qquad\qquad\qquad e_{12} \in A$

$\Rightarrow \qquad\qquad\qquad e_{12} b e_{22} \in A$

$\Rightarrow \qquad\qquad\qquad b e_{12} \in A$

$\Rightarrow \qquad\qquad\qquad c e_{22} \in A$

Then, either $c = 0 \qquad$ or $\qquad e_{22} \in A$.

Therefore, A is either generated by e_{12} and e_{22} or by e_{12} alone.

$\Rightarrow \quad A$ is finitely generated.

Hence, R is right Noetherian.

REMARKS

☞ The ring defined in Example (2), R is right Artinian but not left Artinian.

☞ The ring of integers $\mathbf{Z}$ is Noetherian but not Artinian. Because for any positive integer n, the strictly descending chain

$$< n > \supset < 2n > \supset < 4n > \supset... \text{ of ideals of } \mathbf{Z} \text{ is infinite.}$$

☞ Let F be the ring of all real valued functions on R. For any real number $r > 0$, we define $I_r = \{f \in F : f(x) = 0, \text{ for all } -r < x < r \}$. Then, I_r is an ideal of F. The strictly descending chain of ideals $I_1 \supseteq I_2 \supseteq ...$ and the strictly ascending chain of ideals $I_1 \subseteq I_2 \subseteq I_{1/3} \subseteq ...$ never terminate. Hence, F is neither Noetherian nor Artinian.

☞ $\mathbf{Q}$, being a field, is Artinian. $\mathbf{Z}$ is a subring of $\mathbf{Q}$, but $\mathbf{Z}$ is not Artinian.

☞ Let F be a field. Then F is Artinian. Consider $F[x]$. Then strictly descending chain $<x> \supseteq < x^2 > \supseteq < x^3 > \supseteq...$ of ideals of $F[x]$ is infinite. Hence, $F[x]$ is not Artinian.

Example 3. *Let I be a proper ideal of a Noetherian ring R. Then, show that*

$$\bigcap_{n=1}^{\infty} I^n = \{r \in R : (1-a) r = 0, \text{ for some } a \in I\}$$

Solution. Let us write $\qquad S = \{r \in R : (1 - a)r = 0, \text{ for some } a \in I\}$.

Now, if $r \in S$, so that $(1 - a) r = 0$, for some $a \in I \Rightarrow r = ar$

Similarly, we have $r = ar = a^2r = ... = a^nr = ...$

Therefore, $r \in I^n$ for every integer n and hence $r \in \bigcap_{n=1}^{\infty} I^n$.

Thus, $\qquad\qquad S \subseteq \bigcap_{n=1}^{\infty} I^n$ ...(1)

Further, let us put $J = \bigcap_{n=1}^{\infty} I^n$.

Now, consider the primary decomposition of the ideal IJ such that $IJ = \bigcap_i I_i$, where each I_i is primary. We shall show that $IJ = J$.

Now, since $IJ \subseteq J$, therefore, it is enough to show that $J \subseteq IJ$ for each i. Now, $IJ \subseteq I_i$. Then, we have either $J \subseteq I_i$; or else $(\sqrt{I})^n \subseteq I_i$ for some positive integer.

But if $(\sqrt{I})^n \subseteq I_i$, then $J \subseteq I^n \subseteq (\sqrt{I})^n \subseteq I_i$

$\Rightarrow$ $\qquad J \subseteq I_i$ for each i.

$\Rightarrow$ $\qquad IJ = J$.

Then, there exists an element $a \in I$ such that $(1-a)J = [0]$.

Therefore, we have

$$J = \bigcap_{n=1}^{\infty} I^n \subseteq S \qquad\qquad ...(2)$$

Thus, from (1) and (2), we conclude that

$$\bigcap_{n=1}^{\infty} I^n = S = [r \in R : (1-a)^r = 0, \text{ for some } a \in I\,]$$

Example 4. *If I and J are two ideals of the ring R, with I finitely generated. If IJ = I, then show that there exists an element $r \in J$ such that $(1-r)\,I = [0]$.*

Solution. Suppose that I is generated by the elements $a_1, a_2,..., a_n$. Let I_i denote the ideal $(a_i, a_{i+1},..., a_n)$ and put $I_{n+1} = [0]$.

By induction hypothesis on i, we shall prove the existence of an element $r_i \in J$ such that $(1-r_i)\,I \subseteq I_1 (i = 1, 2,..., n+1)$.

Particularly, $r_{n+1} \in J$. When $i = 1$, the ideal $I_1 = I$, then take $r_1 = 0$.

Now, using induction hypothesis that $(1-r_i)\,I \subseteq I_i$ for some $r_i \in J$ with the fact that $I \subseteq IJ$, we have $(1-r_i)\,I_i \subseteq (1-r_i)\,IJ \subseteq I_1 J$.

Now, since each $a_i \in I$, then $(i-r_i)a_i \in I_i J$ and therefore

$$(1-r_i)\,a_i = \sum_{k=1}^{n} b_{ik}a_k \Rightarrow (1-r_i-b_{ii})\,a_i = \sum_{k=i+1}^{n} b_{ik}a_k \in I_{i+1}$$

Now, we take $1-r_{i+1} = (1-r_i)\,(1-r_i-b_{ii})$. Clearly, $r_{i+1} \in J$

Now, $\qquad (1+r_{i+1})I = (1-r_i-b_{ii})I \subseteq (1-r_i-b_{ii})I_i$

$$= (1-r_i-b_{ii})\,(a_i, I_{i+1}) \subseteq I_{i+1}.$$

Example 5. *Let F be the ring of a real valued functions on R, the set of real numbers. For any positive real number r define $I_r = \{f \in F : f(x) = 0 \text{ for } -r \le x \le r\,\}$. Show that I_r, is an ideal of F such that $... \subseteq I_{1/3} \subseteq I_{1/2} \subseteq I_1 \subseteq I_2 \subseteq I_3 ...$. Also. Show that F is not a Noetherian ring.*

Solution. We have $\qquad I_r = \{f \in F : f(x) = 0, -r \le x \le r\}$

Let $f, g \in I_r$, then $f(x) = 0, g(x) = 0$. For $-r \le x \le r$

$\Rightarrow \qquad f(x) - g(x) = 0 \qquad\qquad \Rightarrow \qquad f - g \in I_r$

Further, let $t(x) \in F$ be arbitraiy and $f(x) \in I_r$

Then, $\qquad f(x)\,t(x) = 0; \ t(x) = 0 \quad$ and $\quad t(x)\,f(x) = t(x).0 = 0$

Therefore, $\qquad f(x) \in I_r, t(x) \in F \Rightarrow f(x)\,t(x) \in I_r$ and $t(x)\,f(x) \in I_r$.

Thus, I_r is an ideal of F.

Further, for $r = 1/3, 1/2, 1, 2,3, ...,$ we have

$$I_{1/3} = \{f \in F : f(x) = 0 ; -1/3 \le x \le 1/2\}$$
$$I_{1/2} = \{f \in F : f(x) = 0 ; -1/2 \le x \le 1/2\}$$
$$I_1 = \{f \in F : f(x) = 0 ; \quad -1 \le x \le 1\}$$
$$I_2 = \{f \in F : f(x) = 0 ; \quad -2 \le x \le 2\}$$
$$I_3 = \{f \in F : f(x) = 0 ; \quad -3 \le x \le 3\}$$

Now, since ... $\begin{bmatrix} -\dfrac{1}{3} & \dfrac{1}{3} \end{bmatrix} \subset \begin{bmatrix} -\dfrac{1}{2} & \dfrac{1}{2} \end{bmatrix} \subset [-1\ 1] \subset [-2\ 2] \subset [-3\ 3] \subset ....$

Therefore, $... \subseteq I_{1/3} \subseteq I_{1/2} \subseteq I_1 \subseteq I_2 \subseteq I_3 ...$ is a non-stationary ascending chain of ideals of F. Hence, F is not Noetherian.

EXERCISE 6.4

1. Show that in a unique factorization domain, ascending chain condition holds on principal ideals.

2. Show that every commutative Artinian ring possesses a finite number of proper prime ideals.

3. Show that every ideal of a commutative Noetherian ring contains a product of prime ideals.

4. If in a Noetherian ring R, every ideal generated by two elements is principal, show that R is a principal ideal ring.

5. If R is a right Noetherian (Artinian), show that R_n, the ring of $n \times n$ matrices over R is right

Noetherian (Artinian).

6. Show that every Noetherian ring R with unity has a maximal ideal.

7. If R is a commutative ring such that $R[x]$ is Noetherian, show that R contains unity.

8. If I be a non-zero ideal of a principal ideal domain R, show that R/I is both Artinian and Noetherian.

9. Show that the ring of real valued continuous functions on the closed interval [0, 1] is not Noetherian.

10. Show that every finite ring is Noetherian.

Chapter Review: | *A competitive Approach*

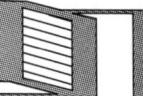

Selected Terms and Results

Terms

- **Radical :** Let R be a commutative ring and I be an ideal of R. The radical of I is said to be the ideal $\cap\, P$, where the intersection is taken over all prime ideals P which contains I.

- **Primary Ideal:** Let R be a commutative ring. An ideal $I(\neq R)$ of R is said to be primary ideal if for any $a, b \in R \Rightarrow a, b \in I$ and $a \in I \Rightarrow b^n \in I$ for some $n \geq 1$.

- **Algebraic variety :** Let S be a finite subset of $F(x)$. The algebraic variety $V(s)$ in F^n is the set of all common zero in F^n of the polynomial in S.

- **Multiplicatively closed subset :** A non-empty subset S of a ring R is said to be multiplicatively closed if $a, b \in S \Rightarrow a.b \in S$.

- **Ring of fractions :** Let R be a ring and S be a multiplicative subset of R. Then the set of equivalence classes of $R \times S$ defined by R_S forms a ring under addition and multiplication. The ring R_S is called ring of fractions.

- **ORE domian :** An integral domain R is said to be right (left) ORE domain if for every pair of non-zero elements $a, b \in R$ there exist non-zero elements x, y of R such that
$$ax = by \ (xa = yb)$$

- **Ascending chain conditions :** A ring R satisfies the ascending chain conditions for ideals if given any sequence of ideals $I_1, I_2, ...,$ of R with $I_1 \subseteq I_2 \subseteq I_3 \subseteq ... \subseteq I_n \subseteq ...$ there exists an integer m such that $I_m = I_n\ \forall\ m \geq n$.

- **Notherian Ring :** A commutative ring with unity in which ascending chain condition holds for right as well as left ideals is called a Notherian ring.

- **Finitely generated ideal :** A right ideal I of ring R is said to be finitely generated if it is generated by a finite subset of R.

- **Artinian Ring :** A ring in which every strictly descending chain of right (left) ideals is finite is called a right (left) Artinian ring.

Results

- Radical of I is also known as prime radical of R.

- Every prime ideal is clearly primary.

- Every prime ideal is primary and belongs to itself.

- The radical of an ideal is also an ideal.

- Let R be a ring. If I be an ideal of R. Then $\mathrm{Rad}(\mathrm{Rad}\, I) = \mathrm{Rad}.\, I$.

- The radical of a primary ideal is a prime ideal.

- Let R be a commutative ring and I be an ideal of R if I have a primary decomposition, then I has a reduced primary decomposition.

- If I_1 and I_2 are primary ideal that have the same radical β. Then $I_1 \cap I_2$ is primary.

- Let D be primary, β its associated prime and let S be any ideal not contains in β then $D : S = D$.

- Let S_1 and S_2 are ideals in a Noetherian ring $S_1 : S_2 = S_1$ if an only if I_2 is not contained in any of the associated prime of S_1.

- The importance of Gröbner basis in applications is due to the fact that they are machine computable. They have applications to engineering and Computer Science as well as Mathematics.

- Let R be a ring and S be a multiplicative subset of R. Then the set of equivalence classes of $R \times S$ defined by R_S forms a ring under addition and multiplication, this ring R_S is called ring of fractions.

- If R is a non-zero ring with a no zero divisors and $0 \notin S$, then R_S is an integral domain.

- Let R be a non-zero ring with no zero divisors and S is the set of all non-zero elements of R, then R_S is a field.

- Let R be a ring with unity and S be a multiplicative subset containing unity of R, then there exists a natural homomorphism $\phi : R \to R_s$ defined by $\phi(a) = a/1$.

- Let R be a commutative principal ideal ring with unity and S be a mulitplicative subset of R such that $1 \in S$. Then R_S is also a commutative principal ideal ring with unity.

- Let R be unique factorisation domain (UFD) and S be a multiplicative subset of R containing the unity of R, then R_S is also a UFD.
- Let R be a commutative Noetherian ring and let S be a multiplicative subset of R. Then, R_S is Noetherian.
- Let R be a commutative ring with unity and S be a mulitplicative subset of R. Then there is one-to-one correspondence between the prime ideals of R which are disjoint from S and the set of prime ideals of R_S.
- An integral domain is right ORE domain if and only if every pair of non-zero right ideals has a non-zero intersection.
- The equivalence relation makes the partition of H into disjoint classes.
- A finite division ring is necessarily a commutative field. This is known as Wedderburn's theorem on finite division ring.
- Jacobian's theorem actually holds for any ring R satisfying $a^{n(a)} = a$ for every $a \in R$, not just for division ring.
- A homomorphic image of a right Noetherian ring is right Noetherian.
- If I be an ideal of R, then R is right Noetherian iff both I and R/I are right Noetherian.
- If R is a right Noetherian ring with unity, then $R[x]$, the rings of polynomials over R is right Noetherian.
- An ideal I of ring R is said to be irrducible if it cannot be expressed as a finite intersection of ideals of R properly containing I.
- Every irreducible ideal in a Noetherian ring is primary.
- Every ideal of a Noetherian ring can be represented as a finite intersection of primary ideals.
- A commutative ring R with unity is Noetherian if every prime ideal of R is finitely generated.
- Every ideal of a right Artinian ring is right Artinian.
- A right Artinian ring having more than one element and having no proper divisors is a division ring.

Review Questions and Project Work

1. For an integral domain, show that the relation $a \sim b \Rightarrow$ if a is an associate of b is an equivalence relation on D.
2. Show that in a PID, every ideal is contained in a maximal ideal.
3. Let D be a UFD, show that a non-constant divisor of a primitive polynomial in $D[x]$ is again a primitive polynomial.
4. Show that a principal ideal ring is a unique factorization domain.
5. Show that $\mathbf{Z}$ is a Noetherian ring but not an Artinian ring.
6. If R is a Noetherian ring, show that $ab = 1$, $a, b \in R$ if and only if $ba = 1$.
7. Show that any right ideal of the ring
$$R = \begin{pmatrix} \mathbf{Z} & \mathbf{Q} \\ O & \mathbf{Q} \end{pmatrix}$$
can be generated by at most two elements then R is right Noetherian.
8. Show that the intersection of all prime ideals in a Noetherian ring is nilpotent.
9. Let R be a left Artinian integral domain with more than one element. Show that R is a division ring.
10. Let R be a finite dimensional algebra over an algebraically closed field F. Suppose R has non-zero nil ideals. Show that R is isomorphic to the direct sum of matrix rings over F.

Objective type Questions

Fill in the Blanks

1. Every prime ideal is clearly _____.
2. Every prime ideal is primary and belongs to _____.
3. Let R be a ring. If I is an ideal of R, then Rad (Rad I) =
4. If I is an ideal of a ring R and m is any positive integer then Rad (I^m) = _____
5. The radical of a primary ideal is _____ ideal.
6. Let R be a commutative ring and I is an ideal of R. If I have a primary decomposition then I has a _____ primary decomposition.
7. If R is a non-zero ring with no zero divisors and $0 \notin S$. Then R_S is _____.

8. Let R be a non-zero ring with no zero divisors and S is the set of all non-zero elements of R then R_S is a _____.

9. Let R be a commutative Noetherian ring and let S be a multiplicative subset of R. Then R/S

10. An integral domain is a right *ORE* domain if and only if every pair of non-zero right ideals has a _____ intersection.

True/False

Write 'T' for true and 'F' for false statement.

1. If R is a principal ideal domain then R_S is also a principal ideal domain. **(T/F)**

2. If p is prime in an integral domain, then p is reducible. **(T/F)**

3. If R is an integral domain with unity in which each right ideal is of the form aR, $a \in R$ then R is a right ORE domain. **(T/F)**

4. A finite division ring is not necessarily a commutative field. **(T/F)**

5. Every finite ring is clearly both right and left Noetherian. **(T/F)**

6. Every PID is a Noetherian ring. **(T/F)**

7. Every ideal of right Artinian ring is right Artinain. **(T/F)**

8. The ring of integers is Noetherian but not artinian. **(T/F)**

9. Every Noetherian ring R with unity has a maximal ideal. **(T/F)**

10. The ring of real valued continuous functions on the closed interval [0, 1] is Noetherian. **(T/F)**

Multiple Choice Questions

Choose the most appropriate one :

1. Every finite ring is :
   (a) Noetherian
   (b) not necessarily Noetherian
   (c) not Noetherian
   (d) none of the above

2. If R is a commutative ring such that $R[x]$ is Noetherian. Then R contains :
   (a) all zero
   (b) all unity
   (c) both (a) and (b) are true.
   (d) none of the above

3. Every ideal of a commutative Noetherian ring contains $\subset$ product of :
   (a) Maximal ideals
   (b) prime ideals
   (c) both (a) and (b) are true.
   (d) none of the above

4. The ring of integers is :
   (a) Noetherian
   (b) Not Artinian
   (c) both (a) and (b) are true.
   (d) none of the above

5. An Artinian integral domain with at least two elements is :
   (a) field
   (b) integral domain
   (c) both (a) and (b) are true.
   (d) none of the above

=========== **Answers** ===========

Fill in the Blanks

1. primary	**2.** itself	**3.** Rad (I)	**4.** Rad (I)	**5.** prime
6. reduced	**7.** Integral domain	**8.** field	**9.** Noetherian	**10.** non-zero

True/ False

1. T	**2.** F	**3.** T	**4.** F	**5.** T	**6.** T	**7.** T	**8.** T
9. T	**10.** F						

Multiple choice questions

1. (a)   **2.** (b)   **3.** (b)   **4.** (c)   **5.** (c)

● ● ●

7 Vector Spaces

Outlines

7.1 INTRODUCTION

We are familiar with the concept of semigroup, a group and a ring. A group was obtained from a semigroup by imposing certain restrictions on the composition of a semigroup. A ring was obtained by defining a certain composition on a group structure and by giving rules that connected the group composition with the new composition. We shall now discuss another algebraic structure called a vector space or linear space which is going to involve a group structure, a ring structure and an operation connecting the elements of these two structures.

In order to discuss a vector space we need two basic things. One of them is the set of vectors and the other is the set of scalars. Therefore, to define a vector space we need a field F. The elements of field F are called the scalars. In addition, we need two binary operations. One of them is internal composition and the other is external composition. Now we distinguish the internal and external compositions as follows:

(i) **Internal Composition.** Let R be any set. If $a * b \in R$ for all $a, b \in R$ and $a*b$ is a unique, then '*' is known as the *internal composition*. That is, a binary operation defined over the vectors is called *internal composition (vector addition)*.

(ii) **External Composition.** Let V be the set of vectors and F be a field. Then a binary operation defined between the vectors and scalars is called external composition. That is, if $a \circ \alpha \in V$ for all $\alpha \in V$ and $a \in F$ and $a \circ \alpha$ is unique, then o is called an external composition or (scalar multiplication).

7.2 VECTOR SPACES

Definition. *Let V be a non-empty set of vectors and F be a field. Then an algebraic structure (V, +, .) together with two binary operations vectors addition and scalar multiplication is said to be vector space over F if this structure satisfies the following conditions :*

(i) *(V, +) is an abelian group*

(ii) *$a(\alpha + \beta) = a\alpha + a\beta, \forall \alpha, \beta \in V$ and $\forall a \in F$*

(iii) $(a + b)\alpha = a\alpha + b\alpha, \; \forall \; \alpha \in V$ and $\forall \; a,b \in F$

(iv) $(ab)\alpha = a(b\alpha), \; \forall \; a \in V$ and $\forall \; a,b \in F$

(v) $I\alpha = \alpha, \; \forall \; a \in V$ and $1 \in F$

This vector space V over F is denoted by $V(F)$.

For example, if $F = \mathbf{R}$, the field of real numbers, then $V(R)$ is a vector space and it is called a real vector space.

■ Illustrations

(i) Let $\mathbf{R}^2 = \{(a_1, a_2): a_1 \in \mathbf{R}, a_2 \in \mathbf{R})$. The set $\mathbf{R}^2$ is a vector space over $\mathbf{R}$ with addition and scalar multiplication defined as follows :

$$(a_1, a_2) + (b_1, b_2) = (a_1+b_1, a_2+b_2)$$

$$c(a_1, a_2) = (ca_1, ca_2), \; \forall \; a_1, a_2, b_1, b_2, c \in \mathbf{R}$$

(ii) Vector in 3-dimensional space form a vector space over $\mathbf{R}$ with respect to addition and scalar multiplication of vectors.

(iii) Let $\mathbf{R}^n$ be the set of n-tuples of real numbers, *i.e.*,

$$\mathbf{R}^n = \{(a_1, a_2,...,a_n : a_i \in \mathbf{R}\}$$

Then $\mathbf{R}^n$ is a vector space over R with pointwise addition and scalar multiplication as defined in (1).

(iv) Let $\mathbf{C}^n$ be the set of all ordered n-tuples of complex numbers. Then $\mathbf{C}^n$ is a vector space over $\mathbf{C}$ with addition and scalar multiplication.

7.3 ELEMENTARY PROPERTIES OF VECTOR SPACES

<u>Theorem 1.</u> *Let $V(F)$ be a vector space over a field F and 0 be the zero (null) vector of V. Then*

(i) $a0=0, \; \forall \; a \in F$

(ii) $0\alpha = 0, \; \forall \; \alpha \in V$

(iii) $a(-\alpha) = -(a\alpha), \; \forall \; a \in F, \alpha \in V$

(iv) $(-a)\alpha = -(a\alpha) \; \forall \; a \in F, \alpha \in V$

(v) $a(\alpha - \beta) = a\alpha - a\beta, \; \forall \; a \in F, \; \alpha, \beta \in V$

(vi) $a\alpha = 0 \Rightarrow a = 0$ or $\alpha = 0$

(vii) $a\alpha = a\beta \Rightarrow \alpha = \beta, \; \forall \; a \in F, \alpha, \beta \in V, a \neq 0$

(viii) $a\alpha = b\alpha \Rightarrow a = b \; \forall \; a, b \in F, \alpha \in V$ and $\alpha \neq 0$.

Proof. (i) we have $\qquad\qquad a0 = a\,(0 + 0) \qquad\qquad\qquad [\because 0+0=0]$

$\qquad\qquad\qquad\qquad\qquad\quad = a0 + a0 \qquad\qquad$ [By property of vector space]

$\qquad\qquad$ or $\qquad\qquad 0 + a0 = a0 + a0 \qquad\qquad\qquad [\because \; 0 + a0 = a0]$

$\qquad\qquad$ or $\qquad\qquad\qquad 0 = a0$

(ii) We have $\qquad\qquad 0\alpha = (0+ 0)\alpha \qquad\qquad\qquad [\because 0 + 0 = 0, 0 \in F]$

$\qquad\qquad\qquad\qquad\qquad\quad = 0\alpha + 0\alpha \qquad\qquad$ [By definition of $V(F)$]

$\qquad\qquad$ or $\qquad\qquad 0 + 0\alpha = 0\alpha + 0\alpha \qquad\qquad\quad [\because \; 0\alpha \in V \therefore 0+ 0\alpha = 0\alpha]$

$\qquad\qquad$ or $\qquad\qquad\qquad 0 = 0\alpha$

(iii) To prove $\qquad a\,(-\alpha) = -(a\alpha)$

We have,

$$a0 = 0 \qquad\qquad\qquad\qquad \text{[From (i)]}$$

or $\qquad a(-\alpha + \alpha) = 0 \qquad\qquad\qquad [\because -\alpha + \alpha = 0]$

or $\qquad a(-\alpha) + a\alpha = 0 \qquad\qquad\qquad \text{[By the definition of } V(F)]$

(iv) To prove $\qquad (-a)\,\alpha = -(a\alpha)$

Since, we have

$$0\alpha = 0 \qquad\qquad\qquad\qquad [\text{ From (ii)}]$$

or $\qquad (-a + a)\alpha = 0 \qquad\qquad\qquad [\because -a + a = 0, a \in F]$

or $\qquad (-a)\alpha + a\alpha = 0 \qquad\qquad\qquad \text{[By the definition of } V(F)]$

or $\qquad (-a)\,\alpha = -\,(a\alpha)$

(v) To prove $\qquad a(\alpha - \beta) = a\alpha - a\beta$

We have,

$$a(\alpha - \beta) = a\,[\alpha + (-\beta)\,]$$
$$= a\alpha + a(-\beta) \qquad\qquad \text{[By definition of } V(F)]$$

$\therefore \qquad a(\alpha - \beta) = a\alpha - a\beta \qquad\qquad [\because a(-\beta) = -(\alpha\beta)]$

(vi) To prove $\qquad a\alpha = 0 \Rightarrow a = 0 \quad$ or $\quad \alpha = 0$

Suppose $0 \neq a \in F$, then a^{-1} exists in F.

Now $\qquad\qquad a\alpha = 0 \qquad\qquad\qquad\qquad\qquad \text{[given]}$

$\Rightarrow \qquad\qquad a^{-1}a\alpha = a^{-1}0$

$\Rightarrow \qquad\qquad (a^{-1}a)\alpha = 0 \qquad\qquad\qquad\qquad [\because a^{-1}0 = 0]$

$\Rightarrow \qquad\qquad 1\alpha = 0 \qquad\qquad\qquad\qquad\qquad [aa^{-1} = 1]$

$\Rightarrow \qquad\qquad \alpha = 0 \qquad\qquad\qquad\qquad\qquad [\because 1\alpha = \alpha]$

Suppose $\alpha \neq 0$, then to prove $a = 0$, let us assume that $a \neq 0$, then a^{-1} exists. Since we have

$$a\alpha = 0$$

$\Rightarrow \qquad\qquad a^{-1}\,(a\alpha) = a^{-1}\,0$

$\Rightarrow \qquad\qquad (a^{-1}a)\alpha = 0 \qquad\qquad\qquad\qquad [\because a^{-1}a = 1]$

$\Rightarrow \qquad\qquad 1\alpha = 0$

$\Rightarrow \qquad\qquad \alpha = 0 \qquad\qquad\qquad\qquad\qquad [\because 1\alpha = \alpha]$

$\Rightarrow \qquad$ This gives a contradiction because we have taken $\alpha \neq 0$.

Hence, $\qquad\qquad a = 0$.

(vii) We have,

$$a\alpha = a\beta \Rightarrow a\alpha - a\beta = 0$$

$\Rightarrow \qquad\qquad a(\alpha - \beta) = 0$

$\Rightarrow \qquad\qquad \alpha - \beta = 0 \qquad\qquad\qquad [\because a \neq 0 \text{ and from (vi) }]$

$\Rightarrow \qquad\qquad \alpha = \beta.$

Hence, $\qquad\qquad a\alpha = a\beta \Rightarrow$ for all $a \neq 0 \in F$ and $\alpha, \beta \in V$

(viii) We have,

$$a\alpha = b\alpha \Rightarrow a\alpha - b\alpha = 0$$
$$\Rightarrow \qquad (a-b)\alpha = 0$$
$$\Rightarrow \qquad a - b = 0 \qquad\qquad [\because \alpha \neq 0 \text{ and from (vi) }]$$
$$\Rightarrow \qquad a = b$$

Hence, $\qquad a\alpha = b\alpha \Rightarrow a = b.$

7.4 VECTOR SUBSPACES: VECTOR SPACE WITHIN VECTOR SPACE

Just like a subgroup and a subring, we do have the concept of a vector subspace which is generally addresses as a subspace.

Definitions. *Let W be a non-empty subset of V, where V is a vector space over a field F. Then W is said to be a vector subspace of V(F) if W is itself a vector space aver F with respect to the same operations as defined on V.*

For example: The set $W = \{ (a, 0, b) : a, b \in \mathbf{R}\}$ is a subspace of $R^3(\mathbf{R})$.

7.5 ELEMENTARY PROPERTIES OF VECTOR SUBSPACES

Theorem 1. *The necessary and sufficient conditions for a non-empty subset W of V(F) to be a subspace are that:*

(i) $\alpha \in W, b \in W \Rightarrow \alpha - \beta \in W$

(ii) $a \in F, \alpha \in W \Rightarrow a\alpha \in W.$

Proof. Suppose W is a subspace of a vector space $V(F)$. Then if

$$\beta \in W \Rightarrow -\beta \in W$$
$$\therefore \qquad \alpha \in W, -\beta \in W \Rightarrow a + (-\beta) \in W$$

$$[\because W \text{ is closed under vector addition.}]$$

$$\Rightarrow \qquad \alpha - \beta \in W$$

and $\qquad a \in F, \alpha \in W \Rightarrow a\alpha \in W \quad [\because W \text{ is closed under scalar multiplication.}]$

Conversely, Suppose W is a subset of V and

(i) $\qquad \alpha \in W, \ \beta \in W \Rightarrow \alpha - \beta \in W.$

(ii) $\qquad a \in F, \ \alpha \in W \Rightarrow a\alpha \in W.$

Now we have to show that W is a subspace. For this purpose we proceed as follows:

$$\alpha \in W, \alpha \in W \Rightarrow \alpha - \alpha \in W. \qquad\qquad \text{[From (i)]}$$
$$\Rightarrow 0 \in W \Rightarrow \text{identity exists.}$$

and $\qquad 0 \in W, \alpha \in W \Rightarrow 0 - \alpha \in W. \qquad\qquad \text{[From (i)]}$

$$\Rightarrow -\alpha \in W \Rightarrow \text{inverse exists.}$$

Now, $\qquad \alpha \in W, -\beta \in W \Rightarrow \alpha - (-\beta) \in W.$

$$\Rightarrow \quad \alpha + \beta \in W \qquad\qquad \text{[From (i)]}$$

$\therefore \quad$ W is closed under vector addition.

Also vectror addition is always associative and commutative. Thereore $(W,+)$ is an abelian group.

From (ii) it is obvious that W is closed under multiplication. Space V is a vector space over F, therefore remaining properties will also hold in W. Hence W is a vector space and Hence W is itself a subspace.

Theorem 2. *The necessary and sufficient condition for a non-empty subset of W of a vector space V(F) to be a subspace of V is*

$$a,b \in F, \alpha,\beta \in W \Rightarrow a\alpha + b\beta \in W$$

Proof.   Suppose W is a subspace of a vector space $V(F)$. Then W is closed under vector addition and multiplication, therefore we have

$$a \in F, \alpha \in W \Rightarrow a\alpha \in W$$

and $b \in F, \beta \in W \Rightarrow b\beta \in W$

∴ $a\alpha \in W, b\beta \in W \Rightarrow a\alpha + b\beta \in W.$

Conversely, Suppose W is a subset of $V(F)$ and

$$a,b \in F, \alpha,\beta \in W \Rightarrow a\alpha + b\beta \in W \text{ is given.}$$

Then we have to show that W is a subspace of $V(F)$.

Now taking $a=1, b=1$, then

$$1 \in F; \alpha,\beta \in W \Rightarrow 1\alpha + 1\beta \in W$$

$\Rightarrow$ $\alpha + \beta \in W$ $[1\alpha = \alpha, 1\beta = \beta]$

∴ W is closed under vector addition.

and taking $a=0, b=-1$, we have

$$0\alpha + (-1)\beta \in W$$

$\Rightarrow$ $0 + (\beta) \in W$ $[\because (-1)\beta = -\beta]$

$\Rightarrow$ $-\beta \in W$

∴   Additive inverse exists in W.

Again, taking $a=0, b=0$, we have

$$0\alpha + 0\beta \in W$$

$\Rightarrow$ $0 \in W$ $[\because 0\alpha = 0.\text{Similarly}, 0\beta = 0]$

Since $W \subseteq V$, theerfore vector addition is associative and commutative.

Thus, W is an abelian group under vector addition.

Further, taking $\beta = 0$, we have

$$a\alpha + b0 \in W$$

$\Rightarrow$ $a\alpha \in W$ $[\because b0 = 0 \text{ and } a\alpha + 0 = a\alpha]$

∴ W is closed under scalar multiplication.

The rest properties will hold in W because $W \subseteq V$ and these properties hold in V.

Hence W is a vector space and consequently W is a subspace of $V(F)$.

7.6 ALGEBRA OF SUBSPACES

Theorem 1. *The intersection of any two subspaces of a vector space is a subspace .*

Proof.   Let $V(F)$ be a vector space over F and W_1, W_2 be two subspaces of $V(F)$. Then, we have to show that $W_1 \cap W_2$ is a subspace of $V(F)$

Let $\alpha,\beta \in W_1 \cap W_2 \Rightarrow \alpha,\beta \in W_1$ and $\alpha, \beta \in W_2$

Since, W_1 and W_2 are subspaces of V, so we have

$$a, b \in W \text{ and } \alpha, \beta \in W_1 \Rightarrow a\alpha + b\beta \in W_1$$ ...(1)

and, $a, b \in F$ and $\alpha, \beta \in W_2 \Rightarrow a\alpha + b\beta \in W_2$ ...(2)

From (1) and (2), we get

if $a,b \in F$ and $\alpha,\beta \in W_1 \cap W_2 \Rightarrow a\alpha + b\beta \in W_1 \cap W_2$

Hence, $W_1 \cap W_2$ is a subspace of V.

Theorem 2. *The intersection of an arbitrary collection of a subspaces of a vector space is also a subspace.*

Proof. Let $\{W_\lambda : \lambda \in \Lambda\}$ be an arbitrary collection of subspaces of a vcector space V(say). Then we have to show that $\cap \{W_\lambda : \lambda \in \Lambda\}$ is a subspace of V.

Let $$\alpha, \beta \in \cap \{W_\lambda : \lambda \in \Lambda\}$$

$\Rightarrow$ $\qquad\qquad\qquad \alpha, \beta \in W_\lambda$ for each $\lambda \in \Lambda$.

Since, each W_λ is a subspace of V, then for any two scalars $a, b \in F$, we have

$$a\alpha + b\beta \in W_\lambda \text{ for each } \lambda \in \Lambda$$

$\Rightarrow$ $\qquad\qquad\qquad a\alpha + b\beta \in \cap \{W_\lambda : \lambda \in \Lambda\}$

Hence, $\cap \{W_\lambda : \lambda \in \Lambda\}$ is a subspace of V.

Theorem 3. *The union of two subspaces of a vector space is not necessarily a subspace.*

Proof. Let W_1, W_2 be two subspaces of a vector space V and suppose that

$$W_1 = \{(a_1, a_2, 0) : a_1, a_2 \in F\}$$

and $$W_2 = \{(a_1, 0, a_3) : a_1, a_3 \in F\}$$

Obviously, W_1 and W_2 are subspaces of $R^3(\mathbf{R})$. By definition of W_1 and W_2, we have $W_1 \cup W_2$ containing all triads i.e., 3-tuples of the form $(a_1, a_2, 0)$ and those of the form $(a_1, 0, a_3)$.

Now, if we consider the elements $\alpha = (1,2,0)$ and $\beta = (3,0,5)$ of $W_1 \cup W_2$, then for scalars $a = 1$ and $b = 2$.

$$a\alpha + b\beta = 1(1,2,0) + 2(3,0,5) = (1,2,0) + (6,0,10)$$

$$= (7,2,10) \notin W_1 \cup W_2$$

Thus, if $\alpha \in W_1 \cup W_2$ and $\beta \in W_1 \cup W_2$, then it is not necessarily implied that $a\alpha + b\beta \in W_1 \cup W_2$ for some $a, b \in F$

Hence, $W_1 \cup W_2$ is a subspace of $\mathbf{R}^3(\mathbf{R})$.

Theorem 4. *The union of two subspaces of a vector space is a subspace iff one is contained in the other.*

Proof. Let $V(F)$ be a vector space and W_1, W_2 be two subspaces of V.

Suppose $W_1 \subseteq W_2$ or $W_2 \subseteq W_1$. Then we have to show that $W_1 \cup W_2$ is a subspace of V.

Now $W_1 \cup W_2 = W_2$ if $W_1 \subseteq W_2$ and W_2 is a subspace, therefore $W_1 \cup W_2$ is subspace. Also, $W_1 \cup W_2 = W_1$ if $W_1 \subseteq W_2$ and since W_1 is a subspace, therefore $W_1 \subseteq W_2$ is a subspace V.

Conversely, Suppose $W_1 \cup W_2$ is a subspace of V. Then we have to show that

$$W_1 \subseteq W_2 \text{ or } W_2 \subseteq W_1.$$

Let us assume that W_1 is not a subset of W_2 and W_2 is not a subset of W_1.

Now, W_1 is not a subset of W_2, this implies that there exists an element α in W_1 which is not in W_2.

Also, W_2 is not a subset of W_1, therefore there exists an element β in W_2 which is not in W_1. But we have $\alpha \in W_1 \cup W_2$ and $\beta \in W_1 \cup W_2$ and since $W_1 \cup W_2$ is a subspace of V, we have.

$$a, b \in F, \alpha, \beta \in W_1 \cup W_2 \Rightarrow a\alpha + b\beta \in W_1 \cup W_2$$

Now taking $a = 1, b = 1$, we have

$$1\alpha + 1\beta \in W_1 \cup W_2 \qquad\qquad [\because 1\alpha \in W_1 \cup W_2 \subseteq V \therefore 1\alpha = \alpha]$$

$\Rightarrow$ $\qquad\qquad \alpha + \beta \in W_1 \cup W_2$

$\Rightarrow$ $\qquad\qquad \alpha + \beta \in W_1 \text{ or } \alpha + \beta \in W_2$

Suppose $\alpha + \beta \in W_1$ and $\alpha \in W_1$, then $(\alpha + \beta) - \alpha \in W_1$, because W_1 is a subspace of V. Therefore $\beta \in W_1$ and this gives a contradiction.

Now, suppose $\alpha + \beta \in W_2$ and $\beta \in W_2$, then

$$(\alpha + \beta) - \beta \in W_2$$

$\Rightarrow \qquad\qquad\qquad \alpha \in W_2 \qquad\qquad\qquad$ [$\because W_2$ is a subspace.]

Hence, either $\qquad W_1 \subseteq W_2$ or $W_2 \subseteq W_1$

7.7 LINEAR SUM OF TWO SUBSPACES

Let W_1 and W_2 be two subspaces of a vector $V(F)$. Then the linear sum of W_1 and W_2 is the set of all those elements each one of which is expessible as the sum of an element of W_1 and an element of W_2. The linear sum of W_1 and W_2 can be written as $W_1 + W_2$. That is

$$W_1 + W_2 = \{\alpha + \beta : \alpha \in W_1, \beta \in W_2\}$$

REMARK

☞ If $a \in W_1$, then $a = a + 0$ with $0 \in W_1$ and $0 \in W_2$ so $W_1 \subseteq W_1 + W_2$, Similarly, $W_2 \subseteq W_1 + W_2$.

Theorem 1. *The linear sum of two subspaces of a vector space is also a subspace.*

Proof. Let W_1 and W_2 be two subspaces of a vector space $V(F)$. Then we have to show that $W_1 + W_2$ is a subspace of $V(F)$.

Let α, β be any two arbitrary elements of $W_1 + W_2$,

Then, $\qquad\qquad \alpha, \beta \in W_1 + W_2$

$\Rightarrow \qquad\qquad\qquad \alpha = \alpha_1 + \alpha_2$ and $\beta = \beta_1 + \beta_2$, where $\alpha_1, \beta_1 \in W_1$ and $\alpha_2, \beta_2 \in W_2$.

Since, $\qquad \alpha_1, \alpha_2, \beta_1, \beta_2 \in V$

$\therefore \qquad\qquad W_1 + W_2 \subseteq V.$

Since, $\qquad W_1$ and W_2 are subspaces of V. Then

$\qquad\qquad\qquad \alpha_1, \beta_1 \in W_1 \qquad\qquad \Rightarrow a\alpha_1 + b\beta_1 \in W_1$ for some $a, b \in F$

and $\qquad\qquad \alpha_2, \beta_2 \in W_2 \qquad\qquad \Rightarrow a\alpha_2 + b\beta_2 \in W_2$

Now, $\qquad a\alpha_1 + b\beta_1 \in W_1 \qquad$ and $\; a\alpha_2 + b\beta_2 \in W_2$

$\qquad\qquad\qquad\qquad\qquad\qquad \Rightarrow (a\alpha_1 + b\beta_1) + (a\alpha_2 + b\beta_2) \in W_1 + W_2$

$\qquad\qquad\qquad\qquad\qquad\qquad \Rightarrow a(\alpha_1 + \alpha_2) + b(\beta_1 + \beta_2) \in W_1 + W_2$

$\qquad\qquad\qquad\qquad\qquad\qquad \Rightarrow a\alpha + b\beta \in W_1 + W_2$

$\therefore \qquad\qquad \alpha, \beta \in W_1 + W_2 , a, b \in F \Rightarrow a\alpha + b\beta \in W_1 + W_2$

Hence, $W_1 + W_2$ is a subspace.

7.8 DIRECT SUM OF VECTOR SUBSPACES

Definition. *Let W_1 and W_2 be two subspaces of a vector space V. Then V is said to be the direct sum of W_1 and W_2 if each element of V can be uniquely expressed as the sum of an element of W_1 and an element of W_2. If V is direct sum of W_1 and W_2, then it can be written as $V = W_1 \oplus W_2$.*

In general, if V is the direct sum of $W_1, W_2, ... W_n$, then

$$V = W_1 \oplus W_2 \oplus ..... \oplus W_n$$

Here $W_1, W_2, ..., W_n$ are called complementary spaces.

Theorem 1. *The necessary and sufficient condition for a vector space V to be the direct sum of two of its subspaces W_1 and W_2 are:*

    (i) $V=W_1+W_2$

    (ii) $W_1 \cap W_2=\{0\}$

Proof.   **Condition is necessary:**

Suppose V is the direct sum of W_1 and W_2, then each element of V can be uniquely experessed as sum of an element of W_1 and an element of W_2, so in particular each element of V is expressible as the sum of an element of W_1 and element of W_2, this conclude that

$$V=W_1+W_2$$

Next, we shall show that $W_1 \cap W_2 =\{0\}$, for this let, if possible, there be a non-zero vector in $W_1 \cap W_2$ and let it be $\alpha \in W_1 \cap W_2$. Then we may write

$$\alpha=\alpha+0 \text{ with } \alpha \in W_1 \text{ and } 0 \in W_2.$$

and $\alpha=0+\alpha \text{ with } 0 \in W_1 \text{ and } \alpha \in W_2.$

Since $W_1+W_2 \in V$, so $\alpha \in V$ and $V=W_1 \oplus W_2$ therefore, α can be uniquely expressed as the sum of an element of W_1 and an element of W_2. Thus contains only zero vector. This implies $W_1 \cap W_2 =\{0\}$.

Condition is Sufficient:

Suppose the conditions:

    (i) $V=W_1+W_2$

    (ii) $W_1 \cap W_2=\{0\}$

hold then we shall show that $V=W_1 \oplus W_2$.

From (i) we conclude that each element of V can be expressed as the sum of an element of W_1 and an element of W_2. Therefore, we shall only show that this representation is unique.

Let, if possible, an element $\alpha \in V$ has two representations, that is,

$$\alpha=\alpha_1+\alpha_2 \text{ with } \alpha_1 \in W_1 \text{ and } \alpha_2 \in W_2.$$

and $\alpha'=\alpha'_1+\alpha'_2 \text{ with } \alpha'_1 \in W_1 \text{ and } \alpha'_2 \in W_2.$

$\Rightarrow$ $\alpha_1+\alpha_2=\alpha'_1+\alpha'_2$

$\Rightarrow$ $\alpha_1-\alpha'_1=\alpha'_2-\alpha_2$

           $\alpha_1,\alpha'_1 \in W_1 \Rightarrow \alpha_1-\alpha'_1 \in W_1$

and $\alpha_2,\alpha'_2 \in W_2 \Rightarrow \alpha'_2-\alpha_2 \in W_2$ $[\because \ W_1 \text{ and } W_2 \text{ are subspaces}]$

$\therefore$ $\alpha_1-\alpha'_1=\alpha'_2-\alpha_2 \in W_1 \cap W_2.$

But $W_1 \cap W_2=\{0\}, \text{ this implies}$

$$\alpha_1-\alpha'_1=0=\alpha'_2-\alpha_2$$

$\Rightarrow$ $\alpha_1 = \alpha'_1 \text{ and } \alpha'_2 = \alpha_2.$

This shows that each element of V can be uniquely expressed as the sum of an element of W_1 and an element of W_2. Hence, $V=W_1 \oplus W_2$ *i.e.*, V is the direct sum of W_1 and W_2.

Solved Examples

▶ An algebraic structure $(V,+,.)$ is said to be vector space over a field F if
   (i) $(V,+)$ is an abelian group.
   (ii) $a(\alpha+\beta) = a\alpha+a\beta \ \forall \ \alpha,\beta \in V, a \in F$
   (iii) $(a+b)\alpha = a\alpha+b\alpha \ \forall \ \alpha \in V, \ a,b \in F$
   (iv) $(ab)\alpha = a(b\alpha) \ \forall \ \alpha \in V, \ a,b \in F$
   (v) $1.\alpha = \alpha \ \forall \ \alpha \in V, \ 1 \in F$

▶ The necessary and sufficient conditions for a non-empty subset W of $V(F)$ to be a subspace is that
$$a,b \in F, \ \alpha,\beta \in W \Rightarrow a\alpha+b\beta \in W$$

▶ $W_1+W_2=\{\alpha+\beta : \alpha \in W_1, \beta \in W_2\}$

▶ The necessary and sufficient condition for a vector space V to be the direct sum of two of its subspaces W_1 and W_2 are
   (i) $V= W_1+W_2$ (ii) $W_1 \cap W_2 = \{0\}$

Example 1. *In $V=R^3$. Let W_1 be the xy-plane and let W_2 be the z-plane given by*
$$W_1=\{(x, y, 0): x, y \in \mathbf{R}\}$$
and $W_2=\{(0,y,z): y, z \in \mathbf{R}\}$
Show that $V =\{W_1 \oplus W_2\}$

Solution.   Let $(x, y, z) \in V$, then this element can be written as the sum of an element of $\mathbf{W_1}$ and an element of W_2 on one and only one way *i.e.,*
$$(x,y,z)= (x, y, 0)+ (0,0, z)$$
Accordingly, V is the direct sum of W_1 and W_2, that is $V=\{W_1 \oplus W_2\}$

Example 2. *In $V=R^3$ and W_1 be the xy-plane and let W_2 be the yz-plane:*
$$W_1=\{(x,y,0): x, y \in \mathbf{R}\}$$
and $W_2=\{(0,y,z): y, z \in \mathbf{R} \}$
then V is not the direct sum of W_1 and W_2.

Solution.   Let (a,b,c) be any element of $\mathbf{R}^3$, then
$$(a,b,c)= (x ,y, 0)+ (0,y, z)= (x,2y,z)$$
$\Rightarrow$ $a=x, b=2y, c=z$
$\Rightarrow$ $a,b,c \in \mathbf{R}$ $[\because x, y, z \in \mathbf{R}]$
$\Rightarrow$ every element of V can be written as the sum of an element of W_1 and an element of W_2.

But such sums are not unique, for example:
Let $(3,5,7) \in \mathbf{R}^3$, then
$$(3,5,7)=(3,2,0)+ (0,3,7), (3,2,0) \in W_1 \text{ and } (0,3,7) \in W_2.$$
Also $(3,5,7)=(3,1,0)+ (0,4,7), (3,1,0) \in W_1 \text{ and } (0,4,7) \in W_2.$
Hence, V is not the direct sum of W_1 and W_2.

Example 3. *If $V_3(\mathbf{R})$ is a vector space and $W_1=\{(a,0,c):a, c \in \mathbf{R}\}$ and $W_2=\{(0,b,c):b,c \in \mathbf{R}\}$ are two subspaces of $V_3(\mathbf{R})$, then show that $V = W_1 + W_2$ and $V \neq W_1 \oplus W_2$.*

Solution.   Let (x,y,z) be an arbitrary element of $V_3(\mathbf{R})$, then
$$(x,y,z)= (a,0,c)+ (0,b,c)= (a,b,2c)$$

$$\Rightarrow \qquad\qquad x=a, y=b,\ z=2c$$

$$\Rightarrow \qquad\qquad x,y,z \in \mathbf{R} \qquad\qquad\qquad [\because a, b, c \in \mathbf{R}]$$

$\Rightarrow$ every element of $V_3(\mathbf{R})$ can be written as the sum of an element of W_1 and an element of W_2.

$$\Rightarrow \qquad\qquad V=W_1+W_2$$

But such representations is not unique, for example:

Let $(3,5,6) \in V_3(\mathbf{R})$, then

$$(3,5,6)=(3,0,2)+ (0,5,4);\ (3,0,2) \in W_1 \text{ and } (0,5,4) \in W_2.$$

Also $\qquad (3,5,6)=(3,0,5)+ (0,5,1);\ (3,0,5) \in W_1 \text{ and } (0,5,1) \in W_2.$

So here $(3,5,6)$ can be written as the sum of elements of W_1 and W_2 in two ways.

Hence, $\qquad\qquad V \neq W_1 \oplus W_2 .$

Example 4. *Let V be a vector space of all functions from* **R** *into* **R***; let V_e be the subset of even functions, such that $f(-x)= f(x)$; let V_0 be the subset of odd functions $f(-x)= -f(x)$. Prove that*

(a) V_e *and* V_0 *are subspaces of V.*

(b) $V_e + V_0 = V.$

(c) $V_e \cap V_0 = \{0\}.$

Solution. (a) Since V is a vector space of all functions, therefore

$$V_e \subseteq V, V_0 \subseteq V.$$

Let $f(x), g(x) \in V_e$, so that $f(-x)= f(x)$ *and* $g(-x)= g(x)$ and let $a,b \in \mathbf{R}$, then consider,

$$h(x) = af(x)+bg(x)$$

Now $\quad h(-x) = af(-x)+bg(-x)$

or $\qquad h(-x) = af(x)+bg(x) \qquad\qquad [f(-x)= f(x)\ ,\ g(-x)= g(x)]$

$$= h(x)$$

$$\Rightarrow \qquad\qquad h(x) \in V_e$$

$$\Rightarrow af(x)+bg(x) \in V_e$$

Consequently, if $f(x),g(x) \in V_e$, then $af(x)+bg(x) \in V_e$. Hence, V_e is a subspace of V.

Similarly, we can prove that V_0 is a subspace of V.

(b) Let $f(x)$ be any element of V.

Consider $f(x)= \dfrac{1}{2}\ [f(x)+f(-x)]+\dfrac{1}{2}[f(x) - f(-x)].$

Let $\qquad \alpha(x)=\dfrac{1}{2}\ [f(x)+f(-x)] \text{ and } \beta(x)=\dfrac{1}{2}[f(x)+f(-x)]$

$\therefore \qquad f(x)= \alpha(x)+ \beta(x).$

Also $\quad \alpha(-x)=\dfrac{1}{2}\ [f(-x)+f(x)]= \alpha(x)$

$\therefore \qquad \alpha(x)= V_e.$

and $\qquad \beta(-x) = \frac{1}{2}[f(-x)+f(x)] = \frac{1}{2}[f(x)+f(-x)] = \beta(x)$

$\therefore \qquad\qquad \beta(x) = V_0.$

Consequently, $f(x) = \alpha(x) + \beta(x)$ where, $\alpha(x) = V_e$ and $\beta(x) = V_0$. Hence every element of V can be expressed as the sum of an element of V_e and an element of V_0. That is,

$$V = V_e + V_0.$$

(c) $V_e \cap V_0 = \{0\}$.

Let if possible, there exist a non-zero function $f(x)$, which belongs to $V_e \cap V_0$

Therefore, if $f(x) \in V_e$, then $f(-x) = f(x)$ and $f(x) \in V_0$, then $f(-x) = -f(x)$. So that,

$$f(x) = -f(x)$$

$\Rightarrow \qquad 2\ f(x) = 0 \Rightarrow f(x) = 0$

This gives a contradiction, because $f(x)$ is assumed to be non-zero function. Hence every function of $V_e \cap V_0$ is a zero function. Consequently, $V_e \cap V_0 = \{0\}$.

Example 5. *If W_1 and W_2 are subspaces of a vector space $V(F)$, then show that $W_1 + W_2$ is also a subspace of $V(F)$.*

Solution. Let us define

$$W_1 + W_2 = \{\alpha_1 + \alpha_2; \alpha_1 \in W_1, \alpha_2 \in W_2\}.$$

We have to show that $W_1 + W_2$ is a subspace of V.

Let $\quad \alpha \in W_1 + W_2 \Rightarrow \alpha = \alpha_1 + \alpha_2$ for some $\alpha_1 \in W_1$ and $\alpha_2 \in W_2$

$\qquad\qquad\qquad \Rightarrow \alpha = \alpha_1 + \alpha_2$ for some $\alpha_1, \alpha_2 \in V \Rightarrow \alpha \in V$

$\therefore \qquad\quad W_1 + W_2 \subseteq V.$

Now, let $\alpha_1 + \alpha_2 \in W_1 + W_2$ and $\beta_1 + \beta_2 \in W_1 + W_2$, where, $\alpha_1 + \alpha_2 \in W_1$ and $\beta_1 + \beta_2 \in W_2$.

Since, W_1 and W_2 are subspaces of V, then

$$a, b \in F, \alpha_1, \beta_1 \in W_1 \Rightarrow a\alpha_1 + b\beta_1 \in W_1$$
$$a, b \in F, \alpha_2, \beta_2 \in W_0 \Rightarrow a\alpha_2 + b\beta_2 \in W_2$$
$$a\alpha_1 + b\beta_1 \in W_1, a\alpha_2 + b\beta_2 \in W_2$$
$$\Rightarrow (a\alpha_1 + b\beta_1) + (a\alpha_2 + b\beta_2) \in W_1 + W_2$$
$$\Rightarrow a(\alpha_1 + \alpha_2) + b(\beta_1 + \beta_2) \in W_1 + W_2$$

Since, $\quad \alpha_1 + \alpha_2 \in W_1 + W_2$ and $\beta_1 + \beta_2 \in W_1 + W_2$. Thus we have

$$a, b \in F, \alpha_1 + \alpha_2, \beta_1 + \beta_2 \in W_1 + W_2$$
$$\Rightarrow a(\alpha_1 + \alpha_2) + b(\beta_1 + \beta_2) \in W_1 + W_2$$

Hence, $W_1 + W_2$ is subspace of $V(F)$.

Example 6. *Let $\mathbf{R}$ be the field of real numbers, show that the set $W = \{(x, 2y, 3z) : x, y, z \in \mathbf{R}\}$ is a subspace of $V_3(\mathbf{R})$.*

Solution. Since $W = \{(x, 2y, 3z) : x, y, z \in \mathbf{R}\}$.

Let $\alpha, \beta \in W$, where $\alpha = \{x_1, 2y_1, 3z_1\}$ and $\beta = \{x_2, 2y_2, 3z_2\}$ and $x_1, y_1, z_1, x_2, y_2, z_2 \in \mathbf{R}$,

If $a, b \in \mathbf{R}$, then

$$a\alpha + b\beta = a(x_1, 2y_1, 3z_1) + b(x_2, 2y_2, 3z_2)$$
$$= (ax_1, 2ay_1, 3az_1) + (bx_2, 2by_2, 3bz_2)$$
$$= (ax_1 + bx_2), 2(ay_1 + by_2), 3(az_1 + bz_2)$$

$\therefore$ $\qquad a\alpha + b\beta \in W,$

Because $ax_1 + bx_2, ay_1 + by_2, az_1 + bz_2 \in \mathbf{R}.$

Hence, W is a subspace of $V_3(\mathbf{R})$.

Example 7. *Show that the set of all real valued continuous functions defined on* [0,1] *is a vector space over field of reals.*

Solution. Let V be the set of all real valued continuous functions defined on [0,1]. Now we have to show that V is a vector spsce over $\mathbf{R}$ (field of real numbers) under vector addition and scalar multiplication which is defined as follows:

$$(f+g)(x) = f(x) + g(x), \forall f, g \in V$$

and $\qquad (af)(x) = af(x), \forall f \in V$ and $a \in \mathbf{R}.$

First we shall show that $(V, +)$ is an abelian group.

Let $f, g \in V$, then

$$(f+g)(x) = f(x) + g(x), \forall f, g \in V$$

$\therefore$ $\qquad f+g \in V.$ Thus V is closed under vector addition.

Now let $0(x) \in V$, we have

$$f(x) + 0(x) = (f+0)(x) = f(x), \forall f \in V$$

$\therefore$ $\qquad 0(x)$ is the additive identity in V.

Let $-f \in V$, then we have

$$-f(x) + f(x) = (-f+f)(x) = 0(x), \forall f \in V$$

$\therefore$ $\qquad -f$ is the additive inverse of V.

Since vector addition is always associative as well as commutative, consequently $(V, +)$ is an abelian group.

Further, since V is closed under scalar multiplication therefore af is a real valued continuous function defined on [0,1].

(i) If $a \in \mathbf{R}$ and $f, g \in V$, then we have

$$a[(f+g)x] = a[f(x) + g(x)]$$
$$= af(x) + ag(x) = (af + ag)(x)$$

$\therefore$ $\qquad a(f+g) = af + ag.$

(ii) If $a, b \in \mathbf{R}$ and $f \in V$, then we have

$$[(a+b)f](x) = (a+b)f(x) = af(x) + bf(x) = (af + bf)(x)$$

$\therefore$ $\qquad (a+b)f = af + bf.$

(iii) If $a, b \in \mathbf{R}$ and $f \in V$, then we have

$$[(ab)f](x) = (ab)f(x) = a[bf(x)] = [a(bf)](x).$$

$\therefore$ $\qquad (ab)f = a(bf).$

(iv) If $1 \in \mathbf{R}$ and $f \in V$, then we have

$$(1f)(x) = 1f(x) = f(x)$$

$$\therefore \qquad\qquad 1f = f, \ \forall f \in V$$

Hence, V is a vector space over **R**.

Example 8. *Show that $R^2(\mathbf{R})$ is not a vector space when addition and scalar multiplication composition are defined by*

$$(a_1, a_2) + (b_1, b_2) = (a_1 + b_1, a_2 + b_2)$$

and $\qquad a(a_1, a_2) = (aa_1, a_2), \ \forall \ a, a_1, a_2, b_1, b_2 \in \mathbf{R}.$

Solution. Suppose $a=1$ and $b=2$ and $(a_1, a_2) = (3,4)$, then using the given compositions, we have

$$(a+b)(a_1, a_2) = (1+2). \ (3+4)$$
$$= 3.(3,4) = (3.3,4) = (9,4)$$

and $a.(a_1, a_2) + b.(a_1, a_2) = 1.(3,4) + 2.(3,4)$
$$= (3.1,4) + (3.2,4)$$
$$= (3,4) + (6,4)$$
$$= (3+6, 4+4) = (9,8)$$

$$\therefore \qquad (a+b).(a_1, a_2) \neq a . (a_1, a_2) + b . (a_1, a_2).$$

Hence, $R^2(\mathbf{R})$ is not a vector space.

Example 9. *Show that* the set

$$W = \{(a,b,c): a - 3b + 4c = 0\}$$

is a subspace of 3-tuple space $R^3(\mathbf{R})$.

Solution. Let $\alpha = (a_1, b_1, c_1)$ and $\beta = (a_2, b_2, c_2)$ be any two elements of W, such that
$$a_1 - 3b_1 + 4c_1 = 0 \text{ and } a_2 - 3b_2 + 4c_2 = 0$$

For $a, b \in \mathbf{R}$, we have

$$a\alpha + b\beta = a(a_1, b_1, c_1) + b(a_2, b_2, c_2)$$
$$= (aa_1, ab_1, ac_1) + (ba_2, bb_2, bc_2)$$
$$= (aa_1 + ba_2, ab_1 + bb_2, ac_1 + bc_2)$$

Now, $\qquad (aa_1 + ba_2) - 3(ab_1 + bb_2) + 4(ac_1 + bc_2)$
$$= (aa_1 - 3ab_1 + 4ac_1) + (ba_2 - 3bb_2 + 4bc_2)$$
$$= a(a_1 - 3b_1 + 4c_1) + b(a_2 - 3b_2 + 4c_2)$$
$$= a.0 + b.0 = 0$$

So, $\qquad\qquad a\alpha + b\beta \in W.$

Thus $\quad \alpha \in W, \beta \in W \Rightarrow a\alpha + b\beta \in W \ \forall \ a, b \in \mathbf{R}.$

Hence, W is a subspace of $R^3(\mathbf{R})$.

Example 10. *Prove the solution set W of the differential equation*

$$2\frac{d^2 y}{dx^2} - 9\frac{dy}{dx} + 2y = 0$$

is a subspace of vector space of all real valued functions of $\mathbf{R}$.

Solution. Let $W = \left\{ y : 2\dfrac{d^2 y}{dx^2} - 9\dfrac{dy}{dx} + 2y = 0 \right\}$, be the set of all solutions of the given

differential equation where, $y = f(x)$.

Now if we define a real valued function denoted by 0 on R by $0(x)=0$, $\forall\, x \in \mathbf{R}$, then $0(x)$ satisfies the given differential equation, so that $0(x) \in W$.

Let $y_1 = f(x)$ and $y_2 = g(x)$ be any two elements of W, then we have

$$2\frac{d^2 f(x)}{dx^2} - \frac{df(x)}{dx} + 2f(x) = 0 \qquad \qquad \dots\text{ (1)}$$

and

$$2\frac{d^2 g(x)}{dx^2} - \frac{dg(x)}{dx} + 2g(x) = 0 \qquad \qquad \dots\text{ (2)}$$

Let a, b be any two scalars.

Now, multiplying (1) by a and (2) by b and then adding, we get

$$2\frac{d^2}{dx^2}[af(x) + bg(x)] - 9\frac{d}{dx}[af(x) + bg(x)] + 2[af(x) + bg(x)] = 0$$

which shows that $af(x) + bg(x)$ is also the solution of the given differential equation.

So that $\qquad\qquad [af(x) + bg(x)] \in W.$

$\therefore \qquad\qquad\qquad f(x) \in W, g(x) \in W \Rightarrow af(x) + bg(x) \in W\ \forall\ a,\, b \in \mathbf{R}.$

Hence W is a subspace of a vector space of all real valued functions of $\mathbf{R}$.

Example 11. *Let V be the vector space of all functions from the real field $\mathbf{R}$ into $\mathbf{R}$. Show that the set $W = \{f : f(7) = 2 + f(1) \}$ is not a subspace of V.*

Solution. Let f and g be any two elements of W, i.e.,

$$f(7) = 2 + f(1) \text{ and } g(7) = 2 + g(1).$$

Then $\qquad\qquad (f + g)(7) = f(7) + g(7)$

$$= 2 + f(1) + 2 + g(1)$$
$$= 4 + f(1) + g(1)$$
$$= 4 + (f + g)(1) \neq 2 + (f + g)(1).$$

Hence $f + g \neq W$, and so W is not a subspace of V.

Example 12. *Let V be the vector space of all square $n \times n$ matrices over a field of reals $\mathbf{R}$. Show that W is a subspace of V, where*

(i) W consists of the symmetric matrices.

(ii) W consists of all matrices which commute with a given matrix M, i.e.,

$$W = \{A \in V : AM = MA\}.$$

Solution. (i) The null matrix $0 \in W$, as all its entries being zero and it is symmetric.

Let $A = [a_{ij}]$ and $B = [b_{ij}]$ be any two elements of W.

Then $a_{ij} = a_{ji}$ and $b_{ij} = b_{ji}$.

For any scalars $a, b \in \mathbf{R}$, we have

$$aA + bB = a[a_{ij}] + b\,[b_{ij}] = [aa_{ij} + bb_{ij}] = [c_{ij}] = C$$

where, $\qquad\qquad c_{ij} = aa_{ij} + bb_{ij}$

Now, $\qquad\qquad c_{ij} = aa_{ij} + bb_{ij} = aa_{ji} + bb_{ji} = c_{ji}$

Thus $aA + bB$ is symmetric so it belong to W. Hence W is subspace of V.

(ii) The null matrix $0 \in W$ as $0M = M0$.

Now suppose A and B be any two elements of W then, $AM = MA$ and $BM = MB$.

For $a, b \in \mathbf{R}$, we have

$$(aA + bB)M = (\alpha A)M + (bB)M = a(AM) + b(BM)$$
$$= a(MA) + b(MB) = M(aA + bB)$$
$$\therefore \qquad aA + bB \in W, A B \in W \text{ and } \forall\, a, b \in \mathbf{R}.$$

Hence W is a subspace of V.

Example 13. *let $V = R^3$. Show the set $W = \{(a, b, c) : a^2 + b^2 + c^2 \leq 1\}$ is not a subspace of* **V**.

Solution.   Let $\alpha = (1, 0, 0) \in W$, $\beta = (0, 1, 0) \in W$. But we have

$$\alpha + \beta = (1, 0, 0) + (0, 1, 0)$$
$$= (1, 1, 0) \notin W \text{ as } 1^2 + 1^2 + 0^0 = 2 > 1.$$

Hence W is not a subspace of V.

Example 14. *Let V be the vector space of all 2×2 matrices over the real field* **R**. *Show that W is not a subspace of V, where*

  (i) *W consists of all matrices with zero determinant,*

  (ii) *W consists of all matrices A from which $A^2 = A$.*

Solution.   (i) Let $A = \begin{pmatrix} 1 & 0 \\ 0 & 0 \end{pmatrix}$ and $B = \begin{pmatrix} 0 & 0 \\ 0 & 1 \end{pmatrix}$ be two elements of W. But,

$$A + B = \begin{pmatrix} 1 & 0 \\ 0 & 0 \end{pmatrix} + \begin{pmatrix} 0 & 0 \\ 0 & 1 \end{pmatrix}$$

$$= \begin{pmatrix} 1 & 0 \\ 0 & 1 \end{pmatrix} \notin W \text{ as } |A + B| = 1 \neq 0$$

  $\therefore$ W is not a subspace of V.

  (ii) The unit matrix $I = \begin{pmatrix} 1 & 0 \\ 0 & 1 \end{pmatrix} \notin W$ as $I^2 = I$.

  But $2I = \begin{pmatrix} 2 & 0 \\ 0 & 2 \end{pmatrix} \notin W$ as

$$(2I)^2 = \begin{pmatrix} 2 & 0 \\ 0 & 2 \end{pmatrix}\begin{pmatrix} 2 & 0 \\ 0 & 2 \end{pmatrix} = \begin{pmatrix} 4 & 0 \\ 0 & 4 \end{pmatrix} \neq 2I.$$

Hence W is not a subspace of V.

EXERCISE 7.1

1. Show that a field K can be regarded as a vector space over any subfield F of K.

2. Show that the complex field **C** is a vector space over the field **R** of reals.

3. Let V be the set of all ordered pairs (x, y) of reals and let F be the field of real numbers, then show that V is not a vector space over F with respect to addition and multiplication defined as

$(x, y) + (x_1, y_1) = (x + x_1, y + y_1)$ and $c(x, y) = (cx, cy)$

4. Let V be the set of all pairs of real numbers and let F be the field of real numbers and define

$(x, y) + (x_1, y_1) = (3y + 3y_1, -x - x_1)$ and $c(x, y) = (3cy, -cx)$.

Show that V is a vector space over F.

5. Show that the set $W = \{(a_1, a_2, 0) : a_1, a_2 \in F\}$ is a subspace of $V_3(F)$.

6. Show that the set W of the elements of the vector space $V_3(\mathbf{R})$ of the form

$(x + 2y, y, -x + 3y)$

where $x, y \in \mathbf{R}$ is a subspace of $V_3(\mathbf{R})$.

7. Prove that the set of all solutions (a, b, c) of the equation $a + b + 2c = 0$ is a subspace of vector space $V_3(\mathbf{R})$.

8. Prove that the arbitrary intersection of subspaces of a vector space is a subspace.

9. Let $V = R^3$. Show that the set $W = \{(a,b,c) : a, b, c \in \mathbf{Q}]$ is not a subspace of R^3.

HINTS TO SELECTED PROBLEMS

2. Let $\mathbf{C}$ be the set of vectors and $\mathbf{R}$ the set of scalars, Since $\mathbf{C}$ is a field so that $a\alpha \in \mathbf{C}$, this show that $\mathbf{C}$ is closed under scalar multiplication. $1 \in \mathbf{R}$ so $1 \in \mathbf{C}$.

Also for $\qquad a, b \in \mathbf{C}$ and $a, b \in \mathbf{R}$,

$\qquad\qquad a, b \in \mathbf{R} \Rightarrow a, b \in \mathbf{C}$ $\qquad\qquad\qquad (\because \mathbf{R} \subseteq \mathbf{C})$

$\therefore \qquad\qquad a(\alpha + \beta\} = a\alpha + a\beta$ and $(a + b)\alpha = a\alpha + b\alpha$.

also, $\qquad\qquad (ab)\alpha = a(b\alpha), \quad 1 . \alpha = \alpha, \forall \alpha \in C$. Hence $\mathbf{C}$ is a vector space over $\mathbf{R}$.

3. Let $\alpha = (x, y)$ and $a, b \in F$, then

$\qquad\qquad (a + b)\alpha = (a + b)(x, y) \quad = ((a+b) x, y) = \{ax + bx, y\}$

and $\qquad\qquad a\alpha + b\alpha = a(x, y) + b(x, y) = (ax, y) + (bx, y) = (ax + bx, 2y)$.

$\therefore \qquad\qquad (a + b)\alpha \neq a\alpha + b\alpha$. Hence V is not a vector space.

6. $W = ((x+2y, y, -x+3y) : x, y \in \mathbf{R})$, then show that $a\alpha + b\beta \in W \ \forall \ \alpha, \beta \in W$ and $a, b \in \mathbf{R}$.

$a\alpha + b\beta = a(x_1 + 2y_1, y_1, -x_1 + 3y_1] + b(x_2 + 2y_2, y_2, -x_2 + 3y_2]$

$\qquad = (ax_1 + 2ay_1, ay_1, -ax_1 + 3ay_1) + (bx_2 + 2by_2, by_2, -bx_2 + 3by_2)$

$\qquad = (ax_1 + bx_2 + 2(ay_1 + by_2), ay_1 +, by_2, -(ax_1 + bx_2) + 3(ay_1 + by_2)) \in W$

$\therefore \ a\alpha + b\beta \in W \Rightarrow W$ is a subspace.

9. $(a, b, c) \in W, (a, b, -c) \in W$ as $-c \in \mathbf{Q}$.

Now $(a, b, c) + (a, b, -c) = (2a, 2b, 0) \in W$. Hence W is not a subspace.

7.9 LINEAR COMBINATION OF VECTORS

Definition. *Let V be a vector space over a field F and $\alpha_1, \alpha_2, \ldots, \alpha_n \in V$, then any vector $\alpha \in V$ can be expressed as below:*.

$$\alpha = a_1\alpha_1 + a_2\alpha_2 + \ldots + a_n\alpha_n$$

where $a_1, a_2, \ldots, a_n \in F$, is said to be the linear combination of vectors $\alpha_1, \alpha_2, \ldots, \alpha_n$.

Definition. *Let $V(F)$ be a vector space aver F and let S be any non-empty subset of V, then the set of all linear combination of finite elements of S, is called the linear span of S. It is denoted by $L(S)$. Therefore, we have*

$$L(S) = \{a_1\alpha_1 + a_2\alpha_2 + \ldots + a_n\alpha_n : a_1, a_2, \ldots, a_n \in F\}$$

and $\alpha_1, \alpha_2, \ldots, \alpha_n$ are a finite elements of S\}

Theorem 1. *The linear span $L(S)$ of a non-empty subset S of a vector space $V(F)$ is the smallest subspace of V containing S.*

Proof. By definition of $L(S)$, we have

$$L(S) = \{a_1\alpha_1 + a_2\alpha_2 + \ldots + a_n\alpha_n : a_i \in V\}$$

[Let $\alpha \in S$, then $\alpha = 1.\alpha, 1 \in F$, so $\alpha \in L(S)$

$\therefore \qquad\qquad S \subseteq L(S)$

Now, we shall show that $L(S)$ is a subspace.

Let α, β be any two arbitrary elements of $L(S)$, then

$$\alpha = a_1\alpha_1 + a_2\alpha_2 + \ldots + a_n\alpha_n; \text{ for } \alpha_1, \alpha_2, \ldots, \alpha_n \in S \text{ and for } a_1, a_2, \ldots, a_n \in F$$

Also, $\quad \beta = b_1\beta_1 + b_2\beta_2 + \ldots + b_m\beta_m \quad$ for all $a, b \in F$, we have

$$a\alpha + b\beta = a(a_1\alpha_1 + a_2\alpha_2 + \ldots + a_n\alpha_n) + b(b_1\beta_1 + b_2\beta_2 + \ldots + b_m\beta_m)$$

$$= (aa_1)\alpha_1 + (aa_2)\alpha_2 + \ldots + (aa_n)\alpha_n + (bb_1)\beta_1 + (bb_2)\beta_2 + \ldots + (bb_m)\beta_m$$

This implies that $a\alpha + b\beta$ is a linear combination of finite numbers of elements of S, so $a\alpha + b\beta \in L(S)$. Hence $L(S)$ is a subspace of V.

Next, we shall show that $L(S)$ is the smallest subspace containing S.

For this there is a subspace W of V containing S. Let $\alpha_1, \alpha_2, \ldots, \alpha_t \in S \subset W$ and W being a subspace, then

$$a_1\alpha_1 + a_2\alpha_2 + \ldots + a_t\alpha_t \in W; \text{ for all } a_i \in F$$

This implies that W contains all linear combinations of finite elements of S, therefore $L(S) \in W$.

Hence $L(S)$ is the smallest sub space of V containing S.

Theorem 2. *If S, T are two subsets of a vector space V, then*

    (i) $S \subseteq T \Rightarrow L(S) \subseteq L(T)$.

    (ii) $L(S \cup T) = L(S) + L(T)$

    (iii) $L[L(S)] = L(S)$

Proof.   Let α be an arbitrary element of $L(S)$, then

$$\alpha \in L(S) \Rightarrow \alpha = a_1\alpha_1 + a_2\alpha_2 + \ldots + a_n\alpha_n$$

where, $\alpha_1, \alpha_2, \ldots, \alpha_n \in S$ and $a_1, a_2, \ldots, a_n \in F$.

    (i) Since $S \subseteq T$, so that $\alpha_1, \alpha_2, \ldots, \alpha_n \in T$, therefore α is also the linear combination of finite elements of T. This implies $\alpha \in L(T)$.

        Thus $\alpha \in L(S) \Rightarrow \alpha \in L(T)$

        Hence $L(S) \subseteq L(T)$ if $S \subseteq T$.

    (ii) Since $S \subseteq S \cup T$ and $T \subseteq S \cup T$, then from (i), we have

$$L(S) \subseteq L(S \cup T)$$

        and $L(T) \subseteq L(S \cup T)$

        $\Rightarrow$ $L(S) + L(T) \subseteq L(S \cup T)$ ...(1)

Next, let α be an arbitrary element of $L(S \cup T)$, then α is a linear combination of finite elements of $S \cup T$. This implies that some of $\alpha_i \in S$ or some of $\alpha_j \in T$. This shows that α is a linear combination of finite elements of S and finite elements of T, therefore $\alpha \in L(S) + L(T)$. Thus,

$$L(S \cup T) \subseteq L(S) + L(T) \qquad\qquad\qquad ...(2)$$

        From (1) and (2), we get

$$L(S \cup T) \subseteq L(S) + L(T) \qquad\qquad\qquad ...(3)$$

    (iii) Since $S \subseteq T(S)$, then from (i), we have

$$L(S) \subseteq L[(S)],$$

Next, let α be any arbitrary element of $L[L(S)]$, then α is a linear combination of finite elements of $L(S)$. Suppose, we have

$$\alpha = b_1\beta_1 + b_2\beta_2 + ... + b_n\beta_n = \sum_{i=1}^{n} b_i\beta_i \qquad ...(2)$$

where, each $\beta_i \in L(S)$ for all $b_1, b_2, ..., b_n \in F$. Also, each β_i is a linear combination of finite elements of S, so that

$$\beta_1 = a_{11}\alpha_1 + a_{12}\alpha_2 + ... + a_{1m}\alpha_m$$
$$\beta_1 = a_{21}\alpha_1 + a_{22}\alpha_2 + ... + a_{2t}\alpha_t$$

....

On putting the values of $\beta_1, \beta_2, ...,$ etc. in (2), we see that α is a linear combination of finite elements of S. Thus $\alpha \in L(S)$

$$\therefore \qquad L[L(S)] \subseteq L(S) \qquad ...(3)$$

From (1) and (3), we get $L(S) = L[L(S)]$.

Theorem 3. *The linear sum of two subspaces W_1 and W_2 of a vector space $V(F)$ is generated by their union. That is, $W_1 + W_2 = L(W_1 \cup W_2)$.*

Proof. We have already proved that the linear sum of two subspaces is also a subspace and linear span of a subset of a vector space is also a subspace.

Therefore, $W_1 + W_2$ and $L(W_1 \cup W_2)$ are subspaces of $V(F)$.

Let α be any arbitrary element of $W_1 + W_2$, then

$$\alpha \in W_1 + W_2$$
$$\Rightarrow \qquad \alpha = \alpha_1 + \alpha_2 \text{ for some } \alpha_1 \in W_1 \text{ and } \alpha_2 \in W_2$$

Since $\alpha_1 \in W_1$ and $\alpha_2 \in W_2$, so, $\alpha_1 + \alpha_2 \in W_1 + W_2$

Also, we may write $\alpha = \alpha_1 + \alpha_2 = 1. \alpha_1 + 1.\alpha_2$. This implies that α is a linear combination of finite elements namely α_1 and α_2 of $W_1 \cup W_2$, so that $\alpha \in L(W_1 + W_2)$

$$\therefore \qquad \alpha \in W_1 + W_2 \Rightarrow a \in L(W_1 \cup W_2)$$

Thus, $\qquad W_1 + W_2 \subseteq L(W_1 \cup W_2)$. $\qquad ...(1)$

But $L(W_1 \cup W_2)$ being the smallest subspace containing $W_1 \cup W_2$, and since $W_1 + W_2$ is a subspace containing $W_1 \cup W_2$, therefore

$$L(W_1 \cup W_2) \subseteq W_1 \cup W_2 \qquad ...(2)$$

From (1) and (2), we get

$$W_1 + W_2 = L(W_1 \cup W_2)$$

7.10 LINEAR DEPENDENCE AND INDEPENDENCE OF VECTORS

In this section, we shall discuss the concept of linear dependence or linear independence which lays the foundations for the key notions (viz., dimensions) of the theory of vector spaces.

Definition 1. *Let $V(F)$ be a vector space over a field F. Then a finite set $\{\alpha_1, \alpha_2, ..., \alpha_n\}$ of vectors of V is said to be linearly dependent if there exists scalars $a_1, a_2, ..., a_n$ not all of them equal to zero such that*

$$a_1\alpha_1 + a_2\alpha_2, ..., + a_n\alpha_n = 0$$

Definition 2. *Let $V(F)$ be a vector space over F. Then a finite set of vectors $\{\alpha_1, \alpha_2, ..., \alpha_n\}$ of V is said to linearly independent if for every expressions of the type.*

$$a_1\alpha_1 + a_2\alpha_2 + ... + a_n\alpha_n = 0$$

where $a_1, a_2, ..., a_n \in F$ implies $a_1 = 0, a_2 = ... = a_n$.

REMARKS

☞ S is linearly independent $\Leftrightarrow$ S is not linearly dependent

$\Leftrightarrow$ no finite subset of S is linearly dependent.

$\Leftrightarrow$ each finite subset of S is linearly independent.

$$\Leftrightarrow \text{ whenever } \sum_{i=1}^{n} a_i\alpha_i = 0, a_i \in F, \alpha_i \in S, i = 1, 2,...,n$$

and $n \in N$, then each $a_i = 0$.

☞ Any infinite set is linearly independent if it's every finite subset is linearly independent otherwise it is linearly dependent.

Theorem 1. *If $\alpha_1, \alpha_2,...,\alpha_n \in V$ are linearly independent, then every element in their linear span has a unique representation in the form $a_1\alpha_1 + a_2\alpha_2, ..., +a_2\alpha_n$ with $a_i \in F$.*

Proof. By definition of linear span, we know that every element in the linear span is of the form $a_1\alpha_1 + a_2\alpha_2 + ... + a_n\alpha_n$. Therefore, we only show the uniqueness of the representation.

Let if possible,

$$a_1\alpha_1 + a_2\alpha_2 + ... + a_n\alpha_n \text{ and } b_1\alpha_1 + b_2\alpha_2 + ... + b_n\alpha_n$$

be two forms of an element in linear span of $\alpha_1, \alpha_2,.., \alpha_n$, then

$$a_1\alpha_1 + a_2\alpha_2 + ... + a_n\alpha_n = b_1\alpha_1 + b_2\alpha_2 + ... + b_n\alpha_n$$

$$\Rightarrow \quad (a_1 - b_1)\alpha_1 + (a_2 - b_2)\alpha_2 + ... + (a_n - b_n)\alpha_n = 0$$

Since, $\alpha_1, \alpha_2,..., \alpha_n$ are linearly independent so, we have

$$a_1 - b_1 = 0, \quad a_2 - b_2 = 0,..., a_n - b_n = 0$$

$$\Rightarrow \quad a_1 = b_1, \quad a_2 = b_2, ..., \quad a_n = b_n .$$

Hence, every element in the linear span of linearly independent vectors $\alpha_1, \alpha_2, ..., \alpha_n$ has a unique form of $a_1\alpha_1 + a_2\alpha_2 + ... + a_n\alpha_n$.

Theorem 2. *If $\alpha_1, \alpha_2,..., \alpha_n \in V$ (are in V, then either they are linearly independent or some α_k is a linear combination of preceding ones $\alpha_1, \alpha_2,..., \alpha_{k-1}$.*

Proof. $\alpha_1, \alpha_2,...,\alpha_n \in V$ are linearly independent, then nothing is to prove. So assume that $\alpha_1, \alpha_2,...,\alpha_n$ are not linearly independent. Then, there are some $a_i \in F$ which are non-zero such that

$$a_1\alpha_1 + a_2\alpha_2 + ... + a_n\alpha_n = 0$$

Let k be the largest integer for which $a_k \neq 0$. Since $a_i = 0$ for $i > k$, and

$$a_1\alpha_1 + a_2\alpha_2 + ... + a_k\alpha_k = 0$$

$$\Rightarrow \quad a_k = a_k^{-1}(-a_1\alpha_1 - a_2\alpha_2 - .... - a_{k-1}\alpha_{k-1}) \qquad [\because a_k \neq 0]$$

$$\Rightarrow \quad a_k = (-a_k^{-1}a_1)\alpha_1 + (-a_k^{-1}a_2)\alpha_2 + ... + (-a_k^{-1}a_{k-1})\alpha_{k-1}$$

$$\Rightarrow \quad a_k \text{ is a linear combination of preceding ones } \alpha_1, \alpha_2,..., \alpha_{k-1}.$$

7.11 BASIS OF A VECTOR SPACE

Definition. *Let V be a vector space over a field F and S be any non-empty subset of V. Then S is said to be a basis of V if*

(i)   *S is linearly independent.*

(ii) $L(S) = V$, *i.e., every element of V is a linear combination of finite elements of S.*

For example: The set $S = \{(1, 0, 0), (0, 1,0), (0, 0, 1)\}$ forms a basis of $V_3(\mathbf{R})$, and is called usual basis.

REMARKS

- ☞ The zero space has no basis.
- ☞ Every finitely generated vector space has a basis.
- ☞ Every non-zero vector space has a basis.
- ☞ A vector space many have more than one basis.

7.12 FINITE DIMENSIONAL VECTOR SPACE

Definition. *Let V(F) be a vector space over a field F and let S be any non-empty subset of V, then V(F) is said to be finite dimensional if S is finite subset of V such that $L(S) = V$. If this set contains n elements, then the dimension of V is n.*

Theorem 1.   *If $S = \{\alpha_1, \alpha_2, \ldots, \alpha_n\}$ is the basis of a vector space V(F), then each element of V is uniquely expressible as a linear combination of elements of S.*

Proof.   Since S is the basis of a vector space $V(F)$, then by the definition of basis, each element of V is a linear combination of elements of S. Thus, we only show the uniqueness. Let there be two different sets $\{a_1, a_2,\ldots,a_n\}$ and $\{b_1, b_2,\ldots,b_n\}$ of scalars corresponding to an element $\alpha \in V$ such that

$$\alpha = a_1\alpha_1 + a_2\alpha_2 + \ldots + a_n\alpha_n$$

and $$\alpha = b_1\alpha_1 + b_2\alpha_2 + \ldots + b_n\alpha_n \qquad \ldots(1)$$

$\Rightarrow$ $a_1\alpha_1 + a_2\alpha_2 + \ldots + a_n\alpha_n = b_1\alpha_1 + b_2\alpha_2 + \ldots + b_n\alpha_n$

$\Rightarrow$ $a_1\alpha_1 - b_1\alpha_1 + a_2\alpha_2 - b_2\alpha_2 + \ldots + a_n\alpha_n - b_n\alpha_n = 0$

$\Rightarrow$ $(a_1 - b_1)\alpha_1 + (a_2 - b_2)\alpha_2 + \ldots + (a_n - b_n)\alpha_n = 0$

Since the set $S = \{\alpha_1, \alpha_2,\ldots, \alpha_n\}$ is linearly independent so that

$$a_1 - b_1 = 0, \ a_2 - b_2 = 0,\ldots, a_n - b_n = 0$$

$\Rightarrow$ $a_1 = b_1, \quad a_2 = b_2, \ldots, \quad a_n = b_n.$

Hence, the expression (1) is unique.

Theorem 2. (Existence Theorem). *Every finitely generated vector space has a finite basis.*

Step Outlines : To make the proof easier, use the following steps :

Step 1. *If S is linearly independent, that theorem is obvious.*

Step 2. *Assume S is linearly dependent and obtained the set $S_1 = (\alpha_1, \alpha_2, \ldots \alpha_{k-1}, \alpha_{k+1}, \ldots \alpha_n)$ by eliminating α_k from S.*

Step 3. *Again if S_1 is linearly independent, result is obvious. And if S_1 is linearly dependent, proceed same as in step-2.*

Proof.   Let V be a vector space over F which is generated by a finite set (say) $S = \{\alpha_1, \alpha_2,\ldots,\alpha_n\}$ of vectors of V. Without loss of any generality we may assume that all the elements in S are non-zero, because zero vector in the linear combination of elements of S is zero.

If S is linearly independent, then S forms a finite basis for V and in this case theorem is proved. So, we assume that S is linearly dependent, then there exists some α_k $(2 < k \leq n)$ in S such that α_k is a linear combination of preceding vectors $\alpha_1, \alpha_2, \ldots \alpha_{k-1}$. Therefore,

$$\alpha_k = a_1\alpha_1 + a_2\alpha_2 + ... + a_{k-1}\alpha_{k-1} = \sum_{i=1}^{k-1} a_i\alpha_i \qquad ...(1)$$

for some scalars $a_i's \in F$.

But S generates V so that an arbitrary element $\alpha \in V$ is expressible as a linear combination of elements of S.

$\therefore \qquad \alpha = b_1\alpha_1 + b_2\alpha_2 + ... + b_k\alpha_k + ... + b_n\alpha_n$

$$= \sum_{i \neq k} b_i\alpha_i + b_k\alpha_k, \text{for some } b_i's \in F$$

$$= \sum_{i \neq k} b_i\alpha_i + b_k \sum_{i=1}^{k-1} a_i\alpha_i$$

$$= (b_1 + b_k a_1)\alpha_1 + (b_2 + b_k a_2)\alpha_2 + ...$$
$$+ (b_{k-1} + b_k a_{k-1})\alpha_{k-1} + (b_{k+1})\alpha_{k+1} + ... + b_n\alpha_n$$

$\Rightarrow \alpha$ is a linear combination of $a_1\alpha_1 + a_2\alpha_2 + ... + a_{k-1}\alpha_{k-1}$

Thus the set,

$$S_1 = \{\alpha_1, \alpha_2, ..., \alpha_{k+1}..., \alpha_k\}.$$

obtained by eliminating some α_k from S also generates V.

If S_1 in linearly independent, then S_1 will form a basis of V and the theorem is proved in this case. If S_1 is linearly dependent, then by above process, we obtain a new set

$$S_2 = \{\alpha_1, \alpha_2, ..., \alpha_{k-1}, \alpha_{k+1}, ..., \alpha_{i-1}, \alpha_{i+1}, ..., \alpha_n\}$$

by eliminating some α_i $(i > k)$ from S_1, which generates V. If S_2 is linearly independent, then S_2 will form a basis. If S_2 is linearly dependent, then we continue the above process, till after a finite number of steps we obtain a linearly independent set which generates V. At the most by repeating the above process we may obtain a singleton set which is always linearly independent and it generating V and will form a basis of V. Hence, in every finitely generated vector space there exists a finite basis.

Theorem 3. *If V is a finite-dimensional vector space and if $a_1, a_2, ..., a_m$ span V, then some subset of $\alpha_1, \alpha_2, ..., \alpha_m$ forms a basis of V.*

Proof. Since a finite-dimensional vector space has a basis of V containing a finite number of elements. Let these vectors be $\alpha_1, \alpha_2, ..., \alpha_n$. Thus every element in V has a unique representation of the form

$$a_1\alpha_1 + a_2\alpha_2 + a_n\alpha_n; \text{ for } a_1, a_2, ..., a_n \in F.$$

If $\alpha \in V$, then

$$\alpha = a_1\alpha_1 + a_2\alpha_2 + ... + a_n\alpha_n$$

Now define a map ϕ from V into $F^{(n)}$ by

$$\phi(a_1\alpha_1 + a_2\alpha_2 + ... + a_n\alpha_n) = (a_1, a_2, ..., a_n)$$

Since $a_1\alpha_1 + a_2\alpha_2 + ... + a_n\alpha_n$ is a unique representation so that ϕ is well defined, one-to-one and onto and also preserves the composition.

Thus V is isomorphic to $F^{(n)}$ for some n, where n is the number of elements in some basis of V over F. If some other basis of V has m elements, then V would be isomorphic to $F^{(m)'}$. Since both $F^{(n)}$ and $F^{(m)}$ are isomorphic to V, therefore, $F^{(n)}$ and $F^{(m)}$ are isomorphic to each other. This implies $n = m$. Hence the theorem.

Theorem 4. *If $\{\alpha_1, \alpha_2, ..., \alpha_n\}$ is a basis of $V(F)$ and if $\beta_1, \beta_2, ..., \beta_m \in V$, are linearly independent over F, then $m \leq n$.*

Proof. Since the set $\{\alpha_1, \alpha_2, ..., \alpha_n\}$ is a basis of $V(F)$, then every element in V is a linear combination of the elements of basis, so in particular, β_m is a linear combination of $\alpha_1, \alpha_2, ..., \alpha_n$. Therefore, the set $\{\beta_m, \alpha_1, \alpha_2, ..., \alpha_n\}$ is linearly dependent. Also, this set spans V because $\alpha_1, \alpha_2, ..., \alpha_n$ span V. Thus proper subset of the set $\{\beta_m, \alpha_1, \alpha_2, ..., \alpha_n\}$ forms a basis of V. Let this proper set be $\{\beta_m, \alpha_1, \alpha_2, ..., \alpha_n\}$ with $k \leq n - 1$. In forming this new basis at least one α_i is replaced by some one β_j. Repeat this procedure with the set $\{\beta_{m-1}, \alpha_{i_1}, \alpha_{i_2}, ..., \alpha_{i_k}\}$ which is obviously linearly dependent, so we can extract a basis of the form $\{\beta_{m-1}, \beta_m, \alpha_{j_1}, \alpha_{j_2}, ..., \alpha_{j_s}\}$ with $s \leq n - 2$. Continuing this procedure with the set $\{\beta_2, \beta_3, ..., \beta_{m-1}, \beta_m, \alpha_x, \alpha_y\}$. Since β_1 is not a linear combination of $\beta_2, \beta_3, ..., \beta_{m-1}, \beta_m, \alpha_x, \alpha_y\}$, therefore the basis $\{\beta_2, \beta_3, ..., \beta_{m-1}, \beta_m, \alpha_x, \alpha_y\}$ must contain some α's. To get this basis. We have introduced $m-1$ β's and each such introduction costs at least one α's and yet there is an 'α' left. Thus $m - 1 \leq n - 1$ implying $m \leq n$.

Theorem 5. *If V is a finite-dimensional vector space over F, then any two bases of V have the same number of elements.*

Proof. Let S_1 and S_2 be any two bases of $V(F)$ and let
$$S_1 = \{\alpha_1, \alpha_2, ..., \alpha_m\} \text{ and } S_2 = \{\beta_1, \beta_2, ..., \beta_n\}.$$
Then we shall have to show that $m = n$.

Since S_1 forms a basis of V, so that every element of V is uniquely expressible as a linear combination of the elements of S_1. In particular β_1 is uniquely expressible as a linear combination of S_1. Thus the set $S_3 = \{\beta_1, \alpha_1, \alpha_2, ..., \alpha_{k-1}\}$ is now linearly dependent.

Therefore, there exists an element α_k in S_s which is linear combination of proceeding ones, $\beta_1, \alpha_1, \alpha_2, ..., \alpha_{k-1}$. But every element of V can be expressed as a linear combination of $\alpha_1, \alpha_2, ..., \alpha_k, ..., \alpha_m$. Also, α_k is a linear combination of $\beta_1, \alpha_1, \alpha_2, ..., \alpha_{k-1}$. This implies that each element of V is expressible as a linear combination of $\{\beta_1, \alpha_1, \alpha_2, ..., \alpha_{k-1} ... \alpha_m\}$. Thus the set
$$S_4 = \{\beta_1, \alpha_1, \alpha_2, ..., \alpha_{k-1}, \alpha_{k+1}, ..., \alpha_m\}$$
generates V. This set is obtained by adjoining β_1 to S_3 and eliminating α_k and S_3. Since $\beta_2 \in V$ so it is the linear combination of elements of S_4, therefore, the set is obtained by adjoining β_2 to S_4 and eliminating α_1 as before. Thus, the set $S_5 = \{\beta_2, \beta_1, \alpha_1, \alpha_2, ..., \alpha_{k-1}, \alpha_{k+1}, ..., \alpha_{l-1}, \alpha_{l+1}, ..., \alpha_m\}$ generates V.

Continuing the above manner, we observe that each step consists of an inclusion of one β' s and the exclusion of an α's and the resulting set generates V. Since all the α's can not be exhausted before the β's. If it is so then a proper subset of S_2 generates V which is a contradiction of linear independence of S_2. Hence $m < n$. Similarly, if we change the role of S_1 and S_2, we obtain $n < m$. Hence $m = n$.

Theorem 6 (Extension Theorem). *If $V(F)$ is a finite dimensional vector space, then every linearly independent subset of V is either a basis of V or can be extended to form a basis of V.*

Proof. Suppose the dimension of $V = n$. Let $S = \{a_1, a_2, ..., a_n\}$ be a basis of V and let $S_1 = \{\beta_1, \beta_2, ..., \beta_m\}$ be a linearly independent subset of V.

Since S is a basis of V so that every element of V is expressible as a linear combination of elements of S, in particular, β_m is a linear combination of elements of S, therefore, the set obtained by adjoining β_m to S is linearly dependent. Since, the superset of a linearly dependent set is linearly dependent, it follows that the set,

$$S_2 = \{\beta_1, \beta_2, ..., \alpha_1, \alpha_2, ..., \alpha_n\}$$

is linearly dependent.

Since the set S_1 is linearly independent, then there exists an element α_k in S_2 which can be expressed as a linear combination of preceding ones $\beta_1, \beta_2, ..., \beta_m, \alpha_1, \alpha_2 , ..., \alpha_{k-1}$, therefore each element in V can be expressed as a linear combination of $\beta_1, \beta_2, ..., \beta_m, \alpha_1, \alpha_2, ..., \alpha_{k-1}, \alpha_{k+1}, ..., \alpha_n$. Thus the set

$$S_3 = \{\beta_1, \beta_2, ..., \beta_m, \alpha_1, \alpha_2, ..., \alpha_{k-1}, \alpha_{k+1}, ..., \alpha_n.\}$$

is obtained be eliminating α_k from S_2 which generates V. If S_3 is linearly independent, then S_3 forms a basis of V containing S_1. Thus, in this case the theorem is proved.

If S_3 is linearly dependent, then repeat the above process of adjoining and eliminating, till after a finite number of steps we obtain a linearly independent set which generates V and contains S_1. At the most by repeating the above process we may obtain the set S_1 itself which generates V and being linearly independent will form a basis of V.

Hence, either S_1 is a basis of V or can be extended to form a basis of V.

Theorem 7. *Let V be a finite-dimensional vector space and let dim. $V = n$. Then*

    (i) any subset of V which contains more than n vectors is linearly dependent.

    (ii) no subset of V which contains less than n vectors can span V.

Proof.   (i) Since $V(F)$ is n-dimensional, every basis of V will contain n vectors. Let S be any subset of V which contains more than n vectors. Let, if possible, S be linearly independent, then by previous theorem either S is a basis of V or can be extended to form a basis of V. But in both cases the basis of V contains more than n elements which is contradictory to the fact that V is n-dimensional. Hence S is linearly dependent.

    (ii) Let S be a subset of V which contains less than n elements and which can span V. Then every element of V can be expressed as a linear combination of elements of S. If S is linearly independent, then S forms a basis of V which show that dim. $V < n$ which is contradictory to the fact that dim. $V = n$. On the other hand it S is linearly dependent, then S cannot span V which is again a contradiction because we have assumed that S spans V. Hence in both cases such subset S cannot exist. Consequently no subset of V containing less than n vectors can span V.

7.13 DIMENSION OF A SUBSPACE OF VECTOR SPACE

Theorem 1. *Let S be a linearly independent subset of a vector space V. Suppose b is a vector in V which is not in the subspace spanned by S. Then the set obtained by adjoining b to S is linearly independent.*

Proof.   Let $S = \{\alpha_1, \alpha_2, ... \alpha_n\}$ be a linearly independent subset of V. Then we shall show that the set

$$S_1 = \{\beta, \alpha_1, \alpha_2, ...\alpha_n\}$$

obtained by adjoining β to S is also linearly independent where $\beta \in V$ but not in the subspace of V which is spanned by S.

Since $\alpha_1, \alpha_2, \ldots, \alpha_n$ are distinct vectors in S such that

$$a_1\alpha_1 + a_2\alpha_2 + \ldots + a_n\alpha_n + b\beta = 0 \qquad \ldots(1)$$

where, all $a's$. are zero.

We actually show that $b = 0$. Let, if possible, $b \neq 0$. Then from (1), we have

$$\beta = \left(-\frac{a_1}{b}\right)\alpha_1 + \left(-\frac{a_2}{b}\right)\alpha_2 + \ldots + \left(-\frac{a_n}{b}\right)\alpha_n$$

$\Rightarrow$ β is a linear combination of $\alpha_1, \alpha_2, \ldots, \alpha_n$.

$\Rightarrow$ β is in the subspace of V spanned by $\alpha_1, \alpha_2, \ldots, \alpha_n$.

But it is contradictory to the hypothesis that β is not in the subspace spanned by S. Hence $b = 0$. Consequently, the set S_1 is linearly independent.

Theorem 2. *If W is a subspace of a finite-dimensional vector space V, every linearly independent subset of W is finite and is a part of a (finite) basis for W.*

Proof. Let S be a linear independent subset of W and let S_1 be a linearly independent subset of W containing S. Then S_1 contains not more than dim. V elements.

If S spans W, then S is a basis for W and in this case, theorem is proved. If S does not span W, then we find a vector β_1 in W such that the set

$$S_2 = S \cup \{\beta_1\}$$

is linearly independent (By theorem 1). If S_2 spans W, it will form a basis for W, If S_2 does not span W, then again we find a vector β_2 in W such that the set

$$S_3 = S \cup \{\beta_1, \beta_2\}$$

is linearly independent.

Continuing above process up to finite number of steps less than dim V, we get a set ,

$$S_m = S \cup \{\beta_1, \beta_2, \ldots, \beta_{m-1}\}$$

which is linearly independent and is a basis for W or a part of basis for W.

Theorem 3. *If $V(F)$ be a finite-dimensional vector space and W be a subspace of V, then W is finite dimensional and dim. $W \leq$ dim V. In particulars, if W is a proper subspace of V, then dim $W <$ dim V. Also $V = W$ if and only if dim $V =$ dim W.*

Proof. Let dim $V = n$. Then every basis of V will contain n vectors of V, therefore, every subset having vectors more than n will be linearly dependent. Thus, a linearly independent set of vectors in W contains at most n-elements. Let $S = \{\alpha_1, \alpha_2, \ldots, \alpha_m\}$, with $m \geq n$ being a maximal linearly independent set in W. If α is an arbitrary element in W, then the set

$$S_1 = \{\alpha_1, \alpha_2, \ldots, \alpha_m\}$$

will be linearly dependent because S being a maximal linearly independent set. Therefore, α is a linear combination of $\alpha_1, \alpha_2, \ldots, \alpha_m$ which shows that S spans V. Hence W is finite-dimensional. Also, dim. $W = m \leq n =$ dim V.

$\therefore$ dim $W <$ dim V.

Next, if $V = W$, then every basis of V is also the basis of W which shows that dim. $V =$ dim W. On the other hand, if dim. $V =$ dim W, then every basis of W will contain the vectors equal to dim V so it will also generate V. Thus each one of V and W is generated by some basis. Hence $V = W$.

Theorem 4. **(Existence of Complementary Subspace).** *Every subspace of a finite dimensional vector space has a complement.*

Proof. Let $V(F)$ be a finite dimensional vector space and let W_1 be its subspace.

Then our aim is to find out a subspace W_2 of V such that $V = W_1 \oplus W_2$. Since $V(F)$ is a finite dimensional vector space so that W_1 will be finite dimensional.

Let $S_1 = [\alpha_1, \alpha_2 ,..., \alpha_n\}$ be the basis of W_1, then by extension theorem, we have, S_1 can be extended to form a basis of V. Let this extended set be

$$S_2 = \{\alpha_1, \alpha_2, ...,\alpha_n, \beta_1, \beta_2, ...,\beta_m\} \text{ be the basis of } V.$$

Let us suppose that the set $\{\beta_1, \beta_2, ...,\beta_m\}$ generates a subspace and let this subspace be W_2.

Now we shall show that $V = W_1 \oplus W_2$ or equivalently $V = W_1 + W_2$ and $W_1 \cap W_2 = \{0\}$.

Let γ be an arbitrary element of V and S_2 being the basis for V, there exist scalars a_1, $a_2, ..., a_n, b_1, b_2, ..., b_m$ such that

$$\gamma = a_1\alpha_1 + a_2\alpha_2 + ... + a_n\alpha_n + b_1\beta_1 + b_2\beta_2 + ... + b_m\beta_m = \alpha + \beta$$

Where, $\quad \alpha = \sum_{i=1}^{n} a_i\alpha_i \text{ and } \beta = \sum_{j=1}^{m} b_j\beta_j$

Since $\{\alpha_1, \alpha_2, ...,\alpha_n\}$ generates W_1 and $\{\beta_1, \beta_2, ...,\beta_m\}$ generates W_2, therefore $a \in W_1$ and $\beta \in W_2$

Thus every element of V is expressible as the sum of an element of W_1 and an element of W_2.

$\therefore \qquad V = W_1 + W_2$

Again, $\quad \alpha = \sum_{i=1}^{n} a_i\alpha_i \in W_1 \text{ and } \beta = \sum_{j=1}^{m} b_j\beta_j \in W_2$

Let if possible $W_1 \cap W_2 \neq \{0\}$. Then three exists a non-zero element which belong to both W_1 and W_2. Let it be x. Then $x \in W_1$ and $x \in W_2$.

$\therefore \qquad x = \sum_{i=1}^{n} a_i'.\alpha_i \text{ and } x = \sum_{j=1}^{m} b_j'.\beta_j$

$\Rightarrow \qquad \sum_{i=1}^{n} a_i'.\alpha_i = \sum_{j=1}^{m} b_j'.\beta_j$

$\Rightarrow \qquad \sum_{i=1}^{n} a_i'.\alpha_i + \sum_{j=1}^{m} (-b_j').\beta_j = 0$

Since the set S_2 is linearly independent, so that

$$a_i' = 0, \text{ for each } i = 1, 2 ,.., n$$

and $\quad b_j' = 0,, \text{ for each } i = 1, 2 ,.., m$

$\Rightarrow \qquad x = 0$

which is a contradiction, because we have taken $x \neq 0$.

Thus the contradiction arises by assuming that $W_1 \cap W_2 \neq \{0\}$

Hence, $W_1 \cap W_2 = \{0\}$

Consequently
$$V = W_1 \oplus W_2.$$

REMARK

☞ Here W_2 is the subspace complementary to the subspace W_1 of finite dimensional vector space V.

Theorem 5. *If W_1 and W_2 are two finite-dimensional subspaces of a vector space $V(F)$, then $W_1 + W_2$ is finite dimensional and $dim. W_1 + dim. W_2 = dim(W_1 \cap W_2) + dim. (W_1 + W_2)$*

Proof. Since W_1 and W_2 are subspaces of V so that $W_1 \cap W_2$ will be a subspace of V and its dimension is finite. Let dim. $W_1 = m$, dim. $W_2 = n$ and dim.$(W_1 \cap W_2) = r$.

Let $\{\alpha_1, \alpha_2, ..., \alpha_r\}$ be a basis of $W_1 \cap W_2$. Therefore, we can extend this basis to a basis of W_1 and also to a basis of W_2.

Let, $\qquad S_1 = \{\alpha_1, \alpha_2, ... \alpha_r, \beta_1, \beta_2, ... \beta_{m-r}\}$

and $\qquad S_2 = \{\alpha_1, \alpha_2, ... \alpha_r, \gamma_1, \gamma_2, ... \gamma_{n-r}\}$

be the basis of W_1 and W_2 respectively. Consider the set

$$S = \{\alpha_1, \alpha_2, ... \alpha_r, \beta_1, \beta_2, ... \beta_{m-r}, \gamma_1, \gamma_2, ... \gamma_{n-r}\}$$

Now we have to show that S will form a basis for $W_1 + W_2$. For this, we shall show that S is linearly independent and spans $W_1 + W_2$. For this, suppose

$$\sum a_i \alpha_i + \sum b_j \beta_j + \sum c_k \gamma_k = 0 \text{ for } a_i's, b_j's, c_k' s \in F.$$

Then $\sum c_k \gamma_k = \sum a_i \alpha_i + \sum b_j \beta_j \Rightarrow \sum c_k \gamma_k \in W_1$

Also, $\sum c_k \gamma_k \in W_2$. It follows that $\sum c_k \gamma_k \in W_1 \cap W_2$ and we have

$\sum c_k \gamma_k \in d_i \alpha_i$ for some scalars $d_1, d_2, ..., d_r$. Since the set $\{\alpha_1, \alpha_2, ... \alpha_r, \gamma_1, \gamma_2, ... \gamma_{n-r}\}$ is linearly independent hence all the scalars $c_1 = 0 = c_2 = ... = c_{n-r}$.

Thus $\sum a_i \alpha_i + \sum b_j \beta_j = 0$

and since the set $\{\alpha_1, \alpha_2, ... \alpha_r, \beta_1, \beta_2, ... \beta_{m-r}\}$ is also linearly independent,
then

$$a_1 = 0 = a_2 = ... = a_r$$

and $\qquad b_1 = 0 = b_2 = ... = b_{m-r}$

Thus the set $S = \{\alpha_1, \alpha_2, ... \alpha_r, \beta_1, \beta_2, ... \beta_{m-r}, \gamma_1, \gamma_2, ... \gamma_{n-r}\}$ is linearly independent. Now we shall show that S spans $W_1 + W_2$.

Let α be an arbitrary element of $W_1 + W_2$ then it can be written as $\alpha = \beta + \gamma$ with $\beta \in W_1$ and $\gamma \in W_2$. Now S_1 and S_2 being the basis of W_1 and W_2 respectively, β and γ can be expressed uniquely in the form of

$$\beta = \sum_{i=1}^{r} a_i \alpha_i + \sum_{j=1}^{n-r} b_j \beta_j \text{ , for some } a_i's \text{ and } b_j's.$$

and $\qquad \gamma = \sum_{i=1}^{r} e_i \alpha_i + \sum_{j=1}^{n-r} c_j \gamma_j \text{ , for some } e_i's \text{ and } c_j's.$

$\therefore \qquad \alpha = \beta + \gamma = \sum_{i=1}^{r} (a_i + e_i) \alpha_i + \sum_{j=1}^{m-r} b_j \beta_j + \sum_{j=1}^{m-r} c_j \gamma_j.$

$\Rightarrow \qquad \alpha$ is a linear combination of elements of S.

$\Rightarrow \qquad S$ spans $W_1 + W_2$.

Hence, S is basis of W_1+W_2 so that W_1+W_2 is finite-dimensional, with dimensional $(m+n-r)$.

Finally,
$$\dim W_1 + \dim W_2 = m+n = r+(m+n-r)$$
$$= \dim .(W_1 \cap W_2) + \dim .(W_1+W_2).$$

Theorem 6. *If a finite dimensional vector space $V(F)$ be the direct sum of its two subspaces W_1 and W_2, then* $\dim.V = \dim. W_1 + \dim.W_2$.

Proof. Since V is finite-dimensional, therefore W_1 and W_2 are also finite-dimensional. Let
$$\dim.W_1=m, \ \dim.W_2= n$$
Also $V=W_1 \oplus W_2,$ *implying that*

(i) $V=W_1+W_2$

(ii) $W_1 \cap W_2=\{0\}.$

Let $S_1=\{a_1,a_2, ..., a_m\}$ be a basis of W_1 and the set $S_2=\{b_1, b_2,..., b_n\}$ is a basis of W_2.

Now consider a set
$$S_3=\{a_1,a_2,...a_m, b_1,b_2,..., b_n\}$$
We claim that S_3 forms a basis of V.

For some scalars $a_1,a_2,...,a_m, b_1,b_2,..., b_n \in F$, we have
$$a_1\alpha_1+a_2\alpha_2+...+a_m\alpha_m+b_1\beta_1+b_2\beta_2+.....+b_n\beta_n=0$$
$\Rightarrow$ $a_1\alpha_1+ a_2\alpha_2+...+a_m\alpha_m= - (b_1\beta_1+b_2\beta_2+.....+b_n\beta_n)=0$

$\Rightarrow$ $a_1\alpha_1+a_2\alpha_2+...+a_m\alpha_m \in W_2$ as $b_1\beta_1+b_2\beta_2+.....+b_n\beta_n \in W_2$

and $b_1\beta_1+b_2\beta_2+...+b_n\beta_n \in W_1$ as $a_1\alpha_1+a_2\alpha_2+...+a_m\alpha_m \in W_1$

$\Rightarrow$ $a_1\alpha_1+a_2\alpha_2+...+a_m\alpha_m \in W_1 \cap W_2$

$\Rightarrow$ $b_1\beta_1+b_2\beta_2+...+b_n\beta_n \in W_1 \cap W_2$

But from (i) $W_1 \cap W_2=\{0\}.$

$\Rightarrow$ $a_1\alpha_1+a_2\alpha_2+...+a_m\alpha_m=0$

and $b_1\beta_1+b_2\beta_2+...+b_n\beta_n=0$

Since S_1 and S_2 both are linearly independent, therefore
$$a_1=0=a_2= ....=a_m , \ b_1=0=b_2= ....=b_n.$$
$\Rightarrow$ S_1 is linearly independent.

Next, let γ be an arbitrary element of V, then
$$\gamma=\alpha+\beta, \ \alpha \in W_1, \ \beta \in W_2 \quad\quad\quad\quad\quad\quad\quad\quad [\because V=W_1+W_2]$$

Since $\alpha \in W_1 \Rightarrow \alpha \in a_1\alpha_1+a_2\alpha_2+...+a_m\alpha_m$ for some $a_i s \in F.$

and $\beta \in W_2 \Rightarrow \beta \in b_1\beta_1+b_2\beta_2+...+b_m\beta_m$ for some $b_i s \in F.$

$\therefore$ $\gamma= \alpha+\beta = a_1\alpha_1+a_2\alpha_2+...+a_m\alpha_m+b_1\beta_1+b_2\beta_2+.....+b_n\beta_n$

$\Rightarrow$ S_3 generates V, thus S_3 forms a basis of V.

Accordingly, $\dim.V = m+n = \dim.W_1 + \dim.W_2.$

REMARK

☞ It should be noted that in case of finite dimensional spaces, since a basis is a maximal linearly independent subset of the vector space, so the dimension of a finite dimensional vector space may be regarded as the maximum of numbers of elements in all linearly independent subsets. If this is adopted as definition of the dimension of a vector space, is n iff it has a basis consisting of n elements.

7.14 COSETS

Let V be a vector space and W be its any subspace, then the sets,
$$\alpha+W =\{\alpha+\beta: \forall \beta \in W\}$$
and
$$W+\alpha=\{\beta+\alpha: \forall \beta \in W\},$$
where α is any arbitrary element of V, are called left and right cosets respectively.

Since addition is commutative in V, so
$$\alpha+W=W+\alpha \ \forall \alpha \in V$$

Therefore, there is no matter we take right coset or left coset.

REMARKS

☞ If $W+\alpha=W+\beta$, then $\alpha-\beta \in W$ and conversely.

☞ Any two cosets are either disjoint or identical.

☞ Since W is an additive subgroup of V, so we take additive cosets.

7.15 ADDITION AND MULTIPLICATION OF TWO COSETS

Let V be a vector space and W be its any subspace and let α, β be any two arbitrary elements of V, $\alpha+W$ and $\beta+W$ are two cosets of W in V, then
$$(\alpha+W) + (\beta+W)=(\alpha+\beta)+W.$$
and
$$(\alpha+W)(\beta+W)=(\alpha\beta)+W.$$

REMARKS

☞ Zero elements of the set $\alpha+W=\{\alpha+\beta : \beta \in W\}$ is W.

☞ Unit elements of this set $1+W$.

7.16 QUOTIENT SPACES

Let V be a vector space over a field F and let W be any subspace of V. Let $\alpha \in V$.

Then, $W+\alpha = \{w+\alpha : w \in W\}$ is called the right coset of W in V.

Similarly $\alpha+W = \{\alpha+w : w \in W\}$ is called the left coset of W in V.

Clearly, the sets $W+\alpha$ and $\alpha+W$ are both subsets of the vector space $V(F)$. Now since $(V,+)$ is an abelian group, therefore $W+\alpha$ and $\alpha+W$ are equal.

Thus, we can say that $W+\alpha$ is a coset of W in V generated by α. Let V/W represents the set of all cosets of W in V, i.e.,
$$V/W=\{W+\alpha: \alpha \in V\}$$

Further, we define vector addition and scalar multiplication of V/W as follows :
$$(W+\alpha)+ (W+\beta)= W+(\alpha+\beta)$$
and
$$a(W+\alpha)= W+a\alpha \ \forall \ \alpha,\beta \in V, a \in F$$

Then, V/W is a vector space over F for these compositions. This vector space is known as *quotient space* or *factor space*.

Theorem 1. *If V is a vector space over a field F and if W is a subspace of V, then the set*
$$V/W=\{\alpha+W : \forall \alpha \in V\}$$
is a vector space over the field F with respect to the linear compositions.

(i) $(\alpha+W)+(\beta+W)=(\alpha+\beta)+W, \ \forall \ \alpha,\beta \in V$

(ii) $a(\alpha+W)=a\alpha+W, \ \forall \ \alpha \in V \ and \ a \in F$

Proof. First of all, we shall show that the above compositions are well defined. For this, let

$$\alpha+W=\alpha'+W \ \text{and} \ \beta+W=\beta'+W, \text{then we have}$$

$$\alpha+W=\alpha'+W \ \text{and} \ \beta+W=\beta'+W$$

$\Rightarrow \qquad \alpha-\alpha' \in W \ \text{and} \ \beta-\beta' \in W$

$\Rightarrow \qquad (\alpha-\alpha') + (\beta-\beta') \in W$

$\Rightarrow \qquad (\alpha+\beta) - (\alpha'+\beta') \in W$

$\Rightarrow \qquad (\alpha+\beta) + W = (\alpha'+\beta') + W$

$\Rightarrow \qquad (\alpha+W)+(\beta+W) = (\alpha'+W)+(\beta'+W)$

$\Rightarrow$ The composition given in (i) is well defined.

Also, $\qquad\qquad\qquad \alpha+W=\alpha'+W$

$\Rightarrow \qquad\qquad \alpha-\alpha' \in W$

$\Rightarrow \qquad\qquad a(\alpha-\alpha') \in W \ \text{for some} \ a \in F.$

$\Rightarrow \qquad\qquad (a\alpha-a\alpha') \in W$

$\Rightarrow \qquad\qquad (a\alpha+W)=(a\alpha'+W).$

$\Rightarrow$ The composition given in (ii) is well defined.

(i) Next, $\alpha+W, \ \beta+W$ and $\gamma+W$ be any three elements of V/W, then

$$[(\alpha+W)+(\beta+W)]+ (\gamma+W) = [(\alpha+\beta)+W]+ (\gamma+W).$$
$$= [(\alpha+\beta)+\gamma]+(W).$$
$$= [\alpha+(\beta+\gamma)]+(W).$$
$$= (\alpha+W)+[(\beta+\gamma)+(W)].$$
$$= (\alpha+W)+[(\beta+W)+(\gamma+W)].$$

$\therefore$ The first composition given in (i) is associative in V/W.

(ii) Let $\alpha+W$ be an element of V/W, then

$$(\alpha+W)+(0+W) = (\alpha+0)+W = \alpha + W.$$

and $\quad (0+W)+(\alpha+W) = (0+\alpha)+W = \alpha + W.$

Thus, $0+W$ is an additive identity in V/W.

(iii) Since, if $\alpha \in V$ and $-\alpha \in V$, so $\alpha + W \in V$ and $(-\alpha) + W \in V/W$. Then

$$(\alpha+W)+[(-\alpha)+W] = [\alpha+(-\alpha)]+W = 0 + W=W.$$

and $\quad (0+W)+(\alpha+W) = (0+\alpha)+W = \alpha + W.$

Thus, $(-\alpha)+W$ is an additive inverse of $\alpha + W$ in V/W.

(iv) Let $\alpha+W, \ \beta+W$ be any two elements of V/W, then

$$(\alpha+W)+(\beta+W) = (\alpha+\beta)+W = (\beta+\alpha)+W$$
$$[\because \text{ addition is commutative in } V.]$$
$$= (\beta+W)+(\alpha+W).$$

Thus addition is commutative in V/W.

(v) Let $\alpha+W, \ \beta+W$ be any elements of V/W and $a \in F$, then

$$a.[(\alpha+W)+(\beta+W)] = a.[(\alpha+\beta)+W]$$
$$= a.(\alpha+\beta)+W$$

$$= (a.\alpha + a.\beta) + W$$
$$= (a\alpha + W) + (a\beta + W)$$
$$= a(\alpha + W) + a(\beta + W).$$

(vi)   Let $\alpha + W$ be an arbitrary element of V/W and $a, b \in F$, then
$$(a + b)(\alpha + W) = (a + b)\alpha + W = (a\alpha + b\alpha) + W$$
$$= (a\alpha + W) + (b\alpha + W)$$
$$= a(\alpha + W) + b(\alpha + W)$$

(vii)   Let $\alpha + W$ be an arbitrary element of V/W and $a, b \in F$, then
$$(ab)(\alpha + W) = (ab)\alpha + W$$
$$= a(b\alpha) + W \quad \text{[By elementary property of (v)]}$$
$$= a (b\alpha + W) = a(b(\alpha + W)).$$

(viii)   Let $\alpha + W$ be an arbitrary element of V/W and $1 \in F$, then
$$1 . (\alpha + W) = (1. \alpha) + W = \alpha + W. \qquad [\because 1. \alpha = 1]$$

Hence V/W is a vector space and this vector space is known as Quotient space.

Theorem 4.   *If V is finite-dimensional and if W is a subspace of V, then W is finite-dimensional and dim. (V/W) = dim.V – dim. W.*

Proof.   Let dim. $V = n$, then any $n+1$ elements in V are linearly dependent, in particular any $n+1$ elements in W are also linearly dependent. Thus we can find a maximal linearly independent subset of W.

Let, $\qquad S = \{\alpha_1, \alpha_2,..., \alpha_m\}$ with $m \le n$.

If $\alpha \in W$, then the set $\{\alpha_1, \alpha_2,..., \alpha_m\}$ is linearly dependent such that
$$a\alpha + a_1\alpha_1 + a_2\alpha_2 +... + a_m\alpha_m = 0$$
where, not all of the $a_i's$ are zero, If $a = 0$, then we get each $a_i = 0$ as S being linearly independent, which contradicts that $a = 0$. Thus $a \ne 0$ so that
$$\alpha = - a^{-1} (a_1\alpha_1 + a_2\alpha_2 +... + a_m\alpha_m)$$
$\Rightarrow \qquad \alpha_1, \alpha_2,...,\alpha_m$ span W

$\Rightarrow W$ is finite dimensional with dim . $W = m$.

Since S forms a basis of W so it can be extended to form a basis of V. Let the extended set $S_1 = (\alpha_1, \alpha_2,..., \alpha_m, \beta_1, \beta_2,..., \beta_{n-m}$ be the basis of V, as n being the dim. V. Consider a set $S_2 = \{ W + \beta_1, W + \beta_2,...,W+\beta_{n-m}\}$ of cosets in V/W. Now, we have to prove that S_2 forms a basis of V/W.

For some scalars $b_1, b_2,...,b_{n-m}$, we have
$$b_1 (W + \beta_1) + b_2(W + \beta_2)+...+ b_{n-m}(W+\beta_{n-m}\} = W + 0$$
where, $W + 0$ being the zero element of V/W.
$$\Rightarrow \qquad (W + b_1\beta_1) + (W + b_2 \beta_2)+...+ (W+ b_{n-m} \beta_{n-m}\} = W + 0$$
$$\Rightarrow \qquad W + (b_1\beta_1 + b_2 \beta_2+...+ b_{n-m} \beta_{n-m}) = W + 0$$
$$\Rightarrow \qquad b_1\beta_1+ b_2 \beta_2+...+ b_{n-m} \beta_{n-m} \in W$$
Since S generates W so that for some scalars $c_1, c_2,...,c_m$ we have
$$b_1\beta_1+ b_2 \beta_2+...+ b_{n-m} \beta_{n-m} = c_1\alpha_1+ c_2 \alpha_2+...+ c_m \alpha_m$$

$\Rightarrow \qquad c_1\alpha_1 + c_2\alpha_2 + ... + c_m\alpha_m + (-b_1)\beta_1 + (-b_2)\beta_2 + ... + (-b_{n-m})\beta_{n-m} = 0$

$\Rightarrow \qquad c_1 = 0 = c_2 = ... = c_m\,, b_1 = 0 = b_2 = ... = b_{n-m}\,,$

$$[\because S_1 \text{ is linearly independent.}]$$

$\Rightarrow \qquad$ each $b_i = 0$ for $i = 1, 2, ..., n-m$

$\Rightarrow \qquad S_2$ is linearly independent.

Also, if $W + \beta$ is an arbitrary element of V/W, then $\beta \in V$ and S_1 being the basis of V, therefore, β can be expressible as a linear combination of elements of S_1 . That is, for some scalars $d_1, d_2,...,d_m,\ e_1, e_2, ..., e_{n-m}$

$$\beta = d_1\alpha_1 + d_2\alpha_2 + ... + d_m\alpha_m + e_1\beta_1 + e_2\beta_2 + ... + e_{n-m}\beta_{n-m}$$

$$= \sum_{i=1}^{m} d_i\alpha_i + \sum_{j=1}^{n-m} e_j\beta_j$$

$$\therefore \qquad W + \beta = W + \left(\sum_{i=1}^{m} d_i\alpha_i + \sum_{j=1}^{n-m} e_j\beta_j \right)$$

$$W + \beta = \left(W + \sum_{i=1}^{m} d_i\alpha_i \right) + \left(W + \sum_{j=1}^{n-m} e_j\beta_j \right) \qquad \left[\because \sum_{i=1}^{m} d_i\alpha_i \in W \right]$$

$$= W + \sum_{i=1}^{n-m} e_i\beta_i = W + (e_1\beta_1 + e_2\beta_2 + ... + e_{n-m}\beta_{n-m})$$

$$= (W + e_1\beta_1) + (W + e_2\beta_2) + ... + (W + e_{n-m}\beta_{n-m})$$

$$= e_1(W + \beta_1) + e_2(W + \beta_2) + ... + e_{n-m}(W + \beta_{n-m}).$$

$\Rightarrow S_2$ generates V/W.

Thus, S_2 forms a basis of V/W and having $(n-m)$ elements so that

$$\text{dim. } V/W = n - m$$

Accordingly,

$$\text{dim .}V/W = n - m = \text{dim. } V - \text{dim. } W.$$

7.17 ISOMORPHISM

Let U and V be too vector spaces over the same field $F.$ Then a mapping $T : U \to V$ which associates to each element $\alpha \in U$ to a unique element $T(\alpha) \in V$ such that

$$T(a\alpha + b\beta) = aT(\alpha) + bT(\beta), \forall\ \alpha, \beta \in U, a, b \in F$$

is called a linear transformation of U into $V.$

A linear transformation T of U onto V is called an isomorphism if it is one-one. Therefore, we can say that U is isomorphic to V and write this as $U \cong V$.

Theorem 1. *Every n-dimensional vector space V(F) is isomorphic to F^n (F).*

Proof. Since the vector space V is n-dimensional. Let the set

$$S = \{\alpha_1, \alpha_2,...,\alpha_n\} \text{ form the basis for } V.$$

Then every vector V can be expressed as the linear combination of the elements of S. Let a be any arbitrary vector in V, then there exist a unique order set $\{a_1, a_2,...,a_n\}$ of scalars such that

$$\alpha = \sum_{i=1}^{n} a_i \alpha_i .$$

Consider a mapping

$$T : V \to F^n \text{ by } \quad T(\alpha) = (a_1, a_2,...,a_n) \ \forall \ \alpha \in V.$$

(i) T is linear.

Let $\qquad \alpha = \sum_{i=1}^{n} a_i \alpha_i$ and $\beta = \sum_{i=1}^{n} b_i \alpha_i$ be any two elements of V,

then for all $a, b \in F$, we have

$$T(a\alpha + b\beta) = T\left(a\sum_{i=1}^{n} a_i \alpha_i + b\sum_{i=1}^{n} b_i \alpha_i \right) = T\left(\sum_{i=1}^{n} (a_i \alpha_i + b_i \alpha_i)\alpha_i \right)$$

$$= (aa_1 + bb_1, aa_2 + bb_2, ..., aa_n + bb_n)$$
$$= (aa_1, aa_2,..., aa_n) + (bb_1, bb_2, ..., bb_n)$$
$$= a(a_1, a_2,...,a_n) + b(b_1, b_2,...,b_n) = a\,T(\alpha) + bT(\beta).$$

$\therefore$ T is linear.

(ii) T is one-one.

Suppose that

$$T(\alpha) = T(\beta)$$

$$\Rightarrow \qquad T\left(\sum_{i=1}^{n} a_i \alpha_i \right) = T\left(\sum_{i=1}^{n} b_i \alpha_i \right)$$

$$\Rightarrow \qquad (a_1, a_2,...,a_n) = (b_1, b_2, ..., b_n)$$
$$\Rightarrow \qquad a_i = b_i, \text{ for each } i = 1, 2, 3, ..., n$$

$$\Rightarrow \qquad \sum_{i=1}^{n} a_i \alpha_i = \sum_{i=1}^{n} b_i \alpha_i \qquad \Rightarrow \qquad \alpha = \beta$$

$\therefore$ T is one-one.

(iii) T is onto

Since, corresponding to each element $(aa_1, aa_2,...,aa_n) \in F^n$, there exists a vector

$$\sum_{i=1}^{n} a_i \alpha_i \in V \text{ such that } T\left(\sum_{i=1}^{n} a_i \alpha_i \right) = (a_1, a_2,.....,a_n)$$

$\Rightarrow$ T is onto.

Hence, V is isomorphic to F^n, i.e., $V \cong F^n$.

Theorem 2. *If W_1 and W_2 are the complementary subspaces of a vector space $V(F)$, then the correspondance that assigns to each $\alpha \in W_2$, the coset $W_1 + \alpha$ is an isomorphism between W_2 and V/W_1.-*

Proof.   Since W_1 and W_2 are the complementary subspaces of V, then $V=W_1 \oplus W_2$, or we have

$$V=W_1+W_2 \text{ and } W_1 \cap W_2 = \{0\}$$

Consider a mapping

$$T: W_2 \to V/W_1$$

defined by $T(\alpha) = W_1+\alpha, \forall \alpha \in W_2$

First to show that T is linear.

Let α,β be any two elements in W_2, then for all $a,b \in F$, we have

$$T(a\alpha+b\beta) = W_1+(a\alpha+b\beta) = (W_1+a\alpha)+(W_1+b\beta)$$
$$= a(W_1+\alpha)+b(W_1+\beta) = aT(\alpha)+bT(\beta)$$

$\therefore$ T is linear.

Second, to show that T is one-one.

Suppose that

$$T(\alpha)=T(\beta)$$
$\Rightarrow$ $W_1+\alpha = W_2+\beta \Rightarrow \alpha-\beta \in W_1$
$\Rightarrow$ $\alpha-\beta \in W_1 \cap W_2$ as $\alpha-\beta \in W_2$
$\Rightarrow$ $\alpha-\beta = 0$ $[\because W_1 \cap W_2=\{0\}]$
$\Rightarrow$ $\alpha = \beta.$

$\therefore$ T is one-one.

Finally, to show that T is onto.

Let $W_1+\gamma$ be an arbitrary element of V/W_1 so that $\gamma \in V$. But V is the direct sum of W_1 and W_2, therefore, γ can be uniquely expressed as the sum of an element of W_1 and an element of W_2. So there exists $\alpha \in W_1$ and $\beta \in W_2$ such that

$$\gamma = \alpha+\beta$$
$\Rightarrow$ $W_1+\gamma = W_1+(\alpha+\beta)$
$$= (W_1+\alpha)+(W_1+\beta)$$
$$= (W_1+0)+(W_1+\beta)   [\because \alpha \in W_1 \Rightarrow W_1+\alpha=W_1=W_1+0]$$
$$= (W_1+\beta) = T(\beta)$$

which shows that to each element $W_1+\gamma$ in V/W_1, there exists a unique element $\beta \in W$ such that

$$T(\beta) = W_1+\gamma$$

$\therefore$ T is onto.

Hence, W_2 is isomorphic to V/W_1.

Theorem 3.   *Any two finite dimensional vector spaces over the same field are isomorphic if and only if they are of same dimension.*

Proof.   Let $U(F)$ and $V(F)$ be two spaces over same field F. Suppose that both are of same dimension n (say).

Let $S_1=\{\alpha_1,\alpha_2,.....,\alpha_n\}$ and $S_2=\{\beta_1,\beta_2,.....,\beta_n\}$ be the basis for U and V respectively. So, corresponding to each element $\alpha \in U$, there exists unique scalars $a_1,a_2,.....,a_n \in F$ such that

$$a=a_1\alpha_1+ a_2\alpha_2 + ...+ a_n\alpha_n = \sum_{i=1}^{n} a_i\alpha_i$$

Consider a mapping $T : U \to V$.

by $\qquad T(\alpha) = \sum_{i=1}^{n} a_i \beta_i \;, \forall \; a_i \in F$ and for all $\alpha \in U$

First we show that T is linear.

Let α, β be any two elements of U, then $\alpha = \sum_{i=1}^{n} a_i \alpha_i$ and $\beta = \sum_{i=1}^{n} b_i \alpha_i$.

For $a, b \in F$, we have

$$T(a\alpha + b\beta) = T\left(a\sum_{i=1}^{n} a_i \alpha_i + b\sum_{i=1}^{n} b_i \alpha_i \right)$$

$$= T\left(\sum_{i=1}^{n}(aa_i + bb_i)\alpha_i \right) = \sum_{i=1}^{n}(aa_i + bb_i)\beta_i$$

$$= \sum_{i=1}^{n} aa_i \beta_i + \sum_{i=1}^{n} bb_i \beta_i$$

$$= a \sum_{i=1}^{n} a_i \beta_i + b \sum_{i=1}^{n} b_i \beta_i$$

$$= aT(\alpha) + bT(\beta)$$

$\therefore \qquad T$ is a linear.

Now, we show that T is one-one.

Suppose $\qquad$ for $\qquad \alpha, \beta \in U$

$\Rightarrow \qquad \sum_{i=1}^{n} a_i \beta_i = \sum_{i=1}^{n} b_i \beta_i$

$\Rightarrow \qquad \sum_{i=1}^{n}(a_i - b_i)\beta_i = 0$ for each $i = 1, 2, 3, ..., n$

$\qquad\qquad\qquad\qquad\qquad\qquad\qquad$ [$\because S_2$ is linearly indepenmdent.]

$\Rightarrow \qquad \sum_{i=1}^{n} a_i \beta_i = \sum_{i=1}^{n} b_i \beta_i \Rightarrow \alpha = \beta.$

$\therefore \qquad T$ is one-one.

Finally, we show that T is onto.

Let γ be any element of V, then there exist scalars $c_1, c_2, ..., c_n$ of F such that

$$g = \sum_{i=1}^{n} c_i \beta_i \; ,$$

Now, $\qquad T\left(\sum_{i=1}^{n} c_i \alpha_i \right) = \sum_{i=1}^{n} c_i \beta_i = \gamma$

$\Rightarrow \qquad T\text{-image of } \sum_{i=1}^{n} c_i \alpha_i \in U$ is $\gamma = \sum_{i=1}^{n} c_i \beta_i$

$\therefore \qquad T$ is a onto.

Hence T is an isomorphism, i.e., $U \cong V$

Conversely, suppose that $U \cong V$ and let T be the corresponding isomorphism.

Let $\qquad$ $S_1=\{\alpha_1,\alpha_2,.....,\alpha_n\}$ be the basis of U.

We claim that the set

$$S_2=\{T(\alpha_1),\ T(\alpha_2),.....,T(\alpha_n)\}$$

is the basis for V.

First we show that S_2 is linearly independent.

For some scalars $a_1,a_2,.....,a_n \in F$, the relation

$$a_1T(\alpha_1)+ a_2T(\alpha_2)+..... a_nT(\alpha_n)= 0$$

$\Rightarrow \qquad\qquad T(a_1\alpha_1+ a_2\alpha_2+..... a_n\alpha_n)=0 \qquad\qquad\qquad$ [$\because$ T is a linear]

$$T\left(\sum_{i=1}^{n}a_i\alpha_i\right) = T(0) \qquad\qquad\qquad [\because T(0)=(0)]$$

$$\sum_{i=1}^{n}a_i\alpha_i = 0 \qquad\qquad\qquad [\because T \text{ is one-one}]$$

$\Rightarrow \qquad\qquad a_i=0$ for each $i=1, 2, 3,.....,n.$ $\qquad$ [$\because S_1$ is linearly independent.]

$\therefore \quad$ S_2 is linearly independent.

Secondly, we show that S_2 spans V.

Let γ be any element of V. Since T is one-one onto mapping, then there exists a unique

vector $\alpha=\displaystyle\sum_{i=1}^{n}a_i\alpha_i \in U$ such that

$$\gamma=T(\alpha)=T\left(\sum_{i=1}^{n}a_i\alpha_i\right) = \sum_{i=1}^{n}a_iT(\alpha_i) \qquad\qquad [\because T \text{ is a linear.}]$$

Thus, $\gamma \in V$ is expressible as a linear combinations of elements of S_2, therefore S_2 spans V.

Consequently dim $V = n = $ dim U.

Hence, both the vectors $U(F)$ and $V(F)$ are finite dimensional of dimension n.

REMARK

☞ The dimension of the solution space W of the homogeneous system of linear equations $AX=O$ is $(n-r)$ where, n is the number of unknowns and r is the rank of the coefficient matrix A.

Solved Examples
Based on the following Results

▶ A finite set of vectors $\{\alpha_1, \alpha_2,...,\alpha_n\}$ of vectors of V is said to be linearly dependant if $\exists$ scalars $a_1,a_2,...,a_n$ not all them equal to zero such that $a_1\alpha_1+ a_2\alpha_2+...+ a_n\alpha_n=0$ and it is said to be linearly independent if every expression of the form $a_1\alpha_1+ a_2\alpha_2,+...+ a_n\alpha_n=0$ implies $a_1=0,a_2=0,....,a_n=0$

▶ A non-empty subset S of $V(F)$ is said to be a basis if S is linearly independent and $L(S)=V$.

▶ $\dim(W_1)+ \dim(W_2)= \dim(W_1 \cap W_2)+ \dim(W_1 + W_2)$

▶ If $V(F)$ is the direct sum of its two subspaces W_1 and W_2. Then dim $V = $ dim W_1+dim W_2

▶ **Left Cosets** $\quad \alpha+W=\{\alpha+\beta, \forall\ b\in W\}$

▶ **Right Cosets** $\quad W+\alpha=\{\beta+\alpha, \forall\ b\in W\}$

Example 1. *Prove that if two vectrors are linearly dependent, one of them is a scalar multiple of the other.*

Solution. Let V be a vector space over F and α, β be two vectors of V. These vectors are linearly dependent, then there exists scalars $a,b \in F$ not both equal to zero such that

$$a\alpha + b\beta = 0.$$

If $a \neq 0$, then a^{-1} exists in F, we have

$$a^{-1}(a\alpha) + a^{-1}(b\beta) = a^{-1}0$$

$$\Rightarrow \qquad (a^{-1}a)\alpha + (a^{-1}b)\beta = 0 \qquad\qquad [\because a^{-1}0 = 0]$$

$$\Rightarrow \qquad 1\alpha + a^{-1}b\beta = 0 \qquad\qquad [\because aa^{-1} = 1]$$

$$\Rightarrow \qquad \alpha + (a^{-1}b)\beta = 0 \qquad\qquad [\because 1\alpha = \alpha]$$

$$\Rightarrow \qquad \alpha = -(a^{-1}b)\beta \qquad\qquad [\because 1\alpha = \alpha]$$

$$\Rightarrow \qquad \alpha \text{ is a scalar multiple of } \beta.$$

Similarly, if $b \neq 0$, then we obtain β as a scalar multiple of α. Hence, one of α or β is scalar multiple of the other.

Example 2. *Show that the set $S = \{(1,2,1),(3,1,5),(3,-4,7)\}$ is linearly dependent where $S \subseteq V_3(\mathbf{R})$*

Solution. Let $a, b, c \in \mathbf{R}$ such that

$$a(1,2,1) + b(3,1,5) + c(3,-4,7) = (0,0,0)$$

or $\qquad (a+3b+3c, 2a+b-4c, a+5b+7c) = (0,0,0)$

or $\qquad \left.\begin{array}{l} a + 3b + 3c = 0 \\ 2a + b - 4c = 0 \\ a + 5b + 7c = 0 \end{array}\right\}$ ...(1)

Equation (1) is a system of linear homogeneous equations.

$\therefore$ Coefficient matrix is given by

$$A = \begin{bmatrix} 1 & 3 & 3 \\ 2 & 1 & -4 \\ 1 & 5 & 7 \end{bmatrix}$$

Now, $\qquad \det(A) = 1(7+20) - 3(14+4) + 3(10-1)$

$$= 27 - 54 + 27 = 0$$

$\therefore$ Rank of A is less than 3, *i.e.*, rank of A is less than the number of variables a,b and c so there exist a non-zero solutions. Thus, a,b,c are all not equal to zero. Hence, the given set S is linearly dependent.

Example 3. *Show that the vectors $(1,1,2,4)$, $(2,-1,-5,2)$, $(1,-1,-4,0)$ and $(2,1,1,6)$ are linearly dependent in R^4.*

Solution. Let $a, b, c, d \in \mathbf{R}$ such that

$$a(1,1,2,4) + b(2,-1,-5,2) + c(1,-1,-4,0) + d(2,1,1,6) = (0,0,0,0)$$

or $\qquad (a+2b+c+2d, a-b-c+d, 2a-5b-4c+d, 4a+2b+6d) = (0,0,0,0)$

or $\qquad \left.\begin{array}{l} a + 2b + c + 2d = 0 \\ a - b - c - d = 0 \\ 2a - 5b - 4c + d = 0 \\ 4a + 2b + 6d = 0 \end{array}\right\}$ ...(1)

Therefore, equation (1) represents a system of linear homogeneous equation.

Now the coefficient matrix of these equation is

$$A = \begin{bmatrix} 1 & 2 & 1 & 2 \\ 1 & -1 & -1 & 1 \\ 2 & -5 & -4 & 1 \\ 4 & 2 & 0 & 6 \end{bmatrix}$$

performing $R_2 \rightarrow R_2 - R_1$, $R_3 \rightarrow R_3 - 2R_1$, $R_4 \rightarrow R_4 - 4R_1$, we get

$$\sim \begin{bmatrix} 1 & 2 & 1 & 2 \\ 0 & -3 & -2 & -1 \\ 0 & -9 & -6 & -3 \\ 0 & -6 & -4 & -2 \end{bmatrix}$$

performing $R_3 \rightarrow R_3 - 3R_2$, $R_4 \rightarrow R_4 - 2R_2$

$$\sim \begin{bmatrix} 1 & 2 & 1 & 2 \\ 0 & -3 & -2 & -1 \\ 0 & 0 & 0 & 0 \\ 0 & 0 & 0 & 0 \end{bmatrix}$$

Thus, this matrix is in Echeleon form and having two non-zero rows so the rank of A is 2 which is less than the number of unknown. Therefore, the system of equation has non-zero solution. That is a, b, c, d are not all zero, Hence the given vectors are linearly dependent.

Example 4. *Show that the vectors* $(1,1,2)$, $(1,2,5)$, $(5,3,4)$ *do not form a basis of* R^3.

Solution. Since we know that the set $(1,0,0)$, $(0,1,0)$, $(0,0,1)$ forms a basis of R^3 so dimension of $R^3 = 3$.

Let $a,b,c \in \mathbf{R}$ such that

$$a(1,1,2) + b(1,2,5) + c(5,3,4) = (0,0,0)$$

or $\qquad (a+b+5c, a+2b+3c, 2a+5b+4c) = (0,0,0)$

or $\qquad \begin{matrix} a + b + 5c = 0 \\ a + 2b + 3c = 0 \\ 2a + 5b + 4c = 0 \end{matrix} \Bigg\}$ ...(1)

The equation (1) represents the system of linear homogeneous equations. Then the coefficient matrix of these equation is given by

$$A = \begin{bmatrix} 1 & 1 & 5 \\ 1 & 2 & 3 \\ 2 & 5 & 4 \end{bmatrix}$$

$\therefore \qquad |(A)| = 1(8 - 15) - 1(4 - 6) + 5(5 - 4) = -7 + 2 + 5 = 0$

Thus A is a singular matrix and therefore the rank of A is less that of 3, *i.e.*, less than the number of unknown a, b, c. Then the scalars a, b, c are all not equal to zero. Hence, the given get of vectors is not linearly independent, and hence this given set does not form a basis of R^3.

Example 5. *Show that the vectors* $(2,1,4)$, $(1,-1,2)$, $(3,1,-2)$ *form a basis of* R^3.

Solution. Since we know that the set $\{(1,0,0), (0,1,0), (0,0,1)\}$ forms a basis of R^3. Then dim of $R^3 = 3$. Therefore, if the given set $\{(2,1,4), (1,-1,2), (3,1,-2)\}$ is linearly independent, then it will form the basis of R^3.

Let $a,b,c \in \mathbf{R}$ such that

$$a(2,1,4) + b(1,-1,2) + c(3,1,-2) = (0,0,0)$$

or $\quad\quad (2a+b+3c, a-b+c, 4a+2b-2c) = (0,0,0)$

or $\quad\quad\begin{aligned} 2a + b + 3c &= 0 \\ a - b + c &= 0 \\ 4a + 2b - 2c &= 0 \end{aligned}\Bigg\}$ ...(1)

The equation (1) represents a system of three linear homogeneous equation. Then the coefficient matrix of these equation is given by

$$A = \begin{bmatrix} 2 & 1 & 3 \\ 1 & -1 & 1 \\ 4 & 2 & -2 \end{bmatrix}$$

Now $\quad\quad |(A)| = 2(2-2) - 1(-2-4) + 3(2+4)$
$$= 0 + 6 + 18 = 24 \neq 0$$

∴ The matrix A is a non-singular matrix and thus the rank of A is 3 which is equal to the number of unknowns a, b, c. Hence, the system of equations has only zero solution, i.e., $a=0$, $b=0$, $c=0$. Consequently the given set is linearly independent and hence forms a basis of R^3.

Example 6. *Find the dimension of the solution W of the system of linear equations.*

$$x+2y-4z+3r-s=0;\ x+2y-2z+4r+s=0;\ 2x+4y-2z+3r+4s=0$$

Solution. Above equation can be written as $AX=O$

where, $A = \begin{bmatrix} 1 & 2 & -4 & 3 & -1 \\ 1 & 2 & -2 & 4 & 1 \\ 2 & 4 & -2 & 3 & 4 \end{bmatrix}$ and $X = \begin{bmatrix} x \\ y \\ z \\ r \\ s \end{bmatrix}$, $O = \begin{bmatrix} 0 \\ 0 \\ 0 \end{bmatrix}$.

Now reduce A to Echelon form as follows:

$$A \sim \begin{bmatrix} 1 & 2 & -4 & 3 & -1 \\ 0 & 0 & 2 & 1 & 2 \\ 0 & 0 & 6 & 3 & 6 \end{bmatrix} \quad\quad \begin{bmatrix} R_2 \to R_2 - R_1 \\ R_3 \to R_3 - 2R_1 \end{bmatrix}$$

$$\sim \begin{bmatrix} 1 & 2 & -4 & 3 & -1 \\ 0 & 0 & 2 & 1 & 2 \\ 0 & 0 & 0 & 0 & 0 \end{bmatrix} \quad\quad \begin{bmatrix} R_3 \to R_3 - 3R_2 \end{bmatrix}$$

Echelon form of A has two non-zero rows, so that rank of $A=2$ and number of unknowns$= 5$. Hence dim.$W = 5 - 2 = 3$.

Example 7. *Prove that a set of non-zero vectors $\{x_1, x_2, ..., x_n\}$ is linearly dependent if some of these vectors say x_i is a linear combination of the preceding vectors $x_1, x_2, ..., x_{i-1}$ and conversely.*

Solution. Let us first assume that x_i can be expressed as a linear combination of $x_1, x_2, ..., x_{i-1}$. Then we have

$$x_i = a_1 x_1 + a_2 x_2 + ... + a_{i-1} x_{i-1} \quad\quad ...(1)$$

For some scalars $a_1, a_2, ..., a_{i-1} \in F$.

Now equation (1) can be written as:

$$a_1 x_1 + a_2 x_2 + ... + a_{i-1} x_{i-1} + (-1) x_i = 0$$

or $\quad\quad a_1 x_1 + a_2 x_2 + ... + a_{i-1} x_{i-1} + (-1)x_i + 0\,x_{i+1} + 0\,x_{i+2} + ... + 0\,x_n = 0$

This equation shows that this relation has at least one non-zero coefficient (scalar) of

x_i which is -1. It shows that the set $\{x_1, x_2, ..., x_n\}$ is linearly dependent. Then we have a relation.

$$a_1 x_1 + a_2 x_2 + ... + a_n x_n = 0$$

in which not all a_i's are zero.

If i is the greatest positive integer less than or equal to n such that $a_i \neq 0$, so that $a_{i+1} = 0 = a_{i+2} = ... = a_n$, then we have

$$a_1 x_1 + a_2 x_2 + ... + a_{i-1} x_{i-1} + a_i x_i + 0. x_{i+1} + 0. x_{i+2} + ... + 0. x_n = 0$$

$\Rightarrow$ $x_i = - a_i^{-1}(a_1 x_1 + a_2 x_2 + ... + a_{i-1} x_{i-1})$.

$\Rightarrow x_i$ is the linear combination of preceding vectors $x_1, x_2, ..., x_{i-1}$.

Example 8. *Prove that the vector* (α_1, α_2), (β_1, β_2) $\in$ $\mathbf{R} \times \mathbf{R}$ *are linearly dependent iff* $\alpha_1 \beta_1 - \alpha_2 \beta_2 = 0$

Solution. The vectors (α_1, α_2) and (β_1, β_2) are linearly dependent if and only if there exist scalar $a, b \in \mathbf{R}$ such that

$$a(\alpha_1, \alpha_2) + b(\beta_1, \beta_2) = (0,0)$$

where, $(0,0)$ is the vector in $\mathbf{R} \times \mathbf{R}$

$\Leftrightarrow$ $(a\alpha_1, a\alpha_2) + (b\beta_1, b\beta_2) = (0,0)$

$\Leftrightarrow$ $(a\alpha_1 + b\beta_1, a\alpha_2 + b\beta_2) = (0,0)$

$\Leftrightarrow$ $a\alpha_1 + b\beta_1 = 0, a\alpha_2 + b\beta_2 = 0$

The equations $a\alpha_1 + b\beta_1 = 0$ and $a\alpha_2 + b\beta_2 = 0$ have non-trival (non-zero) solutions iff the determinant of coefficients.

$$\begin{vmatrix} \alpha_1 & \beta_1 \\ \alpha_2 & \beta_2 \end{vmatrix} = 0, \text{ i.e., } \alpha_1 \beta_2 - \beta_1 \alpha_2 = 0.$$

Hence, the vectors (α_1, α_2) and (β_1, β_2) are linearly dependent if $\alpha_1 \beta_2 - \beta_1 \alpha_2 = 0$.

Example 9. *Prove that the vectors* $(2, i, -i)$, $(2i, -1, 1)$, $(1, 2, 3)$ *are linearly independent in* $V_3(\mathbf{R})$ *but linearly dependent in* $V_3(\mathbf{C})$.

Solution. Let $\alpha_1 = (2, i, -i)$, $\alpha_2 = (2i, -1, 1)$ and $\alpha_3 = (1, 2, 3)$ and let $\alpha_1, \alpha_2, \alpha_3$ be three scalars, Then

$\Rightarrow$ $a_1 \alpha_1 + a_2 \alpha_2 + a_3 \alpha_3 = 0$

$\Rightarrow$ $a_1 (2, i, -i) + a_2(2i, -1, 1) + a_3(1, 2, 3) = (0, 0, 0)$

$\Rightarrow$ $(2a_1 + 2ia_2 + a_3, ia_1 - a_2 + 2a_3, -ia_1 + a_2 + 3a_3) = (0,0,0)$

$\therefore$ $2a_1 + 2ia_2 + a_3 = 0$ ...(1)

           $ia_1 - a_2 + 2a_3 = 0$ ...(2)

           $- ia_1 - a_2 + 3a_3 = 0$ ...(3)

Now adding (2) and (3), we get

$$5a_3 = 0 \Rightarrow a_3 = 0.$$

Putting $a_3 = 0$. in any one of above equations, we get

$$ia_2 - a_2 = 0 \text{ or } a_2 = ia_1. \quad\quad\quad ...(4)$$

Case I: In $V_3(\mathbf{R})$, equation (4) is satisfied when $a_1 = 0$, $a_2 = 0$.

Thus in case of $V_3(\mathbf{R})$, equation (4) is satisfied when $a_1 = 1$, $a_2 = i$.

$\therefore$ $1(2, i, -i) + i(2i, -1, 1) + 0(1, 2, 3) = (0,0,0)$.

Thus in case of $V_3(\mathbf{C})$, the vectors α_1, α_2 and α_3 are linearly dependent.

Example 10. *Prove that the vectors* (1, 1, 0) (3, 1, 3) *and* (5, 3, 3) *are linearly dependent.*

Solution.   Let $\alpha_1 = (1,1,0), \alpha_2 = (3,1,3)$ and $\alpha_3 = (5,3,3)$ and let α_1, α_2 and α_3 be some scalars such that

$$a_1\alpha_1 + a_2\alpha_2 + a_3\alpha_3 = 0$$

$\Rightarrow$ $\qquad a_1(1,1,0) + a_2(3,1,3) + a_3(5,3,3) = (0,0,0)$

$\Rightarrow$ $\qquad (a_1 + 3a_2 + 5a_3, a_1 + a_2 + 3a_3, 3a_2 + 3a_3) = (0,0,0)$

$\Rightarrow$ $\qquad\qquad a_1 + 3a_2 + 5a_3 = 0$ ...(1)

$\qquad\qquad\qquad a_1 + a_2 + 3a_3 = 0$ ...(2)

$\qquad\qquad\qquad 3a_2 + 3a_3 = 0$ ...(3)

Coefficient matrix of above equation is $A = \begin{bmatrix} 1 & 3 & 5 \\ 1 & 1 & 3 \\ 0 & 3 & 3 \end{bmatrix}$

Reduce this matrix into Echelon from as follows:

$$A \sim \begin{bmatrix} 1 & 3 & 5 \\ 0 & -2 & -2 \\ 0 & 3 & 3 \end{bmatrix} \qquad\qquad \left[R_2 \rightarrow R_2 - R_1 \right]$$

$$\sim \begin{bmatrix} 1 & 3 & 5 \\ 0 & -2 & -2 \\ 0 & 0 & 0 \end{bmatrix} \qquad\qquad \left[R_3 \rightarrow R_3 + \frac{3}{2}R_2 \right]$$

This is the Echelon from and having two non-zero rows so that rank of $A=2$. Thus the equation (1), (2) and (3) have non-zero solutions. Hence the vectors (1, 1, 0), (3, 1, 3) and (5, 3, 3) are linearly dependent.

Example 11. *Under what conditions on the scalar 'a' are the vectors* (a, 1, 0), (1, a, 1) *and* (0, 1, a) *in* R^3 *linearly dependent?*

Solution.   Let $\alpha_1 = (a, 1, 0), \alpha_2 = (1, a, 1)$ and $\alpha_3 = (0, 1, a)$ and let a_1, a_2 and a_3 be three scalars such that

$$a_1\alpha_1 + a_2\alpha_2 + a_3\alpha_3 = 0$$

$\Rightarrow$ $\quad a_1(a, 1, 0) + a_2(1,a,1) + a_3(0,1,a) = (0,0,0)$

$\Rightarrow$ $\qquad (a_1a + a_2, a_1 + a_2a + a_3, a_2 + a_3a) = (0,0,0)$

$\Rightarrow$ $\qquad\qquad a_1a + a_2 \quad = 0$ ......(1)

$\qquad\qquad\qquad a_1 + a_2a + a_3 = 0$ ......(2)

$\qquad\qquad\qquad a_2 + a_3a \quad = 0$ ......(3)

Let A be the coefficient matrix of the equations (1), (2), (3)

$\therefore$ $$A = \begin{bmatrix} a & 1 & 0 \\ 1 & a & 1 \\ 0 & 1 & a \end{bmatrix}$$

For non-trivial solution of (1), (2) and (3), we must have

$$\begin{vmatrix} a & 1 & 0 \\ 1 & a & 1 \\ 0 & 1 & a \end{vmatrix} = 0$$

$\Rightarrow$ $\qquad\qquad a[a^2-1] - 1[a-0] = 0 \Rightarrow a(a^2 - 2) = 0$

$\Rightarrow$ $\qquad\qquad\qquad a = 0, a = \pm\sqrt{2}$

Hence the vectors (a, 1, 0), (1, a, 1) and (0, 1, a) are linearly dependent when $a = 0; \pm\sqrt{2}$.

Example 12. *For what value of m, the vector (m, 3, 1) a linear combination of e_1 = (3, 2, 1) and e_2 =(2, 1, 0) .*

Solution : For some scalars a_1 and a_2 such that

$$(m, 3, 1) = a_1e_1 + a_2 e_2 = a_1(3,2,1) + a_2(2,1,0)$$
$$= (3a_1+2a_2, 2a_1+a_2, a_1)$$

$$\Rightarrow \quad \begin{aligned} 3a_1 + 2a_2 &= m \\ 2a_1 + a_2 &= 3 \\ a_1 &= 1 \end{aligned} \qquad \begin{aligned} &......(1) \\ &......(2) \\ &......(3) \end{aligned}$$

Putting the value of $a_1=1$ in (2), we get $2(1) + a_2 = 3$, $a_2 =3 - 2 = 1$.

Now putting the values of a_1 and a_2 in (1),we get m = 3(1) +2(1) = 3 + 2 = 5

Example 13. *Show that the set {(1,2,1), (2,1,0), (I, –1,2)} forms a basis for $V_3(\mathbf{R})$.*

Solution. Since we know that the set {(1, 0, 0), (0, 1, 0),(1, –1,2)} is linearly independent, then it will form the basis of $V_3(\mathbf{R})$.

Let $a, b, c \in R$ such that $a(1,2,1) + b(2,1,0) + c(1,–1,2)=(0,0,0)$

$$\Rightarrow \quad (a + 2b + c, 2a + b - c, a + 2c) = (0, 0, 0)\ 0$$

$$\Rightarrow \quad \begin{aligned} a + 2b + c &= 0 \\ 2a + b - c &= 0 \\ a + 2c &= 0 \end{aligned} \qquad \begin{aligned} &...(1) \\ &...(2) \\ &...(3) \end{aligned}$$

The equation (1), (2), (3) represent a system of three linear homogeneous equations. Then the coefficient matrix of these equations is given by :

$$A = \begin{bmatrix} 1 & 2 & 1 \\ 2 & 1 & -1 \\ 1 & 0 & 2 \end{bmatrix}$$

Now, $|A| = 1(2 - 0) - 2(4 + 1) + 1(0 - 1) = 2 - 10 - 1 = -9 \neq 0.$

Therefore, matrix A is non-singular and thus the rank of A is 3 which is equal to the number of unknowns a, b, c. Hence the system of equations has only zero solution, *i.e.,* $a = 0, b = 0, c = 0$. Consequently, the given set is linearly independent and hence forms a basis of $V_3(\mathbf{R})$.

Example 14. *Under what conditions on the scalar 'a', do the vectors (1, 1, 1) and (1,a, a^2) form a basis of $C^3(\mathbf{C})$.*

Solution. Since dim. $C^3(\mathbf{C})$ = 3 so that every basis of $C^3(\mathbf{C})$ must have three vectors. Therefore, a must have two values. Let a_1 and a_2 be the two values of a such that $a_1^2 = a_2^2$. In other words, we are to find the condition on the scalar 'a' such that the vectors (1,1,1), (1, a, a^2), (1,– a, a^2) form a basis of $C^3(\mathbf{C})$. Since dim. $C^3(\mathbf{C})$ = 3. therefore, we find the condition such that (1,1,1), (1, a, a^2) and (1, – a, a^2) are linearly independent.

For some scalar $a_1, a_2, a_3 \in \mathbf{C}$, we have

$$a_1 (1, 1, 1) + a_2(1,a,a^2) + a_3(1,–a, a^2) = (0, 0, 0)$$
$$\Rightarrow \quad (a_1 + a_2 + a_3 , a_1 +aa_2 - aa_3 , a_1 + a^2a_2 +a^2a_3) = (0, 0, 0)$$

$$\Rightarrow \quad \begin{aligned} a_1 + a_2 + a_3 &= 0 \\ a_1 + aa_2 - aa_3 &= 0 \\ a_1 + a^2a_2 + a^2a_3 &= 0 \end{aligned} \qquad \begin{aligned} &...(1) \\ &...(2) \\ &...(3) \end{aligned}$$

This system of homogeneous equations has a trivial solution if and only if

$$\begin{vmatrix} 1 & 1 & 1 \\ 1 & a & -a \\ 1 & a^2 & a^2 \end{vmatrix} \neq 0$$

$$\Rightarrow \quad 1(a^3+a^3) - 1(a^2 + a) + 1(a^2 - a) \neq 0$$

$$\Rightarrow \quad 2a^3 - 2a \neq 0 \Rightarrow 2a(a^2 - 1) \neq 0 \Rightarrow a \neq 0, a \neq \pm 1.$$

Hence the vectors $(1,1,1)$ and $(1, a, a^2)$ form a basis of $C^3(\mathbf{C})$ if any only if $a \neq 0$, $a \neq \pm 1$.

EXERCISE 7.2

1. Show that the set $S = \{(1, 0, 0), (0, 1, 0), (0, 0, 1)\}$ spans the vector space $V_3(\mathbf{R})$.

2. Prove that every superset of a linearly dependent set is linearly dependent.

3. Show that the set $S = \{(1, 2, 4), (1, 0, 0), (0, 1, 0), (0, 0, 1)\}$ is linearly dependent subset of the vector space $V_3(\mathbf{R})$.

4. Show that the set $\{1, x, 1 + x + x^2\}$ is linearly independent set of vectors in the vector space of all polynomials over real numbers field.

5. Show that the vectors $(1, 1, -1)$, $(2, -3, 5)$, $(-2, 1, 4)$ of R^3 are lineally independent.

6. Show that the vectors $(1, 3, 2)$, $(1, -7, -8)$, $(2, 1, -1)$ of $V_3(\mathbf{R})$ are linearly dependent.

7. Is the vector $(3, -1, 0, 1)$ in the subspace of R^4 spanned by the vectors $(2,-1,3,2)$, $(-1,1,2, -3)$, $(1, 1, 9, -5)$?

8. Is the vector $(2, -5, 3)$ in the subspace of R^3 spanned by the vectors $(1, -3, 2)$, $(2, -4, -1)$, $(1, -5, 7)$?

9. If F is the field of complex numbers, prove that the vectors (a_1, a_2) and (b_1, b_2) in $V(F)$ are linearly dependent iff $a_1 b_2 - a_2 b_1 = 0$.

10. Show that every subset of a linearly independent set of vectors is linearly independent.

11. Examine each of the following sets of vectors for linear dependent in the vector space $V_3(\mathbf{R})$
    (i) $\{(1, 2, 0), (0, 3, 1), (-1, 0, 1)\}$
    (ii) $\{(-1, 2, 1), \{3, 0, -1), (-5, 4, 3)\}$

7.18 LINEAR TRANSFORMATION

The concept of homomorphism is easily carried to vector spaces. To begin with linear transformation (vector space homomorphism) is a function from one vector space to another. Like a ring homomorphism, it is supposed to preserve both the vector space operations. The process of taking functional values and performing the vector space operation should be commutative. This requires that the scalar field in case of either space should be the same.

Definition. *Let U and V be two vector spaces over the same field F. A mapping $T:U \to V$ is said to be a linear transformation from U into V which associates to each element α of U to a unique element $T(\alpha)$ of V such that*

$$T(a\alpha + b\beta) = aT(\alpha) + bT(\beta)$$

for all α and β in U and all scalars a, b in F.

REMARKS

☞ Linear transformation is also known as vector space homomorphism.

☞ If a linear transformation is onto, then it is known as isomorphism.

For example :

(1) If V is any vector space over F, then the identity transformation I, defined by $I(\alpha) = \alpha$, $\forall \alpha \in V$ is a linear transformation from V into V. Also the zero transformation 0 denoted by $0(\alpha) = 0$, is a linear transformation.

(2) Let F be a field of real numbers and let V be the vector space of all polynomials, then a mapping

$$D : V \to V$$

given by $D[f(x)] = \dfrac{d}{dx} [f(x)]$, $\forall f(x) \in V$ is a linear transformation.

Since for any $f(x)$ and $g(x) \in V$ and $a, b \in F$

$$D[af(x) + bg(x)] = \frac{d}{dx}[af(x) + bg(x)] = \frac{d}{dx}[af(x)] + \frac{d}{dx}[bf(x)]$$

$$= a\frac{d}{dx}[f(x)] + b\frac{d}{dx}[g(x)] = aD[f(x)] + bD[g(x)]$$

(3) Let $\mathbf{R}$ be the field of real numbers and let V be the vector space of all functions from $\mathbf{R}$ into $\mathbf{R}$ which are continuous. Then a mapping $T : V \to V$ given by

$$T[f(x)] = \int_0^x f(t)\,dt$$

is a linear transformation.

For any $f(x)$, $g(x) \in V$ and $a, b \in \mathbf{R}$

$\therefore \qquad T[af(x) + bg(x)] = \int_0^x [af(t) + \int_0^x bg(t)]\,dt$

$$= \int_0^x af(t)\,dt + \int_0^x bg(t)\,dt$$

$$= a\int_0^x f(t)\,dt + b\int_0^x g(t)\,dt = aT[f(x)] + bT[g(x)]$$

(4) Let V be the vector space of all $m \times n$ matrices over a field F and let P be a fixed $m \times n$ matrix and Q be a fixed matrix of order $n \times n$.

Then a mapping $T : V \to V$ given by $T(A) = PAQ$, $\forall A \in V$ is a linear transformation.

For any two matrices, $A, B \in V$ and $a, b \in F$

$$T(aA + bB) = P(aA + bB)Q = (aPA + bPB)Q$$
$$= aPAQ + bPBQ = aT(A) + bT(B).$$

Theorem 1. *Let U and V be two finite-dimensional vector spaces over the same field F and let $\{\alpha_1, \alpha_2, ..., \alpha_n\}$ be an ordered basis for U and let $\{\beta_1, \beta_2, ..., \beta_n\}$ be an ordered set in V. Then there is precisely one linear transformation T from U into V such that $T(\alpha_j) = \beta_j$, $j = 1, 2, 3, ...n$.*

Proof. Since the set $\{\alpha_1, \alpha_2, ..., \alpha_n\}$ is a basis of $U(F)$, then for each $\alpha \in U$, there are some scalars $a_1, a_2, ..., a_n$ such that

$$\alpha = a_1\alpha_1 + a_2\alpha_2 + ... + a_n\alpha_n = \sum_{i=1}^{n} a_i\alpha_i$$

For this vector α we define $T : U \to V$ given by

$$T(\alpha) = a_1\beta_1 + a_2\beta_2 + ... + a_n\beta_n = \sum_{i=1}^{n} a_i\beta_i$$

Then T is well defined for each vector α in U and a vector $T(\alpha) \in V$. From the definition it is clear that $T(\alpha_j) = \beta_j$ for each j.

Now we shall show that T is linear. For this if $\alpha = \sum\limits_{i=1}^{n} a_i \alpha_i$ and $\beta = \sum\limits_{i=1}^{n} b_i \alpha_i$ are any two vectors in U, then for all $a, b \in F$, we have

$$T(a\alpha + b\beta) = \left[a \sum_{i=1}^{n} a_i \alpha_i + b \sum_{i=1}^{n} b_i \alpha_i \right] = T\left(\sum_{i=1}^{n} aa_i \alpha_i + \sum_{i=1}^{n} bb_i \alpha_i \right)$$

$$= T\left(\sum_{i=1}^{n} aa_i + \sum_{i=1}^{n} bb_i \right) \alpha_i = \sum_{i=1}^{n} (a_i a_i + b_i b_i) \beta_i$$

$$= a \sum_{i=1}^{n} a_i \beta_i + b \sum_{i=1}^{n} b_i \beta_i$$

$$= aT\left(\sum_{i=1}^{n} a_i \alpha_i \right) + bT\left(\sum_{i=1}^{n} b_i \alpha_i \right) = aT(\alpha) + bT(\beta)$$

Now, we shall show the uniqueness of T.

Let if possible, T_1 be another linear transformation from U into V such that $T_1(\alpha_j) = B_j, j = 1, 2, ..., n$.

Then for any vector $\alpha = \sum\limits_{i=1}^{n} a_i \alpha_i$, we have

$$T_1(\alpha) = T_1\left(\sum_{i=1}^{n} a_i \alpha_i \right) = \sum_{i=1}^{n} a_i T_1(\alpha_i) \qquad\qquad [\because T_1 \text{ is linear}]$$

$$= \sum_{i=1}^{n} a_i \beta_i \qquad\qquad [\because T_1(\alpha_i) = \beta_i]$$

$$= T\left(\sum_{i=1}^{n} a_i \alpha_i \right) = T(\alpha)$$

$\Rightarrow \qquad\qquad T_1 = T \qquad\qquad [\because \alpha \text{ is an arbitrary vector.}]$

Hence, T is unique.

7.19 ALGEBRA OF LINEAR TRANSFORMATIONS

<u>**Theorem 1.**</u> *Let U and V be two vector spaces over the field F. Let T_1 and T_2 be two linear transformations from U into V. then the function $(T_1 + T_2)$ defined by*

$(T_1 + T_2)(\alpha) = T_1(\alpha) + T_2(\alpha), \forall \alpha \in U$

is a linear transformation from U into V. If c is any element of F, then the function (cT) defined by

$(cT)(\alpha) = cT(\alpha)$

is a linear transformation from U into V.

The set of all transformations $L(U, V)$ from U into V, together with the addition and scalar multiplication defined above, is a vector space over the field F.

Proof.   For $\alpha, \beta \in U$ and $a, b \in F$, we have

$$(T_1 + T_2)(a\alpha + b\beta) = T_1(a\alpha + b\beta) + T_2(a\alpha + b\beta) \qquad \text{[By definition]}$$

$$= [aT_1(\alpha) + bT_1(\beta)] + [aT_2(\alpha) + bT_2(\beta)]$$

$$[\because T_1 \text{ and } T_2 \text{ are linear transformations.}]$$

$$= [aT_1(\alpha) + aT_2(\alpha)] + [bT_1(\beta) + bT_2(\beta)]$$

$$= a(T_1 + T_2)(\alpha) + b(T_1 + T_2)(\beta)$$

$\therefore$ $T_1 + T_2$ is a linear transformation.

Again, T is linear transformation and c is any scalar, then for $\alpha, \beta \in U$ and $a, b \in F$, We have

$$(cT)(a\alpha + b\beta) = c[T(a\alpha + b\beta)] \qquad \text{[By definition]}$$

$$= c[aT(\alpha) + bT(\beta)] \qquad [\because T \text{ is linear transformation.}]$$

$$= c[aT(\alpha)] + c[bT(\beta)] = (ca)T(\alpha) + (cb) T(\beta)$$

$$= (ac)T(\alpha) + (bc)T(\beta)]$$

$$[\because \text{Multiplication is commutative in } F.]$$

$$= a(cT)(a) + b(cT)(\beta)$$

$\therefore$ cT is a linear transformation.

Now we shall show that the set of all linear transformations $L(U, V)$ from U into V forms a vector space with respect to the above defined compositions. First we show that $(L(U, V), +)$ is an abelian group :

(i) Closure Property.

If $T_1, T_2 \in L(U, V)$, then we have already proved that $T_1 + T_2$ is linear transformation, so that $T_1, T_2 \in L(U, V)$.

(ii) Associative Property.

For all $T_1, T_2, T_3 \in L(U, V)$ and for all $\alpha \in U$, we have

$$[(T_1 + T_2) + T_3](\alpha) = (T_1 + T_2)(\alpha) + T_3(\alpha)$$

$$= [T_1(\alpha) + T_2(\alpha)] + T_3(\alpha)$$

$$= T_1(\alpha) + [T_2(\alpha) + T_3(\alpha)]$$

$$[\because \text{Addition is associative in } V.]$$

$$= T_1(\alpha) + (T_2 + T_3)(\alpha) = [T_1 + (T_2 + T_3)](\alpha)$$

$\therefore$ $(T_1 + T_2) + T_3 = T_1 + (T_2 + T_3)$.

(iii) Commutative Property.

For all $T_1, T_2, \in L(U, V)$ and $\alpha \in U$, we have

$$(T_1 + T_2)(\alpha) = T_1(\alpha) + T_2(\alpha)$$

$$= T_2(\alpha) + T_1(\alpha) \qquad [\because \text{Addition is commutative in } V.]$$

$$= (T_2 + T_1)(\alpha)$$

$\therefore$ $T_1 + T_2 = T_2 + T_1$

(iv) Existence of Identity .

The zero transformation, denoted by 0 and defined by $0(\alpha) = 0$, $\forall \alpha \in U$ is a linear transformation.

Also if $T \in L(U, V)$, then
$$(T + 0) = 0 + T = T, \text{ for all } T.$$
$\therefore \quad 0 \in L(U, V)$ and is identity transformation.

(v) Existence of Inverse.

For each $T \in L(U, V)$, there exists $(-T) \in L(U,V)$, defined by
$$(-T)(\alpha) = -T(\alpha), \forall \alpha \in U, (-T) \text{ is linear and } T + (-T) = (-T) + T = 0.$$
$\therefore \quad (-T)$ is the additive inverse of T.

(vi) Distributive Property.

For all $T_1, T_2 \in L(U, V)$, $\alpha \in U$ and $a \in F$,
$$a[(T_1 + T_2)](\alpha) = a(T_1 + T_2)(\alpha) = a[(T_1(\alpha) + T_2)(\alpha)]$$
$$= aT_1(\alpha) + aT_2(\alpha) = (aT_1 + aT_2)(\alpha)$$
$\therefore \qquad a(T_1 + T_2) = aT_1 + aT_2$

Also, for all $T \in L(U, V)$ and $\alpha \in U$, $a, b \in F$,
$$[(a+b)T](\alpha) = (a+b)T(\alpha) = aT(\alpha) + bT(\alpha) = (aT + bT)(\alpha)$$
$$\Rightarrow \qquad (a+b)T = aT + bT$$

(vii) For all $T \in L(U,V)$, $\alpha \in U$ and $a, b \in F$,
$$[(ab)T](\alpha) = (ab)T(\alpha) = a[bT(\alpha)] = a(bT)(\alpha)$$
$$(ab).T = a.(bT)$$

(viii) If 1 is the unity in F, then for all $T \in L(U, V)$ and $\alpha \in U$
$$(1.T)(\alpha) = 1.T(\alpha) = T(\alpha)$$
$\therefore$ Hence, $L(U, V)$ is a vector space.

REMARK

☞ The vector space $L(U, V)$ is also denoted by Hom. (U, V), *i.e.*, (the set of all homomorphism from U into V).

Theorem 2. *Let U be an m-dimensional vector space over the field F, and let V be an n-dimensional vector space over F. Then the vector space L(U, V) is finite-dimensional and has dimension mn.*

Proof. Since U and V both are finite-dimensional vector spaces of dimensions m and n respectively, therefore, let
$$\beta = \{\alpha_1, \alpha_2, \dots \alpha_m\} \text{ and } \beta' = \{\beta_1, \beta_2, \dots \beta_n\}$$
be the ordered basis of U and V respectively.

For each pair of integers (i, j) with $1 \le i \le m$ and $1 \le j \le n$, we define a linear transformation T_{ij} from U into V by
$$T_{ij}(\alpha_k) = \begin{cases} 0; & \text{if } k \ne j \\ \beta_i; & \text{if } k = j \end{cases}$$

The existence and uniqueness of above linear transformations follows from preceding theorem. It is obvious that there are mn linear transformations of the type T_{ij}, so we claim that these mn transformations form a basis of $L(U, V)$.

(i) For mn scalars a_{ij}, we have
$$\sum_{i=1}^{n} \sum_{j=1}^{m} a_{ij} T_{ij} = 0 \qquad \text{[zero transformation]}$$

$$\Rightarrow \quad \sum_{i=1}^{n}\sum_{j=1}^{m} a_{ij}T_{ij}(\alpha_k) = 0(\alpha_k), \forall \alpha_k \in U, 1 \le k \le n$$

$$\Rightarrow \quad \sum_{i=1}^{n}\sum_{j=1}^{m} a_{ij}T_{ij}(\alpha_k) = 0$$

$$\Rightarrow \quad \sum_{i=1}^{n}\left(\sum_{j=1}^{m} a_{ij}T_{ij}(\alpha_k)\right) = 0$$

$$\Rightarrow \quad \sum_{i=1}^{n}[a_{i1}T_{i1}(\alpha_k) + a_{i2}T_{i2}(\alpha_k) + \dots + a_{im}T_{im}(\alpha_k)] = 0$$

$$\Rightarrow \quad \sum_{i=1}^{n}[a_{i1}T_{i1}(\alpha_k) + \sum_{i=1}^{n} a_{i2}T_{i2}(\alpha_k) + \dots + \sum_{i=1}^{n} + a_{im}T_{im}(\alpha_k)] = 0$$

$$\Rightarrow \quad a_{11}T_{11}(\alpha_k) + a_{21}T_{21}(\alpha_k) + \dots + a_{n1}T_{n1}(\alpha_k)$$
$$+ a_{12}T_{12}(\alpha_k) + a_{22}T_{22}(\alpha_k) + \dots + a_{n2}T_{n2}(\alpha_k)$$
$$+ \dots\dots\dots\dots\dots\dots$$
$$+ a_{1m}T_{1m}(\alpha_k) + a_{2m}T_{2m}(\alpha_k) + \dots + a_{nm}T_{nm}(\alpha_k) = 0$$
$$= a_{11}\beta_1 + a_{21}\beta_2 + \dots + a_{n1}\beta_n + a_{12}\beta_1 + a_{22}\beta_2 + \dots + a_{n2}\beta_n$$
$$+ a_{1m}\beta_1 + a_{2m}\beta_2 + \dots + a_{nm}\beta_n = 0$$

$$\dots\dots\dots\dots\dots\dots\dots\dots\dots$$

$$\begin{pmatrix} \because T_{ij}(\alpha_k) = 0, (j \ne k) \\ T_{ij}(\alpha_k) = \beta_i, (j = k) \end{pmatrix}$$

Since $\beta' = \{\beta_1, \beta_2, \dots, \beta_n\}$ is a basis of V, therefore it is linearly independent so that
$$a_{11} = 0 = a_{21} = \dots = a_{n1}$$
$$a_{12} = 0 = a_{22} = \dots = a_{n2}$$
$$\dots\dots\dots\dots$$
$$\dots\dots\dots\dots$$
$$a_{1m} = 0 = a_{2m} = \dots = a_{nm}$$
Thus, $\{T_{ij} : 1 \le i \le m, i \le j \le n\}$ is linearly independent.

(ii)　Now we show that $\{T_{ij} : 1 \le i \le m, i \le j \le n\}$ spans $L(U, V)$. For this, let T be an arbitrary linear transformation from U into V, i.e., $T \in L(U, V)$.

For $\alpha_j \in U$, $T(\alpha_j) \in V$ and $\beta' = \{\beta_1, \beta_2, \dots, \beta_n\}$ is a basis of V so that
$$T(\alpha_j) = a_{1j}\beta_1 + a_{2j}\beta_2 + \dots + a_{nj}\beta_n; 1 < j < n$$
where $a_{1j}, a_{2j}, \dots, a_{nj}$ are the coordinates of vector $T(\alpha_j)$ in β'.

$$T(\alpha_j) = \sum_{i=1}^{n} a_{ij}\beta_i = \sum_{i=1}^{n}\sum_{j=1}^{m} a_{ij}T_{ij}(\alpha_j)$$

$$\Rightarrow \quad T = \sum_{i=1}^{n}\sum_{j=1}^{m} a_{ij}T_{ij}$$

$\Rightarrow \quad \{T_{ij} : 1 \leq i \leq m, 1 \leq j \leq n\}$ generates $L(U, V)$.

$\Rightarrow \quad \{T_{ij} : 1 \leq i \leq m, 1 \leq j \leq n\}$ is a basis of $L(U, V)$.

Hence, $L(U, V)$ is finite-dimensional and dim. $L(U, V) = mn$.

Theorem 3. *Let U, V and W be vector spaces aver the field F. Let T_1 be a linear transformation from U into V and T_2 be a linear transformations from V into W, then the composed function T_2T_1 is defined by*

$$(T_2T_1)(\alpha) = T_2[T_1(\alpha)], \text{ for all } \alpha \in U$$

is a linear transformation from U into W.

Proof. For $\alpha, \beta \in U$ and $a, b \in F$, we have

$$(T_2T_1(a\alpha + b\beta)) = T_2[T_1(a\alpha + b\beta)] = T_2[aT_1(\alpha) + bT_1(\beta)) \qquad [\because T_1 \text{ is linear.}]$$
$$= a(T_2T_1)(\alpha) + b(T_2T_1)(\beta) \qquad [\because T_2 \text{ is linear.}]$$

$\therefore \quad T_2T_1$ is a linear transformation from U into W.

7.20 LINEAR OPERATOR

Definition. *If V is a vector space over the field F, then a linear transformation from V into V is called a linear operator.*

In case of above theorem if U, V and W are replaced by V, then T_1 and T_2 are linear operators on the space V and $T_2 T_1$ is also a linear operator on V. Thus the vector space $L(V, V)$ has a 'multiplication' defined on it by composition. In this case the operator $T_1 T_2$ is also defined but in general $T_2T_1 \neq T_1T_2$. Therefore, if T is a linear operator on V, then we can compose T with T as follows :

$$T^2 = TT$$
$$T^3 = TTT$$

in general, $\quad T^n = T \, T...T$ (*n* times) for $n = 1, 2, 3, ...$

REMARK

☞ If $T \neq 0$, then we define $T^0 = 1$ (Identity transformation).

7.21 ALGEBRA OF LINEAR OPERATORS

Theorem 1. *Let V be a vector space over the field F and let T, T_1, , T_2 and T_3 be linear operators on V and let c be an element in F, then*

(i) $IT = TI = T$, *I being an identity operator.*

(ii) $T_1 (T_2 + T_3) = T_1T_2 + T_1T_3; (T_2 + T_3) T_1 = T_2 T_1 + T_3 T_1$

(iii) $T_1 (T_2T_3) = (T_1T_2)T_3$

(iv) $c(T_1T_2) = (cT_1)T_2 = T_1(cT_2)$

(v) $T0 = 0T = 0$, *0 being zero linear operator.*

Proof. (i) For $\alpha \in V$

$$(IT)(\alpha) = I[T(\alpha)] = T(\alpha) \qquad [\because I(\alpha) = \alpha]$$
$$IT = T$$

Also, $\quad (TI)(\alpha) = T[I(\alpha)] = T(\alpha)$

$\Rightarrow \qquad TI = T$

Thus, $\qquad IT = TI = T$

(ii) For any $\alpha \in V$

$$[T_1(T_2 + T_3)](\alpha) = T_1[(T_2+T_3)(\alpha)] = T_1[T_2(\alpha) + T_3(\alpha)]$$
$$= (T_1T_2)(\alpha) + (T_1T_3)(\alpha) = (T_1T_2+T_1T_3)(\alpha)$$

$\therefore \quad T_1(T_2+T_3) = T_1T_2+T_1T_3$

Similarly,

$$(T_2+T_3)T_1 = T_2T_1+T_3T_1$$

(iii) For any $\alpha \in V$

$$[T_1(T_2T_3)](\alpha) = T_1[(T_2T_3)(\alpha)] = T_1[T_2(T_3(\alpha))]$$
$$= (T_1T_2)[T_3(\alpha)] = [(T_1T_2)T_3](\alpha)$$

$\therefore \quad T_1(T_2T_3) = (T_1T_2)T_3$

(iv) For any $\alpha \in V$, $c \in F$

$$[c(T_1T_2)](\alpha) = c[(T_1T_2)(\alpha)] = c\,[T_1(T_2(\alpha))]$$
$$= (cT_1)[T_2(\alpha)] = [(cT_1)T_2](\alpha)$$

$\therefore \quad c(T_1T_2) = (cT_1)T_2$

Also, $[c(T_1T_2)](\alpha) = (cT_1)[T_2(\alpha)] = T_1[cT_2(\alpha)] = T_1[(cT_2)](\alpha)$

$\therefore \quad c(T_1T_2) = T_1(cT_2)$

Thus, $\quad c(T_1T_2) = (cT_1)T_2 = T_1(cT_2)$

(v) For any $\alpha \in V$,

$$(T0)(\alpha) = T[0(\alpha)] = T(0) \qquad\qquad [\because 0(\alpha) = 0]$$
$$= 0$$

Similarly, $\quad 0T = 0$.

7.22   RANGE AND NULL SPACE OF A LINEAR TRANSFORMATION

(i) **Range space of a linear transformation.** If T is a linear transformation from U into V, then the range of T is a subspace of V. Let R_T be the range of T, that is, the set of all vectors β in V such that $T(\alpha) = \beta$ for some $\alpha \in U$,

i.e., $R_T = \{\beta \in V : T(\alpha) = \beta,\ \text{for some } \alpha \in U\}$.

If U is finite-dimensional, then the dimension of range of T is called rank of T and is denoted by $\rho(T)$.

(ii) **Null space of a linear transformation.** If T is a linear transformation from a vector space U into a vector space V, then the null space of T denoted by $N(T)$ is the set of all vectors α in U such that $T(\alpha) = 0$, where 0 is the zero vector in V, i.e.,

$N(T) = \{\alpha \in U : T(\alpha) = 0\}$.

If U is finite-dimensional, then the dimension of null space $N(T)$ is called nullity of T and is denoted by $n(T)$.

REMARK

☞ Kernel of T is also known as null space of T.

Theorem 1. *Let U and V be vector spaces over the field F and let T be a linear transformation from U into V. Suppose U is finite-dimensional. Then*

$$rank\ (T) + nullity\ (T) = dim.\ U$$

i.e., $\qquad \rho(T) + n(T) = dim.\ U$

Proof. Let $\{\alpha_1, \alpha_2, ..., \alpha_k\}$ be the basis of N_T, the null space of T. Let the dimension of U be n, so that $\alpha_{k+1}, \alpha_{k+2}, ..., \alpha_n \in U$ such that $\{\alpha_1, \alpha_2, ..., \alpha_n\}$ forms a basis of U. Therefore, dim. N_T, $= k$ and dim. $U = n$.

We claim that $[T(\alpha_{k+1}), T(\alpha_{k+2}), ..., T(\alpha_n)]$ is a basis of range of T.

For scalars $a_i \in F$ we have

$$a_{k+1}T(\alpha_{k+1}) + a_{k+2}\,T(\alpha_{k+2}) + ... + a_n\,T(\alpha_n) = 0$$

$$\Rightarrow \quad \sum_{i=k+1}^{n} a_i T(\alpha_i) = 0 \qquad \Rightarrow \quad T\left(\sum_{i=k+1}^{n} a_i\alpha_i\right) = 0$$

$$\Rightarrow \quad \sum_{i=k+1}^{n} a_i\alpha_i \in N_T$$

Since $\{\alpha_1, \alpha_1, ..., \alpha_k\}$ is the basis of N_T, so that for some scalars $b_1, b_2, ..., b_k$, we have

$$\sum_{i=k+1}^{n} a_i\alpha_i = b_1\alpha_1 + b_2\alpha_2 + ... + b_k\alpha_k$$

$$\Rightarrow \quad b_1\alpha_1 + b_2\alpha_2 + ... + b_k\alpha_k - \sum_{i=k+1}^{n} a_i\alpha_i = 0$$

$$\Rightarrow \quad b_1\alpha_1 + b_2\alpha_2 + ... + b_k\alpha_k + (-a_{k+1})\alpha_{k+1} + ... + (-a_n)\alpha_n = 0$$

Since $\alpha_1, \alpha_2, ..., \alpha_n$ are linearly independent, we must have

$$b_1 = 0 = b_2 = ... = b_k = a_{k+1} = a_n.$$

$\therefore \quad [T(\alpha_{k+1}), T(\alpha_{k+2}), ..., T(\alpha_n)]$ is linearly independent.

Now, we shall show that $T(\alpha_{k+1}), T(\alpha_{k+2}), ..., T(\alpha_n)]$ spans range of T.

For this, let $T(\alpha) \in R_T$ (range of T) for some $\alpha \in U$.

Since $\{\alpha_1, \alpha_2, ..., \alpha_n\}$ spans U so that

$$\alpha = a_1\alpha_1 + a_2\alpha_2 + ... + a_n\alpha_n$$

For $a_i' s \in F$, we have

$$\begin{aligned}
T(\alpha) &= T(a_1\alpha_1 + a_2\alpha_2 + ... + a_n\alpha_n) \\
&= T(a_1\alpha_1) + T(a_2\alpha_2) + ... + T(a_n\alpha_n) && [\because T \text{ is linear.}]\\
&= a_1T(\alpha_1) + a_2T(\alpha_2) + ... + a_kT(\alpha_k) + a_{k+1}T(\alpha_{k+1}) + ... + a_nT(\alpha_n) \\
&= a_{k+1}T(\alpha_{k+1}) + a_{k+2}T(\alpha_{k+2}) + ... + a_nT(\alpha_n) \\
&&& [\because T(\alpha_i) = 0,\ 1 \le i \le k]
\end{aligned}$$

Thus, $\qquad T(\alpha_{k+1})...T(\alpha_n)$ spans R_T.

Hence, $\qquad [T(\alpha_{k+1}), ..., T(\alpha_n)]$ is a basis of R_T.

Accordingly, $\qquad$ dim. $R_T = n - k = $ dim .$U -$ dim. N_T

$\therefore \qquad$ dim.$R_T +$ dim.$N_T = $ dim. U

Hence, rank$(T) + $ nullity$(T) = $ dim. U.

7.23 INVERTIBLE LINEAR TRANSFORMATION

Definition. *A linear transformation T from a vector space U(F) into V(F) is called invertible or regular if there exists a unique linear transformation T^{-1} (called the inverse of T) from V(F) into U(F) such that (T^{-1}) is the identity linear transformation on U and (TT^{-1}) is the identity transformation on V.*

Furthermore, T is invertible iff

(i) *T is one-to-one*

(ii) *T is onto, i.e., R(T) = V*

Theorem 1. *Let U and V be vector spaces over the same field F and left T be a linear transformation from U into V. If T is invertible, then T^{-1} is a linear transformation from V into U.*

Proof.   Since T is invertible, so for each $\beta \in V$, there is a unique $\alpha \in U$ such that

$$T(\alpha) = \beta \Leftrightarrow T^{-1}(\beta) = \alpha$$

Now, we shall show that T^{-1} is linear.

For $\alpha_1, \alpha_2 \in U$ and $a, b \in F$

$$T(a\alpha_1 + b\alpha_2) = aT(\alpha_1) + bT(\alpha_2) \qquad [\because T \text{ is linear.}]$$

But for β_1 and β_2 in V, there are unique $\alpha_1, \alpha_2 \in U$ respectively, such that

$$T(\alpha_1) = \beta_1 \Leftrightarrow T^{-1}(\beta_1) = \alpha_1$$

and $$T(\alpha_2) = \beta_2 \Leftrightarrow T^{-1}(\beta_2) = \alpha_2$$

Thus, we have

$$T(a\alpha_1 + b\alpha_2) = a\beta_1 + b\beta_2$$

$\Rightarrow \qquad a\alpha_1 + b\alpha_2 = T^{-1}(a\beta_1 + b\beta_2) \quad [\because a\alpha_1 + b\alpha_2 \text{ is unique in } V.]$

$\Rightarrow \quad aT^{-1}(\beta_1) + b\, T^{-1}(\beta_2) = T^{-1}(a\beta_1 + b\beta_2)$

Hence, T^{-1} is a linear transformation.

Theorem 2. *Let T_1 be an invertible linear transformation from U(F) into V(F) and T_2 an invertible linear transformation from V(F) into W(F). Then $T_1 T_2$ is invertible and $(T_2 T_1)^{-1} = T_1^{-1} T_2^{-1}$*

Proof.   To show $T_2 T_1$ is invertible, we shall show that it is one-one and onto.

If $\alpha_1, \alpha_2 \in U$ such that $(T_2 T_1)(\alpha_1) = (T_2 T_1)(\alpha_2)$, then

$$(T_2 T_1)(\alpha_1) = (T_2 T_1)(\alpha_2) \Rightarrow T_2[T_1(\alpha_1)] = T_2[T_1(\alpha_2)]$$

$$\Rightarrow T_1(\alpha_1) = T_1(\alpha_2) \qquad\qquad [\because T_2 \text{ is one-one.}]$$

$$\Rightarrow \alpha_1 = \alpha_2 \qquad\qquad [\because T_1 \text{ is one-one.}]$$

Thus, $T_2 T_1$ is one-one.

Also, T_1 and T_2 being onto, then for each $\beta \in V$, there exists a unique $\alpha \in U$ such that

$$T_1(\alpha) = \beta$$

and for each $\gamma \in W$, there exists a unique $\beta \in V$ such that $T_2(\beta) = \gamma$.

Thus, $\gamma \in W \Rightarrow$ there exists $\beta \in V : \gamma = T_2(\beta)$.

$\Rightarrow \quad$ there exists $\quad \alpha \in U : \gamma = T_2(T_1(\alpha)) \qquad\qquad [\because T_1(\alpha) = \beta]$

$\Rightarrow \quad$ there exists $\quad \alpha \in U : \gamma = (T_2 T_1)(\alpha)$.

Therefore $(T_2 T_1)$ is onto. Hence, $(T_2 T_1)$ is invertible.

Also, $(T_2 T_1)(T_1^{-1} T_2^{-1}) = T_2(T_1 T_1^{-1})T_2^{-1} = (T_2 I)T_2^{-1} = T_2 T_2^{-1} = I$

Similarly,

$$(T_1^{-1}T_2^{-1})(T_2T_1) = T_1^{-1}(T_2^{-1}T_2)T_1 = T_2^{-1}(IT_1) = T_1^{-1}T_1 = I$$

Hence,

$$(T_2T_1^{-1}) = T_1^{-1}T_2^{-1}$$

7.24 NON-SINGULAR LINEAR TRANSFORMATIONS

Definition. *Let U and V be vector spaces over the field F. Then a linear transformation T from U into V is called non-singular if the null space of T is* (0).

Thus, if T is non-singular, then

$$T(\alpha) = 0 \qquad \Rightarrow \alpha = 0$$

Also, when T is non-singular and $\alpha, \beta \in U$

$$T(\alpha) = T(\beta) \qquad \Rightarrow T(\alpha) - T(\beta) = 0$$

$\Rightarrow$ $\qquad\qquad\qquad T(\alpha-\beta) = 0$ $\qquad\qquad\qquad\qquad$ [∵ *T is linear.*]

$\Rightarrow$ $\qquad\qquad\qquad \alpha-\beta = 0$ $\qquad\qquad\qquad\qquad$ [∵ *T is non-singular.*]

$\qquad\Rightarrow$ $\qquad\qquad\qquad \alpha = \beta$

Hence, when T is non-singular, implies that T is one-one.

Theorem 1. *Let T be a linear transformation from U(F) into V(F). Then T is non-singular if and only if T carries each linearly independent subset of U onto a linearly independent subset of V.*

Proof. Let us first suppose that T is non-singular. Now, let

$$S = \{\alpha_1, \alpha_2,...,\alpha_k\}$$

be an arbitrary linearly independent subset of U. Then we have to show that the set

$$S_1 = \{T(\alpha_1), T(\alpha_2),...,T(\alpha_k)\}$$

is linearly independent subset of V.

For scalars $a_1, a_2,... a_k \in F$ we have

$$a_1 T(\alpha_1) + a_2T(\alpha_2) +...+ a_kT(\alpha_k)\} = 0$$

$\Rightarrow$ $\qquad T(a_1\alpha_1) + T(a_2\alpha_2) +...+ T(a_k\alpha_k)\} = 0$ $\qquad$ [∵ *T is linear.*]

$\Rightarrow$ $\qquad\quad T(a_1\alpha_1 + a_2\alpha_2 +...+ a_k\alpha_k) = 0$ $\qquad\quad$ [∵ *T is linear.*]

$\Rightarrow$ $\qquad\qquad a_1\alpha_1 + a_2\alpha_2 +...+ a_k\alpha_k = 0$ $\qquad$ [∵ *T is non-singular.*]

$\Rightarrow$ $\qquad\qquad\quad a_1 = a_2 = ... = a_k = 0$ $\qquad$ [∵ *S is linear independent.*]

Hence, S_1 is linearly independent.

Conversely, suppose that T carries each linearly independent subset of U into a linearly independent subset of V. Let α be a non-zero vector in U, then $\{\alpha\}$ is linearly independent so is $\{T(\alpha)\}$. Consequently, $T(\alpha) \ne 0$ because the set consisting of the zero vector alone is dependent. Therefore, the null space of T is the zero space and hence T is non-singular.

Theorem 2. *Let U and V be finite-dimensional vector spaces over the field F such that dim. U= dim. V. If T is a linear transformation from U into V, then the following are equivalent:*

(i) T is invertible.

(ii) T is non-singular.

(iii) T is onto, that is, the range of T is V.

(iv) If $\{\alpha_1, \alpha_2,..., \alpha_n\}$ is a basis of U, then $\{T(\alpha_1), T(\alpha_2),..., T(\alpha_n)\}$ is a basis of V.

Proof. **(i)** $\Rightarrow$ **(ii)** : Since T is invertible, so it is one-one and onto, therefore, T is non-singular.

(ii) $\Rightarrow$ **(iii):** Let T be non-singular and let $\{\alpha_1, \alpha_2, ..., \alpha_n\}$ be the basis of U, then the set $\{T(\alpha_1), T(\alpha_2),..., T(\alpha_n)\}$ is linearly independent subset of V, but dim. U = dim. V, therefore $\{T(\alpha_1), ..., T(\alpha_n)\}$ is a basis for V.

For any $\beta \in V, a_1, a_2, ..., a_n \in F$, we have

$$\beta = a_1 T(\alpha_1) + a_2 T(\alpha_2) + ... + a_n T(\alpha_n)$$
$$\beta = T(a_1 \alpha_1 + a_2 \alpha_2 + ... + a_n \alpha_n) \qquad [\because T \text{ is linear.}]$$
$$\Rightarrow \qquad \beta \in R_T$$

Thus, $\qquad V \subseteq R_T$, but $R_T \subseteq V$

$\therefore \qquad\qquad R_T = V$

i.e., the range of $T = V$

(iii) $\Rightarrow$ **(iv)** : Suppose range of $T = V$. Let the set $\{\alpha_1, \alpha_2, ..., \alpha_n\}$ be a basis of U so that an arbitrary element $\alpha \in U$ is expressible as linear combination of $\alpha_1, \alpha_2, ..., \alpha_n$.

$\therefore \qquad \alpha = b_1 \alpha_1 + b_2 \alpha_2 + ... + b_n \alpha_n$ for some scalars, $b_1, b_2, ..., b_n \in F$

$\Rightarrow \qquad T(\alpha) = T(b_1 \alpha_1 + b_2 \alpha_2 + ... + b_n \alpha_n)$

$\qquad\qquad = b_1 T(\alpha_1) + b_2 T(\alpha_2) +, ..., + b_n T(\alpha_n) \qquad [\because T \text{ is linear.}]$

This shows that each element of range of T is expressible as a linear combination of $\{T(\alpha_1) + T(\alpha_2), ..., T(\alpha_n)\}$. Thus, the set

$$\{T(\alpha_1) + T(\alpha_2), ..., T(\alpha_n)\}.$$

spans R_T. Since $R_T = V$. Also, dim. U = dim. $V = n$.

Hence, $\{T(\alpha_1), T(\alpha_2), ..., T(\alpha_n)\}$ forms a basis of V.

(iv) $\Rightarrow$ **(i):** Let $\{\alpha_1, \alpha_2,..., \alpha_n\}$ be a basis of U such that $\{T(\alpha_1), T(\alpha_2), ..., T(\alpha_n)\}$ is a basis of V.

Let α be an arbitrary element of U, then for $b_1, b_2, ..., b_n \in F$, we have

$$\alpha = b_1 \alpha_1 + b_2 \alpha_2 + ... + b_n \alpha_n$$

Now, $\qquad T(\alpha) = 0$

$\Rightarrow \qquad T(b_1 \alpha_1 + b_2 \alpha_2 + ... + b_n \alpha_n) = 0$

$\Rightarrow \qquad b_1 T(\alpha_1) + b_2 T(\alpha_2) + ... + b_n T(\alpha_n) = 0 \qquad [\because T \text{ is linear.}]$

$\Rightarrow \qquad b_1 = b_2 = ... = b_n = 0$

$\qquad\qquad [\because \{T(\alpha_1), T(\alpha_2), ..., T(\alpha_n)\} \text{ is linearly independent.}]$

$\Rightarrow b_1 \alpha_1 + b_2 \alpha_2 + ... + b_n \alpha_n = 0 \Rightarrow \alpha = 0$

Hence, T is non-singular and therefore T is one-one.

Also, $\{T(\alpha_1), T(\alpha_2), ..., T(\alpha_n)\}$ spans V and range of T is V. Consequently, T is one-one and hence T is invertible.

7.25 CO-ORDINATE VECTOR

Let V be a finite-dimensional vector space over a field F and let dim. $V = n$, then $B = \{\alpha_1, \alpha_2, ..., \alpha_n\}$ is a basis of V and for $\alpha \in V$, suppose that

$$\alpha = a_1 \alpha_1 + a_2 \alpha_2 + ... + a_n \alpha_n$$

for $a_i's \in F$. Then the coordinate vector of α relative to β, which we write as a column vector unless otherwise specified or implied, is

$$[\alpha]_B = \begin{bmatrix} a_1 \\ a_2 \\ \vdots \\ a_n \end{bmatrix}$$

7.26 MATRIX REPRESENTATION OF A LINEAR TRANSFORMATION

Let U be an m-dimensional vector space over a field F and let V be an n-dimensional vector space over the field F. Let $B = \{\alpha_1, \alpha_2,...,\alpha_m\}$ and $B' = \{\beta_1, \beta_2 ,.., \beta_n)$ be the basis of U and V respectively. If T is a linear transformation from U into V, then $T(\alpha_1), T(\alpha_2), ...,T(\alpha_m)$ are vectors in V. Since $B'= \{\beta_1, \beta_2 ,.., \beta_n\}$ is a basis of V so that each $T(\alpha_i)$ is a linear combination of the elements of B'. For $a_{ij} \in F, 1 \le i \le m, 1 \le j \le n$, we have

$$T(\alpha_1) = \{a_{11}\beta_1 + a_{12}\beta_2 + ,.., + a_{1n}\beta_n)$$
$$T(\alpha_2) = \{a_{21}\beta_1 + a_{22}\beta_2 + ,.., + a_{2n}\beta_n)$$

$$\cdots\cdots\cdots\cdots\cdots\cdots\cdots\cdots$$

$$T(\alpha_m) = \{a_{m1}\beta_1 + a_{m2}\beta_2 + ,.., + a_{mn}\beta_n)$$

Definition. *The transpose of the above matrix of coefficients, denoted by $[T]_B$ is called the matrix representation of T relative to the ordered basis B.*

Thus,

$$[T]_B = \begin{bmatrix} a_{11} & a_{21} & \cdots & a_{m1} \\ a_{12} & a_{22} & \cdots & a_{m2} \\ \cdots & \cdots & \cdots & \cdots \\ a_{1n} & a_{2n} & \cdots & a_{mn} \end{bmatrix}_{n \times m}$$

For example : Let V be the vector space of polynomials in 't' over the field of reals **R,** of degree ≤ 3 , and let

$$D : V \to V$$

be the differential operator defined by $D [p(t)] = \dfrac{d}{dt}[p(t)]$

We compute the matrix of D in the basis $B = [1, t, t^2 , t^3]$ as follows

$$D(1) = 0 = 0 + 0t + 0t^2 + 0t^3$$
$$D(t) = 1 = 1 + 0t + 0t^2 + 0t^3$$
$$D(t^2) = 2t = 0 + 2t + 0t^2 + 0t^3$$
$$D(t^3) = 3t^2 = 0 + 0t + 3t^2 + 0t^3$$

Thus, the matrix of D relative to B is given by

$$[D]_B = \begin{bmatrix} 0 & 1 & 0 & 0 \\ 0 & 0 & 2 & 0 \\ 0 & 0 & 0 & 3 \\ 0 & 0 & 0 & 0 \end{bmatrix}$$

Theorem 1. *Let U be an m-dimensional vector space aver the field F and V an n-dimensional vector space over the field F. Let B be an ordered basis for U and B' an ordered basis for V. Let T be any linear transformation from U into V. Then for any vector $\alpha \in U$.*

$$[T]_B \, [\alpha]_B = [T(\alpha)]_{B'}$$

Proof. Let $B = \{\alpha_1, \alpha_2,..., \alpha_m\}$ be an ordered basis for U and $B' = \{\beta_1, \beta_2,..., \beta_n\}$ an ordered basis for V. T is linear transformation from U into V, then T is determined by its action on the vectors α_j, $1 \le i \le m$. Each of m vectors $T(\alpha_i)$ is uniquely expressible as a linear combination of elements of B' :

$$T(\alpha_i) = \sum_{j=1}^{n} a_{ij} \beta_j \qquad \qquad ...(1)$$

where $a_{i1}, a_{i2},..., a_{in}$ are the coordinates of $T(\alpha_i)$ in the ordered basis B'.
If α be any vector in U, then

$$\alpha = a_1\alpha_1 + a_2\alpha_2 + \ldots + a_m\alpha_n$$

$$\therefore \qquad [\alpha]_B = \begin{bmatrix} a_1 \\ a_2 \\ \vdots \\ a_m \end{bmatrix}$$

Now, $T(\alpha) = T(a_1\alpha_1 + a_2\alpha_2 + \ldots + a_m\alpha_m)$

$$= a_1 T(\alpha_1) + a_2 T(\alpha_2) + \ldots + a_m T(\alpha_m)$$

$$= a_1 \sum_{j=1}^{n} a_{1j}\beta_j + a_2 \sum_{j=1}^{n} a_{2j}\beta_j + \ldots + \sum_{j=1}^{n} a_{mj}\beta_j \qquad \text{[using (1)]}$$

$$= \sum_{j=1}^{n} a_1 a_{1j}\beta_j + \sum_{j=1}^{n} a_2 a_{2j}\beta_j + \ldots + \sum_{j=1}^{n} a_m a_{mj}\beta_j$$

$$= a_1 a_{11}\beta_1 + a_1 a_{12}\beta_2 + \ldots + a_1 a_{1n}\beta_n + a_2 a_{21}\beta_1 + a_2 a_{22}\beta_2 + \ldots + a_2 a_{2n}\beta_n$$

$$\cdots\cdots\cdots\cdots\cdots\cdots\cdots$$

$$+ a_m a_{m1}\beta_1 + a_m a_{m2}\beta_2 + \ldots + a_m a_{mn}\beta_n$$

$$\therefore \qquad [T(\alpha)]_{B'} = \begin{bmatrix} a_1 a_{11} + a_2 a_{21} + \ldots + a_m a_{m1} \\ a_1 a_{12} + a_2 a_{22} + \ldots + a_m a_{m2} \\ \cdots\cdots\cdots\cdots\cdots\cdots \\ a_1 a_{1n} + a_2 a_{2n} + \ldots + a_m a_{mn} \end{bmatrix}_{n \times m}$$

$$= \begin{bmatrix} a_{11} & a_{21} & \cdots & a_{m1} \\ a_{12} & a_{22} & \cdots & a_{m2} \\ \cdots & \cdots & \cdots & \cdots \\ a_{1n} & a_{2n} & \cdots & a_{mn} \end{bmatrix}_{n \times m} \cdot \begin{bmatrix} a_1 \\ a_2 \\ \vdots \\ a_m \end{bmatrix}_{m \times 1}$$

$$[T(\alpha)]_{B'} = [T]_B [\alpha]_B .$$

Theorem 2. *Let U, V and W be vector spaces over the field F of respective dimensions n, m and p. Let T_1 be a linear transformation from U into V and T_2 a linear transformation from V into W. If B, B' and B'' are the ordered bases for the spaces U, V and W respectively, if A is the matrix of T_1, relative to the pair B, B' and B is the matrix of T_2 relative to the pair B' and B'', then the matrix of $(T_2 T_1)$ relative to the pair B, B'' is the product matrix C = BA.*

Proof. Let $B = \{\alpha_1, \alpha_2,..., \alpha_n\}$, $B' = \{\beta_1, \beta_2, ... , \beta_m\}$ and $B'' = \{\gamma_1, \gamma_2, ..., \gamma_p]$ be the bases of U, V and W respectively. If α is any vector in U, then

$$[T_1\{\alpha)]_{B'} = [T_1]_B[\alpha]_B \qquad\qquad \text{[By above theorem]}$$
$$= A[\alpha]_B \qquad\qquad\qquad [\because A = [T_1]_B]$$

and $\quad [T_2(T_1(\alpha))]_{B''} = [T_2]_{B'} [T_1(\alpha)]_{B'} = B[T_1(\alpha)]_{B'} \qquad [\because B = [T_2]B']$

$\therefore \qquad [(T_2 T_1](\alpha)]_{B''} = BA[\alpha]_B.$

Hence, by the definition and uniqueness of the representing matrix, we must have C = BA as the matrix of (T_2T_1) relativeto B, B''.

Theorem 3. *Let V be an n-dimensional vector space over the field F and B be an ordered basis of V. If T_1 and T_2 are linear operators from V into V, then*

(i) $[T_1+T_2]_B = [T_1]_B+[T_2]_B$

(ii) $[cT_1]_B = c[T_1]_B$ *for* $c\in F$

(iii) $[T_2 T_1]_B = [T_1]_B[T_2]_B.$

Proof. Let $\{a_1, a_2 ,..., a_n\}$ be the basis of V, then for $a_{ij}\in F$ and $b_{ij}\in F$, $1\le i \le n$, $1 \le j \le n$, we have

$$T_1 (\alpha_1) = a_{11}\alpha_1 +a_{12}\alpha_2 +....+a_{1n}\alpha_n$$
$$T_1 (\alpha_2) = a_{21}\alpha_1 +a_{22}\alpha_2 +....+a_{2n}\alpha_n$$

$$\cdots\cdots\cdots\cdots$$

$$T_1 (\alpha_n) = a_{n1}\alpha_1 +a_{n2}\alpha_2 +....+a_{nn}\alpha_n$$

$$\therefore \qquad [T_1]_B = \begin{bmatrix} a_{11} & a_{21} & \cdots & a_{n1} \\ a_{12} & a_{22} & \cdots & a_{n2} \\ \cdots & \cdots & \cdots & \cdots \\ a_{1n} & a_{2n} & \cdots & a_{nn} \end{bmatrix}$$

Also,

$$T_2 (\alpha_2) = b_{11}\alpha_1 +b_{12}\alpha_2 +....+b_{1n}\alpha_n$$
$$T_2 (\alpha_2) = b_{21}\alpha_1 +b_{22}\alpha_2 +....+b_{2n}\alpha_n$$

$$\cdots\cdots\cdots\cdots$$

$$T_2 (\alpha_n) = b_{n1}\alpha_1 +b_{n2}\alpha_2 +....+b_{nn}\alpha_n$$

$$[T_2]_B = \begin{bmatrix} b_{11} & b_{21} & \cdots & b_{n1} \\ b_{12} & b_{22} & \cdots & b_{n2} \\ \cdots & \cdots & \cdots & \cdots \\ b_{1n} & b_{2n} & \cdots & b_{nn} \end{bmatrix}$$

$\therefore$

(i) $\quad (T_1 + T_2)(\alpha_1) = (T_1)(\alpha_1)+ (T_2)(\alpha_1)$

$$= (a_{11}+ b_{11})a_1+ (a_{12}+ b_{12})a_2+ ... + (a_{1n}+ b_{1n})a_n$$

$$(T_1 + T_2)(\alpha_2) = (T_1)(\alpha_2) + (T_2)(\alpha_2)$$
$$= (a_{21} + b_{21})\alpha_1 + (a_{22} + b_{22})\alpha_2 + \ldots + (a_{2n} + b_{2n})\alpha_n$$
$$(T_1 + T_2)(\alpha_n) = (T_1)(\alpha_n) + (T_2)(\alpha_n)$$
$$= (a_{n1} + b_{n1})\alpha_1 + (a_{n2} + b_{n2})\alpha_2 + \ldots + (a_{nn} + b_{nn})\alpha_n$$

$$[T_1 + T_2]_B = \begin{bmatrix} (a_{11} + b_{11}) & (a_{21} + b_{21}) & \cdots & (a_{n1} + b_{n1}) \\ (a_{12} + b_{12}) & (a_{22} + b_{22}) & \cdots & (a_{2n} + b_{2n}) \\ \cdots & \cdots & \cdots & \cdots \\ (a_{1n} + b_{1n}) & (a_{2n} + b_{2n}) & \cdots & (a_{nn} + b_{nn}) \end{bmatrix}$$

$$= \begin{bmatrix} a_{11} & a_{21} & \cdots & a_{n1} \\ a_{12} & a_{22} & \cdots & a_{n2} \\ \cdots & \cdots & \cdots & \cdots \\ a_{1n} & a_{2n} & \cdots & a_{nn} \end{bmatrix} + \begin{bmatrix} b_{11} & b_{21} & \cdots & b_{n1} \\ b_{12} & b_{22} & \cdots & b_{2n} \\ \cdots & \cdots & \cdots & \cdots \\ b_{1n} & b_{2n} & \cdots & b_{nn} \end{bmatrix}$$

$$= [T_1]_B + [T_2]_B$$

(ii) $(cT_1)(\alpha_1) = cT_1(\alpha_1) = ca_{11}\alpha_1 + ca_{12}\alpha_2 + \ldots + ca_{1n}\alpha_n$
$(cT_1)(\alpha_2) = cT_1(\alpha_2) = ca_{21}\alpha_1 + ca_{22}\alpha_2 + \ldots + ca_{2n}\alpha_n$

....

$(cT_1)(\alpha_n) = cT_1(\alpha_n) = ca_{n1}\alpha_1 + ca_{n2}\alpha_2 + \ldots + ca_{nn}\alpha_n$

$$\therefore (cT_1)_B = \begin{bmatrix} ca_{11} & ca_{21} & \cdots & ca_{n1} \\ ca_{12} & ca_{22} & \cdots & ca_{n2} \\ \cdots & \cdots & \cdots & \cdots \\ ca_{1n} & ca_{2n} & \cdots & ca_{nn} \end{bmatrix} = c \begin{bmatrix} a_{11} & a_{21} & \cdots & a_{n1} \\ a_{12} & a_{22} & \cdots & a_{n2} \\ \cdots & \cdots & \cdots & \cdots \\ a_{1n} & a_{2n} & \cdots & a_{nn} \end{bmatrix} = c[T_1]_B$$

(iii) $(T_2T_1)(\alpha_1) = T_2(T_1(\alpha_1))$
$$= T_2(a_{11}\alpha_1 + a_{12}\alpha_2 + \ldots + a_{1n}\alpha_n)$$
$$= a_{11}T_2(\alpha_1) + a_{12}T_2(\alpha_2) + \ldots + a_{1n}T_2(\alpha_n) \qquad [\because T_2 \text{ is linear.}]$$
$$= a_{11}(b_{11}\alpha_1 + b_{12}\alpha_2 + \ldots + b_{1n}\alpha_n)$$
$$+ a_{12}(b_{21}\alpha_1 + b_{22}\alpha_2 + \ldots + b_{2n}\alpha_n)$$

....

$$+ a_{1n}(b_{n1}\alpha_1 + b_{n2}\alpha_2 + \ldots + b_{nn}\alpha_n)$$
$$= (a_{11}b_{11} + a_{12}b_{21} + \ldots + a_{1n}b_{n1})\alpha_1$$
$$+ (a_{11}b_{12} + a_{12}b_{22} + \ldots + a_{1n}b_{n2})\alpha_2$$

....

$$+ (a_{11}b_{1n} + a_{12}b_{2n} + \ldots + a_{1n}b_{nn})\alpha_n$$
$$(T_2T_1)(\alpha_2) = T_2(T_1(\alpha_2)) = T_2(a_{21}\alpha_1 + a_{22}\alpha_2 + \ldots + a_{2n}\alpha_n)$$
$$= a_{21}T_2(\alpha_1) + a_{22}T_2(\alpha_2) + \ldots + a_{2n}T_2(\alpha_n)$$
$$= a_{21}(b_{11}\alpha_1) + b_{12}\alpha_2 + \ldots + b_{1n}\alpha_n)$$
$$+ a_{22}(b_{21}\alpha_1 + b_{22}\alpha_2 + \ldots + b_{2n}\alpha_n)$$

....

$$+ a_{2n}(b_{n1}\alpha_1 + b_{n2}\alpha_2 + \ldots + b_{nn}\alpha_n)$$

$$= a_{21}b_{11} + a_{22}b_{21} + \ldots + a_{2n}b_{n1})\alpha_1$$
$$+ (a_{21}b_{12} + a_{22}b_{22} + \ldots + a_{2n}b_{n2})\alpha_2$$

$$\ldots\ldots\ldots\ldots\ldots\ldots\ldots\ldots\ldots$$

$$+ (a_{21}b_{1n} + a_{22}b_{2n} + \ldots + a_{2n}b_{nn})\alpha_n$$

Similarly,

$$(T_2T_1)(\alpha_n) = T_2(T_1(\alpha_n)) = T_2(a_{n1}\alpha_1 + a_{n2}\alpha_2 + \ldots + a_{nn}\alpha_n)$$
$$= a_{n1}T_2(\alpha_1) + a_{n2}T_2(\alpha_2) + \ldots + a_{nn}T_2(\alpha_n)$$
$$= a_{n1}(b_{11}\alpha_1 + b_{12}\alpha_2 + \ldots + b_{1n}\alpha_n)$$
$$+ a_{n2}(b_{21}\alpha_1 + b_{22}\alpha_2 + \ldots + b_{2n}\alpha_n)$$

$$\ldots\ldots\ldots\ldots\ldots\ldots\ldots\ldots\ldots$$

$$+ a_{nn}(b_{n1}\alpha_1 + b_{n2}\alpha_2 + \ldots + b_{nn}\alpha_n)$$
$$= (a_{n1}b_{11} + a_{n2}b_{21} + \ldots + a_{nn}b_{n1})\alpha_1$$
$$+ (a_{n1}b_{12} + a_{n2}b_{22} + \ldots + a_{nn}b_{n2})\alpha_2$$

$$\ldots\ldots\ldots\ldots\ldots\ldots\ldots\ldots\ldots$$

$$+ (a_{n1}b_{1n} + a_{n2}b_{2n} + \ldots + a_{nn}b_{nn})\alpha_n$$

$$[T_2T_1]_B = \begin{bmatrix} a_{11}b_{11} + a_{12}b_{21} + \ldots + a_{1n}b_{n1} \\ a_{11}b_{12} + a_{12}b_{22} + \ldots + a_{1n}b_{n2} \\ \ldots\ldots\ldots\ldots\ldots\ldots\ldots\ldots \\ a_{11}b_{1n} + a_{12}b_{2n} + \ldots + a_{1n}b_{nn} \end{bmatrix}$$

$$\therefore$$

$$\begin{bmatrix} a_{21}b_{11} + \ldots + a_{2n}b_{n1}, \ldots a_{n1}b_{11} + \ldots + a_{nn}b_{n1} \\ a_{21}b_{12} + \ldots + a_{2n}b_{n2}, \ldots a_{n1}b_{12} + \ldots + a_{nn}b_{n2} \\ \ldots\ldots\ldots\ldots\ldots\ldots\ldots\ldots\ldots\ldots\ldots\ldots \\ a_{21}b_{1n} + \ldots + a_{2n}b_{nn} \ldots a_{n1}b_{1n} + \ldots + a_{nn}b_{nn} \end{bmatrix}$$

$$= \begin{bmatrix} b_{11} & b_{21} & \cdots & b_{n1} \\ b_{12} & b_{22} & \cdots & b_{n2} \\ \cdots & \cdots & \cdots & \cdots \\ b_{1n} & b_{2n} & \cdots & b_{nn} \end{bmatrix} \begin{bmatrix} a_{11} & a_{21} & \cdots & a_{n1} \\ a_{12} & a_{22} & \cdots & a_{n2} \\ \cdots & \cdots & \cdots & \cdots \\ a_{1n} & a_{2n} & \cdots & a_{nn} \end{bmatrix} = [T_2]_B[T_1]_B$$

7.27 CHANGE OF BASIS

It has been shown that we can represent vectors by tuples (column vectors) and linear operators by matrix once we have selected a basis.

In this section we will see how the representation of matrix of linear transformation changes if we take another basis?

Let $\{\alpha_1, \alpha_2, \ldots, \alpha_n\}$ be a basis of V and let $\{\beta_1, \beta_2, \ldots, \beta_n\}$ be another basis of V and suppose

$$\beta_1 = a_{11}\alpha_1 + a_{12}\alpha_2 + \ldots + a_{1n}\alpha_n$$
$$\beta_2 = a_{21}\alpha_1 + a_{22}\alpha_2 + \ldots + a_{2n}\alpha_n$$

$$\ldots\ldots\ldots\ldots\ldots\ldots\ldots$$

$$\beta_n = a_{n1}\alpha_1 + a_{n2}\alpha_2 + \ldots + a_{nn}\alpha_n$$

Then the transpose of the coefficient matrix of above equation is called the transition matrix from the basis $\{\alpha_1, \alpha_2,..., \alpha_n\}$ to the basis $\{\beta_1, \beta_2,..., \beta_n\}$

$$P = \begin{bmatrix} a_{11} & a_{21} & \cdots & a_{n1} \\ a_{12} & a_{22} & \cdots & a_{n2} \\ \cdots & \cdots & \cdots & \cdots \\ a_{1n} & a_{2n} & \cdots & a_{nn} \end{bmatrix}$$

REMARK

☞ P is invertible and its P^{-1} is the transition matrix from new basis to old basis.

For example: Let $\{(1,0),(0,1)\}$ and $\{(1,1),(-1,0)\}$ be two bases of R^2, then $(1,1) = 1.(0,1) + 1(0,1)$ and $(-1,0) = -1(1,0) + 0.(0,1)$

$$\therefore \qquad P = \begin{bmatrix} 1 & -1 \\ 1 & 0 \end{bmatrix}$$

Theorem 1. *Let P be the transition matrix from a basis B to a basis B' in a vector space V. Then for any vector $a \in V$, $P[\alpha]_{B'} = [\alpha]_B$ and $[\alpha]_{B'} = P^{-1}[a]_B$.*

Proof. Let V be an n-dimensional vector space and let,

$$B = \{\alpha_1, \alpha_2,..., \alpha_n\} \text{ and } B' = \{\beta_1, \beta_2,..., \beta_n\}$$

be two bases of V and let P be the transition matrix from B to B'. Then we have,

$$\beta_1 = a_{11}\alpha_1 + a_{12}\alpha_2 + ... + a_{1n}\alpha_n$$
$$\beta_2 = a_{21}\alpha_1 + a_{22}\alpha_2 + ... + a_{2n}\alpha_n$$

$$\cdots\cdots\cdots\cdots\cdots\cdots\cdots\cdots$$

$$\beta_n = a_{n1}\alpha_1 + a_{n2}\alpha_2 + ... + a_{nn}\alpha_n; \text{ for } \alpha_{ij} \in F$$

$$\therefore \qquad P = \begin{bmatrix} a_{11} & a_{21} & \cdots & a_{n1} \\ a_{12} & a_{22} & \cdots & a_{n2} \\ \cdots & \cdots & \cdots & \cdots \\ a_{1n} & a_{2n} & \cdots & a_{nn} \end{bmatrix}.$$

Now suppose $\alpha \in V$ such that

$$\alpha = b_1\beta_1 + b_2\beta_2 + ... + b_n\beta_n.$$

Substituting β's from above, we obtain,

$$\alpha = b_1(a_{11}\alpha_1 + a_{12}\alpha_2 + ... + a_{1n}\alpha_n) + b_2(a_{21}\alpha_1 + a_{22}\alpha_2 + ... + a_{2n}\alpha_n) +$$

$$\cdots\cdots\cdots\cdots\cdots\cdots$$

$$+ b_n(a_{n1}\alpha_1 + a_{n2}\alpha_2 + ... + a_{2n}\alpha_n)$$

$$= (b_1a_{11} + b_2a_{12} + ... + b_na_{1n})\alpha_1 + (b_1a_{12} + b_2a_{22} + ... + b_na_{2n})\alpha_2$$

$$\cdots\cdots\cdots\cdots\cdots\cdots$$

$$+ (b_1a_{1n} + b_2a_{2n} + ... + b_na_{2n})\alpha_n$$

Thus, $[\alpha]_{B'} = \begin{bmatrix} b_1 \\ b_2 \\ \cdots \\ b_n \end{bmatrix}$ and $[\alpha]_B = \begin{bmatrix} b_1a_{11} + b_2a_{21} + ... + b_na_{n1} \\ b_1a_{12} + b_2a_{22} + ... + b_na_{n2} \\ \cdots\cdots\cdots\cdots\cdots\cdots\cdots \\ b_1a_{1n} + b_2a_{2n} + ... + b_na_{nn} \end{bmatrix}$

Accordingly,

$$P[\alpha]_{B'} = \begin{bmatrix} a_{11} & a_{21} & \cdots & a_{n1} \\ a_{12} & a_{22} & \cdots & a_{n2} \\ \cdots & \cdots & \cdots & \cdots \\ a_{1n} & a_{2n} & \cdots & a_{nn} \end{bmatrix} \cdot \begin{bmatrix} b_1 \\ b_2 \\ \cdots \\ b_n \end{bmatrix} = \begin{bmatrix} b_1 a_{11} + b_2 a_{21} + \cdots + b_n a_{n1} \\ b_1 a_{12} + b_2 a_{22} + \cdots + b_n a_{n2} \\ \cdots\cdots\cdots\cdots\cdots\cdots\cdots\cdots\cdots \\ b_1 a_{1n} + b_2 a_{2n} + \cdots + b_n a_{nn} \end{bmatrix}$$

$$= [\alpha]_B$$

Furthermore, since P is invertible , hence

$$P[\alpha]_{B'} = [\alpha]_B \qquad \Rightarrow \quad P^{-1}P[\alpha]_{B'} = P^{-1}[\alpha]_B$$

$$\Rightarrow \quad I[\alpha]_{B'} = P^{-1}[\alpha]_B \quad \Rightarrow \qquad [\alpha]_{B'} = P^{-1}[\alpha]_B$$

Theorem 2. *Let P be the transition matrix from a basis to a basis B' in a vector space V. Then for any linear operator T on V,*

$$[T]_{B'} = P^{-1}[T]_B P.$$

Proof. Let α be any vector in V. then we have

$$[T]_B[\alpha]_B = [T(\alpha)]_B \qquad\qquad\qquad ...(1)$$

and

$$P[\alpha]_{B'} = (\alpha)_B \qquad\qquad\qquad ...(2)$$

$$\Rightarrow \qquad [T]_B P[\alpha]_{B'} = [T]_B[\alpha]_B = [T(\alpha)]_B \qquad\qquad \text{[using (1)]}$$

$$\Rightarrow \qquad P^{-1}[T]_B P[\alpha]_{B'} = P^{-1}[T(\alpha)]_B = [T(\alpha)]_{B'} \qquad \text{[By theorem (1)]}$$

$$= [T]_{B'}[\alpha]_{B'} \qquad\qquad \text{[using (1)]}$$

$$\Rightarrow \qquad P^{-1}[T]_B P = [T]_{B'} \qquad\qquad [\because [\alpha]_{B'} \in F \text{ are arbitrary}]$$

Solved Examples

▶ A mapping $T:U \to V$ is said to be linear transformation from U into V if $T(a\alpha + b\beta) = aT(\alpha) + bT(\beta)$ $\forall\ \alpha, \beta \in U$, $a, b \in F$.

▶ Rank (T) + nulltiy (T) = dim(U).

▶ T is invertible iff it is bijective.

▶ T is non-singular if $T(\alpha) = 0 \Rightarrow \alpha = 0$.

Example 1. *Show that the mapping T defined by*

$$T(a,b) = (\alpha+\beta, \alpha-\beta, \beta),\ \forall (\alpha,\beta) \in V_2(\mathbf{R})\ \text{is a linear transformation.}$$

Solution. Obviously, T is a mapping from $V_2(\mathbf{R})$ into $V_3(\mathbf{R})$, because

$$(\alpha+\beta, \alpha-\beta, \beta) \in V_3(\mathbf{R})\ \forall (\alpha,\beta)$$

For each $a,b \in F$ and $(\alpha_1,\beta_1), (\alpha_2,\beta_2) \in V_2(\mathbf{R})$, we have

$$T[a(\alpha_1,\beta_1) + b(\alpha_2,\beta_2)] = T(a\alpha_1 + b\alpha_2,\ a\beta_1 + b\beta_2)$$

$$= [(a\alpha_1 + b\alpha_2) + (a\beta_1 + b\beta_2), (a\alpha_2 + b\alpha_2) - (a\beta_1 + b\beta_2), (a\beta_1 + b\beta_2)]$$

$$= a(\alpha_1 + \beta_1, \alpha_1 - \beta_1, \beta_1) + b(\alpha_2 + \beta_2, \alpha_2 - \beta_2, \beta_2)$$

$$= aT(\alpha_1, \beta_1) + bT(\alpha_2, \beta_2)$$

Hence, T is linear.

Example 2. *Which of the following function T from $\mathbf{R}^2$ into $\mathbf{R}^2$ are linear transformation?*

(i) $T(x_1, x_2) = (x_1^2, x_2)$

(ii) $T(x_1, x_2) = (\sin x_1, x_2)$

            (iii) $T(x_1, x_2) = (x_1{}^2 - x_2, 0)$.

Solution.   (i)   Let (x_1, x_2) and (y_1, y_2) be any two vectors in $\mathbf{R}^2$ and $a, b \in \mathbf{R}$, then

$$T[a(x_1, x_2) + b(y_1, y_2)] = T[(ax_1 + by_1), (ax_2 + by_2)]$$
$$= [(ax_1 + by_1)^2, (ax_2 + by_2)] \qquad [\because T(x_1, x_2) = (x_1{}^2, x_2)]$$
$$= [a^2 x_1{}^2 + b_2 y_1{}^2 + 2abx_1 y_1, \, ax_2 + by_2]$$
$$\neq aT(x_1, x_2) + bT(y_1 + y_2)$$

$\therefore$ T is not a linear transformation.

   (ii)   Let $(x_1, x_2), (y_1, y_2) \in \mathbf{R}^2$ be an arbitrary vector and let $a, b \in \mathbf{R}$, then

$$T[a(x_1, x_2) + b(y_1, y_2)] = T(ax_1 + by_1), (ax_2 + by_2)]$$
$$= [\sin(ax_1 + by_1), \, ax_2 + by_2] \, \{\because T\{x_1, x_2) = (\sin x_1, \, x_2)\}$$

$\therefore$ T is not a linear transformation.

   (iii)   Let $(x_1, x_2), (y_1, y_2) \in \mathbf{R}^2$ be an arbitrary vector and let $a, b \in \mathbf{R}$, then

$$T[a(x_1, x_2) + b(y_1, y_2)] = T[(ax_1 + by_1), (ax_2 + by_2)]$$
$$= (ax_1 + by_1) - (ax_2 + by_2), 0] \qquad [\because T(x_1, x_2) = (x_1 - x_2, 0)]$$
$$= [a(x_1 - x_2) + b(y_1 - y_2), 0]$$
$$= aT(x_1, x_2) + bT(y_1, y_2)$$

Hence, T is a linear transformation.

Example 3.   *Show that the mapping* $T : \mathbf{R}^3 \to \mathbf{R}^2$, *defined by*

$$T(\alpha, \beta, \gamma) = (\alpha, \beta), \, \forall (\alpha, \beta, \gamma) \in \mathbf{R}^3$$

is a homomorphism (linear transformation) of the vector space $R^3(\mathbf{R})$ *onto* $R^2(\mathbf{R})$.

Solution.   Let $(\alpha_1, \beta_1, \gamma_1), (\alpha_2, \beta_2, \gamma_2) \in R^3(\mathbf{R})$ be an arbitrary vector and let $a, b \in \mathbf{R}$, then

$$T[a(\alpha_1, \beta_1, \gamma_1) + b(\alpha_2, \beta_2, \gamma_2)]$$
$$= T[(a\alpha_1 + b\alpha_2, a\beta_1 + b\beta_2, \, a\gamma_1 + b\gamma_2]$$
$$= (a\alpha_1 + b\alpha_2, \, a\beta_1 + b\beta_2) \qquad [\because T(\alpha, \beta, \gamma) = (\alpha, \beta)]$$
$$= a(\alpha_1, \beta_1) + b(\alpha_2, \beta_2) = aT(\alpha_1, \beta_1, \gamma_1) + bT(\alpha_2, \beta_2, \gamma_2)$$

Thus, T is a homomorphism.

Further, if $(\alpha, \beta) \in \mathbf{R}^2$, then $(\alpha, \beta, \gamma) \in \mathbf{R}^3$ and we have $T(\alpha, \beta, \gamma) = (\alpha, \beta)$.

Hence, T is onto.

Example 4.   *Let F be the field of complex numbers and let T be the function from* $\mathbf{R}^3$ *onto* $\mathbf{R}^3$ *defined by* $T(a_1, a_2, a_3) = (a_1 - a_2 + 2a_3, 2a_1 + a_2 - a_3, -a_1 - 2a_2)$. *Verify that T is a linear transformation. Describe the null space of T.*

Solution.   Let $\alpha = (a_1, a_2, a_3)$ and $\beta = (b_1, b_2, b_3)$ be any two vectors in $\mathbf{R}^3$ and $a, b \in \mathbf{R}$, then

$$a\alpha + b\beta = a(a_1, a_2, a_3) + b(b_1, b_2, b_3)$$
$$= (aa_1, aa_2, aa_3) + (bb_1, bb_2, bb_3)$$
$$= (aa_1 + bb_1, aa_2 + bb_2, aa_3 + bb_3)$$

Now, $T(a\alpha + b\beta) = T(aa_1 + bb_1, aa_2 + bb_2, aa_3 + bb_3)$

$$= [(aa_1 + bb_1) - (aa_2 + bb_2) + 2(aa_3 + bb_3), 2(aa_1 + bb_1) + (aa_2 + bb_2)$$
$$\qquad\qquad\qquad - (aa_3 + bb_3), -(aa_1 + bb_1) - 2(aa_2 + bb_2)]$$

$$= [a(a_1 - a_2 + 2a_3) + b(b_1 - b_2 + 2b_3), a(2a_1 + a_2 - a_3)$$
$$\qquad\qquad\qquad + b(2b_1 + b_2 - b_3), a(-a_1 - 2a_2) + b(-b_1 - 2b_2)]$$

$$= [a(a_1-a_2+2a_3, 2a_1+a_2-a_3,-a_1-2a_2)$$
$$+b(b_1-b_2+2b_3,2b_1+b_2-b_3,-b_1-2b_2)]$$

Thus, T is a linear transformation.

Next, By the definition of null space of T, we have

$$N_T = \{\alpha \in \mathbf{R}^3 : T(\alpha) = 0 = (0,0,0)\}$$

Let $\qquad \alpha = (a_1,a_2,a_3) \in \mathbf{R}^3$

$\therefore \qquad T(\alpha) = T(a_1,a_2,a_3) = (0,0,0)$

$\Rightarrow \quad (a_1-a_2+2a_3, 2a_1+a_2-a_3, -a_1-2a_2) = (0,0,0)$

$$\Rightarrow \quad \left.\begin{array}{l} a_1 - a_2 + 2a_3 = 0 \\ 2a_1 + a_2 - a_3 = 0 \\ -a_1 - 2a_2 = 0 \end{array}\right\} \qquad \dots(1)$$

Now, we find the solution of system of equation (1).

Let A be the coefficient matrx of (1), then we have,

$$A = \begin{bmatrix} 1 & -1 & 2 \\ 2 & 1 & -1 \\ -1 & -2 & 0 \end{bmatrix}$$

Performing $R_2 \rightarrow R_2 - 2R_1$, $R_3 \rightarrow R_3 + R_1$, we get

$$A = \begin{bmatrix} 1 & -1 & 2 \\ 0 & 3 & -5 \\ 0 & -3 & 2 \end{bmatrix}$$

Performing $R_3 \rightarrow R_3 + R_2$, we get

$$\sim \begin{bmatrix} 1 & -1 & 2 \\ 0 & 3 & -5 \\ 0 & 0 & -3 \end{bmatrix}$$

This matrix is in Echelon form and having three non-zero rows, thus its rank=3, which is equal to the number of unknowns. Hence, the system of equations (1) has only trival solution,*i.e.*, $a_1 = 0$, $a_2 = 0$, $a_3 = 0$. Consequently, $N_T = \{(0, 0, 0)\}$.

Example 5. *Describe explicitly the linear transformation $T : \mathbf{R}^2 \rightarrow \mathbf{R}^3$ such that $T(2, 3) = (4, 5)$ and $T(1, 0) = (0, 0)$.*

Solution : Let $\alpha = (2, 3)$, $\beta = (1,0)$ and let $a,b \in \mathbf{R}$, then

$$a\alpha + b\beta = 0(0,0) \qquad\qquad \Rightarrow a(2,3) + b(1,0) = (0,0)$$

$\Rightarrow \qquad (2a+b, 3a) = (0,0)$

$\Rightarrow \qquad 2a + b = 0$

$\qquad\qquad 3a = 0$

$\Rightarrow \qquad a = 0, b = 0$

$\Rightarrow$ the set $\{\alpha,\beta\} = \{(2,3),(1,0)\}$ is linearly independent.

Also, dim. $\mathbf{R}^2 = 2$, thus $\{(2, 3), (1, 0)\}$ forms a basis of $\mathbf{R}^2$.

Let (x, y) be any element of $\mathbf{R}^2$ and for some scalars p and q in $\mathbf{R}$, we have

$$(x,y) = p\alpha + q\beta = p(2,3) + q(1,0) = (2p+q, 3p)$$

$$\Rightarrow \qquad x = 2p+q, y = 3p \Rightarrow p = \frac{y}{3}, q = \frac{3x-2y}{3}$$

$$\therefore \qquad (x, y) = \frac{y}{3}(2,3) + \frac{3x - 2y}{3}(1,0)$$

Now, $$T(x, y) = T\left[\frac{y}{3}(2,3) + \frac{3x - 2y}{3}(1,0)\right]$$

$$= \frac{y}{3}T(2,3) + \frac{3x - 2y}{3}T(1,0) \qquad\qquad [\because T \text{ is linear}]$$

$$= \frac{y}{3}(4,5) + \frac{3x - 2y}{3}(0,0) = \left(\frac{4y}{3}, \frac{5y}{3}\right).$$

Example 6. *If a map $T : V_2(\mathbf{R}) \to V_3(\mathbf{R})$ defined by $T(a, b) = (a+b, a-b, b)$ is a linear transformation. Find the range, rank, null-space and nullity of T.*

Solution. (i) *Determination of range of T, i.e., R_T and rank :*

Since the ordered set $\{(1, 0), (0, 1)\}$ forms a basis of $V_2(\mathbf{R})$. Then by definition of T, we have,

$$T(1,0) = (1+0, 1-0, 0) = (1,1,0)$$
and $$T(0,1) = (0+1, 0-1, 1) = (1,-1,1).$$

Since $(1,0), (0,1)$ generates $V_2(\mathbf{R})$. Therefore

$T(1,0), T(0,1)$ will generate $T(V_2(\mathbf{R})) = R_T$

$\Rightarrow \qquad (1,1,0), (1,-1,1)$ generates R_T

Also, for some scalars $a, b \in \mathbf{R}$, such that

$$a\,(1,1,0) + b\,(1,-1,1) = (0,0,0)$$
$\Rightarrow \qquad\qquad (a+b, a-b, b) = (0,0,0)$
$\Rightarrow \qquad\qquad a+b = 0,\ a-b = 0,\ b = 0$
$\Rightarrow \qquad\qquad a = 0,\ b = 0.$

$\therefore \qquad \{(1,1,0), (1, -1, 1)\}$ is linearly independent and spans R_T, so it forms a basis of R_T. Hence, dim. $R_T = 2$.

(ii) *Determination of null space and nullity of T.*

Since T is a linear transformation from $V_2(\mathbf{R})$ into $V_3(\mathbf{R})$. Therefore,

$$\text{dim. } R_T + \text{dim. } N_T = \text{dim. } V_2(\mathbf{R})$$
$$2 + \text{dim. } N_T = 2 \ \Rightarrow\ \text{dim. } N_T = 0$$

Thus, nullity of $T = 0$.

Since dim. $N_T = 0$

$\Rightarrow$ Null space of T, i.e., N_T is a zero space.

$\Rightarrow \qquad\qquad N_T = \{(0, 0)\}.$

Example 7. *Let T be a linear operator on a vector space $V(F)$. If $T_2 = 0$, what can you say about the relation of the range of T to the null space of T? Give an example of a linear operator on $V_2(\mathbf{R})$ such that $T^2 = 0$ but $T \neq 0$.*

Solution. Since $T^2 = 0$, then for $\alpha \in V$

$$T^2(\alpha) = \mathbf{0}(\alpha)$$
$\Rightarrow \qquad\qquad T[T(\alpha)] = 0$

$\Rightarrow$ $\qquad\qquad\qquad$ $T(\alpha) \in N_T$ $\qquad\qquad$ [By definition of null space]

But $T(\alpha) \in R_T \; \forall \; \alpha \in V$

$\therefore$ $\qquad\qquad\qquad\qquad$ $R_T \subset N_T$

Hence, when $T^2 = 0$, the range of T is contained in null space of T.

Next, let T be a linear map from $V_2(\mathbf{R})$ into $V_2(\mathbf{R})$ such that

$$T(a,b) = (0,a), \forall \; (a,b) \in V_2(\mathbf{R})$$

Obviously, $T \neq 0$.

Also, $\qquad\qquad$ $T^2(a, b) = T[T(a, b)] = T[(0, a)] = \{0,0\} = 0 \; (a, b)$

$\Rightarrow$ $\qquad\qquad\qquad\qquad$ $T^2 = 0$.

Example 8. *Find a linear transformation $T : R^2 \to R^2$ such that $T \; (1, \; 0) = (1,1)$ and $T(0, \; 1) = (-1, \; 2)$. Prove that T maps the square with vertices $(0, \; 0)$, $(1, \; 0)$, $(1, \; 1)$ and $(0,1)$ into a parallelogram.*

Solution. Since the ordered set $\{\{1, 0), (0, 1)\}$ forms a basis of R^2, so that for some scalars p and q in $\mathbf{R}$ and $(x, y) \in R^2$ such that

$$(x,y) = p(1,0) + q(0,1) \qquad\qquad ....(1)$$

$\Rightarrow$ $\qquad\qquad$ $T(x,y) = T[p(1,0) + q(0,1)]$

$$= pT(1,0) + qT(0,1) = p(1,1) + q(-1,2) = (p-q, p + 2q) \; ....(2)$$

From (1), we have

$$(x. y) = (p,q) \qquad \Rightarrow \qquad p = x, q = y.$$

From (2), we get

$$T(x, y) = (x-y, x + 2y). \qquad\qquad ...(3)$$

This is the required linear transformation.

Next, let A, B, C and D be the vertices of a square with $A \; (0, 0), B \; (1, 0), C \; (1, 1)$ and $D \; (0, 1)$ and let P, Q, R and S be the T-images of A, B, C and D respectively. Then we have,

$$P = T(A) = T \; (0, 0) = (0, 0) \qquad\qquad \text{[using (3)]}$$
$$Q = T(B) = T(1,0) = (1,1) \qquad\qquad \text{[using (3)]}$$
$$R = T(C) = T(1,1) = (0,3) \qquad\qquad \text{[using (3)]}$$
and $\qquad$ $S = T(D) = T(0,1) = (-1, 2) \qquad\qquad \text{[using (3)]}$

Now, $PQ = $ Distance between $(0, 0)$ and $(1, 1) = \sqrt{(1-0)^2 + (1-0)^2} = \sqrt{2}$

$PS = $ Distance between $(0, 0)$ and $(-1, 2) = \sqrt{(-1-0)^2 + (2-0)^2} = \sqrt{1+4} = \sqrt{5}$

$RS = $ Distance between $(0, 3)$ and $(-1, 2) = \sqrt{(-1-0)^2 + (2-3)^2} = \sqrt{1+1} = \sqrt{2}$

and

$QR = $ Distance between $(1, 1)$ and $(0, 3) = \sqrt{(0-1)^2 + (3-1)^2} = \sqrt{1+4} = \sqrt{5}$

Hence, $PQRS$ is a parallelogram.

Example 9. *Let $T:V_3(\mathbf{R}) \to V_3(\mathbf{R})$ be a linear transformation defined by*

$$T(a,b,c) = (3a, a-b, 2a + b + c), \; \forall \; a,b,c \in \mathbf{R}.$$

Prove that T is invertible and then find T^{-1}. Also prove that $(T^2 - I)(T - 3I) = 0$.

Solution. Let $\alpha = (a_1, b_1, c_1)$ and $\beta = (a_2, b_2, c_2)$ be two vectors in $V_3(\mathbf{R})$. Suppose that

$$T(\alpha) = T(\beta) \Rightarrow T(a_1, b_1, c_1) = T(a_2, b_2, c_2)$$

$$\Rightarrow (3a_1, a_1 - b_1, 2a_1 + b_1 + c_1) = (3a_2, a_2 - b_2, 2a_2 + b_2 + c_2)$$

$$\Rightarrow \left. \begin{array}{r} 3a_1 = 3a_2 \\ a_1 - b_1 = a_2 - b_2 \\ 2a_1 + b_1 + c_1 = 2a_2 + b_2 + c_2 \end{array} \right\}$$

$$\Rightarrow \qquad a_1 = a_2, \, b_1 = b_2 \text{ and } c_1 = c_2 \Rightarrow \alpha = \beta.$$

Thus, T is one-one.

Since, V_3 $(\mathbf{R})$ is a 3-dimensional vector space, then T is onto also. Thus, T is one-one onto mapping. Hence T is invertible.

Determination of T^{-1} .

Let $\qquad T(a, b, c) = (p, q, r)$, then $T^{-1}(p, q, r) = (a, b, c)$

$$T(a, b, c) = (p, q, r) \Rightarrow (3a, a - b, 2a + b + c) = (p, q, r)$$

$$\Rightarrow \qquad \begin{cases} p = 3a \\ q = a - b \\ r = 2a + b + c \end{cases}$$

$$\Rightarrow \qquad a = \frac{p}{3}, \, b = \frac{p}{3} - q, c = r - p + q$$

$$\therefore \qquad T^{-1}(p, q, r) = \left(\frac{p}{3}, \frac{p}{3} - q, r - p + q \right)$$

To prove that $(T^2 - I)(T - 3I) = 0$

$$(T - 3I)(a, b, c) = T(a, b, c) - 3I(a, b, c)$$

$$= T(a, b, c) - 3(a, b, c) \quad [\because I \text{ is identity transformation.}]$$

$$= (3a, a - b, 2a + b + c) - 3(a, b, c)$$

$$= (3a, a - b, 2a + b + c) - (3a, 3b, 3c)$$

$$= (0, a - 4b, 2a + b - 2c)$$

$$[(T^2 - I)(T - 3I)](a, b, c) = (T^2 - I)[(T - 3I)(a, b, c)]$$

$$= (T^2 - I)(0, a - 4b, 2a + b - 2c)$$

$$= T^2(0, a - 4b, 2a + b - 2c) - I(0, a - 4b, 2a + b - 2c)$$

$$= T[T(0, a - 4b, 2a + b - 2c)] - (0, a - 4b, 2a + b - 2c)$$

$$= T[T(0, -a + 4b, 0 + a - 4b + 2a + b - 2c)]$$

$$\qquad\qquad - (0, a - 4b, 2a + b - 2c)$$

$$= T[0, -a + 4b, 3a - 3b - 2c)] - (0, a - 4b, 2a + b - 2c)$$

$$= (0, 0 - (-a + 4b), 0 + (-a + 4b) + (3a - 3b - 2c)$$

$$\qquad\qquad - (0, a - 4b, 2a + b - 2c)$$

$$= (0, a - 4b, 2a + b - 2c) - (0, a - 4b, 2a + b - 2c)$$

$$= (0, 0, 0) = 0 = 0 (a, b, c)$$

$$\Rightarrow \qquad (T^2 - I)(T - 3I) = 0.$$

Example 10. *If T is a linear transformation on a vector space V such that $T^2 - T + I = 0$, then show that T is invertible.*

Proof.   Since $T^2 - T + I = 0$, then

$$T^2 = T - I$$

For every $\alpha_i \in V$, we have

$$T^2(\alpha_i) = (T - I)(\alpha_i) \implies T[T(\alpha_i)] = T(\alpha_i) - I(\alpha_i)$$

$$\implies \qquad T[T(\alpha_i)] = T(\alpha_i) - \alpha_i.$$

Now, for some $\beta_i \in V$, such that $T(\alpha_i) = \beta_i$

$$\implies \qquad T(\beta_i) = \beta_i - \alpha_1 \qquad \qquad ...(1)$$

To show that T is one-one :

For $\beta_1, \beta_2 \in V$, suppose that

$$T(\beta_1) = T(\beta_2) \implies \beta_1 - \alpha_1 = \beta_2 - \alpha_1 \qquad \text{[using (1)]}$$

$$\implies \qquad \beta_1 = \beta_2$$

$\therefore$ T is one-one.

To show T is onto :

For every $\beta_i \in V$, there exists $\beta_i - \alpha_i \in V$ such that $T(\beta_i) = \beta_i - \alpha_i$.

Thus T is also onto. Hence, T is one-one and onto. Hence, T is invertible.

Example 11. *Lei $T : R^2(\mathbf{R}) \to R^2(\mathbf{R})$, where for any $(x,y) \in R^2, T(x, y) = \left(2x, \dfrac{1}{2}y\right)$. Find the matrix associated with T w.r.t. the ordered basis $\{(1,0), (0,1)\}$.*

Solution.   Let $B = \{(1,0), (0,1)\}$ be an ordered basis of $R^2(\mathbf{R})$ and $T(x, y) = \left(2x, \dfrac{1}{2}y\right)$, then

$$T(1, 0) = (2, 0) \text{ and } T(0, 1) = \left(0, \dfrac{1}{2}\right)$$

Now, $T(1,0) = (2,0) = 2(1,0) + 0(0,1) \text{ and } T(0, 1) = \left(0, \dfrac{1}{2}\right) = 0(1,0) + \dfrac{1}{2}(0,1)$

Hence, the matrix associated with T w.r.t., B is $[T]_B = \begin{bmatrix} 2 & 0 \\ 0 & \dfrac{1}{2} \end{bmatrix}$

Example 12. *Find the matrix representation of a linear map $T : R^3 \to R^3$ defined by $T(x,y,z) = (z, y+z, x+y+z)$ relative to the basis $\{(1,0,1),(-1,2,1),(2,1,1)\}$.*

Solution.   Suppose $B = \{(1,0,1),(-1,2,1),(2,1,1)\}$ is the basis of R^3 and $T(x, y, z) = (z, y+z, x+y+z)$ then

$$\left. \begin{array}{l} T(1,0,1) = (1,1,2) \\ T(-1,2,1) = (1,3,0) \\ T(2,1,1) = (1,2,4) \end{array} \right\} \qquad \qquad ...(1)$$

For some $a, b, c \in \mathbf{R}$ and $(x, y, z) \in R^3$ we have

$$(x, y, z) = a(1,0,1) + b(-1,2,1) + c(2,1,1) = (a - b + 2c, 2b + c, a + b + c)$$

$\Rightarrow$ $x = a - b + 2c,\ y = 2b + c,\ z = a + b + c$

$\Rightarrow$ $a = \dfrac{1}{4}(-x - 3y + 5z),\ b = \dfrac{1}{4}(-x + y + z),\ c = \dfrac{1}{4}(2x + 2y - 2z)$

So $(x, y, z) = \dfrac{1}{4}(-x - 3y + 5z)(1,0,1) + \dfrac{1}{4}(-x + y + z)(-1,2,1)$

$$+ \dfrac{1}{4}(2x + 2y - 2z)(2,1,1) \quad\dots(2)$$

Putting $x = 1, y = 1, z = 1$ in (2) and using (1), we get

$$T(1,0,\,1) = (1,1,2) = \dfrac{3}{2}(1,0,1) + \dfrac{1}{2}(-1,2,1) + 0(2,1,1) \qquad\dots(3)$$

Putting $x = 1, y = 3, z = 0$ in (2) and using (1), we get

$$T(-1,2,1) = (1,3,0) = 0.(1,0,1) + 1.(-1,2,1) + 1.(2,1,1) \qquad\dots(4)$$

Putting $x = 1, y = 2, z = 4$ in (2) and using (1), we get

$$T(2,\,1,\,1) = (1,\,2,\,4) = \dfrac{13}{4}(1,0,1) + \dfrac{5}{4}(-1,2,1) - \dfrac{1}{2}(2,1,1) \qquad\dots(5)$$

Now, the matrix of coefficients of equations (3), (4) and (5) is,

$$\begin{bmatrix} \dfrac{3}{2} & \dfrac{1}{2} & 0 \\ 0 & 1 & 1 \\ \dfrac{13}{4} & \dfrac{5}{4} & -\dfrac{1}{2} \end{bmatrix}$$

Thus, the matrix representation of T relative to B is the transpose of above matrix:

$$[T]_B = \begin{bmatrix} \dfrac{3}{2} & 0 & \dfrac{13}{4} \\ \dfrac{1}{2} & 1 & \dfrac{5}{4} \\ 0 & 1 & -\dfrac{1}{2} \end{bmatrix}$$

Example 13. *Let T be linear operator in R^3 defined by*

$$T(x_1, x_2, x_3) = (3x_1 + x_3, -2x_1 + x_2, -x_1 + 2x_2 + 4x_3)$$

Find the matrix of T in the ordered basis $\{\alpha_1, \alpha_2, \alpha_3\}$, where

$$\alpha_1 = (1,0,1),\ \alpha_2 = (-1,2,1)\ and\ \alpha_3 = (2,1,1)$$

Solution.   Suppose $B = \{\alpha_1, \alpha_2, \alpha_3\}$ is the basis of R^3 where, $\alpha_1 = (1,0,1)$, $\alpha_2 = (-1,2,1)$ and $\alpha_3 = (2,1,1)$ also $T : R^3 \to R^3$ defined by

$$T(x_1, x_2, x_3) = (3x_1 + x_3, -2x_1 + x_2, -x_1 + 2x_2 + 4x_3)$$

Then, we get

$$\left.\begin{array}{l} T(\alpha_1) = T(1,0,1) = (4,-2,3) \\ T(\alpha_2) = T(-1,2,1) = (-2,4,9) \\ T(\alpha_3) = T(2,1,1) = (7,-3,4) \end{array}\right\} \qquad\dots(1)$$

Let $(x, y, z) = a\alpha_1 + b\alpha_2 + c\alpha_3$ for some $a, b, c \in \mathbf{R}$. Then

$(x, y, z) = a(1, 0, 1) + b(-1, 2,1) + c(2,1,1) = (a - b + 2c,\ 2b + c,\ a + b + c)$

$\Rightarrow$ $x = a - b + 2c,\ y = 2b + c,\ z = a + b + c$

$\Rightarrow \qquad a = \dfrac{1}{4}(-x-3y+5z),\ b = \dfrac{1}{4}(-x+y+z),\ c = \dfrac{1}{4}(2x+2y-2z)$

$\therefore \qquad (x,y,z) = \dfrac{1}{4}(-x-3y+5z)\alpha_1 + \dfrac{1}{4}(-x+y+z)\alpha_2 + \dfrac{1}{4}(2x+2y-2z)\alpha_3$ ...(2)

Putting $x = 4, y = -2, z = 3$ in (2) and using (1), we get

$$T(\alpha_1) = (4,-2,3) = \dfrac{17}{4}\alpha_1 - \dfrac{3}{4}\alpha_2 - \dfrac{1}{2}\alpha_3 \qquad\qquad ...(3)$$

Putting $x = -2, y = 4, z = 9$ in using (1), we get

$$T(\alpha_2) = (-2,4,9) = \dfrac{35}{4}\alpha_1 + \dfrac{15}{4}\alpha_2 - \dfrac{7}{2}\alpha_3 \qquad\qquad ...(4)$$

Putting $x = 7, y = -3, z = 4$ in (2) and using (1), we get

$$T(\alpha_3) = (7,-3,4) = \dfrac{11}{2}\alpha_1 - \dfrac{3}{2}\alpha_2 + 0.\alpha_3 \qquad\qquad ...(5)$$

Now, the coefficient matrix of the system of equation (3),(4) and (5) is

$$\begin{bmatrix} 17/4 & -3/4 & 11/2 \\ 35/4 & 15/4 & -7/2 \\ 11/2 & -3/2 & 0 \end{bmatrix}$$

Thus, the matrix of T relative to B is obtained by taking the transpose of above coefficeint matrix :

$$[T]_B = \begin{bmatrix} 17/4 & 35/4 & 11/2 \\ -3/4 & 15/4 & -3/2 \\ -1/2 & -7/2 & 0 \end{bmatrix}.$$

Example 14. *If the matrix of a linear transformation T on a vector space $V_2(\mathbf{C})$ w.r.t. the ordered basis $B = \{(1, 0, (0, -1)\}$ is* $\begin{bmatrix} 1 & 1 \\ 1 & 1 \end{bmatrix}$ *what is the matrix w.r.t. the ordered basis $B' = \{(1, 1), (1, -1)\}$.*

Solution. Since $B = \{(1, 0, (0, -1)\}$ and $B' = \{(1,1), (1, -1)\}$ and

$$[T]_B = \begin{bmatrix} 1 & 1 \\ 1 & 1 \end{bmatrix}$$

Determination of $[T]_{B'}$.

Let $\qquad\qquad \alpha_1 = (1,0), \alpha_2 = (0,-1)$

Since, $\qquad [T]_B = \begin{bmatrix} 1 & 1 \\ 1 & 1 \end{bmatrix}$ is the matrix of T w.r.t. B, then

$$T(\alpha_1) = T(1,0) = 1.\alpha_1 + 1.\alpha_2$$
$$= 1.(1,0) + 1.(0,-1) = (1,0) + (0,-1) = (1,-1)$$
$$T(\alpha_2) = T(0,-1) = 1.(\alpha_1) + 1.\alpha_2$$
$$= 1.(1,0) + 1.(0,-1) = (1,-1)$$

If $(a, b) \in V_2(\mathbf{C})$, then we can write, for some $p, q \in \mathbf{C}$

$\Rightarrow \qquad\qquad (a,b) = p(1, 0) + q(0,-1) = (p,-q)$

$\Rightarrow$ $\qquad\qquad\qquad\qquad p = a, q = -b$

Now, $\qquad\qquad T(a,b) = T(p\alpha_1 + q\alpha_2) = pT(\alpha_1) + qT(\alpha_2)$

$\therefore$ $\qquad\qquad\quad T(a,b) = (a-b, a+b).$ ...(1)

Further, since $\quad B' = \{(1,1), (1, -1)\}$ is another basis of $V_2(\mathbf{C})$, let

$$\beta_1 = (1,1), \beta_2 = (1,-1)$$

$T(\beta_1) = T(1,1) = (0,0)$ $\qquad\qquad\qquad$ [using (1)]

$T(\beta_2) = T(1,-1) = (0,0)$ $\qquad\qquad\qquad$ [using (2)] ...(2)

Let $(x,y) \in V_2(\mathbf{C})$ such that

$$(x, y) = p_1\beta_1 + q_1\beta_2, \text{ for some } p_1, q_1 \in C$$
$$= p_1(1,1) + q_1(1,-1) = (p_1 + q_1, p_1 - q_1)$$

$\Rightarrow$ $\qquad\qquad x = p_1 + q_1, y = p_1 - q_1 \Rightarrow \qquad p_1 = \dfrac{x+y}{2}, q_1 = \dfrac{x-y}{2}$

$\therefore$ $\qquad (x,y) = \left(\dfrac{x+y}{2}\right)\beta_1 + \begin{bmatrix} 0 & 0 \\ 0 & 2 \end{bmatrix} \beta_2 = \dfrac{x+y}{2}(1,1) + \dfrac{x-y}{2}(1,-1)$ ...(3)

Putting $x = 0, y = 0$ in (3) and using (2), we get

$$T(1,1) = (0, 0) = 0.(1,1) + 0.(1,-1)$$ ...(4)

Putting $x = 2, y = -2$ in (3) and using (2), we get

$$T(1, -1) = (2, -2) = 0.(1, 1) + 2(1, -1)$$ ...(5)

Now, coefficient matrix of the system of equations (4) and (5) is

$$\begin{bmatrix} 0 & 0 \\ 0 & 2 \end{bmatrix}$$

Thus, the matrix of T relative to B' is the transpose of above coefficient matrix :

$$[T]_B = \begin{bmatrix} 0 & 0 \\ 0 & 2 \end{bmatrix}$$

Example 15. *Let $B = \{(1,0), (0,1)\}$ and $B' = \{(1,2), (2,3)\}$ be any two bases of R^2*

(i) Find the transition matrices P from B to B'

(ii) Verify that $[\alpha]_B = P^{-1}[\alpha]_{B'}, \forall \alpha \in R^2$

(iii) Verify that $P^{-1}[T]_B P = [T]_{B'}$ where $T(x, y) = (2x - 3y, x + y).$

Solution. Let $\alpha_1 = (1, 0), \alpha_2 = (0, 1), \alpha'_1 = (1, 2)$ and $\alpha'_2 = (2, 3).$

Since $T(x, y) = (2x - 3y, x + y)$

(i) $\qquad\qquad\qquad \alpha'_1 = (1,2) = 1.(1,0) + 2(0,1) = 1.\alpha_1 + 2\alpha_2$ ...(1)

and $\qquad\qquad\quad \alpha'_2 = (2,3) = 2.(1,0) + 3(0,1) = 2.\alpha_1 + 3\alpha_2$ ...(2)

The coefficient matrix of the system of equation (1) and (2) is given by

$$\begin{bmatrix} 1 & 2 \\ 2 & 3 \end{bmatrix}.$$

$\therefore$ The transition matrix P is the transpose of the coefficient matrix is given by

$$P = \begin{bmatrix} 1 & 2 \\ 2 & 3 \end{bmatrix}.$$

Let α be an arbitrary element of R^2, then
$$\alpha = a_1\alpha_1 + a_2\alpha_2, \text{ for some } a_1, a_2 \in \mathbf{R}.$$
$$= a_1(1, 0) + a_2(0, 1) = (a, a)$$

If $\alpha = \{x, y\}; x, y \in \mathbf{R}$, then
$$(x, y) = (a_1, a_2) \Rightarrow \quad x = a_1, y = a_2$$
$$\therefore \qquad (x, y) = xa_1 + ya_2$$

Thus, $[\alpha]_B = \begin{bmatrix} x \\ y \end{bmatrix}$, which is the co-ordinate vector of a w.r.t. B.

Again, $\qquad (x, y) = a'_1\alpha'_1 + a'_2\alpha'_2$
$$= a'_1(1, 2) + a'_2(2, 3) = (a'_1 + 2a'_2,\ 2a'_1 + 3a'_2)$$
$$\Rightarrow \qquad x = a'_1 + 2a'_2,\ y = 2a'_1 + 3a'_2$$
$$\Rightarrow \qquad a'_1 = 2y - 3x,\ a'_2 = 2x - y.$$
$$\therefore \qquad (x, y) = (2y - 3x)a'_1 + (2x - y)a'_2 \qquad \qquad \qquad ...(3)$$

Thus, $\qquad [\alpha]_{B'} = \begin{bmatrix} 2y - 3x \\ 2x - y \end{bmatrix}$

$$\therefore \qquad P[\alpha]_{B'} = \begin{bmatrix} 1 & 2 \\ 2 & 3 \end{bmatrix} \begin{bmatrix} 2y - 3x \\ 2x - y \end{bmatrix}$$

$$= \begin{bmatrix} 1(2y - 3x) + 2(2x - y) \\ 2(2y - 3x) + 3(2x - y) \end{bmatrix} = \begin{bmatrix} x \\ y \end{bmatrix} = [\alpha]_B$$

(ii) Since $\qquad P = \begin{bmatrix} 1 & 2 \\ 2 & 3 \end{bmatrix}$

$$|P| = (1 \times 3 - 2 \times 2) = -1.$$

Matrix of co-factors of element of P is
$$\begin{bmatrix} 3 & -2 \\ -2 & 1 \end{bmatrix} \text{ so, } adj\ P = \begin{bmatrix} 3 & -2 \\ -2 & 1 \end{bmatrix}$$

$$\therefore \qquad P^{-1} = \frac{1}{P}\begin{bmatrix} 3 & -2 \\ -2 & 1 \end{bmatrix}$$

$$= \frac{1}{P}\begin{bmatrix} 3 & -2 \\ -2 & 1 \end{bmatrix} = -1\begin{bmatrix} 3 & -2 \\ -2 & 1 \end{bmatrix} = \begin{bmatrix} -3 & 2 \\ 2 & -1 \end{bmatrix}$$

Since $\qquad T(x, y) = (2x - 3y, x + y)$
$$\therefore \qquad T(\alpha_1) = T(1,0) = (2,1) = 2(1,0) + 1.(0, 1) = 2\alpha_1 + \alpha_2 \qquad ...(4)$$
$$T(\alpha_2) = T(0,1) = (-3,1) = -3(1,0) + 1.(0, 1) = -3\alpha_1 + \alpha_2 \qquad ...(5)$$
$\therefore$ $[T]_B$ = Transpose of the coefficient matrix of the system of equations (4) and (5)
$$= \begin{bmatrix} 2 & -3 \\ 1 & 1 \end{bmatrix}$$

Also, $\qquad \begin{aligned} T(\alpha'_1) &= T(1,2) = (-4,3) \\ T(\alpha'_2) &= T(2,3) = (-5,5) \end{aligned} \Big\} \qquad \qquad ...(6)$

From (3), we have

$$(x, y) = (2x - 3x)\, \alpha'_1 + (2x - y)\, \alpha'_2 \qquad \qquad ...(7)$$

Putting $x = -4, y = 3$ in (7) and using (6), we get

$$T(\alpha'_1) = (-4, 3) = 18\alpha'_1 - 11\alpha'_2 \qquad \qquad ...(8)$$

Putting $x = -5, y = 5$ in (7) and using (6), we get

$$T(\alpha'_2\} = (-5, 5) = 25\alpha'_1 - 15\alpha'_2 \qquad \qquad ...(9)$$

$\therefore$ $[T]_{B'}$ = Transpose of coefficient matrix of the system of equations (8) and (9)

$$= \begin{bmatrix} 18 & -11 \\ 25 & -15 \end{bmatrix} = \begin{bmatrix} 18 & 25 \\ -11 & -15 \end{bmatrix}$$

So, $\qquad P^{-1}[T]_B P = \begin{bmatrix} -3 & 2 \\ 2 & -1 \end{bmatrix} \begin{bmatrix} 2 & -3 \\ 1 & 1 \end{bmatrix} \begin{bmatrix} 1 & 2 \\ 2 & 3 \end{bmatrix}$

$$= \begin{bmatrix} -4 & 11 \\ 3 & -7 \end{bmatrix} \begin{bmatrix} 1 & 2 \\ 2 & 3 \end{bmatrix} = \begin{bmatrix} 18 & 25 \\ -11 & -15 \end{bmatrix} = [T]_{B'}$$

Example 6. *If T_1, T_2 and T_3 are linear transformations on a vector space $V(F)$ such that $T_1 T_2 = T_3 T_1 = I$, then T_1 is invertible and $T_1^{-1} = T_2 = T_3$.*

Solution. First we shall show that T_1 is one-one :

For $\alpha, \beta \in V$, suppose that

$$T_1(\alpha) = T_1(\beta) \Rightarrow T_3\,(T_1(\alpha)) = T_3(T_1(\beta))$$

$\Rightarrow \qquad (T_3 T_1)(\alpha) = (T_3 T_1)(\beta) \Rightarrow \qquad I(\alpha) = I(\beta) \qquad [\because T_3 T_1 = I]$

$\Rightarrow \qquad \alpha = \beta.$

Thus, T_1 is one-one.

Secondly, we shall show that T_1 is onto : Let $\beta \in V$ be an arbitrary vector.

Since $T_2 : V \to V,$ then $T_2\,(\beta) \in V$.

Let us take $\qquad T_2(\beta) = \alpha \in V$

Now, $\qquad T_2(\beta) = \alpha \qquad \Rightarrow T_1(T_2(\beta)) = T_1(\alpha)$

$\Rightarrow \qquad (T_1 T_2)(\beta) = T_1(\alpha) \Rightarrow I(\beta) = T_1(\alpha) \qquad [\because T_1 T_2 = 1]$

$\therefore$ For any β, there exists $\alpha \in V$ such that $T_1\,(\alpha) = \beta$. Hence T_1 is onto.

Thus, T_1 is one-one and onto and hence T_1 is invertible.

Next, since $\qquad T_1 T_2 = T_3 T_1 = I$

$\therefore \qquad \qquad T_1 T_2 = I \qquad \Rightarrow T_1^{-1}(T_1)T_2 = T_1^{-1}I$

$\Rightarrow \qquad (T_1^{-1}T_1)T_2 = T_1^{-1} \qquad \Rightarrow \qquad \qquad IT_2 = T_1^{-1}I$

$\Rightarrow \qquad \qquad T_2 = T_1^{-1}$

Also, $\qquad \qquad T_3 T_1 = I \qquad \Rightarrow (T_3 T_1)T_1^{-1} = I\,T_1^{-1}$

$\Rightarrow \qquad T_3(T_1 T_1^{-1}) = T_1^{-1} \qquad \Rightarrow \qquad \qquad T_3 I = T_1^{-1} \Rightarrow T_3 = T_1^{-1}.$

Hence, $\qquad \qquad T_1^{-1} = T_2 = T_3.$

Example 7. *Let T be an invertible linear operator on a vector space $V(F)$. Then show that*

(i) aT is also an invertible linear operator, when $a \neq 0$ and $a \in F$

(ii) $(aT)^{-1} = \left(\dfrac{1}{a}\right)T^{-1}$, where $a \neq 0$ and $a \in F.$

(iii) T^{-1} is invertible and $(T^{-1})^{-1}=T$.

Solution. Since T is invertible so that it is one-one and onto.

(i) Let $\alpha, \beta \in V$ and $0 \neq a \in F$ and suppose that

$$(aT)(\alpha) = (aT)(\beta) \qquad \Rightarrow a(T(\alpha)) = a(T(\beta))$$
$$\Rightarrow \qquad T(\alpha)=T(\beta) \qquad \Rightarrow \alpha = \beta \qquad [\because T \text{ is one-one.}]$$

$\therefore \qquad aT$ is one-one.

Let β be any vector in V, then there exists a vector α in V such that

$$T(\alpha) = \beta \qquad\qquad [\because T \text{ is onto.}]$$
$$\Rightarrow \qquad a(T)(\alpha)=a(\beta) \qquad \Rightarrow (aT)(\alpha) = a\beta$$
$$\Rightarrow \qquad (aT)(\alpha) = \gamma \quad \text{for some } \gamma = a\beta \in V.$$

$\therefore \qquad aT$ is onto.

Thus, aT is one-one and onto and hence aT is invertible.

(ii) Consider,

$$(aT)\left(\frac{1}{a}T^{-1}\right) = a.\frac{1}{a}\left(TT^{-1}\right)$$

$$= 1.(TT^{-1})=I \qquad\qquad [\because T \text{ is invertible.}]$$

Also, $\qquad \left(\frac{1}{a}T^{-1}\right)(aT)= I$

Hence, $\qquad (aT)^{-1}= \left(\frac{1}{a}\right)T^{-1}$

(iii) Since $\qquad T^{-1}T=TT^{-1}=I$, this implies that inverse of T^{-1} is T,

i.e., $\qquad (T^{-1})^{-1}=T$.

$\therefore \qquad (T^{-1})^{-1}T^{-1}=T^{-1}(T^{-1})^{-1}=I$

$\Rightarrow \qquad T^{-1}$ exists.

Hence, T^{-1} exists and $(T^{-1})^{-1} =T$.

EXERCISE 7.3

1. Show that the following mappings are linear:
   (i) $T: \mathbf{R}^2 \to \mathbf{R}^2$ defined by $T(x, y) = (2x-y, x)$.
   (ii) $T: \mathbf{R}^3 \to \mathbf{R}^2$ defined by
   $T(x, y, z) = (z, x+y)$.
   (iii) $T: \mathbf{R} \to \mathbf{R}^2$ defined by $T(x) = (2x, 3x)$.
   (iv) $T: \mathbf{R}^2 \to \mathbf{R}^2$ defined by
   $T(x, y) = (ax+by, cx+dy)$
   where $a, b, c, d \in \mathbf{R}$.

2. Show that the following mappings are linear:
   (i) $T: \mathbf{R}^2 \to \mathbf{R}^2$ defined by $T(x, y) = (x^2, y^2)$.
   (ii) $T: \mathbf{R}^3 \to \mathbf{R}^2$ defined by
   $T(x, y, z) = (x+1, y+z)$.
   (iii) $T: \mathbf{R}^2 \to \mathbf{R}^2$ defined by $T(x, y) = |x-y|$.

3. Show that the map $T : \mathbf{R}^2 \to \mathbf{R}^3$ defined by $T(a,b) = (a-b, b-a, -a)$ is linear transformation. Find the range, rank, null space and nullity of T.

4. Let $F : \mathbf{R}^3 \to \mathbf{R}^2$ be a map given by $F(a, b, c) = (a, b), \forall (a, b, c) \in \mathbf{R}^3$. Prove that F is a (homomorphism) linear tranformation. Also, find the kernel of F (null space of F).

5. Give an exmple of a linear tranformation T on $V_3(\mathbf{R})$ such that $T \neq 0$, $T^2 \neq 0$ but $T^3= 0$ where 0 is the zero tranformation.

6. Let there be a linear operator on $\mathbf{R}^3$ given by $T(x, y, z)= (2x, 2x-5y, 2y + z)$. Find T^{-1}.

7. Let $T : \mathbf{R}^3 \to \mathbf{R}^2$ be a linear map defind by $T(x, y, z) = (x+ 2y-z, y+z, x+y-2z)$. Find basis and dimension of (i) range of T (ii) null space of T.

8. Prove that a linear transformation T on a finite-dimensional vector space $V(F)$ is invertible if and only if T is nonsingular.

9. Describe explicitly a linear transformation from $V_3(\mathbf{R})$ into $V_3(\mathbf{R})$ which has its range subspace spanned by $(1, 0, -1)$ and $(1, 2, 2)$.

10. Let T be the linear oeprator on $\mathbf{R}^2$ defined by $T(x, y) = (4x-2y, 2x + y)$. Compute the matrix of T relative to the basis $\{\alpha_1, \alpha_2\}$, where $\alpha_1 = (1, 1)$, $\alpha_2 = (-1, 0)$.

11. Let $V = R^3$ and $T : V \to V$ be a linear map defined by $T(x, y, z) = (x+z, -2x + y, -x+2y+z)$. What is matrix of T w.r.t. basis $B = \{(1, 0, 1), (-1, 1, 1), (0, 1, 1)\}$?

12. If the matrix of a linear transformation on $V_3(\mathbf{C})$ with respect to the basis $B = \{(1, 0\ 0), (0, 1, 0), (0, 0, 1)\}$ is
$$\begin{bmatrix} 0 & 1 & 1 \\ 1 & 0 & -1 \\ -1 & -1 & 0 \end{bmatrix}.$$

What is the matrix of T w.r.t. basis $B' = \{(0,1,-1), (1, -1, 1), (-1, 0, 1)$ and $B' = (1, 1, -1), (-1, 0, 1) (1, 2, 1)\}$

13. Find the matirx representation of the linear mappings relative to the usual basis $B = \{(1, 0, ... 0), (0, 1, ...,1), (0, 0,..., 1)\}$.

For $\mathbf{R}^n$
(i) $T : \mathbf{R}^3 \to \mathbf{R}^2$ defined by $T(x, y, z)$
$$= (2x-4y + 9z, 5x + 3y-2z)$$
(ii) $T : \mathbf{R} \to \mathbf{R}^2$ defined by $T(x) = (3x, 5x)$.

14. Let T be a linear operator on $\mathbf{R}^3$ defined by $T(x, y) = (2y, 3x -y)$. Find the matrix representation of T relative to the basis $B = \{(1, 3), (2, 5)\}$

15. Let F be a linear operator on $\mathbf{R}^3$ defined by $F(x, y, z) = (2y+z, x-4y, 3x)$
(i) Find the matrix of F in the basis $B' = \{(1, 1, 1), (1, 1, 0), (1, 0, 0)\}$
(ii) Verify that
$$[F]_B = [F(\alpha)]_{B'} = [F(\alpha)]_{B'}, \forall \alpha \in \mathbf{R}^3.$$

16. Let $B = \{(1, 0, 0), (0, 1, 0), (0, 0, 1)\}$ and $B' = \{(1, 1, 1), (1, 1, 0), (1, 0, 0)\}$ be two basis of $\mathbf{R}^3$
(i) Show that $P[\alpha]_{B'} = [\alpha]_{B'} \forall \alpha \in \mathbf{R}^3$.
(ii) Show that $[T]_{B'} = P^{-1}[T]_B P$, where $T(x, y, z) = (2y+z, x-4y, 3x)$.

17. Find two linear tranformations T and S on a vector space $\mathbf{R}^2(\mathbf{R})$ such that $TS=0$ but $ST \neq 0$.

1. (i) Let $\alpha = (x_1, y_1, z_1)$, $\beta = (x_2, y_2, z_2) \in \mathbf{R}^3$ and $a, b \in \mathbf{R}$.
Then $a\alpha + b\beta = a(x_1, y_1, z_1) + b(x_2, y_2, z_2)$
$= (ax_1 + bx_2, ay_1 + by_2, az_1 + bz_2)$
$\therefore T(a\alpha + b\beta)$
$= T(ax_1 + bx_2, ay_1 + by_2, az_1 + bz_2)$
$= (az_1 + bz_2, ax_1 + bx_2 + ay_1 + by_2)$
$= \{az_1 + bz_2, a(x_1 + y_1) + b(x_2 + y_2)]$
$= \{az_1, a(x_1 + y_1)\} + \{bz_2, b(x_2 + y_2)\}$
$= a(z_1, x_1 + y_1) + b(z_2, x_2 + y_2)$
$= aT(\alpha) + bT(\beta)$
Hence T is linear.

2. (iii) Let $\alpha = (x_1, y_1,)$, $\beta = (x_2, y_2) \in \mathbf{R}^2$ and $a, b \in \mathbf{R}$.
Then $a\alpha + b\beta = a(x_1, y_1) + b(x_2, y_2)$
$= (ax_1+bx_2, ay_1+by_2)$
$\therefore T(a\alpha + b\beta) = T(ax_1 + bx_2, ay_1 + by_2)$
$= |(ax_1+bx_2) - ay_1 + by_2)|$
$= |(a(x_1-y_1)+b(x_2-y_2)|$
$\leq |a(x_1-y_1)| + |b(x_2-y_2)|$
(by triangular inequality)
$\leq aT(\alpha+bT(\beta)$.
$\therefore T(a\alpha + b\beta) \neq aT(\alpha)+bT(\beta)$.
Hence T is not linear.

3. $B = \{(1, 0) (0, 1)\}$ be the basis of $\mathbf{R}^2$. Then $T(1, 0) = (1, -1, -1)$ and $T(0, 1) = (-1, 1, 0)$. Range space is generated by the set $B_1 = \{(1, -1, -1), (-1, 1, 0)\}$.
Clerly B_1 is linearly independent.
Thus rank $(T) = 2$.
For Null space.
$N(T) = \{\alpha \in R^2 : T(\alpha) = 0\}$
$= \{(x, y) \in R^2 : T(x, y) = 0\}$
Now, $T(x, y) = (x-y, y-x, -x) = (0, 0, 0)$
$\Rightarrow \quad x-y = 0, y-x=0, -x= 0$
$\Rightarrow \quad x = 0, y = 0, -x = 0$
$\therefore \quad$ Null space of T = $\{0\}$ and nullity $(T) = 1$.

4. Let $\alpha = (a_1, b_1, c_1)$, $\beta = (a_2, b_2, c_2)$ and $a, b, c \in \mathbf{R}$.
Then
$a\alpha + b\beta = (aa_1 + ba_2, ab_1 + bb_2, ac_1 + bc_2)$
$\therefore F(a\alpha+\beta) = (aa_1 + ba_2, ab_1+bb_2)$
$= (aa_1 + ab_1) + (ba_1 + bb_2)$
$= a(a_1 + b_1) + b(a_2 + b_2)$
$= aF(a_1, b_1, c_1) + bF(a_2, b_2, c_2)$
$= aF(\alpha) + bF(\beta)$
$\therefore F$ is linear.
Now Ker $F = \{\alpha \in \mathbf{R}^3 : F(\alpha) = \mathbf{0}\}$ $\{(a, b, c) \in R^3 : F(a, b, c) = \mathbf{0}\}$
$= \{(a, b, c)\} \in R^3 : (a, b) = (0, 0)\}$

$$= \{(a, b, c)\} \in R^3 : a = 0, b = 0\}$$
$$= \{(0, 0, c) : c \in \mathbf{R}\}.$$

6. Since $T(x, y, z,) = (2x, 2x - 5y, 2y + z)$

$$(x, y, z) = T^{-1}(2x, 2x - 5y, 2y + z)$$

Let $p = 2x$, $q = 2x - 5y$, $r = 2y + z$

so that $x = \dfrac{p}{2}$, $y = \dfrac{p-q}{5}$ and $z = \dfrac{5r - 2p + 2q}{5}$

Then $T^{-1}(p, q, r) = \left(\dfrac{p}{2}, \dfrac{p-q}{5}, \dfrac{5r - 2p + 2q}{5} \right)$

9. Since range space is generated by $(1, 0, -1)$
and $(1, 2, 2)$.
So, we take $T(1, 0, 0) = (1, 0, -1)$,
$T(0, 1, 0) = (1, 2, 2)$ and $T(0, 0, 1) = (0, 0, 0)$.
Let $(x, y, z) \in V_3 (\mathbf{R})$, then

$(x, y, z) = x(1, 0, 0) + y(0, 1, 0) + z(0,0,1)$
$T(x, y, z) = xT(1,0,0) + yT(0, 1, 0) + zT(0,0,1)$
$\qquad = x(1, 0, -1) + y(1, 2, 2) + z(0, 0, 0)$
$\qquad = (x+y, 2y, -x+2y).$

10. $T : \mathbf{R}^2 \to \mathbf{R}^2$ given by $T(x, y) = (4x - 2y, 2x + y)$
$T(\alpha_1) = T(1,1) = (2,3) = a(1,1) + b(-1, 0);$
$T(\alpha_2) = T(-1,0) = (-4,-2) = c(1, 1) + d(-1, 0).$
$\therefore a - b = 2$, $a = 3$ and $c - d = -4$. $c = -2$,
i.e., $a = 3$, $b = 1$ and $c = -2$, $d = 2$

Hence, $[T]_{(\alpha_1, \alpha_2)} = \begin{bmatrix} 3 & -2 \\ 1 & 2 \end{bmatrix}$

12. $B = (1, 0, 0), (0, 1, 0), (0, 0, 1)$

and $[T]_{B'} = \begin{bmatrix} 1 & 0 & 0 \\ 0 & 0 & 0 \\ 0 & 0 & -1 \end{bmatrix}$

$\therefore T(1, 0, 0) = (0, 1, -1);$ $T(0,1,0) = (1,0,-1);$
$T(0, 0, 1) = (1, -1, 0).$
Let $(x, y, z) \in V_3(\mathbf{C})$ Then
$(x, y, z) = x(1, 0, 0) + y(0, 1, 0) + z(0, 0, 1).$
$\therefore T(x, y, z) = x T(1, 0, 0) + yT(0, 1, 0)$
$\qquad\qquad + zT(0, 0, 1)$
$\qquad = x(0, 1, -1) + yT(1, 0, -1) + z(1, -1, 0)$
$\qquad = (y + z, x - z, -x - y).$
Now for $B_1 = \{(0, 1, -1, (1, -1, 1), (-1, 0, 1)\}$

$T(0, 1, -1) = (0. 1, -1) = a_1(0, 1, -1)$
$\qquad + b_1 (1, -1, 1) + c_1 (-1, 0, 1)$
$T(1, -1, 1) = (0, 0, 0) = a_2(0, 1, -1)$
$\qquad + b_1 (1, -1, 1) + c_2 (-1, 0, 1)$
$\therefore \quad b_1 - c_1 = 0$, $a_1 - b_1 = 1$, $-a_1 + b_1 + c_1 = -1$
$\qquad b_2 - c_2 = 0$, $a_2 - b_2 = 0$, $-a_2 + b_2 + c_2 = 0$
and $\quad b_3 - c_3 = 1$, $a_3 - b_3 = -2$, $-a_3 + b_3 + c_3 = 1$.
Solving these equations, we get $a_1 = 1$, $b_1 = 0$,
$c_1 = 0$; $a_2 = 1$, $b_2 = 0$, $c_2 = 0$; $a_3 = 0$, $b_3 = 0$,
$c_3 = -1$.

$\therefore \quad [T]_{B'} = \begin{bmatrix} 1 & 0 & 0 \\ 0 & 0 & 0 \\ 0 & 0 & -1 \end{bmatrix}$

15 (ii) From (i)

$$[F]_{B'} = \begin{bmatrix} 3 & 3 & 3 \\ -6 & -6 & -2 \\ 6 & 5 & -1 \end{bmatrix}$$

Let $a \in \mathbf{R}^3$ so taking $\alpha = (a, b, c)$
$\therefore \quad (a, b, c) = p_1 (1, 1, 1) + q_1 (1, 1, 0)$
$\qquad\qquad + r_1 (1, 0, 0)$
$\Rightarrow \qquad p_1 = c$, $q_1 = b - c$, $r_1 = a - b$.

$\therefore \quad [\alpha]_{B'} = \begin{bmatrix} p_1 \\ q_1 \\ r_1 \end{bmatrix} = \begin{bmatrix} c \\ b-c \\ a-b \end{bmatrix}$

$F(\alpha) = F(a, b, c) = (2b+c, a - 4b, 3a)$
$\qquad = x(1, 1, 1) + y(1, 1, 0) + z(1, 0, 0)$
$\therefore \quad x = 3a$, $y = -2a - 4b$, $z = -a + 6b + c$

So that $F[\alpha]_{B'} = \begin{bmatrix} x \\ y \\ z \end{bmatrix} = \begin{bmatrix} 3a \\ -2a - 4b \\ -a + 6b + c \end{bmatrix}$

Now

$[F]_{B'}[\alpha]_{B'} = \begin{bmatrix} 3 & 3 & 3 \\ -6 & -6 & -2 \\ 6 & 5 & -1 \end{bmatrix} \begin{bmatrix} c \\ b-c \\ a-b \end{bmatrix}$

$\qquad = \begin{bmatrix} 3a \\ -2a - 4b \\ -a + 6b + c \end{bmatrix} = [F(\alpha)]_{B'}$

Answers

5. Kernel of $F = \{(0, 0, 0) \in R^3 : T(0, 0, c) = (0, 0)\}$

5. $T : V_3 (\mathbf{R}) \to V_3(\mathbf{R})$ defined by $T(a, b, c) = (0, a, b)$ such that $T \neq 0$, $T^2 \neq 0$ and $T^3 \neq 0$.

6. $T^{-1}(x, y, z) = \left[\dfrac{x}{2}, \dfrac{x - y}{5}, \dfrac{-2x + 2y + 5z}{5} \right]$

7. (i) $\{(1, 0, 1), (2, 1, 1)\}$ is a basis of R_T and dim. $R_T = 2$

(ii) $\{(3, -1, 1\}$ is a basis of N_T and dim. $N_T = 1$.

9. $T(a,b,c) = (a+b, 2b, 2b-a)$

10. Matrix of T relative to $\{\alpha_1, \alpha_2\}$ is $\begin{bmatrix} 3 & -2 \\ 1 & 2 \end{bmatrix}$
11. $[T]_B = \begin{bmatrix} 2 & 1 & 2 \\ 0 & 1 & 1 \\ -2 & 2 & 0 \end{bmatrix}$

12. $[T]_B = \begin{bmatrix} 1 & 0 & 0 \\ 0 & 0 & 0 \\ 0 & 0 & -1 \end{bmatrix}$
13. (i) $[T]_B = \begin{bmatrix} 2 & -4 & 9 \\ 5 & 3 & -2 \end{bmatrix}$
(ii) $[T]_B = \begin{bmatrix} 3 \\ 5 \end{bmatrix}$

14. $[T]_B = \begin{bmatrix} -30 & -48 \\ 18 & 29 \end{bmatrix}$
15. $[T]_{B'} = \begin{bmatrix} 3 & 3 & 3 \\ -6 & -6 & -2 \\ 6 & 5 & -1 \end{bmatrix}$
16. (i) $P = \begin{bmatrix} 1 & 1 & 1 \\ 1 & 1 & 0 \\ 1 & 0 & 0 \end{bmatrix}$

17. $T(a, b) = (2a, 0)$ and $S(a, b) = (0, 2a)$

7.28 LINEAR FUNCTIONAL

We shall study the linear mappings from a vector space V into its field F of scalars. Naturally all the theorems and results for arbitrary linear mappings on V hold for this special case. However, such type of mappings are treated separately because of their fundamental importance and because of the relationship between V and F, but all the theorems and results do not apply in the general case.

Definition. *Let V be a vector space over a field F. Then a linear transformation $\phi : V \to F$ is called a linear functional (or linear form) if for every $\alpha, \beta \in V$ and every $a, b \in F$.*

$$\phi(a\alpha + b\beta) = a\phi(\alpha) + b\phi(\beta).$$

For example:

(1) Let V be the vector space of polynomials in t over **R**. Then, the integral operator $\phi : V \to \mathbf{R}$ defined by

$$\phi[p(t)] = \int_0^1 p(t)\,dt \text{ is a linear functional.}$$

(2) Let V be the vector space of n-square matrices over the field F. Let $T : V \to F$ be the trace mapping defined by $T(A) = a_{11} + a_{12} + \ldots + a_{1n}$ where $A = [a_{ij}]$. This mapping is a linear functional.

(3) If V is a vector over the field F, then a mapping $T : V \to F$ defined by

$$T(\alpha) = 0, \ \forall \ a \in V$$

is a linear functional. This functional is also known as zero functional.

7.29 DUAL SPACES

Definition. *The set of linear functionals on a vector space V over the field F is also a vector space over F with addition and scalar multiplication defined by,*

(i) $(\phi_1 + \phi_2)(\alpha) = \phi_1(\alpha) + \phi_2(\alpha), \ \forall \ \alpha \in V$

(ii) $(c\phi)(\alpha) = c\phi(\alpha), \ c \in F, \ \alpha \in V$

where ϕ_1, ϕ_2, ϕ are linear functionals on V.

This space is called the dual space (conjugate space) of V and is denoted by V^*. We also write $V^* = L(V, F)$.

Theorem 1. *Let V be a finite-dimensional vector space over the field F. Then, dim, V* = dim. V.*

Proof. Let V be n-dimensional vector space. Let V^* be its dual space, so that

$$\text{dim. } V = n, L(V, F) = V^*, \text{dim. } F = 1.$$

Since we know that

$$\text{dim. } L(U, V) = (\text{dim. } U) (\text{dim. } V)$$

$$\Rightarrow \qquad \text{dim. } L(V, F) = (\text{dim.}V) (\text{dim. } F)$$

$$\Rightarrow \qquad \text{dim. } V^* = (\text{dim.}V) .1 \qquad \Rightarrow \quad \text{dim. } V^* = \text{dim. } V.$$

Theorem 2. *Let V be a finite dimensional vector space and a ≠ 0 in V, then there is an element f∈V* such that f(a)≠ 0.*

Proof. Let V be n-dimensional vector space over the field F and let $\alpha \neq 0$ be arbitrary non-zero vector of V and let $\{\alpha_1, \alpha_2,.., \alpha_n\}$ be the basis of V. Then there exists unique scalars $a_i \in F$ such that

$$\alpha = a_1\alpha_1 + a_2\alpha_2 + ... + a_n\alpha_n = \sum_{j=1}^{n} a_j\alpha_j$$

Suppose $\{\phi_1, \phi_2,..., \phi_n\}$ is dual basis of V^*, then

$$\phi_i(\alpha_j) = \begin{cases} 1, & i = j \\ 0, & i \neq j \end{cases} = \delta_{ij}.$$

Now,

$$\phi_i(\alpha) = \phi_i\left(\sum_{j=1}^{n} a_j\alpha_j \right)$$

$$= \sum_{j=1}^{n} a_j\phi_i(\alpha_j) = \sum_{j=1}^{n} a_j\delta_{ij} = a_i$$

$$\Rightarrow \qquad \phi_1(\alpha) = a_1, \phi_2(\alpha) = a_2,.., \phi_n(\alpha) = a_n$$

Since all the scalars a_i are not zero, this implies that there exists a linear functional $\phi \in V^*$ such that $\phi(\alpha) \neq 0$.

7.30 DUAL BASIS

Let $B = \{\alpha_1, \alpha_2,..., \alpha_n\}$ be a basis for V, then there exists a unique linear functional f on V for each i such that

$$\phi_i(\alpha_j) = \begin{cases} 1, & \text{if } i = j \\ 0, & \text{if } i \neq j \end{cases} = \delta_{ij} \text{(Kronecker delta)}$$

Thus we obtain from B, a set of n distinct linear functionals $\phi_1, \phi_2 ..., \phi_n$ on V. These functionals are also linearly independent and generate V^*.

Therefore, the set $B^* = \{\phi_1, \phi_2,.., \phi_n\}$ foms a basis for V^*. This basis is called dual basis of B.

Theorem 1. *If $\{\alpha_1, \alpha_2,..., \alpha_n\}$ is a basis of a vector space V over the field F. Let $\phi_1, \phi_2,..., \phi_n \in V^*$ be the linear functionals defined by*

$$\phi_i(\alpha_j) = \delta_{ij} = \begin{cases} 1, & \text{if } i = j \\ 0, & \text{if } i \neq j \end{cases}$$

Then $\{\phi_1, \phi_2,..., \phi_n\}$ is a basis of V^*

Proof. First we show that $\{\phi_1, \phi_2 ,..., \phi_n\}$ is linarly independent.

For $a_i \in F$ such that $a_1\phi_1 + a_2\phi_2 ,..., a_n\phi_n = 0$.

Applying both sides to α_1, we get

$$(a_1\phi_1 + a_2\phi_2 ,..., a_n\phi_n)(\alpha_1) = 0(\alpha_1) = 0$$

$$\Rightarrow \quad (a_1\phi_1(\alpha_1) + a_2\phi_2(\alpha_1) ,..., a_n\phi_n(\alpha_1) = 0$$

$$\Rightarrow \quad a_1.1 + a_2.0 +...+ a_n.0 \quad \Rightarrow \quad a_1 = 0$$

Similarly, for $i = 2, 3, ...,n$ we have

$$a_1\phi_1(\alpha_i) + a_2\phi_2(\alpha_i) +...+ a_i\phi_i(\alpha_i) + ...+ a_n\phi_n(\alpha_i) = 0 \Rightarrow a_i = 0. \text{ Thus}$$

$$a_1 = 0, a_1 = 0, ... a_n = 0. \text{ Hence } \{\phi_1, \phi_2 ,..., \phi_n\} \text{ is linearly independent.}$$

Now we show that $\{\phi_i\}$ spans V^*.

For this, let ϕ be an arbitrary element of V^* and suppose that

$$\phi(\alpha_1) = c_1, \phi(\alpha_2) = c_2 ,..., \phi(\alpha_n) = c_n$$

and set

$$\psi = c_1\phi_1 + c_2\phi_2 + ... + c_n\phi_n \text{ , then}$$

$$\psi(\alpha_1) = c_1\phi_1(\alpha_1) + c_2\phi_2(\alpha_1) + ... + c_n\phi_n(\alpha_1) = c_1.$$

Similarly, for $i = 2, 3, ..., n$. We have

$$\psi(\alpha_i) = c_i.$$

Thus, $\phi(\alpha_i) = \psi(\alpha_i)$ for $i = 1, 2, 3, ..., n$. Since ϕ and ψ both agree on the basis vectors of V.

$$\therefore \qquad\qquad \phi = \psi = c_1\phi_1 + c_2\phi_2 + ... + c_n\phi_n$$

$$\Rightarrow \quad \{\phi_1, \phi_2 ,..., \phi_n\} \text{ spans } V^*.$$

Hence, $\{\phi_1, \phi_2 ,..., \phi_n\}$ forms a basis of V^*.

Theorem 2. *Let $B = \{\alpha_1, \alpha_2,..., \alpha_n\}$ be a basis of V and let $B^* = \{\phi_1, \phi_2 ,..., \phi_n\}$ be the dual basis of V^*. Then, for any vector $\alpha \in V$, such that*

$$\alpha = \phi(\alpha)\alpha_1 + \phi(\alpha)\alpha_2 +...+ \phi_n(\alpha)\alpha_n$$

and for any linear functional $\phi \in V^*$, $\phi = \phi(\alpha)\phi_1 + \phi(\alpha)\phi_2 +...+ \phi_n(\alpha)\phi_n$.

Proof. Since $B = \{\alpha_1, \alpha_2,..., \alpha_n\}$ is a basis for V and $B^* = \{\phi_1, \phi_2 ,..., \phi_n\}$ is a basis for V^*, then there is a unique ϕ_j for each j such that

$$\phi_j(\alpha_i) = \delta_{ij} \qquad\qquad\qquad ... (1)$$

Since $\{\phi_1, \phi_2 ,..., \phi_n\}$ generates V^*, then for some scalar $a_i \in F$ and for $\phi \in V^*$ such that

$$\phi = a_1\phi_1 + a_2\phi_2 + ... + a_n\phi_n = \sum_{j=1}^{n} a_j\phi_j \qquad\qquad ...(2)$$

Then

$$\phi(\alpha_i) = \sum_{j=1}^{n} a_j\phi_j(\alpha_i) = \sum_{j=1}^{n} a_j\delta_{ij} \qquad\qquad \text{[using (1)]}$$

$$\phi(\alpha_i) = a_i, i = 1, 2,...,n.$$

Thus (2) becomes

$$\phi = \phi(\alpha_1)\phi_1 + \phi(\alpha_2)\phi_2 +...+ \phi(\alpha_n)\phi_n.$$

Similarly, $B = (\alpha_1, \alpha_2,..., \alpha_n)$ generates V, then for $\alpha \in V$ some scalars $b_i \in F$ such that

$$\alpha = b_1\alpha_1 + b_2\alpha_2 +...+ b_n\alpha_n = \sum_{i=1}^{n} b_i\alpha_j \qquad\qquad .... (3)$$

Then, $$\phi_j(\alpha) = \sum_{i=1}^{n} b_i \alpha_j(\alpha_i) = \sum_{i=1}^{n} b_i \delta_{ij}$$ [using (1)]

$$=b_j \quad \text{for } j = 1, 2, ..., n.$$

Hence, (3) becomes

$$\alpha = \phi_1(\alpha)\alpha_1 + \phi_2(\alpha)\alpha_2 + ... + \phi_n(\alpha)\alpha_n.$$

7.31 SECOND DUAL SPACE : BIDUAL SPACE

It has been shown that every vector space V has a dual space V^* which consists of all the linear functionals on V. Therefore, V^* itself has a dual space V^{**}, this dual space of V^* is called the second dual of V.

Furthermore, $\quad \text{dim. } V = \text{dim. } V^* = \text{dim. } V^{**}.$

7.32 NATURAL MAPPING

Definition. *Let V be an n-dimensional vector space, V^* its dual and V^{**} the dual of V^*. Then a mapping $\upsilon \rightarrow \bar{\upsilon}, \upsilon \in V, \bar{\upsilon} \in V^{**}$, where, $\bar{\upsilon}$ is a linear functional on V^* defined by $\bar{\upsilon}(\phi) = \phi(\upsilon), \phi \in V^*$ is an isomorphism. This mapping is called a natural mapping.*

REMARK

☞ If V is not finite dimensional, then the natural mapping can never be onto V^{**}. However, it is always linear and one-one.

7.33 ANNIHILATOR

Definition. *Let V be a vector space over the field F and V^* its dual. Let W be a subset of V which is not necessarily a subspace. Then a linear functional $\phi \in V^*$ is called an annihilator of W if $\phi(\alpha) = 0$ for every $\alpha \in W$, which is denoted by W^0.*

That is, the set of all linear functional ϕ on V such that $\phi(\alpha)=0, \forall \alpha \in V$, i.e., $\phi(W)=\{0\}$ is called annihilator of W.

Also, $\quad W^0 = \{\phi \in V^*: \phi(\alpha) = 0, \forall \alpha \in V\}$

REMARKS

☞ Annihilator of V is the zero functional on V.

☞ $\{0\}^0 = V^*$.

Theorem 1. W^0 *is a subspace of V^*.*

Proof. By definition of W^0, it is clear that $0 \in W^\circ$ and $W^\circ \subseteq V^*$.

Now suppose $\phi_1, \phi_2, \in W^0$ and for any scalars $a, b \in F$ and for any $a \in W$.

$$(a\phi_1 + b\phi_2)(\alpha) = a\phi_1(\alpha) + b\phi_2(\alpha) = a.0 + b.0 \quad [\because \phi_1, \phi_2 \in W^0]$$

$$= 0$$

$\Rightarrow \quad a\phi_1 + b\phi_2 \in W^0$

Hence, W^0 is a subspace of V^*.

Theorem 2. *Let V be a finite-dimensional vector space over the field F and let W be a subspace of V. Then $\text{dim.} W + \text{dim.} W^0 = \text{dim.} V$.*

Proof. Let V be n-dimensional vector space over the field F. Let Dim. $W = m$.

Let W be a subspace of V. Then W^0 is a subspace of V^*(dual of V).

Since W is a subspace of V so that

$$\dim . W \le \dim . V, \quad i.e., \quad m \le n$$

Let $\{a_1, a_2, ..., a_n\}$ be a basis of W so it can be extended to form a basis of V, therefore choose vectors $\{\alpha_{m+1}, \alpha_{m+2}, ... , \alpha_n\}$ in V such that

$$B = \{\alpha_1, \alpha_2, ..., \alpha_m, \alpha_{m+1}, ... , \alpha_n)$$

is a basis of V. Let $\{\phi_1, \phi_2, .., \phi_n\}$ be the basis of V^* which is the dual of B.
Now we claim that $\{\phi_{m+1}, \phi_{m+2}, ..., \phi_n)$ is a basis of W^0. Obviously, $\phi_i \in W^0$ for $i \ge m+1$, because

$$\phi_i(\alpha_j) = \delta_{ij} = \begin{cases} 0, & i \ne j \\ 1, & i = j \end{cases}$$

and $\qquad \qquad \delta_{ij} = 0$ if $i \ge m+1$ and $j \le m$.

Since $\{\phi_{m+1}, \phi_{m+2}, ..., \phi_n\}$ is a subset of linearly independent set $\{\phi_1, \phi_2, ..., \phi_n\}$ hence $\{\phi_{m+1}, \phi_{m+2}, ..., \phi_n\}$ is linearly independent. Now we shall show that $\{\phi_{m+1}, \phi_{m+2}, ..., \phi_n\}$ spans W^0.
Let $\phi \in W^0$ be an arbitrary linear functional, so that

$$\phi(\alpha_i) = 0, \text{ for } 1 \le i \le m \qquad \qquad ...(1)$$

Since $W^0 \subseteq V^*$, then $\phi \in V^*$
But $\{\phi_1, \phi_2, ..., \phi_n\}$ generates V^*, therefore, we have

$$\phi = \sum_{i=1}^{n} \phi(\alpha_i)\phi_i$$

$\qquad \qquad \qquad \qquad \qquad \qquad \qquad \qquad \qquad$ [By theorem]

$$= \phi(\alpha_1)\phi_1 + \phi(\alpha_2)\phi_2 + ... + \phi(\alpha_m)\phi_m + \phi(\alpha_{m+1})\phi_{m+1}$$
$$+ \phi(\alpha_{m+2})\phi_{m+2} + ... + \phi(\alpha_n)\phi_n$$

$$= \phi(\alpha_{m+1})\phi_{m+1} + \phi(\alpha_{m+2})\phi_{m+2} + ... + \phi(\alpha_n)\phi_n = \sum_{i=1}^{n} \phi(\alpha_i)\phi_i$$

This shows that $\{\phi_{m+1}, \phi_{m+2}, ..., \phi_n\}$ spans W^0.
Thus $\{\phi_{m+1}, \phi_{m+2}, ..., \phi_n\}$ forms a basis of W^0.
Accordingly, $\qquad \qquad \dim. W^0 = n - m = \dim. V - \dim. W$
Hence, $\qquad \dim. W + \dim. W^0 = \dim. V$.

Corollary. *If W and W_1 are two subspaces of a vector space V which are annihilated by the subspace W^0, then dim. $W = $ dim. W_1.*

Proof. Since W and W_1 are both annihilated by the W^0, and both are subspaces of V, then by above theorem we have

$$\dim.W + \dim.W^0 = \dim. V \qquad \qquad ....(1)$$

and $\qquad \qquad \dim.W_1 + \dim.W^0 = \dim. V \qquad \qquad ....(2)$

On using (1) and (2), we get dim. $W = $ dim.W_1

Theorem 3. *Let W_1 and W_2 be subspaces of a finite dimensional vector space over the field F, then $W_1 = W_2$ if and only if $W_1^0 = W_2^0$,*

Proof. If $W_1 = W_2$, then obviously, $W_1^0 = W_2^0$. Conversely, if $W_1^0 = W_2^0$, then we can show that $W_1 = W_2$.
Let if possible, $W_1 \ne W_2$, then there is at least one vector in W_1 which is not in W_2.
Suppose $\alpha \in W_2$ and $\alpha \ne W_1$. then there is a linear functional ϕ such that $\phi(\beta) = 0$

$\forall \beta \in W$ but $\phi(\alpha) \neq 0$. This implies that $\phi \in W_1^0$ but $\phi \notin W_2^0$ and thus $W_1^0 \neq W_2^0$.

Hence if $W_1^0 = W_2^0$, then $W_1 = W_2$.

7.34 ANNIHILATOR OF AN ANNIHILATOR

Definition. *Let V be a vector space over the field F. V* its dual space and let V** be the dual space of V*.*

Let W be a subset of a vector space $V(F)$. Then W^0 is a subspace of V^*. Therefore,

$$(W^0)^0 = W^{00} = \{\psi \in V^{**} : \psi(\phi) = 0, \forall \phi \in W^0\}$$

Thus W^{00} is called an annihilator of W^0.

Since V^* is natural isomorphism to V^{**}, then we can write W^{00} as follows

$$W^{00} = \{\alpha \in V : \phi(\alpha) = 0 \ \forall \phi \in W^0\}$$

Theorem 1. *If W is a subspace of a finite-dimensional vector space V over the field F, then $W = W^{00}$.*

Proof. In order to prove $W = W^{00}$, we prove that

(i) $W \subset W^{00}$ (ii) $W^{00} \subset W$

By definition, we have

$$W^0 = \{\phi \in V^* : \phi(\alpha) = 0, \forall \phi \in W^0\} \qquad \ldots (1)$$

and $\qquad W^{00} = \{\psi \in V^{**} : \psi(\phi) = 0, \forall \phi \in W^0\}$

or $\qquad W^{00} = \{\alpha \in V : \phi(\alpha) = 0, \text{ for all } \phi \in W^0\} \qquad \ldots (2)$

Let $\alpha \in W$ and $\phi \in W^0$ be arbitrary. Then

$$\alpha \in W \Rightarrow \phi(\alpha) = 0 \ \forall \phi \in W^0 \Rightarrow \alpha \in W^{00} \qquad \text{[from (2)]}$$

$\therefore \qquad W \subset W^{00}$

Further, we know that

$$\dim.W + \dim.W^0 = \dim.V \qquad \ldots (3)$$

and $\qquad \dim.W^0 + \dim.W^{00} = \dim.V^*. \qquad \ldots (4)$

From (3) and (4), we get $\dim.W = \dim.W^{00}$ $\qquad [\because \dim.V = \dim.V^*]$

As $W \subset W^{00} \Rightarrow W$ is a subspace of W^{00} with the property that $\dim W = \dim W^{00}$.

Hence $W = W^{00}$

Solved Examples

Based on the following Results

▶ A linear transformation $\phi : V \to F$ is called linear functional if for every $\alpha, \beta \in V$ and $a, b \in F$, $\phi(a\alpha + b\beta) = a\phi(\alpha) + b\phi(\beta)$

▶ $\dim.(W) + \dim(W^0) = \dim.V$

▶ The set of linear functionals on a vector space V over the field F is also a vector space such that $(\phi_1 + \phi_2) = \phi_1(\alpha) + \phi_2(\alpha) \forall \alpha \in V$ and $\phi(c\alpha) = c\phi(\alpha)$, $c \in F$, where ϕ_1, ϕ_2 are linear functional on V. This space is called the dual space or conjugate space.

Example 1. *If $B = \{(-1, 1, 1), (1, -1, 1), (1, 1, -1)\}$ is a basis of $V_3(\mathbf{R})$, then find the dual basis of B.*

Solution. Let $\qquad \alpha_1 = (-1, 1, 1), \alpha_2 = (1, -1, 1), \alpha_3 = (1, 1, -1)$.

$\therefore \qquad B = \{\alpha_1, \alpha_2, \alpha_3\}$ is a basis of $V_3(\mathbf{R})$.

Let $B^* = (\phi_1, \phi_2, \phi_3)$ be the dual of B such that

$$\phi_i(\alpha_j) = \begin{cases} 1 \text{ if } & i = j \\ 0 \text{ if } & i \neq j \end{cases} \qquad ...(1)$$

Now write,

$$\phi_1(x, y, z) = a_1 x + b_1 y + c_1 z$$
$$\phi_2(x, y, z) = a_2 x + b_2 y + c_2 z$$
$$\phi_3(x, y, z) = a_3 x + b_3 y + c_3 z$$

Determination of ϕ_1.

$$\phi_1(\alpha_1) = \phi_1(-1, 1, 1) = -a_1 + b_1 + c_1$$
$$\phi_1(\alpha_2) = \phi_1(1, -1, 1) = a_1 - b_1 + c_1$$
$$\phi_1(\alpha_3) = \phi_1(1, 1, -1) = a_1 + b_1 - c_1$$

But using (1), we have $\phi_1(\alpha_1) = 1, \phi_1(\alpha_2) = 0, \phi_1(\alpha_3) = 0$,

$$\therefore \qquad \left. \begin{aligned} -a_1 + b_1 + c_1 &= 1 \\ a_1 - b_1 + c_1 &= 1 \\ a_1 + b_1 - c_1 &= 1 \end{aligned} \right\} \qquad ... (2)$$

From last two equations of (2), we have

$$\frac{a_1}{(-1)(-1) - (1)(1)} = \frac{b_1}{(1)(1) - (1)(-1)} = \frac{c_1}{(1)(1) - (-1)(1)} = k$$

$$\Rightarrow \qquad \frac{a_1}{1-1} = \frac{b_1}{1+1} = \frac{c_1}{1+1} = k$$

$$a_1 = 0, b_1 = 2k, c_1 = 2k.$$

Putting the values of a_1, b_1 and c_1 in the first equation of (2), we get

$$0 + 2k + 2k = 1 \Rightarrow 4k = 1 \Rightarrow k = \frac{1}{4}$$

$$\therefore \qquad a_1 = 0, b_1 = \frac{1}{2}, c_1 = \frac{1}{2}$$

Thus

$$\phi_1(x, y, z) = \frac{1}{2}(y + z)$$

Determination of ϕ_2.

$$\phi_2(\alpha_1) = \phi_2(-1, 1, 1) = -a_2 + b_2 + c_2$$
$$\phi_2(\alpha_2) = \phi_2(1, -1, 1) = a_2 - b_2 + c_2$$
$$\phi_2(\alpha_3) = \phi_2(1, 1, -1) = a_2 + b_2 - c_2$$

Again, $\phi_2(\alpha_1) = 0, \phi_2(\alpha_2) = 1, \phi_2(\alpha_3) = 0$.

So that above equations become:

$$-a_2 + b_2 + c_2 = 0$$
$$a_2 - b_2 + c_2 = 1$$
$$a_2 + b_2 - c_2 = 0$$

Solving these equations, we get

$$a_2 = \frac{1}{2}, b_2 = 0, c_2 = \frac{1}{2}$$

$$\therefore \qquad \phi_2(x,y,z) = \frac{1}{2}(x+z)$$

Determination of ϕ_3.

$$\phi_3(u_1) = \phi_3(-1, 1, 1) = -a_3+b_3+c_3$$
$$\phi_3(\alpha_2) = \phi_3(1, -1, 1) = a_3-b_3+c_3$$
$$\phi_3(\alpha_3) = \phi_3(-1, 1, -1) = a_3+b_3-c_3$$

Again, $\phi_3(\alpha_1) = 0, \phi_3(\alpha_2) = 1, \phi_3(\alpha_3) = 1.$

So that above equations become:

$$-a_3+b_3+c_3 =0$$
$$a_3-b_3+c_3 =0$$
$$a_3+b_3-c_3 =1$$

Solving these equations, we get $a_3 = \frac{1}{2}, b_3 = \frac{1}{2}, c_3 = 0$

$$\therefore \qquad \phi_3(x,y,z) = \frac{1}{2}(x+y)$$

Hence $\{\phi_1, \phi_2, \phi_3\}$ is the dual basis,

where $\phi_1(x,y,z) = \frac{1}{2}(y+z), \phi_2(x,y,z) = \frac{1}{2}(x+z), \phi_3(x,y,z) = \frac{1}{2}(x+y).$

Example 2.   *If $\{\alpha_1= (1, -2, 3), \alpha_2= (1, -1, 1), \alpha_3= (2, -4, 7)\}$ is a basis of $\mathbf{R}^3$, then find the dual basis $\{\phi_i\}$.*

Solution.   Let $\{\phi_1, \phi_2, \phi_3\}$ be the dual to the basis $\{\alpha_1, \alpha_2, \alpha_3\}$ such that

Let $$\phi_i(\alpha_j) = \begin{cases} 1 \text{ if } & i=1 \\ 0 \text{ if } & i \ne j \end{cases} \qquad\qquad \text{...(1)}$$

$$\phi_1(x, y, z) = a_1x+b_1y+c_1z$$
$$\phi_2(x, y, z) = a_2x+b_2y+c_2z$$
$$\phi_3(x, y, z) = a_3x+b_3y+c_3z$$

Determination of ϕ_1.

$$\phi_1(\alpha_1) = \phi_1(1, -2, 3) = a_1- 2b_1+3c_1$$
$$\phi_2(\alpha_2) = \phi_1(1, -1, 1) = a_1- b_1+c_1$$
$$\phi_3(\alpha_3) = \phi_1(2, -4, 7) = 2a_1-4b_1- 7c_1 \qquad\qquad \text{[Using (1)]}$$

But $\phi_1(\alpha_1) = 1, \phi_1(\alpha_2) =0, \phi_1(\alpha_3) =0,$

So that above equations become:

$$a_1-2b_1+3c_1 =1$$
$$a_1-b_1+c_1 =0$$
$$2a_1- 4b_1-7c_1 =0$$

Solving these equations, we get

$$a_1=-3, b_1=-5, c_1 =2.$$

$$\therefore \qquad \phi_1(x, y, z) = -3x - 5y - 2z.$$

Determination of ϕ_2.

$$\phi_2(\alpha_1) = \phi_2(1, -2, 3) = a_2- 2b_2+3c_2$$
$$\phi_2(\alpha_2) = \phi_2(1, -1, 1) = a_2 - b_2+c_2$$

$$\phi_1 (\alpha_3) = \phi_3 (2, -4, 7) = 2a_2 - 4b_2 + 7c_2$$

Since, $\phi_2 (\alpha_1) = 0$, $\phi_2 (\alpha_2) = 1$, $\phi_2 (\alpha_3) = 0$, the above equation become

$$a_2 - 2b_2 + 3c_2 = 0$$
$$a_2 - b_2 + c_2 = 1$$
$$2a_2 - 4b_2 + 7c_2 = 0$$

Solving these equations, we get

$$a_2 = 2, b_2 = 1, c_2 = 0$$

$$\therefore \quad \phi_2(x, y, z) = 2x + y$$

Determination of ϕ_3.

$$\phi_3 (\alpha_1) = \phi_3 (1, -2, 3) = a_3 - 2b_3 + 3c_3$$
$$\phi_3 (\alpha_2) = \phi_3 (1, -1, 1) = a_3 - b_3 + c_3$$
$$\phi_3 (\alpha_3) = \phi_3 (2, -4, 7) = 2a_3 - 4b_3 + 7c_3$$

Again, $\phi_3 (\alpha_1) = 0$, $\phi_3 (\alpha_2) = 0$, $\phi_3 (\alpha_3) = 1$, the above equations become

$$a_3 - 2b_3 + 3c_3 = 0$$
$$a_3 - b_3 + c_3 = 0$$
$$2a_3 - 4b_3 + 7c_3 = 0$$

Solving the equations, we get

$$a_3 = 1, b_3 = 2, c_3 = 1$$

$$\therefore \quad \phi_3 (x, y, z) = x + 2y + z.$$

Hence, $\{\phi_1, \phi_2, \phi_3\}$ is the dual basis, where

$$\phi_1 (x, y, z) = -(3x + 5y + 2z), \phi_2 (x, y, z) = 2x + y, \phi_3 (x, y, z) = 2x + 2y + 2z$$

Example 3. *Find the dual basis of the set*

$$B = \{(1, -1, 3), (0, 1, -1), (0, 3, -2)\} \text{ for } V_3 \text{ } (\mathbf{R})$$

Solution. Let us suppose $\alpha_1 = (1, -1, 3)$, $\alpha_2 = (0, 1, -1)$ and $\alpha_3 = (0, 3, -2)$ and let $B^* = \{(\phi_1, \phi_2, \phi_3)\}$ be the dual to the basis B such that

$$\phi_i(\alpha_j) = \begin{cases} 1 \text{ if } i = j \\ 0 \text{ if } i \neq j \end{cases} \qquad \text{... (1)}$$

Suppose

$$\phi_1 (x, y, z) = a_1 x + b_1 y + c_1 z$$
$$\phi_2 (x, y, z) = a_2 x + b_2 y + c_2 z$$
$$\phi_3 (x, y, z) = a_3 x + b_3 y + c_3 z$$

Determination of ϕ_1.

$$\phi_1 (\alpha_1) = \phi_1 (1, -1, 3) = a_1 - b_1 + 3c_1$$
$$\phi_1 (\alpha_2) = \phi_1 (0, 1, -1) = b_1 - c_1$$
$$\phi_1 (\alpha_3) = \phi_1 (0, 3, -2) = 3b_1 - 2c_1$$

Since $\phi_1 (\alpha_1) = 1$, $\phi_1 (\alpha_2) = 0$, $\phi_1 (\alpha_3) = 0$, [Using (1)]

then above equations become:

$$a_1 - b_1 + 3c_1 = 1$$
$$b_1 - c_1 = 0$$
$$3b_1 - 2c_1 = 0$$

Solving these equations, we get

$$a_1 = 1, b_1 = 0, c_1 = 0.$$

$$\therefore \qquad \phi_1\,(x,\,y,\,z) = x$$

Determination of ϕ_2.

$$\phi_2\,(\alpha_1) = \phi_2\,(1,\,-1,\,3) = a_2 - b_2 + 3c_2$$
$$\phi_2\,(\alpha_2) = \phi_2\,(0,\,1,\,-1) = b_2 - c_2$$
$$\phi_2\,(\alpha_3) = \phi_3\,(0,\,3,\,-2) = 3b_2 - 2c_2$$

Since $\qquad \phi_2\,(\alpha_1) = 0,\,\phi_2\,(\alpha_2) = 1,\,\phi_2\,(\alpha_3) = 0,$

the above equation become

$$a_2 - b_2 + 3c_2 = 0$$
$$b_2 - c_2 = 1$$
$$3b_2 - 2c_2 = 0$$

Solving these equations, we get

$$a_2 = 7,\,b_2 = -2,\,c_2 = -3$$
$$\therefore \qquad \phi_2(x,\,y,\,z) = 7x - 2y - 3z$$

Determination of ϕ_3.

$$\phi_3\,(\alpha_1) = \phi_3\,(1,\,-1,\,3) = a_3 - b_3 + 3c_3$$
$$\phi_3\,(\alpha_2) = \phi_3\,(0,\,1,\,-1) = b_3 - c_3$$
$$\phi_3\,(\alpha_3) = \phi_3\,(0,\,3,\,-2) = 3b_3 - 2c_3$$

Since, $\qquad \phi_3\,(\alpha_1) = 0,\,\phi_3\,(\alpha_2) = 0,\,\phi_3\,(\alpha_3) = 1,$

the above equations become

$$a_3 - b_3 + 3c_3 = 0$$
$$b_3 - c_3 = 0$$
$$3b_3 - 2c_3 = 0$$

Solving the equations, we get

$$a_3 = -2,\,b_3 = 1,\,c_3 = 1$$
$$\therefore \qquad \phi_3\,(x,\,y,\,z) = -2x + y + z.$$

Hence, $\{\phi_1,\,\phi_2,\,\phi_3\}$ is the dual basis, where

$$\phi_1\,(x,\,y,\,z) = x,\,\phi_2\,(x,\,y,\,z) = 7x - 2y - 3z,\,\phi_3\,(x,\,y,\,z) = -2x + y + z$$

Example 4. *A basis of the vector space $R^3(R)$ is*

$$B = \{\alpha_1 = (1,1,0),\,\alpha_2 = (1,0,1),\,\alpha_3 = (0,1,1)\}$$

and f is a linear functional on R^3 such that $f(\alpha_1) = 1,\,f(\alpha_2) = -1,\,f(\alpha_3) = 3$, then find f when $\alpha = (1,\,-1,3)$.

Solution. Since $B = \{\alpha_1 = (1,1,0),\,\alpha_2 = (1,0,1),\,\alpha_3 = (0,1,1)\}$ is a basis of $R^3(R)$.

Let $B^* = \{\phi_1,\,\phi_2,\,\phi_3\}$ be the basis. Then

$$\phi_i(\alpha_j) = \begin{cases} 1 & \text{if } i = j \\ 0 & \text{if } i \neq j \end{cases} \qquad \qquad ...(1)$$

Now write

$$\phi_1(x,\,y,\,z) = a_1x + b_1y + c_1z$$
$$\phi_2(x,\,y,\,z) = a_2x + b_2y + c_2z$$
$$\phi_3(x,\,y,\,z) = a_3x + b_3y + c_3z$$

Determination of ϕ_1.

$$\phi_1(\alpha_1) = \phi_1(1,1,0) = a_1 + b_1$$

$$\phi_1(\alpha_2) = \phi_1(1,0,1) = a_1 + c_1$$
$$\phi_1(\alpha_3) = \phi_1(0,1,1) = b_1 + c_1$$

Since $\qquad \phi_1(\alpha_1) = 1, \phi_1(\alpha_2) = 0, \phi_1(\alpha_3) = 0 \qquad\qquad$ [using (1)]

Then, the above equations become

$$a_1 + b_1 = 1$$
$$a_1 + c_1 = 0$$
$$b_1 + c_1 = 0.$$

Solving these equations, we get

$$a = \frac{1}{2}, b_1 = \frac{1}{2}, c_1 = -\frac{1}{2}$$

$$\therefore \qquad \phi_1(x,y,z) = \frac{1}{2}(x + y - z).$$

Determination of ϕ_2.

$$\phi_2(\alpha_1) = \phi_2(1,1,0) = a_2 + b_2$$
$$\phi_2(\alpha_2) = \phi_2(1,0,1) = a_2 + c_2$$
$$\phi_2(\alpha_3) = \phi_3(0,1,1) = b_2 + c_2.$$

Using (1) we have $\phi_2(\alpha_1) = 1, \phi_2(\alpha_2) = 1, \phi_2(\alpha_3) = 0$,
the above equations become

$$a_2 + b_2 = 0$$
$$a_2 + c_2 = 1$$
$$b_2 + c_2 = 0.$$

Solving these equations, we get

$$a_2 = \frac{1}{2}, b_2 = -\frac{1}{2}, c_2 = \frac{1}{2}.$$

$$\therefore \qquad \phi_2(x,y,z) = \frac{1}{2}(x - y + z).$$

Determination of ϕ_3.

$$\phi_3(\alpha_1) = \phi_3(1,1,0) = a_3 + b_3$$
$$\phi_3(\alpha_2) = \phi_3(1,0,1) = a_3 + c_3$$
$$\phi_3(\alpha_3) = \phi_3(0,1,1) = b_3 + c_3$$

Since $\qquad \phi_3(\alpha_1) = 0, \phi_3(\alpha_2) = 0, \phi_3(\alpha_3) = 1.$

Then, the above equations become

$$a_3 + b_3 = 0$$
$$a_3 + c_3 = 0$$
$$b_3 + c_3 = 1$$

Solving these equations, we get

$$a_3 = -\frac{1}{2}, b_3 = \frac{1}{2}, c_3 = \frac{1}{2}$$

$$\phi_3(x,y,z) = \frac{1}{2}(-x+y+z)$$

Since f is a linear functional $\mathbf{R}^3$ such that $f(\alpha_1) = 1, f(\alpha_2) = -1, f(\alpha_3) = 3$. Then by theorem, we have

$$f = f(\alpha_1)\phi_1 + f(\alpha_2)\phi_2 + f(\alpha_3)\phi_3$$
$$f = \phi_1 - \phi_2 + 3\phi_3.$$

$\therefore$
$$\phi(x, y, z) = \phi_1(x, y, z) - \phi_2(x, y, z) + 3\phi_3(x, y, z).$$

$$= \frac{1}{2}(x+y-z) - \frac{1}{2}(x-y+z) + \frac{3}{2}(-x+y+z)$$

$$= -\frac{3}{2}x + \frac{5}{2}y + \frac{1}{2}z.$$

Determination of f when $\alpha = (1, -1, 3)$

$$f(\alpha) = f(1,-1,3) = -\frac{3}{2}(1) + \frac{5}{2}(-1) + \frac{1}{2}(3).$$

$$= -\frac{3}{2} - \frac{5}{2} + \frac{3}{2} = -\frac{5}{2}.$$

Example 5. *If W_1 and W_2 are subspaces of a vector space V over a field F and $W_1 \subset W_2$, show that $W_2^0 \subset W_1^0$.*

Solution. Since W_1 and W_2 are subspace of V with the condition that $W_1 \subset W_2$. Then we shall show that $W_2^0 \subset W_1^0$.

Let $\phi \in W_2^0$ be an arbitrary linear functional.

Then, we have $\phi \in W_2^0 \Rightarrow \phi(\alpha) = 0, \forall \alpha \in W_1$

$$\Rightarrow \phi(\alpha) = 0, \forall \alpha \in W_2 \qquad\qquad [\because W_1 \subset W_2]$$

$$\Rightarrow \phi \in W_1^0 .$$

Hence, $\qquad\qquad W_2^0 \subset W_1^0.$

Example 6. *If W is a subset of a vector space $V(F)$, then*
  (i) $W^0 = [L(W)]^0$ (ii) $W^{00} = L(W)$.

Solution. (i) Let $W = \{\alpha_1, \alpha_2, \dots \alpha_n\}$ be a subset of a vector space $V(F)$. We shall show that
$$W^0 = [L(W)]^0$$

By the definition of W^0, we have
$$W^0 = \{\phi \in V^* : \phi(\alpha) = 0, \alpha \in W\}.$$

Since $W \subseteq L(W)$, therefore
$$[L(W)]^0 \subset W^0 \qquad\qquad\qquad \dots(1)$$

Now, $\qquad \phi \in W^0 \Rightarrow \phi(\alpha) = 0, \forall \alpha \in W$

$$\Rightarrow \phi(\alpha) = 0, \forall \alpha \in L(W) \qquad\qquad [\because W \subset L(W)]$$

$$\Rightarrow \phi \in [L(W)]^0.$$

$\therefore \qquad\qquad W^{00} \subset [L(W)]^0 \qquad\qquad\qquad \dots(2)$

From (1) and (2), we get

$$W^0 = [L(W)]^0$$

(ii) Since we know that

$$W^{00} = W. \qquad \qquad \qquad \text{...(3)}$$

$$\therefore \qquad W^{00} = (W^0)^0 = [[L(W)]^0]^0 \qquad \qquad [\because W^0 = [L(W)]^0]$$

$$= L(W) \qquad \qquad \qquad \text{[using (3)]}$$

Hence, $\qquad W^{00} = L(W)$.

Example 7. *Let W_1 and W_2 be subspaces of a finite-dimensional vector space V aver the field F. Prove that*

(i) $(W_1 + W_2)^0 = W_1^0 \cap W_2^0$

(ii) $W_1^0 + W_2^0 = (W_1 \cap W_2)^0$

Solution. (i) To prove that : $\qquad (W_1 + W_2)^0 = W_1^0 \cap W_2^0$.

Since we know that : $\quad W_1 \subset W_2 \Rightarrow W_2^0 \subset W_1^0$.

Obviously, $\qquad \qquad W_1 \subset W_1 + W_2 \text{ and } \quad W_2 \subset W_1 + W_2$

$$\therefore \qquad \qquad (W_1 + W_2)^0 \subset W_1^0 \text{ and } (W_1 + W_2)^0 \subset W_2^0$$

$$\Rightarrow \qquad \qquad (W_1 + W_2)^0 \subset W_1^0 \cap W_2^0 \qquad \qquad \text{...(1)}$$

Now, let $\phi \in W_1^0 \cap W_2^0$ be an arbitrary linear functional, then

$$\phi \in W_1^0 \cap W_2^0 \quad \Rightarrow \phi \in W_1^0 \text{ and } \phi \in W_2^0$$

$$\Rightarrow \phi(\alpha) = 0, \forall \alpha \in W_1 \text{ and } \phi(\beta) = 0, \forall \beta \in W_2$$

Let $\gamma \in W_1 + W_2$ be an arbitrary vector so that

$$y = \alpha + \beta, \text{ for some } \alpha \in W_1, \beta \in W_2$$

$$\therefore \qquad \qquad \phi(\gamma) = \phi(\alpha + \beta) = \phi(\alpha) + \phi(\beta) \qquad [\because \phi \text{ is linear.}]$$

$$= 0 + 0 = 0$$

$$\Rightarrow \qquad \qquad \phi(\gamma) = 0, \text{ for all } \gamma \in W_1 + W_2$$

$$\Rightarrow \qquad \qquad \phi \in (W_1 + W_2)^0$$

Thus, $\qquad \qquad W_1^0 \cap W_2^0 \subset (W_1 + W_2)^0 \qquad \qquad \text{... (2)}$

From (1) and (2), we get

$$(W_1 + W_2)^0 = W_1^0 \cap W_2^0$$

(ii) Above result (i) is taken for the vector space $V^*(F)$ in place of $V(F)$, then we

have $\qquad \qquad (W_1^0 + W_2^0)^0 = W_1^{00} \cap W_2^{00}$

$$\Rightarrow \qquad \qquad (W_1^0 + W_2^0)^0 = W_1 \cap W_2 \qquad \qquad [\because W^{00} = W]$$

$$\Rightarrow \qquad \qquad (W_1^0 + W_2^0)^{00} = (W_1 \cap W_2)^0$$

$$\Rightarrow \qquad \qquad W_1^0 + W_2^0 = (W_1 \cap W_2)^0 \qquad \qquad [\because W^{00} = W]$$

EXERCISE 7.4

1. Let $\phi = \mathbf{R}^2 \to \mathbf{R}$ and $\psi = \mathbf{R}^2 \to \mathbf{R}$ be the linear functionals defined by $\phi(x, y) = x + 2y$ and $\psi(x,y) = 3x - y$. Find

   (i) $\phi + \psi$ (ii) 4ϕ (iii) $2\phi - 5\psi$

2. Let $\phi = \mathbf{R}^3 \to \mathbf{R}$ and $\psi = \mathbf{R}^3 \to \mathbf{R}$ be the linear functionals defined by

   $$\phi(x, y, z) = 2x - 2y + z$$

   and $\psi(x, y, z) = 4x - 2y + 3z$. Find
   (i) $\phi + \psi$ (ii) 3ϕ (iii) $2\phi - 5\psi$

3. Let ϕ be the linear functional on $\mathbf{R}^2$ defined by $\phi(2,1) = 15$ and $\phi(1, -2) = -10$. Find $\phi(x, y)$ and, in particular, find $\phi(-2, 7)$.

7.35   EIGEN VALUES AND EIGEN VECTORS OF A LINEAR TRANSFORMATION

Let V be a finite-dimensional vector space over a field F and let T be a linear operator on V, then a scalar $\lambda \in F$ is called an eigenvalue of T if there exists a non-zero vector $\alpha \in V$ such that

$$T(\alpha) = \lambda\alpha.$$

Also, each such non-zero vector $\alpha \in V$ is called an eigenvector of T corresponding to λ.
The set E_λ of all such vectors is a subspace of V called the eigenspace of λ.

Theorem 1. *Non-zero eigenvectors belonging to distinct eigenvalues are linearly independent.*

Proof.   Let $T : V \to V$ be a linear operator and let $\alpha_1, \alpha_2, \dots, \alpha_n$ be non-zero eigenvectors of T corresponding to distinct eigenvalues $\lambda_1, \lambda_2, \dots, \lambda_n$. Then we have to show that $\alpha_1, \alpha_2, \dots, \alpha_n$ are linearly independent.

We shall prove by induction on n. If $n = 1$ then α_1 is linearly independent as $\alpha_1 \neq 0$, so assume that for $n > 1$, we have $\alpha_1, \alpha_2, \dots, \alpha_{n-1}$ are linearly independent.

Suppose $a_1\alpha_1 + a_2\alpha_2 + \dots + a_n\alpha_n = 0$ ...(1)
for $a_1, a_2, \dots, \in F$.

Applying T to (1), we get

$$T(a_1\alpha_1 + a_2\alpha_2 + \dots + a_n\alpha_n) = T(0) = 0$$

$\Rightarrow$ $a_1 T(\alpha_1) + a_2 T(\alpha_2) + \dots + a_n T(\alpha_n) = 0$ $\{\because T \text{ is linear.}\}$ ...(2)

But by given hypothesis

$$T(\alpha_i) = \lambda_i\alpha_i, \forall i.$$

Therefore, (2) becomes

$$a_1\lambda_1\alpha_1 + a_2\lambda_2\alpha_2 + \dots + a_n\lambda_n\alpha_n = 0 \qquad ...(3)$$

Now multiplying (1) by λ_n, we get

$$a_1\lambda_n\alpha_1 + a_2\lambda_n\alpha_2 + \dots + a_n\lambda_n\alpha_n = 0 \qquad ...(4)$$

Subtracting (4) from (3), we get

$$a_1(\lambda_1 - \lambda_n)\alpha_1 + a_2\{\lambda_2 - \lambda_n\}\alpha_2 + \dots + a_{n-1}\{\lambda_{n-1} - \lambda_n\}\alpha_{n-1} = 0 \qquad ...(5)$$

Since $\alpha_1, \alpha_2, \dots, \alpha_{n-1}$ are linearly independent, also λ_i are distinct, therefore from (5) we obtain

$$a_1 = a_2 = \dots = a_{n-1} = 0.$$

Substituting these values of $a_1, a_2 \dots a_{n-1}$ into (1), we get

$$a_n\alpha_n = 0 \Rightarrow a_n = 0 \qquad [\because \alpha_n \neq 0]$$

So, $a_1 = a_2 = \dots = a_n = 0$, hence $\alpha_1, \alpha_2, \dots, \alpha_n$ are linearly independent.

Theorem 2. *Let T: V → V be a linear operator on a finite dimensional vector space over a field F. Then λ ∈ F is an eignvalue of T if and only if the operator (λI − T) is singular. The eigenspace of X is then the kernel of (λI − T).*

Proof. Suppose λ ∈ F is an eigenvalue of T, then we shall show that (λI − T) is singular.

Let α be any non-zero vector of V, then by definition of eigenvalue of T, we have

$$T(\alpha) = \lambda\alpha \text{ or } \lambda\alpha - T(\alpha) = 0$$

$\Rightarrow$ $\quad (\lambda I)(\alpha) - T(\alpha) = 0 \Rightarrow (\lambda I - T)(\alpha) = 0$

$\Rightarrow$ $\quad |(\lambda I - T)| = 0$

$\Rightarrow$ $\quad \lambda I - T = 0 \Rightarrow \lambda I - T$ is singular.

Conversely, suppose (λI−T) is singular, then we shall show that λ is an eigenvalue of T.

Let α ≠ 0 be an element of V so it is linearly independent.

Since (λI − T) is singular so that

$$|(\lambda I - T)| = 0 \Rightarrow (\lambda I - T)(\alpha) = 0 \quad [\because \alpha \text{ is linearly independent.}]$$

$\Rightarrow$ $\quad (\lambda I)(\alpha) - T(\alpha) = 0 \Rightarrow \lambda I(\alpha) - T(\alpha) = 0 \quad\quad\quad [\because I(\alpha) = \alpha]$

$\Rightarrow$ $\quad\quad T(\alpha) = \lambda\alpha$

$\Rightarrow$ $\quad$ λ is an eigenvalue of T.

Also, $\quad\quad (\lambda I - T)(\alpha) = 0 \Rightarrow \alpha \in$ kernel of (λI − T).

Definition. *Let T be a linear operator on a n-dimensional vector space V over a field F. If B is an ordered basis of V, then $|(xI - T) = (xI - [T]_B)|$ is called a monic polynomial of degree n in F[x]. The monic polynomial $|(xI - T)|$ in the variable x is also called characteristic polynomial of T, it is denoted by Δ(x).*

By above theorem, λ is an eigenvalue of T if and only if λ is a root of this polynomial in F.

REMARK

☞ The degree of characteristic polynomial of T is exactly equal to n, then n-dimension of V, T cannot have more than n eigenvalues, counted with multiplicity.

Theorem 3.(Cayley Hamilton Theorem). *If T be a linear transformation on n-dimensional vector space and Δ(λ) be its characteristic polynomial then Δ(T) = 0.*

Proof. Let $T : V(F) \to V(F)$ be a linear operator and let B be an ordered basis of V and the matrix of T corresponding to B be V.

$$[T]_B = A = [a_{ij}]_{n \times n} \text{ for } a_{ij}\text{'s} \in F$$

where, $\quad\quad T(\alpha_j) = \sum_{i=1}^{n} a_{ij}\alpha_i .$

If I is an identity matrix of order n×n and λ an indeterminate scalar, then the characteristic polynomial of T is given by

$$\Delta(\lambda) = |T - \lambda I| = 0 = |[T]_B - \lambda I| = 0 = |A - \lambda I| = 0$$

$$\therefore \quad \Delta(\lambda) = \begin{vmatrix} a_{11} - \lambda & a_{12} & \cdots & a_{1n} \\ a_{21} & a_{21} - \lambda & \cdots & a_{2n} \\ \cdots & \cdots & \cdots & \cdots \\ a_{n1} & a_{n2} & \cdots & a_{nn} - \lambda \end{vmatrix}$$

$$= b_n\lambda^n + b_{n-1}\lambda^{n-1} + \dots b_1\lambda + b_0 \text{ (say)}.$$

Thus the characteristic polynomial of A is

$$\Delta(\lambda) = 0, \text{ i.e., } b_n\lambda^n + b_{n-1}\lambda^{n-1} + \dots b_1\lambda + b_0 = 0. \qquad \dots(1)$$

Since the elements of $A - \lambda I$ are polynomials of degree at most one in λ so that each element of adj$(A-\lambda I)$ is the polynomial at most of degree $n-1$, in λ. Therefore,

$$\text{adj } (A - \lambda I) = B_0 + B_1\lambda + \dots + B_{n-1}\lambda^{n-1} \qquad \dots(3)$$

where B_i's are square matrices of degree $n \times n$.

Since $(A - \lambda I_n)$ adj $(A - \lambda I_n) = |A - \lambda I_n| I_n$

or $\quad (A - \lambda I)(B_0 + B_1\lambda + B_2\lambda^2 + \dots + B_{n-1}\lambda^{n-1}) = (b_n\lambda^n + b_{n-1}\lambda^{n-1} + \dots + b_0) I.$

Comparing the coefficients of like powers in λ, on both sides, we have

$$AB_0 = b_0 I$$
$$AB_1 - B_0 = b_1 I$$
$$AB_2 - B_1 = b_2 I$$
$$\dots \quad \dots \quad \dots \quad \dots$$
$$AB_{n-1} - B_{n-1} = b_{n-1}I - B_{n-1} = b_n I.$$

Multiplying above equation by $I, A, A^2, \dots, A^n$ respectively and then adding, we get

$$0 = b_0 I + b_1 A + b_2 A^2 + \dots + b^n A^n$$

or $\qquad\qquad\qquad \Delta(A) = 0 \qquad\qquad\qquad$ [using (1)]

or $\qquad\qquad\qquad \Delta(T) = 0.$

Theorem 4. *If $\lambda \in F$ is a characteristic root (eigenvalue) of T, then for any polynomial $p(x) \in F[x]$, $p(\lambda)$ is a characteristic root of $p(T)$.*

Proof. Since $\lambda \in F$ is a characteristic root of T, there exists a non-zero vector a in V such that

$$T(\alpha) = \lambda\alpha. \qquad \dots(1)$$

Now, $\qquad\qquad T^2(\alpha) = T[T(\alpha)] = T(\lambda\alpha) \text{ [using (1)]}$

$$= \lambda T(\alpha) \qquad\qquad\qquad [\because T \text{ is linear.}]$$

$$= \lambda^2\alpha. \qquad\qquad\qquad [\text{using (1)}]$$

Continuing in this way, we get

$$T^k(\alpha) = \lambda^k\alpha \qquad \dots(2)$$

for all positive integer k.

Let $\qquad\qquad p(x) = a_0 x^n + a_1 x^{n-1} + \dots + a_{n-1}x + a_n, a_i \in F$

Then $\qquad\qquad p(T) = a_0 T^n + a_1 T^{n-1} + \dots + a_{n-1}T + a_n I.$

Now $\qquad [p(T)](\alpha) = (a_0 T^n + a_1 T^{n-1} \dots + a_{n-1}T + a_n I)(\alpha)$

$$= a_0 T^n (\alpha) + a_1 T^{n-1} (\alpha) + \dots + a_{n-1}T (\alpha) + a_n I (\alpha)$$

$$= a_0\lambda^n\alpha + a_1\lambda^{n-1}\alpha + \dots + a_{n-1}\lambda \alpha + a_n \alpha$$

$$= (a_0\lambda^n + a_1\lambda^{n-1} + \dots + a_{n-1}\lambda + a_n) (\alpha)$$

$$= [p(\lambda)] (\alpha)$$

$\Rightarrow \qquad\qquad [p(\lambda)I - p(T)](\alpha) = 0$

$\Rightarrow \qquad\qquad [p(\lambda)I - p(T)] = 0$

$\Rightarrow \qquad p(\lambda)$ is a characteristic root of $p(T)$.

7.36 MINIMAL POLYNOMIAL

Definition. *A monic polynomial $m_T(x) \in F[x]$ of least degree, is said to be a minimal polynomial of T if $m_T(T) = 0$.*

Theorem 1. *If $\lambda \in F$ is a characteristic root of T then λ is a root of the minimal polynomial of T. In particular, T only has a finite number of characteristic roots in F.*

Proof. Let $m(x) \in F[x]$ be a minimal polynomial of T, then we shall show that λ is a root of $m(x) = 0$.

If $\lambda \in F$ is a characteristic root of T, then there exists a non-zero vector α in V such that

$$T(\alpha) = \lambda\alpha.$$

By above theorem,

$$[m(T)](\alpha) = [m(\lambda)](\alpha) = m(\lambda)\,\alpha.$$

But $m(T) = 0$ as $m(x)$ is a minimal polynomial of T.

$$\therefore \qquad 0\,\alpha = m(\lambda)\alpha$$

$$\Rightarrow \qquad m(\lambda)\,\alpha = 0 \qquad\qquad \text{[By elementary property of vector space]}$$

Since $\alpha \neq 0$ so that $m(\lambda) = 0$.

Thus λ is a root of $m(x)$.

Also, V is a finite dimensional vector space; if dim $V = n$ then degree of $m(x) \leq n^2$, therefore $m(x)$ has atmost n^2 roots implying that $m(x)$ has only a finite number of roots in F, thus there can only be finite number of characteristic roots of T in F.

Theorem 2. *If T and S are two linear operators with S invertible, then T and STS^{-1} have the same minimal polynomial.*

Solution. We have $\qquad (STS^{-1})^2 = (STS^{-1})((STS^{-1}) = ST(S^{-1}S)TS^{-1}$

$$= ST^2S^{-1} \qquad\qquad [S^{-1}S = I\,]$$

Continuing in this way, we get

$$(STS^{-1})^k = ST^kS^{-1} \qquad\qquad\qquad ...(1)$$

Let $m_T(x)$ be a minimal polynomial of T, thus $m_T(T) = 0$ and let

$$m_T(x) = a_0x^n + a_1x^{n-1} + ... + a_{n-1}x + a_n.$$

Then, $\qquad a_0T^n + a_1T^{n-1} + ... + a_{n-1}T + a_nI = 0.$

Now, $\qquad m_T(STS^{-1}) = a_0(STS^{-1})^n + a_1(STS^{-1})^{n-1} + ... + a_n(STS^{-1})$

$$= a_0ST^nS^{-1} + a_1ST^{n-1}S^{-1} + ... + a_nSTS^{-1} \qquad \text{[using (1)]}$$

$$= S(a_0T^n + a_1ST^{n-1} + ... + a_nI\,)S^{-1}$$

$$m_T(STS^{-1}) = S\,.\,0\,.S^{-1}. \qquad\qquad\qquad\qquad \text{[using (2)]}$$

$$= 0.$$

Hence, $m_T(x)$ is also the minimal polynomial of STS^{-1}.

Theorem 3. *Let V be a finite dimensional vector space over a field F and let T be a linear operator on V. Then there exists a vector α_j in V such that $m_T^{\alpha_j}(x) = m_T(x)$.*

Proof. For any vector $\beta \in V$, the polynomial $m_T^\beta(x)$ divides $m_T(x)$. Thus each $m_T^\beta(x)$ is a factor of $m_T(x) \in F[X]$. Since V is finite-dimensional so for all $\beta \in V$, we get only finite number of polynomials. Let these polynomials be $p_1(x), p_2(x), ..., p_k(x)$ corresponding to the vectors $\alpha_1, \alpha_2,...,\alpha_k$, that is

$$m_T^{\alpha_i}(x) = p_i(x) \qquad \text{for } i = 1, 2, ..., k.$$

Let $V_i = \ker p_i(T)$.

Then $V \subseteq \bigcup_{i=1}^{k} V_i$ and so $V = V_j$ for some j, therefore,

$$(p_j(T))_\beta = 0, \forall \beta \in V$$

$$\Rightarrow \qquad p_j(T) = m_T^{\alpha_i}(x) = m_T(x).$$

7.37 INVARIANCE OF LINEAR OPERATOR

Definition. *Let $T : V \to V$ be a linear operator. A subspace W of V is said to be T-invariant or invariant under T if T maps W into itself, that is, if $a \in W$ implies $T(\alpha) \in W$. In this case T restricted to W defines a linear operator on W; that is T induces a linear operator $\hat{T}(\alpha) = T(\alpha), \forall \alpha \in W$.*

Theorem 1. *Let $T : V \to V$ be a linear operator and let $p(x)$ be any polynomial. Then the kernel of $p(T)$ is T-invariant*

Proof. Let $\alpha \in \ker p(T)$, then $(p(T))(\alpha) = 0$.

Now we shall show that $T(\alpha) \in \ker p(T)$.

Since $\qquad p(x)x = x\,p(x)$ so that $p(T)T = Tp(T)$.

$\therefore \qquad (p(T)T)(\alpha) = T(p(T)(\alpha)) = T(0) = 0$

$\Rightarrow \qquad p(T)(T(\alpha)) = 0.$

Thus, $\qquad T(\alpha) \in \ker p(T)$.

Theorem 2. *Let $T : V \to V$ be a linear operator, and suppose that $p(x) = f(x)\,g(x)$ are polynomials such that $p(T) = 0$ and $f(x)$ and $h(x)$ are relatively prime. Then V is the direct sum of the T-invariant subspace W_1 and W_2 where, $W_1 = \ker f(T)$ and $W_2 = \ker g(T)$.*

Proof. Since $f(x)$ and $g(x)$ are relatively prime, then there exist two polynomials $r(x)$ and $s(x)$ such that

$$r(x)f(x) + s(x)g(x) = I.$$

So, for the operator T, we get

$$r(T)f(T) + s(T)g(T) = I. \qquad ...(1)$$

Let $a \in V$, then form (1), we have

$$\alpha = [r(T)f(T)](\alpha) + [s(T)g(T)](\alpha) \qquad ...(2)$$

Since $\qquad [g(T)(r(T)f(T))](\alpha) = r(T)[f(T)g(T)(\alpha)]$

$$= r(T)[p(T)(\alpha)]f = r(T)0.\,\alpha \qquad [\because p(T) = 0]$$

$$= 0.$$

$\therefore \qquad [r(T)f(T)](\alpha) \in W_2 = \ker g(T).$

Similarly, $[s(T)g(T)](\alpha) \in W_1 = \ker f(T).$

Thus, α is the sum of an element of W_1 and an element of W_2. Hence

$$V = W_1 + W_2.$$

Now, to show that $V = W_1 \oplus W_2$, we must show that a sum $\alpha = \beta + \gamma$ with $\beta \in W_1$ and

$\gamma \in W_2$, is uniquely determined by α.

Applying the operator $r(T)f(T)$ to $\alpha = \beta + \gamma$ and using $f(T)(\beta) = 0$, we get

$$[r(T)f(T)](\alpha) = [r(T)f(T)](\beta) + [r(T)f(T)](\gamma) = [r(T)f(T)](\gamma) \qquad ...(3)$$

Also, from (1), we get

$$[r(T)f(T)](\gamma) + [s(T)g(T)]\gamma = \gamma$$

$$\Rightarrow \qquad\qquad \gamma = [r(T)f(T)](\gamma) \qquad\qquad [\because (g(T))(\gamma) = 0] \qquad\qquad ...(4)$$

From (3) and (4), we get

$$\gamma = [r(T)f(T)](\alpha)$$

which shows that γ is uniquely determined by α.

Similarly, β is uniquely determined by α. Hence $V = W_1 \oplus W_2$.

Theorem 3. *If T_1 is the restriction of T to W_1 and T_2 is the restriction of T to W_2, where, $T : V \rightarrow V$ is a linear operator and $p(x) = f(x)g(x)$ such that $p(x)$ is the minimal polynomial of T and $f(x)$ and $g(x)$ are monic. Then $f(x)$ and $g(x)$ are minimal polynomials of T_1 and T_2 respectively.*

Proof.   Let $m_1(x)$ and $m_2(x)$ be the minimal polynomials of T_1 and T_2 respectively, then we shall show that $m_1(x) = f(x)$ and $m_2(x) = g(x)$.

Since $W_1 = \ker f(x)$ and $W_2 = \ker g(x)$.

$\therefore$ $f(T_1) = 0$ and $g(T_2) = 0$.

Thus, $m_1(x)$ divides $f(x)$ and $m_2(x)$ divides $g(x)$.

Let $p(x)$ be the least common multiple of $m_1(x)$ and $m_2(x)$. But $m_1(x)$ and $m_2(x)$ are relatively prime. Accordingly, we have

$$p(x) = m_1(x)\, m_2(x).$$

But we have,

$$p(x) = f(x)g(x).$$

Since $m_1(x)$ is monic and divides $f(x)$ and $m_2(x)$ is also monic and divides $g(x)$.

Thus $f(x) = m_1(x)$ and $g(x) = m_2(x)$.

Theorem 4. (Primary Decomposition Theorem). *Let $T : V \rightarrow V$ be a linear operator with minimal polynomial*

$$m(x) = p_1(x)^{n_1} p_2(x)^{n_2} ... p_r(x)^{n_r}$$

where, the $p_i(x)$ are distinct monic irreducible polynomials. Then V is the direct sum of T-invariant subspaces $W_1, W_2, ..., W_r$, where $W_i = \ker p_i(T)^{n_i}$. Moreover, $p_i(x)^{n_i}$ is the minimal polynomial of the restriction of T to W_i.

Solution.   We shall prove the theorem by induction on r. If $r = 1$, then the theorem is trivially true.

Suppose, that the theorem is true for $r - 1$.

By theorem 1, we may write V as the direct sum of T- invariant subspaces W_1 and V_1 where $W_1 = \ker p_1(T)^{n_1}$ and $V_1 = \ker(p_2(T)^{n_2} ... p_r(T)^{n_r})$.

Now by theorem 1, the minimal polynomials of the restrictions of T to W_1 and V_1 are respectively $p_1(x)^{n_1}$ and $p_2(x)^{n_2} ... p_r(x)^{n_r}$.

Let T_1 be the restriction of T to V_1. Then by induction hypothesis, V_1 is the direct sum of subspaces $W_2, W_3, \ldots, W_r$ such that

$$W_i = \ker p_i(T)^{n_i}$$

where $p_i(x)^{n_i}$ is the minimal polynomial for the restriction of T_1 to W_i.

But $\quad \ker p_i(T)^{n_i} \subseteq V_1$ for $i = 2, 3,\ldots, r$.

Since $\quad p_i(x)^{n_i} \mid p_2(x)^{n_2}\ldots p_r(x)^{n_r}$. Thus the $\ker p_i(T)^{n_i}$ is the same as the $\ker p_i(T_1)^{n_i} = W_1$. Also the restriction of T to W_i is the same as the restriction of T_1 to $W_i (i = 2,3,\ldots,r)$. Hence $p_i(x)^{n_i}$ is also the minimal polynomial for the restriction of T to W_i.

Hence, $\qquad V = W_1 \oplus W_2 \oplus \ldots \oplus W_r.$

Solved Examples
Based on the following Results

▶ A scalar $\lambda \in F$ is called an eigen value of a linear operator T if their exists a non-zero vector $\alpha \in V$ such that $T(\alpha) = \lambda \cdot \alpha$.

▶ λ is an eigen value of T if and only if λ is the root of characteristic polynomial.

▶ Let T be a linear transformation on n-dimensional vector space and $\Delta(\lambda)$ be its characteristic equation then $\Delta(T) = 0$ [Cayley-Hamilton theorem]

Example 1. *Let $I : V \to V$ be the identity mapping on any non-zero vector space V. Show that $\lambda=1$ is an eigenvalue of I. What is the eigenspace E_1 of $\lambda = 1$?*

Solution. Let $\alpha \neq 0$ be any vector of V, then
$$I(\alpha) = \alpha = 1\alpha.$$
$\Rightarrow \quad \lambda = 1$ is an eigenvalue of I.

Also, every vector in V is an eigenvector corresponding to 1, therefore $E_1=V$.

Example 2. *Show that 0 is an eigenvalue of T if and only if T is singular.*

Solution. Let α be a non-zero vector of V.

0 is an eigenvalue of $T \Leftrightarrow$ there exists a non-zero vector such that
$$T(\alpha) = 0 \cdot \alpha$$
$\Leftrightarrow \qquad T(\alpha) = 0$
$\Leftrightarrow \qquad T$ is singular.

Example 3. *Let λ be an eigenvalue of a linear operator $T : V \to V$. Let E_λ be the eigenspace of A. Show that E_λ is a subspace of V.*

Solution. In order to show E_λ to be subspace of V, we shall show that

(i) if $\alpha \in E_\lambda$ then $k\alpha \in E_\lambda$ for any scalar $k \in F$,

(ii) if $\alpha, \beta \in E_\lambda$, then $\alpha + \beta \in E_\lambda$

(i) Since $\alpha \in E_\lambda$, so we have
$$T(\alpha) = \lambda\alpha$$
then $\qquad T(k\alpha) = kT(\alpha) = k(\lambda x) = \lambda(k\alpha).$
$$k\alpha \in E_\lambda.$$

(ii) Since $\alpha, \beta \in E_\lambda$, we have
$$T(\alpha) = \lambda\alpha, \quad T(\beta) = \lambda\beta.$$

Then, $\quad T(\alpha+\beta) = T(\alpha) + T(\beta) = \lambda\alpha + \lambda\beta = \lambda(\alpha+\beta).$

$\therefore \qquad \alpha + \beta \in E_\lambda.$

Hence, E_λ is a subspace of V.

Example 4. *Let V be the vector space of differentiable functions on $\mathbf{R}$ and $D: V \to V$ be the differential operator. Show that functions $e^{a_1 t}, e^{a_2 t}, ..., e^{a_n t}$, where $a_1, a_2, ..., a_n$ are distinct non-zero scalars are eigenvectors of D, to which eigenvalue λ_i does $e^{a_i t}$ belong ?*

Solution. Since $e^{a_i t} \neq 0$ for all $a_i \neq 0.$

Then, we have
$$D(e^{a_i t}) = a_i e^{a_i t}, \quad \forall\, i = 1, 2, 3, ..., n.$$

This equation shows that $e^{a_i t}$ is the eigenvector of D corresponding to $\lambda_i = a_i$.

Example 5. *Suppose λ is an eigenvalue of an invertible operator T. Show that λ^{-1} is an eigenvalue of T^{-1} .*

Solution. Since T is invertible so that it is non-singular, therefore, $\lambda \neq 0$. Also λ is an eigenvalue of T, then there exists a non-zero vector $\alpha \in V$ such that
$$T(\alpha) = \lambda\alpha.$$

Multiplying both sides by T^{-1} , we get
$$T^{-1}[T(\alpha)] = T^{-1}(\lambda\alpha)$$

$\Rightarrow \qquad (T^{-1}T)(\alpha) = T^{-1}(\lambda\alpha) \quad \Rightarrow \quad I(\alpha) = \lambda[T^{-1}(\alpha)]$

$\Rightarrow \qquad \alpha = \lambda\, T^{-1}(\alpha) \Rightarrow T^{-1}(\alpha) = \dfrac{1}{\lambda}\,\alpha = \lambda^{-1}\alpha.$

Hence, λ^{-1} is an eigenvalue of T^{-1}.

Example 6. *Suppose α is a none-zero eigenvector of linear maps S and T . Show that α is an eigenvector of $S + T$.*

Solution. Since α is an eigenvector of S and T . Let λ_1 and λ_2 be eigenvalues of S and T corresponding to α respectively.

Then, $\qquad S(\alpha) = \lambda_1\alpha$ and $T(\alpha) = \lambda_2\alpha.$

Now, $\qquad (S + T)(\alpha) = S(\alpha) + T(\alpha) = \lambda_1\alpha + \lambda_2\alpha$

$\qquad\qquad (S + T)(\alpha) = (\lambda_1 + \lambda_2)\alpha.$

Thus, α is an eigenvector $S + T$ corresponding to the eigenvalue $\lambda_1 + \lambda_2$.

Example 7. *Let A be a square matrix of order $n\times n$ with its elements belong to F. The left multiplication by A defines a linear operator*
$$T_A : F^{n\times m} \to F^{n\times m}$$

such that $\qquad T_A(B) = AB.$

A scalar $\lambda \in F$ is an eigenvale of T_A if and only if λ is an eigenvalue of A .

Solution. Suppose λ is an eigenvalue of A , then there exists a non-zero vector $\alpha \in F^n$ such that
$$A(\alpha) = \lambda\alpha.$$

Let B be an $n \times m$ matrix whose first column is α and all other columns zero, then B is a non-zero matrix and

$$T_A (B) = AB = \lambda B \qquad\qquad [\because A(\alpha) = \lambda\alpha]$$

Thus, λ is an eigenvalue of T_A .

Conversely, if λ, is eigenvalue of T_A, then there exists a non-zero matrix $B \in F^{n\times m}$ such that

$$T_A (B) = \lambda B = AB.$$

Let α be one of the non-zero columns of B, then

$$AB = \lambda B.$$

$$\therefore \qquad\qquad A(\alpha) = \lambda\alpha.$$

Thus, λ is an eigenvalue of A.

7.38 DIAGONALIZATION

Let V be a finite dimensional vector space over a field F. Then a linear operator $T : V \to V$ is said to be diagonalizable if V has a basis consisting of eigenvectors of T only. Equivalently, T has n linearly independent eigenvectors, if V is n-dimensional vector space.

For example: A scalar multiple of identity operator is diagonalizable.

Theorem 1. *A linear operator $T : V \to V$ can be represented by a diagonal matrix A iff V has a basis consisting of eigenvectors of T . In this case the diagonal elements of A are the corresponding eigenvalues.*

Proof. Let V be n-dimensional vector space. Let $B = \{\alpha_1, \alpha_2,...,\alpha_n\}$ be an ordered basis of V such that each α_j is the eigenvector of T.

If $\lambda_1, \lambda_2,..., \lambda_n$ are the corresponding eigenvalue of T, then

$$\left.\begin{array}{l} T(\alpha_1) = \lambda_1\alpha_1 \\ T(\alpha_2) = \lambda_2\alpha_2 \\ \cdots\ \cdots\ \ \cdots\ \cdots \\ T(\alpha_n) = \lambda_n\alpha_n \end{array}\right\} \qquad ...(1)$$

System (1) can also be rewritten as

$$T(\alpha_1) = \lambda_1\alpha_1 + 0\alpha_2 + 0\alpha_3 +...+ 0\alpha_n$$
$$T(\alpha_2) = 0\alpha_1 + \lambda_2\alpha_2 + 0\alpha_3 +...+ 0\alpha_n$$

$$T(\alpha_n) = 0\alpha_1 + 0\alpha_2 + 0\alpha_3 +...+ \lambda_n\alpha_n$$

Thus, the matrix representation of T, with respect to B is given by

$$A = [T]_B = \begin{vmatrix} \lambda_1 & 0 & 0 \\ 0 & \lambda_2 & 0 \\ \vdots & \vdots & \vdots \\ 0 & 0 & \lambda_n \end{vmatrix}.$$

Hence, T can be represented by a diagonal matrix A whose elements are the eigenvalues of T.

Conversely, suppose T can be represented by a diagonal matrix A given by

$$A = \begin{vmatrix} \lambda_1 & 0 & \cdots & 0 \\ 0 & \lambda_2 & \cdots & 0 \\ \vdots & & & \\ 0 & 0 & \cdots & \lambda_n \end{vmatrix}.$$

Since V is n-dimensional, there exists a basis $\{\alpha_1, \alpha_2,...,\alpha_n\}$ of V for which

$$T(\alpha_i) = \lambda_i \alpha_i, \quad \forall \, i = 1, 2,...,n.$$

This equation shows that each α_i in the basis is an eigenvector of T corresponding to each λ_i.

REMARK

☞ Let T be a linear operator on an n-dimensional vector space V over a field F. If T has n distinct eigenvalues, then T is diagonalizable.

Definition. *Let T be a linear operator on V and let λ be an eigenvalue of T. Then $E_\lambda = ker(T - \lambda I)$ is called the eigenspace of T corresponding to λ.*

The dimension of E_λ is called the geometric multiplicity of λ and the multiplicity of λ as a root of the characteristic polynomial is called the algebraic multiplicity of λ.

Theorem 2. *Let λ be an eigenvalue of a linear operator $T : V \to V$. Then the geometric multiplicity of λ does not exceed its algebraic multiplicity.*

Proof. Suppose the geometric multiplicity of λ be r. Then there are r linearly independent eigenvector of T corresponding to λ. Let these be $\alpha_1, \alpha_2,...,\alpha_r$.

If r = dim. V, then $\quad B = \{\alpha_1, \alpha_2,...,\alpha_r\}$ is an ordered basis of V and

$$[T]_B = \lambda I_r.$$

Thus $\quad |x I_r - [T]_B| = (x - \lambda)^r$ so that the algebraic multiplicity of λ is r.

If r < dim.$V = n$, then the set $B = \{\alpha_1, \alpha_2,...,\alpha_r\}$ can be extended to form a basis of V. Let the extension of B is given by

$$B_1 = \{\alpha_1, \alpha_2,...,\alpha_r, \beta_{r+1}, \beta_{r+2},...,\beta_n\}$$

We have

$$T(\alpha_1) = \lambda \alpha_1$$
$$T(\alpha_2) = \lambda \alpha_2$$
$$\cdots \quad \cdots \quad \cdots$$
$$T(\alpha_r) = \lambda \alpha_r$$
$$T(\beta_{r+1}) = a_{11}\alpha_1 +... + a_{1r}\alpha_r + a_{1(r+1)}\beta_{r+1} +...+ a_{1n}\beta_n$$
$$T(\beta_{r+2}) = a_{21}\alpha_1 +... + a_{2r}\alpha_r + a_{2(r+1)}\beta_{r+1} +...+ a_{2n}\beta_n$$
$$\cdots \quad \cdots \quad \cdots \quad \cdots \quad \cdots \quad \cdots \quad \cdots \quad \cdots$$
$$T(\beta_n) = a_{n1}\alpha_1 +... + a_{nr}\alpha_r + a_{n(r+1)}\beta_{r+1} +...+ a_{nn}\beta_n.$$

Thus the matrix of T in the basis B_1, is given by

$$[T]_{B_1} = \begin{vmatrix} \lambda & 0 & \dots & 0 & a_{11} & a_{21} & \dots & a_{n1} \\ 0 & \lambda & \dots & 0 & a_{12} & a_{22} & \dots & a_{n2} \\ \dots & \dots & \dots & \dots & \dots & \dots & \dots & \dots \\ 0 & 0 & \dots & \lambda & a_{1r} & a_{2r} & \dots & a_{nr} \\ 0 & 0 & \dots & 0 & a_{1(r+1)} & a_{2(r+1)} & \dots & a_{n(r+1)} \\ 0 & 0 & \dots & 0 & a_{1(r+2)} & a_{2(r+2)} & \dots & a_{n(r+2)} \\ \dots & \dots & \dots & \dots & \dots & \dots & \dots & \dots \\ 0 & 0 & \dots & 0 & a_{1n} & a_{2n} & \dots & a_{nn} \end{vmatrix}$$

$$= \begin{bmatrix} \lambda I_r & \vdots & A \\ \dots & & \dots \\ O & \vdots & B \end{bmatrix}$$

Since $[T]_{B_1}$ is a block triangular matrix, the characteristic polynomial of λI_r is $(x-\lambda)^r$, which must divide the characteristic polynomial of $[T]_{B_1}$ and hence T. Hence the algebraic multiplicity of λ for the operator T is at least r.

Theorem 3. *Let T be a linear operator on an n- dimensional vector space V over a field F and let $\lambda_1, \lambda_2,...,\lambda_k$ be all the distinct eigenvalues of T. Then the following statements are equivalent :*

   (i) T is diagonalizable

   (ii) $V = E_{\lambda_1} \oplus E_{\lambda_2} \oplus ... \oplus E_{\lambda_k}$, $E_{\lambda_i} = \ker(T - \lambda_i I)$

   (iii) the characteristic polynomial of T splits over F, and the algebraic multiplicity of each eigenvalue equals its geometric multiplicity

   (iv) the minimal polynomial of T is $(x-\lambda_1)(x-\lambda_2)...(x-\lambda_k)$.

Proof.(i) $\Rightarrow$ **(ii).** If T is diagonalizable, then V has an ordered basis

$$B = \{\alpha_1, \alpha_{12},..., \alpha_{1n}, \alpha_{21}, \alpha_{22},..., \alpha_{2n},..., \alpha_{k1},..., \alpha_{kn_k}\}$$

Such that each α_{i_m} is an eigenvector of T corresponding to eigenvalue λ_{i_m}. Obviously $n_1 + n_2 + ... + n_k = n$. Thus it is enough to show that for each m, the set $\{\alpha_{m1}, \alpha_{m2},....,\alpha_{mn_m}\}$ is a basis of E_{λ_m}.

Let if possible $\{\alpha_{m1}, \alpha_{m2},...., \alpha_{mn_m}\}$ is not a basis of $E_{\lambda m}$ for some $m = 1, 2, ...,$ k. Then there exists vector $\beta_m \in E_{\lambda m}$ such that $\{\alpha_{m1}, \alpha_{m2},....,\alpha_{mn_m}, \beta_m\}$ is a linealry independent subset of $E_{\lambda m}$. But $B \cup \{\beta_m\}$ is a linearly independent subset of V, which gives a contradiction as B spans V, therefore, $\{\alpha_{m1}, \alpha_{m2},...., \alpha_{mn_m}\}$ is a basis of $E_{\lambda m}$.

(ii) $\Rightarrow$ **(i)** It is obvious.

(i) $\Rightarrow$ **(iii)** If T is diagonalizable then there is an order basis B on V such that $[T]_B$ is a diagonal matrix with its diagonal elements as eigenvalues of T, so that characteristic ploynomial split over F. Since.

$$V = E_{\lambda_1} \oplus E_{\lambda_2} .... \oplus E_{\lambda_k}$$

so, if B_i is a basis of E_{λ_i} then

$$B = \bigcup_{i=1}^{k} B_i$$

is a basis of V and

$$[T]_B = \text{diag.} \, ([T_1]_{B_1} \cdots [T_k]_{B_k})$$

where T_i is the linear operator of E_{λ_i} induced by T.

Let dim. $E_{\lambda_i} = n_i$, since $T_i \, (\alpha) = T(\alpha) = \lambda_i$ for all $\alpha \in E_{\lambda_i}$) so that

$$[T_i]B_i = (\lambda_i I_{n_i}) \, .$$

Hence the characteristic polynomial is

$$(x - \lambda_1)^{n_1} (x - \lambda_2)^{n_2}, ..., (x - \lambda_k)^{n_k}$$

which shows that the algebraic multiplicity of each λ_i equals geometric multiplicity.

(iii) $\Rightarrow$ **(iv)**　Since characteristic polynomial of T splits over F and it is given by

$$(x - \lambda_1)^{n_1} (x - \lambda_2)^{n_2}, ..., (x - \lambda_k)^{n_k}$$

Also the geometic multiplicity and the algebraic multiplicity of each eigen values are equal, so for each λ_i, there are n_i linearly independent eignvalues of T which are $\alpha_{i1}, \alpha_{i2}, \cdots \alpha_{in_i}$.

Therefore, $B = \{\alpha_{11}, ... \alpha_{1n_1}, \alpha_{21}, ... \alpha_{2n_2}, ... \alpha_{11}, ... \alpha_{kn_k}\}$ is a basis of V and the linear operator $(T - \lambda_1 I)(T - \lambda_2 I) .... (T - \lambda_k I)$ maps all the basis elements of B to the zero vector.

Thus $(T - \lambda_1 I)(T - \lambda_2 I) .... (T - \lambda_k I) = 0$, hence $(x - \lambda_1)(x - \lambda_2) ... (x - \lambda_k)$ is minimal polynomial of T.

Theorem 4.　*A linear operator $T : V \rightarrow V$ has a diagonal matrix respresentation if and only if its minimal polynomial m (x) is product of distinct linear polynomials.*

Proof.　Suppose $m \, (x)$ is a product of distinct linear polynomials, let

$$m \, (x) = (x - \lambda_1) \, (x - \lambda_2) .... \, (x - \lambda_n) \qquad \qquad ...(1)$$

where $\lambda_1, \lambda_2, ..., \lambda_n$, are distinct scalars.

By primary decomposition theorem V is direct sum of subspaces $W_1, W_2, ..., W_n$, where,

$$W_1 = \ker \, (T - \lambda_1 I), \; W_2 = \ker \, (T - \lambda_2 I), \; W_n = \ker \, (T - \lambda_n I).$$

If $\alpha \in W_i$, then

$$(T - \lambda_i I) \, (\alpha) = 0 \Rightarrow T(\alpha) = \lambda_i \alpha$$

$\Rightarrow$ every vector in W_i is an eigen vector corresponding to the eigenvalue λ_i.

Thus, the union of bases for $W_1, W_2 ..... W_n$, form a basis of V. This basis consists of eigen vectors and so T is a diagonlizable

Conversely, suppose T has a diagonal matrix representation, implying that V has a basis consisting of eigevectors of T.

Let $\lambda_1, \lambda_2, ... \lambda_r$ be the distinct eigenvalus of T, then

$$P(T) = (T - \lambda_1 I) \, (T - \lambda_2 I) .... \, (T - \lambda_r I).$$

Since $(T - \lambda_1 I)$ maps each bsais vector to 0 so that $p(T) = 0$.

Thus, the minimal polynomial $m(x)$ of T divides the polynomial $P(x) = (x - \lambda_1)(x - \lambda_2) ... (x - \lambda_r)$. According $m(x)$ is the product of distinct linear polynomials.

Solved Examples

Example 1. *If T is a linear operator on R^3 which is represented in the standard ordered basis by*

$$A = \begin{bmatrix} 5 & -6 & -6 \\ -1 & 4 & 2 \\ 3 & -6 & -4 \end{bmatrix}$$

then show that T is diagonalizable.

Solution. The characteristic polynomial of A is

$$\Delta(x) = |xI - A| = \begin{vmatrix} x-5 & 6 & 6 \\ 1 & x-4 & -2 \\ -3 & 6 & x+4 \end{vmatrix}$$

$$= (x-5)\{(x-4)(x+4) + 12\} - 6\{x + 4 - 6\} + 6\{6 + 3(x-4)\}$$
$$= (x-5)(x^2 - 4) - 6x + 12 + 18x - 36$$
$$= x^3 - 5x^2 - 4x + 20 - 6x + 12 + 18 - 36$$
$$= x^3 - 5x^2 + 8x - 4$$
$$\Delta(x) = (x-1)(x-2)^2$$

Now we find the minimal polynomial of T.

The possible form of the minimal polynomial $m(x)$ of T are :

(i) $m_1(x) = (x-1)(x-2)$

(ii) $m_2(x) = (x-1)(x-2)^2$

Since $m_2(A) = 0$ may be the minimal polyonmial of T of $m_1(A) \neq 0$, but if $m_1(A) = 0$, then the minimal polynomial will be $m_1(x)$ but not the $m_2(x)$. So we test whether $m_1(A) = 0$, or $m_1(A) \neq 0$.

Since, $m_1(A) = (A - I)(A - 2I)$.

Now $A - I = \begin{bmatrix} 5-1 & -6 & -6 \\ -1 & 4-1 & 2 \\ 3 & -6 & -4-1 \end{bmatrix} = \begin{bmatrix} 4 & -6 & -6 \\ -2 & 3 & 2 \\ 3 & -6 & -5 \end{bmatrix}$

And $A - 2I = \begin{bmatrix} 5-2 & -6 & -6 \\ -1 & 4-2 & 2 \\ 3 & -6 & -4-2 \end{bmatrix} = \begin{bmatrix} 3 & -6 & -6 \\ -1 & 2 & 2 \\ 3 & -6 & -6 \end{bmatrix}$

So $(A-I)(A-2I) = \begin{bmatrix} 4 & -6 & -6 \\ -2 & 3 & 2 \\ 3 & -6 & -5 \end{bmatrix} \begin{bmatrix} 3 & -6 & -6 \\ -1 & 2 & 2 \\ 3 & -6 & -6 \end{bmatrix} = \begin{bmatrix} 0 & 0 & 0 \\ 0 & 0 & 0 \\ 0 & 0 & 0 \end{bmatrix} = 0$

$m_1(A) = 0$

Thus the minimal polynomial of T is $m_1(x) = (x-1)(x-2)$
which is the product of monic linear polynomial over F. Hence T is diagonalizable.

Example 2. *Let $T : R^3 \rightarrow R^3$ be defined by $T(x, y, z) = (2x + 3y - 2z, 5y + 4z, x - z)$. Find the characteristic polynomial $\Delta(t)$ of T.*

Solution. Let $B = \{(1, 0, 0), (0, 1, 0), (0, 0, 1)\}$ be the standard basis of R^3. So that

$$T(1, 0, 0) = (2, 0, 1)$$
$$T(0, 1, 0) = (3, 5, 0)$$

$$T (0, 0, 1) = (-2, 4, -1)$$

Thus, $\qquad [T]_B = \begin{bmatrix} 2 & 3 & -2 \\ 0 & 5 & 4 \\ 1 & 0 & -1 \end{bmatrix}$

The characteristic polynomial of T is given by

$$\Delta(t) = \begin{vmatrix} t - 2 & -3 & 2 \\ 0 & t - 5 & -4 \\ -1 & 0 & t + 1 \end{vmatrix}$$

$$= (t - 2) \{(t - 5) (t + 1) - 0\} + 3 (0 - 4)\} + 2 \{0 + t - 5\}$$
$$= (t - 2) (t^2 - 4t - 5) - 12 + 2t - 10$$
$$= t^3 - 2t^2 + 4t^2 + 8t - 5t + 10 - 12 + 2t - 10$$
$$= t^3 - 6t^2 + 5t - 12$$

Example 3. *If A be a square matrix given by* $A = \begin{bmatrix} 3 & 0 & 0 \\ 0 & 2 & -5 \\ 0 & 1 & -2 \end{bmatrix}$

then find all the eigenvalues of A viewed as matrics over (i) Real field **R** *(ii) Complex field* **C**. *Also, find in which case the matrix A is diagonalizable.*

Solution. The characteristic polynomial of A is given by

$$\Delta(t) = | tI - A | = \begin{vmatrix} t - 3 & 0 & 0 \\ 0 & t - 2 & 5 \\ 0 & -1 & t + 2 \end{vmatrix}$$

$$= (t - 3)\{(t - 2) (t + 2) + 5\}$$
$$\Delta (t) = (t - 3) (t^2 + 1) = (t - 3) (t - i) (t + i)$$

The roots of $\Delta t = 0$ are $t = 3, -i, i$.

(i) If A is a matrix over the field of reals, then A has only one eigenvalue which is 3. Thus, A can not be diagonalized.

(ii) If A is a matrix over the field of complex numbers, then A has three distinct eigenvalues, so that the minimal polynomial of A is equal to $\Delta (t)$ which is the product of linear polynomials. Hence A is digonalizable.

Example 4. *If T be a linear tranformation on V (F). Then the following are equivalent :*

   (i) λ *is the characteristic value of T*

   (ii) *the transformation* $T - \lambda I$ *is singular*

   (iii) $|(T - \lambda I)| = 0$.

Solution. **(i)** $\Rightarrow$ **(ii).** If λ is a characteristic value of T, then there exists a non-zero vector $\alpha \in V$ such that

$$T (\alpha) = \lambda \alpha \Rightarrow T(\alpha) = \lambda I(\alpha) \qquad\qquad [\because I(\alpha) = \alpha]$$
$$\Rightarrow T (\alpha) - \lambda I (\alpha) = 0 \Rightarrow (T - \lambda I) (\alpha) = 0 \text{ with } \alpha \neq 0.$$

$\therefore \qquad T - \lambda I$ is singular.

(ii) $\Rightarrow$ **(iii).** Since $(T - \lambda I)$ is singular, *i.e.*, it is not invertible so that $|(T - \lambda I)| = 0$

(iii) $\Rightarrow$ **(i).** Since $|(T - \lambda I)| = 0$, therefore $(T - \lambda I)$ is not invertible, *i.e.*,

$$(T - \lambda I)\alpha = 0 \Rightarrow T(\alpha) - \lambda I(\alpha) = 0$$

$$\Rightarrow \qquad T(\alpha) - \lambda\alpha = 0 \Rightarrow T(\alpha) = \lambda\alpha$$

∴ λ is an eigenvalue of T.

Example 5. *Show that the matrix* $A = \begin{bmatrix} 1 & 2 \\ 0 & 1 \end{bmatrix}$ *is diagonalizable over the field of complex numbers.*

Solution. The characteristic polynomial of A is

$$\Delta(t) = | tI - A | = \begin{vmatrix} t-1 & -2 \\ 0 & t-1 \end{vmatrix} = (t-1)^2$$

Thus A has only one eignvalue which is I.

The eigenvectors of A of $\lambda = 1$ are given by the solution of the homogeneous system

$$(I - A)\begin{bmatrix} x_1 \\ x_2 \end{bmatrix} = \begin{bmatrix} 0 \\ 0 \end{bmatrix} \text{ or } \begin{bmatrix} 0 & -2 \\ 0 & 0 \end{bmatrix}\begin{bmatrix} x_1 \\ x_2 \end{bmatrix} = \begin{bmatrix} 0 \\ 0 \end{bmatrix} \text{ or } -2x_2 = 0$$

Above equation gives $x_2 = 0$ and x_1 is assigned let $x_2 = 1$, thus A has only one linearly independent eigenvector $\begin{bmatrix} 1 \\ 0 \end{bmatrix}$. Hence A is not diagonlizable.

Example 6. *Let V be the vector space of functions which have B = {sinθ , cosθ} as a basis and let D be the differential operator on V. Find the characterstic polynomial Δ(t) of D.*

Solution. First, we find the matrix A which represents D in the basis B :

$$D(\sin\theta) = \cos\theta = 0(\sin\theta) + 1.(\cos\theta)$$
$$D(\cos\theta) = -\sin\theta = (-1)\sin\theta + 0(\cos\theta)$$

Thus, $\qquad [D]_B = A\begin{bmatrix} 0 & 1 \\ -1 & 0 \end{bmatrix}$

Now, $\qquad \Delta(t) = \begin{bmatrix} t & -1 \\ 1 & t \end{bmatrix} = t^2 + 1$

This is the required characteristic polynomial of A.

Example 7. *What is the algebraic and geometric multiplicity of λ = −2, where λ = −2 is one of the eigenvalue of the matrix*

$$A = \begin{bmatrix} -3 & 1 & -1 \\ -7 & 5 & -1 \\ -6 & 6 & -2 \end{bmatrix}$$

Solution. The characteristic polynomial of A is given by

$$\Delta(t) = | tI - A | = \begin{vmatrix} t+3 & -1 & 1 \\ 7 & t-5 & 1 \\ 6 & -6 & t+2 \end{vmatrix}$$

$$= (t + 3)\{(t - 5)(t + 2) + 6\} + 1\{7(t + 2) - 6\} + 1\{-42 - 6(t - 5)\}$$
$$= (t + 3)\{t^2 - 3t - 10 + 6\} + 7t + 8 - 42 - 6t + 30$$
$$= (t + 3)(t^2 - 3t - 4) + t - 4\}$$
$$= t^3 + 3t^2 - 3t^2 - 9t - 4 - 12 + t - 4$$
$$\Delta(t) = t^3 - 12t - 16 = (t + 2)^2 - (t - 4).$$

Since the factor $(t + 2)$ occurs twice in $\Delta(t)$, so that the algebaric multiplicity of $\lambda = -2$ is two.

Now we find a basis of the eigenspace of $\lambda = -2$.

Let $X = \begin{bmatrix} x_1 \\ x_2 \\ x_3 \end{bmatrix}$ be non- zero eigenvector corresponding to $\lambda = -2$, then

$$(\lambda I - A)X = 0 \text{ or } (-2I - A)X = 0$$

or $\begin{bmatrix} 1 & -1 & 1 \\ 7 & -7 & 1 \\ 6 & -6 & 0 \end{bmatrix} \begin{bmatrix} x_1 \\ x_2 \\ x_3 \end{bmatrix} = \begin{bmatrix} 0 \\ 0 \\ 0 \end{bmatrix}$

or
$$\begin{aligned} x_1 - x_2 + x_3 &= 0 \\ 7x_1 - 7x_2 + x_3 &= 0 \\ 6x_1 - 6x_2 &= 0 \end{aligned}$$

or
$$\begin{aligned} x_1 - x_2 + x_3 &= 0 \\ 7x_1 - 7x_2 + x_3 &= 0 \\ x_1 - x_2 &= 0. \end{aligned}$$

This system has only one independent solution, *i.e.*, $x_1 = 1$, $x_2 = 1$ and $x_3 = 0$.
Thus $\alpha = (1, 1, 0)$ forms, a basis of the eigenspace E_2. Hence the geometirc multiplicity of $\lambda = -2$ is one as dim $E_2 = 1$.

EXERCISE 7.5

1. Let α and β be eigenvectors of T corresponding to two eigenvalues λ and μ respectively. Show that for non-zero scalars a and b, $a\alpha + b\beta$ is not an eigenvector. of T.

2. Suppose $T : V \to V$ is a linear operator on a vector space V with dimension n. Define the characteristic polynomial $\Delta(t)$ of T.

3. Suppose a linear map $T : V \to V$ may have matrix representations, is it possible for T to have many characteristic polynomials?

4. Let $T: R^2 \to R^2$ be the linear operator which rotates each vector $\alpha \in R^2$ by an angle $\theta = \pi/2$. Show geometrically that T has no eigenvalues and hence no eigenvectors.

5. Let λ be an eigenvalue of a linear operator $T: V \to V$. Let E_λ be the set of all eigenvectors of T belonging to λ. Show that E_λ is a subspace of V.

6. Let λ be an eigenvalue of a linear operator $T: V \to V$. Define the algebraic multiplicity and the geometric multiplicity of λ .

7. Let A and B be n–square matrices. Show that AB and BA have the same eigenvalues.

8. Show that if λ is an eigenvalue of T and $p(x) \in F[x]$, then $p(\lambda)$ is an eigenvalue of $p(T)$.

9. Find eigenvalues and eigenvectors for the following matrices over **C**, the field of complex numbers:

   (*i*) $A = \begin{bmatrix} 6 & -1 & 2 \\ 4 & 1 & 2 \\ -10 & 0 & 3 \end{bmatrix}$ (*ii*) $A = \begin{bmatrix} 4 & 2 & 2 \\ 3 & 3 & 2 \\ -3 & -1 & 0 \end{bmatrix}$

   (*iii*) $A = \begin{bmatrix} 0 & 0 & 1 \\ 1 & 0 & -1 \\ 0 & 1 & 1 \end{bmatrix}$.

10. Prove that if every non-zero is an eigenvector of T, then T is a scalar multiple of the identity operator.

11. Prove the if char $F \neq 2$, then T can be expressed as the sum of two invertible linear operators.

12. If $V_2(R)$ is a vector space and T is a linear transformation on $V_2(R)$ whose matrix relative to the basis of $V_2(R)$ is $A = \begin{bmatrix} 1 & 2 \\ 3 & 2 \end{bmatrix}$. Find the dimension of the eigenspace of T.

13. Find the minimal polynomial for real matrix

   $A = \begin{bmatrix} 7 & 4 & -1 \\ 4 & 7 & -1 \\ -4 & -4 & 4 \end{bmatrix}$.

14. Find the minimal polynomial for real matrix
$$A = \begin{bmatrix} 1 & 1 & 1 \\ 1 & 1 & 1 \\ 1 & 1 & 1 \end{bmatrix}.$$

15. Find the invertible matrix P such that $P^{-1}AP$ is a diagonal matrix, where $A = \begin{bmatrix} 1 & 4 \\ 2 & 3 \end{bmatrix}$.

16. If T be the linear transformation on $V_3(\mathbf{R})$ which is represented in the standard ordered basis by the matrix
$$\begin{bmatrix} -9 & 4 & 4 \\ -8 & 3 & 4 \\ -16 & 8 & 7 \end{bmatrix}.$$

Prove that T is diagonalizable.

17. Show that the matrix $\begin{bmatrix} 1 & 1 \\ 0 & 1 \end{bmatrix}$ is not diagonalizable.

18. If T is linear operator on $V_3(\mathbf{R})$ which is represented in the standard ordered basis by the matrix:

14. (right column)
$$A = \begin{bmatrix} 5 & -6 & -6 \\ -1 & 4 & 2 \\ -3 & -6 & -4 \end{bmatrix}.$$ Find the eigenvalues of A and prove that T is diagonalizable.

19. Let λ be an eigenvalue of $T \in L(V)$ with algebraic multiplicity m. Prove that T is not diagonalizable if rank $(T-\lambda I)>n-m$, where $n=\dim. V$.

20. A linear operator $T:\mathbf{R}^3 \to \mathbf{R}^3$ defined by $T(x,y,z) = \{2x + y, y-z, 2y+4z\}$.
   (i) Find the characteristic polynomial $\Delta(t)$ of T
   (ii) Find the eigenvalues of T.
   (iii) Is T diagonalizable ?

21. Suppose α is a non-zero eigenvector of T. Show that, for any $k \in F$, $k\alpha$ is an eigenvector of kT.

22. Suppose λ is an eigenvalue of a linear operator T:
   (i) Show that λ^2 is an eigenvalue of T^2
   (ii) Show that λ^n is an eigenvalue of T^n for $n \geq 1$.

Hints to Selected Problems

1. Since α, β are the eigenvectors of T corresponding to two distinct eigenvalues λ and μ. Then we have
$$T(\alpha) = \lambda\alpha \text{ and } T(\beta) = \mu\beta. \text{ Now for } a, b \text{ (non-zero) scalars,}$$
$$T(a\alpha + b\beta) = aT(\alpha) + bT(\beta) = a\lambda\alpha + b\mu\beta \neq v(a\alpha + b\beta), \text{ for some scalar } v.$$
$\therefore$ $a\alpha + b\beta$ is not an eigenvector of T.

5. $E_\lambda = \{\alpha \in V: T(\alpha) = \lambda\alpha\}$. Let $\alpha, \beta \in E_\lambda$ and $a, b \in F$, then
$$T(a\alpha + b\beta) = aT(\alpha) + bT(\beta) = a\lambda a + b\mu\beta \qquad [\because T(\alpha) = \lambda\alpha, T(\beta) = \lambda\alpha]$$
$$= \lambda(a\alpha + b\beta).$$
$\therefore$ $a\alpha + b\beta$ is also an eigenvector of T corresponding to λ.
$\therefore$ $a\alpha + b\beta \in E_\lambda$. Hence E_λ is subspace of V.

8. Since λ is an eigenvalue of T so that for a non-zero vector $\alpha \in V, T(\alpha) = \lambda\alpha$. For $p(x) = F[x]$.
We have $(p(T))(\alpha) = p(T)(\alpha) = p(\lambda\alpha)$ $\qquad [\because T(\alpha)=\lambda\alpha]$
$$= p(\lambda)\alpha.$$
This show that $p(\lambda)$ is an eigenvalue of $p(T)$.

10. Let α be any non-zero vector and if α is an eigenvector of T, then $T(\alpha) = \lambda\alpha$, for some scalar λ
$$(T(\alpha)) = \lambda\alpha = \lambda I(\alpha) \qquad\qquad [\because I(\alpha)=\alpha]$$
$$(T-\lambda I)(\alpha) = 0, \forall \, \alpha \in V$$
$\Rightarrow$ $T=\lambda I$

12. Since matrix representation of T is given by $A = \begin{bmatrix} 1 & 2 \\ 3 & 2 \end{bmatrix}$.

Eigenvalues of A are the root of the equation $\Delta(t) = |A-tI| = 0$ or $\begin{vmatrix} 1-t & 2 \\ 3 & 2-t \end{vmatrix} = 0$

or $(1-t)(2-t)-6 = 0$ or $t^2-3t-4 = 0$ or $(t-4)(t + 1) = 0$ or $t = -1, 4$.

Let $X_1 = \begin{bmatrix} x_1 \\ x_2 \end{bmatrix}$ be an eigenvector corresponding to $\lambda = -1$

$\therefore$ $(A–\lambda I)X_1=0$ or $\begin{bmatrix} 2 & 2 \\ 3 & 3 \end{bmatrix}\begin{bmatrix} x_1 \\ x_2 \end{bmatrix} = \begin{bmatrix} 0 \\ 0 \end{bmatrix}$.

or $2x_1 +2x_2 = 0$, $3x_1 +3x_2 =0$, i.e., $x_1 + x_2 = 0$.
So we have one equation in two variables, then set $x_2 =-1, x_1 =1$.

$\therefore$ $X_1 = \begin{bmatrix} 1 \\ -1 \end{bmatrix}$. Again $X_2 = \begin{bmatrix} x_1 \\ x_2 \end{bmatrix}$ be an eigenvector corresponding to $\lambda = 4$.

$\therefore$ $(A–\lambda I)X_1 = 0$ or $\begin{bmatrix} -3 & 2 \\ 3 & -2 \end{bmatrix}\begin{bmatrix} x_1 \\ x_2 \end{bmatrix} = \begin{bmatrix} 0 \\ 0 \end{bmatrix}$.

or $-3x_1 + 2x_2 = 0$, $3x_1 - 2x_2 = 0$
$\therefore$ We have one equation in two variables so set $x_2 = 3$, and $x_1 = 2$.

$\therefore$ $X_2 = \begin{bmatrix} 2 \\ 3 \end{bmatrix}$.

Hence eigenspace of T is generated by two vectors so the dimension of eigenspace of T is 2.

16. The characteristic polynomial of the matrix of T is given by $\Delta(t) = \begin{vmatrix} -9-t & 4 & 4 \\ -8 & 3-t & 4 \\ -16 & 8 & 7-t \end{vmatrix}$

$= (-9 - t)\{(3 - t)(7 - t) - 32\} + 4\{-64 +(- 8(7- t)\} + 4\{-64 + 16(3 - t)\}$
$= -(9 + t)\{21 - 10t + t^2 - 32\} + 4\{-8 - 8t\} + 4\{-16 - 16t\}$
$= -(9 + t)(t_2-10t-11)-32-32t-64-64t$
$= -(t^3 - 10t^2 - 11t + 9t^2 - 90t - 99) - 96 - 96t = -t^3 + t^2 + 101t + 99 - 96 - 96t$
$\therefore$ $\Delta(t)= -t^3 + t^2 + 5t + 3 = (t+1)^2(3–t)$.
There are two polynomials $m_1(t) = (t + 1)^2(3–1)$ and $m_2(t) = (t + 1)(3 - t)$ one of them may be minimal polynomial.

Now $(A + I)(3I–A) = \begin{bmatrix} -8 & 4 & 4 \\ -8 & 4 & 4 \\ -16 & 8 & 8 \end{bmatrix}\begin{bmatrix} 12 & -4 & -4 \\ 8 & 0 & -4 \\ 16 & -8 & -4 \end{bmatrix} = \begin{bmatrix} 0 & 0 & 0 \\ 0 & 0 & 0 \\ 0 & 0 & 0 \end{bmatrix} = \mathbf{O}$. $\therefore m_2(A)=O$

Thus $m_2(t) = (t+ 1)(3 - t)$ is a minimal polynomial of A which is a product of linear polynomials. Hence the matrix of T is diagonalizable,

21. Let λ be an eigenvalue of T corresponding to which α is its eigenvector. Thus $T(\alpha) = \lambda\alpha$.
For $k \in F$. $(kT)(\alpha) = k(T(\alpha)) = k\lambda\alpha = (k\lambda)\alpha$.
Hence α is also an eigenvector of T corresponding to eigenvalue $k\lambda$.

22. Let $\alpha \neq 0$ be an eigenvector of T corresponding to an eigenvalue k. Then $T(\alpha) = \lambda\alpha$.
(i) $(T^2)(\alpha) = T(T(\alpha)) = T(\lambda\alpha) = \lambda(T(\alpha)) =\lambda(\lambda\alpha) = \lambda^2\alpha$.
$\therefore$ λ^2 is an eigenvalue of T^2. Similarly we may prove that λ^n is an eigenvalue of T^n.

=============================**Answers**=============================

3. No.   **9.** (ii) 1,2,4;[0, 1, –1]',[1, 1, –2]'[1, 1, –1]'.   (iii) [1,0,1]';[1–(1+i)i]',[1–(1–t)–1]'.

12. Two.   **13.** $m(x)=x^2–15+36$. **14.** 0,3;[10–1]',[aaa]', $a\in F$. **15.** $p=\begin{bmatrix} 1 & 2 \\ 1 & -1 \end{bmatrix}$.

18. 1,2,2.   **20.** (i) $\Delta(t)=t^3–7t^2+16t–12$.   (ii) 2,2,3.   (iii) No

7.39 INNER PRODUCT SPACE

Definition 1. *Let V be a vector space over F(**R** or **C**). An inner product on V is a function* $(.,.): V \times V \to F$ *which satisfies the following properties:*

(i) $(\alpha, \alpha) > 0$ *for all non-zero vectors* α *in V.*

(ii) $(\alpha, \beta) = \overline{(\beta, \alpha)}, \forall \alpha, \beta \in V$

(iii) $(a\alpha + b\beta, \gamma) = a(\alpha, \gamma) + b(\beta, \gamma) \ \forall \ \alpha, \beta, \gamma \in V$ *and* $a, b \in F.$

For example:

(1) On F^n, we may defined an inner product, known as standard inner product as:

Let $\alpha = (a_1, a_2, ..., a_n)$ and $\beta = (b_1, b_2, ..., b_n)$ be two elements in F^n. Then standard inner product is defined by

$$(\alpha, \beta) = \sum_{i=1}^{n} a_i \overline{b_i} \ , \text{ if } a_i, b_i \ \in F = \mathbf{C}.$$

and

$$(\alpha, \beta) = \sum_{i=1}^{n} a_i b_i \ , \text{ if } a_i, b_i \ \in F = \mathbf{R}.$$

(2) Let V be the vector space of all continuous complex-valued functions on the unit interval $0 \le t \le 1$.

Let

$$(f, g) = \int_0^1 f(t)\overline{g(t)}d(t).$$

Then $(., .)$ is an inner product on $V.$

For example: let V and W be vector spaces over F and suppose that $\{., .\}$ is an inner product on W. If T is a non-singular linear transformation from V into W, then $p_T(\alpha, \beta) = (T(\alpha), T(\beta))$ defines an inner product p_T on $V.$

REMARKS

☞ In case of reals, the standard inner product is often called the dot or scalar product and it is denoted by α, β .

☞ $(\alpha, a\beta + b\gamma) = \overline{(a\beta + b\gamma, \alpha)} = \overline{a} \, \overline{(\beta, \alpha)} + \overline{b} \, \overline{(\gamma, \alpha)} = \overline{a}(\alpha, \beta) + \overline{b}(\alpha, \gamma)$.

Definition 2. *A vector space V together with an inner product is called an inner product space.*

Definition 3. *The norm or length of a vector* $\alpha \in V$ *is defined by* $\|\alpha\| = \sqrt{\alpha, \alpha}$.

Definition 4. *A vector* $a \in V$ *is called a unit vector, if* $\|a\| = 1.$

Definition 5. *A finite-dimensional real inner product space is often called a Euclidean space.*

Definition 6. *A complex inner product space is often called a Unitary space.*

REMARKS

☞ The distance between two vectors α and β in an inner product space V is $d(\alpha, \beta) = \|\alpha - \beta\|$.

☞ If θ be the angle between two vectors α and β in an inner product space V, then $\cos \theta = \dfrac{(\alpha, \beta)}{\|\alpha\|\|\beta\|}$.

Theorem 1. *Let V be an inner product space over F and let α, β ∈ V. Then*

(i) $\|\alpha \pm \beta\|^2 = \|\alpha\|^2 + 2\mathrm{Re}.(\alpha, \beta) + \|\beta\|^2$

where Re.(α, β) denotes the real parts of (α, β).

(ii) $\|\alpha + \beta\|^2 + \|\alpha - \beta\|^2 = 2\|\alpha\|^2 + 2\|\beta\|^2$ [Parallelogram law]

(iii) $\|k\alpha\| = |k|\,\|\alpha\|$ $\forall\, k \in F$

(iv) $4(\alpha,\beta) = \begin{cases} \|\alpha+\beta\|^2 - \|\alpha-\beta\|^2 & \text{, if } F = \mathbf{R} \\ \|\alpha+\beta\|^2 - \|\alpha-\beta\|^2 + i\|\alpha+i\beta\|^2 - i\|\alpha-i\beta\|^2 & \text{, if } F = \mathbf{C} \end{cases}$

                                     [Polarization identities]

(v) $|(\alpha,\beta)| \le \|\alpha\|\,\|\beta\|$ [Cauchy-Schwarz Inequality]

(vi) $\|\alpha \pm \beta\| \le \|\alpha\| + \|\beta\|$ [Triangle inequality]

(vii) $|\,\|\alpha\| - \|\beta\|\,| > \|\alpha - \beta\|$

Proof.   (i)

$$\|\alpha \pm \beta\|^2 = (\alpha\pm\beta, \alpha\pm\beta) = (\alpha, \alpha\pm\beta) \pm (\beta, \alpha\pm\beta)$$
$$= (\alpha,\alpha) \pm (\alpha,\beta) \pm (\beta,\alpha) + (\beta,\beta)$$
$$= \|\alpha\|^2 \pm (\alpha,\beta) + \overline{(\alpha,\beta)} + \|\beta\|^2$$
$$= \|\alpha\|^2 + 2\mathrm{Re}.(\alpha,\beta) + \|\beta\|^2.$$

(ii) From (i), we have

$$\|\alpha + \beta\|^2 = \|\alpha\|^2 + 2\mathrm{Re}.(\alpha,\beta) + \|\beta\|^2$$

and $\|\alpha - \beta\|^2 = \|\alpha\|^2 - 2\mathrm{Re}.(\alpha,\beta) + \|\beta\|^2.$

Adding these equations, we get

$$\|\alpha + \beta\|^2 + \|\alpha - \beta\|^2 = 2\|\alpha\|^2 + 2\|\beta\|^2.$$

(iii) $\|k\alpha\|^2 = (k\alpha, k\alpha) = k(\alpha, k\alpha) = k\bar{k}\,(\alpha,\alpha) = |k|^2\|\alpha\|^2.$

∴ $\|k\alpha\| = |k|\,\|\alpha\|.$

(iv) From (i), we have

$$\|\alpha + \beta\|^2 = \|\alpha\|^2 + 2\mathrm{Re}.(\alpha,\beta) + \|\beta\|^2$$

and $\|\alpha - \beta\|^2 = \|\alpha\|^2 - 2\mathrm{Re}.(\alpha,\beta) + \|\beta\|^2.$

On subtracting

$$\|\alpha + \beta\|^2 - \|\alpha - \beta\|^2 = 4\mathrm{Re}.(\alpha,\beta)$$

If $F = \mathbf{R}$, then

$$\mathrm{Re}.(\alpha,\beta) = (\alpha,\beta)$$

∴ $\|\alpha + \beta\|^2 - \|\alpha - \beta\|^2 = 4(\alpha,\beta).$

This proves the first result.

Also, if $F = \mathbf{C}$.

From (i), we have

$$\|\alpha + \beta\|^2 - \|\alpha - \beta\|^2 = 2(\alpha,\beta) + 2(\beta,\alpha) \qquad \text{...(1)}$$

Now, $\|\alpha + i\beta\|^2 = (\alpha + i\beta, \alpha + i\beta) = (\alpha, \alpha + i\beta) + (i\beta, \alpha + \beta)$

$$= (\alpha,\alpha) + i(\alpha,\beta) + (i\beta,\alpha) - i(i\beta,\beta)$$

$$= \|\alpha\|^2 - i(\alpha,\beta) + i(\beta,\alpha) - i.i(\beta,\beta)$$

$$= \|\alpha\|^2 - i(\alpha,\beta) + i(\beta,\alpha) + \|\beta\|^2$$

$$\therefore \quad i\|\alpha + i\beta\|^2 = i\|\alpha\|^2 + (\alpha,\beta) - (\beta,\alpha) + i\|\beta\|^2. \qquad \text{...(2)}$$

Similarly, $-i\|\alpha - i\beta\|^2 = -i\|\alpha\|^2 + (\alpha,\beta) - (\beta,\alpha) - i\|\beta\|^2. \qquad \text{...(3)}$

Adding (1), (2) and (3), we get

$$\|\alpha + \beta\|^2 - \|\alpha - \beta\|^2 + i\|\alpha + i\beta\|^2 - \|\alpha - i\beta\|^2 = 4(\alpha,\beta).$$

This proves the second result.

(v) If $\beta = 0$, then the statement is trivially true. If $\beta \neq 0$, then for a, $b \in F$, using (i) and (iii), we have

$$\|a\alpha + b\beta\|^2 \geq 0.$$

$$\Rightarrow \quad |a|^2\|\alpha\|^2 + 2 \operatorname{Re}.a\bar{b}\,(\alpha,\beta) + |b|^2\|\beta\|^2 \geq 0$$

in particular, $a = \|\beta\|^2$ and, $b = -(\alpha,\beta)$, then

$$\Rightarrow \quad \|\alpha\|^2\|\beta\|^4 + 2\operatorname{Re}.\|\beta\|^2\{-\overline{(\alpha,\beta)}\,(\alpha,\beta)\} + |(\alpha,\beta)|^2\|\beta\|^2 \geq 0$$

$$\Rightarrow \quad \|\alpha\|^2\|\beta\|^4 - 2|(\alpha,\beta)^2|\|\beta\|^2 + |(\alpha,\beta)|^2\|\beta\|^2 \geq 0$$

$$\Rightarrow \quad \|\alpha\|^2\|\beta\|^4 - |(\alpha,\beta)|^2\|\beta\|^2 \geq 0$$

$$\Rightarrow \quad \|\alpha\|^2\|\beta\|^2 - |(\alpha,\beta)|^2 \geq 0 \qquad [\because \beta \neq 0]$$

$$\therefore \quad |(\alpha,\beta)| \leq \|\alpha\| + \|\beta\|$$

(vi) From (i), we have

$$\|\alpha \pm \beta\|^2 = \|\alpha\|^2 \pm 2\operatorname{Re}.(\alpha,\beta) + \|\beta\|^2$$

$$\leq \|\alpha\|^2 \pm 2|\operatorname{Re}.(\alpha,\beta)| + \|\beta\|^2$$

$$\leq \|\alpha\|^2 \pm 2|\alpha,\beta| + \|\beta\|^2$$

$$\leq \|\alpha\|^2 \pm 2\|\alpha\|\,\|\beta\| + \|\beta\|^2 \qquad \text{[Using v]}$$

$$\leq (\|\alpha\| + \|\beta\|)^2.$$

$$\therefore \quad \|\alpha \pm \beta\| \leq \|\alpha\| + \|\beta\|.$$

(vii) From (vi), we have

$$\|\alpha\| = \|(\alpha - \beta) + \beta\| \leq \|\alpha - \beta\| + \|\beta\|$$

so that $\qquad \|\alpha\| - \|\beta\| \leq \|\alpha - \beta\| \qquad$ Now interchanging α and β, we get

$$\|\beta\| - \|\alpha\| \leq \|\beta - \alpha\| = \|\alpha - \beta\|$$

$$\therefore \qquad +(\|\alpha\| - \|\beta\|) \leq \|\alpha - \beta\| \quad \text{or} \quad |\,\|\alpha\| - \|\beta\|\,| \leq \|\alpha - \beta\|.$$

7.40 ORTHOGONALITY AND ORTHONORMALITY

Definition 1. *Let V be an inner product space. Then any two vectors* $\alpha, \beta \in V$ *are said to be orthogonal to each other if* $(\alpha, \beta) = 0$.

For example:

(1) The zero vector is orthogonal to every vector of V.

    As $(0, \alpha) = (O\alpha, \alpha)$ $[\because 0\alpha = 0]$

                       $= 0 \, (\alpha, \alpha) = 0$.

(2) The vector (a, b) in R^2 is orthogonal to $(-b, a)$ with respect to the standard inner product, for $((a, b), (-b, a)) = -ab + ba = 0$.

(3) The standard basis of either R^n or C^n is an orthogonal with respect to the standard inner product.

Definition 2. *Let V be an inner product space. If S and T are the subsets of V, then S is said to be orthogonal to T, denoted by* $S \perp T$, *if for each vector* $s \in S$ *and* $t \in T$, $(s, t) = 0$.

Definition 3. *Let V be an inner product space, and S is a subset of V. Then a set of all those vectors of V which are orthogonal to each vector of S, is called the orthogonal complement of S, denoted by* $S^\perp$

i.e., $S^\perp = \{\alpha \in V : (\alpha, s) = 0, \forall \, s \in S\}$.

It can easily be verified that $S^\perp$ is a subspace of V, and if S is a subspace of V, then

$$S \cap S^\perp = \{0\}.$$

Also, if $S \perp T$, then $T \subseteq S^\perp$.

Definition 4. *The set of vectors in V is called an orthonormal set if*

    (i) $\|\alpha\| = 1, \forall \, \alpha \in V$

    (ii) $(\alpha, \beta) = 0, \forall \, \alpha, \beta \in V$.

For example : The standard basis of either $\mathbf{R}^n$ or $\mathbf{C}^n$ is an orthonormal set with respect to the standard inner product.

Definition 5. *A maximal orthonormal set of vectors in an inner product space V is called a complete orthonormal set.*

Definition 6. *If V be a finite dimensional inner product space, then a basis of V which is also orthonormal set (orthogonal set) is called an* orthonormal basis (orthogonal basis).

Theorem 1. *An orthogonal set of non-zero vectors is linearly independent.*

Proof.   Let S be a finite or infinite orthogonal set of non-zero vectors of an inner product space $V(F)$. Suppose $\alpha_1, \alpha_2, ..., \alpha_n$ are distinct vectors in S.

Let $\displaystyle\sum_{i=1}^{n} a_i \alpha_i = 0$ for $a_i \in F$.

Then, for any $k = 1, 2, 3, \ldots, n$ suppose $\alpha_k = 0$.

So, $\left(\displaystyle\sum_{i=1}^{n} a_i \alpha_i, \, \alpha_k \right) = (0, \alpha_k) = 0.$

or $(a_1 \alpha_1 + a_2 \alpha_2 + \ldots + a_k \alpha_k + \ldots + a_n \alpha_n, \, \alpha_k) = 0$

or $a_1(\alpha_1, \alpha_k) + a_2(\alpha_2, \alpha_k) + \ldots + a_k(\alpha_k, \alpha_k) + \ldots + a_n(\alpha_n, \alpha_k) = 0$

or $\qquad a_k(\alpha_k, \alpha_k) = 0 \qquad\qquad\qquad [\because (\alpha_j, \alpha_j) = 0 \text{ for } i \neq j]$

or $\qquad\qquad a_k = 0 \text{ as } (\alpha_k, \alpha_k) = 1$.

$\therefore \qquad\qquad a_k = 0 \text{ for } k = 1, 2, 3, ..., n$.

Hence, S is a linearly independent subset of V.

Theorem 2. *In a finite-dimensional inner product space a complete orthonormal set is a basis.*

Proof. Let S be a complete orthonormal subset of a finite dimensional inner product space V. Since V is finite dimensional so that S is linearly independent.

Also, S is finite set.

Let $\qquad\qquad S = \{\alpha_1, \alpha_2,..., \alpha_m\}$.

Then, for some $\alpha \in V$, we have

$$\beta = \alpha - \sum_{i=1}^{m} (\alpha, \alpha_i)\alpha_i \ .$$

Let $\alpha_k \neq 0$ for $1 \leq k \leq m$. Then

$$(\beta, \alpha_k) = \left(\alpha - \sum_{i=1}^{m}(\alpha, \alpha_i)\alpha_i, \alpha_k \right) = (\alpha, \alpha_k) - \sum_{i=1}^{m}(\alpha, \alpha_k)(\alpha_i, \alpha_k)$$

$$= (\alpha, \alpha_k) - (\alpha, \alpha_k)(\alpha_k, \alpha_k) \quad [\because (\alpha_i, \alpha_k) = 0 \text{ for } i \neq k]$$

$$= (\alpha, \alpha_k) - (\alpha, \alpha_k) \qquad\qquad\qquad [\because (\alpha_k, \alpha_k) = 1]$$

$\therefore \qquad\qquad (\beta, \alpha_k) = 0, \quad \forall k = 1, 2, 3,...,m.$

Thus, $\beta \perp S$, Since $\alpha_k \neq 0$, so that

$$\beta = 0 \Rightarrow \alpha = \sum_{i=1}^{m}(\alpha, \alpha_i)\alpha_i.$$

Hence, S forms a basis of V.

Theorem 3.(Gram-Schmidt Orthogonalization). *Let V be an inner product space and let $\beta_1, \beta_2,.., \beta_n$ be any linearly independent vectors in V. Then one may construct orthogonal vectors $\alpha_1, \alpha_2,...,\alpha_n$ in V such that for each $k = 1, 2, 3,...,n$, the set $\{\alpha_1, \alpha_2,..., \alpha_k\}$ forms a basis for the subspace spanned by $\beta_1, \beta_2 .....\beta_k$.*

Proof. Let $S_k = \{\beta_1, \beta_2 .....\beta_k\}$, for $k = 1, 2, 3, ..., n$. We inductively construct orthogonal vectors $\{\alpha_1, \alpha_2,...\}$ such that $\{\alpha_1, \alpha_2,..., \alpha_k\}$ is a basis of the subspace spanned by S_k.

First let $\alpha_1 = \beta_1$. The other vectors are then defined inductively as follows :

Suppose $\alpha_1, \alpha_2,..., \alpha_m$ for $m = 1, 2, 3, ..., n$ have been constructed so that for each $k = 1, 2, 3, ..., m$, $\{\alpha_1, \alpha_2,..., \alpha_k\}$ is an orthogonal basis of the subspace of V which is spanned by $\{\beta_1, \beta_2,..., \beta_k\}$.

Now, we construct α_{m+1}. Let,

$$\alpha_{m+1} = \beta_{m+1} - \sum_{k=1}^{m} \frac{(\beta_{m+1}, \alpha_k)}{\|\alpha_k\|^2} \alpha_k. \qquad\qquad ...(1)$$

Then, $\alpha_{m+1} \neq 0$, for otherwise β_{m+1} will be a linear combination of $\alpha_1, \alpha_2,..., \alpha_m$ and so a linear combination of $\{\beta_1, \beta_2,..., \beta_n\}$, which will give a contradiction because

$\{\beta_1, \beta_2,..., \beta_n\}$ is linear independent.

Also, if $1 \le j \le m$, then

$$(\alpha_{m+1}, \alpha_j) = \left(\beta_{m+1} - \sum_{k=1}^{m} \frac{(\beta_{m+1}, \alpha_k)}{\|\alpha_k\|^2}(\alpha_k, \alpha_j)\right)$$

$$= (\beta_{m+1}, \alpha_j) - \sum_{k=1}^{m} \frac{(\beta_{m+1}, \alpha_k)}{\|\alpha_k\|^2}(\alpha_k, \alpha_j)$$

$$= (\beta_{m+1}, \alpha_j) - (\beta_{m+1}, \alpha_j) = 0$$

Thus, $\{\alpha_1, \alpha_2,..., \alpha_{m+1}\}$ is an orthogonal set consisting $m+1$ non-zero vectors in a subspace spanned by $\{\beta_1, \beta_2,..., \beta_{m+1}\}$. Hence, it is basis for this subspace.

REMARK

☞ Every finite-dimensional inner product space has an orthonormal basis.

Solved Examples

Example 1. *Decide which of the following functions define an inner product on* $\mathbf{R}^2$. *For* $\alpha = [a_1, b_1]^t$, $\beta = [a_2, b_2]^t$

   (i) $(\alpha, \beta) = a_1b_2 + a_2b_1$ (ii) $(\alpha, \beta) = a_1b_1 + a_2b_2$

   (iii) $(\alpha, \beta) = a_1a_2b_1b_2$ (iv) $(\alpha, \beta) = a_1a_2 - a_1b_2 - a_2b_1 + 3b_1b_2$

Solution.   (i) Since $(\alpha, \alpha) = a_1b_1 + a_1b_1 = 2a_1b_1$.

   ∴ (α, α) may or may not be positive.

So that the function, defined by $(\alpha, \beta) = a_1b_2 + a_2b_1$ is not an inner product.

(ii)   Similarly, the function defined by $(\alpha, \beta) = a_1b_1 + a_2b_2$ is not an inner product.

(iii)   (a) $(\alpha, \alpha) = a_1^2 b_1^2 > 0, \forall \alpha \in R^2$

   (b) $(\beta, \alpha) = b_1b_2a_1a_2$ [By definition]

   $= a_1a_2b_1b_2 = \bar{a}_1\bar{a}_2\bar{b}_1b_2$ [∵ a's and b's are real]

   $= \overline{a_1a_2b_1b_2} = \overline{(\alpha, \beta)}$

(c)   Let $\lambda, \mu \in \mathbf{R}$ and let $\alpha = [a_1, b_1]^t$, $\beta = [a_2, b_2]^t$ and $\gamma = [a_3, b_3]^t$, then

   $\lambda\alpha + \mu\beta = \lambda[a_1b_1]^t + \mu[a_2b_2]^t = [\lambda a_1 + \mu a_2, \lambda b_1 + \mu b_2]^t$

Now, $(\lambda\alpha + \mu\beta, \gamma) = (\lambda a_1 + \mu a_2)a_3 + (\lambda b_1 + \mu b_2)b_3$

and, $\lambda(\alpha, \gamma) + \mu(\beta, \gamma) = \lambda(a_1a_3b_1b_3] + \mu(a_2a_3b_2b_3)$.

   ∴ $(\lambda\alpha + \mu\beta, \gamma) \ne \lambda(\alpha, \gamma) + \mu(\beta, \gamma)$.

Thus the function defined by

   $(\alpha, \beta) = a_1a_2b_1b_2$

is not an inner product.

(iv)   We verify the three axioms of an inner product:

(a) When $a_1 \ne 0, b_1 \ne 0$,

$$(\alpha, \alpha) = a_1^2 - 2a_1b_1 + 3b_1^2 = a_1^2 - 2a_1b_1 + b_1^2 + 2b_1^2$$

$$= (a_1 - b_1)^2 + 2b_1^2 > 0$$

$\therefore$ $(\alpha, \alpha) > 0, \forall 0 \neq \alpha \in \mathbf{R}^2.$

(b) $(\beta, \alpha) = a_2 a_1 - a_2 b_1 - b_2 a_1 + 3b_2 b_1 = a_1 a_2 - a_1 b_2 - a_2 b_1 + 3b_1 b_2$

$\qquad\qquad = (\alpha, \beta) = \overline{(\alpha, \beta)}$ $[\because \alpha, \beta \in \mathbf{R}^2]$

$\therefore$ $(\beta, \alpha) = \overline{(\alpha, \beta)}, \forall \alpha, \beta \in \mathbf{R}^2.$

(c)   Let $\gamma = [a_3, b_3]^t$, and $\lambda, \mu \in \mathbf{R}$, then

$\qquad\qquad \lambda\alpha + \mu\beta = \lambda[a_1, b_1]^t + \mu[a_2, b_2]^t$

$\qquad\qquad (\lambda\alpha + \mu\beta) = [\lambda a_1 + \mu a_2, \lambda b_1 + \mu b_2]^t$

$\qquad$ Now, $(\lambda\alpha + \mu\beta, \gamma) = ([\lambda a_1 + \mu a_2, \lambda b_1 + \mu b_2]^t, [a_3, b_3]^t)$

$\qquad\qquad\qquad = (\lambda a_1 + \mu a_2)a_3 - (\lambda a_1 + \mu a_2)b_3 - (\lambda b_1 + \mu b_2)a_3 + 3(\lambda b_1 + \mu b_2)b_3$

$\qquad\qquad\qquad = \lambda(a_1 a_3 - a_1 b_3 - a_3 b_1 + 3b_1 b_3) + \mu(a_2 a_3 - a_2 b_3 - a_3 b_2 + 3b_2 b_3)$

$\qquad\qquad\qquad = \lambda(\alpha, \gamma) + \mu (\beta + \gamma)$

Hence, the function defined in (iv) is an inner product.

Example 2.   *Let V be a vector space of real continuous functions on the interval $0 \leq t \leq 1$ with inner product defined by $(f, g) = \int_0^1 f(t)g(t)\, dt$ and the polynomials $f(t) = t+2$, $g(t) = 3t - 2$ and $h(t) = t^2 - 2t - 3$. Find $(f, g), (f, h), \|f\| and \|g\|$.*

Solution.   (i) $(f,g) = \int_0^1 f(t)g(t)dt = \int_0^1 (t + 2)(3t - 2)dt$

$$= \int_0^1 (3t^2 - 4t - 4)dt = [t^3 + 2t^2 - 4t]_0^1 = (1 + 2 - 4) = -1$$

(ii) $(f,h) = \int_0^1 f(t)h(t)dt = \int_0^1 (t + 2)(t^2 - 2t - 3)dt$

$$= \int_0^1 (t^3 - 7t - 6)dt = \left[\frac{t^4}{4} - \frac{7}{2}t^2 - 6t\right]_0^1$$

$$= \left(\frac{1}{4} - \frac{7}{2} - 6\right) = -\frac{37}{4}$$

(iii) $(f,f) = \int_0^1 f(t)f(t)dt = \int_0^1 (t + 2)^2 dt$

$$= \int_0^1 (t^2 + 4t + 4)dt = \left[\frac{t^3}{3} + 2t^2 + 4t\right]_0^1$$

$$= \left(\frac{1}{3} + 2 + 4\right) = \frac{19}{3}$$

$$\|f\| = \sqrt{(f,f)} = \sqrt{\frac{19}{3}} = \frac{1}{3}\sqrt{57}$$

$\therefore$

(iv) $(g, g) = \int_0^1 g(t)g(t)dt = \int_0^1 (3t - 2)^2 dt$

$$= \int_0^1 (9t^2 - 12t + 4)dt = [3t^2 - 6t^2 + 4t]_0^1$$

$$= (3 - 6 + 4) = 0$$

$\therefore$ $\|g\| = \sqrt{g, g} = \sqrt{1} = 1.$

Example 3. *Apply Gram-Schmidt orthogonalization process to the vectors $\beta_1 = (1, 0, 1)$, $\beta_2 = (1, 0, -1)$ and $\beta_3 = (0, 3, 4)$ to obtain an orthonormal basis $(\alpha_1, \alpha_2, \alpha_3)$ for $\mathbf{R}^3$ with standard inner product.*

Solution.   Let $\alpha_1 = \dfrac{\beta_1}{\|\beta_1\|} = \dfrac{(1,0,1)}{\sqrt{1^2 + 0^2 + 0^2}} = \dfrac{(1,0,1)}{\sqrt{2}} = \left(\dfrac{1}{\sqrt{2}}, 0, \dfrac{1}{\sqrt{2}}\right)$

Let $\gamma_2 = \beta_2 - (\beta_2, \alpha_1)\alpha_1.$

Now, $(\beta_2, \alpha_1) = (1,0,-1) \cdot \left(\dfrac{1}{\sqrt{2}}, 0, \dfrac{1}{\sqrt{2}}\right) = \dfrac{1}{\sqrt{2}} + 0 - \dfrac{1}{\sqrt{2}} = 0$

$\therefore$ $\displaystyle\sum_{i=1}^{n} a_i\alpha_i = b_1\alpha_1 + b_2\alpha_2 + ... + b_k\alpha_k$

Now $(\beta_2, \alpha_2) = (1,0,-1) \cdot \left(\dfrac{1}{\sqrt{2}}, 0, \dfrac{1}{\sqrt{2}}\right) = \dfrac{1}{\sqrt{2}} + 0 - \dfrac{1}{\sqrt{2}} = 0$

Again, let $\gamma_3 = \beta_3 - (\beta_3, \alpha_2)\alpha_2 - (\beta_3, \alpha_1)\alpha_1.$

Now $(\beta_3, \alpha_2) = (0,3,4) \cdot \left(\dfrac{1}{\sqrt{2}}, 0, -\dfrac{1}{\sqrt{2}}\right) = (0 + 0 - 2\sqrt{2}) = -2\sqrt{2}.$

and $(\beta_3, \alpha_1) = (0,3,4) \cdot \left(\dfrac{1}{\sqrt{2}}, 0, \dfrac{1}{\sqrt{2}}\right) = (0 + 0 + 2\sqrt{2}) = 2\sqrt{2}.$

$\therefore$ $\gamma_3 = (0,3,4) - (-2\sqrt{2})\left(\dfrac{1}{\sqrt{2}}, 0, -\dfrac{1}{\sqrt{2}}\right) - 2\sqrt{2}\left(\dfrac{1}{\sqrt{2}}, 0, \dfrac{1}{\sqrt{2}}\right)$

$$= (0, 3, 4) + (2, 0, -2) - (2, 0, 2)$$

$$= (0, 3, 4) + (0, 0, -4) = (0, 3, 0)$$

So, $\phi_i(\alpha_j) = \begin{cases} 1, & \text{if } i = j \\ 0, & \text{if } i \neq j \end{cases} = \delta_{ij} \text{(Kronecker delta)} \cdot$

Thus the required orthonormal basis is

$$\left\{\left(\dfrac{1}{\sqrt{2}}, 0, \dfrac{1}{\sqrt{2}}\right), \left(\dfrac{1}{\sqrt{2}}, 0, -\dfrac{1}{\sqrt{2}}\right), (0,1,0)\right\}.$$

Example 4. *Let $\beta_1 = (3, 0, 4)$, $\beta_2 = (-1, 0, 7)$, $\beta_3 = (2, 9, 11)$ be vectors in $\mathbf{R}^3$ equipped with the standard inner product. Obtain an orthogonal basis.*

Solution. Let $\alpha_1 = \beta_1 = (3, 0, 4)$, then

$$\|\alpha_1\|^2 = 3^2 + 0^2 + 4^2 = 25$$

Now $\alpha_2 = \beta_2 - \dfrac{(\beta_2, \alpha_1)}{\|\alpha_1\|^2} \alpha_1 = (-1, 0, 7) - \dfrac{\{(-1, 0, 7).(3, 0, 4)\}}{25} (3, 0, 4)$

$$= (-1, 0, 7) - \dfrac{(-3 + 0 + 28)}{25} (3, 0, 4)$$

$$= (-1, 0, 7) - \dfrac{25}{25} (3, 0, 5) = (-1, 0, 7) - (3, 0, 4)$$

$$\alpha_2 = (-4, 0, 3) \text{ and } \|\alpha_2\|^2 = 16 + 0 + 9 = 25$$

Also, $\alpha_3 = \beta_3 - \dfrac{(\beta_3, \alpha_1)}{\|\alpha_1\|^2} \alpha_1 - \dfrac{(\beta_3, \alpha_1)}{\|\alpha_2\|^2} \alpha_2$

$$= (2, 9, 11) - \dfrac{\{(2, 9, 11).(3, 0, 4)\}}{25} (3, 0, 4) \dfrac{-\{(2, 9, 11).(-4, 0, 3)\}}{25} (-4, 0, 3)$$

$$= (2, 9, 11) - \dfrac{(6 + 0 + 44)}{25} (3, 0, 4) - \dfrac{(-8 + 0 + 33)}{25} (-4, 0, 3)$$

$$= (2, 9, 11) - 2(3, 0, 4) - (-4, 0, 3)$$

$$= (2, 9, 11) - (6, 0, 8) - (4, 0, 3) = (0, 9, 0).$$

Thus the required orthogonal basis is $\{(3, 0, 4), (-4, 0, 3), (0, 9, 0)\}$.

EXERCISE 7.6

1. Show that $(\alpha, \beta) = x_1 y_1 + x_2 \cdot y_2 - x_3 y_3$ is not an inner product on $\mathbf{R}^3$ where $\alpha = (x_1, x_2, x_3)$ and $\beta = (y_1, y_2, y_3)$.

2. Decide which of the following functions define an inner product on $\mathbf{R}^2$ for $\alpha = [x_1. x_2]^t$, $\beta = [y_1. y_2]^t$:

   (i) $(\alpha ; \beta) = x_1 y_2 - x_2 y_1$

   (ii) $(\alpha, \beta) = 2x_1 y_1 + x_1 y_2 + x_2 y_1 + 2x_2 y_2$.

3. Decide which of the following functions define an inner product on $\mathbf{C}^2$. For

   $$\alpha = [x_1 x_2]^t, \ \beta = [y_1 y_2]^t$$

   (i) $(\alpha, \beta) = x_1 \bar{y}_2$ (ii) $(\alpha, \beta) = x_1 \bar{y}_1 + x_2 \bar{y}_2$

   (iii) $(\alpha, \beta) = 2x_1 \bar{y}_1 + i(x_2 \bar{y}_1 - x_1 \bar{y}_2) + 2x_2 \bar{y}_2$

4. Let α and β be vectors in an inner product space such that $\|\alpha + \beta\| = 8, \|\alpha - \beta\| = 6, \|\alpha\| = 7$, find $\|\beta\|$.

5. Prove that for α, β and γ in an inner product space V.

   $$\|\alpha - \beta\| \|\gamma\| \le \|\beta - \gamma\| \|\alpha\| + \|\gamma - \alpha\| \|\beta\|.$$

6. Expand $(3\alpha_1 + 2\alpha_2, 5\beta_1 - 6\beta_2 + 4\beta_3)$.

7. Let $S = \{\alpha_1, \alpha_2, \alpha_3, \alpha_4\}$ be an orthonomal basis of $\mathbf{R}^4$ and let α, $\beta \in V$ be represented by $(1, 2, 3, -1)$ and $(2, 4, -1, 1)$ respectively. Compute inner product (α, β).

8. Find the orthonormal basis of $V_3(\mathbf{R})$ with standard inner product using Gram-Schmidt orthogonalization process to the vectors $\beta_1 - (1, 0, 1)$, $\beta_2 = (1, 2, -2)$ and $\beta_3 = (2, -1, 1)$.

9. Orthonormalize the set of linearly independent vectors $\{(1, 0, 1, 1), (-1, 0, -1, 1), (0, -1, 1, 1)\}$ of $V_4(\mathbf{R})$.

10. Find an orthonormal basis of the subspace of $V_2(\mathbf{C})$ with standard inner product spanned by $\alpha_1 = (1, 0, i), \ \alpha_2 = (2, 1, 1+i)$.

11. Let W be a subspace of the inner product space V spanned by $\{(0, 1, 1, 0), (0, 5, -3, -2), (-3, -3, 5, -7)\}$ Find an orthonormal basis for W.

Hints to Selected Problems.

1. Let $\alpha = (3, 4, 5)$, then $(\alpha, \alpha) = 3 \times 3 + 4 \times 4 - 5 \times 5 = 9 + 16 - 25 = 0$

$\therefore \qquad (\alpha, \alpha) = 0$ where $\alpha = 0$.

Hence $(\alpha, \beta) = x_1 y_1 + x_2 y_2 - x_3 y_3$ is not an inner product on $\mathbf{R}^3$.

4. Since we know that $\|\alpha + \beta\|^2 + \|\alpha - \beta\|^2 = 2\|\alpha\|^2 + 2\|\beta\|^2$.

$\therefore \qquad\qquad\qquad\qquad 8^2 + 6^2 = 2(7)^2 + 2\|\beta\|^2$

$\therefore \qquad\qquad\qquad\qquad \|\beta\| = 1.$

6. $(3\alpha_1 + 2\alpha_2, 5\beta_1 - 6\beta_2 + 4\beta_3) = 3(\alpha_1, 5\beta_1 - 6\beta_2 + 4\beta_3) + 2(\alpha_2, 5\beta_1 - 6\beta_2 + 4\beta_3)$
$$= 15(\alpha_1, \beta_1) - 18(\alpha_1, \beta_2) + 12(\alpha_1, \beta_3) + 10(\alpha_2, \beta_1) - 12(\alpha_2, \beta_2) + 8(\alpha_2, \beta_3).$$

7. $(\alpha, \beta) = ((1, 2, 3, -1), (2, 4, -1, 1) = 1 \times 2 + 2 \times 4 + 3 \times (-1) + (-1) \times 1 = 2 + 8 - 3 - 1 = 6.$

8. Let $\gamma_1 = \beta_1$, and let $\alpha_1 = \dfrac{\gamma_1}{\|\gamma_1\|} = \dfrac{\beta_1}{\|\beta_1\|} = \dfrac{(1,0,1)}{\sqrt{1^2 + 0^2 + 1^2}} = \dfrac{1}{\sqrt{2}}(1,0,1).$

Now $\gamma_2 = \beta_2 - (\beta_2, \alpha_1)\alpha_1$

$$= (1, 2, -2) - \frac{1}{\sqrt{2}}(1 \times 1 + 2 \times 0 + (-2) \times 1)\left(\frac{1}{\sqrt{2}}, 0, \frac{1}{\sqrt{2}}\right)$$

$$= (1, 2, -2) + \frac{1}{\sqrt{2}}\left(\frac{1}{\sqrt{2}}, 0, \frac{1}{\sqrt{2}}\right) = (1, 2, -2) + \left(\frac{1}{2}, 0, \frac{1}{2}\right) = \left(\frac{3}{2}, 2, -\frac{3}{2}\right)$$

So, $\qquad \alpha_2 = \dfrac{\gamma_2}{\|\gamma_2\|} = \dfrac{\left(\frac{3}{2}, 2, \frac{3}{2}\right)}{\sqrt{\frac{9}{4} + 4 + \frac{9}{4}}} = \left(\dfrac{3}{\sqrt{34}}, \dfrac{3}{\sqrt{34}}, -\dfrac{3}{\sqrt{34}}\right)$

Now $\qquad \gamma_3 = \beta_3 - (\beta_3, \alpha_2)\alpha_2 - (\beta_3, \alpha_1)\alpha_1$

$(\beta_3, \alpha_2) = \dfrac{1}{\sqrt{34}}(2 \times 3 + (-1) \times 4 + 1 \times (-3)) = \dfrac{1}{\sqrt{34}}(6 - 4 - 3) = \dfrac{1}{\sqrt{34}}$

$(\beta_3, \alpha_1) = \dfrac{1}{\sqrt{2}}(2 \times 1 + (-1) \times 0 + 1) = \dfrac{1}{\sqrt{2}}(2 + 1) = \dfrac{3}{\sqrt{2}}.$

$\therefore \qquad \gamma_3 = (2, -1, 1) + \dfrac{1}{\sqrt{34}}\left(\dfrac{3}{\sqrt{34}}, \dfrac{4}{\sqrt{34}}, -\dfrac{4}{\sqrt{34}}\right) - \dfrac{3}{\sqrt{2}}\left(\dfrac{1}{\sqrt{2}}, 0, \dfrac{1}{\sqrt{2}}\right)$

$$= (2, -1, 1) + \frac{1}{34}(3, 4 - 3) - \frac{3}{2}(1, 0, 1) = \left(\frac{10}{17}, \frac{15}{17}, \frac{10}{17}\right) = \frac{5}{17}(2, -3, -2).$$

$\therefore \qquad \alpha_3 = \dfrac{\gamma_3}{\|\gamma_3\|} = \dfrac{\frac{5}{17}(2, -3, -2)}{\frac{5}{17}\sqrt{4 + 9 + 4}} = \left(\dfrac{2}{\sqrt{17}}, \dfrac{-3}{\sqrt{17}}, \dfrac{-2}{\sqrt{17}}\right).$

Answers

2. (i) Not inner product   (ii) Inner product

3. (i) Not inner product   (ii) Inner product   (iii) Not inner product

4. $\|\beta\| = 1$

6. $(\alpha_1, \beta_2) - 18(\alpha_1, \beta_2) + 12(\alpha_1, \beta_3) + 10(\alpha_2, \beta_1) - 12(\alpha_2, \beta_2) + 8(\alpha_2, \beta_3)$

7. 6   **8.** $\left\{\left(\dfrac{1}{\sqrt{2}},0,\dfrac{1}{\sqrt{2}}\right),\left(\dfrac{3}{\sqrt{34}},\dfrac{4}{\sqrt{34}},\dfrac{-3}{\sqrt{34}}\right),\left(\dfrac{2}{\sqrt{17}},\dfrac{-3}{\sqrt{17}},\dfrac{-2}{\sqrt{17}}\right)\right\}$

9. $\left\{\left(\dfrac{1}{\sqrt{3}},0,\dfrac{1}{\sqrt{3}},\dfrac{1}{\sqrt{3}}\right),\left(-\dfrac{1}{\sqrt{6}},0,-\dfrac{1}{\sqrt{6}},\dfrac{2}{\sqrt{6}}\right),\left(-\dfrac{1}{\sqrt{6}},-\dfrac{2}{\sqrt{6}},\dfrac{1}{\sqrt{6}},0\right)\right\}$

10. $\left\{\dfrac{1}{\sqrt{2}}(1,0,i),\dfrac{1}{\sqrt{2}}\left(\dfrac{1+i}{2},1,\dfrac{1-i}{2}\right)\right\}$ **11.** $\left\{\dfrac{1}{\sqrt{2}}(0,1,1,0),\dfrac{1}{3}(0,2,-2,-1),\dfrac{1}{9}(-3,-2,2,-8)\right\}$

7.41 THE ADJOINT OF A LINEAR TRANSFORMATION

Definition 1. *Let U and V be finite-dimensional inner product space and $T : U \to V$ be a linear transformation. Then there exists a unique linear mapping $T : V \to U$ such that for all $\alpha \in V$ and $\beta \in U$*

$(T(\alpha), \beta) = (\alpha, T^*(\beta))$ *or* $(T\alpha, \beta) = (\alpha, T^*\beta)$

The mapping T^* is called the adjoint of T.

Definition 2. *If $A \in F^{m \times n}$, then A^* denotes the conjugate transpose of the matrix A. If we take standard inner product spaces F^n and F^m, and consider A as a linear transformation from F^n to F^m, as $\alpha \to A(\alpha)$, then for $\alpha \in F^n$ and $\beta \in F^m$*

$(A(\alpha), \beta) = \beta * (A(\alpha)) = (A^*(\alpha))^* \beta = (\beta, A^*(\alpha)).$

Thus $A^* = (\bar{A})^t$ is called adjoint of linear transformation A between standard inner product space.

7.42 PROPERTIES OF THE ADJOINT

Theorem 1. *Let U and V be finite-dimensional inner product spaces over the same field F, then*

(i) $(T_1+T_2)^* = T_1^* + T_2^*$, *for* $T_1, T_2 \in L(U, V)$

(ii) $(\lambda T)^* = \bar{\lambda} T^*$, *for* $T \in L(U, V)$ *and* $\lambda \in F$

(iii) $(T_1 T_2)^* = T_2^* \cdot T_1^*$, *for* $T_1, T_2 \in L(U, V)$

(iv) $(T^*)^* = T$, $T \in L(U, V)$

(v) *If* $T \in L(U, V)$ *and T is invertible, then* $(T^*)^{-1} = (T^{-1})^*$.

Proof.   (i) Let $\alpha \in U$ and $\beta \in V$

$(\alpha, T_1 +T_2)^*(\beta)) = ((T_1 +T_2)(\alpha), (\beta))$
$= (T_1 (\alpha) +T_2(\alpha), \beta) = (T_1(\alpha),\beta) +(T_2(\alpha),\beta)$
$= (\alpha, T_1^* (\beta)) +(\alpha, T_2^* (\beta)) = (\alpha, (T_1^* + T_2^*)(\beta)).$

Therefore, by uniqueness of adjoint mapping, $(T_1+T_2)^* = T_1^* + T_2^*$.

(ii) Let $\alpha \in U$ and $\beta \in V$, then

$(\alpha,(\lambda T)^*(\beta)) = (\lambda T (\alpha), \beta) = \lambda(T(\alpha),\beta)$
$= \lambda(\alpha,T^* (\beta)) = (\alpha,(\bar{\lambda} T^*)(\beta)).$

$\Rightarrow$ $(\lambda T)^* = \bar{\lambda} T^*$

(iii) Let $\alpha \in U$ and $\beta \in V$ then

$(\alpha, (T_1 T_2)^* (\beta)) = ((T_1 T_2)(\alpha), \beta) = (T_1(T_2(\alpha), \beta)$
$= (T_2(\alpha), T_1^*(\beta)) = (\alpha, T_2^* T_1^* (\beta)).$

$\therefore$ $(T_1 T_2)^* = T_2^* T_1^*.$

(iv) Let $\alpha \in U$ and $\beta \in V$, then

$$(\alpha, (T^*) * (\beta)) = (T * (\alpha), \beta) = \overline{(\beta, T * (\alpha))}$$

$$= \overline{(T(\beta), \alpha)} = (\alpha, T(\beta)).$$

$$\therefore \qquad (T^*)^* = T$$

(v) Let $\alpha \in U$ and $\beta \in V$, then as

$$(\alpha, \beta) = (I(\alpha), \beta) \qquad\qquad [\because I(\alpha) = \alpha \text{ as } I \text{ an identity operator}]$$

$$= (\alpha, I^*(\beta))$$

$$\Rightarrow \qquad I^* = I$$

Since T is invertible, then

$$TT^{-1} = I \Rightarrow (TT^{-1})^* = I^*$$

$$\Rightarrow \quad (T^{-1})^* T^* = I \Rightarrow (T^*)^{-1} = (T^{-1})^*.$$

Theorem 2. *Let U and V be finite dimensional inner product spaces over F and let $T \in L(U, V)$. If B and B' are ordered orthonormal bases of U and V respectively, then the matrix representation of T^* with respect to these bases is the conjugate transpose of the matrix representation of T with respect to these bases, that is,*

$$[T^*]_{B'} = [T]^*_{B}.$$

Proof. Let $B = \{\alpha_1, \alpha_2, ..., \alpha_n\}$ and $B' = \{\beta_1, \beta_2, ...\beta_m\}$ be ordered orthonormal bases of U and V respectively. Let $[T]_B = A \in F^{m \times n}$ and let $[T]_{B'} = C \in F^{n \times m}$.

Then
$$T(\alpha_j) = \sum_{k=1}^{m} a_{kj} \beta_k \qquad\qquad ... (1)$$

for $j = 1, 2, 3, ..., n$ and so

$$\left(T(\alpha_j), (\beta_i) \right) = \sum_{k=1}^{m} a_{kj} (\beta_k, \beta_i) = a_{ij} \qquad [\because B' \text{ is oirthonormal}]$$

$\therefore$ $(T(\alpha_j), \beta_i)$ is (i, j)th entry of $m \times n$ matrix $[T]_B$.

If follows that

$$[C]_{ij} = [T * (\beta_j), \alpha_i] = \overline{[\alpha_i, T * (\beta_j)]}$$

$$= \overline{[T(\alpha_i), \beta_j]} = [A]_{ji} = [A^*]_{ij}$$

$$\Rightarrow \qquad C = A^* \Rightarrow [T^*]_{B'} = [T]^*_B$$

REMARK

☞ Let T be a linear operator on a finite-dimensional inner product space V over F. Then $|T^*| = |\overline{T}|$ and $tr\, T * = \overline{tr.T}$ where, $tr.$ stands for trace of a matrix.

7.43 SELF-ADJOINT TRANSFORMATION

Definition. *A linear transformation T on an inner product space V(F) is said to be self-adjoint if $T^* = T$ and T is said to be skew-adjoint if $T^* = -T$.*

In an Euclidean space, the self-adjoint transformation is called symmetric and in an unitary space it is called Hermitian.

Theorem 1. Let T_1 and T_2 be self-adjoint transformations on an inner product space $V(F)$. Then

   (i) $T_1 + T_2$ is self-adjoint

   (ii) $T_1 T_2$ is self-adjoint if and only if $T_1 T_2 = T_2 T_1$

   (iii) T^{-1} is self-adjoint if T is invertible

   (iv) λT is self-adjoint iff λ is real for $T \neq 0$ and $\lambda \neq 0$.

Proof.   It is given that $T_1^* = T_1$ and $T_2^* = T_2$

   (i) $(T_1 + T_2)(\alpha), (\beta)) = (T_1(\alpha) + T_2(\alpha), \beta) = (T_1(\alpha), \beta) + (T_2(\alpha), \beta)$

$$= (\alpha, T_1^*(\beta) + (\alpha, T_2^*(\beta)) = (\alpha, T_1(\beta) + (\alpha, T_2^*(\beta))$$

$$= (\alpha, (T_1 + T_2)(\beta)) \qquad \dots (1)$$

Since, $((T_1 + T_2)(\alpha), (\beta)) = (\alpha, (T_1 + T_2)^*(\beta)).$ $\qquad \dots (2)$

From (1) and (2), we get

$$(\alpha, (T_1 + T_2)^*(\beta)) = (\alpha, (T_1 + T_2)(\beta))$$

$\Rightarrow \qquad (T_1 + T_2)^* = (T_1 + T_2).$

Thus $T_1 + T_2$ is self-adjoint.

   (ii) Suppose $T_1 T_2$ is self-adjoint, then

$(T_1 T_2)^* = T_1 T_2 \quad \Rightarrow T^*_2 T^*_1 = T_1 T_2$ [By reversal rule]

$\Rightarrow \quad T_2 T_1 = T_1 T_2$ $\qquad [\because T_1^* = T_1 \text{ and } T_2^* = T_2]$

Conversely, suppose $T_2 T_1 = T_1 T_2$, then we have to show that $T_1 T_2$ is self-adjoint.

Now, $\qquad (T_1 T_2)^* = T_2^* T_1^* = T_2 T_1 = T_1 T_2$ $\qquad [\because T_2 T_1 = T_1 T_2]$

Thus $T_1 T_2$ is self-adjoint.

   (iii) Since T is invertible, so

$$T T^{-1} = T^{-1} T = I \Rightarrow (T T^{-1})^* = (T^{-1} T)^* = I^*$$

$\Rightarrow \qquad (T^{-1})^* T^* = T^* (T^{-1})^* = I$ $\qquad [\because I^* = I]$

$\Rightarrow \qquad (T^{-1})^* T = T(T^{-1})^* = I$ $\qquad [\because T^* = T]$

$\therefore \qquad (T^{-1})^* = T^{-1}.$ Thus T^{-1} is self- adjoint.

   (iv) For each $\alpha, \beta \in V$, we have

$$((\lambda T)(\alpha), \beta) = (\alpha, (\lambda T)^* (\beta)) \qquad \dots (3)$$

Also, $\qquad ((\lambda T)(\alpha), \beta) = (\lambda T(\alpha), \beta) = \lambda (T(\alpha), \beta)$

$$= \lambda(\alpha, T^*(\beta)) = \lambda(\alpha, T(\beta)) \qquad [\because T^* = T]$$

$$= (\alpha, (\bar{\lambda} T)(\beta)) \qquad \dots (4)$$

From (3) and (4) , we get

$$(\alpha, (\lambda T)^* (\beta)) = (\alpha, (\bar{\lambda} T) (\beta)) \Rightarrow (\lambda T)^* = \bar{\lambda} T$$

$\Rightarrow \qquad (\lambda T)^* = \lambda T$ iff $\bar{\lambda} = \lambda.$

Thus λT is self-adjoint iff λ is real.

Theorem 2. If T is a self-adjoint transformation on an inner product space $V(F)$, then S^*TS is self-adjoint for all S, also if S is invertible and $S^* T S$ is self-adjoint, then T is self-adjoint.

Proof.   Let T be self-adjoint, so that

$$T^* = T.$$

Now, $\qquad (S^*TS)^* = S^*T^*(S^*)^*$ [By reversal rule]

$$= S^* TS \qquad [\because T^* T \text{ and } (S^*)^* = S]$$

$\therefore$ S^* T S is self-adjoint

Secondly, let S be invertible, so that
$$SS^{-1} = S^{-1}S = I$$

Also S^* is invertible , then
$$S^* (S^*)^{-1} = (S^*)^{-1} S^* = I. \qquad \qquad ... (2)$$

Since S^*TS is self-adjoint so.
$$(S^*TS)^* = S^*TS \Rightarrow S^* T^*(S^*)^* = S^*TS$$

$\Rightarrow$ $\qquad S^*T^*S = S^*TS$

$\Rightarrow$ $\qquad (S^*)^{-1} S^*T^*S = (S^*)^{-1} S^*TS$ [By left inverse of S^*]

$\Rightarrow$ $\qquad IT^*S = ITS$ [using (2)]

$\Rightarrow$ $\qquad T^*S = TS \Rightarrow (T^*S)S^{-1} = (TS)S^{-1}$ [By right inverse of T]

$\Rightarrow$ $\qquad T^*(SS^{-1}) = T(SS^{-1}) \Rightarrow T^*I = T I$ [using (1)]

$\Rightarrow$ $\qquad T^* = T.$

Thus T is self-adjoint.

Theorem 3. *If T is a linear transformation on an inner product space V(F). Then T^* is self-adjoint$\Leftrightarrow [T] = [T^*]$ with respect to the orthonomal basis B of V(F).*

Proof. Let the matrix of T be
$$[T]_B = [a_{ij}] \qquad \qquad ...(1)$$
relative to B.

Then the matrix of T^* relative to basis B^* the dual of B, is
$$[T^*]_B = [\overline{a_{ij}}] \qquad \qquad ...(2)$$
First, suppose that T is self-adjoint, so $T^* = T$.

Since orthonormal basis is self dual *i.e.*, $B^* = B$, then $a_{ij} = \overline{a_{ij}} \Rightarrow [T] = [T^*]$.

Conversely, suppose that
$$[T] - [T^*] \ i.e., \ a_{ij} = \overline{a_{ij}} \qquad \qquad ...(3)$$
Let $B = \{\alpha_1, \alpha_2,...,\alpha_n\}$ be an ordered orthonomal basis of V. Then we have
$$T(\alpha_j) = \sum_{i=1}^{n} a_{ij}\alpha_i \qquad \qquad ...(4)$$

where, $\qquad [T] = [a_{ij}]$

Also from (2), we have $[T^*] = [\overline{a_{ji}}]$

where, $\qquad T^*(\alpha_j) = \sum_{i=1}^{n} \overline{a_{ji}}\alpha_i \quad$ for $j = 1, 2,..., n.$ $\qquad ...(5)$

Let $\alpha \in V$ be any vector. Then for $b_j \in F$, we have $\alpha = \sum_{j=1}^{n} b_j\alpha_i$ [$\because$ B spans V]

so, $\quad T(\alpha) = T\left(\sum_{j=1}^{n} b_j\alpha_j\right) = \sum_{j=1}^{n} b_jT(\alpha_j) = \sum_{j=1}^{n} b_j \sum_{j=1}^{n} a_{ij}\alpha_i$ [using (4)]

$\qquad \qquad = \sum_{i=1}^{n}\left(\sum_{i=1}^{n} \overline{a}_{ji}b_j\right)\alpha_i = \sum_{i=1}^{n}\left(\sum_{j=1}^{n} a_{ji}b_j\right)\alpha_i$ [using (3)]

$$= \sum_{j=1}^{n} b_j \left(\sum_{i=1}^{n} \overline{a}_{ji}\alpha_i \right) = \sum_{j=1}^{n} b_j . T^*(\alpha_j) \qquad \text{[using (5)]}$$

$$= T^* \left(\sum_{j=1}^{n} b_j \alpha_j \right)$$

$$\therefore \qquad T(\alpha) = T^*(\alpha).$$

Since α is an arbitrary vector of V. Thus, $T^* = T$ and hence T is self-adjoint.

Theorem 4. *If T be a self-adjoint linear transformation on an inner product space $V(F)$, then $T = 0 \Leftrightarrow (T(\alpha), \alpha) = 0, \forall \alpha \in V$.*

Proof. Let $T = 0$, then for $\alpha \in V$

$$(T(\alpha), \alpha) = (0(\alpha), \alpha) = (0, \alpha) = 0.$$

Conversely, let $(T(\alpha), \alpha) = 0 \ \forall \ \alpha \in V$.

Now consider the identity

$$(T(\alpha), \beta) + (T(\beta), \alpha) = (T(\alpha+\beta), (\alpha + \beta)) - (T(\alpha), \alpha) - (T(\beta), \beta), \text{ for } \forall \ \alpha, \beta \in V.$$

Then, $(T(\alpha), \beta) + (T(\beta), \alpha) = 0 - 0 - 0 = 0 \quad$ or $\quad (T(\alpha), \beta) + (\beta, T^*(\alpha)) = 0$

or $(T(\alpha), \beta) + \overline{(T^*(\alpha), \beta)} = 0 \quad$ or $\quad (T(\alpha), \beta) + \overline{(T(\alpha), \beta)} = 0 \quad [\because T \text{ self-adjoint}]$

$\Rightarrow \qquad 2 \text{ Real } \{(T(\alpha), \beta)\} = 0 \ \Rightarrow \qquad \text{Real } (T(\alpha), \beta) = 0.$ ...(1)

There arises two cases :

Case I. If $V(F)$ is an Euclidean space, then $(T(\alpha), \beta)$ is real, so from (1)

$$(T(\alpha), \beta) = 0, \forall \ \alpha, \beta \in V \ \Rightarrow \quad T = 0.$$

Case II. If $V(F)$ is an inner product over a field of complex numbers, then

$$(T(i\alpha), \beta) = \text{Real } i \ (T(\alpha), \beta) = 0 = |(T(\alpha), \beta)| = 0$$

$$\Rightarrow \qquad |(T(\alpha), \beta)| = 0 \ \Rightarrow \ T = 0.$$

Solved Examples

Based on the following Results

▶ Let $T : U \to V$ be a linear transformation. Then $T^* : V \to U$ is said to be adjoint of T if $(T(\alpha), \beta) = (\alpha, T^*(\beta)), \alpha \in U, \beta \in V$.

▶ $T^* = T \Leftrightarrow T$ is self- adjoint.

▶ $T^* = -T \Leftrightarrow T$ is skew-adjoint.

▶ If T_1 and T_2 are self-adjoint then $T_1 + T_2$ is self-adjoint and $T_1.T_2$ is self-adjoint iff $T_1 T_2 = T_2 T_1$.

Example 1. *If T is a linear transformation on $V_3(F)$, so that*

$$T(a, b, c) = (a + b, b, a + b + c)$$

for arbitrary $(a, b, c) \in V_3$. Find $T^*(a_1, b_1, c_1)$.

Solution. Let $(a_1, b_1, c_1) \in V_3(F)$, then by definition of T^*

$$((a, b, c), T^*(a_1, b_1, c_1)) = (T(a, b, c), (a_1, b_1, c_1))$$

$$= (a + b, b, a + b + c), (a_1, b_1, c_1))$$

$$= (a + b)\overline{a}_1 + b\overline{b}_1 + (a + b + c)\overline{c}_1$$

$$= a(\bar{a}_1 + \bar{c}_1) + b(\bar{b}_1 + \bar{c}_1 + \bar{a}_1) + c\bar{c}_1$$

$$= \overline{a(a_1 + c_1)} + \overline{b(b_1 + c_1 + a_1)} + c\bar{c}_1$$

$$= ((a, b, c), (a_1 + c_1, b_1 + c_1 + a_1, c_1))$$

Since (a, b, c) and (a_1, b_1, c_1) are arbitrary elements in $V_3(F)$, so we have

$$T^*(a_1, b_1, c_1) = (a_1 + c_1, b_1 + c_1 + a_1, c_1).$$

Example 2. *Let T be the linear operator on $\mathbf{C}^3$ defined by*

$$T(x, y, z) = (2x + (1-i)y, (3 + 2i)x - 4iz, 2ix + (4 - 3i)y - 3z).$$

Find $T^*(x,y,z)$.

Solution.   Let $B = \{\{1,0,0),(0,1,0),(0,0,1)\}$ be the standard basis of $\mathbf{C}^3$. Then the matrix of T relative to B is

$$[T] = \begin{bmatrix} 2 & 1-i & 0 \\ 3+2i & 0 & -4i \\ 2i & 4-3i & -3 \end{bmatrix}.$$

Now, the conjugate transpose of $[T]$ is

$$[T^*]' = \begin{bmatrix} 2 & 3-2i & -2i \\ 1+i & 0 & 4+3i \\ 0 & 4i & -3 \end{bmatrix}.$$

Thus, $T^*(x, y, z) = (2x + (3-2i)y - 2iz, (1+i)x + (4+3i)z, 4iy - 3z).$

Example 3. *Let T be the linear operator on $\mathbf{C}^3$ defined by*

$$T(x,y, z) = (2x + iy, y - 5iz, x + (1 - i)y + 3z)$$

Find $T^*(x,y,z)$.

Solution.   Let $B = \{(1,0,0),(0,1,0),(0,0,1)\}$ be the standard basis of $\mathbf{C}^3$. Then

$$T(1,0,0) = (2,0,1)$$
$$T(0,1,0) = (i, 1,1-i)$$
$$T(0,0,1) = (0, -5i, 3).$$

The matrix of T is

$$[T] = \begin{bmatrix} 2 & i & 0 \\ 0 & 1 & -5i \\ 1 & 1-i & 3 \end{bmatrix}.$$

Now the conjugate transpose of $[T]$ is

$$[T^*] = \begin{bmatrix} 2 & 0 & 1 \\ -i & 1 & 1+i \\ 0 & 5i & 3 \end{bmatrix}.$$

Thus, $T^*(x, y, z) = (2x + z, -ix + y + (1+i)z, 5iy + 3z).$

Example 4. *Let T be a linear operator on V, let W be a T-invariant subspace of V. Show that $W^\perp$ is invariant under T^*.*

Solution.   Let $\alpha \in W^\perp$. Since W is T-invariant subspace of V, then $T(\beta) \in W$ for $\beta \in W$.

Now $\alpha \in W^\perp \Rightarrow (T(\beta), \alpha) = 0 \quad \forall T((\beta)x \in W$

$$\Rightarrow (\beta, T^*(\alpha)) = 0 \Rightarrow T^*(\alpha) \perp \beta \; \forall \beta \in W$$

$$\Rightarrow T^*(\alpha) \in W^{\perp}.$$

Thus $W^{\perp}$ is invariant under T^*.

Example 5. *Use the definition of adjoint to show that* $0^* = 0$.

Solution.   For every $\alpha, \beta \in V$,

$$(0(\alpha), \beta) = (0, \beta) = 0 = (\alpha, 0) = (\alpha, 0(\beta))$$

$$\Rightarrow \qquad (\alpha, 0^*(\beta)) = (a, 0(\beta)) \quad \Rightarrow \quad 0^* = 0.$$

Example 6.   *If T be a linear transformation on an inner product space V(F) satisfying* $T^2 = T$, *then show that T is self-adjoint* $\Leftrightarrow$ $T^*T = TT^*$.

Solution.   It is given that $\qquad T^2 = T$. Suppose that T is self-adjoint, *i.e.*, $T^* = T$.

Then, for $\alpha, \beta \in V$, we have

$$(T(\alpha), \beta) = (\alpha, T^*(\beta)) = (\alpha, T(\beta)) \qquad\qquad [\because T^* = T]$$

$$= (\alpha, T^2(\beta)) \qquad\qquad [\because T = T^2]$$

$$= (\alpha, TT(\beta)) = (\alpha, T^*T(\beta)) \qquad\qquad [\because T = T^*]$$

$$\therefore \qquad (T(\alpha), \beta) = (\alpha, T^*T(\beta))$$

$$\Rightarrow \qquad T^* = T^*T \Rightarrow T = TT^* \qquad\qquad\qquad ...(1)$$

Also, $\qquad (T(\alpha), \beta) = (\alpha, TT^*(\beta))$

$$\Rightarrow \qquad T^* = TT^* \Rightarrow T = T^*T. \qquad\qquad\qquad ...(2)$$

From (1) and (2), $\quad TT^* = T^*T$.

Conversely, suppose that

$$T^*T = TT^* = T^2T^* \qquad\qquad\qquad [\because T^2 = T]$$

$$= T(TT^*) = T(T^*T)$$

$$T^*T = (TT^*)T$$

$$\Rightarrow \qquad T^* = TT^* \Rightarrow T = (TT^*)^* = T^*T = TT^* = T^*$$

$$\therefore \qquad T^* = T.$$

Thus T is self-adjoint.

Example 7.   *If* $V_2(\mathbf{C})$ *be an inner product space having standard basis* $B = \{(1, 0), (0,1)\}$ *and T be a linear transformation on* $V_2(\mathbf{C})$ *defined by*

$$T(0,1) = (1, -2), \ T(0, 1) = (i, -1).$$

Then, find T(x, y) and T^*(x, y).

Solution.   Since T is linear, so that

$$T(x, y) = T[x(1, 0) + y(0, 1)] = xT(1, 0) + yT(0,1)$$

$$= x(1, -2) + y(i, -1) = (x + iy, -2x - y).$$

The matrix of T relative to B is

$$[T] = \begin{bmatrix} 1 & i \\ -2 & -1 \end{bmatrix}.$$

If T^* is the adjoint of T, then the matrix of T^* relative to B^* (the dual basis of B)is

$$[T^*] = \begin{bmatrix} 1 & -2 \\ -i & -1 \end{bmatrix}.$$

Thus, $T^*(x, y) = (x - 2y, -ix, -y)$

Example 8. *If T be a linear transformation on an inner product space $V_2(\mathbf{C})$ having*
$B = \{(1,0), (0,1)\}$ *as basis and if T is defined by*
$$T(1, 0) = (1+ i, 2), \quad T(0, 1) = (i, i) \quad then \ find \ the \ matrix \ of \ T^*$$
relative to B. Does T^* commute with T?

Solution.   Let $(x, y) \in V_2(\mathbf{C})$. Then
$$(x, y) = x(1, 0) + y(0, 1).$$
$$\therefore \qquad T(x, y) = xT(1, 0) + yT(0,1) = (1+i, 2)+ y(i, i)$$
$$= ((1+i) x+iy, 2x + iy). \qquad\qquad ...(1)$$
The matrix of T relative to B is
$$[T] = \begin{bmatrix} 1+i & i \\ 2 & i \end{bmatrix}.$$
$\therefore$ The matrix of T^* relative to B is
$$[T^*] = [\overline{T}]^t = \begin{bmatrix} 1-i & 2 \\ -i & -i \end{bmatrix}.$$
Thus, we have
$$T^*(x, y) = ((1-i)x + 2y, -ix - iy). \qquad\qquad ...(2)$$
Now, $\qquad\quad TT^*(x, y) = T(T^*(x,y)) = T((1-i)x+ 2y, -ix-iy) \qquad\quad$ [using (2)]
$$= ((1+i)\{(1- i)x+2y\}+i(-ix-iy), 2\{(1-i)x+2y\}+i(-ix-iy)\}$$
$$= ((1- i^2x + 2(1+ i)y+ x+ y, 2(1-i)x+4y+x+y)$$
$$TT^*(x, y) = (3x+(3 + 2i)y, (3-2i)x + 5y). \qquad\qquad ...(3)$$
Also, $\qquad\quad T^*T(x, y) = T^*[(x, y)] = T^*((1 + i)x + iy, 2x + iy) \qquad\quad$ [using (1)]
$$= ((1- i)\{(1+i)x+iy\} + 2(2x+iy), -i(1+i)x+y -2ix + y)$$
$$= (2x+i(1-i)y +4x+2iy, (1- 3i)x+2y) \qquad\qquad ...(4)$$
$$T^*T(x, y) = (6x + (1 + 3i)y, (1 - 3i)x + 2y).$$
From (3) and (4), we get $TT^* \neq T^*T$.

Hence, T^* does not commute with T.

Example 9. *If T is a self-adjoint operator on V. Let λ be an eigenvalue of T. Then show that λ is*
real.

Solution.   Let α be a non-zero eigenvector of T belonging to λ , then
$$T(\alpha) = \lambda\alpha. \qquad\qquad ...(1)$$
Since $\alpha \neq 0$, so (α, α) is positive.

Now, $\qquad\qquad \lambda\{\alpha,\alpha\} = (\lambda\alpha, \alpha) = (T(\alpha), \alpha) \qquad\qquad$ [using (1)]
$$= (\alpha, T^*(\alpha)) \qquad\qquad\qquad \text{[By the definition of adjoint]}$$
$$= (\alpha, T(\alpha)) \qquad\qquad\qquad [\because T^* = T]$$
$$= (\alpha, \lambda\alpha)$$
$$\lambda(\alpha,\alpha) = \overline{\lambda}\,(\alpha, \alpha).$$
Since $(\alpha, \alpha) \neq 0$ so that $\lambda = \overline{\lambda}$, this shows that λ is real.

Example 10. *If T is a self-adjoint operator on V and $T^2 = 0$, show that $T = 0$.*

Solution.   For any $\alpha \in V$, we have
$$\|T(\alpha)\|^2 = (T(\alpha),T(\alpha)) = (\alpha,T^*T(\alpha)) = (\alpha,T^2(\alpha)) \qquad\qquad [\because T^* = T]$$
$$= (\alpha, 0(\alpha)) \qquad\qquad\qquad\qquad\qquad [\because T^2 = 0]$$

$$= (\alpha, 0) = 0$$

$$\Rightarrow \quad \|T(\alpha)\| = 0 \quad \Rightarrow \quad T(\alpha) = 0 \ \forall \ \alpha \in V$$

$$\Rightarrow \quad T = 0.$$

Example 11. *If T is skew-symmetric transformation on an Euclidean space V, then* $(T(\alpha), \alpha) = 0$
for $\alpha \in V$. *Is converse true?*

Solution. Since T is skew-symmetric so that

$$T^* = -T \qquad \qquad ...(1)$$

then for $\alpha \in V$, we have

$$(T(\alpha), \alpha) = (\alpha, T^*(\alpha)) = (\alpha(-T)(\alpha)) \qquad \text{[using (1)]}$$

$$= -(\alpha, T(\alpha)) = -\overline{(T(\alpha), \alpha)}$$

$$= -(T(\alpha), \alpha) \qquad \qquad [\because V \text{ is Euclidean space.}]$$

$$\Rightarrow \quad 2(T(\alpha), \alpha) = 0 \quad \Rightarrow \quad (T(\alpha), \alpha) = 0.$$

Conversely, if $(T(\alpha), \alpha) = 0 \ \forall \ \alpha \in V$

$$\Rightarrow \qquad \qquad T = 0.$$

Thus, converse is not true.

EXERCISE 7.7

1. Let $T : \mathbf{R}^3 \rightarrow \mathbf{R}^3$ defined by $T(x, y, z)$ $= (3x + 4y - 5z, 2x - 6y + 7z, 5x - 9y + z)$. Find $T(x, y, z)$.

2. Let V be an inner product space. For each $\alpha \in V$, there is a mapping $\phi : V \rightarrow F$ defined by $\phi(\beta) = (\beta, \alpha), \beta \in V$. Show that ϕ is linear.

3. Show that the product of two self-adjoint operators is self-adjoint if and only if they commute each other.

4. Let λ be an eigen value of a linear operator T on V,
   (i) If $T^* = T^{-1}$, then $|\lambda| = 1$.
   (ii) If $T^* = -T$, then λ, is purely imaginary.
   (iii) If $T = S^*S$ with S non-singular, then λ, is real and positive.

5. If T is a self-adjoint operator on V, then eigenvectors of T belonging to distinct eigenvalues are orthogonal.

6. Show that $T + T^*$ is self adjoint for any operator T on V.

7. Show that $T - T^*$ is self adjoint for any operator T on V.

8. Show that any operator T on V can be expressed as the sum of a self-adjoint operator and a skew-adjoint operator.

9. Show that $T^*T - I$ is self-adjoint for any operator T on V.

10. Let T be the linear operator on $\mathbf{R}^2$ defined by $T(x, y) = (y, -x)$. Then $(T(\alpha), \alpha) = 0$ for every $\alpha \in V$, but $T \neq 0$.

Hints to Selected Problems

6. $(T + T^*) = T^* + (T^*)^* = T^* + T = T + T^*$.

   $\therefore \qquad T + T^*$ is self-adjoint. Similarly we can prove that $T - T^*$ is skew.

8. Suppose $T = P + Q$, where $P = \dfrac{T + T^*}{2}, Q = \dfrac{T - T^*}{2}$.

   Now $P^* = \left(\dfrac{T + T^*}{2}\right)^* = \dfrac{1}{2}(T + T^*)^* = \dfrac{1}{2}(T + T^*)^* = P \qquad \Rightarrow \quad P$ is self-adjoint.

   and $Q^* = \dfrac{1}{2}(T - T^*)^* = \dfrac{1}{2}(T^* - T) = -\dfrac{1}{2}(T - T^*) = -Q \qquad \Rightarrow \quad Q$ is skew-adjoint.

   Hence T can be expressed as the sum of a self-adjoint and a skew-adjoint operators.

11. The matrix of T relative to B is

$$[T] = \begin{bmatrix} i & 1 & -i & 1 \\ 1 & i & 1 & -i \\ -1 & 1 & i & 1 \\ 1 & -i & 1 & i \end{bmatrix}$$

$$[T^*] = [\bar{T}]^t = \begin{bmatrix} -i & 1 & i & 1 \\ 1 & -i & 1 & i \\ -1 & 1 & -i & 1 \\ 1 & i & 1 & -i \end{bmatrix}^t = \begin{bmatrix} -i & 1 & -1 & 1 \\ 1 & -i & 1 & i \\ i & 1 & -i & 1 \\ 1 & i & 1 & -i \end{bmatrix}.$$

13. Since $T^* = -T$.

Now $(T^2)^* = (TT)^* = (T^*T) = ((-T)(-T)) = T^2$

$\Rightarrow \qquad T^2$ is self-adjoint and $(T^3)^* = (T^2T)^* = T^*(T^2)^* = (-T)(T^2) = -T^3$

$\Rightarrow \qquad T^3$ is skew-adjoint.

Answers

1. $T^*(x,y,z) = (3x+2y+5z, 4x-6y-9z, -5x+7y+z)$

2. $[T^*] = \begin{bmatrix} -i & 1 & -1 & 1 \\ 1 & -i & 1 & i \\ i & 1 & -i & 1 \\ 1 & i & 1 & -i \end{bmatrix}$ **10.** $[T^*] = \begin{bmatrix} 1 & 1 \\ 2 & -1 \end{bmatrix}$ **13.** T^2 is self-adjoint, T^3 is skew.

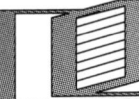
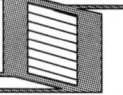

Chapter Review: *A competitive Approach*

Selected Terms and Results

Terms

- **Vector space:** An algebraic structure $(V, +.,)$ is said to be vector space over a field F if
  - (i) $(V, +)$ is an abelian group
  - (ii) $a(\alpha + \beta) = a\alpha + a\beta \; \forall \alpha, \beta \in V, a \in F$
  - (iii) $(a+b)\alpha = a\alpha + b\alpha \; \forall \alpha \in V, a, b \in F$
  - (iv) $(ab)\alpha = a(b\alpha) \; \forall \alpha \in V, a, b \in F$
  - (v) $1. \alpha = \alpha \forall \in V$

- **Vector subspace:** A non-empty subset W of a vector space $V(F)$ which itself is a vector space is called vector subspace of $V(F)$

- **Linear sum:** The linear sum of two subspaces W_1 and W_2 is the set of all those elements each one of which is expressible as the sum of an element of W_1 and an element of W_2.

- **Direct sum:** V is said to be direct sum of W_1 and W_2 if each element of V can be uniquely expressed as the sum of an element of W_1 and an element of W_2.

- **Linear combination of vectors:** Let $V(F)$ be a vector space and $\alpha_1, \alpha_2, ...\alpha_n \in V$. Then any vector $\alpha \in V$ can be expressed as $\alpha = a_1\alpha_1 + a_2\alpha_2 + ... + a_n\alpha_n$, where $a'_i s \in F$ is said to be linear combination of vectos $\alpha_1, \alpha_2, ..., \alpha_n$.

- **Linear span:** Let $V(F)$ be a vector space and S be any non-empty subset of V, then set of all linear combination of finite elements of S is called the linear span of S.

- **Linearly dependent vectors:** A finite set $\{\alpha_1, \alpha_1, ..., \alpha_n\}$ of vectors of V is said to be linearly dependent if there exists scalars $\alpha_1, \alpha_2, ..., \alpha_n$ not all of them equal to zero such that
  $$a_1\alpha_1 + a_2\alpha_2 + , ..., a_n\alpha_n = 0$$

- **Linearly Independant vectors:** A finite set of vectors $\{\alpha_1, \alpha_2, ..., \alpha_n\}$ of the vector space $V(F)$ is said to be linearly independent if for every expression of the type $a_1\alpha_1 + a_2\alpha_2 + ... + a_n\alpha_n = 0$, $a_i's \in F$ implies $a_1 = a_2 = ... = a_n = 0$.

- **Basis of a vector space:** A non-empty subset

S of a vector space $V(F)$ is said to be basis if
  - (i) S is linearly independent.
  - (ii) $L(S) = V$

- **Finite-dimensional vector space:** Let S be a non-empty subset of a vector space $V(F)$, then $V(F)$ is said to be finite dimensional if S is finite subset of V such that $L(S) = V$.

- **Cosets:** Let W be a subspace of a vector space $V(F)$ then the set $\alpha + W = \{\alpha + \beta \; \forall \beta \in W\}$ and $W + \alpha = \{\beta + \alpha \forall \beta \in W\}$ are called left and right cosets respectively.

- **Quotient space:** Let $V/W = \{ W + \alpha : \alpha \in V\}$ be the set of all cosets of W in V such that $(W+\alpha)+(W+\beta) = W+(\alpha+\beta)$ and $a(W+\alpha) = W+a\alpha$. Then vector space V/W is called quotient space.

- **Linear transformation:** Let U and V be two vector spaces over the same field F, then a mapping $T : U \to V$ which associates to each element $\alpha \in U$ to a unique element $T(\alpha) \in V$ such that $T(a\alpha + b\beta) = aT(\alpha) + bT(\beta)$ is called a linear transformation of U into V.

- **Isomorphism:** A linear transformation $T : U \to V$ which is one-one and onto.

- **Linear operator:** Let $V(F)$ be a vector space, then a linear transformation T from V to V is called linear operator.

- **Range of a linear Transformation** $R_T = \{\beta \in V : T(\alpha)\beta$ for some $\alpha \in U\}$

- **Null space of a linear transformation:** $N_T = \{\alpha \in U : T(\alpha) = 0\}$

- **Non-singular linear transformation:** $T(\alpha) = 0 \Rightarrow \alpha = 0$

- **Linear functionals:** Let $V(F)$ be a vector space. Then a linear transformation $f : V \to F$ is called linear functional if ;
  $$f(a\alpha + b\beta) = af(\alpha) + bf(\beta) \; \forall \alpha, \beta \in V, a, b \in F.$$

- **Dual space:** The set of linear functionals on a vector space $V(F)$ is also a vector space, called dual space.

- **Annhilators:** The set of all linear functional ϕ on V such that $\phi(\alpha) = 0 \forall \ \alpha \in V$ is called an annihilator of W.

- **Eigen values of linear transformation:** Let $V(F)$ be finite dimensional vector space and T be a linear operator on V then a scalar $\lambda \in F$ is called an eigen value of T if there exists a non-zero vector $\alpha \in V$ such that $T(\alpha) = \lambda \alpha$.

- **Diagonalization :** A linear operator $T : V \to V$ is said to be diagonalizable if V has a basis consisting of eigen vectors of T only.

- **Inner product space:** An inner product on V is a function $(,) : V \times V \to F$ satisfy the following conditions
  (i) $(\alpha, \alpha) > 0 \ \forall \ \alpha \in V$
  (ii) $(\alpha, \beta) = (\bar{\beta}, \bar{\alpha}) \forall \alpha, \beta \in V$
  (iii) $(a\alpha + b\beta, \gamma) = a(\alpha, \gamma) + b(\beta, \gamma) \ \forall \alpha, \beta, \gamma \in V,$
  $a, b \in F$
  further a vector space V together with an inner product is called an inner product space.

- **Euclidean space:** A finite-dimensional real inner product space is called a Euclidean space.

- **Unitary space:** A complex inner product space is called unitary space.

- **Orthogonal vectors:** Any two vectors α, β of an inner product space $V(F)$ are said to be orthogonal if $(\alpha, \beta) = 0$.

- **Orthonormal set of vectors:** The set of vectors in V is called an orthonormal set if
  (i) $\|\alpha\| = 1 \forall \alpha \in V$ (ii) $(\alpha, \beta) = 0 \forall \alpha, \beta \in V$

- **Complete orthonormal set:** A maximal orthonormal set of vectors in an inner product space V is called a complete orthonormal set.

- **Orthonormal basis:** Let V be a finite dimensional inner product space then a basis of V which is also orthonormal is called an orthonormal basis.

- **Adjoint of a linear transformation:** Let $T : U \to V$ be a linear transformation then there exists a unique linear mapping $T^* : V \to U$ such that for all $\alpha \in V$, $\beta \in U$ such that $(T\alpha, \beta) = (\alpha, T^*\beta)$. Then T^* is called adjoint of T.

- **Self-adjoint transformation** $T^* = -T$

Results

- The intersection of any two subspaces of a vector space is a subspace.

- The union of two subspaces of a vector space is a subspace iff one is contained in the other.

- The linear span $L(S)$ of a non-empty subset S of a vector space $V(F)$ is the smallest subspace of V containing S.

- The zero space has no basis.

- Every finitely generated vector space has a finite basis **(Existence theorem)**.

- Every non-zero vector space has a basis. A vector space may have more than one basis.

- Every linearly independent subset of a vector space V is either a basis of V or can be extended to form a basis of V **(Extension theorem)**.

- Every subspace of a finite-dimensional vector space has a compliment.

- Any two cosets are either disjoint or identical.

- Every n-dimensional vector space $V(F)$ is isomorphic to $F^n(F)$.

- Any two finite-dimensional vector spaces over the same field are isomorphic if and only if they are of same dimension.

- If two vectors are linearly dependent, then one of them is a scalar multiple of other.

- A set of non-zero vectors $[x_1, x_2,..., x_n]$ is linearly dependent if some of these vectors, say x_i is a linear combination of the preceding vectors $x_1, x_2,..., x_{i-1}$ and conversely.

- If W_1 and W_2 are subspaces of a vector space $V(F)$, then $W_1 + W_2$ is also a subspace of $V(F)$.

- The set of all real valued continuous function defined in $[0, 1]$ is a vector space over field of reals.

- The complex field **C** is a vector space over the field of reals.

- Arbitrary intersection of subspaces of a vector space is a subspace.

- Linear transformation is also known as vector space homomorphisrn.

- Let V be an m-dimensional vector space over the field F and let V be an n-dimensional vector space over F. Then the vector space $L(U, V)$ is finite dimensional and has dimension mn.

- If V is a vector space over a field F, then a linear transformation from V into V is called a linear operator.

- Kernel of a linear transformation is also known as null space of T.

- A linear transformation T is said to be invertible if T is one-one and onto.

- A linear transformation T from U into V is called non-singular if the null space of T is $\{0\}$.

- If T is non-singular, then T is one-one.

- If V is not finite dimensional, then the natural mapping can never be onto V^{**}. However, it is always linear and one-one.

- Non-zero eigen vectors belonging to distinct

eigen values are linearly independent.

■ 0 is the eigen value of T if and only if T is singular.

■ If λ is an eigen value of T, then λ^{-1} is an eigen value of T^{-1}.

■ A vector space V together with an inner product is called an inner product space.

■ A finite dimensional real inner product space is called an Euclidean space.

■ A complete inner product space is called a unitary space.

■ Any two vectors α, $\beta \in V$ is said to be orthogonal

to each other if $(\alpha, \beta) = 0$.

■ An orthonormal set of non-zero vectors is linearly independent.

■ A maximal orthonormal set of vectors in an inner product space V is called a complete orthonormal set.

■ In a finite dimensional inner product space, a complete orthonormal set is basis.

■ Every finite dimensional inner product space has an orthonormal basis.

■ A linear transformation T on an inner product space $V(F)$ is said to be self-adjoint if $T^* = T$ and skew symmetric if $T^* = -T$.

Review Questions and Project Work

1. If V is a vector space over an infinite field F then show that it is not possible to write V as union of a finite number of proper subspaces.

2. Show that $L(S)$ is the smallest subspace of V containing S.

3. Show that following vectors are linearly dependent :

   (i) $(1, 1, 2)$, $(-3, 1, 0)$, $(1, -1, 1)$ $(1, 2, -3)$ in $R^3(\mathbf{R})$

   (ii) $(1, -1, 2, 0)$, $(3, 0, 0, 1)$, $(2, 1, -1, 0)$, $(1, -1, 2, 0)$ in $R^4(\mathbf{R})$

4. Show that following vectors are linearly independent :

   (i) $(1, 1, 0)$, $(1, 0, 1)$, $(0, 1, 1)$ in $R^3(\mathbf{R})$

   (ii) $(1, 0, 0)$, $(1, 1, 1)$, $(1, 2, 3)$ in $R^3(\mathbf{R})$

5. Show that the vectors (v_1, v_2) and (w_1, w_2) in $\mathbf{C}$ are linearly dependent if $v_1 w_2 = v_2 w_1$

6. Let S be a finite subset of a vector space such that S is linearly independent and every proper superset of S in V is linearly dependent, show that S is a basis of V.

7. Verify that following is an inner product on R^2.

   $(u, v) = x_1 y_1 - 2x_1 y_2 - 2x_2 y_1 + 5x_2 y_2$

8. If W is a subspace of V and $v \in V$ satisfies

   $(v, w) + (w, v) \le (w, w) \forall\ w \in W$

   prove that $(v, w) = 0\ \forall w \in W$, where V is an inner product space over F.

9. Let T be a linear operator on V and let Rank T^2 = Rank T then show that

Range $(T) \subseteq$ Ker $(T) = \{0\}$

10. Show that a necessary and sufficient condition for the map $T : F^2 \to F^2$ such that $T(x_1, x_2)$ $= (\alpha x_1 + \beta x_2,\ \gamma x_1 + \delta x_2)$, $(\alpha, \beta, \gamma, \delta$, are some fixed element of $F)$ to be an isomorphism is

    that $\begin{vmatrix} \alpha & \beta \\ \gamma & \delta \end{vmatrix} \ne 0$

11. Let A be $n \times n$ matrix over F. Show that A is invertiable if and only if rows of A are linearly independent over F.

12. Let $u, v \in V$ and that $f(u) = 0 \Rightarrow f(v) = 0$ for all $f \in V^*$. Show that $v = \alpha u$ for some scalar α.

13. Let T be a linear operator on R^2 which is represented in the standard ordered basis by

    the matrix $A = \begin{bmatrix} 0 & -1 \\ 1 & 0 \end{bmatrix}$, show that T has no

    eigen values in R.

14. Let V be the vector space of all real valued continuous functions. Define $T : V \to V$ by

    $Tf(x) = \int_0^x f(t)dt$ show that T has no eigen values.

15. Let a, b, c be elements of the field F and

    $A = \begin{bmatrix} 0 & 0 & c \\ 1 & 0 & b \\ 0 & 1 & a \end{bmatrix}$

    Prove that the characteristic polynomial of A is same as that of its minimal polynomial.

Objective type Questions

Fill in the blanks

1. In a vector space $(V, '+', '.')$ the external   composition is also known as _____ .

2. In a vector space $V(F)$, the vector addition is also known as_____ .

3. The elements of a field F for $V(F)$ are called_____ .

4. In a vector space $(V, '+', '.')$, $(V, +)$ must be _____ .

5. The additive identity for the vector space $V(F)$ is _____ .

6. If $a.\ \alpha \in V$ for $a \in F$ and $\alpha \in V$, then V is closed under _____ .

7. If $a \in F$ and $\alpha \in V$, then $(ab)\alpha = $_____ .

8. If $\alpha,\ \beta,\ \gamma \in V$ and $\alpha + \beta = \gamma$, then $\alpha + \beta - \gamma = $_____ .

9. If F is any field, then F is a vector space over_____ .

10. If $V(F)$ is a vector space and $0 \in V$, then _____ $= 0\ \forall\ \alpha \in V$.

11. If $V(F)$ is a vector space and $a \in F,\ \alpha \in V$, then $a\alpha = 0 \Rightarrow a = $ _____ or $\alpha = $_____ .

12. If W be a subset of a vector space $V(F)$ and $a\alpha + b\beta \in W$, for all $a, b \in F$ and $\alpha, \beta \in W$, then W is a _____ .

13. If $W = \{(a_1, a_2, 0): a_1, a_2 \in F\}$, then W is a _____ of vector space.

14. $R(C)$ is _____ .

15. For any non-empty subset W of $V(F)$, if $\alpha - \beta \in W$ and $a\alpha \in W$ for all $a \in F$ and $\alpha, \beta \in W$, then W is _____ .

16. If W_1 and W_2 are two subspaces of $V(F)$, then $W_1 \cap W_2$ is a _____ .

17. If W_1 and W_2 are two subspaces of $V(F)$ and either $W_1 \subseteq W_2$ or $W_2 \subseteq W_1$, then _____ is a subspace of $V(F)$.

18. $L(\phi) = $ _____ .

19. If $S = \{(1,0,0), (0,1,0), (0,0,1)\}$ is a subset of $V_3(F)$, then $L(S) = $_____ .

20. If $a \ne 0,\ b \ne 0 \in F$ and $a\alpha + b\beta = 0$ for $\alpha, \beta \in V$, then α, β are _____ .

21. The vectors $(1, 0, 0), (0, 1, 0), (0, 0, 1)$ are linearly _____ .

22. For any subset S of $V(F)$, $L(S) = V$, then S is a basis of $V(F)$ if S is linearly _____ .

23. Every superset of a linearly dependent set of vectors is linearly _____ .

24. The vectors in a basis are linearly_____

25. Every vector space has a_____ .

26. Let $S = \{\alpha\}$ and $\alpha \ne 0 \in V(F)$, then S is always linearly_____ .

27. Any infinite set of vectors of V is linearly independent if its every finite subset is linearly _____ .

28. If a basis of vector space has 4 elements, then dimension of the vector space is _____ .

29. The set $\{(1, 0, 0), (0, 1, 0), (0, 0, 1)\}$ forms a .... of $V_3(F)$.

30. The set $S = \{(1, 0), (0, 1)\}$ is a basis of $V_n(F)$ for $n = $ _____ .

31. Any subset containing $(n+1)$ vectors of an n-dimensional vector space is linearly _____ .

32. M and N are two subspaces of a vector space V, then $V = M \oplus N$ if
(i) _____ and (ii) _____

33. If V^* is a dual space of V, then dim. $V = $ _____ .

34. Let V be a vector space and V^* be its dual space. Let W be a non-empty subset of V, then W^* is a subspace of _____ .

35. If W is a subspace of a finite-dimensional vector space $V(F)$, then W^{00} _____ .

36. M is an m-dimensional subspace of a n-dimensional vector space V and M' is an annihilator of $M°$, then dim. $M° = $_____ .

37. If W is a subset of a vector space $V(F)$, then $[L(W)]° = $_____ .

38. If W_1 and W_2 are subsets of a vector space V with, $W_1 \subset W_1$, then _____ .

39. V is a finite-dimensional vector space, V^* is its dual space, $x, y \in V,\ x \ne y$ then there is an $f \in V^*$ such that_____ .

40. Let W_1 and W_2 be subspace of a vector space over a field F, then $W_1° + W_2° = $_____ .

True/False

Write 'T' for true and 'F' for false statement.

1. In a vector space $V(F)$, the vector addition is also called an internal composition.   **(T/F)**

2. The elements of V are scalars.   **(T/F)**

3. The elements of F are vectors.   **(T/F)**

4. Let $V(F)$ be a vector space. Then the zero space $\{0\}$ is called a trivial subspace.   **(T/F)**

5. The set $W = \{(a, 0, b) : a, b \in R\}$ is not a subspace of $R^3(R)$.   **(T/F)**

6. Let $V(F)$ be a vector space and $\alpha \in V,\ a \in F$, then $a0 = 0$ and $0\alpha = 0$.   **(T/F)**

7. If K is a field and $F \subseteq K$, then $K(F)$ is a vector space.   **(T/F)**

8. The field of complex numbers is not a vector space over a field of real numbers. **(T/F)**

9. The field of real numbers is a vector space over a field of complex numbers. **(T/F)**

10. If any subset W of V is closed under addition and scalar multiplication in V, then W is a subspace of V. **(T/F)**

11. The intersection of two subspaces of a vector space is also a subspace. **(T/F)**

12. If $\{(1, 0), (0, 1)\} \subseteq V_2(F)$, then $L\{1, 0), (0, 1)\} = F^2$. **(T/F)**

13. $L(\phi) = \{0\}$ **(T/F)**

14. The single non-zero vector is always linearly dependent. **(T/F)**

15. If W_1 and W_2 are two subspaces of $V(F)$, then $W_1 + W_2$ is also a subspace of $V(F)$. **(T/F)**

16. In a vector space, every subset of a linearly independent set is linearly dependent. **(T/F)**

17. The set containing the zero vector is linearly dependent. **(T/F)**

18. The vectors $(1 + i, 2i), (1, 1 + i)$ in $C^2(\mathbf{C})$ are linearly dependent but in $C^2(\mathbf{R})$ are linearly independent. **(T/F)**

19. For a non-empty subset S of $V(F)$, $L(S) = V$ and S is linearly independent, then S is a basis of $V(F)$. **(T/F)**

20. Every vector space has a finite basis. **(T/F)**

21. The vectors in a basis are linearly independent. **(T/F)**

22. Every vector space has a basis. **(T/F)**

23. Let V^* be the dual space of V and V^{**} be the dual of V^*, then
$$\dim. V^* < \dim. V^{**} \quad \textbf{(T/F)}$$

24. If W is any subset of a finite-dimensional vector space V, then $W^{\circ\circ} = L(W)$. **(T/F)**

25. If W_1 and W_2 are subspace of $V(F)$ with $W_1 \subset W_2$, then $W_1^{\circ} \subset W_2^{\circ}$ **(T/F)**

26. Every linear functional is a linear transformation. **(T/F)**

27. If W is a subset of V with $W = V$, then $W^{\circ\circ} = \{0\}$. **(T/F)**

28. If W is a subset of V with $W = \{0\}$, then $W^{\circ} = V$. **(T/F)**

29. If W_1 and W_2 are two subspaces of $V(F)$, then $W_1 = W_2$ iff $W_1^{\circ} = W_2^{\circ}$ **(T/F)**

30. If W is a subspace of a finite dimensional vector space $V(F)$, then $W = W^{\circ\circ}$ **(T/F)**

Multiple choice Questions

Choose the most appropriate option :

1. In a vector space $V(F)$, $a.0$ equals :
   (a) 0
   (b) **0**
   (c) a
   (d) 1

2. In a vector space $V(F)$, $\alpha \in V$ and $a, b \in \mathbf{F}$, then $(ab) \alpha$ equals :
   (a) $a(b\alpha)$
   (b) ab
   (c) a
   (d) $\alpha(ab)$

3. If W_1 and W_2 are two subspaces of $V(F)$, then $W_1 \cup W_2$ is a subspace if:
   (a) $W_1 - W_2$
   (b) $W_1 \subseteq W_2$
   (c) $W_1 \cap W_2$
   (d) None of these

4. $L(\phi)$ equals :
   (a) 0
   (b) ϕ
   (c) $\{0\}$
   (d) None of these

5. If $S = \{(1, 0), (0, 1) \subseteq V_2(\mathbf{R})\}$, then $L(S)$ equals :
   (a) $\mathbf{R}$
   (b) S
   (c) R^3
   (d) R^2

6. If $\alpha = k\beta$ for $\alpha, \beta \in V$ and $k \in F$, then $\{\alpha, \beta\}$ is linearly :
   (a) dependent
   (b) independent

   (c) None of these

7. The set $S = \{(1, 0, 0), (0, 1, 0), (0, 0, 1)\}$ forms a basis for $V_n(\mathbf{R})$ if 'n' equals :
   (a) 2
   (b) 3
   (c) 4
   (d) 1

8. The dimension of a vector space $R^3(\mathbf{R})$ is :
   (a) 2
   (b) 4
   (c) 1
   (d) 3

9. Which of the sets is linearly dependent ?
   (a) $\{0\}$
   (b) $\{\phi\}$
   (c) $\{1\}$
   (d) $\{\alpha\}$

10. A subset S of $V(F)$ forms a basis of $V(F)$ if S is linearly independent and :
    (a) $L(S) = S$
    (b) $L(S) = V$
    (c) $L(S) = F$
    (d) None of these

11. Which of the following is not a vector space ?
    (a) $\mathbf{R(R)}$
    (b) $\mathbf{C(C)}$
    (c) $\mathbf{R(C)}$
    (d) $\mathbf{C(R)}$

12. Condition that vectors (a_1, a_2) and (b_1, b_2) are linearly dependent is :
    (a) $a_1b_1 + a_2b_2 = 0$
    (b) $a_1b_2 + a_2b_1 = 0$
    (c) $a_1b_2 - a_2b_1 = 0$
    (d) $a_1b_1 - a_2b_2 = 0$

13. If W_1 and W_2 are subsets of a vector space V such that $W_2 \subset W_1$ then:

(a) $W_1^{\circ} \subset W_2^{\circ}$

(b) $W_1^{\circ} = W_2^{\circ}$

(c) $W_2^{\circ} \subset W_1^{\circ}$

(d) None of the above

14. If V is a finite dimensional vector space and V^* is its dual and V^{**} is the dual of V^*, then:

(a) V is isomorphic to V^* and but not to V^{**}

(b) V is isomorphic to V^{**} but not to V^*

(c) V is not isomorphic to both V^* and V^{**}

(d) None of the above

15. If W is a subspace of a finite-dimensional vector space $V(F)$, then dim. V + dim. W° is equal to:

(a) dim. V

(b) 0

(c) 1

(d) None of the above

16. If W_1 and W_2 are subspace of a vector space $V(F)$, then $(W_1 \cap W_2)^{\circ}$ is equal to:

(a) $W_1^{\circ} + W_2^{\circ}$ (b) $W_1^{\circ} \cup W_2^{\circ}$

(c) $W_1^{\circ} \cap W_2^{\circ}$ (d) None of these

17. If W_1 and W_2 are subspaces of a vector space $V(F)$, then $(W_1 + W_2)^{\circ}$ is equal to:

(a) $W_1^{\circ} \cup W_2^{\circ}$

(b) $W_1^{\circ} \cap W_2^{\circ}$

(c) $W_1^{\circ} + W_2^{\circ}$

(d) None of the above

18. If W_1 and W_2 are subspace of a vector space $V(F)$, such that $V = W_1 \oplus W_2$, then $W_1^{\circ} \oplus W_2^{\circ}$ is equal to:

(a) V^*

(b) V

(c) V^{**}

(d) None of the above

19. If W_1 and W_2 are subspaces of a vector space $V(F)$ which are annihilated by the subspace W°, then dim. W_1 + dim.W_2 is equal to:

(a) 2 (dim. V – dim. W°)

(b) dim. V – dim. W°

(c) $\dfrac{1}{2}$ (dim. V – dim. W°)

(d) 2 (dim. W° – dim. V)

Answers

Fill in the Blanks

1. scalar multiplication **2.** internal composition **3.** scalars **4.** abelian group **5.** zero vectors
6. scalar multiplication. **7.** $a(b\alpha)$ **8.** zero vector *i.e.*, 0 **9.** F **10.** 0α **11.** 0, 0 **12.** subspace **13.** subspace
14. not a vector space **15.** subspace **16.** subspace **17.** $W_1 \cup W_2$ **18.** {0} **19.** V_3 **20.** linearly dependent
21. independent **22.** independent **23.** dependent **24.** independent **25.** basis **26.** independent
27. independent **28.** 4 **29.** basis **30.** 2 **31.** dependent **32.** $V = M+N, M \cup N = \{0\}$ **33.** div V^* **34.** V^*

35. W **36.** $(n - m)$ **37.** W^* **38.** $W_2^{\circ} \subset W_1^{\circ}$ **39.** $f(x) \neq f(y)$ **40.** $(W_1 \cap W_2)^*$

True/ False

1. T	2. F	3. F	4. T	5. F	6. F	7. T	8. F	9. F
10. T	11. T	12. T	13. T	14. T	15. T	16. F	17. F	18. T
19. T	20. F	21. T	22. T	23. F	24. T	25. F	26. T	27. T
28. F	29. T	30. T						

Multiple choice questions

1. (b)	2. (a)	3. (b)	4. (c)	5. (d)	6. (a)	7. (b)	8. (d)	9. (a)
10. (b)	11. (c)	12. (c)	13. (a)	14. (c)	15. (a)	16. (a)	17. (b)	18. (a)
19. (a)								

COMPETITION CORNER

for JRF, NET/SET, GATE Aspirants

Some Fascinating Facts

1. In the definition of a vector space V over a field F, the condition $f(x)=x \; \forall \; x \in V$ can be replaced by the conditon '$\lambda x = 0$' holds only if $\lambda = 0$ or $x = 0$, $\lambda \in F$, $x \in V$.

2. If **R** is considered as a vector space over **Q**, then the necessary and sufficient condition that the vector 1 and x in **R** be linearly independent is that the real number x be irrational.

3. In a vector space every subset of a linearly independent set is linearly independent and every superset of a linearly dependent set is linearly dependent.

4. If W is a subspace of a vector space V then there is a one-one correspondence between subspaces of V which contain W and subspaces of V/W.

5. Let W be a subspace of a vector space $V(F)$ for $a,b \in W$ we define $a \equiv b (\bmod W)$ if and only if $a-b \in W$ and for all $\alpha \in F$, $\alpha(a-b) \in W$. Also congruence module W is an equivalence relation on V.

6. Every subspaces of a finite-dimensional vector space has a complement.

7. If U is a subspace of a finite-dimensional vector space V then every complement of U has equal to dim V–dim U.

8. All complement of a subspace are of same dimension.

9. Complement of a subspace is not unique.

10. If $V(F)$ be a vector space of dimension n. If V_1 and V_2 are subspaces each of dimensions strictly greater than $\dfrac{n}{2}$ then $V_1 \cap V_2 \neq \{0\}$.

11. Every inner product space is a metric space.

12. If x,y are vectors in a Euclidean space such that $\|x\| = \|y\|$ then $x+y$ is orthogonal to $x-y$.

13. Schwarz inequality implies that cosine of an angle is of absolute value atmost 1.

14. A linear transformation $T:U \to V$ is one-one if and only if it maps every linearly independent subset of U into a linearly independent subset of V.

15. If U and V be two subspaces of a vector space W. Then $(U+V)|U \cong V|(V \cap U)$.

16. If diagonal operator has only the eigen value 0 and 1, then it is a projection.

17. If T is a linear operator on a finite-dimensional space V over F and minimal polynomial $p(x)$ of T is a product of distinct linear factors, then T is diagonalisable.

18. A linear opearator T on a finite-dimensional vector space V is diagonalisable if and only if V is a direct sum of one dimensional T-invariant subspace of V.

19. A linear transformation is nothing but a vector space homomorphism.

20. Every linear transformation is a group homomorphism.

21. Elements of a quotient space V/W are actually subsets of V and therefore $V/W \subset P(V)$.

22. A linear transformation from V into W forms a vector space with vector addition and scalar multiplication.

23. A linear transformation $T:V \to W$ is an isomorphism iff the null space of T consists of a single element.

Some Important Illustrations

1. A field F itself is a vector space over the field F.

2. **C** is a vector space over **R**.

3. The set V of all $m \times n$ matrices with their elements as real numbers is a vector space over the field F of real numbers *w.r.t.* addition of matrices as addition of vectors and multiplication of a matrix by a scalar as scalar multilication.

4. **R** is not a vector space over **C**.

5. **Q** is not a vector space over **R**.

6. **R** is a vector space over **Q**.

7. The set of all polynomials in x of degree less than equal to 2 is a vector space.

8. Let **R**^{n} be the set of all n-types of real numbers,*i.e.*,

$$\mathbf{R}^n = \{a_1, a_2, ..., a_n : a_i's \in R\}$$

Then **R**^{n} is a vector space over **R** with additon and scalar multiplication defined as follows:

$$(a_1,a_2,...,a_n)+(b_1,b_2,...,b_n)$$
$$=(a_1+b_1,a_2+b_2,...,a_n+b_n)$$
$$c(a_1,a_2,...,a_n)=(ca_1,ca_2,...,ca_n)$$

9. The set V of all real valued functions of $(0,1)$ is a vector space over $\mathbf{R}$ with respect to addition and scalar multiplication of functions.

10. The set V of all those polynomials functions over $\mathbf{R}$ with coefficeint in $\mathbf{R}$ which are of degree $\leq n$ together with the zero function in $\mathbf{R}$ is a vector space over $\mathbf{R}$.

11. The set of all Hermitian matrices of order n is a vector space over $\mathbf{R}$ with respect to matrix addition and multiplication of a matrix by a scalar.

12. Let S be the set of all matrix of the form

$$\begin{pmatrix} a & b \\ -b & a \end{pmatrix}, a,b \in \mathbf{C}$$

Then S is a vector space over $\mathbf{C}$.

13. The set of all odd functions from $\mathbf{R}$ to itself is a vector space *w.r.t.* addition and scalar multiplication of functions.

14. The set of all 4×4 complex matrices *w.r.t.*

matrix addition and multiplication of a matrix by a scalar is a vector space over $\mathbf{C}$.

15. The set of all $n\times n$ matrices over $\mathbf{Q}$ is a vector space over $\mathbf{Q}$ *w.r.t.* matrix addition and multiplication of a matrix by a scalar.

16. The set of all n-rowed real skew-symmetric matrices over $\mathbf{Q}$ *w.r.t.* matrix addition and multiplication by a scalar is a vector space over $\mathbf{R}$.

17. The set of all skew-hermitian matrices of order n is a vector space over $\mathbf{R}$ *w.r.t.* matrix addition and multiplication of a matrix by a scalar.

18. Each of the following set of matrices (form given below) is a vector space over $\mathbf{C}$ *w.r.t.* matrix addition and matrix multiplication by a scalar.

(i) $\begin{pmatrix} x & y \\ z & 0 \end{pmatrix} : x, y, z \in \mathbf{C}$

(ii) $\begin{pmatrix} x & 0 \\ 0 & y \end{pmatrix} : x, y \in \mathbf{C}$

(iii) $\begin{pmatrix} x & 0 \\ 0 & 0 \end{pmatrix} : x \in \mathbf{C}$

Self Assessment Test

1. In $V_3(\mathbf{R})$, examine each of the following set of vectors for linear dependence :
   (i) $\{(1,2,0),\ (0,3,1),\ (-1,0,1)\}$
   (ii) $\{(-1,2,1),\ (3,0,-1),\ (-5,4,3)\}$
   (iii) $\{(1,3,2),\ (1,-7,-8),\ (2,1,-1)\}$
   (iv) $\{(1,1,-1),\ (2,-3,5),\ (-2,1,4)\}$

2. Prove that the four vectors $(1,0,0)$, $(0,1,0)$, $(0,0,1)$, $(1,1,1)$ in $V_3(\mathbf{C})$ form a linearly dependent set but any of them are linearly independent.

3. If α, β and γ are vectors such that $\alpha+\beta+\gamma=0$ then show that α and β span the same subspaces as β and γ.

4. In the vector space R^3, let $\alpha=(1,2,1)$, $\beta=(3,1,5)$, $\gamma=(3,-4,7)$. Show that the subspace spanned by $S=\{\alpha,\beta\}$ and $T=\{\alpha,\beta,\gamma\}$ are the same.

5. Let f be a linear transformation from a vector space U into a vector space V. If S is a subspaces of U, show that $f(S)$ will be a subspaces of V.

6. Show that the space of all real functions is the direct sum of the subspaces of odd functions and even functions.

7. Show that the vectors S and α of the vector space $\mathbf{R}$ over $\mathbf{Q}$ are linearly independent iff α is an irrational number but that the same is not true in the real vector space of $\mathbf{R}$.

8. Show that $\{a+ib, c+id\}$ forms a basis of the vector space of complex numbers over the field of real numbers.

9. Let $W = (1,2,3)$ be a vector in Euclidean space R^3. Find an orthonormal basis of $W^\perp$.

10. Let P be orthogonal, prove that $\|P_u\|=\|u\|$ for -every $u \in V$.

11. If A is orthogonally equivalent to B. Show that B is orthogonally equivalent to A.

12. Find an orthonormal basis for the subspaces U of R^4 spanned by $V_1=(1,1,1,1)$, $V_2=(1,2,4,5), V_3=(1,-3,-4,-2)$.

13. Let V be a vector space of polynomials $f(t)$ with inner product $(f,g)=\int\limits_{-1}^{1}f(t)g(t)\,dt$. Apply the Gram-schmidt algorithm to the set $\{1, t, t^2, t^3\}$ to obtain an orthonormal set $\{f_0, f_1, f_2, f_3\}$.

14. Find the matrix A which represents the given inner product on R^2 with respect to usual basis $\{(1,0),\ (0,1)\}$ of R^2.

15. Find the matrix relative to the basis $\{1+i, 1+2i\}$.

16. Let $F : R^2 \rightarrow R^2$ be defined by $F(1,0)=(2,4)$ and $F(0,1)=(5,8)$.
    Find the matrix A representing F with respect to the usual basis for R^2.

17. Let V be the finite dimensional and T be a linear operator on V. Show that T is invertible if and only if T is non-singular.

18. If a vector space has one basis that contains infinitely many elements, prove that every basis contains infinitely many elements.

19. Let T be a linear transformation from V to W. Prove that the image of V under T is a subspacce of W.

20. If $\{u,v,w\}$ is a linearly independent subset of a vector space, show that $\{u, u+v, u+v+w\}$ is also linearly independent.

● ● ●

8

Structure Theory of Vector Spaces
(Some Miscellaneous Concepts)

■ **Outlines**

- Bilinear Forms
- Orthogonal Diagonalization
- Canonical Forms
- Nilpotent Transformation
- Rational Canonical Form

- Quadratic Form
- Hermitian Forms
- Similarity of Linear Transformation
- Jordan Canonical Form

8.1 INTRODUCTION

Let V be a finite-dimensional inner product space over a field F and if T be a linear operator on V, then a function f defined on $V \times V$ by

$$f(\alpha,\beta) = (T(\alpha),\, \beta), \text{ for } \alpha,\, \beta \in V$$

may be regarded as a kind of substitute for T and it can be easily seen that f determines the linear mapping T.

For example, if $B = \{\alpha_1,\alpha_2,.....\alpha_n\}$ be an orthonormal basis of V, then the elements of the matrix of T relative to B are given by

$$a_{ij} = f(\alpha_j,\alpha_i).$$

In this chapter we extend the notion of inner product to vector space over arbitrary field F and we shall discuss three forms namely Bilinear, Quadratic and Hermitian forms and their properties.

8.2 BILINEAR FORMS

Definition 1. *A bilinear form on a vector space $V(F)$ is a mapping $f : V \times V \to F$ which satisfies the following properties*

(i) $f(a\alpha+b\beta,\gamma) = af(\alpha,\gamma)+bf(\beta,\gamma)$

(ii) $f(\alpha,\ a\beta+\beta\gamma) = af(\alpha,\beta)+bf(\alpha,\gamma)$ *for all $a,b \in F$ and all $\alpha,\beta,\gamma \in V$.*

We express condition (i) by saying f is linear in its first variable (co-ordinate) and condition (ii) by saying f is linear in its second variable (co-ordinate).

For example.

(1) *Let f be the dot product on $\mathbf{R}^n$, then f is bilinear form of $\mathbf{R}^n$.*

(2) *Let f be the dot product on $\mathbf{C}^n$, that is*

$$f(\alpha,\beta) = a_1b_1 + a_2b_2 +... + a_nb_n$$

where $\alpha = (a_1, a_2,..., a_n)$, $\beta = (b_1, b_2,..., b_n)$, then f is bilinear.

Definition 2. *A vector space V together with a bilinear form f defined above is called a bilinear space which is denoted by (V, f) of B(V, f).*

Definition 3. *Let f be a bilinear form on V over F. Then a vector $\alpha \in V$ is said to be orthogonal to $\beta \in V$ with respect to f if $f(\alpha, \beta) = 0$.*

Definition 4. *A bilinear form f is called reflexive if orthogonality relation with respect to f is symmetric. That is for $\alpha, \beta \in V$, $f(\beta, \alpha) = 0$ whenever $f(\alpha, \beta) = 0$.*

Definition 5. *A bilinear form f is called symmetric if $f(\alpha, \beta) = f(\beta, \alpha)$ for all $\alpha, \beta \in V$.*

Definition 6. *A bilinear form f is called skew-symmetric if $f(\alpha, \beta) = -f(\beta, \alpha)$ for all $\alpha, \beta \in V$.*

Definition 7. *A bilinear form f is called alternating if $f(\alpha, \alpha) = 0$ for all $\alpha \in V$.*

Theorem 1. *A bilinear form is reflexive if and only if it is either symmetric or alternating.*

Proof. Let f be a reflexive bilinear form on a vector space V over F. Then for $\alpha, \beta, \gamma \in V$, we have

$$f(\alpha, f(\alpha,\beta)\gamma - f(\alpha, \gamma)\beta) = 0$$

$$\Rightarrow \quad f(f(\alpha,\beta)\gamma - f(\alpha, \gamma)\beta, \alpha) = 0$$

$$\Rightarrow \quad f(\alpha,\beta)f(\gamma,\alpha) - f(\alpha,\gamma)f(\beta,\alpha) = 0 \qquad \qquad ...(1)$$

In particular for $\gamma = \alpha$, we get

$$f(\alpha,\beta)f(\alpha,\alpha) - f(\alpha,\alpha)f(\beta,\alpha) = 0$$

or $\qquad \qquad f(\alpha,\alpha).(f(\alpha,\beta) - f(\beta,\alpha)) = 0 \qquad \qquad ...(2)$

Now assume that f is not symmetric, then we shall prove that f is alternating.

If for $u \in V$, there exists $v \in V$ such that

$$f(u, v) \neq f(v, u)$$

Then from (2), we get

$$f(u, u) = 0$$

Let $u \in V$ be such that $f(\alpha, u)$ for all $\alpha \in V$. Now, choose $v, w \in V$ such that

$$f(v, w) \neq f(w, v).$$

Then from (1), we have

$$f(u, v) f(w, u) - f(u, w) f(v, u) = 0$$

$$\Rightarrow \quad f(u, v) f(w, u) - f(u, w) f(u, v) = 0 \qquad \qquad [\because f(u, v) = f(v, u)]$$

$$\Rightarrow \quad f(u, v)(f(w, u) - f(u, w)) = 0$$

so that $\qquad f(u, v) = 0 = f(v, w)$

Similarly,

$$f(u,w) = 0 = f(w,u).$$

Also, $\qquad f(v, u+w) = f(v,u) + f(v,w).$

$$= f(v,w) + f(w,v) = f(u+w,v)$$

so that $f(u +w, u+w) = 0$. Hence, $f(u, u) = 0, \forall u \in V$.

Thus f is alternating.

Conversely, let f be symmetric, then $f(u, v) = f(v, u) \forall u, v \in V$.

When $f(u, v) = 0$, then $f(v, u) = 0$, this shows that f is reflexive.

REMARK

☞ A bilinear form is reflexive iff it is either symmetric or skew-symmetric.

Theorem 2. *Let V be a vector space of dimension n over F. Let $\{f_1, f_2, ..., f_n\}$ be a basis of the dual space V^*. Then $\{f_{ij} : i, j = 1, 2, .., n\}$ is a basis of $B(V, f)$ where f_{ij} is defined by $f_{ij}(\alpha, \beta) = f_i(\alpha) f_j(\beta)$. Thus in particular dim $B(V, f) = n^2$.*

Proof. Let $\{\alpha_1, \alpha_2 ..., \alpha_n\}$ be the basis of V dual to $\{f_i\}$. We first show that $\{f_{ij}\}$ spans $B(V, f)$. Let $f \in B(V, f)$ and suppose that $f(\alpha_i, \alpha_j) = a_{ij}$. Then we shall show that $f = \sum a_{ij} f_{ij}$.

It is sufficient to show that $f(\alpha_k, \alpha_m) = (\sum a_{ij} f_{ij})(\alpha_k, \alpha_m)$ for $k, m = 1, 2, ..., n$. We have $(\sum a_{ij} f_{ij})(\alpha_k, \alpha_m) = \sum a_{ij} f_{ij}(\alpha_k, \alpha_m)$

$$= \sum a_{ij} f_i(\alpha_k) f_j(\alpha_m) \qquad \text{[By the definition of } f_{ij}]$$
$$= \sum a_{ij} \cdot \delta_{ik} \delta_{jm} \qquad [\because f_i(\alpha_j) = \delta_{ij}]$$
$$= a_{km} \qquad \left[\because \delta_{ij} = \begin{cases} 1, i = j \\ 0, i \neq j \end{cases} \right]$$
$$= f(\alpha_k, \alpha_m)$$

Hence, $\{f_{ij}\}$ spans $B(V, f)$.

Now, we shall show that $\{f_{ij}\}$ is linearly independent.

Let $\sum a_{ij} f_{ij} = 0$ for $i, j = 1, 2, ..., n$.

Then for $k, m = 1, 2, ..., n$, we have

$$\Rightarrow \quad (\sum a_{ij} f_{ij})(\alpha_k, \alpha_m) = 0(\alpha_k, \alpha_m)$$
$$\Rightarrow \quad \sum a_{ij} f_{ij}(\alpha_k, \alpha_m) = 0 \Rightarrow a_{km} = 0$$

Thus, $\{f_{ij}\}$ is linearly independent.

Hence, $\{f_{ij}\}$ forms a basis of $B(V, f)$. Further, i and j take all the values from 1 to n therefore there are n^2 elements in $\{f_{ij}\}$, which shows that

dim. $B(V, f) = n^2$.

8.3 BILINEAR FORM AND MATRICES

Definition 1. *Let f be a bilinear form on $V(F)$ and let $S = \{\alpha_1, \alpha_2, ..., \alpha_n\}$ be a basis of V. Then a matrix $A = \{a_{ij}\}$, where $a_{ij} = f(\alpha_i, \alpha_j)$ is called the matix representation of f with respect to the basis S, this matrix A is denoted by $[f]_s$. The above matrix A represents f in another way as follows :*

Let $\alpha, \beta \in V$, then for $a_1, a_2, ..., a_n, b_1, b_2, ..., b_n \in F$, we have

$$\alpha = a_1 \alpha_1 + a_2 \alpha_2 + ... a_n \alpha_n$$
$$\beta = b_1 \alpha_1 + b_2 \alpha_2 + ... b_n \alpha_n$$

Then, $f(\alpha, \beta) = f(a_1\alpha_1 + a_2\alpha_2 + ... a_n\alpha_n, b_1\alpha_1 + b_2\alpha_2 + ... b_n\alpha_n)$

$$= a_1 b_1 f(\alpha_1, \alpha_1) + a_1 b_2 f(\alpha_1, \alpha_2) + ... + a_n b_n f(\alpha_n, \alpha_n)$$

$$= [a_1, a_2, ..., a_n] \, A \begin{bmatrix} b_1 \\ b_2 \\ \vdots \\ b_3 \end{bmatrix} = [\alpha]^t{}_n \, A [\beta]_s$$

Definition 2. *A matix B is said to be congruent to A if there exists an invertible matix P such that*

$$B = P'AP.$$

<u>**Theorem 1.**</u> *Let P be the transition matrix from one basis of V to another. Let A be the matix of bilinear form f in the oiginal basis. Then B = P′AP is the matrix of f in the new basis.*

<u>**Proof.**</u> Let $S=\{\alpha_1,\alpha_2,..,\alpha_n\}$ and $S'=\{\beta_1,\beta_2,..,\beta_n\}$ be two ordered basis of V. Let T be a linear operator on V defined by

$$T(\alpha_i) = \beta_i \text{ for } i=1, 2,..., n. \qquad \qquad ...(1)$$

Obviously T is a mapping from S to S′ so it is invertible.

Let $P = [p_{ij}]_{n\times n}$, be the matrix of T relative to the basis S so P is also invertible.

Also, $\qquad \qquad T(a_j) = \sum P_{ij}\,\alpha_i = \beta_i \qquad \qquad ...(2)$

For any $\alpha, \beta \in V$, we have

$$[\alpha]_s = P[\alpha]_s \qquad \qquad ...(3)$$

and $\qquad \qquad [\beta]_s = P[\beta]_s \qquad \qquad ...(4)$

$\therefore \qquad \qquad [\alpha]^t{}_s = (P[[\alpha]_{s'})^t = [\alpha]^t{}_s . P^t. \qquad \qquad ...(5)$

Now, from definition of t, we have

$$f(\alpha,\beta) = [\alpha]^t{}_s\, A[\beta]_s = [\alpha]^t{}_s\, P^t AP = [\beta]_{s'} \quad \text{[Using (4) and (5)]}$$

Since α and β are arbitrary elements of V. Thus P^tAP is the matrix of f in the basis S′. If it be B, then $B= P^tAP$ or $[f]_{s'} = P^t[f]_s, P$

REMARKS

☞ The matrix P defined in above theorem is called transition matrix from ordered basis S to S′.

☞ The above theorem indicates one main difference between bilinear forms and linear forms operator, both of which can be represented by square matrix, namely, if B and A represent the same linear opearator, then B is similar to A, that is $B=P^{-1}AP$, where P is the change of basis matrix but if B and A represent the same linear form then $B= P^tAP$

Definition 3. *Let f be a bilinear form on V, then rank of a bilinear form f, denoted by* rank(f) *is defined to be the rank of any matrix representation of f.*

Definition 4. *A bilinear form f on V is said to be degenerate if rank (f)< dim. V and it is said to be* non-degenerate *if rank (f) = dim. V.*

Definition 5. *Let f and g be two bilinear forms on V and on W respectively over the field F. Then a linear mapping T : V → W is called an isometry if it preserves the bilinear forms, that is, for all α,β ∈ V.*

$$f(\alpha,\beta) = g(T(\alpha),T(\beta)).$$

Also, if T is isomorphism then it is called an isometric isomorphism. If such a mapping exists, then we can say that the bilinear spaces B(V, f) and B(W, g) are isomorphic.

Definition 6. *A mapping q:V→F is called a quadratic mapping if $q(\alpha) = f(\alpha, \alpha)$, where f is a symmetric bilinear form on V.*

Definition7. *If in a field F, we have $1+1\neq0$, then f is obtainable from q from the following identity f(α, β) $= \dfrac{1}{2}\,(q(\alpha+ \beta) - q(\alpha) - q(\beta)).$*

This form is called polar form *of f.*

Theorem 1. Let $B(V, f)$ be a fnite dimensional bilinear space over F and let $T : B(V, f) \to B(W, g)$ be an isometic isomorphism, where $B(W, g)$ is other bilinear space over F. Then $[f]_s$ and $[g]_s$, are congruent matices, where S and S' are ordered bases of V and W respectively.

Proof. For any $\alpha, \beta \in V$, we have

$$f(\alpha, \beta) = g(T(\alpha), T(\beta)) \qquad \qquad ...(1)$$

Also, $\qquad \qquad f(\alpha, \beta) = [\alpha]^t_s \, [f]_s \, [\beta]_{s'} \qquad \qquad ...(2)$

Similarly, $g(T(\alpha), T(\beta)) = [T(\alpha)]^t_{s'} [g]_{s'} \, [T(\beta)]_{s'}$

$$= ([T]_s[\alpha]_s)^t \, [g]_{s'} ([T]_s \, [\beta]_s)$$

$$= ([\alpha]^t_s[T]^t_s) \, [g]_{s'} ([T]_s \, [\beta]_s)$$

$$= [\alpha]^t_s ([T]^t_s \, [g]_{s'} ([T]_s) [\beta]_{s'} \qquad \qquad ...(3)$$

From (1) and (2), we get

$$g(T(\alpha), T(\beta)) = [\alpha]^t_s [f]_s [\beta]_s \qquad \qquad ...(4)$$

Now from (3) and (4), we have

$$[f]_s = [T]^t_s \, [g]_{s'} \, [T]_{s'}$$

Hence, $[f]_s$ and $[g]_{s'}$ are congruent.

Theorem 2. Finite dimensional bilinear spaces $B(V, f)$ and $B(W, g)$ over F are isomorphic if and only if f and g admit the same matix representation with respect to a suitably chosen ordered bases of V and W.

Proof. Let $T : B(V, f) \to B(W, g)$ be an isometric isomorphism and let

$S = \{\alpha_1, \alpha_2 ...\alpha_n\}$ be an ordered basis of V.

Then $S' = \{T(\alpha_1), T(\alpha_2), .., T(\alpha_n)\}$ is an ordered basis of W and $[T]_s = I_n$,

where $I_{n'}$ is a unit matrix of order n. Therefore from above theorem, we have

$$[f]_s = [g]_{s'}.$$

Thus, f and g admit the same matrix representation.

Conversely, suppose f and g have the same matrix representation, with respect to some ordered bases S and S' of V and W respectively, and dim. $V =$ dim. W.

Let $S = \{\alpha_1, \alpha_2, ..., \alpha_n\}$ and $S' = \{\beta_1, \beta_2, ..., \beta_n\}$ be ordered basis of V and W such that

$$[f]_s = [g]_{s'}.$$

Let $T : V \in W$ be a mapping defined by $T(a_i) = b_i$, $i = 1, 2, ...,n$.

Then, T is obviously isomorphism.

Now, for $\alpha, \beta \in V$, we have

$$\alpha = a_1\alpha_1 + a_2\alpha_2 + ... a_n\alpha_n; \quad \text{for } a\text{'s} \in F$$

and $\qquad \qquad \beta = b_1\alpha_1 + b_2\alpha_2 + ... b_n\alpha_n; \quad \text{for } b\text{'s} \in F$

Then, $\qquad \displaystyle f(\alpha, \beta) = f\left(\sum_i a_i\alpha_i, \sum_j b_j\alpha_j\right) = \left(\sum_i \sum_j a_i b_j f(\alpha_i, \alpha_j)\right)$

$$= \left(\sum_i \sum_j a_i b_j g(\beta_i, \beta_j)\right) = \left(\sum_i \sum_j a_i b_j g(T(\alpha_i), (T(\alpha_j))\right)$$

$$= g\left(\sum_i a_i T(\alpha_i), \sum_j b_j T(\alpha_j) \right) = g(T(\alpha).T(\beta))$$

Hence, T is an isometry.

Theorem 3. *Let f be an alternating bilinear form on V. Then there exists a basis of V in which f is represented by a matrix of the form.*

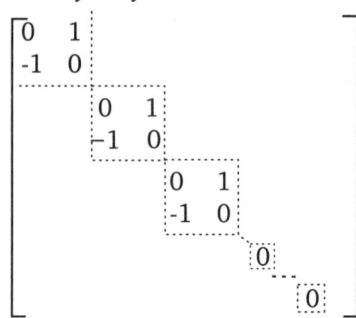

Moreover, the number of $\begin{bmatrix} 0 & 1 \\ -1 & 0 \end{bmatrix}$ *is uniquely determined by f (because it is equal to* $\dfrac{1}{2}$ *rank (f)).*

Proof.
Suppose $f = 0$, then the theorem is obvious. Also, if dim. $V = 1$, then for $\alpha \in V$ and $a, b \in F$, we have

$$f(a\alpha, b\alpha) = abf(\alpha, \alpha) = 0 \text{ and so } f = 0.$$

Therefore, we may assume that $\dim V > 1$ and $f \neq 0$.

Since $f \neq 0$, so there exist non-zero vectors $\alpha_1, \alpha_2 \in V$ such that $f(\alpha_1, \alpha_2) \neq 0$. In fact multiplying α_1 by an appropriate factor, we may assume that $f(\alpha_1, \alpha_2) = 1$ and so $f(\alpha_2, \alpha_1) = -1$.

Now, α_1 and α_2 are linearly independent. Let if possible α_1 and α_2 are dependent, i.e., $\alpha_2 = k\alpha_1$, then $f(\alpha_1, \alpha_2) = f(\alpha_1, k\alpha_1) = kf(\alpha_1, \alpha_1) = 0$, which is not possible, thus α_1, α_2 must be linearly independent.

Let U be a subspace spanned by α_1 and α_2 that is $U = \text{span}(\alpha_1, \alpha_2)$. Thus, we have

(i) The matrix representation of the restriction of f to U in the basis $\{\alpha_1, \alpha_2\}$ is

$$\begin{bmatrix} 0 & 1 \\ -1 & 0 \end{bmatrix}.$$

(ii) If $u \in U$, so $u = a\alpha_1 + b\alpha_2$, then

$$f(u, \alpha_1) = f(a\alpha_1 + b\alpha_2, \alpha_1) = af(\alpha_1, \alpha_1) + bf(\alpha_2, \alpha_1)$$
$$= a0 + b(-1) = -b$$

$$f(u, \alpha_2) = f(a\alpha_1 + b\alpha_2, \alpha_2) = af(\alpha_1, \alpha_2) + bf(\alpha_2, \alpha_2)$$
$$= a(1) + b0 = a$$

Let W consist vector $w \in V$ such that

$$f(w, \alpha_1) = 0 \text{ and } f(w, \alpha_2) = 0.$$

We now claim that $V = U \oplus W$. From the definition of U and W, it is clear that $U \cap W = \{0\}$. Therefore, we only have to show that $V = U + W$.

Let $v \in V$ and setting

$$u = f(v,\alpha_2)\alpha_1 - f(v,\alpha_1)\alpha_2 \Big\rbrace \qquad \qquad ...(1)$$

and $\qquad w = v - u$

Since u is a linear combination of α and β, then we show that $w \in W$.

Now, $\qquad f(u, \alpha_1) = f(f(v, \alpha_2)\alpha_1 - f(v, \alpha_1)\alpha_2, \alpha_1)$

$$= f(v,\alpha_2)f(\alpha_1, \alpha_1) - f(v, \alpha_1) f(\alpha_2, \alpha_1)$$

$$= 0 - (-1) f(v, \alpha_1) = f(v, \alpha_1).$$

So that $\qquad f(w, \alpha_1) = f(v - u, \alpha_1) \qquad \qquad$ [Using (1)]

$$= f(u,\alpha_1) - f(u,\alpha_1) = 0.$$

This implies that $w \in W$ and then by (1),

$$v = u + w$$

where, $u \in U$ and $w \in W$.

Thus, every element of V can be expressed as the linear sum of an element of U and an element of W, i.e., $V = U + W$. Hence, $V = U \oplus W$.

Now, the restriction of f to W is an alternating bilinear form on W. By induction hypothesis, there exists a basis $\alpha_3, \alpha_4, ..., \alpha_n \in W$ in which the matrix representation of f restricted to W has the above desired form. Hence, $\alpha_1, \alpha_2, ..., \alpha_n$, is a basis of V in which the matrix representing f has the desired form.

Theorem 4. *Let f be a symmetric bilinear form on V over F. Then V has a basis $\{\alpha_1, \alpha_2, ..., \alpha_n\}$ in which f is represented by a diagonal matix, i.e., $f(\alpha_i, \alpha_j) = 0$ for $i \neq j$.*

Proof. If $f = 0$ or if dim.$V = 1$, then the theorem is evidently true. Thus, we may assume that $f \neq 0$ and dim.$V = n > 1$. If for every $\alpha \in V$, $q(\alpha) = f(\alpha, \alpha)$, then the polar form of f gives that $f = 0$.

Hence, we may assume that there is a vector $\alpha_1 \in V$ such that $f(\alpha_1, \alpha_1) \neq 0$.

Let W_1 be a subspace of V spanned by α_1 and let W_2 consist of those vectors $\alpha \in V$ for which $f(\alpha_1, \alpha) = 0$. We now claim that $V = W_1 \oplus W_2$.

(i) We first show that $W_1 \cap W_2 = \{0\}$.

If possible $W_1 \cap W_2 \neq \{0\}$, then there exists a non-zero vector β such that $\beta \in W_1$ and $\beta \in W_2$.

Now, $\qquad \beta \in W_1 \Rightarrow \beta = k\alpha_1$ for some $k \in F$.

Also, $\qquad \beta \in W_2 \Rightarrow f(\beta, \beta) = 0 \Rightarrow f(k\alpha_1, k\alpha_1) = 0$

$$\Rightarrow k^2 f(\alpha_1, \alpha_1) = 0 \Rightarrow k = 0 \qquad \qquad [\because f(\alpha_1, \alpha_1) \neq 0]$$

$$\Rightarrow \beta = 0.$$

Thus, $\qquad W_1 \cap W_2 = \{0\}$.

(ii) Now, to show that $V = W_1 + W_2$.

Let $\beta \in V$ and set

$$y = \beta - \frac{f(\alpha_1, \beta)}{f(\alpha_1, \alpha_1)}\alpha_1 \qquad \qquad ...(1)$$

Then, $f(\alpha_1, g) = f\left(\alpha_1, \beta - \dfrac{f(\alpha_1, \beta)}{f(\alpha_1, \alpha_1)} \alpha_1\right) = f(\alpha_1, \beta) - \dfrac{f(\alpha_1, \beta)}{f(\alpha_1, \alpha_1)} f(\alpha_1, \alpha_1)$

$$= f(\alpha_1, \beta) - f(\alpha_1, \beta) = 0$$

Thus, $\gamma \in W_2$, then from (1), we have

$$\beta = \gamma + \frac{f(\alpha_1, \beta)}{f(\alpha_1, \alpha_1)} \alpha_1.$$

This shows that $\beta \in V$ is the sum of an element of W_1, and an element of W_2, thus

$$V = W_1 + W_2 \qquad \qquad ...(2)$$

Hence, from (1) and (2), we have

$$V = W_1 \oplus W_2$$

Now, f restricted to W_2, is a symmetric bilinear form on W_1. Also,

$$\dim.W_2 = \dim.V - \dim.W_1 = n - 1$$

Thus, by induction, we may assume that there is a basis $\{\alpha_2, \alpha_3, ..., \alpha_n\}$ of W_2 such that $f(\alpha_i, \alpha_j) = 0$ for $i \neq j$ and $2 \leq i, i \leq n$. But by definition of W_2, $f(\alpha_i, \alpha_j) = 0$ for $i = 2, 3, ..., n$. Therefore, we obtain the basis $\{\alpha_1, \alpha_2, ..., \alpha_n\}$ of V such that $f(\alpha_i, \alpha_j) = 0$, for all $i \neq j$.

Theorem 5. *If f is a non-degenerate bilinear form on $V(F)$. Then the set G of all linear transformation T on V which preserves f, is a group under the product of transformations.*

Proof. It is given that $G = \{T: f(T(\alpha), T(\beta)) = f(\alpha, \beta), \forall \alpha, \beta \in V\}$. Now to show that G is a group under the product of transformations.

(i) **Closure property.** For each $T_1, T_2 \in G$, we have

$$f((T_1 T_2)(\alpha), (T_1 T_2)(\beta)) = f(T_1(T_2(\alpha)), T_1(T_2(\beta)))$$
$$= f(T_2(\alpha), T_2(\beta)) \qquad [\because T_1 \text{ preserves } f.]$$
$$= f(\alpha, \beta). \qquad [\because T_2 \text{ preserves } f.]$$

Thus, $T_1 T_2$ preserves f so that $T_1 T_2 \in G$.

(ii) **Associativity.** Multiplication of linear transformations is always associative.

(iii) **Existence of Identity.** Since $I(\alpha) = \alpha$, $I(\beta) = \beta$.

[By definition of identity mapping]

Now, $f(I(\alpha), I(\beta)) = f(\alpha, \beta), \forall \alpha, \beta \in V$.

This implies that $I \in G$ which is the identity toansformation.

(iv) **Existence of inverse.** For each $T \in G$. Let α be a vector in the null space of T, so that

$$T(\alpha) = 0, \forall \alpha \in V.$$

Then, $f(T(\alpha), T(\beta)) = f(0, T(\beta)) = 0$.

Since T preserves f, i.e., $f(\alpha, \beta) = f(T(\alpha), T(\beta))$.

$$f(\alpha, \beta) = 0, \quad \forall \beta \in V.$$

Also, f is non-degenerate, therefore $f(\alpha, \beta) = 0$, $\forall \beta \in V$ is only possibility when $\alpha = 0$.

$\therefore$ $\qquad\qquad\qquad T(\alpha)=T(0)=0.$

This implies T is non-singular. Now, as V is finite dimensional so T is invertible.

Also, $\qquad f(T^{-1}(\alpha), T^{-1}(\beta)) = f(T(T^{-1}(\alpha)), T(T^{-1}(\beta)))$

$\qquad\qquad\qquad\qquad\qquad = f((TT^{-1})(\alpha), (TT^{-1})(\beta))$

$\qquad\qquad\qquad\qquad\qquad = f(I(\alpha), I(\beta)) = f(\alpha, \beta)$

$\therefore$ $\quad T^{-1}$ preserves f.

Thus, $T^{-1} \in G$. This shows that every element in G has its inverse in G. Hence G is a group.

Definition 8. *Let f be a reflexive bilinear form on $V(F)$ and S be a subspace of V. Define a set $S^{\perp}=\{\beta \in V : f(\alpha, \beta) = 0, \forall\ \alpha \in S\}$. Clearly, $S^{\perp}$ is a subspace of V. The subspace S is called isotropic if $S \subseteq S^{\perp} \neq \{0\}$, otherwise, anisotropic. Also, S is called totally isotropic if $S \subseteq S^{\perp}$.*

The subspace $V^{\perp}$ is called the radical of f, denoted by rad (f). Thus

$$\text{rad } (f) = \{\beta \in V : f(\alpha, \beta) = 0, \forall\ \alpha \in V\}$$

If rad $(f) = \{0\}$, then f is non-degenerate, otherwise degenerate.

Theorem 6. *Let f be a reflexive bilinear form on a finite dimensional vector space V over F. Then f is non-degenerate if and only if the matix of f with respect to an ordered basis of V is invertible.*

Proof. Let B be an ordered basis of V. Let if possible $[f]_B$ is not invertible, *i.e.*, $\det.[f]_B = 0$, then the system of equations (matrix form)

$$[f]_B X = 0$$

has a non-zero solution. Let this solution be $X_0 \in F$, where $n = \dim.V$ and let $\beta \in V$ such that $\qquad\qquad X_0 = [b]_B.$

Then for any $\alpha \in V$, we have

$$f(\alpha, \beta) = [\alpha]_B^t\ [f]_B\ [\beta]_B = [\alpha]_B\ [f]_B X_0 = [\alpha]_B\ 0 = 0.$$

$\Rightarrow \qquad\qquad\qquad \beta \in \text{rad}(f) - \{0\}.$

Thus, $\text{rad}(f) \neq \{0\}$. Hence, f is degenerate, therefore if $[f]_B$ is invertible, then f is non-degenerate.

Conversely, let us suppose that f is degenerate, then there exists a non-zero vector $\beta \in V$ such that

$$f(\alpha, \beta) = 0, \forall\ \alpha \in V.$$

Let $B = \{\alpha_1, \alpha_2, ..., \alpha_n\}$ be an ordered basis of V, so

$$\beta = a_1\alpha_1 + a_2\alpha_2 + ... + a_n\alpha_n, \text{ for } a_i \in F.$$

Then $[\beta]_B = [a_1, a_1, ..., a_n]^t$ is a non-zero element of F^n.

Now, for $i = 1, 2, ..., n$, then i-th entry of $[f]_B\ [\beta]_B$ is equal to

$$[\alpha_i]_B^t [f]_B [\beta]_B = f(\alpha_i, \beta) = 0.$$

Hence, $[f]_B$ is not invertible. Therefore, if f is non-degenerate, then $[f]_B$ is invertible.

Theorem 7. *Let f be a reflexive bilinear form on a vector space $V(F)$. Let W_1 and W_2 be subspaces of V, then*

(i) If $W_1 \subseteq W_2$, then $W_2^{\perp} \subseteq W_1^{\perp}$ $\qquad$ *(ii) $(W_1 + W_2)^{\perp} = W_1^{\perp} \cap W_2^{\perp}$*

(iii) $W_1 \subseteq W_1^{\perp\perp}$ (iv) $W_1^{\perp} \subseteq W_1^{\perp\perp\perp}$.

Proof.   (i) Let $\beta \in W_2^{\perp}$, then $\beta \in V$ such that

$$f(\alpha, \beta) = 0, \forall \ \alpha \in W_2.$$

Since every element of W_1 is also the element of W_2, therefore

$$f(\alpha, \beta) = 0 \ \forall \ \alpha \in W_1 \Rightarrow \beta \in W_1^{\perp}.$$

$$\therefore \qquad W_2^{\perp} \subseteq W_1^{\perp}.$$

(ii) Since W_1 and W_2 are subspaces of V, then $W_1 + W_2$ is also a subspace of V.

Let $\alpha \in W_1 + W_2$, then

$$\alpha = \alpha_1 + \alpha_2 \text{ where } \alpha_1 \in W_1 \text{ and } \alpha_2 \in W_2.$$

Now, $\beta = (W_1 + W_2)^{\perp} \Leftrightarrow f(\alpha, \beta) = 0 \ \forall \ \alpha \in W_1 + W_2$

$\Leftrightarrow f(\alpha_1 + \alpha_2, \beta) = 0 \ \forall \ \alpha_1 \in W_1 \text{ and } \alpha_2 \in W_2$

$\Leftrightarrow f(\alpha_1, \beta) + f(\alpha_2, \beta) = 0$

$\Leftrightarrow f(\alpha_1, \beta) = 0 \ \forall \ \alpha_1 \in W_1 \text{ and } f(\alpha_2, \beta) = 0 \ \forall \ \alpha_2 \in W_2$

$\Leftrightarrow \beta \in W_1^{\perp} \text{ and } \beta \in W_2^{\perp} \Leftrightarrow \beta \in W_1^{\perp} \cap W_2^{\perp}$

$$\therefore \qquad (W_1 + W_2)^{\perp} = W_1^{\perp} \cap W_2^{\perp}.$$

In a similar way we may prove result (iii) and (iv).

Theorem 8. **(Riesz representation theorem)**. *Let f be a non-degenerate reflexive bilinear form on a finite dimensional vector space V over F. If ϕ is a linear form on V, then there exists a unique vector $\beta \in V$ such that $\phi(\alpha) = f(\alpha, \beta)$ for all $\alpha \in V$.*

Proof.   Let $\psi: V \to V^*$ be a mapping defined by $\psi(y) = \phi(\alpha, y)$, where V^* is dual to V.

Obviously, ψ is well defined homomorphism of vector space. If $\gamma \in \ker.\psi$, then $f(\alpha, \gamma) = 0, \forall \ \alpha \in V$. Since f is non-degenerate so that $\gamma = 0$.

Therefore, ψ is one to one. Also, V is finite dimensional so that $\dim V = \dim V^*$. Thus the mapping ψ is surjective.

Hence, if $\phi \in V^*$, then there exists a unique vector $\beta \in V$ such that

$$\phi(\beta) = f(\alpha, \beta), \forall \ \alpha \in V.$$

Theorem 9. *Let f be a non-degenerate reflexive bilinear form on a finite dimensional vector space V over F. Then for a subspace W of V, $W^{\perp\perp} = W$.*

Proof.   It is known that

$$W \subseteq W^{\perp\perp} \qquad\qquad\qquad ......(1)$$

Suppose that the inclusion is proper, then there exists a linear map ϕ on V such that $\phi = 0$, for all $\alpha \in W$ and $\phi \neq 0$, for all $\beta \in W^{\perp\perp}$.

Let $\gamma \in V$ such that $\phi(\alpha) = f(\alpha, \gamma)$ for all $\alpha \in V$.

Since $f(\alpha, \gamma) = 0$, for all $\alpha \in W$ and $f(\alpha, \gamma) \neq 0$, for some $\alpha \in W^{\perp\perp}$, this implies $\gamma \in W^{\perp} \sim W^{\perp\perp}$, which gives a contradiction as $W^{\perp} = W^{\perp\perp}$.

Hence, $W = W^{\perp\perp}$.

Theorem 10. *Let f be a reflexive bilinear form on a vector space V over F. If S is a finite dimensional anisotropic subspace of V, then $V = S \oplus S^{\perp}$.*

Proof.   Let $\beta \in V$. The mapping $\phi: S \to F$ defined by $\phi(\alpha) = f(\alpha, \beta)$, for all $\alpha \in S$. Obviously

ϕ is linear on S. The bilinear form $\hat{f} = f|_{S \times S}$ is non-degenerate on S. Since if $\gamma \in S$

such that $f(\alpha,\gamma)=0$, for all $\alpha \in S$, then $\gamma \in S \cap S^{\perp} = \{0\}$. Therefore, by Riesz

representation theorem there exists $\gamma \in S$ such that $\phi(\alpha) = \hat{f}(\alpha,\gamma)$ for all $\alpha \in S$.

Thus for all $\alpha \in S$.

$$f(\alpha,\beta) = \hat{f}(\alpha,\gamma) = f(\alpha,\gamma) \Rightarrow f(\alpha, \beta - \gamma) = 0, \text{ for all } \alpha \in S$$

$$\Rightarrow \quad \beta - \gamma \in S^{\perp} \quad \Rightarrow \quad \beta = \gamma + \beta - \gamma \in S + S^{\perp}.$$

Since β is an arbitrary element of V. Hence $V = S \oplus S^{\perp}$.

Theorem 11. *A symmetric bilinear form on a finite dimensional vector space V over a field F of characteristic not equal to 2 is diagonolizable.*

Proof. We shall prove the theorem by induction on dim. V.

If dim. $V = 1$, then every non-zero vector of V is an orthogonal basis. Now assume that induction hypothesis holds for all symmetric bilinear forms on vector spaces of dimension less than dim V.

Let f be a symmetric bilinear form on V. If $f(\alpha,\alpha)=0$, $\forall \alpha \in V$. Then for $\beta, \gamma \in V$.

$$0 = f(\beta+\gamma, \beta+\gamma) = f(\beta,\beta) + 2f(\beta,\gamma) + f(\gamma,\gamma) \Rightarrow f = 0 \text{ as characteristic of } F \neq 2.$$

This shows that every basis of V is orthogonal.

Assume that for some $\alpha_1 \in V, f(\alpha_1,\alpha_1) \neq 0$.

Let W be a subspace spanned by α_1, then $V = W \oplus W^{\perp}$. If $f_1 = f \mid_{W^{\perp} \times W^{\perp}}$,

then f_1 is a symmetric bilinear form on $W^{\perp}$.

Since dim. $W^{\perp} = $ dim.$V - 1$. By induction hypothesis $W^{\perp}$ has an orthogonal basis B_1. Hence, $B = \{\alpha_1\} \cup B_1$ is an orthogonal basis of V relative to f such that f is of the form

$$[f]_B = \text{diag}\left(\begin{bmatrix} 0 & 1 \\ -1 & 0 \end{bmatrix}, \ldots, \begin{bmatrix} 0 & 1 \\ -1 & 0 \end{bmatrix}, 0, \ldots, 0\right)$$

Theorem 12. *Let f be a symmetric and non-alternating bilinear form on a finite dimensional vector space V over a field F of characteristic 2. Then f is diagonalizable.*

Proof. Let V_1 be a subspace of V complementary to rad(f). Then $V = V_1 \oplus \text{rad } f$ and

$\hat{f} = f \mid_{V_1 \times V_1}$ is a non-degenerate, non-alternating and symmetric bilinear form on

V_1. If V_1 has an orthogonal basis B_1 with respect to $\hat{f}$ and B_2 any basis of rad(f), then $B = B_1 \cup B_2$ is an orthogonal basis of V. Therefore, we can assume that f is also non-degenerate.

Now, we shall prove the main theorem by induction on dim$V$.

If dim $V = 1$, obviously the theorem is true in this case. It is also assumed that the hypothesis hold for all non-degenerate, non-alternating and symmetric bilinear forms on vector spaces of dimension less than dim. V.

Since f is non-alternating, then there exists a vector $\alpha \in V$ such that $f(\alpha,\alpha)=k \neq 0$.

Let W be a subspace of V spanned by the vector α. Then $V = W \oplus W^{\perp}$. Let $f_0 = f \mid_{W^{\perp} \times W^{\perp}}$. If f_0 is non-altenating, then by induction hypothesis, $W^{\perp}$ has an orthogonal basis B_0 relative to f_0 and so $B = \{\alpha\} \cup B_0$ is an orthogonal basis of V relative to f.

Since f_0 is non-degenerate and if f_0 is alternating then there is a basis

$$B' = \{\alpha_2, \alpha_3, \ldots, \alpha_n\} \text{ such that}$$

$$[f_0]_{B'} = \text{diag}\left(\begin{bmatrix} 0 & 1 \\ 1 & 0 \end{bmatrix}, \dots\dots, \begin{bmatrix} 0 & 1 \\ 1 & 0 \end{bmatrix} \right)$$

Let $W_1 = (\alpha, \alpha_2, \alpha_3)$ and let $f_1 = f|_{W_1 \times W_1}$. Then clearly f_1 is a non-degenerate bilinear form on W_1. Also, the vectors $\beta_1 = \alpha + \alpha_2 + \alpha_3$, $\beta_2 = \alpha + k\alpha_2$, $\beta_3 = \alpha + (1+k)\alpha_2 + \alpha_3$, are mutually orthogonal and the matrix of f_1 relative to the ordered basis $\{\beta_1, \beta_2, \beta_3\}$ is equal to dig.(k, k, k).

Now, let W_2 be a subspace spanned by the vectors $\beta_3, \alpha_4, \dots, \alpha_n$.

Since β_3 is a linear combination of vectors α, α_2 and α_3 and $f(\beta_3, \beta_3) = k$, the bilinear form $f_2 = f|_{W_2 \times W_2}$ is non-degenerate, non-alternating and symmetric. Also

dim $W_2 <$ dim V, then by induction hypothesis, W_2 has an orthogonal basis B''. Hence, $B = \{\beta_1, \beta_2\} \cup B''$ is an orthogonal basis of V.

Theorem 13. *Let V be a finite-dimensional vector space over the field of complex numbers. Let f be a symmetric bilinear form on V which has rank r. Then there is an ordered basis $B = \{\beta_1, \beta_2, \dots, \beta_n\}$ for V such that*

(i) the matix of f in the ordered basis B is diagonal.

(ii) $f(\beta_j, \beta_j) = \begin{cases} 1, j = 1, 2, \dots, r \\ 0, j > r \end{cases}$

Proof. By theorem 4, there is an ordered basis $\{\alpha_1, \alpha_2, \dots, \alpha_n\}$ of V such that $f(\alpha_i, \alpha_j) = 0$ for $i \neq j$.

Since f has a rank r, therefore the matrix of f is relative to the ordered basis $\{\alpha_1, \alpha_2, \dots, \alpha_n\}$.

Hence, $f(\alpha_i, \alpha_j) \neq 0$ for exactly r values of j. Now, by reordering the vectors α_j, we may assume that $f(\alpha_j, \alpha_j) \neq 0$, $j = 1, 2, \dots, r$.

Also, the field is the field of complex numbers, and if $\sqrt{f(\alpha_j, \alpha_j)}$ denotes complex square root of $f(\alpha_j, \alpha_j)$ and if we put

$$\beta_j = \begin{cases} \dfrac{1}{\sqrt{f(\alpha_j, \alpha_j)}} \alpha_j, 1, 2, \dots, r \\ \alpha_j, \qquad\qquad j > r \end{cases}$$

Then the basis $\{\beta_1, \beta_2, \dots, \beta_n\}$ satisfies the given hypothesis (i) and (ii).

Definition 9. A symmetric bilinear form f is said to be positive definite if for $0 \neq \alpha \in V$, $f(\alpha, \alpha) > 0$.

Theorem 14. *Let V be an n-dimensional vector space over the field of real numbers and let f be a symmetric bilinear form on V which has rank r. Then there is an ordered basis $\{\beta_1, \beta_2, \dots, \beta_n\}$ for V in which the matrix of f is diagonal and such that*

$$f(\beta_j, \beta_j) = \pm 1, j = 1, 2, \dots, r$$

Furthermore, the number of basis vectors β_j for which $f(\beta_j, \beta_j) = 1$ is independent of the choice of basis.

Proof. By theorem 4, there is an ordered basis $\{\alpha_1, \alpha_2, \dots, \alpha_n\}$ such that $f(\alpha_i, \alpha_j) = 0, i \neq j$.

Therefore, $f(\alpha_j, \alpha_j) \neq 0, 1 \leq j \leq r$

and $\qquad f(\alpha_j, \alpha_j) = 0, j > r$

Let us assume that

$$\beta_j = \begin{cases} \dfrac{1}{\sqrt{|f(\alpha_j,\alpha_j)|}}\alpha_j, & 1 \le j \le r \\ \alpha_j, & j > r \end{cases}$$

Then, $\{\beta_1,\beta_2,...,\beta_n\}$ forms a basis with the given properties.

Now, assume that there are k number of basis vectors β_j for which $f(\beta_j,\beta_j) = 1$. Then we show that the number k is independent of this basis.

Let V^+ be the subspace of V spanned by the basis vectors β_j for which $f(\beta_j,\beta_j)=1$. and let V^- be the subspace of V spanned by the basis vectors β_j. For which $f(\beta_j,\beta_j)=-1$. Now $k= \dim V^+$, so it is the uniqueness of the dimension of V^+ which we must show. Clearly, if α is a non-zero vector in V^+, then $f(\alpha,\alpha)<0$ implying f is negative definite on V^-.

Now, Iet $V^\perp$ be the subspace spanned by the vector β_j for which $f(\beta_j,\beta_j)=0$, so if $\alpha \in V^\perp$, then $f(\alpha,\beta) = 0$, $\forall\ \beta \in V$.

Since $\{\beta_1, \beta_2,...,\beta_n\}$ is a basis of V, then we have

$$V=V^+ \oplus V^- \oplus V^\perp.$$

Next, we claim that if W is any subspace of V on which f is positive definite, then the subspaces W, V^- and $V^\perp$ are independent.

For, suppose $\alpha \in W$, $\beta \in V^-$ and $\gamma \in V^\perp$ and $\alpha+\beta+\gamma=0$,then

$$\left.\begin{array}{l} 0 = f(\alpha,\alpha+\beta+\gamma) = f(\alpha,\alpha) + f(\alpha,\beta) + f(\alpha,\gamma) \\ 0 = f(\beta,\alpha+\beta+\gamma) = f(\beta,\alpha) + f(\beta,\beta) + f(\beta,\gamma) \end{array}\right\} \qquad ...(1)$$

Since, $\gamma \in V^\perp$ so $f(\alpha, \gamma) = f(\beta,\gamma) = 0$, and since f is symmetric, then (1) reduces to
$$0 = f(\alpha, \alpha) + f(\alpha, \beta) \text{ and } 0 = f(\beta, \beta) + f(\alpha, \beta)$$
From these equations, we obtain that $f(\alpha, \alpha) = f(\beta, \beta)$.

But $f(\alpha, \alpha) \ge 0$ and $f(\beta, \beta) \le 0$, therefore it follows that
$$f(\alpha, \alpha) = f(\beta, \beta) = 0.$$
Also, f is positive definite on W and negative definite on V^-, then we obtain
$$\alpha=\beta=0$$
and hence
$$\gamma=0$$
Since $V = V^+ \oplus V^- \oplus V^\perp$ and W, V^-, $V^\perp$ are independent so that
$$\dim.W \le \dim V^+.$$

If β_1 is another ordered basis of V which satisfies the conditions of the theorem, then we shall have corresponding subspaces $V_1{}^+$, $V_1{}^-$ and $V_1{}^\perp$ and using above process, we have $\dim V_1{}^+ \le \dim V^+$. Now, interchanging the role of $V_1{}^+$ and V^+, we have $\dim.V^+ \le \dim V_1{}^+$.

Hence, $\dim V^+ = \dim V_1{}^+$.

Definition 10. *Let f be a symmetic bilinear form on V and let V^+ ,V^- be the subspaces of V such that $f(\beta_j,\beta_j)=1$ on V^+ and $f(\beta_j,\beta_j)=-1$. on V^-, where $\{\beta_1, \beta_2,...., \beta_n\}$ is an ordered basis of V, then the number $(\dim V^+ + \dim V^-)$ is called the rank of f and the number $(\dim V^+ - \dim V^-)$ is called the signature of f.*

Theorem 15. *Let V be an n-dimensional vector space over a subfield of the complex numbers, and let f be a skew-symmetric bilinear form on V. Then the rank r = 2k, there is an ordered basis for V in which ihe matix of f is the direct sum of the $(n-r) \times (n-r)$ zero matrix and k copies of the 2×2 matrix* $\begin{bmatrix} 0 & 1 \\ -1 & 0 \end{bmatrix}$.

Proof. Let $\alpha_1, \beta_1, \alpha_2, \beta_2, ..., \alpha_k, \beta_k$ be the vectors having the following properties:

(i) $f(\alpha_j, \beta_j) = 1$ for $j = 1, 2, ..., k$.

(ii) $f(\alpha_i, \beta_j) = f(\beta_i, \beta_j) = f(\alpha_i, \beta_j) = 0, i \neq j$.

(iii) If W_j is the two dimensional subspace spanned by α_j and β_j, then
$$V = W_1 \oplus W_2 \oplus ,..., \oplus W_k \oplus W_0$$
where every vector in W_0 is orthogonal to all α_j and β_j and the restriction of f to W_0 is the zero form.

Assuming that $[\gamma_1, \gamma_2, ..., \gamma_m]$ be any ordered basis for the subspace W_0, then
$$B = \{\alpha_1, \beta_1, ..., \alpha_k, \beta_k, \gamma_1, \gamma_2, ..., \gamma_n\}$$
forms an ordered basis for V.

From conditions (i), (ii) and (iii), it is clear that the matrix of f in the ordered basis B is the direct sum of the $(n - 2k) \times (n - 2k)$ zero matrix and k copies of the 2×2 matrix

$$\begin{bmatrix} 0 & 1 \\ -1 & 0 \end{bmatrix}$$

It is also clear that the rank of this matrix of f is $2k$.

Solved Examples

▶ A mapping $f : V \times V \rightarrow F$ is said to be bilinear form on a vector space if
(i) $f(a\alpha + b\beta, \gamma) = af(\alpha, \gamma) + bf(\beta, \gamma)$
(ii) $f(\alpha, a\beta + b\gamma) = af(\alpha, \beta) + bf(\alpha, \gamma) \ \forall \ a, b \in F, \ \alpha, \beta, \gamma \in V$

▶ Let f be a bilinear form on V over F. Then a vector $\alpha \in V$ is said to be orthogonal to $\beta \in V$ with respect to f if $f(\alpha, \beta) = 0$

▶ Let f be a bilinear form on $V(F)$ and let $S = \{\alpha_1, \alpha_2, ..., \alpha_n\}$ be a basis of V. Then a matrix $A = \{a_{ij}\}$ where $a_{ij} = f(\alpha_i, \alpha_j)$ is called the matrix representation of f w.r.t. the basis S.

Example 1. *Let A be any square matix of order $n \times n$ over F. Then the map $f(X, Y) = X^T A Y$ is a bilinear form on F^n.*

Solution. For any $a, b \in F$ and any $X_i, Y_i \in F^n$
$$f(aX_1 + bX_2, Y) = (aX_1 + bX_2)^T AY = (aX_1^T + bX_2^T) A \in V$$
$$= aX_1^T AY + bX_2^T AY = af(X_1, Y) + bf(X_2, Y).$$
This shows that f is linear in first variable.

Also, $f(X, aY_1 + bY_2) = X^T A(aY_1 + bY_2)$
$$= aX^T AY_1 + bX^T AY_2 = af(X, Y_1) + bf(X, Y_2)$$

This shows that f is linear in second variable.

Hence, f is bilinear form on F^n.

Example 2. *Let ϕ and ψ be any linear functionals on a vector space V. Let $f : V \times V \to F$ be defined by $f(\alpha, \beta) = \phi(\alpha)\psi(\beta)$. Show that f is bilinear form.*

Solution. For any $a, b \in F$ and $\alpha_i, \beta_i \in V$, we have

$$f(a\alpha_1 + b\alpha_2, \beta) = \phi(a\alpha_1 + b\alpha_2)\psi(\beta) = (a\phi(\alpha_1) + b\phi(\alpha_2))\psi(\beta)$$

$$= a\phi(\alpha_1)\psi(\beta) + b\phi(\alpha_2)\psi(\beta) = af(\alpha_1, \beta) + bf(\alpha_2, \beta)$$

Also, $f(\alpha, a\beta_1 + b\beta_2) = \phi(\alpha) + \psi(a\beta_1 + b\beta_2) = \phi(\alpha)a(\psi(\beta_1) + b\psi(\beta_2))$

$$= a\phi(\alpha)\psi(\beta_1) + b\phi(\alpha)\psi(\beta_2) = af(\alpha, \beta_1) + bf(\alpha, \beta_2)$$

Hence, f is bilinear form.

Example 3. *Let f be a bilinear form on $\mathbf{R}^2$ defined by $f((x_1, x_2), (y_1, y_2)) = 2x_1y_1 - 3x_1y_2 + x_2y_2$. Find the matrix of f in the basis $\{\alpha_1 = (1,0), \alpha_2 = (1,1)\}$.*

Solution. Let $A = \{a_{ij}\}$ be the matrix of f in the given basis, where $a_{ij} = f(\alpha_i \alpha_j)$, $1 \le i, j \le 2$.

Now, $a_{11} = f(\alpha_1, \alpha_1) = f(1,0), (1,0)$

$$= 2(1)(1) - 3(1)(0) + (0)(0) = 2$$

$a_{12} = f(\alpha_1, \alpha_2) = f(1,0), (1,1)$

$$= 2(1)(1) - 3(1)(1) - 0(1) = 2 - 3 = -1$$

$a_{21} = f(\alpha_1, \alpha_1) = f(1,0), (1,0)$

$$= 2(1)(1) - 3(1)(0) + (0)(0) = 2$$

$a_{22} = f(\alpha_2, \alpha_2) = f(1,1), (1,1)$

$$= 2(1)(1) - 3(1)(1) + (1)(1) = 2 - 3 + 1 = 0$$

Thus, $A = \begin{bmatrix} 2 & -1 \\ 2 & 0 \end{bmatrix}$ is the matrix of f in the basis $\{\alpha_1, \alpha_2\}$.

Example 4. *Let $[f]$ denote the matix representation of a bilinear form f on V relative to a basis $S = \{\alpha_1, \alpha_2, ..., \alpha_n\}$ of V. Show that the mapping $\phi : f \to [f]$ is an isomorphism of $B(V, f)$ on the vector space of n-square matrices.*

Solution. Since f is completely determined by the scalars $f(\alpha_i, \alpha_j)$, therefore the mapping ϕ is one-one and onto. Now to show that ϕ is homomorphism.

For any $a, b \in F, f_1, f_2 \in B(V, f)$,

$$(af_1 + bf_2)(\alpha_i, \alpha_j) = af_1(\alpha_i, \alpha_j) + bf_2(\alpha_i, \alpha_j)$$

$$\Rightarrow \qquad [af_1 + bf_2] = a[f_1] + b[f_2].$$

Hence, the result.

Example 5. *Which of the following mappings $f : \mathbf{R}^2 \times \mathbf{R}^2 \to \mathbf{R}$ are bilinear form : if for $\alpha = (x_1, x_2)$, $\beta = (y_1, y_2)$*

(i) $f(\alpha, \beta) = x_1x_2 + y_1y_2 - x_2y_1 - x_1y_2$

(ii) $f(\alpha, \beta) = (x_1 - y_1)^2 + x_2y_2$

(iii) $f(\alpha, \beta) = x_1y_2 \pm x_2y_1$

(iv) $f(\alpha, \beta) = x_1y_1 + 1$

Solution. Let $\alpha = (x_1, x_2)$, $\beta = (y_1, y_2)$ and $\gamma = (z_1, z_2)$.

(i) For $a, b \in \mathbf{R}$

$$f(a\alpha + b\beta, \gamma) = f(a(x_1, x_2) + b(y_1, y_2), (z_1, z_2))$$
$$= f((ax_1 + by_1, ax_2 + by_2), (z_1, z_2))$$
$$= (ax_1 + by_1)(ax_2 + by_2) + z_1 z_2 - (ax_2 + by_2)z_1 - (ax_1 + by_1)z_2$$
$$= a^2 x_1 x_2 + abx_1 y_2 + aby_1 x_2 + b^2 y_1 y_2 + z_1 z_2$$
$$\quad - ax_2 z_1 - by_2 z_1 - ax_1 z_2 - by_1 z_2$$

and $af(\alpha, \gamma) + bf(\beta, \gamma) = af((x_1, x_2), (z_1, z_2)) + bf((y_1, y_2), (z_1, z_2))$
$$= ax_1 x_2 + az_1 z_2 - ax_2 z_1 - ax_1 z_2 + by_1 y_2 + bz_1 z_2 - by_2 z_1 + by_1 z_2$$

Clearly, $f(a\alpha + b\beta, \gamma) \neq af(\alpha, \gamma) + bf(\beta, \gamma)$.

Thus f is not bilinear form.

(ii) For $a, b \in \mathbf{R}$;

$$f(a\alpha + b\beta, \gamma) = f((ax_1 + by_1, ax_2 + by_2), (z_1, z_2))$$
$$= (ax_1 + by_1 - z_1)^2 + (ax_2 + by_2) z_2$$

and $af(\alpha, \gamma) + bf(\beta, \gamma) = af((x_1, x_2), (z_1, z_2)) + bf((y_1, y_2), (z_1, z_2))$
$$= a[(x_1 - z_1)^2 + x_2 z_2] + b[(y_1 - z_1)^2 + y_2 z_2]$$

Clearly $f(a\alpha + b\beta, \gamma) \neq af(\alpha, \gamma) + bf(\beta, \gamma)$.

Thus f is not bilinear form.

(iii) For $a, b \in \mathbf{R}$

$$f(a\alpha + g\beta, \gamma) = f((ax_1 + by_1, ax_2 + by_2), (z_1, z_2)) = (ax_1 + by_1)z_2 \pm (ax_2 + by_2)z_1$$
$$= a(x_1 z_2 + x_2 z_1) + b(y_1 z_2 + y_2 z_1)$$
$$= af((x_1, x_2), (z_1, z_1)) + bf((y_1, y_2), (z_1, z_2)) = af(\alpha, \gamma) + bf(\beta, \gamma)$$

Also, $f(\alpha, a\beta + b\gamma) = f((x_1, x_2)(ay_1 + bz_1, ay_2 + bz_2))$
$$= x_1(ay_2 + bz_2) \pm x_2(ay_1 + bz_1)$$
$$= a(x_1 y_2 \pm x_2 y_1) + b(x_1 z_2 \pm x_2 z_1)$$
$$= af((x_1, x_2), (y_1, y_2)) + bf((x_1, x_2), (z_1, z_2))$$
$$= af(\alpha, \beta) + bf(\alpha, \gamma)$$

Hence, f is bilinear form.

(iv) For $a, b \in \mathbf{R}$

$$f(a\alpha + b\beta, \gamma) = f((ax_1 + by_1, ax_2 + by_2), (z_1, z_2))$$
$$= (ax_1 + by_1)z_1 + 1 = a(x_1, z_1) + b(y_1, z_1) + 1$$

and $af(\alpha, \gamma) + bf(\beta, \gamma) = af((x_1, x_2), (z_1, z_2)) + bf((y_1, y_2), (z_1, z_2))$
$$= a[x_1 z_1 + 1] + b[y_1 z_1 + 1] = a(x_1, z_1) + b(y_1, z_1) + a + b$$

Clearly, $f(a\alpha + b\beta, \gamma) \neq af(\alpha, \gamma) + bf(\beta, \gamma)$.

Thus, f is not bilinear form.

Example 6. *Prove that if f is a bilinear form on $V(F)$, then*

(i) $f(-\alpha, \beta) = -f(\alpha, \beta) = f(\alpha, -\beta)$, $\forall\ \alpha, \beta \in V$

(ii) $f(\alpha,0) = 0 = f(0,\alpha), \forall\ \alpha \in V$

Solution.   (i) For any $a \in F$ and any $\alpha, \beta \in V$.

$$f(a\alpha,\beta) = af(\alpha,\beta)$$

taking $a = -1$, we get

$$f(-\alpha,\beta) = -f(\alpha,\beta)$$

Similarly, $f(\alpha,a\beta) = af(\alpha,\beta)$

again taking $a = -1$, we get

$$f(\alpha,-\beta) = -f(\alpha,\beta).$$

$$f(-\alpha,\beta) = -f(\alpha,\beta) = f(\alpha,-\beta).$$

(ii) $f(\alpha,0)\ = f(\alpha,0.0)$ $[\because\ 0.0 = 0]$

$$= 0f(\alpha,0) = 0.$$

Similarly, $f(0,\alpha) = 0.$

$\therefore$ $f(\alpha,0) = 0 = f(0,\alpha), \forall\ \alpha \in V.$

Example 7. *Let F be a field of chnracteristic 2. Verify that the mapping $\phi: F \times F \to F$ defined by $\phi(x,y) = xy$ is a skew-symmetric bilinear form which is not alternating.*

Solution.   First we show that ϕ is bilinear.

For $a, b \in F$ and $x, y, z \in F$, we have $(ax + by, z) \in F \times F$ so,

$$\phi(ax + by, z) = (ax + by)z$$

$$= axz + byz = a\phi(x, z) + b\phi(x, z)$$

Also, $\phi(x, ay + bz) = x(ay + bz) = axy + bxz = a\phi(x, y) + b\phi(x, z).$

Thus, Q is bilinear form.

Since F is of characteristic 2 so that $2xy = 0, \forall\ x, y \in F$.

$\Rightarrow$ $xy + xy = 0$ $\Rightarrow\ xy = -xy$

$\Rightarrow$ $\phi(x,y) = -\phi(x, y), \forall x, y \in F.$Thus, ϕ is skew-symmetric.

Now, ϕ is skew-symmetric so that

$$\phi(x,x) = -\phi(x,x), \forall\ x \in F \text{ or } 2\phi(x,x) = 0$$

$\therefore$ $\phi(x,x) \neq 0$ as characteristic of $F = 2.$

This shows that ϕ is not alternating form.

Example 8. *If f is a bilinear form on V(F) and S be a subspace of V, then the set*
$$S^{\perp} = \{\beta \in V: f(\alpha,\beta) = 0, \forall\ \alpha \in S\} \text{ is a subspace of V.}$$

Solution.   Let $\alpha \in S$ so $f(\alpha,0) = 0$. Therefore $0 \in S^{\perp}$. Suppose $\alpha_1, \alpha_2 \in S^{\perp}$ and $k \in F$. Then $f(\alpha,\alpha_1) = 0$ and $f(\alpha,\alpha_2) = 0.$

Thus, $f(\alpha, \alpha_1 + \alpha_2) = f(\alpha,\alpha_1) + f(\alpha,\alpha_2) = 0 + 0 = 0$

and $f(\alpha, k\alpha_1) = kf(\alpha,\alpha_1) = k.0 = 0.$

This implies $\alpha_1 + \alpha_2 \in S^{\perp}$ and $k\alpha_1 \in S^{\perp}$.

Hence, $S^{\perp}$ is a subspace of V.

REMARKS

☞ The subspace S is called isotropic if $S \cap S^\perp \neq \{0\}$ otherwise anisotropic.

☞ S is called totally isotropic if $S \subseteq S^\perp$.

☞ The subspace $V^\perp$ is called the radical of f, i.e., rad $f = \{\beta \in V : f(\alpha, \beta) = 0 \ \forall \alpha \in V\}$

Example 9. *Show that a bilinear form f is symmetric if and only if any matrix A representing f is symmetric.*

Solution. Suppose f is symmetric and A represents f. For X, $X^T A Y$, is a scalar therefore, it is equal to its transpose. Then

$$f(X,Y) = X^T A Y = (X^T A Y)^T = Y^T A^T X$$

Since f is symmetric, i.e., $f(X, Y) = f(Y, X)$

So, $\qquad Y^T A^T X = Y^T A X, \ \forall \ X, Y$

$\Rightarrow \qquad\qquad A^T = A \qquad \Rightarrow \quad A$ is symmetric.

Conversely, suppose A is symmetric. Then

$$f(X, Y) = X^T A Y = (X^T A Y)^T = Y^T A^T X = Y^T A X \qquad\qquad [\because A^T = A]$$

$$= f(Y, X)$$

$\therefore f$ is symmetric.

Example 10. *Let V be a finite-dimensional vector space and L_1, L_2 linear functionals on V. Show that the equation $f(\alpha, \beta) = L_1(\alpha) L_2(\beta) - L_1(\beta) L_2(\alpha)$ defnes a skew-symmetric bilinear form on V. Show that $f = 0$ if and only if L_1, L_2 are linearly dependent.*

Solution. First we show f is linear form on V.

Let $\alpha, \beta, \gamma \in V$ and $a, b \in F$, we have

$$f(\alpha, a\beta + b\gamma) = L_1(\alpha) L_2(a\beta + b\gamma) - L_1(a\beta + b\gamma) L_2(\alpha)$$

$$= L_1(\alpha)\{aL_2(\beta) + bL_2(\gamma)\} - \{aL_1(\beta) + bL_1(\gamma)\} L_2(\alpha)$$

$$[\because L_1, L_2 \text{ are linear.}]$$

$$= aL_1(\alpha) L_2(\beta) + bL_1(\alpha) L_2(\gamma) L_2(\alpha) - aL_1(\beta) L_2(\alpha) - bL_1(\gamma) L_2(\alpha)$$

$$= a\{L_1(\alpha) L_2(\beta) - L_2(\beta) L_2(\alpha)\} + b\{L_1(\alpha) L_2(\gamma) - L_1(\gamma) L_2(\alpha)\}$$

$$= af(\alpha, \beta) + bf(\alpha, \gamma).$$

$\therefore \quad f$ is also linear in second coordinate.

Thus f is a bilinear form on V.

Secondly, we show that f is skew-symmetric.

For any $\alpha, \beta \in V$, we have

$$f(\alpha, \beta) = L_1(\alpha) L_2(\beta) - L_1(\beta) L_2(\alpha)$$

$$= -\{L_1(\beta) L_2(\alpha) - L_1(\alpha) L_2(\beta)\}$$

$$= -f(\beta, \alpha)$$

$\therefore \qquad\qquad f(\alpha, \beta) = -f(\beta, \alpha) \ \forall \ \alpha, \beta \in V$

Hence, f is skew-symmetric.

Next, we show that $f = 0$ iff L_1, L_2 are linearly dependent.

Suppose L_1, L_2 are linearly dependent then

$$f(\alpha, \beta) = L_1(\alpha) L_2(\beta) - L_1(\beta) L_2(\alpha)$$

$$= (kL_2)(\alpha)L_2(\beta) - (kL_2)(\beta)L_2(\alpha)$$
$$= kL_2(\alpha)L_2(\beta) - kL_2(\beta)L_2(\alpha)$$
$$= k(L_2(\alpha)L_2(\beta) - L_2(\beta)L_2(\alpha))$$
$$= k.0$$
$$= 0.$$

Since α, β are arbitrary element of V so that $f = 0$.

Conversely, suppose $f = 0 \ \forall \ \alpha < \beta \in V$, then

$\Rightarrow$ $L_1(\alpha) L_2(\beta) - L_1(\beta)L_2(\alpha) = 0.$

This is possible only if either $L_1 = kL_2$ or $L_2 = kL_1$.

Hence, L_1, L_2 are linearly dependent.

Example 11. *Let n be a positive integer, and let V be the space of all $n \times n$ matrices over the field of complex numbers. Show that the equation*

$$f(A, B) = n \ tr(AB) - tr(A)tr(B)$$

defines a bilinear form f on V. Is it true that $f(A, B) = f(B, A)$ for all A, B ?

Solution.   Let $A, B, D \in V$ and $a, b \in \mathbf{C}$ (complex field).

Then, $f(A, aB+bD) = n \ tr[A(aB+bD)] - tr(A)tr(aB+bD)$
$$= n \ tr(aAB+bBD) - tr(A)\{tr(aB)+tr(bD)\}$$
$$= na \ tr(AB) + nb \ tr(BD) - a \ tr(A) \ tr(B) - tr(A)tr(bD)$$
$$= na \ tr(AB) + nb \ tr(BD) - a \ tr(A) \ tr(B) - b \ tr(A) \ tr(D)$$
$$= a[n \ tr \ (AB) - tr(A)tr(B)] + b[n \ tr(BD) - tr(A) \ tr \ (D)]$$
$$= af(A,B) + bf(A,D)$$

$\therefore f$ is linear in one coordinate.

Also, $f(aB+bD, A) = n \ tr((aB+bD)A) - tr(aB+bD)tr(A)$
$$= n \ tr(aBA+bDA) - tr(aB) \ tr(A) - tr(bD) \ tr(A)$$
$$= na \ tr(BA) + nb \ tr(DA) - a \ tr(B) \ tr(A) - b \ tr(D) \ tr(A)$$
$$= a[n \ tr(DA) - tr(B) \ tr(A)] + b[n \ tr(DA) - tr(D) \ tr(A)]$$
$$= af(B,A) + bf(D,A).$$

$\therefore f$ is linear in second coordinate.

Hence, f is a bilinear form on V.

Further, since $tr(AB) = tr(BA)$ for all A and B, then

$$n \ tr(AB) - tr(A) \ tr(B) = n \ tr(BA) - tr(B)tr(A)$$

$\Rightarrow$ $f(A, B) = f(B, A)$ for all A and B.

EXERCISE 8.1

1. Decide which of the following mappings $f : \mathbf{R}^2 \times \mathbf{R}^2 \to \mathbf{R}$ are bilinear, if for $\alpha = [x_1, x_2]^t, \beta = [y_1, y_2]^t$ in $\mathbf{R}^2$:

(i) $f(\alpha, \beta) = (x_1 + x_2)(y_1 + y_2)$

(ii) $f(\alpha, \beta) = (x_1 + x_2)/(y_1 + y_2)$

(iii) $f(\alpha, \beta) = x_1 x_2 y_1 y_2$

(iv) $f(\alpha, \beta) = (3x_1 + 5)y_1 + (3y_2 + 5)x_1$

(v) $f(\alpha, \beta) = x_1 y_2 + x_2 y_1^2$

(vi) $f(\alpha, \beta) = (x_1 + y_1)^2 - (x_1 - y_1)^2$

2. Let f be the bilinear form on $\mathbf{R}^2$ defined by $f((x_1,y_1),(x_2,y_2))=x_1y_1+x_2y_2$.
Find the matrix of f in each of the following bases:

(i) $\{(1,0),(0,1)\}$ (ii) $\{(1,-1),(1,1)\}$
(iii) $\{(1,2),(3,4)\}$

3. Let V be the vector space of all 2 x 3 matrices over $\mathbf{R}$, and let f be the bilinear form on V defined by $f(X,Y)=\text{trace}(X^tAY)$ where

$$A=\begin{bmatrix} 1 & 2 \\ 3 & 4 \end{bmatrix}.$$

Find the matrix of f in the ordered basis $\{E^{11}, E^{12}, E^{13}, E^{21}, E^{22}, E^{23}\}$, where E^{ij} is the matrix whose non-zero entry is 1 in $(i,j)^{th}$ position.

4. Let $\alpha=(x_1, x_2, x_3)$ and $\beta=(y_1, y_2, y_3)$ and let
$$f(\alpha,\beta)=3x_1y_1-2x_1y_2+5x_2y_1$$
$$+7x_2y_2-8x_2y_3+4x_3y_2-x_3y_3.$$
Express f in matrix notation.

5. Let f and g be bilinear forms on V. Show that the sum $f+g$ defined by
$(f+g)(\alpha,\beta)=f(\alpha,\beta)+g(\alpha,\beta)$ is bilinear.

6. Let f be a bilinear form on V and let $k \in F$. show that the map kf, defined by
$(kf)(\alpha,\beta)=kf(\alpha,\beta)$ is bilinear.

7. Let $B(V, f)$ denote the collection of all bilinear forms on V. Show that $B(V, f)$ is a vector space with respect to addition $f + g$ and scalar multiplication kf defined by
$(f + g)(\alpha,\beta)= f(\alpha,\beta)+g(\alpha,\beta)$
$(kf)(\alpha,\beta)=kf(\alpha,\beta) \; \forall \alpha,\beta \in V$

8. Let f be the bilinear form on $\mathbf{R}^2$ defined by
$f(\alpha,\beta)=3x_1y_1-2x_1y_2+ 4x_2y_1-x_2y_2$
where $\alpha=(x_1,x_2),\beta=(y_1, y_2)$.
(i) Express f in matrix form with respect to an ordered basis $B = \{(1, 0), (0, 1)\}$.
(ii) Express f in matrix form with respect to an ordered basis $B = \{(1, 1), (1,2)\}$.

(iii) Find the matrix P such that $[f]_{B'}=P^T[f]_B P$.

9. Prove that if char. $F\neq 2$, then a bilinear form on V over F is skew-symmetric if and only if it is altenating.

10. Prove that if char. $F\neq 2$ and f is a symmetric and altenating bilinear form on $V(F)$, then $f=0$.

11. Prove that in a reflexive bilinear space if $\alpha_1,\alpha_2,...,\alpha_n$ are mutually orthogonal anisotropic vectors, then $\{\alpha_1,\alpha_2,...,\alpha_n\}$ are linearly independent.

12. Prove that a diagonalizable bilinear form is symmetric.

13. Let $B(V,f)$ and $B(W,g)$ be isomorphic bilinear spaces. Prove that f is reflexive if and only if g is reflexive.

14. If f is reflexive, non-degenerate bilinear form on a vector space $V(F)$, then prove that for $\alpha, \beta \in V$
(i) if $f(\alpha,\gamma)= f(\beta,\gamma)$ for all $\gamma \in V$, then $\alpha=\beta$.
(ii) if $f(\gamma,\alpha)=f(\gamma,\beta)$ for all $\gamma \in V$, then $\alpha=\beta$.

15. Let f be any bilinear form on a finite-dimensional vector space V. Let W be the subspace of all β such that $f(\alpha, \beta)= 0$ for every $\alpha \in V$. Show that
$$\text{rank}(f)=\dim V- \dim W$$

16. Let f, g be bilinear forms on a finite-dimensional vector space $V(F)$. Suppose g is non-singular. Show that there exist unique operators T_1,T_2 on V such that
$$f(\alpha,\beta)=g(T_1(\alpha),\beta)=g(\alpha,T_2(\beta))$$
for all $\alpha,\beta \in V$.

17. Let f be a symmetric bilinear form on $\mathbf{C}^n$ and g a skew-symmetric bilinear form on $\mathbf{C}^n$. Suppose $f+g=g$. Show that $f=g=0$.

18. Let m and n be positive integers and F a field. Let V be the vector space of all $m \times n$ matrices over F. Let A be a fixed $m \times m$ matrix over F. Define $f_A(X,Y)=\text{trace}(X^tAY)$. Then show that f_A is bilinear.

Hints to Selected Problems

1. Since $\alpha= (x_1,x_2),\beta=(y_1,y_2)$, $\gamma=(z_1,z_2)$.

$\therefore$ For $a,b \in F, \; a\alpha+b\beta=a(x_1, x_2) +b(y_1,y_2) = (ax_1 +by_1, ax_2+by_2)$

$f(a\alpha + b\beta, \gamma)= f((ax_1 + by_1, ax_2 + by_2),(z_1, z_2))$

$\qquad = (ax_1 +by_1 +ax_2+by_2)(z_1+z_2)$ $[\because f(\alpha,\beta)=(x_1 +x_2)(y_1 +y_2)]$

$\qquad = (ax_1+ax_2)(z_1+z_2)+(by_1+by_2)(z_1+z_2)$

$\qquad = a(x_1 +x_2)(z_1+z_2)+b(y_1+y_2)(z_1+z_2)$

$\qquad = af(\alpha,\beta)+bf(\alpha, \gamma)$

Now, $f(\gamma, a\alpha + b\beta) = f((z_1, z_2), (ax_1 + by_1, ax_2 + by_2))$
$$= (z_1 + z_2)(ax_1 + by_1 + ax_2 + by_2)$$
$$= a(z_1 + z_2)(x_1 + x_2) + b(z_1 + z_2)(y_1 + y_2) = af(\gamma, \alpha) + bf(\gamma, \beta)$$

Hence, f is bilinear.

2. (i) Since $f(x_1, y_1), (x_2, y_2)) = x_1 y_1 + x_2 y_2$

$B = \{(1,0), (0,1)\}$. Let $\alpha_1 = (1,0), \alpha_2 = (0,1)$

$\alpha_{11} = f(\alpha_1, \alpha_1) = f((1, 0, (1, 0)) = 1 \times 0 + 1 \times 0 = 0$
$\alpha_{12} = f(\alpha_1, \alpha_2) = f((1, 0), (0, 1)) = 1 \times 0 + 0 \times 1 = 0$
$\alpha_{21} = f(\alpha_2, \alpha_1) = f(0, 1), (1, 0)) = 0 \times 1 + 1 \times 0 = 0$
$\alpha_{22} = f(\alpha_2, \alpha_2) = f((0, 1), (0, 1)) = 0 \times 1 + 0 \times 1 = 0$

$\therefore$ The matix of f relative to $B = \begin{bmatrix} 0 & 0 \\ 0 & 0 \end{bmatrix}$.

3. $E^{11} = \begin{bmatrix} 1 & 0 & 0 \\ 0 & 0 & 0 \end{bmatrix}, E^{12} = \begin{bmatrix} 0 & 1 & 0 \\ 0 & 0 & 0 \end{bmatrix}, E^{13} = \begin{bmatrix} 0 & 0 & 1 \\ 0 & 0 & 0 \end{bmatrix}$

$E^{21} = \begin{bmatrix} 0 & 0 & 0 \\ 1 & 0 & 0 \end{bmatrix}, E^{22} = \begin{bmatrix} 0 & 0 & 0 \\ 0 & 1 & 0 \end{bmatrix}, E^{23} = \begin{bmatrix} 0 & 0 & 0 \\ 0 & 0 & 1 \end{bmatrix}$

Now, $(E^{11})^t A(E^{11}) = \begin{bmatrix} 1 & 0 \\ 0 & 0 \\ 0 & 0 \end{bmatrix} \begin{bmatrix} 1 & 2 \\ 3 & 4 \end{bmatrix} \begin{bmatrix} 1 & 0 & 0 \\ 0 & 0 & 0 \end{bmatrix} = \begin{bmatrix} 1 & 2 \\ 0 & 0 \\ 0 & 0 \end{bmatrix} \begin{bmatrix} 1 & 0 & 0 \\ 0 & 0 & 0 \end{bmatrix} = \begin{bmatrix} 1 & 0 & 0 \\ 0 & 0 & 0 \\ 0 & 0 & 0 \end{bmatrix}$

$\therefore$ trace $(E^{11})^t A(E^{11}) = 1$

$\therefore$ $a^{11} = f(E^{11}, E^{11}) = 1$

and $(E_{11})^t A(E^{12}) = \begin{bmatrix} 1 & 0 \\ 0 & 0 \\ 0 & 0 \end{bmatrix} \begin{bmatrix} 1 & 2 \\ 3 & 4 \end{bmatrix} \begin{bmatrix} 0 & 1 & 0 \\ 0 & 0 & 0 \end{bmatrix} = \begin{bmatrix} 1 & 2 \\ 0 & 0 \\ 0 & 0 \end{bmatrix} \begin{bmatrix} 0 & 1 & 0 \\ 0 & 0 & 0 \end{bmatrix} = \begin{bmatrix} 0 & 1 & 0 \\ 0 & 0 & 0 \\ 0 & 0 & 0 \end{bmatrix}$

$\therefore$ trace$[(E^{11})^t AE^{12}] = 0$

$\therefore$ $a_{12} = f(E^{11}, E^{12}) = 0$

and $(E_{11})^t A(E^{13}) = \begin{bmatrix} 1 & 2 \\ 0 & 0 \\ 0 & 0 \end{bmatrix} \begin{bmatrix} 0 & 0 & 1 \\ 0 & 0 & 0 \end{bmatrix} = \begin{bmatrix} 0 & 0 & 1 \\ 0 & 0 & 0 \\ 0 & 0 & 0 \end{bmatrix}$

$\therefore$ trace$[(E^{11})^t AE^{13}] = 0$, so $a_{13} = 0$

and $(E^{11})^t A(E^{21}) = \begin{bmatrix} 1 & 2 \\ 0 & 0 \\ 0 & 0 \end{bmatrix} \begin{bmatrix} 0 & 0 & 0 \\ 1 & 0 & 0 \end{bmatrix} = \begin{bmatrix} 2 & 0 & 0 \\ 0 & 0 & 0 \\ 0 & 0 & 0 \end{bmatrix}$

$\therefore$ trace$[(E^{11})^t AE^{21}] = 0$, so $a_{14} = 0$

and $(E^{11})^t A(E^{22}) = \begin{bmatrix} 1 & 2 \\ 0 & 0 \\ 0 & 0 \end{bmatrix} \begin{bmatrix} 0 & 0 & 0 \\ 0 & 1 & 0 \end{bmatrix} = \begin{bmatrix} 0 & 2 & 0 \\ 0 & 0 & 0 \\ 0 & 0 & 0 \end{bmatrix}$

$\therefore$ trace$[(E^{11})^t AE^{22}] = 0$, so $a_{15} = 0$

and $(E^{11})^t A(E^{23}) = \begin{bmatrix} 1 & 2 \\ 0 & 0 \\ 0 & 0 \end{bmatrix} \begin{bmatrix} 0 & 0 & 0 \\ 0 & 0 & 1 \end{bmatrix} = \begin{bmatrix} 0 & 0 & 2 \\ 0 & 0 & 0 \\ 0 & 0 & 0 \end{bmatrix}$

$\therefore$ trace$[(E^{11})^t AE^{23}] = 0$, so $a_{16} = 0$

Hence first row of the matrix of f relative to the given basis is given by $a_{11}, a_{12}, a_{13}, a_{14}, a_{15}$ and $a_{16}, i.e.,$ 1,0,0,2,0,0.

In the similar manner, we can also find other rows.

6. For $\alpha, \beta, \gamma \in V$, $a, b \in F$,

$$(kf)\ (a\alpha + b\beta,\ \gamma) = k[f(a\alpha + b\beta, \gamma)] = k\ [af(\alpha, \gamma) + bf(\alpha,\ \gamma)]$$
$$= a(kf)(\alpha, \gamma) + b(kf)(\alpha, \gamma).$$

Similarly, $(kf)(\gamma, a\alpha + b\beta) = a(kf)(\gamma, \alpha) + b(kf)(\gamma, \beta).$

Hence kf is bilinear.

8. $f(\alpha, \beta) = 3x_1 y_1 - 2x_1 y_2 + 4x_2 y_1 - x_2 y_2$

Since, $\alpha = (x_1, x_2), \beta = (y_1, y_2)$

For $B = ((1, 0), (0, 1))$. Let $\alpha_1 = (1, 0), \alpha_2 = (0, 1)$

$$a_{11} = f(\alpha_1, \alpha_1) = f((1, 0)\ (1, 0)) = 3$$
$$a_{12} = f(\alpha_1, \alpha_2) = f((1, 0)\ (0, 1)) = -2$$
$$a_{13} = f(\alpha_2, \alpha_1) = f((0, 1)\ (1, 0)) = 4$$
$$a_{14} = f(\alpha_2, \alpha_2) = f((0, 1)\ (0, 1)) = -1$$

$\therefore$ $[f]_B = \begin{bmatrix} 3 & -2 \\ 4 & -1 \end{bmatrix}$

Similarly, for $B' = \{(1, 1), (1, 2)\}$

$$[f]_{B'} = \begin{bmatrix} 4 & 1 \\ 7 & 3 \end{bmatrix}$$

Let P be the matrix whose columns are the basis vectors in B'.

$\therefore$ $P = \begin{bmatrix} 1 & 1 \\ 1 & 2 \end{bmatrix}$

Then, $P^t [f]_B\ P = \begin{bmatrix} 1 & 1 \\ 1 & 2 \end{bmatrix} \begin{bmatrix} 3 & -2 \\ 4 & -1 \end{bmatrix} \begin{bmatrix} 1 & 1 \\ 1 & 2 \end{bmatrix} = \begin{bmatrix} 4 & 1 \\ 7 & 3 \end{bmatrix} = [f]_{B'}$

9. Since char. $F \neq 2$, i.e., $1 + 1 \neq 0$. Let f be a bilinear form.

Suppose f is alternating.

Let $\alpha, \beta \in V$, then $0 = f(\alpha + \beta, \alpha + \beta)$ [$\because f$ is alternating.]

$$= f(\alpha,\ \alpha) + f(\alpha,\ \beta) + f(\beta, \alpha) + f(\beta, \beta)$$
$$= f(\alpha, \beta) + f(\beta, \alpha)$$ [$\because f(\alpha, \alpha) = 0,\ f(\beta, \beta) = 0$]

$\Rightarrow$ $f(\alpha, \beta) = -f(\beta, \alpha)$

$\Rightarrow$ f is skew-symmetric.

Conversely, suppose f is skew-symmetric, then

$$f(\alpha, \alpha) = -f(\alpha, \alpha)\ \forall\ \alpha \in V$$

$\Rightarrow$ $2f(\alpha, \alpha) = 0$

$\Rightarrow$ $f(\alpha, \alpha) = 0\ \forall \alpha \in V$ [as $1 + 1 \neq 0, i.e., 2 \neq 0$]

$\Rightarrow$ f is alternating.

12. Let f be a diagonalizable bilinear form. Then $f(\alpha_i, \alpha_j) = 0$ for $i \neq j$.

$\therefore B = \{\alpha_1, \alpha_2, ..., \alpha_n\}$ forms an orthogonal basis with respect to f.

Let $\alpha, \beta \in V$, then

$$\alpha = \sum_i^n a_i \alpha_i,\ \beta = \sum_{j=1}^n b_j \beta_j\ \text{ for } a_i, b_j \in F$$

$$f(\alpha, \beta) = f\left(\sum_{i=1}^n a_i \alpha_i, \sum_{j=1}^n b_j \alpha_j \right) = \sum_i \sum_j a_i b_j f(\alpha_i \alpha_j)$$

$$= \sum_{i}^{n} a_i b_i f(\alpha_i, \alpha_i) \qquad\qquad [\because f(\alpha_i, \alpha_j) = 0, i \neq j]$$

and $$f(\beta, \alpha) = f\left(\sum_{j=1}^{n} b_j \alpha_j, \sum_{i=1}^{n} a_i \alpha_i\right) = \sum_i \sum_j b_j a_i f(\alpha_j \alpha_i) = \sum_{i=1}^{n} b_i a_i f(\alpha_i, \alpha_i)$$

Since $a_i b_i = b_i a_i.$

$\therefore$ $f(\alpha, \beta) = f(\beta, \alpha) \ \forall \alpha, \beta \in V.$

Hence, f is symmetric.

14. Since f is reflexive and non-degenerate. Then for $\alpha, \beta, \gamma \in V$

         $f(\alpha, \gamma) = 0,$ whenever $f(\gamma, \alpha) = 0$ and $f(0, \gamma) = 0 \ \forall \gamma \in V.$

(i) $f(\alpha, \gamma) = f(\beta, \gamma).$

$\therefore$ $f(\alpha - \beta, \gamma) = f(\alpha, \gamma) - f(\beta, \gamma) = 0 \ \forall \gamma \in V$

$\Rightarrow$ $\alpha - \beta = 0$ $\therefore$ $\alpha = \beta.$

(ii) $f(\gamma, \alpha - \beta) = f(\gamma, \alpha) - f(\gamma, \beta), \ \forall \gamma \in V$

         $= 0 \ \forall \gamma \in V$

$\Rightarrow$ $f(\alpha - \beta, \gamma) = 0$

$\therefore$ $\alpha - \beta = 0, f$ is non degenerate.

$\Rightarrow$ $\alpha = \beta.$

Answers

1. No one is bilinear..

2. (i) $\begin{bmatrix} 0 & 0 \\ 0 & 0 \end{bmatrix}$ (ii) $\begin{bmatrix} -2 & 0 \\ 0 & 2 \end{bmatrix}$ (iii) $\begin{bmatrix} 4 & 14 \\ 14 & 24 \end{bmatrix}$

3. $\begin{bmatrix} I_3 & 2I_3 \\ 2I_3 & I_3 \end{bmatrix}$, where I_3 is a unit matrix of order 3×3.

4. $\begin{bmatrix} 3 & -2 & 0 \\ 5 & 7 & -8 \\ 0 & 4 & -1 \end{bmatrix}$ 8. (i) $\begin{bmatrix} 3 & -2 \\ 4 & -1 \end{bmatrix}$ (ii) $\begin{bmatrix} 4 & 1 \\ 7 & 3 \end{bmatrix}$ (iii) $P = \begin{bmatrix} 1 & 1 \\ 1 & 2 \end{bmatrix}$

8.4 QUADRATIC FORMS

Definition 1. *A mapping $q : V \to F$ is called a quadratic form if $q(\alpha) = f(\alpha, \alpha)$ for some bilinear form f on V.*

Alternatively, a quadratic form is a polynomial $q(X) = X^t A X$, where

$X^t = (x_1, x_2, \ldots, x_n)$ and A is a symmetric matrix. That is

$$q(x) = (x_1, x_2, \ldots, x_n) \begin{bmatrix} a_{11} & a_{12} & \cdots & a_{1n} \\ a_{21} & a_{22} & \cdots & a_{2n} \\ \cdots & \cdots & \cdots & \cdots \\ a_{n1} & a_{n2} & \cdots & a_{nn} \end{bmatrix} \begin{bmatrix} x_1 \\ x_2 \\ \vdots \\ x_n \end{bmatrix}$$

$$= \sum_i \sum_j a_{ij} x_i x_j \quad \text{for } 1 \leq i \leq n, \ 1 \leq j \leq n$$

$$q(x) = a_{11}x_1^2 + a_{12}x_2^2 + \ldots + a_{nm}x_n^2 + 2\sum_{i<j} a_{ij}\, x_i\, x_j$$

Clearly, $q(X)$ is a polynomial in which every term has degree two.

Definition 2. *Let V be a real inner-product space and suppose that A is a real symmetric matrix linear transformation on V. Then a real valued function $Q(\alpha)$ defined by $Q(\alpha) = (A(\alpha),\alpha)$, is called the quadratic form associated with A.*

For example.

(1) $5x^2 - 6xy + 8y^2$ *is a real quadratic form in two variables x and y.*

(2) $2x^2 - y^2 + 2z^2 - 2yz - 4zx + 6xy$ *is a real quadratic form in three variables x, y and z.*

Definition 3. *If the matrix A is diagonal, then the corresponding quadratic form q has diagonal representation as*

$$q(X) = X^t A X = a_{11}x_1^2 + a_{22}x_2^2 + \ldots + a_{nn}x_n^2.$$

8.5 REAL SYMMETRIC BILINEAR AND QUADRATIC FORMS: LAW OF INERTIA

In this section, we shall discuss symmetric bilinear forms and quadratic forms on vector spaces over the field of reals. These forms appear in many branches of mathematics and physics.

Definition 1. *If $q(X) = X^t A X$ is a real quadratic form, then the rank of q is defined as the rank of the matix A.*

Definition 2. *Let q be a real quadratic form and f be a bilinear form, then the signature of f and of q are defned by sig.$(f) = $sig.$(q) = p - N$, where p is the number of positive entries and N the number of negative entries in any diagonal representation of f and q.*

Theorem 1.(Sylvester's Theorem or Law of Inertia). *Let f be a symmetic bilinear form on V over R. Then there is a basis of V in which f is represented by a diagonal matrix, every other diagonal representation has the same number p of positive entries and the same number N of negative entries.*

Proof. By theorem 4, there is a basis $\{\alpha_1, \alpha_2, \ldots, \alpha_n\}$ of V such that $f(\alpha_i, \alpha_j) = 0$ for $i \neq j$. This shows that f is represented by a diagonal matrix, say, with p positive and N negative entries.

Now assume that $\{\beta_1, \beta_2, \ldots, \beta_n\}$ is another basis of V in which f is represented by a diagonal matrix say, with p_1 positive and N_1 negative entries. Without any loss of generality, we may assume that the positive entries in each matrix representation appear first.

Since rank$(f) = p + N = p_1 + N_1$. Now we only prove that $p = p_1$.

Let U be a subspace of V spanned by the vectors $\alpha_1, \alpha_2, \ldots, \alpha_p$ and let W be a subspace of V spanned by $\beta_{p_1+1} + \beta_{p_1+2} + \ldots + \beta_n$. Then $f(\alpha,\alpha) > 0$ for every non-zero $\alpha \in U$ and $f(\alpha,\alpha) \leq 0$ for every non-zero vector $\alpha \in W$.

Obviously, $U \cap W = \{0\}$. Also, dim.$U = p$ and dim.$W = n - p_1$. Thus

$$\dim(U+W) = \dim U + \dim W - \dim(U \cap W)$$
$$= p + n - p_1 - 0 = p - p_1 + n.$$

But $\dim(U+W) \leq \dim V = n$.

∴ $\dim(U+W) \leq n$ or $p - p_1 + n \leq n$ or $p \leq p_1$.

Similarly, $p_1 \leq p$, and therefore $p = p_1$. Hence the theorem.

Theorem 2. If $q(x) = X^t A X$ be a real quadratic form of rank r in n variables, then there exists a real orthogonal transformation $X = PY$ which transform $q(X)$ to the form

$$\lambda_1 y_1^2 + \lambda_2 y_2^2 + \dots + \lambda_r y_r^2$$

where λ_1, λ_2,...λ_r are the r non-zero eigenvalues of A and $n-r$ eigenvalues of A being equal to zero and $Y = (y_1 y_2, \dots y_r)^t$.

Proof. Since A is a real symmetric matrix of rank r, therefore there exists a real orthogonal matrix P such that

$$P^{-1}AP = \text{diag.}[\lambda_1, \lambda_2,\dots\lambda_r,0,0,\dots0].$$

Also, P is orthogonal so $P^{-1} = P^t$.

$\therefore \quad P^{-1}AP = P^t A P = \text{diag.}[\lambda_1, \lambda_2,\dots\lambda_r,0,0,\dots0].$

Now, consider the real orthogonal transformation $X = PY$, then we have

$$q(X) = X^t A X = (PY)^t A (PY)$$
$$= Y^t P^t A P Y = Y^t \text{ diag.}[\lambda_1, \lambda_2,\dots\lambda_r,0,0,\dots0]Y$$
$$= \lambda_1 y_1^2 + \lambda_2 y_2^2 + \dots + \lambda_r y_r^2.$$

Theorem 3. *Every real quadratic form*

$$q(x_1,x_2,\dots,x_n) = X^t A X = y_1^2 + y_2^2 + y_p^2 - y_{p+1}^2 - \dots - y_r^2.$$

where r is the rank of A and p is the number of positive eigenvalues of A.

Proof. From above theorem, we have

$$Q^{-1}AQ = Q^t A Q = \text{diag.}[\lambda_1, \lambda_2,\dots\lambda_r,0,0,\dots0].$$

where Q is a real orthogonal matrix.

Let $\lambda_1,\lambda_2,\dots\lambda_p$ be positive eigenvalues and $\lambda_{p+1},\lambda_{p+2},\dots,\lambda_r$ be negative. Let D be $n \times n$ real matrix diagonal elements as

$$\frac{1}{\sqrt{\lambda_1}}, \frac{1}{\sqrt{\lambda_2}}, \dots, \frac{1}{\sqrt{\lambda_p}}, \frac{1}{\sqrt{(-\lambda_{p+1})}}, \frac{1}{\sqrt{(-\lambda_{p+2})}}, \dots, \frac{1}{\sqrt{(-\lambda_p)}}, 1,1,\dots,1$$

Then D is a non-singular matrix and also $D' = D$.

If we take $P = QD$, then P is also a real non-singular matrix, so that

$$P'AP = (QD)'A(QD) = D'Q'AQD$$
$$= D' \text{ diag.}[\lambda_1, \lambda_2,\dots\lambda_r,0,0,\dots0]D$$
$$= \text{diag.}[1,1,\dots1,-1,-1,\dots,-1,0,0,\dots,0].$$

It is noted that here 1 and -1 appear p and $r - p$ times respectively.

Now consider a real non-singular linear transformation $X = PY$ which reduces $X'AX$ to the form $Y'P'APY$ as

$$q(X) = y_1^2 + y_2^2 + \dots + y_p^2 - y_{p+1}^2 - y_{p+1}^2 - \dots - y_r^2.$$

8.6 ORTHOGONAL DIAGONALIZATION OF THE QUADRATIC FORM

Let q be a quadratic form on a vector space V over F determined by a symmetric bilinear form f. Let $B = \{\alpha_1,\alpha_2,\dots\alpha_n\}$ be an ordered orthogonal basis of V with respect to f. Then $[f]_B = \text{diag.}(\lambda_1, \lambda_2,\dots\lambda_n)\lambda_i = f(\alpha_i,\alpha_i)$.

For $\beta \in V$, we may write

$$\beta = a_1\alpha_1 + a_2\alpha_2 + \dots + a_n\alpha_n, a_i \in F$$

$$= \sum_{i=1}^{n} a_i \alpha_i$$

so that $q(\alpha) = f(\alpha, \alpha)$ [By definition]

$$= f\left(\sum_{i=1}^{n} a_i \alpha_i, \sum_{j=1}^{n} a_j \alpha_j \right)$$

$$= a_1^2 f(\alpha_1, \alpha_1) + a_2^2 f(\alpha_2, \alpha_2) + \ldots\ldots + a_n^2 f(\alpha_n, \alpha_n) \qquad [\because f(\alpha_i, \alpha_j) = 0, i \neq j]$$

$$= \lambda_1 a_1^2 + \lambda_2 a_2^2 + \ldots + \lambda_n y_n^2$$

Since, $[\beta]_B = [a_1 a_2 \ldots a_n]'$ [By definition of co-ordinate of a vector]

$\therefore \qquad q(\alpha) = [\beta]_B^t \, \text{diag.}(\lambda_1, \lambda_2, \ldots \lambda_n) [\beta]_B.$

This is called the diagonalizatiion.

Definition 1. *Let V be an n-dimensional vector space over R. Then a quadratic form q on V is called positive semidefinite if $q(\alpha) \geq 0$ for all $\alpha \in V$.*

Definition 2. *A quadratic form q on n-dimensional vector space V is said to be positive definite if $q(\alpha) > 0$ for all $\alpha \in V - \{0\}$.*

Definition 3. *A quadratic form q is said to be negative semidefinite if $q(\alpha) \leq 0$ for all $\alpha \in V$.*

Definition 4. *A quadratic form q is said to be negative definite if $q(\alpha) < 0$ for all $\alpha \in V - \{0\}$.*

If the quadratic form is none of above, then it is called indefinite.

WORKING PROCEDURE

To describe the algorithm which diagonalizes quadratic form $q(X)$ on R^n by means of an orthogonal change of co-ordinates $X = PY$. We proceed as follows :

Step 1. Find the symmetic matix A which represents q and find its characteristic polynomial $\Delta(t)$.

Step 2. Find the eigenvalues of A which are the roots of the equation $\Delta(t) = 0$.

Step 3. For each eigenvalue λ of A in step 2. Find an orthogonal basis of its eigenspace.

Step 4. Normalize all eigen vectors which are obtained in step 3 which then form an orthogonal basis of R^n.

Step 5. Find the matrix P whose columns are the normalized eigenvectors obtained in step 4.

Then $X = PY$ is the required orthogonal change of co-ordinates and the diagonal entries in $P'AP$ will be the eigenvalues of A which correspond to the columns of P.

To descibe the algoithm which gives the matrix P such thnt $P'AP$ is diagonal.

(i) First form the matix $M = (A : I)$

(ii) Then apply the row and column operations to M is such a way that the row operation will change both halves of M, but the column operations will only change the left half of M. Then algorithm will finally transform M into the form $M' = (D:Q)$ where D is diagonal matrix. Then $P = Q'$, and $P'AP = D$.

Solved Examples (Based on the Quadratic Forms)

Example 1. *Find the quadratic form $q(x,y)$ corresponding to the symmetic matrix*

$$A = \begin{bmatrix} 5 & -3 \\ -3 & 8 \end{bmatrix}$$

Solution.
$$q(x, y) = (x,y) \begin{bmatrix} 5 & -3 \\ -3 & 8 \end{bmatrix} \begin{bmatrix} x \\ y \end{bmatrix} = (5x - 3y, -3x + 6y) \begin{bmatrix} x \\ y \end{bmatrix}$$

$$= x(5x - 3y) + y(-3x + 8y) = 5x^2 - 3xy - 3yx + 8y^2$$
$$= 5x^2 - 6xy + 8y^2$$

Example 2. *Find the quadratic form $q(x_1, x_2, x_3)$ corresponding to the symmetric matrix*

$$A = \begin{bmatrix} 1 & 2 & -4 \\ 2 & 3 & 5 \\ -4 & 5 & -7 \end{bmatrix}.$$

Solution.   Let $X = \begin{bmatrix} x_1 \\ x_2 \\ x_3 \end{bmatrix}$,

$$\text{Then } q(X) = X'AX = [x_1 x_2 x_3] \begin{bmatrix} 1 & 2 & -4 \\ 2 & 3 & 5 \\ -4 & 5 & -7 \end{bmatrix} \begin{bmatrix} x_1 \\ x_2 \\ x_3 \end{bmatrix}$$

$$= [x_1 + 2x_2 - 4x_3, 2x_1 + 3x_2 + 5x_3, -4x_1 + 5x_2 - 7x_3] \begin{bmatrix} x_1 \\ x_2 \\ x_3 \end{bmatrix}$$

$$= x_1(x_1 + 2x_2 - 4x_3) + x_2(2x_1 + 3x_2 + 5x_3) + x_3(-4x_1 + 5x_2 - 7x_3)$$
$$= x_1^2 + 2x_1x_2 - 4x_1x_3 + 2x_2x_1 + 3x_2^2 + 5x_2x_3 - 4x_3x_1 + 5x_3x_2 - 7x_3^2$$
$$= x_1^2 + 3x_2^2 - 7x_3^2 + 4x_1x_2 - 8x_1x_3 + 10x_2x_3$$

Example 3. *The following expressions defne quadratic forms q on $\mathbf{R}^2$. Find the symmetric bilinear form f corresponding to each q.*

(i) ax_1^2 (ii) bx_1x_2 (iii) $x_1^2 + 9x_2^2$

(iv) $2x_1^2 - \dfrac{1}{3} x_1x_2$ (v) $3x_1x_2 - x_2^2$ (vi) $4x_1^2 + 6x_1x_2 - 3x_2^2$.

Solution.   Let $X = \begin{bmatrix} x_1 \\ x_2 \end{bmatrix}$. By definition of quadratic form, we have

(i) $f((x_1, x_2),(x_1, x_2)) = q(x_1,x_2) = ax_1^2$

(ii) $f((x_1, x_2),(x_1, x_2)) = bx_1x_2$

(iii) $f((x_1, x_2),(x_1, x_2)) = x_1^2 + 9x_2^2$

(iv) $f((x_1, x_2),(x_1, x_2)) = 2x_1^2 - \dfrac{1}{3} x_1x_2$

(v) $f((x_1, x_2),(x_1, x_2)) = 3x_1x_2 - x_2^2$

(vi) $f((x_1, x_2),(x_1, x_2)) = 4x_1^2 + 6x_1x_2 - 3x_2^2$

Example 4. *Find the symmetric matrix A corresponding to the quadratic form*
$$q(x, y, z)=3x^2+4xy-y^2+8xz-6yz+z^2.$$

Solution. Since the symmetric matrix $A= [a_{ij}]$ representing $q(x_1, x_2,..., x_n)$ has the diagonal entry a_{ij} equal to the coefficient of x_i^2, and has the entries a_{ij} and a_{ji} each equal to half the coefficient of x_ix_j.

Thus
$$A = \begin{bmatrix} 3 & 2 & 4 \\ 2 & -1 & -3 \\ 4 & -3 & 1 \end{bmatrix}$$

Aliter. $q(x, y, z)$ may be rewitten as
$$q(x, y, z)= (3x^2 +2xy+4xz)+(2yx-y^2-3yz)+(4zx-3zy+z^2)$$
$$= x(3x +2y+4z)+y(2x-y-3z)+z(4x-3y+z)$$
$$q(x,y,z) = (x,y,z)\begin{bmatrix} 3 & 2 & 4 \\ 2 & -1 & -3 \\ 4 & -3 & 1 \end{bmatrix}\begin{bmatrix} x \\ y \\ z \end{bmatrix}$$

Thus
$$A = \begin{bmatrix} 3 & 2 & 4 \\ 2 & -1 & -3 \\ 4 & -3 & 1 \end{bmatrix}$$

Example 5. *Find the symmetric matrix A which corresponds to*
$$q(x, y, z)= 4xy +5y^2.$$

Solution. Since $q(x, y, z)$ indicates that there are three variables, so $q(x, y, z)$ may be written as
$$q(x, y, z) = 0. x^2 +4xy+5y^2 +0 .xz+0. yz+0.z^2$$
$$= (0. x^2 +2xy+0 .xz)+(2yx+5y^2 +0 . yz)+(0. 2z+0. zy+0.z^2)$$
$$= x (0. x +2y+0 .z)+y(2x+5y +0 .z)+z(0. x+0. y+0.z)$$
$$= (x,y,z)\begin{bmatrix} 0 & 0 & 0 \\ 2 & 5 & 0 \\ 0 & 0 & 0 \end{bmatrix}\begin{bmatrix} x \\ y \\ z \end{bmatrix}$$

Thus,
$$A = \begin{bmatrix} 0 & 2 & 0 \\ 2 & 5 & 0 \\ 0 & 0 & 0 \end{bmatrix}$$

Example 6. *Consider the quadratic form*
$$q(x, y, z)= x^2 +4xy +3y^2-6xz+10yz+7z^2.$$
Find a non-singular linear substitution expressing the variables x, y, z in terms of variables r, s, t such that q(r, s, t) is diagonal.

Solution.
$$A = \begin{bmatrix} 1 & 2 & -3 \\ 2 & 3 & 5 \\ -3 & 5 & 7 \end{bmatrix}$$

Now, $M= (A|I).$

$$\therefore \quad M = \begin{pmatrix} 1 & 2 & -3 & \vdots & 1 & 0 & 0 \\ 2 & 3 & 5 & \vdots & 0 & 1 & 0 \\ -3 & 5 & 7 & \vdots & 0 & 0 & 1 \end{pmatrix}$$

Apply the row operations $R_2 \to R_2 - 2R_1$ and $R_3 \to R_3 + 3R_1$ to M and then the corresponding column operations $C_2 \to C_2 - 2C_1$ and $C_3 \to C_3 + 3C_1$ to A, we get

$$\begin{pmatrix} 1 & 2 & -3 & \vdots & 1 & 0 & 0 \\ 0 & -1 & 11 & \vdots & -2 & 1 & 0 \\ 0 & 11 & -2 & \vdots & 3 & 0 & 1 \end{pmatrix}$$

and then

$$\begin{pmatrix} 1 & 0 & 0 & \vdots & 1 & 0 & 0 \\ 0 & -1 & 11 & \vdots & -2 & 1 & 0 \\ 0 & 11 & -2 & \vdots & 3 & 0 & 1 \end{pmatrix}$$

Next apply the row operations $R_3 \to R_3 + 11R_2$ and then the corresponding olumn operation $C_3 \to C_3 + 11C_2$, we finally obtain,

$$\begin{pmatrix} 1 & 0 & 0 & \vdots & 1 & 1 & 0 \\ 0 & -1 & 0 & \vdots & -2 & 1 & 0 \\ 0 & 0 & 119 & \vdots & -19 & 11 & 1 \end{pmatrix}$$

Thus

$$P = \begin{pmatrix} 1 & -2 & -19 \\ 0 & 1 & 11 \\ 0 & 0 & 1 \end{pmatrix} \quad \text{and} \quad P'AP = \begin{pmatrix} 1 & 0 & 0 \\ 0 & -1 & 0 \\ 0 & 0 & 119 \end{pmatrix}$$

Thus the linear substitutions are given by

$$\begin{bmatrix} x \\ y \\ z \end{bmatrix} = P \begin{bmatrix} r \\ s \\ t \end{bmatrix}$$

$$\therefore \quad x = r - 2s - 19t, \ y = s + 11t, \ z = t$$

Hence $q(r,s,t) = (r,s,t)P'AP \begin{bmatrix} r \\ s \\ t \end{bmatrix}$

$$= (r,s,t) \begin{bmatrix} 1 & 0 & 0 \\ 0 & -1 & 0 \\ 0 & 0 & 119 \end{bmatrix} \begin{bmatrix} r \\ s \\ t \end{bmatrix} = r^2 - s^2 + 119r^2$$

Example 7. *Give an example of a quadratic form q on $\mathbf{R}^2$ such that $q(\alpha) = 0$ and $q(\beta) = 0$ for some $\alpha, \beta \in \mathbf{R}^2$, but $q(\alpha + \beta) \neq 0$.*

Solution. Let $q(x, y) = x^2 - y^2$ and $\alpha = (1, 1), \ \beta = (1, -1)$

Then $q(\alpha) = q(1, 1) = 1^2 - 1^2 = 0$

And $q(\beta) = q(1, -1) = 1^2 - (-1^2) = 0$

But $q(\alpha + \beta) = q((1, 1) + (1, -1)) = q(2, 0) = 2^2 - 0^2 = 4 \neq 0.$

Example 8. *Find the signature of the quadratic form $q(x, y, z)$*

$$q(r, y, z) = x^2 + 4xy + 3y^2 - 6xz + 10yz + 7z^2.$$

Solution. From example 6, the equivalent diagonal form of $q(x, y, z)$ is
$$q(r, s, t) = x^2 - s^2 + 119t^2.$$

Obviously, this quadratic form has $p = 2$ positive entries on the diagonal and $N = 1$ negative entry on the diagonal. Thus sig.$(q) = p - N = 2 - 1 = 1$.

Example 9. Let $q(x, y, z) = x^2 + 2y^2 - 4xz - 4yz + 6z^2$. Is q positive definite?

Solution. The symmetric matrix corresponding to the quadratic form is

$$A = \begin{bmatrix} 1 & 0 & -2 \\ 0 & 2 & -2 \\ -2 & -2 & 7 \end{bmatrix}$$

Now convert A into diagonal form by applying $R_3 \to R_3 + 2R_1$, and $C_3 \to C_3 + 2C_1$ and then $R_3 \to R_3 + R_2$ and $C_3 \to C_3 + C_2$,

$$A = \begin{bmatrix} 1 & 0 & -2 \\ 0 & 2 & -2 \\ -2 & -2 & 7 \end{bmatrix} - \begin{bmatrix} 1 & 0 & 0 \\ 0 & 2 & -2 \\ 0 & -2 & 3 \end{bmatrix} - \begin{bmatrix} 1 & 0 & 0 \\ 0 & 2 & 0 \\ 0 & 0 & 1 \end{bmatrix}.$$

The diagonal representation of q only contains positive entries, 1,2 and 1 on the diagonal. Hence, q is positive definite.

Example 10. Show that $q(x, y)' = ax^2 + byx + cy^2$ is positive definite if and only if $b^2 - 4ac < 0$.

Solution. Suppose $\alpha = (x, y) \neq 0$ say $y \neq 0$. Let $t = \dfrac{x}{y}$. Then,

$$q(\alpha) = y^2 \left[a \left(\frac{x}{y} \right)^2 + b \left(\frac{x}{y} \right) + c \right] = y^2 (ar^2 + bt + c)$$

Let $s = ar^2 + bt + c$, thus s lies above the t-axis i.e., s is positive for every value of t if and only if $b^2 - 4ac < 0$. Hence q is positive definite if and only if $b^2 - 4ac < 0$.

Example 11. Find an orthogonal change of co-ordinates which diaonalizes the real quadratic form $q(x, y) = 2x^2 - 4xy + 5y^2$.

Solution. The matrix A corresponding to the given quadratic form is

$$A = \begin{bmatrix} 2 & -2 \\ -2 & 5 \end{bmatrix}$$

Then its characteristic polynomial $\Delta(t)$ is given by

$$\Delta(t) = |tI - A| = \begin{bmatrix} t - 2 & -2 \\ 2 & t - 5 \end{bmatrix} = (t - 6)(t - 1)$$

$\therefore$ The eigenvalues of A are 6 and 1.

Let $X_1 = \begin{bmatrix} x_1 \\ x_2 \end{bmatrix} \neq 0$ be an eigenvector corresponding to $t = 6$, then

$$(tI - A)X_1 = 0 \Rightarrow \qquad (6I - A)X_1 = 0$$

$$\Rightarrow \begin{bmatrix} 4 & 2 \\ 2 & 1 \end{bmatrix} \begin{bmatrix} x_1 \\ x_2 \end{bmatrix} = \begin{bmatrix} 0 \\ 0 \end{bmatrix} \qquad \Rightarrow 4x_1 + 2x_2 = 0 \text{ and } 2x_1 + x_2 = 0$$

A non-zero solution of these equation is

$$X_1 = \begin{bmatrix} 1 \\ -2 \end{bmatrix}$$

Now, let $X_2 = \begin{bmatrix} x_1 \\ x_2 \end{bmatrix} \neq 0$ be an eigenvector corresponding to $t = 1$, then

$$(tI - A)X_2 = 0 \text{ or } (I - A)X_2 = 0 \begin{bmatrix} -1 & 2 \\ 2 & -4 \end{bmatrix} \begin{bmatrix} x_1 \\ x_2 \end{bmatrix} = \begin{bmatrix} 0 \\ 0 \end{bmatrix}$$

$\Rightarrow \quad -x_1 + 2x_2 = 0$ and $2x_1 - 4x_2 = 0$

The non-zero solution of these equations is

$$X_2 = \begin{bmatrix} 2 \\ 1 \end{bmatrix}$$

Normalizer X_1 and X_2 to obtain the orthogonal basis

$$\{\alpha_1 = (1/\sqrt{5}, -2/\sqrt{5}), \alpha_2 = (2/\sqrt{5}, 1/\sqrt{5})\}$$

Then

$$P = \begin{bmatrix} 1/\sqrt{5} & 2/\sqrt{5} \\ -2/\sqrt{5} & 1/\sqrt{5} \end{bmatrix} \text{ and } P'AP = \begin{bmatrix} 6 & 0 \\ 0 & 1 \end{bmatrix}$$

Thus the required orthogonal change of co-ordinates is

$$\begin{bmatrix} x \\ y \end{bmatrix} = P \begin{bmatrix} x' \\ y' \end{bmatrix} \text{ i.e., } x = \frac{x'}{\sqrt{5}} + \frac{2y'}{\sqrt{5}}, y = \frac{2x'}{\sqrt{5}} + \frac{y'}{\sqrt{5}}$$

Under this change of co-ordinates q is transformed into the diagonal form

$$q(x', y') = 6x'^2 + y'^2$$

Example 12. *Let q be the quadratic form on R^2 given by*

$$q(x_1 + x_2) = ax_1^2 + 2bx_1x_2 + cx_2^2, a \neq 0$$

Find an invertible linear operator U on R^2 such that

$$(U^* q)(x_1, x_2) = ax_1^2 + \left(c + \frac{b^2}{a}\right)x^2.$$

Solution.   Diagonalize q by the method known as "completing the square".

$$\therefore \quad q(x_1, x_2) = a\left(x_1^2 + \frac{2b}{a}x_1x_2\right) + cx_2^2$$

$$= a\left(x_1^2 + \frac{2b}{a}x_1x_2 + \frac{b^2}{a^2}x_2^2\right) - \frac{b^2}{a}x_2^2 + cx_2^2$$

$$= a\left(x_1 + \frac{b}{a}x_2\right)^2 + \left(c - \frac{b^2}{a}\right)x_2^2.$$

Let $s = x_1 + \dfrac{b}{a}x_2, t = a_2,$ then this linear substitution yields the quadratic form as

$$q'(s,t) = as^2 + \left(c - \frac{b^2}{a}\right)t^2$$

or $$(U^* q)(s,t) = as^2 + \left(c - \frac{b^2}{a}\right)t^2$$

But $$q(U(s,t)) = (U^* q)(s,t) = as^2 + \left(c - \frac{b^2}{a}\right)t^2$$

Now if we put $x_1 = s - \dfrac{b}{a}t, x_2 = t$ in given quadratic, then

$$q\left(s - \frac{b}{a}t,t\right) = a\left(s - \frac{b}{a}t\right)^2 + 2b\left(s - \frac{b}{a}t\right)t + ct^2$$

$$= as^2 + \left(c - \frac{b^2}{a}\right)t^2$$

$\therefore$ $$q(U(s,t)) = q\left(a - \frac{b}{a}t,t\right)$$

$\therefore$ $$U(s,t) = \left(s - \frac{b}{a}t,t\right)$$

or $$U^{-1}(s,t) = \left(s + \frac{b}{a}t,t\right)$$

This is the required invertible linear operator.

Example 13. *Let q be the quadratic form on R^2 given by $q(x_1, x_2) = 2bx_1x_2$. Find an invertible linear operator U on R^2 such that*

$$(U^* q)(x_1,x_2) = \frac{b}{2}x_1^2 - \frac{b}{2}x_2^2.$$

Solution.   Making the complete square in q.

$$q(x_1,x_2) = 2bx_1x_2$$

$$q(x_1,x_2) = \frac{b}{2}[(x_1 - x_2)^2 - (x_1 - x_2)^2]$$

Let $s = x_1 + x_2, t = x_1 - x_2,$ then this linear substitution yields the quadratic form as

$$q'(s, t) = bs^2 - bt^2$$

or $\qquad (U^* \, q)(s,t) = \dfrac{b}{2}s^2 - \dfrac{b}{2}t^2$

But we know that

$$(U^* \, q)(s,t) = q(U(s,t)) = \frac{b}{2}s^2 - \frac{b}{2}t^2.$$

Now if we put $x_1 = \dfrac{s+t}{2}, x_2 = \dfrac{s-t}{2}$ in the given quadratic then.

$$q\left(\frac{s+t}{2},\frac{s-t}{2}\right) = 2b\left(\frac{s+t}{2}\right)\left(\frac{s-t}{2}\right)$$

$$= \frac{b}{2}(s^2 - t^2) = \frac{b}{2}s^2 - \frac{b}{2}t^2$$

$\therefore \qquad q(U(s,t)) = q\left(\dfrac{s+t}{2},\dfrac{s-t}{2}\right)$

$\therefore \qquad U(s,t)) = \left(\dfrac{s+t}{2},\dfrac{s-t}{2}\right)$

This is the required invertible linear operator.

8.7 HERMITIAN FORMS

In this section we assume that V is a vector space over the complex field **C**.

Definition. *If $A = [a_{ij}]$ is an $n \times n$ matrix over **C**, then we write $\bar{A}$ for the matrix obtained by taking the complex conjugate of every entry of A, i.e., $\bar{A} = [\overline{a_{ij}}]$. We also write $A^\theta = (\bar{A})^t = (\overline{A^t})$. Thus A^θ is the conjugate transpose of A.*

For example: If $A = \begin{bmatrix} 2+3i & 5+4i \\ 6-7i & 1+9i \end{bmatrix}$, $A^\theta = (\bar{A})^t = \begin{bmatrix} 2-3i & 6+7i \\ 5-4i & 1-9i \end{bmatrix}$

Definition 2. *A matrix A is said to be Hermitian if $A^\theta = A$.*

Definition 3. *A matrix A is said to be skew-Hermitian if $A^\theta = -A$.*

Definition 4. *Let V be a vector space over the complex field **C**, then a mapping $f:V \times V \to \mathbf{C}$ is said to Hermitian form which satisfies the following properties*

(i) $f(a\alpha + b\beta,\gamma) = af(\alpha,\gamma) + bf(\alpha,\gamma)$

(ii) $f(\alpha,\beta) = \overline{f(\beta,\alpha)}$, where $a,b \in \mathbf{C}$ and $\alpha,\beta \in V$.

If f is Hermitian on V, then $f(\alpha, a\beta + b\gamma) = \bar{a}f(\alpha,\beta) = \bar{b}f(\alpha,\gamma)$

or

If $\mathbf{A} = [a_{ij}]_{nxn}$ is a Hermitian matrix of order $n \times n$, then

$$X^\theta A X = \sum_{i=1}^{n}\sum_{j=1}^{n} a_{ij}\bar{x}_i x_j$$

is called a Hermitian form in n variables $x_1, x_2, ..., x_n$. The matrix A is called the matrix of this Hermitian form.

Definition 5. *Let f be a Hermitian form on V. Then a mapping $q: V \to R$ defined by $q(\alpha) = f(\alpha, \alpha)$ is called the Hermitian form or complex quadratic form associated with the Hermitian form f.*

Moreover, one can obtain f from q by the following identity, called polar form of

$$f : f(\alpha, \beta) = \frac{1}{4}(q(\alpha+\beta) - q(\alpha-\beta)) + (q(\alpha+i\beta) - q(\alpha-i\beta)).$$

Definition 6. *A Hermitian form f and its quadratic form q are said to be non-negative semidefinite if $q(\alpha) = f(\alpha, \alpha) > 0$ for every $\alpha \in V$, and are said to be positive definite if $q(\alpha) = f(\alpha, \alpha) > 0$ for every $\alpha \neq 0$.*

8.8 MATRIX REPRESENTATION OF A HERMITIAN FORM

Let V be an n-dimensional vector space over C and let $B = \{\alpha_1, \alpha_2, ..., \alpha_n\}$ of V. Then the matrix $A = [a_{ij}]$, where $a_{ij} = f(\alpha_i, \alpha_j)$ is called the matrix representation of f relative to the basis B.

Since $f(\alpha_i, \alpha_j) = \overline{f(\alpha_i, \alpha_j)}$, thus A is Hermitian and in particular the diagonal entries of A are all real.

Theorem 1. *Let f be a Hermitian form on V. Let A be the matrix of f relative to the basis $B = \{\alpha_1, \alpha_2, ..., \alpha_n\}$ of V. Then $f(\alpha, \beta) = [\alpha]^t_B A[\bar{\beta}]_B$, for all $\alpha, \beta \in V$.*

Proof. Since $\alpha, \beta \in V$ so that

$$\alpha = a_1\alpha_1 + a_2\alpha_2 + ... + a_n\alpha_n \quad \text{and} \quad \beta = b_1\alpha_1 + b_2\alpha_2 + ... + b_n\alpha_n, \text{ for } a_i, b_i \in C$$

Then, $f(\alpha, \beta) = f(a_1\alpha_1 + a_2\alpha_2 + ... + a_n\alpha_n, b_1\alpha_1 + b_2\alpha_2 + ... + b_n\alpha_n)$

$$= \sum_{i=1}^{n}\sum_{j=1}^{n} a_i \bar{b}_j \cdot f(\alpha_i, \alpha_j) = (a_1, a_2, ..., a_n) A \begin{bmatrix} \bar{b}_1 \\ \bar{b}_2 \\ \vdots \\ \bar{b}_n \end{bmatrix} = [\alpha]^t_B A[\bar{\beta}]_B$$

Theorem 2. *Let P be the change of basis matrix from a basis B of V to a new basis B'. Let A be the matrix of a Hermitian form f in the original basis B. Then $P'AP = Q^\theta AQ$, where $Q = \bar{P}$ is the matrix of f in the new basis B'.*

Proof. Let $\alpha, \beta \in V$. Since P is the change of basis matrix from B to B', we have

$$P[\alpha]_{B'} = [\alpha]_B \quad \text{and} \quad P[\beta]_{B'} = [\beta]_B$$

$$\therefore \quad [\alpha]^t_B = [\alpha]^t_{B'} P' \quad \text{and} \quad [\bar{\beta}]_B = \overline{P[\beta]_B}$$

Then by above theorem 1, we have

$$f(\alpha, \beta) = [\alpha]^t_B = A[\bar{\beta}]_B = [\alpha]^t_{B'} P'A\bar{P}[\beta]_B$$

Since α, β are arbitrary elements of V. Hence $P'A\bar{P}$ is the matrix of f in the new basis B'.

REMARK

☞ Let f be a Hermitian form on V. Then there exists a basis $\{\alpha_1, \alpha_2, ..., \alpha_n\}$ of V in which f is represented by a diagonal matrix, i.e., $f(\alpha_i \alpha_j) = 0$ for $i \neq j$. Moreover, every diagonal representation of f has the same number p of positive entries and the same number N of negative entries. The difference $p - N$ is called signature of f.

Solved Examples

Example 1. *If f is a Hermitian form on V. Show that*

$$f(\alpha, a\beta + b\gamma) = \bar{a}\, f(\alpha, \beta) + \bar{b}\, f(\alpha, \gamma)$$

Solution. By definition of Hermitian form

$$f(a, \alpha\beta + b\gamma) = \overline{f(a\beta + b\gamma, \alpha)} = \overline{af(\beta, \alpha) + bf(\gamma, \alpha)}$$

$$= \overline{af(\beta, \alpha)} + \overline{bf(\gamma, \alpha)} = \bar{a}\, f(\alpha, \beta) + \bar{b}\, f(\alpha, \gamma)$$

Example 2. *Let C^n be the set all square matrices of order $n \times n$ and A be a Hermitian matrix. Then show that f is Hermitian form on C^n where f is defined by $f(X, Y) = X^t A \bar{Y}$.*

Solution. For $a, b \in C$ and all $X_1, X_2, Y \in C^n$,

$$f(aX_1 + bX_2, Y) = (aX_1 + bX_2)^t A\bar{Y} = (aX_1' + bX_2')A\bar{Y}$$

$$aX_1' A\bar{Y} + bX_2' A\bar{Y} = af(X_1, Y) + bf(X_2, Y)$$

Also
$$\overline{f(X, Y)} = \overline{(XA\bar{Y})} = \overline{(\overline{XAY})^t}$$

$$= \overline{(Y^t A X)} = Y^t A^\theta \bar{X} = Y^t A \bar{X} \qquad\qquad [\because A^\theta = A]$$

$$\overline{f(X, Y)} = f(Y, X)\ .\ \text{Hence } f \text{ is Hermitian form.}$$

Example 3. Let $A = \begin{bmatrix} 1 & 1+i & 2i \\ 1-i & 4 & 2-3i \\ -2i & 2+3i & 7 \end{bmatrix}$, *a Hermitian matrix. Find a non-singular matrix P*

such that $P'A\bar{P}$ is diagonal.

Solution. First form the matrix $M = (A : I)$ as

$$M = \begin{bmatrix} 1 & 1+i & 2i & : & 1 & 0 & 0 \\ 1-i & 4 & 2-3i & : & 0 & 1 & 0 \\ -2i & 2+3i & 7 & : & 0 & 0 & 1 \end{bmatrix}$$

Apply the row operations $R_2 \rightarrow R_2 + (-1+i)R_1$ and $R_3 \rightarrow R_3 + 2iR_1$ to M and corresponding Hermitian column operations $C_2 \rightarrow C_2 + (-1-i)C_1$ and $C_3 \rightarrow C_3 - 2iC_1$ to A we get

$$\begin{pmatrix} 1 & 1+i & 2i & : & 1 & 0 & 0 \\ 0 & 2 & -5i & : & -1+i & 1 & 0 \\ 0 & 5i & 3 & : & 2i & 0 & 1 \end{pmatrix}$$

and then $\begin{pmatrix} 1 & 0 & 0 & : & 1 & 0 & 0 \\ 0 & 2 & -5i & : & -1+i & 1 & 0 \\ 0 & 5i & 3 & : & 2i & 0 & 1 \end{pmatrix}$

Now apply the row operation $R_3 \rightarrow 2R_3 - 5iR_2$ and the corresponding Hermitian column operation $C_3 \rightarrow 2C_3 + 5iC_2$, we get

$$\begin{pmatrix} 1 & 0 & 0 & : & 1 & 0 & 0 \\ 0 & 2 & -5i & : & -1+i & 1 & 0 \\ 0 & 0 & -19 & : & 5+9i & -5i & 2 \end{pmatrix}$$

and then
$$\begin{pmatrix} 1 & 0 & 0 & : & 1 & 0 & 0 \\ 0 & 2 & 0 & : & -1+i & 1 & 0 \\ 0 & 0 & -38 & : & 5+9i & -5i & 2 \end{pmatrix}$$

Thus A is diagonalized.

Setting $P = \begin{bmatrix} 1 & -1+i & 5+9i \\ 0 & 1 & -5i \\ 0 & 0 & 2 \end{bmatrix}$. Then, $P^t A \bar{P} = \begin{bmatrix} 1 & 0 & 0 \\ 0 & 2 & 0 \\ 0 & 0 & -38 \end{bmatrix}$.

EXERCISE 8.2

1. Find the quadratic form $q(x,y,z)$ corresponding to the following matrices:

(i) $A = \begin{bmatrix} 3 & 0 & 0 \\ 0 & -4 & 0 \\ 0 & 0 & 6 \end{bmatrix}$

(ii) $A = \begin{bmatrix} 2 & -5 & 1 \\ -5 & -6 & -7 \\ 1 & -7 & 9 \end{bmatrix}$

(iii) $A = \begin{bmatrix} 2 & 1 & 5 \\ 1 & 3 & -2 \\ 5 & -2 & 4 \end{bmatrix}$

(iv) $A = \begin{bmatrix} 1 & 2 & 3 \\ 2 & 0 & 3 \\ 3 & 3 & 1 \end{bmatrix}$

2. Find the real symmetric matrices corresponding to the following quadratic forms:

(i) $q(x,y) = ax^2 + 2hxy + by^2$

(ii) $q(x,y,z) = 2xy + 6xz - 4yz$

(iii) $q(x,y) = 4x^2 + 5xy - 7y^2$

(iv) $q(x,y,z) = 2x^2 - 10xy - 6y^2 + 2xz - 14yz + 9z^2$

(v) $q(x,y,z) = x^2 - 2yz + xz$.

3. Let $q(x,y,z) = x^2 + 4xy + 3y^2 - 8xz - 12yz + 9z^2$. Find a non-singular linear substitution expressing the variables x,y,z in terms of the variables r,s,t so that $q(r,s,t)$ is diagonal. Also find the signature of q.

4. Show that $q(0) = 0$ for any quadratic q on V.

5. Suppose $q(\alpha) = 0$ for a quadratic form q on $V(F)$. Show that $q(k\alpha) = 0$ for any $k \in F$.

6. Show that rank $(f) = \text{rank}(q) = p + N$, where p is the number of positive entries and N the number of negative entries in any diagonal representation of f and q.

7. Let $q(x,y,z) = x^2 + y^2 + 2xz + 4yz + 3z^2$. Is q positive definite?

8. Let $q(x,y) = x - 4xy + 5y^2$. Show that q is positive definite.

9. Let $q(x,y) = x^2 - 6xy + 3y^2$. Show that q is not positive definite.

10. Let $q(x,y) = x^2 + 4xy + y^2$. Find an orthogonal change of co-ordinates which diagonalizes q. Also find its signature.

11. Let $q(x,y) = 3x^2 - 6xy + 11y^2$. Find an orthogonal change of co-ordinates which diagonalizes q. Also find its signature and rank.

12. The following expressions define quadratic forms q on R^2. Find the symmetric bilinear form corresponding to each q.

(i) cx_2^2

(ii) $3x_1 x_2 - 2x_2^2$

(iii) $2x_1^2 - \frac{1}{3}x_1 x_2$.

13. Suppose f is a Hermitian form on V. Show that $f(\alpha,\alpha)$ is real for any $\alpha \in V$.

14. Define a non-negative semidefinite and a positive definite Hermitian form.

15. Let f be the dot product on C^n. Let $f(\alpha,\beta) = x_1 \bar{y}_1 + x_2 \bar{y}_2 + \dots + x_n \bar{y}_n$ for $\alpha - (x_1, x_2, \dots, x_n)$ $\beta = (y_1, y_2, \dots, y_n) \in C^n$. Is f a Hermitian? Is f positive definite?

16. Show that the Hermitian form assumes one real values for all complex n-vectors X.

17. Every Hermitian form $X^\theta A X$ is unitarily equivalent to the form

$$\lambda_1 \bar{y}_1 y_1 + \lambda_2 \bar{y}_2 y_2 + \dots + \lambda_n \bar{y}_n y_n$$

where $\lambda_1, \lambda_2, \dots, \lambda_n$ are the eigenvalues of the Hermitian matrix A, and $X = PY, P$ the unitary.

1. (ii) Let $X = \begin{bmatrix} x_1 \\ x_2 \\ x_3 \end{bmatrix}$. Then we have,

$$q(X) = X'AX = [x_1, x_2, x_3] \begin{pmatrix} 2 & -5 & 1 \\ -5 & -6 & -7 \\ 1 & -7 & 9 \end{pmatrix} \begin{bmatrix} x_1 \\ x_2 \\ x_3 \end{bmatrix}$$

$$= [2x_1 - 5x_2 + x_3, \ -5x_1 - 6x_2 - 7x_3, \ x_1 - 7x_2 + 9x_3] \begin{bmatrix} x_1 \\ x_2 \\ x_3 \end{bmatrix}$$

$$= x_1(2x_1 - 5x_2 + x_3) + x_2(-5x_1 - 6x_2 - 7x_3) + x_3(x_1 - 7x_2 + 9x_3)$$
$$= 2x_1^2 - 10x_1x_2 + 2x_1x_3 - 6x_2^2 - 14x_2x_3 + 9x_3^2$$
$$= 2x_1^2 - 6x_2^2 + 9x_3^2 - 10x_1x_2 + 2x_1x_3 - 14x_2x_3$$

2. (iv) $q(x,y,z) = 2x^2 - 10xy - 6y^2 + 2xy - 14yz + 9z^2$

$\qquad q(x,y,z) = 2x^2 - 5xy + xz - 5yx - 6y^2 - 7yz + zx - 7zy + 9z^2$

$$= x(2x - 5y + z) + y(-5x - 6y - 7z) + z(x - 7y + 9z)$$

$$= [x, y, z] \begin{pmatrix} 2 & -5 & 1 \\ -5 & -6 & -7 \\ 1 & -7 & 9 \end{pmatrix} \begin{bmatrix} x \\ y \\ z \end{bmatrix}.$$

Hence the matrix corresponding to the given quadratic form is $\begin{pmatrix} 2 & -5 & 1 \\ -5 & -6 & -7 \\ 1 & -7 & 9 \end{pmatrix}$

4. Since we have $q(\alpha) = f(\alpha, \alpha)$, $q(0) = f(0,0) = f(0\alpha, 0) = f(\alpha, 0) = 0$.

7. $q(x,y,z) = x^2 + y^2 + 2xz + 4yz + 3z^2$.

$$A = \begin{bmatrix} 1 & 0 & 1 \\ 0 & 1 & 2 \\ 1 & 2 & 3 \end{bmatrix}$$

Reduce A to diagonal form as follows :

$$A = \begin{bmatrix} 1 & 0 & 0 \\ 0 & 1 & 2 \\ 0 & 2 & 2 \end{bmatrix} = \begin{bmatrix} 1 & 0 & 0 \\ 0 & 1 & 0 \\ 0 & 0 & -2 \end{bmatrix}.$$

There is a negative entry -2 in the diagonal representation of q. Hence q is not positive definite.

10. $q(x, y) = x^2 + 4xy + y^2$, the matrix of q is $A = \begin{bmatrix} 1 & 2 \\ 2 & 1 \end{bmatrix}$, then

$$\Delta(t) = |tI - A| = \begin{vmatrix} t-1 & 2 \\ -2 & t-1 \end{vmatrix} = t^2 - 2t - 3 = (t-3)(t+1)$$

Thus eigenvalues of A are 3 and -1. Substitute into the matrix $tI - A$ to obtain the corresponding homogeneous system of linear equations $2x - 2y = 0$, $-2x + 2y = 0$. A non-zero solution is $\alpha_1 = (1,1)$. Next substitute $t = -1$ into the matrix $tI - A$ to obtain the corresponding homogeneous system of linear equation $-2x - 2y = 0$, $-2x - 2y = 0$. A non-zero solution is $\alpha_2 = (1,-1)$. Normalize α_1, α_2 to obtain orthonormal basis.

$$\left\{ \beta_1 = \left(1\sqrt{2}, 1/\sqrt{2}, 1/\sqrt{2}\right), \beta_2 = \left(-1/\sqrt{2}, 1/\sqrt{2}\right) \right\}$$

Let P be the matrix whose columns are β_1 and β_2 respectively, then

$$\begin{bmatrix} x \\ y \end{bmatrix} = P \begin{bmatrix} x' \\ y' \end{bmatrix} \qquad \text{or} \qquad x = \frac{x' - y'}{\sqrt{2}}, \ y = \frac{x' + y'}{\sqrt{2}}.$$

Under this change of co-ordinates, q is transformed into the diagonal form $q(x', y')=3x'^2-y'^2$ $\text{sig}(q)=1-1=0$.

13. Since f is a Hermitian form, then by definition of Hermitian form, $f(\alpha,\alpha)=\overline{f(\alpha,\alpha)}$. This shows that $f(\alpha,\alpha)$ is real.

15. Let $\alpha=(x_1, x_2,..., x_n)$, $\beta=(y_1, y_2,...y_n)$, $z=(z_1, z_2,..., z_n) \in \mathbf{C}^n$

For $a, b \in \mathbf{C}$, we have

$$a\alpha+b\beta=(ax_1+by_1, ax_2+by_2,...,ax_n+by_n)$$
$$f(a\alpha+b\beta, \gamma) = (ax_1+by_1)\,\overline{z}_1 + (ax_2+by_2)\,\overline{z}_2 +...+(ax_n+by_n)\,\overline{z}_n$$

$$=a(x_1\overline{z}_1 +x_2\overline{z}_2 +...+x_n\overline{z}_n)+b(y_1\overline{z}_1 +y_2\overline{z}_2 +...+y_n\overline{z}_n)$$
$$=af(\alpha,\gamma)+bf(\beta,\gamma)$$

Also, $\qquad f(\alpha,\beta)= x_1\overline{y}_1 + x_2\overline{y}_2 +......+ x_n\overline{y}_n$

$$= \overline{(y_1\overline{x}_1 + y_2\overline{x}_2 +...+ y_n\overline{x}_n)} = \overline{f(\beta,\alpha)}$$

Thus, f is Hermitian.

Let $\qquad \alpha \neq 0$

$$f(\alpha,\alpha) = x_1\overline{x}_1 + x_2\overline{x}_2 +...+ x_n\overline{x}_n =| x_1 |^2 +| x_2 |^2 +...+| x_n |^2 > 0$$

Hence f is positive definite.

Answers

1. (i) $q(x,y,z)=3x^2-4y^2+6z^2$

(ii) $q(x,y,z)=2x^2 - 6y^2 + 9z^2 - 10xy + 2xz - 14yz$

(iii) $q(x,y,z) = 2x^2 + 3y^2 + 4z^2 +2xy+10xz-4yz$

(iv) $q(x,y,z) = x^2 +z^2 +4xy+6xz+6yz$

2. (i) $A=\begin{bmatrix} a & h \\ h & b \end{bmatrix}$ (ii) $A=\begin{bmatrix} 0 & 1 & 3 \\ 1 & 0 & -2 \\ 3 & -2 & 0 \end{bmatrix}$ (iii) $A=\begin{bmatrix} 4 & 5/2 \\ 5/2 & -7 \end{bmatrix}$

(v) $A=\begin{bmatrix} 2 & -5 & 1 \\ -5 & -6 & -7 \\ 1 & -7 & 9 \end{bmatrix}$ (vi) $A=\begin{bmatrix} 1 & 0 & 1/2 \\ 0 & 0 & -1 \\ 1/2 & -1 & 0 \end{bmatrix}$

3. $x=r-2s, y=s+2t, z=t, \text{sig.}(q)=-1$

7. q is not positive definite, change of co-ordinates is $x = \dfrac{x'}{\sqrt{2}} - \dfrac{y'}{\sqrt{2}}, y = \dfrac{x'}{\sqrt{2}} + \dfrac{y'}{\sqrt{2}}$.

$\therefore \qquad q(x', y')=3x'^2- y^2, \text{sig.}(q)=0$.

11. Orthogonal change of co-ordinates is $x = \dfrac{3x' - y'}{\sqrt{10}}, y = \dfrac{x' + 3y'}{\sqrt{10}}$.

so $\qquad q(x', y') = 3x'^2 - y'^2, \text{sig.}(q)=2, \text{rank}(q)=0$.

12. (i) $f((x_1, x_2),(x_1,x_2))=cx_2^2$.

(ii) $f((x_1, x_2),(x_1,x_2))=3x_1x_2-2x_2^2$.

(iii) $f((x_1, x_2),(x_1,x_2))= 2x_1^2 -\dfrac{1}{3} x_1x_2$.

8.9 CANONICAL FORM

The relation of similarity (in matrices) arise when we study the various matrix representation of linear transformation of vector space into itself. Under similarity, the rank of matrix A is invariant since two similar matrices are certainly equivalent and rank is even invariant under equivalence.

Now, to check the similarity of two linear transformation, we have to compute a particular canonical form for each and check if these are same.

Here, first we define the concept of similarity of matrices and similarity of linear transformation.

8.10 SIMILARITY OF MATRICES

For two square matrices, A and B, the matrix B is said to be similar to A if there exists an invertible matrix P such that

$$B = P^{-1}AP$$

REMARK

☞ We can prove easily that similarity of matrices is an equivalence relation, *i.e.*, similarity of matrices is reflexive, symmetric and transitive.

8.11 SIMILARITY OF LINEAR TRANSFORMATIONS

Definition 1. *Let $V(F)$ be the n-dimensional vector space over the field F and $A(V)$ be the set of all linear transformation from V to V. Then two linear transformation S, $T \in A(V)$ are said to be similar if $\exists$ an invertible linear transformation $C \in A(V)$ such that*

$$T = CSC^{-1}$$

The similarity of linear transformation is also an equivalence relation. Thus, we can decompose $A(V)$ into equivalence classes, each such classes is called similarity class.

Definition 2. *The special form of matrix representation (in some basis of V) of linear transformation in each similarity class is called* canonical forms.

Therefore, as we earlier said, in order to check the similarity of two linear transformations, we have to form a particular canonical form for each matrix representation of linear transformation and then we have to verify if these are the same.

8.12 INVARIANT SUBSPACES

Definition. *Let $T:V \to V$ be a linear transformation. Then a subspace W of V is said to be invariant under T if $T(W) \subset W$, i.e., $\forall \alpha \in W, T(\alpha) \in W$.*

Theorem 1. *If W is a subspace invariant under $T \in A(V)$, then T induces a linear transformation T_q on quotient space V/W defined by $T_q(\alpha + W) = T(\alpha) + W$.*

Further, if T satisfies the polynomial $q(x) \in F[x]$, then so is T_q. Thus the minimal polynomial of T_q divide the minimal polynomial of T.

Proof. We have to show that T_q is well defined and also T_q is linear.

   (i) T_q is well defined.

   Take two elements $\alpha + W$ and $\beta + W$ of $V|W$ such that

$$\alpha + W = \beta + W, \text{ this implies } \alpha - \beta \in W$$

Now, $T(\alpha-\beta)= T(\alpha)-T(\beta) \in W$ [W is T-invariant.]

so, $T(\alpha)+W = T(\beta) + W$

$\Rightarrow$ $T_q(\alpha + W) = T_q(\beta + W)$

Thus, T is well-defined.

(ii) T_q is linear transformation. For $(\alpha + W)$, $(\beta + W) \in V/W$

We have $T_q\{(\alpha + W) +(\beta+W)\} = T_q(\alpha+\beta+W)$

$\qquad\qquad\qquad\qquad = T(\alpha+\beta)+W$

$\qquad\qquad\qquad\qquad = T(\alpha)+T(\beta)+W$ [$\because T$ is linear.]

$\qquad\qquad\qquad\qquad = T(\alpha)+W+T(\beta)+W$

$\qquad\qquad\qquad\qquad = T_q(\alpha+W)+T(\beta+W)$

Also, $T_q\{C(\alpha+W)\} = T_q(C\alpha+W)$

$\qquad\qquad\qquad\qquad = T(C\alpha)+W = CT(\alpha)+W$ [$\because T$ is linear.]

$\qquad\qquad\qquad\qquad = C\{T(\alpha)+W\}$

$\qquad\qquad\qquad\qquad = CT_q(\alpha+W)$

So, T_q is linear.

Again $\alpha + W \in V/W$, then

$\qquad T_q^2(\alpha+W) = T^2(\alpha)+W=T(T(\alpha))+W$

$\qquad\qquad\qquad\quad = T_q(T(\alpha)) + W = T_q (T_q(\alpha+W))$

$\qquad\qquad\qquad\quad = T_q^2(\alpha+W)$

Similarly, we can prove $(T_q^n)=(T_q)^n; \forall n \geq 0$

Now, for a polynomial $q(x) \in F[x]$ where

$\qquad q(x)=a_n x^n+a_{n-1}x^{n-1}+......+a_0$

$q(T_q)(\alpha+W)=q(T)(\alpha)+W$

$\qquad\qquad\quad = a_n T^n(\alpha)+a_{n-1}T^{n-1}(\alpha)+......+a_0 I(\alpha)+W$

$\qquad\qquad\quad = \Sigma a_i T^i(\alpha)+W= \Sigma a_i(T^i(\alpha)+W)$

$\qquad\qquad\quad = \Sigma a_i T_q^i(\alpha+W)$

$\qquad\qquad\quad = a_i(T_q)^i(\alpha+W)$

$\qquad\qquad\quad = q(T_q)(\alpha+W)$

Hence, T_q satisfy the polynomial $q(x)=0$, i.e., $q(T_q)=0$. Thus, T_q is root of $q(x)=0$.

8.13 INVARIANT DIRECT-SUM DECOMPOSITIONS

Definition. *Let $T:V \to V$ be a linear transformation such that V is the direct sum of T-invariant subspaces $W_1,W_2,......,W_r$, i.e., $V = W_1 \oplus W_2 \oplus......\oplus W_r$, where $T(W_i) \subset W_i$ for $i \in \mathbf{N}$ of T_i in the linear transformation restricted to W_i. Then T is said to be decomposable into operator T, so T can be written as $T = T_1 \oplus T_2 ......\oplus T_r$ and subspace $W_1, W_2,......W_r$ are said to be T-invariant direct sum decomposition of V.*

Theorem 1. *If $V = W_1 \oplus W_2 \oplus......\oplus W_r$ where n_i is dimension of each subspace W_i and every subspace is invariant under $T \in A(V)$, then a basis of V can be found so that the matrix of T in this basis is of the form*

$$\begin{bmatrix} A_1 & 0 & \cdots & 0 \\ 0 & A_1 & \cdots & 0 \\ \vdots & \vdots & \cdots & \vdots \\ 0 & 0 & \cdots & A_r \end{bmatrix}$$

where each A_i is an $n_i \times n_i$ matrix of linear transformation induced by T on W_r.

Proof. Let $\left\{\alpha_1^{(1)}, \alpha_2^{(1)}, ..., \alpha_n^{(1)}\right\}, \left\{\alpha_1^{(2)}, \alpha_2^{(2)}, ..., \alpha_n^{(2)}\right\} ... \left\{\alpha_1^{(r)}, \alpha_2^{(r)}, ..., \alpha_n^{(r)}\right\}$ be the basis of $W_1, W_2, ..., W_r$ respectively.

Since $V = W_1 \oplus W_2 \oplus ...... \oplus W_r$, therefore

$$\left\{\alpha_1^{(1)}, \alpha_2^{(1)}, ..., \alpha_n^{(1)}, \; \alpha_1^{(2)}, \alpha_2^{(2)}, ..., \alpha_n^{(2)}, ..., \alpha_1^{(r)}, \alpha_2^{(r)}, ..., \alpha_n^{(r)}\right\}$$

form a basis of V. Also, each W_i is T-invariant, so that $T(\alpha_j^i) \in W_i$ and it is linear combination of $\alpha_1^{(i)}, \alpha_2^{(i)}, ..., \alpha_n^{(i)}$,i.e.,

$$T\left(\alpha_j^{(i)}\right) = a_1^{(i)}\alpha_1^{(i)} + a_2^{(i)}\alpha_2^{(i)} + ... + a_n^{(i)}\alpha_n^{(i)} \qquad ...(1)$$

for every $i = 1, 2,..., r$ and $j = 1, 2,..., n$.

Thus, matrix representation of T with respect to basis V is obtained by (1) which is

$$\begin{bmatrix} A_1 & 0 & \cdots & 0 \\ 0 & A_1 & \cdots & 0 \\ \vdots & \vdots & \cdots & \vdots \\ 0 & 0 & \cdots & A_r \end{bmatrix}$$,where A_i is the matrix of T_i induced on W_i by T.

Now, we are in position to discuss some important canonical form for checking the similarity of two linear transformations. They are of following types:

(i) Normal form

(ii) Triangular form

(iii) Jordan form

(iv) Rational form

8.14 NORMAL FORM

A matrix A is said to be in normal form if it can be written as $A = \begin{pmatrix} I_r & 0 \\ 0 & 0 \end{pmatrix}$, where I_r is square identity matrix of order r.

Theorem 1. *Let $T: U \rightarrow V$ be a linear transformation and rank(T) =r. Then there exist bases of U and V such that matrix representation of T has the form $A = \begin{pmatrix} I_r & 0 \\ 0 & 0 \end{pmatrix}$,$I_r \rightarrow$ represents the identity matrix of order r.*

Proof. Let the dim $U = m$ and dim $V = n$. Let W be kernel of T. Now rank of T is r so the dimension of kernel space of T is $m-r$. Consider $\{\alpha_1, \alpha_2,..., \alpha_{m-r}\}$ be then basis of W. By extension theorem, it can be extended to form the basis of U. Let this extension be $\{v_1, v_2,..., v_n, \alpha_1, \alpha_2,..., \alpha_{m-r}\}$. By setting a transformation $T(v_i) = u_i$, then set $\{u_1, u_2,...,u_r\}$ form a basis of image (T) and thus base can be extended to form the basis of V. Let this base be

$$\{u_1, u_2,...,u_r, u_{r+1},....,u_n\}$$

Here, we can observe that every v_i under T can be written as the linear combination of u_i as

$$T(v_1)=u_1=1u_1+0u_2+...+0u_r+0u_n$$
$$T(v_2)=u_2=0u_1+1u_2+...+0u_r+0u_n$$

... ...

$$T(v_r)=u_r=0u_1+0u_2+...+1u_r+0u_n$$
$$T(\alpha_1)=0=0u_1+0u_2+...+0u_r+0u_{r+1}+...+0u_n$$
$$T(\alpha_2)=0=0u_1+0u_2+...+0u_r+0u_{r+1}+...+0u_n$$

... ...

$$T(\alpha_{m-r})=0=0u_1+0u_2+...+0u_r+0u_{r+1}+...+0u_n$$

so the matrix representation of T is given by

$$A=\begin{bmatrix} 1 & 0 & 0 & \cdots & 0 & 0 & \cdots & 0 \\ 0 & 1 & 0 & \cdots & 0 & 0 & \cdots & 0 \\ \cdots & \cdots & \cdots & \cdots & \cdots & \cdots & \cdots & \cdots \\ 0 & 0 & 0 & \cdots & 1 & 0 & \cdots & 0 \\ 0 & 0 & 0 & \cdots & 0 & 0 & \cdots & 0 \\ \vdots & \vdots & \vdots & \cdots & \vdots & \vdots & \cdots & \vdots \\ 0 & 0 & 0 & \cdots & 0 & 0 & \cdots & 0 \end{bmatrix}_{n\times n}$$

Hence, $A=\begin{pmatrix} I_r & 0 \\ 0 & 0 \end{pmatrix}.$

8.15 TRIANGULAR FORM

Definition. *Let $T:V{\to}V$ be a linear transformation on V over F, then the matrix of T in the basis $\{\alpha_1,\alpha_2......\alpha_n\}$ of V is triangular if*

$$T(\alpha_1)=a_{11}\alpha_1$$
$$T(\alpha_2)=a_{21}\alpha_1+a_{22}\alpha_2$$
$$T(\alpha_3)=a_{31}\alpha_1+a_{32}\alpha_2+a_{33}\alpha_3$$

... ...

$$T(\alpha_n)=a_{n1}\alpha_1+a_{n2}\alpha_2+a_{nn}\alpha_n$$

Theorem 1. *If $T \in A(V)$ has all its characteristic root in F, then there is a basis of V in which matrix representation of T is triangular.*

Proof.   This result can be proved by induction on the dimension of V

(i)   If dim $V=1$, then every matrix representation is a matrix of order 1×1 which is trivially triangular.

(ii) Let this result hold good for all vector space over F of dimension $n-1$. Let dim $V=n > 1$. If $\lambda_1 \in F$ be a characteristic root of T, then $\exists$ a non-zero eigen vector α_1 corresponding to λ_1 such that $T(\alpha_1)=a_{11}\alpha_1$. It is due to the fact that T has all its characteristic roots in F. Let W be the one dimensional subspace of V spanned by α_1 and T invariant, then quotient space $V_q =V/W$, then

$$\text{dim } V_q = \text{dim } V-\text{dim } W = n-1.$$

Now, T induces a linear transformation T_q on V_q, whose minimal polynomial divides the minimal polynomial or T so all roots of minimal polynomial of T_q are also the roots of minimal polynomial of T and hence all roots lie in F. Thus V and T satisfy the hypothesis of the theorem.

Now, dim of $V_q = n-1$. Then by the hypothesis of induction, there is a basis for V_q as $\left\{ \bar{\alpha}_2, \bar{\alpha}_3, ..., \bar{\alpha}_n \right\}$ of V_q such that

$$T_q\left(\bar{\alpha}_2\right) = a_{22}\bar{\alpha}_2$$
$$T_q\left(\bar{\alpha}_3\right) = a_{32}\bar{\alpha}_2 + a_{33}\bar{\alpha}_3$$
$$... \quad ... \quad ... \quad ......$$
$$T_q\left(\bar{\alpha}_n\right) = a_{n2}\bar{\alpha}_2 + a_{n3}\bar{\alpha}_3 + ... + a_{nn}\bar{\alpha}_n$$

Now, elements $\{\alpha_2, \alpha_3, ..., \alpha_n\}$ being the elements of V, also belong to cosets $\bar{\alpha}_2, \bar{\alpha}_3, ..., \bar{\alpha}_n$ respectively, i.e., $\bar{\alpha}_i = \alpha_i + W$.

Now, $\qquad T_q\left(\bar{\alpha}_2\right) = a_{22}\bar{\alpha}_2$

$\Rightarrow \qquad T_q(\alpha_2 + W) = a_{22}(\alpha_2 + W) \Rightarrow T(\alpha_2) + W = a_{22}\alpha_2 + W$

$\Rightarrow \qquad T(\alpha_2) - a_{22}\alpha_2 \in W$

But W is spanned by α_1 so

$$T(\alpha_2) - a_{22}\alpha_2 = a_{21}\alpha_1$$
$\Rightarrow \qquad\qquad T(\alpha_2) = a_{21}\alpha_1 + a_{22}\alpha_2$

Similarly, for $\bar{\alpha}_3, \bar{\alpha}_4, ..., \bar{\alpha}_n$ we have $T(\alpha_i) = a_{i1}\alpha_1 + a_{i2}\alpha_2 + ... + a_{in}\alpha_n$.

In this way, we get

$$T(\alpha_1) = a_{11}\alpha_1$$
$$T(\alpha_2) = a_{21}\alpha_1 + a_{22}\alpha_2$$
$$... \quad ... \quad ... \quad ... \quad ... \quad ....$$
$$T(\alpha_n) = a_{n1}\alpha_1 + a_{n2}\alpha_2 + a_{nn}\alpha_n$$

Hence, the matrix of T in the basis $\{\alpha_1, \alpha_2 ...... \alpha_n\}$ is triangular.

REMARKS

☞ The above theorem can be restated as : "If a square matrix A has all its characteristic roots in F, then A is similar to a triangular matrix, i.e., there exists an invertible matrix P such that $P^{-1}AP$ is triangular."

☞ If any linear transformation T is represented by a triangular matrix

$$A = \begin{bmatrix} a_{11} & a_{12} & \cdots & a_{1n} \\ 0 & a_{22} & \cdots & a_{2n} \\ \vdots & \vdots & \cdots & \vdots \\ 0 & 0 & \cdots & a_{nn} \end{bmatrix}$$

The characteristic polynomial of T is a product of linear factors and is given by
$$\Delta(x) = |A-x| = (x-a_{11})(x-a_{22})......(x-a_{nn}).$$

Theorem 2. *If dim $V = n$ and if $T \in A(V)$ has all its characteristic roots in F, then T satisfy a polynomial of degree n over F.*

Proof.   Let $\lambda_1, \lambda_2, \ldots\ldots, \lambda_n$ be the characteristic roots of T. Since T has all its characteristic root in F so there exists a basis $\{\alpha_1, \alpha_2, \ldots, \alpha_n\}$ of V such that

$$T(\alpha_1) = \lambda_1 \alpha_1$$
$$T(\alpha_2) = a_{21}\alpha_1 + \lambda_2 \alpha_2$$
$$T(\alpha_3) = a_{31}\alpha_1 + a_{32}\alpha_2 + \lambda_3 \alpha_3$$
$$\ldots \quad \ldots \quad \ldots \quad \ldots \quad \ldots \quad \ldots$$
$$T(\alpha_n) = a_{n1}\alpha_1 + a_{n2}\alpha_2 + \ldots + \lambda_n \alpha_n$$

The all above relations can be rewritten as

$$(T - \lambda_1 I)(\alpha_1) = 0$$
$$(T - \lambda_2 I)(\alpha_2) = a_{21}\alpha_1$$
$$(T - \lambda_3 I)(\alpha_3) = a_{31}\alpha_1 + a_{32}\alpha_2$$
$$\ldots \quad \ldots \quad \ldots \quad \ldots \quad \ldots \quad \ldots$$
$$(T - \lambda_n I)(\alpha_n) = a_{n1}\alpha_1 + a_{n2}\alpha_2 + \ldots + \alpha_{n(n-1)}\alpha_{n-1}$$

Now, $(T - \lambda_2 I)(T - \lambda_1 I)(\alpha_2) = (T - \lambda_1 I)(T - \lambda_2 I)(\alpha_2)$

$$= (T - \lambda_1 I)a_{21}\alpha_1 \qquad \text{[From above relations]}$$
$$= a_{21}(T - \lambda_1 I)(\alpha_1)$$

and $(T - \lambda_3 I)(T - \lambda_2 I)(T - \lambda_1 I)(\alpha_3)$

$$= (T - \lambda_2 I)(T - \lambda_1 I)(T - \lambda_3 I)(\alpha_3)$$
$$= (T - \lambda_2 I)(T - \lambda_1 I)(a_{31}\alpha_1 + a_{32}\alpha_2) \qquad \text{[From above relations]}$$
$$= (T - \lambda_2 I)(T - \lambda_1 I)(a_{31}\alpha_1) + (T - \lambda_2 I)(T - \lambda_1 I)(a_{32}\alpha_2)$$
$$= a_{31}(T - \lambda_2 I)(T - \lambda_1 I)(\alpha_1) + a_{32}(T - \lambda_2 I)(T - \lambda_1 I)(\alpha_2)$$
$$= 0 + 0 = 0.$$

Proceeding in the same way, we get

$$(T - \lambda_n I)(T - \lambda_{n-1} I)(T - \lambda_{n-2} I)\ldots(T - \lambda_1 I)(\alpha_n) = 0$$

If $(T - \lambda_n I)(T - \lambda_{n-1} I)\ldots(T - \lambda_1 I)$ is represented by S, then we have $S(\alpha_1) = S(\alpha_2) = \ldots\ldots S(\alpha_n) = 0$. Thus S, being the annihilator of base of V, annihilates all elements of V.

i.e., $S = 0 \Rightarrow (T - \lambda_n I)(T - \lambda_{n-1} I), \ldots (T - \lambda_1 I) = 0$

Hence, T satisfy the polynomial $q(x) = (x - \lambda_n)(x - \lambda_{n-1}), \ldots (x - \lambda_1)$ in $F[x]$ of degree n.

Theorem 3. *If a subspace W of V is T-invariant, then T has a matrix representation $\begin{pmatrix} A & B \\ 0 & C \end{pmatrix}$, where A is matrix representation of the restricted T_q of T to W.*

Proof.   The proof of this result can be showed simply by matrix representation of T_q. Let $\{\beta_1, \beta_2, \ldots\ldots, \beta_r\}$ be the basis of W then by extension theorem, it can be extended to the basis $\{\beta_1, \beta_2, \ldots\ldots, \beta_r, \alpha_1, \alpha_2, \ldots\ldots, \alpha_s\}$ of V.

Since W is T invariant, so

$$T_q(\beta_i) = T(\beta_i), \text{ for } i = 1, 2, \ldots, r. \text{ Now we have}$$

$$T_q(\beta_1) = T(\beta_1) = a_{11}\beta_1 + \ldots\ldots + a_{1r}\beta_r$$
$$T_q(\beta_2) = T(\beta_2) = a_{21}\beta_1 + \ldots\ldots + a_{2r}\beta_r$$

$$T_q(\beta_r) = T(\beta_r) = a_{r1}\beta_1 + a_{r2}\beta_2 + \ldots.. + a_{rr}\beta_r$$

and $\qquad T(\alpha_1) = b_{11}\beta_1 + \ldots\ldots + b_{1r}\beta_r + c_{11}\alpha_1 + \ldots\ldots + c_{1s}\alpha_s$

Continuing in this way, we get

$$T(\alpha_1) = b_{s1}\beta_1 + \ldots\ldots + b_{sr}\beta_r + c_{s1}\alpha_1 + \ldots\ldots + b_{ss}\alpha_s$$

$$\begin{bmatrix} a_{11} & a_{21} & \cdots & a_{r1} & b_{11} & b_{21} & \cdots & b_{s1} \\ \vdots & \vdots & \cdots & \vdots & \vdots & \vdots & \cdots & \vdots \\ a_{1r} & a_{2r} & \cdots & a_{r2} & b_{1r} & b_{2r} & \cdots & b_{sr} \\ \cdots & \cdots & \cdots & \cdots & \cdots & \cdots & \cdots & \cdots \\ 0 & 0 & \cdots & 0 & C_{11} & C_{21} & \cdots & C_{s1} \\ \vdots & \vdots & \cdots & \vdots & \vdots & \vdots & \cdots & \vdots \\ 0 & 0 & \cdots & 0 & C_{1s} & C_{2s} & \cdots & C_{ss} \end{bmatrix}$$

$$= \begin{pmatrix} A & B \\ 0 & S \end{pmatrix}$$

where, $\qquad A = \begin{bmatrix} a_{11} & a_{21} & \cdots & a_{r1} \\ \vdots & \vdots & \vdots & \vdots \\ a_{1r} & a_{2r} & \cdots & a_{rr} \end{bmatrix}_{r \times r} \qquad B = \begin{bmatrix} b_{11} & b_{21} & \cdots & b_{s1} \\ \vdots & \vdots & \vdots & \vdots \\ b_{1r} & b_{2r} & \cdots & b_{sr} \end{bmatrix}_{r \times s}$

$$C = \begin{bmatrix} c_{11} & c_{21} & \cdots & c_{r1} \\ \vdots & \vdots & \vdots & \vdots \\ c_{1s} & c_{2s} & \cdots & c_{ss} \end{bmatrix} \qquad O = \begin{bmatrix} 0 & 0 & \cdots & 0 \\ \vdots & \vdots & \vdots & \vdots \\ 0 & 0 & \cdots & 0 \end{bmatrix}_{s \times r}$$

8.16 NILPOTENT TRANSFORMATION

Definition. *A linear transformation T: V→V is said to be nilpotent if $T^n = 0$ for some least positive integer n.*

or

Any $T \in A(V)$ is nilpotent then for some $k \in \mathbf{Z}^+$, $T^k = 0$ but $T^{k-1} \neq 0$, where k is index of nilpotency.

REMARK

☞ The characteristic root of nilpotent transformation are zero, So they belong to F and hence all characteristic roots of nilpotent transformation belong to F. Then, we can say that nilpotent linear transformation can always be brought to triangular form over F.

8.17 JORDAN CANONICAL FORM

Definition. *The matrix of the form*

$$J = \begin{bmatrix} \lambda & 1 & 0 & \cdots & 0 & 0 \\ 0 & \lambda & 1 & \cdots & 0 & 0 \\ \cdots & \cdots & \cdots & \cdots & \cdots & \cdots \\ 0 & 0 & 0 & \cdots & \lambda & 1 \\ 0 & 0 & 0 & \cdots & 0 & \lambda \end{bmatrix}$$

is called Jordan block matrix belonging to λ. In this matrix A, λ's are on the diagonal and 1's are on the superdiagonal and other elements are zero.

Theorem 1. *Let $T : V \to V$ be a linear operator whose characteristic and minimal polynomial are respectively.*

$$\Delta(x) = (x - \lambda_1)^{n_1} (x - \lambda_2)^{n_2} \ldots\ldots (x - \lambda_r)^{n_r}$$

and $\quad m(x) = (x - \lambda_1)^{m_1} (x - \lambda_2)^{m_2} \ldots\ldots (x - \lambda_r)^{m_r}$

where, λ_i are different scalars, then T has a block diagonal matrix representation

$$J = \begin{bmatrix} J_1 & & & \\ & J_2 & & \\ & & \ddots & \\ & & & J_r \end{bmatrix}$$

where $\quad J_i = \begin{bmatrix} \lambda_i & 1 & 0 & \cdots & 0 & 0 \\ 0 & \lambda_i & 1 & \cdots & 0 & 0 \\ \cdots & \cdots & \cdots & \cdots & \cdots & \cdots \\ 0 & 0 & 0 & \cdots & \lambda_i & 1 \\ 0 & 0 & 0 & \cdots & 0 & \lambda_i \end{bmatrix}$

For each λ_i, the corresponding block J_i have the following properties :

1. *There is at least one J_i of order m, and all other J_i are of order less than or equal to m_i.*

2. *The sum of the orders of J_i is n_i.*

3. *The number J_i equals the geometric multiplicity of λ_i.*

4. *The number J_i of each possible order is uniquely determined by T.*

Proof. We can write T by primary decomposition theorem as

$$T = T_1 \oplus T_2 \oplus \ldots\ldots \oplus T_r.$$

where $(x - \lambda_i)^{m_i}$ is the minimal polynomial of T_i.

Since the minimal polynomial is satisfied by the operator, therefore we have

$$\left(T_i - \lambda_i I\right)^{m_i} = 0 \quad \text{for } i = 1,2,3,\ldots\ldots,r$$

Now taking $\quad N_i = T_i - \lambda_i I$ for $i = 1,2,3,\ldots,r$

$\Rightarrow \qquad\qquad T_i = N_i + \lambda_i I$ and $N_i^{m_i} = 0$

This implies that N_i is nilpotent of index m_i and T_i is the sum of N_i, and scalar operator $\lambda_i I$.

Now, N_i being the nilpotent of index m_i, we can select a basis in which N_i is represented by a canonical form as

$$\begin{bmatrix} 0 & 1 & 0 & \cdots & 0 & 0 \\ 0 & 0 & 1 & \cdots & 0 & 0 \\ \cdots & \cdots & \cdots & \cdots & \cdots & \cdots \\ 0 & 0 & 0 & \cdots & 0 & 1 \\ 0 & 0 & 0 & \cdots & 0 & 0 \end{bmatrix}$$

In this basis $T_i = N_i + \lambda_i I$ can be reduced to a block diagonal matrix whose block are block Jordan matrix J_i. T is direct sum of $T_1, T_2, \ldots\ldots T_r$ therefore the direct sum of matrix representation of T_i gives the block diagonal matrix representation of T whose diagonal block are matrices J_i which have the following properties:

1. N_i is the nilpotent of index m_i so there is at least one J_i of order m_i.

2. Since T and the block diagonal matrix representation of T have the same characteristic polynomial so that the sum of orders of J_i is n_i.

3. Since the nullity of N_i is equal to the geometric multiplicity of eigen value λ_i because characteristic equation of N_i is $(x - \lambda_i)^{m_i} = 0$. Hence the number of J_i is equal to the geometric multiplicity of λ_i.

4. Since T_i and N_i are uniquely determined by T. Hence number of J_i of each possible order is uniquely determined by T.

8.18 RATIONAL CANONICAL FORM

The Jordan canonical form is exerted when the minimal polynomials cannot be factored into linear polynomial while in rational canonical form, the minimal polynomial is taken as the product of linear polynomial.

Theorem 1. *Let $T : V \to V$ be a linear operator with minimal polynomial*

$$b_1(x) = q_1(x)^{l_1} . q_2(x)^{l_2} ... q_k(x)^{l_k}, \text{ where } q_1(x), q_2(x), ..., q_k(x) \text{ are distinct monic}$$

irreducible polynomial. Then T has a unique block diagonal matrix representation

$$\begin{bmatrix} C_1 & & & & \\ & C_2 & & & \\ & & C_3 & & \\ & & & \ddots & \\ & & & & C_k \end{bmatrix}$$

where C_i is the companion matrix of polynomial $q_i(x)^{l_{ij}}$ where

$$l_1 = l_{11} \geq l_{12} \geq ... \geq l_{1r}...$$
$$l_k = l_{k_1} \geq l_{k_2} \geq .... \geq l_{k_r}$$

Proof. By primary decomposition theorem, V can be decomposed as $V = V_1 \oplus V_2 \oplus ...... \oplus V_k$, where each V_i is T invariant and the minimal polynomial of T_i, linear transformation induced by T on V_i has minimal polynomial $q_i(x)^{l_{ij}}$ .....

The matrix representation of T_i in some of V_i is companion matrix C_i. But $V = V_1 \oplus V_2 \oplus ...... \oplus V_k$.

Thus the matrix of T is

$$\begin{bmatrix} C_1 & & & & \\ & C_2 & & & \\ & & C_3 & & \\ & & & \ddots & \\ & & & & C_k \end{bmatrix}$$

This matrix representation is called rational canonical form and polynomial

$$q_1(x)^{l_{11}}, q_1(x)^{l_{12}}, ..., q_1(x)^{l_{1r}} ... q_k(x)^{l_{k1}} ... q_k(x)^{l_{k2}} ... q_k(x)^{l_{kr}}$$

are called the elementary divisor of T.

Solved Examples

Example 1. *Suppose W is invariant under S:V→V and T : V→V. Show that W is also invariant under S+T and ST.*

Solution.   Since W is S-invariant and T-invariant, if $\alpha \in W$, then $S(\alpha) \in W$ and $T(\alpha) \in W$.

Now, if $\alpha \in W$, then

$$(S+T)(\alpha) = S(\alpha) + T(\alpha)$$

Since $S(\alpha), T(\alpha) \in W$, so $S(\alpha) + T(\alpha) \in W$

$$(S + T)(\alpha) \in W$$

∴ W is invariant under $S + T$.

Also, for $\alpha \in W$.

$$(ST)(\alpha) = S(T(\alpha))$$

Since $T(\alpha) \in W$ and W is invariant under S, so that

$$S(T(\alpha)) \in W$$

⇒ $(ST)(\alpha) \in W$

∴ W is invariant under ST.

Example 2. *Determine all invariant subspaces of $A = \begin{bmatrix} 2 & -5 \\ 1 & -2 \end{bmatrix}$ viewed as an operator on R^2.*

Solution.   Since R^2 and $\{0\}$ are invariant subspace of A. If A has any other invariant subspaces, then it must be one-dimensional. The characteristic polynomial of A is

$$\Delta(x) = |xI - A|$$

$$= \begin{vmatrix} x-2 & 5 \\ -1 & x+2 \end{vmatrix}$$

$$= (x-2)(x+2) + 5$$

$$= x^2 + 1$$

Clearly, A has no eigenvalue in R so A has no eigenvectors in R^2. Hence, $R^2 + \{0\}$ are the only subspace invariant under A.

Example 3. *Suppose T: V→V is linear and suppose $T = T_1 \oplus T_2$, with respect to a T-invariant direct sum decomposition $V = V_1 \oplus V_2$, Show that*

(i) *$m(x)$ is the least common multiple of $m_1(x)$ and $m_2(x)$ where $m(x), m_1(x)$ and $m_2(x)$ are the minimal polynomials of T, T_1 and T_2, respectively.*

(ii) *$\Delta(x) = \Delta_1(x)\Delta_2(x)$ where $\Delta(x), \Delta_1(x)$ and $\Delta_2(x)$ are the characteristic polynomials of T, T_1 and T_2 respectively.*

Solution.   (i) Since T_1 is induced of T on V_1 and T_2 is induced of T on V_2, therefore the minimal polynomials $m_1(x)$ of T_1 and $m_2(x)$ of T_2 each divides $m(x)$.

Suppose $p(x)$ is a multiple of both $m_1(x)$ and $m_2(x)$, then

$$(P(T_1))(V_1) = 0 \quad \text{and} \quad (P(T_2))(V_2) = 0$$

Let $\alpha \in V$, then $\alpha = \alpha_1 + \alpha_2$, where $\alpha_1 \in V_1$ and $\alpha_2 \in V_2$

Now, $(P(T))(\alpha) = (P(T))(\alpha_1) + (P(T))(\alpha_2)$

$$= 0 + 0$$

$$= 0$$

T is the root of $P(x)$. Hence $m(x)|p(x)$. So that $m(x)$ is the least common multiple.

(ii) Since $T=T_1 \oplus T_2$, then the matrix representation of T is

$$M= \begin{bmatrix} A & 0 \\ 0 & B \end{bmatrix}$$

where A and B are the matrix representations of T_1 and T_2 respectively.

$$\Delta(x) = |xI - M|$$
$$= \begin{vmatrix} xI - A & 0 \\ 0 & xI - B \end{vmatrix} = (xI-A)(xI-B)$$

$$= \Delta_1(x)\Delta_2(x)$$

Example 4. *Let $T: V \to V$ be linear and let W be the eigenspace belonging to an eigen value λ of T. Show that W is T-invariant.*

Solution. By the definition of eigenspace, we have

$$W = \text{kernel}(T-\lambda I)$$

If $\alpha \in W$, then

$$(T-\lambda I)(\alpha) = 0 \Rightarrow T(\alpha) - \lambda I(\alpha)=0$$
$$\Rightarrow \qquad T(\alpha)= \lambda\alpha \qquad\qquad\qquad \text{...(1)}$$

Since W is a subspace of W, so for any scalar $\lambda \in F$ and $\alpha \in W$, we have

$$\lambda\alpha \in W \Rightarrow T(\alpha) \in W \qquad\qquad \text{[Using (1)]}$$

Hence, W is T-invariant.

Example 5. *If $\{W_i\}$ is a collection of T-invariant subspaces of a vector space V. Show that the intersection $W = \cap W_i$ is also T-invariant.*

Solution. Since each W_i is T-invariant, then for $\alpha \in W_i$, we have $T(\alpha) \in W_i$ for every i.

$$\therefore \qquad T(\alpha) \in \cap W_i \text{ for every } i.$$
$$\Rightarrow \qquad T(\alpha) \in W.$$

Hence, W is T-invariant.

Example 6. *Let A be a square matrix over the complex field $\mathbf{C}$. Suppose λ is an eigenvalue of A^2. Show that $\sqrt{\lambda}$ or $-\sqrt{\lambda}$ is an eigen value of A.*

Solution. Since, A is similar to a triangular matrix

$$B= \begin{bmatrix} u_1 & b_{12} & \cdots & b_{1n} \\ 0 & u_2 & \cdots & b_{2n} \\ \cdots & \cdots & \cdots & \cdots \\ 0 & 0 & \cdots & u_n \end{bmatrix}$$

Thus, A^2 is similar to the matrix.

$$B^2= \begin{bmatrix} u_1^2 & u_1 b_{12} + b_{12}u_2 & \cdots & u_1 b_{1n} +\ldots+ u_n b_{1n} \\ 0 & u_2^2 & \cdots & u_2 b_{2n} +\ldots+ u_n b_{2n} \\ \cdots & \cdots & \cdots & \cdots \\ 0 & 0 & \cdots & u_n^2 \end{bmatrix}$$

Since similar matrices have the same eigenvalues, so $\lambda = u_i^2$ for some i.

$$\Rightarrow \quad u_i= \sqrt{\lambda} \quad \text{ or } \quad u_i = -\sqrt{\lambda}.$$

Hence, $\sqrt{\lambda}$ or $-\sqrt{\lambda}$ is an eigenvalue of A.

Example 7. *Show that similar matrices have the same eigenvalues.*

Solution.   If a matrix A is similar to B, then there exists an invertible matrix P such that
$A = P^{-1}BP$.

So, $|xI{-}A| = |xI{-}P^{-1}BP|$
$$= |P^{-1}(xI{-}B)P| = |P^{-1}|\,|xI{-}B|\,|P|$$
$$= |xI{-}B| \qquad\qquad\qquad [\,|P^{-1}|\,|P| = 1\,]$$

This shows that similar matrices have the same characteristic polynomials.
Hence, they have the same eigenvalues.

Example 8. *Let the matrix A is given by*

$$A = \begin{bmatrix} 0 & 1 & 1 & 0 & 1 \\ 0 & 0 & 1 & 1 & 1 \\ 0 & 0 & 0 & 0 & 0 \\ 0 & 0 & 0 & 0 & 0 \\ 0 & 0 & 0 & 0 & 0 \end{bmatrix}$$

Show that it is nilpotent and find its index of nilpotency. Also, find the nilpotent matrix m canonical form, which is similar to A.

Solution.   We have, $A^2 = \begin{bmatrix} 0 & 0 & 1 & 1 & 1 \\ 0 & 0 & 0 & 0 & 0 \\ 0 & 0 & 0 & 0 & 0 \\ 0 & 0 & 0 & 0 & 0 \\ 0 & 0 & 0 & 0 & 0 \end{bmatrix} \neq 0$

$A^3 = A^2 A$

$$= \begin{bmatrix} 0 & 0 & 1 & 1 & 1 \\ 0 & 0 & 0 & 0 & 0 \\ 0 & 0 & 0 & 0 & 0 \\ 0 & 0 & 0 & 0 & 0 \\ 0 & 0 & 0 & 0 & 0 \end{bmatrix} \begin{bmatrix} 0 & 1 & 1 & 0 & 1 \\ 0 & 0 & 1 & 1 & 1 \\ 0 & 0 & 0 & 0 & 0 \\ 0 & 0 & 0 & 0 & 0 \\ 0 & 0 & 0 & 0 & 0 \end{bmatrix}$$

$$= \begin{bmatrix} 0 & 0 & 0 & 0 & 0 \\ 0 & 0 & 0 & 0 & 0 \\ 0 & 0 & 0 & 0 & 0 \\ 0 & 0 & 0 & 0 & 0 \\ 0 & 0 & 0 & 0 & 0 \end{bmatrix} = 0$$

Thus, A is a nilpotent matrix of index 2.

Since A is nilpotent of index, thus we can say that M contains the diagonal block matrix of order less than or equal to 2.

Clearly, the rank of $A = 2$ and the matrix A is of order 3. So that nullity of $A = 3$.

Thus, M will contain 3 diagonal block matrices, in which 2 diagonal block of order 2 each and 1 diagonal block of order 1.

$$M = \begin{bmatrix} M_2 & & \\ & M_2 & \\ & & M_1 \end{bmatrix}$$

where, $M_2 = \begin{bmatrix} 0 & 1 \\ 0 & 0 \end{bmatrix}$, $M_1 = [0]$

Thus, we have

$$M = \begin{bmatrix} 0 & 0 & 0 & 0 & 0 \\ 0 & 0 & 0 & 0 & 0 \\ 0 & 0 & 0 & 1 & 0 \\ 0 & 0 & 0 & 0 & 0 \\ 0 & 0 & 0 & 0 & 0 \end{bmatrix}$$

Example 9. *If T is nilpotent of index k, show that $T^n, n > 1$ is nilpotent of index k.*

Solution.   Since $T^k = 0$ but $T^{k-1} \neq 0$, then

$$(T^n)^k = (T^k)^n = 0^n = 0 \qquad\qquad [\because T^{k-1} \neq 0]$$

$$(T^n)^{k-1} = (T^{k-1})^n \neq 0$$

and thus T^n is nilpotent of index $\leq k$.

Example 10. *Suppose S and T are nilpotent operators which commutes, i.e., $ST = TS$, Show that S+T and ST are also nilpotent.*

Solution.   Since S and T are nilpotent, so we have $S^m = 0$ and $T^n = 0$ for some positive integers m and n. Since S and T commutes, then

$$(S+T)^{m+n} = \sum_{r=0}^{m+n} C_r T^{m+n-r} \cdot S^r \qquad\qquad \ldots(1)$$

(i) If $r \geq m$, then $S^r = 0$. So from (1), we get

$$(S+T)^{m+n} = 0.$$

(ii) If $r < m$, so $m+n-r \geq n$, then $T^{m+n-r} = 0$.

Therefore, from (1), we get

$$(S+T)^{m+n} = 0$$

$\Rightarrow$ $S+T$ is nilpotent.

Now to show that ST is nilpotent.

Suppose $S^m = 0$ and $T^n = 0$ and $m \leq n$.

Then, $(ST)^n = S^n T^n = S^n = 0 = 0$

If $m \geq n$, then

$$(ST)^m = S^m T^m = 0.T^m = 0$$

Thus ST is nilpotent.

Example 11. *Determine all possible Jordan canonical forms for a linear operator $T: V \to V$ whose characteristic polynomial is $\Delta(x) = (x-2)^3 (x-5)^2$.*

Solution.   Since $x - 2$ has exponent 3 in $\Delta(x)$ and $x-5$ has exponent 2 in $\Delta(x)$, therefore Jordan canonical form will be the matrix of order 5×5.

We may write $\Delta(x)$ as

$$\Delta(x) = (x - \lambda_1)^3 (x - \lambda_2)^2, \text{ where } \lambda_1 = 2 \text{ and } \lambda_2 = 5.$$

Now, $\lambda_1 = 2$ must appear a three times on the main diagonal and $\lambda_2 = 5$ must appear two times. Hence, the possible Jordan canonical forms are

(i)
$$\begin{bmatrix} 2 & 1 & 0 & 0 & 0 \\ 0 & 2 & 1 & 0 & 0 \\ 0 & 0 & 2 & 0 & 0 \\ 0 & 0 & 0 & 5 & 1 \\ 0 & 0 & 0 & 0 & 5 \end{bmatrix}$$

(ii)
$$\begin{bmatrix} 2 & 1 & 0 & 0 & 0 \\ 0 & 2 & 1 & 0 & 0 \\ 0 & 0 & 2 & 0 & 0 \\ 0 & 0 & 0 & 5 & 1 \\ 0 & 0 & 0 & 0 & 5 \end{bmatrix}$$

(iii)
$$\begin{bmatrix} 2 & 1 & 0 & 0 & 0 \\ 0 & 2 & 1 & 0 & 0 \\ 0 & 0 & 2 & 0 & 0 \\ 0 & 0 & 0 & 5 & 1 \\ 0 & 0 & 0 & 0 & 5 \end{bmatrix}$$

(iv)
$$\begin{bmatrix} 2 & 1 & 0 & 0 & 0 \\ 0 & 2 & 1 & 0 & 0 \\ 0 & 0 & 2 & 0 & 0 \\ 0 & 0 & 0 & 5 & 1 \\ 0 & 0 & 0 & 0 & 5 \end{bmatrix}$$

(v)
$$\begin{bmatrix} 2 & 1 & 0 & 0 & 0 \\ 0 & 2 & 1 & 0 & 0 \\ 0 & 0 & 2 & 0 & 0 \\ 0 & 0 & 0 & 5 & 1 \\ 0 & 0 & 0 & 0 & 5 \end{bmatrix}$$

(vi)
$$\begin{bmatrix} 2 & 1 & 0 & 0 & 0 \\ 0 & 2 & 1 & 0 & 0 \\ 0 & 0 & 2 & 0 & 0 \\ 0 & 0 & 0 & 5 & 1 \\ 0 & 0 & 0 & 0 & 5 \end{bmatrix}$$

Example 12. *If A is a complex 5×5 matrix with characteristic polynomial*

$$\Delta(x) = (x-2)^3(x+7)^2$$

and minimal polynomial $m(x) = (x-2)^2(x+7)^2$. What is the Jordan form for A?

Solution. Here, in $\Delta(x)$, the exponent of $(x-2)$ is 3 and that of $(x+7)$ is 2. Thus, A will be of order 5×5. Also, in $m(x)$, the exponent of $(x-2)$ is 2 and that of $(x+7)$ is 1. Therefore, Jordan form will have one block of order 2 and other two blocks must be of the order 2 or 1. Hence the required Jordan form of A is

$$\begin{bmatrix} 2 & 1 & 0 & 0 & 0 \\ 0 & 2 & 0 & 0 & 0 \\ 0 & 0 & 2 & 1 & 0 \\ 0 & 0 & 0 & 2 & 0 \\ 0 & 0 & 0 & 0 & -7 \end{bmatrix}$$

Example 13. *Determine all possible Jordan canonical forms for a matrix of order 5 whose minimal polynomial is $m(x) = (x-2)^2$.*

Solution. Since the minimal polynomial of a matrix of order 5×5 is $(x-2)^2$, therefore, its characteristic polynomial will be $(x-2)^5$.

Thus, Jordan canonical form must have a Jordan block matrix of order 2 and other must be of order 2 or 1.

Hence, the possible Jordan canonical forms are :

(i)
$$\begin{bmatrix} 2 & 1 & 0 & 0 & 0 \\ 0 & 2 & 0 & 0 & 0 \\ 0 & 0 & 2 & 1 & 0 \\ 0 & 0 & 0 & 2 & 0 \\ 0 & 0 & 0 & 0 & 2 \end{bmatrix}$$

(ii)
$$\begin{bmatrix} 2 & 1 & 0 & 0 & 0 \\ 0 & 2 & 0 & 0 & 0 \\ 0 & 0 & 2 & 1 & 0 \\ 0 & 0 & 0 & 2 & 0 \\ 0 & 0 & 0 & 0 & 2 \end{bmatrix}$$

Example 14. *Suppose* $T : V \to V$ *has characteristic polynomial*

$$\Delta(x) = (x + 6)^4 (x - 2)^3$$

and minimal polynomial

$$M(x) = (x + 6)^3 (x - 1)^2$$

Find the Jordan canonical form of the matrix representation of T.

Solution. Since degree of $\Delta(x)$ is 7 so that Jordan form will be a matrix of order 7×7 , in which -8 will be repeated 4 times on the diagonal and 1 will be three times on the diagonal. Also, $(x + 8)$ has the exponent 3 in $m(x)$, therefore Jordan form will have one block of order 3×3 belonging to -8 and $(x - 1)$ has the exponent 2 in $m(x)$. So, there must be one block of order 2×2 belonging to 1.

Hence, the required Jordan canonical form is

$$\begin{bmatrix} -8 & 1 & 0 & 0 & 0 & 0 & 0 \\ 0 & -8 & 1 & 0 & 0 & 0 & 0 \\ 0 & 0 & -8 & 0 & 0 & 0 & 0 \\ 0 & 0 & 0 & -8 & 0 & 0 & 0 \\ 0 & 0 & 0 & 0 & 1 & 1 & 0 \\ 0 & 0 & 0 & 0 & 0 & 1 & 0 \\ 0 & 0 & 0 & 0 & 0 & 0 & 1 \end{bmatrix}$$

Example 15. *Find all possible rational canonical forms for* 6×6 *matrices with minimal polynomial* $m(x) = (x + 1)^3$.

Solution. Let $T : V \to V$ be a linear with minimal polynomial $m(x) = (x + 1)^3$ and dim $V = 6$, then T is one of the following direct sum of companion matrices:

(i) $C((x + 1)^3) \oplus C((x + 1)^3)$

(ii) $C((x + 1)^3) \oplus C((x + 1)^2) \oplus C(x + 1)$

(iii) $C((x + 1)^3) \oplus C(x + 1) \oplus C(x + 1) \oplus C(x + 1)$

$$\text{Now,} \quad C((1 + x)^3) = C(x^3 + 3x^2 + 3x + 1) = \begin{bmatrix} 0 & 0 & -1 \\ 1 & 0 & -3 \\ 0 & 1 & -3 \end{bmatrix}$$

$$C((1 + x)^2) = C(x^2 + 2x + 1) = \begin{bmatrix} 0 & -1 \\ 1 & -2 \end{bmatrix}$$

$$C(1 + x) = [-1].$$

Thus, the rational canonical form of T is one of the following matrices :

(i)
$$\begin{bmatrix} 0 & 1 & -1 & 0 & 0 & 0 \\ 1 & 0 & -3 & 0 & 0 & 0 \\ 0 & 1 & -3 & 0 & 0 & 0 \\ 0 & 0 & 0 & 0 & 0 & -1 \\ 0 & 0 & 0 & 1 & 0 & -3 \\ 0 & 0 & 0 & 0 & 1 & -3 \end{bmatrix}$$

(ii)
$$\begin{bmatrix} 0 & 0 & -1 & 0 & 0 & 0 \\ 1 & 0 & -3 & 0 & 0 & 0 \\ 0 & 1 & -3 & 0 & 0 & 0 \\ 0 & 0 & 0 & 0 & -1 & 0 \\ 0 & 0 & 0 & 1 & -2 & 0 \\ 0 & 0 & 0 & 0 & 1 & -1 \end{bmatrix}$$

$$(iii) \begin{bmatrix} 0 & 0 & -1 & 0 & 0 & 0 \\ 1 & 0 & -3 & 0 & 0 & 0 \\ 0 & 1 & -3 & 0 & 0 & 0 \\ 0 & 0 & 0 & -1 & 0 & 0 \\ 0 & 0 & 0 & 0 & -1 & 0 \\ 0 & 0 & 0 & 0 & 0 & -1 \end{bmatrix}$$

Example 16. *Let A be a 4×4 matrix with minimal polynomial* $M(x) = (x^2 + 1)(x^2 - 3)$. *Find the rational canonical form for A if A is a matrix over (i) the rational field* **F**, *(ii) the real field* **R**, *(iii) the complex field* **C**.

Solution. (i) If the field is rational **Q**. The rational canonical form of A is the following direct sum of companion matrices :

(a) $\qquad C(x^2 + 1) \oplus C(x^2 - 3)$

Now, $\quad C(x^2 + 1) = \begin{bmatrix} 0 & -1 \\ 1 & 0 \end{bmatrix}$

$$C(x^2 - 3) = \begin{bmatrix} 0 & 3 \\ 1 & 0 \end{bmatrix}$$

Thus the rational canonical form is

(i) $\begin{bmatrix} 0 & -1 & 0 & 0 \\ 1 & 0 & 0 & 0 \\ 0 & 0 & 0 & 3 \\ 0 & 0 & 1 & 0 \end{bmatrix}$

(ii) If the field is a field of reals **R**, then the rational canonical form of A is the direct sum of the companion matrices

$$C(x^2 + 1) \oplus C(x - \sqrt{3}) \oplus C(x + \sqrt{3})$$

Now, $\quad C(x - \sqrt{3}) = [\sqrt{3}] \quad$ and $\quad C(x + \sqrt{3}) = [-\sqrt{3}]$

Thus the required rational canonical form is

$$\begin{bmatrix} 0 & -1 & 0 & 0 \\ 1 & 0 & 0 & 0 \\ 0 & 0 & \sqrt{3} & 0 \\ 0 & 0 & 0 & -\sqrt{3} \end{bmatrix}$$

(iii) If the field is field of complex numbers **C**, then the rational canonical form of A is the direct sum of the companion matrices

$$C(x - i) \oplus C(x + i) \oplus C(x - \sqrt{3}) \oplus C(x + \sqrt{3})$$

Thus the required rational canonical form is

$$\begin{bmatrix} i & 0 & 0 & 0 \\ 0 & -i & 0 & 0 \\ 0 & 0 & \sqrt{3} & 0 \\ 0 & 0 & 0 & -\sqrt{3} \end{bmatrix}$$

Example 17. *Let V be a vector space of dimension 6 over* **R** *and let T be a linear operator whose minimal polynomial is* $m(x) = (x^2 - x + 3)(x- 2)^2$. *Find the rational canonical form of T.*

Solution. Since $\dim V = 6$, then T is one of the following direct sums of companion matrices:

(i) $C(x^2 - x + 3) \oplus C(x^2 - x+3) \oplus C((x - 2)^2)$

(ii) $C(x^2 - x + 3) \oplus C((x - 2)^2) \oplus C((x - 2)^2)$

(iii) $C(x^2 - x + 3) \oplus C((x - 2)^2) \oplus C(x - 2) \oplus C(x - 2)$

where $C(q(x))$ is the companion matrix of $q(x)$.

Now $C(x^2 - x + 3) = \begin{bmatrix} 0 & -3 \\ 1 & 1 \end{bmatrix}$

$$C((t - 2)^2) = C(t^2 - 4t + 4) = \begin{bmatrix} 0 & -4 \\ 1 & 4 \end{bmatrix}$$

$$C((x-2)) = [2]$$

Thus the canonical forms of T is one of the following matrices :

(i) $\begin{bmatrix} 0 & -3 & 0 & 0 & 0 & 0 \\ 1 & 1 & 0 & 0 & 0 & 0 \\ 0 & 0 & 0 & -3 & 0 & 0 \\ 0 & 0 & 1 & 1 & 0 & 0 \\ 0 & 0 & 0 & 0 & 0 & -4 \\ 0 & 0 & 0 & 0 & 1 & 4 \end{bmatrix}$ (ii) $\begin{bmatrix} 0 & -3 & 0 & 0 & 0 & 0 \\ 1 & 1 & 0 & 0 & 0 & 0 \\ 0 & 0 & 0 & -4 & 0 & 0 \\ 0 & 0 & 1 & 4 & 0 & 0 \\ 0 & 0 & 0 & 0 & 0 & -4 \\ 0 & 0 & 0 & 0 & 1 & 4 \end{bmatrix}$

(iii) $\begin{bmatrix} 0 & -3 & 0 & 0 & 0 & 0 \\ 1 & 1 & 0 & 0 & 0 & 0 \\ 0 & 0 & 0 & -4 & 0 & 0 \\ 0 & 0 & 1 & 4 & 0 & 0 \\ 0 & 0 & 0 & 0 & 2 & 0 \\ 0 & 0 & 0 & 0 & 0 & 2 \end{bmatrix}$

EXERCISE 8.3

1. Prove that the relation of similarity is an equivalence relation in $A(V)$.

2. If A is triangular $n \times n$ matrix with entries λ_1, λ_2, ... λ_n on the diagonal, then
$$(A - \lambda_1 I) (A - \lambda_2 I) \dots (A - \lambda_n I) = 0$$

3. Suppose $T : V \to V$ is linear. Show that each of the following is invariant under T :
   (i) kernel of T (ii) {0}
   (iii) Image of T (iv) V

4. Determine the invariant subspace of
$$A = \begin{bmatrix} 2 & -4 \\ 5 & -2 \end{bmatrix}$$ viewed as linear operator on:

   (i) R^2 (ii) C^2

5. Suppose A is super triangular matrix (all entries below the main diagonal are 0). Show that A is nilpotent.

6. What is the minimal polynomial of nilpotent matrix A of index k?

7. Show that following matrix are nilpotent :
$$A = \begin{bmatrix} -2 & 1 & 1 \\ -3 & 1 & 2 \\ -2 & 1 & 1 \end{bmatrix}; A = \begin{bmatrix} 1 & -3 & 2 \\ 1 & -3 & 2 \\ 1 & -3 & 2 \end{bmatrix}$$
Find also the index of nilpotency in each matrix.

8. Find the canonical nilpotent form of the matrix

$$A = \begin{bmatrix} -2 & 1 & 1 \\ -3 & 1 & 2 \\ -2 & 1 & 1 \end{bmatrix}$$

9. If matrix A and B are similar, then show that A is nilpotent of index k, if and only if B is nilpotent of index k.

10. If F is a field of characteristic zero and if S and T in $A(V)$ are such that $ST - TS$ commutes with S, then $ST - TS$ is nilpotent.

11. If $T : V \rightarrow V$ is linear operator on V over a field F of characteristic zero such that tr. $T^k = 0$ for all $k \geq P$, then show that T is nilpotent.

12. Find all possible Jordan canonical forms for those matrix whose characteristic polynomial $\Delta(x)$ and minimal polynomial $m(x)$ are as follows :
(i) $\Delta(x) = (x - 2)^4 \cdot (x - 3)^2$,
$m(x) = (x - 2)^2 (x - 3)^2$
(ii) $\Delta(x) = (x - 2)^7, m(x) = (x - 2)^2$
(iii) $\Delta(x) = (x - 3)^4 (x - 5)^4$,
$m(x) = (x - 3)^2 (x - 5)^2$

13. How many possible Jordan forms are there for 6×6 complex matrix with characteristic polynomial $\Delta(x) = (x + 2)^4(x- 1)^2$?

14. Find all possible Jordan forms 8×8 matrices having $x^2(x - 1)^3$ as minimal polynomial.

15. Show that every complex matrix is similar to its transpose.

16. Prove that the matrix $\begin{bmatrix} 1 & 1 & 1 \\ -1 & -1 & -1 \\ 1 & 0 & 0 \end{bmatrix}$ is nilpotent and find its invariant and Jordan form.

17. Determine the Jordan canonical form for the matrices

(i) $\begin{bmatrix} 1 & 1 & 1 & 1 \\ 0 & 1 & 1 & 1 \\ 0 & 0 & 1 & 1 \\ 0 & 0 & 0 & 1 \end{bmatrix}$ (ii) $\begin{bmatrix} 1 & 0 & 1 & 1 \\ 0 & 1 & 0 & 1 \\ 0 & 0 & 1 & 0 \\ 0 & 0 & 0 & 1 \end{bmatrix}$

18. Show that all complex matrices of order n for which $A^n = I$ are similar.

19. Find all possible rational canonical forms for 6×6 matrix with minimal polynomial $m(x) = (x^2+2)^2 (x + 3)^2$.

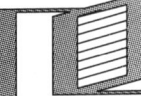 **Chapter Review:** | *A competitive Approach*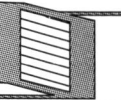

Selected Terms and Results

Terms

- **Bilinear form:** A bilinear form on a vector space $V(f)$ is a mapping $f : V \times V \rightarrow f$ which satisfies the following properties:
  - (i) $f(a\alpha+b\beta,\gamma)=af(\alpha,\gamma)+bf(\beta,\gamma)$
  - (ii) $f(\alpha,a\beta+b\gamma)=af(\alpha,\beta)+bf(\alpha,\gamma)$
  - for all $a,b \in F$, $\alpha,\beta,\gamma \in V$.

- **Bilinear space :** A vector space V together with a bilinear form f defined above is called a bilinear space.

- **Rank of a bilinear form:** Let f be a bilinear form on V, then rank of a bilinear form f, denoted by rank(f) is defined to be the rank of any matrix representation of f.

- **Degenerate and non-degenerate bilinear form:** A bilinear form f on V is said to be degenerate if rank $(f)<$dim V and it is said to be non-degenerate if rank $(f)=$dim V

- **Quadratic Mapping:** A mapping $q:V\rightarrow f$ is called a quadratic mapping if $q(\alpha)=f(\alpha,\alpha)$, where f is a symmetric bilinear form on V.

- **Quadratic form:** A mapping $q:V\rightarrow f$ is called a quadratic form if $q(\alpha)= f(\alpha,\alpha)$ for some bilinear form f on V.

- **Signature:** Let q be a real quadratic form and f be a bilinear form then the signature of f and of q are defined by sig$(f)=$ sig$(q)= p-N$ where p is the number of positive entries and N is the number of negative entries in any diagonal representation of f and q.

- **Hermitian form :** A matrix A is said to be Hermitian if $A^\theta=A$ and skew-hermitian if $A^\theta=-A$.

- **Similar martices:** For two square matrices A and B, the matrix B is said to be similar to A if there exist an invertible matrix P such that $B= P^{-1}AP$.

- **Invariant subspace:** Let $T:V\rightarrow V$ be a linear transformation. Then a subspace W of V is said to be invariant under T if $T(W) \subset W \; \forall \; \alpha \in W$, $T(\alpha) \in W$.

- **Normal form:** A matrix A is said to be in normal form if it can be written as $A= \begin{pmatrix} I & 0 \\ 0 & 0 \end{pmatrix}$ where I is the square identity matrix of order r.

- **Nilpotent Transformation :** A linear transformation $T: V \rightarrow V$ is said to be nilpotent if $T^n=0$ for some least positive integer n.

Results

- A bilinear form f is called symmetric if $f(\alpha,\beta)=f(\beta,\alpha)$ for all $\alpha, \beta \in V$.
- A bilinear form f is called skew-symmetric if $f(\alpha,\beta)= -f(\beta.\alpha)$ for all $\alpha,\beta \in V$.
- A bilinear form f is called alternating if $f(\alpha,\alpha)=0$ for all $\alpha \in V$.
- A bilinear form is reflexive if and only if it is either symmetric or alternating.
- A bilinear form is reflexive if it is either symmetric or skew-symmetric.
- A symmetric bilinear form on a finite dimensional vector space V over a field F of characteristic not equal to 2 is diagonalizable.
- A mapping $q:V\rightarrow F$ is called a quadratic form if $q(\alpha)=f(\alpha,\alpha)$ for some bilinear form f on V.

- Similarity of matrices is an equivalence relation.
- If dim $V=n$ and if $T \in A(V)$ has all its characteristic roots in F, then T satisfy a polynomial of degree n over F.
- A linear transformation $T:V\rightarrow V$ is said to be nilpotent if $T^n=0$ for some least positive integer n.
- The characteristic root of nilpotent transformation are zero, so they belong to the concerned field F and hence, all characteristic roots of nilpotent transformation belongs to F.
- The Jordan canonical form is exerted when the minimal polynomial cannot be factored into

linear polynomial while in rational canonical form, the minimal polynomial is taken as the product of linear polymomial.

▪ If $T:V{\to}V$ is linear and W be the eigenspace belonging to an eigenvalue λ of T, then W is T-invariant.

▪ Similar matrices have the same eigenvalues.

▪ If T is nilpotent of index k, then $T^n, n>1$ is nilpotent of same index.

▪ If S and T are nilpotent operators such that $ST=TS$, then $S+T$ and ST are nilpotent.

Review Questions and Project Work

1. If V is an n-dimenstonal vector space over the field F, then show that dim $L[V,V,F]=n^2$

2. Let F be a subfield of complex numbers and A be a symmetric $n{\times}n$ matrix over F. Show that there exist an invertible $n{\times}n$ matrix P over F such that $P'AP$ is diagonal.

3. Describe explicity all symmetric bilinear form S on $\mathbf{R}^3$.

4. Let V be an n-dimensional vector space over the field of complex numbers and f be a skew-symmetric bilinear form on V. Prove that rank of f is even.

5. Let A be $n \times n$ matrix over the field F, show that

$$f(x,y)=X'AY$$

is a bilinear form on F^n.

6. Show that the symmetric matrix of a real quadratic form is unique.

7. Let $\alpha=(x_1,x_2,x_3)$, $\beta=(y_1,y_2,y_3){\in}V_3(\mathbf{C})$
Define a function $f: V_3(\mathbf{C}){\times}V_3(\mathbf{C}){\to}\mathbf{C}$ by

$$f(\alpha,\beta)= ix_1\bar{y}_2-ix_2\bar{y}_1+x_3\bar{y}_3$$

Show that f is a Hermitian form on $V_3(\mathbf{C})$.

8. Find all invariant subspace of $A = \begin{pmatrix} 2 & -4 \\ 5 & -2 \end{pmatrix}$ viewed as a linear operator $\mathbf{R}^2$.

9. Let $T:V{\to}V$ be linear and suppose $T=T_1{\oplus}T_2$ w.r.t. a T-invariant direct sum decomposition $V=V \oplus W$. Let $m(t)$, $m_1(t)$ and $m_2(t)$ denote respectively the minimum polynomials of T,T_1 and T_2. Show that $m(T)$ is the least common multiple of $m_1(t)$ and $m_2(t)$.

10. Show that a Jordan block J may be written as the sum of a scalar matrix and a canonical

nilpotent block N ,i.e., $J=\lambda I+N$.

11. Find all possible Jordan canonical forms for a linear map $T:V{\to}V$ whose characteristic polynomial is $\Delta(t)=(t-7)^5$ and whose minimal polynomial is $m(t)=(t-7)^2$.

12. Find all possible rational canonical forms with $m(t)=(t-2)^3$ and $\Delta(t)=(t-2)^5$.

13. Show that $f(t)=t^3-3t^2+3t+2$ is irreducible over the rational field $\mathbf{Q}$.

14. Show that congurent matrices have the same rank.

15. Let $A= \begin{pmatrix} 0 & 1 & 1 \\ 1 & -2 & 2 \\ 1 & 2 & -1 \end{pmatrix}$ be a symmetric matrix. Find a non-singular matrix P such that P^TAP is diagonal. Also find the matrix P^TAP.

16. Show that the quadratic form $q(x,y,z)$ corresponding to $\begin{pmatrix} 2 & -5 & 1 \\ -5 & -6 & -7 \\ 1 & -7 & 9 \end{pmatrix}$ is given by

$$q(x,y,z)=2x^2-10xy-6y^2+2xz-14yz+9z^2.$$

17. Consider the quadratic form $q(x,y,z)=x^2 + 4xy + 3y^2 - 6xz + 10yz + 7z^2$. Find a non-singular linear substituion expressing the variables x,y,z in terms of variables r,s,t such that $q(r,s,t)$ is diagonal.

18. Diagonalize $q(x,y) = 3x^2 - 12xy + 7y^2$ by completing the square.

19. Let $q(x,y,z)=x^2+y^2+2xz+4yz+3z^2$. Show that q is not positive definite.

20. Show that $q(x,y)=ax^2+bxy+cy^2$ is positive definite iff $b^2-4ac<0$.

Objective type Questions

Fill in the Blanks

1. A bilinear form f is reflexive if for $\alpha,\beta{\in}V, f(\beta,\alpha)=0$ whenever _____ .

2. A bilinear form f is _____ if $f(\alpha,\beta)=f(\beta,\alpha)$ $\forall \alpha,\beta{\in}V$.

3. A bilinear form f is _____ if $f(\alpha,\beta)= -f(\beta,\alpha)$ $\forall \alpha,\beta{\in}V$.

4. A bilinear form f is _____ if $f(\alpha,\alpha)=0$ $\forall \alpha{\in}V$.

5. A bilinear form f is non-degenerate if rank $(f) = $ _____

6. A bilinear form f is alternating if $f(\alpha, \alpha) = $ _____ $\forall \alpha \in V$.

7. Let f be a bilinear form on a vector space V over F. Then the quadratic form on V associated with the bilinear form f is the function $q : V \to F$ defined by $q(\alpha) = $ _____ $\forall \alpha \in V$.

8. Every bilinear form on the vector space V over the field of complex numbers can be uniquely expressed as the _____ of a symmetric and skew-symmetric bilinear forms.

9. If f is a Hermitian form on a vector space V over the complex field **C**. Then $f(\alpha, a\beta + b\gamma)$ _____ .

10. The sum of Hermitian form is _____ .

True/False

Write 'T' for true and 'F' for false statement.

1. A bilinear form f is alternating if $f(\alpha, \alpha) = 0$ $\forall \alpha \in V$ **(T/F)**

2. A symmetrical bilinear form f is positive definite for $0 \neq \alpha \in V, f(\alpha, \alpha) > 0$ **(T/F)**

3. Let F be a symmetric form on V, then signature$(f) = \dim V^+ + \dim V^-$.   **(T/F)**

4. A quadratic form f is negative semidefintie if $q(\alpha) \neq 0 \ \forall \alpha \in V$.   **(T/F)**

Multiple Choice Questions

Choose the most appropriate one :

1. Let f be a symmetric bilinear form on V, then sig(f) is equal to :
   (a) $\dim V^+ + \dim V^-$
   (b) $\dim V^+ - \dim V^-$
   (c) $\dim V^+$
   (d) none of the above

2. Let f be a symmetric bilinear form on V, then rank(f) is equal to :
   (a) $\dim V^+ + \dim V^-$
   (b) $\dim V^+ - \dim V^-$
   (c) $\dim V^+$
   (d) none of the above

3. A bilinear form f on a vector space $V(F)$, is called degenerate if:
   (a) for each $0 \neq \alpha \in V; f(\alpha, \beta) = 0 \ \forall \beta \in V$
   (b) for each $0 \neq \beta \in V; f(\alpha, \beta) = 0 \ \forall \alpha \in V$
   (c) both (a) and (b) are true.
   (d) none of the above

4. If a linear transformation $X = BY$ is applied to a quadratic form $A(x, x)$, then the resultant is a quadratic form $C(y, y)$ whose matrix C can be expressed as :
   (a) $C = B'AB$ (b) $A = B'CB$
   (c) $A = B^{-1}CB$ (d) none of these

5. When a quadratic form $q = X'AX$ is transformed to a quadratic form $Y'CY$ by a non-singular transformation $X = BT$, then :
   (a) $\rho(C) < \rho(A)$
   (b) $\rho(C) > \rho(A)$
   (c) $\rho(C) \leq \rho(A)$
   (d) none of the above

6. The quadratic form corresponding to the diagonal matrix diag$(\lambda_1, \lambda_2, ..., \lambda_n)$ is :
   (a) $x_1^2 + x_2^2 + ... + x_n^2$
   (b) $\lambda_1 x_1^2 + \lambda_2 x_2^2 + ... + \lambda_n x_n^2$
   (c) $\lambda_1^2 x_1^2 + \lambda_2^2 x_2^2 + ... + \lambda_n^2 x_n^2$
   (d) none of the above

7. The corresponding matrix of the quadratic form $q = ax^2 + 2hxy + by^2$ is :
   (a) $\begin{bmatrix} a & h \\ -h & b \end{bmatrix}$

   (b) $\begin{bmatrix} a & h \\ h & b \end{bmatrix}$

   (c) $\begin{bmatrix} a & -h \\ -h & b \end{bmatrix}$

   (d) none of the above

8. The corresponding matrix of the quadratic form
   $q = ax^2 + by^2 + cz^2 + 2fyz + 2gzx + 2hxy$ is :
   (a) $\begin{bmatrix} a & h & g \\ h & b & f \\ g & f & c \end{bmatrix}$

   (b) $\begin{bmatrix} a & h & -g \\ -h & -b & f \\ g & f & c \end{bmatrix}$

   (c) both (a) and (b) are true.
   (d) none of the above

=================================== Answers ===================================

Fill in the Blanks

1. $f(\beta,\alpha) = 0$ **2.** symmetric   **3.** Skew-symmetric   **4.** alternating

5. $\dim V^+ + \dim V^-$ **6.** 0   **7.** $f(\alpha,\alpha)$ **8.** sum

9. $\bar{a}f(\alpha,\beta) + \bar{b}f(\alpha,\gamma)$ **10.** Hermitian

True/False

1. T   **2.** T   **3.** F   **4.** T

Multiple Choice Questions

1. (b)   **2.** (b)   **3.** (c)   **4.** (a)   **5.** (b)   **6.** (b)   **7.** (b)   **8.** (a)

● ● ●

Chapter

9 Modules

9.1 INTRODUCTION

Module is a generalized concept of vector space. Basically vector space is generalization of abelian group. By this fact, the Module can be considered as the generalization of abelian group. In module, the scalar element can be taken from a ring while in vector space scalar element belong to the field.

9.2 MODULES

Definition 1. *Let R be a ring. A R-module (left) is an additive abelian group M together with a function $R \times M \to M : \forall r,s \in R$ and $a,b \in M$ with following conditions :*

(i) $r(a + b) = ra + rb$

(ii) $(r + s)a = ra + sa$

(iii) $r(sa) = (rs)a$

Definition 2. *(Unitary R-module). If R has an identity element, i.e., $1 \in R$ and $1a = a$ $\forall a \in M$, then M is said to be a unitary R-module. If R is a field then unitary R-module is called a vector space over the field R.*

Similarly we can define a right R-module by putting the elements of ring, in right to the element of abelian group in the all conditions mentioned above. If R is commutative then every left R-module M can be given the structure of a right R-module by defining $ar = ra$ for $r \in R$ and $a \in M$ the difference between left R-module and right R-module is merely that of notations therefore theory of both module (left of right) can be developed in the same manner. Here, we shall develop the theory of left R-modules and omit the replaced use of adjective (left) in this way left R-module is written now, simply R-module.

Solved Examples

Example 1. *Every abelian group G is a module over the ring of integers **Z**.*

Solution. To prove that G is module over the ring of integer **Z**, we have to show that, for all

$m,n \in \mathbf{Z}$ and $a,b \in G$

(i) $m(a + b) = ma + mb$ (ii) $(m + n)a = ma + na$ (iii) $m(na) = (mn)a$

All these conditions can be proved by using the properties of elements in vector space. In fact the concept of module is generalisation of that of a vector space.

Example 2. *Every left ideal M of ring R is an R-module.*

Solution. M, being left ideal, is an additive abelian group. Also, by definition of left ideal of R, $rm \in M$, $\forall r \in R$ and $m \in M$, further

     (i) $r(m_1 + m_2) = rm_1 + rm_2$, $\forall r \in R$ and $m_1, m_2 \in M$.

        This is hold by left distributions in R.

     (ii) $(r_1 + r_2)m = r_1 m + r_2 m$ $\forall r_1, r_2 \in R$ and $m \in M$ [By right distribution in R]

     (iii) $r(sm) = (rs)m$ $\forall r, s \in R$ and $m \in M$. *This is hold by associativity.*

        Hence, M is an R-module.

Example 3. *Every ring is a R module over itself.*

        (if we take $M = R$ then this result is a special case of example 2)

Example 4. *Let R be a ring then set of R^n of all n-tuples is an R-module with an internal and external composition defined by*

$$(a_1, a_2, ..., a_n) + (b_1, b_2, ..., b_n) = (a_1 + b_1, a_2 + b_2, ..., a_n + b_n)$$

and $a(a_1, a_2, ..., a_n) = (aa_1 + aa_2, ..., aa_n)$

9.3 COSET R-MODULE

Let R be a ring and S be a left ideal of R. Let, $M = \{a + S : a \in R\}$ be the set of all cosets of S in R. Then, M is a R-module with the composition defined by

$$(a + S) + (b + S) = (a + b) + S; \text{ and } r(a + S) = (ra + S)$$

9.4 GENERAL PROPERTIES OF MODULES

Let M be a module over a ring R. Then

    (i) $r0 = 0, \forall r \in R$

   (ii) $0a = 0, \forall a \in M$

  (iii) $(-r)a = r(-a) = -(ra) \forall r \in R$ and $a \in M$

  (iv) $(-r)(-a) = ra, \forall r \in R$ and $a \in M$

   (v) $r(a-b) = (ra-rb), \forall r \in R$ and $a, b \in M$

  (vi) $(r-s)a = (ra-sa), \forall r, s \in R$ and $a \in M$

The proof of all these properties may be given in the same way as in vector space.

9.5 SUBMODULES

Definition. *A non empty subset S of a R-module M is said to be its submodule if*

               (i) S is an additive subgroup of M;

and   *(ii) $r \in R$; and $a \in S = ra \in S$*

9.5.1 Improper Submodules of M

     If M is module over a ring R then M and $\{0\}$ are improper submodules of M.

9.5.2 Proper Submodules of M

     Any submodule of M other than M and $\{0\}$ is called the proper submodule of M.

9.5.3 Irreducible R-Module

A R-module of M is said to be irreducible if its only submodule are M and {0}.

Theorem 1. (*Intersection of two submodules*): *If A and B one two submodules of module M over a ring R, then $A \cap B$ is also a submodule of M.*

Proof. We have

(i) $A \cap B$ is an additive subgroup of M [$\because$ *A and B are subgroups.*]

(ii) For $r \in R$ and $a \in A \cap B \Rightarrow ra \in A \cap B$

Now, since A and B are the submodule of M.

Therefore, $r \in R$, $a \in A \Rightarrow ra \in A$

Also, $r \in R$, $a \in B \Rightarrow ra \in B$

So $ra \in A \cap B$ [since $ra \in A$ and $ra \in B$]

Hence $A \cap B$ is a submodule of M.

Theorem 2. *Arbitrary intersection of submodules is a submodule.*

Proof. Let $\pi = \{\cap A_\lambda : A_\lambda$ is a submodule for $\lambda = 1,2,...,n\}$. Then we have

(i) $\pi = \cap A_\lambda$ is an additive subgroup of M [$\because$ each λ is a subgroup of $M.$]

(ii) For $r \in R$ and $a \in \pi$

$\Rightarrow$ $r \in R$ and $a \in A_\lambda$ for each $\lambda = 1,2,...,n$

Now since each A_λ is submodule of M

Therefore; $r \in R, a \in A_\lambda \Rightarrow ra \in A_\lambda$ for each $\lambda = 1,2,...,n$

$\Rightarrow ra \in \pi \cap A_\lambda$ for each $\lambda = 1,2,...,n$ $\Rightarrow ra \in \pi$

Hence π (intersection of n submodules) is a submodule.

9.5.4 Submodule Generated by a Set

Let M be a module over a ring R and A be a non empty subset of M. Then a submodule S of M containing A is called the submodule generated by A; if S is contained in every submodule of M containing A.

Theorem 1. *Let S be a non-empty subset of M and M be the unital R-module, then the set of all linear combinations of elements of S is a submodule of M, generated by S.*

Proof. Let $a_1, a_2, a_3, ..., a_n$ be arbitrary finite subsets of S, then take $L(S) = r_1 a_1 + r_2 a_2 + ... + r_n a_n$ where $L(S)$ denote the set of all linear combination of elements of S and $r_1, r_2, ..., r_n$ is any arbitrary finite subset of the ring R. Now, if a, b be two elements of S then

$$a = r_1 a_1 + r_2 a_2 + ... + r_n a_n; \quad b = s_1 b_1 + s_2 b_2 + ... + s_n b_n$$

where $r_i's, s_i's \in R$; $a_i's$ and $b_i's \in S$.

So, $a - b = r_1 a_1 + r_2 a_2 + ... + r_n a_n + (-s_1)b_1 + (-s_2)b_2 + ... + +(-s_n)b_n$

which is linear combination of some element of S.

So, $a - b \in L(S)$. Here $a, b \in L(S) \Rightarrow a - b \in L(S)$. Hence, $L(S)$ is an additive subgroup of M.

Again for $r \in R$, we have

$$ra = r(r_1 a_1 + r_2 a_2 + ... + r_n a_n)$$
$$= r(r_1 a_1) + r(r_2 a_2) + ... + r(r_n a_n)$$
$$= (rr_1)a_1 + (rr_2)a_2 + ... + (rr_n)a_n$$

$$\Rightarrow \quad ra \in L(S) \qquad \left[\begin{array}{l}\text{Since } rr_1, rr_2 \ldots rr_n \in R \\ \text{so, } ra \text{ is linear combination or some element or } S\end{array}\right]$$

Therefore $L(S)$ is submodule of M.

Also each $a_i \in S$ can be expressed as $a_i = 1.a_1 \in L(S)$.So, $S \subset L(S)$.

So $L(S)$ is submodule containing S.

Now we have to show $L(S) \subset T$ (any submodule of m containing S)

Let $a = r_1 a_1 + r_2 a_2 + \ldots + r_n a_n \in L(S)$, where a_i's $\in S$ so a_i's $\in T$ since $S \subset T$

Then $a \in T$ (because T is also a module). Therefore, $L(S) \subset T$

Hence, $L(S)$ is the smallest submodule of M containing S of $L(S)$ is submodule generalised by S.

9.6 LINEAR SUM OF TWO MODULES

Definition. *Let M_1 and M_2 be two submodules of an R-module then linear sum of M_1 and M_2 denoted by $M_1 + M_2$ is defined as*

$$M_1 + M_2 = [a + b : a \in M_1, b \in M_2)$$

Since M_1 and M_2 are submodule of M so $M_1 + M_2$ is a subset of M.

Theorem 1. *The linear sum of two submodules of an R-module is also a submodule of the R module.*

Proof. Let M_1 and M_2 be two submodules of a R-module then $a = M_{11} + M_{21}, b = M_{12} + M_{22}$ be the two elements of $M_1 + M_2$ where $m_{11}, m_{12} \in M_1$ and $m_{12}, m_{22} \in M_2$

Now $\quad a - b = (m_{11} + m_{21}) - (m_{12} + m_{22}) = (m_{11} - m_{12}) + (m_{21} - m_{22})$

Since M_1 is an additive subgroup so $m_{11} + m_{21} \in M_1$

Similarly $m_{21} - m_{22} \in M_2$.

Therefore $(m_{11} - m_{12}) + (m_{21} - m_{22}) \in M_1 + M_2$

Therefore $M_1 + M_2$ is additive subgroup of M.

Now for any $r \in R$

$$ra = r((m_{11} + m_{21}) = rm_{11} + rm_{21} \qquad \ldots(1)$$

Since M_1 is submodule so $r \in R$, $m_{11} \in M_1 \Rightarrow rm_{11} \in M_1$

Similarly $rm_{21} \in M_2$ thus $rm_{11} + rm_{21} \in M_1 + M_2$. Also from(1) $ra \in M_1 + M_2$

Hence, $M_1 + M_2$ is a submodule of M.

9.6.1 Direct Sum Of Submodules

A R-module M is a direct sum of two of its submodules M_1 and M_2 if each element of M can be uniquely expressed as sum of an element of M_1 and element of M_2. It is denoted as $M = M_1 \oplus M_2$.

This definition of direct sum can be extended to n submodules of M. Let $M_1, M_2 \ldots M_n$ be n submodule of M.

Then M is said to be the direct sum of $M_1, M_2 \ldots M_n$ of each element of $a \in M$ can be uniquely written as $a = a_1 + a_2 + \ldots + a_n$ where $a_1 \in M_1, a_2 \in M_2 \ldots a_n \in M_n$

Theorem 2. *The necessary and sufficient conditions for a modules to be a direct sum of its two submodules M_1 and M_2 are that*

$$(i)\ M = M_1 + M_2 \qquad (ii)\ M_1 \cap M_2 = \{0\}$$

Proof.   Let us first suppose M is the direct sum of M_1 and M_2, i.e., $M = M_1 \oplus M_2$. By definition of direct sum, every element of M can be uniquely expressed as the sum of the element of M_1 and the element of M_2. If $m \in M$ be arbitrary, then

$$m = \text{element of } M_1 + \text{element of } M_2.$$

$\Rightarrow$ $M = M_1 + M_2$

Further, we shall show that $M_1 \cap M_2 = (0)$

Let if possible $0 \neq x \in M_1 \cap M_2$. Then clearly x is a non-zero element common to both M_1 and M_2. Therefore, we can write

$$x = x + 0 \in M_1 + M_2, \quad x \in M_1 \text{ and } 0 \in M_2$$

and $\qquad x = 0 + x \in M_1 + M_2, \quad 0 \in M_1 \text{ and } x \in M_2$

$\Rightarrow$ $x \in M$ may be expressed as the sum of element of M_1 and M_2 in two ways which contradicts the fact that each element of M is uniquely expressed as sum of an element of M_1 and an element of M_2.

Therefore, 0 is the only element common to both M_1 and M_2. Hence, $M_1 \cap M_2 = \{0\}$.

Conversely, suppose that (*i*) and (*ii*) conditions are satisfied. We have to prove that $M = M_1 \oplus M_2$. For this we shall prove that each element of M is uniquely expressed as the sum of an element of M_1 and an element of M_2. Now, from (*i*) we have $M = M_1 + M_2$

$\Rightarrow$ Every element of M is expressed as the sum of an element of M_1 and an element of M_2.

Let $x \in M$ be arbitrary then $x = x_1 + y_1, x_1 \in M, y_1 \in M_2$

We shall show that the representation $x = x_1 + y_1$ is unique.

Let if possible $\qquad\qquad x = x_1 + y_1$ with $x_1 \in M_1, y_1 \in M_2$

and $\qquad\qquad\qquad x = x_2 + y_2$ with $x_2 \in M_1, y_2 \in M_2$ be two representations.

Then, clearly we have $\quad x = x_1 + y_1 = x_2 + y_2$

$\Rightarrow$ $x_1 - x_2 = y_2 - y_1 \in M_1 \cap M_2$

$$[\because x_1 - x_2 \in M_1 \text{ and } y_2 - y_1 \in M_2 \text{ and } x_1 - x_2 = y_2 - y_1]$$

But, we have assumed that $M_1 + M_2 = \{0\}$

Therefore $\qquad\qquad x_1 - x_2 = y_2 - y_1 = 0$

$\Rightarrow$ $x_1 = x_2$ and $y_1 = y_2$

$\Rightarrow$ The presentation $x = x_1 + y_1$ is unique.

$\Rightarrow$ $M = M_1 \oplus M_2$

Hence, M is the direct sum of M_1 and M_2.

9.7 HOMOMORPHISM OF MODULES (LINEAR TRANSFORMATIONS)

Definition. *A mapping f from a R-module M to other R-module N; is said to be module homomorphism (R-homomorphism)if*

(*i*) $f(m_1 + m_2) = f(m_1) + f(m_2), \forall m_1 m_2 \in M$

and   (*ii*) $f(rm) = r f(m), \forall r \in R$ and $m \in M$

The *R*-module *N* is said to be homomorphic image of *M* under *f*.

9.7.1 Isomorphism of Modules

A mapping f from a R-module M to other R-module N is said to isomorphism (R-isomorphism) if

(i) f is one-one and onto

(ii) $f(m_1 + m_2) = f(m_1) + f(m_2)$, $\forall m_1, m_2 \in M$

(iii) $f(rm) = rf(m)$, $\forall r \in R$ and $m \in M$

REMARKS

☞ From the definitions of homomorphism and isomorphism it is clear that **i**somorphism is a particular case of homomorphism.

☞ If f be a homomorphism of a R-module M into this R-module N. Then,

(i) $f(0) = 0$

(ii) $f(-m) = -f(m)$, $\forall\, m \in M$

(iii) $f(m_1 - m_2) = f(m_1) - f(m_2)$ $\forall m_1, m_2 \in M$

9.7.2 Kernel of a homomorphism

Let f be a homomorphism of an R-module M into an R-module N then Kernel of f denoted by $K(f)$ is set of all those elements of M which are mapped to 0 (identity element of N), i.e.,

$K(f) = \{m \in M : f(m) = 0\}$ where 0 is identity element of additive group N.}

Theorem 1.   *The Kernel of a homomorphism is a submodule.*

Proof.   Let f be a homomorphism from R-module M into an R-module N then kernel of f is given by $K(f) = \{m \in M : f(m) = 0\}$. Since $f(0) = 0$.

Therefore at least $0 \in K(f)$. So $K(f)$ is non empty subset of M

Now for any $m_1, m_2 \in K(f)$.

Then $f(m_1) = 0$ and $f(m_2) = 0$ and $f(m_1 - m_2) = f(m_1) - f(m_2) = 0 - 0 = 0$ therefore $m_1 - m_2 \in K(f)$. Thus $K(f)$ is an additive subgroup of M.

Again for second condition $r \in R$ and $m \in K(f)$ such that $f(m) = 0$ We have $f(rm) = rf(m) = rf(m) = r0 = 0$ therefore $rm \in K(f)$. Hence $K(f)$ is submodule of M.

Theorem 2.   *If f be a module homomorphism. Then f is an isomorphism iff $K(f) = 0$, i.e., $Kr(f) = 0$.*

Proof.   Let f be a homomorphism of an R-module M onto a R-module N. First we shall prove that if $K(f) = 0$ then f is an isomorphism. For this take any two elements $m_1, m_2 \in M$ such that

$$f(m_1) = f(m_2) \qquad \rightarrow f(m_1) - f(m_2) = 0$$
$$\Rightarrow \quad f(m_1 - m_2) = 0 \qquad \Rightarrow m_1 - m_2 \in K(f)$$
$$\Rightarrow m_1 - m_2 = 0 \qquad [\ker (f) = 0 \text{ gives}]$$
$$\Rightarrow \qquad m_1 = m_2$$

Therefore f is one-one. Hence f is isomorphism.

Conversely. Suppose f be an isomorphism. Then f is one-one.

Now for any $m \in \ker(f)$, we have

$$\ker(f) = f(m) = 0 = f(0) \Rightarrow m = 0.$$

Hence, $K(f)$ or $\ker(f) = 0$.

Theorem 3. *The range of module-homomorphism is submodule.*

Proof. Let f be a homomorphism of an R-module M into a R-module M and let $I(f)$ denote the range of T. Then $I(f) = \{f(m): m \in M\}$. To prove that $I(f)$ is an additive subgroup of N. Let $f(m_1)$ and $f(m_2)$ be the elements of $I(f)$ then $f(m_1) - f(m_2) = f(m_1 - m_2) \in I(f)$ therefore $I(f)$ is an additive subgroup of N.

Again for any $r \in R$ and $f(m) \in I(f)$ we have
$$rf(m) = f(rm) \in I(f) \text{ since } rm \in M.$$
Hence, $I(f)$ is a submodule of N.

9.8 QUOTIENT MODULE

Let M be an R-module and A be its any submodule. Then by definition A is a subgroup of the additive group M for $m \in M$, $A + m$ is a coset of A in M. Such set of all cosets of A in M is denoted by M/A and the structure M/A is called quotient modules of M relative to the submodule A such that $M/A = \{A+m : m \in M)$, the set of all coset of A in M with respect to the composition defined by
$$(A + m_1) + (A + m_2) = A + (m_1 + m_2), \forall r \in R \text{ and } m \in M$$

REMARK

☞ Let M be the R-module and A be its submodule then the mapping $f : m \to M/A :$ $f(m) = W + m, \forall m \in M$ is a homomorphism of M onto (M/W) with $\ker(f) = W$.

Theorem 4. *(Fundamental theorem on homomorphism of modules).*

If f be is a homomorphism of a R-module M onto a R-module N with $\ker(f) = A$ then N is isomorphic to M/A, i.e., $N \cong M/A$.

Proof. Let f be a homomorphism of an R-module onto R-module N then $\ker(f)$ is submodule of M. But $\ker(f) = A$ (Given) so A is submodule of M and M/A is a quotient module which is given by
$$M/A = \{m+A : m \in M\}$$
Also, for all $m_1 \in A$, we have $f_1(m_1) = 0$ and range of $T(N) = N$ (Since f is onto).

Let us define a mapping $T : N \to M/A$ such that
$$T(f(m)) = (m + A), \forall \in N, f(m) \in N \text{ when } m \in M.$$
Obviously T is well defined mapping. To show $N \cong M/A$. First, take $m_1, m_2 \in M$ so that $f(m_1), f(m_2) \in N$ such that
$$T(f(m_1)) = m_1 + A, T(f(f(m_2))) = m_2 + A$$
$$T(f(m_1)) = T(f(m_2)) \Rightarrow m_1 + A = m_2 + A$$
$$\Rightarrow m_1 - m_2 \in A \Rightarrow f(m_1 - m_2) = 0$$
$$\Rightarrow f(m_1) + f(m_2) = 0 \Rightarrow f(m_1) = f(m_2)$$
Hence $T(f(m_1)) = T(f(m_2)) = T(f(m_1 + m_2)) = (m_1 + m_2) + A$
$$= (m_1 + A) + (m_2 + A) = T(f(m_1)) + T(f(m_2))$$
Let $r \in R$ then
$$T(rf(m)) = T(f(rm)) \qquad \text{[By definition of } f]$$
$$= rm + A \qquad \text{[By definition of } f]$$

$$= r(m + A) = rT(f(m))$$

Thus T is also linear. Hence T is isomorphism.

9.9 CYCLIC MODULE

A R-module M is said to be a cyclic module generated by an element $a \in M$ if each $m \in M$ is expressible as $m = ra$ for some $r \in R$.

The element a is called generator of M and we write it as $M = (a)$

REMARK

☞ Let M be a unital R-module and $W = \{ra : r \in R$ and $a \in M)$. Then W is a cyclic submodule of m, generated by a.

Theorem 1. *An irreducible R-module is cyclic.*

Proof.   Let M be a unital, irreducible R-module. Then only submodule of M are $\{0\}$ and M itself.

Case (i) If $M = \{0\}$, it is clearly cyclic.

Case (ii) If $M \neq \{0\}$ Take $a \in M$ such that $a \neq 0$.

Let $W = \{ra : r \in R\}$, Also $a = 1. a = 1(R$ is unital$) = a \in W$

So $W \neq \{0\}$ this W is non zero submodule of m.

Hence $W = m$ [$\because$ only non zero submodule of M is m.]

Also W is generated by $a \in M$ so w is cyclic. Hence M is cyclic.

9.9.1 Finitely Generated Module

A R-module M is called finitely generated if there exist finite number of elements $m_1, m_2, ...,m_n \in M$ such that every $m \in M$ is expressible as

$$m = \sum_{i=1}^{n} r_i m_i, \text{ for some } r_i \in R.$$

Theorem 2.   *(Fundamental theorem of finitely generated modules over Euclidean Rings).*

If R is an Euclidean Ring, then any finitely generated R-module M is the direct sum of a finite number of cyclic modules.

Proof.   Let M be a R-module which is finitely generated and R be a Euclidean rings whose generating sets have the minimum element as minimal generating sets. The number of elements in such minimal generating is said to be the rank of M. We shall prove the theorem by method of mathematical induction on rank of M.

If $n = 1$, then M is generated by single element which is generator of M and M is cyclic. Hence the result is true. Now let the result is true for all module of rank $n = k - 1$.

Let M be R-module of rank k, we shall prove that the result is true for this M. For any given minimal generating set $\{a_1, a_2,...,a_n\}$ of M and relation of the form $r_1 a_1 + r_2 a_2 + ... + r_k a_k$ for $r_i \in R$ implies that $r_1 a_1 = r_2 a_2 = ... = r_k a_k = 0$ then M clearly is the direct sum of $M_1, M_2,...M_k$, where each M_i is cyclic submodule of M generated by a_i. Thus

$$M = M_1 \oplus M_2 \oplus ... \oplus M_k.$$

Hence by mathematical induction the theorem is true for every n.

So for any given minimal generated set $\{b_1, b_2,...,b_k\}$ of M, there must be elements $r_1, ...r_k \in R$ such that

$$r_1 b_1 + r_2 b_2 + ... + r_k b_k = 0$$

in which not all of $r_1 b_1, r_2 b_2, ... , r_k b_k$ are equal to zero.

Among all such possible relations, for all minimal generating sets, there is an element $s_i \in R$ whose d-value $d(s_i)$ is minimal. Let the set $\{C_1, C_2, ...,C_k\}$ be the generating set such that

$$\sum_{i=1}^{n} s_i C_i = 0 \qquad \qquad ... (1)$$

Now R is an Euclidean ring so for $s_1, r_1 \in R$, $\exists\ m, t \in R$

such that $\qquad \qquad r_1 = m s_1 + t \qquad \qquad ... (2)$

where $t = 0$ or $\quad d(t) < d(s_1)$.

Let $\sum_{i=1}^{k} r_i c_i = 0$ for some $r_1 \in R$ $\qquad \qquad ...(3)$

Multiply (1) by m and subtracting from (3) we get

$$\sum_{i=1}^{k} (r_i - m s_i) c_i = 0$$

$$\Rightarrow \quad (r_1 - m s_1) c_1 + \sum_{i=1}^{k} (r_i - m s_i) c_i = 0$$

$$\Rightarrow \qquad \qquad t c_1 + \sum_{i=1}^{k} (r_i - m s_i) c_i = 0 \qquad \text{[from (2)]} \qquad ...(4)$$

If $t \neq 0$, then $d(t) < d(S_1)$ which contrary to the fact $d(s_1)$ is minimal. Hence $t = 0$ which implies $r_1 = m s_1$ by (2) . This means s_1 is a divisor of r_1 . Now we shall show that s_1 is divisor of s_i for $i = 2, 3, ... k$.

For $s_1, s_2 \in R \exists\ m_2$, $t_2 \in R$ such that

$$s_2 = m_2 s_1 + t_2$$

where either $t_2 = 0$ or $d(t_2) < d(s_1)$.

Now if $c_1^1 = c_1 + m_2 c_2$, then $\{c_1^1, c_2, c_3 ....ck\}$ is also a generating set of M

So $\quad s_1 c_1^1 + t_2 c_2 + s_3 c_3 +...+ s_k c_k = s_1(c_1 + m_2 c_2) + t_2 c_2 + s_3 c_3 +...+ s_k c_k$

$$= s_1 c_1 + (s_1 m_2 + t_2) c_2 + s_3 c_3 +...+ s_k c_k$$

$$= s_1 c_1 + (s_2 c_2 +...+ s_k c_k = 0 \qquad \qquad ... (5)$$

If $t_2 \neq 0$ then $d\ (t_2) < d(s_1)$ and in this case solution (5) contradict for our choice of s_1 so $t_2 = 0$ then $s_2 = m s_1$, which means s_1 is divisor of s_2.

If we proceed in this manner, we can show that s_1 is a divisor of s_i for $i = 3, 4,...., k$.

Therefore $\qquad s_2 = m_2 s_1, s_3 = m_3 s_1 .... = m_k s_1$

Again let

$$c^*_1 = c_1 + m_2 c_2 + ... + m_k c_k \qquad \qquad ... (A)$$

Let M_1 (generted by $c_1, c_2, ...c_k$) be cyclic module of M generated by c_1 and M_2 be generated by $\{c_1, c_2, ..., c_k\}$

$\therefore$ $\qquad\qquad M = M_1 + M_2$ ...(6)

Now for any $\alpha \in M_1 \cap M_2 \Rightarrow \alpha \in M_1, \alpha \in M_2$

$\Rightarrow$ $\qquad\qquad \alpha = r_1, c^*_1, \alpha = r_2 c_2 + r_3 c_3 + ... + r_k c_k$

Since M is generated by a^*_1, therefore

$$r_1 c^*_1 = r_2 c_2 + r_3 c_3 + ... + r_k c_k$$

$\Rightarrow$ $\quad r_1(c_1 + m_2 c_2 + ... + m_k c_k) - r_2 c_2 - r_3 c_3 ... - r_k c_k = 0$

$\Rightarrow$ $\qquad r_1 c_1 + (r_1 m_2 - r_2) c_2 + ... + (r_1 m_k - r_k) c_k = 0$ ...(7)

In this relation coefficent of c_i is r_i. Therefore s_1 is divisor of r_1 and hence $r_1 = p s_1$, where $p \in R$.

Thus $\qquad\qquad \alpha = r_1 c^*_1 = p s_1 c^*_1 = p (s_1 c^*_1)$

$\therefore$ $\qquad\qquad \alpha = p[s_1(c_1 + m_2 + c_2 + ... m_k c_k)]$ $\qquad$ [By (A)]

$\qquad\qquad\qquad = p[s_1 c_1 + s m_2 c_2 + ... s_1 m_k c_k]$

$\qquad\qquad\qquad = p[s_1 c_1 + s_2 c_2 + ... s_k c_k]$ $\qquad$ [$\because s_i = m_i c_i + 1 = 2, 3, ...k$]

$\qquad\qquad\qquad = p . 0 = 0$ $\qquad$ [By (3)]

Thus $\qquad\qquad \alpha \in M_1 \cap M_2$ and $\alpha = 0$

$\qquad\qquad M_1 \cap M_2 = \{0\}$

Finaly, we have $M = M_1 \oplus M_2$ and $M_1 \cap M_2 = \{0\}$

Therefore, $\qquad\qquad M = M_1 \oplus M_2$

Now $\qquad\qquad M_1 = \{a_1^*\} \Rightarrow$ rank of $M_1 = 1 \Rightarrow M_1$ is cyclic.

$\qquad\qquad M_2 = \{a_2, ...a_k\} \Rightarrow$ rank of M_2 is at most $k - 1$.

Hence by induction, M_2 is direct sum of cyclic modules.

Therefore, M is direct sum of cyclic modules.

EXERCISE 9.1

1. Show that the polynomial ring $R (x)$ over ring R is a R–module.

2. If M and N are R – modules and if $f : M \to N$ is homomorphism then

   (i) $f(0) = 0$

   (ii) $f(-m) - -f(m), \forall, m \in M$

3. Show that a subring S of a ring R is a module over the whole ring only if S is an ideal of R.

4. If M is the set of $m \times n$ matrices over a ring R. Then show that M is a module over R.

5. If M, N, Q are R-module of an R-module M such that $M \supset N$, Show that

   $M \cap (N + Q) = N + (M \cap Q)$

6. If R is a non-zero commutative ring with unity. Show that $R (x)$ is not a finitely generated R-module.

7. Show that, the set of rational numbers $\mathbf{Q}$ is not finitely generated $\mathbf{Z}$-module.

8. If K is a submodule of a finitely generated R-module M, then show that M/K is also a finitely generated R-module.

9. Let $f : M \to N$ be R-module homomorphism and let the kernel contains a submodule M_1 of M. Show that the mapping $\phi : M / M_1 \to N$ defined by $\phi (x + M_1) = f(x)$ is an R-module homomorphism.

10. If R is a ring with unity. Show that a R-module M is cyclic if and only if $M \cong R / I$ for some left ideal I or R.

9.10 SIMPLE AND SEMI-SIMPLE MODULES

Notation

(1) In this section, all rings will be rings with unity and all R-modules will be unital.

(2) R-modules means left "R-module."

Definition 1. *An R-module M is said to be simple if it has no submodules other than itself and the zero submodule*

For example :

(1) An abelian group $G \neq 0$ is a simple Z-module if and only if it is cyclic of prime order.

(2) A vector space over a division ring R is a simple R-module if and only if it has dimension 1.

(3) If M is the left R-module formed by the addtive group of a ring R, then a submodule $N \neq 0$ of M is simple if and only if it is a minimal left ideal of M.

Definition 2. *A homomorphism of a ring R into the ring End_R (M) of all endomorphism of an abelian group M is a called representation of R.*

Definition 3. *A left R-module M is said to be faithful if and only if the representaion of R associated with M is injective.*

Definition 4. *An R-module M is said to be semi-simple if it can be expressed as a direct sum of simple submodules.*

REMARK

☞ If M is a left R-module, then a mapping $f : R \to End_R$ (M), $f \to I_r$ is the representation of R associated with M, where for each $r \in R$, the mapping $I_r(x) = rx$ is an endomorphism of M.

Theorem 1. (*Schur's Theorem*). *If M is a simple R-module and N is any R-module then*

      (i) *every non-zero homomorphism $f : M \to N$ is injective (Monomorphism)*

      (ii) *every non-zero homomorphism $f : M \to N$ is surjective (Epimorphism)*

      (iii) *End_R is a division ring, where End_R $(M) = Hom_R$ (M, M)*

Proof.   (i) Since, f is a homomorphism, therefore, ker(f) is a submodule of M. Now, since M is simple, thus either Ker $(f) = [0]$ or Ker $(f) = M$. But $f \neq 0$, therefore

$$\text{Ker } (f) \neq M$$

$$\Rightarrow \quad \text{Ker } (f) = [0]$$

Henec, f is surjective.

(ii) Since M is a non-zero R-module homomorphism from M to N. But Im (f) is also a non-zero submodule of M and M is simple. Thus

$$\text{Im } (f) = M$$

Hence, f is surjective.

(iii) Suppose, f is a nor-zero R-module homomorphism from M to M. Now, since, M is simple, then using (i) and (ii), f is an automorphism.

$\Rightarrow$ f is unit in the ring End_R (M). Hence, End_R (M) is a division ring.

Theorem 2. *If R is a ring and I is a minimal left ideal of R. Also, if M is a faithful simple left R-module, then M is isomorphic to the left R-module I.*

Proof.   Let R be a ring and I be a minimal left ideal of R.

Let $a_0 \in I$, $a_0 \neq 0$

Since M is faithful, thus there is some $x \in R$ such that $a_0 x \neq 0$ and the mapping $f_x : I \to R$ defined by $f_x(a) = ax$, $a \in I$ is a non-zero homomorphism. By Schur's lemma, f_x is an epimorphism because the R-module M is simple.

Hence, f_x is an isomorphism.

Theorem 3. *Let M be an R-module and $\{ M_i : i \in \Delta \}$ be a family of simple submodules of M such that $M = \sum_{i \in \Delta} M_i$, then for each submodule N of M, there is a Δ' of Δ such that*

$$M = \bigoplus_{i \in \Delta'} M_i \oplus N \ .$$

Proof. In case of $M = N$, $\Delta' = \phi$, thus therom is trivially true. Suppose $M \neq N$, then for some $k \in \Delta$, $M_K \not\subset N$. Now, since M_k is simple $M_k \cap N = [0]$ and $N + M_k$ is a direct sum Therefore, the set of all families $F = \{M_i : i \in \Delta'' : \Delta'' \subset \Delta\}$ for which the sum $\sum_{i \in \Delta''} M_i + N$ is direct, is non empty. By Zorn's lemma, F contains a maximal subfamily

$\{M_i : i \in \Delta'\}$. Now, suppose $\bigoplus_{i \in \Delta'} M_i \oplus N . \neq M$, then for some $j \in \Delta$, $M_j \not\subset \bigoplus_{i \in \Delta'} M_i \oplus N$.

Because M_j is simple, therefore $M_j \cap [\bigoplus_{i \in \Delta'} M_i \oplus N] = 0$ and $M_j \cap [\bigoplus_{i \in \Delta'} M_i \oplus N]$ is a

direct sum. Then, we have

$$[M_i : i \in \Delta' \cup [J]] \in F$$

which contridicts the maximality of $\{M_i : i \in \Delta'\}$

Hence, we have

$$M = \bigoplus_{i \in \Delta'} M_i \oplus N$$

Theorem 4. *Let M be an R-module. Then following conditions are equivalent :*

    (i) M is a sum of simple submodules.

    (ii) M is a semi-simple module.

    (iii) Every submodule of M is a direct summand of M.

Proof. (i) $\Rightarrow$ **(ii)** Let us suppose $\{M_i : i \in \Delta\}$ be a family of simple submodule of M and if $M = \sum_{i \in \Delta} M_i$, then using previous theorem 3, with $N = 0$, there is a subset Δ'

of Δ such that

$$M = \bigoplus_{i \in \Delta'} M_i$$

Hence, M is a sum of simple submodule.

(ii) $\Rightarrow$ **(iii)** Let N be a submodule of M and let

$$M = \bigoplus_{i \in \Delta'} M_i$$

$$M = \bigoplus_{i \in \Delta} M_i \qquad\qquad [\because M \text{ is semi simple}]$$

Then again by theorem-3, there is a subset Δ' of Δ such that

$$M = \bigoplus_{i \in \Delta'} M_i \oplus N.$$

(iii) ⇒ (i) Let us suppose that every submodule of M is a direct summand of M. Firstly, we shall prove that M has simple submodule.

Let $a \in M$, $a \ne 0$ and let $N = Ra$. Then clearly, N is a finitely generated R-module, therefore N has a maximal submodule L. Thus, N/P is a simple R-module. By our hypothesis, L has a complementary submodule L' such that

$$M = L \oplus L'$$

Thus $\qquad N = L \oplus (N \cup L')$.

⇒ $N \cup L'$ is a submodule of M which is isomorphic to N/L .

⇒ $N \cup L'$ is simple.

Now, since every non-zero submodule of M contains the cyclic submodule generated by each of its non-zero elements. Thus, every non-zero submodule contains a simple submodule.

Further, let P be the sum of the simple submodules of M. Then, $M = P \oplus P'$ for some submodule P . If P' is non-zero, then it contains a simple submodule which is not possible, since all the simple submodules are contained in P.

Therefore $\qquad M = P$.

Hence, M is a sum of simple submodules.

Theorem 5. *Every submodule and every homomorphic image of a semi-simple module is semi-simple.*

Proof. Let M be a semi-simple R-module and let N be a submodule of M and L be a submodule of N. Then, there exists a submodule L' of M such that $M = L \oplus L'$. In this case, $L \oplus (L' \cap N) = N$.

Then, by previous theorem, every submodule of N is direct summand of N.

Hence, N is semi-simple.

Finally, since every homomorphic image of R-module M is isomorphic to M / N for some submodule N of M. But $M / N \cong N'$.

Hence, M is semi-simple.

Theorem 6. *Let M be a semi simple R-module and it is equal to the direct sum of a family $[M_i : i \in \Delta]$ of simple submodules, then every simple submodule of M is isomorphic to one of the M_i.*

Proof. Let M be a semi-simple R-module and N is a simple submodule of M, then N is a direct summand of M. Therefore, there is an epimorphism

$$\phi : M \to N \text{ and } N = \sum_{i \in \Delta} \phi(M_i).$$

Now, since $N = \phi (M_k)$ for $k \in \Delta$. But M_k is simple. Hence, by Schur's lemma, ϕ is an isomorphism.

9.11 FREE MODULES

Definition 1. *Let M. be a module aver a ring R with unity and let S be a subset of M, then S is said to be basis for M. if*

(i) *S generates M.*

(ii) *S is linearly independent.*

Definition 2. *An R-module M is said to be free module if there exists a subset S of M such that S generates M and S is linearly independent set.*

For example. If $M = (0)$, then the set ϕ is its basis. Therefore, it is free module.

REMARK

☞ If S is a basis for M, then, in particular $M \neq [0]$ and if $R \neq [0]$, then every element of M can be expressed uniquely as a linear combination of the elements of S.

Theorem 1. *Let R be a ring and M be a module over R. Also let S be a non-empty set and let $\{x_i : i \in S\}$ be a basis of M. If N is an R-module and $\{y_i : i \in S)$ be a family of elements of N, then, there exists a unique homomorphisrn $f : M \to N$ such that $f(x_i) = y_i, \forall\, i \in S$*

Proof. It is given that $\{x_i : i \in S\}$ is a basis of M, then every element of M can be expressed uniquely as linear combination of elements of the set $\{x_i : i \in S\}$. Let $x \in M$. Then, there exists a unique family $\{r_i : i \in S\}$ of elements of R such that

$$x = r_1 x_1 + r_2 x_2 + \dots + r_k x_k \dots = \sum_{i \in S} r_i x_i \qquad \dots(1)$$

Define a mapping $f : M \to N$ such that

$$f(x) = \sum_{i \in S} r_i y_i, \forall\, x \in M, \forall y_i \in N \qquad \dots(2)$$

Clearly, f is a homomorphism and it is unique, because

$$f(x) = \sum_{i \in S} r_i f(x_i) \qquad \dots(3)$$

From (2) and (3), we conclude that $f(x_i) = y_i$, $\forall\, i \in S$.

Theorem 2. *Let M be a free R-module with a basis $[e_i : i \in N]$, where e_i is n-tuples in which the ith entry is 1 and rest are zero. Then $M \cong R^n$.*

Proof. Let us define a mapping $\phi : M \to R^n$ such that

$$\phi(x) = \sum_{i=1}^{n} r_i\, y_i \text{ where } x \in M \text{ and } y_i \in R^n.$$

Thus,

$$x = \sum_{i=1}^{n} r_i\, e_i$$

Let x, y be any two elements of M and suppose that

$$x = y \Rightarrow \sum_{i=1}^{n} r_i\, e_i = \sum_{i=1}^{n} s_i\, e_i$$

Now, since $\{e_1, e_2, \dots, e_n)$ is linearly independent.

$\Rightarrow \qquad\qquad r_i - s_i = 0$

$\Rightarrow \qquad\qquad r_i = s_i$

$\Rightarrow \qquad\qquad \displaystyle\sum_{i=1}^{n} r_i y_i = \sum_{i=1}^{n} s_i y_i$

$\Rightarrow \qquad\qquad \phi(x) = \phi(y)$

Thus, ϕ is well defined.

Again, if $\qquad\qquad x = \displaystyle\sum_{i=1}^{n} r_i e_i, y = \sum_{i=1}^{n} s_i e_i$ and $r \in R$. Then

$$\phi\ (x+y)\ =\ \phi\left(\sum_{i=1}^{n} r_i e_i + \sum_{i=1}^{n} s_i e_i\right) = \phi\left(\sum_{i=1}^{n} (r_i + s_i) e_i\right)$$

$$= \sum_{i=1}^{n} (r_i + s_i)y_i = \sum_{i=1}^{n} r_i y_i + \sum_{i=1}^{n} s_i y_i$$

$$= \phi\ (x) + \phi\ (y)$$

Also $$\phi\ (rx)\ =\ \phi\left(r\sum_{i=1}^{n} r_i e_i\right) = \phi\left(r\sum_{i=1}^{n} r_i e_i\right)$$

Therefore, ϕ is a homomorphism.

Finally, if $\phi\ (x) = 0$, then $\sum\limits_{i=1}^{n} r_i \cdot y_i\ = 0$

$\Rightarrow \quad r_1 = 0 = r_2 = \ ......= r_n \Rightarrow x = 0$

$\Rightarrow \quad \phi$ is one-one

Also, ϕ is onto.

Hence, ϕ is an isomorphism. Thus, $M \cong R^n$.

Theorem 3. *Every finitely generated module is a homomorphic image of a finitely generated free module.*

Proof. Let M be an R-module generated by the set $[x_1, x_2, ..., x_n]$, *i.e.*, $M = \{x_1, x_2, ... , x_n\}$.

Now, let $e_i = (0, 0, ..., 1, 0......0\}$, then the set $\{e_1, e_2, ...e_n\}$ is linearly independent over R and generates a free module R^n.

Now, define a mapping $\phi : R^n \to M$ given by

$$\phi(x) = \sum_{i=1}^{n} r_i\ x_i,\ x \in R^n \text{ so that } x = \sum_{i=1}^{n} r_i e_i$$

Now, since every element of R^n has a unique representation as $x = \sum\limits_{i=1}^{n} r_i e_i$. Thus, ϕ

is well defined.

Let $\quad x\ =\ \sum\limits_{i=1}^{n} r_i e_i$ and $y\ =\ \sum\limits_{i=1}^{n} s_i e_i$ be any two elements of R^n. Then, we have

$$\phi\ (x+y)=\phi\left(\sum_{i=1}^{n} (r_i + s_i)e_i\right) = \sum_{i=1}^{n} (r_i + s_i)x_i$$

$$= \sum_{i=1}^{n} r_i\ x_i + \sum_{i=1}^{n} s_i\ x_i = \phi(x) + \phi(y)$$

Also $\quad \phi\ (rx) = \phi\ = \left(\sum\limits_{i=1}^{n} rr_i e_i\right) = \sum\limits_{i=1}^{n} r_i r x_i = r\sum\limits_{i=1}^{n} r_i x_i = r\phi(x)$

Hence, ϕ is a homomorphism of R^n into M.

Theorem 4. *Let R be a ring and M be a free module with basis B. If N is any R-module and F : B → N is any mapping, then there exists a unique R-module homomorphism $\phi : M \to N$ such that $\phi_{1B} = f$.*

Proof. Let $B = \{x_i : i \in \Lambda\}$ be a basis of M. For any $x \in M$,

we have $x = \sum_{i \in \Lambda} r_i x_i, r_i \in R$ [By definition of basis]

Define a mapping $\phi : M \to N$ be $\phi\,(x) = \sum_{i \in \Delta} r_i f(x_i)$

If $x, y \in M$, then for $r_i \in R$ and $s_i \in R$, we have

$$x = \phi \sum_{i \in \Delta} r_i x_i, \quad y = \sum_{l \in \Delta} s_i x_i$$

Consider, $\phi\,(x+y) = \phi \left(\sum_{i \in \Delta} r_i x_i + \sum_{i \in \Delta} s_i x_i \right)$

$$= \phi \left(\sum_{i \in \Delta} (r_i + s_i) x_1 \right) = \sum_{i \in \Delta} (r_i + s_i) f(x_i)$$

$$= \sum_{i \in \Delta} r_i f(x_i) + \sum_{i \in \Delta} s_i f(x_i) = \phi(x) + \phi(y)$$

Now, for $r \in R, x \in M$,

$$\phi(rx) = \sum_{i \in \Delta} r_i f(x_i) = r \sum_{i \in \Delta} r_i f(x_i) = r\phi(x)$$

Hence, ϕ is an R-module homomorphism and $\phi_{1B} = f$.

Theorem 5. *Let M be a finitely generated free module over a commutative ring R. Then, all basis of M are finite..*

Proof. Since M is finitely generated free module, let $M = (x_1, x_2.., x_M)$ and $\{e_i : i \in S\}$ be a basis for M.

Then, for any $x_j \in M$, we have

$$x_j = \sum_i a_{ij} \text{ where } a_{ij} \in R \qquad ...(1)$$

Now, since finite number of a_{ij} are zero, therefore the set of those e_i's which occur in all expression of the type (1) for $j = 1, 2, ..., n$ is finite.

Theorem 6. *If R is a commutative ring and M is a finitely generated free module over R, then all basis of M have the same number of elements.*

Proof. It is known that a free module with a basis $[e_1, e_2, ...-, e_n]$ is isomorphic to R^n. So, let R^m and R^n be any two basis of M. We have to prove that if $R^m \cong R^n$, then $m = n$.

Let us suppose that $m < n$. Define a homomorphism $\phi : R^m \to R^n$ such that $\phi = \phi^{-1}$

Suppose $[e_1, e_2, ...e_m]$ and $[f_1, f_2, ...f_n]$ be the ordered basis of R^m and R^n respectively. Now, since ϕ is an isomorphism from R^m to R^n and g is an isomorphism from R^n to R^m. Thus, we have

$$\left.\begin{array}{l}\phi(e_1) = a_{11}f_1 + a_{21}f_2 + \ldots\ldots + a_{n1}f_n\\ \phi(e_2) = a_{12}f_1 + a_{22}f_2 + \ldots\ldots + a_{n2}f_n\\ \cdots\quad\cdots\quad\cdots\quad\cdots\\ \phi(e_m) = a_{1m}f_1 + a_{2m}f_2 + \ldots\ldots + a_{nm}f_n\end{array}\right] \quad\ldots\ldots(1)$$

and

$$\left.\begin{array}{l}g(f_1) = b_{11}e_1 + b_{21}e_2 + \ldots\ldots + b_{m1}e_m\\ g(f_2) = b_{12}e_1 + b_{22}e_2 + \ldots\ldots + b_{m2}e_m\\ \cdots\quad\cdots\quad\cdots\quad\cdots\\ g(f_n) = b_{1n}e_1 + b_{2n}e_2 + \ldots\ldots + b_{mn}e_m\end{array}\right] \quad\ldots\ldots(2)$$

Let $A = [a_{ij}]_{n \times m}$ and $B = [b_{kj}]_{m \times n}$ be two matrices.
Then, we, can write

$$g(\phi(e_i)) = g(a_{11}f_1 + a_{21}f_2 + \ldots + a_{n1}f_n)$$

$$= a_{11}g(f_1) + a_{21}g(f_2) + \ldots + a_{n1}g(f_n)$$

Similarly

$$\left.\begin{array}{l}g(\phi(e_1)) = a_{11}b_{11}e_1 + a_{21}b_{21}e_2 + \ldots + a_{n1}b_{m1}e_m\\ g(\phi(e_2)) = a_{12}b_{12}e_1 + a_{22}b_{22}e_2 + \ldots + a_{n2}b_{m1}e_m\\ \cdots\ \cdots\ \cdots\ \cdots\ \cdots\ \cdots\ \cdots\ \cdots\\ g(\phi(e_m)) = a_{1m}b_{1m}e_1 + a_{2m}b_{2m}e_2 + \ldots + a_{nm}b_{mn}e_m\end{array}\right] \quad\ldots(3)$$

Now, equation (3) can be rewritteen as

$$g(\phi(e_i)) = \sum_{k=1}^{m}\sum_{j=1}^{n} b_{kj}a_{ji}a_{ji}e_k \text{ for } 1 \le i \le m$$

Also, since e_i's are linearly independent and $g = \phi^{-1}$, then

$$\sum_{k=1}^{m}\sum_{j=1}^{n} b_{kj}a_{ji} = \delta_{ki} = \begin{cases} 1 & ; & k = i\\ 0 & ; & k \ne i\end{cases} \quad\ldots(4)$$

which gives

$$\begin{bmatrix} b_{11} & b_{12} & \cdots & b_{1n}\\ b_{21} & b_{22} & \cdots & b_{2n}\\ \vdots & \vdots & \vdots & \vdots\\ b_{m1} & b_{m2} & \cdots & b_{mn}\end{bmatrix}\begin{bmatrix} a_{11} & a_{12} & \cdots & a_{1n}\\ a_{21} & a_{22} & \cdots & a_{2n}\\ \vdots & \vdots & \vdots & \vdots\\ a_{m1} & a_{m2} & \cdots & a_{mn}\end{bmatrix} = \begin{bmatrix} 1 & 0 & \cdots & 0\\ 0 & 1 & \cdots & 0\\ \vdots & \vdots & \vdots & \vdots\\ 0 & 0 & \cdots & 1\end{bmatrix}$$

$\Rightarrow \quad BA = I_m$, where I_m is the unit matrix of order $m \times n$.
Similarly, we may get
$$AB = I_n$$

Now, let

$$A_1 = [A : 0] \text{ and } B_1 = \begin{bmatrix} B\\ \vdots\\ 0\end{bmatrix}$$

be augmented matrices, where 0 blocks are zero matrices.

Then, we have

$$A_1 B_1 = I_n \quad \text{and} \quad B_1 A_1 = \begin{bmatrix} i_m & 0 \\ 0 & 0 \end{bmatrix}$$

$$\Rightarrow \qquad |A_1 B_1| = |I_n| = 1 \quad \text{and} \quad |B_1 A_1| = 0 \qquad \qquad \dots(5)$$

Also, since R is commutative, therefore $|A_1 \cdot B_1| = |B_1 A_1|$, which is a contradiction by virtue of (5).

Thus $m \nless n$ so that $m \geq n$. If the role of $[e_i]$ and $[f_i]$ are interchanged, then we get

$n \geq m$. Hence, $m = n$.

EXERCISE 9.2

1. Show that every ideal of **Z** is free as **Z**-module.

2. Show that every left ideal is an integral domain R with unity is free as a left R-module.

3. Show that the **Z**-module **Q** is not free.

4. If M and N are free modules over a commutative ring with unity such that both can be freely generated by n elements, show that $M = N$.

5. Let R be a non-zero commutative ring with unity. Consider R as an R-module. Show that

every submodule of R is free if and only if R is PID.

6. Let M be a semi-simple module and let $K \neq M$ be a submodule of M. Show that M/K is semi-simple module.

7. Show that, the polynomial ring $R[x]$ is a free R-module with basis $[X^n : n \in \mathbf{N}]$.

8. Show that a simple R-module M is free if and only if R is a division ring and rank $(M) = 1$.

9.12 NOETHERIAN AND ARTINIAN MODULES

Definition 1. (*Modules with Chain Conditions*)

Let R be a ring and M be the R-module. Also $S = [M_\alpha : \alpha \in \Delta]$ is the set of all submodules of M. Then S is said to satisfy

(i)   *the ascending chain condition if and only if every increasing chain of submodules*

$$M_1 \subseteq M_2 \subseteq M_3 \subseteq \dots\dots$$

in S is stationary, i.e., there exists a positive integer n such that

$$M_n = M_{n+1} = M_{n+2} = \dots\dots$$

(*ordered by the inclusion relation* $\subseteq$)

(ii) *the descending chain condition if and only if every decreasing chain of submodules*

$$M_1 \supseteq M_2 \supseteq M_3 \supseteq \dots\dots$$

in S is stationary, i.e., there exists a positive integer n such that

$$M_n = M_{n+1} = \dots\dots$$

(*S is ordered by* $\supseteq$)

Definition 2. *An R-module is said to be Noetherian module if it satisfy the ascending chain condition.*

Definition 3. *An R-module is said to be Artinian if it satisfies the descending chain condition.*

Definition 4. *An R-module M is said to be co-generated if for each family* $[M_\alpha : \alpha \in A]$ *of submodules of M,* $\bigcap_{\alpha \in \wedge} M_\alpha = 0$ *implies that* $\bigcap_{\alpha \in \Delta'} M_\alpha = 0$ *for some finite subset* Δ *'of* Δ.

REMARK

☞ The finitely co-generated modules satisfies the certain chain condition on their submodules.

Definition 5. (*Uniform Module*). *A non-zero R-module is said to be uniform if any two non-zero submodules of M have non-zero intersection.*

Definition 6. (*Subisomorphic submodule*). *Let M and N be two uniform modules, then M is said to be subisomorphic to N, divided by M ∼ N if M find N contain non-zero isomorphic submodules.*

REMARK

☞ Clearly ∼ is an equivalence relation and [M] denotes the equivalence class of M, i.e., [M] = [N ∼ M , for all uniform modules N].

Definition 7. (*Primary Module*). *An R-module M is said to be primary if each non-zero submodule of M has uniform submodule and any two uniform submodules of M are subisomorphic.*

Definition 8. *Let R be a commutative Noetherian ring and P be the prime ideal of R. Then, P is said to be associated with the module M if P = r(x), for some x ∈ M, where r(x) = [a ∈ R: xa = 0] is the annihilator of x or if R/P embeds in M.,*

Definition 9. *An R-module M is said to be P-primary for some prime ideal P if P is the only prime ideal associated with M.*

Theorem 1. *Let M be the R-module. Then following are equivalent :*

     (i) *M is Noetherian.*

     (ii) *Every submodule of M is of finitely generated.*

     (iii) *Every non-empty set S of submodules of M has a maximal element.*

Proof.(i) $\Rightarrow$ **(ii).** Let M be Noetherian module. Let us suppose that every submodule N of M is finitely generated.

Let if possible N is not finitely generated. Then for any positive integer r if
$$a_1, a_2, ..., a_r \in N.$$
Then, $N \neq (a_1, a_2, ..., a_r)$

Now, let $a_{r+1} \in N$, then $a_{r+1} \notin (a_1, a_2, ..., a_r)$.

Thus we obtain an infinite proper ascending chain.

$(a_1) \subset (a_1, a_2) \subset (a_1, a_2, a_3) \subset ..... \subset (a_1, a_2, ..., a_r) \subset (a_1, a_2, ..., a_{r+1}) \subset ...$ of submodules of M. Since, M satisfies a finite ascending chain of its submodules, therefore we arrive at a contradiction, because N is taken as not to be finitely generated. Hence, N is finitely generated.

(ii)$\Rightarrow$(iii). Let us suppose every submodule of M is finitely generated. We have to show that every non-empty set of submodules of M has a maximal element.

Let N_0 be an element of S. Suppose that N_0 is not maximal. In this case, there is a submodule N_1 in S such that
$$N_0 \subset N_1 \qquad ...(1)$$
If N_1 is not maximal, then similarly as above, there is a submodule N_2 in S such that
$$N_1 \subset N_2 \qquad ...(2)$$

From (1) and (2), we conclude that
$$N_0 \subset N_1 \subset N_2$$
Proceeding in the similar manner, we obtain an infinite property ascending chain of submodules
$$N_0 \subset N_1 \subset N_2 \,.....\subset N_i \subset N_{i+1} \subset \,...... \qquad ...(3)$$

Let us write $N = \bigcup_i N_i$, then N is a submodule of M, being the arbitrary union of property containing submodules. But by our assumption, every submodule of M is finitely generated, therefore N is finitely generated. Thus, there exist $a_1, a_2, ..., a_n \in N$ such that
$$N = (a_1, a_2, ..., a_n)$$

Now, since $N = \bigcup_i N_i$, then there must exist $N_k \subset N$, such that $a_1, a_2, ..., a_n \in N_k$.

Further, N is the smallest submodule containing $a_1, a_2, ..., a_n$.

Then, we get $\qquad N_k = N$.

Therefore, $\qquad N_k = N_{k+1} = ...$

which is a contradiction because we have assumed that S has no maximal element. Hence, S must have a maximal element.

(iii)$\Rightarrow$(i) Let us suppose that every non-empty set of submodules of M has maximal element. We have to show that M is Noetherian. Let us consider an ascending chain
$$N_1 \subset N_2 \subset N_3 \subset ...... \qquad ...(4)$$
of submodule of M, but we have assumed that every non-empty set of submodules has a maximal element. Therefore, the chain (4) has a maximal element say N_i, but then
$$N_i = N_{i+1} = N_{i+2} = \,......$$

$\Rightarrow$ (4) is a finite ascending chain of submodules of M. Hence, M is Noetherian module.

Theorem 2. *Let M be a R-module. Then following are equivalent :*

    (*i*) *M is Artinian.*

    (*ii*) *Every quotient module of M is finitely co-generated.*

    (*iii*) *Every non-empty set of submodules of M has a minimal element.*

Proof. The proof is consequence of Theorem 1 and left for the reader.

Theorem 3. *Every submodule of a Noetherian (Artinian) module is Noetherian (Artinian).*

Proof. Let M be a Noetherian (Artinian) module and N be its submodule. We have to prove that N is Noetherian (Artinian). Since M is Noetherian (Artinian), then it satisfying ascending (descending) chain condition of submodules. Also, every non-empty set of submodules of M has a maximal (minimal) element.

$\Rightarrow$ N is Noetherian (Artinian).

Hence, every submodule of M is Noetherian (Artinian).

Theorem 4. *Every homomorphic image of a Noetherian (Artinian) module is Noetherian (Ertinian).*

Proof. Let f be a homomorphism of a Noetherian (Artinian) module M. We have to show that $f(M)$ is Noetherian (Artinian).

Since M is Noetherian, then we have an ascending chain of submodules of M.

$$M_1 \subset M_2 \subset M_3 \subset ..... \subset M_n \subset ...$$

such that $M_k = M_{k+1} = ...$ for some positive integer k. Also, in case of Artinian, we have a descending chain of submodules of M.

$$N_1 \supset N_2 \supset N_3 \supset ..... \supset N_n \supset ...$$

such that $N_r = N_{r+1} = ...$, for some positive integer r.

Now, in case of Noetherian

$$f(M_1) \subset f(M_2) \subset ....f(M_n) \subset ... \text{ such that } f(M_k) = f(M_{k+1}) = ...$$

Also, in case of Artinian, we have

$$f(N_1) \supset f(N_2) \supset ....\supset f(N_n) \supset ....$$

such that $\qquad f(N_r) = f(N_{r+1}) = ...$

$\Rightarrow \qquad f(M)$ is Noetherian (Artinian).

Hence, every homomorphic image of a Noetherian (Artinian) module is Noetherian (Artinian).

Theorem 5. *Let M be an R-module and N be an R-submodule of M, then M is Noetherian (Artinian) if and only if both N and M/N are Noetherian (Artinian).*

Proof. Let us first suppose that N and M/N are both Noetherian. Also, let L be a submodule of M, then $(L + N)/N$ is a submodule of M/N. But M/N is Noetherian, therefore $(L + N)/N$ is finitely generated. Therefore, there exist $\bar{a}_1, \bar{a}_2, ..., \bar{a}_n \in L/N \cap L$ such that

$$L/N \cap L = (\bar{a}_1, \bar{a}_2, ..., \bar{a}_n)$$

which implies that

$$L = (a_1, a_2, ..., a_n) + (N \cap L), \forall a_i \in L$$

Since, N is Noetherian and $N \cap L$ is a submodule of N. Therefore, $N \cap L$ is finitely generated. Then, there exist $b_1, b_2, ..., b_m \in N \cap L$ such that

$$N \cap L = (b_1, b_2, ..., b_m)$$

$\Rightarrow \quad L = (a_1, a_2, ..., a_n) + (b_1, b_2, ..., b_m)$

$\Rightarrow \quad L$ is finitely generated and L is any submodule of M.

$\Rightarrow \quad M$ is Noetherian.

Conversely, let M be Noetherian. Then, its every submodule is Noetherian.

$\Rightarrow \quad N$ is Noetherian.

Now, let $f : M \to M/N$ be a canonical homomorphism and let $\overline{M_1} \subset \overline{M_2} \subset .....$be an ascending chain of submodule of M/N.

If $M_i = f^{-1}(\overline{M_i})$, then we have $M_1 \subset M_2 \subset M_3 \subset .....$is an ascending chain of submodules of M.

Since, M is Noetherian, therefore there exists a positive integer r such that

$$M_r = M_{r+1} = ...\text{Then, we have}$$

$$f(M_i) = \overline{M_i}, \forall i \geq r \Rightarrow \overline{M_r} = \overline{M_{r+1}} = ...$$

$\Rightarrow \quad M/N$ is Noetherian.

The proof for Artinian module is similar as above.

Theorem 6. *Let M be a module and N, N ' be submodules. If M = N + N ' and if both N, N' are Noetherian, then M is Noetherian. Further, a finite direct sum of Noetherian module is Noetherian.*

Proof. We observe that the direct product $N \times N'$ is Noetherian because it contains N as a submodule whose factor module is isomorphic to N'. Then, applying previous theorem, we obtain a surjective homomorphism $N \times N' \rightarrow M$ such that the pair (x, x'), $x \in N, x' \in N'$ maps on $x + x'$.

Therefore, M is Noetherian.

Further, by using induction, we can prove the result of finite direct sum.

REMARK

☞ A ring is called Noetherian if it is Noetherian as a left module over itself, *i.e.*, every left ideal is finitely generated.

Theorem 7. *Let A be a Noetherian ring and M be a finitely generated module. Then,M is Noetherian.*

Proof. Let $x_1,..., x_n$ be the generators of M.

Then, there exists a homomorphism $f : A \times A \times ...\times A \rightarrow M$ of the product of A with itself n times such that

$$f(a_1,...,a_n)= a_1x_1+...+a_nx_n$$

Since, this homomorphism is surjective, then by previous theorem, the product is Noetherian. Also, we know that every homomorphic image of a Noetherian module is Noetherian. Hence, M is Noetherian module.

Theorem 8. *Let M be a Noetherian module. Then, each non-zero submodule of M contains a uniform module.*

Proof. Let M be a Noetherian R-module. We have to show that each non-zero submodule of M contains a uniform module. For this, we shall prove that μR contains a uniform submodule for non-zero element μ of M.

If R is Noetherian ring, then μR is also Noetherian. Also, if M is Noetherian, then the submodule μR is also Noetherian. Further, let N be a non-zero submodule of M and if $N \cap L \neq [0]$ for all non-zero submodules L of M.

Now, let S be a family of all submodules of μR which are not large. Since, μR is Noetherian, then every non-empty set of submodules of μR has a maximal element, thus S has a maximal element, say L.

But L is not large, so $L \cap L = [0]$ for some non-zero submodule K of μR.

We shall prove that K is uniform.

Suppose K_1 and K_2 be any two non-zero submodules of K. Let if possible,

$$K_1 \cap K_2 = [0]$$

If $u \in (L \oplus K_1) \cap K_2$, then we have

$$u \in (L \oplus K_1 \quad \text{and} \quad u \in K_2$$

$\Rightarrow \qquad u = a + a_1 = a_2,$ for some $a \in L, a_1 \in K_1, a_2 \in K_2$

$\Rightarrow \qquad a = u - a_1 \in L \cap K$

$\Rightarrow \qquad a = 0 \quad \text{and} \quad a_1 = a_2$

$\Rightarrow \qquad a_1 = a_2 \in K_1 \cap K_2 =[0]$

$$\Rightarrow \qquad\qquad a_1 = a_2 = 0$$

$$\Rightarrow \qquad\qquad u = 0$$

which is a contradiction, therefore $K_1 \cap K_2 \neq [0]$. Hence, K is uniform.

Theorem 9. *Let R be a commutative Noetherian ring and M be a uniform R- module. Then, M contains a submodule isomorphic to R/P for exactly one prime ideal of P.*

Proof. Let $x \in M$, $x \neq 0$ and S be the family of annihilators ideals $r(x)$. Further, since R is Noetherian, S will contain a maximal element say $r(x)$. We have to show that $P = r(x)$ is prime ideal.

Suppose $a.b \in r(x)$ with $a \notin r(x)$.

By definition of $r(x)$, $\qquad xa \neq 0$ and $r(x) \subset r(xa)$.

But $r(x)$ is maximal in S, therefore $r(x) = r(xa)$.

Therefore, $\qquad\qquad xab = 0$, then $b \in r(x)$

$\Rightarrow \quad P$ is prime,

and xR is isomorphic to $R/r(x) = R/P$. Therefore R/P embeds in M.

Now, it remains to prove that P is unique.

Let if possible, Q be an other prime ideal of R such that R/Q embeds in M.

Also, $yR \simeq R/Q$ and $[R/P] = [R/Q]$.

Then, there exist two cyclic modules xR of R/P and yR of R/Q such that $xR \simeq yR$.

Therefore, $\qquad\qquad R/P \simeq R/Q$

$\Rightarrow \qquad\qquad P = Q$

REMARK

☞ Let M be a non-zero finitely generated module over a commutative Noetherian ring R, then there are only a finite number of prime associated with M.

Theorem 10.(NOETHER-LASKER THEOREM)

Let M be a finitely generated module over a commutative Noetherian ring R. Then, there exists a finite family $\{N_1, N_2, ..., N_m\}$ of submodules of M such that

(i) $\displaystyle\bigcap_{j=1}^{m} N_j = 0$ and $\displaystyle\bigcap_{\substack{j=1 \\ j \neq r}}^{m} N_j \neq 0$ for all $1 \leq r \leq m$

(ii) Each quotient module M/N_j is a P_j-primary for some prime ideal P_j associated with M.

(iii) The P_j are all distinct $1 \leq j \leq m$.

(iv) The primary components N_j is unique if and only if P_j does not contain P_k for any $k \neq j$.

9.13 FILTERED AND GRADED MODULES

Definition 1 (*Filtration*). *Let R be a commutative ring and E be a module. By a filtration of E, we mean a sequence of submodules.*

$$E = E_0 \supset E_1 \supset E_2 \supset ... \supset E_n \supset$$

For example: Let I be an ideal of a ring R and E an R-module. Also, let $E_n = I^n E$. Then, the sequence of submodules $< E_n >$ is a filtration.

REMARK

☞ The filtration defined above is known as descending filtration.

Definition 2. *Let $< E_n >$ be any filtration of a module E, It is said to be an I-filtration of $IE_n \subset E_{n+1}$ for all n.*

Definition 3. *An l-filtration is said to be I-stable or stable if $IE_n = E_{n+1}$ $\forall n$, where n is sufficiently large.*

Definition 4 (Graded Ring). *A ring R is called graded (by the natural numbers) if we can write R as a direct sum (as abelian group).*

$$A = \bigoplus_{n=0}^{\infty} A_n$$

such that for all integers, $m, n \geq 0$, we have $R_n R_m \subset R_{n+m}$.

REMARK

☞ R_0 is a subring and each component R_n is an R_0-module.

Definition 5 (Graded Module). *Let R be a graded ring. A module E is called a graded module if E can be expressed as a direct sum (as abelian groups)*

$$E = \bigoplus_{n=0}^{\infty} E_n$$

such that $R_n E_m \subset E_{n+m}$.

In particular, E_n is an R_0-module and elements of En are then called homogeneous of degree n.

REMARK

☞ Using above definition, any element E can be expressed uniquely as a finite sum of homogeneous elements.

Theorem 1. *Let $<E_n>$ and $< E'_n >$ be stable I-filtration of E, then there exists a positive integer d such that $E_{n+d} \subset E'_n$ and $E'_{n+d} \subset E_n$ for all $n \geq 0$.*

Proof. Here, it is sufficient to prove the proposition when $E'_n = I^n E$. Since

$IE_n \subset E_{n+1}$ for all n, we have $I^n E \subset E_n$.

By stability hypothesis, there exists d such that

$$E_{n+d} = I^n E_d \subset I^n E$$

$$\Rightarrow \qquad E_{n+d} \subset E'_n.$$

Theorem 2. *Let R be a graded ring. Then R is Noetherian if and only if R_0 is Noetherian and R is finitely generated as R_0-Algebra.*

Proof. Clearly, a finitely generated algebra over a Noetherian ring is Noetherian because it is a homomorphic image of the polynomial ring in finitely many variables and we can then apply Hilbert's theorem.

Conversely, suppose R is Noetherian, the sum

$$p_n^+ = \bigoplus_{n=0}^{\infty} R_n$$

is an ideal of R, whose residue class ring is R_0 , which is thus a homomorphic image of R, and is therefore Noetherian. Also, R^+ has a finite number of generators $x_1, \ldots x_s$ by hypothesis. Expressing each generator as a sum of homogeneous elements, we may assume without loss of generality that these generators are homogeneous, say of degrees d_1 , $\ldots\ldots d_s$ respectively with all $d_i > 0$. Let B be the subring of R generated over R_0 by $x_1, \ldots, x_s$. We claim that $R_n \subset B$ for all n. This is certainly true for $n = 0$. Let $n > 0$. Let x be homogeneous of degree n, then there exists some elements $a_i \in R_{n-d}$ such that

$$x = \sum_{i=1}^{s} a_i x_i$$

Since $d_i > 0$, by induction, each a_i is in $R_0 [x_1, \ldots, x_s] = B$.

$\Rightarrow \qquad\qquad\qquad x \in B$

Hence, the theorem.

9.13.1 Construction of Graded Rings from Filtration

Let R be a ring and I an ideal. We view R is a filtered ring by the powers I^n . We define the first associated graded ring to be

$$S_I(A) = S = \bigoplus_{n=0}^{\infty} I^n$$

Similarly, if E is an R-module and E is filtered by an I-filtration, we define

$$E_s = \bigoplus_{n=0}^{\infty} E_n$$

Then, we can easily verify that E_s is a graded S-module.

REMARK

☞ If R is Noetherian and I is generated by elements $x_1 , \ldots, x_s$, then S is generated as an R-algebra also by $x_1, \ldots, x_s$ and is therefore also Noetherian.

Theorem 3. *Let R be a Noetherian ring and E a finitely generated module with an I-filtration. Then E_s is finite over S if and only if tlie filtration of E is I-stable.*

Proof. Let $F_n = \bigoplus_{i=0}^{n} E_i$ and let $G_n = E_0 \oplus \ldots \oplus E_n \oplus IE_n \oplus I^2 E_n \oplus I^3 E_n \oplus \ldots$ Then, G_n is a

S-submodule of E_s and is finite over S, since F_n is finite over R.

We have $G_n \subset G_{n+1}$ and $\cup G_n = E_s$

Since S is Noetherian, we get

$\qquad E_s$ is finite over S

$\Rightarrow \quad E_s = G_N$ for some N.

$\Rightarrow \quad E_{N+M} = I^m E_n$ for all $m \geq 0$.

Hence, filtration of E is I-stable.

Theorem 4 (Artin-Ress). *Let R be a Noetherian ring and I an ideal E a finite R-module with a stable I-filtration. Let F be a siibmodule and let $F_n = F \cap E_n$. Then $<F_n>$ is a stable I-filtration of F.*

Proof. We have $I[F \cap E_n] \subset IF \cap IE_n \subset F \cap E_{n+1}$

Therefore, $< F_n >$ is an I-filtration of F. Then, we can form the associated graded S-module F_s, which is a submodule of E_s and is finite over S, since S is Noetherian. Finally, using the previous theorem, we conclude that $< F_n >$ is a stable I-filtration of F.

REMARK

☞ Artin-Ress theorem can be restated as follows : "Let R be a Noetherian ring, E is a finite R-module and F a submodule. Let I be an ideal. There exists an integer i such that for all integers $n \geq i$, we have $I^n E \cap F = I^{n-i}(a^i E \cap F)$

Theorem 5 (Krull Theorem). *Let R be a Noetherian ring and I be an ideal contained in every maximal ideal of R. Let E be a finite R-module. Then*

$$\bigcap_{n=1}^{\infty} I^n E = 0$$

Proof. Write $F = \cap I^n E$. Then apply Nakayama's lemma, whose statement is given below.

Nakayama's Lemma

Let I be an ideal of R which is contained in every maximal ideal of R. Let E be a finitely generated R-module. Suppose that $IE = E$. Then $E = [0]$.

9.14 PROJECTIVE AND INJECTIVE MODULES

Every free module is projective and arbitrary projective modules (which need not be free) have some of the same properties as free modules. Projective modules are useful in categorical setting since they are defined solely in terms of modules and homomorphisms. Injectivity is the dual notion to projectivity.

Definition 1 (*Projective Module*). *A module P over a ring R is said to be projective if given any diagram of R-module homomorphism*

$$P$$
$$\downarrow f$$
$$A \xrightarrow{g} B \longrightarrow 0$$

with buttom row exact (i.e., g is an epimorphism) there exists an R-module homomorphism $h : P \to A$ such that the diagram

$$P$$
$$h \swarrow \quad \downarrow f$$
$$A \xleftarrow{g} B \longrightarrow 0$$

is commutative (i.e., $gh = f$).

Definition 2 (*Injective Module*) *A module J over a ring R is said to be injective if given any diagram of R-module homomorphism*

$$0 \longrightarrow A \xrightarrow{g} B$$
$$f \downarrow \qquad \diagup$$
$$J \swarrow$$

with top row exact (i.e., *g is a monomorphism.) there exists an R-module homomorphism*
$h : B \rightarrow J$ *such that the diagram*

$$0 \longrightarrow A \xrightarrow{g} B$$
$$f \downarrow \qquad \diagup$$
$$J \swarrow$$

is commutative (i.e., hg = f).

Theorem 1. *Every free module M. over a ring R wiih identity is projective.*

Proof.   We may assume that we are given a diagram of homomorphisms of unitary R-module.
F if

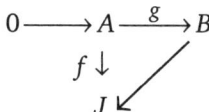

$$F$$
$$\downarrow f$$
$$\bar{A} \xrightarrow{g} B \longrightarrow 0$$

with g an epimorphism and M a free module on the set X $(t : X \rightarrow M)$. For each $x \in X$,
$f(t(x)) \in B$. Since g is an epimorphism, there exists $a_x \in A$ with $g(a_x) = f(t(x))$. Now,
since F is free the map $h(t(x)) = a_x \forall X \in X$. Thus $ght(x) = g(a_x) = f(t(ax)) \forall x \in X$ so
that $ght = ft : X \rightarrow B$. Then, $gh = f$.

Hence, F is projective.

REMARKS

☞ Every module M over a ring R is the homomorphic image of a projective R-module.

☞ Let R be a ring. Then following conditions on an R-module P are equivalent:
  (i) P is projective.
  (ii) every short exact sequence $0 \rightarrow A \xrightarrow{f} B \xrightarrow{g} P \rightarrow 0$ is split exact.   Thus $B \cong A \oplus P$.

  (iii)   there is a free module F and an R-module K such that $F \cong K \oplus P$.

☞ Let R be a ring. A direct sum of R-modules $\sum_{i \in I} P_i$ is projective if and only if each P_i is projective.

☞ A direct product of R-module $\prod_{i \in I} J_i$ is injective if and only if J_i is injective for every $i \in I$.

☞ Let R be a ring with identity. A unitary R-module J is injective if and only if for every left ideal L of

  R any R-module homomorphism $L \rightarrow J$ may be extended to an R-module homomorphism $R \rightarrow J$.

☞ Every unitary module M over a ring R with identity may be embedded in an injective R-module.

☞ Let R be a ring with identity. The following conditions on a unitary R- module J are equivalent:
  (i) J is injective.
  (ii) every short exact sequence $0 \rightarrow J \xrightarrow{f} B \xrightarrow{g} C \rightarrow 0$ is split exact. Hence, $B \cong J \oplus C$.
  (iii) J is a direct summand of any module B of which it is a submodule.

9.15 SMITH NORMAL FORM OVER A PID AND RANK

In this section, we shall consider the ring R as a principal ideal domain with unity, so in particular, R may be a field.

Definition 1 (*Row Module and Column Module*). *Let R be a ring and A be $n \times n$ matrix of order $n \times n$ over R. The submodule of R^n generated by the m rows of A is called the rows module of A and the submodule of R^m generated by the n columns of A and the submodule of R^m generated by the n columns of A is known as column module of A.*

Definition 2 (*Rank*). *Let A be an $m \times n$ matrix over R. The rank of the row module (column module) is known as the row rank (column rank) of A.*

Definition 3. *Let A be an $m \times n$ matrix over R. The following types of operations on the rows (column) of A are known as elementary row (column) operations*

   (i)   *interchanging two rows (columns)*

   (ii) *multiplying the element of one row (column) by a non-zero element of R*

   (iii) *adding to the elements of one row (column) x times the corresponding elements of a different row (column) where $x \in R$.*

Definition 4. We define the following matrices

   (1) E_{ij} = *the matrix obtained from the identity matrix by interchanging i^{th} and i^{th} rows (columns).*

   (2) $L_i(x)$ = *the matrix obtained from the identity matrix by multiplying the ith row (column) by a non-zero $\alpha \in R$.*

   (3) $M_{ij}(\alpha)$ = *the matrix obtained from the identity matrix by adding to the elements of the ith row (column) α times the corresponding elements of the jth row (column).*

Then the matrices E^{ij}, $L_i (\alpha)$ and $M_{ij} (\alpha)$, obtained from identity matrix by elementary operations are known as elementary matrices.

Definition 5. *Two $m \times n$ matrices A and B over R are said to be equivalent if there exists an invertible matrix $P \in R^m$ and an invertible matrix $Q \in R^n$ such that $B = PAQ$.*

Definition 6 (*Smith Normal Form*). *Let A be $m \times n$ matrix over a principal ideal domain R, then A is equivalent to a matrix of the diagonal form*

$$\begin{bmatrix} q_1 & & & & & \\ & q_1 & & & & \\ & & \ddots & & & \\ & & & q_1 & & \\ & & & & 0 & \\ & & & & & \ddots \\ & & & & & & 0 \end{bmatrix}_{m \times n} , \text{with } q_i \neq 0 \text{ and } q_1 \mid q_2 \mid q_3 \cdots$$

This form of matrix A is called Smith normal form, denoted by $S(A)$.

Definition 7. *The non-zero diagonal elements of Smith normal form are called the invariant factors of A, where $S(A)$ is the Smith normal form.*

Definition 8. *Let A be an $m \times n$ matrix over a PID R. The common value of the row rank of A and the common rank of A is called the rank of A.*

Definition 9. *Let R be a PID, and an invariant factor*

$$q_k = u_k p_1^{\alpha_{k_1}} p_2^{\alpha_{k_2}}, \ldots, p_m^{\alpha_{k_m}}, \; k = 1, 2, \ldots, r$$

where $p_1, p_2, \ldots, p_m$ are distinct primes in the complex system of non-associates in R, α_{kj} are non-negative.

Integers $j = 1, 2, \ldots, m$, $k = 1, 2, \ldots r$ and u_k are units.

From $q_k | q_{k+1}$ it follows that $\alpha_{k_i} \subseteq \alpha_{(k+1)_j}$, $k = 1, 2, \ldots, r-1$, $j = 1, 2, \ldots, m$.

A prime factor $p_i^{\alpha_{kj}}$ in which $\alpha_{kj} > 0$ is called an elementary divisor of h.

Solved Examples

Example1. *Find the Smith normal form and rank of the matrix*

$$A = \begin{bmatrix} 1 & 2 & 3 \\ 4 & 5 & 0 \end{bmatrix}$$

Solution. The given matrix is $A = \begin{bmatrix} 1 & 2 & 3 \\ 4 & 5 & 0 \end{bmatrix}$

Performing elementary row and column operation

Performing $R_2 \to R_2 - 4R_1$, we get

$$\begin{bmatrix} 1 & 2 & 3 \\ 0 & -3 & -12 \end{bmatrix}$$

Performing $C_2 \to C_2 - 2C_1$ and $C_3 \to C_3 - 3C_1$, we get

$$\begin{bmatrix} 1 & 0 & 0 \\ 0 & -3 & -12 \end{bmatrix}$$

$C_3 \to C_3 - 4C_2$

$$\begin{bmatrix} 1 & 0 & 0 \\ 0 & -3 & 0 \end{bmatrix}$$

$R_2 \to (-1)R_2$

$$A \sim \begin{bmatrix} 1 & 0 & 0 \\ 0 & 3 & 0 \end{bmatrix}$$

Therefore Smith normal form

$$S(A) = \begin{bmatrix} 1 & 0 & 0 \\ 0 & 3 & 0 \end{bmatrix}$$

with $q_1 = 1$, $q_2 = 3$

Clearly $q_1 | q_2$. Hence, rank $(A) = 2$.

Example 2. *Reduce the matrix*

$$A = \begin{bmatrix} -x & 4 & -2 \\ -3 & 8-x & 3 \\ 4 & -8 & -2-x \end{bmatrix}$$

Over the ring Q(x) to smith normal form. Also, find the rank.

$$A = \begin{bmatrix} -x & 4 & -2 \\ -3 & 8-x & 3 \\ 4 & -8 & -2-x \end{bmatrix} \sim \begin{bmatrix} 4 & -8 & -2-x \\ -3 & 8-x & 3 \\ -x & 4 & -2 \end{bmatrix}$$

$$[\text{Performing } R_1 \leftrightarrow R_3]$$

$$\sim \begin{bmatrix} 4 & -8 & -(2+x) \\ 0 & 2-x & \frac{1}{4}(6-3x) \\ 0 & 4-2x & -\frac{1}{4}(x^2+2x+8) \end{bmatrix} \quad [R_2 \rightarrow R_2 + \frac{3}{4}R_1, R_3 \rightarrow R_3 + \frac{x}{4}R_1]$$

$$\sim \begin{bmatrix} 4 & -8 & -(2+x) \\ 0 & 2-x & \frac{1}{4}(6-3x) \\ 0 & 0 & -\frac{1}{4}(x^2+4x+20) \end{bmatrix} \quad [R_3 \rightarrow R_3 \rightarrow 2R_2]$$

$$\sim \begin{bmatrix} 4 & 0 & -(2+x) \\ 0 & 2-x & \frac{1}{4}(6-3x) \\ 0 & 0 & -\frac{1}{4}(x^2+4x+20) \end{bmatrix} \quad [C_2 \rightarrow C_2 \rightarrow 2C_1]$$

$$\sim \begin{bmatrix} 4 & 0 & -(2+x) \\ 0 & 2-x & 0 \\ 0 & 0 & -\frac{1}{4}(x^2+4x+20) \end{bmatrix} \quad [C_3 \rightarrow C_3 - \frac{3}{4}C_2]$$

$$\sim \begin{bmatrix} 1 & 0 & -(2+x) \\ 0 & 2-x & 0 \\ 0 & 0 & -\frac{1}{4}(x^2+4x+20) \end{bmatrix} \quad [C_1 \rightarrow \frac{1}{4}C_1]$$

$$\sim \begin{bmatrix} 1 & 0 & 0 \\ 0 & 2-x & 0 \\ 0 & 0 & x^2+4x+20 \end{bmatrix} \qquad [C_3 \to C_3 + (2+x)\,C_3 \to -4C_3]$$

In a similar manner, we apply some more transformation, we get

$$\sim \begin{bmatrix} 1 & 0 & 0 \\ 0 & 1 & 0 \\ 0 & 0 & (x-2)(x^2+4x+20) \end{bmatrix}$$

Then, we have $q_1 = 1,\ q_2 = 1,$

$$q_3 = (x-2)\,(x^2 - 4x + 20)$$

$$q_1 \ne 0,\ q_2 \ne 0,\ q_3 \ne 0 \text{ if } x \ne 2$$

Also $q_1 \mid q_2$ and $q_2 \mid q_3$

Therefore, Smith normal form is given by

$$\begin{bmatrix} 1 & 0 & 0 \\ 0 & 1 & 0 \\ 0 & 0 & (x-2)\!\left(x^2 + 4x + 20\right) \end{bmatrix}$$

Hence, rank $(A) = 3$

Example 3. *If $D = diag\ (3^2 - 7^3.\ 24)$, then find its Smith normal form over* **Z**.

Solution. Clearly, the elementary divisors of D are $2^3.\ 3.\ 3^2.\ 7^3$.

Then, the list of invariant factors is given by

$$q_3 = 2^3.3^2.7^3 = 24696$$

$$q_2 = 3$$

$$q_1 = 1$$

Hence, ths Smith normal of D is given by

$$S\ (D) = diag.\ (1,\ 3,\ 24969)$$

Example 4. *Let $2^3,\ 2^4,\ 2^4,\ 3,\ 3^3,\ 3^3,\ 5^2,\ 5^5, 7\ 7^2$ be the complete list of elementary divisors of $A \in \mathbf{Z}^5 \times \mathbf{Z}^7$, i.e., A is 5×7 matrix over* **Z** *and rank of $A = 4$. Find the Smith normal form of A.*

Solution. It is given that rank of A is 4.

Thus, there are four invariant factors $q_1,\ q_2,\ q_3,\ q_4$.

Also, q_4 must be the product of all the highest powers of the distinct primes.

Therefore, $q_4 = 2^4.\ 3^3.\ 5^5.\ 7^2$.

Further, q_3 must be the product of all the highest powers of the distinct primes in the list after q_4 has been constructed, i.e., $q_3 = 2^4.\ 3^3.\ 5^2.7$

In a similar manner, we get $q_2 = 2^3.\ 3$ and q_1 must be a unit, so $q_1 = 1$.

Hence, Smith normal form is given by

$$S(A) = \begin{bmatrix} q_1 & 0 & 0 & 0 & 0 & 0 & 0 \\ 0 & q_2 & 0 & 0 & 0 & 0 & 0 \\ 0 & 0 & q_3 & 0 & 0 & 0 & 0 \\ 0 & 0 & 0 & q_4 & 0 & 0 & 0 \\ 0 & 0 & 0 & 0 & 0 & 0 & 0 \end{bmatrix}$$

9.16 FINITELY GENERATED MODULES OVER A PID

Definition 1. (*Torsion and Torsion Free Element*). *Let R be a principal ideal domain and M be an R-module. An element $x \in M$ is called torsion element if $rx = 0$ for some non-zero $r \in R$. If x is not a torsion element, then it is known as torsion free element.*

Definition 2. (*Torsion and Torsion Free Module*). *Let R be a principal ideal domain and M be an R-module. If T(M) is a set of all torsion element of M, then T(M) is a submodule of M. Also, if T(M) = M, then M is known as torsion module. On the other hand, if T(M) = 0, then M is called torsion free module.*

Theorem 1 (Decompositon Theorem). *If M is a non-zero finitely generated module over a PID R. If n is the minimal number of elements required to generate M, then M is the direct sum of cylic submodules.*

$$M = Ra_1 \oplus Ra_2 \oplus \dots \oplus Ra_n \oplus = \sum_{i=1}^{n} Ra_i$$

such that Ann. $(a_{i+1}) \subseteq$ Ann.(a_i) for all $i = 1, 2, \dots, n-1$, where Ann. $(a_1) \neq R$ and $A_{nn}. (a_n) = A_{nn}. (M)$

Proof. As per given, M is a non-zero finitely generated module over a *PID R*. Since M is finitely generated, therefore, there exist $\alpha_1, \alpha_2, \dots, \alpha_n \in R$ such that

$$M = (\alpha_1, \alpha_2, \dots, \alpha_n)$$

Define a mapping $f : R^n \to M$ such that $f(c_1, c_2, \dots, c_n) = c_1\alpha_1 + c_2\alpha_2 + \dots + c_n\alpha_n$

Clearly, f is an R-module epimorphism.

Now, if $K = \text{Ker}(f)$, then K is a free submodule of rank m (say) $m \leq n$.

Further, choose a basis $[\beta_1, \beta_2, \dots, \beta_n]$ of R^n and non-zero elements $d_1, d_2, \dots, d_m \in R$ such that $[d_1 \beta_1, d_2 \beta_2, \dots, d_m\beta_m]$ form a basis of K and for each i, $d_i \mid d_{i+1}$.

Now, since f is epimorphism, then for each $i = 1, 2, \dots, n$.

$$a_i = f(\beta_i)$$

Thus, $[a_1, a_2, \dots, a_n]$ will generate M.

Since, for any $a \in M$, there is $b \in R^n$ such that $a = f(b)$ but

$$b = c_1\beta_1 + c_2\beta_2 + \dots + c_n\beta_n \text{ for some } c_1, c_2, \dots, c_n \in \mathbf{R}$$

such that

$$a = c_1a_1 + c_2a_2 + \dots + c_na_n = \sum_{i=1}^{n} c_i a_i$$

We want to show that $M = Ra_1 \oplus Ra_2 \oplus \dots \oplus Ra_n$.

Now, suppose that $a = 0$, *i.e.*, $\sum\limits_{i=1}^{n} c_i a_i = 0$, for $a_i \in R$.

Then, $f(b) = 0$

$\Rightarrow$ $b \in \text{Ker}(f) = K$

$\Rightarrow$ $C_1\beta_1 + C_2\beta_2 + ... + c_n\beta_n \in K$

$\Rightarrow$ $\sum\limits_{i=1}^{n} c_i\beta_i \in K$

But $[d_1\beta_1, d_2\beta_2,..., d_m b_m]$ is the basis of K so that

$$\sum_{i=1}^{n} c_i\beta_i = \sum_{j=1}^{m} e_i d_j\beta_j, \text{ for some } e_i \in R$$

$\Rightarrow$ $$\sum_{i=1}^{m}(c_i - e_i d_i)\beta_i = \sum_{i=m+1}^{n} c_i\beta_i = 0$$

Now, since $[\beta_i : i=1, 2, ..., m, m+1, ..., n]$ is linearly independent so that $c_i - e_i d_i = 0$ for $i = 1, 2, ..., m$.

and $c_i = 0$ for $i = m+1, m+2, ..., n$.

For $i = 1, 2, ..., m$, we have $a_i = f(\beta_i)$

So that $c_i a_i = f(C_i \beta_i) = f(e_i d_i \beta_i) = 0$

Therefore, $C_i a_i = 0$, for all i. Thus, we get

$$M = Ra_1 \oplus Ra_1 \oplus ... \oplus Ra_n$$

We have to show that Ann. $(a_{i+1}) \subseteq$ Ann. (a_i) for $i = 1, 2, ..., n-1$

For $\alpha \in$ Ann. (a_i) $\alpha a_i = 0$ and $a_i = f(\beta_i)$ so that

$$f(\alpha\beta_i) = \alpha a_i = 0$$

$\Rightarrow$ $\alpha\beta_i \in K$

If $i > m$, then $\alpha = 0$, thus Ann. $(a_i) = [0]$. If $i \le m$, then $\alpha\beta_i = \sum\limits_{j=1}^{m} e_j d_j\beta_j$ for some $e_1, e_2,..., e_m \in R$, so that $\alpha = e_i d_i$

$\Rightarrow$ $d_i | a$

Therefore, Ann. $(a_i) \subseteq d_i$

But $d_i a_i = d_i f(\beta_i) = f(d_i \beta_i) = 0$

because $d_i \beta_i \in K$ so that $d_i \in Ann.(a_i)$.

$\Rightarrow$ $(d_i) \subseteq$ Ann.$[a_i]$

Thus, we get Ann. $(a_i) = d_i$

Since $d_i | d_{i+1}$ for each i, therefore

$$\text{Ann. } (a_{i+1}) \subseteq \text{Ann } (a_i)$$

It remains to prove that Ann. $(a_n) =$ Ann. (M).

If $m < n$, then Ann. $(a_n) = [0] =$ Ann. $[M]$

If $m = n$, then M is torsion module and $d_i | d_n$ for all i

$$\Rightarrow \qquad d_i M = [0]$$

Therefore, $\quad$ Ann. $(a_n) = (d_n) =$ Ann. (M)

Theorem 2 (Uniqueness of Decompositon Theorem). *Let M be a finitely generated module over a PID R. If M has cyclic decompositions given by*

$$M = Ra_1 \oplus Ra_2 \oplus ... \oplus Ra_m$$

and $\qquad\qquad M = Rb_1 \oplus Rb_2 \oplus ... \oplus Rb_m$

such that $\quad$ Ann. $(a_{i+1}) \subseteq$ Ann. (a_i), *for* $i = 1, 2, ..., m-1$

With $\qquad\qquad$ Ann. $(a_i) \neq R$ *and* Ann. $(a_n) =$ Ann. (M)

and $\qquad\qquad$ Ann. $(b_{j+1}) \subseteq$ Ann. (b_j), *for* $j = 1, 2, ..., n-1$

With $\qquad\qquad$ Ann. $(b_1) \neq R$ *and* Ann. $(b_n) =$ Ann. (M), *then* $m = n$.

and $\qquad\qquad$ Ann. $(a_i) =$ Ann. (b_j) *for all* $i = 1, 2, ..., m$.

Proof. $\qquad$ We have

$$\text{Ann. } (a_m) = \text{Ann. } (M) = \text{Ann. } (b_n)$$

So that, $\qquad$ Ann. $(a_m) =$ Ann. (b_n)

Let us assume that $\qquad m \geq n$.

Let Ann. $(a_i) = (r_i)$ for all $i = 1, 2,...,m$, $r_i \in R$. Also, let p be a prime element in R such that $p \mid r$.

Then, $\qquad\qquad$ Ann. $(a_i) \subseteq (p)$ $\forall i = 1, 2, ..., m$

Clearly, pM is a submodule of R and M/pM is a module of R/pR with scalar multiplication defined by

$$(r + pR)(a + pM) \Rightarrow ra + pM, \text{ for } r \in R, a \in M \qquad\qquad ...(1)$$

But R/pR is a field, so that $M|pM$ can be treated as a vector space over R/pR. Thus, M/pM is a free module. Now, we shall prove that

$$S = [a_1 + pM, a_2 + pM, ...a_m + pM] \text{ is a basis of } M/pM$$

Since, M is finitely generated. Therefore, $M = \{a_1, a_2, ..., a_m\}$ and if $x \in M$, then

$$x = \alpha_1 a_1 + \alpha_2 a_2 + ... + \alpha_m a_m$$

for $\alpha_1, \alpha_1, ..., \alpha_m \in R$.

Therefore,

$$\begin{aligned} x + pM &= (\alpha_1 a_1 + \alpha_2 a_2 + ... + \alpha_m a_m) + pM \\ &= (\alpha_1 a_1 + \alpha_2 a_2 + ... + \alpha_m a_m) + pM \\ &= (\alpha_1 a_1 + pM) + (\alpha_2 a_2 + pM) + ... + (\alpha_m a_m + pM) \end{aligned}$$

Thus, $\qquad\qquad x = \sum_{i=1}^{m} (\alpha_i + pR)(a_i + pM)$ $\qquad\qquad$ [Using (1)]

Therefore S spans M/pM because $a_i + pM \neq 0$. Further, we shall show that S is linearly independent. Suppose

$$\sum_{i=1}^{m} (\alpha_i + pR)(a_i + pM) = pM \qquad\qquad [\because pM \text{ is the zero element of } M/pM.]$$

$$\Rightarrow \qquad \sum_{i=1}^{m} (a_i \alpha_i + pM) = pM, \qquad\qquad\qquad i.e., \sum_{i=1}^{m} \alpha_i a_i \in pM, \forall a_i \in R$$

$$\Rightarrow \qquad \sum_{i=1}^{m} \alpha_i a_i = p \sum_{i=1}^{m} \beta_i a_i, \text{ for some } \beta_i \in R$$

$$\Rightarrow \qquad \sum_{i=1}^{m} (\alpha_i - p\beta_i) a_i = 0$$

$$\Rightarrow \qquad (\alpha_i - p\beta_i) a_i = 0 \;\; \forall \; i = 1, 2, ..., m$$

$$\Rightarrow \qquad \alpha_i - p\beta_i \in A_{nn}(a_i) \subseteq (p) \Rightarrow a_i \in (p)$$

$$\Rightarrow \qquad \alpha_i + pR = pR \qquad\qquad [\because pR \text{ is the zero element of } R/pR]$$

$\Rightarrow$ S is linearly independent.

Thus, S form a basis of M/pM over R/pR of dimension m.

Also, $\qquad\qquad\qquad M = (b_1, b_2, ..., b_n).$ Then

$$S_1 = (b_1 + pM, b_2 + pM, ..., b_n + pM)$$

gernerates M/pM, thus we get $m \leq n$.

$$\Rightarrow \qquad\qquad m = n.$$

It remains to prove that

Ann.$(a_i)=$ Ann.(b_i), for $\quad i=1,2,3,...,m.$

We know that if I is an ideal of R, then $I=(x)$.Then, $l(x)$ denotes the primes appearing in the prime factorization of X.

Now, we apply induction on $l(\text{Ann.}(M))$. If $l(\text{Ann}(m))=1$ then Ann $(M) = (p)$

But $\quad$ Ann.$(a_m) =$ Ann.$(M) =$ Ann.$(b_m) = (p)$ and p is the maximal ideal of R.

Therefore, $\qquad\qquad$ Ann.$(a_i) = (p) =$ Ann.$(b_i) \; \forall \; i=1,2,....,n$

$\Rightarrow \quad$ Theorem is true for $l(A_{nn}, (M))=1$

Further, assume that theorem is true for all finitely generated module N over R with

$$l(A_{nn}, (N)) < l(A_{nn}, (M))$$

Thus, we have

$$pM = pRa_1 \oplus pRa_2 \oplus ....\oplus pRa_m = pRa_{s+1} \oplus pRa_{s+2} \oplus ....\oplus pRa_m$$

and $\qquad\qquad pM = pRb_1 \oplus pRb_2 \oplus ....\oplus pRb_m = pRb_{t+1} \oplus pRb_{t+2} \oplus ....\oplus pRb_m$

where, $\quad$ Ann.$(a_i) =$ Ann.$(a_2) = ...=$ Ann.$(a_s) = (p)$

Also, Ann.$(a_{s+1}) \neq (p)$ and Ann.$(b_1) =$ Ann. $(b_2) = ...=$ Ann.$(b_t) = (p)$

$\qquad\qquad$ Ann.$(b_{t+1}) \neq (p)$

Now, $\;$ Ann.$(pM) = (r|p)$, if Ann.$(M) = (r)$

Therefore,

$\qquad l(\text{Ann.}(p(M)) = l(\text{Ann.}(p(M)) - 1$

Then by induction

$$m-s = m-t \Rightarrow \quad s = t$$

and $\quad$ Ann.$(pa_i) =$ Ann.(pb_i) for $i=s+1,......,m$

Hence, $\;$ Ann.$(a_i) =$ Ann. $(b_i) \; \forall \; i=1,2,...,n$

EXERCISE 9.3

1. Let M be a finitely generated module over a PID R, then show that $M = \text{Tor}(M) \oplus F$, where F is a free module of finitely rank.

2. Find the Smith normal form and rank of the following matrices over a PID,

   (i) $\begin{bmatrix} 0 & 2 & -1 \\ -3 & 8 & 3 \\ 2 & -4 & -1 \end{bmatrix}$, $R = \mathbf{Z}$

   (ii) $\begin{bmatrix} -x-3 & 2 & 0 \\ 1 & -x & 1 \\ 1 & -3 & -x-2 \end{bmatrix}$, $R = \mathbf{Q}$

3. Obtain the invariant of the following matrix over $\mathbf{Q}(x)$.

$$\begin{bmatrix} 5-x & 1 & -2 & 4 \\ 0 & 5-x & 2 & 2 \\ 0 & 0 & 5-x & 3 \\ 0 & 0 & 0 & 4 \end{bmatrix}$$

Answers

2. (i) $\begin{bmatrix} 1 & 0 & 0 \\ 0 & 1 & 0 \\ 0 & 1 & 10 \end{bmatrix}$, rank$=3$

   (ii) $\begin{bmatrix} 1 & 0 & 0 \\ 0 & 1 & 0 \\ 0 & 0 & (x+1)^2(x+3) \end{bmatrix}$, rank$=3$

 **Chapter Review:** | *A competitive Approach*

Selected Terms and Results

Terms

■ **Module :** Let R be a ring. A R-module (left) is an additive abelian group M together with a function $R \times M \to M$ $\forall r, s \in R, a,b \in M$ with the following conditions.

(i) $r(a+b) = ra+rb$

(ii) $(r+s)(a) = ra +sa$

(iii) $r(sa) = (rs)(a)$

■ **Unitary R-module :** If R has identily element, i.e., $1 \in R$ and $1.a = \forall a \in m$ Then M is said to be a unitary R-module.

■ **Coset R-module :** Let R be a ring and S be a left ideal of R. Let $M = \{a+S : a \in R\}$ be the set of all cosets of S in R. Then M is a R-Module with the composition defined by

$(a+s) + (b+s) = (a+b)+s$ and $r(a+S)=ra+S$

■ **Submodules :** A non-empty subset S of a R-module M is said to be submodule if

(i) S is an additive subgroup of M

and (ii) $r \in R, a \in S \Rightarrow ra \in S$

■ **Module homomorphism :** A mapping f from a R-module M to other R-module N is said to be module homomorphism if

(i) $f(m_1 + m_2) = f(m_1) + f(m_2) \forall m_1, m_2 \in M$

(ii) $f(rm) = r.f(m) \forall r \in R, m \in M.$

■ **Modulcs Isomorphism :** A mapping f from a R-module M to other R-module N is said to be an isomorphism if

(i) f is a homomorphism.

(ii) f is one-one and onto.

■ **Kernel of a Module Homomorphism:** Let f be homomorphism of an R-module M into an R-module N then kernal of f denoted by $K(f)$ is the set of all those elements which are mapped to 0(identity element of N),

i.e., $K(f) = \{m \in M : f(m) =0$, where 0 is the identity element of additive group N.$\}$

■ **Quotient Module :** $M/A = \{A+m : m \in M\}$

■ **Cyclic Module :** A R-module M is said to be simple if it has no submodules other than it self and zero submodule.

■ **Semi-simple Module :** An R-module M is said to be semi-simple if it can be expressed as a direct sum of simple submodules.

■ **Basis of Module:** Let M be a module over a ring R with unity and let S be a subset of M then S is said to be basis for m if

■ (i) S generates M

■ (ii) S is linearly independent.

■ **Free Module :** An R-module M is said to be free module if there exists a subset S of M such that S generates M and S is linearly independent set.

■ **Noetherian Module:** An R-module is said to be Noetherian module if it satisfy the ascending chain condition.

■ **Artinian Module:** An R-module is said to be Artinian if it satisfies the descending chain condition.

■ **Uniform Module:** A non-zero R-module is said to be uniform if any two non-zero submodules of M have non-zero intersection.

■ **Primary Module:** An R module M is said to be primary if each non-zero submodule of M has uniform submodule and any two uniform submodules of M are subisomorphic.

■ **Row and Column Module:** Let R be a ring and A be $n \times n$ matrix over R. The submodule of R^n generated by the m rows of A is called the row module of A and the submodule of R^m generated by the n-columns of A and the submodule of R^m generated by the n-columns of A is known as column module of A.

■ **Rank of a Module:** Let A be a $m \times n$ matrix over R. The rank of the row module (column module) is known as the row rank (column rank) of A.

■ **Invariant Factor of a Smith Normal form:** The non-zero diagonal elements of Smith normal form are called the invariant factors of A.

■ **Torison and Torsion free Element:** Let R be a principal ideal domain and M be an R-module. An element $x \in M$ is called torsion element if $rx = 0$ for some non-zero $r \in R$. If

x is not a torsion element, then it is called torsion free element.

■ **Torsion and Torsion free module:** Let R be a principal ideal domain and M be an R-module. If $T(m)$ is a set of all torsion element of M, then $T(M)$ is a submodule of M. Also if $T(M) = M$, then M is called as torsion module and if $T(m)$, then M is called torsion free module.

Results

■ Module is a generalised concept of vector space.

■ Every abelian group is a module over the ring of integers.

■ Every left ideal M of a ring R is an R-module.

■ Every ring is a R-module itself.

■ Arbitrary intersection of submodules is again a submodule.

■ The linear sum of two submodules of an R-module is also a submodule of the R-module.

■ The kernel of a homomorphism is a submodule.

■ Let f be a module homomorphism. Then f is an isomorphism iff $\mathrm{Ker}(f) = 0$.

■ The range of a module homomorphism is a submodule.

■ An irreducible R-module is cyclic.

■ If R is an Euclidean ring, then any finitely generated R-module M is the direct sum of a finite number of cyclic modules (*Fundamental theorem*).

■ Polynomial ring $R(x)$ over a ring R is a R-module.

■ A subring S of a ring R is a module over the whole ring only if S is an ideal of R.

■ The set of rational numbers $\mathbf{Q}$ is not finitely generated Z-module.

■ If R is a ring with unity, then R-module M is cyclic if and only if $M \cong R/I$ for some left ideal I of R.

■ Every submodule and every homomorphic image of a semi-simple module is semi simple.

■ If R is a ring and I is a minimal left ideal of R. Also, if M is a faithful simple left R-module, then M is isomorphic to the left R-module I.

■ Every finitely generated module is a homomorphic image of a finitely generated free module.

■ If M is a finitely generated free module over a commutative ring R. Then, all basis of M are finite.

■ If R is a commutative ring and M is a finitely generated free module over R, then all basis of M have the same number of elements.

■ The finitely co-generated modules satisfies the certain chain condition on their submodules.

■ Every submodule of a Noetherian (Artinian) module is Noetherian (artinian).

■ Every homomorphic image of a Noetherian (artinian) module is Noetherian (artinian).

■ A ring is called Noetherian if it is Noetherian as a left module over itself, *i.e.*, every left ideal is finitely generated.

■ Every free module M over a ring R with identity is protective.

■ Every module M over a ring R is the homomorphic image of protective R-rnodule.

■ Every unitary module M over a ring R with identity may be embedded in an injeclive R-module

Review Questions and Project Work

1. Let M be the set of all $m \times n$ matrix over a ring R. Show that M is a module over R.

2. Let G be a multiplicative abelian group. Define $ng = g^n$ for $g \in G$ and $n \in \mathbf{Z}$. Show that G is a Z-module.

3. Let M be an additive abelian group. Show that there is one way of making it a Z-module.

4. If M is a R-module and $x \in M$, prove that the set
$$K = \{rx + nx : r \in R, n \in \mathbf{Z}\}$$
is a R-module of M containing x. Further if R has unity then prove that $K = Rx$.

5. Let R be a commutative ring over I, a finitely generated ideal in R with $I = I^2$. Show that I is a direct summand of R. Also, give an example to show that if I is not finitely generated then I need not be a summand.

6. If A and B are R-submodule of an R-module M. Show that $(A + B)|B \cong A|A \cap B$.

7. Let M be a completely reducible module and let K be a non-zero submodule of M. Show that K is completely reducible. Also show that K is a direct summand of M.

8. Show that every finitely generated module is a homomorphic image of a finitely generated free module.

9. Show that every principal left ideal is an integral domain R with unity is free as a left R-module.

10. Let R be a commutative noetherian ring and let S be a multiplicative subset of R. Show that the ring of fraction R_S is also Noetherian.

11. Let R be a left Artinian integral domain with more than one element. Show that R is a division ring.

12. Let R be a prime left Artinian ring with unity. Show that R is isomorphic to the $n \times n$ ring over a division ring. Hence, show that a prime ideal is an Artinian ring is maximal.

13. Show that an infinite dimensional vector space is neither artinian nor noetherian.

14. Show that the abelian group generated by x_1 and x_2 subject to $x_1 + x_2 = 0$ is isomorphic to $\mathbf{Z}$.

Objective type Questions

Fill in the Blanks

1. Every abelian group G is a module over the _____ .

2. Let M be an additive abelian group. Then there is _____ way of making it a Z-module.

3. Let $f : M \rightarrow N$ be a R-module of N. Then the range of homomorphism f is an _____ of N.

4. An anti-homomorphism which is one-one and onto is called _____ .

5. Any minimal R-submodule N of an R-module M is _____ .

6. An R-submodule N of M is _____ in M if and only if M/N is simple.

7. Any one-dimensional vector space is _____ .

8. A R-module M is called _____ if it is a direct sum of a family of minimal submodule.

9. An R-module M is called _____ if M admits a basis.

10. Let M be a finitely generated free module over a commutative ring R. Then all basis of M are _____ .

True/False

Write 'T' for true and 'F' for false statement.

1. If R is a field then the rank of M is called dimension of the vector space. **(T/F)**

2. Every ideal of $\mathbf{Z}$ is not necesarily is free as Z-module. **(T/F)**

3. Every principal left ideal is an integrated domain R with unity is free as a left R-module. **(T/F)**

4. Every finite abelian group is free as a module over Z. **(T/F)**

5. A module which has only finitely many submodules is Artinian. **(T/F)**

6. Infinite cycle group are Artinian. **(T/F)**

7. If M is Artinian then every quotient module of M is finitely generated. **(T/F)**

8. The ring of integers is Noetherian and as well as Artinian. **(T/F)**

9. Every homomorphic image of a Noetherian module is a Noetherian. **(T/F)**

10. The subring of a Noetherian ring is Noetherian. **(T/F)**

Multiple Choice Questions

Choose the most apporiate one.

1. The ring of integers is :
   (a) Noetherian
   (b) Artinian
   (c) both (a) and (b) are true
   (d) none of the above

2. The subring of Artinian ring is :
   (a) Artinian
   (b) need not be Artinian
   (c) never Artinian
   (d) none of the above

3. If R is a Noetherian ring then $ab = 1, a, b \in R$ if and only if :
   (a) $a = 1$
   (b) $b = 1$
   (c) $ba = 1$
   (d) none of the above

4. If R is right Noetherian then :
    (a) maximum condition holds for right ideal of R
    (b) every right ideal of R is finitely generated
    (c) both (a) and (b) are true
    (d) none of the above

5. The set of integers **Z** as a **Z**-module is :
    (a) uniform module
    (b) primary module
    (c) both (a) and (b) are true.
    (d) none of the above

6. Let R be a PID with unity then :
    (a) every irreducible element is prime in R.
    (b) every non-zero prime ideal is maximal.
    (c) both (a) and (b) are true.
    (d) none of the above

7. If $rx = 0$, $r \in R$ $r \neq 0$ then element x is called :
    (a) torsion element
    (b) torsion free element
    (c) Free element
    (d) None of the above

8. Which one of the following is not module ?
    (a) An additive abelian group
    (b) A ring R with property $a, m \in R \Rightarrow am \in R$
    (c) The polynomial ring $R[x]$ over a ring R
    (d) All are true.

9. An R-module M is called semi-simple if :
    (a) it is a direct sum of a family of minimal submodule.
    (b) it is a direct sum of simple submodules.
    (c) both (a) and (b) are true.
    (d) none of the above

10. A boolean Noetherian ring is :
    (a) finite
    (b) a finite direct product of fields with two elements
    (c) both (a) and (b) are true.
    (d) none of the above

Answers

Fill in the Blanks

1. ring of integers   **2.** one   **3.** R-submodules   **4.** anti-isomorphism   **5.** simple   **6.** semi-simple
7. simple   **8.** semi-simple   **9.** free module   **10.** finite

True/False

1. T   **2.** F   **3.** T   **4.** T   **5.** T   **6.** F   **7.** T   **8.** F   **9.** T
10. F

Multiple Choice Questions

1. (a)   **2.** (b)   **3.** (c)   **4.** (c)   **5.** (c)   **6.** (c)   **7.** (b)   **8.** (d)   **9.** (c)
10. (c)

● ● ●

10 Extension Fields and Galois Theory

Outlines

- Field extension (Simple and algebraic)
- Multiple roots
- Algebraically closed fields
- Finite Fields
- Solvability by radicals
- Construction with ruler and compass

- Splitting or decomposition field
- Normal and separable extension
- Galois group
- Galois field
- Cyclic extensions

10.1 INTRODUCTION

It is known that a commutative ring with unity in which every non-zero element having a multiplicative inverse is known as field. If F is a field and $a \neq 0$, $a \in F$, $b \in F$, then $a^{-1}b$ can also be written as $a|b$. Thus $a|b$ is an element of F obtained on dividing b by a. Therefore, a field is a commutative ring in which we can divide by any non-zero element. In this chapter, we shall discuss the theory of finite field extensions and related properties.

10.2 FIELD EXTENSIONS

Definition 1. *Let F be a field. Then, a field K is said to be an extension of F if F is a subfield of K.*

For example: R is an extension field of Q, and C is an extension field of R and Q.

REMARK

☞ If F is a subfield of a field K, then K can be regarded as a vector space over F under the ordinary field operations in K.

Definition 2. *Let K be an extension of the field F. The dimension of K as a vector space over F, i.e., the dimension of the vector space $K(F)$ is known as the degree of K over F. The degree of K over F is denoted by $[K : F]$.*

Definition 3. *(Finite Field Extension). Let F be a field and K be an extension of F. Then, K is said to be a finite extension of F if the degree of K over F is finite.*

REMARKS

☞ K is finite extension of F if the vector space $K(F)$ is finite dimensional.

☞ A field extension of a field F is an ordered pair (K, ϕ) where K is a field and ϕ is a monomorphism of F into K.

For example : If K is a field and F is a subfield of K. The injective mapping $i:F \to K$ defined by $i(x) = x \ \forall x \in F$ is a monomorphism. Thus (K, i) a field extension of F.

Definition 4. *A field extension K is said to be finite or infinite extension of F according as the degree of K(F) is finite or infinite.*

For example :

(1) The field **C** of complex numbers is a finite extension of the field **R** of real numbers. Also, [**C** : **R**] = 2.

(2) Since a field *F* can be regarded as a subfield of *F*. Thus, *F* is an extension of *F*. Therefore, dimension of the vector space *F(F)* is one. The unit element 1 of *F* is a basis of this vector space. Hence, the degree of *F* over *F* is one, *i.e.*, [*F*:*F*]=1.

Theorem 1. *(Transitivity of Finite Extensions): Let L be a finite extension of K and if K is finite extension of F, then L is a finite extension of F, i.e.,* [*L*:*F*] = [*L*: *K*][*K*:*F*].

Step Outlines : To make the proof easier, use the following steps :

Step 1. *Suppose* [*L* : *K*] =*m and* [*K* : *F*] =*n and let* $\alpha_1, \alpha_2, ..., \alpha_m \in L$ *and* $\beta_1, \beta_2, ..., \beta_n \in K$

Step 2. *Prove that the set* (α_i, β_j) *generates L over F.*

Step 3. *Prove that the set* (α_i, β_j) *is linearly independent.*

Proof. Let *L* be a finite extension of *K*. Let *K* be a subfield of *L* and *F* be a subfield of *K*, *i.e.,*

$$F \subseteq K \subseteq L$$

Let[*L* : *K*] = *m* and [*K*:*F*] = *n*.

Now, suppose that $\alpha_1, \alpha_2, ...\alpha_m$ is a basis of *L* over *K* and $\beta_1, \beta_2, ...\beta_n$ is a basis of *K* over *F*.

Then, $\alpha_1, \alpha_2, ...\alpha_m \in L$ and $\beta_1, \beta_2, ...\beta_n \in K$.

Since $K \subset L$, therefore, $\beta_1, \beta_2, ...\beta_n \in L$.

Thus the *mn* elements $\alpha_i \beta_j$ where $i = 1, ...m, j = 1, ..., n$ are all in *L*.

We have to prove that the set of these *mn* elements forms a basis of *L* over *F*, *i.e.*, we shall have [*L* : *F*] = *mn*.

(i) *Set* $[\alpha_i, \beta_j]$ *generates L over F :*

Let *l* be any element of *L*. Now, since $[\alpha_1, \alpha_2, ...\alpha_m]$ is a basis of *L(K)*, thus *l* can be expressed as a linear combinations of $\alpha_1, ...\alpha_m$ over the elements in *K*.

Therefore, we have

$$l = \sum_{i=1}^{m} k_i \alpha_i, k_i \in K \qquad \qquad ...(1)$$

Now, $k_i \in K$ and $[\beta_1, ..., \beta_n]$ is a basis of *K(F)*. Then we have

$$k_i = \sum_{i=1}^{m} f_{ij} \beta_i, f_{ij} \in F \qquad \qquad ...(2)$$

From (1) and (2), we conclude that

$$l = \sum_{i=1}^{m} \left(\sum_{j=1}^{n} f_{ij} \beta_j \right) \alpha_i = \sum_{i=1}^{m} \sum_{j=1}^{n} f_{ij} \left(\alpha_i \beta_j \right), f_{ij} \in F.$$

l is a linear combination of the elements $\alpha_i \beta_j$ elements over *F*. Thus, the set of *mn* elements $\alpha_i \beta_j$ generates the vector space *L(F)*.

(ii) *The set* $[\alpha_i, \beta_j]$ *is linearly independent :*

Here, we shall show that the set $[\alpha_i, \beta_j]$ is linearly independent over F.

We have

$$\sum_{i=1}^{m} \sum_{j=1}^{n} f_{ij} \left(\alpha_i \beta_j \right) = 0, f_{ij} \in F$$

$$\Rightarrow \qquad \sum_{i=1}^{m} \left(\sum_{j=1}^{n} f_{ij} \beta_j \right) \alpha_i = 0$$

$$\Rightarrow \qquad \sum_{j=1}^{n} f_{ij} \beta_j = 0, \qquad \text{for } i = 1, \ldots, m$$

$$(\because [\alpha_1, \ldots, a_m] \text{ is a basis of } L(K) \text{ and each } f_{ij} \beta_j \in K)$$

$$\Rightarrow \qquad f_{ij} = 0, \text{ for } i = 1, \ldots, m, \quad j = 1, \ldots, n$$

$$(\because [\beta_1, \ldots, \beta_n] \text{ is a basis of } K(F) \text{ and each } f_{ij} \in F)$$

$$\Rightarrow \qquad [\alpha_i \beta_j] \text{ is linearly independent over } F.$$

Thus, the set of these mn elements forms a basis of L over F.

$$\Rightarrow \qquad [L : M] = mn$$

$$\Rightarrow \qquad L \text{ will also be a finite extension of } F.$$

Hence, $\qquad [L{:}F] = [L{:}K][K{:}F]$

Theorem 2. *Let L be a finite extension of F and if K is a subfield of L which contains F, then $[K{:}F] \mid [L{:}F]$, i.e., $[K : F]$ is a divisor of $[L{:}F]$.*

Proof.   Let L, K and F be three fields such that $F \subseteq K \subseteq L$. Now, suppose that $[L : F]$ is finite and is equal to n. Also, let $[\alpha_1, \ldots, a_n]$ be a basis of L over F. Then, by definition $[\alpha_1, \ldots, a_n]$ generates L over F. Since $K \subseteq F$. Thus, any linear combination of $\alpha_1, \ldots, \alpha_n$ over F will also be a linear combination of $\alpha_1, \ldots, \alpha_n$ over K. Then, the set $[\alpha_1, \ldots, \alpha_n]$ also generates L over K. This set may not be linearly independent over K. Since, $L(K)$ is generated by a finite set, thus it is finite dimensional vector space and $[L : K]$ is finite. Also, $K(F)$ is a subspace of $L(F)$. Also, since $[L : F]$ is finite, thus $[K : F]$ is finite. We know that each subspace of a finite dimensional vector space is also finite dimensional. Using previous theorem, we have

$$[L{:}F] = [L{:}K][K{:}F]$$

$$\Rightarrow \quad [K{:}F] \text{ is a divisor of } [L{:}F].$$

REMARK

☞ If $[L : F]$ is a prime number, then there can be no field K properly contained between L and F, i.e., if $[L{:}F]$ is prime and K is any subfield of L containing F, then either we have $K = L$ or $K = F$.

Theorem 3. *If K is an extension of F, then $K = F$ if and only if $[K : F] = 1$.*

Proof.   If $K = F$, then $[K : F] = [K : K] = 1$.

If $[K : F] = 1$, let $[a]$ be a basis of K over F. Thus, $1 \in K \Rightarrow 1 = \alpha a, \alpha \in F, \alpha \neq 0$ as $1 \neq 0$

$$\Rightarrow \qquad a = \alpha^{-1} \in F.$$

Now, let $b \in K \Rightarrow b = \beta a, \beta \in F, a \in F.$

Therefore, $b \in F \Rightarrow K \subseteq F.$ Hence, $K = F.$

Theorem 4. *Let L be an extension of F and [L : F] is a prime number p, then there is no field K such that $F \subset K \subset L$.*

Proof. Let if possible, there exists a field K such that $F \subseteq K \subseteq L$.

Then, we have

$$p = [L : F] = [L : K] \, [K : F]$$

$\Rightarrow$ $\qquad [L{:}K] = 1$ or $[K{:}F] = 1$

$\Rightarrow$ $\qquad\qquad K = L$ or $K = F$, which is a contradiction.

Hence, there is a no field K such that $F \subset K \subset L$.

REMARK

☞ If K is an extension of F of prime degree then for every $a \in K$, $F(a) = F$ or $F(a) = K$.

10.3 FIELD ADJUNCTIONS

Let K be an extension of a field F. Let $a \in K$. Let C be the collection of all subfields of K containing both F and a. Clearly C is not empty because at least K itself belongs to C. Also, the intersection of an arbitrary collection of subfields of K is also a subfield of K. Further if $F(a)$ denote the intersection of all those subfield of K which are members of C. Then, clearly $F(a)$ is a subfield of K. Then $F(a)$ containing both F and a because each member of C contained both F and a.

$\Rightarrow$ $\qquad\qquad F(a)$ is a member of C.

Now, if E is any subfield of K containing both F and a, then $F(a)$ will be contained in E because $F(a)$ is the intersection of all the members of C and E is a member of C. Therefore $F(a)$ is a subfield of K containing both F and a and itself is contained in any subfield of K containing F and a. Thus, $F(a)$ is the smallest subfield of K containing both F and a. This $F(a)$ is known as subfield obtained by adjoining a to F. Also, a has been adjoined to F. This process is called field adjunction.

Construction of F(a). Let K be an extension of a field F. If $a \in K$ and

$$U = \left\{ \frac{k_0 a^n + k_1 a^{n-1} + \dots\dots\dots + k_n}{l_0 a^m + l_1 a^{m-1} + \dots\dots\dots + l_m} : k_i, l_k \in F \right\}$$

Here, $l_0 a^m + l_1 a^{m-1} + \dots\dots + l_m$ is not equal to the zero element of K and n, m are any non-negative integers. Then, we can easily verify that

(i) $\alpha, \beta \in U \Rightarrow \alpha - \beta \in U$

(ii) $\alpha \in U, 0 \neq \beta \in U \Rightarrow a | \beta \in U$

Therefore, U is a subfield of K. Also, $U = F(a)$.

10.4 SIMPLE EXTENSION OF A FIELD

Let K be an extension of a field F. Let $a \in K$. Also, let $T = F(a)$. Since $F(a)$ is the subfield of K. Then K is an extension of $F(a)$. If W is the subfield of K obtained by adjoining b to $F(a)$. Then, $W = (F(a))(b)$, we write this by $F(a,b)$.

Now, $\qquad\qquad\qquad F(a,b) = (F(a))(b)$

$\qquad\qquad\qquad\qquad\quad =$ smallest subfield of K containing both $F(a)$ and b

$\qquad\qquad\qquad\qquad\quad =$ smallest subfield of K containing F,a and b.

Similarly, $\qquad F(b, a)$ = smallest subfield of K containing F, a and b.

We know that the subfield of K obtained by adjoining both a and b to F. Similarly, if $a_1, a_2, ..., a_n$ ∈ K, then $F(a_1, a_2, ..., a_n)$ will be described as the subfield of K generated by $F, a_1, a_2, ..., a_n$. Hence, $F(a_1, a_2, ..., a_n)$ will be the smallest subfield of K containing K as well as $a_1, a_2, ..., a_n$.

Definition 1. *An extension K of a field F is called a simple extension of F if $K=F(a)$ for some a in K.*

The element a is known as primitive *element of K over F.*

Definition 2. *A field K is said to be finitely generated over F if there exists a finite number of elements say $a_1, ..., a_n$ ∈ K such that $K = F(a_1, a_2, ..., a_n)$.*

10.5 ALGEBRAIC EXTENSION OF A FIELD

Let $F(x)$ be the ring of polynomials in x over F. Let $g(x)$ ∈ $F(x)$ such that

$$q(x) = a_0 x^m + a_1 x^{m-1} + ... + a_m$$

If b ∈ K, where K is any extension of F. Then

$$q(b) = a_0 b^m + a_1 b^{m-1} + ... + a_m$$

Here, $q(b)$ is called the value of $q(x)$ for $x = b$. The element b is said to satisfy $q(x)$ if $q(b)=0$. In this case b is a root of $q(x)$.

Definition 1. *(Algebraic Element). Let K be an extension of a field F. An element a ∈ K is said to be algebraic over F, if there is a non-zero polynomial $p(x)$ ∈ $F(x)$ for which $p(a)=0$.*

or

An element a ∈ K is said to be algebraic over F if there exists elements $\beta_0, \beta_1, ..., \beta_n$ in F, not all 0, such that $\beta_0 a^n + \beta_1 a^{n-1} + ... + \beta_n = 0$.

Definition 2. *(Transcendental Element). Let K be an extension of a field F. An element a ∈ K is said to be transcendental over F if it is not algebraic over F.*

Definition 3. *(Algebraic Number). A complex number is said to be an algebraic number if it is algebraic over the field of rational numbers.*

Definition 4. *(Transcendental Number). A complex number which is not algebraic is called Transcendental number.*

For example, the number e is transcendental number.

Definition 5. *(Minimal Polynomial). Let K be an extension of a field F and a ∈ K is algebraic over F. If $p(x)$ is a polynomial over F of lowest positive degree satisfied by a, then $p(x)$ is known as minimal polynomial for a over F.*

Definition 6. *(Monic Polynomial). A non-zero polynomial $f(x)$ ∈ $F(x)$ is said to be monic over F, if the leading coefficient, i.e., coefficient of highest power of x in $f(x)$ is 1 in F.*

REMARK

☞ If the degree of minimal polynomial of a over F is n, then a is said to be an algebraic element of degree n.

Definition 7. *(Algebraic Extension of a Field). Let K be an extension of a field F. Then, K is said to be an algebraic extension of F if every element of K is algebraic over F.*

Definition 8. *If there exists a ∈ K such that a is not algebraic over F, then K is known as transcendental extension of F.*

For example:

(1) The field of complex numbers is an algebraic extension of the field of real numbers.

(2) The field of real numbers is not an algebraic extension of the field of rational numbers. Because π is an element of **R** which is not algebraic over **Q**. Hermite proved that there exists no non-zero polynomial with rational coefficients satisfied by π.

REMARK

☞ If F is any field, then F is always an algebraic extension of F.

Theorem 1. *Let F be a field and a $\in$ K be algebraic over F. Then, any two minimal polynomials for a over F are equal.*

Proof. Consider two monic polynomials

$$x^n + \alpha_1 x^{n-1} + ... + \alpha_n \text{ and } x^n + \beta_1 x^{n-1} + ... + \beta_n$$

for a over F. Then, we have

$$a^n + \alpha_1 a^{n-1} + ... + \alpha_n = 0 = a^n + \beta_1 a^{n-1} + ... + \beta_n$$

$\Rightarrow \qquad (\alpha_1 - \beta_1)a^{n-1} + (\alpha_2 - \beta_2)a^{n-2} + ... + (\alpha_n - \beta_n) = 0$

$\Rightarrow \quad a$ satisfies the polynomials $q(x) = (\alpha_1 - \beta_1)x^{n-1} + ... + (\alpha_n - \beta_n)$ belonging to $F(x)$.

$\Rightarrow \quad q(x)$ must be the zero polynomial because minimal polynomial for a over F is of degree n while $q(x)$, if it is not the zero polynomial, is of degree less than n.

$\Rightarrow \qquad \qquad a_1 - \beta_1 = 0 ......, \alpha_n - \beta_n = 0$

$\Rightarrow \qquad \qquad a_1 = \beta_1 ...... \alpha_n = \beta_n.$

Hence, $x^n + \alpha_1 x^{n-1} + ... + \alpha_n = x^n + \beta_1 x^{n-1} + ... + \beta_n$

Theorem 2. *Let K be an extension of a field F and let $\alpha \in K$ be an algebraic element over F. Then α is a root of a unique monic polynomial f(x) of positive degree over F is satisfied by α, then $f(x) | g(x)$.*

Step Outlines: To make the proof easier, use the following steps :

Step 1. *Show that there is a non-zero polynomial in F(x) which is satisfied by α.*

Step 2. *Show that f(x) is a monic polynomial of positive degree over F satisfied by α.*

Step 3. *Prove the uniquness of f(x).*

Proof. Let F be a field and $\alpha \in K$ is an algebraic element. Then, there is a non-zero polynomial in $F(x)$, which is satisfied by α.

If $h(x) = a_0 + a_1 x + a_2 x^2 + ... + a_r x^n$, $a_0 \neq 0$ is a non-zero polynomial of $F(x)$ such that $h(\alpha) = 0$.

Then $\qquad \qquad h(\alpha) = 0 \qquad \Rightarrow \qquad a_0 + a_1 \alpha + a_2 \alpha^2 + ... + a_n \alpha^n = 0$

Because $a_0 \neq 0$, then $n \neq 0$. Therefore, $n > 0$

$\Rightarrow \qquad \qquad \deg (h(x)) > 0$

$\Rightarrow \qquad \qquad h(x) = a_0 + a_1 x + a_2 x^2 + ...$ with $a_n \neq 0$, for some n.

Define $\qquad \qquad f(x) = \dfrac{1}{a_n} h(x)$

Then, we have

$$f(x) = \frac{1}{a_n} (a_0 + a_1 x + a_2 x^2 + ... + a_{n-1} x^{n-1}) + x^n$$

which is a monic polynomial of degree n over F.

Further, $$f(\alpha) = \frac{1}{a_n} h(\alpha) = 0$$

$\Rightarrow$ $f(x)$ is a monic polynomial of positive degree over F satisfied by α.

Let $g(x)$ be any other polynomial over F satisfied by α. Then, by division algorithm, there exist two polynomials $q(x)$ and $r(x)$ of $F(x)$ such that

$$g(x) = g(x)f(x) + r(x) \qquad \qquad \text{...(1)}$$

where either $r(x) = 0$ or deg. $r(x) < \deg f(x)$.

But we have $g(\alpha) = 0$

Then, $g(\alpha) = 0 \Rightarrow g(\alpha)f(\alpha) + r(\alpha) = 0$

$\Rightarrow$ $r(\alpha) = 0$ $[\because f(\alpha) = 0]$

Now, since no non-zero polynomial of degree less than $f(x)$ is satisfied by α and $\deg(r(x)) < \deg f(x)$.

Thus, $r(x) = 0$

$\Rightarrow$ $g(x) = q(x)f(x)$ [Using (1)]

$\Rightarrow$ $f(x) | g(x)$

Now, it remains to prove that $f(x)$ is unique.

Let if possible $f_1(x)$ be another monic polynomial of positive degree satisfied by α, then $f_1(\alpha) = 0$.

Also, $f_1(x)$ is a divisor of any other non-zero polynomial satisfied by α. Thus,

$$f_1(x) | f(x) \qquad \qquad [\because f(\alpha) = 0]$$

Thus, there exists a polynomial $p(x) \in F[x]$ such that

$$f(x) = p(x)f_1(x)$$

But $f(x)$ is monic polynomial of smallest degree over F satisfied by a, so that $p(x)$ is a non-zero constant polynomial whose coefficient is 1.

i.e., $p(x) = 1 \Rightarrow f(x) = f_1(x)$.

Hence, $f(x)$ is unique.

Theorem 3. *Let F be a field and $a \in K$ be algebraic over F. If $p(x)$ is a minimal polynomial for a over F. Then $p(x)$ is irreducible over F.*

Proof.   Let $p(x)$ be a polynomial in $F(x)$ of smallest positive degree such that $p(a) = 0$.

Let if possible $p(x)$ is not irreducible over F. Then, by definition $p(x)$ can be resolved into non-trivial factors.

Let $p(x) = f(x) g(x)$, where $f(x)$ and $g(x)$ are polynomials of positive degree in $F(x)$ and each of them is of degree less than that of $p(x)$.

Now, we have

$$p(a) = f(a)g(a)$$

$\Rightarrow$ $0 = f(a) g(a)$ $[\because p(a) = 0 \text{ because } a \text{ satisfies } p(x)]$

$\Rightarrow$ $f(a) = 0 \text{ or } g(a) = 0$

$\Rightarrow$ a satisfies $f(x)$ or $g(x)$.

$\Rightarrow$ $P(x)$ is not a minimal polynomial for a over F because

$$\deg f(x) < \deg (p(x))$$

and $\deg(g(x)) < \deg (p(x))$

Further, since $p(x)$ is a minimal polynomial for a over F, which contradicts the fact that $p(x)$ is not irreducible over F.

Theorem 4. *Let K be an extension of a field F and $a \in K$ be algebraic of degree n over F. Then*

$$F(a) = [\beta_0 + \beta_1 a + \beta_2 a^2 + \dots\dots + \beta_{n-1} a^{n-1} : \beta_0, \beta_1, \dots, \beta_{n-1} \in K]$$

Further, the expression for each element of $F(a)$ in the form

$$\beta_0 + \beta_1 a + \dots\dots + \beta_{n-1} a^{n-1} \text{ is unique.}$$

Proof.   In case of $n = 1$, we get $F(a) = F$. Then, theorem is true.

If $n > 1$. Since it is given that $a \in K$ is algebraic of degree n over F. Thus the minimal polynomial for a over F is of degree n.

Now, let $p(x) = x^n + \alpha_1 x^{n-1} + \dots\dots + \alpha_n$ be the minimal polynomial for a over F. Then

$$a^n + \alpha_1 a^{n-1} + \dots\dots + \alpha_n = 0$$

$\Rightarrow$ $a^n = - (\alpha_1 a^{n-1} + \dots\dots + \alpha_n)$ ...(1)

$\Rightarrow$ $a^{n+1} = - (\alpha_1 a^n + \dots\dots + \alpha_{n+1})$

$\Rightarrow$ $= - (-\alpha_1(\alpha_1 a^{n-1} + \dots\dots + \alpha_n) + \alpha_2 a^{n-1} + \dots\dots + \alpha_n . a)$ [Using (1)

$\Rightarrow$ a^{n+1} is a linear combination of the elements $1, a, \dots, a^{n-1}$ over F.

Continuing this process, we can easily show that a^{n+k}, for $k \geq 0$ is a linear combination over F of $1, a, \dots a^{n-1}$.

Let us write

$$T = [\beta_0 + \beta_1 a + \dots\dots + \beta_{n-1} a^{n-1} : \beta_0, \beta_1, \dots, \beta_{n-1} \in F]$$

We have to show that $T = F(a)$. For this we shall prove that T is a subfield of K.

Let $u = \beta_0 + \beta_1 a + \dots\dots + \beta_{n-1} a^{n-1}$

$v = \gamma_0 + \gamma_1 a + \dots\dots + \gamma_{n-1} a^{n-1}$

be any two elements of T. Then, we have

$$u - v = (\beta_0 - \gamma_0) + (\beta_1 - \gamma_1)a + \dots\dots + (\beta_{n-1} - \gamma_{n-1})a^{n-1} \in T$$

Now, let

$$0 \neq u = \beta_0 + \beta_1 a + \dots\dots + \beta_{n-1} a^{n-1} \in T$$

Also, let

$$q(x) - (\beta_0 + \beta_1 x + \dots\dots + \beta_{n-1} x^{n-1}) \in F(x)$$

Then, $q(a) = u \neq 0$

We want to show that $q(x)$ is not a divisor of $p(x)$. If $q(x)$ is a divisor of $p(x)$, then we must have

$$p(x) = (a_0 x + a_1) q(x)$$ ...(2)

where $a_0 x + a_1$ is a non-zero polynomial in $F(x)$.

From (2), we get

$$p(a) = (a_0 a + a_1) q(a)$$

$\Rightarrow$ $0 = (a_0 a + a_1) q(a)$

$\Rightarrow$ $a a_0 + a_1 = 0$

$\Rightarrow$ a satisfies a polynomial $a_0 x + a_1$ of degree 1 over F because deg $p(x)=n$ which is greater than 1.

$\Rightarrow$ $q(x)$ is not a divisor of $p(x)$. Further, since $p(x)$ is irreducible over F, $q(x)$ and $p(x)$ must be relatively prime. Thus, we can find polynomials $s(x)$ and $t(x)$ in $F(x)$, such that

$$p(x)s(x) + q(x)t(x) = 1$$

$\Rightarrow$ $\qquad\qquad p(a)s(a) + q(a)t(a) = 1$

$\Rightarrow$ $\qquad\qquad q(a)\, t(a) = 1$ $\hfill [\because p(a)=0 \text{ and } q(a)=u.]$

$\Rightarrow$ $\qquad\qquad\quad ut(a) = 1$

$\Rightarrow$ $t(a)$ is the inverse of u.

Now in $t(a)$, all powers of a higher than $n-1$ can be replaced by linear combinations of $1, a, ..., a^{n-1} \in F$.

$\Rightarrow$ $\qquad\qquad\quad t(a) \in T$

$\Rightarrow$ $\qquad\qquad t(a) - u^{-1} \in T$ $\hfill [\because t(a) \text{ is the inverse of } u]$

Further, in the product of $u^{-1}v$ all powers of a higher than $n-1$ can be replaced by linear combinations of $1, a, ..., a^{n-1}$ over F.

Thus, $0 \neq u, \quad v \in T \quad \Rightarrow \quad u^{-1}v \in T$

$\Rightarrow$ T is a subfield of K.

Obviously both F and a are in T. Since K is closed under addition and multiplication, therefore, any subfield of K which contains both F and a must contain T.

$\Rightarrow$ $\qquad\qquad T$ is the smallest subfield of K containing both F and a.

Therefore, $\qquad\quad T = F(a)$

Further, let $u \in T$. Now, let

$$u = \beta_0 + \beta_1 a + ... + \beta_{n-1} a^{n-1}$$

and $\qquad\qquad v = \gamma_0 + \gamma_1 a + ... + \gamma_{n-1} a^{n-1}$

Then clearly, we have

$$\beta_0 + \beta_1 a_1 + ... + \beta_{n-1} a^{n-1} = \gamma_0 + \gamma_1 a + ... + \gamma_{n-1} a^{n-1}$$

$\Rightarrow$ $\qquad (\beta_0 - \gamma_0) + (\beta_1 - \gamma_1)a + ... (\beta_{n-1} - \gamma_{n-1})\, a^{n-1} = 0$

$\Rightarrow$ a satisfies the polynomial.

$$h(x) = (\beta_0 - \gamma_0) + (\beta_1 - \gamma_1)x + ... (\beta_{n-1} - \gamma_{n-1})\, x^{n-1} \in F(x)$$

$\Rightarrow$ $h(x)$ must be the zero polynomial, because otherwise a will not be of degree n over F.

$\Rightarrow$ $\qquad\quad \beta_0 - \gamma_0 = 0, \beta_1 - \gamma_1 = 0 = ... \beta_{n-1} - \gamma_{n-1} = 0$

$\Rightarrow$ $\qquad\qquad \beta_0 = \gamma_0, \beta_1 = \gamma_1 = ...\beta_{n-1} = \gamma_{n-1}$

Hence, the expression for u in the form $\beta_0 + \beta_1 a + ...+ \beta_{n-1} a^{n-1}$ is unique.

Theorem 5. *Let F be a field and K be an extension of F. Let $a \in K$ be algebraic over F. If a satisfies an irreducible polynomial $p(x)$ in $F(x)$, then $p(x)$ must be a minimal polynomial for a over F.*

Proof. Define $\qquad M = \{f(x) \in F(x): f(x) = 0\}$.

Firstly, we shall prove that M is an ideal of $F[x]$.

Let $f(x), g(x) \in M$. Then, $f(a) = 0, g(a) = 0$

Define $\qquad d(x) = f(x) - g(x)$

$\Rightarrow \qquad\qquad d(a) = f(a) - g(a) = 0$

$\Rightarrow \qquad\qquad d(x) \in M$

Further, let $f(x) \in M$ and $h(x) \in F[x]$, then $f(a) = 0$.

Let $\qquad\qquad t(x) = f(x). h(x)$

$\Rightarrow \qquad\qquad t(a) = f(a). h(a) = 0 . h(a) = 0$

$\Rightarrow \qquad\qquad t(x) \in M$

$\Rightarrow \qquad M$ is an ideal of $F[x]$.

Also, $\qquad\qquad M \neq F[x]$.

$$[\because \text{ if } f(x) = 1 \in F[x], \text{ then } f(a) = 1 \neq 0 \Rightarrow f(x) = 1 \text{ is not in } M]$$

Further, $p(x)$ is an irreducible polynomial in $F[x]$. Thus, the ideal $N = (p(x))$ of $F[x]$ generated by $p(x)$ is a maximal ideal because $F[x]$ is an Euclidean ring.

Now, we have $p(x) \in M$, because we have given $p(a) = 0$.

If $n(x) = m(x) p(x)$ is any element of $(p(x))$, then

$$n(a) = m(a)p(a) = 0$$

$\Rightarrow \qquad\qquad n(x) \in M$

$\Rightarrow \qquad\qquad N \subset M$

$\Rightarrow \qquad M$ is an ideal of $F[x]$ contained between N and $F[x]$.

$\Rightarrow \qquad\qquad N \subseteq M \subseteq F[x]$

Now, it remains to prove that $p(x)$ is minimal. Let if possible, $p(x)$ is not minimal for a over F.

If $s(x)$ be a polynomial in $F[x]$ of degree less than that of $p(x)$ and satisfied by a. Now, since $s(a) = 0 \Rightarrow s(x) \in M$.

$\Rightarrow \qquad\qquad s(x) = p(x) r(x) \text{ , for some } r(x) \in F[x]$

which is not possible because

$$deg(s(x)) < deg \ (p(x)).$$

Hence, $p(x)$ must be a minimal polynomial for a over F.

Theorem 6. *Let F be a field and K be an extension of F. An element $a \in K$ is algebraic over F if and only if $F(a)$ is a finite extension of F.*

Proof. Let us first suppose $F(a)$ is a finite extension of F. We have to prove that a is algebraic over F.

Let $\qquad [F(a) : F] = m$

Since $F(a)$ is a field and $a \in F(a)$, then all $m + 1$ elements $1, a, a^2 , ..., a^{m-1} , a^m$ are all in $F(a)$.

Now, since the dimension of the vector space $F(a)$ over F is m, thus, these $m+1$ elements of $F(a)$ are linearly dependent over F. Therefore, there exist elements α_0, $\alpha_1, \alpha_2, ..., \alpha_m \in K$, not all 0 such that

$$\alpha_0 .1 + \alpha_1 a + \alpha_2 a^2 + ... + \alpha_m a^m = 0$$

which shows that a satisfies a non-zero polynomial

$$f(x) = \alpha_0 + \alpha_1 x + \dots + \alpha_m x^m \in F(x).$$

Thus, a is algebraic over F.

Conversely, let $a \in K$ is algebraic over F. We have to prove that $F(a)$ is a finite extension of F.

Let $s(x)$ be a polynomial over F of lowest degree satisfied by a.

Let degree of $s(x)$ be n.

$\Rightarrow$ a is algebraic of degree n over F.

Thus, $F(a) = (\beta_0 + \beta_1 a + \beta_2 a^2 + \dots + \beta_{n-1} a^{n-1} : \beta_0, \beta_1 \dots, \beta_{n-1} \in F\}$

$\Rightarrow$ $F(a)$ is a vector space over F spanned by the element $1, a, a^2 , \dots a^{n-1}$.

Also, these elements of $F(a)$ are linearly independent over F. For this, consider $c_0 1 + c_1 a + c_2 a^2 + \dots + c_{n-1} a^{n-1} = 0$ with $c_i \in F$

$\Rightarrow$ a satisfies a polynomial $q(x) = c_0 + c_1 x + \dots + c_{n-1} x^{n-1} \in F(x)$

$\Rightarrow$ $q(x)$ must be the zero polynomial otherwise $s(x)$ is not a polynomial of lowest positive degree satisfied by a, because $\deg s(x) = n$, while $\deg. q(x) < n$ if $q(x) \neq 0$.

$\Rightarrow$ $c_0 = 0, c_1 = 0, c_2 = 0, \dots, c_{n-1} = 0$

$\Rightarrow$ $1, a, a^2 , \dots a^{n-1} \in F(a)$ are linearly independent over F.

which shows that the n elements $1, a, a^2 , \dots a^{n-1}$ forms a basis for $F(a)$ over F.

$\Rightarrow$ $[F(a) : F] = n$

Hence, $F(a)$ is a finite extension of F.

REMARK

☞ If degree of a over F is n, then we have $[F(a) ; F] = n$. Thus $m = n$. Therefore, if $[F(a): F] = m$, then a is algebraic of degree m over F.

Theorem 7. *Every finite extension of a field F is algebraic.*

Proof.   Let F be a field and K be an algebraic extension of F. Also, let the degree of K over F be n, i.e., $K(F)$ is n-dimensional vector space. To prove that K is algebraic extension. For this, we shall prove that every element of K is algebraic element over F. Let α be any arbitrary element of K, then K being a field $\alpha, \alpha^2, \dots, \alpha^n$ are all belongs to K. Also, $1 \in K$. Thus, the set $[1, \alpha, \alpha^2, \dots, \alpha^n]$ of $n+1$ elements of K is linearly dependent because dimension of K over F is n.

For the set $[1, \alpha, \alpha^2, \dots, \alpha^n]$, there exist elements $a_0, a_1, \dots a_n \in F$ not all zero such that

$$a_0.1 + a_1 \alpha + a_2 \alpha^2 + \dots + a_n \alpha^n = 0$$

i.e., $a_0 + a_1 \alpha + a_2 \alpha^2 + \dots + a_n \alpha^n = 0$

$\Rightarrow \alpha$ is a root of a non-zero polynomial $a_0 + a_1 x + a_2 x^2 + \dots + a_n x^n \in F(x)$.

Therefore, α is an algebraic element of K over F.

Since, α is arbitrary, therefore, every element of K is algebraic over F. Hence, K is an algebraic extension of F.

Theorem 8. *Let K be an extension of F and let $a_1, a_2, \dots, a_n$ be n elements in K algebraic over F. Then $F(a_1, a_2, \dots, a_n)$ is a finite extension of F and hence an algebraic extension of F.*

Proof.   It is given that K is an extension of F.

We know that

$$F(a) \subseteq F\{a_1\} \subseteq F(a_1, a_2) \subseteq \ .....\subseteq F(a_1, a_2,..., a_n) \subseteq K.$$

Now, since a_k is algebraic over F, so it is also algebraic over $F(a_1, a_2,...,a_{k-1})$ which is a subfield of F. It is known that any non-zero polynomial over F is also a non-zero polynomial over $F(a_1, a_1,...,a_{k-1})$.

Since a_K is algebraic over $F(a_1, a_2,...,a_{k-1})$, therefore

$$(F(a_1, a_2,..., a_{k-1})) (a_k) \text{ is a finite extension of } F(a_1, a_2,..., a_{k-1}).$$

$\Rightarrow$ $F(a_1, a_2,..., a_{k-1}, a_k)$ is a finite extension of $F(a_1, a_2,..., a_{k-1})$.

$\Rightarrow$ $[F(a_1, a_2,..., a_K): F(a_1, a_2,...,a_{K-1})]$ is finite, say λ_k .

Further,

$$[F(a_1,a_2,,...,a_n): F] = [F(a_1, a_2,,...,a_n): F(a_1, a_2,,...,a_{n-1})]$$
$$[F(a_1, a_2,,...,a_{n-1}) : F(a_1, a_2,,...,a_{n-2})].....[F(a_1): K]$$
$$= \lambda_n \lambda_{n-1} ....\lambda_2 \lambda_1$$
$$= \text{finite, since each } \lambda \text{ is finite.}$$

Thus, $F(a_1, a_1......a_n)$ is a finite extension of F. Hence, $F(a_1 a_2......a_n)$ is an algebraic extension of F.

Theorem 9. *If α, $\beta \in K$ are algebraic over F, then $\alpha \pm \beta$, $\alpha\beta$ and $\alpha|\beta$ ($\beta \neq 0$) are all algebraic over F. Hence, the elements in K which are algebraic over F form a subfield K.*

Proof. Let F be a field. Since $\alpha \in K$ is algebraic over F. Then, $[F(\alpha): F]$ is finite. Further $\beta \in K$ is algebraic over F, therefore it satisfies a non-zero polynomial over F, but $F(\alpha)$ is a subfield of F. Thus, every non-zero polynomial over F also a non-zero polynomial over $F(\alpha)$. Therefore, β satisfies a non-zero polynomial over $F(a)$.

So, $(F(\alpha))(\beta) = F[\alpha, \beta]$ is a field generated by β over $F(\alpha)$, so that $F(\alpha, \beta)$ is a finite extension of $F(\alpha)$, *i.e.,* $[F(\alpha, \beta): F(\alpha)]$ is finite.

Using transitivity of finite extension of fields, we have

$$[F(\alpha,\beta): F] = [F(\alpha,\beta)) : F(\alpha)][F(\alpha):F]$$

Since, $[F(\alpha, \beta), F(\alpha)]$ and $(F(\alpha), F)$ are finite, therefore $[F(a, b): F]$ is finite.

Thus, $F(\alpha, \beta)$ is a finite extension of F, therefore, it is an algebraic extension.

Thus, every element of $F(\alpha, \beta)$ is algebraic over F. Since $F(\alpha, \beta)$ is a field so

$$\alpha, \beta \in F(\alpha,\beta) \Rightarrow \alpha \pm \beta \in F(\alpha,\beta), \alpha\beta \in F(\alpha,\beta)$$

Now, $$\beta \neq 0 \Rightarrow \beta^{-1} \in F(\alpha, \beta)$$

$\Rightarrow$ $$\alpha|\beta \in F(\alpha, \beta)$$

Hence, $$\alpha \pm \beta, \ \alpha\beta \text{ and } \alpha| \beta \ (\beta \neq 0) \text{ are algebraic over } F.$$

Thoerem 10. *Let F be a field and K be an extension of F. If α and β in K are algebraic over F of degree m and n respectively, then $\alpha \pm \beta$, $\alpha\beta$ and $\alpha|\beta(\beta \neq 0)$ are algebraic over F of degree atmost mn.*

Proof. It is given that $\alpha \in K$ is algebraic over F of degree m. Thus $[F(\alpha): F] = m$.

Further, $\beta \in K$ is an algebraic element of degree n over F.

$\Rightarrow$ β satisfies a minimal polynomial of degree n over F.

But $F(\alpha)$ is a finite extension of F, so that F will satisfy a minimal polynomial of degree atmost n over F.

Thus, $[F(\alpha, \beta): F(\alpha)] < n$

Now, by the transitivity of field extension, we have

$$[F(\alpha, \beta): F] = [F(\alpha, \beta): F(\alpha)] \, [F(\alpha) : F] \subseteq mn$$

$\Rightarrow$ $F(\alpha, \beta)$ is a finite extension of F.

$\Rightarrow$ $F(\alpha, \beta)$ is an algebraic extension of F of degree atmost mn.

$\Rightarrow$ Every element of $F(a, \beta)$ is algebraic of degree atmost mn.

Further, $\beta \neq 0 \Rightarrow \beta^{-1}$ exists and $\beta^{-1} \in F(\alpha, \beta)$

Then, we have $\alpha\beta^{-1} \in F(\alpha, \beta)$

Hence, $\alpha \pm \beta$, $\alpha\beta$ and α/β $(\beta \neq 0)$ are algebraic over F of degree atmost mn.

Theorem 11. *(Transitivity of Algebraic Field Extension). Let L be an algebraic extension of K and if K is an algebraic extension of F, then L is an algebraic extension of F.*

Proof.   Let $\alpha \in L$ be arbitrary. We have to show that α will satisfy a non-zero monic polynomial over F.

Since L is an algebraic over K and $a \in L$, then α will satisfy a non-zero monic polynomial

$$p(x) = x^n + a_1 x^{n-1} + a_2 x^{n-2} + \dots + a_{n-1} + a_n$$

over K, i.e., $a_1, a_2, \dots, a_n \in K$.

But K is an algebraic extension of F, thus each elements $a_1, a_2, \dots, a_n$ is algebraic over F. So an extension $K = F(a_1, a_2, \dots, a_n)$ is a finite extension of F. Now, since α satisfies the polynomial $p(x)$ whose coefficients are in K. Thus, α is algebraic over K so that $K(\alpha)$ is a finite extension of K.

Using the transitivity of finite extension fields, we have

$$[K(\alpha):F] = [K(\alpha): K][K : F]$$

$\Rightarrow$ $[K(\alpha): K]$ is finite, because $[K(\alpha): K]$ and $[K; F]$ both are finite.

$\Rightarrow$ $K(a)$ is a finite extension of F.

$\Rightarrow$ α is algebraic over F.

Since α is arbitrary, thus every element of L is algebraic over F.

Hence, L is an algebraic extension of F.

Theorem 12. *Let F be a field and K be an extension of a field F. Then, the mapping $\phi: F[x] \rightarrow F(a)$ defined by $h(x)\phi = h(a)$ is a homomorphism ($h(x)\phi$ is the image of $h(x)$ under the mapping ϕ).*

Proof.   Let $h(x), g(x) \in F(x)$.

Then, $h(x) \, \phi = h(a)$ and $g(x) \, \phi = g(a)$

Also, let

$$s(x) = h(x) + g(x) \text{ and } \quad t(x) = h(x). \, g(x)$$

$\Rightarrow$ $s(a) = h(a) + g(a) \text{ and } \quad t(a) = h(a) \, g(a)$

We have

$$[h(x) + g(x)]\phi = s(x)\phi = s(a)$$
$$= h(a) + g(a) = h(x)\phi + g(x) \, \phi$$

and $\qquad [h(x)\,g(x)]\,\phi = t(a)\phi = t(a)$
$$= h(a)g(a) = [h(x)\phi][g(x)\phi].$$

Hence, ϕ is a homomorphism from $F[x]$ to $F(a)$.

Theorem 13. *Every finite extension K of a field F is algebraic and may be obtained from F by the adjunction of finitely many algebraic dements.*

Proof. Let F be a field and K be an extension of F. If $[K : F] = 1$. Then $K = F$. In this case, result is trivially true.

Let $[K : F] = n > 1$

Let $a \in K$ be arbitrary such that $a \in F$. Now, since K is a field and $a \in K$, thus $n + 1$ elements $1, a, a^2,..., a^{n-1}, a^n$ are all in K. Since, the dimension of the vector space $K(F)$ is n, so these $n + 1$ elements of K are linearly dependent over F. Therefore, there exist elements $\alpha_0, \alpha_1,......\alpha_n \in F$ not all 0, such that

$$\alpha_0.1 + \alpha_1 a + \alpha_2 a^2 + ...+ \alpha_n a^n = 0$$

$\Rightarrow a$ is algebraic over F.

$\Rightarrow K$ is an algebraic extension of F.

Now, $\qquad\qquad F \subset F(a) \subseteq K$.

Further $n = [K : F] = [K : F(a)]\,[F(a): F]$

If $[F(a): F] = n$, then $[K : F(a)] = 1$

$\Rightarrow \qquad\qquad\qquad K = F(a)$

In this case result is true.

If $[F(a): F] \neq n$, then suppose $[F(a): F] = m < n$

Since $a \in F$, so $\qquad\qquad m > 1.$

From (I), $\qquad [K : F(a)] > 1.$

Let $b \in K$ such that $\qquad b \notin F(0)$. We can easily shown that b is algebraic over F.

Since a and b are algebraic over F. Thus, $F(a, b)$ is a finite extension of F.

If $[F(a, b): F] = p$ where $p > m$, because $F(a)$ is a proper subset of $F(a, b)$.

Further

$$n = [K : F] = [K : F(a, b)]\,[F(a, b): F]$$
$$= [K : F(a,b)]p$$

If $p = n$, then $\qquad F(a,b) = K.$

In this case result is true.

If $p \neq n$, then continue the above process a finite number of times, we get

$$[F(a,b,......,K):F] = n$$

Then, we have

$$M = [k : F] = [K : F(a, b,......k)]\,[\,K(a, b,......k): F]$$
$$= [k : F\,(a, b,......., k)]n$$

Thus, $\quad [K : F\,(a, b,......k)] = 1$

Therefore, $k = F(a, b,......., k)$, where $a, b, ..., k$ are algebraic over F.

Theorem 14. *Let K be an extension of a field F and let $a \in K$ be algebraic over F. Then F(a) is isomorphic to F[x]/V, where V is ideal of F(x) generated by the minimal polynomial for a over F.*

Proof. Let f be a field and K be an extension of F. If $p(x)$ is the minimal polynomial for a over F. Then, clearly $p(x)$ is irreducible over F, i.e., $p(x)$ is a prime element of $F(x)$.

Define $V = [h(x) \in F[x] : h(a) = 0]$

Then, clearly V is an ideal of $F(x)$. Now $F(x)$ is a principal ideal ring and $p(x)$ is an element of lowest degree in the ideal V of $F[x]$. Thus, V is the ideal of $F[x]$ generated by $p(x)$. Since $F(x)$ is irreducible, thus V is a maximal ideal of $F[x]$.

So, $F[x]/V$ is a field.

Now, define a mapping $\phi : F[x] \rightarrow F(a)$

such that

$$f(x)\phi = f(a) \text{ for any } f(x) \in F[x].$$

Clearly, ϕ is a homomorphic mapping from the ring $F[x]$ into the field $F(a)$. The ideal V of $F[x]$ will be the kernel of ϕ. Then, by fundamental theorem on ring homomorphism, $F[x]/V$ is isomorphic to $F[x]\phi$. which is the image of $F[x]$ under ϕ. Also, $F[x] \phi$ is a subset of $F(a)$. Since $F[x]$ is a field, therefore, $F[x]\phi$ is a subfield of $F(a)$.

Thus, $a \in F(x)\phi$

Further, if $\alpha \in F$, then $\alpha \in F[x]$.

Now, by definition of ϕ, we have $\alpha\phi = \alpha$, therefore $\alpha \in F[x] \phi$.

$\Rightarrow$ $F[x] \phi$ is a subfield of $F(a)$ and it contains both F and a. Since $F(a)$ is the smallest subfield of K containing both F and a, therefore, we must have

$$F[x] \phi = F(a)$$

Hence, $F[x]/V$ is isomorphic to $F(a)$.

Theorem 15. *Let F be a field and let F[x] be the ring of polynomials in x over F. Let $g(x) \in F(x)$ of degree n and let $V = (g(x))$ be the ideal generated by g(x) in F[x]. Then F[x] / V is an n-dimensional vector space over F.*

Proof. Define $\qquad V = \{f(x).g(x): f(x) \in F[x]\}$

and $\qquad F[x] / V = \{V + f(x): f(x) \in F[x]\}$

We can define addition in F[x]/V as follows

Let $V + f_1(x), V + f_2(x) \in F[x] / V$, then we can define

$$[V + f_1(x)] + [V + f_2(x)] = V + f_1(x) + f_2(x)$$

Further, the scalar multiplication in $K[x] / V$ over F can be defined as follows:

Let $\qquad a \in F$ and $V + f(x) \in F[x]/V$

Then, we can define

$$a[V + f(x)] = V + af(x)$$

Clearly F[x]/V is an abelian group with respect to addition t. Also, the residue class V is the zero vector.

Now, let $a, b \in F$ and $f_1(x), f_2(x) \in F[x]$. Then

(i) $(a + b) [V + f_1(x)] = V + (a + b)f_1(x)$

$$= V + af_1(x) + bf_1(x)$$
$$= [V + af_1(x)] + [V + bf_1(x)] = a[V + f_1(x)] + b[V + f_1(x)]$$

(ii) $a[[V + f_1(x)] + [V + f_2x)]] = a[V + f_1(x) + f_2(x)]$

$$= V + a[f_1(x) + f_2(x)] = V + af_1(x) + af_2(x)$$
$$= [V + af_1(x)] + [V + af_2(x)] = a[V + f_1(x)] + a[V + f_2(x)]$$

(iii) $a[b[V + f_1(x)]] = a[V + bf_1(x)] = V + (ab)f_1(x)$

$$= ab[V + f_1(x)]$$

(iv)   1. $[V + f_1(x)] = V + 1.f_1(x) = V + f_1(x)$

Therefore, $F[x]/V$ is a vector space over F.

If $g(x)$ is of degree n. Then, we have to show that $F[x]/V$ is of dimension n over F. For this, we shall show that $V + 1, V + x, V + x^2, \ldots\ldots V + x^{n-1}$ constitute a basis of $F[x]/V$ over F.

(i) $V + 1, V + x, \ldots, V + x^n$ are linearly independent

We have

$$a_0(V + 1) + a_1(V + x) + a_2(V + x^2) + \ldots + a_{n-1}(V + x^{n-1}) = V, a_i \in F$$

$\Rightarrow \quad (V + a_0) + (V + a_1x) + (V + a_2x^2) + \ldots + (V + a_{n-1}x^{n-1}) = V$

[Here V is the zero vector and $V + f(x) = V \Leftrightarrow f(x) \in V$]

$\Rightarrow \quad (V + a_0) + (V + a_1x) + (V + a_2x^2) + \ldots + (V + a_{n-1}x^{n-1}) = V$

$\Rightarrow \quad V + a_0 + a_1x + a_2x^2 + \ldots + a_{n-1}x^{n-1} = V$

$\Rightarrow \quad a_0 + a_1x + a_2x^2 + \ldots + a_{n-1}x^{n-1} \in V$

$\Rightarrow \quad a_0 + a_1x + a_2x^2 + \ldots + a_{n-1}x^{n-1} = f(x).g(x)$, for some $f(x) \in F[x]$.

$\Rightarrow \qquad f(x) = 0$

[Because if $f(x) \neq 0$, then $\deg f(x) g(x) \geq \deg g(x) = n$, so we cannot have

$$f(x) g(x) = a_0 + a_1x + a_2x + \ldots + a_{n-1}x^{n-1}]$$

$\Rightarrow \quad a_0 + a_1x + a_2x^2 + \ldots + a_{n-1}x^{n-1} = 0$

$\Rightarrow \quad a_0 = 0, a_1 = 0, \ldots, a_{n-1} = 0$

Thus, $V + 1, V + x, V + x^2, \ldots\ldots V + x^{n-1}$ are linearly independent over F.

(ii) $V + 1, V + x, V + x^2, \ldots\ldots V + x^{n-1}$ generates $F[x] / V$ over F.

Let $V + F(x)$ be any element in $F(x)/V$. Then $f(x) \in F[x]$.

By division algorithm, there exists $q(x), r(x) \in F[x]$ such that

$$f(x) = q(x) g(x) + r(x)$$

where either $r(x) = 0$ or $\deg r(x) < \deg g(x)$

Also, $V + f(x) = V + q(x) g(x) + r(x)$

$$= [V + q(x)g(x)] + [V + r(x)]$$
$$= V + V + r(x) = V + r(x)] \qquad\qquad [\therefore V \text{ is zero vector}]$$
$$= V + a_0 + a_1x + a_2x_2 + a_{n-1}x^{n-1}$$

Where, $a_0, a_1, \ldots a_{n-1} \in K$ [$\therefore r(x) = 0$ or $\deg r(x) < n$, i.e., $\deg g(x)]$

$\Rightarrow \quad V + 1, V + x, V + x^2 \ldots + V + x^{n-1}$ generates $F[x]/V$ over F.

Hence, $\dim F[x] / V$ over $F = n$.

10.6 ROOTS OF A POLYNOMIAL

Definition 1. *Let F be a field and let $p(x) \in F[x]$. Then an element a lying in some extension field of F is called a root of $p(x)$ if $p(a) = 0$.*

Definition 2. *Let F be a field and let $p(x) \in F[x]$. If K is an extension of F, then $a \in K$ is said to be a root of $p(x)$ of multiplicity m if in $K[x]$, we have "$(x-a)^m$ is a divisor of $p(x)$ whereas $(x-a)^{m+1}$ is not a divisor of $p(x)$".*

Definition 3. *A root of multiplicity 1 is called a simple root and a root of multiplicity greater than one is called a multiple root.*

Definition 4. *Let F be a field and let $F[x]$ be the ring of polynomials in x over F. Clearly $F[x]$ is an integral domain with unity containing F as a proper subring. Then a polynomial $f(x) \in F[x]$ is called irreducible over F if whenever $f(x) = p(x)\, q(x)$, with $p(x), q(x) \in F[x]$, then either $p(x)$ or $q(x)$ has degree zero, i.e., either $p(x) \in F$ or $q(x) \in F$.*

REMARK

☞ If $f(x)$ is not irreducible, then it is known as reducible.

Theorem 1. (*Remainder Theorem*). *If $f(x) \in F[x]$ and if K is an extension of F, then for any element $c \in K$, $f(x) = (x-c)\, q(x) + f(c)$, where $q(x) \in K[x]$ and where $\deg q(x) = \deg p(x) - 1$.*

Proof.   Since K is an extension of F, therefore

$$F \subseteq K$$

$$F[x] \subseteq K[x] \Rightarrow f(x) \in F[x] \subseteq K[x] \Rightarrow f(x) \in K[x]$$

Now, $f(x)$ and $x - c$ are both in $K[x]$, then by division algorithm, there exists polynomials $q(x)$ and $r(x)$ in $K[x]$, such that

$$f(x) = (x-c)q(x) + r(x)$$

where either $r(x) = 0$ or $\deg r(x) < \deg (x-c)$. But since $\deg(x-c) = 1$,

therefore either $r(x) = 0$ or $\deg r(x) = 0$.

$\Rightarrow$ $r(x)$ is a constant polynomial in $K[x]$.

$\Rightarrow$ $r(x)$ is simply an element say r in K.

$\Rightarrow$ $\qquad\qquad f(x) = (x - c)q(x) + r$

$\Rightarrow$ $\qquad\qquad f(c) = (c - c)q(c) + r$

$\Rightarrow$ $\qquad\qquad f(c) = 0 \cdot q(c) + r$

$\Rightarrow$ $\qquad\qquad f(c) = r$

$\Rightarrow$ $\qquad\qquad f(x) = (x - c)q(x) + f(c)$ $\qquad\qquad\qquad$ ....(1)

Further, let us suppose $\deg f(x) = n$ and $\deg q(x) = m$. Then, degree of the polynomial on RHS of (1) is then $m + 1$. By equality of two polynomials, we have

$$n = m + 1.$$

$\Rightarrow$ $\qquad\qquad m = n - 1$, i.e., $\deg q(x) = \deg f(x) - 1$.

Theorem 2   (*Factor Theorem*). *If $a \in K$ is a root of $f(x) \in F[x]$, where $F \subseteq K$, then in $K[x]$, $(x-a) \mid f(x)$.*

Proof.   Let F be a field and K be an extension of F such that $F \subseteq K$.

Let $f(x) \in F[x]$ and let $a \in K$, where K is an extension field of F. Using remainder theorem, we have

$$f(x) = (x–a)q(x) + f(a)$$

$$\Rightarrow \qquad f(x) = (x - a)\, q(x) + 0 \qquad\qquad [\because a \text{ is a root or } f(x).]$$

$$= (x–a)q(x)$$

Thus, in $K[x]$, we have $(x - a)$ is a divisor of $f(x)$. Hence, $(x–a) \mid f(x)$.

Theorem 3. *A polynomial of degree n over a field can have at most n roots is any extension field.*

Proof. Let K be an extension of a field F and let $f(x) \in F[x]$ be a polynomial of degree n over F.

We shall prove this theorem by induction hypothesis on n.

Step (i). If $n = 1$, then we have $f(x) = a_0 x + a_1$, $a_0 \neq 0$, $a_0, a_1 \in K$.

Now, $x = - a_1/a_0$ is a unique root of $f(x) \in F[x]$ which belongs to K.

$\Rightarrow$ Theorem is true in this case.

Step (ii). Let $\deg f(x) = n > 1$, then by induction, we may assume that the theorem is true for each polynomial of degree less than n. If $f(x)$ has no root in K, then theorem is trivially true. Now, there is a root of multiplicity m, $\alpha \in K$ of $f(x)$. Then, there exists a polynomial $q(x) \in K[x]$ such that

$$f(x) = (x - \alpha)^m q(x), \text{ with } q(\alpha) \neq 0 \qquad\qquad \text{... (1)}$$

Clearly, $\deg (q(x)) = n - m < n$. Then by induction hypothesis $q(x)$ will have at most $n - m$ roots in K.

Since $q(\alpha) \neq 0$, therefore α is not a root of $q(x)$. Also, any root of $f(x)$ in K other than α is a root of $q(x)$.

Let $\beta \neq \alpha$ be any root of $f(x)$ in K, then from (1), we have

$$f(\beta) = (\beta - \alpha)^m q(\beta)$$

But $f(\beta) = 0$, then $(\beta - \alpha)^m q(\beta) = 0$ so that $q(\beta) = 0$ because $(\beta - \alpha)^m \neq 0$.

$\Rightarrow \quad \beta$ is a root of $q(x)$ in K.

$\Rightarrow q(x)$ has at most $n - m$ roots in K other than α.

$\Rightarrow \quad f(x)$ has at most $n - m$ roots in K other than α.

Hence, $f(x)$ has at most $(n - m) + m = n$ roots in K.

Theorem 4. *(Kroncker's Theorem). Let $f(x)$ be an irreducible polynomial of positive degree in $F[x]$. Then, there exists an extension K of F in which $f(x)$ has a root. Also $[K : F] = \deg f(x)$.*

Proof. Let $\deg f(x) = n$. It is given that $f(x)$ is irreducible in $F[x]$. Then $f(x)$ is a maximal ideal of $F[x]$. Thus, $F(x)/f(x)$ is a field which is an extension of F.

Now, let $K = F[x] /(f(x))$.

and $K' = [f(x) + a : a \in F]$ is a subset of K.

Firstly, we shall show that K' is a subfield of K.

Let $(f(x)) + a$ and $(f(x)) + b$ be any two-elements of K' such that $(f(x))+b \neq (f(x))$. Now, since $K' \subset K$ and $f(x) + b \in K$.

Then, $((f(x)) + b)^{-1} \in K$

$\Rightarrow \qquad (f(x)) + b^{-1} \in K$

$\Rightarrow$ $\qquad\qquad\qquad\qquad b^{-1} \in F$ $\qquad\qquad\qquad\qquad\qquad$ [$\because b \neq 0$]

Therefore,

$$((f(x)) + a)((f(x)) + b)^{-1} = ((f(x)) + a)((f(x)) + b^{-1}) = (f(x)) + ab^{-1} \in K^{-1}$$

$\Rightarrow$ K' is a subfield of K.

Now, define a mapping $\phi : F \to K'$ such that $\phi(a) = (f(x)) + a$, $\forall\ a \in F$. We have to show that ϕ is one-one onto homomorphism.

(i) ϕ **is one-one.** Let $a, b \in F$

Suppose $\qquad\qquad \phi(a) = \phi(b)$

$\Rightarrow$ $\qquad\qquad (f(x)) + a = ((f(x)) + b$

$\Rightarrow$ $\qquad\qquad a - b \in (f(x))$

$\Rightarrow$ $\qquad\qquad a - b = f(x)\, g(x)$, for some $g(x) \in F[x]$

Since $f(x)$ is irreducible, so its product with any other non-zero polynomial cannot be equal to a constant polynomials.

Thus, $\qquad\qquad g(x) = 0\ = a - b = 0$

$\Rightarrow$ $\qquad\qquad a = b$

$\Rightarrow$ $\qquad \phi$ is one-one.

(ii) ϕ **is onto.** Let $(f(x)) + a \in K'$ such that $a \in K$, then, we have

$$\phi\,(a) = (f\,(x)) + a,\ \forall a \in F$$

$\Rightarrow$ $\qquad \phi$ is onto.

(iii) ϕ **preserves the structure.**

Let $a, b \in F$. Then, we have

$$\phi(a + b) = (f(x)) + (a + b)$$
$$= ((f(x)) + a) + ((f(x)) + b)$$
$$= \phi(a) + \phi\,(b)$$

and $\qquad \phi(ab) = (f(x)) + ab = ((f(x)) + a)\,((f(x)) + b))$
$$= \phi(a)\phi(b)$$

Therefore, K' is a subfield of K-isomorphic to F. From above isomorphism, we can say that every element of K' is corresponding element of K so that F can be regarded as a subfield of K. Therefore K is an extension of F. Now, we claim that K is a finite extension of F. For this, we shall prove that n elements $V+1, V+x, V + x^2, ..., V+x^{n-1}$ form of a basis of K over F (For this proof see Theorem 15 of previous section).

$\therefore$ $\qquad\qquad [K : F] = n = \deg f(x)$

Finally, we shall show that $f(x)$ has a root in K.

Let $f(x) = a_0 + a_1 x + a_2 x^2 + ... + a_n x^n$, $a_n \neq 0$, $a_0, a_1, ...a_n \in F$

Now, since $f(x) \in (f(x))$

$\Rightarrow$ $\quad ((f(x)) + (f(x) = (f(x))$

$\Rightarrow$ $\quad ((f(x)) + (a_0 + a_1 x + ... + a_n x^n) = (f(x))$

$\Rightarrow$ $\quad ((f(x) + a_0)((f(x)) + a_1 x) + ... + ((f(x)) + a_n x^n) = (f(x))$

$\Rightarrow$ $\quad a_0\,((f(x) + 1) + a_1(f(x)) + x) + ... + a_n((f(x)) + x^n) = (f(x))$

$\Rightarrow$ $\quad a_0\,((f(x) + x)^0 + a_1((f(x)) + x)^1 + ... + a_n((f(x)) + x)^n = (f(x))$

But $(f(x))$ is the zero element of $F[x]/(f(x)) = K$.

Therefore, each element $(f(x)) + x$ in K satisfies the polynomial $f(x)$.

Hence, $f(x)$ has a root in K.

Theorem 5. *If $f(x) \in F[x]$, then there is a finite extension K of F in which $f(x)$ has a root and $[K:F] \le \deg f(x)$.*

Proof. If $f(x)$ is irreducible polynomial, then using above theorem, this theorem is also true.

If $f(x)$ is not irreducible, then it can be expressed as a product of two polynomials $g(x)$ and $h(x)$ where $g(x)$ and $h(x) \in F[x]$ and $g(x)$ is irreducible factor of $f(x)$.

i.e., $$f(x) = g(x).h(x)$$

$\Rightarrow$ $$\deg g(x) \le \deg . f(x)$$

Further, since $g(x)$ is irreducible over F, again by above theorem, there exists an extension K of F such that $g(x)$ has a root in K and

$$[K : F] = \deg g(x) \le \deg f(x)$$

If α is a root of $g(x)$ which is in K, then $g(\alpha) = 0$, therefore

$$f(\alpha) = g(\alpha).h(\alpha) = 0$$

$\Rightarrow$ α is a root of $f(x)$.

Hence, there exists an extension K of F such that $f(x)$ has a root in K and $[K : F] \le \deg f(x)$.

Theorem 6. *Let $f(x) \in F[x]$ be of positive degree, say n. Then there is a finite extension K of F of degree at most $n!$ in which $f(x)$ has n roots.*

Proof. Let $$\deg f(x) = n.$$

Now, we shall prove this theorem by induction.

Step (i). If $\deg f(x) = 1$, then $f(x) = a_0 x + a_1$ where, $a_0, a_1 \in F, a_0 \ne 0$.

Now, F itself is an extension of F and $[F : F] = 1$.

Also, $-\dfrac{a_1}{a_0} \in F$ is a root of $a_0 x \ne a_1$.

Therefore, if $\deg f(x) = 1$, then there is a finite extension F of F of degree at most $1!$ $= 1$, in which $f(x)$ has one root.

Step (ii). Let us assume that theorem is true in any field for all polynomials of degree less than n. Let $f(x)$ be a polynomial of degree n over a field F. Using previous theorem, there is an extension K_0 of F with $[K_0 : F] \le n$ in which $f(x)$ has a root, say α . Then by factor theorem in $K_0(x)$, $f(x)$ factored as

$$f(x) = (x - \alpha)q(x)$$

where $$\deg q(x) = \deg f (x) - 1 = n - 1$$

Now, $q(x)$ is a polynomial over E_0 of degree $n-1$. Now, since $\deg q(x)$ is less than n, then by our induction hypothesis there is an extension K of K_0 of degree at most $(n-1)!$ in which $q(x)$ has $n-1$ roots.

Now, since any root of $f(x)$ is either a or a root of $q(x)$. Thus, we find in K all n roots of $f(x)$.

Further, K is an extension of K_0 and K_0 is an extension of F implies that K is an extension of F. Thus, we have

$$[K : F] = [K : K_0][K_0 : F]$$
$$\leq (n - 1)!.n = n!$$

Hence, E is a finite extension of F of degree at most $n!$ in which $f(x)$ has n roots.

Solved Examples

▶ An element a lying in some extension field of F is called a root of $p(x)$ if $p(a) = 0$

▶ A root of mutiplicity 1 is called simple root and a root of multiplicity greater than one is called a multiple root.

▶ If $f(x) \in F(x)$ and if K is an extension of F then for any element $c \in K$, $f(x) = (x-c)\, g(x) + f(c)$, $g(x) \in K(x)$ and where $\deg g(x) = \deg p(x) - 1$ (Remainder theorem)

▶ If $a \in K$ is a root of $f(x) \in F(x)$ where $F \subset K$ then in $K[x]$, $(x-a)\,|\,f(x)$. (Factor theorem)

Example 1.　*Let* **R** *and* **Q** *be the field of real and rational numbers, respectively. In* **R** *,* $\sqrt{2}$ *and* $\sqrt{3}$ *are both algebraic over* **Q**

(i) *Exhibit a polynomial of degree 4 over* **Q** *satisfied by* $\sqrt{2} + \sqrt{3}$

(ii) *Show that* $\mathbf{Q}(\sqrt{2}, \sqrt{3}) = \mathbf{Q}(\sqrt{2} + \sqrt{3})$

(iii) *Show that* $\deg(\sqrt{2} + \sqrt{3})$ *over* **Q** *is 4.*

Solution.　(i) Here, $\sqrt{2}$ satisfies a polynomial $f(x) = x^2 - 2$ over **Q** and $f(x)$ is irreducible over **Q** of degree 2.

$\Rightarrow$　　$\sqrt{2}$ is algebraic of degree 2 over **Q**.

$\Rightarrow$　　$[\mathbf{Q}\sqrt{2} : \mathbf{Q}] = 2$.

Again, $\sqrt{3}$ satisfies a polynomial $g(x) = x^2 - 3$ of degree 2 over **Q**, it is also irreducible over **Q** .

$\Rightarrow$　　$\sqrt{3}$ is algebraic over **Q** of degree 3.

　　　$[\mathbf{Q}\sqrt{3} : \mathbf{Q}] = 2$

Further, let $x = \sqrt{2} + \sqrt{3}$. Then

$$x^2 = (\sqrt{2} + \sqrt{3})^2 = 5 + 2\sqrt{6} \quad \Rightarrow \quad = (x^2 - 5)^2 = (2\sqrt{6})^2$$

$\Rightarrow$　　　　$x^4 - 10x^2 + 1 = 0$

Therefore, $\sqrt{2} + \sqrt{3}$ satisfies a polynomial $x^4 - 10x^2 + 1$ of degree 4.

(ii) $\mathbf{Q}(\sqrt{2}, \sqrt{3}) = (a + b\sqrt{2} + c\sqrt{3} + d\sqrt{2}\sqrt{3}; a, b, c, d \in \mathbf{Q})$

and $\mathbf{Q}(\sqrt{2} + \sqrt{3}) = [a + b(\sqrt{2} + \sqrt{3}) : a, b \in \mathbf{Q}]$

Now,　$\sqrt{3} + \sqrt{2} \in \mathbf{Q}(\sqrt{2}, \sqrt{3})$ so that

$$\mathbf{Q}(\sqrt{2} + \sqrt{3}) \subseteq \mathbf{Q}(\sqrt{2}, \sqrt{3}) \quad\quad\quad …(i)$$

Now, since $\mathbf{Q}(\sqrt{2} + \sqrt{3})$ is a field, therefore, every non-zero element of $\mathbf{Q}(\sqrt{2} + \sqrt{3})$ has its multiplicative inverse. Therefore,

$$\sqrt{2} + \sqrt{3} \in \mathbf{Q}(\sqrt{2} + \sqrt{3})$$

$$\Rightarrow \qquad (\sqrt{2} + \sqrt{3})^{-1} \in \mathbf{Q}(\sqrt{2} + \sqrt{3})$$

$$\Rightarrow \qquad (\sqrt{3} - \sqrt{2}) \in \mathbf{Q}(\sqrt{2} + \sqrt{3})$$

Therefore,

$$\sqrt{3} = \frac{1}{2}(\sqrt{2} + \sqrt{3} + \sqrt{3} - \sqrt{2}) \in \mathbf{Q}(\sqrt{2} + \sqrt{3})$$

and

$$\sqrt{2} = \frac{1}{2}[(\sqrt{2} + \sqrt{3}) - (\sqrt{3} - \sqrt{2})] \in \mathbf{Q}(\sqrt{2} + \sqrt{3})$$

$$\Rightarrow \qquad (\sqrt{2}, \sqrt{3}) \in \mathbf{Q}(\sqrt{2} + \sqrt{3})$$

$$\Rightarrow \qquad \mathbf{Q}(\sqrt{2}, \sqrt{3}) \subseteq \mathbf{Q}(\sqrt{2} + \sqrt{3})$$

From (1) and (2), we conclude that

$$\mathbf{Q}(\sqrt{2}, \sqrt{3}) = \mathbf{Q}(\sqrt{2} + \sqrt{3})$$

(iii) From (1), we have

$$[\mathbf{Q}\sqrt{2} : \mathbf{Q}) = 2$$

Also, $\sqrt{3}$ satisfies a polynomial $x^2 - 3$ over $\mathbf{Q}$ but $x^2 - 3$ is also reducible over $\mathbf{Q}(\sqrt{2})$.

Thus $[(\mathbf{Q}(\sqrt{2}))(\sqrt{3}):\mathbf{Q}(\sqrt{2})] = 2$

or $[\mathbf{Q}(\sqrt{2}, \sqrt{3}) : \mathbf{Q}(\sqrt{2})] = 2$

Using transitivity of algebraic extension of fields, we have

$$[\mathbf{Q}(\sqrt{2}, \sqrt{3}) : \mathbf{Q}] = [\mathbf{Q}(\sqrt{2}, \sqrt{3}):\mathbf{Q}(\sqrt{2})][\mathbf{Q}(\sqrt{2}) : \mathbf{Q}] = 2 \times 2 = 4$$

or $[\mathbf{Q}(\sqrt{2} + \sqrt{3}) : \mathbf{Q}] = 4$ [By using $(\sqrt{2}, \sqrt{3}) = \mathbf{Q}(\sqrt{2} + \sqrt{3})$]

Hence, the degree of $\sqrt{2} + \sqrt{3}$ over $\mathbf{Q}$ is 4.

Example 2. *If a is any algebraic number, show that there exists a positive integer n such that na is an algebraic integer.*

Solution. It is given that a is an algebraic number. Then by definition, a is algebraic over the field of rationals $\mathbf{Q}$. Therefore, there exists a non-zero monic polynomial $f(x) \in \mathbf{Q}(x)$ such that $f(a) = 0$.

Let $f(x) = x^m + a_1 x^{m-1} + \ldots + a_{m-1}x + a_m, \ a_i \in \mathbf{Q}$

Putting $a_i = \dfrac{p_i}{q_i}, p_i, q_i (\neq 0)$ are integers in $f(a) = 0$, we get

$$a^m + \frac{p_1}{q_1} a^{m-1} + \ldots + \frac{p_{m-1}}{q_{m-1}} a + \frac{p_m}{q_m} = 0$$

Further, let $n = q_1 \ldots q_m$. Then clearly n is a positive integer and

$$na^m + p_1 q_2 \ldots q_m a^{m-1} + \ldots + p_m q_1 \ldots q_{m-1} = 0$$

$$\Rightarrow \quad n^m a^m + p_1 q_2 \ldots q_m a^{m-1} n^{m-1} + \ldots + p_m q_1 \ldots q_{m-1} n^{m-1} = 0$$

$$\Rightarrow \quad na \text{ satisfies the polynomial.}$$

$$x^m + p_1 q_2 \ldots q_m \cdot x^{m-1} + \ldots + p_m q_1 \ldots q_{m-1} n^{m-1} = 0$$

Hence, na is an algebraic integer.

Example 3. *Find the inverse of* $3 - \sqrt{2} \in \mathbf{Q}\sqrt{2}$.

Solution. We wish to find $\dfrac{1}{\mathbf{Q}\sqrt{2}}$ where $q(x) = 3 - x$.

Now, since $\sqrt{2}$ satisfies a minimal polynomial $\phi(x) = x^2 - 2$, thus we have

$$\phi(x) = x^2 - 2$$

If we divide $f(x) = x^2 - 2$ by $q(x) = 3 - x$, we get the quotient $-x - 3$ with remainder 7.

Therefore, we can write

$$7 = 1(x^2 - 2) + (x - 3)(3 - x)$$

$$\Rightarrow \qquad 1 = \frac{1}{7}(x^2 - 2) + \frac{1}{7}(x - 3)(3 - x)$$

Then, we can express 1, as *gcd* of $\phi(x)$ and $q(x)$ in the form

$$1 = g(x)\,\phi(x) + h(x)q(x)$$

where $\quad g(x) = \dfrac{1}{7}. h(x) = \dfrac{x + 3}{7}$

But $\quad \phi(\sqrt{2}) = 0 \;\Rightarrow\; 1 = h(\sqrt{2})q(\sqrt{2})$

$$\Rightarrow \qquad \frac{1}{q(\sqrt{2})} = h(\sqrt{2})$$

$$= \frac{1}{7}(\sqrt{2} + 3) = \frac{3}{7} + \frac{1}{7}\sqrt{2}$$

Hence, the inverse of $3 - \sqrt{2}$ is $\dfrac{3}{7} + \dfrac{1}{7}\sqrt{2}.$

Example 4. *Find the inverse of* $1 - (2)^{1/3} + (4)^{1/3} \in \mathbf{Q}((2)^{1/3})$

Solution. Clearly $(2)^{1/3}$ satisfies a minimal polynomial $\phi(x) = x^3 - 2$.

Here, we wish to find $\dfrac{1}{q((2)^{1/3})}$, where $q(x) = 1 - x + x^2$.

If we divide $\phi(x) = x^3 - 2$ by $q(x) = 1 - x + x^2$, we get the quotient $x + 1$ with a remainder of -3. Therefore, we have

$$x^3 - 2 = (x+1)(x^2 - x + 1) - 3$$

$$\Rightarrow \qquad 3 = (-1)(x^3 - 2) + (x+1)(x^2 - x + 1)$$

$$\Rightarrow \qquad 1 = -\frac{1}{3}(x^3 - 2) + \frac{1}{3}(x+1)(x^2 - x + 1)$$

Then, we can express 1 as *gcd* of $\phi(x)$ and $q(x)$ in the form

$$1 = g(x)\,\phi(x) + h(x)q(x)$$

where $\quad g(x) = -\dfrac{1}{3}, h(x) = \dfrac{x+1}{3}$

Also, $\quad \phi((2)^{1/3}) = 0$

$$\Rightarrow \qquad\qquad 1 = h((2)^{1/3})\, q((2)^{1/3})$$

$$\Rightarrow \qquad \frac{1}{q((2)^{1/3})} = h((2)^{1/3})$$

$$\Rightarrow \qquad \frac{1}{q((2)^{1/3})} = h((2)^{1/3}) = \frac{(2)^{1/3}+1}{3}$$

Hence, $\dfrac{(\sqrt[3]{2}+1)}{3}$ is the inverse of $1-(2)^{1/3}+(4)^{1/3}$.

Example 5. *If the rational number r is also an algebraic integer, show that r must be an ordinary integer.*

Solution.   Let $\qquad r = \dfrac{p}{q}, q > 0, (p,q) = 1$

Now, since r is an algebraic integer so

$$r^m + \alpha_1 r^{m-1} + \dots + \alpha_{m-1} r + \alpha_m = 0 \text{ ; where } \alpha_i\text{'s are integers.}$$

Thus, $\dfrac{p^m}{q^m} + \alpha_1 \dfrac{p^{m-1}}{q^{m-1}} + \dots + \alpha_{m-1}\dfrac{p}{q} + \alpha_m = 0$

So, $\qquad\qquad\qquad p^m + q \text{ (an integer)} = 0$

$\Rightarrow \qquad\qquad\qquad q$ divides p^m, but $(p, q) = 1$

Thus, $\qquad\qquad\qquad q\,|1 \Rightarrow q = 1 \Rightarrow r = p =$ integer.

Hence, r must be an ordinary integer.

Example 6. *Find the degree of the following expressions by finding bases*

$$\mathbf{Q}(\sqrt[3]{2},\sqrt{2}) \text{ over } \mathbf{Q}(\sqrt[3]{2}), \text{ over } \mathbf{Q}(\sqrt{2}), \text{ over } \mathbf{Q}.$$

Solution.   It is known that

$\mathbf{Q}(\sqrt[3]{2},\sqrt{2})$ is an extension of $\mathbf{Q}(\sqrt[3]{2})$ and $\mathbf{Q}(\sqrt[3]{2})$ of $\mathbf{Q}$.

By transitivity of extension we get

$$[\mathbf{Q}(\sqrt[3]{2}, \sqrt{2}): \mathbf{Q}] = [\mathbf{Q}(\sqrt[3]{2}, \sqrt{2}): \mathbf{Q}(\sqrt{2})]\,[\mathbf{Q}(\sqrt{2}): \mathbf{Q}]$$

$$= [\mathbf{Q}(\sqrt[3]{2},\sqrt{2}): \mathbf{Q}(\sqrt[3]{2})][\mathbf{Q}(\sqrt[3]{2}): \mathbf{Q}]$$

$\Rightarrow$ $[\mathbf{Q}(\sqrt[3]{2}, \sqrt{2}): \mathbf{Q}]$ is a multiple of 2 and 3 and hence a multiple of 6.

Since $\sqrt[3]{2}$ is algebraic of degree ≤ 3 over every field containing $\mathbf{Q}$, we have

$$[\mathbf{Q}(\sqrt[3]{2}, \sqrt{2}): \mathbf{Q}] = [\mathbf{Q}(\sqrt[3]{2}, \sqrt{2}): \mathbf{Q}(\sqrt{2})]\,[\mathbf{Q}(\sqrt{2}): \mathbf{Q}] \le 6$$

$\Rightarrow \qquad [\mathbf{Q}(\sqrt[3]{2}, \sqrt{2}): \mathbf{Q}] = 6$

Thus, $[\mathbf{Q}(\sqrt[3]{2}, \sqrt{2}): \mathbf{Q}(\sqrt{2})] = 3$ and $[\mathbf{Q}(\sqrt[3]{2}, \sqrt{2}): \mathbf{Q}(\sqrt[3]{2})] = 2$

Now, a basis of $\mathbf{Q}(\sqrt[3]{2}, \sqrt{2})$ over $\mathbf{Q}$ can be obtained from the bases $[1,\sqrt[3]{2},\sqrt[3]{4}]$ of $\mathbf{Q}(\sqrt[3]{2})$ over $\mathbf{Q}$ and $[1,\sqrt{2}]$ of $\mathbf{Q}(\sqrt[3]{2}, \sqrt{2})$ over $\mathbf{Q}(\sqrt[3]{2})$.

Therefore, we may get

$$[1, \sqrt[3]{2}, \sqrt[3]{4}, \sqrt{2}, 2^{5/6}, 2^{7/6})$$

Since, $2^{7/6} = 2 \times 2^{1/6}$, a simple basis is given by $[2^{k/6} : 0 \leq k \leq 5]$.

Hence, $\mathbf{Q}(\sqrt[3]{2}, \sqrt{2}) = \mathbf{Q}(\sqrt[6]{2})$

Example 7. *Prove that an element $\alpha \in K$, the finite extension of a field F generates the whole extension if and only if $[K : F] = [\alpha : F] = degree$ of α over F.*

Solution.   It is known that

$$(F(\alpha) : F) = degree\ of\ \alpha\ over\ F = [\alpha : F] \qquad \qquad ...(i)$$

Let us suppose α generates the whole extension K, then we have

$$K = F(\alpha)$$

Thus, $[K : F] = [F(\alpha) : F] = [\alpha : F]$

Conversely, suppose that $[K : F] = [\alpha : F]$. We have to prove that α generates the whole extension K, *i.e.,* $K = F(\alpha)$.

Using transitivity of extensions, we get

$$[K : F] = [K : F(\alpha)][F(\alpha) : F]$$

$\Rightarrow$ $[K : F] = [K : F(\alpha)][\alpha : F]$ [From (1)]

$$= [K : F(\alpha)][K : F]$$

$\Rightarrow$ $1 = [K : F(\alpha)]$

$\Rightarrow$ $K = F(\alpha)$

Example 8. *Let F be a field and let α, $\beta \in K$ be algebraic over F of degree m and n respectively and let m, n be relatively prime. Then, prove that $F(\alpha, \beta)$ is of degree mn over F.*

Solution.   We have

$$[F(\alpha) : F] = m \text{ and } [F(\beta) : F] = n$$

We have to prove that $[F(a, \beta) : F] = mn$.

Let $[F(\alpha, \beta) : F] = k$, then we shall show that $k = mn$.

Consider $[F(\alpha, \beta) : F] = [F(\alpha, \beta) : F(\alpha)]\ [F(\alpha) : F]$ [By transitivity]

$\Rightarrow$ $k = [F(\alpha, \beta) : F(\alpha)]\ m$

$\Rightarrow$ $m | k$

Also, $[F(\alpha, \beta) : F] = [F(\alpha, \beta) : F(\beta)]\ [F(\beta) : F]$

$\Rightarrow$ $k = [F(\alpha, \beta) : F(\beta)]n$

$\Rightarrow$ $n | k$

Since m and n are relatively prime, so that $m | k$ and $n | k$

$\Rightarrow$ $mn | k$ $\Rightarrow$ $k \geq mn$ ....(1)

Also, we have

$$[F(\alpha, \beta) : F] \leq [F(\alpha) : F]\ [F(\beta) : F]$$

$\Rightarrow$ $k \leq mn$ ...(2)

Using (1) and (2), we conclude that

$$k = mn$$

Hence, $F(\alpha, \beta)$ is an algebraic extension of degree mn over F.

Example 9. Find a basis of $\mathbf{Q}(\sqrt{3}, \sqrt{5})$ over $\mathbf{Q}$.

Solution.   We have

$$[\mathbf{Q}(\sqrt{3}, \sqrt{5}) : \mathbf{Q}] = [\mathbf{Q}(\sqrt{3})(\sqrt{5}) : \mathbf{Q}]$$

$$= [\mathbf{Q}(\sqrt{3})(\sqrt{5}) : \mathbf{Q}(\sqrt{3})][\mathbf{Q}(\sqrt{3}) : \mathbf{Q}]$$

$$= [L(\sqrt{5}) : L][\mathbf{Q}(\sqrt{3}) : \mathbf{Q}], \text{ where } L = \mathbf{Q}(\sqrt{3})$$

$$= \deg. \text{Irr}(L, \sqrt{5}) \times \deg. \text{Irr}(\mathbf{Q}, \sqrt{3})$$

$$= \deg(x^2 - 5) \times \deg(x^2 - 3)$$

$$= 2 \times 2 = 4$$

which shows that basis has 4 elements.

Further, if $[(F(a) : F)] = n$, then $1, a, a^2, ..., a^{n-1}$ is a basis of $F(a)$ over F and therefore

basis of $L(\sqrt{5})$ over L is $[1, \sqrt{5}]$

and   basis of $\mathbf{Q}(\sqrt{3})$ over $\mathbf{Q}$ is $[1, \sqrt{3}]$

Therefore, basis of $[L(\sqrt{5}) : L][\mathbf{Q}(\sqrt{3}) : \mathbf{Q}] = [\mathbf{Q}(\sqrt{3}, \sqrt{5}) : \mathbf{Q}]$ is given by

$$1.1, 1.\sqrt{3}, 1.\sqrt{5}, \sqrt{3}.\sqrt{5}$$

i.e., $1, \sqrt{3}, \sqrt{5}, \sqrt{15}$

Example 10. *Let F be a field and K be an extension of F. If a is an algebraic element of odd degree over F, show that α^2 is algebraic over F and that $F(\alpha) = F(\alpha^2)$.*

Solution.   It is given that α is an algebraic element of odd degree over F, so $F(\alpha)$ is the extension of F of odd degree.

Now, $F(\alpha)$ is a field, therefore, $\alpha \in F(\alpha)$ implies that $\alpha^2 \in F(\alpha)$.

$\Rightarrow \alpha^2$ is algebraic over F and $F(\alpha^2) \subseteq F(\alpha)$.

Since α is a root of the polynomial $x^2 - \alpha^2$ with coefficients in $F(\alpha^2)$,

therefore α is algebraic of degree at most 2 over $F(\alpha^2)$. i.e, $[F(\alpha) : F(\alpha^2)] \leq 2$

Also, $[F(\alpha) : F] = [F(\alpha) : F(\alpha^2)][F(\alpha^2) : F]$

$\Rightarrow$ $[F(\alpha) : F(\alpha^2)] \,|\, [F(\alpha) : F]$

But $[F(\alpha) : F(\alpha^2)] \leq 2$ and $[F(\alpha) : F]$ is odd, therefore, we must have

$$[F(\alpha) : F(\alpha^2)] = 1$$

Hence, $F(\alpha) = F(\alpha^2)$

Example 11. *Show that $\sin m°$ is an algebraic number for every integer m.*

Solution.   We know that

$$e^{\pi m i / 180} = \cos \frac{\pi m}{180} + i \sin \frac{\pi m}{180}$$

$\Rightarrow$ $(e^{\pi m i / 180})^m = \cos m\pi + i \sin m\pi = \pm 1$

Therefore,

$e^{\pi m i / 180}$ is a root of $x^{180} = \pm 1$.

$\Rightarrow$ $e^{\pi m i / 180}$ is an algebraic number for all integers m.

$\Rightarrow$ $\cos \dfrac{m\pi}{180} + i \sin \dfrac{m\pi}{180}$ is an algebraic number.

$\Rightarrow$ $\cos \dfrac{m\pi}{180} - i \sin \dfrac{m\pi}{180}$ is an algebraic number.

$\Rightarrow$ $2 \cos \dfrac{m\pi}{180}$ is an algebraic number for all integers m.

$\Rightarrow$ $\cos \dfrac{m\pi}{180}$ algebraic number for all integers m.

$\Rightarrow$ $\cos m°$ is algebraic number for all integers m.

Further, since $\cos \dfrac{m\pi}{180}$ and $\cos \dfrac{m\pi}{180} + i \sin \dfrac{m\pi}{180}$ are algebraic numbers, therefore

$i \sin \dfrac{m\pi}{180}$ is algebraic number.

$\Rightarrow \sin \dfrac{m\pi}{180}$ is algebraic number.   [$\because$ i also an algebraic number]

Hence, $\sin m°$ is an algebraic number.

Example 12. *Let F be a field and α be an algebraic element of degree n and of minimal polynomial $p(x)$ over F. Let β be an algebraic element over F, whose degree m is relatively prime to n. Find the degree of $F(\alpha, \beta)$ over F and show that $p(x)$ is irreducible over $F(\beta)$. Also, find $F(\alpha) \cap F(\beta)$.*

Solution.   It is given that

$$[F(\alpha) : F] = n \quad \text{and} \quad [F(\beta) : F] = m$$

Therefore, $[F(\alpha, \beta) : F] \le [F(\alpha) : F]\,[F(\beta) : F]$

$\Rightarrow$ $[F(\alpha, \beta) : F] < mn$ ...(1)

Further, $[F(\alpha, \beta) : F] = [F(\alpha, \beta) : F(\alpha)][F(\alpha) : F]$

$\Rightarrow$ $[F(\alpha) : F]$ divides $[F(\alpha, \beta) : F]$

Also, $[F(\alpha, \beta): F] = [F(\alpha, \beta): F(\beta)]\,[F(\beta): F]$

$\Rightarrow$ $[F(\beta): F]$ divides $[F(\alpha, \beta): F]$

Since m and n are relatively prime so that

$$[F(\alpha, \beta) : F] \ge mn$$ ...(2)

From (1) and (2), we conclude that

$$[F(\alpha, \beta): F] = mn$$

Now, since α satisfies a minimal polynomial $p(x) \in F[x]$, therefore $p(x)$ will not be expressed as a product of irreducible factors over $F[x]$. Thus, $p(x)$ is irreducible over $F(\beta)$.

Now, since $[F(\alpha) \cap F(\beta): F]$ divides both $[F(\alpha): F]$ and $[F(\beta): F]$ and $[F(\beta): F]$ are relatively prime, therefore

$$[F(\alpha) \cap F(\beta) : F] = 1$$

Hence, $F(\alpha) \cap F(\beta) = F$

Example 13. *Let F be a field and K be an extension of F of prime degree, then show that any element in K but not in F generates the whole extension K.*

Solution.   Let us suppose $[K : F] = p$, where p is prime.

Let $\alpha \in K$ such that $\alpha \notin F$. Then, we have to prove that $K = F(\alpha)$.

Now, since $[K : F] = [K : F(\alpha)]\,[F(\alpha): F]$

$\Rightarrow$ $[K : F(\alpha)]\,[F(\alpha) : F] = p$

Since $\alpha \notin F$, therefore $[F(\alpha): F] \neq 1$

$\Rightarrow$ $[F(\alpha) : F] > 1$

But p is prime, then we can say that one of the factor must be 1 and the other must be p.

Further, $[F(\alpha): F\,] > 1$

$\Rightarrow$ $[K : F(\alpha)] = 1$

Hence, $K = F(\alpha)$.

Example 14. *Show that every quadratic extension of α is of the form $\mathbf{Q}(\sqrt{a}\,)$ where a is a square free relative integer.*

Solution.   It is known that a quadratic extension of $\mathbf{Q}$ is an extension by a number of the form $(x \pm \sqrt{y}\,)\,/\,2$, with x, y relations.

Therefore, it is an extension of the form $\mathbf{Q}(\sqrt{y}\,)$.

If $y = \dfrac{m}{n}$, $m, n \in Q$, then $\mathbf{Q}(\sqrt{y}\,) = \mathbf{Q}\!\left(\sqrt{\dfrac{m}{n}}\right) = \mathbf{Q}\!\left(\dfrac{\sqrt{mn}}{n}\right) = \mathbf{Q}(\sqrt{mn})$

If $mn = k^2 a$ for a square free integer a, then we have

$$\mathbf{Q}\!\left(\sqrt{mn}\right) = \mathbf{Q}\!\left(k\sqrt{a}\right) = \mathbf{Q}(\sqrt{a})$$

Example 15. *Let K be an extension of a field F and $\alpha \in K$ algebraic over F. Then show that F(a) is isomorphic to F(x)/V, where V is an ideal of F[x] generated by minimal polynomial for α over F.*

Solution.   Let F be a field and K be an extension of F. Let $p(x)$ be the minimal polynomial for α over F, then $p(x)$ is irreducible over F.

Thus, $V = [q(x) \in F(x): q(\alpha) = 0]$ ...(1)

Clearly V is an ideal of $F[x]$ generated by the minimal polynomial for α over F. Also, $p(x) \in V$ is of lowest degree so that V is an ideal generated by $p(x)$, i.e., $V = (p(x))$. Further, $p(x)$ is irreducible so that V is maximal ideal of $F[x]$. Therefore $F[x]/V$ is a field.

Define a mapping

$\phi : F[x] \to F(\alpha)$ such that $\phi(f(x)) = f(\alpha),\ \forall\, f(x) \in F[x]$

Then, $\phi[\,f(x) + g(x)] = f(\alpha) + g(\alpha) = \phi\,(f(x)) + \phi(g(x))$

$\Rightarrow$ ϕ is a homomorphism.

Also, ϕ is onto.

Therefore, $\text{Ker } \phi = [q(x) \in F[x] : \phi(q(x)) = 0]$

$$= [q(x) \in F[x] : q(\alpha) = 0] = V$$

Hence, by fundamental theorem of homomorphism, we have

$$F(\alpha) \cong F[x]/V$$

Example 16. *Let K be an extension of a field F of finite degree n and α be an element of K which is algebraic over F of minimal polynomial p(x) over F. Show that deg p(x) divides n.*

Proof.   Let $\text{deg. } p(x) = m$

Since $\alpha \in K$, then $F \subset F(\alpha) \subset K.$

Also, since $\alpha \in K$ is algebraic element over F of minimal polynomial $p(x) \in F[x]$, then

$$[F(\alpha) : F] = m$$

Further, $[K : F] = n$, then we have

$$[K : F] = [K : F(\alpha)][F(\alpha) : F]$$

$\Rightarrow$ $n = [K : F(\alpha)]\, m$

$\Rightarrow$ $m \mid n$

Hence, deg $p(x)$ divides n.

Example 17. *Let F[x] be the field of rational functions in an indeterminate x. Show that every element of F(x) which is not in K is transcendental over F.*

Proof.   Let $0 \neq \dfrac{f}{g} \in F[x]$ such that $\dfrac{f}{g} \notin F$ and $(f, g) = 1$.

Let if possible f/g is not transcendental over F. Then, f/g is algebraic over F.

Therefore $F\left(\dfrac{f}{g}\right) = F\left[\dfrac{f}{g}\right].$

Now, consider $\dfrac{g}{f} \in F\left[\dfrac{f}{g}\right] = F\left(\dfrac{f}{g}\right)$

Becaus $0 \neq \dfrac{f}{g} \in F\left[\dfrac{f}{g}\right]$ and $F\left[\dfrac{f}{g}\right]$ is a field, $\dfrac{g}{f} \in F\left(\dfrac{f}{g}\right) = F\left[\dfrac{f}{g}\right]$

Further, $\dfrac{g}{f} = \alpha_0 + \alpha_1\left(\dfrac{f}{g}\right) + ... + \alpha_n\left(\dfrac{f}{g}\right)^n$, $\alpha_i \in F$

Therefore, $g^{n+1} = (\alpha_0 g^n + \alpha_1 f g^{n-1} + ... + \alpha_n f^n) f$

As per given, $(f, g) = 1,\ f \mid g^{n+1} \Rightarrow f \mid g \Rightarrow f = \text{unit}$

$\Rightarrow$ $g = \text{unit}$

$\Rightarrow$ $f \mid g = \text{unit} \in F$, which is a contradiction.

Hence, $f \mid g$ is transcendental over F.

EXERCISE 10.1

1. If a is an algebraic integer and m is an ordinary integer, show that
   (i) $a + m$ is an algebraic integer.
   (ii) ma is an algebraic integer.

2. Let a field L be a finite field extension of a field K. Define the degree $[L : K]$ of L over K. Let **Q** denote the field of rational numbers $K = \mathbf{Q}(\sqrt{2})$, $L = \mathbf{Q}(\sqrt{3})$. Show that $[L : K]=2$ and $[K : \mathbf{Q}] = 2$.

3. If $a \in K$ is algebraic over F of odd degree. Show that $F(a) = F(a^2)$.

4. Show that every finite extension K of a field F is algebraic and may be obtained by the adjunction of finitely many algebraic element.

5. If $a, b \in K$ are algebraic over F of degrees m and n respectively and if $(m, n) = 1$, show that $F(a, b)$ is of degree mn over F.

6. Find the degree and a basis for the following field extensions :
   (i) $\mathbf{Q}(\sqrt{2},\sqrt{3},\sqrt{18})$ over **Q**
   (ii) $\mathbf{Q}(\sqrt{2},\sqrt[3]{2})$ over **Q**
   (iii) $\mathbf{Q}(\sqrt{2},\sqrt{3})$ over **Q**
   (iv) $\mathbf{Q}(\sqrt[3]{2},\sqrt[3]{6},\sqrt[2]{24})$ over **Q**
   (v) $\mathbf{Q}(\sqrt{2},\sqrt{6}+\sqrt{10})$ over $\mathbf{Q}(\sqrt{3}+\sqrt{5})$ **Q**

7. For each of the given numbers $\alpha \in \mathbf{C}$, show that α is algebraic over **Q** by finding $f(x) \in \mathbf{Q}(x)$ such that $f(\alpha) = 0$
   (i) $1 + \sqrt{2}$　　　　　(ii) $\sqrt{2} + \sqrt{3}$
   (iii) $1 + i$　　　　　(iv) $\sqrt{1 + \sqrt[3]{2}}$
   (v) $\sqrt{\sqrt[3]{2} - i}$

8. Show that degree of $\sqrt{2}+\sqrt{3}$ over **Q** is 4 and degree of $\sqrt{2}+\sqrt{5}$ over **Q** is 6.

9. For each of the given algebraic number $\alpha \in \mathbf{C}$, find irreducible polynomial of α over **Q** and its degree.
   (i) $\sqrt{3-\sqrt{6}}$　(ii) $\sqrt{\frac{1}{3}+\sqrt{7}}$　(iii) $\sqrt{2}+i$

10. Show that the sum and product of two algebraic integers is an algebraic integer.

11. Find a suitable number α such that
    (i) $\mathbf{Q}(\sqrt{2}+\sqrt{5}) = \mathbf{Q}(\alpha)$
    (ii) $\mathbf{Q}(\sqrt{3},i) = \mathbf{Q}(\alpha)$

12. Find a basis of $\mathbf{Q}(\sqrt{2},\sqrt{3})$ over **Q** .

13. Let $w = \cos\dfrac{2\pi}{n} + i\sin\dfrac{2\pi}{n}$ and $u = \cos\dfrac{2\pi}{n}$. Show that $[\mathbf{Q}(w): \mathbf{Q}(u)] = 2$.

14. If K is an extension of F, $c \in K$, $a, b \in F$, $a \neq 0$, show that $F(c) = F(ac + b)$.

Answers

6. (i) $4,[1,\sqrt{3},\sqrt{2},\sqrt{6}]$　(ii) $6,[1,\sqrt{2},\sqrt[3]{2},\sqrt{2}\sqrt[3]{2},(\sqrt[3]{2})^2,\sqrt{2}(\sqrt[3]{2})^2]$　　(iii) $2,[1,\sqrt{6}]$
   (iv) $9,[1,\sqrt[3]{2},\sqrt[3]{4},\sqrt[3]{3},\sqrt[3]{6},\sqrt[3]{12},\sqrt[3]{9},\sqrt[3]{18},\sqrt[3]{36}]$　(v) $2,[1,\sqrt{2}]$

7. (i) $x^2 - 2x - 1$　　　(ii) $x^4 - 10x^2 + 1$　(iii) $x^2 - 2x - 2$　　(iv) $x^6 - 3x^4 + 3x^2 - 3$
   (v) $x^2 - 3x^8 - 4x^6 + 3x^4 + 12x^2 + 5$

9. (i) $x^4 - 6x^2 + 3$, degree = 4　　(ii) $x^4 - \frac{2}{3}x^2 - \frac{62}{9}$, , degree = 4　(iii) $x^4 - 2x^2 + 9$, degree = 4

12. $[1,\sqrt{2},\sqrt{3},\sqrt{6}]$

10.7 SPLITTING OR DECOMPOSITION FIELD

Definition 1 (*Root Fields*). *Let F be a field and K be a simple extension of F. Then K is said to be a root field of an irreducible polynomial $f(x) \in F[x]$ if K contains a root of $f(x)$.*

For example: The field $\mathbf{Q}(\sqrt{2}) = [a + b\sqrt{2} : a, b \in \mathbf{Q}]$ is a simple extension of **Q** and contains a root $\sqrt{2}$ of an irreducible polynomial $x^2 - 2$ over the field **Q** .

Definition 2 (*Splitting Field*). *Let F be a field and K be an extension of F. If f(x) ∈ F[x], then K is said to be splitting field over K for f(x) if over K, but not over any proper subfield K, f(x) can be factored as a product of linear factors.*

or

K is said to be a splitting field of a polynomial f(x) ∈ F[x] if it contains all its roots and there is no proper subfield of K which contains all the roots of f(x).

or

An extension K of a field F is said to be splitting field of f(x)∈F[x] if f(x)∈K[x] is expressible as f(x) = a(x − α₁)(x − α₂)...(x−αₙ), where a is a non-zero element of F, $\alpha_1, \alpha_2,... \alpha_n$ ∈ K and K = F (α₁, α₂,...αₙ).

REMARKS

☞ The field F defined above is called the base field or the initial field.

☞ The splitting field is also known as decomposition field.

☞ Splitting field of a polynomial over a field depends on both the polynomials as well as the field and that is why it is essential to mention splitting field of $f(x)$ over F.

Definition 3. Let K and L be the finite extensions of the fields F_1 and F_2 respectively, then an isomorphic mapping $\phi : F_1^i \to F_2$ is said to be continuation of the isomorphic mapping $\psi : K \to L$ if $\phi(\alpha) = \psi(\alpha)$, $\forall \alpha \in K$.

Definition 4. (*F-Isomorphism*). Let K_1 and K_2 be two extensions of a field F. Then an isomorphism $\sigma: K_1 \to K_2$ is said to be F-isomorphism if $\sigma(a) = a$, $\forall a \in F$.

Definition 5. (*F-Automorphism*). Let K be an extension of F. Then an F-isomorphism $\sigma : K \to K$ is said to be F-automorphism.

Theorem 1. (*Existence of a Decomposition Field*).

There exists a splitting field for every $f(x) \in F[x]$.

Proof.   Let $f(x)$ be a polynomial of positive degree say n over a field F.

We shall prove this theorem by induction on n.

If $n = 1$, i.e., $f(x)$ is a polynomial of degree 1.

Then $f(x) = ax + b$, $a, b \in F, a \neq 0$

Now, $x = -\dfrac{b}{a}$ ∈ K is a root of $f(x) = 0$.

So, we have

$$f(x) = a\left(x - \frac{b}{a}\right)$$

⇒ F is the decomposition field of $f(x) \in F[x]$.

⇒ Theorem is true for each polynomial of degree 1 over F.

Assume that theorem is true for each polynomial of degree less than n over F. Let $f(x)$ be a polynomial of degree $n > 1$. Also, let

$$f(x) = f_1(x)f_2(x)....f_m(x) \qquad\qquad ...(1)$$

where each polynomial in RHS of (1) is irreducible over F. Further, if each $f_i(x)$ in (1) is of degree 1, then F itself will be a decomposition field of $f(x)$. Suppose that at least one of $f_i (x)$, say $f_1(x)$ is of degree ≥ 2 . Then there is an extension $F(\alpha_1)$ containing a root α_1 of $f_1(x)$. So in the field $F(\alpha_1)$, we have

$$f(x) = (x - a_1)\, g(x)$$

where $\deg g(x) = n - 1$ and $g(x) \in F(\alpha_1)[x]$

Since $\deg g(x) < n$, then by induction hypothesis, there exists a decomposition field

$$(F(\alpha_1))(\alpha_2, \alpha_3,..\alpha_n) = F(\alpha_2, \alpha_3,..\alpha_n)$$

of $g(x)$ and thus of $f(x)$.

Hence, by induction, theorem is true.

Theorem 2. *Decomposition fields are algebraic, extensions.*

Proof. Let K be the decomposition field of a polynomial $f(x)$ over a field F. Let us suppose $\alpha_1, \alpha_2,..., \alpha_n$ be the roots of $f(x)$.

We have to prove that K is an algebraic extension of F.

As per given $\alpha_1, \alpha_2, \alpha_3,..\alpha_n$ are the roots of $f(x) = 0$, therefore

$$K = F(\alpha_1, \alpha_2, \alpha_3,..\alpha_n)$$

Then, $\qquad\qquad K_1 = F(\alpha_1)$

$$K_2 = K_1(\alpha_2) = F(\alpha_1, \alpha_2)$$

$$K_3 = K_2(\alpha_3) = F(\alpha_1, \alpha_2, \alpha_3)$$

$$\cdots \quad \cdots \quad \cdots \quad \cdots \qquad \cdots$$

$$K = K_n = K_{n-1}(\alpha_n) = F(\alpha_1, \alpha_2,..\alpha_n)$$

Here, each of the elements $\alpha_1, \alpha_2,..\alpha_n$ is the root of a non-zero polynomial over F, therefore $\alpha_1, \alpha_2,..\alpha_n$ are all algebraic elements over F. Further, since each of the fields $F, K_1, K_2,..., K_n = K$ can be obtained on adjoining an algebraic element to its predecessor field extensions so that each of the degrees

$$[K_1 : F], [K_2 : K_1],......[K : K_{n-1}] \text{ is finite.}$$

Then, by transitivity of extension fields

$$[K : F] = [K : K_{n-1}][K_{n-1}: K_{n-2}]...[K_2 : K_1][K : F]$$

is finite.

$\Rightarrow$ K is a finite extension of F.

Since we know that every finite extension of a field is an algebraic extension.

Hence, K is an algebraic extension of F.

Theorem 3. *Let ψ be an isomorphism of a field F_1 onto a field F_2 such that $\alpha\psi = \alpha'$ for every $\alpha \in f_1$. Then there is an isomorphism ϕ of $F_1[x]$ onto $F_2[t]$ with the property $\alpha\phi = \alpha\psi = \alpha'$, for each $\alpha \in F_1$.*

Proof. Let $f(x) = \alpha_0 + \alpha_1 x + ... + \alpha_n x^n$ be an arbitrary polynomial of $F_1[x]$ with $\alpha_0, \alpha_1,..\alpha_n \in F_1$.

Define a mapping

$$\phi : F_1(x) \to F_2[t]$$

such that

$$f(x)\,\phi = (\alpha_0 + \alpha_1 x + ... + \alpha_n x^n)\,\phi$$

$$= \alpha_0\psi + \alpha_1\psi t + ...+ \alpha_n\psi t^n$$

$$= \alpha'_0 + \alpha'_1 t + ...+ \alpha'_n t_n \quad = f'(t) \text{ (say)}$$

(i) ϕ *is one-one.*

Let $\quad\quad f(x) = \alpha_0 + \alpha_1 x + ... + \alpha_n x^n$

and $\quad\quad g(x) = \beta_0 + \beta_1 x + ... + \beta_m x^m$

be any two elements of $F_1[x]$. Consider

$$f(x)\phi = g(x)\phi$$

$$\Rightarrow (\alpha_0 + \alpha_1 x + ... + \alpha_n x^n)\phi = (\beta_0 + \beta_1 x + ... + \beta_m x^m)\phi$$

$$\Rightarrow \alpha'_0 + \alpha'_1 t + ... + \alpha'_n t^n = \beta'_0 + \beta'_1 t + ... + \beta'_m t^m$$

$$\Rightarrow \quad\quad n = m \text{ and } \alpha'_i = \beta'_i \text{ for each } i = 0, 1, 2, ... \, n \, .$$

$$\Rightarrow \quad\quad n = m \text{ and } \alpha_i \psi = \beta_i \psi \text{ for each } i.$$

$$\Rightarrow \quad\quad n = m \text{ and } \quad \alpha_i = \beta_i \text{ for each } i \quad\quad\quad\quad [\because \psi \text{ is one-one.}]$$

$$\Rightarrow \quad\quad f(x) = g(x)$$

$$\Rightarrow \phi \text{ is one-one.}$$

(ii) ϕ *is onto.*

Let $b'_0 + b'_1 t + ... + b'_n \, t^n$ be any element of $F_2[x]$ where $b'_0, b'_1, ... b'_n \in F_2$. Now, since ψ is onto , thus $\exists \, b_0, b_1, ..., b_n \in F_2$ such that

$$b_0 \psi = b'_0, b_1 \psi = b'_1, ..., b_n \psi = b'_n$$

Now, $\quad\quad b_0 + b_1 x + ... + b_n x^n \in F[x]$

Also, $\quad [b_0 + b_1 x + ... + b_n x^n]\phi = b'_0 + b'_1 t + ... + b'_n \, t^n$

$\Rightarrow \quad$ The mapping ϕ is onto.

(iii) ϕ *preserves addition of polynomials.*

Let $\quad\quad f(x) = a_0 + a_1 x + ... + a_n x^n, \quad g(x) = b_0 + b_1 x + ... + b_m x^m$

be any two elements of $F_1[x]$. Let us assume $n \geq m$.

Firstly, consider the case when $n > m$, we have

$$[f(x) + g(x)]\phi = [(a_0 + b_0) + (a_1 + b_1)x + ... + (a_m + b_m)x^m$$
$$+ \, a_{m+1} x^{m+1} + ... + a_n x^n]\phi$$
$$= (a_0 + b_0)\psi + (a_1 + b_1)\psi t + ... + (a_m + b_m)\psi t^m$$
$$+ \, a_{m+1} \psi t^{m+1} + ... + a_n \psi t^n$$
$$= (a'_0 \psi + b'_0 \psi) + (a'_1 \psi + b'_1 \psi)t + ... + (a'_m \psi + b'_m \psi)t^m$$
$$+ \, a'_{m+1} \psi t^{m+1} + ... + a'_n \psi t^n$$
$$= (a'_0 + a'_1 t + ... + a'_m t^m + a'_{m+1} t^{m+1} + ... + a'_n t^n)$$
$$+ (b'_0 + b'_1 t + ... + b'_m t_m]$$
$$= f(x)\phi + g(x)\phi$$

Similarly if $\quad n = m$, then we can easily show that

$$[f(x) + g(x)]\phi = f(x)\phi + g(x)\phi$$

(iv) ϕ *preserves multiplication of polynomials.*

Consider

$$[f(x) \cdot g(x)]\phi = [(a_0 + a_1 x + ... + a_n x^n)(b_0 + b_1 x + ... + b_m x^m)]\phi$$
$$= [a_0 b_0 + (a_0 b_1 + a_1 b_0)x + ... + a_n b_m x^{n+m}]\phi$$
$$= (a_0 b_0)\psi + (a_0 b_1 + a_1 b_0)\,\psi \, t + ... + (a_n b_m)\,\psi \, t^{n+m}$$

$$= a'_0 b'_0 + (a'_0 b'_1 + a'_1 b'_0) \; t + \ldots + (a'_n b'_m) \; t^{n+m}$$
$$= (a'_0 + a'_1 t + \ldots + a'_n t^n) \; (b'_0 + b'_1 t + \ldots + b'_m t^m)$$
$$= [f(x)\phi][g(x)\phi]$$

Therefore, ϕ is an isomorphism of $F_1[x]$ and $F_2[t]$. Also, if $f(x) \in F_1[x]$ be simply taken as α, where $\alpha \in F_1$. Then by definition of ϕ, we have

$$\alpha\phi = \alpha\psi = \alpha'$$

Theorem 4. *Let ψ be an isomorphism of a field F_1 onto a field F_1 defined by $a\psi = a'$ for $a \in F_1$ for an arbitrary polynomial $f(x) = a_0 + a_1 x + \ldots + a_n x^n \in F_1[x]$. Let us define $f'(t) = a'_0 + a'_1 t + \ldots + a'_n t^n \in K_2(t)$. If $f(x)$ is irreducible in $F_2[x]$, then there is an isomorphism θ of $F_1[x] / (f(x))$ onto $F_2(t) / (f^k(t))$ with the property that for every $\theta \in F, a\theta = a\psi = a'$.*

Proof. Define a function $\phi : F_1[x] \to F_2[t]$ such that

$$f(x)\phi = f'(t) \text{ for every } f(x) \in F[x] \; .$$

Using previous theorem, we can say that ϕ is an isomorphism of $F[x]$ onto $F_2[t]$.

Now, let $f(x)$ is irreducible in $F_1[x]$. Then $f'(t)$ will be irreducible in $F_2(t)$. If $V = (f(x))$ be the ideal of $F_1[x]$ generated by $f(x)$ and $V' = (f'(t))$ be the ideal of $F_2(t)$ generated by $f_2(t)$. Both V and V' are maximal ideals because $f(x)$ and $f'(t)$ are irreducible. Thus, $F_1[x]/V$ and $F_2[t]/V'$ are both fields.

Now, define a mapping $\theta : F_1 [x] \xrightarrow{\text{into}} F_2 [t]/V'$

such that

$$[V + g(x)]\theta = V' + g(x)\phi = V' + g'(t), \text{ for every } g(x) \in F_1[x].$$

(1) θ *is well defined.*

Let $\qquad V + g(x) = V + h(x)$

$\Rightarrow \qquad g(x) - h(x) \in V$

$\Rightarrow \qquad g(x) - h(x) = K(x) \, f(x), \text{ for some } K(x) \in F[x]$

$\Rightarrow \qquad [g(x) - h(x)]\phi = [K(x)f(x)]\phi$

$\Rightarrow \qquad g(x)\phi - h(x)\phi = [K(x)\phi] \, [f(x)\phi] \qquad\qquad [\because \phi \text{ is an isomorphism.}]$

$\Rightarrow \qquad g'(t) - h'(t) = K'(t)f'(t)$

$\Rightarrow \qquad g'(t) - h'(t) \in V$

$\Rightarrow \qquad V' + g'(t) = V' + h'(t)$

$\Rightarrow \qquad [V + g(x)]\theta = [V + h(x)]\theta$

Thus, θ is well defined.

(ii) θ *is one-one.*

Let $g(x), h(x) \in F_1[x]$

Consider,

$$[V + g(x)]\theta = [V + h(x)]\theta$$

$\Rightarrow \qquad V' + g'(t) = V' + h'(t)$

$\Rightarrow \qquad g'(t) - h'(t) \in V'$

$\Rightarrow \qquad g'(t) - h'(t) = K'(t) \, f'(t), \text{ for some } K'(t) \in F_2(t)$

$\Rightarrow \qquad g(x)\phi - h(x)\phi = [K(x)\phi][f(x)\phi]$

$\Rightarrow \quad [g(x) - h(x)]\phi = [K(x)f(x)]\phi$

$\Rightarrow \quad g(x) - h(x) = K(x)f(x)$ $\qquad\qquad\qquad$ [$\because \phi$ is one-one]

$\Rightarrow \quad g(x) - h(x) \in V$

$\Rightarrow \quad V + g(x) = V + h(x)$

$\Rightarrow \quad \theta$ is one-one.

(iii) θ *is onto.*

Since the mapping ϕ is onto, thus corresponding to any polynomial $g'(t)$ in $F_2[t]$, we have a polynomial $g(x)$ in $F_1[x]$. Thus

$$V' + g'(t) \in F_2(t) | V'$$

$\Rightarrow$ there exists $V + g(x) \in F_1[x] | V$ such that $[V + g(x)]\theta = V' + g'(t)$

(iv) θ *preserves additions and multiplications.*

Let $g(x), h(x) \in F_1[x]$

Consider

$$[\{V + g(x)\} + \{V + h(x)\}]\theta = [V + g(x) + h(x)]\theta$$
$$= V' + [g(x) + h(x)]\phi = V' + g(x)\phi + h(x)\phi]$$
$$= V' + g'(t) + h'(t) = [V + g'(t)] + [V + h'(t)]$$
$$= [V + g(x)]\theta + [V + h(x)]\theta$$

Further

$$[V + g(x)][V + h(x)]\theta = [V + g(x)h(x)]\theta$$
$$= V' + [g(x) + h(x)]\phi = V + [g(x)\phi][h(x)\phi]$$
$$= V' + g'(t)h'(t) = [V' + g'(t)][V' + h'(t)]$$
$$= [\{V + g(x)\}\theta][\{V + h(x)\}\theta]$$

Therefore, θ is an isomorphism of $F_1[x] / V$ onto $F_2(t) / V'$.

Using previous theorem, we conclude that the field F can be imbedded in the field $F_1[x]/V$ by identifying the element $\alpha \in F$ with the residue class (coset) $V + \alpha$ in $F_1(x)/V$. In a similar way, we can consider F_2 to be contained in $F_2[t]/V'$.

Therefore, for any $\alpha \in F$, we have

$$\alpha\theta = (V + \alpha)\theta$$
$$= V' + \alpha\phi$$
$$= V' + \alpha' = \alpha'.$$

REMARKS

☞ Let ψ be an isomorphism of a field F_1 onto a field F_2 such that $\alpha\psi = \alpha'$ for every $\alpha \in F_1$. Let $f(x) = \alpha_0 + \alpha_1 x + \dots + \alpha_n x^n$ be an irreducible polynomial in $F[x]$ which is mapped onto the irreducible polynomial $f'(t) = \alpha'_0 + \alpha'_1 + \dots + \alpha'_n t^n$ in $F_2[t]$. If V is a root of $f(x)$ in some extension field of F_1 and w is a root of $f'(t)$ in some extension field of F_2, then the field $F_1(V)$ is isomorphic to $F_2(W)$ by an isomorphism σ such that

(i) $V\sigma = w$

(ii) $\alpha\sigma = \alpha\psi = \alpha'$, for every $\alpha \in F_1$

☞ Let $f(x) \in F_1[x]$ be irreducible and if a, b be any two roots of $f(x)$, then $F_1[a]$ is isomorphic to $F_1[b]$ by an isomorpliism which takes a onto b and which leaves every element of F_1 fixed.

Theorem 5. *Let ψ be an isomorphism of a field F_1 defined by $\alpha\psi = \alpha'$ for every $\alpha \in F_1$. Corresponding to a polynomial*

$$f(x) = a_0 + a_1x + \ldots + a_nx^n \text{ in } F_1[x]$$

Let $f'(t) = a'_0 + a'_1t + \ldots + a'_nt^n$ be a polynomial in $F_2[t]$. Then, splitting fields K_1 and K_2 of $f(x) \in F_1[x]$ and $f'(t) \in F_2(t)$ respectively are isomorphic by an isomorphism ϕ with the property that

$$\alpha\phi = \alpha\psi = \alpha' \text{ for every } \alpha \in F_1.$$

Proof.

We shall prove this theorem by induction on the degree of splitting field over the initial field.

Let $[K:F_1] = 1$. Then, $K = F$. So $f(x)$ resolves into a product of linear factors over F_1 itself. Now by Theorem 3, $f'(t)$ must be also resolves into a product of linear factors over F_2 itself. So F_2 is a splitting field for $f'(t)$, i.e., $F_2 = K'$. Thus $\phi = \psi$.

$\Rightarrow$ We get an isomorphism of K onto K' coinciding with ϕ on F_1.

Further, assume that the theorem is true for any field F_0 and any polynomial $g(x) \in F_0[x]$ provided the degree of some splitting field K_0 of $g(x)$ is less than n.

i.e., $(K_0:F_0) < n$

Further, let $[K:F_1] = n > 1$, where K is a splittng field of $f(x)$ over F_1. If $f(x)$ resolves into a product of linear factors over F_1. Then, F_1 will be a splitting field for $f(x)$ and so we can not have $[K:F_1] > 1$.

$\Rightarrow$ $f(x) \in F_1[x]$ must have an irreducible factor $p(x) \in F_1[x]$ of degree $r > 1$.

Now, let $p(t)$ be the corresponsing irreducible factor of $f'(t)$.

Since K is a splitting field for $f(x) \in F_1[x]$.

$\Rightarrow$ a full compliment of roots of $f(x)$ are in K.

$\Rightarrow$ a full compliment of roots of $p(x)$ are in K.

$\Rightarrow$ there exist $V \in K$ such that $p(r) = 0$

$\Rightarrow$ $[F_1(V) : F_1] = r = \deg p(x)$

In a similar manner, there is a $w \in K'$ such that $p'(w) = 0$. Using the result of remark given just above the theorem, there is an isomorphism σ of $F_1(V)$ onto $K_2(w)$ such that

$$\alpha\sigma = \alpha\psi = \alpha', \text{ for every } \alpha \in F$$

Now, $[K : F_1] = [K : F_1(V)] [F_1(V) : F]$

$\Rightarrow$ $[K:F_1(V)] = \dfrac{[K : F_1]}{[K(V):F_1]} = \dfrac{n}{V} < n$

If $F_0 = F_1(V)$ and $F'_0 = F_2(w)$.

Now, since $F_1 \subset F_0$, thus $f(x) \in F_1[x]$ can also be regarded as $f(x) \in F_0(x)$. We want to show that K is a splitting field for $f(x) \in F_0[x]$. Clearly $f(x) \in F_0[x]$ resolves into a product of linear factors over K. No proper subfield of K containing F_0 and so F_1 can split $f(x)$ into linear factors. Thus, K is a splitting field for $f(x) \in F_0[x]$. Similarly K' is a splitting field for $f'(t) \in F_0[t]$.

Finally σ is an isomorphism of F_0 onto F'_0. K is splitting field for $f(x) \in F_0[x]$ and K' is a splitting field for $f'(t) \in F'_0[t]$.

Now, since

$$[K : F_0] = [K:F_t(V)] < n$$

Then, by our induction hypothesis, there is an isomorphism ϕ of K onto K such that

$$a\phi = a\sigma \text{ for all } a \in F_0$$

Also, $\alpha \in F_1 \Rightarrow \alpha \in F_0$. So for every $\alpha \in F_1$ we have

$$a\phi = a\sigma = a\psi = \alpha'.$$

Theorem 6. (*Uniqueness of Splitting Field*). *Any two splitting fields of the same polynomial over a given field F are isomorphic by an isomorphism leaving every element of F fixed.*

Proof. In the above theorem, let us take $F_2 = F_1$ and take ψ as the identity mapping of F, i.e., $\alpha\psi = \alpha \ \forall \ \alpha \in F$.

Also, ψ is an isomorphism of F onto F and it leaves every element of F fixed. If K_1 and K_2 are two splitting fields for $f(x) \in F(x)$. Then take $K_1 = K$ and $K_2 = K'$. Then K_1 and K_2 are isomorphic by an isomorphism leaving every element of F fixed.

Theorem 7. *Let K be an extension of F and if $f(x) \in F[x]$ and if ϕ is an automorphism of K leaving every element of F fixed, then ϕ must take a root of $f(x)$ lying in K into a root of $f(x)$ in K.*

Proof. Let $f(x) = a_0 + a_1x + ... + a_nx_n$, where $a_0, a_1, ..., a_n \in F$. Since K is an extension of F and ϕ is an automorphism of E leaving every element of F fixed. Therefore, $\phi(a) = a, \ \forall \ a \in F$.

If α is a root of $f(x)$ in E, then $a_0 + a_1\alpha + ... + a_n\alpha^n = 0$

We have to show that $\phi(a)$ is also a root of $f(x)$ in E, Clearly, $\phi(\alpha) \in K$, because ϕ is a mapping from K onto K.

Next, let $\phi(\alpha) = \beta$.

If r is any positive integer, so we have

$$\phi(\alpha^r) = \phi(\alpha.\alpha......r \text{ times})$$
$$= \phi(\alpha) \phi(\alpha).....r \text{ times} \qquad \text{(By structure preserving property)}$$
$$= \beta.\beta.\beta .....r \text{ times}$$
$$= \beta^r$$

Further, since ϕ will map 0 onto 0 because it is an automorphism. So $\phi(0) = 0$.

Consider, $a_0 + a_1\alpha + ... + a_n\alpha^n = 0$

$\Rightarrow$ $\phi(a_0 + a_1\alpha + ... + a_n\alpha^n) = \phi(0)$

$\Rightarrow$ $\phi(a_0) + \phi(a_1).\phi(\alpha) + ... + \phi(a_n).\phi(\alpha^n) = 0$

$\Rightarrow$ $a_0 + a_1\beta + a_2\beta^2 + ... + a_n\beta^n = 0$

Hence, β is a root of $f(x)$.

10.8 MULTIPLE ROOTS

Definition1. (*Derivative of a Polynomial over a Field*). *Let $f(x) = a_0 + a_1x + a_2x^2 + ... + a_{n-1}x^{n-1}$ be a polynomial over a field F. The derivative of $f(x)$ denoted by $f'(x)$ is defined by the polynomial.*

$$f'(x) = a_1 + 2a_2x + ... + (n-1)a_{n-1}x^{n-2} + na_nx^{n-1} \in F[x].$$

REMARKS

☞ If F is a field of finite characteristic p, then the derivative of the polynomial x^p is 0.

☞ The derivative of a non-constant polynomial can be zero if field F is a field of finite characteristics.

Definition 2. *(Multiple Roots). Let K be a splitting field of a polynomial $f(x) \in F[x]$; If α is a root of $f(x)$, then $(x-\alpha)\,|f(x)$ over K. If m is the largest positive integer for which $(x-\alpha)^m\,|f(x)$ in K[x], then m is called the multiple of α.*

REMARK

☞ If $m = 1$, then α is called a simple root.

Definition 3. *(Primitive Element). Let K be an extension of a field F, then an element $\alpha \in K$ is said to be primitive if $K = F(\alpha)$.*

Theorem 7. *Let $f(x)$ and $g(x)$ be any two polynomials over a field F and $a, b \in F$, then $(af(x) + bg(x))' = af'(x) + bg'(x)$.*

Proof. Consider two polynomials $f(x)$ and $g(x)$ such that

$$f(x) = a_0 + a_1 x + a_2 x^2 + \ldots\ldots + a_m x^m.$$

and

$$g(x) = b_0 + b_1 x + b_2 x^2 + \ldots\ldots + b_n x^n$$

without loss of any generality, we may assume that $m \geq n$. Then, for $a, b \in F$.

$$af(x) + bg(x) = (aa_0 + bb_0) + (aa_1 + bb_1)x + (aa_2 + bb_2)x^2 + \ldots.$$
$$+ (aa_n + bb_n)x^n + aa_{n+1}x^{n+1} + \ldots.. + aa_m x^m$$

So, $(af(x) + bg(x))' = (aa_1 + bb_1) + 2(aa_2 + bb_2)x + \ldots\ldots + n(aa_n + bb_n)x^{n-1}$
$$+ aa_{n+1}(n+1)x^n \ldots\ldots + aa_m m x^{m-1}$$

$$= a(a_1 + 2a_2 x + \ldots\ldots + na_n x^{n-1} + (n+1)a_{n+1}x^n + \ldots\ldots$$
$$+ ma_m x_{m-1}) + b[b_1 + 2b_2 x + 3b_3 x^2 + \ldots\ldots + nb_n x^{n-1}]$$

$$= af'(x) + bg'(x)$$

REMARK

☞ For any two polynomials $f(x)$ and $g(x)$ in $F[x]$, $[f(x)\,g(x)]' = f'(x)\,g(x) + f(x)\,g'(x)$.

Theorem 2. *Let K be an extension of a field F and f (x) be a polynomial of positive degree over F. Then $\alpha \in K$ is a multiple root of f (x) if and only if α is a common root of $f(x)$ and $f'(x)$.*

Proof. If $f(x)$ be a polynomial of positive degree over F. If α is a multiple root of $f(x)$ of multiplicity m greater than equal to α. Then we have

$$f(x) = (x-\alpha)^m g(x), g(\alpha) \neq 0$$
$$g(x) \in K[x]$$

Further,

$$f(x) = m(x-2)^{m-1}g(x) + (x-\alpha)^m g'(x)$$

$\Rightarrow \qquad f'(\alpha) = 0$

$\Rightarrow \quad \alpha$ is a common root of $f(x) = 0$ and $f'(x) = 0$.

Conversely, suppose that $f(x)$ and $f'(x)$ have a common root say α. We have to show that α is a multiple root of $f(x)$.

Let if possible, α is the simple root of $f(x)$, then
$$f(x) = (x-\alpha)g(x), \quad g(x) \in K[x] \quad \text{and} \quad g(\alpha) \neq 0$$
Also,
$$f'(x) = g(x) + (x-\alpha) g'(x)$$
$$\Rightarrow \qquad f(\alpha) = g(\alpha) + 0 = g(\alpha) \neq 0$$
$\Rightarrow \quad \alpha$ is not a common root of $f(x) = 0$ and $f'(x) = 0$, which is a contradiction.

$\Rightarrow \quad \alpha$ is not a simple root.

Hence, α is a multiple root of $f(x)$.

Theorem 3. *If $f(x)$ is an irreducible polynomial over a field F, then $f(x)$ has a multiple root in some field extension if and only if $f'(x) = 0$.*

Proof. Let F be a field and $f(x)$ be an irreducible polynomial over F. Let us first suppose α is a multiple root of $f(x) \in F[x]$ in some field extension K of F. Then, using previous theorem α is also a root of $f'(x)$.

Now, since $f(x)$ is an irreducible polynomial over F such that $f(\alpha) = 0$ and $f'(x)$ is another polynomial over F such that $f'(\alpha) = 0$.
$$\Rightarrow \qquad f(x) | f'(x)$$
Further if $f'(x) \neq 0$, then $\deg f'(x) < \deg f(x)$
$$\Rightarrow \qquad f(x) \nmid f'(x)$$
which is a contradiction
$$\Rightarrow \qquad f'(x) = 0.$$
Conversely, suppose $f(x)$ is an irreducible polynomial of degree n over F and let $f'(x) = 0$. We have to prove that $f(x)$ has a multiple root in some extension field of F.

Let if possible, $f(x)$ has no multiple root, then
$$f(x) = a(x-\alpha_1)(x-\alpha_2)......(x-\alpha_n)$$
$$= a \prod_{i=1}^{n} (x - \alpha_i)$$
where all $\alpha_i's$ are distinct roots in K.

But then
$$f'(x) = a[(x-\alpha_2)(x-\alpha_3)......(x-\alpha_n) + (x-\alpha_1)(x-\alpha_3)......(x-\alpha_n)$$
$$+ ....(x-\alpha_1)(x-\alpha_2)......(x-\alpha_{n-1})]$$
For each $i = 1, 2, 3,.......n$
$$f'(\alpha_i) = a \prod_{j \neq 1} (\alpha_i - \alpha_j) \neq 0$$

$\Rightarrow \quad f'(x) = 0$ holds if one of the roots $\alpha_1, \alpha_2,...,\alpha_n$ is a multiple root of $f(x)$. Hence, $f(x)$ has a multiple root of $f(x)$ in K.

Theorem 4. *Let $f(x)$ be an irreducible polynomial over $F[x]$, Then.*
(i) if the characteristic of F is 0, $f(x)$ has no multiple roots.
(ii) if the characteristic of F is $p \neq 0$, $f(x)$ has a multiple root only if it is of the form $f(x) = g(x^p)$.

Proof. (i) Consider an irreducible polynomial $f(x)$ of degree $n \geq 1$ given by
$$f(x) = a_0 + a_1 x + ... + a_0 x^n , \quad a_n \neq 0$$

over a field of characteristic 0.

Then, we have

$$f'(x) = a_1 + 2a_2x + \ldots + na_nx^{n-1}$$

Now, since F is of characteristic 0 and $a_n \neq 0$, therefore $na_n \neq 0$

$$\Rightarrow \qquad f'(x) \neq 0$$

Again, $\deg f'(x) < \deg f(x)$

We have to show that $f(x)$ has no multiple roots. Let if possible, $f(x)$ has a multiple root, say α in some extension field of F, then clearly $f(x) \,|\, f'(x)$.

Since $f'(x) \neq 0$ and $f(x)$ and $f'(x)$ both are irreducible with $\deg f'(x) \subset \deg f(x)$

$$\Rightarrow \qquad f(x) \text{ cannot divide } f'(x).$$

$$\Rightarrow \qquad \alpha \text{ is not a root of } f'(x), \text{ which is a contradiction.}$$

$$\Rightarrow \qquad \alpha \text{ is not a multiple root of } f(x).$$

(ii) As per given, the characteristic of F is p. Let us suppose α is a multiple root of $f(x)$. Then, clearly we have

$$f'(x) = 0$$

$$\Rightarrow \qquad a_1 + 2a_2x + 3a_3x^2 + \ldots + na_nx^{n-1} = 0$$

$$\Rightarrow \qquad\qquad a_1 = 0, 2a_2 = 0, 3a_3 = 0 \ldots na_n = 0$$

$$\Rightarrow \qquad\qquad ra_r = 0 \quad \forall\; 1 \leq r \leq n$$

Since, F is of characteristic $p \neq 0$, then

$$ra_r = 0 \qquad\qquad \Rightarrow \text{ either } a_r = 0 \text{ or } r \text{ is a multiple of } p.$$

$$\Rightarrow \qquad\qquad \text{either } a_r = 0 \text{ or if } a_r \neq 0, \text{ then } r = k \cdot p, k \in \mathbf{Z}^+$$

$$\Rightarrow \text{ The term } a_rx^r \text{ in } f(x) \text{ with } a_r \neq 0 \text{ will be of the form}$$

$$a_rx^r = a_{kp}\, x^{\,kp} = a_{kp}\, (x^p)^k$$

Hence, $f(x)$ will be a polynomial in x^p,

i.e., $\qquad\qquad f(x) = g(x^p) \text{ for some } g(x) \in F[x]$.

REMARK

☞ If $f(x)$ and $g(x)$ in $F[x]$ have a non-trivial common factor in $K[x]$, where K is some extension of F, then they must have a non-trivial common factor in $F[x]$.

Theorem 5. *Let F be a field of characteristic $p \neq 0$, then the polynomial*

$$x^{p^n} - x \in F[x], \text{ for } n \geq 1 \text{ has distinct roots.}$$

Proof. Let us write

$$f(x) = x^{p^n} - x$$

$$\Rightarrow \qquad\qquad f'(x) = p^n x^{p^n - 1} - 1$$

By $p \in F$, we mean $1 + 1 + 1 \ldots$ upto p times. Since F is of characteristic p, so the order of 1 as an element of the additive group of F is p. Therefore $p = 1 + 1 + \ldots$ upto p times $= 0$.

$$\Rightarrow \qquad\qquad p^n = 0$$

$$\Rightarrow \qquad\qquad f'(x) = -1$$

We observe that $f(x)$ and $f'(x)$ have no non-trivial common factor. Hence, using the result just given before this theorem, $f(x)$ has no multiple root.

Theorem 6. *Let F be a field of characteristic 0 and if a, b are algebraic over F, then there exists an element $c \in F[a, b]$ such that*

$$F[a, b] = F(c)$$

i.e., F[a, b] is a simple extension of F.

Proof. Consider two elements a and b which are algebraic over F. Let $f(x)$ and $g(x)$ be the irreducible polynomial over F, which are satisfied by a and b respectively.

Let $\deg f(x) = m$ and $\deg g(x) = n$

If K is an extension of F such that both $f(x)$ and $g(x)$ split completely. Further, since $f(x)$ is irreducible and characteristic of F is zero. Thus, by Theorem 4, all the roots of $f(x)$ are distinct. In a similar way, we can show that all the roots of $g(x)$ are distinct.

Now, let $a_1, a_2, ..., a_m$ be the roots of $f(x)$ and $b_1, b_2, ..., b_n$ be the roots of $g(x)$.

For simplicity, let us write $a_1 = a, b_1 = b$.

Now, in K, we have

$$a_i + \lambda b_j = a + \lambda b, i = 1, ...m, j = 2, ...n$$

Then, $$\lambda = \frac{a_i - a}{b - b_j}$$

which is a unique element of K.

$\Rightarrow$ For each pair of values of i and j ($j \neq 1$), the equation $a_i + \lambda b_j = a + \lambda b$ has only one solution in K.

Therefore, there are only a finite number of elements of E such that

$$a_i + \lambda b_i = a + \lambda b, i = 1, ..., m, j = 2, ..., n$$

Since, F has an infinite number of elements and characteristic of F is 0, each element of F is in K.

Thus, we must have an element $t \in F$ such that

$$a_i + t b_j \neq a + tb, \forall i, \forall j \neq 1$$

Let $c = a + tb$ and $a, b, t \in F[a, b]$

$\Rightarrow$ $c \in F[a, b]$

We have to show that $F(a, b) = F(c)$

Further, since $c \in F(a, b) \Rightarrow F(c) \subseteq F(a, b)$ ....(1)

Now, let $a, b \in F(a, b)$

Since b satisfies the polynomial $g(x)$ over F. The polynomial $g(x)$ over F can be regarded as a polynomial over $F(c)$.

Let $F(c) = K$

and $h(x) = f(c - tx)$

Since $c \in K$ and $t \in F \Rightarrow t \in K$

$\Rightarrow$ $h(x)$ is a polynomial in $K[x]$.

$\therefore$ $$h(b) = f(c - tb)$$
$$= f(a) = 0 \qquad\qquad [\because a \text{ is a root of } f(x).]$$

$\Rightarrow$ b satisfies both the polynomials $g(x)$ and $h(x)$ in $K[x]$.

$\Rightarrow$ $x - b$ is a common factor of $g(x)$ and $h(x)$ in some extension E of K.

We wish to show that $x - b$ is the g.c.d. of $g(x)$ and $h(x)$ in $K[x]$. Let $b_j \neq b$ be another root of $g(x)$. Then, we have $h(b_j) = f(c - tb_j) \neq 0$

$\Rightarrow$ b_j is not a root of $h(x)$.

$\Rightarrow$ Any factor of $g(x)$ in $E(x)$ other than $x - b$ is not a factor of $h(x)$. Further, $g(x)$ has all distinct roots, therefore

$$(x - b)^r, \quad r \geq 2 \text{ is not a divisor of } g(x).$$

$\Rightarrow$ $(x - b)$ is the g.c.d. of $g(x)$ and $h(x)$ in some extension E of K. Now, since $g(x)$ and $h(x)$ have a non-trivial factor over some extension of K. Thus, they must have a non-trivial common factor over K.

Therefore, they must have a non-trivial g.c.d. over K which must be a divisor of $x - b$. But degree of $x - b$ is one, so $x - b$ itself is the g.c.d. of $g(x)$ and $h(x)$ in $K[x]$.

$\Rightarrow$ $\qquad\qquad b \in K = F(c)$

Also $\qquad\quad c, t, b \in F(c) \Rightarrow c - tb \in F(c)$

$\Rightarrow$ $\qquad\qquad a \in F(c) \Rightarrow a, b \in F(c)$

Now, $\qquad\quad a, b \in F(c) = F(a, b) \subseteq F(c)$

$\Rightarrow$ $\qquad\qquad F[a, b] \subseteq F(c)$ $\qquad\qquad\qquad\qquad\qquad$ ...(2)

From (1) and (2), we conclude that $F(a, b) = F(c)$.

Theorem 7. *Any finite extension of a field of characteristic 0 is a simple extension.*

Proof. Let F be a field of characteristic 0 and K be a finite extension. Then, K is an algebraic extension of F. So, K can be obtained by adjoining a finite number of algebraic elements of F. Let $K = F[\alpha_1, \alpha_2, ...\alpha_n]$

To prove there is an element $c \in K$ such that $K = F(c)$.

Consider

$$K = F(\alpha_1, \alpha_2, ...\alpha_{n-2}, \alpha_{n-1}, \alpha_n)$$
$$= (F(\alpha_1, \alpha_2, ...\alpha_{n-2}))(\alpha_{n-1}, \alpha_n)$$
$$= (F(\alpha_1, \alpha_2, ...\alpha_{n-2}))$$

where, $\qquad\qquad d \in F(\alpha_1, \alpha_2, ...\alpha_{n-2}))(\alpha_{n-1}, \alpha_n)$
$$= (F(\alpha_1, \alpha_2, ...\alpha_{n-2}, \alpha_n) = K$$

$\qquad\qquad\qquad\qquad\qquad\qquad\qquad\qquad\qquad$ [Using previous theorem]

$$= F(\alpha_1, \alpha_2, ...\alpha_{n-2}, d)$$

Applying previous theorem by a finite number of times, we get

$$K = F(c), \text{ where } c \in F(c) = K.$$

Hence, K is a simple extension of F.

REMARK

☞ The above theorem is known as "Primitive element theorem".

10.9 NORMAL AND SEPARABLE EXTENSIONS OF A FIELD

Definition 1 (*Normal Extension*). *Let F be a field and K be an algebraic extension of F. Then K is said to be normal extension of F, if the splitting field of the minimal polynomial $f(x) \in F[x]$ for each element of K is the splitting field of some polynomial over F.*

Definition 2 (*Separable Polynomial*). *An irreducible polynomial $f(x) \in F[x]$ is said to be separable over F if $f(x)$ has no multiple roots in its splitting field, i.e., the roots of $f(x)$ in its splitting field are simple.*

Definition 3. *A polynomial which is not separable, is called* inseparable.

Definition 4. *Let F be a field and K be an algebraic extension of F. Then an element $\alpha \in K$ is said to be separable element over F if the minimal polynomial for α over F is separable.*

If α is not separable, then it is said to be inseparable element of F.

Definition 5 (*Separable Extensions*). *Let F be afield. An algebraic extension K of F is said to be separable extensions of F if every element of K is separable over F. If K is not a separable extension, then it is said to be* inseparable extension.

Definition 6 (*Perfect Field*). *A field F is said to be perfect if each of its extension is separable.*

<u>**Theorem 1.**</u> *A finite algebraic extension of K of a field F is normal over F if and only if K is the splitting field of some polynomial over F.*

<u>**Proof.**</u> Let F be a field and $K = F(\alpha_1, \alpha_2, ...,\alpha_n)$ be a finite algebraic extension of F. Here each α_i's is algebraic over F. Let $f_1(x), f_2(x),... f_n(x)$ be the minimal polynomial of $(\alpha_1, \alpha_2, ...,\alpha_n)$ over F respectively.

Let us first suppose that K is normal over F, then the splitting field of each $f_1(x)$, $f_2(x)....,f_n(x)$ is contained in K. Therefore, K is a splitting field of a polynomial.

$$f(x) = f_1(x), f_2(x).....f_n(x) \text{ over } F.$$

Conversely, let K be a splitting field of some polynomial $f(x) \in F[x]$ and let $\alpha_1, \alpha_2, ...,\alpha_m$ be the roots of $f(x)$ in K, then, we have

$$K = F(\alpha_1, \alpha_2, ..., \alpha_m)$$

To prove K is normal over F. For this, we shall prove that the splitting field of the minimal polynomial for each element of K is contained in K.

Let $\alpha \in K$ be an arbitrary element and let $p(x)$ be the minimal polynomial of α over F.

Let if possible $p(x)$ do not have all of its roots in K and suppose β is a root of $p(x)$. which is not in K.

Then clearly $F(\alpha)$ is isomorphic to $F(\beta)$ such that α is mapped on β and each element of F remains fixed. Further, since $\alpha \in K$, the field K is the splitting field of $f(x)$ over $F[\alpha]$. Also, $K(\beta)$ is generated by the root of $f(x)$ over $F(\beta)$.

Here $K(\alpha)$ is the splitting field of $f(x)$ over $F(\beta)$.

$\Rightarrow$ An isomorphism of $F[\alpha]$ onto $F(\beta)$ is extended to an isomorphism of K onto $K(\beta)$ such that each element of F remains fixed under this isomorphism.

$\Rightarrow$ K and $K[\beta]$ will have same degree over F.

$\Rightarrow$ $K[\beta]$ cannot be a proper extension of K, which is a contradiction, because $K \subseteq K[\beta]$

Hence, K is a normal extension of F.

Theorem 2. *Let F be a field and K be a normal extension of F. If L is an immediate field so that $F \subseteq L \subseteq K$, then K is also a normal extension of L.*

Proof. It is given that F be a field and K is a normal extension of F and L is also a field such that $F \subseteq L \subseteq K$. We have to show that K is a normal extension of L.

For this, we shall prove that K is the splitting field of a minimal polynomial for every element of K over L is contained in K.

Suppose $\alpha \in K$ is any element and let $f(x)$ and $g(x)$ be the minimal polynomial for α over F and L respectively.

Now, since

$$f(x) \in F[x]$$
$$\Rightarrow \qquad f(x) \in L[x] \qquad\qquad\qquad [\because F \subseteq L]$$

Further, $g(x)$ is the minimal polynomial of α over L, then we have

$$g(x) \,|\, f(x)$$

$\Rightarrow$ Every root of $g(x)$ is the root of $f(x)$.

But every root of $f(x)$ is in K, therefore every root of $g(x)$ in $L[x]$ is in K.

Hence, K is normal extension of F.

Theorem 3. *Every field of characteristic zero is perfect.*

Proof. Let F be a field of characteristic 0 and K be any extension of F. To prove F is perfect. For this, we shall prove that all the finite extension of F are separable, *i.e.,* every element of K is separable, or the minimal polynomial for each element of K over F is separable.

Let $a \in K$ be arbitrary. Also, let

$$f(x) = a_0 + a_1x + a_2x^2 + \dots + a_nx^n, \quad a_n \neq 0 \qquad\qquad \dots(1)$$

and $a_i's \in F$ be a minimal polynomial for a over F.

Now, $\qquad f'(x) = 0 \Rightarrow a_1x + 2a_2x + \dots + na_nx^{n-1} = 0 \qquad$ [Using (1)]

Further, suppose $f(x)$ has a multiple root in K, then $f'(x) = 0$.

$$\Rightarrow \qquad a_1 + 2a_2x + \dots + na_nx^{n-1} = 0$$
$$\Rightarrow \qquad a_1 = 0, 2a_2, \dots, na_n = 0 \qquad\qquad \dots (2)$$

Since F is of characteristic zero, then (2) is possible only if $a_1 = 0, = a_2 = \dots = a_n$.

$\Rightarrow$ $f(x) = a_0$, a constant polynomial having a multiple root, which is a contradiction, because no constant polynomial has a root.

$\Rightarrow$ $f(x)$ has no multiple root.

Thus, a is separable over F.

Since a is arbitrary, so every element of K is separable over F.

$\Rightarrow$ K is a separable extension of F.

Also, since K is an arbitrary finite extension of F, so we conclude that all finite extension of F are separable. Hence, F is perfect.

Theorem 4. *An irreducible polynomial $f(x)$ over a field F of characteristic $p > 0$ is inseparable if and only if $f(x) = g(x^p)$, i.e., $f(x)$ is a polynomial in x^p.*

Proof. Consider an irreducible polynomial

$$f(x) = a_0 + a_1x + \dots + a_nx^n, a_n \neq 0$$

over a field F of characteristic $p > 0$

Then
$$f'(x) = 0 \Rightarrow a_1 + 2a_2 x + \ldots + na_n x^{n-1}$$

Let us first suppose $f(x)$ is inseparable, then $f(x)$ has at least one multiple root in its splitting field over F'.

Then $\qquad\qquad f'(x) = 0 \Rightarrow a_1 + 2a_2 x + \ldots + na_n x^{n-1} = 0$

$\Rightarrow \qquad\qquad\qquad a_1 = 0, 2a_2 = 0, .., na_n = 0$

i.e., $\qquad\qquad\qquad ra_r = 0, 0 \le r \le n$

Now, since F is a field of characteristic $p > 0$, then
$$ra_r = 0$$

$\Rightarrow$ either $\qquad a_r = 0$ or p/r

$\Rightarrow$ either $\qquad a_r = 0$ or if $a_r \ne 0$ then $r = k/p, k \in \mathbf{Z}^+$

$\Rightarrow \qquad\qquad f(x) = b_0 + b_1 x^p + b_2 x^{2p} + \ldots b_m x^{m/p}$, where $b_j = a_{jp}$

$\Rightarrow$　$f(x)$ is a polynomial in x^p over F of characteristic $p > 0$

Then $\qquad\qquad f'(x) = 0.$

Conversely, let if possible $f(x)$ has no multiple root, then $f'(x) = 0$ which is a contradiction.

$\Rightarrow$　$f(x)$ has multiple root in its splitting field.

Hence, $f(x)$ is inseparable.

Theorem 5. *A field of characteristic $p \ne 0$ such that each element of the field is the pth power of some member of the same is perfect.*

Proof. Let F be field of characteristic $p \ne 0$ such that
$$a_i = b_0^p, \text{ for } a_i, b_i \in F$$

To prove F is perfect. For this we shall prove that all the finite extensions of F are separable extension of F. We have to show that K is separable. Since F is a field of characteristic $p \ne 0$, then using above theorem we can say that the only irreducible polynomial, which are inseparable are of the form $f(x) = f(x^p)$.

Then, we have
$$f(x^p) = a_0 + a_1 x^p + a_2 x^{2p} + \ldots + a_m x^{np}, a_i \in F$$

But $\qquad\qquad a_0 = b_0^p, a_1 = b_1^p, \ldots, a_n = b_n^p$

$\Rightarrow \qquad\qquad f(x^p) = b_0^p + b_1^p x^p + b_2^p x^{2p} + \ldots + b_n^p x^{np}$

$$= (b_0 + b_1 x + b_2 x^2 + \ldots + b_n x^n)^p \quad [\because b_i^p = 0, \forall 0 = b_i \in F]$$

$\Rightarrow$　$f(x^p)$ is not irreducible.

$\Rightarrow$　No irreducible polynomial over F is inseparable.

$\Rightarrow$　Every irreducible polynomial over F is separable. Hence, F is perfect.

Theorem 6. *Every finite field is perfect.*

Proof. Let F be a finite field. It is known that its characteristics $p \ne 0$, will be the prime number.

To prove F is perfect. For this we shall prove that every element of F is the pth power of the element.

Define a mapping $\phi : F \to F$ such that $\phi(a) = a^p$, $\forall a \in F$

We have to show that ϕ is one-one

Let $a, b \in F$.

Consider $\qquad \phi(a) = \phi(b) \Rightarrow a^p = b^p \qquad\qquad \Rightarrow a^p - b^p = 0$

$\Rightarrow \qquad\qquad (a-b)^p = 0 \qquad\qquad\qquad [\because x^p = 0 \text{ for all non-zero } x \in F]$

$\Rightarrow \qquad\qquad a = b$

Therefore, ϕ is one-one.

Also, since ϕ is one-one from F, so that ϕ is certainty onto.

$\Rightarrow$ Every element of F is the pth power of some other element of K.

Hence, using above theorem, F is perfect.

AN IMPORTANT NOTATION: E/K is normal $\Rightarrow E$ is a normal extension of K.

Theorem 7. Let $F \subseteq K \subseteq E$ be a tower of fields. If E/F is normal then E/K is normal.

Proof. Let F be field such that $F \subseteq K \subseteq E$.

Since E/F is normal $\qquad \Rightarrow E/F$ is algebraic $\Rightarrow E/K$ is algebraic.

Let $\alpha \in E$ and let $p(x)$ be any irreducible polynomial over K of α, i.e.,

$$p(x) = \text{Irr.}(F, \alpha).$$

Similarly let $\qquad g(x) = \text{Irr.}(F, \alpha)$.

Then $\qquad f(x) \in F(x) \subseteq K(x)$

$\Rightarrow \quad g(x) \in K(x)$ and $f(\alpha) = 0$

$\Rightarrow \quad p(x)$ divides $g(x)$ in $K(x)$

Now, since E/K normal and $a \in E$, $g(x)$ splits in $E(x)$

$\Rightarrow \quad P(x)$ splits in $E(x)$. Hence, E/K is normal.

Theorem 8. *A minimal splitting field of a non-constant polynomial $f(x) \in K(x)$ over K is normal extension of K.*

Proof. Let F be a field and E be a minimal splitting field of $f(x)$ over K . Then E/K algebraic and finite.

Consider a polynomial

$$f(x) = \alpha_0(x-\alpha_1)...(x-\alpha_n), \alpha_i \in E.$$

Then $\qquad\qquad E = K(\alpha_1, \alpha_2, ..., \alpha_n).$

Let $\qquad\qquad \alpha \in E, p(x) = \text{Irr.}(K, \alpha) \in K(x) \subseteq E(x).$

Then $p(x)$ splits in some extension of E.

Let b be a root of $p(x)$ in some extension of E. To show $b \subset E$.

Now, a, b are roots of $p(x)$, therefore, there exists a K-isomorphism

$$\sigma : K(\alpha) \to K(b)$$

such that $\qquad\qquad \sigma(\alpha) = b$

Then, a minimal splitting field of f over $K(\alpha)$ is $K(\alpha)(\alpha_1... \alpha_n)$

$$= K(\alpha_1... \alpha_n)(\alpha_n)$$

$$= E(\alpha)$$

$$= E \qquad\qquad\qquad\qquad\qquad [\because \alpha \in E]$$

Further, a minimal splitting field of $\sigma(f) = f$ over $K(b)$ is giving by

$$K(b) = (\alpha_1, \alpha_2 \ldots \alpha_n)$$
$$= K(\alpha_1, \alpha_2, \ldots \alpha_n)(b)$$
$$= E(b)$$

Therefore, there exists an isomorphism $\theta : E \rightarrow E(b)$

such that $\theta(a) = \sigma(a), \forall a \in K(\alpha)$.

$\Rightarrow$ $\qquad\qquad\qquad\qquad \theta(\alpha) = \sigma(\alpha) = b$

Also, $\qquad\qquad\qquad K \subseteq K(\alpha) \subseteq E \subseteq E(b)$

$\Rightarrow$ $\qquad\qquad [E : K(\alpha)] = [\theta(E) : \theta(K(\alpha))]$
$$= [E(b) : \sigma(K(\alpha)]$$
$$= [E(b) : (K(b)]$$

Therefore

$$[E(b) : K] = [E(b) : K(b)][K(b) : K]$$
$$= [E : K(\alpha)]\deg. p(x)$$
$$= [E : K(\alpha)][K(a) : K]$$
$$= [E : K] \qquad\qquad\qquad\qquad \text{[By transitivity]}$$

Finally, since $E \subseteq E(b)$ and $E, E(b)$ as vector space over K have same dimension $E = E(b)$

$\Rightarrow$ $\qquad\qquad\qquad\qquad b \in E$

$\Rightarrow$ $\qquad\qquad\qquad\qquad p(x)$ splits E.

Hence, E/K is normal .

Theorem 9. *A finite normal extension is a minimal splitting filed of some polynomial.*

Proof. Let E / K be a finite normal extension.

Since E/K finite, therefore, $E = K(\alpha_1, \alpha_2, \ldots \alpha_n)$.

Let $\qquad\qquad\qquad p_i(x) = \text{Irr}(K, \alpha_i)$.

Further, since $\alpha_i \in E$ and E / K is normal, each $p_i(x)$ splits in E

Let $\qquad\qquad\qquad f = p_1 p_2 \ldots p_n \in K(x)$

Then a minimal splitting field of f, over K is given by

$$K(\alpha_1, \alpha_2 \ldots a_n, \text{ roots of } p_i\text{'s in } E) = E$$

Hence, E is a minimal splitting field of f over K.

Theorem 10. *Let $K \subseteq E_1 \subseteq E, K \subseteq E_2 \subseteq E$ be towers of fields such that $E_1 / K, E_2 / K, i.e., E_1$ and E_2 are normal extension of K then $E_1 E_2$, the smallest subfield of E containing $E_1 \cup E_2$, is finite normal extension of K.*

Proof. Since E_1 is finite over K. so $E_1 = K(\alpha_1 \ldots \alpha_n)$

Therefore $\qquad\qquad E_1 E_2 = K(\alpha_1 \ldots \alpha_n) E_2$
$$= E_2(\alpha_1 \ldots \alpha_n) \qquad\qquad [\because K \subseteq E_2 \Rightarrow KE_2 = E_2]$$

Consider

$$(E_1 E_2 : E_2) = (E_2(\alpha_1 \ldots \alpha_n) : E_2)$$
$$= [E_2(\alpha_1 \ldots \alpha_n) : E_2(\alpha_1 \ldots \alpha_{n-1})] \ldots (E_2(\alpha_1) : E_2)$$

$$\le [K(\alpha_1... \alpha_n) : K(\alpha_1... \alpha_{n-1})] ...[K(\alpha_1) : K]$$
$$= [K(\alpha_1... \alpha_n) : K]$$
$$= [E_1 : K]$$
$$\Rightarrow \qquad (E_1 E_2 : K) = (E_1 E_2 : E_2)(E_2 : K)$$
$$\le (E_1 : K)(E_2 : K) = \text{finite}$$

Now E_1 /K is finite normal $\Rightarrow E_1$ is a minimal splitting field of f_1 over K and E_2/K is finite normal

$\Rightarrow \qquad E_2$ is a minimal splitting field of f_2 over K

Let $\qquad f = f_1 f_2, E_1 = K(a_1...a_r), E_2 = K(b_1...b_s)$.

Then, a minimal splitting field of f over K is

$$K(a_1...a_r, b_1...b_s)$$
$$= E_1 (b_1...b_s)$$
$$= E_1 K (b_1...b_s) \qquad\qquad [\because E_1 K = E_2]$$
$$= E_1 E_2$$

$\Rightarrow E_1 E_2$ finite normal extension over K.

REMARKS

☞ An irreducible polynomial is separable if and only if $f'(x) \ne 0$.

☞ Every non-zero polynomial over a field of characteristic zero is, separable.

☞ If F is a field of characteristic zero, then any algebraic extension of F is separable.

☞ If E is a finite extension of K. Then E is separable over K if and only if each element of E is separable over K.

☞ Let E be an algebraic extension of K generated by a family of elements $\{\alpha_i : i \in \mathbf{Z}\}$. If each α_i is separable over K. then, E is separable over K.

☞ Let $E \supset F \supset K$ be a tower of field. Then $[E : K]_s = [E : F]_s [F : K]_s$. Also, if E is finite over K, then $[E : K]$ is finite and $[E : K]_s \le [E ; K]$.

Example 1. *If p is prime then show that $f(x) = x^{p-1} \in \mathbf{Q}(x)$ has splitting field $\mathbf{Q}(\alpha)$ where $\alpha \ne 1$ and $\alpha^p = 1$. Also show that $(\mathbf{Q}(\alpha) : \mathbf{Q}) = p - 1$.*

Solution. It is given that

$$f(x) = x^p - 1$$
$$= (x-1)(x^{p-1} + x^{p-2} + ...x + 1)$$

Let $\qquad g(x) = (x^{p-1} + x^{p-2} + ...x + 1)$

Then $g(x) \in \mathbf{Q}(x)$ is clearly irreducible over $\mathbf{Q}$. Let α be a root of $g(x)$ in the splitting field of $f(x)$ over $\mathbf{Q}$.

Then $\qquad g(\alpha) = 0$.

So $\qquad f(\alpha) = (\alpha - 1)g(\alpha) = 0 \qquad\qquad [\because g(\alpha) = 0 \text{ and } \alpha \ne 1]$

$\Rightarrow \qquad \alpha^p - 1 = 0 \Rightarrow \alpha^p = 1$

We wish to show that $1, \alpha, \alpha^2, ..., \alpha^{p-1}$ are p-distinct roots of $f(x)$. Let α^p be any one of them.

SInce $\alpha^p = 1$, thus, $(\alpha^i)^p = (\alpha^p)^i = (1)^i = 1$, for all non-negative integers i.

It remains to prove that they all are distinct. If m is the least positive integers such that $\alpha^m = 1$. Then m/p because $\alpha^p = 1$, but p is prime so $m = p$.

Therefore $1, \alpha, \alpha^i, ..., \alpha^{p-1}$ are p distinct roots of $f(x)$ over $\mathbf{Q}$.

So the splitting field of $x^p - 1 \in \mathbf{Q}(x)$ is $\mathbf{Q}(x)$.

Finally, the minimal polynomial of α is $g(x)$ whose degree is $p-1$.

Hence, $(\mathbf{Q}(\alpha):\mathbf{Q}) = p-1$

Example 2. *Let F be a field of characteristic p. Let b be a root of $f(x) = x^p - x - a \in F(x)$. Prove that splitting field of $f(x)$ over F is $f(b)$.*

Solution. As per given, b is a root of $f(x)$

Therefore $\qquad b^p - b = a \in K$

and $\quad (b+1)^p - (b+1) = b^p + 1 - b - 1 = a \Rightarrow (b+1)$ is a root of $f(x)$.

In a similar way, $b+2, ..., b+(p-1)$ are the roots of $f(x)$.

Therefore $f(x) = (x-b)(x-b-1)...(x-b-p+1)$

Hence, splitting field of $f(x)$ over F is given by

$$F(b, b+1), ..., b+(b-1) = F(b).$$

Example 3. *Obtain the degree of the splitting field of the polynomial $x^3 - 2$ over the field $\mathbf{Q}$ of rational numbers.*

Solution. Let $f(x) = x^3 - 2$

The roots of $f(x) = 2^{1/3}, 2^{1/3}\omega$ and $2^{1/3}\omega^2$, where ω is the cube roots of unity, *i.e.*,

$\omega = \dfrac{-1+\sqrt{3}i}{2}$. We observe that out of all the roots of $f(x)$, $2^{1/3}$ belongs to $\mathbf{Q}$, thus $f(x)$

is an irreducible polynomial of degree 3 over $\mathbf{Q}$.

Since $2^{1/3}$ is a root of $f(x)$.

$\Rightarrow$ it is algebraic over $\mathbf{Q}$ of degree 3

So $\qquad\qquad (\mathbf{Q}(2^{1/3}):\mathbf{Q}) = 3$

Further, let K be the splitting field of $f(x)$ over $\mathbf{Q}$. Then $\mathbf{Q}(2^{1/3}) \subset K$.

Now, since the roots $2^{1/3}\omega$ and $2^{1/3}\omega^2$ of $f(x)$ are in K but not in $\mathbf{Q}(2^{1/3})$.

Thus $\quad [\mathbf{K:Q}] > [\mathbf{Q}(2^{1/3}):\mathbf{Q}] = 3$

But we have

$$[\mathbf{K:Q}] \leq 3 = 6$$

Then, by the transitivity of extensions, we have

$[\mathbf{K:Q}] = [\mathbf{K:Q}(2^{1/3})][\mathbf{Q}(2^{1/3}):\mathbf{Q}] = [\mathbf{K:Q}(2^{1/3})] \cdot 3$

$\Rightarrow$ 3 is a divisor of $(\mathbf{K:Q})$.

$\Rightarrow$ $[\mathbf{K:Q}] \leq 6, 3[\mathbf{K:Q}] > 3$ and 3 divides $(\mathbf{K:Q})$

Hence $\qquad\qquad (\mathbf{K:Q}) = 6$.

Example 4. *Let $x^4 + x^2 + 1 \in \mathbf{Q}(x)$. Prove the splitting field of $f(x)$ over $\mathbf{Q}$ is $\mathbf{Q}(\omega)$ and $(\mathbf{Q}(\omega):\mathbf{Q}) = 2$.*

Solution. It is known that

$$1 + \omega + \omega^2 = 0$$
$$\omega^4 + \omega^3 + \omega^2 = 0$$
$\Rightarrow \qquad\qquad 1 + \omega^2 + \omega^4 = 0 \qquad\qquad\qquad\qquad [\because \omega^3 = 1]$

$\Rightarrow$ ω is a root of $x^4 + x^2 + 1$.

$\Rightarrow$ ω^2 is also a root of $x^4 + x^2 + 1$ Thus, we can write

$$f(x) = x^4 + x^2 + 1$$
$$= (x^2 - \omega^2)(x^2 - \omega) = (x^2 - \omega^2)(x^2 - \omega^4) \qquad\qquad [\because \omega = \omega^4]$$
$$= (x - \omega)(x + \omega)(x - \omega^2)(x + \omega^2)$$

Hence, splitting field of $x^4 + x^2 + 1$ over $\mathbf{Q}$ is

$$\mathbf{Q}(\omega, -\omega, \omega^2, -\omega^2) = \mathbf{Q}(\omega)$$

Example 5. *Let $F = z/2$ show that the splitting field of $f(x) = x^3 + x^2 + 1 \in F(x)$ is a finite field with eight elements.*

Solution. It is known that

$$F = z/2 = \{(2) + a : a \in \mathbf{Z}\}$$
$$= \{(2) + 0, (2) + 1\}$$

where

$$(2) + 0 = \{...-6, -4, 0, 2, 4, 6, ...\}$$
$$(2) + 1 = \{...-3, -1, 0, 1, 3, 5...\}$$

It is clear that $f(0) \neq 0$ and $f(1) \neq 0$. So neither 0 nor 1 is a root of $f(x)$ over E.

Therefore $f(x) = x^3 + x^2 + 1$ is irreducible over F.

Now, let α be a root of $f(x)$ in some extension of F, then $f(\alpha) = 0$

$\Rightarrow$ $\alpha^3 + \alpha^2 + 1 = 0$

$$f(x) = x^3 + x^2 + 1 = x^3 + x^2 - \alpha^3 - \alpha^2$$
$$= (x - \alpha)(x^2 + \alpha^2 + \alpha x) + (x - \alpha)(x + \alpha)$$

$\Rightarrow$ $\qquad f(x) = (x - \alpha)(x^2 + (1 + \alpha)x + \alpha + \alpha^2) = (x - \alpha)g(x)$

where $\qquad g(x) = x^2 + (1 + \alpha)x + \alpha + \alpha^2 \in F(\alpha)(x)$.

Further, since every element of $F(\alpha)$ is of the form $a_0 + a_1\alpha + a_2\alpha^2$ for $a_i = 0,1$

So $F(x)$ has the following elements

$$0, 1, \alpha, \alpha^2, \alpha + 1, \alpha^2 + 1, \alpha^2 + \alpha, \alpha^2 + \alpha + 1$$

Now, $\qquad g(x) = x^2 + (1 + \alpha)x + \alpha + \alpha^2$

$$= (x - \alpha^2)(x - (1 + \alpha + \alpha^2)).$$

Therefore the roots of $g(x)$ are also in $F(x)$.

Hence, $F(x)$ is the splitting field of $f(x) = x^3 + x^2 + 1$ over F containing 8 elements.

Example 6. *Show that the splitting field of $x^4 + 1 \in \mathbf{Q}$ is $\mathbf{Q}(\sqrt{2}, i)$ whose degree over $\mathbf{Q}$ is 4.*

Solution. It is known that, roots of $x^4 + 1$ are given by

$$x = (-1)^{1/4}$$
$$= (\cos(2r+1)\pi + i\sin(2r+1)\pi)^{1/4}$$
$$= (\cos(2r+1)\frac{\pi}{4} + i\sin(2r+1)\frac{\pi}{4}) \qquad\qquad (r = 0,1,2,3)$$
$$= \frac{1}{\sqrt{2}}(1+i), \frac{-1}{\sqrt{2}}(1-i), -\frac{1}{\sqrt{2}}(1+i), \frac{1}{\sqrt{2}}(1-i)$$

Thus, splitting field K of $x^4 + 1$ over $\mathbf{Q}$ is

$$K=\mathbf{Q}\left(\pm\frac{1}{\sqrt{2}}(1+i),\pm\frac{1}{\sqrt{2}}(1-i)\right)$$

We have to show that $\quad K = \mathbf{Q}(V\sqrt{2},i)$

$$\frac{1+i}{\sqrt{2}},\frac{1-i}{\sqrt{2}}\in K.$$

$\Rightarrow \qquad\qquad \frac{1+i}{\sqrt{2}}\pm\frac{1-i}{\sqrt{2}}\in K$

$\Rightarrow \qquad\qquad \sqrt{2},\sqrt{2}\ i\in K$

$\Rightarrow \qquad\qquad \sqrt{2},i\in K$

Further $\qquad\qquad \mathbf{Q}\subseteq K \Rightarrow \mathbf{Q}(\sqrt{2},i)\subseteq K$

Also, $\qquad\qquad \sqrt{2},i\in \mathbf{Q}(\sqrt{2},i)$

$\Rightarrow \qquad \pm\frac{1}{\sqrt{2}}(1+i),\pm\frac{1-i}{\sqrt{2}}\in \mathbf{Q}(\sqrt{2},i)$

$\Rightarrow \qquad\qquad K\subseteq \mathbf{Q}(\sqrt{2},i)$ $\qquad\qquad\qquad$ $[\because \mathbf{Q}\subseteq\mathbf{Q}(\sqrt{2},i)]$

$\Rightarrow \qquad\qquad K=\mathbf{Q}(\sqrt{2},i)$

Now, $x^2+1\in \mathbf{Q}(\sqrt{2})(x)$ is irreducible over $\mathbf{Q}(\sqrt{2})$

$\Rightarrow \qquad (\mathbf{Q}(\sqrt{2},i):\mathbf{Q}(\sqrt{2}))=\deg\ \text{Irr.}(\mathbf{Q}\sqrt{2},i)$

$\qquad\qquad\qquad\qquad = 2$ $\qquad\qquad$ $[\because\ i\ \text{satisfies}\ x^2+1=0]$

But $\qquad (\mathbf{Q}(\sqrt{2}:\mathbf{Q})=\deg.\text{Irr.}(\mathbf{Q},\sqrt{2})=\deg(x^2-2)=2$.

So $\qquad\qquad (K:\mathbf{Q})=[K:\mathbf{Q}(\sqrt{2})][\mathbf{Q}(\sqrt{2}):\mathbf{Q})$

$\qquad\qquad\qquad\qquad$ [By transitivity of Extensions]

$\qquad\qquad\qquad\qquad = 2.2 = 4.$

Example 8. *Show that the splitting field of $f(x) = x^4-2 \in \mathbf{Q}(x)$ over $\mathbf{Q}$ is $\mathbf{Q}(2^{1/x},i)$ and its degree of splitting field is 8.*

Solution. Here, we have

$$f(x) = x^4-2 = (x^2-\sqrt{2})(x^2+\sqrt{2})$$

$$=(x-2^{1/4})(x+2^{1/4})(x^2+\sqrt{2})$$

$$=(x-2^{1/4})(x+2^{1/4})(x+i2^{1/4})(x-i2^{1/4})$$

So, $\qquad\qquad$ roots of $f(x)=0$ are $2^{1/4},-2^{1/4},2^{1/4}i,-2^{1/4}i$

Clearly $f(x)$ is irreducible over $\mathbf{Q}$.

Also, $\mathbf{Q}(2^{1/4})$ contains $2^{1/4}$ and $-2^{1/4}$, so $(\mathbf{Q}(2^{1/2}):\mathbf{Q})=4$.

$\qquad\qquad\qquad$ $[\because 2^{1/4}\ \text{satisfies the minimal polynomial}\ x^4-2 \in \mathbf{Q}(x)]$

Further, since $2^{1/4}i$ and $-2^{1/4}i$ do not belong to $\mathbf{Q}(2^{1/4})$ but $\mathbf{Q}(2^{1/4},i)$ also contains $2^{1/4}i$ and $-2^{1/4}i$. Therefore $\mathbf{Q}(2^{1/4},i)$ is the splitting field of $f(x) = x^4-2$ over $\mathbf{Q}$.

Also, $2^{1/4}i$ and $-2^{1/4}i$ satisfies the minimal polynomial $x^2 + \sqrt{2} \in \mathbf{Q}(2^{1/4})$ of degree 2 so that $(\mathbf{Q}(2^{1/4},i):\mathbf{Q}(2^{1/4})) = 2$.

Using the transitivity of extensions, we get
$$[\mathbf{Q}(2^{1/4}, i):\mathbf{Q}] = [\mathbf{Q}(2^{1/4}i):\mathbf{Q}\,(2^{1/4})][\mathbf{Q}(2^{1/4}):\mathbf{Q}] = 2.4 = 8.$$
Hence, the degree of the splitting field of $f(x) = x^4 - 2$ over $\mathbf{Q}$ is 8.

Example 9. *Find the degree of a minimal splitting field of $x^6 + 1$ over* $\mathbf{Q}$.

Solution. The roots of $f(x) = x^6 + 1$ are given by
$$\frac{\sqrt{3}+i}{2}, \frac{-\sqrt{3}+i}{2}, \frac{-\sqrt{3}-i}{2}, \frac{\sqrt{3}-i}{2}, i, -i$$

Let E be a minimal splitting field of $f(x)$ over α. Then $E = \mathbf{Q}(\sqrt{3}, i)$
and $(E:\mathbf{Q}) = [\mathbf{Q}(\sqrt{3}, i):\mathbf{Q}]$

$$= [\mathbf{Q}(\sqrt{3}, i): \mathbf{Q}(\sqrt{3})]\,[\mathbf{Q}(\sqrt{3}): \mathbf{Q}] \qquad \text{[By transitivity of extensions]}$$

$$= \deg Irr.(\mathbf{Q}\sqrt{3}, i) \times \deg Irr.(\mathbf{Q}, \sqrt{3})$$

$$\le 2.2 \qquad\qquad [\because \deg Irr(\mathbf{Q}, \sqrt{3}) = \deg x^2 - 3 = 2]$$

Also i satisfies $x^2 + 1$ over $\mathbf{Q}(\sqrt{3})$. Thus $[\mathbf{Q}(\sqrt{3}, i):\mathbf{Q}(\sqrt{3}, i)] = 1$ or 2.

If $(\mathbf{Q}(\sqrt{3}, i):\mathbf{Q}(\sqrt{3})] = 1$

Then $(\mathbf{Q}(\sqrt{3}) = \mathbf{Q}(\sqrt{3}, i)$, which is not true because $i \notin \mathbf{Q}\sqrt{3}$.

$\Rightarrow [\mathbf{Q}(\sqrt{3}, i):\mathbf{Q}(\sqrt{3})] = 2.$

Hence $[E:\mathbf{Q}] = 4.$

Example 10. *Show that the polynomial $f(x) = x^2 + x + 1$ and $g(x) = x^2 + 3x + 3$ over F have the same splitting field.*

Solution. We have $f(x) = x^2 + 3x + 3$

$$= (x-\omega)(x-\omega^2), \text{ where } \omega = \frac{-1+i\sqrt{3}}{2}.$$

Then, $F(\omega)$ is the splitting field of $f(x)$.

Further $g(x) = (x-(\omega - 1))(x-(\omega^2 - 1))$. Then $\omega = 1$ and $\omega^2 = 1$ are the roots of $g(x)$ which belongs to $F(\omega)$. Hence, $f(x)$ and $g(x)$ have the same splitting field.

Example 11. *Determine the splitting field of $ax^2 + bx + c$ with $a \ne 0$ over $P = \mathbf{Q}(a,b,c)$.*

Solution. Let $f(x) = ax^2 + bx + c$

$$\Rightarrow f(x) = a\left(x^2 + \frac{b}{a}x + \frac{c}{a}\right) = a(x-\alpha)(x-\beta)$$

where, $\alpha + \beta = -\dfrac{b}{a}$ and $\alpha\beta = \dfrac{c}{a}$

$$\Rightarrow \alpha = \frac{-b+\sqrt{b^2 - 4ac}}{2a}, \quad \beta = \frac{-b-\sqrt{b^2 - 4ac}}{2a}$$

Also, if $b^2 - 4ac$ is or is not a square in $\mathbf{Q}[a, b, c]$, then the splitting field of $ax^2 + bx + c$ is either $\mathbf{Q}[a,b,c]$ or the quadratic extension of $\mathbf{Q}[a,b,c]$ generated by a root of $b^2 - 4ac$.

Example 12. *Find necessary and sufficient conditions on a and b so that the splitting field of irreducible polynomial $x^3 + ax + b$ has degree 3 over $\mathbf{Q}$.*

Solution. Let $f(x) = x^3 + ax + b \in \mathbf{Q}[x]$

Suppose K is the splitting field of $f(x)$ over $\mathbf{Q}$.

Also, let $\qquad f(x) = (x - \alpha_1)\,(x - \alpha_2)\,(x - \alpha_3)$

Then, $\qquad\qquad K = \mathbf{Q}[\alpha_1, \alpha_2, \alpha_3]$

Further, $\alpha_1 + \alpha_2 + \alpha_3 = 0$, $\quad \alpha_1\alpha_2 + \alpha_2\alpha_3 + \alpha_3\alpha_1 = a$

and $\qquad\qquad \alpha_1\alpha_2\alpha_3 = -b$

Write $\qquad\qquad D = [(\alpha_1 - \alpha_2)(\alpha_2 - \alpha_3)(\alpha_3 - \alpha_1)]^2$

Then, we can easily verify that

$$D = -4a^3 - 27b^2$$

Now, we have to show that $\mathbf{Q}(\sqrt{D}, \alpha_3) \subseteq K$

For this, consider

$$\alpha_1, \alpha_2, \alpha_3 \in K$$

$\Rightarrow \qquad \alpha_1 - \alpha_2, \alpha_2 - \alpha_3, \alpha_3 - \alpha_1 \in K$

$\Rightarrow \quad (\alpha_1 - \alpha_2)\,(\alpha_2 - \alpha_3)\,(\alpha_3 - \alpha_1) \in K$

$\Rightarrow \qquad\qquad \sqrt{D} \in K$. Also, $\alpha_3 \in K$

$\Rightarrow \qquad\quad \mathbf{Q}(\sqrt{D}, \alpha_3) \subseteq K \qquad\qquad\qquad\qquad\qquad \dots(1)$

Then, $\qquad \sqrt{D} = (\alpha_1 - \alpha_2)[\alpha_3(\alpha_1 + \alpha_2) - \alpha_3{}^2 - \alpha_1\alpha_2] \in \mathbf{Q}(\sqrt{D}, \alpha_3)]$

Since, $\qquad \alpha_1\alpha_2 = -\dfrac{b}{\alpha_3} \in \mathbf{Q}\,(\sqrt{D}, \alpha_3)$

and $\qquad \alpha_1 + \alpha_2 = -\alpha_3 \in \mathbf{Q}(\sqrt{D}, \alpha_3)$

$\qquad\qquad\qquad \alpha_1 - \alpha_2 \in \mathbf{Q}(\sqrt{D}, \alpha_3)$

$\Rightarrow \qquad\qquad \alpha_1, \alpha_2 \in \mathbf{Q}(\sqrt{D}, \alpha_3)$

Thus, $\qquad\qquad K \subseteq \mathbf{Q}(\sqrt{D}, \alpha_3) \qquad\qquad\qquad\qquad\qquad \dots(2)$

From (1) and (2), we conclude that $K = \mathbf{Q}(\sqrt{D}, \alpha_3)$

Now, suppose $\quad \sqrt{D} \in \mathbf{Q}$, then $K = \mathbf{Q}(\alpha_3)$

So, $\qquad\qquad [\mathbf{K} : \mathbf{Q}] = [\mathbf{Q}(\alpha_3) : \mathbf{Q}]$

$$= \deg \text{ Irr. } (\mathbf{Q}, \alpha_3)$$

$$= \deg f(x) = 3$$

Conversely, let $\ [K, \mathbf{Q}] = 3$

Also, let $\sqrt{D} \notin \mathbf{Q}$, then

$$\mathbf{Q} \subset \mathbf{Q}(\sqrt{D}) \subseteq \mathbf{Q}(\sqrt{D}, \alpha_3) = K$$

But $\sqrt{D}$ satisfies $x^2 - D \in \mathbf{Q}[x]$

$\Rightarrow \qquad [\mathbf{Q}(\sqrt{D}) : \mathbf{Q}] = 2 \qquad\qquad\qquad\qquad [\because x^2 - D \text{ is irreducible over } \mathbf{Q}$

$\qquad\qquad\qquad\qquad\qquad\qquad\qquad\qquad [K : \mathbf{Q}] = [K : \mathbf{Q}(\sqrt{D})][\mathbf{Q}(\sqrt{D}) : \mathbf{Q}]$

$\qquad\qquad\qquad\qquad\qquad\qquad\qquad\qquad 3 = [K : \mathbf{Q}(\sqrt{D})].\, 2 \text{ a contradiction}]$

$\Rightarrow \qquad\qquad\qquad \sqrt{D} \in \mathbf{Q}$

Hence, a necessary and sufficient condition for the splitting field of irreducible cubic $x^3 + ax + b$ over $\mathbf{Q}$ to have degree 3 is $\sqrt{D} \in \mathbf{Q}$.

Example 13. *Find the degree of a minimal splitting field of $x^4 + 2$ over* **Q**.

Solution. Let $$f(x) = x^4 + 2.$$

Then, clearly, roots of $f(x)$ are given by

$$2^{1/4}\left(\frac{1}{\sqrt{2}} + \frac{i}{\sqrt{2}}\right), \ 2^{1/4}\left(-\frac{1}{\sqrt{2}} + \frac{i}{\sqrt{2}}\right), \ 2^{1/4}\left(-\frac{1}{\sqrt{2}} - \frac{i}{\sqrt{2}}\right), 2^{1/4}\left(\frac{1}{\sqrt{2}} - \frac{i}{\sqrt{2}}\right)$$

Let E be a minimal splitting field of $f(x)$ over **Q**. Then, we have

$$E = \mathbf{Q}(2^{1/4}, i) \qquad\qquad [\because (2^{1/4})^2 = \sqrt{2}\,]$$

Consider

$$
\begin{aligned}
[E : \mathbf{Q}] &= [\mathbf{Q}(2^{1/4}, i) : \mathbf{Q}(2^{1/4})] \ [\mathbf{Q}(2^{1/4}) : \mathbf{Q}] \\
&= [\mathbf{Q}(2^{1/4}, i) : \mathbf{Q}(2^{1/4})] \ \text{deg Irr. } (\mathbf{Q}, 2^{1/4})] \\
&= [\mathbf{Q}(2^{1/4}, i) : \mathbf{Q}(2^{1/4})] \ \deg(x^4 - 2) \\
&= [\mathbf{Q}(2^{1/4}, i) : \mathbf{Q}(2^{1/4})] \cdot 4
\end{aligned}
$$

$\Rightarrow$ 4 divides $[E : \mathbf{Q}]$.

Again, $$
\begin{aligned}
[E : \mathbf{Q}] &= [\mathbf{Q}(2^{1/4}, i) : \mathbf{Q}(i)] \ [\mathbf{Q}(i) : \mathbf{Q}] \\
&= [\mathbf{Q}(2^{1/4}, i) : \mathbf{Q}(i)] \ \text{deg Irr.}(\mathbf{Q}. i) \\
&= [\mathbf{Q}(2^{1/4}, i) : \mathbf{Q}(i)] \deg(x^2 + 1) \\
&\leq 4 \times 2 = 8 \qquad\qquad [\because 2^{1/4} \text{ satisfies } x^4 - 2 \text{ over } \mathbf{Q}(i)]
\end{aligned}
$$

$\Rightarrow$ 4 divides $[E : \mathbf{Q}]$ and $[E : \mathbf{Q}] \leq 8$.

Thus, $[E : \mathbf{Q}] = 4$ or 8

If $[E : \mathbf{Q}] = 4$, then we have

$[\mathbf{Q}(2^{1/4}, i) : \mathbf{Q}(2^{1/4})] = 1$

$\Rightarrow \qquad\qquad \mathbf{Q}(2^{1/4}) = \mathbf{Q}(2^{1/4}, i)$

$\Rightarrow \qquad\qquad i \in \mathbf{Q}(2^{1/4})$, which is not possible because $\mathbf{Q}(2^{1/4})$ is a subfield of real numbers. Hence, $[E : \mathbf{Q}] = 8$.

Example 14. *Show that the field of real number* **R** *is not a normal extension* **Q**, *the field of rational numbers.*

Solution. Let $f(x) = x^3 - 2 \in \mathbf{Q}[x]$ be any polynomial.

Then, $f(x)$ is irreducible over **Q** and has a root $2^{1/3}$ in **R**.

But it does not split into linear factors in **R**, because other two roots are complex.

Hence, **R** is not a normal extension of **Q**.

Example 15. *Find the normal extensions of* **Q** *generated by* $\sqrt{3} + \sqrt{2}, \sqrt{3} + i, 2^{1/3} + \sqrt{2}, \sin\dfrac{2\pi}{5}$,

$t = \dfrac{2\pi i}{n}$ *for integers n.*

Solution. It is known that

$$\mathbf{Q}(\sqrt{3} + \sqrt{2}) = \mathbf{Q}(\sqrt{3}, \sqrt{2})$$

$$\mathbf{Q}(\sqrt{3} + i) = \mathbf{Q}(\sqrt{3}, i)$$

and $$\mathbf{Q}(2^{1/3} + \sqrt{2}) = \mathbf{Q}(2^{1/3}, \sqrt{2})$$

Clearly, $\mathbf{Q}(\sqrt{3}+\sqrt{2})$ and $\mathbf{Q}(\sqrt{3}+i)$ are normal extensions, since they are the splitting fields of $(x^2-2)(x^2-3)$ and $(x^2+1)(x^2-3)$.

Also, $2^{1/3}+\sqrt{2}$ generates the normal extensions $\mathbf{Q}(2^{1/3}, \sqrt{2}, i)$ which is the splitting field of $(x^2-2)(x^3-2)$.

Further, if $a = 4\sin\dfrac{2\pi}{5} = \sqrt{10+2\sqrt{5}}$ and that the minimal polynomial of a over $\mathbf{Q}$ is

$x^4 - 20x^2 + 80$ and the conjugate of a over $\mathbf{Q}$ are thus $\pm\sqrt{10+2\sqrt{5}}$.

Now $\sqrt{10-2\sqrt{5}} = 4\sqrt{5}\sqrt{10+2\sqrt{5}}$ belongs to $\mathbf{Q}\left(\sin\dfrac{2\pi}{5}\right)$ because $\sqrt{5} = \dfrac{a^2-10}{2}$

belong to it. Therefore $\mathbf{Q}\left(\sin\dfrac{2\pi}{5}\right)$ is a normal extension of $\mathbf{Q}$.

In case of $t = \dfrac{2\pi i}{n}$, for an integer n, $\mathbf{Q}(t)$ is the splitting field of x^n-1 over $\mathbf{Q}$.

Hence, it is a normal extension of $\mathbf{Q}$.

Example 16. *If $a = \cos\dfrac{\pi}{4}+i\sin\dfrac{\pi}{4}$, then show that $\mathbf{Q}(a)$ is a normal extension of $\mathbf{Q}$.*

Solution.   Let $a = \cos\dfrac{\pi}{4}+i\sin\dfrac{\pi}{4}$

Then, clearly a is a root of $x^4 + 1 \in \mathbf{Q}[x]$

Consider, $x^4 + 1 = (x^2 - i)(x^2 + i)$

$$= (x-\sqrt{i})(x+\sqrt{i})(x-i^{3/2})(x+i^{3/2})$$

Since $a = \cos\dfrac{\pi}{4}+i\sin\dfrac{\pi}{4}$, so $a^2 = i \Rightarrow \sqrt{i} = a$, $i^{3/2} = a^3$

$$x^4+1 = (x-a)(x+a)(x-a^3)(x+a^3)$$

$\Rightarrow$ $\mathbf{Q}(a)$ is the splitting field of $x^4 + 1 \in \mathbf{Q}[x]$.

Hence, $\mathbf{Q}(a)$ is the normal extension of $\mathbf{Q}$.

Example 17. *Show that a minimal splitting field over K for a polynomial of degree n is generated over K by any of (n— 1) of its zeroes.*

Solution.   Let $f(x) \in K[x]$ such that $\deg f(x) = n$

Further, let $f(x) = a_n x^n + a_{n-1}x^{n-1}+ ... + a_1x + a_0$, $a_i \in K$

Also, let $E = K[\alpha_1, ..., \alpha_n]$ be a minimal splitting field of f over K.

Now, $\Sigma\alpha_i = -\dfrac{a_{n-1}}{a_n} \in K \subseteq K[\alpha_1,...,\alpha_n] = E'$

So, $\alpha_1 + \alpha_2 + ...+ \alpha_n \in E$ and $\alpha_i \in E$

Let us denote $E' = K[\alpha_1......\alpha_{i-1}, \alpha_{i+1}......\alpha_n]$

Now, $\alpha_i = -\dfrac{a_{n-1}}{a_n} - (\alpha_1......\alpha_{i-1}, \alpha_{i+1}.....\alpha_n) \in E'$

$\Rightarrow$ $E \subseteq E'$

But $E' \subseteq E$

Therefore, $E = E'$, which is generated by $n-1$ zeroes of $f(x)$ over K.

Example 18. *If K is a quadratic extension of a field F. Show that F is necessarily normal.*

Solution. It is given that K is a quadratic extension of F. Therefore, we must have $K = F[a]$ $\forall\ a \in K$, which is not in F. Also, since they are all of degree 2 over F.

Further, let b be the second root of the minimal polynomial of a over F. Since $a+b \in F$, therefore we have $\quad K = F\,[a,\,b]$, so K is a splitting field. Hence, F is normal.

Example 19. *Check which of the following polynomials are separable over* **Q**.

$\quad$ (i) $x^2 - 5x + 6$ $\qquad$ (ii) $x^3 + 3x^2 - x + 3$.

Solution. (i) $f(x) = x^2 - 5x + 6 = (x-3)\,(x-2)$

$\qquad$ Clearly, $2, 3 \in \mathbf{Q}$, so $f(x)$ has no multiple root in $\mathbf{Q}$. Hence, $f(x)$ is separable over $\mathbf{Q}$.

$\qquad$ (ii) Here, $\qquad f(x) = x^3 + 3x^2 - x + 3 \in \mathbf{Q}[x]$

$\qquad$ Clearly, $f(x)$ is non-constant irreducible over $\mathbf{Q}$.

$\qquad$ Now, $\qquad f'(x) = 3x^2 + 6x - 1$

$\qquad \Rightarrow f(x)$ and $f'(x)$ are relatively prime.

$\qquad$ Hence, $f(x)$ is separable over $\mathbf{Q}$.

Example 20. *Let a, b, c be the roots of $x^3 - 3x + 1$ in* **C**. *Express b and c in the basis* $[1, a, a^2]$.

Solution. Since a, b, c are the roots of $x^3 - 3x + 1$. Therefore

$$b + c = -a \ \text{ as } c = -1$$

$$\text{Discriminant} = [(a-b)\,(b-c)\,(c-a)]^2$$

But discriminant of $x^3 - 3x + 1 = 81$

So we have

$$(a-b)(b-c)(c-a) = \pm 9$$

Without loss of any generality, we may take positive sign.

i.e., $(a-b)(b-c)(c-a) = 9$

$$\Rightarrow \quad (b-c) = \frac{9}{(a-b)(c-a)} = \frac{9}{-a^2 + ac - bc + ab} = \frac{9}{-a^2 - (-1/a) + a(b+c)}$$

$$\Rightarrow \quad b - c = \frac{9}{-a^2 + \dfrac{1}{a} + a(-a)} = \frac{9}{-2a^2 + \dfrac{1}{a}}$$

Using successive Euclidean division, we have

$$b - c = \frac{9}{-2a^2 + \dfrac{1}{a}} = 4 - a - 2a^2$$

But $\ b + c = -a$

Hence, we have

$$b = 2 - a - a^2$$

and $\qquad c = a^2 - 2$.

EXERCISE 10.2

1. Show that $f(x) = x^6 - 2ax^3 + a$ is inseparable over $Z_3(a)$ with a is any extension field of Z_3.

2. Show that the degree of splitting field of $x^5 - 3x^3 + x^2 - 3$ over Q is 4.

3. Show that the splitting field of $f(x) = x^4 - 2 \in Q[x]$ over Q is $Q(2^{1/4}, i)$ and its degree of extension is 8.

4. Find the splitting field of $x^p - 1$ over Q, p being a prime number.

5. Let $f(x) = x^4 - 2x^2 - 2 \in Q[x]$. Find the roots α, β of $f(x)$ such that $Q(\alpha) = Q(\beta)$. What is the splitting field of $f(x)$?

6. Find the splitting fields and their degrees of the following polynomials over Q

   (i) $x^6 + 1$ (ii) $x^4 - 2$

   (iii) $x^5 - 1$ (iv) $(x^2 - 2)(x^3 - 2)$

7. Construct splitting fields over Q for the following polynomials:

   (i) $x^3 - 1$ (ii) $x^4 + 1$

   (iii) $x^6 - 1$ (iv) $(x^2 - 2)(x^3 - 3)$

8. Find the degree of a minimal splitting field of $x^6 + 1$ over F_2, the field of $[0, 1]$ mod 2.

9. Find the smallest normal extension (upto isomorphism) of $Q(2^{1/4}, 3^{1/4})$ in Q.

10. Show that every extension of Q is separable.

11. Does the fact that $\sqrt[3]{2}$ is an element of $Q(\sqrt[3]{2})$ whereas $\omega\sqrt[3]{2}$ and $\omega^2\sqrt[3]{2}$ do not imply that $Q(3/2)$ is a not a normal extension of itself.

12. Show that $F(3\sqrt{2})$, where F is the field of rationals, has no automorphisms other than the identity automorphism.

Answers

1. $\alpha = \sqrt{1 + \sqrt{3}}$, $\beta = \sqrt{1 - \sqrt{3}}$, splitting field of $f(x) = Q(\sqrt{1 + \sqrt{3}})$.

6. (i) $Q(\sqrt{3}, i), 4$ (ii) $Q(2^{1/4}, i), 8$ (iii) $Q(e^{2\pi i/5}), 5$ (iv) $Q(\sqrt{2}, 2^{1/3}, i), 12$

7. (i) $Q(\sqrt{3}\,i), 2$ (ii) $Q(\sqrt{2}\,i), 4$ (iii) $Q(\sqrt{3}\,i), 2$ (iv) $Q(\sqrt{2}, 3\sqrt{3}\,\omega), 12$

8. 2

9. The smallest normal extensioin of $Q(2^{1/4}, 3^{1/4})$ is the splitting field of $(x^4 - 2)(x^4 - 3)$ namely $Q(2^{1/4}, 3^{1/4}, i)$.

10.10 ALGEBRAICALLY CLOSED FIELDS AND ALGEBRAIC CLOSURE

Let S be a set of polynomials over K. If $f \in S$ splits in a field E containing K, then E is called a splitting field of S over K and K is known as minimal splitting field of S over K. Let E_1 be a minimal splitting field of f_1 over K, E_2 be a minimal splitting field of f_2 over E_1 and so on. E_n be a minimal splitting field of f_n over E_n. Then $E_1 \subseteq E_2 \subseteq \ldots \subseteq E_n$ and each f_i splits in $E_i \subseteq E_n$ which implies S splits in E_n. Thus, K is a minimal splitting field of S over K. It is also a minimal splitting field of $f = f_1 f_2 \ldots f_n$, over K.

    Definition 1. *A field K is called algebraically closed if every polynomial f over K splits in K.*

or

    A field K is said to be algebraically closed if every polynomial in $K[x]$ of degree greater than or equal to 1 has a root in K.

    Definition 2. *Let K be a field. An extension E of K is called algebraic closure of K if*

    (i) *E is algebraic over K.*

    (ii) *E is algebraically closed.*

Theorem 1.   *A field is algebraically closed if and only if every irreducible polynomial in $F[x]$ is of degree 1.*

Proof.   Let us first suppose F be algebraically closed and let $f(x) \in F[x]$ be an irreducible

polynomial in $F[x]$ of degree n. Then, there exists a finite extension K of F such that
$$[K : F] = n.$$
But F is algebraically closed so $\quad K = F$.

$\Rightarrow \qquad\qquad\qquad [K : F] = 1, i.e., n = 1$

Hence, $\qquad\qquad\qquad \deg f(x) = 1.$

Conversely, let us suppose every irreducible polynomial in $F[x]$ is of degree 1.

To show F is algebraically closed.

Let K be any algebraic extension of F and $a \in K$, then α is a root of a irreducible polynomial of $F[x]$. Since the degree of every irreducible polynomial in $F[x]$ is of degree 1, so $\alpha \in F$.

$\Rightarrow \qquad\qquad\qquad [K : F] = 1$

$\Rightarrow \qquad\qquad\qquad K = F$

Hence, F is algebraically closed.

Theorem 2. *A field F is algebraically closed if and only if every polynomial in $F[x]$ of positive degree factors in $F[x]$ into linear factors.*

Proof. Let us first suppose F be algebraically closed field and $f(x)$ be a polynomial in $F[x]$.

Clearly, $f(x)$ has a root in $F[x]$.

If this root is α, then $(x - \alpha)$ is a factor of $f(x)$. Then, $f(x) = (x - \alpha)\, g(x)$.

If $g(x) \in F[x]$ is of positive degree, then it will be a root, say β in F such that
$$g(x) = (x - \beta)h(x)$$
$\Rightarrow \qquad\qquad\qquad f(x) = (x - \alpha)(x-\beta)h(x)$

Proceeding in the same way, we get a factorization of $f(x)$ in $F[x]$ into linear factors.

Conversely, let us suppose that every polynomial in $F[x]$ of positive degree factors in $F[x]$ into linear factors. We have to show that F is algebraically closed.

If $(ax - b)$, $a \neq 0$ is a linear factor of $f(x)$, then b/a is a root of $f(x)$, which belongs to F. Hence, F is algebraically closed.

Theorem 3. *A field F is algebraically closed if and only if every algebraic extension of F is F itself.*

Proof. Let us first suppose F be algebraically closed. Let K be an algebraic extension of F.

Let $\alpha \in K$ and $p(x)$ be any irreducible polynomial of α in F,

i.e., $\qquad\qquad\qquad p(x) = \text{Irr } (F, \alpha)$

Then, by theorem 1, $\qquad \deg p(x) = 1$

$\Rightarrow \qquad\qquad\qquad p(x) = x - \alpha \in F[x]$

$\Rightarrow \qquad\qquad\qquad \alpha \in K \Rightarrow F = K$

Conversely, let $f \in F[x]$. Let K be a minimal splitting field of f over F.

Then, K is algebraic over F.

By our hypothesis, $\qquad\qquad\qquad F = K.$

Therefore, $f(x)$ splits in $F[x] \Rightarrow F$ is algebraically closed.

Theorem 4. *Let F be an algebraically closed field such that K is an extension of F. Let $S = [a \in K: a$ is algebraic over F $]$. Then S is an algebraic closure of F.*

Proof.   It is known that
$$K \subseteq S \subseteq K \text{ is a tower of fields.}$$
By definition of S, S is algebraic over F.

Let $f \in S(x)$. Then, $f \in K[x]$. Since K is algebraically closed, f splits in K.

Let $f = \alpha(x - \alpha_1)\ldots\ldots(x - \alpha_n), \quad \alpha_i \in K$

Now, since α_i is algebraic over S, $S(\alpha_i)$ is algebraic over S for all i.

Also, S is algebraic over F

$\Rightarrow$ $S(\alpha_i)$ is algebraic over F for all i.

$\Rightarrow$ $\alpha_i \in K$ is algebraic over F.

$\Rightarrow$ $\alpha_i \in S$

$\Rightarrow$ f splits in S.

$\Rightarrow$ S is algebraically closed.

$\Rightarrow$ S is algebraic closure of F.

$\Rightarrow$ $S = [a \in \mathbf{C} : a \text{ is algebraic over } \mathbf{Q}]$ is an algebraic closure of $\mathbf{Q}$.

Theorem 5. *Let F be a field, $K = F(\alpha)$ an algebraic extension of F and $\sigma : F \to L$ an embedding (F-homomorphism) of F into an algebraically closed field L. Then σ can be extended to $\eta : E \to L$, an embedding of E into L and the number of such extensions is equal to the number of distinct roots of the minimal polynomial of α .*

Proof.   Consider a minimal polynomial
$$f(x) = a_0 + a_1 x + \ldots + x^n$$
of x over F. Also, let
$$F^\sigma(x) = \sigma(a_0) + \sigma(a_1)x + \ldots + x^n$$
be a polynomial in $L[x]$.

Since L is algebraically closed, $f^\sigma(x)$ has a root in L. Let β be a root of $f^\sigma(x)$ in L. Now, since a satisfies $f(x)$, then an element of $F(\alpha)$ can be expressed as the linear combination of elements $1, \alpha, \alpha^2, \ldots\ldots \alpha^m$ with $m < n$. Thus, for some $b_0, b_1, \ldots, b_m \in F$, we define a mapping
$$\eta : F[x] \to L \text{ by}$$
$$\eta(b_0 + b_1\alpha + b_2\alpha^2 + \ldots + b_m\alpha^m) = \sigma(b_0) + \sigma(b_1)\beta + \ldots + \sigma(b_m)\beta^m$$
Obviously, η is well defined and homomorphism, therefore η is an embedding of $F(\alpha)$ into L.

Clearly, to each α , the root of minimal polynomial $f(x)$, we get an embedding of $F(\alpha)$ into L. Thus, there is one-one correspondence between the set of distinct roots of $f^\sigma(x)$ in L and set of embeddings η of $F(\alpha)$ into L that extend σ. Hence, the number of such η is equal to the number of distinct roots of the minimal polynomial of α.

Theorem 6. *Let K be an algebraic closure of F. Then K is a minimal splitting field of the set S of all polynomials over F.*

Proof.   As per given, K be an algebraic extension of F.

$\Rightarrow$ K is algebraically closed.

$\Rightarrow$ Each $f \in S$ splits in K.

Let $K' = F \text{ (zeros of } f \in S \text{ in } f) \subseteq K$.

Let $\alpha \in F$, then α is algebraic over F because K is algebraic over F. Let $p(x)$ be an irreducible polynomial of α over F, i.e., $p(x) = \text{Irr}(F, \alpha)$.

Then, α is a zero of $p(x) \in S$ in K.

Therefore, $\qquad \alpha \in K' \Rightarrow K \subseteq K'$

So, $K' = K \Rightarrow K$ is a minimal splitting field of K' of the set of all polynomials over F.

Theorem 7. *Any two algebraic closures of a field are isomorphic.*

Proof. Let F be a field and K_1, K_2 be algebraic closures of F. Then K_1, K_2 are minimal splitting fields of the set of all polynomials over F. Therefore, K_1, K_2 are isomorphic because any two minimal splitting fields of a set of polynomials over F are isomorphic.

Theorem 8. *Let K be an algebraic extension of F and $\sigma : F \to L$ an embedding of F into an algebraically closed field. Then there exists an extension of σ to an embedding of K into L. If K is algebraically closed and L is algebraic over $\sigma(F)$, then any such extension of σ is an isomorphism of K into L.*

Proof. Define a set S as follows

$$S = [(E, \phi): E \text{ is a subfield of } K \text{ containing } F \text{ and } \phi \text{ is an extension of } \sigma$$
$$\text{to an embedding of } E \text{ into } L]$$

If (E, ϕ) and (E', ϕ') are any two elements of S, then we may write

$$(E, \phi) \subseteq (E', \phi').$$

If $E \subset E$ and $\phi'/E = \phi$, i.e., ϕ' restricted to E is ϕ, it is clear that $S \neq \phi$ because it contains (F, σ) and ordered inductively.

If $[(E_i, \phi_i)]$ is totally ordered subset of S, we set $E = \cup E_i$ and define ϕ on E to be equal to ϕ_i on each E_i, i.e., $\phi(\alpha) = \phi_i(\alpha)$ for $\alpha \in E$ and $\alpha \in E_i$ for some i. Then (E, ϕ) is an upper bound for the totally ordered subset $[(E_i, \phi_i)]$.

Then, by Zorn's lemma, let (E, η) be a maximal elements in S. Then η is an extension of σ. So $E = K$ is not possible. Otherwise there exist $\alpha \in E$, $\alpha \notin K$. Then by Theorem 5, the embedding η has an extension to $K(\alpha)$, which contradict the maximality of (K, η). Thus, there exists an extension to $K(\alpha)$.

Next, if K is algebraically closed and L is algebraic over $\sigma(F)$, then $\eta(K)$ is algebraically closed and L is algebraic over $\eta(K)$. Thus $L = \eta(K)$.

Hence, η is an isomorphism of K onto L.

Theorem 9. *Algebraic closure of a countable field is countable.*

Proof. Let F be a countable field. Then, for each integer $n \geq 1$, there is a countable set of polynomials of degree n over F. So the set S of all polynomials over F is countable.

Let $\quad S = [f_1, f_2, \ldots\ldots f_n, \ldots\ldots]$

Let $E_0 = F$ and E_1 be a minimal splitting field of f_1 over $E_0 = F$. Let E, be a minimal splitting field of f_i over E_{i-1}. Then, $E_{n-1} \subseteq E_n$ for all n.

Therefore,

$$E = \bigcup_n E_n \text{ is a field.}$$

$\Rightarrow \quad$ Each f_1 splits in E.

$\Rightarrow \quad$ E is a splitting fields of S over F.

If $K = F$ [zeroes of f_i in E] $\subseteq E$

Then $F \subseteq K \subseteq E$ is a tower of fields and K is a minimal splitting field of S over F. Therefore, K is an algebraic closure of F.

⇒ K is algebraically closed.

⇒ K is not finite.

Since E is countable, K is also countable. Hence, any algebraic closure K' of F being isomorphic to K is also countable.

Theorem 10. *Let K and K' be algebraic closures of a field F, then there exists an isomorphism $\phi:K \to K'$ of K onto K' inducing the identity on F.*

Proof.   Define an identity mapping $\theta: F \to K$ such that $\theta(\alpha) = \alpha, \forall \ \alpha \in F$. Using theorem 8, θ is extended to an embedding $\phi:K \to K'$. Now, $\phi(K) \cong K$. Since K is algebraic closure of F, thus $\phi(K)$ is algebraically closed field containing F. Also, K' is an algebraic extension of F, therefore, K' is also an algebraic extension of $\phi(K)$, which lies between F and K'. Then $\phi(K) = K'$. Hence, ϕ is an isomorphism of K onto K'.

EXERCISE 10.3

1. Let K be an extension of a field F and K be an algebraically closed field, then show that the algebraic closure $\overline{F}$ of F in K is an algebraically closed field.

2. Show that the field **C** of complex numbers is an algebraically closed field.

3. Let F be a field. Then show that following are equivalent:
   (i) F is algebraically closed.
   (ii) Every irreducible polynomial over F has degree one.
   (iii) Every algebraic extension over F is F itself.

4. Show that a finite field is not algebraically closed.

5. Let S be a set of polynomials over a field F. Then show that there is a minimal splitting field of S over F.

6. Let E be an algebraic extension of F. Then show that E is algebraically closed if and only if E has no algebraic extension other than E itself.

7. Show that any two minimal splitting fields of a set of polynomials over K are isomorphic.

8. Let S be the set of all polynomials over a field F. Then show that a minimal splitting field of S over F is an algebraic closure of **R**.

10.11 GALOIS THEORY

The Galois theory gives a beautiful interplay of group and field theory, *i.e.*, this theory is an excellent composite of the theory of groups with the theory of algebraic field extensions. It also has a very important applications in the theory of equations.

We shall start by recalling the following results:

(i)   Let $F \subseteq E \subseteq \overline{F}$, $\alpha \in E$ and β be a conjugate of α over F, *i.e.*, F is irreducible over α (irr. (F, α)) has β as a zero also. Then there is an isomorphism $\psi_{\alpha,\beta}$ mapping $F(\alpha)$ onto $F(\beta)$ that leaves F fixed and maps α onto β.

(ii)   If $F \subseteq E \subseteq \overline{F}$ and $\alpha \in E$, then an automorphism σ of $\overline{F}$ that leaves F fixed must map α onto some conjugate of α over F.

(iii)   If $F \subseteq E$, the collection of all automorphisms of E leaving F fixed forms a group $G(E/F)$. For any subset S of $G(E/F)$, the set of all elements of E left fixed by all elements of S is a field E_s. Also $F \subseteq E_{G(E/F)}$.

(iv)   If E is a finite extension of F and is a separable splitting field over F. Then $|G(E|F)| = [E : F] = [F : E]$

10.12 AUTOMORPHISM AND GROUP OF AUTOMORPHISMS OF FIELDS

Definition 1. *A one-one mapping σ of a field F onto itself is called an automorphism of F if*
$$\sigma(a+b) = \sigma(a)+\sigma(b)$$
and $$\sigma(ab) = \sigma(a)\sigma(b), \quad \forall \; a, b \in F.$$

Definition 2. *Two automorphisms σ_1 and σ_2 of F are said to be equal if $\sigma_1(a)=\sigma_2(a)$, $\forall$ $a \in F$.*

Also, σ_1 and σ_2 will be distinct if $\sigma_1(a) \neq \sigma_2(a)$ for some element a in K.

Definition 3. *(Fixed Field). Let G be a subgroup of the automorphism group of a field F. Let F_0 be the subset of F consisting of those elements of F which are left fixed by every automorphism in G, i.e., $F_0 = \{a \in F: \sigma(\alpha) = a, \forall \sigma \in G\}$ Then F_0 is known as the fixed field of G.*

Definition 4. *Let K be a subfield of a field F. Then the group of automorphisms of F relative to K, written as G(F,K) is the set of all those automorphisms of F which leaves every element of K fixed.*

REMARKS

☞ If σ is an automorphism of a field F, then

(i) $\sigma(0) = 0$ (ii) $\sigma(-a) = -\sigma(a)$ (iii) $\sigma(1) = 1$

(iv) $\sigma(-1) = -1$ (v) If $a \neq 0$, then $\sigma(a^{-1})=[\sigma(a)]^{-1}$

☞ Automorphism is a special kind of isomorphism.

☞ The definition of fixed field will be sensible only if we are able to show that all elements $a \in F$ such that $\sigma(a) = a$ for all $\sigma \in G$ actually form a subfield of F.

☞ $G(F, K)$ is a subgroup of the groups of all automorphisms of F. If $\alpha \in F$ be such that $\alpha \in K$, then $f(\alpha)=\alpha$ for all $f \in G(F, K)$. So the field of $G(F, K)$ must contain F.

☞ Every rational number is left fixed by every automorphism of F.

Theorem 1. *Let F be a field and A(F) be the collection of all automorphisms of F. Then A(F) is a group with respect to the operation known as composite of two functions.*

Proof.   Let F be a field and $A(F)$ be the collection of all automorphisms of a field F, i.e.,
$$A(F)=\{f : f \text{ is an automorphism of } F\}.$$

We have to prove that $A(F)$ is a group with respect to composite of mapping as operation

(i) Closure property.

Let $f, g \in A(F)$

$\Rightarrow$ f and g are automorphisms.

$\Rightarrow$ f and g are one-one mapping of F onto itself.

$\Rightarrow$ gf also a one-one mapping of F onto itself.

Let a, b be any two elements of F, then we have
$$(gf)(ab)=g(f(ab)) = g[f(a)f(b)]$$
$$=g(f(a))g(f(b))$$
$$=(gf(a))(gf(b)).$$

Again $(gf)(a+b)=g(f(a+b))=g(f(a)+f(b))$
$$=g(f(a))+g(f(b))$$

$$= gf(a) + gf(b)$$

$\Rightarrow$　　　gf is also an automorphism of F.

$\Rightarrow$　　　　　$gf \in A(F)$

$\Rightarrow$　　　$A(F)$ is closed with respect to composite composition.

(ii) Associativity. It is known that composite of arbitrary mappings is associative. Thus, composite of automorphism is also associative.

(iii) Existence of Identity. The identity mapping I of F is also an automorphism of F. Clearly, I is one-one-onto and if $a, b \in F$

$$I(ab) = ab = I(a)\,I(b)$$

and　　　$I(a + b) = a + b = I(a) + I(b)$

Therefore, $I \in A(F)$ and if $f \in A(F)$, we have $If = f = fI$.

(iv) Existence of Inverse. Let $f \in A(F)$. Now, since f is one-one mapping of F onto itself

$\Rightarrow$　　　f^{-1} exists and f^{-1} is one-one mapping of F onto itself.

Further, it remains to prove that f^{-1} is also an automorphism.

Let $a, b \in F$. Then there exists $a', b' \in F$ such that

$$f^{-1}(a) = a'$$

$\Leftrightarrow$　　　　$f(a') = a$

and　　　$f^{-1}(b) = b' \Leftrightarrow f(b') = b$.

Consider

$$f^{-1}(ab) = f^{-1}[f(a')f(b')] \ = f^{-1}[f(a'b')] = a'b'$$
$$= f^{-1}(a)f^{-1}(b).$$

Also $f^{-1}(a + b) = f^{-1}[f(a') + f(b')] = f^{-1}[f(a' + b')] = a' + b'$
$$= f^{-1}(a) + f^{-1}(b)$$

Therefore f^{-1} is an automorphism of F and so

$$f \in A(F) \Rightarrow f^{-1} \in A(F).$$

Since f is arbitrary, so each element of $A(F)$ possesses inverse. Hence, $A(F)$ is a group with respect to composite compositions.

Theorem 2. *Let F be a field and if $\sigma_1, \sigma_2, ..., \sigma_n$ are distinct automorphism of F, then we cannot find elements $a_1, a_2, ..., a_n$ not all 0 in F such that*

$$\sum_{i=1}^{n} a_i \sigma_i(b) = 0, \forall\, b \in F$$

Proof.　　　Let if possible, we can find a set of element $a_1, a_2, ..., a_m$ not all zero in F such that

$$\sum_{i=1}^{m} a_i \sigma_i(b) = 0, b \in F \qquad\qquad\qquad \text{... (1)}$$

If $m = 1$, then (1) reduces to

$$a_1 \sigma_1(b) = 0, \forall b \in F$$

$\Rightarrow$　　　　$a_1 \sigma_1(1) = 0,\, 1 \in F$

$\Rightarrow$　　　　　$a_i 1 = 0$　　　　　　　　　　　$[\because \sigma_1(1) = 1]$

$$\Rightarrow \qquad a_i = 0$$

which is contrary to our assumption. Therefore, we must have $n>1$. Now the automorphisms σ_1 and σ_m are distinct, so $\exists$ an element $c \in F$ such that $\sigma_1(c) \neq \sigma_m(c)$.

Further, $\qquad c \in F, b \in F \Rightarrow cb \in F.$

$\Rightarrow \qquad$ Relation (1) holds good for cb.

$\Rightarrow \qquad a_1\sigma_1(cb) + a_2\sigma_2(cb) + \ldots + a_m\sigma_m(cb) = 0, \; \forall \, b \in F$

$\Rightarrow \qquad a_1\sigma_1(c)\sigma_1(b) + a_2\sigma_2(c)\sigma_2(b) + \ldots + a_m\sigma_m(c)\sigma_m(b) = 0, \; \forall \, b \in F \qquad \ldots \text{(2)}$

From (1) and (2), we can find

$$a_2[\sigma_2(c) - \sigma_1(c)]\sigma_2(b) + \ldots + a_m[\sigma_m(c) - \sigma_1(c)]\sigma_m(b) = 0 \qquad \ldots \text{(3)}$$

Let $\qquad u_i = a_i(\sigma_i(c) - \sigma_1(c)) \qquad$ for $\qquad i = 2, \ldots, m$

Now, since a_i, $\sigma_i(c)$, $\sigma_1(c)$ are all belong to K, so each u_i is in F.

and $\qquad u_m = a_m[\sigma_m(c) - \sigma_1(c) \neq 0] \qquad [\because a_m \neq 0 \text{ and } \sigma_m(c) - \sigma_1(c) = 0]$

So, from (3), we get

$$u_2\sigma_2(b) + \ldots + u_m\sigma_m(b) = 0 \; \forall b \in K \qquad \ldots \text{(4)}$$

Clearly relation (4) has $m-1$ terms. Since $u_m \neq 0$, then from (4), we can find a relation of the type (4) having less than m non-zero terms, which contradicts the fact that (1) is a minimal relation.

Theorem 3. *Let G be a subgroup of the group of all automorphisms of a field F. Then the fixed field of G is a subfield of F.*

Proof. Let G be the subgroup of the group of all automorphisms of a field F.

Also, let

$$H = \{a \in F : \sigma(a) = a, \forall \sigma \in G\}$$

We have to prove that H is a subfield of F.

It is clear that H is not empty because at least 0 and 1 are in H and $\sigma(0) = 0$, $\sigma(1) = 1$, $\forall \sigma \in G$.

Let $a, b \in H$, then $\qquad \sigma(a) = a, \sigma(b) = b, \forall \, \sigma \in G$

Now, for any $\sigma \in G$, $\quad \sigma(a - b) = \sigma(a) - \sigma(b)$

$$= a - b$$

Therefore, $\qquad a, b \in H \Rightarrow a - b \in H.$

Also, let $a \in H, a \neq b \in H$.

Then $\qquad \sigma(a) = a, \sigma(b) = b, \forall \, \sigma \in G.$

and $\qquad 0 \neq b \Rightarrow b^{-1}$ exists.

Now, for any $\qquad \sigma \in G$

$$\sigma(ab^{-1}) = \sigma(a)\,\sigma(b^{-1})$$
$$= [\sigma(a)][\sigma(b)]^{-1}$$
$$= ab^{-1}$$

Therefore, $a \in H, 0 \neq b \in H \Rightarrow ab^{-1} \in H$, which is the required necessary and sufficient condition for subfield. Hence, H is a subfield of F.

Theorem 4. *Let K be a subfield of a field F. Let G(F, K) be the set of all those automorphisms of F which leave every element of K fixed, i.e., the automorphism f of F is in G(F, K)* *if and only if $f(\alpha) = \alpha, \forall \ \alpha \in K$. Then $G \ (F, K)$ is a subgroup of the group of all automorphisms of F.*

Proof.   Since at least the identity automorphism I of F is in $G(F, K)$. So, $G(F, K)$ is not empiy.

Also, I is the identity mapping of F, therefore $I(\alpha) = \alpha, \forall \alpha \in K$

Further, let

$f, g \in G(F, K)$,   then $f(\alpha) = \alpha, \forall \alpha \in K$

and $g(\alpha) = \alpha$ for each $\alpha \in K$.

For arbitrary $\alpha \in K$, we have

$$(gf^{-1})(\alpha) = g(f^{-1}(\alpha))$$
$$= g(\alpha) \qquad\qquad [\because f(\alpha) = \alpha \Rightarrow f^{-1}(\alpha)=\alpha]$$
$$= \alpha$$

$\Rightarrow$ $gf^{-1} \in G(F, K)$

i.e., $g, f \in G(F, K) \qquad \Rightarrow \ gf^{-1} \in G(F, K),$

which is the required necessary and sufficient condition for subgroups.

Hence, $G(F, K)$ is a subgroup of the group of all automorphism of F.

Theorem 5. *Let K be a finite extension of a field F. Then $G(K, F)$ is a finite group and its order $O(G(K, F))$ satisfies the relation*

$$O(G(K, F \,) \subseteq [K : F]$$

Proof.   Let F be a field and K be a finite extension of F.

Let $[K : F] = n$

Further, let $b_1, b_2, ...b_n$ be a basis of the vector space $K(F)$, then every

element $\alpha \in K$ is of the form

$$\alpha = b_1\alpha_1 + b_2\alpha_2 +.....+ b_n\alpha_n \text{ for } b_i \in F$$

Suppose we can find $(n+1)$ distinct automorphism $\sigma_1, \sigma_2, ..., \sigma_n, \sigma_{n+1}$ in $G \ [K: F]$.

Now, consider the following system of homogeneous linear equations.

$$\sigma_1(b_1) \, x_1 + \sigma_2(b_1) \, x_2 +.....+ \sigma_{n+1}(b_1)x_{n+1}=0$$
$$\sigma_1(b_2) \, x_1 + \sigma_2(b_2) \, x_2 +.....+ \sigma_{n+1}(b_2)x_{n+1}=0$$
$$\text{.......} \quad \text{....} \quad\quad \text{....} \quad\quad \text{....} \quad\quad \text{....} \quad\quad \text{... ... ...}$$
$$\sigma_1(b_i) \, x_1 + \sigma_2(b_i) \, x_2 +.....+ \sigma_{n+1}(b_i)x_{n+1}=0$$
$$\text{.......} \quad \text{....} \quad\quad \text{....} \quad\quad \text{....} \quad\quad \text{... ... ...}$$
$$\sigma_1(b_n) \, x_1 + \sigma_2(b_n) \, x_2 +.....+ \sigma_{n+1}(b_n)x_{n+1}=0$$

In the above system of linear homogeneous equations, the number of equations is less than the number of unknowns, so this system must have a non-trivial solution (not all 0) $x_1 = a_1 , x_2 = a_2, ...x_{n+1} = a_{n+1}$ in K.

Thus, we have

$$a_1\sigma_1(b_i) + a_2\sigma_2(b_i) + ...+ a_{n+1}\sigma_{n+1}(b_i) = 0 \qquad\qquad ...(1)$$

Since every element in F is left fixed by each σ_i, so that for $j = 1, 2, ..., n+1$, we have

$$\sigma_j(b) = \sigma_j \, (x_1\alpha_1 + x_2\alpha_2 +.....+ x_n\alpha_n)$$
$$= \sigma_j \, (x_1\alpha_1) + \sigma_j \, (x_2\alpha_2) +.....+ \sigma_j \, (x_n\alpha_n)$$

$$= \sigma_j(x_1)\sigma_j(\alpha_1) + \sigma_j(x_2)\sigma_j(\alpha_2) + \dots + \sigma_j(x_n)\,\sigma_j(\alpha_n)$$
$$= x_1\,\sigma_j(\alpha_1) + x_2\sigma_j(\alpha_2) + \dots + x_n\,\sigma_j(\alpha_n) \qquad \dots(2)$$

Multiplying, first, second, ...equation of (1) by $x_1, x_2,...$ respectively and adding and then using (2), we get

$$a_1\,\sigma_1(\alpha) + a_2\sigma_2(\alpha) + \dots + a_{n+1}\sigma_{n+1}(\alpha) = 0 \qquad \dots(3)$$

which contradicts the facts that the distinct automorphism $\sigma_1, \sigma_2, ..., \sigma_{n+1}$ of K, it is not possible to find the elements $a_1, a_2, ., ..., a_{n+1}$ in K such that

$$a_1\,\sigma_1(\alpha) + a_2\sigma_2(\alpha) + \dots + a_{n+1}\sigma_{n+1}(\alpha) = 0, \forall \alpha \in K$$

Hence, $O(G(K, F)) < n = [K : F]$

Theorem 6. *Let G be a finite group of automorphisms of a field F and F_0 be the set of all elements of F which are left fixed by each element of G. Then F_0 is a field and $[F : F_0] = o(G)$.*

Proof. Let us suppose $o(G) = n$ such that $G = (\sigma_1, \sigma_2, ..., \sigma_n)$. Firstly assume that $[F : F_0] < n$ and let $[F : F_0] = m < n$.

Since we have assumed that $[F : F_0] = m$, let $\{b_1, b_2, ...b_m\}$ be a basis for F over F_0. It is known that a system of homogeneous linear equations with more unknowns than equations has a non-trivial solution, so that there exists elements $x_1, x_2, ..., x_n$ of F not all 0 such that

$$\sum_{i=1}^{n} \sigma_i(b_j)x_k = 0, j = 1, 2, ..., m \qquad \dots(1)$$

If a be any element of F, then there exist $a_1, a_2, ..., a_m$ of F_0 such that

$$a = a_1 b_1 + a_2 b_2 + ... + a_m b_m \qquad \dots(2)$$

and

$$\sigma_1(a_j) = a_j = \sigma_2(a_j) \dots = \sigma_n(a_j) \qquad \dots(3)$$

$$\Rightarrow \qquad \sigma_1(a_j b_j)x_1 + \sigma_2(a_j b_j)x_2 + \dots + \sigma_n(a_j b_j)x_n = 0 \qquad \dots(4)$$

On expanding (4) from $j = 1$ to m and then adding, we get

$$\sigma_1(a)x_1 + \sigma_2(a)x_2 + \dots + \sigma_n(a)x_n = 0 \qquad \dots(5)$$

Equation (5) is true for every $a \in F$, so there exists a set containing $n+1$ elements of $F[b_1(b_2, ..., b_{n+1}]$ which is linearly dependent over F_0.

Further, using the same argument of non-trivial solution as above, we can say there exists elements $x_1, x_2, ..., x_{n+1}$ of F, not all zero, such that

$$\sum_{i=1}^{n+1} \sigma_i(b_i)x_i = 0, j = 1, 2, ..., n \qquad \dots(6)$$

Suppose the elements $x_1, x_2, ..., x_{n+1}$ are so chosen that as few as possible are non-zero. Thus, equation (6) has a non-trivial solution $x_1, x_2, ..., x_{n+1}$ in F. Let us suppose that at least one of the x's is not in F_0. Suppose each $x_i \in F_0$. If σ_1 is the identity element of G, then first equation of (6) is

$$\sum_{i=1}^{n+1} x_i b_i = 0$$

which is a contradiction because $b_1, b_2, ..., b_{n+1}$ are linearly independent over F.

Among all the solutions of (6), select one with the smallest number of non-zero terms; we may assume that $x_i \neq 0$ for $i = 1, 2, ..., r$ and $x_i = 0$ for $i = r+1, r+2, ...n+1$.

Then equation (6) reduces to

$$\sum_{i=1}^{r} \sigma_j(b_i)x_i = 0, \quad j = 1, 2, ..., n$$

$$\Rightarrow \quad \sigma_j(b_1)x_1 + \sigma_j(b_2)x_2 + ... + \sigma_j(b_r)x_r = 0 \qquad ...(7)$$

Now, dividing by x_r and setting $y_i = x_i/x_r$, $i = 1, 2, ... r-1$ then for $j = 1, 2, ... n$, we have

$$\sigma_j(b_1)y_1 + \sigma_j(b_2)y_2 + ... + \sigma_j(b_{r-1})y_{r-1} + \sigma_j(b_r) = 0 \qquad ...(8)$$

Since σ_1 is the identity automorphism, then from (8), we have

$$b_1y_1 + b_2y_2 + ... + b_{r-1}y_{r-1} + b_r = 0 \qquad ...(9)$$

$\Rightarrow \quad y_1, y_2, ... y_{r-1}$ do not belong to F_0.

$$[\because [b_1, b_2, ..., b_{n+1}] \text{ is linearly independent over } F_0]$$

Further, suppose $y_1 \notin F_0$

$\Rightarrow$ It is not left fixed by all the automorphisms in G.

$\Rightarrow$ $\exists$ an automorphism σ_2 (say) in G such that $\sigma_2(y_1) \neq y_1$

Apply σ_2 to all the equations of (8), we have

$$\sigma_2[\sigma_j(b_1)y_1 + \sigma_j(b_2)y_2 + ... + \sigma_j(b_{r-1})y_{r-1} + \sigma_j(b_r)] = 0$$

$$\Rightarrow \quad (\sigma_2\sigma_j)(b_1)\sigma_2y_1 + (\sigma_2\sigma_j)(b_2)\sigma_2y_2 + ...$$
$$+ (\sigma_2\sigma_j)(b_{r-1})\sigma_2y_{r-1} + (\sigma_2\sigma_j)b_r = 0 \qquad ...(10)$$

But the set $[\sigma_2\sigma_1, \sigma_2\sigma_2, ..., \sigma_2\sigma_n]$ coincides with the set $[\sigma_1, \sigma_2, ... \sigma_n]$ though the order of the elements will be different. Thus, equation (10) may be rewritten as

$$\sigma_j(b_1)\sigma_2(y_1) + \sigma_j(b_2)\sigma_2(y_2) + ... + \sigma_j(b_{r-1})\sigma_2(y_{r-1}) + \sigma_j(b_r) = 0 \ ...(11)$$

Subtracting (11) from (8), we get

$$\sigma_j(b_1)[y_1 - \sigma_2(y_1)] + \sigma_j(b_2)[y_2 - \sigma_2(y_2)] + ... + \sigma_j(b_{r-1})[y_{r-1} - \sigma_2(y_{r-1})] = 0$$

Set $z_i = y_i - \sigma_2(y_i)$, for $i = 1, 2, ..., r-1$ and $z_i = 0$ for $i = r, r+1, ... n+1$,

where

$$y_i = x_l/x_r \quad [l = 1, 2, ... r], \text{ then } z_i \neq 0$$

$\Rightarrow \quad z_1, z_2, ... z_{n+1}$ are elements of F not all zero satisfying (7) but with fewer than r non-zero elements among them.

which contradict the choice of the set $[x_1, x_2, ... x_{n+1}]$

$\Rightarrow \qquad [F : F_0] \not> n$

Hence, $\qquad [F : F_0] = n = o(G)$.

10.13 NORMAL EXTENSIONS AND ELEMENTARY SYMMETRIC FUNCTIONS OF THE ELEMENTS OF A FIELD

Definition 1 (*Normal Extension*). *Let F be a field and K be a finite extension of F. Then K is said to be normal extension of F if the fixed field G(K, F) is F.*

Definition 2. *Let $a_1, a_2, ..., a_n$ be n elements of a field F. Then, $y_1, y_2, ..., y_n$ are known as elementary functions of $a_1, a_2, ..., a_n$ if*

$$y_1 = a_1 + a_2 + ... + a_n = \sum_{i=1}^{n} a_i$$

$$y_2 = a_1 a_2 + a_2 a_3 + ... = \sum_{i<1} a_i a_j$$

$$y_3 = a_1 a_2 a_3 + a_2 a_3 a_4 + ... = \sum_{i<j<k}^{n} a_i a_j a_k \qquad ... (1)$$

$$\vdots$$

$$y_n = a_1 a_2 ... a_n$$

From equation (1), it is clear that $a_1, a_2,..., a_n$, are the roots of the polynomial equation

$$y^n - x_1 y^{n-1} + x_2 y^{n-2} - ... + (-1)^n y_n = 0$$

REMARK

☞ From the definition of normal extension, it is clear that if a is any element of K which is not in F, then we must have some automorphism σ in $G(K, F)$ such that $\sigma(a) \neq a$

Theorem 1. *Let F be a field and K be a normal extension of F. If H is a subgroup of G(K, F). Define*

$$K_H = [a \in K : \sigma(a) = a, \forall \sigma \in H] \qquad ...(1)$$

i.e., K_H is the fixed field of H. Then

(i) $[K : K_H] = o(H);$ (ii) $H = G(K, K_H)$

Proof.   (i) Let F be a field and H be a subgroup of $G(K, F)$ so that

$$H \subseteq G(K, K)$$

Now, K_H is a subfield of K. Also, since K is normal extension of F, therefore, it is a finite extension of F. Then we have already proved that

$$o(G(K, K_H)) \subseteq [K : KH] \qquad ...(2)$$

Now, $G(K, K_H) = \{\sigma \in A(K): \sigma(a) = a, \forall a \in K_H\}$ ...(3)

If σ is any automorphism in H, then we have

$$\sigma \in H \Rightarrow \sigma(b) = b, \forall b \in K_H$$

$\Rightarrow$ $\sigma \in G(K, K_H)$

So, $H \subset G(K, K_H)$

$\Rightarrow$ H is a subgroup of $G(K, K_H)$, then we have

$$o(H) \subseteq o(G(K, K_H)) \qquad ...(4)$$

Using (2) and (4), we conclude that

$$o(H) \subseteq [K, K_H]$$

Further, we shall prove that $o(H) \geq [K : K_H]$.

Suppose $o(H) = m$ and $[K : K_H] = n$

Now, since $K \subset K_H$ and K is a normal extension of F

$\Rightarrow$ K is a normal extension of K_H

$\Rightarrow$ K is a simple extension of K_H.

$\Rightarrow$ $\exists$ an element $a \in K$ such that $K = K_H(a)$.

We also have $[K : K_H] = n$

$\Rightarrow$ $\quad$ a will satisfy an irreducible polynomial over K_H of degree n.

Now, let $\sigma_1, \sigma_1,..., \sigma_m$ be the elements of H, where σ_1 is the identity automorphism of K.

Let $x_1, x_2,.., x_m$ be the symmetric functions of the elements $\sigma_1(a), \sigma_2(a)......\sigma_m(a)$ of K. Then

$$x_1 = \sigma_1(a) + \sigma_2(a) + ... + \sigma_m(a) = \sum_{i=1}^{m} \sigma_i(a)$$

$$x_2 = \sum_{i<j} \sigma_i(a)\sigma_i(a)$$

$$\vdots$$

$$x_m = \sigma_1(a)\sigma_2(a) + ... + \sigma_m(a)$$

Further, we shall show that each x_i is invariant under every element of H. Let $\sigma \in H$ be arbitrary, then m products $\{\sigma\sigma_1, \sigma\sigma_2,...\sigma\sigma_m\}$ belongs to H.

Now, $\qquad \sigma(x_1) = \sigma\{\sigma_1(a) + \sigma_2(a) + ... + \sigma_m(a)\}$

$$= (\sigma\sigma_1(a) + (\sigma\sigma_2)(a) + ... + (\sigma\sigma_m)(a)$$

$$= \sum_{i=1}^{m} (\sigma\sigma_i(a) = \sum_{i=1}^{m} \sigma_i(a)$$

$$= x_1$$

In a similar way, we may get

$$\sigma(x_2) = \sigma\left[\sum_{i<j} \sigma_i(a)\sigma_j(a) \right] = \sum_{i<j} \sigma[\sigma_i(a)\sigma_j(a)]$$

$$= \sum_{i<j} [(\sigma\sigma_i)(a)(\sigma\sigma_j)(a)]$$

$$= \sum_{i<j} [\sigma_i(a)\sigma_j(a)] = x_2$$

In general, we get $\quad \sigma(x_i) = x_i$

$\Rightarrow$ $\quad$ Each x_i is invariant under every $\sigma \in H$

$\Rightarrow$ $\qquad\qquad x_i \in K_H$

Now, consider a polynomial.

$$p(x) = (x - \sigma_1)(a)) \, (x - \sigma_2)(a)) \, ...(x - \sigma_m)(a))$$
$$= x^m - y_1 x^{m-1} + y_2 x^{m-2} - ....+ (-1)^m y_m$$

Since $\qquad\qquad \sigma_1(a) = a$

$\Rightarrow$ $\quad$ a is a root of $p(x)$ over K_H and a cannot satisfy a non-zero polynomial over K_H of lowest degree.

So, we have

$$m \geq n \Rightarrow o(H) \geq [K : K_H] \qquad ....(6)$$

From (5) and (6), we conclude that
$$o(H) = (K : K_H)$$

(ii) Using (2) and (4), we get
$$o(H) \leq o[G(K, K_H)] \leq [K : K_H]$$

Using (7), we get
$$o(H) \leq o[G(K, K_H)] \leq o(H)$$
$$\Rightarrow \qquad o(H) = o[G(K, K_H)]$$

Also, H is a subgroup of $G(K, K_H)$.

Also, H is a subgroup of $G(K, K_H)$, hence we have $H = G(K, K_H)$.

Theorem 2. *Let K be a normal extension of a field F of characteristic 0. Then $[K : F] = o[G(K, F)]$.*

Proof. Let F be a field of characteristic 0 and K be a normal extension of F. Since K is a normal extension of F, therefore, the fixed field of $G(K, F)$ is F itself.

Since, we know that every normal extension is a finite extension

$\Rightarrow \qquad K$ is a finite extension of F.

Using theorem 1 by taking $H = G[K, F]$, we get
$$K_H = \text{the fixed field of } G[K, F] = F.$$

Hence, $\qquad o[G(K, F)] = [K : F]$.

Theorem 3. *Let K be the splitting field of $f(x) \in F[x]$ and let $p(x)$ be an irreducible factor of $f(x)$ in $F[x]$. If the roots of $p(x)$ are $a_1, a_2,..., a_r$, then for each i, there exists an automorphism α_i in $G[K, F]$ such that $\sigma_i(\alpha_1) = \alpha_i$.*

Proof. Let K be the splitting field of $f(x) \in F[x]$ and $p(x)$ be an irreducible factor of $f(x) \in F[x]$.

$\Rightarrow \qquad$ Every root of $p(x)$ is a root of $f(x)$, therefore it lies K.

Suppose that $\qquad F_1 = F(\alpha_1) \quad \text{and} \quad F_1' = F(\alpha_i)$

where α_1 and α_i are any two roots of $p(x)$.

Since $p(x)$ is irreducible in $F[x]$, then there exists an isomorphism
$$\phi : F(\alpha_1) \to F(\alpha_i)$$

such that $\phi(\alpha_1) = \alpha_i$ and $\phi(a) = a \ \forall \ a \in F$

Since $\qquad F \subset F_1 \Rightarrow$ if $f(x) \in F[x]$. Then $f(x) \in F[x]$

$\Rightarrow \qquad K$ is the splitting field of $f(x) \in F_1[x]$.

Similarly, K is the splitting field of $f(x) \in F'_1[x]$. Therefore, there exists an isomorphism.
$$\sigma_i : K \to K \text{ such that } \sigma_i(a) = \phi(a) \ \forall \ a \in F_1$$

But $F_1 = F(\alpha_1)$, therefore $\alpha_1 \in F_1$, then we have
$$\sigma_i(\alpha_1) = \phi(\alpha_1) = \alpha_1$$

Since α_i coincides with ϕ on F_1 and therefore on F

$\Rightarrow \qquad F_i(\alpha) = \alpha \ \forall \ \alpha \in F$

Hence, $\qquad \sigma_i \in G[K, F]$

Theorem 4. *K is a normal extension of a field F of characteristic 0 if and only if K is the splitting field of some polynomial over F.*

Proof. Let us first suppose K is a normal extension of F. We have to prove that K is the splitting field of some polynomial over F.

Since K is a normal extension of F.

$\Rightarrow$ K is a finite extension of F.

Since, characteristic of F is 0, so K is a normal extension of F.

$\Rightarrow$ $\exists$ an element a in K such that $K = F(a)$.

Since the group $G[K, F]$ is of finite order, say n.

Let $\sigma_1, \sigma_2, ...\sigma_m$ be the distinct elements of $G[K, F]$.

Further, let σ_1 be the identity of $G[K, F] \Rightarrow \sigma_1$ is the identity automorphism of K.

Let $\sigma_1, \sigma_2, ..., \sigma_n$ be the elementary symmetric functions with respect to $\sigma_1(a), ...\sigma_n(a) \in K$, i.e.,

$$\alpha_1 = \sum_{i=1}^{n} \sigma_i(a)$$

$$\alpha_2 = \sum_{i<j}^{n} \sigma_i(a)\sigma_j(a)$$

$$... \quad ... \quad ... \quad ... \quad ... \quad ...$$

$$\alpha_n = \sigma_1(a)\sigma_2(a)...\sigma_n(a)$$

Clearly α_i is invariant under every $\sigma \in G[K, F]$

Since K is a normal extension of F, so F is the fixed field of $G[K, F]$.

$\Rightarrow$ $\alpha_i \in F$ for each i.

Now, consider the polynomial

$$P(x) = [x - \sigma_1(a)][x - \sigma_2(a)]....[x - \sigma_n(a)] \text{ over } K \qquad ....(1)$$

$\Rightarrow$ $\qquad\qquad P(x) = x^n - \alpha_1 x^{n-1} + ...+ (-1)^n \alpha_n$

Now, since each $\qquad \alpha_i \in F \Rightarrow p(x) \in F[x]$.

From (1), we can say that K splits $p(x) \in F[x]$ into a product of linear factors.

Again from (1), $\sigma_1(a) = a$, a root of $p(x)$ in K. Now, since a generates K over F

$\Rightarrow$ a can be in no proper subfield of K which contains F.

$\Rightarrow$ K is the splitting field of $p(x)$ over F.

Conversely, let K be the splitting field of the polynomial $f(x) \in F[x]$.

We have to show that K is a normal extension of F.

Since K is a normal extension of $F \Rightarrow K$ is a finite extension of F.

Let $[K : F] = n$. Then, we shall apply the induction on n.

If $[K : F] = 1$, then $K = F$

$\qquad\qquad \Rightarrow F$ is a normal extension of F.

Assume that if K_1 is the splitting field over F_1 of a polynomial in $F_1(x)$ and if $[K_1 : F_1] < n$, then K_1 is a normal extension of F_1 .

Let $[K : F] = n > 1$ and K is the splitting field over F of $f(x)$ in $F[x]$. If $f(x)$ splits into linear factors over F, then $K = F$.

$\Rightarrow \quad [K : F] = 1$

Thus, assume that $f(x)$ has an irreducible factor $p(x) \in F[x]$ of degree $r > 1$. Now, since characteristic of F is 0 and $p(x)$ is irreducible, so $p(x)$ cannot have multiple roots. Also, $p(x)$ is a factor of $f(x)$ and K is the splitting field of $f(x)$, so $p(x)$ has a full compliment of roots in K. Let $\alpha_1, \alpha_2, ..., \alpha_r \in K$ be the r distinct roots of $p(x)$.

Now, since $p(x)$ is irreducible over F and $\deg p(x) = r$, then

$$[F(\alpha_1) : F] = r$$

Also, by transitivity of extension, we have

$$[K : F] = [K : F(\alpha_1)] [F(\alpha_1) : F]$$

$\Rightarrow \qquad\qquad [K : F(\alpha_1)] = \dfrac{[K : F]}{[F(\alpha_1) : F]} = \dfrac{n}{r} < n$

Further, K is also the splitting field of $f(x)$ considered as a polynomial over $F(\alpha_1)$.

Again, since $[K : F(\alpha_1)] < n$, then by induction hypothesis, K is a normal extension of $F(\alpha_1)$.

Further, let $\theta \in K$ be left fixed by every automorphism in $G[K, F]$. We wish to show that θ is in F. In this case F will be the fixed field of $G[K, F]$ and K will be a normal extension of F.

$\qquad$ Since $\qquad\qquad\qquad \sigma \in G(K, F(\alpha_1))$

$\Rightarrow \quad \sigma$ leaves every element of $F(\alpha_1)$ fixed.

$\Rightarrow \quad \sigma$ leaves every element of F fixed.

$\Rightarrow \quad \sigma \in G[K, F]$

$\Rightarrow \quad \sigma(\theta) = \theta \qquad$ [Because θ is left fixed by every automorphism in $G[K, F]$]

$\Rightarrow \quad \sigma(\theta) = \theta, \forall \sigma \in G[K, F(\alpha_1)]$.

Since K is a normal extension of $F(\alpha_1)$ is the fixed field of $G[K, F(\alpha_1)]$

$\Rightarrow \quad \theta$ is in $F(\alpha_1)$.

Therefore, we can write

$$\theta = \lambda_0 + \lambda_1\alpha_1 + \lambda_2 \alpha_1^2 + ... + \lambda_{r-1} \alpha_1^{r-1}, \text{ where } \lambda_0, ..., \lambda_{r-1} \in F \quad ....(2)$$

Now, for each $i = 1, 2, ...r$, there is an automorphism σ_i of K, $\sigma_i \in G[K, F]$ such that $\sigma_i(\alpha_1) = \alpha_i$.

Since θ is left fixed by every automorphism in $G[K, F]$

$\Rightarrow \quad \sigma_i(\theta) = \theta$

$\qquad$ Also, $\quad \lambda_0, \lambda_1, ...\lambda_{r-1} \in F$

$\qquad \Rightarrow \quad \sigma_i(\lambda_0) = \lambda_0, \qquad \sigma_i(\lambda_1) = \lambda_1... \sigma_i(\lambda_{r-1}) = \lambda_{r-1}$

Further operating σ_i on both sides of (2), we get

$$\sigma_i(\theta) = \sigma_i(\lambda_0 + \lambda_1\alpha_1 + \lambda_2\alpha_1^2 + ... + \lambda_{r-1}\alpha_1^{r-1})$$

$\Rightarrow \qquad\qquad \theta = \sigma_i(\lambda_0) + \sigma_i(\lambda_1)\sigma_i(\alpha_1) + \sigma_i(\lambda_2)\sigma_i(\alpha_1)^2 + ... + \sigma_i(\lambda_{r-1})[\sigma_i(\alpha_1)]^{r-1}$

$\Rightarrow \qquad\qquad \theta = \lambda_0 + \lambda_1\alpha_i + \lambda_2\alpha_i^2 + ... + \lambda_{r-1}\alpha_i^{r-1}$

$\Rightarrow \quad \lambda_{r-1}\alpha_i^{r-1} + \lambda_{r-2}\alpha_i^{r-2} + ... + \lambda_i\alpha_i + (\lambda_0 - \theta) = 0$ ...(3)

Now, since $\alpha_1, \alpha_2, ... \alpha_r$ are all distinct, then from (2), the polynomial

$q(x) = \lambda_{r-1}x^{r-1} + \lambda_{r-2}x^{r-2} + ... + \lambda_1 x + (\lambda_0 - \theta)$

In $K[x]$ of degree at most $r-1$, has r distinct roots $(\alpha_1, \alpha_2 ..., \alpha_r)$, which is possible only when all the coefficients of $q(x)$ are 0.

In particular $\lambda_0 - \theta = 0 \Rightarrow \theta = \lambda_0$

$\Rightarrow \quad \theta$ is in F.

Hence, K is a normal extension of F.

Theorem 5. *Let F be a field of characteristic 0 and K be a normal extension of F. If T is a subfield of K containing F, then T is a normal extension of F if and only if*

$$\sigma(T) \subseteq T, \forall \sigma \in G[K, F].$$

Proof. Let F be a field. Since K is a normal extension of $F \Rightarrow K$ is a finite extension of F.

Let T be a subfield of K containing F.

$\Rightarrow \quad T$ is also a finite extension of F, a field with characteristic 0.

$\Rightarrow \quad T$ is a simple extension of F.

$\Rightarrow \quad \exists$ an element $a \in T$ such that $T = F(a)$.

Let us first suppose that $\sigma(T) \subseteq T, \forall \sigma \in G[K, F]$. We have to show that T is a normal extension of F. Since K is a normal extension of F.

$\Rightarrow \quad G[K, F]$ is a finite group, say of order n.

Now, let $\sigma_1, \sigma_2, ..., \sigma_n$ be all the n elements of $G[K, F]$ and let σ_1 be the identity of $G[K, F]$. Then

$$\sigma \in T \text{ and } \sigma(T) \subseteq T, \forall \sigma \in G[K, F]$$

$\Rightarrow \quad \sigma_1(a), \sigma_2(a), ..., \sigma_n(a)$ are all in T.

Now, consider the polynomial over T.

$$p(x) = [x - \sigma_1(a)][x - \sigma_2(a)]....[x - \sigma_n(a)] \quad ...(1)$$

$$= x^n - \alpha_1 x^{n-1} + \alpha_2 x^{n-2} .... + (-1)^n \alpha_n$$

where $\alpha_1, \alpha_2 ..., \alpha_n$ an are elementary symmetric functions of the elements $\sigma_1(a), \sigma_2(a), ... \sigma_n(a)$ of T.

We observe that $\alpha_1, \alpha_2, ... \alpha_n$ are each invariant with respect to every $\sigma \in G[K, F]$.

$\Rightarrow \quad$ Each α_i, for $i = 1, 2, ..., n$ must be an element of the fixed field $G[K, F]$.

$\Rightarrow \quad$ Fixed field of $G[K, F]$ is F because K is a normal extension of K.

$\Rightarrow \quad$ Each α_i is in F.

$\Rightarrow \quad p(x) \in F[x]$

From (1), we can say that T splits the polynomial $p(x) \in F[x]$ into a product of linear products.

Also, $\sigma_1(a) = a$ is a root of $p(x)$ in $T = F(0)$.

Further, since a generates T over F, so a can be in no proper subfield of T which contains F.

⇒   *T* is the splitting fields of $p(x) \in F[x]$.

Hence, using previous theorem, *T* is a normal extension of *F*.

Conversely, suppose *K* is a normal extension of *F*. We have to prove that $\sigma(T) \subseteq T$ $\forall \sigma \in G[K,F]$.

Since *T* is a normal extension of *F*, so $G[T,F]$ is a finite group, say of order *m*. Also, let $\psi_1, \psi_2, \ldots \ldots \psi_m$ be all the *m* elements of $G[T,F]$ and let ψ_1 be the identity of $G[T,F]$. Now since $a \in T = F(a)$, so $\psi_1(a), \psi_2(a), \ldots \ldots, \psi_m(a)$ are all in *T*.

Now consider the polynomial over *T*.

$$q(x) = [x-\psi_1(a)][x-\psi_2(a)] \ldots \ldots [x-\psi_m(a)] \qquad \ldots\ldots(2)$$
$$= x^m - \beta_1 x^{m-1} + \beta_2 x^{m-2} - \ldots\ldots + (-1)^m \beta_m,$$

where, β_i's are elementary symmetric functions of the element $\psi_1(a), \psi_2(a), \ldots\ldots \psi_m(a)$ of *T* .

We observe that each β_i is invariant w.r.t. each ψ in $G[T,F]$. Now, since *T* is a normal extension of *F*. Each β_i must be in *F*. Therefore, $q(x) \in F[x]$.

Also, from (2), we can say that $q(x)$ has all its roots in *T* and $\psi_1(a) = a$ is a root of $q(x)$ in *T* and hence in *K*.

Let σ be any element of $G(F,K)$. Then, since σ is an automorphism of *K* leaving every element of *F* fixed and *a* is a root of $q(x) \in F(x)$ in *K*.

⇒ $\sigma(a)$ is also a root of $q(x)$ in *K*.

Now since all roots of $q(x)$ are in *T*   ⇒ $\sigma(a)$ must be an element of *T*.

Now, $T = F(a)$

If $[T : F] = r$, then $t \in T$ can be written as

$$t = \lambda_0 + \lambda_1 a + \ldots\ldots \lambda_{r-1} a^{r-1}, \ \lambda_0, \lambda_1, \ldots\ldots \lambda_{r-1} \in F.$$
$$\Rightarrow \quad \sigma(t) = \sigma(\lambda_0 + \lambda_1 a + \ldots\ldots \lambda_{r-1} a^{r-1})$$
$$= \sigma(\lambda_0) + \sigma(\lambda_1)\sigma(a) + \ldots\ldots \sigma(\lambda_{r-1})[\sigma(a)]$$
$$= \lambda_0 + \lambda_1 \sigma(a) + \ldots\ldots \lambda_{r-1}[\sigma(a)]^{r-1}$$
$$= \text{an element of } T, \text{ since } \sigma(a) \in T \text{ and } \lambda_0, \lambda_1, \ldots\ldots, \lambda_{r-1} \in F$$

For every $\sigma \in G(K, F)$ and for every $t \in T$, we have $\sigma(t) \in T$.

Hence, $\sigma(T) \subseteq T \ \ \forall \ \sigma \in G(K, F)$

10.14 GALOIS GROUP

Definition 1. *Let F be a field and K be a finite extension of F. Then the group G[K, F] of all F automorphism of K is known as Galois group of K over F. Here, K is known as Galois extension of F.*

Definition 2. *Let K be a finite extension of F, then two elements α and β of K are said to be conjugate over F, if they have the same minimal polynomial over F.*

<u>**Theorem 1.**</u> *An element of K which remains invariant under each member of the group G(K,F) of K over F is necessarily a member of F.*

<u>**Proof.**</u>   Let α be an arbitrary element of *K* which remains invariant under every member of

$\sigma \in G(K,F)$, *i.e.,* $\sigma(\alpha) = \alpha$, $\forall\ \sigma \in G(K, F)$. Then, we shall show that $\alpha \in F$.

Since $G(K, F)$ is Galois group, so that K is a finite normal extension of K, therefore it is the splitting field of some polynomial $f(x) \in F[x]$.

Let $p(x)$ be the minimal polynomial of α over F. Since K is normal over F and one root α of $p(x)$ is in K, therefore each root of $p(x)$ belongs to K.

Consider a polynomial $p(x) \in F[x]$, then K is the splitting field of $F(x)\ p(x)$.

Let if possible deg $p(x) \geq 2$.

Then, K is also separable over F and all the roots of $p(x)$ are distinct, then there exists an element $\beta \in K$ with $\beta \neq \alpha$ such that P is a root of $p(x)$ over F.

So, α and β are two distinct roots of an irreducible polynomial $p(x) \in F[x]$, so there exists an F-isomorphism $\sigma : F(\alpha) \to F(\beta)$ such that $\sigma(\alpha) = \beta$. Thus, isomorphism can be contained to an F-automorphism of K. Therefore, there exists a member $\sigma \in G(K,F)$ which maps α qn $\beta \neq \alpha$. Thus, we arrive at a contradiction.

Hence, the deg $p(x) \ngeq 2$, accordingly deg $p(x) = 1$, *i.e.,* $p(x) = x-\alpha$, where $\alpha \in F$ because $p(x) \in F[x]$ and $p(\alpha) = 0$.

Theorem 2. *Let K be a finite normal extension of a field F. If α, β are any two elements of K conjugate over F, then their exists an F-automorphism ϕ of K such that $\phi(\alpha) = \beta$.*

Proof. Let K be a finite normal extension of F, therefore it is the splitting field of some polynomial $f(x) \in F[x]$. Also, α and β are conjugate over F, so they will have the same minimal polynomial over F and are algebraic over F. Thus, there exists an isomorphism $\psi : F(\alpha) \to F(\beta)$ such that $\psi(\alpha) = \beta$. Now K the splitting field of $f(x)$ over $F(\alpha)$ and over $F(\beta)$ because $F \subset F(\alpha)$ and $F \subset F(\beta)$ and $\alpha, \beta \in K$, therefore there exists an automorphism ϕ of K which is an extension of ψ.

Also, $\phi(\alpha) = \psi(\alpha) = \beta$ and for any element $\gamma \in F$, we have $\phi(\gamma) = \psi(\gamma) = \gamma$.

Hence, ϕ is an F-automorphism of K such that $\phi(\alpha) = \beta$.

Theorem 3. *Let K be any extension of a field F and let an element α of K be algebraic over F. Then for any F-automorphism ϕ of K, $\phi(\alpha)$ is a conjugate of α over F.*

Proof. We shall prove that $\phi(\alpha)$ and α will have the same minimal polynomial over F.

Let $p(x) = x^n + a_1 x^{n-1} + \ldots + a_{n-1}x + a_n$ be the minimal polynomial of α over F, then $p(\alpha) = 0$, *i.e.,* $\alpha^n + a_1 \alpha^{n-1} + \ldots + a_{n-1}\alpha + a_n = 0$.

Since ϕ is F-automorphism, so that $\phi(0) = 0$

$\Rightarrow\quad \phi(\alpha^n + a_1\alpha^{n-1} + \ldots + a_{n-1}\alpha + a_n) = 0$

$\Rightarrow\quad \phi(\alpha^n) + \phi(a_1\alpha^{n-1}) + \ldots + \phi(a_{n-1}\alpha) + \phi(a_n) = 0$

$\Rightarrow\quad \phi(\alpha^n) + \phi(a_1)\phi(\alpha^{n-1}) + \ldots + \phi(a_{n-1})\phi(\alpha) + \phi(a_n) = 0$

$\Rightarrow\quad \phi(\alpha^n) + a_1\phi(\alpha^{n-1}) + \ldots + a_{n-1}\phi(\alpha) + a_n = 0$ $[\because \phi(a_i) = J_i, \forall\ a \in F]$

$\Rightarrow\quad [\phi(\alpha)]^n + a_1[\phi(\alpha)]^{n-1} + \ldots + a_{n-1}\phi(\alpha) + a_n = 0$

             $[\because \phi \text{ preserves the multiplication in } K.]$

This shows that $\phi(\alpha)$ is also a root of $p(x)$.

Hence, $\phi(\alpha)$ and α have the same minimal polynomial, accordingly $\phi(\alpha)$ is conjugate of α over F.

Theorem 4. *The order of the Galois group $G(K,F)$ is equal to $[K:F]$.*

Proof.   Let K be a finite separable extension of F, therefore it is simple extension of F. So, there exists an element $\alpha \in K$ such that

$$K = F(\alpha).$$

Let $p(x)$ be a minimal polynomial of α over F and let $\deg p(x) = n$, then we have

$$[K:F] = n.$$

Since K is separable over F, the roots of $p(x) = 0$ are all simple. Let these roots be $\alpha = \alpha_1, \alpha_2, ..., \alpha_n$ which are distinct because degree of $p(x)$ is n. Now each F-automorphism of K maps α to another root of $p(x)$. Evidently $K = F(\alpha_i)$ for $i = 1, 2, 3, ..., n$.

We know that the F-mapping which maps α_1 and α_i actually determines a F-automorphism ϕ_i of $F(\alpha_1)$ on $F(\alpha_i)$ such that $\phi(\alpha_1) = \alpha_i$. Each ϕ_i is unique because α_i generates K over F.

Hence, the Galois group $G(K,F)$ consists of $\phi_1, \phi_2, ...\phi_n$.

Consequently, $o(G(K,F)) = n = [K:F]$

Theorem 5. *Let K be a finite separable, normal extension of a field F of characteristic zero, then the fixed field of Galois group $G(K, F)$ is F itself.*

Proof.   It is known that every finite separable extension is a simple extension, then K is a simple extension of F, i.e., $K = F(\alpha)$ for some $\alpha \in K$.

Let $p(x) \in F[x]$ be a minimal polynomial of α and let E be the splitting field of $p(x)$. Since K is normal extension of F, so the splitting field of every polynomial over F is K.

Thus, we have $E \subseteq K$ ...(1)

Also, $\alpha \in E$, because E being the splitting field of $p(x)$ for which $p(\alpha) = 0$.

Now, $a \in E$ $\Rightarrow$ $F(\alpha) \subseteq E \Rightarrow K \subseteq E$ $[\because K = F(\alpha)]$ ...(2)

From (1) and (2), we conclude that

$$E = K$$

$\Rightarrow$ K is the splitting field of the minimal polynomial $p(x)$ of α.

If $\deg . p(x) = n$, then $[K:F] = n$.

Let $\alpha = \alpha_1, \alpha_2, ...\alpha_n$ be the conjugate of α over F.

Then, $K = F(\alpha_i)$ for $i = 1, 2, ..., n$. Now, for each i, there exists an F-automorphism ϕ of K such that $\phi(\alpha_1) = \alpha_i$ for some α_1 with $\phi(\alpha_1)$ is conjugate of α_1.

Hence, $G(K, F)$ consists of $\phi_1, \phi_2, ...,\phi_n$.

If the fixed field of $G(K, F)$ is denoted by $K_{G(K,F)}$, then

$$[K:K_{G(K,F)}] = O(G(K,F)) = n = [K:F] \Rightarrow K_{G(K,F)} = F$$

Hence, the fixed field under $G(K, F)$ is F itself.

Theorem 6. *(Fundamental Theorem of Galois Theory). Let K be a separable normal extension of finite degree over a subfield F, then there exists a one-one correspondence between the family of sub-fields of K containing F and the family of all subgroups of $G(K,F)$. Further, if E is any subfield of K, such that $F \subset E \subset K$, then*

   (i) $[K:E] = o(G(K, E))$ and $[E:F] = $ index of $G(K, E)$ in $G(K, F)$

   (ii) E is a normal extension of F if and only if $G(K, E)$ is a normal subgroup of $G(K, F)$.

(iii) If E is a normal extension of F, then $G(E, F) = G(K, F)/G(K, E)$.

Proof. First, we shall show that there exists a one-one correspondence between the family of subfields of K containing F and the family of all subgroups of $G(K, F)$.

Let E be any subfield of K containing F and let $G(K, E)$ be the group of all E-automorphisms of K.

Since $F \subset E \subset K$, so that $G(K, E) \subseteq G(K, F)$.

Also, $G(K, E)$ and $G(K, F)$ are the subgroups of the group of all automorphisms of K, therefore $G(K, E)$ is a subgroup of $G(K, F)$. Thus, for each subfield E of K, we can find a subgroup $G(K, E)$ of $G(K, F)$.

Consider a mapping ψ of the set of all subfields of K containing F into the set of all subgroups of $G(K, F)$, defined by $\psi(E) = G(K, E)$ for all subfields E of K containing F.

Now, we show that ψ is bijective.

(i) **ϕ is one-one.** Let E_1 and E_2 be any two subfields of K containing F and suppose that

$$\psi(E_1) = \psi(E_2) \Rightarrow G(K,E_1) = G(K,E_2).$$

$\therefore$ The fixed field of $G(K, E_1) = $ The fixed field of $G(K, E_2)$

$\Rightarrow$ $E_1 = E_2$ [$\because K$ is finite normal extension of F.]

Thus, ψ is one-one.

(ii) **ψ is onto.** Let H be an arbitrary subgroup of $G(K, F)$, then the fixed field of H denoted by K_H is given by

$$K_H = \{a \in K : \phi(a) = a, \forall \phi \in H\}$$

which shows that ϕ is a K_H-automorphism of K, so that

$$\phi \in H \Rightarrow \phi \in G(K,K_H)$$

$\therefore$ $H \subseteq G(K,K_H)$

Also, $o(H) = [K:K_H]$.

Since K is a normal extension of F and K_H is a subgroup of K containing F, so that K is also normal extension of K_H . Therefore, K_H is the fixed field under the Galois group $G(K, K_H)$.

$$[K:K_H] = o[G(K,K_H)].$$

Now, from (1) and (2), we conclude that

$$o(H) = o[G(K,K_H)]$$

Also, $H \subseteq G(K,K_H)$

$\therefore$ $H = G(K,K_H)$.

$\Rightarrow$ Each subgroup of $G(K, F)$ is of the form $G(K, K_H)$ such that $F \subseteq K_H \subseteq K$, and so corresponding to this subgroup $G(K, K_H)$, there exists a sub-field K_H of K containing such that $\psi(K_H) = G(K,K_H)$. Hence, ψ is onto.

(i)   Since K is a normal extension of F and E is a subfield of K such that $F \subseteq E \subseteq K$, then K is a normal extension of E, therefore, we have

$$[K:F] = o[G(K,F)] \text{ and } [K:E] = o[G(K,E)].$$

Moreover,

$$[K:F] = [K:E][E:F]$$

$\Rightarrow \quad o[G(K,F)] = o[G(K,E)][E:F]$

$\Rightarrow \qquad [E:F] = \dfrac{o[G(K,F)]}{o[G(K,E)]} = $ index of $G(K, E)$ in $G(K, F)$.

(ii) Let us suppose E is a normal extension of F, then we shall show that $G(K,F)$ is a normal subgroup of $G(K,F)$. For any $\sigma \in G(K, F)$ and $\psi \in G(K, E)$, we actually show that $\sigma^{-1}\psi\sigma \in G(K, E)$.

Let α be an arbitrary element of E. Since E is a normal extension of F, so that the splitting field of the minimal polynomial of α over F is contained in E and every conjugate of α is therefore in E.

Since $\sigma(\alpha)$ is conjugate of α for any $\sigma \in G(K,F)$, then $\sigma(\alpha) \in E$. Therefore, for automorphism $\psi \in G(K, E)$, we have $\psi(\sigma(\alpha)) = \sigma(\alpha)$.

Now, $(\sigma^{-1}\psi\sigma)(\alpha) = \sigma^{-1}[\psi(\sigma(\alpha))] = \sigma^{-1}(\sigma(\alpha)) = \alpha$

$\Rightarrow \qquad \sigma^{-1}\psi\sigma \in G(K, E), \forall \sigma \in G(K, F), \psi \in G(K, E).$

Hence, $G(K, E)$ is a normal subgroup of $G(K, F)$.

Conversely, now suppose that $G(K, E)$ is a normal subgroup of $G(K, F)$, then we show that E is normal extension of F.

Let $\alpha \in E$ be an arbitrary element and let $p(x)$ be the minimal polynomial of α over F. Let L be the splitting field of $p(x)$. Since K is normal extension of F so that $L \subseteq K$.

If β is any root of $p(x)$ in L, then β is conjugate of α over F, therefore there exists an F-automorphism σ of K such that $\sigma(\alpha) = \beta$.

Now, $G(K, E)$ is a normal subgroup of $G(K, F)$, then for $\sigma \in G(K, F)$ and $\psi \in G(K, E)$, we have

$\sigma^{-1}\psi\sigma \in G(K,E) \Rightarrow (\sigma^{-1}\psi\sigma)(\alpha) = \alpha$

$\Rightarrow (\sigma^{-1}\psi)(\sigma(\alpha)) = \alpha \Rightarrow (\sigma^{-1}\psi)(\beta) = \alpha \quad [\because \sigma(\alpha)=\beta]$

$\Rightarrow \psi(\beta) = \beta \qquad \Rightarrow \beta \in E$ so $\beta \in L \Rightarrow \beta \in E.$

$\therefore \qquad\qquad L \subseteq E.$

This shows that every splitting field of $p(x)$ of $\alpha \in E$ is contained in E.

Hence, E is the normal extension of F.

(iii) Let E be the normal extension of F.

Let σ be any element of $G(K,F)$. Define a mapping σ' of E into K be splitting $\sigma'(\alpha) = \sigma(\alpha), \forall \alpha \in E$.

Since σ is an F-automorphism of K and E is normal extension of F, so that $E = F(\alpha)$, therefore, σ' is F-automorphism of E, i.e., $\sigma' \in G(E, F)$.

Therefore, $\sigma(E) = \sigma'(E) = E$.

Consider a mapping $\phi : G(K, F) \to G(E, F)$ by setting $\phi(\alpha) = \sigma'$, $\forall \sigma \in G(K,F)$. This mapping ϕ is a group homomorphism. If σ_1 and σ_2 are any two elements of $G(K, F)$ and $\alpha \in E$, then

$(\phi(\sigma_1\sigma_2))(\alpha) = ((\sigma_1\sigma_2)')(\alpha)$ [By definition of ϕ]

$= (\sigma_1\sigma_2)(\alpha)$ $[\because \sigma'(\alpha)=\sigma(\alpha), \forall \alpha \in E]$

$= \sigma_1(\sigma_2(\alpha))$

and $\quad (\phi(\sigma_1)\phi(\sigma_2))(\alpha) = \phi(\sigma_1)\phi(\sigma_2)(\alpha) = (\phi(\sigma_1))(\sigma_2'(\alpha))$

$$= (\phi(\sigma_1))(\sigma_2(\alpha)) = \sigma_1'(\sigma_2(\alpha)) = \sigma_1(\sigma_2(\alpha)).$$

So, $\quad\quad\quad \phi(\sigma_1\sigma_2) = \phi(\sigma_1)\phi(\sigma_2)$.

Consider any $\psi \in G(E, F)$, then $\psi(\alpha)$ is conjugate of α over F, so there exists an F-automorphism σ of K such that $\sigma(\alpha) = \psi(\alpha)$. Also, σ and ψ are both identity of F and E and $E = F(\alpha)$, so that $\sigma(a) = \psi(a)$, $\forall a \in F(\alpha) = E$.

$\therefore \quad\quad\quad\quad \psi = \sigma' = \phi(\sigma)$.

Hence, f is onto.

Also, $\quad$ kernel $\phi = \{\sigma \in G(K,F) : \phi(\sigma) = I$, the identity of $G(E, F)\}$

$$= \{\sigma \in G(K,F) : \sigma' = I\}$$

$$= \{\sigma \in G(K,F) : \sigma'(\alpha) = I(\alpha) = \alpha, \forall \alpha \in E\}$$

$$= \{\sigma \in G(K,F) : \sigma(\alpha) = \alpha, \forall \alpha \in E\} = G(K, E).$$

Hence, by the fundamental theorem on homomorphism of groups, we conclude that

$$G(E,F) \cong \frac{G(K,F)}{G(K,E)} \quad \cdot \cdot$$

Theorem 7. *Let F be a field and S be a set of automorphism of K. Then fixed field of S and that of $\overline{S}$, the subgroup of the group of all automorphism of K generated by S are identical.*

Proof. Let F be a field and S be a set of automorphism of F. Also, $\overline{S}$ denotes the subgroup of the group of all automorphisms of F generated by S. Further, suppose that F_1 and F_2 are two fixed fields of S and $\overline{S}$ respectively. We have to prove that $F_1 = F_2$.

Consider $a \in L_2 \Rightarrow \sigma(\alpha) = a, \forall a \in \overline{S}$

$\Rightarrow \quad \sigma(\alpha) = a \forall \sigma \in S \Rightarrow \quad\quad a \in L \quad\quad\quad\quad\quad\quad\quad\quad\quad\quad [\because S \subseteq \overline{S}]$

Since a is arbitrary

$\Rightarrow \quad\quad L_2 \subseteq L_1 \quad\quad\quad\quad\quad\quad\quad\quad\quad\quad\quad\quad\quad\quad\quad\quad\quad\quad\quad ......(1)$

If σ is an automorphism of F, then we have $\sigma(a) = a$ if and only if $\sigma^{-1}(a) = a$. Therefore, if $\sigma(a) = a$, then $\sigma^n(a) = a$, where n is any integer.

Also, $\sigma^2(a) = \sigma[\sigma(a)] = \sigma(a) = a$.

If $a \in L_1$ is arbitrary, then $\sigma(a) = a, \forall \sigma \in S$.

Further, let f be any element of S, then $\overline{S}$ is a subgroup of the group of all automorphism of F generated by S.

$\Rightarrow \quad f$ can be expressed as a product of positive and negative integral powers of a finite number of elements of S.

Let $\sigma_1^{P_1}\sigma_2^{P_2}........\sigma_m^{P_m}$, $\sigma_1,......\sigma_m \in S$ and $p_1,.....p_m$ are integers. Then, we have

$$f(a) = \left(\sigma_1^{P_1}\sigma_2^{P_2}........\sigma_m^{P_m}\right)(a)$$

$$= a \quad\quad\quad\quad\quad\quad\quad\quad\quad\quad\quad\quad [\because \sigma_i(a) = a, \forall i = 1,........,m]$$

Therefore $a \in L_1 \Rightarrow f(a) = a \forall f \in \overline{S} \Rightarrow a \in L_2$

Since a is arbitrary, therefore $L_1 \subseteq L_2 \quad\quad\quad\quad\quad\quad\quad\quad\quad\quad\quad\quad .....(2)$

From (1) and (2), we conclude that $L_1 = L_2$.

Theorem 8. *Let K be an extension of the field of rational numbers* **Q**. *Then, any automorphism of K must leave every element of* **Q** *fixed.*

Proof. Let σ be an automorphism of K. Let p be any integer. We have to show that $\sigma(p)=p$.

Now, there are following two cases :

(i) If p is 0, then $\sigma(p) = \sigma(0) = 0 = p$

(ii) If p is a positive integer, therefore, we have

$$p = 1 + 1 + 1 + ...\text{upto } p \text{ times.}$$
$$\Rightarrow \qquad \sigma(p) = \sigma(1 + 1 + ...\text{upto } p \text{ times})$$
$$= \sigma(1) + \sigma(1) + ...\text{upto } p \text{ times.}$$
$$= 1 + 1 + ...\text{upto } p \text{ times.}$$
$$= p.$$

(iii) If p is a negative integer, say $p = -q$, where q is a positive integer.

$$\text{Consider } \sigma(p) = \sigma(-q) = \sigma((-1)+(-1)+.....\text{upto } q \text{ times})$$
$$= \sigma(-1) + \sigma(-1) + ...\text{upto } q \text{ times}$$
$$= (-1) + (-1) + ...\text{upto } q \text{ times.}$$
$$= -q = p$$

Further, let $\alpha \in F$, *i.e.*, if α is any rational number. Then $\alpha = \dfrac{m}{n}$, where m and n are integers with $n \neq 0$.

If σ is any automorphism of K, then we have

$$\sigma(\alpha)=\sigma(m\,|\,n)= \sigma(mn^{-1}) = \sigma(m)\,\sigma(n^{-1}) = m(\sigma(n))^{-1}$$
$$=mn^{-1} = m\,|\,n = \alpha.$$

Hence, in each case σ leaves every element of **Q** fixed.

Theorem 9. *Let* **C** *be the field of complex numbers and* **R** *be the field of real numbers, then* **C** *is a normal extension of* **R**.

Proof. Define **C** $= [a + ib : a, b \in \mathbf{R}]$

Clearly **C** is an extension of **R**.

The elements 1, i of **C** form a basis of **C** over **R**. Thus [**C** : **R**] $= 2$ and **C** is a finite extension of **R**. Let σ be any automorphism of **C**, then

$$[\sigma(i)]^2 = \sigma(i)\,\sigma(i) = \sigma(i \,.\, i) = \sigma(i^2) = \sigma(-1) = -\sigma(1) = -1$$

Let $\sigma \in G(\mathbf{C}, \mathbf{R})$, *i.e.*, σ leaves every real numbers fixed if $a + ib$ is any element of **C**, $a, b \in \mathbf{R}$, then

$$\sigma(a + ib) = \sigma(a) + \sigma(i)\,\sigma(b)$$
$$= a \pm ib$$

If $\sigma \in G[\mathbf{C}, \mathbf{R}]$, then we must have $\sigma(a + ib) = a \pm ib$

Clearly each of the mapping

$\sigma_1(a + ib) = a + ib$ and $\sigma_2(a + ib) = a - ib$ defines an automorphism of **C**, where σ_1 is the identity automorphism and σ_2 is its complex conjugate.

$\Rightarrow$ The group $G[\mathbf{C}, \mathbf{R}]$ consists of two elements σ_1, σ_2.

$\Rightarrow$ $o[G(\mathbf{C},\mathbf{R})] = 2$

Further, let us find the fixed field $G[\mathbf{C},\mathbf{R}]$. The fixed field of $G[\mathbf{C}, \mathbf{R}]$ necessarily contains $\mathbf{R}$.

Let $a + ib \in$ the fixed field of $G[\mathbf{C}, \mathbf{R}]$. Then, we must have

$$\sigma(a + ib) = a + ib, \forall\, a \in G[\mathbf{C}, \mathbf{R}]$$

$$\Rightarrow \qquad \sigma_2(a + ib) = a + ib$$

$$\Rightarrow \qquad\qquad a - ib = a + ib$$

$$\Rightarrow \qquad\qquad 2ib = 0$$

$$\Rightarrow \qquad\qquad b = 0$$

$$\Rightarrow \qquad\qquad a + ib = a$$

$$\Rightarrow \qquad\qquad a + ib \in \mathbf{R}$$

Therefore, the elements of $\mathbf{R}$ are the only elements of $\mathbf{C}$ which belong to the fixed field of $G[\mathbf{C}, \mathbf{R}]$

$\Rightarrow$ Fixed field of $G[\mathbf{C}, \mathbf{R}]$ is exactly $\mathbf{R}$.

Hence, $\mathbf{C}$ is a normal extension of $\mathbf{R}$.

10.15 GALOIS GROUP OF A SEPARABLE POLYNOMIAL

Let $f(x) \in F[x]$ be a separable polynomial over F and let K be the splitting field of $f(x)$ over F. Then, the group $G(K, F)$ of all F-automorphism whose fixed field is F is known as the Galois group of $f(x)$.

10.16 GALOIS GROUP OF A POLYNOMIAL REPRESENTED AS A GROUP OF PERMUTATION OF ITS ROOTS

Let $f(x) \in F[x]$ be a separable polynomial of degree n and let $K=F(\alpha_1, \alpha_2,...,\alpha_n)$ be the decomposition field of $f(x)$ over F, where $\alpha_1, \alpha_2,...,\alpha_n$ are the roots of $f(x) = 0$. Suppose that $f(x)$ has no multiple root. Each element of K is obtained as the result of a finite number of operations of addition, subtraction, multiplication and division performed on the elements of F and $\alpha_1, \alpha_2,...,\alpha_n$.

Each F-automorphism of K maps a root of $f(x)$ on other root of the $f(x)$ such that distinct roots are mapped separately. Thus each F-automorphisrn determines a permutation of the roots and is itself determined by it. Hence, there exists a group of permutations of the roots of $f(x)$ which is isomorphic to the Galois group of $f(x)$.

REMARK

☞ Let K and E be the extension of a field F with $[K : F]$ finite and assume that both K and E are subfields of some larger field. If K is a Galois extension of F, then KE is a Galois extension of E and
$$G(KE, E)=G(K, K \cap E).$$

10.17 FINITE FIELDS

For every prime number p, there exists a field with p elements, namely the field Z_p of residue classes of integers modulo p. Also, every field consisting of p elements is isomorphic to Z_p. Thus, for each prime number p, there exists one and only one field with p elements. The characteristic of a field having a finite number of elements is always a prime number.

Definition 1. *A field having a finite number of elements is known as finite field.*

Definition 2. *A field is called prime if it has no proper subfield.*

For example: **Q** and **Z**/p where p is the prime, are prime fields.

Theorem 1. *The characteristic of a finite field is necessarily non-zero but converse is not true.*

Proof. Let F be a field and suppose that the characteristic of F is zero, then F will have a subfield with an infinite number of elements since the field of characteristic zero is isomorphic to the field of rational numbers which is in fact an infinite field.

Hence the characteristic of a finite field is non-zero. But the converse is not true, for example, the field $I_p[x]$ of polynomial quotients in an indeterminate x over I_p has an infinite number of elements because $I_p[x]$ has non-zero characteristic p.

Theorem 2. *The number of elements in a finite field of characteristic p is of the form p^n, where n being some positive integer.*

Proof. Let K be a finite field of characteristic p and let F be the prime field in K, then F contains p elements and K is finite extension of F. Therefore, K forms a finite-dimension vector space over F.

Suppose $[K : F] = n$, then K will have a basis of n elements over F.

Let $[\alpha_1, \alpha_2,..., \alpha_n]$ be a basis of K. Then each element of K is uniquely expressed as

$$a_1\alpha_1 + a_2\alpha_2 + ... + a_i\alpha_i + ... + a_n\alpha_n \qquad ...(1)$$

where each $\alpha_i \in F$.

Thus , the number of elements in K is the number of the element of the form (1).

Since each α_i can have p values. Hence, K must clearly have p^n elements

Theorem 3. *Each element of a finite field with F with p^n elements satisfies the equation $x^{p^n} = x$.*

Proof. The zero element of F trivially satisfies the equation $x^{p^n} = x$.

Therefore, consider the set of non-zero elements of F.

The non-zero elements of F form a multiplicative group with $p^n - 1$ elements, then for each element $x \in F$, we have

$$o(x) = p^n - 1, \quad i.e., \quad x^{p^n} - 1 = 1$$

Multiplying this relation by x, we get

$$x^{p^n} = x.$$

Theorem 4. *If a finite field F has p^n elements, then F is the splitting field of the polynomial $f(x) = x^{p^n} - x$.*

Proof. We observe that each element of a finite field F with p^n elements satisfies the equation $x^{p^n} = x$. Since the polynomial $f(x) = x^{p^n} - x$ is of degree p^n so that $f(x) = 0$ will have at most p^n roots, these roots are nothing but the elements of F. Hence, F is the splitting field of the polynomial $f(x) = x^{p^n} - x$.

Theorem 5. *(Existence and Uniqueness Theorem). For every prime integer p and every positive integer n, there exists a finite field with p^n elements and field is unique apart from isomorphism.*

Proof. Let F be a prime field of characteristic p.

Consider a polynomial:

$$f(x) = x^{p^n} - x \in F[x].$$

Clearly $f(x)$ is a polynomial of degree p^n , so it will have at most p^n elements. Let K be its splitting field over F.

Now, $f'(x) = p^n x^{p^n - 1} - 1 = -1$ $[\because F$ is of characteristic $p.]$

Thus, $f(x)$ and $f'(x)$ do not have a non-trivial common factor of positive degree, this implies that $f(x)$ has all its roots simple in its splitting field. Let $F = \{a \in K : a^{p^n} = a\}$. Now, we claim that F is a field with p^n elements. Let a, b be any two elements of F, then

$$a^{p^n} = a \text{ and } b^{p^n} = b.$$

Since F is of characteristic p, so that

$$(a - b)^{p^n} = a^{p^n} - b^{p^n} = a - b \Rightarrow a - b \in F.$$

Again, let $a, b \in F$ with $b \neq 0$, then

$$(b^{p^n})^{-1} = b^{-1} \Rightarrow (b^{-1})^{p^n} = b^{-1} \Rightarrow b^{-1} \in F.$$

Also, $(ab^{-1})^{p^n} = a^{p^n} (b^{-1})^{p^n} = ab^{-1} \Rightarrow ab^{-1} \in F.$

Hence, F is a field with p^n elements because F is the set of all p^n roots of $f(x)$.

$\Rightarrow$ There exists a field with p^n elements.

Now, we shall prove the uniqueness.

Since a field F with p^n elements is a splitting field of the polynomial $f(x) = x^{p^n} - x$ over the field of characteristic p and F consists, the p^n roots of $f(x) = 0$.

Now, any two splitting fields of a polynomial are isomorphic and any two prime fields of characteristic p are isomorphic.

Theorem 6. *Let F be a finite field with q elements and suppose $F \subset K$, where K is also a finite field, then K has q^n elements where $n = [K : F]$.*

Proof.   It is known that every field can be regarded as a vector space over a subfield, so K can be regarded as a vector space over the field F. Also, number of elements in K is finite, thus the vector space $K(F)$ is finite dimensional. Let the dimension of the vector space $K(F)$ be n, i.e., $[K : F] = n$.

Let $[b_1, b_2,..., b_n]$ be a basis of vector space $K(F)$. Then every element of K can be uniquely expressed in the form of $a_1 b_1 + a_2 b_2 + ... + a_n b_n$, $a_i's \in F$.

Then, the number of elements in K is the number of $a_1 b_1 + a_2 b_2 + ... + a_n b_n$ as the a_1, $a_2 ... a_n$ range over F. Since F has q elements, so each of the n coefficients $a's$ can have q values. Hence, K must have qn elements.

Theorem 7. *Any two finite fields having the same number of elements are isomorphic.*

Proof.   Let F be a finite field of characteristic p. Further, let the number of elements in F be equal to $-q = p^n$. Firstly, we shall prove that F can be regarded as the splitting field of the polynomial $x^q - x \in Z_p (x)$, where Z_p is the field of integers modulo p. Now, F is a field of finite characteristic p, so it contains a subfield F_0 isomorphic to the field Z_p. Thus, we may assume that F can be regarded as an extension of the field Z_p.

Now, since the field F contains q elements. Then we have already proved that

$$a^q = a, \forall a \in F.$$

$\Rightarrow$ Every a in F satisfies the polynomial $x^q - x \in Z_p(x)$.

The degree of the polynomial $x^q - x$ is q so it can have at most q roots in any extension field of $\mathbf{Z}_p$. But all the q elements of F are the roots of $x^q - x$. Thus,

we have
$$x^q - x = \prod_{\lambda \in K} (x - \lambda)$$

$\Rightarrow$ The polynomial $x^q - x \in I_p[x]$ splits in the field F. But this polynomial cannot split in any smaller field for that field would have to have all the roots of this polynomial and so would have to have at least q elements. Also, roots of this polynomial are distinct. Therefore, F is the splitting field of $x^q - x \in \mathbf{Z}_p[x]$.

Further, suppose that F^* is any other finite field having $q = p^n$ elements. Then proceeding same as above, F^* is also the splitting field of $x^q - x \in \mathbf{Z}_p[x]$. It is known that any two splitting fields of $x^q - x \in \mathbf{Z}_p[x]$ must be isomorphic. Hence, $F^* \cong F$.

10.18 GALOIS FIELD

Definition. *For any prime integer p and for any integer n, a field with p^n elements is known as Galois field. It is denoted by $GF(p^n)$.*

Theorem 1. *The multiplicative group of a Galois field is cyclic.*

Proof. Let F be a finite field with $q = p^n$ elements and let F^* denote the set of $q - 1$ non-zero elements so that F forms a multiplicative group of finite order $q - 1$, i.e., $o(F^*) = q - 1$.

Now, the elements of F^* are the roots of the polynomial
$$f(x) = x^{p^n - 1} - 1 \text{ or } \quad f(x) = x^{q-1} - 1$$

We shall show that F^* is cyclic. For this, we shall show that there exists an element of F^* whose order is $q - 1$.

Write
$$q - 1 = p_1^{m_1} p_2^{m_2} \cdots p_r^{m_r}$$
where $p_i \neq p$ and all p_i are distinct primes.

Let $m = \text{l.c.m.} \{p_1^{m_1}, p_2^{m_2}, ..., p_r^{m_r}\}$. Since F^* is finite so that the order of each element of F^* is finite.

Let α be an element of F^* with $o(\alpha) = m$.

Also, $\qquad \alpha \in F^*$.

$\Rightarrow$ α satisfies a polynomial $f(x) = x^{q-1} - 1$.

$\Rightarrow \qquad \alpha^{q-1} - 1 = 0 \Rightarrow \quad \alpha^{q-1} = 1 \Rightarrow q - 1 \text{ divides } m.$ $\qquad [\because o(\alpha) = m]$

$\Rightarrow \qquad q - 1 = m$ $\qquad\qquad\qquad\qquad\qquad\qquad [\because m \leq q - 1]$

Hence, F^* is cyclic.

Theorem 2. *Every finite fields of characteristic p has an automorphism $a \Leftrightarrow a^p$.*

Proof. Let F be a finite field of characteristic p. Define a mapping
$$\phi : F \to F \text{ such that } \phi(a) = a^p, \forall\, a \in F.$$

(i) ϕ is one-one. Let $a, b \in F$. Then consider
$$\phi(a) = \phi(b)$$
$$\Rightarrow \qquad\qquad a^p = b^p$$

$\Rightarrow$ $\qquad\qquad a^p - b^p = 0$

$\Rightarrow$ $\qquad\qquad (a–b)^p = 0$

$\qquad\qquad$ [$\because$ In a field of characteristic p, we have $(a–b)^p = a^p–b^p$]

$\Rightarrow$ $\qquad\qquad a–b = 0$

$\Rightarrow$ $\qquad\qquad a = b$

Therefore, ϕ is one-one.

(ii) ϕ **is onto.** Since the set F is finite and we have already proved that ϕ's one-one, so ϕ is always onto.

(iii) ϕ **preserves the compositions.**

Let a, b $\in$ F, then

$\qquad\qquad \phi(a + b) = (a + b)^p$ $\qquad\qquad$ [By definition of ϕ]

$\qquad\qquad\qquad = a^p + b^p$

$\qquad\qquad$ [$\because$ In a field of characteristic p, we have $(a + b)^p = a^p+b^p$]

$\qquad\qquad\qquad = \phi(a) + \phi(b)$

Further, $\qquad \phi(ab) = (ab)^p = a^p\, b^p\ = \phi(a)\, \phi(b)$

Hence, ϕ is an automorphism of the field F.

REMARK

☞ In a finite field of characteristic p, every element has a pth root.

Theorem 3. *Let F be a finite field and $\alpha \neq 0$, $\beta \neq 0$ are two elements of F, then we can find elements a and b in F such that $1 + \alpha a^2 + \beta b^2 = 0$.*

Solution. Here, we shall discuss the following two cases :

Case I. Let the characteristic of F be 2. Then clearly F has 2^m elements and it is known that every element x in F satisfies $x^{2m} = x$.

$\Rightarrow$ Every element in F is a square of some element in F.

In particular

$\qquad\qquad \alpha^{-1} = c^2$ for some $c \in F$.

Taking $a = c$ and $b = 0$. Then, we have

$\qquad 1 + \alpha a^2+\beta b^2 = 1 + \alpha a^2+\beta.0 = 1 + \alpha\alpha^{-1} = 1 + 1 = 0$

$\qquad\qquad$ [Because if the characteristic of F is 2, then $1 + 1 = 0$]

Case II. Let us suppose that the characteristic of F is a prime number p other than 2. Then, F has p^m elements.

Define $\qquad\qquad W_\alpha = \{1 + \alpha x^2 : x \in F\}$.

Let us find the number of distinct elements in W_α.

We have $\qquad 1 + \alpha x^2 = 1 + \alpha y^2$

$\Rightarrow$ $\qquad\qquad \alpha x^2\ = \alpha y^2$

$\Rightarrow$ $\qquad\qquad x^2 = y^2$ $\qquad\qquad\qquad\qquad$ [$\because$ $\alpha \neq 0$]

$\Rightarrow$ $\qquad\qquad x = \pm y$.

So, for $x \neq 0$, we get from each pair x and $-x$, one element in U_α and for $x = 0$, we get $1 \in W_\alpha$. Thus, W_α has

$$1 + \frac{p^m - 1}{2} = \frac{p^m + 1}{2} \text{ distinct elements.}$$

In a similar way, we can show that

$$W_\beta = [-\beta x^2 : x \in F]$$

has $\dfrac{p^m + 1}{2}$ distinct elements.

Therefore, W_α and W_β are subsets of F and each of them has more than half the elements of F. Thus the intersection of W_α and W_β cannot be empty.

Let $\qquad\qquad t \in W_\alpha \cap W_\beta$

$\Rightarrow \qquad\qquad t \in W_\alpha$ and $t \in W_\beta$

Now, $\qquad\qquad t \in W_\alpha \Rightarrow t = 1 + \alpha a^2$, for some $a \in F$.

Also, since $t \in W_\beta$, therefore $t = -\beta b^2$ for some $b \in F$.

Therefore, there exists $a, b \in F$, such that $1 + \alpha a^2 = -\beta b^2$.

Hence, $\qquad\qquad 1 + \alpha a^2 + \beta b^2 = 0$.

10.19 CONSTRUCTION OF GALOIS FIELD AND ITS SUBFIELDS

Theorem 1. *For each positive integer n, there exists an irreducible polynomial f(x) of degree n over* $\mathbf{Z}_p$.

Proof. Using Theorem 1 of previous section, we can say that there exists $\alpha \in GF(p^n)$ such that $\qquad\qquad GF(p^n) = \mathbf{Z}_p(\alpha)$.

Now, let $f(x)$ be the minimal polynomial for α over $\mathbf{Z}_p$, then we have

$$n = [\mathbf{Z}_p(\alpha) : \mathbf{Z}_p] = \deg [f(x)]$$

Hence, the irreducible polynomial $f(x)$ is of degree n over $\mathbf{Z}_p$.

Theorem 2. *Let K be a subfield of the Galois field GF(p^n), then there exists an integer m such that K contains p^m elements and m is a divisor of n.*

Proof. It is known that the characteristic of $GF(p^n)$ is p. Therefore, K is a field of characteristic p.

Then, $o(K) = p^m$, for some positive integer m. Further, since $K \subseteq GF(p^n)$,

then $[GF(p^n) : K]$ is finite. If $[GF(p^n): K] = r$, then $[\alpha_1, \alpha_2,..., \alpha_r]$ forms a basis of $GF(p^n)$ over K. Thus, each element of $GF(p^n)$ can be expressed as a linear combination of $\alpha_1, \alpha_2,..., \alpha_r$.

But each α_i assumes the m values, because $o(K) = p^m$.

$\Rightarrow$ The number of elements contained in $GF(p^n)$ is mr.

But $\qquad\qquad\qquad mr = n \Rightarrow m \mid n$

Hence, m is a divisor of n.

Solved Examples

Example 1. *Let Q be the field of rational numbers and let $K = \mathbf{Q}(2^{1/3})$, where $2^{1/3}$ is the real cube root of 2. Show that the only automorphism of K is the identity automorphism. Is K is a normal extension of F ?*

Solution. By definition of $\mathbf{Q}(2^{1/3})$, we have

$$K = \mathbf{Q}(2^{1/3}) = \{a+b(2^{1/3}) + c(2^{2/3}) : a, b, c \in \mathbf{Q}\}$$

Clearly, K is an extension of $\mathbf{Q}$ such that

$$\sigma(x) = x \ \forall \, x \in \mathbf{Q}$$

where $\sigma \in A(K)$, i.e., $\sigma : K \to K$.

Further, since $2^{1/3} \in K$, then $\sigma(2^{1/3}) \in K$.

Now, $[\sigma(2^{1/3})]^3 = \sigma(2^{1/3})\sigma(2^{1/3})\sigma(2^{1/3}) = \sigma(2^{1/3}.\,2^{1/3}.\,2^{1/3}) = \sigma(2)$

$$= 2 \qquad\qquad [\because \alpha(x) = x \ \forall \, x \in \mathbf{Q}]$$

$\therefore$ $\sigma(2^{1/3})$ is a cube root of 2 lying in K.

Since there is only one real cube roots of 2, i.e., $2^{1/3}$ and K is a subfield of real numbers

$\therefore$ $\sigma(2^{1/3}) = 2^{1/3}$.

Let x be any element of K, so that

$$x = a + b.(2^{1/3}) + c.(2^{2/3}), a, b, c \in \mathbf{Q}.$$

Then, $\sigma(x) = \sigma(a + b.2^{1/3} + c.2^{2/3})$

$$= \sigma(a) + \sigma(b.2^{1/3}) + \sigma(c.2^{2/3})$$

$$= \sigma(a) + \sigma(b)\,\sigma(2^{1/3}) + \sigma(c)\sigma(2^{2/3})$$

$$= a + b.2^{1/3} + c.\sigma(2^{1/3}.2^{1/3}) \qquad [\because \sigma(\alpha) = a \ \forall \, a \in \mathbf{Q}]$$

$$= a + b.2^{1/3} + c.\sigma(2^{1/3})\sigma(2^{1/3})$$

$$= a + b.2^{1/3} + c.2^{1/3}.2^{1/3} = a + b.2^{1/3} + c.2^{2/3} = x.$$

$\therefore$ $\sigma(x) = x, \ \forall \, x \in K$.

Thus, σ is an identity automorphism of K

i.e., $\sigma = I$

where $I : K \to K$ is an identity automorphism.

$\therefore$ $G(K, F) = \{I\}$.

Now, by definition of fixed field, we have

Fixed field of $G(K, F) = \{x \in K : \sigma(x) = x, \ \forall \, \sigma \in G(K, F)\}$

$$= \{x \in K : \mathbf{I}(x) = x \text{ as } G(K, F) = \{I\}\} = K \neq F.$$

Hence, K is not a normal extension of F.

Example 2. *Let $K = \mathbf{Q}(2^{1/3}, \omega)$, where $\omega^3 = 1, \omega \neq 1$. Let σ_1 be the identity automorphism of K, and let σ_2 be an automorphism of K such that $\sigma_2(\omega) = \omega^2$ and $\sigma^2(2^{1/3}) = 2^{1/3}\omega$. If $G(K, \mathbf{Q}) = \{\sigma_1, \sigma_2\}$, then show that fixed field of $G(K, \mathbf{Q}) = \mathbf{Q}(2^{1/3}\omega^2)$.*

Solution. Since $(1, \omega)$ is the basis of $\mathbf{Q}(2^{1/3}, \omega)$ over $\mathbf{Q}(2^{1/3})$ and $\{1, 2^{1/3}, 2^{2/3}\}$ is the basis of $\mathbf{Q}(2^{1/3}, \omega) = K$ over $\mathbf{Q}$.

Then $\{1, 2^{1/3}, 2^{2/3}, \omega, 2^{1/3}\omega, 2^{2/3}\omega\}$ will be a basis of $\mathbf{Q}(2^{1/3}, \omega) = K$ over $\mathbf{Q}$.

Now, by definition,

$$K = \mathbf{Q}(2^{1/3}, \omega) = \{x_1 + x_2(2^{1/3}) + x_3(2^{2/3}) + x_4(\omega) + x_5(2^{1/3}\omega) + x_6(2^{1/3}\omega) :$$
$$x_1, x_2, x_3, x_4, x_5, x_6 \in \mathbf{Q}\}.$$

Let α be an arbitrary element of K, then

$$\alpha = x_1 + x_2(2^{1/3}) + x_3(2^{2/3}) + x_4(\omega) + x_5(2^{1/3}\omega) + x_6(2^{2/3}\omega)$$

and $\sigma_2(\alpha) = \sigma_2(x_1) + \sigma_2(x_2)\sigma_2(2^{1/3}) + \sigma_2(x_3)\sigma_2(2^{2/3}) + \sigma_2(x_4)\sigma_2(\omega)$

$$+ \sigma_2(x_5)\sigma_2(2^{1/3})\sigma_2(\omega) + \sigma_2(x_6)\sigma_2(2^{1/3})\sigma_2(\omega)$$

$$= x_1 + x_2(2^{1/3}\omega) + x_3(2^{2/3}\omega^2) + x_4\omega^2 + x_5(2^{1/3}\omega)(\omega^2) + x_6(2^{2/3}\omega^2)(\omega)$$

$$= x_1 + x_2(2^{1/3}\omega) + x_3(-1-\omega)2^{2/3} + x_4(-1-\omega) + x_5(2^{1/3}) + x_6(2^{2/3})$$

$$= (x_1-x_4) + x_5(2^{1/3}) - x_3(2^{2/3}) - x_4(\omega) + x_2(2^{1/3}\omega) - (x_3-x_6)(2^{2/3}\omega)$$

$$[\because 1 + \omega + \omega^2 = 0 \text{ and } \omega^3 = 1]$$

For the fixed field, $\sigma_2(\alpha) = \alpha$, then comparing both sides, we get

$$x_1 = x_1 - x_4, \ x_2 = x_5, \ x_3 = -x_3, x_4 = -x_4, x_5 = x_2 \text{ and } x_6 = -x_3 + x_6.$$

Solving above equations, we obtain $x_3 = 0, x_4 = 0, x_2 = x_5$ and x_1, x_6 are arbitrary.

$$\therefore \qquad \alpha = x_1 + x_2 2^{1/3}(1 + \omega) + x_6(2^{2/3}\omega) = x_1 - x_2(2^{1/3}\omega^2) + x_6(2^{1/3}\omega^2)^2$$

$$\Rightarrow \qquad \alpha \in \mathbf{Q}(2^{1/3}\omega^2).$$

Hence, the fixed field of $G(K, \mathbf{Q})$ is $\mathbf{Q}(2^{1/3}\omega^2)$.

Example 3. *The group $G[\mathbf{Q}(\alpha), \mathbf{Q}]$, where $\alpha^5 = 1$ and $\alpha \neq 1$, is isomorphic to cyclic group of order 4.*

Solution. Since we have

$$\alpha^5 = 1 \Rightarrow \alpha^5 - 1 = 0$$

$$\Rightarrow \quad (\alpha-1)(1 + \alpha + \alpha^2 + \alpha^3 + \alpha^4) = 0$$

$$\Rightarrow \qquad 1 + \alpha + \alpha^2 + \alpha^3 + \alpha^4 = 1$$

Thus α is a root of a polynomial.

$$f(x) = 1 + x + x^2 + x^3 + x^4 \in \mathbf{Q}[x].$$

Since $f(x)$ is irreducible over $\mathbf{Q}$, then $[\mathbf{Q}(\alpha) : \mathbf{Q}] = 4$.

Also, all the roots of the polynomial $g(x) = x^5 - 1 \in \mathbf{Q}[x]$ are $1, \alpha, \alpha^2, \alpha^3, \alpha^4$. So, $\mathbf{Q}(\alpha)$ is the splitting field of $g(x) = x^4 - 1$ over $\mathbf{Q}$, hence $\mathbf{Q}(\alpha)$ is normal extension of $\mathbf{Q}$, therefore

$$o[G(\mathbf{Q}(\alpha), \mathbf{Q})] = [\mathbf{Q}(\alpha) : \mathbf{Q}] = 4.$$

This shows that, there are four $\mathbf{Q}$-automorphism of $\mathbf{Q}(\alpha)$.

Since $\{1, \alpha, \alpha^2, \alpha^3\}$ is the basis of $\mathbf{Q}(\alpha)$ over $\mathbf{Q}$, then

$$\mathbf{Q}(\alpha) = \{a + b\alpha + c\alpha^2 + d\alpha^3 : a, b, c, d \in \mathbf{Q}\}$$

Let $G(\mathbf{Q}(\alpha), \mathbf{Q}) = \{\sigma_1, \sigma_2, \sigma_3, \sigma_4\}$ with σ_1 an identity automorphisms.

The four $\mathbf{Q}$-automorphisms of $\mathbf{Q}(\alpha)$ are as follows :

$$\sigma_1(a + b\alpha + c\alpha^2 + d\alpha^3) = a + b\alpha + c\alpha^2 + d\alpha^3$$

$$\sigma_2(a + b\alpha + c\alpha^2 + d\alpha^3) = a + b\alpha^2 + c\alpha^4 + d\alpha$$

$$\sigma_4(a + b\alpha + c\alpha^2 + d\alpha^3) = a + b\alpha^4 + c\alpha^3 + d\alpha^2.$$

Clearly, $\{\sigma_1, \sigma_2, \sigma_3, \sigma_4\}$ forms a cyclic group of order 4 generated by σ_2 and σ_3.

Example 4. *Let n be a positive integer and let F be a field containing all the roots of unity. Let K be the splitting field of $x^n - a \in F[x]$ over F. Then $K = F(\alpha)$, where $\alpha^n = a$, and the Galois group $G(K, F)$ is abelian.*

Solution. Since F is a field of all n^{th} roots of unity, then

$$F = \{1, \omega, \omega^2, ..., \omega^{n-1}\}, \text{ where } \omega = \cos\frac{2\pi}{n} + i\sin\frac{2\pi}{n}$$

Let $f(x) = x^n - a$, then the roots of $f(x) = 0$ are $\alpha, \alpha\omega, \alpha\omega^2, ..., \alpha\omega^{n-1}$. Thus the splitting K field of $f(x)$ over F is $F(\alpha)$, i.e., $K = F(\alpha)$.

Now, we shall show that $G(K, F)$ is abelian.

Let σ_1 and σ_2 be any two elements of $G(K, F)$.

Since $f(\alpha) = 0$, then $f((\sigma_1(\alpha)) = 0$ and $f(\sigma_2(\alpha)) = 0$, i.e., $\sigma_1(\alpha)$ and $\sigma_2(\alpha)$ are also roots of $x^n - a = 0$. Then, they can be written as

$$\sigma_1(\alpha) = \alpha\omega^i \text{ and } \sigma_2(\alpha) = \alpha\omega^j, 0 \le i \le n-1, 0 \le j \le n-1.$$

Then, we have

$$(\sigma_1\sigma_2)(\alpha) = \sigma_1(\sigma_2(\alpha))$$
$$= \sigma_1(\alpha\omega^j) = \sigma_1(\alpha)\omega^j = \alpha\omega^i\omega^j = \alpha\omega^{i+j}.$$

Also, $(\sigma_2\sigma_1)(\alpha) = \sigma_2(\sigma_1(\alpha)) = \sigma_2(\alpha\omega^i)$
$$= \sigma_2(\alpha)\omega^i = \alpha\omega^j\omega^i = \alpha\omega^{i+j}$$

$\therefore$ $\sigma_1\sigma_2 = \sigma_2\sigma_1$

Hence, $G(K, F)$ is abelian.

Example 5. *Let F be a field of characteristic $\ne 2$ or 3. Let $f(x) = x^3 + bx + c$ be separable polynomial over F. If $f(x)$ is irreducible over F, then the Galois group of $f(x)$ is of order 3 or 6. Also, the Galois group of $f(x)$ is S_3 (symmetric group of symbols 3) if and only if $\Delta = -4b^3 - 27c^2$ is not a square of F.*

Solution. If $f(x)$ has a root α in F, then

$$f(x) = (x-\alpha)g(x), \; g(x) \in F[x].$$

Now, we have two cases :

Case I. If $g(x)$ has a root in F, then $f(x)$ splits into linear factors in F, so that F itself will be the splitting field of $f(x)$ over F.

So, $[F : F] = 1 = o(G(F, F))$.

Thus, the Galois group of $f(x)$ is of order 1.

Case II. If $g(x)$ is irreducible over F, then the splitting field of $f(x)$ will not be F. If E is the splitting field of $f(x)$ over F, then

$$[E : F] = 2 = o(G(E, F)).$$

Thus, if $f(x)$ is irreducible over F, then

$$o(G(E, F)) \ne 1 \text{ or } 2.$$

Since $f(x)$ is separable, i.e., all its roots are simple and distinct and if E is the splitting of $f(x)$ over F, then $G(E, F) \subset S_3$ [Because if $f(x) \in F[x]$ has r roots in its splitting field E over F, then the Galois group $G(E, F)$ of $f(x)$ is a subgroup of the symmetric group S_r].

Hence, $o(G(E, F)) = 3$ or 6, because $o(S_3) = 6$.

Next, we find the necessary and sufficient conditions on the coefficients of $f(x)$ such that $G(E, F) \cong S_3$.

Let α, β and γ be the roots $f(x) = x^3 + bx + c$ and let $\delta = (\alpha - \beta)(\beta - \gamma)(\gamma - \alpha)$ and $\Delta = \delta^2$, then for all $\sigma \in G(E, F)$, $\sigma(\delta) = \pm \delta$.

$\therefore \qquad\qquad \sigma(\Delta) = \sigma(\delta^2) = \sigma(\delta)\, \sigma(\delta) = (\pm \delta)(\pm \delta) = \delta^2 = \Delta.$

Thus, Δ is in the fixed field of $G(E, F)$, then $\Delta \in F$.

Now, if $\delta \in F$, then $\sigma(\delta) = \delta \ \forall\ \sigma \in G(E, F)$, therefore, σ cannot be an odd permutation, so that $\sigma \in A_3$, where A_3 is the subgroup of even permutation of S_3 (the group of all permutation).

Conversely, if $\sigma \in A_3$, then $\sigma(\delta) = \delta$.

Thus, $A_3 = G(E, F)$ if and only if $\delta \in F$ and $G(E, F) = S_3$ if and only if $\delta \neq F$, that is the polynomial $x^2 - \Delta \in F[x]$ is irreducible over F because $\Delta = \delta^2 \in F$.

Also, by some computation, we get,

$$\Delta = -4b^3 - 27c^2.$$

Hence, the Galois group of $f(x)$ is S_3 iff $\Delta = -4b^3 - 27c^2$ is not a square in F.

Example 6. *Find the Galois group of the $x^3 - 2 \in \mathbf{Q}[x]$ over $\mathbf{Q}$, the field of rational numbers.*

Solution. Let K be the splitting field of $f(x) = x^3 - 2$ over the field $\mathbf{Q}$ of rational numbers. The roots of $f(x)$ are not all reals, therefore K may be considered as subfield of the field of complex number.

Let $\alpha = 2^{1/3}$. Now, to obtain K, we first adjoin α to $\mathbf{Q}$. In $\mathbf{Q}(\alpha)[x]$, we have

$$f(x) = (x - \alpha)(x^2 + \alpha x + \alpha^2) = (x - \alpha)\, g(x).$$

The field $\mathbf{Q}(\alpha)$ is a real field and $g(x)$ has no real root, so that $g(x)$ is irreducible over $\mathbf{Q}(\alpha)[x]$. The roots of $g(x)$ are $\alpha\omega$ and $\alpha\omega^2$, where $\omega = \dfrac{-1 + \sqrt{3}\, i}{2}$ and $\omega^2 = \dfrac{-1 - \sqrt{3}\, i}{2}$.

Hence, $\qquad K = \mathbf{Q}(\alpha, \sqrt{3}\, i)$.

We have $[K : \mathbf{Q}] = 6$ and therefore, the order of $G(K, \mathbf{Q}) = 6$. Thus, if we can find a distinct $\mathbf{Q}$-automorphisms of K, they will constitute all of $G(K, \mathbf{Q})$, one element of this group is its identity I. Let σ be the $\mathbf{Q}$-automorphism of K which leaves α fixed and maps $\sqrt{3}\, i$ in $-3i$. Let τ be the $\mathbf{Q}$-automorphism of K such that $\tau(\alpha) = \alpha\omega$, $\tau(\sqrt{3}\, i) = -\sqrt{3}\, i$. Thus, we have the following table which gives the images of α and $\sqrt{3}\, i$ under indicative $\mathbf{Q}$-automorphisms of K.

	I	σ	τ	$\sigma\tau$	$\tau\sigma$	$\sigma\tau\sigma$
α	α	α	$\alpha\omega$	$\alpha\omega^2$	$\alpha\omega$	$\alpha\omega^2$
$\sqrt{3}\, i$	$\sqrt{3}\, i$	$-\sqrt{3}\, i$	$-\sqrt{3}\, i$	$\sqrt{3}\, i$	$\sqrt{3}\, i$	$-\sqrt{3}\, i$

Hence, the Galois group of $x^3 - 2 \in \mathbf{Q}[x]$ is $\{I, \sigma, \tau, \tau\sigma, \tau\sigma, \sigma\tau\sigma\}$.

Example 7. *Show that every element in a finite field can be written as the sum of two squares.*

Solution. Let F be a finite field such that $O(F) = p^n$.

Now, we have the following two cases :

Case I. If $p = 2$, define $\theta : F \to F$ such that $\theta(b) = b^2$.

Clearly θ is one-one.

Also, since F is finite and θ is one-one. Therefore θ is onto.

Further, $\theta(b_1 + b_2) = (b_1 + b_2)^2 = b_1^2 + b_2^2 = \theta(b_1) + \theta(b_2)$

and $\theta(b_1 . b_2) = (b_1 b_2)^2 = b_1^2 b_2^2 = \theta(b_1)\theta(b_2)$

Therefore, θ is a homomorphism.

Now, let $a \in F$, then there is $b \in F$ such that $\theta(b) = a$ which implies

$$a = b^2 = b^2 + 0^2 = \text{sum of two squares in } F.$$

CaseII. If $p \neq 2$. Let $a \in F$ and define $X = \{a - x^2 : x \in F\}$.

Then $a - x_1^2 = a - x_2^2 , x_1, x_2 \in F$

$\Rightarrow$ $x_1^2 = x_2^2 \Rightarrow x_1 = -x_2 \text{ if } x_1 \neq x_2$

$\Rightarrow$ $o(X) = \dfrac{p^n - 1}{2} + 1 = \dfrac{p^n + 1}{2}$

Let $Y = [y^2 : y \in F]$. Then proceed same as above, we get

$$o(Y) = \frac{p^n + 1}{2}$$

Since $X, Y \subseteq F$ and $O(F) = p^n, X \cap Y \neq \phi$

Therefore, $a - x^2 = y^2$, for some $x, y \in F$.

$\Rightarrow$ $a = x^2 + y^2 = \text{sum of two squares in } F.$

Example 8. *Prove that for any integer a and prime p, $a^p \equiv a \pmod p$.*

Solution.   Let us write $a = pq + r, 0 \leq r \leq p$. Then, $a = r \pmod p$.

For $0 \leq r \leq p \Rightarrow r \in F_p$

$\Rightarrow$ $rr........r = r$

                 p times

$\Rightarrow$ $r^p - pu = r$

$\Rightarrow$ $r^p \equiv r \pmod p$

$\Rightarrow$ $r^p \equiv a \pmod p$

Therefore, $a \equiv r \pmod p$ $\Rightarrow a^p \equiv r^p \pmod p$

Hence, $a^p \equiv a \pmod p$.

EXERCISE 10.4

1. For each of the following polynomials $f(x)$, find the Galois group G of the splitting field K of $f(x)$ over $\mathbf{Q}$ and also find a subgroup of S_n isomorphic to G, by numbering the roots, whore n is the degree of the polynomial.

  (i) $x^3 - 2$ (ii) $(x^2 - 2)(x^2 - 3)$

  (iii) $x^3 - 1$ (iv) $x^3 + 2$

2. Obtain the Galois group of the following :

  (i) $x^3 - 2x - 1$ over $\mathbf{Q}$ and over $\mathbf{Q}(\sqrt{5})$

  (ii) $x^3 - 2x + 2$ over $\mathbf{Q}$ and over $\mathbf{Q}(i\sqrt{19})$.

3. Find the Galois groups $G[K, \mathbf{Q}]$ of the following K of $\mathbf{Q}$

  (i) $K = \mathbf{Q}(\alpha)$ where $\alpha = \cos\dfrac{2\pi}{3} + i\sin\dfrac{2\pi}{3}$

  (ii) K is the splitting field of $x^4 - 3x^2 + 4 \in \mathbf{Q}(x)$.

4. For each of the following subgroups S_r of the group $G(K, \mathbf{Q})$, find K_{S_r} (the fixed field of $G(K : S_r)$ where $K = \mathbf{Q}(2^{1/3}, \omega]$ be an extension of a field $\mathbf{Q}, \omega^3 = 1, \omega \neq 1.$

  (i) $S_1 = [I, \sigma_2]$,
      where $\sigma_2(2^{1/3}) = 2^{1/3}\omega^2, \sigma_2(\omega) = \omega^2.$

  (ii) $S_2 = [I, \sigma_3]$,
      where $\sigma_3(2^{1/3}) = 2^{1/3}\omega, \sigma_4(\omega) = \omega^2.$

  (iii) $S_3 = [I, \sigma_4]$,
      where $\sigma_4(2^{1/3}) = 2^{1/3}, \sigma_4(\omega) = \omega^2.$

(iv) $S_4 = [I, \sigma_5, \sigma_6]$,

where $\sigma_5(2^{1/3}) = 2^{1/3}\omega, \sigma_5(\omega) = \omega$

and $\sigma_6(2^{1/3}) = 2^{1/3}\omega^2, \sigma_6(\omega) = \omega$.

5. Find the splitting field $x^4 - 1$ over the field **Q** of rational numbers.

6. Let F be the field of rational numbers. Determine the degree of the splitting field of the polynomial $x^7 - 1$ over F.

7. Show that the polynomials $x^2 + 3$ and $x^2 - x + 1$ have the same splitting field over **Q**, the field of rational numbers.

8. If a field F has q elements, then F is a splitting field of $x^q - x$ over its prime subfield.

9. If $f(x) \in F[x]$ has n distinct roots in its splitting field K over F, then show that Galois groups

$G(K, F)$ of $f(x)$ is a subgroup of the symmetric group S_n.

10. Let F be the field of characteristic not equal to 2. Let $x^2 - a \in F[x]$ be an irreducible polynomial over F. Then show that it is a Galois group of order 2.

11. Show that every finite extension of a finite field is Galois.

12. Let F be a finite field. Then show that there exists an irreducible polynomial of any given degree n over K.

13. Show that $x^m - 1$ divides $x^n - 1$ over a field F if and only if m divides n.

14. Show that $x^{p^m} - x$ divides $x^{p^m} - x$, if m divides n.

Answers

1. (i) $G = G(\mathbf{Q}(\sqrt{2}), \mathbf{Q}) = [I, \sigma]$, where $\sigma(\sqrt{2}) = -\sqrt{2}$ and $G = S_2$

(ii) $G = G(\mathbf{Q}(\sqrt{2}, \sqrt{3}), \mathbf{Q}) = [I = \sigma_1, \sigma_2, \sigma_3, \sigma_4]$

where $\sigma_2(\sqrt{2}) = -\sqrt{2}, \sigma_2(\sqrt{3}) = \sqrt{3}, \sigma_3(\sqrt{2}) = \sqrt{2}, \sigma_3(\sqrt{3}) = -\sqrt{3}, \sigma_4(\sqrt{2}) = -\sqrt{2}$

$\sigma_4(\sqrt{3}) = -\sqrt{3}$ and $G \cong S_4$

(iii) $G = G(\mathbf{Q}(\sqrt{3}), \mathbf{Q}) = [I, \sigma]$ where $\sigma(i\sqrt{3}) = -i\sqrt{3}, \sigma_3$ and $G \cong S_2$

(iv) $G = G(\mathbf{Q}(2^{1/3}, \sqrt{3}\,i), \mathbf{Q}) = [I, \sigma, \tau, \sigma\tau, \tau\sigma, \sigma\tau\sigma]$

where $\sigma(2^{1/3}) = 2^{1/3}, \sigma(\sqrt{3}\,i) = -\sqrt{3}\,i, \tau(2^{1/3}) = 2^{1/3}\omega$, and $\tau(\sqrt{3}\,i) = -\sqrt{3}\,i, G \cong S_3$

2. (i) The splitting field of $x^3 - 2x - 1$ over **Q** is $\mathbf{Q}(\sqrt{5})$. Its Galois group over **Q** is isomorphic to $z/2z$ and its Galois group over $\mathbf{Q}(\sqrt{5})$ is trivial.

(ii) $x^3 - 2x + 2$ is irreducible over **Q**. Also, it is irreducible over $\mathbf{Q}(i\sqrt{5})$. Since $D = -76$, which is not a square in **Q** so $G \cong S_3$ in **Q**, but $D = -76 = (2i\sqrt{19})^2$, which is a square in $\mathbf{Q}(i\sqrt{19})$. Thus $G \cong z/3z$ in it.

3. (i) $G(K, Q) = [I, \sigma]$, where $\sigma(\alpha) = \alpha^2$.

(ii) $G[\mathbf{Q}(7, i), \mathbf{Q}] = [I, \sigma_1, \sigma_2, \sigma_3]$, where $\sigma_1(\sqrt{7}) = -\sqrt{7}, \sigma_1(i) = -i$

$\sigma_2(\sqrt{7}) = -\sqrt{7}, \sigma_2(i) = -i, \sigma_3(\sqrt{7}) = \sqrt{7}, \sigma_3(i) = -i$

5. $\mathbf{Q}(i)$ 6. 6.

10.20 THE GALOIS GROUP OF CYCLOTOMIC EXTENSIONS

Definition 1. (*Primitive n^{th} Roots of Unity*). *Let F be a field. The roots $x^n - 1 = 0$ over F are known as the primitive n^{th} roots of unity in its splitting field if $\omega^n = 1$, but $\omega^m \neq 1$ for any positive integer $m < n$.*

REMARK

☞ There are exactly $\phi(n)$ primitive n^{th} roots of unity for every integer n.

Definition 2. *(Cyclotomic Extension). Consider the equation $x^n - 1 = 0$ over the field of rational numbers. Then, the complex number satisfying $x^n - 1 = 0$ divide the circumference of a circle with centre at origin and of unit radius into n equal parts. The equation $x^n - 1 = 0$ is known as* **cyclotomic** *extension over F.*

Definition 3. *(n^{th} Cyclotomic Polynomial). The polynomial*

$$\phi_n(x) = \prod_{i=1}^{\psi(n)} (x - \alpha_i)$$

where α_i are the primitive n^{th} roots of unity in $\overline{F}$, is known as n^{th} cyclotomic polynomial over F.

REMARKS

☞ Since an automorphism of the Galois group $G(K : F)$ must permit the primitive n^{th} roots of unity, $\phi_n(x)$ is left fixed under every element of $G[K : F]$ regarded as extended in the natural way to $K[x]$.

☞ Each nth roots of unity is a primitive d^{th} root of unity for one and only one divisor d of n. Thus we have

$$x^n - 1 = \prod_{d|n} \phi_d(x)$$

Definition 4. *(n^{th} Cyclotomic Field). The splitting field $x^n - 1 \in Q[x]$ which is contained in the field of complex numbers is known as n^{th} cyclotomic field.*

SOME CYCLOTOMIC POLYNOMIALS

(i)   For $n=1$, $\phi_1[x] = x - 1$ (ii) For $n=2$, $\phi_2[x] = x + 1$

(iii) For $n=3$, $\phi_3[x] = x^2 + x + 1$ (iv) For $n=4$, $\phi_4[x] = x^2 + 1$

(v)   For $n=5$, $\phi_5[x] = x^4 + x^3 + x^2 + 1$ (vi) For $n=6$, $\phi_6[x] = x^2 - x + 1$

In general, for any prime p

$$\phi_p[x] = x^{p-1} + x^{p-2} + \ldots\ldots + x + 1$$

Theorem 1. *Let F be a field and n be a positive integer. Then there exists a primitive n^{th} roots of unity in some extension K of F if and only if either characteristic of F is zero or not a divisor of n.*

Proof.   Let us first suppose the characteristic of F is zero or not a divisor of n.

Consider a polynomial $f(x) = x^n - 1$ over F.

Then, $f'(x) = nx^{n-1} \neq 0$.

$\Rightarrow$ $f(x) = 0$ will have n distinct roots in the splitting field. These n roots will form a multiplicative cyclic group. If ω is one of its generators, then $\omega^n = 1$.

Also, no integer m less than n exists such that $\omega^m = 1$.

$\Rightarrow$ ω is a primitive n^{th} roots of unity in the extension of F.

Conversely, let ω is a primitive n^{th} roots of unity is some extension of F.

Clearly, $1, \omega, \omega^2, \ldots, \omega^{n-1}$ are all distinct roots of $x^n - 1 = 0$.

$\Rightarrow$ $f(x)$ has no multiple roots.

$\Rightarrow$ $f'(x) \neq 0$, *i.e.*, $nx^{n-1} = 0$

Characteristic of $F = 0$ or does not divide n.

Theorem 2. *For any positive integer n,* $x^n - 1 = \prod_{d|n} f_d(x)$

Proof. Let us write $t = \cos\frac{2\pi}{x} + i\sin\frac{2\pi}{x}$

The roots of x^n-1 are $1, t, t^2, ..., t^{n-1}$.

Therefore, $x^n-1 = (x-1)(x-t)(x-t^2) ... (x-t^{n-1})$

$$= \prod_{r=1}^{n}(x-t^r)$$

$$[\because t^n = 1]$$

$$= \prod_{d/n}\left[\prod_{1\leq n/d}(x-t^{2d})\right], \text{ with } (q, n/d) = 1$$

Further, $t^d = \left[\cos\frac{2\pi}{n} + i\sin\frac{2\pi}{n}\right]^d = \cos\frac{2\pi d}{n} + i\sin\frac{2\pi d}{n}$

is a primitive $(n|d)^{th}$ roots of unity and

$$x^n - 1 = \prod_{d/n}\phi_{n/d}(x) = \prod_{d/n}\phi_d(x)$$

Theorem 3. *For any positive integer n,* $\phi_d(x)$ *has integer coefficients.*

Proof. If $n = 1$, then $\phi_1(x) = x-1$.

Clearly, the coefficients of $\phi_1(x) = x-1$ are all integers.

$\Rightarrow$ Theorem is true for $n = 1$.

Further, assume $n > 1$ and $\phi_m(x)$ has integer coefficients for $1\leq m \leq n-1$.

Then $\phi_n(x) = \dfrac{x^n - 1}{\prod_{d|n}\phi_d(x)}$

Now, since $\prod_{d|n}\phi_d(x)$ has integer coefficients and x^{n-1} has integer coefficients, then

$\phi_n(x)$ has integer coefficients. Hence the theorem.

Theorem 4. $\Phi_n(x)$ *is irreducible polynomial of degree* $\Phi(n)$ *in* $\mathbf{Z}[x]$ *for any positive integer n.*

Proof. Let $\Phi_n(x)$ be an irreducible polynomial of degree $\Phi(n)$ in $\mathbf{Z}[x]$.

Further, let $f(x)$ be a non-constant monic irreducible factor of $\Phi_n(x)$.

Let t be a primitive n^{th} root of unity, which is also a root of $f(x)$. We shall prove that for any prime p with $p/n, f(t^p) = 0$. Now, since t is a primitive n^{th} root of unity, therefore t^p is also a primitive n^{th} root of unity.

Also, since $f(x)\in\mathbf{Z}[x]$ is a factor of $\Phi_n(x)$, then there exists $g(x)\in\mathbf{Z}[x]$ such that

$$\Phi_n(x) = f(x).g(x)$$

Further, suppose that $f(t^p) \neq 0$, then $g(t^p) = 0$

$\Rightarrow$ t is a root of $g(x^p) = 0$.

Therefore, $f(x)$ and $g(x^p)$ have a common factor over some extension of $\mathbf{Z}$.

Since $f(x)$ is irreducible over $\mathbf{Z}$, so it is also irreducible over $\mathbf{Q}$. Then, we get

$$f(x)\,|\,g(x^p) \Rightarrow g(x^p) = f(x)h(x),\ h(x) \text{ also a monic polynomial over } \mathbf{Z}.$$

Consider a canonical epimorphism $\sigma : \mathbf{Z} \to \mathbf{Z}p$ defined by $\sigma(a) = \bar{a}$, where $\bar{a} = a(\mathrm{mod}\ p)$.

Then, mapping induces an epimorphism $\sigma^* : \mathbf{Z}[x] \to \mathbf{Z}_p[x]$ such that

$$\sigma^*\,(r(x)) = \bar{r}(x),\ \text{where } \bar{r}\,(x) = r(x)(\mathrm{mod}\ p)$$

$\Rightarrow \qquad\qquad \bar{g}(x^p) = \bar{f}(x)\bar{h}(x)$

$\Rightarrow \qquad\qquad \bar{g}(x^p) = (\bar{g}(x))^p$

Therefore, $\qquad g(x^p) = f(x)h(x)$ gives that $\bar{f}(x)$ and $\bar{g}(x)$ have common factor.

Therefore

$$\phi_n(x) = f(x)\bar{g}(x)$$

and $\qquad\qquad \Phi_n(x)\,|\,x^n - 1$

$\Rightarrow\quad x^n - 1$ has no multiple roots because $p \nmid n$.

Further, if $p_1, p_2, ..., p_m$ are primes, not necessarily distinct and $p_i \nmid n,\ \forall\ i = 1, 2,...,$ m. Since $\qquad\quad f(t) = 0 \Rightarrow f(t^{p_1}) = 0$

$\Rightarrow \qquad\qquad f(t^{p_1 p_2}) = 0$ ...and so on.

Then, we have $f(t^{p_1 p_2 \cdots p_m}) = 0$

If ξ, is a primitive n^{th} roots of unity, then $\xi = t^m$, where m is a positive relatively prime to n. So $f(t^m) = 0$.

Further, $\Phi_n(x)$ has no multiple roots, therefore, $\Phi_n(x)\,|\,f(x)$.

But $\quad \Phi_n(t) = 0 \Rightarrow f(x)\,|\,\phi_n\,(x) \Rightarrow \Phi_n(x) = f(x)$.

and $f(x)$ is irreducible over $\mathbf{Z}$. Hence, $\Phi_n(x)$ is irreducible over $\mathbf{Z}$. Finally, $\phi_n(x)$ is a factor of $x^n - 1$ and $\phi_n(x)$ is monic, therefore $\Phi_n[x] \in \mathbf{Z}[x]$. Also, since number of primitive n^{th} roots of unity is $\phi(n)$. Hence, the degree of $\Phi_n(x)$ is $\phi_n(x)$.

10.21 SOLVABILITY BY RADICALS

10.21.1 Solution of Polynomial Equations

Definition 1 (*Radical of Exponents n*). *Let F be a field and an element $a \in F$ such that it is not the nth power of an element of F i.e., $x^n - a \in F[x]$ has no root in F[x]. Then a root of $x^n = a$ denoted by $\sqrt[n]{a}$ is known as radical or exponent n over F.*

Definition 2 (*Reducible or Irreducible Radicals*). *A radical $\sqrt[n]{a}$ of exponent n over F is called reducible or irreducible over F according as $x^n - a$ is reducible or irreducible polynomial of F[x].*

Definition 3 (*Pure Equation and Pure Extension*). *The equation of the form $x^n - a = 0$ is known as pure equation.*

Further, an extension field $F[\sqrt[n]{a}]$ is called pure extension of F.

Definition 4 (*Solvable by Radicals*). *Any field extension which can be reached through a finite series of successive pure extensions is known as solvable by radicals or tower.*

For example, *Let K be an extension of a field F and we have*

$$F = F_0 \subseteq F_1 \subseteq F_2 \subseteq ... \subseteq F_r = K$$

such that each member of the series is a pure extension of its predecessor, then K is known as tower over F or as a solvable by radicals over F.

Definition 5. *Let $f(x) \in F[x]$ be any polynomial, then the polynomial equation $f(x) = 0$ is said to be solvable by radicals over F, if the splitting field K of $f(x)$ is a tower of F.*

Also, if K is a normal extension of F, then tower is said to be normal radical tower *over F.*

10.21.2 Method of Solution by Radicals

Let F be a field and $f(x) \in F[x]$ be any non-constant polynomial.

If $f(x) = x^2 + bx + c$ over F of characteristic not equal to 2. Then solution of

$f(x) = 0$ is given by $-b \pm \sqrt{b^2 - 4ac}$.

$\Rightarrow$ The equation either has solution in F or solutions in $F(\sqrt{b^2 - 4ac})$.

$\Rightarrow f(x)$ can always be solved in some field which is obtained by adjoining a radical of F.

We have to find when an arbitrary polynomial equation $f(x) = 0$ can be solved.

If it is not solved in F, then there exists some field in which $f(x) = 0$ will have the solution. This field is obtained from F by the successive adjunction of radicals.

10.21.3 Solvability by Radicals of Cyclotomic Fields

Definition. *A group G of finite order is said to be solvable if their composition factors are prime.*

Theorem 1. *The cyclotomic field for a prime p over a field of characteristic 0 is solvable by radicals over the same field.*

Proof. Suppose that F is the field of characteristic 0 and K is the splitting field $x^p - 1 \in F[x]$.

The Galois group of $x^p - 1 \in F[x]$ is cyclic of order d (say), where d is a divisor of $p-1$.

Now, let $\phi = p_1^{m_1}, p_2^{m_2}, ..., p_r^{m_r}$, where $p_1, p_2, ..., p_r$ are distinct primes.

We shall prove this theorem by induction on exponents.

If $p = 2$. Then roots of $x^2 - 1$ are ± 1, which already belong to F.

$\Rightarrow$ Theorem is true for the exponent 2.

Further assume that theorem is true for each prime exponent less than p.

By adjoining $p_1^{th}, p_2^{th}, ..., p_r^{th}$ roots of unity successively, we get K.

Then, we arrive at a field E of characteristic zero containing the $p_1^{th}, p_2^{th}, ..., p_r^{th}$ roots of unity.

$\Rightarrow \qquad F \subseteq E \subseteq K$.

It is known that Galois group $G(K, E)$ is a subgroup of $G(K, F)$, where $G(K, F)$ is cycle of order d.

Evidently, the Galois group $G(K, E)$ is cycle of order of the form $p_1^{n_1}, p_2^{n_2}, ..., p_r^{n_r}$ with

$n_1 \leq m_1, n_2 \leq m_2, ..., n_r \leq m_r$. Then by fundamental theorem of Galois theory $\exists$ a finite sequence of fields given by

$$E, E_1, E_2, ...E_{n_1}$$

such that the Galois group $G(E_1, E)$, $G(E_2, E_1)$, ...$G(E_{n_1}, E_{n_1-1})$ are all cyclic of order p_1.

Thus $[E_n : E] = p_n^{n_1}$

Continuing this process, finally we arrive at K of degree $p_1^{n_1}, p_2^{n_2}, ..., p_r^{n_r}$ over E. So, K is a tower over F. Hence, F is solvable by radicals.

Theorem 2(Fundamental Theorem). *Let F be a field of characteristic 0. Then, a polynomial $f(x) \in F[x]$ is solvable by radicals over F if and only if its splitting field K over F has solvable Galois group $G[K, F]$.*

Proof.   Let us first suppose that $G[K, F]$ is solvable. We have to prove that $f(x) = 0$ is solvable by radicals.

Let $[K : F] = o[G(K, F)] = n$.

Let us assume that F contains a primitive n^{th} roots of unity. Then, F will contain a primitive m^{th} root-root of unity for all positive integer m that divides n. Let $G = G(K, F)$ and G is solvable and finite.

Then, $G = G_0 \supseteq G_1 \supseteq G_2 \supseteq .....\supseteq G_s = 1$

is a chain of subgroups of G such that for $i = 1, 2,..., s$, G_i is a normal subgroup of G_{i-1} and $G_{i-1} | G_i$ is cyclic.

If F_i is the fixed field of G_i, then

$$F = F_0 \subseteq F_1 \subseteq F_2 \subseteq ...\subseteq F_s = K \qquad ... (1)$$

If $n_i = [F_i : F_{i-1}]$, then $n_i | n$, therefore F_{i-1} contains a primitive n_i^{th} root of unity. Also, F_i / F_{i-1} is cyclic because $G(F_i, F_{i-1})$ is isomorphic to G_{i-1}/ G_i.

$\Rightarrow$ F_i is a cyclic extension of F_{i-1}.

$\Rightarrow$ F_i is the splitting field of an irreducible polynomial $x^{n_i} - a_i \in F_{i-1}(x)$ and

$$F_i = F_{i-1}(\alpha_i), \text{ where } \alpha_i \text{ is a root of } x^{n_i} - a_i = 0.$$

$\Rightarrow$ (1) is the radical tower over F and F_s is the splitting field of $f(x)$ over F.

$\Rightarrow$ $f(x) = 0$ is solvable by radicals.

Further, let C be an algebraic closure of F such that $K \subseteq C$. Let t be a primitive n^{th} root of unity in C. Then $C(t)$ is the splitting field of $f(x)$ over $F(t)$ and $G[C(t), K]$ is isomorphic to a subgroup of G.

Now, since G is solvable, so $G[C(t), K]$ is solvable.

$C(t)$ is a radical extension of $F(t) \Rightarrow C(t)$ is a radical extension of F.

Thus, $f(x)$ is solvable by radicals.

Conversely, let $f(x) = 0$ is solvable by radicals. Then, there is a normal radical tower of F such that

$$F = K_0 \subseteq K_1 \subseteq K_2 \subseteq ...\subseteq K_r$$

over F such that K_r contains a splitting field K of $f(x)$ over F.

For $i = 1, 2,..., r$, $K_i = K_{i-1}(a_i)$ with $a_i^{n_i} = b_i \in K_{i-1}$.

Now, let $n = n_1 n_2 .... n_r$ and let t be a primitive n^{th} root of unity in some algebraic closure of F which contains K_r.

For $i = 1, 2,..., r$, let $E_i = K_i(t)$, then

$$F(t) = E_0 \subseteq E_1 \subseteq E_2 \subseteq ... \subseteq E_r$$

is a normal radical tower over $F(t)$.

Further, we have $E_i = E_{i-1}(a_i)$ and E_{i-1} contains n_i^{th} root of unity, therefore

$$E_i = E_{i-1} \text{ is cyclic.}$$

If $\qquad\qquad H_i = G(E_r, E_i)$, then

$$G(E_r', F(t)) = H_0 \supseteq H_1 \supseteq ... \supseteq H_r = 1$$

and $H_{i-1} | H_i \cong G(E_i, E_{i-1})$, which is cyclic.

$\Rightarrow \quad G(E_r, F(t))$ is solvable.

We have $G(E_r, F(t)) \cong G(K_r, K_r \cap F(t))$ and $F(t)$ is an abelian extension of F. Then $K_r \cap F(t)$ is an abelian extension of F, therefore $G(K_r \cap F(t), F)$ is solvable. Therefore $G(K_r, F)$ has a solvable normal subgroup $G(K_r \cap F(t), F)$ such that the factor group $G(K_r, F) | G(K_r, K_r \cap F(t)) = G(K_r \cap F(t), F)$ is solvable.

Hence, $G(K, F)$ is solvable.

Theorem 3. *The Galois group of $x^n - 1$ over any field of characteristic zero is abelian.*

Proof. Let F be a field of characteristic 0 and K be the splitting field of x^{n-1} over F. If t is a primitive n^{th} roots of unity, then $K = F(t)$.

It is known that, in every F-automorphism of K, t is mapped on some power t^a of t and such automorphism is determined by its effect on t, i.e.,

$$G(K, F) = [\sigma_a : \sigma_a(t) = t^a]$$

Consider two elements σ_a and σ_b of $G(K, F)$. Then

$$\sigma_a(t) = t^a \text{ and } \sigma_b(t) = t^b$$

Now, $\qquad (\sigma_a \sigma_b)(t) = \sigma_a(\sigma_b(t)) = \sigma_a(t^b) = [\sigma_a(t)]^b$

$$= [t^a]^b = t^{ab}$$

$$= t^{ba} = (t^b)^a = [\sigma_b(t)]^a = \sigma_b(t)^a$$

$$= (\sigma_b \sigma_a)(t)$$

$\Rightarrow \qquad\qquad \sigma_a \sigma_b = \sigma_b \sigma_a$

Hence, $G(K, F)$ is abelian.

Theorem 4. *The Galois group of $x^p - 1$ over the prime field of characteristic zero is the cyclic group of order $p-1$, p is prime integer.*

Proof.   Let F be a prime field of characteristic 0. Now, since

$$x^p - 1 = (x + 1)\Phi_p(x)$$

Clearly, the splitting field K of $x^p - 1 \in F[x]$ is the same as that of

$$\Phi_p(x) = x^{p-1} + x^{p-2} + \dots + x + 1$$

Further, $\Phi_p(x)$ is irreducible over F.

Let t be a root of $\Phi_p(x)$, then $t, t^2, \dots, t^{p-1}$ are $p-1$ roots of $\phi_p(x)$. Also, all roots are distinct. Thus, there are $p-1$ automorphism defined by $\sigma_a(t) = t^a$.

Clearly, $\sigma_a \to a$ is an isomorphism of the Galois group of automorphism of order $p-1$ with the group of $p-1$ integers $1, 2, 3, \dots, p-1$ for multiplication modulo p.

Then, we have $(\sigma_a \sigma_b)(t) = \sigma_a(\sigma_b(t)) = \sigma_a(t^b) = (t^b)^a = t^{ba} = t^{ab} \Rightarrow \sigma_a \sigma_b \to ab$.

Therefore $\{1, 2, \dots, p-1\}$ forms a multiplicative group of prime field I_p of characteristic p, which is cyclic of order $p-1$.

10.22 CYCLIC EXTENSIONS

Definition. *A Galois extension K of a field F is cyclic over F if $G(K, F)$ is a cyclic group.*

For example; Let t be a primitive n^{th} root of unity and $K = \mathbf{Q}(t)$ is the splitting field of $x^n - 1 \in \mathbf{Q}(x)$, then K is a Galois extension of $\mathbf{Q}$ as well as cyclic extension.

Theorem 1. *Let F be a field containing a primitive n^{th} root of unity. If K is a cyclic extension of degree n over F, then there exists $a \in F$, $\alpha \in K$ such that the minimal polynomial of α over F is $x^n - a$ and $K = F(\alpha)$.*

Proof.   Let K be a cyclic extension of degree n over F. Also, let $t \in F$ be primitive n^{th} roots of unity. If σ is a generator of $G[K, F]$. Now, since $t \in F$, therefore, $N_{K/F}(t) = t^{(K:F)} = t^n = 1$.

$\Rightarrow$ $\exists$ a non-zero $\alpha \in K$ such that $t = \alpha | \sigma(\alpha)$

$\Rightarrow$ $\sigma^i(a) = t^{-i} \cdot a$ for $i = 1, 2, \dots, n$ and α has distinct F-conjugate.

Now, since $n = [K : F] \supseteq [F(\alpha): F] \geq n$, $K = F(\alpha)$ and $[F(\alpha): F] = n$.

Also from $t = \sigma \mid \sigma(a)$, we have $1 = t^n = a^n \mid \sigma(a^n)$

$\Rightarrow$ $\sigma^i(\alpha^n) = \alpha^n$, for $i = 1, 2, \dots n$.

$\Rightarrow$ α^n is in the fixed field of $G(K, F)$ and $x^n = a \in F$.

$\Rightarrow$ α is a root of the polynomial $x^n - a \in F[x]$.

Finally, since α is of the degree n over F, hence $x^n - a$ is the minimal polynomial of α over F.

Theorem 2. *Let F be a field which contains a primitive n^{th} root of unity. If $K = F(\alpha)$, where $\alpha^n = a$ over F, then K is a cyclic extension of F, and $[K : F] \mid n$ and*

$$[K : F] = \min \{r \in \mathbf{N} : \alpha^r \in F\}$$

Proof.   If α is a root of $x^n - a$ and $t \in F$ is a primitive of nth root of unity, then

$$[\alpha, t\alpha, t^2\alpha, \dots t^{n-1}\alpha]$$

is a complete set of distinct roots of $x^n - a = 0$. Therefore $K = F(\alpha)$ is a splitting field of separable polynomial over F. Thus, K is a finite Galois extension of F with Galois group $G(K, F)$.

Let $\sigma \in G(K, F)$, then $\sigma(\alpha) = t_\sigma(\alpha)$, where $t_\sigma = t^{m(\sigma)}$

and $m(\sigma)$ is an integer uniquely determined mod n by σ.

Now the mapping $\eta: G(K, F) \to S$ where S is the multiplicative group of nth roots of unity defined by $n(\sigma) = t_\sigma$ is a monomorphism. Then $G(K, F)$ is isomorphic to a subgroup of a cyclic group of order n. Therefore, $G(K, F)$ is cyclic and $(K : F) = O[G(K, F)] \mid n$.

$\Rightarrow K$ is the cyclic extension of F.

Also,
$$N_{K/F}(\alpha) = \prod_{\sigma \in G(K,F)} (t_\sigma \alpha) = \alpha^{[K:F]}. \prod_{\sigma \in G(K,F)} (t_\sigma) \in F \text{ and } t_\sigma \in F.$$

Also, for each $\sigma \in G(K, F)$, $\alpha^{[K:F]} \in F$. If $r < [K : F]$, then $\alpha^r \notin F$. Because of $\alpha^r = b \in F$, α is a root of $x^r - b$ and $[K : F] = [\alpha : F] < r < [K : F]$, which is a contradiction. Hence, $[K : F] = \min [r \in N : \alpha^r \in F]$.

10.23 CONSTRUCTION WITH RULER AND COMPASS

The word 'Ruler' is to be considered as a synonym for straightedge. We shall use field extensions to settle two famous problems of antiquity :

1. it is possible to trisect an arbitrary angle by ruler and compass construction.

2. it is possible via ruler and compass construction to duplicate an arbitrary cube, *i.e.*, to construct the side of a cube having twice the volume of the given cube.

We shall assume as known all the standard ruler and compass construction as presented in almost any plane geometry text. For example, given a straight line L and a point P on L, the unique straight line through P and parallel L is constructive. Here, constructible means constructible by ruler and compass constructions.

Now, we shall use the concept of analytic geometry as follows. We may construct with ruler and compass two perpendicular straight lines. Choose a unit length, then we can construct all points of the plane with integer coordinates.

If F is a subfield of the field **R** of real numbers. The plane of F is the subset of the plane consisting of all points (c, d) with $c \in F$, $d \in F$. If P, Q are distinct points in the plane of F, the unique line through P and Q is called a circle in F.

It is easily verified that every straight line in F has an equation of the form $ax + by + c = 0$, $a, b, c \in F$ and every circle in F an equation of the form $x^2 + y^2 + ax + by + c = 0$, $a, b, c \in F$.

Definition 1. *A point $r \in R^2$ is said to be constructible from P_0 if $\exists$ a finite sequence of points $r_1, r_2, ...,r_n = r$ such that r_i is constructible at one step from $P_0 \cup [r_1 ... r_{i-1}]$. In fact this is known as ruler and compass constructions.*

Definition 2. (*Constructible Number*). *Any real number α is said to be constructible number if by the use of ruler and compass alone, we can construct a line segment of length $|a|$.*

Definition 3. *A point (x, y) is said to be constructible from P_0 if $(x, y) \in P_i$ for some $i = 0$, 1, 2, ...*

Definition 4. *A line I is constructible from P_0 if it is passed through two distinct points in some P_i.*

Definition 5. *A circle c is said to be constructible from P_0 if its centre is in some P_i and it passes through another point lying in P_i.*

Definition 6. *A real number* α *is said to be constructible from* $\mathbf{Q}$ *if point* $(\alpha, 0)$ *is constructible from* $\mathbf{Q} \times \mathbf{Q} \subseteq R^2$.

Definition 7. *An angle* α *is constructible by ruler and compass if the point* $(\cos \alpha, \sin \alpha)$ *is constructible from* $\mathbf{Q} \times \mathbf{Q}$.

REMARKS

☞ The point $(\cos \alpha, \sin \alpha)$ is constructible from $\mathbf{Q} \times \mathbf{Q}$ if and only if $\cos \alpha$ is constructible number or if and only if $\sin \alpha$ is constructible number.

☞ The following are equivalent statements :

    (i) $\alpha \in \mathbf{R}$ is constructible from $\mathbf{Q}$.
    (ii) $(\alpha, 0)$ is constructible from $\mathbf{Q} \times \mathbf{Q}$.
    (iii) (α, α) is constructible from $\mathbf{Q} \times \mathbf{Q}$.
    (iv) $(0, \alpha)$ is constructible from $\mathbf{Q} \times \mathbf{Q}$.

Theorem 1. *If real number a is constructible, then* $(a, 0)$, (a, a) *and* $(0, a)$ *are also constructible.*

Proof.   By definition $(a, 0)$ is constructible.

Consider the circle c whose equation is given by
$$(x - a)^2 + y^2 = a^2$$

Its centre is $(a, 0)$ and passes through $(0, 0)$ which are constructible points. Therefore, c is constructible.

Also, line $L y = x$ joining $(0, 0)$ and $(1,1)$ is constructible.

Therefore $(a, a) = L \cap c$ is constructible.

The line $y = -x$ passing through $(0, 0)$, $(1, -1)$ and circle $x^2 + y^2 = 2a^2$ with centre $(0,0)$ and passing through (a, a) are constructible.

Thus $(-a, a)$, their product of intersection is constructible.

Also, line $y = a$ joining (a, a) and $(-a, a)$ is constructible.

Hence, $(a, 0) = (y = a) \cap (x = 0)$ is constructible.

Theorem 2. *If a is constructible, then* $x = a$ *and* $y = a$ *are constructible lines.*

Proof.   If $a = 0$, then the lines $x = 0$ and $y = 0$ are trivially constructible. So assume that $a \neq 0$. Since a is constructible, so $(a, 0)$, $(0, a)$ and (a, a) are constructible.

Now, the line $x = a$ passes through two constructible points $(a, 0)$ and (a, a), thus $x = a$ is constructible. Similarly the line $y = a$ passes through two constructible point $(0, a)$ and (a, a), thus $y = a$ is constructible.

Hence, $x = a$ and $y = a$ are constructible.

Theorem 3. *If a and b are constructible, then* (a, b) *is constructible.*

Proof.   Since a and b are constructible, then $x = a$ and $y = b$ are constructible and (a, b) is the point of intersection of two constructible points. Hence (a, b) is constructible.

Theorem 4. *If a and b are constructible, then* $a \pm b$, ab *and* a / b *when* $b \neq 0$ *are also constructible.*

Proof.   (i)   Since a and b are constructible, so that (a, b) is constructible.

The circle $(x - a)^2 + y^2 = b^2$ is constructible, because its centre is constructible point $(a, 0)$ and passes through a constructible point (a, b). Also, the line $y = 0$ is constructible. Now, the points $(a \pm b, 0)$ are the intersection points of the constructible circle $(x - a)^2 + y^2 = b$ and the constructible line $y = 0$. Hence $a \pm b$ are constructible.

(ii) Since $(0, b)$ and $(a, b-1)$ are constructible, then the line $ay = -x + ab$ is constructible, because it passes through two points $(0, b)$ and $(a, b-1)$.

Now, the point $(ab, 0)$ is the intersection of two constructible lines $ay = -x + ab$ and $y = 0$. So $(ab, 0)$ is constructible. Hence ab is constructible.

(iii) If $a = 0$, then $\dfrac{a}{b} = 0, b \neq 0$, so in this case $a|b$ is constructible. So assume that $a \neq 0$. Since $(0, a)$ and $(a, a(1-b))$ are constructible. Then the line $bx = a - y$ is constructible,. because it passes two distinct constructible points $(0, a)$ and $(a, a(1-b))$. Now $(a|b, 0)$ is the point of intersection of two constructible lines $bx = a - y$ and $y = 0$, so $(a|b, 0)$ is constructible. Hence, $a|b$ is constructible.

Theorem 5. If $a > 0$ is constructible, then $\sqrt{a}$ is also constructible.

Proof. Since $a > 0$ is constructible, then by theorem 4, $1 + a$ is constructible and $\dfrac{1+a}{2}$ also constructible so far.

Therefore $\left(\dfrac{1+a}{2}, 0\right)$ is constructible.

Now the circle $\left(x - \dfrac{1+a}{2}\right)^2 + y^2 = \left(\dfrac{1+a}{2}\right)^2$ is constructible, because its centre is $\left(\dfrac{1+a}{2}, 0\right)$ and passes through the point $(0, 0)$. Also the line $x = 1$ is constructible. Now the point $(1, \sqrt{a})$ is the point of intersection of above constructible circle and the constructible line $x = 1$, so $(1, \sqrt{a})$ is constructible. Therefore, the circle $(x-1)^2 + (y - \sqrt{a})^2 = a + 1$ is constructible because its centre is $(1, \sqrt{a})$ and passes through a constructible point $(0,0)$.

Now, the point $(0, 2\sqrt{a})$ is the intersection point of a constructible circle $(x-1)^2 + (y - \sqrt{a})^2 = a + 1$ and a constructible line $x = 0$. So that $(0, 2\sqrt{a})$ is constructible. Thus $2\sqrt{a}$ is constructible, and by theorem 4, $\sqrt{a}$ is constructible.

Theorem 6. Let **Q** be the field of rational numbers. Then a real number a is constructible if and only if we can find real numbers $a_1, a_2, ..., a_n$ such that

$$a_1^2 \in \mathbf{Q}, \ a_2^2 \in \mathbf{Q}(a_1), a_3^2 \in \mathbf{Q}(a_1, a_2), ..., a_n^2 \in \mathbf{Q}(a_1, a_2, ..., a_{n-1})$$

and $a \in \mathbf{Q}(a_1, a_2, ..., a_n)$.

Proof. It is clear that the point of intersection of two lines in **Q** is in **Q** , and a line and a circle in the plane of **Q** either intersect in point in the plane of **Q** or in the plane of $\mathbf{Q}(a_1)$, where $a_1^2 \in \mathbf{Q}$ and the two circles in the plane of **Q** either intersect in **Q** or in the plane of $\mathbf{Q}(a_1)$, where $a_1^2 \in \mathbf{Q}$.

Thus if the lines and circles in the plane of **Q** lead to points either in **Q** or in quadratic extension of **Q**.

Now, if the lines and circles are in $\mathbf{Q}(a_1)$, then they intersect in points in the plane of $\mathbf{Q}(a_1, a_2)$, where $a_2^2 \in \mathbf{Q}(a_1)$. Continuing in this way, a point a is constructible from **Q** if we can find real numbers $a_1, a_2, ..., a_n$ such that $a_1^2 \in \mathbf{Q}$, $a_2^2 \in \mathbf{Q}(a_1)$, $a_3^2 \in \mathbf{Q}(a_1, a_2), ..., a_n^2 \in \mathbf{Q}(a_1, a_2, ..., a_{n-1})$ such that a is in the plane of $\mathbf{Q}(a_1, a_2, ..., a_n)$.

Theorem 7. *If a real number a is constructible, then a lies in some extension K of the field of rational numbers of degree, which is some power of 2.*

Proof. Let $\mathbf{Q}$ be a field of rational numbers. Then by above theorem 6, if a real number a is constructible, then we can find real numbers $a_1, a_2 ,....., a_n$ such that $a_1^2 \in \mathbf{Q}$, $a_2^2 \in \mathbf{Q}(a_1)$, $a_3^2 \in \mathbf{Q}(a_1,a_2),......,a_n^2 \in \mathbf{Q}(a_1,a_2,....,a_{n-1})$ and $a \in \mathbf{Q}(a_1,a_2,....,a_n)$. Further, we have

$$[\mathbf{Q}(a_1):\mathbf{Q}] = 1 \text{ or } 2.$$

and $\quad\quad [\mathbf{Q}(a_1,a_2,.....,a_i):\mathbf{Q}(a_1,a_2,.....,a_{i-1}] = 1 \text{ or } 2 \text{ for } i = 1, 2, ...,n.$

Now, by transitivity of extensions, we have

$$[\mathbf{Q}(a_1,a_2,.....,a_i):\mathbf{Q}] = [\mathbf{Q}(a_1,a_2,.....,a_n):\mathbf{Q}(a_1,a_2,.....,a_{n-1})]$$

$$[\mathbf{Q}(a_1,a_2 ......a_{n-1}) : \mathbf{Q}(a_1,a_2, ...a_{n-2})] ... [\mathbf{Q}(a_1,a_2): \mathbf{Q}(a_1)][\mathbf{Q}(a_1): \mathbf{Q}]$$

Since each term in the product of above expression is either 1 or 2

$$\therefore \quad [\mathbf{Q}(a_1,a_2,.....,a_n):\mathbf{Q}] = 2^m \text{ ; where } m \text{ is some non-negative integer.}$$

Hence, $\alpha \in \mathbf{Q}(a_1,a_2,.....,a_n)$, which is finite extension of $\mathbf{Q}$ degree, a power of 2.

Theorem 8. *If a real number a satisfies an irreducible polynomial of degree n over the field $\mathbf{Q}$ of rational numbers and if n is not a power of 2, then a is not constructible.*

Proof. If a satisfies an irreducible polynomial of degree n over $\mathbf{Q}$, then we have

$$[\mathbf{Q}(a):\mathbf{Q}] = n.$$

Let us first suppose n is not a power of 2 and if possible, let a be constructible, then by previous theorem, there is a finite extension K of $\mathbf{Q}$ in which a lies such that $[K:\mathbf{Q}] = 2^m$ for some non-negative integer m.

Now, K contains $\mathbf{Q}$ and a and $\mathbf{Q}(a)$ being the smallest subfield of $\mathbf{R}$ containing both $\mathbf{Q}$ and a, so

$$\mathbf{Q}(a) \subseteq \mathbf{K}.$$

Therefore, $[\mathbf{Q}(a): \mathbf{Q}]$ must divide $[K : \mathbf{Q}]$

$\Rightarrow \quad [\mathbf{Q}(a): \mathbf{Q}]$ must divide 2^m,

$\Rightarrow \quad n \,|\, 2^m$

$\Rightarrow \quad n$ is also a power of 2, which gives a contradiction, this contradiction arises by assuming that a is constructible.

Hence, a is not constructible.

10.23.1 Some General Constructions

(1) **Squaring a Circle.** *It is not always possible to construct by ruler and compass, a square equal in area to the area of a circle.*

Let us assume that the radius of the circle is 1, then its area will be π. If a is the side of the square, whose area is equal to that of this circle, then $a^2 = \pi$.

Since π is not algebraic over the field $\mathbf{Q}$ of rational numbers, so that $\sqrt{\pi}$ is not algebraic over $\mathbf{Q}$ thereby implying that a is not algebraic over $\mathbf{Q}$, thus we have $[\mathbf{Q}(a):\mathbf{Q}] \neq 2^m$ for any non-negative integer.

Hence, a is not constructible by ruler and compass.

(2) **Duplicating a Cube.** *It is not possible to construct by ruler and compass a cube with a volume equal to twice the value of a given cube.*

Let us consider a cube of unit side, then its volume is 1.

Let x be the side of a cube with volume 2 to be constructed.

Then, its volume is x^3. So that we have

$$x^3 = 2 \quad \text{or} \quad x^3 - 2 = 0.$$

Now $x = 2^{1/3}$ is the only real root of the irreducible polynomial x^3-2 of degree 3 over **Q**, therefore, $[\mathbf{Q}(2^{1/3}):\mathbf{Q}]=3\neq 2^m$ for any non-negative integer m.

Thus, $x = 2^{1/3}$ is not constructible by ruler and compass.

(3) **Trisecting an Angle.** *There does not exist an angle that can be trisected by ruler and compass.*

Let us take an angle 60°. If this angle can be trisected by ruler and compass, then the number cos20° is constructible over **Q**.

Let $x = \cos 20°$. Then consider an identity

$$\cos 3\theta = 4\cos^3\theta - 3\cos\theta.$$

Putting $\theta = 20°$, we get

$$\cos 60° = 4\cos^3 20° - 3\cos 20°$$

or $\qquad \dfrac{1}{2} = 4x^3 - 3x$ $\qquad\qquad\qquad\qquad\qquad\qquad$ $[\because x = \cos 20°]$

or $\qquad 8x^3 - 6x - 1 = 0.$

This polynomial is irreducible over **Q** and has a root $x = \cos 20°$, so that

$$[\mathbf{Q}(x):\mathbf{Q}]=3\neq 2^m$$

for any non-negative integer m. Thus, $x=\cos 20°$ is not constructible. Hence, there does not exist an angle that can be trisected by ruler and compass.

Solved Examples

Example 1. *Show that 60° is constructible.*

Solution. Let $O(0, 0)$ be the vertex.

Since 3 is constructible

$\Rightarrow \quad \sqrt{3}$ is constructible.

Therefore, $A(1, \sqrt{3})$ is constructible.

Join O and A, we get OA as constructible line. Then

$$\theta = \angle XOA = 60°$$

As A is constructible, x-axis is constructible. So, line AP through A and perpendicular to x-axis is constructible.

Now, $\qquad OP = 1, PA = \sqrt{3}, OA = \sqrt{3+1} = 2.$

Hence, $\qquad \cos\theta = \dfrac{1}{2} \quad \Rightarrow \quad \theta = 60°.$

Example 2. Determine if an angle 10° can be constructed by a ruler and compass.

Solution. Let $a = 10°$, a is constructible if $\sin a$ is constructible. Therefore, we have to check whether sin 10° is constructible or not.

We know that

$$\sin 3a = 3\sin a - 4\sin^3 a$$

Putting $a = 10°$, we get

$$\sin 30° = 3\sin 10° - 4\sin^3 10°$$

or $\qquad\qquad \dfrac{1}{2} = 3a - 4a^3$, where $a = \sin 10°$

or $\qquad 8a^3 - 6a + 1 = 0.$

Thus a is a root of cubic polynomial $8x^3 - 6x + 1 = 0$ over $\mathbf{Q}$, which is irreducible over $\mathbf{Q}$, then $[\mathbf{Q}(a): \mathbf{Q}] = 3 \neq 2^m$ for any non-negative integer m.

Therefore $b = \sin 10°$ is not constructible. Hence, $a = 10°$ is not constructible by ruler and compass.

Example 3. *Show that regular pentagon is constructible.*

Solution. If we construct $\alpha = \cos\dfrac{2\pi}{5} = 2\cos 72° = 2\sin 180°$, then it would be possible to construct a pentagon.

Now, since $\sin 180° = \dfrac{-1+\sqrt{5}}{5}$, which is constructible.

Hence, it is possible to construct a regular pentagon.

Example 4. A regular n-gon is constructible if and only if $\phi(n)$, Euler function is a power of 2.

Solution. It would be possible to construct a regular n-gon if we construct the angle $\cos\dfrac{2\pi}{n}$. Let

$$t = \cos\dfrac{2\pi}{n} + i\sin\dfrac{2\pi}{n}, \text{ where } t \text{ is a primitive root of units.}$$

Then, $\qquad t + \dfrac{1}{t} = 2\cos\dfrac{2\pi}{n}$

Let $u = \cos\dfrac{2\pi}{n}$. Then, consider $\mathbf{Q} \subseteq \mathbf{Q}(u) \subseteq \mathbf{Q}(t)$.

Then, $\qquad t - \cos\dfrac{2\pi}{n} = i\sin\dfrac{2\pi}{n}$

$$\Rightarrow \quad \left(t - \cos\dfrac{2\pi}{n}\right)^2 = -\sin^2\dfrac{2\pi}{n} = -\left(1 - \cos^2\dfrac{2\pi}{n}\right)$$

$$\Rightarrow \qquad (t-u)^2 = u^2 - 1 \Rightarrow \quad t^2 + u^2 - 2ut = u^2 - 1$$

$$\Rightarrow \qquad t^2 - 2ut + 1 = 0$$

$\Rightarrow \quad t$ satisfies a polynomial $x^2 - 2ux + 1 = 0$ over $\mathbf{Q}(u)$ which is irreducible over $\mathbf{Q}(u)$ of degree α. Therefore, we have

$$[\mathbf{Q}(t):\mathbf{Q}(u)] = 2$$

But we know that $[\mathbf{Q}(t):\mathbf{Q}] = \phi(u)$

Using transitivity of extensions, we get

$$[\mathbf{Q}(t):\mathbf{Q}] = [\mathbf{Q}(t):\mathbf{Q}(u)]\,[\mathbf{Q}(u):\mathbf{Q}]$$

$$\Rightarrow \qquad \phi(n) = 2[\mathbf{Q}(u):\mathbf{Q}]$$

$$\Rightarrow \qquad [\mathbf{Q}(u):\mathbf{Q}] = \dfrac{1}{2}\phi(n)$$

Hence, u is constructible if and only if $\phi(n)$ is a power of 2, because u is constructive if $[\mathbf{Q}(u): \mathbf{Q}] = 2^m$, for some $m > 0$.

10.24 INSOLVABILITY OF THE GENERAL EQUATION OF DEGREE 5 (QUINTIC) BY RADICALS

Consider the field F of radical numbers, let $f(x) \in F[x]$ be a non-constant polynomial. If $f(x) = x^2 + bx + c$ and if characteristic $F \neq 2$, then the solution of $f(x) = 0$ are given by

$$\frac{-b \pm \sqrt{b^2 - 4ac}}{2a}$$

$\Rightarrow$ This equation either has solution in F or in $F\left(\sqrt{b^2 - 4ac}\right)$.

$\Rightarrow f(x) = 0$ can always be solved in a field by obtaining a radical to F.

Similarly, the roots of a polynomial of degree 3 or 4 over F can be expressed in terms of radicals. Abel proved that *a* quintic, *i.e.*, 5th degree polynomial need not be solvable by radicals.

10.24.1 Elementary Symmetric Functions

If $y_1 \in \mathbf{R}$ be transcendental over $\mathbf{Q}$, $y_2 \in \mathbf{R}$ be a transcendental over $\mathbf{Q}(y_1)$ and so on until we get $y_5 \in \mathbf{R}$ such that $y_5 \in \mathbf{Q}(y_1, y_2, y_3, y_4)$.

Let
$$K = \mathbf{Q}(y_1, \ldots, y_5) \text{ and let}$$
$$f(x) = (x - y_1)(x - y_2)(x - y_3)(x - y_4)(x - y_5) \qquad \ldots(1)$$

Then,
$$f(x) \in K[x].$$

The elementary functions of $y_1, \ldots, y_5$ can be defined as follows :

$$s_1 = \sum_{i=1}^{5} y_i ; \quad s_2 = \sum_{i<1} y_i y_j ; \qquad s_3 = \sum_{i<j<k} y_i y_j y_k$$

$$s_4 = \sum_{i<j<k<l} y_i y_j y_k y_l \quad \text{and} \qquad s_5 = y_1 y_2 y_3 y_4 y_5$$

Then, (1) can be written as $f(x) = x^5 - s_1 x^4 + s_2 x^3 - s_3 x^2 + s_4 x - s_5$

10.24.2 Some Important Results

1. Let F be a subfield of real numbers and let $f(x) \in F[x]$ be a polynomial of degree 5 such that the splitting field K of $f(x)$ over F has Galois group isomorphic to S_5.

2. If a subgroup H of S_5 contains a cycle of length 5 and transposition, then $H \approx S_5$.

3. If $f(x)$ is an irreducible polynomial in $Q[x]$ of degree 5 having exactly two complex and three real roots, then Galois group of $f(x)$ over $\mathbf{Q}$ is isomorphic to S_5 . Hence, $f(x)$ is not solvable by radicals.

Solved Examples

Example 1. *Show that the Galois group of $x^4 + x^2 + 1$ is the same as that of $x^6 - 1$ and is of order 2.*

Solution. The primitive n^{th} root of unity is given by $e^{2\pi i/n} = \cos \dfrac{2\pi}{n} + i \sin \dfrac{2\pi}{n}$ and primitive third root of unity is given by

$$e^{2\pi i/3} = \cos \frac{2\pi}{3} + i \sin \frac{2\pi}{3} = -\frac{1}{2} + \frac{i\sqrt{3}}{2}$$

Therefore, $x^4 + x^2 + 1 = y^2 + y + 1$, where $y = x^2$.

But $-\dfrac{1}{2} + \dfrac{i\sqrt{3}}{2}$ is a root of $y^2 + y + 1 = 0$, thus $y^2 + y + 1$ is the minimal polynomial for a primitive third root of unity. So that the splitting field of $x^4 + x^2 + 1$ will contain the square root of $e^{2\pi i/3}$ and $e^{4\pi i/3}$. Thus we need to adjoin

$$e^{\pi i/3} = (e^{2\pi i/3})^{1/2}, \; e^{4\pi i/3} = -e^{\pi i/3} = -(e^{2\pi i/3})^{1/2}$$

$$e^{2\pi i/3} = (e^{4\pi i/3})^{1/2} \text{ and } e^{5\pi i/3} = -e^{2\pi i/3} = -(e^{4\pi i/3})^{1/2}$$

$\Rightarrow$ $K = \mathbf{Q}(\alpha)$, where $\alpha = e^{\pi i/3}$, is the splitting field of $x^4 + x^2 + 1$ over $\mathbf{Q}$.

Now, α is a primitive sixth root of unity, therefore K is also the splitting field of x^6-1 over $\mathbf{Q}$. Then, $G(K, \mathbf{Q}) = (Z/(6))^* = \{T,F\}$ is the group of order 2.

Example 2. *Let K be the splitting field of $x^n - a \in F[x]$. Then show that $G(K, F)$ is a solvable group.*

Solution. Let F contains a primitive n^{th} root of unity, then $G(K, F)$ is abelian and hence it is solvable. Assume that F contains no primitive n^{th} root of unity. Let $t \in \overline{F}$ (the closure of F) be a generator of the cyclic group of the nth root of unity. If α be a root of $x^n - a = 0$. Then αt is also a root of $x^n - a = 0$. Thus $t = \alpha^{-1} (\alpha t)$ is in the splitting field K of $x^n-a \in F[x]$.

Consider $F \subseteq F(t) \in K$. $F(t)$ is a normal extension of F, because $F(t)$ is the splitting field of x^n-1, therefore $G(K,F(t))$ is a normal subgroup of $G(K, F)$.

Since K is the splitting field of $x^n-a \in F[x]$ so that $G(K, F(t))$ is abelian. Then, we have $(e) \in G(K, F(t)) \subseteq G(K, F)$ as a normal series.

Then, by the fundamental theorem of Galois theory, we have

$$G(K,F)/G(K,F(t)) \cong G(F(t),F).$$

Now, since $G(F(t),F)$ is abelian, so that $G(K, F)$ has a normal series with abelian factors. Hence, $G(K, F)$ is solvable.

Example 3. *Find $\phi_8(x)$.*

Solution. It is known that 1, 3, 5, 7 are the integers relatively prime to 8, so $\phi(8) = 4$. If t be a primitive 8th root of unity, then

$$t = \cos \dfrac{2\pi}{8} + i \sin \dfrac{2\pi}{8} = \cos \dfrac{\pi}{4} + i \sin \dfrac{\pi}{4} = \dfrac{1+i}{\sqrt{2}}$$

Now, all the primitive 8th roots of unity are t, t^3, t^5 and t^7.

Now, $$t^3 = \left(\cos \dfrac{\pi}{4} + i \sin \dfrac{\pi}{4} \right)^3 = \cos \dfrac{3\pi}{4} + i \sin \dfrac{3\pi}{4} = \dfrac{-1+i}{\sqrt{2}}$$

Similarly, $$t^5 = \dfrac{-1-i}{\sqrt{2}} \text{ and } t^7 = \dfrac{1-i}{\sqrt{2}}$$

Thus, $$\Phi_8(x) = (x - t)(x - t^3)(x - t^5)(x - t^7)$$

$$= \left(x - \dfrac{1+i}{\sqrt{2}} \right)\left(x - \dfrac{-1+i}{\sqrt{2}} \right)\left(x - \dfrac{-1-i}{\sqrt{2}} \right)\left(x - \dfrac{1-i}{\sqrt{2}} \right)$$

$$= \left[x^2 - \left(\dfrac{1+i}{\sqrt{2}} \right)^2 \right]\left[x^2 - \left(\dfrac{1-i}{\sqrt{2}} \right)^2 \right]$$

$$= (x^2-i)(x^2+i) = x^4+1$$

Example 4. *Compute each of the following for $K = Q(\sqrt{2}, \sqrt{3})$.*

  (i) $N_{K/Q}(\sqrt{2})$ (ii) $N_{K/Q}(\sqrt{2} + \sqrt{3})$

Solution. We know that $K = Q(\sqrt{2}, \sqrt{3})$ and K is normal extension of Q, so that there are four Q-automorphism of K leaving Q fixed.

Now, $G(K, Q) = \{\sigma_1 = I, \sigma_2, \sigma_3, \sigma_4\}$, where σ_1 is the identity map. Therefore

$$\sigma_2(\sqrt{2}) = -\sqrt{2}, \sigma_2(\sqrt{6}) = -\sqrt{6}, \sigma_2(1) = 1, \sigma_2(\sqrt{3}) = \sqrt{3}$$

$$\sigma_3(\sqrt{3}) = -\sqrt{3}, \sigma_3(\sqrt{6}) = -\sqrt{6}, \sigma_3(1) = 1, \sigma_3(\sqrt{2}) = \sqrt{2}$$

$$\sigma_4(\sqrt{2}) = -\sqrt{2}, \sigma_4(\sqrt{3}) = -\sqrt{3}, \sigma_4(1) = 1, \sigma_4(\sqrt{6}) = \sqrt{6}$$

(i) $\qquad N_{K/Q}(\sqrt{2}) = \sigma_1(\sqrt{2}).\sigma_2(\sqrt{2}).\sigma_3(\sqrt{3}).\sigma_4(\sqrt{2})$

$$= (\sqrt{2})(-\sqrt{2})(-\sqrt{2})(-\sqrt{2}) = 4.$$

(ii) $N_{K/Q}(\sqrt{2} + \sqrt{3}) = \sigma_1(\sqrt{2} + \sqrt{3}).\sigma_2(\sqrt{2} + \sqrt{3}).\sigma_3(\sqrt{2} + \sqrt{3}).\sigma_4(\sqrt{2} + \sqrt{3})$

$$= (\sigma_1(\sqrt{2}) + \sigma_1(\sqrt{3})).(\sigma_2(\sqrt{2}) + \sigma_2(\sqrt{3})).(\sigma_3(\sqrt{2})$$

$$+ \sigma_3(\sqrt{3})).(\sigma_4(\sqrt{2}) + \sigma_4(\sqrt{3}))$$

$$= (\sqrt{2} + \sqrt{3})(-\sqrt{2} + \sqrt{3})(\sqrt{2} - \sqrt{3})(-\sqrt{2} - \sqrt{3})$$

$$= (2-3)(2-3) = 1$$

Example 5. *Show that the polynomial $x^5 - 9x + 3$ is not solvable by radicals over Q.*

Solution. Let $f(x) = x^5 - 9x + 3$. Using Descartes's rule of signs, we can say

(i) there are two variation of signs in $f(x)$, so $f(x)$ has 2 positive real roots,

(ii) there are one variation of signs in $f(-x) = -x^5 + 9x + 3$.

$\Rightarrow$ $f(x)$ has one negative real root.

Since $\deg.f(x) = 5$, thus $f(x)$ has exactly two non-real roots.

$\Rightarrow$ The Galois group of $f(x)$ is isomorphic to S_5 and S_5 is not solvable. Hence $f(x)$ is not solvable.

Example 6. *Show that the polynomial $2x^5 - 5x^4 + 5$ is not solvable by radicals over Q.*

Solution. Let $f(x) = 2x^5 - 5x^4 + 5$.

Now, put $x = y/2$, we obtain $g(y) = \dfrac{1}{16}(y^5 - 5y^4 + 80)$.

Let $h(y) = y^5 - 5y^4 + 80$, then the roots of $g(y)$ or those of $h(y)$ are the twice of the roots of $f(x)$. The solvability of $f(x)$ by radicals is equivalent to the solvability by radicals of $h(y)$.

Then, by Eisenstein's criterion, we have $h(y)$ is irreducible over Q, so $f(x)$ is irreducible over Q. By Descartes rule of signs, $h(y)$ will contain at most two positive real roots and at one negative real root, so, $h(y)$ will contain at most three real roots, but by intermediate theorem, $h(y)$ has three roots, one in each of the intervals $(-2, -1)$, $(2, 3)$ and $(4, 5)$. Thus, $h(y)$ has exactly three real roots and two non-real roots of $h(y)$ and so of $f(x)$. Therefore, Galois group of $f(x)$ is isomorphic to S_5. Hence, $f(x)$ is not solvable by radicals.

EXERCISE 10.5

1. Show that in $\mathbf{Q}[x]$, $\Phi_{2n}(x)=\Phi_n(-x)$, for odd integers $n>1$

2. Show that if F is a field of characteristic not equal to 2 and $f(x)=ax^4 + bx^2 + C$, $a\neq0$, then $f(x)$ is solvable by radicals over F.

3. Show that if an irreducible polynomial $f(x) \in F[x]$ over a field F has a root in a radical extension of F, then $f(x)$ is solvable by radicals over K.

4. Show that $\Phi_8(x)$ and $x^8 -1$ have the same Galois group, namely $(\mathbf{Z}/(8))^* =(1,3,5,7)$ has the Klein Four Group (a group in which every element except the identity has order 2 known as Klein Four Group).

5. Show that regular 7-gon is not constructible.

6. Show that regular 9-gon is not constructible.

7. Show that regular 15-gon is not constructible.

8. Show that regular 17-gon is not constructible.

9. Show that it is possible to trisect 59°.

10. Show that it is possible to trisect 72°.

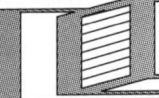

Chapter Review: | *A competitive Approach*

Selected Terms and Results

Terms

- **Field Extension:** A field K is said to be an extension of a field F if F is a subfield of K.

- **Algebraic Element:** Let K be an extension of a field F. An element $a \in K$ is said to be algebraic over F if there is a polynomial $p(x) \in F(x)$ for which $p(a) = 0$

- **Trancendental Element:** Which is not algebraic.

- **Algebraic Number:** A complex number is said to be an algebraic number if it is algebraic over the field of rational numbers.

- **Algebraic Extension of a Field:** Let K be an extension of a field F, then K is said to be an algebraic extension of F if every element of K is algebraic over F.

- **Transcendental Extension of a Field:** If there exist $a \in K$ such that a is not algebraic over F, then K is called as tanscendental extension of F.

- **Simple and Multiple Root:** A root of multicity one is called simple and a root of multiplicity greater than one is called a multiple root.

- **Root Field:** Let F be a field and K be an extension of F. Then K is said to be a root field of an irreducible polynomial $f(x) \in F[x]$ if K contains a root of $f(x)$.

- **Splitting Field:** An extension K of a field F is said to be splitting field of a polynomial $f(x) \in F[x]$ if it contains all it roots and there is no proper subfield of K which contains all the roots of $f(x)$.

- **Primitive Element:** Let K be an extension of a field F then an element $\alpha \in K$ is said to be primitive if $K = F(\alpha)$.

- **Normal Extension:** Let F be a field and K be an algebraic extension of F. Then K is said to be normal extension of F, if the spliting field of the minimal polynomial $f(x) \in F(x)$, each element of K is the splitting field of some polynomial over F.

- **Separable polynomial:** An irreducible polynomial $f(x) \in F[x]$ is said to be algebraic over F if $f(x)$ has no multiple roots in its

splitting field.

- **Separable Element:** Let F be a field and K be an algebraic extension of F, then an element $\alpha \in K$ is said to be separable element over F if the minimal polynomial or α over F is seperable.

- **Separable Extension:** An algebraic extension K of a field F is said to be separable extension of F if every element of K is separable over F.

- **Perfect field:** A field F is said to be perfect if each of its extension is seperable.

- **Algebraically closed field:** A field K is called algebraically closed if every polynomial F over K splits in K.

- **Fixed field:** Let G be a subgroup of the automorphism group of a field F. Let F_0 be the subset of F consisting of those elements of F which are left fixed by every automorphism in G.

- **Galois group:** Let K be a finite extension of a field F. Then the group $G(K : F)$ of all automorphism of K is called Galois group of K over F.

- **Galois field:** For any prime integer P and for any integer n, a field with p^n elements is called Galois field.

- **Primitive n^{th} roots of unity:** Let F be a field. The roots $x^{n-1} = 0$ over F are known as the primitive n^{th} roots of unity in its splitting field if $\omega^n = 1$ but $\omega^m \neq 1$ for any positive integer $m < n$.

- **Cyclotomic Extension:** The complex numbers satisfying $x^n - 1 = 0$ divide the circumference of a circle with centre at origin and of unit radius into n equal parts. Then the equation $x^n - 1 = 0$ is called cyclotomic extension over F.

- **Radicals:** Let F be a field. Let an element $a \in F$ it is not the nth power of an element of F such that $x^n - a \in F[x]$ has no root in $F[x]$. Then a root of $x^n = a$ denoted by $\sqrt[n]{a}$ is called radical or exponent over F.

- **Reducible or Irreducible Radicals:** A radical $\sqrt[n]{a}$ of exponent n over F is called reducible or irreducible over F according as $x^n - a$ is reducible or irreducible polynomial of $F[x]$.
- **Solvable group:** A group G of finite order is said to be solvable if their composition factors are prime.
- **Constructible number:** Any real numbers is said to be constructible if by the use of ruler and compass alone, we can construct a line segment of length $|\alpha|$.

Results

- A field extension K is said to be finite or infinite extension of F according as the degree of $K(F)$ is finite or infinite.
- A complex number is said to be an algebraic number if it is algebraic over the field of rational numbers.
- A complex number which is not algebraic is called Transcendental number.
- Let K be an extension of a field F and $a \in K$ is algebraic over F. If $p(x)$ is a polynomial over F of lowest positive degree satisfied by a, then $p(x)$ is known as minimal polynomial for a over F.
- A non-zero polynomial $f(x) \in F(x)$ is said to be monic over F, if 1 is the leading coefficient, *i.e.,* coefficient of highest power of x in $f(x)$ is 1 in F.
- Every rational number is left fixed by every automorphism of a field K.
- If G is a subgroup of the group of all automorphism of a field F, then the fixed field is a subfield F.
- A finite extension K of a field F is said to be normal if the fixed field $G(K, F)$ is F.
- K is a normal extension of a field F of characteristic 0 if and only if K is the splitting field of some polynomial over F.
- Let K be a finite extension of F, then two elements α and β of K are said to be conjugate over F, if they have the same minimal polynomial over F.
- Let K be an extension of a field of rational numbers $\mathbf{Q}$, then any automorphism of K must leave every element $\mathbf{Q}$ fixed.
- The characteristic of a finite field is necessarily non-zero, but converse is not true.
- The number of elements in a finite field of characteristic p is of the form p^n, where n is some positive integer.
- Any two finite fields having the same number of elements are isomorphic.
- Let F be a finite field with q elements and $F \subseteq K$, K is a finite field. Then K has q^n elements where $n = [K : F]$.
- The multiplicative group of a Galois field is cyclic.
- Every finite extension of a finite field is Galois.
- A group G of finite order is said to be solvable if their composition factors are prime.
- The cyclotomic field for a prime p over a field of characteristic 0 is solvable by radicals over the same field.
- The Galois group of $x^n - 1$ over any field of characteristic zero is abelian.
- If real number a is constructible, then $(a, 0)$, (a, a) and $(0, a)$ are also constructible.
- If real number a is constructible, then a lies in some extension K of the field of rational numbers of degree, which is some power of 2.

Review Questions and Project Work

1. Show that $x^3 + ax^2 + bx + 1 \in \mathbf{Z}(x)$ is reducible over $\mathbf{Z}$ if and only if either $a = b$ or $a + b = -2$.

2. Show that a cubic over a field F is irreducible over F if and only if it has no root in F.

3. Let K be an extension of a field F. Also let $a \in K$ and $f(x) \in F[x]$. Show that $f(x)$ is divisible by $x - a$ in $K[x]$ if and only if a is a root of $f(x)$.

4. Show that the field $\mathbf{C}$ of complex numbers is an algebraic closed field.

5. Find the degree of splitting field of $x^5 - 3x^2 + x^2 - 3$ over $\mathbf{Q}$.

6. Show that over any extension field K of $\mathbf{Q}$ the polynomial $x^3 - 3x + 1 \in \mathbf{Q}(x)$ is either irreducible or splits into linear factors.

7. Show that the field generated by a root of $x^3 - x - 1 \in \mathbf{Q}(x)$ is not normal over $\mathbf{Q}$.

8. Show that a finite field F of p^n elements has exactly one subfield with p^m elements for each divisior m of n.

9. If F is a field of characteristic p, show that each element a of F has a unique p^{th} root of $\sqrt[p]{a}$ in F.

10. Show that in any finite field, any element can be written as the sum of two squares.

11. Let E be an extension of a field F and $\alpha \in E$ be algebraic over F. Show that α is separable over F if and only if $F(\alpha)$ is a separable extension of F.

12. Show that the group $G[\mathbf{Q}(\alpha), \mathbf{Q}]$ where $\alpha^5 = 1$

and $\alpha \neq 1$ is isomorphic to the cyclic group of order 4.

13. Let F be a field of characteristic not equal to 2. Let $x^2 - a \in F[x]$ be an irreducible polynomial over F, show that its Galois group is of order 2.

14. Find the Galois group G of $x^4 - 2 \in \mathbf{Q}[x]$ and illustrate the one-one correspondence between subgroup of G and subfield of the splitting field of $x^4 - 2$.

15. Show that the polynomial $2x^5 - 5x^4 + 5$ is not solvable by radicals.

Objective type Questions

Fill in the Blanks

1. A prime element is _____ element.

2. A comutative integral domain with unity that is a principal ideal ring is called a _____ .

3. An irreducible element in a commutative princpal ideal domain is always _____ .

4. Let $f(x) \in \mathbf{Z}[x]$ be the primitive. Then $f(x)$ is irreducible over $\mathbf{Q}$ if and only if $f(x)$ is reducible over _____ .

5. The polynomial is _____ over $\mathbf{Q}$.

6. If K be an extension of F then $K = F$ if and only if _____ .

7. A polynomial of degree n over a field can have almost _____ roots in any extension field.

8. Every finite extension of a field is _____ .

9. The extension which is not algebraic is called _____ .

10. If every algebraic extension of a field K coincides with K is known as .....

True/False

Write 'T' for true and 'F' for false statement.

1. K is algebrically closed if and only if every irreducible polynomial in $K[x]$ is of degree one. **(T/F)**

2. If $a \in K$ is algebraic of degree n over F then $[F(a):F] = n$ **(T/F)**

3. Let F be a field. Then there exists an algebraically closed field K containing F as a subfield. **(T/F)**

4. The degree of $\mathbf{Q}\sqrt{2}$ over $\mathbf{Q}$ is 3. **(T/F)**

5. Field of complex numbers is not algebrically closed field. **(T/F)**

6. The splitting field of $x^2 + 1 \in \mathbf{R}[x]$ over $\mathbf{R}$ is the field of **C**. **(T/F)**

7. The degree of splitting field of $x^3 - 2 \in \mathbf{Q}(x)$ is 6. **(T/F)**

8. The field of complex numbers is not a splitting field. **(T/F)**

9. The field generated by a root of $x^3 - x - 1 \in \mathbf{Q}(x)$ is not normal over $\mathbf{Q}$. **(T/F)**

10. The field of rational number is a prime field. **(T/F)**

Multiple Choice Questions

Choose the most appropriate one:

1. A field F is called prime if it has :
   (a) proper subfield
   (b) no proper field
   (c) no improper field
   (d) none of the above

2. The multiplicative group of non-zero elements of a finite field is :
   (a) cyclic
   (b) non-cyclic

   (c) may be cyclic
   (d) None of the above

3. Let F be a finite then there exists a polynomial of any given degree n over F which is :
   (a) reducible
   (b) irreducible
   (c) may or may not be reducible
   (d) None of the above

4. An irreducible polynomial $f(x) \in F[x]$ is called a separable polynomial if all its roots are :

(a) simple

(b) zero

(c) ≥ 1

(d) None of the above

5. Let **Q** be the field of rationals then $\mathbf{Q}\left(\sqrt{2},\sqrt{3}\right)$

   (a) $\mathbf{Q}\left(\sqrt{2}-\sqrt{3}\right)$

   (b) $\mathbf{Q}\left(\sqrt{2}.\sqrt{3}\right)$

   (c) $\mathbf{Q}\left(\sqrt{2}+\sqrt{3}\right)$

   (d) None of the above

6. Let E be a finite separable extension of a field F then which is not true :

   (a) E is a normal extension of F.

   (b) F is the field of $G(E|F)$.

   (c) $[E{:}F]=|G(E|F)|$

   (d) All are true.

7. Let E be a Galois extension of F then E is called cyclic extension of F if $G(E|F)$ is a :

   (a) cyclic group

(b) non-cyclic group

(c) may be cyclic group

(d) none of the above

8. Let E be a splitting field of $x^n{-}a\in F[x]$. Then $G(E|F)$ is :

   (a) solvable

   (b) not solvable

   (c) may be solvable

   (d) None of the above

9. The degree of general polynomial which is not solvable by radicals is :

   (a) ≥ 5

   (b) ≥ 6

   (c) ≥ 7

   (d) None of the above

10. If α is a constructible number then :

    (a) $x=a$ is a constructible line.

    (b) $y=a$ is a constructible line.

    (c) both (a) & (b) are true.

    (d) None of the above

=== Answers ===

Fill in the blanks

1. irreducible **2.** PID **3.** prime **4.** **Z**, the set of integers **5.** Irreducible **6.** [K:F]=1 **7.** *n* **8.** algebraic
9. transcendental **10.** algebraically closed

True/False

1. T	**2.** T	**3.** T	**4.** F	**5.** F	**6.** T	**7.** T	**8.** F	**9.** T

10. T

Multiple Choice Questions

1. (b)	**2.** (a)	**3.** (b)	**4.** (a)	**5.** (c)	**6.** (d)	**7.** (a)	**8.** (a)	**9.** (a)

10. (c)

BIBLIOGRAPHY

1.	**Asha Rani Singhal**	*Algebraic Structures* Rastogi and Company, Meerut
2.	**E.Artin**	*Galois Theory* University of Notre Dame Press
3.	**F.M. Hall**	*An Introduction to Abstract Algebra* Cambridge University Press
4.	**F.Stewart**	*Introduction to Linear Algebra* East-West Press, New Delhi
5.	**G. Briefkhoff and S.K. Lane**	*A Survey of Modern Algebra* Macmillan Company, New York
6.	**I.N. Herstein**	*Topics in Algebra* Vani Educational Books
7.	**J.B. Frayleigh**	*A First Course in Abstract Algebra* Addison-Wesley Publishing Company
8.	**J.T. Moore**	*Introduction to Abstract Algebra* Academic Press
9.	**L.R. Goldstein**	*Abstract Algebra: A First Course* Prentice Hall Ind.
10.	**N. Jacobson**	*Lectures in Abstract Algebra* East-West Press Pvt. Ltd., New Delhi
11.	**P.L. Bhatnagar**	*Introductory Lessons in Modern* Mathematical Concepts East-west Press, New Delhi
12.	**P.R. Halmos**	*Finite Dimensional Vector Spaces* Springer Verlag
13.	**S.Lang**	*Algebra* Addison-Wesley Longman
14.	**Surjeet Singh and Qazi Za-meeruddin**	*Modern Algebra* Vikas Publishing House, New Delhi
15.	**Vijay K. Khanna and S.K. Bhambri**	*A Course in Abstract Algebra* Vikas Publishing House, New Delhi

INDEX